国外优秀数学著作
原 版 系 列

对称群的表示论

Representation Theory of Symmetric Groups

（英文）

[法] 皮埃尔-洛伊克·梅利奥特（Pierre-Loïc Méliot） 著

HITP
哈尔滨工业大学出版社
HARBIN INSTITUTE OF TECHNOLOGY PRESS

黑版贸审字 08－2020－204 号

图书在版编目(CIP)数据

对称群的表示论＝Representation Theory of Symmetric Groups: 英文/(法)皮埃尔－洛伊克・梅利奥特著. —哈尔滨: 哈尔滨工业大学出版社, 2023.1
ISBN 978-7-5767-0619-2

Ⅰ. ①对…　Ⅱ. ①皮…　Ⅲ. ①对称群－英文　Ⅳ. ①O152.1

中国国家版本馆 CIP 数据核字(2023)第 030111 号

DUICHENQUN DE BIAOSHILUN

策划编辑　刘培杰　杜莹雪
责任编辑　聂兆慈　李兰静
封面设计　孙茵艾
出版发行　哈尔滨工业大学出版社
社　　址　哈尔滨市南岗区复华四道街 10 号　邮编 150006
传　　真　0451－86414749
网　　址　http://hitpress.hit.edu.cn
印　　刷　哈尔滨博奇印刷有限公司
开　　本　787 mm×1 092 mm　1/16　印张 44.25　字数 640 千字
版　　次　2023 年 1 月第 1 版　2023 年 1 月第 1 次印刷
书　　号　ISBN 978-7-5767-0619-2
定　　价　98.00 元

Contents

Preface v

I Symmetric groups and symmetric functions 1

1 Representations of finite groups and semisimple algebras 3

1.1 Finite groups and their representations 3
1.2 Characters and constructions on representations 13
1.3 The non-commutative Fourier transform 18
1.4 Semisimple algebras and modules 27
1.5 The double commutant theory 40

2 Symmetric functions and the Frobenius–Schur isomorphism 49

2.1 Conjugacy classes of the symmetric groups 50
2.2 The five bases of the algebra of symmetric functions 54
2.3 The structure of graded self-adjoint Hopf algebra 69
2.4 The Frobenius–Schur isomorphism 78
2.5 The Schur–Weyl duality 87

3 Combinatorics of partitions and tableaux 99

3.1 Pieri rules and Murnaghan–Nakayama formula 99
3.2 The Robinson–Schensted–Knuth algorithm 108
3.3 Construction of the irreducible representations 131
3.4 The hook-length formula 140

II Hecke algebras and their representations 147

4 Hecke algebras and the Brauer–Cartan theory 149

4.1 Coxeter presentation of symmetric groups 151
4.2 Representation theory of algebras 161
4.3 Brauer–Cartan deformation theory 173
4.4 Structure of generic and specialized Hecke algebras 183
4.5 Polynomial construction of the q-Specht modules 207

5 Characters and dualities for Hecke algebras 217

5.1 Quantum groups and their Hopf algebra structure 218
5.2 Representation theory of the quantum groups 230
5.3 Jimbo–Schur–Weyl duality 252

5.4 Iwahori–Hecke duality 263
5.5 Hall–Littlewood polynomials and characters of Hecke algebras 272

6 Representations of the Hecke algebras specialized at $q = 0$ 287
6.1 Non-commutative symmetric functions 289
6.2 Quasi-symmetric functions 299
6.3 The Hecke–Frobenius–Schur isomorphisms 306

III Observables of partitions 325

7 The Ivanov–Kerov algebra of observables 327
7.1 The algebra of partial permutations 328
7.2 Coordinates of Young diagrams and their moments 339
7.3 Change of basis in the algebra of observables 347
7.4 Observables and topology of Young diagrams 354

8 The Jucys–Murphy elements 375
8.1 The Gelfand–Tsetlin subalgebra of the symmetric group algebra 375
8.2 Jucys–Murphy elements acting on the Gelfand–Tsetlin basis . . 387
8.3 Observables as symmetric functions of the contents 396

9 Symmetric groups and free probability 401
9.1 Introduction to free probability 402
9.2 Free cumulants of Young diagrams 418
9.3 Transition measures and Jucys–Murphy elements 426
9.4 The algebra of admissible set partitions 431

10 The Stanley–Féray formula for characters and Kerov polynomials 451
10.1 New observables of Young diagrams 451
10.2 The Stanley–Féray formula for characters of symmetric groups 464
10.3 Combinatorics of the Kerov polynomials 479

IV Models of random Young diagrams 499

11 Representations of the infinite symmetric group 501
11.1 Harmonic analysis on the Young graph and extremal characters 502
11.2 The bi-infinite symmetric group and the Olshanski semigroup . 511
11.3 Classification of the admissible representations 527
11.4 Spherical representations and the Gelfand-Naimark-Segal construction 538

12 Asymptotics of central measures 547
12.1 Free quasi-symmetric functions 548
12.2 Combinatorics of central measures 562
12.3 Gaussian behavior of the observables 576

13 Asymptotics of Plancherel and Schur–Weyl measures **595**
13.1 The Plancherel and Schur–Weyl models 596
13.2 Limit shapes of large random Young diagrams 602
13.3 Kerov's central limit theorem for characters 614

Appendix 629

Appendix A Representation theory of semisimple Lie algebras **631**
A.1 Nilpotent, solvable and semisimple algebras 631
A.2 Root system of a semisimple complex algebra 635
A.3 The highest weight theory 641

References **649**

Index **661**

编辑手记 **667**

Preface

The objective of this book is to propose a modern introduction to the representation theory of the symmetric groups. There is now a large literature on the general representation theory of finite groups, see for instance the classical *Linear Representations of Finite Groups* by J.-P. Serre ([Ser77]); and among this literature, a few books are concentrated on the case of symmetric groups, for example *The Symmetric Group: Representations, Combinatorial Algorithms and Symmetric Functions* by B. Sagan (see [Sag01]). The point of view and interest of the present book is the following: we shall show that most of the calculations on symmetric groups can be performed, or at least eased by using some appropriate *algebras of functions*. It is well known since the works of Frobenius and Schur that the algebra of symmetric functions encodes most of the theory of characters of symmetric groups. In this book, we shall use the algebra of symmetric functions *as the starting point* of the representation theory of symmetric groups, and then go forward by introducing other interesting algebras, such as:

- the algebra of *observables of partitions*, originally called "polynomial functions on Young diagrams," and whose construction is due to Kerov and Olshanski.
- the Hopf algebras of *non-commutative symmetric functions*, *quasi-symmetric functions* and *free quasi-symmetric functions*, which contain and generalize the algebra of symmetric functions.

This algebraic approach to the representation theory of symmetric groups can be opposed to a more traditional approach which is of combinatorial nature, and which gives a large role to the famous Young tableaux. The approach with algebras of functions has several advantages:

1. First, if one tries to replace the symmetric group by finite-dimensional algebras related to it (the so-called partition algebras, or the Hecke algebras), then one can still use the algebra of symmetric functions to treat the character theory of these algebras, and in this setting, most of the results related to the symmetric groups have direct analogues. In this book, we shall treat the case of Hecke algebras, which is a good example of this kind of extension of the theory of symmetric groups (the case of partition algebras is treated for instance in a recent book by Ceccherini-Silberstein, Scarabotti and Tolli, see [CSST10]).
2. On the other hand, the algebraic approach leads to a new formula for the irreducible characters of the symmetric groups, due to Stanley and Féray. The

combinatorics underlying this formula are related to several interesting topics, such as free probability theory, or the theory of Riemann surfaces and maps drawn on them.

3. Finally, the approach with algebras is adequate to deal with *asymptotic* representation theory, that is to say representations of symmetric groups $\mathfrak{S}(n)$ with n large, going to infinity. In this setting, a natural question is: what are the typical properties of a representation of $\mathfrak{S}(n)$ with n large, and in particular what is the decomposition of such a large representation in irreducible components? Since the irreducible representations of $\mathfrak{S}(n)$ are labeled by integer partitions of size n, this question leads to the study of certain models of random partitions, in particular the so-called Plancherel measures. There, the algebra of observables of partitions will prove a very powerful tool.

Besides, our approach enables us to present in a book the theory of *combinatorial Hopf algebras*, which is nowadays a quite active field of research in algebraic combinatorics.

Let us now detail more precisely the content of the book, which is split into four parts:

▷ *Part I: Symmetric groups and symmetric functions.*

The first part of the book is devoted to a presentation of the classical theory of representations of symmetric groups, due mainly to Frobenius, Schur and Young. In Chapter 1, we explain the representation theory of finite groups and finite-dimensional semisimple algebras, thereby bringing most of the prerequisites to the reading of the book. One thing that we shall try to do in each chapter is to obtain a big "black box theorem," which summarizes most of the results and allows one to recover at once the remainder of the theory. For the general theory of linear representations of finite groups, one such summarizing statement is the fact that the non-commutative Fourier transform of finite groups is an isomorphism of algebras, of Hilbert spaces and of bimodules (Theorem 1.14). An analogous result holds for finite-dimensional semisimple algebras, the language of algebras and modules being a bit more flexible than the language of groups and representations.

In Chapter 2, we introduce the Hopf algebra of symmetric functions Sym, and we show that the Schur functions correspond to the irreducible representations of the symmetric groups: thus, Sym is isomorphic to the Grothendieck ring formed by these representations (Theorem 2.31). This theorem due to Frobenius and Schur can be used as a starting point to the combinatorics of representations, which are developed in Chapter 3 and rely on Young tableaux, that is numberings of Young diagrams of integer partitions. Two other building blocks of this deep combinatorial theory are the Schur–Weyl duality (Section 2.5), which relates the representations of $\mathfrak{S}(n)$ to the representations of the general linear groups $\mathrm{GL}(N)$;

and the Robinson–Schensted–Knuth algorithm (cf. Section 3.2), which connects the Young tableaux to words or permutations. These two tools will have a pervasive use throughout the book.

▷ *Part II: Hecke algebras and their representations.*

In the second part, we explain how one can extend the theory of symmetric groups to other related combinatorial algebras, namely, the so-called Iwahori–Hecke algebras. These algebras are continuous deformations $\mathfrak{H}_z(n)$ of the group algebras $\mathbb{C}\mathfrak{S}(n)$, the parameter z being allowed to take any value in $\mathbb{C}$; one recovers $\mathbb{C}\mathfrak{S}(n)$ when $z = 1$. In Chapter 4, we show that for almost any value of z, $\mathfrak{H}_z(n)$ is isomorphic to $\mathbb{C}\mathfrak{S}(n)$ and has the same representation theory: its irreducible modules S_z^λ are again labeled by integer partitions of size n, and they have the same dimension as the irreducible representations S^λ of $\mathfrak{S}(n)$ (Theorem 4.67). This chapter can be considered as an introduction to modular representation theory that is focused on a specific example. In Chapter 5, we compute the characters of the Hecke algebras in the generic case, by using an extension of Schur–Weyl duality, in which symmetric groups are replaced by Hecke algebras and linear groups are replaced by quantum groups. We obtain a formula that generalizes the Frobenius–Schur formula and involves the Hall–Littlewood symmetric functions (see Theorem 5.49). In Chapter 6, we consider the case $z = 0$, which is not generic and does not yield a semisimple algebra. In this setting, one can still use combinatorial Hopf algebras to describe the representations of $\mathfrak{H}_0(n)$ (see Theorem 6.18): the algebra of noncommutative symmetric functions NCSym, and the algebra of quasi-symmetric functions QSym, which are in duality. Thus, the extension of the representation theory of symmetric groups to the case of Hecke algebras leads quite naturally to an extension of the theory of symmetric functions to more general functions, which will also appear later in the book (Chapters 10 and 12).

▷ *Part III: Observables of partitions.*

The third part of the book is devoted to what is now known as the *dual combinatorics* of the characters of the symmetric groups. In the first part of the book, the characters of the symmetric groups are introduced as functions $\mathrm{ch}^\lambda : \mathfrak{S}(n) \to \mathbb{C}$ or $\mathfrak{Y}(n) \to \mathbb{C}$ that are labeled by integer partitions λ of size n, and that can be computed with the help of the Frobenius–Schur formula:

$$\mathrm{ch}^\lambda(\mu) = \langle s_\lambda \mid p_\mu \rangle,$$

where $\langle \cdot \mid \cdot \rangle$ is the Hall scalar product on the algebra of symmetric functions Sym. However, one can also consider the quantity $\mathrm{ch}^\lambda(\mu)$ as a function of λ labeled by the conjugacy class $\mu \in \mathfrak{Y}(n)$. This point of view leads one to consider functions of irreducible representations of symmetric groups, and to introduce an algebra $\mathscr{O}$ formed by these functions, which we call the algebra of observables of partitions. Our Chapter 7 presents this algebra and several bases of it, and it explains how the character ch^λ of the symmetric groups is related to the geometry of the Young diagram of the integer partition λ (see in particular Theorems 7.13 and 7.25). In

Chapters 8 and 9, we introduce other observables of partitions, related to the so-called Jucys–Murphy elements or to the theory of free probability. In particular, we present an important algebraic basis $(R_k)_{k\geq 2}$ of $\mathscr{O}$, whose elements are called free cumulants, and whose combinatorics are related to constructions on set partitions and to maps on surfaces. Chapter 10 explores the interactions between the basis of free cumulants $(R_k)_{k\geq 2}$, and the basis of renormalized character values $(\Sigma_k)_{k\geq 1}$ in $\mathscr{O}$. This study relies on a new formula for the characters of the symmetric groups (Theorem 10.11):

$$\mathrm{ch}^{\lambda}(\mu) = \frac{\dim S^{\lambda}}{|\lambda|(|\lambda|-1)\cdots(|\lambda|-|\mu|+1)} \sum_{\rho_\mu=\sigma\tau} \varepsilon(\tau) N^{\sigma,\tau}(\lambda),$$

where the sum runs over factorizations of a permutation ρ_μ with cycle type μ, and where the quantities $N^{\sigma,\tau}(\lambda)$ count certain numberings of the cells of the Young diagram λ. Thus, if instead of Sym one uses the combinatorial algebra $\mathscr{O}$ as the starting point of the representation theory of the symmetric groups, then one gets another totally different formula for the irreducible characters, though to be precise the Stanley–Féray formula sits in a larger algebra $\mathscr{Q} \supset \mathscr{O}$. A careful analysis of this formula leads to an explicit change of basis formula between the symbols R_k and the symbols Σ_k; see Theorem 10.20, which explains how to compute the coefficients of the Kerov polynomials.

▷ *Part IV: Models of random Young diagrams.*

In the last part of the book, we use the results of the previous chapters in order to describe the properties of the representations of large symmetric groups. In Chapter 11, we start with a classification of the extremal characters of the infinite symmetric groups $\mathfrak{S}(\infty)$ (Theorem 11.31). They play with respect to $\mathfrak{S}(\infty)$ a role similar to the irreducible characters of the finite symmetric groups $\mathfrak{S}(n)$, and they allow one to consider coherent families $(\tau_n)_{n\in\mathbb{N}}$ of representations or more generally of traces of these finite groups. The classification involves an infinite-dimensional convex compact space known as the Thoma simplex. For any parameter $t \in T$ in this simplex, one can consider traces $\tau_{t,n}$ on the symmetric groups $\mathfrak{S}(n)$, whose decompositions in irreducible characters yield probability measures $\mathbb{P}_{t,n}$ on the sets $\mathfrak{Y}(n)$ of integer partitions of size n. Thus, the representation theory of $\mathfrak{S}(\infty)$ leads one to study random models of partitions, and this study is performed in Chapters 12 and 13.

In Chapter 12, we show that every family of measures $(\mathbb{P}_{t,n})_{n\in\mathbb{N}}$ (the so-called central measures) satisfies a law of large numbers (Theorem 12.19) and a central limit theorem (Theorem 12.30). To this purpose, we introduce a new combinatorial Hopf algebra FQSym which extends both NCSym and QSym; and a method of joint cumulants of random variables that mixes well with the theory of observables of partitions. In Chapter 13, we study the particular case of Plancherel and Schur–Weyl measures, which have degenerate asymptotics in comparison to the other central measures, and which on the other hand allow one to solve the problem of the longest increasing subsequences in uniform random permutations or uniform random words (Theorem 13.10).

The target audience of this book consists mainly of graduate students and researchers. We tried to make the presentation as self-contained as possible, but there remain inevitably certain prerequisites to the reading. Thus, the reader is supposed to have a good familiarity with the basics of algebra (algebraic structures and related constructions) and of combinatorics (counting arguments, bijections); in the last part of the book, we shall also use arguments from probability theory. One prerequisite that helps understanding certain results and that we did not take for granted is the theory of representations of classical Lie algebras; therefore, Appendix Appendix A is devoted to a short presentation (without proof) of this theory. To be honest, there may be some inconsistencies in the prerequisites that we suppose: for instance, we start the book by recalling what is a group, but later we freely use the language of equivalence of categories. We hope that the long bibliography given at the end of the book will smooth a bit the peaks in difficulty that the reader might encounter.

Regarding the bibliography, each chapter is followed by a section called "Notes and references," where we explain precisely which sources we used in order to write the book. All the credit is due to the authors that are cited in these special sections, and we tried sincerely not to forget anyone, and to attribute each result to the right mathematician (this task can sometimes be very difficult to accomplish). Nonetheless, a few results in this book have proofs that are (to our knowledge) either new, or unpublished until now, or very difficult to find in the literature; this is also explained in the notes and references. We also used these special sections to detail some results that we did not have the courage to treat, but that we still wanted to appear in the book.

▷ *Acknowledgments.*

This book has been built from the contents of my PhD thesis, and from notes of lecture courses that I taught at the University of Zürich in 2012–2013. I am very thankful to my PhD director Philippe Biane for introducing me to the subject of asymptotic representation theory, which is one of the main topics of this work. During the years of preparation of my PhD thesis, I also benefited from the expertise of Jean-Yves Thibon, Jean-Christophe Novelli, Florent Hivert and Alain Lascoux; they introduced me to the theory of combinatorial Hopf algebras, and they showed me how to use them in order to solve many difficult computations.

I am much indebted to my colleague Valentin Féray, who explained to me several points of the theory of Kerov polynomials which he developed with Piotr Sniady and Maciej Dołęga; the discussions that we have are always enlightening. I am grateful to Reda Chhaibi for his explanations on the weight theory and the

Littelmann path theory of Lie groups and algebras, and for his comments on an early version of the manuscript. Many thanks are also due to Ashkan Nikeghbali, who has a profound influence on the mathematics that I am doing, and invited me numerous times to Zürich.

I thank Miklós Bóna for proposing that I write this book, and Bob Ross and José Soto at CRC Press for their assistance and their patience with respect to the numerous small delays that the writing of such a long book caused. I am also very grateful to Karen Simon for supervising the many necessary corrections.

Finally, my greatest thanks go to my family, and especially my fiancee Véronique who is a constant support and source of inspiration.

Pierre-Loïc Méliot

Part I

Symmetric groups and symmetric functions

1

Representations of finite groups and semisimple algebras

In this first chapter, we present the general representation theory of finite groups. After an exposition of Maschke's theorem of complete reducibility of representations (Section 1.1) and of Schur's lemma of orthogonality of characters (Section 1.2), we construct the non-commutative Fourier transform (Section 1.3), which provides a decomposition of the complex group algebra $\mathbb{C}G$ in blocks of endomorphism rings of the irreducible representations of G. It implies that any function $f : G \to \mathbb{C}$ can be expanded uniquely as a linear combination of the matrix coefficients of the irreducible representations of G (Proposition 1.15). This can be seen as a motivation for the study of representations of groups, and on the other hand, the Fourier isomorphism

$$\mathbb{C}G \to \bigoplus_{\lambda \in \widehat{G}} \mathrm{End}(V^\lambda)$$

can be generalized to the case of complex semisimple algebras. This language and theory of algebras and modules is in many situations more flexible than the language of groups and representations, and we devote Section 1.4 to the extension of the theory of representations to this setting. In Section 1.5, this extension allows us to detail the double commutant theory, of which the Frobenius–Schur formula for characters of symmetric groups (see Chapter 2) will be an instance. In the second part of the book, we shall explain the representation theory of some combinatorial algebras that are deformations of the symmetric group algebra $\mathbb{C}\mathfrak{S}(n)$; there, the knowledge of the representation theory of semisimple algebras will also prove necessary.

1.1 Finite groups and their representations

▷ *Finite groups.*

We assume the reader to be familiar with the notions of groups, rings, fields, vector spaces and algebras. Thus, recall that a **group** is a set G endowed with an operation $\cdot_G : G \times G \to G$ (the product of the group), such that

(G1) $\cdot_G$ is associative and admits a neutral element:

$$\forall g,h,i \in G,\ (g \cdot_G h) \cdot_G i = g \cdot_G (h \cdot_G i);$$
$$\exists e \in G,\ \forall g \in G,\ g \cdot_G e = e \cdot_G g = g.$$

The neutral element e is then unique.

(G2) every element of G has a (unique) inverse for the product:

$$\forall g \in G,\ \exists h \in G,\ g \cdot_G h = h \cdot_G g = e.$$

We shall usually omit the notation $\cdot_G$, and just denote $g \cdot_G h = gh$. Also, the inverse of $g \in G$ will be denoted g^{-1}, and the neutral element will be denoted indifferently e_G, e or 1. A group will be called finite if as a set it has finite cardinality. We then write

$$|G| = \mathrm{card}\, G = \text{number of elements of } |G|.$$

Example. Let p be a prime number, and denote $\mathbb{Z}/p\mathbb{Z} = \{[1],[2],\ldots,[p]\}$ the set of classes of integers modulo p, which is a ring (quotient of the ring of integers $\mathbb{Z}$). Endowed with the product of classes

$$[a] \times [b] = [ab],$$

the set $(\mathbb{Z}/p\mathbb{Z})^* = \{[1],[2],\ldots,[p-1]\}$ is a finite group of cardinality $p-1$, with neutral element $e = [1]$. It is **commutative**, which means that for any $g,h \in G$, $gh = hg$.

Example. If S is a set, denote $\mathfrak{S}(S)$ the set of maps $\sigma : S \to S$ that are bijective. This is a group with respect to the operation of composition of maps; the neutral element is the identity

$$\mathrm{id}_S : s \in S \mapsto s,$$

and the inverse of a bijection $\sigma \in \mathfrak{S}(S)$ is the inverse function σ^{-1} with $t = \sigma(s)$ if and only if $s = \sigma^{-1}(t)$. This book is devoted to the study of the groups $\mathfrak{S}(n) = \mathfrak{S}([\![1,n]\!])$, where $[\![1,n]\!]$ is the set of integers $\{1,2,3,\ldots,n-1,n\}$ between 1 and n. We shall say a bit more about them in a moment.

A **morphism** between two groups G and H is a map $\phi : G \to H$ compatible with the products of G and H, i.e., such that

$$\phi(g_1 g_2) = \phi(g_1)\,\phi(g_2)$$

for all $g_1, g_2 \in G$. One speaks of **isomorphism** of groups if ϕ is bijective; then, the inverse map ϕ^{-1} is also a morphism of groups. On the other hand, a **subgroup** H of a group G is a subset of G stable by the operations of product and inverse:

$$\forall h_1,h_2 \in H,\ h_1h_2 \in H \quad ; \quad \forall h \in H,\ h^{-1} \in H.$$

Then, H is a group for the restriction of the product map from $G \times G$ to $H \times H$. In the following, we shall say that a group H *can be seen as a subgroup of* G if there is an injective morphism of groups $\phi : H \to G$, which thus identifies H with a subgroup of G.

Example. Let k be a field, and V be a k-vector space. Then $\mathrm{GL}(V)$, the set of bijective linear maps $\phi : V \to V$, is a group for the operation of composition of maps. Similarly, if $k = \mathbb{C}$ and V is a complex vector space endowed with a scalar product, then the set $\mathrm{U}(V)$ of linear isometries of V is a group for the composition of maps, and it is a subgroup of $\mathrm{GL}(V)$. Going to the matrix point of view, the following sets of matrices are also groups for the product of matrices, the neutral element being the identity matrix $I_n = \mathrm{diag}(1, 1, \ldots, 1)$:

$$\begin{aligned}\mathrm{GL}(n,k) &= \{M \in \mathrm{M}(n,k) \mid \det(M) \neq 0\};\\ \mathrm{SL}(n,k) &= \{M \in \mathrm{M}(n,k) \mid \det(M) = 1\};\\ \mathrm{U}(n,\mathbb{C}) &= \{M \in \mathrm{M}(n,\mathbb{C}) \mid M^*M = MM^* = I_n\}.\end{aligned}$$

If V is a complex vector space of dimension n, then the groups of matrices $\mathrm{GL}(n,\mathbb{C})$, $\mathrm{SL}(n,\mathbb{C})$ and $\mathrm{U}(n,\mathbb{C})$ are isomorphic respectively to $\mathrm{GL}(V)$, $\mathrm{SL}(V)$ and $\mathrm{U}(V)$, the isomorphism being the map which sends a linear map to its matrix in a (unitary) basis. On the other hand, for $\mathrm{GL}(n,k)$ to be finite, we need k to be finite, and then, if $q = \mathrm{card}\, k$, one has

$$\mathrm{card}\,\mathrm{GL}(n,k) = (q^n - 1)(q^n - q)(q^n - q^2)\ldots(q^n - q^{n-1}).$$

Indeed, this is the number of distinct bases of k^n to which an arbitrary basis $(e_1, \ldots, e_n)$ can be sent by an element of $\mathrm{GL}(n,k) = \mathrm{GL}(k^n)$.

▷ *Symmetric groups.*

The **symmetric group** of order n is the group of bijections $\mathfrak{S}(n) = \mathfrak{S}([\![1,n]\!])$. It is a finite group with cardinality

$$|\mathfrak{S}(n)| = n! = 1 \times 2 \times 3 \times \cdots \times n = \prod_{i=1}^{n} i.$$

Indeed, to choose a bijection σ between elements of $[\![1,n]\!]$, one has:

- n possibilities for the image $\sigma(1)$ of 1 (all the integers between 1 and n);
- $n-1$ possibilities for the image $\sigma(2)$ of 2 (all the integers but the one already chosen for $\sigma(1)$);
- in general, assuming the images $\sigma(1), \ldots, \sigma(k)$ already chosen, $n-k$ possibilities for the image $\sigma(k+1)$ of $k+1$.

Multiplying these choices yields $|\mathfrak{S}(n)| = n!$. The elements of $\mathfrak{S}(n)$ are called **permutations of size** n, and we shall denote $\sigma = \sigma(1)\sigma(2)\ldots\sigma(n)$ a permutation given by the list of its values. So for instance, 4132 is the permutation in $\mathfrak{S}(4)$ that sends 1 to 4, 2 to 1, 3 to 3 and 4 to 2.

Let us now list some easy and well-known properties of these symmetric groups:

1. If $n \leq N$, then $\mathfrak{S}(n)$ can be seen naturally as a subgroup of $\mathfrak{S}(N)$. Indeed, a bijection σ between the n first integers can be extended in a bijection $\widetilde{\sigma}$ between the N first integers by setting:
$$\widetilde{\sigma}(k) = \begin{cases} \sigma(k) & \text{if } k \leq n, \\ k & \text{if } n < k \leq N. \end{cases}$$
In the sequel, we shall use these natural imbeddings $\mathfrak{S}(n) \hookrightarrow \mathfrak{S}(N)$ constantly, and unless the distinction is needed, we shall keep the same notation for a bijection $\sigma \in \mathfrak{S}(n)$ and its extension to a larger symmetric group $\mathfrak{S}(N)$.

2. For $n \geq 3$, $\mathfrak{S}(n)$ is a **non-commutative** group, which means that one can find g and h such that $gh \neq hg$. Indeed, using the previous property, it suffices to prove the case $n = 3$, and in this case, if $\sigma = 321$ and $\tau = 213$, then their composition products are
$$\sigma\tau = 231 \quad ; \quad \tau\sigma = 312$$
and they are different.

3. Any finite group can be seen as a subgroup of a finite symmetric group. Indeed, given a finite group G, consider the map
$$\begin{aligned} \phi : G &\to \mathfrak{S}(G) \\ g &\mapsto (\phi(g) : h \mapsto gh). \end{aligned}$$
This is a morphism of groups, which is injective since g can be recovered from the map $\phi(g)$ by the formula $g = \phi(g)(e)$. On the other hand, given two finite sets A and B with the same cardinality and a bijection $\psi : A \to B$, there is an isomorphism between the groups $\mathfrak{S}(A)$ and $\mathfrak{S}(B)$, namely,
$$\begin{aligned} \Psi : \mathfrak{S}(A) &\to \mathfrak{S}(B) \\ \sigma &\mapsto \psi \circ \sigma \circ \psi^{-1}. \end{aligned}$$
Thus, if $n = \operatorname{card} G$, then $\mathfrak{S}(G)$ and $\mathfrak{S}(n)$ are isomorphic, so G can be seen as a subgroup of $\mathfrak{S}(n)$.

A more crucial property of permutations deserves the following proposition. Call **cycle of length k and support** $(a_1, a_2, \ldots, a_k)$ the permutation that sends a_1 to a_2, a_2 to a_3, a_3 to a_4, etc., and a_k to a_1; and that leaves invariant all the other elements of $[\![1, n]\!]$. For instance, the cycle $(1, 4, 2)$ in $\mathfrak{S}(4)$ sends 1 to 4, 4 to 2, 2 to 1, and the remaining element 3 to itself; thus, $(1, 4, 2) = 4132$. If $k \geq 2$, then a cycle of length k and support $(a_1, \ldots, a_k)$ is uniquely determined by the sequence $(a_1, a_2, \ldots, a_k)$, up to a cyclic permutation of this sequence:
$$(a_1, a_2, \ldots, a_k) = (a_2, a_3, \ldots, a_k, a_1) = (a_3, \ldots, a_k, a_1, a_2) = \cdots$$
On the other hand, a cycle of length 1 is just the identity permutation, and can be seen as a way to design a particular fixed point (later, we shall make this idea of marked fixed point more rigorous with the notion of partial permutation).

Proposition 1.1. *Any permutation $\sigma \in \mathfrak{S}(n)$ can be written as a product of cycles with disjoint supports, the sum of the lengths of these cycles being equal to n. This decomposition*

$$\sigma = c_1 \circ c_2 \circ \cdots \circ c_r$$

is unique up to permutation of the cycles $c_1, \ldots, c_r$.

Proof. In a finite group, every element g has for inverse a positive power of itself $g^{k \geq 1}$. As a consequence, the permutation $\sigma \in \mathfrak{S}(n)$ being fixed, the relation on $[\![1,n]\!]$ defined by

$$i \sim_\sigma j \iff \exists k \geq 0,\ j = \sigma^k(i),$$

which is clearly reflexive and transitive, is also symmetric, so it is an equivalence relation. Call **orbit** of σ a class for the equivalence relation $\sim_\sigma$ on $[\![1,n]\!]$; then, the orbit of i, if it has length k, is $\{i, \sigma(i), \sigma^2(i), \ldots, \sigma^{k-1}(i)\}$, and the restriction of σ to this orbit is the cycle

$$c = (i, \sigma(i), \sigma^2(i), \ldots, \sigma^{k-1}(i)).$$

The decomposition of σ in disjoint cycles is then obtained by choosing one representative for each orbit, and the unicity comes from the fact that if $\sigma = c_1 \circ c_2 \circ \cdots \circ c_r$ is a product of cycles with disjoint supports, then these supports are orbits of σ, with the order of elements for each cycle entirely determined by the action of σ on each support. □

Example. Consider the permutation $\sigma = 874312659$ in $\mathfrak{S}(9)$. Its orbits are $\{1,5,8\}$, $\{2,6,7\}$, $\{3,4\}$ and $\{9\}$, and the cycle decomposition of σ is

$$\sigma = (1,8,5)(2,7,6)(3,4)(9).$$

Thus, we get two different writings for a given permutation $\sigma \in \mathfrak{S}(n)$: the notation in line $\sigma = \sigma(1)\sigma(2)\ldots\sigma(n)$, and the cycle decomposition

$$\sigma = (a_1, \ldots, a_r)(b_1, \ldots, b_s) \cdots (z_1, \ldots, z_t).$$

In the cycle decomposition, it will sometimes be convenient to omit the cycles of length 1, since they correspond to the identity. This is in particular the case if σ is itself a single cycle. We say that σ is a **transposition** if it is a cycle of length 2; then, it writes as $\sigma = (i,j) = (j,i)$ and it exchanges i and j.

▷ *Representations of groups.*

If V is a complex vector space, we denote as before $\mathrm{GL}(V)$ the group of complex linear isomorphisms $u : V \to V$. If V is finite-dimensional and if $(e_1, \ldots, e_n)$ is a fixed linear basis of V, we denote $(u_{ij})_{1 \leq i,j \leq n}$ the matrix of the linear map u in this basis, which means that

$$u(e_j) = \sum_{i=1}^{n} u_{ij}\, e_i.$$

Then, the map $\psi : u \in \mathrm{GL}(V) \mapsto (u_{ij})_{1\leq i,j\leq n} \in \mathrm{GL}(n,\mathbb{C})$ is an isomorphism of groups.

Definition 1.2. *A (complex, linear)* ***representation*** *of a group G is given by a complex vector space V, and a morphism of groups*

$$\rho : G \to \mathrm{GL}(V).$$

We shall always assume the space V to be finite-dimensional, and we shall denote $(\rho_{ij}(g))_{1\leq i,j\leq \dim V}$ the matrix of $\rho(g)$ in a fixed basis of the representation. For every $g \in G$, $\rho(g)$ is a linear isomorphism of V, and we can make it act on vectors $v \in V$. Thus, we shall frequently manipulate vectors

$$(\rho(g))(v) \in V \text{ with } g \in G \text{ and } v \in V,$$

and if the representation (V,ρ) is fixed, we shall abbreviate $(\rho(g))(v) = g \cdot v$. Then, to describe a representation of G amounts to giving a formula for $g \cdot v$, with the condition that

$$g \cdot (h \cdot v) = (gh) \cdot v$$

for any $g, h \in G$ and any $v \in V$. Notice then that for any $v \in V$, $1 \cdot v = v$ if 1 denotes the neutral element of G.

Example. For any group G, one has the so-called **trivial representation** of G on $V = \mathbb{C}$, given by $g \cdot v = v$ for any $g \in G$ and any $v \in V$.

Example. Fix a positive integer n, and consider the **permutation representation** of $\mathfrak{S}(n)$ on $\mathbb{C}^n$, given by

$$\sigma \cdot (x_1, \ldots, x_n) = (x_{\sigma^{-1}(1)}, \ldots, x_{\sigma^{-1}(n)}).$$

This is indeed a representation, since

$$\begin{aligned}\sigma \cdot (\tau \cdot (x_1, \ldots, x_n)) &= \sigma \cdot (x_{\tau^{-1}(1)}, \ldots, x_{\tau^{-1}(n)}) \\ &= (x_{\tau^{-1}\sigma^{-1}(1)}, \ldots, x_{\tau^{-1}\sigma^{-1}(n)}) \\ &= (x_{(\sigma\tau)^{-1}(1)}, \ldots, x_{(\sigma\tau)^{-1}(n)}) = (\sigma\tau) \cdot (x_1, \ldots, x_n).\end{aligned}$$

The matrix of $\rho(\sigma)$ in the canonical basis of $\mathbb{C}^n$ is the permutation matrix $(\delta_{i,\sigma(j)})_{1\leq i,j\leq n}$, where $\delta_{a,b}$ denotes the Dirac function, equal to 1 if $a = b$ and to 0 otherwise (this notation will be used throughout the whole book).

Example. Let G be a finite group. We denote $\mathbb{C}G$ the vector space of functions from G to $\mathbb{C}$, and we identify a function f with the formal linear sum

$$f = \sum_{g \in G} f(g)\, g.$$

So for instance, if $G = \mathfrak{S}(3) = \{123, 132, 213, 231, 312, 321\}$, then

$$2(123) - (213) + (1 + \mathrm{i})(321)$$

represents the function which sends 123 to 2, 213 to -1, 321 to $1+\mathrm{i}$, and the other permutations in $\mathfrak{S}(3)$ to 0. With these notations, a basis of $\mathbb{C}G$ is G, an element $g \in G$ being identified with the Dirac function δ_g.

The (left) **regular representation of** G is the representation with space $V = \mathbb{C}G$, and with $g \cdot f(h) = f(g^{-1}h)$, which writes more easily as

$$g \cdot \left(\sum_{h \in G} f(h)\, h \right) = \sum_{h \in G} f(h)\, gh.$$

The morphism underlying this regular representation is the composition of the morphism $G \to \mathfrak{S}(G)$ described in the previous paragraph, and of the permutation representation $\mathfrak{S}(G) \to \mathrm{GL}(\mathbb{C}G)$.

▷ *Irreducible representations and Maschke's theorem.*

A **subrepresentation** of a representation (V, ρ) is a vector subspace $W \subset V$ that is stable by the action of G, which means that

$$\forall g \in G,\ \ \forall v \in W,\ \ g \cdot v \in W.$$

Then, W is a representation of G for the new morphism $\rho_{|W}(g) = (\rho(g))_{|W}$. A representation (V, ρ) of G is said to be **irreducible** if it has positive dimension and if there is no stable subspace (subrepresentation) $W \subset V$ with $W \neq \{0\}$ and $W \neq V$.

As we shall see at the end of this paragraph, any representation of a finite group can be split into smaller irreducible representations. Let us first detail the notions of **morphism of representations** and of **direct sum of representations**. If V_1 and V_2 are two representations of G, then their direct sum is the representation of G with underlying vector space $V_1 \oplus V_2$, and with

$$g \cdot (v_1 + v_2) = g \cdot v_1 + g \cdot v_2$$

for any $g \in G$ and any $(v_1, v_2) \in V_1 \times V_2$. On the other hand, given again two representations V_1 and V_2, a morphism of representations between V_1 and V_2 is a linear map $\phi : V_1 \to V_2$ such that

$$\phi(g \cdot v) = g \cdot \phi(v)$$

for any $g \in G$ and any $v \in V$. Thus, for any element of the group, the following diagram of linear maps is commutative:

$$\begin{array}{ccc} V_1 & \xrightarrow{\rho_1(g)} & V_1 \\ {\scriptstyle \phi}\downarrow & & \downarrow{\scriptstyle \phi} \\ V_2 & \xrightarrow[\rho_2(g)]{} & V_2 \end{array}$$

Example. Consider the permutation representation of $\mathfrak{S}(n)$ on $\mathbb{C}^n$. It admits as stable subspaces

$$V_1 = \{(x_1, \ldots, x_n) \in \mathbb{C}^n \mid x_1 + \cdots + x_n = 0\};$$
$$V_2 = \mathbb{C}(1, 1, \ldots, 1),$$

and $\mathbb{C}^n$ is the direct sum of these two representations: $\mathbb{C}^n = V_1 \oplus V_2$. It is clear for dimension reasons that V_2 is irreducible; we shall see later that V_1 is also irreducible. Notice that V_2 is isomorphic to the trivial representation of $\mathfrak{S}(n)$, the isomorphism being given by $\lambda(1, 1, \ldots, 1) \mapsto \lambda$.

Theorem 1.3 (Maschke). *Let G be a finite group and V be a (finite-dimensional) representation of G.*

1. *There exists a decomposition of V as a direct sum of irreducible representations of G:*
$$V = \bigoplus_{i=1}^{r} V_i, \quad \text{with each } V_i \text{ irreducible representation of } G.$$

2. *Fix an irreducible representation I of G. The number of components V_i of V that are isomorphic to I is independent of the decomposition of V in irreducible representations. Moreover, the regular representation $\mathbb{C}G$ of G has* $\dim I$ *components isomorphic to I.*

Before we prove it, let us restate in a clearer way the consequences of Theorem 1.3. There exists a decomposition of the regular representation

$$\mathbb{C}G = \bigoplus_{\lambda \in \widehat{G}} d_\lambda V^\lambda,$$

where $\widehat{G}$ is a finite set; each V^λ is an irreducible representation appearing with multiplicity $d_\lambda = \dim V^\lambda$; and two representations V^λ and V^μ with $\lambda \neq \mu$ are non-isomorphic. Then, every other representation V of G writes up to an isomorphism of representations as

$$V = \bigoplus_{\lambda \in \widehat{G}} m_\lambda V^\lambda,$$

with the multiplicities $m_\lambda \in \mathbb{N}$ uniquely determined by V.

The proof of Theorem 1.3 relies on the two following lemmas:

Lemma 1.4. *Let (V, ρ) be a representation of a finite group G. There exists a scalar product $\langle \cdot \mid \cdot \rangle$ on V such that $\rho(g) \in \mathrm{U}(V)$ for any $g \in G$:*

$$\forall v_1, v_2 \in V, \ \langle g \cdot v_1 \mid g \cdot v_2 \rangle = \langle v_1 \mid v_2 \rangle .$$

Remark. In this book, every instance of a Hermitian scalar product $\langle \cdot \mid \cdot \rangle$ will be antilinear in the first variable, and linear in the second variable. Thus, if v and w are in V and $a, b \in \mathbb{C}$, then

$$\langle av \mid bw \rangle = \overline{a}b \, \langle v \mid w \rangle .$$

Proof. We start with an arbitrary scalar product $(\cdot|\cdot)$ on V, and consider the new scalar product

$$\langle v_1 \mid v_2 \rangle = \sum_{g \in G} (g \cdot v_1 | g \cdot v_2).$$

Then, $\langle \cdot \mid \cdot \rangle$ is obviously again a scalar product, and

$$\langle h \cdot v_1 \mid h \cdot v_2 \rangle = \sum_{g \in G} (gh \cdot v_1 | gh \cdot v_2) = \sum_{g \in G} (g \cdot v_1 | g \cdot v_2) = \langle v_1 \mid v_2 \rangle$$

for any $h \in G$. □

Lemma 1.5 (Schur). *Given two representations V and W of a finite group G, denote* $\mathrm{Hom}_G(V, W)$ *the vector space of morphisms of representations between V and W. If V and W are irreducible, then*

$$\dim \mathrm{Hom}_G(V, W) = \begin{cases} 1 & \text{if } V \text{ and } W \text{ are isomorphic;} \\ 0 & \text{otherwise.} \end{cases}$$

On the other hand, for any representation V of a finite group G, there is an isomorphism of vector spaces between V and $\mathrm{Hom}_G(\mathbb{C}G, V)$.

Proof. For any morphism of representations $\phi : V \to W$, the kernel and the image of ϕ are subrepresentations respectively of V and of W. Fix then an irreducible representation V, and a morphism of representations $\phi : V \to V$. For any $\lambda \in \mathbb{C}$, $\phi - \lambda \mathrm{id}_V$ is also an endomorphism of representations. Take λ among the non-empty set of eigenvalues of ϕ: then, $\mathrm{Ker}(\phi - \lambda \mathrm{id}_V)$ is a subrepresentation of V, so it is equal to V and $\phi = \lambda \mathrm{id}_V$.

Consider then another irreducible representation W of G. If V and W are isomorphic by $\psi : V \to W$, then for any morphism of representations $\phi : V \to W$, $\psi^{-1} \circ \phi \in \mathrm{Hom}_G(V, V)$, so it is a multiple of id_V, and $\phi = \lambda \psi$ for some scalar λ. So, $\dim \mathrm{Hom}_G(V, W) = 1$ if V and W are isomorphic. If V and W are not isomorphic, then given a morphism of representations $\phi : V \to W$, either its kernel is non-zero, or its image is not equal to W. By irreducibility, this implies that either $\mathrm{Ker}\, \phi = \{0\}$ or $\mathrm{Im}\, \phi = \{0\}$, so $\phi = 0$, and the second case for the computation of $\dim \mathrm{Hom}_G(V, W)$ is treated.

Finally, consider a representation V of G and a vector $v \in V$. The map

$$\begin{aligned} \phi : \mathbb{C}G &\to V \\ \sum_{g \in G} f(g)\, g &\mapsto \sum_{g \in G} f(g) (g \cdot v) \end{aligned}$$

is a morphism of representations between $\mathbb{C}G$ and V, and conversely, given a morphism $\phi \in \mathrm{Hom}_G(\mathbb{C}G, V)$, it is easy to see that it is given by the previous formula for $v = \phi(e)$. Thus, we have a natural identification between V and $\mathrm{Hom}_G(\mathbb{C}G, V)$. □

Proof of Theorem 1.3. Let V be a representation of G, endowed with a G-invariant scalar product as in Lemma 1.4. If V is not itself irreducible, consider a stable subspace $V_1 \subset V$ with $V_1 \neq \{0\}$ and $V_1 \neq V$. The orthogonal $V_2 = (V_1)^\perp$ is also stable: if $v \in V_2$, then for any $g \in G$ and any $w \in V_1$,

$$\langle g \cdot v \mid w \rangle = \left\langle v \,\middle|\, g^{-1} \cdot w \right\rangle = 0 \quad \text{since } g^{-1} \cdot w \in V_1 \text{ and } v \in V_2,$$

so $g \cdot v \in V_2$. Thus, we have the decomposition in stable subspaces $V = V_1 \oplus V_2$, and by induction on the dimension of V, the representation V can be totally split in irreducible representations.

For the second part of the theorem, since $\mathrm{Hom}_G(\cdot,\cdot)$ is compatible with direct sums, if $V = \bigoplus_{i=1}^r V_i$ is a decomposition in irreducible representations of V, then given another irreducible representation I, by Schur's lemma,

$$\dim \mathrm{Hom}_G(V, I) = \text{number of indices } i \text{ such that } I \text{ and } V_i \text{ are isomorphic.}$$

This irreducible representation I is also always a component of $\mathbb{C}G$, with multiplicity

$$\dim \mathrm{Hom}_G(\mathbb{C}G, I) = \dim I > 0$$

by the second part of Lemma 1.5. □

Remark. From the proof of Theorem 1.3, we see that if $V = \bigoplus_{\lambda \in \widehat{G}} m_\lambda V^\lambda$ is a decomposition of V into non-isomorphic irreducible representations (with multiplicities), then $m_\lambda = \dim \mathrm{Hom}_G(V, V^\lambda)$.

Let us reformulate once more the content of Theorem 1.3. We shall always denote $\widehat{G}$ the set of non-isomorphic irreducible representations $\lambda = (V^\lambda, \rho^\lambda)$ appearing as components of the regular representation $\mathbb{C}G$. By the previous discussion, they are also the irreducible components of *all* the representations of G. Now, consider the set $\mathrm{S}(G)$ of classes of isomorphism of representations of G. The operation of direct sum $\oplus$ makes $\mathrm{S}(G)$ into a commutative monoid, with neutral element the class of the null representation $\{0\}$. Denote $\mathrm{R}_0(G)$ the Grothendieck group built from $\mathrm{S}(G)$, that is to say, the set of formal differences $V \ominus W$ of (classes of isomorphism of) representations of G, with $V_1 \ominus W_1 = V_2 \ominus W_2$ if and only if $V_1 \oplus W_2$ and $V_2 \oplus W_1$ are isomorphic, and

$$(V_1 \ominus W_1) \oplus (V_2 \ominus W_2) = (V_1 \oplus V_2) \ominus (W_1 \oplus W_2).$$

We call $\mathrm{R}_0(G)$ the **Grothendieck group of representations of** G.

Proposition 1.6. *For any finite group G,*

$$\mathrm{R}_0(G) = \bigoplus_{\lambda \in \widehat{G}} \mathbb{Z} V^\lambda,$$

and the elements of $\mathrm{R}_0(G)$ with non-negative coefficients correspond to classes of isomorphism of representations of G.

The main result of Chapter 2 will be a description of the Grothendieck groups of representations of the symmetric groups $\mathfrak{S}(n)$.

1.2 Characters and constructions on representations

▷ *Characters and Schur's lemma of orthogonality.*

From the previous paragraph, we know that a linear representation of a finite group G is entirely determined up to isomorphisms by a finite sequence of non-negative numbers $(m_\lambda)_{\lambda \in \widehat{G}}$. However, these numbers have been described so far as dimensions of spaces of morphisms of representations, and one may ask for a simpler way to compute them in terms of V. The theory of characters yields a convenient tool in this setting.

Definition 1.7. *The* ***character*** *of a representation (V, ρ) of G is defined by*

$$\mathrm{ch}^V(g) = \mathrm{tr}(\rho(g)).$$

Thus, if $(\rho_{ij}(g))_{i,j}$ is the matrix of $\rho(g)$ in a basis of V, then the character is $\mathrm{ch}^V(G) = \sum_{i=1}^{\dim V} \rho_{ii}(g)$. In many situations, it will be also useful to deal with the **normalized character** χ^V of a representation: it is defined by

$$\chi^V(g) = \frac{\mathrm{ch}^V(g)}{\mathrm{ch}^V(1)} = \frac{\mathrm{ch}^V(g)}{\dim V}.$$

Example. Consider the regular representation of a finite group G. Its character is

$$\mathrm{ch}^V(g) = \sum_{h \in G} \delta_{h,gh} = \begin{cases} |G| & \text{if } g = 1, \\ 0 & \text{otherwise.} \end{cases}$$

Notice that for any representation V, $\mathrm{ch}^V(g^{-1}) = \overline{\mathrm{ch}^V(g)}$. Indeed, we can write the matrices of representation in a basis of V that is unitary with respect to a G-invariant scalar product on V. Then,

$$\overline{\mathrm{ch}^V(g)} = \overline{\mathrm{tr}\,\rho(g)} = \mathrm{tr}\,(\rho(g))^* = \mathrm{tr}\,(\rho(g))^{-1} = \mathrm{ch}^V(g^{-1}).$$

On the other hand, the characters always have the **trace property**:

$$\forall g,h \in G,\ \mathrm{ch}^V(gh) = \mathrm{tr}(\rho(g)\rho(h)) = \mathrm{tr}(\rho(h)\rho(g)) = \mathrm{ch}^V(hg).$$

We view the characters as elements of $\mathbb{C}G$, and we endow this space of functions on G with the scalar product

$$\langle f_1 \mid f_2 \rangle_G = \frac{1}{|G|} \sum_{g\in G} \overline{f_1(g)}\, f_2(g).$$

For any irreducible representation $\lambda \in \widehat{G}$, we fix a G-invariant scalar product on V^λ, and we denote $(\rho^\lambda_{ij}(g))_{1\le i,j\le d_\lambda}$ the matrix of $\rho(g)$ in a unitary basis of V^λ. We also write ch^λ for the irreducible character of V^λ; thus, $\mathrm{ch}^\lambda = \sum_{i=1}^{d_\lambda} \rho^\lambda_{ii}$, where $d_\lambda = \dim V^\lambda$.

Theorem 1.8 (Schur). *For any irreducible representations λ and μ of G,*

$$\left\langle \rho^\lambda_{ij} \,\middle|\, \rho^\mu_{kl} \right\rangle_G = \frac{1}{\dim V^\lambda}\, \delta_{\lambda,\mu}\, \delta_{i,k}\, \delta_{j,l}.$$

As a consequence, $\left\langle \mathrm{ch}^\lambda \,\middle|\, \mathrm{ch}^\mu \right\rangle_G = \delta_{\lambda,\mu}$.

Proof. Let u be an arbitrary linear map between V^μ and V^λ. We set

$$\phi = \frac{1}{\mathrm{card}\, G} \sum_{g\in G} \rho^\lambda(g) \circ u \circ \rho^\mu(g^{-1}).$$

This map is a morphism of representations between V^μ and V^λ. Indeed, for any $h \in G$,

$$\begin{aligned}\phi \circ \rho^\mu(h) &= \frac{1}{|G|} \sum_{g\in G} \rho^\lambda(g) \circ u \circ \rho^\mu(g^{-1}h) \\ &= \frac{1}{|G|} \sum_{k\in G} \rho^\lambda(hk) \circ u \circ \rho^\mu(k^{-1}) = \rho^\lambda(h) \circ \phi.\end{aligned}$$

Since V^μ and V^λ are irreducible representations, the previous map ϕ is 0 unless $\lambda = \mu$. In this latter case, if one makes the identification $V^\lambda = V^\mu$, then ϕ is a scalar multiple of the identity, and the coefficient of proportionality can be found by taking the trace:

$$\phi = \lambda\, \mathrm{id}_{V^\lambda} \quad \text{with } \lambda\,(\dim V^\lambda) = \mathrm{tr}\,\phi = \mathrm{tr}\, u.$$

Now, let us write the representations in matrix form. The previous computations become:

$$\frac{1}{|G|} \sum_{g\in G} \sum_{j=1}^{d_\lambda} \sum_{k=1}^{d_\mu} \rho^\lambda_{ij}(g)\, u_{jk}\, \rho^\mu_{kl}(g^{-1}) = \begin{cases} \frac{\delta_{i,l}}{\dim V^\lambda} \sum_{j=1}^{d_\lambda} u_{jj} & \text{if } \lambda = \mu, \\ 0 & \text{otherwise.} \end{cases}$$

Both sides of the equation are linear forms in the coefficients of u, so their coefficients must be equal. Therefore, for any coefficients $i,j \in [\![1,d_\lambda]\!]$ and $k,l \in [\![1,d_\mu]\!]$,

$$\left\langle \rho^\lambda_{ij} \,\middle|\, \rho^\mu_{lk} \right\rangle_G = \frac{1}{|G|} \sum_{g\in G} \overline{\rho^\lambda_{ij}(g)}\, \rho^\mu_{lk}(g)$$

$$= \frac{1}{|G|} \sum_{g\in G} \rho^\lambda_{ij}(g^{-1})\, \rho^\mu_{kl}(g) = \begin{cases} \frac{\delta_{i,l}\,\delta_{j,k}}{\dim V^\lambda} & \text{if } \lambda = \mu, \\ 0 & \text{otherwise.} \end{cases}$$

This is the equality stated by the theorem with the roles of k and l exchanged. By taking $i = j$ and $k = l$, and summing over indices $i \in [\![1,d_\lambda]\!]$ and $k \in [\![1,d_\mu]\!]$, we get the orthogonality relation for characters. □

Corollary 1.9. *A representation V of a finite group G is entirely determined (up to isomorphisms) by its character.*

Proof. If $V = \bigoplus_{\lambda\in\widehat{G}} m_\lambda V^\lambda$, then $\mathrm{ch}^V = \sum_{\lambda\in\widehat{G}} m_\lambda \,\mathrm{ch}^\lambda$. By using the orthogonality of characters, one sees then that the multiplicity m_λ of an irreducible representation V^λ in V is given by $m_\lambda = \left\langle \mathrm{ch}^V \,\middle|\, \mathrm{ch}^\lambda \right\rangle_G$. As a by-product, one gets a criterion of irreducibility for a representation of G: V is irreducible if and only if $\left\langle \mathrm{ch}^V \,\middle|\, \mathrm{ch}^V \right\rangle_G = 1$. □

Example. Consider the permutation representation of $\mathfrak{S}(n)$ on $\mathbb{C}^n$, with $n \geq 2$. Its character is

$$\mathrm{ch}^{\mathbb{C}^n}(\sigma) = \sum_{i=1}^{n} \delta_{i,\sigma(i)} = \text{number of fixed points of } \sigma.$$

Let us compute the norm of this character:

$$\left\langle \mathrm{ch}^{\mathbb{C}^n} \,\middle|\, \mathrm{ch}^{\mathbb{C}^n} \right\rangle_{\mathfrak{S}(n)} = \frac{1}{n!} \sum_{\sigma\in\mathfrak{S}(n)} \left(\sum_{i=1}^{n} \delta_{i,\sigma(i)} \right)^2 = \frac{1}{n!} \sum_{i,j=1}^{n} \left(\sum_{\sigma\in\mathfrak{S}(n)} \delta_{i,\sigma(i)}\, \delta_{j,\sigma(j)} \right)$$

$$= \frac{1}{n!} \left(\sum_{i=1}^{n} (n-1)! + \sum_{1\leq i\neq j\leq n} (n-2)! \right) = \frac{n! + n!}{n!} = 2.$$

Indeed, on the second line, if two indices i and j are fixed, then the number of permutations with $\sigma(i) = i$ and $\sigma(j) = j$ is $(n-2)!$; and similarly, if one index i is fixed, then the number of permutations with $\sigma(i) = i$ is $(n-1)!$.

It follows that $\mathbb{C}^n$ is necessarily the direct sum of 2 non-isomorphic irreducible representations; indeed, if $V = \bigoplus_{\lambda\in\widehat{G}} m_\lambda V^\lambda$, then $\left\langle \mathrm{ch}^V \,\middle|\, \mathrm{ch}^V \right\rangle_G = \sum_{\lambda\in\widehat{G}} (m_\lambda)^2$. By the discussion on page 10, these two components are the trivial representation on $\mathbb{C}(1,1,\ldots,1)$, and the representation on $\{(x_1,\ldots,x_n) \in \mathbb{C}^n \mid x_1 + \cdots + x_n = 0\}$.

▷ *Tensor products, induction and restriction of representations.*

Because of Corollary 1.9, each statement on representations has an equivalent statement in terms of characters, and from now on we shall try to give both statements at each time. For instance, let us present some constructions on representations, and their effect on characters. We already used the fact that the direct sum of representations corresponds to the sum of characters:

$$\mathrm{ch}^{(V \oplus W)} = \mathrm{ch}^V + \mathrm{ch}^W.$$

The **tensor product** of two representations V and W of G is the representation with underlying vector space $V \otimes W$, and with

$$g \cdot (v \otimes w) = (g \cdot v) \otimes (g \cdot w).$$

Since simple tensors $v \otimes w$ span linearly $V \otimes W$, the previous formula entirely defines a representation on $V \otimes W$.

Proposition 1.10. *The character of a tensor product of representations is the product of the characters:*

$$\mathrm{ch}^{(V \otimes W)} = \mathrm{ch}^V \times \mathrm{ch}^W.$$

Proof. Fix a basis $(e_1, \ldots, e_m)$ of V, and a basis $(f_1, \ldots, f_n)$ of W. A basis of $V \otimes W$ is $(e_i \otimes f_j)_{i,j}$, and with respect to these bases,

$$\rho^{V \otimes W}_{(i,j)(k,l)}(g) = \rho^V_{ik}(g)\, \rho^W_{jl}(g).$$

Therefore,

$$\begin{aligned} \mathrm{ch}^{V \otimes W}(g) &= \sum_{i,j} \rho^{V \otimes W}_{(i,j)(i,j)}(g) \\ &= \left(\sum_i \rho^V_{ii}(g)\right)\left(\sum_j \rho^W_{jj}(g)\right) = \mathrm{ch}^V(g)\, \mathrm{ch}^W(g). \end{aligned}$$

□

Consider now two finite groups $H \subset G$. There is a canonical way to transform representations of H into representations of G and conversely. First, if (V, ρ^V) is a representation of G, then the **restricted representation** $\mathrm{Res}^G_H(V)$ is the representation of H defined by

$$\forall h \in H,\ \rho^{\mathrm{Res}^G_H(V)}(h) = \rho^V(h).$$

Thus, it has the same underlying vector space as V, and the action of H is just obtained by restriction of the definition of the action of G. In particular,

$$\mathrm{ch}^{\mathrm{Res}^G_H(V)} = \left(\mathrm{ch}^V\right)_{|H}.$$

The converse operation of induction from H to G is a bit more cumbersome to define without the language of algebras and modules. If (V, ρ^V) is a representation of H, denote $\mathrm{Ind}_H^G(V)$ the set of functions $f : G \to V$ such that

$$\forall h \in H,\ \forall g \in G,\ f(hg) = h \cdot f(g).$$

If one fixes a set of representatives $\widetilde{g_1}, \ldots, \widetilde{g_r}$ of $H\backslash G$, then a function in $\mathrm{Ind}_H^G(V)$ is determined by its values on $\widetilde{g_1}, \ldots, \widetilde{g_r}$, so

$$\dim \mathrm{Ind}_H^G(V) = [G : H](\dim V) = \frac{|G|}{|H|} \dim V.$$

We make G act on $\mathrm{Ind}_H^G(V)$ by $(g \cdot f)(g') = f(g'g)$, and we call $\mathrm{Ind}_H^G(V)$ the **induced representation** of V from H to G.

Proposition 1.11. *The character of an induced representation* $\mathrm{Ind}_H^G(V)$ *is given by the formula*

$$\mathrm{ch}^{\mathrm{Ind}_H^G(V)}(g) = \sum_{g_j \in G/H} \mathrm{ch}^V(g_j^{-1} g g_j),$$

where $\mathrm{ch}^V(g) = 0$ *if* g *is not in* H*, and the sum runs over representatives* $g_1, \ldots, g_r$ *of the left cosets* gH.

We postpone the proof of this proposition to the end of Section 1.4, where the framework of algebras and modules will provide a more natural definition of the induction of representations, and a simple explanation of the formula for characters.

▷ *Frobenius' reciprocity.*

An important feature of the operations of induction and restriction of representations is their adjointness in the sense of functors on categories. More concretely, one has:

Proposition 1.12 (Frobenius). *Let* $H \subset G$ *be two finite groups, and* V *and* W *be two representations of* H *and* G*. One has*

$$\left\langle \mathrm{ch}^V \,\middle|\, \mathrm{ch}^{\mathrm{Res}_H^G(W)} \right\rangle_H = \left\langle \mathrm{ch}^{\mathrm{Ind}_H^G(V)} \,\middle|\, \mathrm{ch}^W \right\rangle_G.$$

In particular, if V and W are irreducible, then the multiplicity of V in $\mathrm{Res}_H^G(W)$ is the same as the multiplicity of W in $\mathrm{Ind}_H^G(V)$.

Proof. We compute:

$$\begin{aligned}\left\langle \mathrm{ch}^{\mathrm{Ind}_H^G(V)} \,\middle|\, \mathrm{ch}^W \right\rangle_G &= \frac{1}{|G|}\sum_{g\in G} \overline{\mathrm{ch}^{\mathrm{Ind}_H^G(V)}(g)}\,\mathrm{ch}^W(g)\\ &= \frac{1}{|G|}\sum_{g\in G}\sum_{g_j\in G/H} \overline{\mathrm{ch}^V(g_j^{-1}gg_j)}\,\mathrm{ch}^W(g)\\ &= \frac{1}{|G|}\sum_{k\in G}\sum_{g_j\in G/H} \overline{\mathrm{ch}^V(k)}\,\mathrm{ch}^W(g_jkg_j^{-1})\end{aligned}$$

because when g runs over G, so does $k = g_j^{-1}gg_j$. Now, by the trace property of characters, $\mathrm{ch}^W(g_jkg_j^{-1}) = \mathrm{ch}^W(k)$, so

$$\left\langle \mathrm{ch}^{\mathrm{Ind}_H^G(V)} \,\middle|\, \mathrm{ch}^W \right\rangle_G = \frac{1}{|G|}\sum_{k\in G}\sum_{g_j\in G/H} \overline{\mathrm{ch}^V(k)}\,\mathrm{ch}^W(k) = \frac{1}{|H|}\sum_{k\in G} \overline{\mathrm{ch}^V(k)}\,\mathrm{ch}^W(k).$$

Finally, since $\mathrm{ch}^V(k) = 0$ if k is not in H, the last sum runs in fact over H, and we get indeed $\left\langle \mathrm{ch}^V \,\middle|\, \mathrm{ch}^{\mathrm{Res}_H^G(W)} \right\rangle_H$. □

When we shall deal with representations of symmetric groups (Chapter 2), Frobenius' reciprocity will translate into a property of self-adjointness for the Hopf algebra of symmetric functions.

1.3 The non-commutative Fourier transform

▷ *The Fourier transform and the algebra $\mathbb{C}\widehat{G}$.*

For any group G, Theorem 1.3 ensures that there is an isomorphism of representations

$$\mathbb{C}G \to \bigoplus_{\lambda\in\widehat{G}} d_\lambda V^\lambda,$$

where $\widehat{G}$ is the finite set of all classes of isomorphism of irreducible representations of G. However, this isomorphism is for the moment an abstract one, and one may ask for a concrete realization of it. This realization will be provided by the so-called **non-commutative Fourier transform**, whose properties will allow us to restate and summarize most of the previous discussion.

The vector space $\mathbb{C}G$ can be endowed with a structure of algebra for the **convolution product**:

$$(f_1f_2)(k) = \sum_{gh=k} f_1(g)\,f_2(h).$$

This rule of product is easy to understand if one identifies as before a function f with the formal sum $\sum_{g\in G} f(g)\, g$:

$$f_1 f_2 = \left(\sum_{g\in G} f_1(g)\right)\left(\sum_{h\in G} f_2(h)\right) = \sum_{g,h\in G} f_1(g) f_2(h)\, gh$$
$$= \sum_{k\in G} \left(\sum_{gh=k} f_1(g)\, f_2(h)\right) k.$$

We say that $\mathbb{C}G$ is the **group algebra** of G. On the other hand, we denote $\mathbb{C}\widehat{G}$ the complex algebra which is the direct sum of all the algebras $\mathrm{End}(V^\lambda)$:

$$\mathbb{C}\widehat{G} = \bigoplus_{\lambda\in\widehat{G}} \mathrm{End}(V^\lambda).$$

It is again convenient to see formal sums of endomorphisms in $\mathbb{C}\widehat{G}$ as functions on $\widehat{G}$.

Definition 1.13. *The non-commutative Fourier transform $\widehat{f}$ of a function $f \in \mathbb{C}G$ is the element of $\mathbb{C}\widehat{G}$ defined by*

$$\widehat{f}(\lambda) = \sum_{g\in G} f(g)\, \rho^\lambda(g).$$

Example. Consider the symmetric group $\mathfrak{S}(3)$. We already know two non-isomorphic irreducible representations of $\mathfrak{S}(3)$:

$$V_1 = \{(x_1, x_2, x_3) \in \mathbb{C}^3 \mid x_1 + x_2 + x_3\} \quad ; \quad V_2 = \mathbb{C};$$

the first representation being the (restriction of) the permutation representation, and the second representation being the trivial one. A third irreducible representation is provided by the **signature representation** (cf. Section 2.1)

$$V_3 = \mathbb{C} \quad ; \quad \rho(\sigma) = \varepsilon(\sigma) = (-1)^{\sum_{i<j} \delta_{\sigma(i)>\sigma(j)}} \in \mathbb{C}^\times = \mathrm{GL}(1, \mathbb{C}).$$

It is of dimension 1, hence irreducible, and equal to its character, which is different from the trivial character; so, it is non-isomorphic to V_1 and V_2. Since

$$\mathrm{card}\, \mathfrak{S}(3) = 6 = 2^2 + 1^2 + 1^2 = (\dim V_1)^2 + (\dim V_2)^2 + (\dim V_3)^2,$$

we thus have a complete set of representatives of $\widehat{\mathfrak{S}(3)}$. Denoting $(\rho_{ij}(\sigma))_{i,j}$ the 2×2 matrix of the representation V_1, the Fourier transform of a permutation $\sigma \in \mathfrak{S}(3)$ can thus be seen as the block-diagonal matrix

$$\begin{pmatrix} \rho_{11}(\sigma) & \rho_{12}(\sigma) & & \\ \rho_{21}(\sigma) & \rho_{22}(\sigma) & & \\ & & 1 & \\ & & & \varepsilon(\sigma) \end{pmatrix}.$$

▷ *The fundamental isomorphism.*

The space $\mathbb{C}\widehat{G}$ is a representation of G for the action $g \cdot \sum_{\lambda\in\widehat{G}} u^\lambda = \sum_{\lambda\in\widehat{G}} \rho^\lambda(g)\, u^\lambda$. Also, it admits for G-invariant scalar product

$$\langle u \,|\, v\rangle_{\widehat{G}} = \sum_{\lambda\in\widehat{G}} \frac{d_\lambda}{|G|^2}\, \mathrm{tr}((u^\lambda)^* v^\lambda),$$

the adjoint of an endomorphism in each space $\mathrm{End}(V^\lambda)$ being taken with respect to a G-invariant scalar product on V^λ. In the following, we fix for each λ a unitary basis $(e_i^\lambda)_{1\leq i\leq d_\lambda}$ of each space V^λ, and denote $(e_{ij}^\lambda)_{1\leq i,j\leq d_\lambda}$ the associated basis of $\mathrm{End}(V^\lambda)$:

$$e_{ij}^\lambda(e_k^\lambda) = \delta_{j,k}\, e_i^\lambda.$$

Theorem 1.14. *The Fourier transform $\mathbb{C}G \to \mathbb{C}\widehat{G}$ is an isomorphism of algebras, of representations of G, and of Hilbert spaces. The matrix coefficients of irreducible representations $(\rho_{ij}^\lambda)_{\lambda\in\widehat{G},\,1\leq i,j\leq d_\lambda}$ form an orthogonal basis of $\mathbb{C}G$. If $\eta_{ij}^\lambda(g) = \rho_{ji}^\lambda(g^{-1})$, then this new orthogonal basis (η_{ij}^λ) is sent by the Fourier transform to*

$$\widehat{\eta_{ij}^\lambda} = \frac{|G|}{d_\lambda}\, e_{ij}^\lambda.$$

Proof. We saw in Theorem 1.8 that the matrix coefficients of irreducible representations are orthogonal, and since

$$|G| = \dim \mathbb{C}G = \sum_{\lambda\in\widehat{G}} (d_\lambda)^2,$$

we have the right number of terms to form a basis of $\mathbb{C}G$. On the other hand, the Fourier transform is indeed compatible with the product on each algebra:

$$\begin{aligned}\widehat{f_1 f_2} &= \sum_{\lambda\in\widehat{G}} \sum_{k\in G} (f_1 f_2)(k)\, \rho^\lambda(k) = \sum_{\lambda\in\widehat{G}} \sum_{g,h\in G} f_1(g) f_2(h)\, \rho^\lambda(gh)\\ &= \sum_{\lambda\in\widehat{G}} \left(\sum_{g\in G} f_1(g)\rho^\lambda(g)\right)\left(\sum_{h\in G} f_2(h)\,\rho^\lambda(h)\right) = \sum_{\lambda\in\widehat{G}} \widehat{f_1}(\lambda)\widehat{f_2}(\lambda) = \widehat{f_1}\,\widehat{f_2}.\end{aligned}$$

The compatibility with the action of G is trivial. Suppose that $\widehat{f} = 0$. Then, for any irreducible representation λ and any indices $1 \leq i,j \leq d_\lambda$,

$$\left\langle \overline{f} \,\middle|\, \rho_{ij}^\lambda \right\rangle_G = \sum_{i,j} f(g)\, \rho_{ij}^\lambda(g) = (\widehat{f}(\lambda))_{ij} = 0,$$

so $\overline{f}$ is orthogonal to all the elements of an orthogonal basis, and $f = \overline{f} = 0$. It follows that the Fourier transform is injective, and since $\dim \mathbb{C}\widehat{G} = \dim \mathbb{C}G$, it is an isomorphism of algebras and of representations of G.

To prove that it is also an isomorphism of Hilbert spaces, it suffices to show that for any $g,h \in G$, $\left\langle \delta_g \,\middle|\, \delta_h \right\rangle_G = \left\langle \widehat{\delta_g} \,\middle|\, \widehat{\delta_h} \right\rangle_{\widehat{G}}$; indeed, the functions δ_g form an orthogonal basis of $\mathbb{C}G$. Notice that

$$\left\langle \delta_g \,\middle|\, \delta_h \right\rangle_G = \frac{\delta_{g,h}}{|G|} = \frac{\delta_{e,g^{-1}h}}{|G|} = \frac{\mathrm{ch}^{\mathbb{C}G}(g^{-1}h)}{|G|^2}.$$

However,

$$\begin{aligned}\left\langle \widehat{\delta_g} \,\middle|\, \widehat{\delta_h} \right\rangle_{\widehat{G}} &= \sum_{\lambda \in \widehat{G}} \frac{d_\lambda}{|G|^2} \operatorname{tr}\left((\rho^\lambda(g))^* \rho^\lambda(h)\right) = \sum_{\lambda \in \widehat{G}} \frac{d_\lambda}{|G|^2} \operatorname{tr}\left(\rho^\lambda(g^{-1}h)\right) \\ &= \frac{1}{|G|^2} \sum_{\lambda \in \widehat{G}} d_\lambda \, \mathrm{ch}^{V^\lambda}(g^{-1}h) = \frac{\mathrm{ch}^{\mathbb{C}G}(g^{-1}h)}{|G|^2},\end{aligned}$$

the last identity coming from the isomorphism $\mathbb{C}G = \bigoplus_{\lambda \in \widehat{G}} d_\lambda V^\lambda$.

Finally, we compute the Fourier transform of a matrix coefficient of irreducible representations:

$$\widehat{\eta^\lambda_{ij}} = \sum_{\mu \in \widehat{G}} \sum_{g \in G} \overline{\rho^\lambda_{ij}(g)} \, \rho^\mu(g) = |G| \sum_{\mu \in \widehat{G}} \sum_{1 \leq k,l \leq d_\mu} \left\langle \rho^\lambda_{ij} \,\middle|\, \rho^\mu_{kl} \right\rangle_G e^\mu_{kl} = \frac{|G|}{d_\lambda} e^\lambda_{ij}. \qquad \square$$

▷ *Decomposition of functions on groups.*

An important consequence of Theorem 1.14 is the possibility to expand every function on the group G as a linear combination of matrix coefficients of irreducible representations:

$$\begin{aligned} f(g) &= \sum_{\lambda,i,j} d_\lambda \left\langle \eta^\lambda_{ij} \,\middle|\, f \right\rangle_G \eta^\lambda_{ij}(g) = \sum_{\lambda \in \widehat{G}} \frac{d_\lambda}{|G|} \sum_{1 \leq i,j \leq d_\lambda} \left(\sum_{h \in G} f(h) \rho^\lambda_{ij}(h) \right) \overline{\rho^\lambda_{ij}(g)} \\ &= \sum_{\lambda \in \widehat{G}} \frac{d_\lambda}{|G|} \sum_{1 \leq i,j \leq d_\lambda} \left(\widehat{f}(\lambda)\right)_{ij} (\rho^\lambda_{ji})^*(g) = \sum_{\lambda \in \widehat{G}} \frac{d_\lambda}{|G|} \operatorname{tr}\left(\rho^{\lambda *}(g) \widehat{f}(\lambda)\right). \end{aligned}$$

To get a good intuition of these results, it can be useful to compare this expansion of functions with the usual Fourier theory of functions on a circle (or on a multi-dimensional torus). To this purpose, it is convenient to renormalize a bit the algebra structures on $\mathbb{C}G$ and $\mathbb{C}\widehat{G}$. *These modifications will only hold during this paragraph.* To avoid any ambiguity, the dual elements (irreducible representations, Fourier transforms) will be denoted in this paragraph with a symbol $\widetilde{\cdot}$ instead of $\widehat{\cdot}$. We renormalize the convolution product on $\mathbb{C}G$ by setting

$$(f_1 * f_2)(k) = \frac{1}{|G|} \sum_{gh=k} f_1(g) f_2(h),$$

and we keep the same Hilbert scalar product on $\mathbb{C}G$ as before. We define as before the dual group algebra $\mathbb{C}\widetilde{G} = \bigoplus_{\lambda\in\widetilde{G}} \mathrm{End}(V^\lambda)$, but we change the scalar product of $\mathbb{C}\widetilde{G}$ into

$$\langle u \mid v\rangle_{\widetilde{G}} = \sum_{\lambda\in\widetilde{G}} d_\lambda \, \mathrm{tr}((u^\lambda)^* v^\lambda).$$

We define the Fourier transform of a function f by

$$\widetilde{f}(\lambda) = \frac{1}{|G|} \sum_{g\in G} f(g)\rho^\lambda(g).$$

Then, the new Fourier transform $f \mapsto \widetilde{f}$ is as before an isomorphism of $\mathbb{C}$-algebras, of G-representations, and of Hilbert spaces. Moreover:

Proposition 1.15. *For any function f on the group G,*

$$f(g) = \sum_{\lambda\in\widetilde{G}} d_\lambda \, \mathrm{tr}\left(\rho^{\lambda *}(g)\widetilde{f}(\lambda)\right);$$

$$\langle f \mid f\rangle_G = \left\langle \widetilde{f} \,\middle|\, \widetilde{f}\right\rangle_{\widetilde{G}} = \sum_{\lambda\in\widetilde{G}} d_\lambda \, \mathrm{tr}\left((\widetilde{f}(\lambda))^* \widetilde{f}(\lambda)\right).$$

These formulas are exactly the same as those satisfied by the Fourier series of a square-integrable function f on the circle $\mathbb{T} = \mathbb{R}/(2\pi\mathbb{Z})$:

$$f(\theta) = \sum_{k\in\mathbb{Z}} \widetilde{f}(k)\,\mathrm{e}^{-\mathrm{i}k\theta} \quad ; \quad \frac{1}{2\pi}\int_0^{2\pi} |f(\theta)|^2 \, d\theta = \sum_{k\in\mathbb{Z}} \left|\widetilde{f}(k)\right|^2$$

where $\widetilde{f}(k) = \frac{1}{2\pi}\int_0^{2\pi} f(\theta)\,\mathrm{e}^{\mathrm{i}k\theta}\, d\theta$. The reason for this correspondence is that the formulas of Proposition 1.15 hold in fact for any square-integrable function on a topological compact group, the means

$$\frac{1}{|G|}\sum_{g\in G} \cdot$$

being replaced in this theory by integrals

$$\int_G \cdot \;\mathrm{Haar}(dg)$$

against the Haar measure. For instance, with the circle $\mathbb{T}$, the set of irreducible representations is labeled by $\mathbb{Z}$, each irreducible representation of label $k \in \mathbb{Z}$ being one-dimensional and given by $\rho(\theta) = \mathrm{e}^{ik\theta}$. The Haar measure on $\mathbb{T} = [0, 2\pi]$ is $\frac{d\theta}{2\pi}$ and Proposition 1.15 gives indeed the Fourier series of harmonic functions. We leave as an exercise to the reader (see also the notes at the end of the chapter) the proof that almost all results proved so far for representations and characters of finite groups extend to topological compact groups, the only difference being that the set $\widehat{G}$ of irreducible representations appearing as components of the regular representation $L^2(G, \text{Haar})$ can now be discrete infinite (this being a consequence of the spectral theory of compact operators). In particular, most of the theory exposed before can be applied without big changes to the classical compact Lie groups $\mathrm{SU}(n)$, $\mathrm{SO}(n)$, $\mathrm{USp}(n)$.

The fact that the coefficients of representations yield an expansion of arbitrary functions in orthogonal components can be seen as one of the main motivations for the study of representations of groups. In particular, it enables one to solve evolution problems such as the heat equation on non-commutative groups (instead of the basic setting of the real line). In the next paragraph, we shall give an example of this in the case of finite groups.

▷ *Center of the group algebra.*

In this paragraph, we consider the restriction of the non-commutative Fourier transform to the **center** of the group algebra $\mathbb{C}G$. Write $Z(\mathbb{C}G)$ for the set of functions on G such that $f(gh) = f(hg)$ for any g, h.

Lemma 1.16. *The following assertions are equivalent:*

(Z1) The function f belongs to $Z(\mathbb{C}G)$.

(Z2) The function f commutes with any other function d of $\mathbb{C}G$: $fd = df$.

(Z3) The function f is a linear combination of ***conjugacy classes***

$$C_g = \sum_{g'=h^{-1}gh} g'.$$

Proof. If $f \in Z(\mathbb{C}G)$, then $f(h^{-1}gh) = f(hh^{-1}g) = f(g)$ for any g, h, so f is constant on conjugacy classes, and this proves the equivalence between (Z1) and (Z3). Then, if $f \in Z(\mathbb{C}G)$, one has for any other function d

$$\begin{aligned} fd &= \sum_{g,h \in G} f(g)\,d(h)\,gh = \sum_{g,h \in G} f(h^{-1}gh)\,d(h)\,hh^{-1}gh \\ &= \sum_{h,g' \in G} d(h)\,f(g')\,hg' = df, \end{aligned}$$

so $(Z1) \Rightarrow (Z2)$. Conversely, if f commutes with any other function, then

$$f(gh) = (f\,\delta_{h^{-1}})(g) = (\delta_{h^{-1}}\,f)(g) = f(hg).$$ □

Since $\mathbb{C}G$ and $\mathbb{C}\widehat{G}$ are isomorphic by the Fourier transform, their centers are isomorphic, and the center of an endomorphism algebra $\mathrm{End}(V^\lambda)$ is the one-dimensional space $\mathbb{C}\,\mathrm{id}_{V^\lambda}$. It will be convenient to identify an element of

$$Z(\mathbb{C}\widehat{G}) = \bigoplus_{\lambda \in \widehat{G}} \mathbb{C}\,\mathrm{id}_{V^\lambda}$$

with a $\mathbb{C}$-valued function on $\widehat{G}$, according to the following rule:

$$(k : \widehat{G} \to \mathbb{C}) \quad \text{corresponds to} \quad \sum_{\lambda} k(\lambda)\,\mathrm{id}_{V^\lambda}.$$

Then, the restriction of the scalar product $\langle \cdot \mid \cdot \rangle_{\widehat{G}}$ to $Z(\mathbb{C}\widehat{G})$ is defined on functions $\widehat{G} \to \mathbb{C}$ by

$$\langle k_1 \mid k_2 \rangle_{\widehat{G}} = \sum_{\lambda \in \widehat{G}} \frac{d_\lambda}{|G|^2}\,\mathrm{tr}\left(\overline{k_1(\lambda)}\,k_2(\lambda)\,\mathrm{id}_{V^\lambda}\right) = \sum_{\lambda \in \widehat{G}} \left(\frac{d_\lambda}{|G|}\right)^2 \overline{k_1(\lambda)}\,k_2(\lambda).$$

In the following, for any function f in the center, we set $f^*(g) = f(g^{-1})$. Notice that $(\mathrm{ch}^V)^* = \overline{\mathrm{ch}^V}$ for any character of representation. Theorem 1.14 restricted to $Z(\mathbb{C}G)$ reads now as:

Theorem 1.17. *Redefine the Fourier transform of an element $f \in Z(\mathbb{C}G)$ as the function*

$$\widehat{f}(\lambda) = \sum_{g \in G} f(g)\,\chi^\lambda(g).$$

Then, the Fourier transform is an isometry between $Z(\mathbb{C}G)$ and $Z(\mathbb{C}\widehat{G}) = \mathbb{C}^{\widehat{G}}$. An orthonormal basis of $Z(\mathbb{C}G)$ consists in the irreducible characters ch^λ, and the image of $\mathrm{ch}^{\lambda *}$ by the Fourier transform is the function

$$\widehat{\mathrm{ch}^{\lambda *}} = \frac{|G|}{d_\lambda}\,\delta_\lambda.$$

Proof. The redefinition of the Fourier transform on $Z(\mathbb{C}G)$ is compatible with the definition used in Theorem 1.14:

$$\widehat{f}(\lambda) = \frac{\mathrm{tr}\left(\widehat{f}(\lambda)\right)}{d_\lambda}\,\mathrm{id}_V^\lambda = \left(\sum_{g \in G} f(g)\,\chi^\lambda(g)\right)\mathrm{id}_{V^\lambda}.$$

Hence, the first part is an immediate consequence of Theorem 1.14. Then, we compute

$$\widehat{\mathrm{ch}^{\lambda *}}(\mu) = \sum_{g \in G} \overline{\mathrm{ch}^\lambda(g)}\,\chi^\mu(g) = \frac{|G|}{d_\mu}\left\langle \mathrm{ch}^\lambda \mid \mathrm{ch}^\mu \right\rangle_G = \frac{|G|}{d_\lambda}\,\delta_{\lambda,\mu}.$$ □

Corollary 1.18. *The number of distinct irreducible representations in $\widehat{G}$ is the number of conjugacy classes of the group G.*

Proof. This is the dimension of $Z(\mathbb{C}\widehat{G})$, which is isomorphic to $Z(\mathbb{C}G)$. □

Corollary 1.19. *Any central function f expands on irreducible characters as*

$$f(g) = \sum_{\lambda \in \widehat{G}} \frac{(d_\lambda)^2}{|G|}\,\widehat{f^*}(\lambda)\,\chi^\lambda(g).$$

Proof. Since irreducible characters form an orthonormal basis of $Z(\mathbb{C}G)$,

$$f(g) = \sum_{\lambda \in \widehat{G}} \left\langle \mathrm{ch}^\lambda \,\middle|\, f \right\rangle_G \mathrm{ch}^\lambda(g) = \sum_{\lambda \in \widehat{G}} (d_\lambda)^2 \left\langle \chi^\lambda \,\middle|\, f \right\rangle_G \chi^\lambda(g).$$

Then, $\left\langle \chi^\lambda \,\middle|\, f \right\rangle_G = \frac{1}{|G|} \sum_{g\in G} f(g)\,\chi^\lambda(g^{-1}) = \frac{1}{|G|} \sum_{g\in G} f^*(g)\,\chi^\lambda(g) = \frac{\widehat{f^*}(\lambda)}{|G|}$. □

The last result involves the so-called **Plancherel measure** $\mathrm{Pl}(\lambda) = \frac{(d_\lambda)^2}{|G|}$. This is a probability measure on $\widehat{G}$, and for any central function,

$$f = \int_{\widehat{G}} \widehat{f^*}(\lambda)\, \mathrm{Pl}(\lambda)\, \mathrm{ch}^\lambda.$$

In particular, consider the normalized character of the regular representation of G:

$$\chi^{\mathbb{C}G}(g) = \delta_{e,g}.$$

It has for Fourier transform the constant function equal to 1, since

$$\chi^{\mathbb{C}G} = \sum_{\lambda \in \widehat{G}} \mathrm{Pl}(\lambda)\, \chi^\lambda.$$

Thus, the Plancherel measure corresponds to the decomposition in normalized irreducible characters of the normalized regular trace of the group. The study of this probability measure in the case of symmetric groups will be the main topic of Chapter 13 of this book.

Example. As an application of the results of this section, consider the following random process on the symmetric group $\mathfrak{S}(n)$. We consider a deck of cards that are ordered from 1 to n, and at each time $k \in \mathbb{N}$, we choose at random two independent indices $i, j \in [\![1, n]\!]$, and we exchange the i-th card of the deck with the j-th card of the deck, cards being counted from top to bottom. Each index i or j has probability $\frac{1}{n}$, and it is understood that if $i = j$, then one leaves the deck of cards invariant. The configuration after k random transpositions of cards can be encoded by a permutation σ_k of size n, with $\sigma(1)$ denoting the label of the first card of the deck, $\sigma(2)$ denoting the label of the second card of the deck, etc. For instance, assuming $n = 5$, a possible trajectory of the process up to time $k = 8$ is

k	0	1	2	3	4	5	6	7	8
σ_k	12345	15342	13542	43512	43512]	43521	23514	25314	25314

and there are two steps ($k = 4$ and $k = 8$) where the same index $i = j$ was chosen.

We denote $\mathbb{P}[A]$ the probability of an event A, and $\mathbb{P}[A|B]$ the probability of an event A conditionally to another event B. Consider the law $f_k(\sigma) = \mathbb{P}[\sigma_k = \sigma]$, viewed as an element on $\mathbb{C}\mathfrak{S}(n)$. The rules of the random process are:

$$\mathbb{P}[\sigma_{k+1} = \tau | \sigma_k = \sigma]$$
$$= \begin{cases} \frac{1}{n} & \text{if } \tau = \sigma \text{ (corresponding to choices of indices } i = j); \\ \frac{2}{n^2} & \text{if } \tau = \sigma(i,j) \text{ for some pair } i \neq j. \end{cases}$$

Therefore, we get a recursion formula for f_k:

$$f_{k+1} = \sum_{\tau \in \mathfrak{S}(n)} \mathbb{P}[\sigma_{k+1} = \tau]\, \tau = \sum_{\sigma,\tau \in \mathfrak{S}(n)} f_k(\sigma)\, \mathbb{P}[\sigma_{k+1} = \tau | \sigma_k = \sigma]\, \tau$$
$$= \sum_{\sigma \in \mathfrak{S}(n)} f_k(\sigma)\, \sigma \left(\frac{1}{n} + \frac{1}{n^2} \sum_{1 \leq i \neq j \leq n} (i,j) \right) = f_k \left(\frac{1}{n} + \frac{1}{n^2} \sum_{1 \leq i \neq j \leq n} (i,j) \right).$$

So, if $f = \frac{1}{n} + \frac{1}{n^2} \sum_{1 \leq i \neq j \leq n} (i,j)$, then $f_k = f^k$ for any k, in the sense of convolution in $\mathbb{C}\mathfrak{S}(n)$. It should be noticed that the recursion formula can be rewritten as

$$f_{k+1} - f_k = f_k \left(\frac{1}{n^2} \sum_{1 \leq i \neq j \leq n} (i,j) - 1 \right).$$

Thus, we are looking at the analogue in the setting of the symmetric group of the heat equation $\frac{\partial f}{\partial t} = \frac{1}{2} \Delta f$.

Notice now that f is a linear combination of conjugacy classes in the symmetric group. Indeed, the identity 1 is a conjugacy class on its own, and on the other hand, two transpositions (i,j) and (k,l) are always conjugated:

$$(k,l) = (i,k)(j,l)(i,j)(j,l)^{-1}(i,k)^{-1}$$

and each transposition appears twice in $\sum_{1 \leq i \neq j \leq n} (i,j)$. Moreover, since $1 = (1)^{-1}$ and $(i,j) = (i,j)^{-1}$, $f^* = f$. Therefore, f is in $Z(\mathbb{C}\mathfrak{S}(n))$, and

$$f_k(\sigma) = \sum_{\lambda \in \widehat{\mathfrak{S}}(n)} \mathrm{Pl}(\lambda)\, \widehat{(f^k)^*}(\lambda)\, \chi^\lambda(\sigma) = \sum_{\lambda \in \widehat{\mathfrak{S}}(n)} \mathrm{Pl}(\lambda) \left(\widehat{f}(\lambda)\right)^k \chi^\lambda(\sigma).$$

since the Fourier transform is an isomorphism of algebras. This can be rewritten as:

$$f_k(\sigma) = \sum_{\lambda \in \widehat{\mathfrak{S}}(n)} \mathrm{Pl}(\lambda) \left(\frac{1}{n} + \frac{n-1}{n} \chi^\lambda(1,2) \right)^k \chi^\lambda(\sigma),$$

since $\chi^\lambda(1) = 1$ for any representation, and $\chi^\lambda(i,j) = \chi^\lambda(1,2)$ for any transposition (i,j) and any representation λ. This formula can be used to compute the asymptotics of the laws f_k. In particular, it can be shown that $-1 \leq \chi^\lambda(1,2) < 1$ if λ is not the trivial representation of $\mathfrak{S}(n)$ on $\mathbb{C}$. As a consequence, all the terms of the previous formula go to zero as k grows to infinity, but the term corresponding to the trivial representation, which is

$$\frac{(1)^2}{n!}\left(\frac{1}{n}+\frac{n-1}{n}\right)^k 1=\frac{1}{n!}.$$

So, $\lim_{k\to\infty} f_k(\sigma)=\frac{1}{n!}$, and the laws of the random process converge towards the uniform law on permutations.

1.4 Semisimple algebras and modules

By Theorem 1.14, for any finite group G, the group algebra $\mathbb{C}G$ is isomorphic to a direct sum of matrix algebras $\bigoplus_{\lambda\in\widehat{G}}\mathrm{End}(V^\lambda)$, and if one endows $\mathbb{C}G$ and this sum of matrix algebras with adequate Hermitian structures, then one is able to do many computations on the group, e.g., to decompose any function in elementary orthogonal components. Roughly speaking, the content of this section is the following: the same theory exists for *any* complex algebra that is isomorphic to a direct sum of matrix algebras, and moreover, there exists an abstract criterion in order to ensure that a given algebra is isomorphic to a direct sum of matrix algebras. There are many good reasons to consider this more general framework, and in this chapter, we shall see in particular that

- it makes certain constructions on representations much more natural (in particular, the induction of representations);
- it allows one to develop a theory of duality between groups acting on a vector space (see Section 1.5).

Later, it will also enable the study of combinatorial algebras that are modifications of the symmetric group algebras $\mathbb{C}\mathfrak{S}(n)$, and that are not group algebras. In this setting, we shall give and apply concrete criterions in order to ensure the semisimplicity of the algebras considered. As a matter of fact, we will then also need to know the general representation theory of possibly non-semisimple algebras; this will be explained in Section 4.2, and the present section is an introduction to this more general theory.

▷ *Algebras and modules.*

Though we mostly want to deal with algebras over $\mathbb{C}$, it will be convenient in the beginning to consider algebras over an arbitrary field k. Thus, a field k being fixed, we recall that an **algebra A over the field k** is a k-vector space endowed with a product $\times_A : A\times A\to A$ that is

(A1) associative and with a (unique) neutral element (the unity of the algebra):

$$\forall a,b,c \in A,\ (a \times_A b) \times_A c = a \times_A (b \times_A c);$$
$$\exists 1_A \in A,\ \forall a \in A,\ a \times_A 1_A = 1_A \times_A a = a.$$

(A2) compatible with the external product of k:

$$\forall \lambda \in k,\ \forall a,b \in A,\ \lambda(a \times_A b) = (\lambda a) \times_A b = a \times_A (\lambda b).$$

(A3) distributive with respect to the internal addition:

$$\forall a,b,c \in A,\ a \times_A (b+c) = a \times_A b + a \times_A c \quad \text{and} \quad (a+b) \times_A c = a \times_A c + b \times_A c.$$

In other words, a k-algebra is a **ring** and a k-vector space whose structures are compatible with one another in every possible way that one can think of. *In the two first parts of this book, unless explicitly stated, we shall only work with finite-dimensional algebras*, and denote $\dim_k A$, or simply $\dim A$ the dimension of A as a k-vector space. An algebra is said to be commutative if its product $\times_A$ is commutative: $a \times_A b = b \times_A a$. As before, we shall omit in most cases the product $\times_A$ and write $a \times_A b = ab$. The properties listed above for an algebra ensure that this is a non-ambiguous notation.

Example. Given a finite-dimensional k-vector space V, the set $\mathrm{End}(V)$ of k-linear maps $u : V \to V$ endowed with the product of composition of functions is a finite-dimensional algebra of dimension $(\dim V)^2$. Similarly, the set of matrices $\mathrm{M}(n,k)$ of size $n \times n$ and with coefficients in k is an algebra for the matrix product. If $n = \dim_k V$, then the two algebras $\mathrm{End}(V)$ and $\mathrm{M}(n,k)$ are isomorphic, an isomorphism being given by

$$u \mapsto \mathrm{mat}_{(e_1,\ldots,e_n)}(u),$$

where $(e_1,\ldots,e_n)$ is an arbitrary basis of V.

Example. For any finite group G, the set kG of formal k-linear combinations of elements of G (or, in other words, the set of functions $G \to k$) is a k-algebra for the convolution product defined at the beginning of Section 1.3. It has dimension $\dim_k kG = |G|$.

Example. For any field k, the set $k[X_1,\ldots,X_n]$ of polynomials in n variables with coefficients in k is a commutative k-algebra. It is **graded** by the degree of polynomials, and this gradation is compatible with the algebra structure, meaning that for any elements a and b in the algebra,

$$\deg(ab) \leq \deg a + \deg b.$$

This inequality is an equality as soon as a and b are not zero.

A **left module** M over a k-algebra A is a k-vector space endowed with an external product $\cdot : A \times M \to M$, such that

(M1) $\cdot$ is compatible with the addition and the product in A:

$$\forall a, b \in A,\ \forall m \in M,\ (a +_A b) \cdot m = (a \cdot m) + (b \cdot m);$$
$$(a \times_A b) \cdot m = a \cdot (b \cdot m).$$

(M2) $\cdot$ is compatible with the k-vector space structure on M:

$$\forall a \in A,\ \forall m, n \in M,\ a \cdot (m + n) = a \cdot m + a \cdot n;$$
$$\forall a \in A,\ \forall \lambda \in k,\ \forall m \in M,\ \lambda(a \cdot m) = (\lambda a) \cdot m = a \cdot (\lambda m).$$

(M3) for all $m \in M$, $1_A \cdot m = m$.

Again, *in the two first parts of this book, we shall only deal with finite-dimensional A-modules*, and this assumption holds always implicitly in the following. Dually, one defines a **right module** M over a k-algebra A as a k-vector space endowed with an external product $\cdot : M \times A \to M$ that is compatible with the structures of M and A. Notice that a k-vector space is a left module over the k-algebra k, so one can see the notion of module over a k-algebra as an extension of the notion of vector space over k. Then, the notions of (left or right) A-submodule and of morphism of (left or right) A-modules are defined in the obvious way, thereby generalizing the notions of k-vector subspace and of k-linear map. In the following, when a result holds for both left and right modules, we shall just speak of modules, and usually do the reasoning with left modules.

Remark. Notice that a structure of A-module on a k-vector space M is equivalent to a morphism of k-algebras $A \to \mathrm{End}(M)$.

Example. Let V be a k-vector space. Then, V is a left module over $\mathrm{End}(V)$ for the operation $u \cdot v = u(v)$.

Example. Let G be a finite group, and (V, ρ) a representation of G. Then, V is a left module over $\mathbb{C}G$ for the operation

$$\left(\sum_{g \in G} f(g)\, g\right) \cdot v = \sum_{g \in G} f(g)\, \rho(g)(v).$$

Conversely, any left module V over $\mathbb{C}G$ is a representation of G for the rule $\rho(g)(v) = g \cdot v$, the $\cdot$ denoting the product map $\mathbb{C}G \times V \to V$. With this new point of view, a morphism between two representations V and W of G is a morphism of $\mathbb{C}G$-modules. Therefore, there is an equivalence of categories between complex linear representations of G and left $\mathbb{C}G$-modules.

This reinterpretation already sheds a new light on certain results previously stated. For instance, the regular representation of G is an instance of the regular left module A associated to an algebra A, the action $A \times A \to A$ being given by the product of the algebra. Then, in the second part of Lemma 1.5, the isomorphism of vector spaces between V and $\mathrm{Hom}_G(\mathbb{C}G, V)$ comes from the more general fact that for any k-algebra A and any left A-module M, the module M is isomorphic as a k-vector space to $\mathrm{Hom}_A(A, M)$, the isomorphism being

$$m \mapsto (a \mapsto a \cdot m).$$

If M is a left A-module, then M is also canonically a right A^{opp}-module, where A^{opp} denotes the k-algebra with the same underlying vector space as A, and with product

$$a \times_{A^{\mathrm{opp}}} b = b \times_A a.$$

The right A^{opp}-module structure on a left A-module M is then defined by $m \cdot_{A^{\mathrm{opp}}} a = a \cdot_A m$. In the case where $A = \mathbb{C}G$, there is a simple realization of A^{opp} by using the inverse map. More precisely,

$$\mathbb{C}G \to (\mathbb{C}G)^{\mathrm{opp}}$$
$$\sum_{g \in G} f(g)\, g \mapsto \sum_{g \in G} f(g)\, g^{-1}$$

is an isomorphism of $\mathbb{C}$-algebras. Therefore, any left representation V of G admits a corresponding structure of right representation of G, given by

$$v \cdot g = g^{-1} \cdot v.$$

Example. The permutation representation of $\mathfrak{S}(n)$ on $\mathbb{C}^n$ is more natural when given by a structure of *right* $\mathbb{C}\mathfrak{S}(n)$-module. Indeed, it writes then as

$$(x_1, x_2, \ldots, x_n) \cdot \sigma = (x_{\sigma(1)}, x_{\sigma(2)}, \ldots, x_{\sigma(n)}).$$

Similarly, consider a finite alphabet A, and denote the elements of A^n as words of length n with letters in A:

$$A^n = \{a_1 a_2 a_3 \ldots a_n \mid \forall i \in [\![1, n]\!],\ a_i \in A\}.$$

There is a natural structure of right $\mathbb{C}\mathfrak{S}(n)$-module on the space $\mathbb{C}[A^n]$ of formal linear combinations of these words:

$$(a_1 a_2 a_3 \ldots a_n) \cdot \sigma = a_{\sigma(1)} a_{\sigma(2)} a_{\sigma(3)} \ldots a_{\sigma(n)}.$$

There is also a natural structure of left $\mathbb{C}\mathfrak{S}(A)$-module given by

$$\tau \cdot (a_1 a_2 a_3 \ldots a_n) = \tau(a_1)\tau(a_2)\tau(a_3)\ldots\tau(a_n).$$

This kind of construction justifies the need of both notions of left and right modules on an algebra. We shall study this double action more rigorously in Section 1.5, by introducing the notion of bimodule.

Example. Let A be any k-algebra. Then A is both a left and right A-module, for the actions given by the product of the algebra. The left A-submodules of A are exactly the left **ideals** of A, and similarly on the right.

▷ *Semisimplicity and Artin–Wedderburn theorem.*

Let A be a k-algebra, and M be a (left) module over A that is not the zero module. The module M is said to be **simple** if it is of positive dimension and if its only submodules are $\{0\}$ and M itself. It is said to be **semisimple** if it is a direct sum of simple modules. The notion of simple module is the generalization to the framework of modules and algebras of the notion of irreducible representation, and indeed, a representation V of G is irreducible if and only if it is a simple $\mathbb{C}G$-module.

Proposition 1.20. *A finite-dimensional module M over a k-algebra A is semisimple if and only if, for every submodule $N \subset M$, there exists a complement A-submodule P with $M = N \oplus P$.*

Proof. Suppose that $M = \bigoplus_{i=1}^{r} M_i$ is a direct sum of simple modules, and let N be a submodule of M. We take a subset $I \subset [\![1, r]\!]$ that is maximal among those such that $N \cap \bigoplus_{i \in I} M_i = \{0\}$. By choice, if $P = \bigoplus_{i \in I} M_i$, then $N + P$ is a direct sum. We claim that $N \oplus P = M$. It suffices to show that for every $i \in [\![1, r]\!]$, $N \oplus P$ contains M_i. This is clear if $i \in I$. If $i \notin I$, then $(N \oplus P) \cap M_i$ is not the zero submodule, since otherwise the set I would not be maximal. But M_i is simple, so $(N \oplus P) \cap M_i$ is a non-zero submodule of M_i, hence equal to the whole of M_i. This proves the existence of a complement A-module P of N such that $M = N \oplus P$.

Conversely, suppose that every submodule N of M has a complement submodule P. We can exclude the trivial case $M = \{0\}$. Then, since M is finite-dimensional, there is no infinite descending chain of submodules of M, so M has necessarily a simple submodule M_1. Denote P a complement of M_1:

$$M = M_1 \oplus P \text{ with } M_1 \text{ simple.}$$

To show that M is semisimple, it suffices now to prove that P has the same property as M, that is to say, that every submodule of P has a complement submodule in P. Indeed, an induction on the dimension of M will then allow us to conclude. Fix a submodule $S \subset P$. There is an isomorphism of A-modules

$$\begin{aligned} \psi : P &\to M/M_1 \\ p &\mapsto [p]_{M_1}. \end{aligned}$$

The A-submodule $\psi(S)$ of M/M_1 can be realized as the quotient module $(M_1 \oplus S)/M_1$. By hypothesis, $(M_1 \oplus S)$ has a complement R in M:

$$M = (M_1 \oplus S) \oplus R.$$

Then, if $\pi_{M_1} : M \to M/M_1$ is the canonical projection, $\pi_{M_1}(R)$ is a complement submodule of $\psi(S)$ in M/M_1, and $T = \psi^{-1}(\pi_{M_1}(R))$ is a complement submodule of S in P since ψ is an isomorphism.

$$\begin{array}{ccc} & & M = (M_1 \oplus S) \oplus R \\ & & \Big\downarrow \pi_{M_1} \\ P = S \oplus T & \xleftrightarrow[\psi]{} & M/M_1 = \psi(S) \oplus \pi_{M_1}(R) \end{array}$$

□

Corollary 1.21. *Semisimplicity of modules is kept by looking at submodules, quotient modules, and direct sum of modules.*

Proof. The stability by direct sum is trivial. For the two other properties, let M be a semisimple A-submodule, and P a submodule of M. We saw during the proof of Proposition 1.20 that P has the same property as M, so the stability for submodules is shown. On the other hand, if N is a complement of P in M, then the quotient M/P is isomorphic to N, which is a submodule of a semisimple module, hence semisimple; so the stability for quotient of modules is also proven. □

Definition 1.22. *A finite-dimensional k-algebra A is said to be semisimple if every A-module M is semisimple.*

Proposition 1.23. *A k-algebra A is semisimple if and only if the (left) A-module A is semisimple.*

Proof. If A is a semisimple algebra, then all its modules are semisimple, so A viewed as an A-module is semisimple. Conversely, suppose that A viewed as a module is semisimple, and consider another finite-dimensional A-module M. Since M is finitely generated, it is isomorphic to a quotient of a module $A \oplus A \oplus \cdots \oplus A$. However, semisimplicity is kept for direct sums and quotients, so M is semisimple. □

We leave the reader to check that an easy consequence of this proposition and of Corollary 1.21 is that a quotient or a direct sum of semisimple k-algebras is again semisimple.

We are now ready to classify the semisimple k-algebras. Recall that a **division ring** C over a field k is a (finite-dimensional) k-algebra such that for every non-zero $c \in C$, there exists b with $bc = cb = 1$. The difference with the notion of field extension of k is that we do not ask for the commutativity of the product in C. Given a division ring C, we denote $\mathrm{M}(n, C)$ the space of matrices with coefficients in C; it is a (non-commutative) k-algebra for the product of matrices. A consequence of the possible non-commutativity of a division ring is that the multiplication on the left of C by C is not C-linear. Therefore, $\mathrm{End}_C(C^n)$ is not k-isomorphic to the algebra $\mathrm{M}(n, C)$, but to the algebra $\mathrm{M}(n, C^{\mathrm{opp}})$, where C^{opp} acts on C by multiplication on the right (this is C-linear). This subtlety appears in most of the following discussion.

Theorem 1.24 (Artin–Wedderburn). *Every semisimple k-algebra A is isomorphic to*

$$\bigoplus_{\lambda \in \widehat{A}} \mathrm{M}(d_\lambda, C^\lambda)$$

for some k-division rings C^λ and some multiplicities $d_\lambda \geq 1$.

Lemma 1.25 (Schur). *Let M be a simple A-module. Then,* $\mathrm{End}_A(M)$ *is a k-division ring.*

Proof. The kernel and the image of a morphism between A-modules are A-submodules. Thus, if M is a simple A-module and if $u : M \to M$ is a morphism of modules, then it is either 0 or an isomorphism. Hence, if $u \neq 0$, then u has an inverse v with $uv = vu = \mathrm{id}_M$. For the same reason, if M_1 and M_2 are two simple modules, then either they are isomorphic, or $\dim_k \mathrm{Hom}_A(M_1, M_2) = 0$. □

Proof of Theorem 1.24. We decompose the left A-module A in a direct sum of simple modules (ideals), gathered according to their classes of isomorphism as A-modules:

$$A = \bigoplus_{\lambda \in \widehat{A}} d_\lambda\, M^\lambda,$$

with $\dim \mathrm{Hom}_A(M^\lambda, M^\mu) = 0$ if $\lambda \neq \mu$. For the moment, $\widehat{A}$ denotes the set of non-isomorphic simple modules appearing in A; we shall see hereafter that *every* simple module on A is isomorphic to some $M^\lambda \in \widehat{A}$. We now use the following sequence of isomorphisms of k-algebras:

$$A^{\mathrm{opp}} = \mathrm{End}_A(A) = \bigoplus_{\lambda \in \widehat{A}} \mathrm{End}_A(d_\lambda\, M^\lambda) = \bigoplus_{\lambda \in \widehat{A}} \mathrm{M}(d_\lambda, \mathrm{End}_A(M^\lambda)) = \bigoplus_{\lambda \in \widehat{A}} \mathrm{M}(d_\lambda, D^\lambda)$$

where the D^λ are division rings. Let us detail each identity:

1. An endomorphism of left A-modules on A is necessarily $r_a : b \mapsto ba$ for some $a \in A$. The composition of two endomorphisms reads then as $r_{a_1} \circ r_{a_2} = r_{a_2 a_1}$, so, $a \mapsto r_a$ is an isomorphism of k-algebras between A^{opp} and $\mathrm{End}_A(A)$.

2. One has

$$\mathrm{End}_A(A) = \mathrm{Hom}_A\left(\bigoplus_{\lambda \in \widehat{A}} d_\lambda M^\lambda, \bigoplus_{\mu \in \widehat{A}} d_\mu M^\mu\right) = \bigoplus_{\lambda,\mu \in \widehat{A}} \mathrm{Hom}_A(d_\lambda M^\lambda, d_\mu M^\mu)$$
$$= \bigoplus_{\lambda \in \widehat{A}} \mathrm{End}_A(d_\lambda\, M^\lambda)$$

since two non-isomorphic simple modules M^λ and M^μ have no non-trivial morphism between them. These identities are a priori isomorphisms of k-vector spaces, but the two extremal terms are k-algebras, and it is easily seen that the identification between them is compatible with the product of composition.

3. Using again the multilinearity of $\mathrm{Hom}_A(\cdot,\cdot)$, we know that $\mathrm{End}_A(dM)$ is isomorphic as a k-vector space to d^2 copies of $\mathrm{End}_A(M)$. To make this into an isomorphism of k-algebras, if $u \in \mathrm{End}_A(M_1 \oplus \cdots \oplus M_d)$, denote $u_{ij}(m)$ the i-th component of $u(m_j)$. Here we use indices to denote the different copies $M_1,\dots,M_d$ of M. Then, the elements of $\mathrm{End}_A(dM)$ act indeed as $d\times d$ matrices with coefficients in $\mathrm{End}_A(M)$:

$$\begin{aligned} u(m_1,\dots,m_d) &= \left(\sum_{j=1}^{d} u_{1j}(m_j),\dots,\sum_{j=1}^{d} u_{dj}(m_j)\right) \\ &= (u_{ij})_{1\leq i,j\leq d} \times (m_1,\dots,m_d). \end{aligned}$$

Hence, $\mathrm{End}_A(dM) = \mathrm{M}(d,\mathrm{End}_A(M))$ as k-algebras.

4. Finally, by Schur's lemma, each $\mathrm{End}_A(M^\lambda)$ with M^λ simple is a division ring D^λ.

Taking again the opposites of algebras, we conclude that

$$A = \bigoplus_{\lambda\in\widehat{A}} (\mathrm{M}(d_\lambda, D^\lambda))^{\mathrm{opp}} = \bigoplus_{\lambda\in\widehat{A}} \mathrm{M}(d_\lambda,(D^\lambda)^{\mathrm{opp}}) = \bigoplus_{\lambda\in\widehat{A}} \mathrm{M}(d_\lambda, C^\lambda)$$

since the opposite of a division ring D^λ is also a division ring C^λ, and the opposite of a matrix algebra is the matrix algebra of the opposite. This ends the proof of a decomposition of any semisimple algebra as a direct sum of matrix algebras over division rings. □

▷ *Central idempotents and the Fourier transform for semisimple algebras.*

We now want to show that there is unicity in the Artin–Wedderburn decomposition: up to permutation, the division rings C^λ and the multiplicities d_λ are entirely determined by A. A part of the proof of this unicity relies on the notion of **central idempotent**. Call central idempotent of a k-algebra A an element e such that $e^2 = e$ and $ef = fe$ for all $f \in A$. Notice that if e is a central idempotent, then so is $e' = 1-e$. A central idempotent is called **primitive** if it is non-zero, and if it cannot be written as the sum $e = e_1 + e_2$ of two non-zero central idempotents e_1 and e_2 with $e_1e_2 = 0$. Given two central primitive idempotents e and f, one has either $e = f$ or $ef = 0$. Indeed, ef and $(1-e)f$ are both central idempotents, and

$$f = ef + (1-e)f \qquad ; \qquad ef(1-e)f = (e-e^2)f^2 = 0.$$

Hence, since f is primitive, either $ef = 0$ or $(1-e)f = 0$. Similarly, either $ef = 0$ or $e(1-f) = 0$. Suppose $ef \neq 0$. Then, $(1-e)f = f - ef = 0$ and $e(1-f) = e - ef = 0$, so $f = ef = e$.

If $e = f + g$ is a decomposition of a central idempotent into two other central idempotents that are orthogonal ($fg = 0$), then $eA = fA \oplus gA$. Indeed, if x belongs to $fA \cap gA$, then $x = fx = fgx = 0$. As a consequence, since A is finite-dimensional, when one tries to split a central idempotent into orthogonal parts, one necessarily ends at some point with a sum of *primitive* central idempotents. In particular, there exists a decomposition of 1 into orthogonal central primitive idempotents:

$$1 = e_1 + e_2 + \cdots + e_r, \quad \text{with the } e_i \text{ orthogonal central primitive idempotents.}$$

This decomposition is unique, and $\{e_1, \ldots, e_r\}$ is a complete list of the central primitive idempotents of A. Indeed, if f is another central primitive idempotent, then

$$f = 1f = \sum_{i=1}^{r} e_i f.$$

By the previous discussion, each $e_i f$ is either equal to 0 or to e_i, and since $f \neq 0$, there is one e_i such that $e_i = f$. Then, given another decomposition

$$1 = f_1 + f_2 + \cdots + f_s$$

of 1 into orthogonal central primitive idempotents, each f_j is equal to exactly one e_i, so the decomposition is unique up to a permutation of the terms. One says that $1 = e_1 + \cdots + e_r$ is a **partition of the unity** in the algebra A; by the previous discussion, it is unique. We are then in a situation to prove the unicity in Theorem 1.24:

Lemma 1.26. *Let A a k-algebra isomorphic to $\mathrm{M}(n, C)$ for some k-division ring C. The ring C and the integer n are entirely determined by A.*

Proof. Denote $B = \mathrm{M}(n, C)$. Notice that as a left B-module, B splits into n simple modules all isomorphic to C^n. Indeed, B acts independently by multiplication on each column of a matrix:

$$b \cdot \begin{pmatrix} C_1 & C_2 & \cdots & C_n \end{pmatrix} = \begin{pmatrix} b(C_1) & b(C_2) & \cdots & b(C_n) \end{pmatrix}.$$

Moreover, C^n is a simple B-module, because if $S \subset C^n$ is a non-zero submodule, then it contains a non-zero vector v, and by multiplying by matrices in B one obtains in S all the other vectors of C^n. Therefore, as a B-module, $B = \bigoplus_{i=1}^{n} C^n$, and by a module version of the well-known theorem of Jordan–Hölder, any simple submodule of B is isomorphic to C^n, and any decomposition of B in simple modules contains n copies of C^n.

Thus, if A is isomorphic to B, then the number n is the number of terms in a decomposition of A in simple modules, so it is indeed determined by A. As for the division ring C, notice that $C^{\mathrm{opp}} = \mathrm{End}_B(C^n)$. Indeed, if $u : C^n \to C^n$ is a linear map that commutes with the action of $\mathrm{M}(n, C)$, then it is easy to see that it must write as $(c_1, \ldots, c_n) \mapsto (c_1 c, \ldots, c_n c)$ for some $c \in C$. Therefore, C is the opposite of the endomorphism ring of the unique type of simple submodule of B (or A), and it is uniquely determined. □

Remark. The version of Jordan–Hölder theorem for modules that we are speaking of is quite easy to prove if one assumes the modules to be semisimple; indeed, it is then a simple application of Schur's lemma, with an induction on the number of simple modules in a decomposition. On the other hand, Lemma 1.26 ensures that the converse of Theorem 1.24 is true: any direct sum of matrix algebras over k-division rings is semisimple, since $B = \mathrm{M}(n, C)$ splits into simple B-modules, hence is semisimple.

Proposition 1.27. *In Theorem 1.24, the d_λ's and the C^λ's are uniquely determined by A (up to permutation).*

Proof. Suppose that we have an isomorphism of k-algebras ψ between A and a sum of matrix algebras $B = \bigoplus_{\lambda \in \widehat{A}} \mathrm{M}(d_\lambda, C^\lambda)$. Denote e^λ the central idempotent of $\bigoplus_{\lambda \in \widehat{A}} \mathrm{M}(d_\lambda, C^\lambda)$ given by the matrix $I_{d_\lambda} = \mathrm{diag}(1_{C^\lambda}, \ldots, 1_{C^\lambda})$ in $\mathrm{M}(d_\lambda, C^\lambda)$. By construction, the e^λ's are orthogonal central idempotents in B. We claim that they are primitive. Indeed, otherwise, a decomposition $e^\lambda = f + g$ would correspond to a decomposition of $M = \mathrm{M}(d_\lambda, C^\lambda)$ into two non-trivial two-sided ideals fM and gM. However, it is well known that a matrix algebra over a division ring has no non-trivial two-sided ideal (this can be shown readily by looking at elementary matrices). So,

$$1_B = \sum_{\lambda \in \widehat{A}} e_\lambda$$

is the unique decomposition of 1_B in orthogonal central primitive idempotents. It follows that the number of terms $|\widehat{A}|$ in a decomposition of A is entirely determined by A, since the partition of the unity in an algebra is unique. Then, each block $\psi^{-1}(e_\lambda)(A)$ is isomorphic as a k-algebra to $\mathrm{M}(d_\lambda, C^\lambda)$, and by Lemma 1.26, the multiplicity d_λ and the division ring C^λ is uniquely determined for this block. $\square$

To conclude our analysis of general semisimple algebras, let us present the analogue of the fundamental Fourier isomorphism 1.14. We want to make concrete the isomorphism of k-algebras

$$A \to \bigoplus_{\lambda \in \widehat{A}} \mathrm{M}(d_\lambda, C^\lambda).$$

A prerequisite is a better description of the set $\widehat{A}$:

Lemma 1.28. *Denote $B = \bigoplus_{\lambda \in \widehat{A}} \mathrm{M}(d_\lambda, C^\lambda)$. Any simple module over B is isomorphic to a unique module $(C^\lambda)^{d_\lambda}$, the action of $b = \sum_{\lambda \in \widehat{A}} b^\lambda$ on $(C^\lambda)^{d_\lambda}$ being $b \cdot v^\lambda = b^\lambda(v^\lambda)$.*

Proof. The decomposition of B as a left B-module is $B = \bigoplus_{\lambda \in \widehat{A}} d_\lambda\, C^\lambda$, each matrix space being split into d_λ spaces of column vectors. Fix such a space of columns M, and a simple B-module N. Suppose that $MN \neq \{0\}$. Then, if $n \in N$ is such that $Mn \neq \{0\}$, consider the map

$$\psi : M \to N$$
$$m \mapsto mn.$$

It is a non-zero morphism between two simple B-modules, hence, an isomorphism. So, for any simple submodule M of B, either $MN = \{0\}$, or M and N are isomorphic. Since $N = BN$, there exists at least one simple submodule $M^\lambda = (C^\lambda)^{d_\lambda}$ of B isomorphic to N. Finally, there is unicity, because given two distinct submodules $M^\lambda = (C^\lambda)^{d_\lambda}$ and $M^\mu = (C^\mu)^{d_\mu}$, their product is 0 in B, so they are not isomorphic. □

We can now state the analogue for semisimple algebras of Theorems 1.3 and 1.14. Let A be a semisimple k-algebra, and $\widehat{A} = \{M^\lambda\}$ be a complete family of non-isomorphic simple left ideals of A. Denote

$$k\widehat{A} = \bigoplus_{\lambda \in \widehat{A}} \mathrm{End}_{\mathrm{End}_A(M^\lambda)}(M^\lambda).$$

Let us detail this a bit. An element d^λ of $D^\lambda = \mathrm{End}_A(M^\lambda)$, which is a division ring, acts naturally on the left of M^λ, with the rule

$$\forall a \in A,\ d^\lambda(a \cdot m^\lambda) = a \cdot d^\lambda(m^\lambda).$$

Then, an element of $k\widehat{A}$ is a formal sum $\sum_{\lambda \in \widehat{A}} u^\lambda$ of k-linear maps $u^\lambda : M^\lambda \to M^\lambda$, such that

$$\forall d^\lambda \in D^\lambda,\ d^\lambda(u^\lambda(m^\lambda)) = u^\lambda(d^\lambda(m^\lambda)).$$

Theorem 1.29. *We define the Fourier transform $\widehat{a}$ of $a \in A$ as the element of $k\widehat{A}$ whose λ-component is*

$$\widehat{a}(\lambda) : m^\lambda \mapsto a \cdot m^\lambda.$$

1. *The Fourier transform is an isomorphism of k-algebras and of A-modules between A and $k\widehat{A}$.*

2. *The Grothendieck group of the category of left modules of A is $\bigoplus_{\lambda \in \widehat{A}} \mathbb{Z}M^\lambda$.*

Proof. First, notice that $\widehat{a}(\lambda)$ belongs to $\mathrm{End}_{D^\lambda}(M^\lambda)$ for all $\lambda \in \widehat{A}$. Indeed, this is the relation $d^\lambda(a \cdot m^\lambda) = a \cdot d^\lambda(m^\lambda)$ previously stated. Suppose that $\widehat{a} = 0$. Then, for any $m^\lambda \in M^\lambda$, $a \cdot m^\lambda = 0$. However, A is a sum of copies of the M^λ, so for any $b \in A$, $ab = 0$. In particular, $a = a1_A = 0$; hence, the Fourier transform is injective. Using Theorem 1.24, one sees that $\dim_k A = \dim_k k\widehat{A}$, so the Fourier transform is an isomorphism of vector spaces. It is evident that $\widehat{ab} = \widehat{a}\,\widehat{b}$, so it is even an isomorphism of k-algebras. The compatibility with the action of A is also evident. Remark that the Fourier transform indeed yields a concrete realization of the identification

$$A = \bigoplus_{\lambda \in \widehat{A}} \mathrm{M}(d_\lambda, C^\lambda),$$

since by a previous remark $\mathrm{End}_{D^\lambda}((D^\lambda)^{d_\lambda}) = \mathrm{M}(d_\lambda, (D^\lambda)^{\mathrm{opp}}) = \mathrm{M}(d_\lambda, C^\lambda)$.

As for the second part of the theorem, we saw in the previous lemma that any simple module is isomorphic to some M^λ, so any module over A, which is semisimple, writes as

$$M = \bigoplus_{\lambda \in \widehat{A}} m_\lambda \, M^\lambda$$

for some multiplicities $m_\lambda \in \mathbb{N}$. The multiplicities m_λ are uniquely determined by M, since

$$\dim_k \mathrm{Hom}_A(M, M^\lambda) = m_\lambda \, \dim_k \mathrm{End}_A(M^\lambda) = m_\lambda \, \dim_k(D^\lambda).$$

Therefore, the group of classes of isomorphism of modules over A is indeed $\bigoplus_{\lambda \in \widehat{A}} \mathbb{Z} M^\lambda$. □

Remark. In terms of central primitive idempotents, $\widehat{a}$ is the collection of endomorphisms ae^λ acting on each space $d_\lambda M^\lambda$. Hence, one can see the Fourier transform as a reformulation of the partition of the unity of a semisimple algebra.

Example. As an example of the general theory developed before, let us detail the representation theory of finite groups over an arbitrary field k (not necessarily $k = \mathbb{C}$). This is equivalent to the module theory of the group algebra kG. Assume that $\mathrm{char}(k)$ does not divide $|G|$ (this is in particular the case in characteristic zero). Then, the group algebra kG is semisimple. Indeed, consider a kG-module M and a submodule N. If π is a linear projection $M \to N$, then

$$\widetilde{\pi} = \frac{1}{|G|} \sum_{g \in G} g \pi g^{-1}$$

is a G-equivariant projection $M \to N$. Indeed, since N is a kG-submodule, $\mathrm{Im}\, \widetilde{\pi}$ is a subspace of N. Moreover, on N,

$$\widetilde{\pi}(n) = \frac{1}{|G|} \sum_{g \in G} g \pi g^{-1}(n) = \frac{1}{|G|} \sum_{g \in G} g g^{-1}(n) = n,$$

so $\widetilde{\pi}$ is a projection on N. Finally, for all $h \in G$,

$$h\widetilde{\pi} = \frac{1}{|G|} \sum_{g \in G} h g \pi g^{-1} = \frac{1}{|G|} \sum_{k \in G} k \pi k^{-1} h = \widetilde{\pi} h.$$

Set $P = \mathrm{Ker}\, \widetilde{\pi}$. Then, $M = N \oplus P$, and the G-equivariance of $\widetilde{\pi}$ ensures that P is a kG-submodule of M. Therefore, assuming always $\mathrm{char}(k) \nmid \mathrm{card}\, G$, one has

$$kG = \bigoplus_{\lambda \in \widehat{G}^k} \mathrm{End}_{D^\lambda}(V^\lambda),$$

where $\widehat{G}^k$ is a complete family of representatives of the isomorphism classes of representations of G on a k-vector space; and the $D^\lambda = \mathrm{End}_G(V^\lambda)$ are k-division rings. In particular, if k is algebraically closed, then the same results as in Section 1.1 hold: kG is isomorphic to the direct sum of the matrix algebras $\mathrm{End}_k(V^\lambda) = \mathrm{End}(V^\lambda)$, and $|\widehat{G}^k|$ is the number of conjugacy classes of G.

Over a general field k, one can then develop a character theory, which in particular allows one to decide whether non-trivial k-division rings D appear in the expansion of kG (if this is not the case then k is called a splitting field for G); see the references at the end of the chapter.

▷ *Constructions on the category of modules.*

In this paragraph, we use the language of algebras and modules to revisit some constructions on representations introduced in Section 1.2. The direct sum and the tensor product of two representations V and W of a finite group G are generalized by the direct sum and the tensor product of modules over an algebra

$$M \oplus N \qquad ; \qquad M \otimes_A N.$$

Let $A \subset B$ be two k-algebras. Then, any left module over B can be seen as left module over A for the restriction of the map $B \times M \to M$ to $A \times M \to M$. This gives rise to the restriction functor

$$\mathrm{Res}_A^B : \text{left } B\text{-modules} \to \text{left } A\text{-modules}.$$

Given two groups $H \subset G$, one has an inclusion of algebras $\mathbb{C}H \subset \mathbb{C}G$, and the restricted representation $\mathrm{Res}_H^G(V)$ is with the language of modules $\mathrm{Res}_{\mathbb{C}H}^{\mathbb{C}G}(V)$.

The theory becomes more interesting with induced modules and representations. If M is a left A-module, we can consider the tensor product

$$\mathrm{Ind}_A^B(M) = B \otimes_A M,$$

where B is considered as a right A-module for $b \cdot a = ba \in B$. Thus, we have in $\mathrm{Ind}_A^B(M)$ the rule of calculus $ba \otimes m = b \otimes (a \cdot m)$ for any $a \in A$, $b \in B$ and $m \in M$. Now, the k-vector space $\mathrm{Ind}_A^B(M)$ is a left B-module for the operation $b \cdot (b' \otimes m) = bb' \otimes m$. We say that $\mathrm{Ind}_A^B(M)$ is the induced module of M from A to B, and we have an induction functor

$$\mathrm{Ind}_A^B : \text{left } A\text{-modules} \to \text{left } B\text{-modules}.$$

Remark. More generally, given a k-vector space X which is both a left B-module and a right A-module, $M \mapsto X \otimes_A M$ yields a functor of *generalized* induction. This kind of construction is particularly useful in the representation theory of Lie groups over finite fields, such as $\mathrm{GL}(n, \mathbb{F}_q)$. The usual induction functor Ind_A^B corresponds to the choice $X = B$.

Proposition 1.30. *Given two groups $H \subset G$ and a representation V of H, the induced representation $\mathrm{Ind}_H^G(V)$ as defined in Section 1.2 is equal to $\mathrm{Ind}_{\mathbb{C}H}^{\mathbb{C}G}(V)$ in the sense of induced modules over algebras.*

Proof. We defined $\mathrm{Ind}_H^G(V)$ as the set of functions $f : G \to V$ such that $f(hg) = h \cdot f(g)$. On the other hand, an element of $\mathbb{C}G \otimes_{\mathbb{C}H} V$ can be written as a formal linear sum of elements $(g \otimes f(g^{-1}))$ with the $f(g^{-1})$'s in V. Using the rules of computation in $\mathbb{C}G \otimes_{\mathbb{C}H} V$, we have

$$|H|\,(g \otimes f(g^{-1})) = \sum_{h \in H} gh \otimes h^{-1} \cdot f(g^{-1}).$$

Therefore, an element of $\mathbb{C}G \otimes_{\mathbb{C}H} V$ can be written uniquely as a formal linear sum

$$\sum_{g \in G} g \otimes f(g^{-1})$$

where f satisfies $h^{-1} \cdot f(g^{-1}) = f(h^{-1}g^{-1})$. This yields the isomorphism asked for. □

An application of this point of view is a proof of Proposition 1.11. Let V be a representation of H; G a group containing H; $(e_1, \ldots, e_n)$ of the representation V; and $(g_1, \ldots, g_r)$ a set of representatives of G/H. Since $\mathrm{Ind}_H^G(V) = \mathbb{C}G \otimes_{\mathbb{C}H} V$, a basis of $\mathrm{Ind}_H^G(V)$ consists in the tensors $g_j \otimes e_i$ with $i \in [\![1, n]\!]$ and $j \in [\![1, r]\!]$. Denote $[g_j \otimes e_i](x)$ the coefficient of $g_j \otimes e_i$ in a general tensor $x \in \mathrm{Ind}_H^G(V)$. Then,

$$\begin{aligned}
\mathrm{ch}^{\mathrm{Ind}_H^G(V)}(g) &= \sum_{i=1}^{n} \sum_{j=1}^{r} [g_j \otimes e_i](gg_j \otimes e_i) = \sum_{i=1}^{n} \sum_{j \mid gg_j = g_j h} [g_j \otimes e_i](g_j \otimes h \cdot e_i) \\
&= \sum_{i=1}^{n} \sum_{j \mid g_j^{-1} g g_j = h} [e_i](h \cdot e_i) = \sum_{g_j \in G/H} \mathrm{ch}^V(g_j^{-1} g g_j).
\end{aligned}$$

1.5 The double commutant theory

▷ *Bimodules and the canonical bimodule associated to a left module.*

In the theory of semisimple algebras, given a simple A-module M, we saw in the previous section the importance of $D = \mathrm{End}_A(M)$ and of its opposite $C = D^{\mathrm{opp}}$. One has a right action of C on M given by $m \cdot c = c(m)$, and by definition it commutes with the action of A:

$$a \cdot (m \cdot c) = (a \cdot m) \cdot c.$$

The notion of **bimodule** generalizes this kind of situation. Let A and C be two k-algebras, and B be a k-vector space (here and in the next example we make no assumption of finite dimension). One says that B is a bimodule for (A, C) if it is a left A-module and a right C-module, with the compatiblity rule $a \cdot (b \cdot c) = (a \cdot b) \cdot c$ for all $a \in A$, $b \in B$ and $c \in C$.

Example. Consider the space of tensors $(\mathbb{C}^N)^{\otimes n} = \mathbb{C}^N \otimes \mathbb{C}^N \otimes \cdots \otimes \mathbb{C}^N$. It is endowed with a right action of $\mathbb{C}\mathfrak{S}(n)$ given by

$$(v_1 \otimes v_2 \otimes \cdots \otimes v_n) \cdot \sigma = v_{\sigma(1)} \otimes v_{\sigma(2)} \otimes \cdots \otimes v_{\sigma(n)}.$$

There is also a left action of $\mathrm{GL}(N,\mathbb{C})$, the so-called diagonal action

$$u \cdot (v_1 \otimes v_2 \otimes \cdots \otimes v_n) = u(v_1) \otimes u(v_2) \otimes \cdots \otimes u(v_n).$$

The two structures are compatible and $(\mathbb{C}^N)^{\otimes n}$ is a $(\mathbb{C}\mathrm{GL}(N,\mathbb{C}), \mathbb{C}\mathfrak{S}(n))$-bimodule — here, by $\mathbb{C}\mathrm{GL}(N,\mathbb{C})$ we mean the set of *finite* formal sums of elements of $\mathrm{GL}(N,\mathbb{C})$; it is an infinite-dimensional $\mathbb{C}$-algebra for the convolution product.

In fact, given an algebra A, every left A-module M is canonically a bimodule for a pair (A, C) with C adequately chosen, namely,

$$C = (\mathrm{End}_A(M))^{\mathrm{opp}}.$$

Every structure of bimodule is a sub-structure of this construction, in the following sense: if B is a (A, C)-bimodule, then there is a morphism of algebras $C \to (\mathrm{End}_A(B))^{\mathrm{opp}}$ by definition of the structure of bimodule. Thus, to understand the structures of bimodules, it suffices in a sense to understand the **commutant** of a left A-module, defined as the algebra

$$\mathrm{Com}(A, M) = (\mathrm{End}_A(M))^{\mathrm{opp}}.$$

▷ *The double commutant theorem.*

In the case of (finite-dimensional) semisimple algebras, the commutants have some special properties:

Proposition 1.31. *Let A be a semisimple algebra, and M be a left A-module. The commutant algebra $A' = \mathrm{Com}(A, M)$ is semisimple.*

Proof. As a module over a semisimple algebra A, $M = \bigoplus_{\lambda \in \widehat{A}} m_\lambda M^\lambda$. Then,

$$\mathrm{End}_A(M) = \bigoplus_{\lambda \in \widehat{A}} \mathrm{End}_A(m_\lambda M^\lambda) = \bigoplus_{\lambda \in \widehat{A}} \mathrm{M}(m_\lambda, \mathrm{End}_A(M^\lambda)) = \bigoplus_{\lambda \in \widehat{A}} \mathrm{M}(m_\lambda, D^\lambda).$$

Thus, $\mathrm{End}_A(M)$ is a sum of matrix algebras over k-division rings, and by the discussion of Lemma 1.26, these algebras are semisimple. Therefore, $\mathrm{End}_A(M)$ is a semisimple algebra, and its opposite $\mathrm{Com}(A, M)$ also. □

Theorem 1.32 (Wedderburn). *With the same assumptions, the bicommutant $A'' = (A')'$ is equal to the image of A in $\mathrm{End}_k(M)$ by*

$$a \mapsto (m \mapsto a \cdot m).$$

In other words, if A is a subalgebra of $\mathrm{End}_k(M)$, then $A'' = A$.

In order to prove this, notice first that if $A \subset \mathrm{End}_k(M)$, then in the decomposition $M = \bigoplus_{\lambda \in \widehat{A}} m_\lambda M^\lambda$, all the multiplicities m_λ are ≥ 1. Indeed, consider a simple left ideal M^λ of A. We have to show that $\mathrm{Hom}_A(M^\lambda, M)$ is non-zero. Since $M^\lambda \subset A \subset \mathrm{End}_k(M)$, one can find a vector $v \in M$ such that $M^\lambda v \neq \{0\}$. Then,

$$a \in M^\lambda \mapsto a \cdot v \in M$$

is a non-zero element in $\mathrm{Hom}_A(M^\lambda, M)$.

Lemma 1.33. *Consider the canonical map*

$$\psi : \bigoplus_{\lambda \in \widehat{A}} M^\lambda \otimes_{(\mathrm{End}_A(M^\lambda))^{\mathrm{opp}}} \mathrm{Hom}_A(M^\lambda, M) \to M$$

which sends $m^\lambda \otimes u^\lambda$ to $u^\lambda(m^\lambda)$. This map ψ is an isomorphism of (A, A')-bimodules, and it yields a decomposition of M into non-isomorphic simple A' modules $\mathrm{Hom}_A(M^\lambda, M)$, each with multiplicity $\dim_{C^\lambda}(M^\lambda)$.

Proof. The tensor products are well defined, since $C^\lambda = (\mathrm{End}_A(M^\lambda))^{\mathrm{opp}}$ acts on the right of M^λ, and on the left of $\mathrm{Hom}_A(M^\lambda, M)$ by the composition map

$$\begin{aligned}(\mathrm{End}_A(M^\lambda))^{\mathrm{opp}} \times \mathrm{Hom}_A(M^\lambda, M) &\to \mathrm{Hom}_A(M^\lambda, M)\\ (t, u) &\mapsto u \circ t.\end{aligned}$$

Since M is a direct sum of copies of M^λ's, it is easily seen that $\mathrm{Im}\,\psi$ contains a generating family of M, so ψ is surjective. For dimension reasons, it is an isomorphism of k vector-spaces. The compatibility with the left action of A comes from the fact that right terms in the tensors of $M^\lambda \otimes_k \mathrm{Hom}_A(M^\lambda, M)$ are morphisms of A-modules. As for the right action of A', on the source of ψ, it comes from the map of composition

$$\begin{aligned}\mathrm{Hom}_A(M^\lambda, M) \times (\mathrm{End}_A(M))^{\mathrm{opp}} &\to \mathrm{Hom}_A(M^\lambda, M)\\ (u, v) &\mapsto v \circ u,\end{aligned}$$

and the compatibility of ψ is then quite evident.

Now, we claim that each $\mathrm{Hom}_A(M^\lambda, M)$ is a simple right A'-module. To prove this, consider a non-zero morphism of A-modules $u_1 : M^\lambda \to M$. Since M^λ is simple, u_1 is an imbedding of M^λ into M. We want to show that if $u_2 : M^\lambda \to M$ is another imbedding of M^λ into M, then there exists $v \in \mathrm{End}_A(M)$ with $v \circ u_1 = u_2$; it will ensure the simplicity of $\mathrm{Hom}_A(M^\lambda, M)$ as a right A'-module. Let P_1 and P_2 be complements of $u_1(M^\lambda)$ and of $u_2(M^\lambda)$ in M. Denote $v = u_2 u_1^{-1}$, which is an isomorphism of A-modules between $u_1(M^\lambda)$ and $u_2(M^\lambda)$. One can extend it between $M = u_1(M^\lambda) \oplus P_1$ and $M = u_2(M^\lambda) \oplus P_2$, by setting $v_{|P_1} = 0$. Then, v is indeed a morphism of A-modules such that $v \circ u_1 = u_2$. Suppose that two spaces $\mathrm{Hom}_A(M^\lambda, M)$ and $\mathrm{Hom}_A(M^\mu, M)$ are isomorphic as A'-modules by a map ψ. Let

$u \in \mathrm{Hom}_A(M^\lambda, M)$ be a non-zero element, which is an imbedding of M^λ into M; and $\pi \in A'$ be an A-projection on $u(M^\lambda)$. Then, $\psi(u)$ is an imbedding of M^μ into M, and $\psi(u) = \psi(u \cdot \pi) = \psi(u) \cdot \pi$, so its image is included into a simple module isomorphic to M^λ; hence, $M^\lambda = M^\mu$. As a consequence,

$$M = \bigoplus_{\lambda \in \widehat{A}} M^\lambda \otimes_{C^\lambda} \mathrm{Hom}_A(M^\lambda, M) = \bigoplus_{\lambda \in \widehat{A}} \dim_{C^\lambda}(M^\lambda)\, \mathrm{Hom}_A(M^\lambda, M)$$

is the decomposition of M into non-isomorphic right A' simple modules. □

Proof of Theorem 1.32. Consider the map

$$a \in A \mapsto (m \mapsto a \cdot m) \in A''.$$

It is injective since $A \subset \mathrm{End}_k(M)$, and every non-zero endomorphism of M is non-zero on some vector $m \in M$. Thus, it suffices to show that A and A'' have the same dimensions. Using the previous Lemma, we compute

$$\begin{aligned}\mathrm{End}_{A'}(M) &= \bigoplus_{\lambda \in \widehat{A}} \mathrm{End}_{A'}\left(d_\lambda \,\mathrm{Hom}_A(M^\lambda, M)\right) \\ &= \bigoplus_{\lambda \in \widehat{A}} \mathrm{M}\big(d_\lambda, \mathrm{End}_{A'}(\mathrm{Hom}_A(M^\lambda, M))\big).\end{aligned}$$

Let us identify $\mathrm{End}_{A'}(\mathrm{Hom}_A(M^\lambda, M))$. Let $\phi : \mathrm{Hom}_A(M^\lambda, M) \to \mathrm{Hom}_A(M^\lambda, M)$ be a non-zero endomorphism of right A'-module. It is uniquely determined by its value on a single imbedding $u : M^\lambda \to M$, since $u \cdot A' = \mathrm{Hom}_A(M^\lambda, M)$ by simplicity of the module. Denote $v = \phi(u)$. By using as before an A-projection π on $\mathrm{Im}\, u$, we see that $v = \phi(u) = \phi(u \cdot \pi) = \phi(u) \cdot \pi = v \cdot \pi$;therefore, v has the same image as u. Then, one sees that ϕ is the map

$$\begin{aligned}\mathrm{Hom}_A(M^\lambda) &\to \mathrm{Hom}_A(M^\lambda) \\ w &\mapsto w \circ u^{-1} \circ v,\end{aligned}$$

so it is given by $u^{-1}v \in \mathrm{End}_A(M^\lambda)$. Therefore, $\mathrm{End}_{A'}(\mathrm{Hom}_A(M^\lambda, M))$ is simply equal to $D^\lambda = \mathrm{End}_A(M^\lambda)$, and we conclude that there is an isomorphism of k-algebras

$$A'' = (\mathrm{End}_{A'}(M))^{\mathrm{opp}} = \bigoplus_{\lambda \in \widehat{A}} \mathrm{M}(d_\lambda, (D^\lambda)^{\mathrm{opp}}) = \bigoplus_{\lambda \in \widehat{A}} \mathrm{M}(d_\lambda, C^\lambda) = A.$$ □

If B is an (A, C)-bimodule with $A = C'$ and $C' = A$, we shall say that A and C are in **duality** for their actions on B. If A is semisimple, then it suffices to check that $C' = A$ by the bicommutant theorem. Moreover, in this case, there is a decomposition of B as

$$\bigoplus_{\lambda} M^\lambda \otimes_{C^\lambda} N^\lambda,$$

where M^λ runs over a complete family of simple left A-modules, N^λ runs over a complete family of simple right C-modules, and $C^\lambda = (\mathrm{End}_A(M^\lambda))^{\mathrm{opp}} = \mathrm{End}_C(N^\lambda)$. One also has $N^\lambda = \mathrm{Hom}_A(M^\lambda, B)$ and $M^\lambda = \mathrm{Hom}_C(N^\lambda, B)$. This is a very powerful tool to transform problems in the representation theory of A into problems in the representation theory of C, which might be a simpler algebra than A. In Chapter 5, we shall encounter an important example of this phenomenon, with C a deformation of a symmetric group algebra $\mathbb{C}\mathfrak{S}(n)$.

Example. Let A be a semisimple k-algebra. It acts on itself by multiplication on the left, and the commutant of A is A itself. Thus, one has the decomposition of A as an (A,A)-bimodule:

$$A = \bigoplus_{\lambda \in \widehat{A}} M^\lambda \otimes_{(\mathrm{End}_A(M^\lambda))^{\mathrm{opp}}} \mathrm{Hom}_A(M^\lambda, A).$$

Example. Let $H \subset G$ be two finite groups, and $V = \mathrm{Ind}_H^G(\mathbb{C})$ be the representation of G induced from the trivial representation of H. An element of V can be seen as a function $f = \sum_{g \in G} f(g)\, g \in \mathbb{C}G$ such that $f = fh$ for any $h \in H$. Hence, it is a function on G/H:

$$V = \mathbb{C}[G/H].$$

Notice that $\mathbb{C}[G/H]$ is a subalgebra of $\mathbb{C}G$, and a left ideal of it. Then,

$$\mathbb{C}[G/H] \subset \mathbb{C}G \Rightarrow \mathrm{Com}(\mathbb{C}G, \mathbb{C}[G/H]) \subset \mathrm{Com}(\mathbb{C}[G/H], \mathbb{C}[G/H]) = \mathbb{C}[G/H],$$

so the commutant of $\mathbb{C}G$ with respect to $V = \mathbb{C}[G/H]$ has to be a subalgebra of $\mathbb{C}[G/H]$. We claim that:

$$\begin{aligned} \mathrm{Com}(\mathbb{C}G, C[G/H]) = \mathbb{C}[H\backslash G/H] &= \left\{ \sum f(g)\, HgH \right\} \\ &= \{\text{functions } f \text{ bi-}H\text{-invariant}\}, \end{aligned}$$

the action on $\mathbb{C}[G/H]$ being by multiplication on the right. Indeed, let $u : \mathbb{C}[G/H] \to \mathbb{C}[G/H]$ be a morphism of $\mathbb{C}G$-modules, and denote $f = u(H)$. Since u is a G-morphism, for any element $v = \sum_{g \in G/H} v(g)\, gH$,

$$u(v) = \sum_{g \in G/H} v(g)\, g\, u(H) = vf$$

so u is the multiplication by f on the right. Then, $f = u(H) = u(hH) = h\,u(H) = hf$, so f, which by construction is in $\mathbb{C}[G/H]$, is also left H-invariant, and our claim is proven. The commutant algebra $\mathbb{C}[H\backslash G/H]$ is called the **Hecke algebra** of the pair (G,H). It is semisimple, with a family of classes of isomorphism of modules that is in bijection with the simple G-modules occurring in the decomposition of $\mathbb{C}[G/H]$. In Chapter 5, we shall study an important example of Hecke algebra $\mathfrak{H}(G,H)$ with $G = \mathrm{GL}(n, \mathbb{F}_q)$, and H equal to the subgroup of upper triangular matrices.

▷ *Bitraces and their expansions.*

As in Section 1.2, if A is a (semisimple) k-algebra and M is a left A-module, we call character of M the map

$$\begin{aligned} \mathrm{ch}^M : A &\to k \\ a &\mapsto \mathrm{tr}(\rho^M(a)) \end{aligned}$$

where $\rho^M(a)$ is the k-endomorphism of M defined as $m \mapsto a \cdot m$. Every character is a sum of the simple characters ch^λ, and the double commutant theorem is a very powerful tool in order to compute these simple characters. Suppose to simplify that k is algebraically closed, so that every division ring over k is k itself. Then, if B is an (A,C)-bimodule with A and C in duality, call **bitrace** of a pair (a,c) the trace $\mathrm{btr}^M(a,c)$ of the k-endomorphism of M

$$b \mapsto a \cdot b \cdot c.$$

Writing $B = \bigoplus_\lambda M^\lambda \otimes_k N^\lambda$, one has

$$\mathrm{btr}^M(a,c) = \sum_\lambda \mathrm{ch}^{M^\lambda}(a)\, \mathrm{ch}^{N^\lambda}(c).$$

Thus, if one wants to compute the simple characters ch^{M^λ} of an algebra A, one can proceed as follows:

1. Find a "sufficiently big" module M on which A acts. By sufficiently big, we mean that every class of simple A-module M^λ occurs in M.

2. Identify $C = \mathrm{Com}(A, M)$, and decompose M as the sum of simple (A,C)-bimodules $M^\lambda \otimes_k N^\lambda$.

3. Compute btr^M and the characters of the simple C-modules appearing in M. Then, the characters of the simple A-modules are obtained by the duality formula presented above.

In many situations, it turns out that the computation of the bitrace is easy. So if the character theory of C is already known (using other techniques), then the duality between A and C allows one to obtain the character values of A. In Chapter 2, we shall prove a well-known duality of algebras, namely, the **Schur–Weyl duality** between (the Schur algebra of) $\mathrm{GL}(N,\mathbb{C})$ and (the group algebra of) $\mathfrak{S}(n)$ on the space of tensors $(\mathbb{C}^N)^{\otimes n}$. This will explain the Frobenius formula for characters of the symmetric groups.

Example. Let A be a semisimple algebra over an algebraically closed field k. We have the decomposition of A as an (A,A)-bimodule

$$A = \bigoplus_{\lambda \in \widehat{A}} M^\lambda \otimes_k \mathrm{Hom}_A(M^\lambda, A).$$

We claim that $\mathrm{Hom}_A(M^\lambda, A) = \mathrm{Hom}_k(M^\lambda, k)$ as k-vector spaces. To perform this identification, consider the map

$$\psi : \mathrm{Hom}_A(M^\lambda, A) \to \mathrm{Hom}_k(M^\lambda, \mathrm{End}_A(M^\lambda))$$
$$u \mapsto (m^\lambda \mapsto (n^\lambda \mapsto u(n^\lambda)\, m^\lambda)).$$

If $u \in \mathrm{Hom}_A(M^\lambda, A)$, then for any $m^\lambda, n^\lambda \in m^\lambda$ and any $a \in A$,

$$(\psi(u)(m^\lambda))(an^\lambda) = u(an^\lambda)\, m^\lambda = a\, u(n^\lambda)\, m^\lambda = a\left((\psi(u)(m^\lambda))(n^\lambda)\right),$$

so $\psi(u)(m^\lambda) \in \mathrm{End}_A(M^\lambda)$, and $\psi(u)$ is a k-linear map in $\mathrm{Hom}_k(M^\lambda, \mathrm{End}_A(M^\lambda))$. Our definition thus makes sense. Suppose that $\psi(u) = 0$. Then, for every $m^\lambda \in M^\lambda$, $u(M^\lambda)\, m^\lambda = 0$, hence $u(M^\lambda)\, M^\lambda = 0$. By simplicity of M^λ, using the same reasoning as in Lemma 1.28, this implies that $u(M^\lambda) = 0$, hence, $u = 0$. So, ψ is injective. Since

$$\dim_k \mathrm{Hom}_A(M^\lambda, A) = d_\lambda = \dim_k M^\lambda = \dim_k \mathrm{Hom}_k(M^\lambda, k),$$

it is an isomorphism of k vector spaces (with $k = \mathrm{End}_A(M^\lambda)$ in the last term). Thus, as an (A,A)-bimodule,

$$A = \bigoplus_{\lambda \in \widehat{A}} M^\lambda \otimes_k \mathrm{Hom}_k(M^\lambda, k),$$

where the action of A on the right tensors in $\mathrm{Hom}_k(M^\lambda, k)$ is $(u^\lambda \cdot a)(\cdot) = u^\lambda(a \cdot)$. At the level of characters, this leads to the formula:

$$\mathrm{btr}^A(a, b) = \sum_{\lambda \in \widehat{A}} \mathrm{ch}^\lambda(a)\, \mathrm{ch}^\lambda(b)$$

since the right A-module $\mathrm{Hom}_k(M^\lambda, k)$ corresponds to the left A-module M^λ, hence has the same character.

Remark. The previous example connects the Artin–Wedderburn decomposition $A = \bigoplus_{\lambda \in \widehat{A}} \mathrm{M}(d_\lambda, k)$, and the decomposition as an (A,A)-bimodule, since

$$M^\lambda \otimes_k \mathrm{Hom}_k(M^\lambda, k) = \mathrm{End}_k(M^\lambda) = \mathrm{M}(d_\lambda, k).$$

We leave to the reader the extension of this identification to the case of general fields k, where division rings C^λ can be non-trivial.

Example. Let G be a finite group. We already know the **first orthogonality relation**

$$\forall \lambda, \mu \in \widehat{G},\ \frac{1}{|G|} \sum_{g \in G} \mathrm{ch}^\lambda(g^{-1})\, \mathrm{ch}^\mu(g) = \left\langle \mathrm{ch}^\lambda \,\middle|\, \mathrm{ch}^\mu \right\rangle_G = \delta_{\lambda,\mu}.$$

There is a **second orthogonality relation** coming from the decomposition of the bitrace of $\mathbb{C}G$. Indeed, one has

$$\mathrm{btr}^{\mathbb{C}G}(g^{-1}, h) = \sum_{k \in G} \delta_{k, g^{-1}kh} = \sum_{k \in G} \delta_{k^{-1}gk, h}$$
$$= \begin{cases} \frac{\mathrm{card}\, G}{\mathrm{card}\, C_g} & \text{if } g \text{ and } h \text{ are conjugated;} \\ 0 & \text{otherwise,} \end{cases}$$

where C_g is the conjugacy class of g. Consequently,

$$\forall g,h \in G,\ \sum_{\lambda \in \widehat{G}} \mathrm{ch}^\lambda(g^{-1})\, \mathrm{ch}^\mu(h) = \begin{cases} \frac{|G|}{|C_g|} & \text{if } g \text{ and } h \text{ are conjugated;} \\ 0 & \text{otherwise.} \end{cases}$$

Remark. Let G be a finite group, and V^λ be an irreducible representation of G, or in other words a simple $\mathbb{C}G$-module. We defined before ch^λ and χ^λ as functions on G. The previous discussion allows one to extend these definitions, and to also see the character ch^λ and the normalized character χ^λ as maps $\mathbb{C}G \to \mathbb{C}$, defined by

$$\mathrm{ch}^\lambda\left(\sum_g f(g)\, g\right) = \sum_g f(g)\, \mathrm{ch}^\lambda(g);$$

$$\chi^\lambda\left(\sum_g f(g)\, g\right) = \sum_g f(g)\, \chi^\lambda(g).$$

In particular, if $a \in Z(\mathbb{C}G)$, then one recognizes $\chi^\lambda(a) = \widehat{a}(\lambda)$, where $\widehat{\cdot}$ is the restriction of the Fourier transform to the center of the group algebra, that is to say a map $Z(\mathbb{C}G) \to Z(\mathbb{C}\widehat{G})$. In the remainder of the book, in order to distinguish between the global Fourier transform $\widehat{\cdot} : \mathbb{C}G \to \mathbb{C}\widehat{G}$ and its restriction to the center, we shall use in the latter case the character notation χ^λ. This new notation allows us to avoid the redefinition of $\widehat{\cdot} : Z(\mathbb{C}G) \to Z(\mathbb{C}\widehat{G}) = \mathbb{C}^{\widehat{G}}$ performed in Section 1.3.

Notes and references

For generalities in algebra (the language of groups, vector spaces, modules, etc.), we refer to [Lan02], whose first chapters will fill any possible gap. As for the representation theory of finite groups, the content of this chapter can also be found in [Ser77, CSST08], with a very similar approach but for the module point of view. A more elementary treatment is contained in [JL93]. We do not use the terminology of *intertwiner* for a morphism between two representations: since morphisms of representations of G are morphisms of $\mathbb{C}G$-modules, we did not find it useful to introduce another term to describe them. We also omitted the notion of *simple algebra*, and only spoke of semisimple algebras; for our purpose it will prove sufficient.

Our description of the non-commutative Fourier transform is inspired by the classical textbooks on representations of compact Lie groups, such as [BD85, Var89, Far08]; as explained after Proposition 1.15, one does not need too much additional work in order to extend the theory of representations of finite groups to this setting. Again, we were also inspired by [CSST08], and by the first chapter of [CSST10]. In particular, the heat equation on the symmetric group $\mathfrak{S}(n)$ is studied in much more details in [CSST08, Chapter 10]. We also refer to the papers of Diaconis [DS81, AD86, Dia86] for a proof of the cut-off phenomenon at time $t = \frac{1}{2} n \log n$ of the total variation distance between the law of the random process and the uniform law on $\mathfrak{S}(n)$.

The extension of representation theory to the setting of semisimple algebras is treated in part in [GW09, Chapter 4], and many duality results are exposed in this book. We also followed the beginning of [GS06] for the Artin–Wedderburn theorem. It should be noticed that many deep results in the representation theory of algebras are proved by *abstract non-sense*, that is to say, by using only natural transformations of the spaces considered. This is in particular the case for the sequence of isomorphisms in the proof of Theorem 1.24; for the reasonings on the non-commutative Fourier transform $A \to k\widehat{A}$; and for the identifications $\mathrm{End}_{A'}(\mathrm{Hom}_A(M^\lambda, M)) = \mathrm{End}_A(M^\lambda)$ and $\mathrm{Hom}_A(M^\lambda, A) = \mathrm{Hom}_{C^\lambda}(M^\lambda, C^\lambda)$. Thus, though more abstract, the representation theory of algebras is in some sense more canonical, and we tried to present it accordingly.

In our approach of the bicommutant theorem, our definition of $\mathrm{Com}(A, M)$ as *the opposite of* $\mathrm{End}_A(M)$ is a bit unusual, but it is certainly more adequate for dealing with bimodules. More generally, we tried throughout this chapter to make each k-algebra act on the most natural side of a k-vector space, and this leads to some small differences between our presentation and those of other textbooks. For instance, we always consider tensor products of modules $M \otimes_A N$ with a *right* A-module M and a left A-module N, although the usual definition is with two *left* A-modules. Similarly, we consider more natural the permutation representation of $\mathfrak{S}(n)$ on $\mathbb{C}^n$ when written on the right. Finally, though our book is not concerned with the representation theory of groups over other fields than $\mathbb{C}$, the remarks on page 38 lead eventually to the so-called *modular representation theory*, and we refer to [Ser77, Chapters 12-19] and to [Alp93] for an introduction to this subject. The discussion of Sections 4.2 and 4.3 will also shed a light on this theory.

2

Symmetric functions and the Frobenius–Schur isomorphism

If n is a non-negative integer, we call **(integer) partition of size** n a non-increasing sequence of positive integers

$$\mu = (\mu_1 \geq \mu_2 \geq \cdots \geq \mu_\ell)$$

such that $|\mu| = \sum_{i=1}^{\ell} \mu_i = n$. Denote $\mathfrak{Y}(n)$ the set of all integer partitions of size n; for instance, $\mathfrak{Y}(4) = \{(4),(3,1),(2,2),(2,1,1),(1,1,1,1)\}$. In Section 2.1, we shall see that conjugacy classes in $\mathfrak{S}(n)$ are in bijection with $\mathfrak{Y}(n)$. From the discussion of Chapter 1 (cf. Corollary 1.18), the set $\mathfrak{Y}(n)$ of integer partitions of size n should then also be in bijection with $\widehat{\mathfrak{S}}(n)$, the set of classes of isomorphism of irreducible representations of $\mathfrak{S}(n)$.

Quite surprisingly, if one is only interested in the irreducible characters of the symmetric groups, then there is a way to realize this bijection and to compute these characters *without knowing the actual irreducible representations* S^λ labeled by integer partitions $\lambda \in \mathfrak{Y}(n)$. The trick is to introduce the Grothendieck ring of representations

$$\mathrm{R}_0(\mathfrak{S}) = \bigoplus_{n=0}^{\infty} \mathrm{R}_0(\mathfrak{S}(n))$$

of all the symmetric groups, that is endowed with a product stemming from the inclusions $\mathfrak{S}(m) \times \mathfrak{S}(n) \hookrightarrow \mathfrak{S}(m+n)$. The main result of this chapter is an isomorphism between $\mathrm{R}_0(\mathfrak{S})$ and the algebra of symmetric functions Sym (see Sections 2.2, 2.3 and 2.4). It allows one to interpret and compute the irreducible character values $\mathrm{ch}^\lambda(\sigma)$ of symmetric groups as coefficients relating two bases of Sym, namely, the basis of power sums and the basis of Schur functions. This interpretation can in turn be related to the duality between $\mathrm{GL}(N,\mathbb{C})$ and $\mathfrak{S}(n)$ for their actions on $(\mathbb{C}^N)^{\otimes n}$, and we shall give a detailed account of this Schur–Weyl duality in Section 2.5.

During this chapter, the reader will never need to know anything about the vector spaces S^λ underlying the irreducible representations of $\mathfrak{S}(n)$. We also tried to make our presentation as independent as possible from the combinatorics of tableaux, that are usually pervasive in the theory of symmetric functions. The exposition of these combinatorial properties, and their use in the description of the representation spaces S^λ, are purposedly postponed to Chapter 3.

2.1 Conjugacy classes of the symmetric groups

▷ *Conjugacy classes and integer partitions.*

Fix $n \geq 0$, and let σ be a permutation in $\mathfrak{S}(n)$. Recall that by Proposition 1.1, σ can be written uniquely as a product of disjoint cycles

$$\sigma = (a_1, \ldots, a_{\mu_1})(b_1, \ldots, b_{\mu_2}) \cdots (z_1, \ldots, z_{\mu_\ell}),$$

with $\mu_1 + \mu_2 + \cdots + \mu_\ell = n$. Up to permutation of these cycles, one can suppose their lengths in decreasing order $\mu_1 \geq \cdots \geq \mu_\ell$, so that $\mu = (\mu_1, \mu_2, \ldots, \mu_\ell)$ is a partition of size n. We say that μ is the **cycle type** of σ, and we denote $\mu = t(\sigma)$. Given a partition μ, we shall also denote $\ell(\mu)$ its **length**, which is its number of parts.

Proposition 2.1. *Two permutations σ and τ in $\mathfrak{S}(n)$ are conjugated if and only if they have the same cycle type.*

Proof. We suppose that σ has cycle type μ, and we write

$$\sigma = (a_1, \ldots, a_{\mu_1})(b_1, \ldots, b_{\mu_2}) \cdots (z_1, \ldots, z_{\mu_\ell}).$$

For any $\rho \in \mathfrak{S}(n)$, one has then

$$\rho\sigma\rho^{-1} = (\rho(a_1), \ldots, \rho(a_{\mu_1}))(\rho(b_1), \ldots, \rho(b_{\mu_2})) \cdots (\rho(z_1), \ldots, \rho(z_{\mu_\ell})),$$

and this is again a decomposition in disjoint cycles. Hence, $\rho\sigma\rho^{-1}$ has also cycle type μ, and the cycle type is constant on conjugacy classes in $\mathfrak{S}(n)$. Conversely, if

$$\begin{aligned}\sigma &= (a_1, \ldots, a_{\mu_1})(b_1, \ldots, b_{\mu_2}) \cdots (z_1, \ldots, z_{\mu_\ell}) \\ \tau &= (a'_1, \ldots, a'_{\mu_1})(b'_1, \ldots, b'_{\mu_2}) \cdots (z'_1, \ldots, z'_{\mu_\ell})\end{aligned}$$

have the same cycle type, let $\rho \in \mathfrak{S}(n)$ be the only permutation which sends a_1 to a'_1, a_2 to a'_2, etc. Then, $\tau = \rho\sigma\rho^{-1}$, so σ and τ are conjugated. □

Example. The three conjugacy classes in $\mathfrak{S}(3)$ are the class of the neutral element $\{\mathrm{id}_{[\![1,3]\!]}\}$; the class of transpositions $\{(1,2),(1,3),(2,3)\}$; and the class of 3-cycles $\{(1,2,3),(1,3,2)\}$.

In the following, for $\mu \in \mathfrak{Y}(n)$, we denote C_μ the conjugacy class of $\mathfrak{S}(n)$ that consists in permutations with cycle type μ. On the other hand, for a partition μ, we denote $m_j(\mu)$ the number of parts of μ equal to j. It is then sometimes convenient to denote the partition multiplicatively

$$\mu = 1^{m_1(\mu)} 2^{m_2(\mu)} \ldots s^{m_s(\mu)}.$$

For instance, if $\mu = (3,2,2,2,2,1,1)$, then one will sometimes write $\mu = 1^2\, 2^4\, 3$.

Proposition 2.2. *The cardinality of* C_μ *is* $\frac{n!}{z_\mu}$, *where*

$$z_\mu = \prod_{j\geq 1} j^{m_j(\mu)}\,(m_j(\mu))!\,.$$

Proof. Fix a permutation $\sigma = (a_1,\ldots,a_{\mu_1})\cdots(z_1,\ldots,z_{\mu_\ell})$ of cycle type μ, and consider the map $\rho \in \mathfrak{S}(n) \mapsto \rho\sigma\rho^{-1} \in C_\mu$. If $\rho_1\sigma\rho_1^{-1} = \rho_2\sigma\rho_2^{-1}$, then $\sigma = (\rho_1^{-1}\rho_2)\,\sigma\,(\rho_1^{-1}\rho_2)^{-1}$, so each term in C_μ is attained $z(\sigma)$ times, where

$$z(\sigma) = \mathrm{card}\,\{\rho \in \mathfrak{S}(n) \mid \rho\sigma\rho^{-1} = \sigma\}.$$

Thus, $\mathrm{card}\,C_\mu = \frac{n!}{z(\sigma)}$. However, $\rho\sigma\rho^{-1} = \sigma$ if and only if

$$(\rho(a_1),\ldots,\rho(a_{\mu_1}))(\rho(b_1),\ldots,\rho(b_{\mu_2}))\cdots(\rho(z_1),\ldots,\rho(z_{\mu_\ell}))$$

is a rewriting of $(a_1,\ldots,a_{\mu_1})(b_1,\ldots,b_{\mu_2})\cdots(z_1,\ldots,z_{\mu_\ell})$. By the discussion of Proposition 1.1, such a rewriting corresponds

- to a cyclic permutation of each cycle ($\prod_{i=1}^{\ell}\mu_i$ possibilities),
- and a possible permutation of the cycles ($\prod_{j\geq 1} m_j(\mu)!$ possibilities in order to keep the cycles with lengths of decreasing order).

So,

$$z(\sigma) = z_\mu = \left(\prod_{i=1}^{\ell}\mu_i\right)\left(\prod_{j\geq 1}(m_j(\mu))!\right) = \prod_{j\geq 1} j^{m_j(\mu)}\,(m_j(\mu))!\,. \qquad \square$$

Example. In $\mathfrak{S}(3)$, the class of transpositions is $C_{(2,1)} = \{(1,2),(2,3),(1,3)\}$, and it contains indeed $\frac{6}{2} = 3$ terms, since $z_{(2,1)} = 1^1 2^1 = 2$.

▷ *Signature representation.*

Since the characters of a group are constant on conjugacy classes, if ch^V is a character of a representation of $\mathfrak{S}(n)$ and $\mu \in \mathfrak{Y}(n)$, then we can denote without ambiguity $\mathrm{ch}^V(\mu)$ the value of ch^V on any permutation σ of cycle type μ. One of the main results of this chapter will be Theorem 2.32, which is a formula that allows the calculation of $\mathrm{ch}^{S^\lambda}(\mu)$ for any irreducible representation S^λ of $\mathfrak{S}(n)$. As a warm-up, let us find the characters and representations of dimension 1.

Lemma 2.3. *Any permutation of* $\mathfrak{S}(n)$ *can be written as a product of transpositions* (i,j), *with* $1 \leq i < j \leq n$. *One can even restrict oneself to the* ***elementary transpositions*** $s_i = (i,i+1)$ *with* $i \in [\![1,n-1]\!]$.

Proof. By Proposition 1.1, it suffices to show the result for cycles, and

$$(a_1,a_2,\ldots,a_r) = (a_1,a_2)(a_2,a_3)\cdots(a_{r-1},a_r).$$

Then, each transposition (i,j) is a product of transpositions s_i:

$$(i,j) = s_{j-1}s_{j-2}\cdots s_{i+1}s_is_{i+1}\cdots s_{j-2}s_{j-1}. \qquad \square$$

Proposition 2.4. *For $n \geq 2$, the symmetric group $\mathfrak{S}(n)$ has exactly two one-dimensional representations: the trivial representation $1_n : \sigma \mapsto 1$, and the* ***signature representation***

$$\varepsilon_n : \sigma \mapsto (-1)^{N(\sigma)} = (-1)^{r(\sigma)} = (-1)^{|t(\sigma)|-\ell(t(\sigma))},$$

where:

1. *$N(\sigma)$ is the number of* ***inversions*** *of σ, that is to say the pairs $(i<j)$ such that $\sigma(i) > \sigma(j)$;*
2. *$r(\sigma)$ is the number of elements of a decomposition of σ in a product of transpositions (its parity does not depend on the decomposition chosen).*

Proof. Let $\rho : \mathfrak{S}(n) \to \mathbb{C}^*$ be a morphism of groups. Since $\mathbb{C}^*$ is commutative, ρ is constant on conjugacy classes, and in particular on the class of transpositions. Moreover, given a transposition τ,

$$(\rho(\tau))^2 = \rho(\tau^2) = \rho(1) = 1.$$

so this constant value is a square root of 1, hence $+1$ or -1. If it is $+1$, then $\rho(\sigma) = 1$ for any permutation in $\mathfrak{S}(n)$, because every permutation is a product of transpositions (Lemma 2.3); one looks then at the trivial representation. Suppose now that $\rho(\tau) = -1$ for any transposition; then, $\rho(\sigma) = (-1)^{r(\sigma)}$, where $r(\sigma)$ is the number of transpositions of an arbitrary decomposition $\tau_1\tau_2\cdots\tau_r = \sigma$. To verify that this morphism exists, one has to check that the parity of r does not depend on the chosen decomposition. This follows from the following fact: for any transpositions $\tau_1,\ldots,\tau_r$,

$$N(\tau_1\tau_2\cdots\tau_r) \equiv r \mod 2.$$

We proceed by induction over r, the case $r=0$ being trivial. Suppose the result to be true up to order r, and consider a product $\sigma' = \tau_1\tau_2\cdots\tau_{r+1} = \sigma\tau_{r+1}$, with $\sigma = \tau_1\cdots\tau_r$; by hypothesis, $N(\sigma) \equiv r \mod 2$. We denote $\tau_{r+1} = (i,j)$. If A and B are two parts of $[\![1,n]\!]$, we denote $N(\sigma;A,B)$ the number of inversions (i,j) of σ such that $i \in A$ and $j \in B$. Then,

$$\begin{aligned}N(\sigma) = N(\sigma;[\![1,n]\!],[\![1,i-1]\!]) + N(\sigma;[\![1,n]\!],\{i\}) + N(\sigma;[\![1,n]\!],\{j\})\\ + N(\sigma;[\![1,n]\!],[\![i+1,j-1]\!]) + N(\sigma;[\![1,n]\!],[\![j+1,n]\!])\end{aligned}$$

$$\begin{aligned}
&= N(\sigma;[\![1,i-1]\!],[\![1,i-1]\!]) + N(\sigma;[\![1,i-1]\!],\{i\}) \\
&\quad + N(\sigma;[\![1,i-1]\!],[\![i+1,j-1]\!]) + N(\sigma;\{i\},[\![i+1,j-1]\!]) \\
&\quad + N(\sigma;[\![i+1,j-1]\!],[\![i+1,j-1]\!]) + N(\sigma;[\![1,i-1]\!],\{j\}) \\
&\quad + N(\sigma;\{i\},\{j\}) + N(\sigma;[\![i+1,j-1]\!],\{j\}) \\
&\quad + N(\sigma;[\![1,i-1]\!],[\![j+1,n]\!]) + N(\sigma;\{i\},[\![j+1,n]\!]) \\
&\quad + N(\sigma;[\![i+1,j-1]\!],[\![j+1,n]\!]) + N(\sigma;\{j\},[\![j+1,n]\!]) \\
&\quad + N(\sigma;[\![j+1,n]\!],[\![j+1,n]\!]).
\end{aligned}$$

One shows readily that

$$\begin{aligned}
N(\sigma';[\![1,i-1]\!],\{i\}) &= N(\sigma;[\![1,i-1]\!],\{j\}) \\
N(\sigma';[\![1,i-1]\!],\{j\}) &= N(\sigma;[\![1,i-1]\!],\{i\}) \\
N(\sigma';\{i\},[\![i+1,j-1]\!]) &= (j-i-1) - N(\sigma;[\![i+1,j-1]\!],\{j\}) \\
N(\sigma';\{i\},\{j\}) &= 1 - N(\sigma;\{i\},\{j\}) \\
N(\sigma';[\![i+1,j-1]\!],\{j\}) &= (j-i-1) - N(\sigma;[\![i+1,j-1]\!],\{i\}) \\
N(\sigma';\{i\},[\![j+1,n]\!]) &= N(\sigma;\{j\},[\![j+1,n]\!]) \\
N(\sigma';\{j\},[\![j+1,n]\!]) &= N(\sigma;\{i\},[\![j+1,n]\!])
\end{aligned}$$

and all the other terms in the detailed expansion of $N(\sigma)$ stay the same for $N(\sigma')$. Hence, if one takes the number of inversions of σ' modulo 2, then it differs from the number of inversions of σ modulo 2 by 1, so

$$N(\sigma') \equiv N(\sigma) + 1 \equiv r + 1 \mod 2.$$

The existence of the signature morphism is then established, and using the decomposition of a cycle of length r as a product of $r-1$ transpositions, one sees that a permutation of cycle type μ is a product of $\sum_{j=1}^{\ell(\mu)} \mu_j - 1 = |\mu| - \ell(\mu)$ transpositions; the triple identity for the signature morphism is thus proven. □

▷ *Size of a decomposition in transpositions.*

Actually, there is a deeper connection between Lemma 2.3 and Proposition 2.4:

1. $N(\sigma)$ is the minimal number of terms necessary to write σ as a product of elementary transpositions s_i with $i \in [\![1,n-1]\!]$.

2. $|t(\sigma)| - \ell(t(\sigma))$ is the minimal number of terms necessary to write σ as a product of transpositions (i,j) with $1 \leq i < j \leq n$.

To prove the first claim, notice that by the proof of Proposition 2.4, for any permutation σ and any elementary transposition s_i,

$$N(\sigma s_i) = N(\sigma) \pm 1 = \begin{cases} N(\sigma)+1 & \text{if } (i,i+1) \text{ is not an inversion of } \sigma, \\ N(\sigma)-1 & \text{if } (i,i+1) \text{ is an inversion of } \sigma. \end{cases}$$

Therefore, any decomposition $\sigma = s_{i_1}s_{i_2}\cdots s_{i_r}$ has a number of terms $r \geq N(\sigma)$. Conversely, suppose that σ admits $N(\sigma)$ inversions, and let us show by induction on $N(\sigma)$ that σ writes as a product of $N(\sigma)$ elementary transpositions. It suffices to show that if $N(\sigma) > 0$, then there exists a pair $(k, k+1)$ that is an inversion of σ. Then, $N(\sigma s_k) = N(\sigma) - 1$, and one can apply the induction hypothesis. Thus, consider a permutation that admits at least one inversion (i, j). Then, since $\sigma(j) < \sigma(i)$, one has necessarily a **descent** k between i and j such that $\sigma(k+1) < \sigma(k)$, because otherwise

$$\sigma(i) < \sigma(i+1) < \cdots < \sigma(k) < \sigma(k+1) < \cdots < \sigma(j).$$

Thus, every permutation writes as a product of $N(\sigma)$ elementary transpositions, and this is the minimal possible number.

As for the second claim, consider a decomposition $\sigma = (i_1, j_1)\cdots(i_r, j_r)$ of a permutation of $\mathfrak{S}(n)$ in transpositions. Consider the graph G on $[\![1, n]\!]$ whose edges are the (i_k, j_k)'s. If two elements are in the same orbit for σ, then they must be in the same connected component of G. Therefore, the number of connected components of G is smaller than the number of orbits $\ell(t(\sigma))$ of σ. However, a graph with r edges and n vertices has at most $n - r$ connected components, so $|t(\sigma)| - r \leq \ell(t(\sigma))$, i.e.,

$$r \geq |t(\sigma)| - \ell(t(\sigma)).$$

A decomposition with $r = |t(\sigma)| - \ell(t(\sigma))$ transpositions is obtained by writing each orbit $(a_1, \ldots, a_k)$ of σ as $(a_1, a_2)(a_2, a_3)\cdots(a_{r-1}, a_r)$.

Example. Consider the permutation $\sigma = 35124 = (1,3)(2,5,4)$. It is the product of $3 = 5 - 2$ transpositions, namely, $(1,3)(2,5)(5,4)$, and of 5 elementary transpositions, namely, $s_4 s_2 s_3 s_1 s_2$. And 5 is also the number of inversions: $\{(1,3), (1,4), (2,3), (2,4), (2,5)\}$.

2.2 The five bases of the algebra of symmetric functions

In this section, $X = \{x_1, x_2, \ldots\}$ is an infinite set of independent commuting variables, and we shall present the basic properties of the symmetric functions in these variables. This is for the moment largely independent from the previous discussion, and the connection between this theory and the representation theory of symmetric groups will be explained in Section 2.4.

▷ *Symmetric polynomials and symmetric functions.*

In the sequel, we denote $\mathbb{C}[x_1, x_2, \ldots, x_N]$ the $\mathbb{C}$-algebra of polynomials in N variables. It is infinite-dimensional, and graded by

$$\deg(x_1^{k_1} x_2^{k_2} \cdots x_N^{k_N}) = k_1 + k_2 + \cdots + k_N.$$

In this chapter, *we shall try to always denote the numbers of variables with upper case letters $M, N, \ldots$ and the powers of variables and degrees with lower case letters* $k, n, \ldots$ That said, there is a left action of $\mathfrak{S}(N)$ on $\mathbb{C}[x_1, x_2, \ldots, x_N]$ given by

$$\sigma \cdot P(x_1, \ldots, x_N) = P(x_{\sigma(1)}, \ldots, x_{\sigma(N)}).$$

For instance, if $N = 3$, $\sigma = (1,2)$ and $P(x,y,z) = x^2yz^3$, then $(\sigma \cdot P)(x,y,z) = y^2xz^3$.

Remark. The reader should pay attention to the following fact: $\mathfrak{S}(N)$ acts *on the left* of polynomials in N variables, but *on the right* of sequences of size N. In particular, given a monomial $x^k = (x_1)^{k_1} \cdots (x_N)^{k_N}$, one has $\sigma \cdot (x^k) = x^{k \cdot \sigma^{-1}}$. In the sequel, we shall sometimes make $\mathfrak{S}(N)$ act on polynomials, and sometimes make it act on exponents of monomials viewed as sequences; the place of a permutation σ that acts is then different, but this is only a change of notation.

Definition 2.5. *A* ***symmetric polynomial*** *in N variables is a polynomial P such that $\sigma \cdot P = P$ for any permutation $\sigma \in \mathfrak{S}(N)$.*

The subspace

$$\mathrm{Sym}^{(N)} = \mathbb{C}[x_1, x_2, \ldots, x_N]^{\mathfrak{S}(N)}$$

of symmetric polynomials in $\mathbb{C}[x_1, x_2, \ldots, x_N]$ is in fact a graded subalgebra, since for any polynomials P, Q, one has

$$\sigma \cdot (PQ) = (\sigma \cdot P)(\sigma \cdot Q) \quad ; \quad \deg(\sigma \cdot P) = \deg P.$$

Example. For $k \in [\![0, N]\!]$, set

$$e_k(x_1, \ldots, x_N) = \sum_{1 \le i_1 < i_2 < \cdots < i_k \le N} x_{i_1} x_{i_2} \cdots x_{i_k},$$

with by convention $e_0 = 1$. For instance, $e_2(x_1, x_2, x_3) = x_1x_2 + x_2x_3 + x_1x_3$. The polynomial e_k is called the k-th **elementary symmetric polynomial** of the x_i's, and the e_k's appear in the expansion of the polynomial $\prod_{i=1}^{N}(1 + tx_i)$:

$$E^{(N)}(t) = \prod_{i=1}^{N}(1 + tx_i) = \sum_{k=0}^{N} e_k(x_1, \ldots, x_N)\, t^k.$$

Since $E^{(N)}(t)$ is obviously symmetric in the x_i's, the same holds for its coefficients, so the e_k's are indeed symmetric polynomials, with $\deg e_k = k$.

Example. For $k \ge 0$, set

$$h_k(x_1, \ldots, x_N) = \sum_{1 \le i_1 \le i_2 \le \cdots \le i_k \le N} x_{i_1} x_{i_2} \cdots x_{i_k},$$

with by convention $h_0 = 1$. For example, $h_3(x_1, x_2) = (x_1)^3 + (x_1)^2x_2 + x_1(x_2)^2 +$

$(x_2)^3$. The polynomial is called the k-th **homogeneous symmetric polynomial** of the x_i's, and the h_k's appear in the expansion of the formal series $\prod_{i=1}^N 1/(1-tx_i)$:

$$H^{(N)}(t) = \prod_{i=1}^{N} \frac{1}{1-tx_i} = \sum_{k=0}^{\infty} h_k(x_1,\ldots,x_N)\,t^k.$$

Since $H^{(N)}(t)$ is obviously symmetric in the x_i's, the same holds for its coefficients, so the h_k's are indeed symmetric polynomials, with $\deg h_k = k$. Notice that $H^{(N)}(t)\,E^{(N)}(-t) = 1$.

Example. For $k \geq 1$, set

$$p_k(x_1,\ldots,x_N) = \sum_{i=1}^{N} (x_i)^k = (x_1)^k + (x_2)^k + \cdots + (x_N)^k.$$

Hence, one has for instance $p_4(x_1,x_2,x_3) = (x_1)^4 + (x_2)^4 + (x_3)^4$. Again, the p_k's are symmetric polynomials in the variables $x_1,\ldots,x_n$ with $\deg p_k = k$, and p_k is called the k-th **power sum**. One can compute a generating function for these polynomials:

$$\begin{aligned} P^{(N)}(t) &= \sum_{k\geq 1} p_k(x_1,\ldots,x_N)\,\frac{t^k}{k} = \sum_{i=1}^{N}\sum_{k=1}^{\infty} \frac{(tx_i)^k}{k} \\ &= \sum_{i=1}^{N} \log\left(\frac{1}{1-tx_i}\right) = \log\left(\prod_{i=1}^{N} \frac{1}{1-tx_i}\right) = \log H^{(N)}(t). \end{aligned}$$

In these examples, the number of variables N does not play an important role, and indeed, there is a way to get rid of it and to consider "generic" symmetric polynomials. More precisely, for $M \geq N$, consider the map

$$\begin{aligned} \pi_N^M : \mathbb{C}[x_1,\ldots,x_M] &\to \mathbb{C}[x_1,\ldots,x_N] \\ P(x_1,\ldots,x_M) &\mapsto P(x_1,\ldots,x_N,0,\ldots,0). \end{aligned}$$

The map π_N^M is a morphism of graded algebras. If $\sigma \in \mathfrak{S}(N)$, then $\pi_N^M(\sigma \cdot P) = \sigma \cdot (\pi_M^N(P))$, so it restricts to a morphism of graded algebras $\mathrm{Sym}^{(M)} \to \mathrm{Sym}^{(N)}$.

Definition 2.6. *The algebra of symmetric functions is the projective limit*

$$\mathrm{Sym} = \varprojlim_{N\to\infty} \mathrm{Sym}^{(N)}$$

of the algebras of symmetric polynomials, the limit being taken in the category of graded algebras and with respect to the morphisms $(\pi_N^M)_{M\geq N}$.

Thus, a **symmetric function** is a sequence $(P^{(N)} \in \mathrm{Sym}^{(N)})_{N\in\mathbb{N}}$ of symmetric polynomials bounded in degree, such that $\pi_N^M(P^{(M)}) = P^{(N)}$ for all $M \geq N$. Let us give an alternate and more concrete definition:

Definition 2.7. *A symmetric function is a power series $P \in \mathbb{C}[[x_1, x_2, \ldots]]$, bounded in degree, and such that for any bijection $\sigma : \mathbb{N}^* \to \mathbb{N}^*$,*

$$P(x_1, x_2, \ldots) = P(x_{\sigma(1)}, x_{\sigma(2)}, \ldots).$$

Let us show the equivalence between the two definitions. Given a sequence $k = (k_1, k_2, \ldots, k_N, 0, 0, \ldots)$ of non-negative integers with a finite number of non-zero terms, we denote $x^k = x_1^{k_1} x_2^{k_2} \cdots x_N^{k_N}$. Then, the degree of x^k is $|k| = \sum_{i=1}^{\infty} k_i$, and a formal power series is an infinite series

$$f(x) = \sum_k c_k \, x^k,$$

where the sum runs over sequences $k \in \mathbb{N}^{(\mathbb{N}^*)}$. In this setting we note $c_k = [x^k](f)$. The formal power series f is bounded in degree if there exists $K \geq 0$ such that

$$|k| > K \Rightarrow [x^k](f) = 0.$$

Notice then that f is invariant by any bijection $\sigma : \mathbb{N}^* \to \mathbb{N}^*$ if and only if it is invariant by any element $\sigma \in \varinjlim_{N \to \infty} \mathfrak{S}(N) = \mathfrak{S}(\infty)$, that is to say any *finite* bijection that only modifies a finite number of integers in $\mathbb{N}^*$. Indeed, the first statement is equivalent to

$$\forall k \in \mathbb{N}^{(\mathbb{N}^*)},\ \forall \sigma \in \mathfrak{S}(\mathbb{N}^*),\ [x^k](f) = [x^{k \cdot \sigma}](f),$$

and for a power series bounded in degree by K, one can realize any bijection between k and $k' = k \cdot \sigma$ by a bijection which permutes at most $2K$ elements, hence belongs to $\mathfrak{S}(\infty) = \bigcup_{N=0}^{\infty} \uparrow \mathfrak{S}(N)$.

Now, let $(P^{(N)})_{N \in \mathbb{N}}$ be a sequence in the projective limit $\varprojlim_{N \to \infty} \mathrm{Sym}^{(N)}$. If $k \in \mathbb{N}^{(\mathbb{N}^*)}$, then for N big enough (bigger than the largest index of a non-zero entry of k), x^k belongs to $\mathbb{C}[x_1, \ldots, x_N]$, and then $[x^k](P^{(M)})$ does not depend of $M \geq N$, as

$$[x^k](P^{(N)}) = [x^k](\pi_N^M(P^{(M)})) = [x^k](P^{(M)}).$$

We set $f = \sum_k c_k \, x^k$, where $c_k = [x^k](P^{(N)})$ for N big enough. This defines a power series bounded in degree, since $(P^{(N)})_{N \in \mathbb{N}}$ is itself bounded in degree. For any permutation σ belonging to a finite symmetric group $\mathfrak{S}(N)$, $[x^{k \cdot \sigma}](f) = [x^k](f)$, because this is true for the symmetric polynomials $P^{(M)}$ with $M \geq N$ large enough. By the previous discussion, this shows that f satisfies the hypotheses of Definition 2.6. Conversely, given such a power series f, it defines a sequence of polynomials $(P^{(N)})_{N \in \mathbb{N}}$:

$$P^{(N)}(x_1, \ldots, x_N) = f(x_1, \ldots, x_N, 0, 0, \ldots) = \sum_{k \,|\, k_M = 0 \text{ for } M > N} c_k \, x^k,$$

which is in $\varprojlim_{N \to \infty} \mathrm{Sym}^{(N)}$.

Remark. Notice that Sym *is not* the projective limit of the algebras $\mathrm{Sym}^{(N)}$ in the category of algebras; this object contains more functions than Sym, namely, symmetric functions possibly not bounded in degree. We shall encounter projective limits in the category of *graded* algebras several times in the book, e.g. when constructing the algebras NCSym and QSym in Chapter 6, or the algebra of observables $\mathscr{O}$ in Chapter 7. Each time, this is the graded limit that will be the most convenient object to manipulate, in opposition to the limit in the category of algebras, which is larger.

When we shall need to differentiate symmetric functions from symmetric polynomials, we shall denote a symmetric function $f(X)$, where X denotes the infinite sequence of variables $\{x_1, x_2, \ldots, \}$, also called **alphabet** of variables. The specialization of f in a finite number of variables will then be denoted $f(x_1, x_2, \ldots, x_N)$. Since Sym is a projective limit, any computation that holds between symmetric functions also holds for the corresponding symmetric polynomials in an algebra $\mathbb{C}[x_1, \ldots, x_N]$.

Example. The elementary symmetric functions, homogeneous symmetric functions and power sums are defined by

$$e_k(X) = \sum_{i_1 < i_2 < \cdots < i_k} x_{i_1} x_{i_2} \cdots x_{i_k};$$
$$h_k(X) = \sum_{i_1 \leq i_2 \leq \cdots \leq i_k} x_{i_1} x_{i_2} \cdots x_{i_k};$$
$$p_k(X) = \sum_i (x_i)^k$$

and they correspond as symmetric functions in Sym to the symmetric polynomials previously introduced. Their generating series are given by formal infinite products:

$$E(t) = \sum_{k=0}^{\infty} e_k(X)\, t^k = \prod_i (1 + t x_i);$$
$$H(t) = \sum_{k=0}^{\infty} h_k(X)\, t^k = \prod_i \frac{1}{1 - t x_i} = \frac{1}{E(-t)};$$
$$P(t) = \sum_{k=1}^{\infty} p_k(X)\, \frac{t^k}{k} = \log H(t).$$

Notice that the restriction of a symmetric function $f \neq 0$ to a finite alphabet $\{x_1, \ldots, x_n\}$ can be equal to the zero polynomial:

$$e_k(x_1, \ldots, x_N) = 0 \text{ if } k > N.$$

▷ *Monomial functions and linear bases of* Sym.

If $k \in \mathbb{N}^{(\mathbb{N}^*)}$, then there exists a permutation $\sigma \in \mathfrak{S}(\infty)$ such that $k \cdot \sigma$ is a non-increasing sequence: for instance, starting from $(0, 1, 0, 3, 2, 0, 2, 0, \ldots)$, one can reorder it to get the non-increasing sequence $(3,2,2,1,0,0,\ldots)$. Looking only at the non-zero terms, one gets an integer partition $\lambda(k) = (\lambda_1 \geq \cdots \geq \lambda_\ell)$, uniquely determined by k. Let f be a symmetric function. Then, $[x^k](f) = [x^{\lambda(k)}](f)$, so f is entirely determined by the coefficients $[x^\lambda](f)$ with $\lambda \in \mathfrak{Y} = \bigsqcup_{n=0}^{\infty} \mathfrak{Y}(n)$. For an arbitrary partition λ, set

$$m_\lambda(X) = \sum_{k \text{ reordering of } \lambda} x^k.$$

For example, $m_{2,1}(X) = \sum_{i \neq j} x_i^2 x_j$, and $m_{1,1}(X) = \sum_{i<j} x_i x_j$. Each $m_\lambda(X)$ is a symmetric function, and by the previous discussion, any symmetric function is a linear combination of functions $m_\lambda(X)$, which is unique: if $f(X) = \sum_\lambda c_\lambda \, m_\lambda(X)$, then $c_\lambda = [x^\lambda](f)$. Thus:

Proposition 2.8. *The family of* ***monomial functions*** $(m_\lambda)_{\lambda \in \mathfrak{Y}}$ *is a linear basis of* Sym.

One can define three other linear bases of Sym by using elementary symmetric functions, homogeneous symmetric functions and power sums. If $\lambda = (\lambda_1 \geq \lambda_2 \geq \cdots \geq \lambda_\ell)$ is an integer partition, set

$$e_\lambda(X) = e_{\lambda_1}(X)\, e_{\lambda_2}(X) \cdots e_{\lambda_\ell}(X);$$
$$h_\lambda(X) = h_{\lambda_1}(X)\, h_{\lambda_2}(X) \cdots h_{\lambda_\ell}(X);$$
$$p_\lambda(X) = p_{\lambda_1}(X)\, p_{\lambda_2}(X) \cdots p_{\lambda_\ell}(X).$$

Theorem 2.9. *The families* $(e_\lambda)_{\lambda \in \mathfrak{Y}}$, $(h_\lambda)_{\lambda \in \mathfrak{Y}}$ *and* $(p_\lambda)_{\lambda \in \mathfrak{Y}}$ *are linear bases of* Sym.

The remainder of this paragraph is devoted to a complete proof of Theorem 2.9. We shall use the two following combinatorial properties of the set of integer partitions $\mathfrak{Y}$. Given an integer partition λ, we call **Young diagram** of λ the array of boxes with λ_1 boxes on the first row, λ_2 boxes on the second row, etc. For instance, the Young diagram of $\lambda = (5,2,2,1)$ is

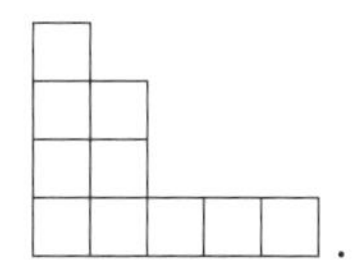

The notion of Young diagrams allows a simple definition of the **conjugate** of a partition λ: this is the partition λ' whose Young diagram is obtained by symmetrizing the Young diagram of λ with respect to the first diagonal. Thus, with the same example $\lambda = (5,2,2,1)$, the conjugate partition is $\lambda' = (4,3,1,1,1)$, with Young

diagram

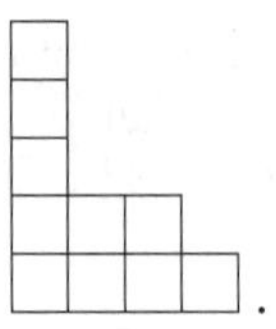

In algebraic terms, it can be checked that $\lambda'_i = \sum_{j \geq i} m_j(\lambda)$. By the geometric interpretation, it is clear that $\lambda \mapsto \lambda'$ is a size-preserving involution of $\mathfrak{Y}$.

On the other hand, one has the following (non-total) order of **dominance** on partitions:

$$\lambda \succeq \mu \iff \forall i \geq 1,\ \lambda_1 + \lambda_2 + \cdots + \lambda_i \geq \mu_1 + \mu_2 + \cdots + \mu_i.$$

This is not a total order, though it is when restricted to partitions of small size. For instance, restricted to $\mathfrak{Y}(5)$, the order $\succeq$ is given by the diagram in Figure 2.1.

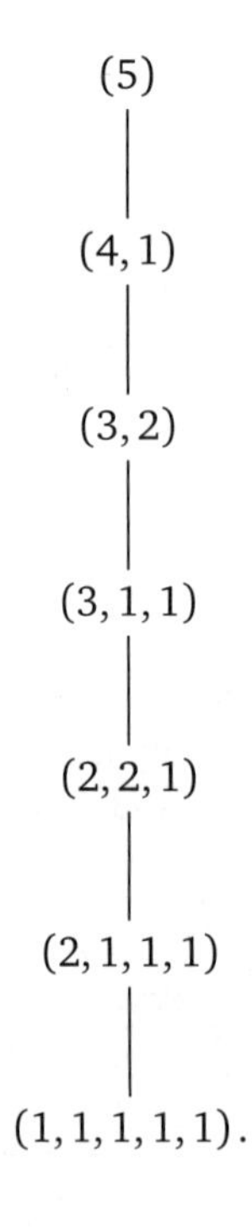

Figure 2.1
Dominance order on $\mathfrak{Y}(5)$.

Lemma 2.10. *Fix an integer partition λ. One has*

$$e_{\lambda'}(X) = m_\lambda(X) + \sum_{\mu \prec \lambda} c_{\lambda\mu}\, m_\mu(X)$$

for some non-negative integer coefficients $c_{\lambda\mu}$. Therefore, $(e_\lambda)_{\lambda \in \mathfrak{Y}}$ is a linear basis of Sym, *and the transition matrix between $(e_{\lambda'})_{\lambda\in\mathfrak{Y}}$ and $(m_\lambda)_{\lambda\in\mathfrak{Y}}$ is upper triangular with 1's on the diagonal.*

Proof. Set $r = \lambda_1 = \ell(\lambda')$. By definition, $e_{\lambda'}(X)$ is the sum of monomials

$$\left(x_{i_{1,1}} x_{i_{1,2}} \cdots x_{i_{1,\lambda_1'}}\right)\left(x_{i_{2,1}} x_{i_{2,2}} \cdots x_{i_{2,\lambda_2'}}\right)\cdots\left(x_{i_{r,1}} x_{i_{r,2}} \cdots x_{i_{r,\lambda_r'}}\right),$$

with $i_{j,1} < i_{j,2} < \cdots < i_{j,\lambda_j'}$ for any $j \in [\![1,r]\!]$. Given such a monomial, we put each label $i_{j,k}$ in the cell of the Young diagram of λ that is placed on the j-th column and on the k-th row:

$$\begin{array}{|c|c|c|c|c|}
\cline{1-1}
i_{1,\lambda_1'} \\
\cline{1-2}
i_{1,3} & i_{2,\lambda_2'} \\
\cline{1-2}
i_{1,2} & i_{2,2} \\
\hline
i_{1,1} & i_{2,1} & i_{3,\lambda_3'} & i_{4,\lambda_4'} & i_{5,\lambda_5'} \\
\hline
\end{array}$$

Since all the columns of this tableau are strictly increasing, all the entries smaller than r must be in the r first rows of the Young diagram.

Consider now an integer partition μ such that x^μ appears in the sum $e_{\lambda'}(X)$. For any $i \geq 1$, the entries 1 with multiplicity μ_1, 2 with multiplicity μ_2, etc. up to i with multiplicity μ_i must appear in the i first rows of the Young diagram of λ, so by counting boxes,

$$\mu_1 + \mu_2 + \cdots + \mu_i \leq \lambda_1 + \lambda_2 + \cdots + \lambda_i.$$

Hence, if $[x^\mu](e_{\lambda'}) \neq 0$, then $\mu \preceq \lambda$. Moreover, if $\mu = \lambda$, then there is exactly one way to fill the Young diagram of λ with the entries $1^{\lambda_1}, 2^{\lambda_2}, \ldots$, namely, with all the 1's on the first row, all the 2's on the second row, etc.; thus, $[x^\lambda](e_{\lambda'}) = 1$. As the coefficients of the expansion of a symmetric function $f(X) = \sum c_\mu\, m_\mu(X)$ in monomial functions are given by

$$c_\mu = [x^\mu](f),$$

the lemma is proven. □

Example. The matrix of change of basis between $(e_\lambda)_{\lambda \in \mathfrak{Y}(4)}$ and $(m_\lambda)_{\lambda \in \mathfrak{Y}(4)}$ is

$$\begin{array}{rlllll} e_4 = & & & & m_{1,1,1,1}; \\ e_{3,1} = & & & m_{2,1,1} & +4\,m_{1,1,1,1}; \\ e_{2,2} = & & m_{2,2} & +2\,m_{2,1,1} & +6\,m_{1,1,1,1}; \\ e_{2,1,1} = & m_{3,1} & +2\,m_{2,2} & +5\,m_{2,1,1} & +12\,m_{1,1,1,1}; \\ e_{1,1,1,1} = m_4 & +4\,m_{3,1} & +6\,m_{2,2} & +12\,m_{2,1,1} & +24\,m_{1,1,1,1}. \end{array}$$

Lemma 2.11 (Newton). *An elementary function e_λ can be written as a linear combination of homogeneous functions h_λ, and conversely. The same result holds for homogeneous functions h_λ and power sums p_λ.*

Proof. Denote as before $E(t)$, $H(t)$ and $P(t)$ the generating series of the e_k's, the h_k's and the p_k's. Notice that $h_1(X) = e_1(X) = p_1(X) = \sum_i x_i$. Since $H(t)\,E(-t) = 1$, for any $k \geq 1$,

$$\begin{aligned} 0 &= \sum_{j=0}^{k} (-1)^{k-j}\, h_j(X)\, e_{k-j}(X) \\ &= h_k(X) - h_{k-1}(X)\, e_1(X) + \cdots + (-1)^{k-1}\, h_1(X)\, e_{k-1}(X) + (-1)^k\, e_k(X). \end{aligned}$$

Therefore, by induction on k, h_k is a linear combination of products of functions e_j, and conversely. This proves the first part of the lemma. For the second part, since $H(t) = \exp P(t)$, one has

$$\forall k \geq 1,\ h_k(X) = [t^k](\exp P(t)) = \sum_{l_1+l_2+\cdots+l_r=k} \frac{p_{l_1}(X) \cdots p_{l_r}(X)}{r!\ (l_1 \cdots l_r)}.$$

If $\mu = (\mu_1, \ldots, \mu_r)$ is a partition of size k, then the number of possible reorderings $(l_1, \ldots, l_r)$ of the parts of μ is $\frac{r!}{\prod_{i \geq 1} m_i(\mu)!}$, so the previous formula rewrites as

$$h_k(X) = \sum_{\mu \in \mathfrak{Y}(k)} \frac{p_\mu(X)}{z_\mu},$$

where z_μ is the quantity introduced in Proposition 2.2. Thus, any homogeneous symmetric function is a linear combination of functions $p_\lambda(X)$, and the converse is true since

$$p_k(X) = k\,[t^k](\log H(t)) = k \sum_{\mu \in \mathfrak{Y}(k)} (-1)^{\ell(\mu)-1}\, (\ell(\mu)-1)!\, \frac{h_\mu(X)}{m_1(\mu)! \cdots m_s(\mu)!}. \qquad \square$$

Remark. If $Q(t) = \sum_{k \geq 1} p_k(X)\, t^k$, then

$$Q(t) = t\,P'(t) = t\,\frac{H'(t)}{H(t)} = t\,\frac{E'(-t)}{E(-t)},$$

which leads to some new relations between the power sums and the elementary and homogeneous symmetric functions:

$$\begin{aligned} k\,h_k(X) &= \sum_{j=1}^{k} p_j(X)\,h_{k-j}(X) \\ &= p_k(X) + p_{k-1}(X)\,h_1(X) + \cdots + p_1(X)\,h_{k-1}(X); \\ k\,e_k(X) &= \sum_{j=1}^{k} (-1)^{j-1}\,p_j(X)\,e_{k-j}(X) \\ &= (-1)^{k-1}\,p_k(X) + (-1)^{k-2}\,p_{k-1}(X)\,e_1(X) + \cdots + p_1(X)\,e_{k-1}(X). \end{aligned}$$

These relations are called the **Newton identities**.

Proof of Theorem 2.9. From Lemma 2.10, we already know that the family of elementary symmetric functions $(e_\lambda)_{\lambda\in\mathfrak{Y}}$ is a basis of Sym. More precisely, denote Sym_n the subspace of Sym that consists in symmetric functions that are homogenenous of degree n. Then, two linear bases of Sym_n are $(m_\lambda)_{\lambda\in\mathfrak{Y}(n)}$ and $(e_\lambda)_{\lambda\in\mathfrak{Y}(n)}$. In particular,

$$\dim \mathrm{Sym}_n = \mathrm{card}\,\mathfrak{Y}(n) < \infty.$$

By Lemma 2.11, $(h_\lambda)_{\lambda\in\mathfrak{Y}(n)}$ and $(p_\lambda)_{\lambda\in\mathfrak{Y}(n)}$ are also linear bases of Sym_n, since

- they span the same vector space as $(e_\lambda)_{\lambda\in\mathfrak{Y}(n)}$;
- they have their cardinality equal to $\dim \mathrm{Sym}_n$.

As $\mathrm{Sym} = \bigoplus_{n=0}^{\infty} \mathrm{Sym}_n$ and $\mathfrak{Y} = \bigsqcup_{n=0}^{\infty} \mathfrak{Y}(n)$, we conclude that $(h_\lambda)_{\lambda\in\mathfrak{Y}}$ and $(p_\lambda)_{\lambda\in\mathfrak{Y}}$ are linear bases of Sym. □

A reformulation of Theorem 2.9 is that $(e_k)_{k\geq 1}$, $(h_k)_{k\geq 1}$ and $(p_k)_{k\geq 1}$ are free families generating the algebra of symmetric functions:

$$\mathrm{Sym} = \mathbb{C}[e_1, e_2, \ldots] = \mathbb{C}[h_1, h_2, \ldots] = \mathbb{C}[p_1, p_2, \ldots].$$

Remark. Notice that all the bases considered have the degree property

$$\deg m_\lambda = \deg e_\lambda = \deg h_\lambda = \deg p_\lambda = |\lambda|.$$

In the following, we call **graded** a linear basis $(f_\lambda)_\lambda$ of Sym that is labeled by integer partitions $\lambda \in \mathfrak{Y}$, with each f_λ homogeneous of degree $|\lambda|$.

Remark. The canonical projection $\pi_N : \mathrm{Sym} \to \mathrm{Sym}^{(N)}$ writes in the basis of monomial symmetric functions as

$$\pi_N(m_\lambda(X)) = \begin{cases} m_\lambda(x_1, \ldots, x_N) & \text{if } \lambda \in \mathfrak{Y}^{(N)}, \\ 0 & \text{otherwise.} \end{cases}$$

In particular, $(m_\lambda(x_1, \ldots, x_N))_{\lambda\in\mathfrak{Y}^{(N)}}$ is a linear basis of the algebra of polynomials

$\mathrm{Sym}^{(N)}$. This implies that $(e_1(x_1,\ldots,x_N),\ldots,e_N(x_1,\ldots,x_N))$ is a free generating family of the algebra $\mathrm{Sym}^{(N)}$. Indeed, we already know that $\pi_N(e_M)=0$ if $M>N$, and that $(e_1,e_2,\ldots)$ is a generating family of the algebra Sym. By applying the projection map π_N, we get that $(e_1(x_1,\ldots,x_N),\ldots,e_N(x_1,\ldots,x_N))$ is a generating family of the algebra Sym. To show the freeness, notice that a polynomial in the $e_{k\leq N}$'s is a linear combination of functions $e_{\lambda'}$ with $\lambda\in\mathfrak{Y}^{(N)}$. Consider thus $P(x_1,\ldots,x_N)=\sum_{\lambda\in\mathfrak{Y}^{(N)}} c_\lambda\, e_{\lambda'}(x_1,\ldots,x_N)$. If one of the c_λ is not equal to 0, then one can choose it such that the partition λ is maximal with respect to the dominance order. Then, by Lemma 2.10, the expansion of $P(x_1,\ldots,x_N)$ in monomial symmetric polynomials has c_λ for coefficient of $m_\lambda(x_1,\ldots,x_N)$. Since $(m_\lambda)_{\lambda\in\mathfrak{Y}^{(N)}}$ is a linear basis, $P(x_1,\ldots,x_N)\neq 0$, and the freeness is shown. Consequently, the projection $\pi_N:\mathrm{Sym}\to\mathrm{Sym}^{(N)}$ writes in the basis of elementary symmetric functions as

$$\pi_n(e_\lambda(X))=\begin{cases} e_\lambda(x_1,\ldots,x_N) & \text{if } N\geq\lambda_1\geq\cdots\geq\lambda_\ell,\\ 0 & \text{otherwise,}\end{cases}$$

and $\mathrm{Sym}^{(N)}=\mathbb{C}[e_1,\ldots,e_N]$.

Remark. An important consequence of the previous reasonings is that a symmetric function f of degree less than N is entirely determined by its specialization $f(x_1,\ldots,x_N)$ on a finite alphabet of size N. Indeed, suppose $f=\sum_{|\lambda|\leq N} c_\lambda\, m_\lambda$. Then,

$$f(x_1,\ldots,x_N)=\sum_{|\lambda|\leq N} c_\lambda\, m_\lambda(x_1,\ldots,x_N),$$

and $c_\lambda=[x^\lambda](f(x_1,\ldots,x_N))$. Thus, the expansion of f in monomial functions is entirely determined by $f(x_1,\ldots,x_N)$.

▷ *Symmetrization and antisymmetrization of polynomials.*

There is a fifth graded linear basis $(s_\lambda)_{\lambda\in\mathfrak{Y}}$ that will play an extremely important role when discussing the connection between Sym and the representation theory of symmetric groups. To construct it, it is convenient to introduce the operations of **symmetrization** and **antisymmetrization** of polynomials in $\mathbb{C}[x_1,\ldots,x_N]$. First, call **antisymmetric** a polynomial $P\in\mathbb{C}[x_1,\ldots,x_N]$ such that

$$\forall\sigma\in\mathfrak{S}(N),\ (\sigma\cdot P)(x_1,\ldots,x_N)=\varepsilon(\sigma)\,P(x_1,\ldots,x_N).$$

Example. The **Vandermonde determinant**

$$\Delta(x_1,\ldots,x_N)=\det((x_i^{j-1})_{1\leq i,j\leq N})=\prod_{1\leq i<j\leq N}(x_j-x_i)$$

is an antisymmetric polynomial, as follows trivially from its determinantal expression.

We denote $A^{(N)}$ the vector space of antisymmetric polynomials in N variables. If $P\in A^{(N)}$ and $Q\in\mathrm{Sym}^{(N)}$, then $\sigma\cdot(PQ)=(\sigma\cdot P)(\sigma\cdot Q)=\varepsilon(\sigma)\,PQ$, so $PQ\in A^{(N)}$.

Proposition 2.12. *The map* $m_\Delta : P \mapsto \Delta P$ *is an isomorphism of vector spaces between* $\mathrm{Sym}^{(N)}$ *and* $A^{(N)}$.

Proof. Since $\Delta \neq 0$ and $\mathbb{C}[x_1, \ldots, x_N]$ is an integral domain, we only have to check that m_Δ is surjective. Let Q be an antisymmetric polynomial, and consider two indices $i < j$. The antisymmetry implies that for any $k \in \mathbb{N}^N$,

$$[x^{k \cdot (i,j)}](Q) = -[x^k](Q).$$

In particular, if $k_i = k_j$, then $[x^k](Q) = 0$. Therefore, one can decompose $Q(x_1, \ldots, x_N)$ as

$$\sum_{\substack{k_l \\ l \neq i,j}} \sum_{k_i < k_j} c(k) \, (x^k - x^{k \cdot (i,j)}).$$

Each monomial in this sum is divisible by $x_j - x_i$, because

$$x_i^{k_i} x_j^{k_j} - x_i^{k_j} x_j^{k_i} = (x_j - x_i) \left(\sum_{t=0}^{k_j - k_i - 1} x_i^{k_j - 1 - t} x_j^{k_i + t} \right).$$

So, for any $i < j$, $x_j - x_i$ divides Q. Now, the ring $\mathbb{C}[x_1, \ldots, x_N]$ is factorial, and the factors $x_j - x_i$ with $i < j$ are prime and therefore prime together. As a consequence, the whole product $\Delta = \prod_{i<j}(x_j - x_i)$ divides Q, and $Q = \Delta P$ for some P in $\mathbb{C}[x_1, \ldots, x_N]$. Then, $P = Q/\Delta$ satisfies the transformation rule

$$\sigma \cdot P = \frac{\sigma \cdot Q}{\sigma \cdot \Delta} = \frac{\varepsilon(\sigma) Q}{\varepsilon(\sigma) \Delta} = \frac{Q}{\Delta} = P,$$

so it is a symmetric polynomial. □

Consider now the two linear operators

$$\mathscr{S}(P) = \frac{1}{N!} \sum_{\sigma \in \mathfrak{S}(N)} \sigma \cdot P \quad ; \quad \mathscr{A}(P) = \frac{1}{N!} \sum_{\sigma \in \mathfrak{S}(N)} \varepsilon(\sigma) (\sigma \cdot P),$$

which are a priori defined from $\mathbb{C}[x_1, \ldots, x_N]$ to itself.

Proposition 2.13. *The linear map* $\mathscr{S}$ *is a projection from the space* $\mathbb{C}[x_1, \ldots, x_N]$ *to* $\mathrm{Sym}^{(N)}$, *and the linear map* $\mathscr{A}$ *is a projection from* $\mathbb{C}[x_1, \ldots, x_N]$ *to* $A^{(N)}$.

Proof. For any polynomial P and any permutation τ, one computes

$$\tau \cdot (\mathscr{S}(P)) = \frac{1}{N!} \sum_{\sigma \in \mathfrak{S}(N)} \tau\sigma \cdot P = \frac{1}{N!} \sum_{\sigma' \in \mathfrak{S}(N)} \sigma' \cdot P = \mathscr{S}(P);$$

$$\begin{aligned} \tau \cdot (\mathscr{A}(P)) &= \frac{1}{N!} \sum_{\sigma \in \mathfrak{S}(N)} \varepsilon(\sigma) (\tau\sigma \cdot P) \\ &= \frac{1}{N!} \sum_{\sigma' \in \mathfrak{S}(N)} \varepsilon(\tau) \, \varepsilon(\sigma') (\sigma' \cdot P) = \varepsilon(\tau) \, \mathscr{A}(P). \end{aligned}$$

So, $\mathscr{S}$ and $\mathscr{A}$ have indeed their images in $\mathrm{Sym}^{(N)}$ and in $A^{(N)}$. It suffices then to verify that $\mathscr{S}_{|\mathrm{Sym}^{(N)}} = \mathrm{id}_{|\mathrm{Sym}^{(N)}}$ and $\mathscr{A}_{|A^{(N)}} = \mathrm{id}_{|A^{(N)}}$:

$$\mathscr{S}(P \in \mathrm{Sym}^{(N)}) = \frac{1}{N!} \sum_{\sigma \in \mathfrak{S}(N)} \sigma \cdot P = \frac{1}{N!} \sum_{\sigma \in \mathfrak{S}(N)} P = P;$$

$$\mathscr{A}(P \in A^{(N)}) = \frac{1}{N!} \sum_{\sigma \in \mathfrak{S}(N)} \varepsilon(\sigma)(\sigma \cdot P) = \frac{1}{N!} \sum_{\sigma \in \mathfrak{S}(N)} P = P. \qquad \square$$

As a consequence, one can construct generating families of $\mathrm{Sym}^{(N)}$ and $A^{(N)}$ by applying the operators $\mathscr{S}$ and $\mathscr{A}$ to generating families of $\mathbb{C}[x_1, \ldots, x_N]$, e.g., to the family of monomials $(x^k)_{k \in \mathbb{N}^N}$. Since every sequence k can be reordered by a permutation $\sigma \in \mathfrak{S}(N)$ in a partition $\lambda(k)$, it even suffices to look at the images of the monomials x^λ with

$$\lambda \in \mathfrak{Y}^{(N)} = \{\text{partitions of length } \ell \leq N\}.$$

Example. The family $(\mathscr{S}(x^\lambda))_{\lambda \in \mathfrak{Y}^{(N)}}$ is a linear basis of $\mathrm{Sym}^{(N)}$. Indeed,

$$\mathscr{S}(x^\lambda) = \frac{\mathrm{Aut}(\lambda)}{N!}\, m_\lambda(x_1, \ldots, x_N),$$

where $\mathrm{Aut}(\lambda)$ is the cardinality of the set of permutations $\sigma \in \mathfrak{S}(N)$ such that $\lambda = \lambda \cdot \sigma$. This combinatorial factor is $\mathrm{Aut}(\lambda) = \prod_{i \geq 1} m_i(\lambda)!$.

▷ *Schur functions and Jacobi–Trudi determinantal formulas.*

An integer N being fixed, we set $\rho = (N-1, N-2, \ldots, 0)$, and we define the addition of integer partitions of length smaller than N:

$$\lambda + \mu = (\lambda_1 + \mu_1 \geq \lambda_2 + \mu_2 \geq \cdots \geq \lambda_N + \mu_N).$$

Proposition 2.14. *The polynomial $\mathscr{A}(x^\mu)$ with $\mu \in \mathfrak{Y}^{(N)}$ is non-zero if and only if $\mu = \lambda + \rho$ for some partition $\lambda \in \mathfrak{Y}^{(N)}$. The family $(\mathscr{A}(x^{\lambda+\rho}))_{\lambda \in \mathfrak{Y}^{(N)}}$ is a linear basis of $A^{(N)}$.*

Proof. We already know that $(\mathscr{A}(x^\mu))_{\mu \in \mathfrak{Y}^{(N)}}$ spans linearly $A^{(N)}$. If μ has two equal coordinates $\mu_i = \mu_j$, then

$$\mathscr{A}(x^\mu) = \mathscr{A}(x^{\mu \cdot (i,j)}) = \varepsilon(i,j)\, \mathscr{A}(x^\mu) = -\mathscr{A}(x^\mu),$$

so $\mathscr{A}(x^\mu) = 0$. Thus, one can restrict oneself to integer partitions $\mu \in \mathfrak{Y}^{(N)}$ that are strictly decreasing, so that write as $\lambda + \rho$ for some partition $\lambda \in \mathfrak{Y}^{(N)}$.

Now, if $P(x_1, \ldots, x_n) = \sum_{\lambda \in \mathfrak{Y}^{(N)}} c_\lambda\, \mathscr{A}(x^{\lambda+\rho})$, notice that the only monomial in $\mathscr{A}(x^{\lambda+\rho})$ that is labeled by an integer partition is $x^{\lambda+\rho}$, with coefficient $\frac{1}{N!}$. Therefore,

$$c_\lambda = N!\,[x^{\lambda+\rho}](P(x_1, \ldots, x_n)),$$

and the family $(\mathscr{A}(x^{\lambda+\rho}))_{\lambda \in \mathfrak{Y}^{(N)}}$ is linearly independent, hence a basis. $\square$

Notice that $\mathscr{A}(x^\rho) = \frac{1}{N!}\Delta(x_1, \ldots, x_N)$. By Proposition 2.12, we thus get a new linear basis $(s_\lambda(x_1, \ldots, x_N))_{\lambda \in \mathfrak{Y}^{(N)}}$ of $\mathrm{Sym}^{(N)}$:

$$\forall \lambda \in \mathfrak{Y}^{(N)},\ s_\lambda(x_1, \ldots, x_N) = \frac{\mathscr{A}(x^{\lambda+\rho})}{\mathscr{A}(x^\rho)}.$$

We say that $s_\lambda(x_1, \ldots, x_N)$ is the **Schur polynomial** of label λ. One has a compatibility between Schur polynomials in different numbers of variables:

$$\begin{aligned}\pi_N^{N+1}(s_\lambda(x_1, \ldots, x_{N+1})) &= s_\lambda(x_1, \ldots, x_N, 0) = \frac{a_{\lambda+\rho}(x_1, \ldots, x_N, 0)}{a_\rho(x_1, \ldots, x_N, 0)} \\ &= \frac{a_{\lambda+\rho}(x_1, \ldots, x_N)\, x_1 x_2 \cdots x_N}{a_\rho(x_1, \ldots, x_N)\, x_1 x_2 \cdots x_N} = \frac{a_{\lambda+\rho}(x_1, \ldots, x_N)}{a_\rho(x_1, \ldots, x_N)} \\ &= s_\lambda^{(N)}(x_1, \ldots, x_N)\end{aligned}$$

where $a_\lambda(x_1, \ldots, x_N) = N!\, \mathscr{A}(x^\lambda) = \sum_{\sigma \in \mathfrak{S}(N)} \varepsilon(\sigma)\, x^{\lambda \cdot \sigma}$. Therefore, there exist limiting objects $s_\lambda(X) \in \mathrm{Sym}$ such that for every N,

$$\pi_N(s_\lambda(X)) = \begin{cases} s_\lambda(x_1, \ldots, x_N) & \text{if } \lambda \in \mathfrak{Y}^{(N)}, \\ 0 & \text{otherwise.} \end{cases}$$

Moreover, the previous discussion shows readily that:

Definition 2.15. *Call **Schur function** of label $\lambda \in \mathfrak{Y}$ the symmetric function $s_\lambda(X)$. The family $(s_\lambda)_{\lambda \in \mathfrak{Y}}$ is a graded linear basis of* Sym.

The combinatorics of Schur polynomials rely deeply on the combinatorics of Young diagrams and their fillings called tableaux: we shall give a detailed account of this theory in Chapter 3. For the moment, we shall only need to express Schur functions in terms of the elementary and homogeneous symmetric functions. The transition matrices are provided by the **Jacobi–Trudi determinants**:

Theorem 2.16 (Jacobi–Trudi). *For any partition $\lambda \in \mathfrak{Y}$,*

$$s_\lambda = \det(h_{\lambda_i - i + j})_{1 \le i,j \le N} = \det(e_{\lambda'_i - i + j})_{1 \le i,j \le M}$$

where $N \ge \ell(\lambda)$ and $M \ge \ell(\lambda') = \lambda_1$ (by convention, $h_{-k} = e_{-k} = 0$ for any $k \ge 1$).

Proof. We start with the first identity, and we prove it for the corresponding symmetric polynomials in N variables $x_1, \ldots, x_N$ (this is sufficient for reasons similar to those stated in the remark on page 64). Notice that the Schur polynomials are naturally defined as quotients of determinants:

$$s_\lambda(x_1, \ldots, x_N) = \frac{a_{\lambda+\rho}(x_1, \ldots, x_N)}{a_\rho(x_1, \ldots, x_N)} = \frac{\det(x_i^{\lambda_j + N - j})_{1 \le i,j \le N}}{\det(x_i^{N-j})_{1 \le i,j \le N}}.$$

Let us compute the determinants in the right-hand side. For i,j less than N, we denote

$$e_{i,j} = e_j(x_1,\dots,x_{i-1},x_{i+1},\dots,x_N)$$

and consider the matrix $M = ((-1)^{N-j}\, e_{i,N-j})_{1\leq i,j\leq N}$. If $H(t) = \sum_{j=0}^{\infty} h_j\, t^j$ and $E_i(t) = \sum_{j=0}^{N-1} e_{i,j}\, t^j$, then

$$E_i(-t)\,H(t) = \left(\prod_{j\neq i} 1 - t x_j\right)\left(\prod_{j=1}^{N} \frac{1}{1-t x_j}\right) = \frac{1}{1-t x_i} = \sum_{j=0}^{\infty} (x_i)^j\, t^j.$$

Thus, for any fixed sequence of coefficients $(k_1,\dots,k_N) \in \mathbb{N}^N$,

$$(x_i)^{k_j} = \sum_{a=0}^{N} (-1)^{N-a}\, e_{i,N-a}\, h_{k_j+a-N} = (M\,H_k)_{i,j},$$

where $H_k = (h_{k_j+i-N})_{1\leq i,j\leq N}$. Therefore, for any partition $\lambda \in \mathfrak{Y}^{(N)}$,

$$\begin{aligned} a_{\lambda+\rho}(x_1,\dots,x_N) &= \det(x_i^{\lambda_j+N-j}) = (\det M)\,(\det H_{(\lambda_1+N-1,\lambda_2+N-2\dots,\lambda_N)}) \\ &= (\det M)\,\det(h_{\lambda_j-j+i})_{1\leq i,j\leq N} = (\det M)\,\det(h_{\lambda_i-i+j})_{1\leq i,j\leq N}. \end{aligned}$$

In particular, for $\lambda = 0$, one gets $a_\rho(x_1,\dots,x_N) = \det M$, because the matrix $H_{(N-1,N-2,\dots,0)}$ is in this case triangular with diagonal coefficients $h_0 = 1$. This proves the first identity.

For the second identity, consider with $k \geq 1$ the two matrices

$$\Psi = \begin{pmatrix} 1 & h_1 & \cdots & h_{k-1} \\ 0 & 1 & \ddots & \vdots \\ \vdots & & \ddots & h_1 \\ 0 & \cdots & 0 & 1 \end{pmatrix} \quad\text{and}\quad \Phi = \begin{pmatrix} 1 & -e_1 & \cdots & (-1)^{k-1}\, e_{k-1} \\ 0 & 1 & \ddots & \vdots \\ \vdots & & \ddots & -e_1 \\ 0 & \cdots & 0 & 1 \end{pmatrix}.$$

Because of the identity $H(t)\,E(-t) = 1$, $\Psi = \Phi^{-1}$. Since these matrices have determinant 1, Ψ is then the transpose of the comatrix of Φ, and more generally, every minor $\det_{I,J}(\Psi^t)$ is equal to the cominor $\det_{[\![1,k]\!]\setminus I,[\![1,k]\!]\setminus J}(\Phi)$. Take $k = N+M$, and consider the sets

$$I = [\![1,N]\!] \quad ; \quad J = \{j + \lambda_{N+1-j}\}_{j\in[\![1,N]\!]}.$$

Then,

$$\begin{aligned} \det_{I,J}(\Psi^t(x_1,\dots,x_N)) &= \det(h_{\lambda_{N+1-i}+i-j}(x_1,\dots,x_N))_{1\leq i,j\leq N} \\ &= \det(h_{\lambda_i-i+j}(x_1,\dots,x_N))_{1\leq i,j\leq N} \\ &= s_\lambda(x_1,\dots,x_N) = \det_{[\![1,k]\!]\setminus I,[\![1,k]\!]\setminus J}(\Phi(x_1,\dots,x_N)). \end{aligned}$$

We claim that the complement of J is $\{N+j-\lambda'_j\}_{j\in[\![1,M]\!]}$. Indeed, if one lists the elements of J in increasing order, they are:

$$\begin{aligned}
&1,2,\ldots,m_0,\\
&(m_0+1)+1,(m_0+2)+1,\ldots,(m_0+m_1)+1,\\
&(m_0+m_1+1)+2,(m_0+m_1+1)+2,\ldots,(m_0+m_1+m_2)+2,\\
&etc.
\end{aligned}$$

each line consisting in consecutive integers, with $m_i = m_i(\lambda) = \mathrm{card}\{j \in [\![1,N]\!] \mid \lambda_j = i\}$. Therefore, the complement of J is

$$\begin{aligned}
&\{m_0+1, m_0+m_1+2, m_0+m_1+m_2+3,\ldots,m_0+\cdots+m_{M-1}+M\}\\
&=\{N-\lambda'_1+1, N-\lambda'_2+2, N-\lambda'_3+3,\ldots,N-\lambda'_M+M\}.
\end{aligned}$$

We conclude that

$$\begin{aligned}
s_\lambda(x_1,\ldots,x_N) &= \det_{[\![N+1,N+M]\!],\{N+j-\lambda'_j\}_{j\in[\![1,M]\!]}}(\Phi(x_1,\ldots,x_N))\\
&= \det(e_{\lambda'_j-j+i}(x_1,\ldots,x_n))_{1\leq i,j\leq M},
\end{aligned}$$

which is the second identity up to an exchange of the indices i,j. □

Example. The Schur functions of degree 4 write as follows in terms of the homogeneous symmetric functions:

$$\begin{aligned}
s_4(X) &= h_4(X);\\
s_{3,1}(X) &= h_{3,1}(X)-h_4(X);\\
s_{2,2}(X) &= h_{2,2}(X)-h_{3,1}(X);\\
s_{2,1,1}(X) &= h_{2,1,1}(X)-h_{2,2}(X)-h_{3,1}(X)+h_4(X);\\
s_{1,1,1,1}(X) &= h_{1,1,1,1}(X)-3h_{2,1,1}(X)+h_{2,2}(X)+2h_{3,1}(X)-h_4(X).
\end{aligned}$$

2.3 The structure of graded self-adjoint Hopf algebra

▷ *The Hopf algebra structure of* **Sym**.

We can now introduce the full **Hopf algebra** structure of Sym. Recall that a Hopf algebra over a field k is a k-vector space A such that:

(H1) A is endowed with a structure of associative algebra. The unity 1_A can then be considered as a morphism $1_A : \lambda \in k \mapsto \lambda 1_A \in A$; and the multiplication $\nabla(a,b) = ab$ can be considered as a linear map $A \otimes_k A \to A$. The following diagrams are commutative:

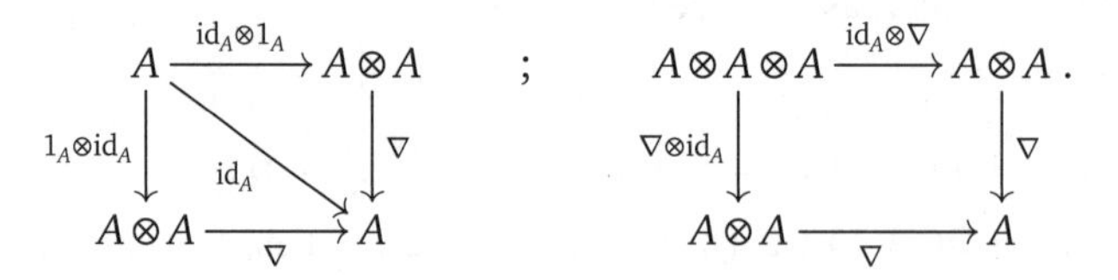

(H2) A is endowed with a structure of associative **coalgebra**, that is to say a linear map $\Delta : A \to A \otimes_k A$ (the **coproduct**) and a morphism of algebras $\eta_A : A \to k$ (the **counity**), such that the following diagrams are commutative:

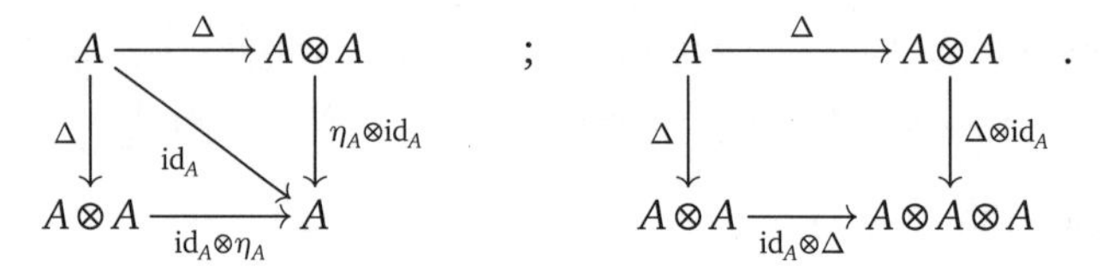

(H3) A is a **bialgebra**, i.e., the structures of algebra and of co-algebra are compatible. Hence, ∇ and 1_A are morphisms of coalgebras, and Δ and η_A are morphisms of algebras, so the following diagrams are commutative:

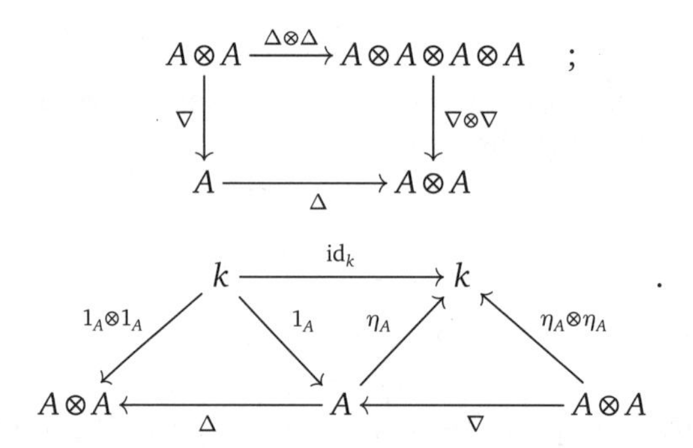

In the diagram on the left-hand side, $\nabla \otimes \nabla$ is the linear extension of the map defined on simple tensors by $a_1 \otimes a_2 \otimes a_3 \otimes a_4 \mapsto (a_1 a_3) \otimes (a_2 a_4)$ (beware of the order of the tensors).

(H4) A is endowed with a linear map $\omega_A : A \to A$ called the **antipode** and which makes the following diagram commutative:

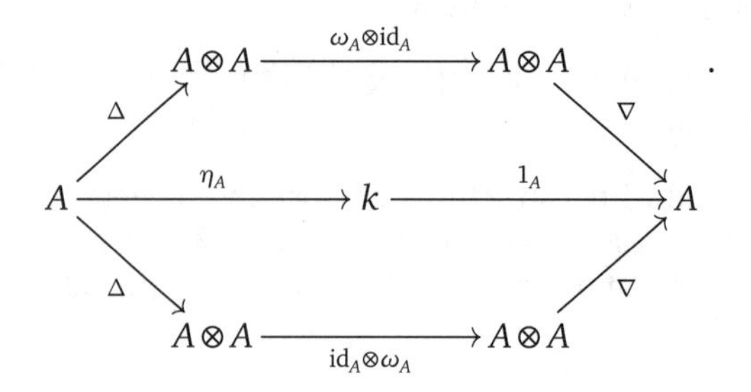

This list of conditions might be a little intimidating for a reader not accustomed to the notion of Hopf algebra. Intuitively, one can understand a product in an algebra A as a way to *combine* two elements a_1 and a_2 in order to get $a = a_1 a_2$. Then, similarly, a coproduct in a coalgebra A is a way to *split* elements $a \in A$ into linear combinations of pairs $a_1 \otimes a_2$. A bialgebra is given by two compatible ways of combining and splitting elements, and a Hopf algebra is endowed on top of the previous structures with a kind of *symmetry*, the antipode. In this book, we shall encounter several combinatorial Hopf algebras, and Sym will be the most important. Let us give before some simpler examples.

Example. Let G be a finite group, and $\mathbb{C}G$ its group algebra. The following maps make it into a Hopf algebra over $\mathbb{C}$:

$$\Delta(g) = g \otimes g \quad ; \quad \eta_{\mathbb{C}G}(g) = 1 \quad ; \quad \omega_{\mathbb{C}G}(g) = g^{-1}.$$

This Hopf algebra is **involutive**, meaning that $\omega^2 = \mathrm{id}$. It is also **cocommutative**, which means that if $\tau : A \otimes A \to A \otimes A$ is the exchange of tensors $a \otimes b \mapsto b \otimes a$, then $\tau \circ \Delta = \Delta$. However, it is not commutative unless G is commutative.

Example. Let G be a finite group, and $\mathbb{C}^G$ the set of functions from $G \to \mathbb{C}$, which for once we do not consider as formal sums $\in \mathbb{C}G$. The pointwise product of functions $(f_1 f_2)(g) = f_1(g) f_2(g)$ makes $\mathbb{C}^G$ into a commutative algebra, and the following maps makes it into a Hopf algebra over $\mathbb{C}$:

$$\Delta(f)(g \otimes h) = f(gh) \quad ; \quad \eta_{\mathbb{C}G}(f) = f(e_G) \quad ; \quad (\omega_{\mathbb{C}G} f)(g) = f(g^{-1}).$$

This Hopf algebra is involutive, and it is usually not cocommutative, unless G is commutative. Thus, we get two different Hopf algebra structures on the set of functions from G to $\mathbb{C}$.

Example. Let V be a vector space over $\mathbb{C}$, and $T(V) = \mathbb{C} \oplus V \oplus V^{\otimes 2} \oplus V^{\otimes 3} \oplus \cdots$ its **tensor algebra**. It is a Hopf algebra for the maps defined on V by

$$\Delta(v) = v \otimes 1 + 1 \otimes v \quad ; \quad \eta_{T(V)}(v) = 0 \quad ; \quad \omega_{T(V)}(v) = -v,$$

and extended uniquely to $T(V)$ by using the conditions of Hopf algebra. This Hopf algebra is involutive and cocommutative.

Remark. There is another definition for the antipode: it is the unique inverse of the identity map $\mathrm{id}_A : A \to A$ with respect to the convolution product of the bialgebra A; see Lemma 5.4. This alternative definition will be used in the theory of quantum groups, and in Chapter 12 in order to define the antipode of the algebra of free quasi-symmetric functions.

Let $X = \{x_1, x_2, \ldots\}$ and $Y = \{y_1, y_2, \ldots\}$ be two infinite alphabets. The **sum of alphabets** $X + Y$ is their reunion $\{x_1, x_2, \ldots, y_1, y_2, \ldots\}$. Notice that

$$p_k(X + Y) = p_k(X) + p_k(Y)$$

for any $k \geq 1$. Since $\mathrm{Sym} = \mathbb{C}[p_1, p_2, \ldots]$, this implies that any symmetric function in $X+Y$ can be expanded into a finite sum of products $g_i(X)\,h_i(Y)$:

$$f(X+Y) = \sum_{i=1}^{I} g_i(X)\,h_i(Y).$$

In this situation, set $\Delta(f) = \sum_{i=1}^{I} g_i \otimes h_i$; we claim that this defines in a non-ambiguous way a coproduct $\Delta : \mathrm{Sym} \to \mathrm{Sym} \otimes_{\mathbb{C}} \mathrm{Sym}$. Indeed, Δ is compatible with products of functions, and on the algebraic basis $(p_k)_{k\geq 1}$ of Sym, the only possibility is

$$\Delta(p_k) = p_k \otimes 1 + 1 \otimes p_k,$$

so Δ is uniquely determined. The previous formula shows readily that Δ is a morphism of graded algebras, the gradation on $\mathrm{Sym} \otimes_{\mathbb{C}} \mathrm{Sym}$ being given by $\deg(g \otimes h) = \deg g + \deg h$. Given a symmetric function f, set also

$$\eta(f) = f(0) = \text{constant term in } f.$$

Then, it is easy to verify that (Δ, η) makes Sym into a graded cocommutative bialgebra, the cocommutativity coming from the identity of alphabets $X+Y = Y+X$, or from the formula of Δ on power sums.

We define the antipode of Sym by linear extension of the formula $\omega(p_\lambda) = (-1)^{\ell(\lambda)}\, p_\lambda$. Then, for $k \geq 1$,

$$\nabla \circ (\omega \otimes \mathrm{id}) \circ \Delta(p_k) = \nabla \circ (\omega \otimes \mathrm{id})(p_k \otimes 1 + 1 \otimes p_k) = \nabla(-p_k \otimes 1 + 1 \otimes p_k) = 0.$$

Since the operations above are all morphisms of algebras, we conclude that for any symmetric function f with graded expansion $f(X) = \sum_{k=0}^{\infty} f_k(X)$,

$$\nabla \circ (\omega \otimes \mathrm{id}) \circ \Delta(f) = f_0 = 1 \circ \eta(f).$$

The same identity holds with $\mathrm{id} \otimes \omega$ instead of $\omega \otimes \mathrm{id}$, so:

Theorem 2.17. *Endowed with the coproduct Δ, the counity η, and the antipode ω,* Sym *is an involutive, commutative and cocommutative graded Hopf algebra.*

Example. Let us compute the coproduct and the antipode of elementary and homogeneous functions. We claim that

$$\begin{aligned}
\Delta(h_k) &= h_k \otimes 1 + h_{k-1} \otimes h_1 + \cdots + h_1 \otimes h_{k-1} + 1 \otimes h_k;\\
\Delta(e_k) &= e_k \otimes 1 + e_{k-1} \otimes e_1 + \cdots + e_1 \otimes e_{k-1} + 1 \otimes e_k;\\
\omega(h_k) &= (-1)^k\, e_k \quad ; \quad \omega(e_k) = (-1)^k\, h_k.
\end{aligned}$$

The two first identities are obvious if one writes

$$h_k(Z) = \sum_{i_1 \leq \cdots \leq i_k} z_{i_1} \cdots z_{i_k} \quad ; \quad e_k(Z) = \sum_{i_1 < \cdots < i_k} z_{i_1} \cdots z_{i_k}.$$

Since $Z = X + Y = \{x_1, x_2, \ldots, y_1, y_2, \ldots\}$, it suffices then to split each monomial in $h_k(X+Y)$ or $e_k(X+Y)$ as a product of a monomial in the x_i's with a monomial in the y_j's. On the other hand, the identities for the antipode can be shown by induction on k and using the Newton identities. For instance, for the homogeneous symmetric functions,

$$\begin{aligned}\omega(h_k(X)) &= \frac{1}{k}\sum_{j=1}^{k} \omega(p_j(X)\, h_{k-j}(X)) \\ &= \frac{1}{k}\sum_{j=1}^{k} (-1)^{k-j+1}\, p_j(X)\, e_{k-j}(X) = (-1)^k\, e_k(X).\end{aligned}$$

In a moment we shall also compute $\Delta(s_\lambda)$ and $\omega(s_\lambda)$ for any partition λ.

Remark. It is convenient to introduce the formal **opposite** $-X$ of an alphabet X, with for any symmetric function f,

$$f(-X) = (\omega(f))(X).$$

Given two alphabets X and Y, one has $f(X - Y) = (\mathrm{id} \otimes \omega) \circ \Delta(f)$, and more generally, one can deal with quantities such as $f(X - Y + Z)$ for three alphabets X, Y, Z.

▷ *Cauchy identities and the Hall scalar product.*

There is an additional structure on Sym that eases many Hopf computations, and is related to the following identities. Let X and Y be two infinite alphabets, and $\Omega(X,Y)$ be the **Cauchy product**

$$\Omega(X,Y) = \prod_{i,j} \frac{1}{1 - x_i y_j}.$$

It is obviously symmetric separately in X and in Y.

Theorem 2.18 (Cauchy). *The Cauchy product satisfies*

$$\Omega(X,Y) = \sum_{\lambda \in \mathfrak{Y}} \frac{1}{z_\lambda}\, p_\lambda(X)\, p_\lambda(Y) = \sum_{\lambda \in \mathfrak{Y}} h_\lambda(X)\, m_\lambda(Y) = \sum_{\lambda \in \mathfrak{Y}} s_\lambda(X) s_\lambda(Y).$$

Proof. Consider the **product alphabet** $XY = \{x_i y_j\}_{i,j \geq 1}$. Notice that $p_k(XY) = p_k(X)\, p_k(Y)$ for any k, and more generally, $p_\lambda(XY) = p_\lambda(X)\, p_\lambda(Y)$ for any λ. Then, by using the expansion of homogeneous functions in power sums (cf. Lemma 2.11),

$$\Omega(X,Y) = \prod_{i,j \geq 1} \frac{1}{1 - x_i y_j} = \sum_{k=0}^{\infty} h_k(XY) = \sum_{\lambda} \frac{p_\lambda(XY)}{z_\lambda} = \sum_{\lambda} \frac{1}{z_\lambda}\, p_\lambda(X)\, p_\lambda(Y).$$

The second form of the Cauchy product can be shown by computing first the product over i, then the product over j. Thus,

$$\Omega(X,Y)=\prod_{i,j\geq 1}\frac{1}{1-x_iy_j}=\prod_{j\geq 1}H(X,y_j)=\sum_k h_k(X)\,y^k=\sum_\lambda h_\lambda(X)\,m_\lambda(Y)$$

where the intermediate sum is over all possible finite sequences of integers $k\in\mathbb{N}^{(\mathbb{N}^*)}$, and the last identity is obtained by gathering the sequences k according to their non-increasing reorderings λ.

For the last identity, one can assume without loss of generality that X and Y have a finite number N of variables, and consider

$$\begin{aligned}a_\rho(X)\,a_\rho(Y)\prod_{1\leq i,j\leq N}\frac{1}{1-x_iy_j}&=a_\rho(X)\,a_\rho(Y)\sum_k h_k(X)\,y^k\\&=a_\rho(X)\sum_{\substack{\sigma\in\mathfrak{S}(N)\\k}}h_k(X)\,\varepsilon(\sigma)\,y^{k+\rho\cdot\sigma}.\end{aligned}$$

Denote $l=k+\rho\cdot\sigma=(k_i+N-\sigma(i))_{1\leq i\leq N}$. The previous expression can now be rewritten as a sum

$$a_\rho(X)\sum_{\substack{\sigma\in\mathfrak{S}(N)\\l}}\varepsilon(\sigma)\,h_{l-\rho\cdot\sigma}(X)\,y^l=a_\rho(X)\sum_l s_{l-\rho}(X)\,y^l=\sum_l a_l(X)\,y^l,$$

where for the first equality we used the first Jacobi–Trudi formula. The sum runs over all sequences such that the number of terms larger than $N-i$ is larger than i. Using the identity $a_{l\cdot\sigma}(X)=\varepsilon(\sigma)\,a_l(X)$ and gathering the terms by non-increasing reorderings of l, one gets finally $\sum_\lambda a_{\lambda+\rho}(X)\,a_{\lambda+\rho}(Y)$, whence the identity by dividing by $a_\rho(X)\,a_\rho(Y)$. □

The **Hall scalar product** on Sym is the unique $\mathbb{C}$-bilinear form for which $(s_\lambda)_{\lambda\in\mathfrak{Y}}$ is an orthonormal basis:

$$\forall\lambda,\mu\in\mathfrak{Y},\ \langle s_\lambda\,|\,s_\mu\rangle_{\mathrm{Sym}}=\delta_{\lambda,\mu}=\begin{cases}1&\text{if }\lambda=\mu,\\0&\text{otherwise.}\end{cases}$$

Remark. Usually, a scalar product on a $\mathbb{C}$-vector space is defined as a sesquilinear form; this is what has been done so far, for instance when manipulating the group algebras $\mathbb{C}G$. However, when working with Sym (or any other combinatorial Hopf algebra), it will be easier to use a (non-degenerate) complex bilinear form. The advantage of this choice is that we shall not have to take care of complex conjugates of scalar coefficients (the complex structure does not mix very well with the Hopf algebra structure, which is purely algebraic). The only place where this choice will lead to a small complication of the arguments is at the beginning of the proof of Proposition 2.26.

The connection between the Hall scalar product and the Cauchy identities is given by:

Proposition 2.19. *Consider two graded bases $(u_\lambda)_{\lambda\in\mathfrak{Y}}$ and $(v_\lambda)_{\lambda\in\mathfrak{Y}}$ of* Sym. *Then, $\Omega(X,Y)=\sum_\lambda u_\lambda(X)\,v_\lambda(Y)$ if and only if $(u_\lambda)_{\lambda\in\mathfrak{Y}}$ and $(v_\lambda)_{\lambda\in\mathfrak{Y}}$ are dual bases with respect to the Hall scalar product:*

$$\forall \lambda,\mu\in\mathfrak{Y},\ \left\langle u_\lambda \,\middle|\, v_\mu\right\rangle_{\mathrm{Sym}}=\delta_{\lambda,\mu}.$$

Proof. Since $(s_\mu)_{\mu\in\mathfrak{Y}}$ is an orthonormal basis, one has the expansions

$$u_\lambda(X)=\sum_{\mu\in\mathfrak{Y}}\left\langle u_\lambda \,\middle|\, s_\mu\right\rangle_{\mathrm{Sym}} s_\mu(X)=\sum_{\mu\in\mathfrak{Y}} U_{\lambda\mu}\, s_\mu(X);$$
$$v_\lambda(X)=\sum_{\mu\in\mathfrak{Y}}\left\langle v_\lambda \,\middle|\, s_\mu\right\rangle_{\mathrm{Sym}} s_\mu(X)=\sum_{\mu\in\mathfrak{Y}} V_{\lambda\mu}\, s_\mu(X).$$

The duality amounts to saying that $\delta_{\lambda,\mu}=\sum_\nu U_{\lambda\nu}V_{\mu\nu}$, or in other words that the transpose of the matrix $(U_{\lambda\mu})_{\lambda,\mu\in\mathfrak{Y}}$ is equal to the inverse of $(V_{\lambda\mu})_{\lambda,\mu\in\mathfrak{Y}}$. But then,

$$\begin{aligned}\sum_\lambda u_\lambda(X)\,v_\lambda(Y)&=\sum_{\lambda,\mu,\nu} U_{\lambda\mu}V_{\lambda\nu}\,s_\mu(X)\,s_\nu(Y)\\&=\sum_{\mu,\nu}(U^tV)_{\mu,\nu}\,s_\mu(X)\,s_\nu(Y)=\sum_\mu s_\mu(X)\,s_\mu(Y)=\Omega(X,Y).\end{aligned}$$

The sequence of identities can be reversed, hence the equivalence announced. In particular, $(h_\lambda)_{\lambda\in\mathfrak{Y}}$ and $(m_\lambda)_{\lambda\in\mathfrak{Y}}$ are dual bases of Sym, and $(p_\lambda)_{\lambda\in\mathfrak{Y}}$ is an orthogonal basis with

$$\left\langle p_\lambda \,\middle|\, p_\mu\right\rangle_{\mathrm{Sym}}=\delta_{\lambda,\mu}\,z_\lambda.$$

□

Proposition 2.20. *The Hopf algebra* Sym *is self-adjoint with respect to its scalar product $\langle\cdot\,|\,\cdot\rangle_{\mathrm{Sym}}$, i.e., for any symmetric functions f,g,h,*

$$\langle f\,g \mid k\rangle_{\mathrm{Sym}}=\langle f\otimes g\mid \Delta(h)\rangle_{\mathrm{Sym}\otimes\mathrm{Sym}}\quad;\quad \langle f\mid \omega(g)\rangle_{\mathrm{Sym}}=\langle \omega(f)\mid g\rangle_{\mathrm{Sym}}\,.$$

Moreover, ω is an isometry of Sym.

Proof. For the formula involving the coproduct, it suffices to prove it with functions in a linear basis of Sym, say the power sums. Since $\Delta(p_k)=p_k\otimes 1+1\otimes p_k$, for any partition $\nu\in\mathfrak{Y}$,

$$\Delta(p_\nu)=\prod_{i=1}^{\ell(\nu)}\Delta(p_{\nu_i})=\sum_{I\subset[\![1,\ell(\nu)]\!]}\left(\prod_{i\in I}p_{\nu_i}\right)\otimes\left(\prod_{i\notin I}p_{\nu_i}\right).$$

If one gathers the terms in the right-hand side of this formula according to the partitions λ and μ obtained by reordering of $\{\nu_i\}_{i\in I}$ and $\{\nu_i\}_{i\notin I}$, one gets:

$$\Delta(p_\nu)=\sum_{(\lambda,\mu)\,|\,\lambda\sqcup\mu=\nu}\frac{z_\nu}{z_\lambda\,z_\mu}\,p_\lambda\otimes p_\mu.$$

Then,

$$\begin{aligned}\left\langle p_\lambda p_\mu \,\middle|\, p_\nu\right\rangle_{\mathrm{Sym}} &= \delta_{\lambda\sqcup\mu,\nu}\, z_\nu;\\ \left\langle p_\lambda \otimes p_\mu \,\middle|\, \Delta(p_\nu)\right\rangle_{\mathrm{Sym}\otimes\mathrm{Sym}} &= \sum_{(\lambda',\mu')\,|\,\lambda'\sqcup\mu'=\nu} \frac{z_\nu}{z_{\lambda'}z_{\mu'}} \left\langle p_\lambda \otimes p_\mu \,\middle|\, p_{\lambda'}\otimes p_{\mu'}\right\rangle_{\mathrm{Sym}}\\ &= \delta_{\lambda\sqcup\mu,\nu}\, z_\nu.\end{aligned}$$

Regarding the antipode, the self-adjointness is obvious again by looking at power sums $f = p_\lambda$ and $g = p_\mu$. Finally, ω being auto-adjoint, since $\omega^2 = \mathrm{id}$, it is an isometry:

$$\langle \omega(f) \mid \omega(g)\rangle_{\mathrm{Sym}} = \left\langle \omega^2(f) \,\middle|\, g\right\rangle_{\mathrm{Sym}} = \langle f \mid g\rangle_{\mathrm{Sym}}\,. \qquad \square$$

▷ *Skew Schur functions and Hopf operations on Schur functions.*

We now have a structure of graded, commutative and cocommutative, involutive and self-adjoint Hopf algebra on Sym. The last ingredient needed before we can explain the connection between Sym and the representation groups of the symmetric groups $\mathfrak{S}(n)$ is a description of the action of Δ and ω on Schur functions. To this purpose, it is convenient to introduce a generalization of Schur functions, the so-called **skew Schur functions**. They are defined as follows. Since $(s_\lambda)_{\lambda\in\mathfrak{Y}}$ is a graded linear basis of Sym, one can expand a product of Schur functions as a linear combination of other Schur functions:

$$s_\mu(X)\,s_\nu(X) = \sum_\lambda c^\lambda_{\mu\nu}\, s_\lambda(X).$$

Since the left-hand side is homogeneous of degree $|\mu| + |\nu|$, so is the right-hand side, and the sum runs over partitions λ with $|\lambda| = |\mu| + |\nu|$.

Definition 2.21. *Fix two partitions λ and μ. The skew Schur function $s_{\lambda\backslash\mu}(X)$ is*

$$s_{\lambda\backslash\mu}(X) = \sum_\nu c^\lambda_{\mu\nu}\, s_\nu(X).$$

It is a symmetric function either equal to 0, or homogeneous of degree $|\lambda| - |\mu|$.

Alternatively, $s_{\lambda\backslash\mu}$ is the unique symmetric function such that

$$\left\langle s_{\lambda\backslash\mu} \,\middle|\, s_\nu\right\rangle_{\mathrm{Sym}} = \left\langle s_\lambda \,\middle|\, s_\mu s_\nu\right\rangle_{\mathrm{Sym}} = c^\lambda_{\mu\nu}$$

for any Schur function s_ν. There is a simple condition in order to guarantee the vanishing of $s_{\lambda\backslash\mu}$, which is in fact a consequence of a formula for $s_{\lambda\backslash\mu}(X)$. If λ and μ are two partitions, we write $\mu \subset \lambda$ if the Young diagram of μ is included in the Young diagram of λ, which is equivalent to $\mu_i \leq \lambda_i$ for all i.

Proposition 2.22. *The skew Schur function $s_{\lambda\backslash\mu}(X)$ is non-zero if and only if $\mu \subset \lambda$, and then, it is given by*

$$s_{\lambda\backslash\mu} = \det(h_{\lambda_i-\mu_j-i+j})_{1\leq i,j\leq N}$$

where $N \geq \ell(\lambda)$.

Proof. We start with the determinantal formula, and we consider two finite alphabets $X = \{x_1, \ldots, x_N\}$ and $Y = \{y_1, \ldots, y_N\}$. One has

$$\begin{aligned}\sum_{\lambda} s_{\lambda\backslash\mu}(X)\, s_\lambda(Y) &= \sum_{\nu,\lambda} c^{\lambda}_{\mu\nu}\, s_\nu(X)\, s_\lambda(Y) = \sum_{\nu} s_\nu(X)\, s_\nu(Y)\, s_\mu(Y)\\ &= \sum_{\nu} h_\nu(X)\, m_\nu(Y)\, s_\mu(Y);\\ \sum_{\lambda} s_{\lambda\backslash\mu}(X)\, a_{\lambda+\rho}(Y) &= \sum_{\nu} h_\nu(X)\, m_\nu(Y)\, a_{\mu+\rho}(Y)\\ &= \sum_{\substack{\sigma\in\mathfrak{S}(N)\\ k}} h_k(X)\, \varepsilon(\sigma)\, y^{k+(\mu+\rho)\cdot\sigma}.\end{aligned}$$

One gets back $s_{\lambda\backslash\mu}(X)$ by taking the coefficient of $y^{\lambda+\rho}$:

$$s_{\lambda\backslash\mu}(X) = \sum_{\substack{\sigma\in\mathfrak{S}(N)\\ k=\lambda+\rho-(\mu+\rho)\cdot\sigma}} h_k(X)\, \varepsilon(\sigma) = \det(h_{\lambda_i-\mu_j-i+j})_{1\leq i,j\leq N}.$$

We now claim that the determinantal formula implies the vanishing of $s_{\lambda\backslash\mu}$ if $\mu \not\subset \lambda$. Indeed, if μ is not included into λ, choose an index k with $\lambda_k < \mu_k$. The matrix $(h_{\lambda_i-\mu_j-i+j})_{1\leq i,j\leq N}$ has then the following form:

$$\begin{pmatrix} h_{\lambda_1-\mu_1} & \cdots & h_{\lambda_1-\mu_k+k-1} & h_{\lambda_1-\mu_{k+1}+k} & \cdots & h_{\lambda_1-\mu_N+N-1}\\ \vdots & & \vdots & \vdots & & \vdots\\ h_{\lambda_{k-1}-\mu_1-k+2} & \cdots & h_{\lambda_{k-1}-\mu_k+1} & h_{\lambda_{k-1}-\mu_{k+1}+2} & \cdots & h_{\lambda_{k-1}-\mu_N+N+1-k}\\ 0 & \cdots & 0 & h_{\lambda_k-\mu_{k+1}+1} & \cdots & h_{\lambda_k-\mu_N+N-k}\\ \vdots & & \vdots & \vdots & & \vdots\\ 0 & \cdots & 0 & h_{\lambda_N-\mu_{k+1}-N+k+1} & \cdots & h_{\lambda_N-\mu_N} \end{pmatrix}.$$

Therefore, it is upper triangular by blocks of size $k-1$, 1 and $N-k$, with the block of coordinates $\{k\} \times \{k\}$ equal to 0. Hence, the determinant is zero. □

In particular, $s_{\lambda\backslash\emptyset}(X) = s_\lambda(X)$, so skew Schur functions generalize Schur functions. One calls **skew partition** a pair of partitions (λ, μ) such that $\mu \subset \lambda$.

Theorem 2.23. *For any skew partition* $\lambda \setminus \mu$,

$$\Delta(s_{\lambda\backslash\mu}) = \sum_{\nu\,|\,\mu\subset\nu\subset\lambda} s_{\lambda\backslash\nu} \otimes s_{\nu\backslash\mu} \quad ; \quad \omega(s_{\lambda\backslash\mu}) = (-1)^{|\lambda|-|\mu|}\, s_{\lambda'\backslash\mu'}.$$

Proof. For the coproduct formula, consider three independent alphabets X, Y and

Z, and the generating series

$$\begin{aligned}\sum_{\lambda,\mu} s_{\lambda\backslash\mu}(X)s_\mu(Y)s_\lambda(Z) &= \sum_{\lambda,\mu,\nu} c^\lambda_{\mu\nu}\, s_\nu(X)s_\mu(Y)s_\lambda(Z)\\ &= \sum_{\mu,\nu} s_\nu(X)s_\mu(Y)s_\nu(Z)s_\mu(Z) = \left(\prod_{i,k}\frac{1}{1-x_iz_k}\right)\left(\prod_{j,k}\frac{1}{1-y_jz_k}\right)\\ &= \Omega(X,Z)\,\Omega(Y,Z) = \Omega(X+Y,Z) = \sum_\lambda s_\lambda(X+Y)s_\lambda(Z).\end{aligned}$$

By identification of the part in X,Y, $s_\lambda(X+Y) = \sum_\mu s_{\lambda\backslash\mu}(X)s_\mu(Y)$, which is the identity stated in the case of a simple (not skew) partition. For a skew partition, one writes then

$$\begin{aligned}\sum_\mu s_{\lambda\backslash\mu}(X+Y)s_\mu(Z) &= s_\lambda(X+Y+Z)\\ &= \sum_\nu s_{\lambda\backslash\nu}(X)s_\nu(Y+Z) = \sum_{\nu,\mu} s_{\lambda\backslash\nu}(X)s_{\nu\backslash\mu}(Y)s_\mu(Z),\end{aligned}$$

whence the coproduct formula in the general case by identification of the part that depends on X and Y.

For the computation of the antipode, the same reasoning as in the proof of Theorem 2.16 gives

$$s_{\lambda\backslash\mu} = \det(e_{\lambda'_i-\mu'_j-i+j})_{1\leq i,j\leq M}.$$

It suffices then to use $\omega(h_k) = (-1)^k\, e_k$ and the fact that ω is a morphism of algebras. □

In Chapter 3, we shall use this **duplication rule** in order to give the expansion of Schur functions in monomials, and to connect them to fillings of Young diagrams.

2.4 The Frobenius–Schur isomorphism

It is now time to make the connection between Sym and the representations of the symmetric groups. We denote $\{S^\lambda\}_{\lambda\in\widehat{\mathfrak{S}}(n)}$ the set of irreducible representations of $\mathfrak{S}(n)$, and we recall that $\mathrm{R}_0(\mathfrak{S}(n))$ is the group of formal sums of these irreducible representations:

$$\sum_{\lambda\in\widehat{\mathfrak{S}}(n)} m_\lambda\, S^\lambda,\quad m_\lambda\in\mathbb{Z}.$$

By tensorization by $\mathbb{C}$, one can even consider formal linear combinations with coefficients in $\mathbb{C}$; we thus set $\mathrm{R}_\mathbb{C}(\mathfrak{S}(n)) = \mathbb{C}\otimes_\mathbb{Z}\mathrm{R}_0(\mathfrak{S}(n))$. There is an isomorphism of vector spaces

$$\begin{aligned}\phi_n : \mathrm{R}_{\mathbb{C}}(\mathfrak{S}(n)) &\to Z(\mathbb{C}\mathfrak{S}(n))\\ \sum_{\lambda \in \widehat{\mathfrak{S}}(n)} c_\lambda S^\lambda &\mapsto \sum_{\lambda \in \widehat{\mathfrak{S}}(n)} c_\lambda \,\mathrm{ch}^\lambda.\end{aligned}$$

Definition 2.24. *The* ***Grothendieck ring of representations of the symmetric groups*** *is*

$$\mathrm{R}_{\mathbb{C}}(\mathfrak{S}) = \bigoplus_{n=0}^{\infty} \mathrm{R}_{\mathbb{C}}(\mathfrak{S}(n)).$$

We shall see in a moment that $\mathrm{R}_{\mathbb{C}}(\mathfrak{S})$ is indeed a ring, and in fact a graded, involutive and self-adjoint Hopf algebra, which will turn out to be isomorphic to Sym.

▷ *Outer tensor products and the Hopf algebra structure on* $\mathrm{R}_{\mathbb{C}}(\mathfrak{S})$.

If M and N are two left modules over k-algebras A and B, then $M \otimes_k N$ is a natural $A \otimes_k B$ module for

$$(a \otimes b) \cdot (m \otimes n) = (a \cdot m) \otimes (b \cdot n).$$

Beware that if $A = B$, then this $A \otimes_k A$-module is very different from the *inner* tensor product $M \otimes_A N$, which is an A-module. We call $M \otimes_k N$ the **outer tensor product** of M and N. In particular, if V and W are representations of finite groups G and H, then their outer tensor product $V \otimes_{\mathbb{C}} W$ is a representation of the Cartesian product $G \times H$, since we have the isomorphism of algebras $\mathbb{C}(G \times H) = \mathbb{C}G \otimes_{\mathbb{C}} \mathbb{C}H$. To make clear that we are dealing with outer tensor products, we shall use the notation $\boxtimes$ for $V \otimes_{\mathbb{C}} W = V \boxtimes W$. By using characters, it is easy to see that an outer tensor product of representations V and W is irreducible if and only if V and W are irreducible. Therefore, for any finite groups G, H, $\mathrm{R}_0(G \times H) = \mathrm{R}_0(G) \otimes_{\mathbb{Z}} \mathrm{R}_0(H)$, and $\mathrm{R}_{\mathbb{C}}(G \times H) = \mathrm{R}_{\mathbb{C}}(G) \otimes_{\mathbb{C}} \mathrm{R}_{\mathbb{C}}(H)$.

Consider now two representations V of $\mathfrak{S}(m)$ and W of $\mathfrak{S}(n)$. The Cartesian product $\mathfrak{S}(m) \times \mathfrak{S}(n)$ can be seen as a subgroup of $\mathfrak{S}(m+n)$: in a pair (σ, τ), the first term acts by permutation of $[\![1, m]\!]$, and the second term acts by permutation of $[\![m+1, m+n]\!]$. We define

$$V \times W = \mathrm{Ind}_{\mathfrak{S}(m) \times \mathfrak{S}(n)}^{\mathfrak{S}(m+n)} (V \boxtimes W).$$

The operation $\times$ is compatible with direct sums, and with isomorphisms of representations; so, it yields a linear map $\mathrm{R}_{\mathbb{C}}(\mathfrak{S}(m)) \otimes_{\mathbb{C}} \mathrm{R}_{\mathbb{C}}(\mathfrak{S}(n)) \to \mathrm{R}_{\mathbb{C}}(\mathfrak{S}(m+n))$. Consequently, $\mathrm{R}_{\mathbb{C}}(\mathfrak{S})$ is a graded $\mathbb{C}$-algebra, with $\deg(S^\lambda) = n$ if S^λ is an irreducible representation of $\mathfrak{S}(n)$.

We endow $\mathrm{R}_{\mathbb{C}}(\mathfrak{S}(n))$ with the following non-degenerate bilinear form:

$$\left\langle S^\lambda \,\middle|\, S^\mu \right\rangle_{\mathrm{R}_{\mathbb{C}}(\mathfrak{S}(n))} = \begin{cases} 1 & \text{if } S^\lambda \text{ and } S^\mu \text{ are irreducible and isomorphic,} \\ 0 & \text{otherwise.} \end{cases}$$

Considering the direct sum in Definition 2.24 to be orthogonal, we thus get a scalar product on $\mathrm{R}_{\mathbb{C}}(\mathfrak{S})$. By Lemma 1.5, for two representations V and W of the same symmetric group $\mathfrak{S}(n)$,

$$\langle V \mid W\rangle_{\mathrm{R}_{\mathbb{C}}(\mathfrak{S})} = \dim \mathrm{Hom}_{\mathfrak{S}(n)}(V,W).$$

We define the coproduct on $\mathrm{R}_{\mathbb{C}}(\mathfrak{S})$ by using the functors of reduction: if V is a representation of $\mathfrak{S}(p)$, we set

$$\Delta(V) = \sum_{m+n=p} \mathrm{Res}^{\mathfrak{S}(p)}_{\mathfrak{S}(m)\times\mathfrak{S}(n)}(V),$$

being understood that a representation in $\mathrm{R}_{\mathbb{C}}(\mathfrak{S}(m)\times\mathfrak{S}(n))$ can be seen uniquely as an element of $\mathrm{R}_{\mathbb{C}}(\mathfrak{S}(m)) \otimes_{\mathbb{C}} \mathrm{R}_{\mathbb{C}}(\mathfrak{S}(n))$.

Proposition 2.25. *Endowed with Δ and the scalar product $\langle\cdot\mid\cdot\rangle_{\mathrm{R}_{\mathbb{C}}(\mathfrak{S})}$, $\mathrm{R}_{\mathbb{C}}(\mathfrak{S})$ is an algebra and a coalgebra whose structures are adjoint of one another. The counity is the projection on $\mathrm{R}_{\mathbb{C}}(\mathfrak{S}(0)) = \mathbb{C}$.*

Proof. Let S^λ, S^μ and S^ν be three irreducible representations of $\mathfrak{S}(p)$, $\mathfrak{S}(m)$ and $\mathfrak{S}(n)$. We compute:

$$\begin{aligned}\left\langle S^\lambda \,\middle|\, S^\mu \times S^\nu\right\rangle_{\mathrm{R}_{\mathbb{C}}(\mathfrak{S})} &= \delta_{p,m+n} \dim \mathrm{Hom}_{\mathfrak{S}(p)}(S^\lambda, \mathrm{Ind}^{\mathfrak{S}(m+n)}_{\mathfrak{S}(m)\times\mathfrak{S}(n)}(S^\mu \boxtimes S^\nu))\\ &= \delta_{p,m+n} \dim \mathrm{Hom}_{\mathfrak{S}(m)\times\mathfrak{S}(n)}(\mathrm{Res}^{\mathfrak{S}(m+n)}_{\mathfrak{S}(m)\times\mathfrak{S}(n)}(S^\lambda), S^\mu \boxtimes S^\nu)\\ &= \left\langle \Delta(S^\lambda) \,\middle|\, S^\mu \otimes S^\nu\right\rangle_{\mathrm{R}_{\mathbb{C}}(\mathfrak{S})\otimes\mathrm{R}_{\mathbb{C}}(\mathfrak{S})}\end{aligned}$$

by using Frobenius' reciprocity. Therefore, Δ is the adjoint of the multiplication map $\nabla : \mathrm{R}_{\mathbb{C}}(\mathfrak{S})\otimes\mathrm{R}_{\mathbb{C}}(\mathfrak{S}) \to \mathrm{R}_{\mathbb{C}}(\mathfrak{S})$. The coassociativity of Δ is then a consequence of the associativity of ∇. Finally, the unity of the algebra $\mathrm{R}_{\mathbb{C}}(\mathfrak{S})$ is $\lambda \in \mathbb{C} \mapsto \lambda S^\emptyset$, where $S^\emptyset$ is the trivial representation of $\mathfrak{S}(0)$. Its adjoint is the projection $\mathrm{R}_{\mathbb{C}}(\mathfrak{S}) \to \mathrm{R}_{\mathbb{C}}(\mathfrak{S}(0))$, and it is the counity for the coalgebra structure. □

The compatibility between the two structures in order to get a bialgebra will be proven later by using the isomorphism with Sym. We finally need a candidate for the antipode of $\mathrm{R}_{\mathbb{C}}(\mathfrak{S})$. Notice that if V is representation of $\mathfrak{S}(n)$, then the inner tensor product $\varepsilon_n \times_{\mathbb{C}\mathfrak{S}(n)} V$ with the signature representation is again a representation of $\mathfrak{S}(n)$, of same dimension; moreover, this operation is involutive, since $\varepsilon_n \otimes_{\mathbb{C}\mathfrak{S}(n)} \varepsilon_n = 1_n$ is the trivial representation of $\mathfrak{S}(n)$. We set:

$$\omega(V) = (-1)^n\, \varepsilon_n \otimes_{\mathbb{C}\mathfrak{S}(n)} V$$

if V is a representation of $\mathfrak{S}(n)$, and we extend this into a linear involution of $\mathrm{R}_{\mathbb{C}}(\mathfrak{S})$. Notice that ω is self-adjoint, and even an involutive isometry. Therefore, $\mathrm{R}_{\mathbb{C}}(\mathfrak{S})$ is a graded algebra and coalgebra, self-adjoint, and endowed with an involution ω.

▷ *The characteristic map and the Frobenius–Schur theorem.*

We identify as before $R_{\mathbb{C}}(\mathfrak{S})$ with $\bigoplus_{n=0}^{\infty} Z(\mathbb{C}\mathfrak{S}(n))$, and we consider the linear maps

$$\Psi_n : C_\mu \in Z(\mathbb{C}\mathfrak{S}(n)) \mapsto \frac{p_\mu(X)}{z_\mu},$$

where μ runs over integer partitions in $\mathfrak{Y}(n)$, and C_μ is the conjugacy class of permutations with cycle type μ. Since the spaces $Z(\mathbb{C}\mathfrak{S}(n))$ and Sym_n have for respective bases $(C_\mu)_{\mu \in \mathfrak{Y}(n)}$ and $(p_\mu)_{\mu \in \mathfrak{Y}(n)}$, each Ψ_n is an isomorphism of vector spaces, and $\Psi = \sum_{n=0}^{\infty} \Psi_n$ yields a linear isomorphism between $R_{\mathbb{C}}(\mathfrak{S})$ and $\mathrm{Sym} = \bigoplus_{n=0}^{\infty} \mathrm{Sym}_n$. We call Ψ the **characteristic map**.

Proposition 2.26. *The characteristic map is a morphism with respect to the structures of graded algebras and coalgebras, and an isometry with respect to the scalar products. Moreover,* $\omega_{\mathrm{Sym}} \circ \Psi = \Psi \circ \omega_{R_{\mathbb{C}}(\mathfrak{S})}$.

Lemma 2.27. *For any finite group G, any $g \in G$ and any representation $\rho^V : G \to \mathrm{GL}(V)$ with character ch^V, $\mathrm{ch}^V(g)$ is an algebraic integer, that is the solution of a monic polynomial equation with coefficients in $\mathbb{Z}$.*

Proof. One can assume the representation to be unitary, in which case $\mathrm{ch}^V(g)$ is the sum of the eigenvalues of the unitary matrix $\rho^V(g)$. These eigenvalues are m-roots of unity, where m is the order of g, that is to say the smallest integer $m \geq 1$ such that $g^m = e_G$ (it is a well-known fact that this order always divides $\mathrm{card}\, G$). The roots of unity are algebraic integers, and a sum of algebraic integers is again an algebraic integer, hence the result. □

Lemma 2.28. *For any irreducible representation S^λ of $\mathfrak{S}(n)$, $\mathrm{ch}^\lambda(\sigma)$ is an integer in $\mathbb{Z}$ for any permutation $\sigma \in \mathfrak{S}(n)$.*

Proof. Notice that since the representations of $\mathfrak{S}(n)$ are direct sums of irreducible representations, the result extends immediately to arbitrary (reducible) representations. Now, the proof of this result relies on an elementary argument from Galois theory. Let σ be a permutation in $\mathfrak{S}(n)$, which is of order $m \geq 1$. For any (irreducible) representation $\rho^\lambda : \mathfrak{S}(n) \to \mathrm{GL}(S^\lambda)$, $\rho^\lambda(\sigma)$ has its order that divides m, hence, the eigenvalues of this unitary transformation of S^λ belong to $\{1, \xi, \xi^2, \ldots, \xi^{m-1}\}$, where $\xi = \mathrm{e}^{\frac{2i\pi}{m}}$. Consequently, the sum of these eigenvalues, which is the character value $\mathrm{ch}^\lambda(\sigma)$, is in the number field $\mathbb{Q}[\xi]$, with ξ primitive m-th root of the unity. Suppose that we can prove that $\mathrm{ch}^\lambda(\sigma)$ belongs in fact to $\mathbb{Q}$. Then, since it is also an algebraic integer, it will be an integer in $\mathbb{Z}$.

The Galois group of the field extension $\mathbb{Q}[\xi] \,|\, \mathbb{Q}$ is the group $(\mathbb{Z}/m\mathbb{Z})^\times$ of invertible elements in the ring $\mathbb{Z}/m\mathbb{Z}$; an invertible class $[k] \in (\mathbb{Z}/m\mathbb{Z})^\times$ acts as an isomorphism of extension field of $\mathbb{Q}[\xi]$ by sending ξ to ξ^k (this is the Galois theory of cyclotomic extensions). In the following, we denote $[k]\cdot x$ the action of $[k]\in(\mathbb{Z}/m\mathbb{Z})^\times$ on $\mathbb{Q}[\xi]$. The field of rationals $\mathbb{Q}$ is characterized in $\mathbb{Q}[\xi]$ by

$$\mathbb{Q}=\{x\in\mathbb{Q}[\xi]\,|\,\forall[k]\in(\mathbb{Z}/m\mathbb{Z})^\times,\ [k]\cdot x=x\}.$$

Thus, we have to prove that

$$[k]\cdot\big(\mathrm{ch}^\lambda(\sigma)\big)=\mathrm{ch}^\lambda(\sigma)$$

for any $[k]$ in the Galois group. However, the left-hand side is also $\mathrm{ch}^\lambda(\sigma^k)$, because if $\rho^\lambda(\sigma)$ has eigenvalues $x_1,\ldots,x_d\in\{1,\xi,\ldots,\xi^{m-1}\}$, then $\rho^\lambda(\sigma^k)$ has eigenvalues $[k]\cdot(x_1),\ldots,[k]\cdot(x_d)$. Then, the lemma follows from the following claim: if k is coprime to the order m of the permutation σ, then σ and σ^k are conjugated in $\mathfrak{S}(n)$, and therefore have the same character values. This is because if $t(\sigma)=(\lambda_1,\lambda_2,\ldots,\lambda_r)$, then

$$m=\gcd(\lambda_1,\lambda_2,\ldots,\lambda_r).$$

Suppose k coprime to any length λ_i of a cycle $c_i=(n_1,\ldots,n_{\lambda_i})$ of σ, that is coprime to m. When raised to the power k, the cycle c_i becomes another cycle $(c_i)^k$ of length λ_i. So, σ^k has the same cycle type as σ, hence is conjugated to it by the discussion of Section 2.1. □

Corollary 2.29. *Denote ϕ_n the isomorphism $\mathrm{R}_\mathbb{C}(\mathfrak{S}(n))\to Z(\mathbb{C}\mathfrak{S}(n))$ previously considered. Its restriction to*

$$\mathrm{R}_\mathbb{R}(\mathfrak{S}(n))=\mathbb{R}\otimes_\mathbb{Z}\mathrm{R}_0(\mathfrak{S}(n))$$

has for image $Z(\mathbb{R}\mathfrak{S}(n))=\mathrm{Span}_\mathbb{R}(\{C_\mu,\ \mu\in\mathfrak{Y}(n)\})$. This restriction is an isometry with respect to:

- *the restriction of the bilinear form of $\mathrm{R}_\mathbb{C}(\mathfrak{S}(n))$ to $\mathrm{R}_\mathbb{R}(\mathfrak{S}(n))$, which is a real scalar product;*
- *the restriction of the Hermitian scalar product of the group algebra $\mathbb{C}\mathfrak{S}(n)$ to the real vector space $Z(\mathbb{R}\mathfrak{S}(n))$.*

Proof. The image of ϕ_n restricted to $\mathrm{R}_\mathbb{R}(\mathfrak{S}(n))$ is the real span of the characters

$$\mathrm{ch}^\lambda=\sum_{\sigma\in\mathfrak{S}(n)}\mathrm{ch}^\lambda(\sigma)\,\sigma=\sum_{\mu\in\mathfrak{Y}(n)}\mathrm{ch}^\lambda(\sigma_\mu)\,C_\mu,$$

where σ_μ is any permutation with cycle type μ. By the previous lemma, it is therefore included in $Z(\mathbb{R}\mathfrak{S}(n))$, and for dimension reasons, it is equal to the center of the real group algebra. The second part of the corollary follows now from the identity $\phi_n(S^\lambda)=\mathrm{ch}^\lambda$ and from

$$\langle S^\lambda\,|\,S^\mu\rangle_{\mathrm{R}_\mathbb{C}(\mathfrak{S}(n))}=\langle\mathrm{ch}^\lambda\,|\,\mathrm{ch}^\mu\rangle_{\mathbb{C}\mathfrak{S}(n)}=\delta_{\lambda\mu}.$$

□

Proof of Proposition 2.26. By using Proposition 2.2, we obtain

$$\langle C_\mu \mid C_\nu \rangle_{\mathbb{C}\mathfrak{S}(n)} = \delta_{\mu,\nu} \frac{\operatorname{card} C_\mu}{n!} = \frac{\delta_{\mu,\nu}}{z_\mu} = \left\langle \frac{p_\mu}{z_\mu} \,\middle|\, \frac{p_\nu}{z_\nu} \right\rangle_{\mathrm{Sym}} = \langle \Psi(C_\mu) \mid \Psi(C_\nu) \rangle_{\mathrm{Sym}},$$

where on the left-hand side, one considers the restriction to $\mathbb{R}\mathfrak{S}(n)$ of the Hermitian scalar product of $\mathbb{C}\mathfrak{S}(n)$. However, by the previous corollary, via the isomorphism ϕ_n,

$$\langle C_\mu \mid C_\nu \rangle_{\mathbb{C}\mathfrak{S}(n)} = \langle C_\mu \mid C_\nu \rangle_{\mathrm{R}_\mathbb{C}(\mathfrak{S}(n))}.$$

This proves the compatibility with scalar products (complex bilinear forms). The compatibility with the gradation is also obvious.

For the compatibility with the algebra structures, we want to show that if ch^V and ch^W are characters of irreducible representations of $\mathfrak{S}(m)$ and $\mathfrak{S}(n)$, then $\Psi_{m+n}(\mathrm{ch}^{V\times W}) = \Psi_m(\mathrm{ch}^V)\,\Psi_n(\mathrm{ch}^W)$. Since irreducible characters form linear bases of the centers of the group algebras, this will indeed suffice. Notice that for any character of a representation V of $\mathfrak{S}(m)$,

$$\Psi_m(\mathrm{ch}^V) = \Psi_m\left(\sum_{\mu\in\mathfrak{Y}(m)} \mathrm{ch}^V(\mu)\, C_\mu\right) = \frac{1}{n!}\sum_{\sigma\in\mathfrak{S}(m)} \mathrm{ch}^V(\sigma)\, p_{t(\sigma)}.$$

Then,

$$\begin{aligned}\Psi_{m+n}(\mathrm{ch}^{V\times W}) &= \frac{1}{(m+n)!}\sum_{\sigma\in\mathfrak{S}(m+n)} \mathrm{ch}^{V\times W}(\sigma)\, p_{t(\sigma)} \\ &= \frac{1}{(m+n)!}\sum_{\sigma\in\mathfrak{S}(m+n)} \sum_{\substack{\tau\in\mathfrak{S}(m+n)/(\mathfrak{S}(m)\times\mathfrak{S}(n)) \\ \tau^{-1}\sigma\tau=(\sigma_1,\sigma_2)\in\mathfrak{S}(m)\times\mathfrak{S}(n)}} \mathrm{ch}^V(\sigma_1)\,\mathrm{ch}^W(\sigma_2)\, p_{t(\sigma_1)}\, p_{t(\sigma_2)}.\end{aligned}$$

If one sums first over pairs (σ_1,σ_2), then $\sigma = \tau(\sigma_1,\sigma_2)\tau^{-1}$ is entirely determined by $\tau \in \mathfrak{S}(m+n)/(\mathfrak{S}(m)\times\mathfrak{S}(n))$, and there are $\frac{(m+n)!}{m!\,n!}$ possibilities for τ. Therefore, the last term is simply

$$\frac{1}{m!\,n!}\sum_{\sigma_1\in\mathfrak{S}(m),\ \sigma_2\in\mathfrak{S}(n)} \mathrm{ch}^V(\sigma_1)\,\mathrm{ch}^W(\sigma_2)\, p_{t(\sigma_1)}\, p_{t(\sigma_2)} = \Psi_m(\mathrm{ch}^V)\,\Psi_n(\mathrm{ch}^W),$$

hence the property that Ψ is a morphism of algebras. Since $\mathrm{R}_\mathbb{C}(\mathfrak{S})$ and Sym are both self-adjoint and Ψ is an isometry, it follows immediately that Ψ is also a morphism of coalgebras. Finally, for the antipodes, we compute for $\mu \in \mathfrak{Y}(n)$

$$\begin{aligned}\Psi \circ \omega_{\mathrm{R}_\mathbb{C}(\mathfrak{S})}(C_\mu) &= (-1)^{|\mu|}\,\varepsilon(\mu)\,\psi(C_\mu) = (-1)^{\ell(\mu)}\,\frac{p_\mu}{z_\mu} \\ &= \omega_{\mathrm{Sym}}\left(\frac{p_\mu}{z_\mu}\right) = \omega_{\mathrm{Sym}} \circ \Psi(C_\mu).\end{aligned}$$

□

Lemma 2.30. *For any partition $\mu \in \mathfrak{Y}$, $s_\mu(X)\,p_1(X)$ is a positive linear combination of Schur functions.*

Proof. A generalization of this fact will be given in Section 3.1. We work with a finite alphabet of size N big enough. Then,

$$\begin{aligned} &a_\rho(x_1,\dots,x_N)\,s_\mu(x_1,\dots,x_N)\,p_1(x_1,\dots,x_n)\\ &= \sum_{\sigma\in\mathfrak{S}(N)}\sum_{j=1}^{N} \varepsilon(\sigma)\,x_j\,(\sigma\cdot(x_1^{\mu_1+N-1}x_2^{\mu_2+N-2}\cdots x_N^{\mu_N}))\\ &= \sum_{\sigma\in\mathfrak{S}(N)}\sum_{j=1}^{N} \varepsilon(\sigma)\,\sigma\cdot(x_j\,(x_1^{\mu_1+N-1}x_2^{\mu_2+N-2}\cdots x_N^{\mu_N}))\\ &= \sum_{j=1}^{N} a_{(\mu+\rho)+(0,\dots,0,1_j,0,\dots,0)}(x_1,\dots,x_N). \end{aligned}$$

Suppose that $\mu_{j-1}=\mu_j$. Then, $a_{(\mu+\rho)+(0,\dots,0,1_j,0,\dots,0)}(x_1,\dots,x_N)=0$, because it corresponds to the antisymmetrization of a monomial with its $(j-1)$-th and its j-th coordinates equal. It follows that the sum can be rewritten as

$$\sum_{\lambda\,|\,\mu\nearrow\lambda} a_{\lambda+\rho}(x_1,\dots,x_N),$$

where the partitions λ are obtained from μ by adding a cell to the Young diagram of μ (we denote this situation by $\mu\nearrow\lambda$). By dividing by $a_\rho(x_1,\dots,x_N)$ we get

$$s_\mu(X)\,p_1(X) = \sum_{\lambda\,|\,\mu\nearrow\lambda} s_\lambda(X). \qquad \square$$

Remark. The expansion over the Schur basis of the product $s_\mu(X)\,p_1(X)$ will be reinterpreted in the next chapter as branching rules between irreducible representations of $\mathfrak{S}(n)$ and $\mathfrak{S}(n+1)$; see Corollary 3.7. An important combinatorial object that encodes this combinatorial rule is the **Young graph**, which is the oriented graph with vertex set $\mathfrak{Y}=\bigsqcup_{n\in\mathbb{N}}\mathfrak{Y}(n)$ and with an edge between μ and λ if and only if $\mu\nearrow\lambda$; see Figure 2.2. The harmonic analysis of this graph will be performed in Chapter 11.

Theorem 2.31 (Frobenius–Schur). *The operations defined on $\mathrm{R}_{\mathbb{C}}(\mathfrak{S})$ make it into a graded self-adjoint involutive Hopf algebra, isomorphic to* Sym *by the characteristic map. Moreover, the images of the irreducible representations S^λ of $\mathfrak{S}(n)$ by the characteristic map Ψ are the Schur functions $(s_\lambda)_{\lambda\in\mathfrak{Y}(n)}$, hence an explicit bijection*

$$\begin{aligned} &\widehat{\mathfrak{S}}(n)\to\mathfrak{Y}(n)\\ &\quad S^\lambda \mapsto \text{unique partition } \lambda \text{ such that } \Psi(S^\lambda)=s_\lambda. \end{aligned}$$

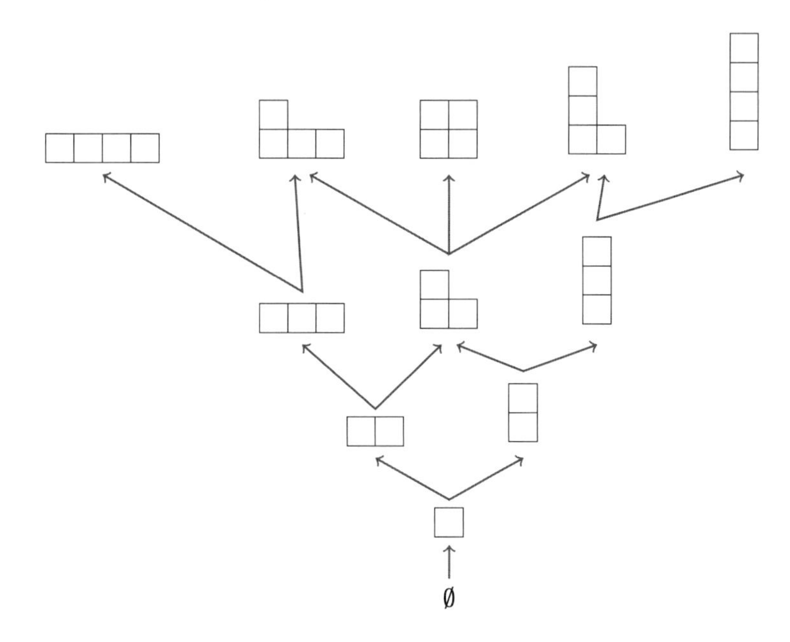

Figure 2.2
The four first levels of the Young graph.

Proof. Since Ψ intertwines all the operations on $\mathrm{R}_{\mathbb{C}}(\mathfrak{S})$ with the operations on Sym, and since Sym is a graded self-adjoint involutive Hopf algebra, the same is true for $\mathrm{R}_{\mathbb{C}}(\mathfrak{S})$. In particular, one gets for free the compatibility between ∇, Δ, 1, η and ω on $\mathrm{R}_{\mathbb{C}}(\mathfrak{S})$, which might have been painful to prove a priori.

The second part of the theorem is a bit more tricky. To begin with, notice that for every n, $h_n(X) = s_n(X)$ is the image by Ψ_n of the trivial representation 1_n of $\mathfrak{S}(n)$:

$$\Psi_n(1_n) = \Psi_n\left(\sum_{\mu \in \mathfrak{Y}(n)} C_\mu\right) = \sum_{\mu \in \mathfrak{Y}(n)} \frac{p_\mu}{z_\mu} = h_n,$$

by the formula in Lemma 2.11. More generally, since Ψ is a ring morphism, for every (non-increasing) sequence of positive integers $n_1, \ldots, n_r$, $h_{n_1,\ldots,n_r}$ is the image by Ψ of a representation, namely, $\mathrm{Ind}_{\mathfrak{S}(n_1)\times\cdots\times\mathfrak{S}(n_r)}^{\mathfrak{S}(n_1+\cdots+n_r)}(1)$.

Using the Jacobi–Trudi equations (Theorem 2.16), it follows that for every partition $\lambda \in \mathfrak{Y}(n)$, $s_\lambda = \det(h_{\lambda_i - i + j})_{i,j}$ is the image by Ψ_n of a linear combination of (irreducible) representations of $\mathfrak{S}(n)$, with integer coefficients (to this point, they might be negative integers):

$$s_\lambda = \Psi_n\left(\sum_{\nu \in \widehat{\mathfrak{S}}(n)} c_\nu^\lambda S^\nu\right), \quad c_\nu^\lambda \in \mathbb{Z}.$$

However, since Ψ_n is an isometry, and since the Schur functions are orthogonal in Sym and the irreducible representations are orthogonal in $\mathrm{R}_\mathbb{C}(\mathfrak{S})$,

$$1 = \langle s_\lambda \mid s_\lambda \rangle_{\mathrm{Sym}} = \sum_{\nu \in \widehat{\mathfrak{S}}(n)} (c_\nu^\lambda)^2.$$

The only possibility is then that $c_\nu^\lambda = \pm 1$ for a unique $\nu \in \widehat{\mathfrak{S}}(n)$, and 0 for the other partitions. Using again the orthogonality of the Schur functions and of the irreducible representations, one concludes that there is a labeling of the irreducible representations of $\mathfrak{S}(n)$ such that for each partition $\lambda \in \mathfrak{Y}(n)$, $\Psi_n(S^\lambda) = \pm_\lambda s_\lambda$. We claim then that the sign $\pm_\lambda$ is always $+1$. Indeed, we can write

$$\pm_\lambda s_\lambda = \Psi_n(S^\lambda) = \sum_{\mu \in \mathfrak{Y}(n)} \mathrm{ch}^\lambda(\mu)\, \frac{p_\mu}{z_\mu},$$

so $\pm_\lambda$ is the sign of $\langle s_\lambda \mid p_{1^n} \rangle_{\mathrm{Sym}}$, because $\mathrm{ch}^\lambda(1^n) = \dim V^\lambda > 0$. But then, $\langle s_\lambda \mid p_{1^n} \rangle_{\mathrm{Sym}}$ is the coefficient of s_λ in the expansion in Schur functions of $(p_1)^n$. By the previous lemma, this is indeed a positive integer. □

From now on, we shall always label irreducible representations of $\mathfrak{S}(n)$ by integer partitions of $\mathfrak{Y}(n)$, being understood that S^λ is the unique irreducible representation of $\mathfrak{S}(n)$ such that $\Psi_n(S^\lambda) = s_\lambda(X)$. Thus, the bijection announced in the introduction of this chapter is realized. It yields readily a formula for the character values $\mathrm{ch}^\lambda(\mu)$:

Theorem 2.32. *For any integer partitions $\lambda, \mu \in \mathfrak{Y}(n)$,*

$$p_\mu(X) = \sum_{\lambda \in \mathfrak{Y}(n)} \mathrm{ch}^\lambda(\mu)\, s_\lambda(X);$$

$$s_\lambda(X) = \sum_{\mu \in \mathfrak{Y}(n)} \mathrm{ch}^\lambda(\mu)\, \frac{p_\mu(X)}{z_\mu}.$$

Proof. Both formulas follow from the fact that $(s_\lambda)_{\lambda \in \mathfrak{Y}}$ and $(p_\mu)_{\mu \in \mathfrak{Y}}$ are orthogonal bases of Sym, and from

$$\langle s_\lambda \mid p_\mu \rangle_{\mathrm{Sym}} = \langle \mathrm{ch}^\lambda \mid z_\mu\, C_\mu \rangle_{\mathrm{R}_\mathbb{C}(\mathfrak{S})} = \delta_{|\lambda|,|\mu|}\, \mathrm{ch}^\lambda(\mu)$$

since Ψ is an isometry. □

Example. With three variables x_1, x_2, x_3, one can easily compute

$$\begin{aligned}
s_3(x_1,x_2,x_3) &= x_1^3 + x_2^3 + x_3^3 + x_1x_2^2 + x_1x_3^2 + x_2x_3^2 + x_2x_1^2 \\
&\quad + x_3x_1^2 + x_3x_2^2 + x_1x_2x_3 \\
&= \frac{p_{1,1,1}(x_1,x_2,x_3)}{6} + \frac{p_{2,1}(x_1,x_2,x_3)}{2} + \frac{p_3(x_1,x_2,x_3)}{3}; \\
s_{2,1}(x_1,x_2,x_3) &= x_1x_2^2 + x_1x_3^2 + x_2x_3^2 + x_2x_1^2 + x_3x_1^2 + x_3x_2^2 + 2x_1x_2x_3 \\
&= \frac{p_{1,1,1}(x_1,x_2,x_3) - p_3(x_1,x_2,x_3)}{3}; \\
s_{1,1,1}(x_1,x_2,x_3) &= x_1x_2x_3 \\
&= \frac{p_{1,1,1}(x_1,x_2,x_3)}{6} - \frac{p_{2,1}(x_1,x_2,x_3)}{2} + \frac{p_3(x_1,x_2,x_3)}{3}.
\end{aligned}$$

From this one deduces the values of the irreducible characters of $\mathfrak{S}(3)$. We identified in Chapter 1 the irreducible representations of $\mathfrak{S}(3)$; so, for instance, one gets that the irreducible character of $\mathfrak{S}(3)$ corresponding to the permutation action on the vector space $V = \{(v_1, v_2, v_3) \in \mathbb{C} \mid v_1 + v_2 + v_3 = 0\}$ is:

$$\mathrm{ch}^V(3) = -1 \quad ; \quad \mathrm{ch}^V(2,1) = 0 \quad ; \quad \mathrm{ch}^V(1^3) = \dim V = 2.$$

2.5 The Schur–Weyl duality

It is a remarkable fact that one can compute the irreducible characters of all the symmetric groups with the help of the algebra Sym, and without ever knowing the matrix forms of the irreducible representations S^λ of these groups. Moreover, Frobenius–Schur Theorem 2.31 yields other information on these representations, e.g., what is the expansion in irreducibles of

$$\mathrm{Ind}_{\mathfrak{S}(n)}^{\mathfrak{S}(n+p)}(S^\lambda) = \Psi^{-1}(s_\lambda\,(s_1)^p).$$

We shall make most of these computations combinatorially concrete in Chapter 3. To conclude this chapter, we want to give another interpretation of the Frobenius formulas of Theorem 2.32. It turns out that one can see $p_\mu(X)$ as a bitrace for the action of the symmetric group and of its commutant on a space of tensors. Then, the Frobenius formula can be interpreted as the expansion of the bitrace corresponding to the decomposition of a bimodule in a sum of tensor products of simple modules. In this framework, the Schur polynomials appear naturally as the characters of polynomial representations of the general linear groups $\mathrm{GL}(N,\mathbb{C})$.

Remark. The whole discussion of this section can be followed without prior knowledge of the representation theory of the general linear groups $\mathrm{GL}(N,\mathbb{C})$, which classically relies on Weyl's highest weight theorem. However, the reader can find in Appendix Appendix A a presentation of this theory, which will be required for an understanding of quantum groups (Chapter 5).

▷ *Spaces of tensors and the commutant algebra of $\mathfrak{S}(n)$.*

In this paragraph, n and N are two fixed positive integers, and we consider as in the beginning of Section 1.5 the following structure of bimodule on $(\mathbb{C}^N)^{\otimes n}$: it is a left $\mathrm{GL}(N,\mathbb{C})$-module for

$$u \cdot (v_1 \otimes v_2 \otimes \cdots \otimes v_n) = u(v_1) \otimes u(v_2) \otimes \cdots \otimes u(v_n),$$

and a right $\mathfrak{S}(n)$-module for the action by permutation of the tensors. Obviously the two actions commute, so one has indeed a bimodule.

To make this structure compatible with the framework of duality between semisimple algebras developed in Section 1.5, we need to replace $\mathbb{C}\mathrm{GL}(N,\mathbb{C})$, which is an infinite-dimensional $\mathbb{C}$-algebra, by its image in $\mathrm{End}((\mathbb{C}^N)^{\otimes n})$. We call this image the **Schur algebra** of order n and of the group $\mathrm{GL}(N,\mathbb{C})$, and it will be denoted $\mathscr{S}_n(N,\mathbb{C})$. This is the set of endomorphisms of $(\mathbb{C}^N)^{\otimes n}$ that can be written as linear combinations of morphisms

$$v_1 \otimes v_2 \otimes \cdots \otimes v_n \mapsto u(v_1) \otimes u(v_2) \otimes \cdots \otimes u(v_n)$$

with $u \in \mathrm{GL}(N,\mathbb{C})$. Since it is a finite-dimensional subalgebra of $\mathrm{End}((\mathbb{C}^N)^{\otimes n})$, it is topologically closed, so, by density of $\mathrm{GL}(N,\mathbb{C})$ in $\mathrm{End}(\mathbb{C}^N)$, $\mathscr{S}_n(N,\mathbb{C})$ is also the set of linear combinations of diagonal morphisms with $u \in \mathrm{End}(\mathbb{C}^N)$ (instead of $\mathrm{GL}(N,\mathbb{C})$).

Remark. Beware that the map $u \mapsto u^{\otimes n}$ *is not* a morphism of algebras from $\mathrm{End}(\mathbb{C}^N)$ to $\mathscr{S}_n(N,\mathbb{C})$: it is compatible with the product, but not at all with sums and multiples by a scalar. Another way to remember this is that $\mathrm{End}(\mathbb{C}^N)$ has a unique class of simple module, namely, $\mathbb{C}^N$; whereas the Schur algebra $\mathscr{S}_n(N,\mathbb{C})$ will be shown to have non-isomorphic simple modules labeled by integer partitions of size n and length smaller than N. The only morphism of algebras in our setting is

$$\begin{aligned} \mathbb{C}\mathrm{GL}(N,\mathbb{C}) &\to \mathscr{S}_n(N,\mathbb{C}) \\ \sum c_g\, g &\mapsto \sum c_g\, (g \otimes g \otimes \cdots \otimes g). \end{aligned}$$

This map can also be defined on the universal enveloping algebra $U(\mathfrak{gl}(N))$; see Appendix Appendix A for the definition of this object.

Theorem 2.33 (Schur–Weyl). *The two algebras $\mathscr{S}_n(N,\mathbb{C})$ and $\mathbb{C}\mathfrak{S}(n)$ are in duality for their actions on $(\mathbb{C}^N)^{\otimes n}$.*

Proof. Denote C the commutant of $\mathbb{C}\mathfrak{S}(n)$ for the action on the space of tensors. We already know that the vector subspace D generated by the image of the map

$$\begin{aligned} \mathrm{End}(\mathbb{C}^N) &\to \mathrm{End}((\mathbb{C}^N)^{\otimes n}) \\ u &\mapsto u^{\otimes n} \end{aligned}$$

is contained in C. To prove that $D = C$, we shall consider a non-degenerate bilinear form on C, such that the orthogonal of D in C is reduced to zero. This is a bit indirect, but easier than proving that every element of C can be obtained by the previous map.

If $u \in \mathrm{End}(\mathbb{C}^N)$, we denote $(u_{ij})_{1\leq i,j\leq N}$ its matrix in the canonical basis $(e_1, \dots, e_N)$ of $\mathbb{C}^N$. A basis of the space of n-tensors $(\mathbb{C}^N)^{\otimes n}$ is

$$e_I = e_{i_1} \otimes e_{i_2} \otimes \cdots \otimes e_{i_n}, \quad I = (i_1, \dots, i_n) \in [\![1, N]\!]^n.$$

For $U \in \mathrm{End}((\mathbb{C}^N)^{\otimes n})$, we denote $(U_{IJ})_{I,J}$ its matrix in this basis. Let us then identify the matrix forms of the elements of D and C. An element of D coming from $u \in \mathrm{End}(\mathbb{C}^N)$ acts on the space of tensors by the matrix

$$(u^{\otimes n})_{IJ} = u_{i_1 j_1} u_{i_2 j_2} \cdots u_{i_n j_n}.$$

On the other hand, if $U \in C$, then in matrix form,

$$\begin{aligned}(U(e_J)) \cdot \sigma &= \left(\sum_I U_{IJ}\, e_I\right) \cdot \sigma = \sum_I U_{IJ}\, e_{I\cdot\sigma} = \sum_I U_{(I\cdot\sigma^{-1})J}\, e_I \\ &= U(e_{J\cdot\sigma}) = \sum_I U_{I(J\cdot\sigma)}\, e_I.\end{aligned}$$

Therefore, $U_{(I\cdot\sigma^{-1})J} = U_{I(J\cdot\sigma)}$, or, equivalently,

$$\forall \sigma \in \mathfrak{S}(n),\ \forall I, J \in [\![1, N]\!]^n,\ U_{IJ} = U_{(I\cdot\sigma)(J\cdot\sigma)}.$$

Consider the bilinear form $b(U, V) = \mathrm{tr}(UV^t) = \sum_{I,J} U_{IJ} V_{IJ}$ on $\mathrm{End}((\mathbb{C}^N)^{\otimes n})$. It is non-degenerate, and by the previous discussion, its restriction to C writes as

$$\sum_{(I,J)\in\Omega} N_{IJ}\, U_{IJ}\, V_{IJ}$$

where Ω is a set of representatives of the orbits for the diagonal action of the symmetric group $\mathfrak{S}(n)$ on $[\![1, N]\!]^n \times [\![1, N]\!]^n$, and N_{IJ} is the cardinality of the orbit of (I, J). Given a pair (I, J), denote x_{IJ} the monomial $x_{i_1 j_1} x_{i_2 j_2} \cdots x_{i_n j_n}$, where the variables x_{ij} belong to the polynomial algebra $\mathbb{C}[x_{11}, \dots, x_{NN}]$. Notice that x_{IJ} is constant on each orbit for the action of $\mathfrak{S}(n)$. The orthogonal of D in C is the set of endomorphisms U such that

$$\sum_{(I,J)\in\Omega} N_{IJ}\, U_{IJ}\, x_{IJ} = 0.$$

To show that this orthogonal $D^\perp$ is reduced to zero amounts to showing that $(x_{IJ})_{(I,J)\in\Omega}$ is a family of linearly independent monomials, or, in other words, that $x_{IJ} = x_{I'J'}$ implies that (I, J) and (I', J') are in the same orbit for the action of

$\mathfrak{S}(n)$. This can be proven by induction on n. Indeed, suppose that $x_{IJ} = x_{I'J'}$. There exists $k \in [\![1,n]\!]$ such that $x_{i_1j_1} = x_{i'_kj'_k}$. We set $\sigma(1) = k$, and consider then

$$\frac{x_{I,J}}{x_{i_1j_1}} = \frac{x_{I',J'}}{x_{i'_kj'_k}}.$$

By induction on n, there is a bijection conjugating the remaining pairs of sequences of indices, which allows one to complete σ in a permutation in $\mathfrak{S}(n)$ such that $(I',J') = (I \cdot \sigma, J \cdot \sigma)$. □

▷ *Schur algebras and representations of linear groups.*

Assume $N \geq n$. Then, the morphism of algebras $\mathbb{C}\mathfrak{S}(n) \to \mathrm{End}((\mathbb{C}^N)^{\otimes n})$ is injective, because the image of $\sigma \in \mathfrak{S}(n)$ is characterized by its action on the tensor

$$e_1 \otimes e_2 \otimes \cdots \otimes e_n \otimes (e_N)^{\otimes N-n},$$

$(e_1,\ldots,e_N)$ being the canonical basis of $\mathbb{C}^N$. Thus, $\mathbb{C}\mathfrak{S}(n)$ can be seen as a subalgebra of $\mathrm{End}((\mathbb{C}^N)^{\otimes n})$, and by Proposition 1.31, the Schur algebra $\mathscr{S}_n(N,\mathbb{C})$ is a semisimple complex algebra, since it is the commutant of (the image of) $\mathbb{C}\mathfrak{S}(n)$. We shall see in a moment that $\mathscr{S}_n(N,\mathbb{C})$ is always semisimple, even when $N < n$. To begin with, let us compute the dimension of $\mathscr{S}_n(N,\mathbb{C})$:

Proposition 2.34. *The complex dimension of $\mathscr{S}_n(N,\mathbb{C})$ is $\binom{N^2+n-1}{n}$.*

Proof. Consider the polynomial algebra $\mathbb{C}[x_{11},\ldots,x_{NN}]$. The subspace of homogeneous polynomials of degree n is of dimension $\binom{N^2+n-1}{n}$. By the proof of Theorem 2.33, there is a non-degenerate pairing between this space and the Schur algebra $\mathscr{S}_n(N,\mathbb{C})$. □

Our goal is now to determine the simple modules and the characters of this Schur algebra, and to prove in general that it is semisimple. To this purpose, it is necessary to relate the modules on $\mathscr{S}_n(N,\mathbb{C})$ to the representations of $\mathrm{GL}(N,\mathbb{C})$. Let V be a finite-dimensional representation of $\mathrm{GL}(N,\mathbb{C})$.

Definition 2.35. *We say that V is a (homogeneous)* ***polynomial representation*** *of $\mathrm{GL}(N,\mathbb{C})$ of degree n if, for any basis of V, the matrix coefficients of $\rho^V(g)$ are homogeneous polynomials of degree n in the coefficients of $g \in \mathrm{GL}(N,\mathbb{C})$.*

More generally, given a morphism of algebras $\rho : \mathbb{C}\mathrm{GL}(N,\mathbb{C}) \to A$ where A is some $\mathbb{C}$-algebra, it is convenient to introduce the **coefficient space** of this morphism: it is the vector space of functions on $\mathbb{C}\mathrm{GL}(N,\mathbb{C})$ that can be written as $\phi \circ \rho$, where ϕ is a linear form on A. Then, a representation V of the group $\mathrm{GL}(N,\mathbb{C})$ is polynomial of degree n if and only if its coefficient space $\mathrm{cf}(V)$ is a subspace of $\mathscr{P}_n(\mathrm{GL}(N,\mathbb{C}))$, the space of homogeneous polynomials of degree n in $g_{11},\ldots,g_{NN}$.

Example. The coefficient space of the morphism $\rho : \mathbb{C}\mathrm{GL}(N,\mathbb{C}) \to \mathscr{S}_n(N,\mathbb{C})$ is $\mathscr{P}_n(\mathrm{GL}(N,\mathbb{C}))$. Indeed, a linear form on $\mathscr{S}_n(N,\mathbb{C})$ is a restriction of a linear form on the space $\mathrm{End}((\mathbb{C}^N)^{\otimes n})$, and these functions yield all the monomials of degree n in the coefficients of $g \in \mathrm{GL}(N,\mathbb{C})$, since we have seen that

$$(g^{\otimes n})_{IJ} = g_{i_1j_1}g_{i_2j_2}\cdots g_{i_nj_n}.$$

Theorem 2.36. *The map $\psi : \mathbb{C}\mathrm{GL}(N,\mathbb{C}) \to \mathscr{S}_n(N,\mathbb{C})$ yields a functorial equivalence between the category of (left, finite-dimensional) modules over the Schur algebra $\mathscr{S}_n(N,\mathbb{C})$, and the category of (finite-dimensional) polynomial representations of $\mathrm{GL}(N,\mathbb{C})$ of degree n, which is a subcategory of the category of representations of this group.*

Proof. Consider the pairing

$$\begin{aligned} b : \mathbb{C}^{\mathrm{GL}(N,\mathbb{C})} \times \mathbb{C}\mathrm{GL}(N,\mathbb{C}) &\to \mathbb{C} \\ f\,,\sum_g c_g g &\mapsto \sum_g c_g\, f(g). \end{aligned}$$

It is a non-degenerate pairing. Indeed, suppose $b(f,\cdot) = 0$. Then, $b(f,g) = f(g)$ for any $g \in \mathrm{GL}(N,\mathbb{C})$, so f is the zero function. Similarly, if one has $b(\cdot,\sum_g c_g f(g)) = 0$, then one can choose functions f that separate points in the support of the finite sum $\sum_g c_g\, g$ to prove that $c_g = 0$ for all g in this support; hence, $\sum_g c_g\, g = 0$.

Let $\rho : \mathbb{C}\mathrm{GL}(N,\mathbb{C}) \to A$ be a morphism of algebras. Then, $(\mathrm{cf}(A,\rho))^{\perp} = \mathrm{Ker}\,\rho$. Indeed, if $\phi \circ \rho(\sum c_g\, g) = 0$ for any linear form ϕ, then $\rho(\sum c_g\, g) = 0$, hence, $\sum c_g\, g \in \mathrm{Ker}\,\rho$. Now, consider a polynomial representation V of $\mathrm{GL}(N,\mathbb{C})$ of degree n. We saw that this means:

$$\mathrm{cf}(V) \subset \mathscr{P}_n(\mathrm{GL}(N,\mathbb{C})) = \mathrm{cf}(\mathscr{S}_n(N,\mathbb{C})).$$

By looking at orthogonals, we have therefore $\mathrm{Ker}\,\rho^V \supset \mathrm{Ker}\,\psi$. Hence, by the factorization theorem for morphisms, there is a unique way to factor ρ^V through ψ:

$$\begin{array}{ccc} \mathbb{C}\mathrm{GL}(N,\mathbb{C}) & \xrightarrow{\psi} & \mathscr{S}_n(N,\mathbb{C}) \\ \downarrow \rho^V & \swarrow \pi^V & \\ \mathrm{End}(V) & & \end{array}$$

So, every polynomial representation V of $\mathrm{GL}(N,\mathbb{C})$ of degree n is endowed with a structure of $\mathscr{S}_n(N,\mathbb{C})$-module.

Conversely, consider a Schur module V, associated to a morphism of algebras $\pi : \mathscr{S}_n(N,\mathbb{C}) \to \mathrm{End}(V)$. The composition $\pi \circ \psi$ yields a representation of

$\mathrm{GL}(N,\mathbb{C})$ on V. Moreover, the coefficient space of this representation is included into $\mathrm{cf}(\mathscr{S}_n(N,\mathbb{C})) = \mathscr{P}_n(\mathrm{GL}(N,\mathbb{C}))$, since if ϕ is a linear form on $\mathrm{End}(V)$, then $\phi\circ(\pi\circ\psi) = (\phi\circ\pi)\circ\psi$ is a linear form on $\mathscr{S}_n(N,\mathbb{C})$ composed with ψ. Hence, this is a polynomial representation of degree n, and the equivalence is constructed. We leave it to the reader to check that our construction is functorial. □

An important consequence of Theorem 2.36 is:

Proposition 2.37. *For any N and n, the Schur algebra $\mathscr{S}_n(N,\mathbb{C})$ is semisimple.*

Proof. By the previous result, this is equivalent to show that any polynomial representation of $\mathrm{GL}(N,\mathbb{C})$ is completely reducible. We shall use the famous **unitarian trick** in order to prove the result. Denote $G = \mathrm{GL}(N,\mathbb{C})$, and $K = \mathrm{U}(N,\mathbb{C})$, which is a compact real Lie group. We assume that the reader is familiar with the basics of Lie group theory, and we denote $\mathfrak{k} \subset \mathfrak{g}$ the real Lie algebras of K and G. They are the spaces of matrices

$$\mathfrak{k} = \{m \in \mathrm{M}(N,\mathbb{C}) \mid m^* = -m\} \qquad ; \qquad \mathfrak{g} = \mathrm{M}(N,\mathbb{C}),$$

endowed with the Lie bracket $[m_1, m_2] = m_1 m_2 - m_2 m_1$. Notice that $\mathfrak{g} = \mathfrak{k} \oplus \mathrm{i}\mathfrak{k}$; one says that $\mathfrak{k}$ is a compact **real form** of $\mathfrak{g}$. Let (V,ρ) be a polynomial representation of G. Assume that it has a non-trivial G-invariant space W. We construct as in Lemma 1.4 a K-invariant scalar product on V: if $(\cdot|\cdot)$ is an arbitrary scalar product, then the integral against the Haar measure of K

$$\langle v \mid w \rangle = \int_K (\rho(k)(v) \,|\, \rho(k)(w))\, dk$$

provides a K-invariant scalar product. It follows that W admits a K-stable complement in V, namely, $W^\perp$ (for the same reasons as in the proof of Theorem 1.3). We claim that $W^\perp$ is also G-stable, and in fact, that a subspace $W \subset V$ is G-stable if and only if it is K-stable. Indeed, suppose W stable under K. Since ρ is polynomial, the map $d_e\rho : \mathfrak{g} \to \mathrm{End}(V)$ is a *complex* linear map, because

$$d_e\rho(m) = \lim_{t\to 0} \frac{\rho(\exp(tm))}{t},$$

and the function that is differentiated is analytic, hence has a $\mathbb{C}$-linear derivative. The vector subspace W is stable for the action of the Lie algebra $\mathfrak{k}$:

$$\begin{aligned} \mathfrak{k} \times W &\to W \\ m, w &\mapsto (d_e\rho(m))(w) = \lim_{t\to 0} \frac{\rho(\exp(tm))(w)}{t}. \end{aligned}$$

Since $d_e\rho$ is $\mathbb{C}$-linear and $\mathfrak{g} = \mathfrak{k} \oplus \mathrm{i}\mathfrak{k}$, W is also stable under $\mathfrak{g}$. Then, W is stable for the action of any element $\exp(tm)$ with $m \in \mathfrak{g}$ and $t \in \mathbb{R}$, and these elements generate G, so it is stable by G. Thus, if V has a non-trivial G-stable subspace W, then there always exists a complement of W that is also G-stable. By induction on the dimension of V, this ensures the complete reducibility of any polynomial representation of G. Actually, our proof works for any analytic representation of a connected complex Lie group that admits a compact real form. □

▷ *Computation of bitraces.*

By combining the Schur–Weyl duality given by Theorem 2.33, and the equivalence between Schur modules and polynomial representations of $\mathrm{GL}(N,\mathbb{C})$ provided by Theorem 2.36, one obtains a classification of all the polynomial representations of $\mathrm{GL}(N,\mathbb{C})$, and a description of their characters. The main tool is the following computation:

Proposition 2.38. *Let $\sigma \in \mathfrak{S}(n)$ and $g \in \mathrm{GL}(N,\mathbb{C})$ acting on $(\mathbb{C}^N)^{\otimes n}$. The bitrace of (g,σ) only depends on the cycle type μ of σ and on the roots $x_1,\ldots,x_N$ of the characteristic polynomial of g. It is then given by the formula*

$$\mathrm{btr}(g,\sigma) = p_\mu(x_1,\ldots,x_N).$$

Proof. The bitrace $\mathrm{btr}(g,\sigma)$ is invariant by conjugation of its two arguments, so in particular it only depends on the cycle type $\mu = (\mu_1,\ldots,\mu_\ell)$ of σ, and we can assume without loss of generality that

$$\sigma = \sigma_\mu = (1,2,\ldots,\mu_1)(\mu_1+1,\ldots,\mu_1+\mu_2)\cdots(n-\mu_\ell+1,\ldots,n).$$

As for g, if $x_1,\ldots,x_N$ are the roots of $\det(g-X)$, then it is well known that there exists for any $\varepsilon > 0$ a matrix $h_\varepsilon \in \mathrm{GL}(N,\mathbb{C})$ such that

$$h_\varepsilon\, g\, h_\varepsilon^{-1} = \begin{pmatrix} x_1 & * & \cdots & * \\ & x_2 & \ddots & \vdots \\ & & \ddots & * \\ & & & x_N \end{pmatrix}$$

with all the upper-diagonal terms smaller than ε in modulus. Since the bitrace is invariant by conjugation and continuous, we conclude that $\mathrm{btr}(g,\sigma) = \mathrm{btr}(\mathrm{diag}(x_1,\ldots,x_N),\sigma)$, so the bitrace only depends on the roots $x_1,\ldots,x_N$, and we can assume without loss of generality that

$$g = d(x) = \mathrm{diag}(x_1,\ldots,x_N).$$

Let us now compute the bitrace of $(d(x),\sigma_\mu)$. The action of this pair on an element $e_I = e_{i_1} \otimes e_{i_2} \otimes \cdots \otimes e_{i_n}$ is $d(x)\cdot e_I \cdot \sigma = x_I\, e_{I\cdot\sigma}$. Therefore,

$$\begin{aligned}\mathrm{btr}(d(x),\sigma) &= \sum_I \delta_{I,I\cdot\sigma}\, x_I = \sum_{j_1,j_2,\ldots,j_\ell} (x_{j_1})^{\mu_1}(x_{j_2})^{\mu_2}\cdots(x_{j_\ell})^{\mu_\ell} \\ &= p_{\mu_1}(x_1,\ldots,x_N)\, p_{\mu_2}(x_1,\ldots,x_N)\cdots p_{\mu_\ell}(x_1,\ldots,x_N) \\ &= p_\mu(x_1,\ldots,x_N).\end{aligned}$$

Indeed, $I = I \cdot \sigma$ if and only if its μ_1 first terms are indices all equal; its μ_2 terms after are indices also all equal; etc. □

Theorem 2.39. *The irreducible polynomial representations of* $\mathrm{GL}(N,\mathbb{C})$ *of degree* n *are labeled by integer partitions of size* n *and length less than* N*. Moreover, one can choose the labeling such that* V^λ *has character given by*

$$\mathrm{ch}^{V^\lambda}(g) = s_\lambda(x_1,\ldots,x_N),$$

where $x_1,\ldots,x_N$ *are the roots of the characteristic polynomial of* g.

Proof. Let A be the image of $\mathbb{C}\mathfrak{S}(n)$ in $\mathrm{End}((\mathbb{C}^N)^{\otimes n})$. We know that A and $\mathscr{S}_n(N,\mathbb{C})$ are in duality for their actions on $(\mathbb{C}^N)^{\otimes n}$, and that $\mathscr{S}_n(N,\mathbb{C})$ is semisimple in all cases, even when $N < n$. Therefore, A is semisimple, and there exists a decomposition

$$(\mathbb{C}^N)^{\otimes n} = \sum_{\lambda\in\Lambda} V^\lambda \otimes S^\lambda,$$

where $\{S^\lambda,\ \lambda\in\Lambda\}$ is a complete collection of non-isomorphic simple right modules over A, and $\{V^\lambda,\ \lambda\in\Lambda\}$ is a complete collection of non-isomorphic simple left modules over the Schur algebra. By Theorem 2.36, $\{V^\lambda,\ \lambda\in\Lambda\}$ is also a complete collection of representatives of the classes of isomorphism of polynomial representations of $\mathrm{GL}(N,\mathbb{C})$ of degree n. Let us identify the modules S^λ. Since they are non-isomorphic simple modules over A, they are also non-isomorphic irreducible representations of $\mathfrak{S}(n)$, so Λ can be seen as a subset of $\widehat{\mathfrak{S}}(n) = \mathfrak{Y}(n)$. Then, the expansion of the bitrace reads as:

$$p_{t(\sigma)}(x_1,\ldots,x_N) = \mathrm{btr}(g,\sigma) = \sum_{\lambda\in\Lambda} \mathrm{ch}^{V^\lambda}(g)\,\mathrm{ch}^\lambda(\sigma).$$

In this expansion, the irreducible characters $\mathrm{ch}^{V^\lambda}(g) = \mathrm{ch}^{V^\lambda}(d(x))$ are polynomials of degree n in the variables $x_1,\ldots,x_N$, and since $d(x)$ and $d(x\cdot\sigma)$ are conjugated for any $\sigma\in\mathfrak{S}(N)$, they are symmetric polynomials in $\mathrm{Sym}^{(N)}$. On the other hand, we know that the symmetric function $p_{t(\sigma)}(X)$ expands over Schur functions, therefore, by Theorem 2.32,

$$p_{t(\sigma)}(X) = \sum_{\lambda\in\mathfrak{Y}(n)} s_\lambda(X)\,\mathrm{ch}^\lambda(\sigma);$$

$$p_{t(\sigma)}(x_1,\ldots,x_N) = \sum_{\substack{\lambda\in\mathfrak{Y}(n)\\ \ell(\lambda)\leq N}} s_\lambda(x_1,\ldots,x_N)\,\mathrm{ch}^\lambda(\sigma).$$

Indeed, the map $\mathrm{Sym}\to\mathrm{Sym}^{(N)}$ vanishes on the Schur functions that are labeled by integer partitions of length strictly larger than N. Finally, in order to identify the two expansions obtained for $p_{t(\sigma)}(x_1,\ldots,x_N)$, we take a scalar product in the symmetric group algebra $\mathbb{C}\mathfrak{S}(n)$ tensorized by $\mathrm{Sym}_n^{(N)}$. Thus, for any partition

$\mu \in \mathfrak{Y}(n)$, one has:

$$\begin{aligned}\frac{1}{n!}\sum_{\sigma\in\mathfrak{S}(n)} p_{t(\sigma)}(x_1,\ldots,x_N)\,\mathrm{ch}^{\mu}(\sigma^{-1}) &= \sum_{\substack{\lambda\in\mathfrak{Y}(n)\\ \ell(\lambda)\leq N}} s_\lambda(x_1,\ldots,x_N)\,\left\langle \mathrm{ch}^{\lambda}\,\middle|\,\mathrm{ch}^{\mu}\right\rangle_{\mathfrak{S}(n)}\\ &= \begin{cases} s_\mu(x_1,\ldots,x_N) & \text{if } \ell(\mu)\leq N;\\ 0 & \text{if } \ell(\mu)>N\end{cases}\\ &= \sum_{\lambda\in\Lambda} \mathrm{ch}^{V^\lambda}(g)\,\left\langle \mathrm{ch}^{\lambda}\,\middle|\,\mathrm{ch}^{\mu}\right\rangle_{\mathfrak{S}(n)}\\ &= \begin{cases} \mathrm{ch}^{V^\mu}(g) & \text{if } \mu\in\Lambda;\\ 0 & \text{if } \mu\notin\Lambda.\end{cases}\end{aligned}$$

This allows us to identify the subset $\Lambda \subset \mathfrak{Y}(n)$ and to compute the characters of the polynomial representations of $\mathrm{GL}(N,\mathbb{C})$. □

▷ *Polynomial and rational representations.*

To conclude this chapter, let us remark that the previous results yield a classification of all the polynomial representations of $\mathrm{GL}(N,\mathbb{C})$ (not only the *homogeneous* ones):

Theorem 2.40. *Any polynomial representation V of $\mathrm{GL}(N,\mathbb{C})$ splits into a direct sum of homogeneous polynomial representations:*

$$V = \bigoplus_{n=0}^{\infty} V_n \quad \textit{with } V_n \textit{ homogeneous polynomial representation of degree } n.$$

Thus, the map $V \mapsto \mathrm{ch}^V$ gives an isomorphism of graded rings between:

- $\mathrm{R}_{\mathrm{pol}}(\mathrm{GL}(N,\mathbb{C}))$*, the ring of polynomial representations of* $\mathrm{GL}(N,\mathbb{C})$ *(endowed with the inner tensor product of representations),*
- *and* $\mathrm{Sym}^{(N)}(\mathbb{Z})$*, the ring of symmetric polynomials in N variables spanned by $\mathbb{Z}$-linear combinations of Schur polynomials.*

Proof. The best proof consists in a reinterpretation of the notion of polynomial representation. Denote $\mathscr{P} = \mathscr{P}(\mathrm{GL}(N,\mathbb{C}))$ the ring of (complex) polynomials in the matrix coefficients $g_{11},\ldots,g_{NN}$. This ring is a coalgebra for the coproduct

$$\Delta f(g,h) = f(gh).$$

The function on the right-hand side is indeed a finite linear combination of polynomials in the coefficients of g and h. Since Δ is a morphism of algebras, it suffices to show this for the generators $g \mapsto g_{ij}$ of $\mathscr{P}$, and indeed,

$$(gh)_{ij} = \sum_{k=1}^{N} g_{ik}\,h_{kj}.$$

The counity of $\mathscr{P}$ is the map $\eta : \mathscr{P} \to \mathbb{C}$ which sends a function f to $f(I_N)$. Now, consider a polynomial representation V of $\mathrm{GL}(N,\mathbb{C})$, that is to say a representation whose coefficient space $\mathrm{cf}(V)$ is included into $\mathscr{P} = \bigoplus_{n=0}^{\infty} \mathscr{P}_n$. This is equivalent to a structure of **comodule** on $\mathscr{P}$, that is to say, a map $\tau : V \to \mathscr{P} \otimes_{\mathbb{C}} V$ such that

$$(\mathrm{id}_{\mathscr{P}} \otimes \tau) \circ \tau = (\Delta \otimes \mathrm{id}_V) \circ \tau;$$
$$(\eta \otimes \mathrm{id}_V) \circ \tau = \mathrm{id}_V.$$

Indeed, given the representation ρ, consider the map $\tau : v \mapsto \rho(\cdot)\, v$. The term $\rho(\cdot)\, v$ is a polynomial function on G with values in V, hence, an element of $\mathscr{P} \otimes V$, so τ is indeed a linear map $V \to \mathscr{P} \otimes V$. Moreover,

$$\begin{aligned}(\mathrm{id}_{\mathscr{P}} \otimes \tau) \circ \tau(v) &= (\mathrm{id}_{\mathscr{P}} \otimes \tau)(\rho(\cdot)\, v) = \rho(\cdot_1) \circ \rho(\cdot_2)\, v \\ &= (\Delta \otimes \mathrm{id}_V)(\rho(\cdot)\, v) = (\Delta \otimes \mathrm{id}_V) \circ \tau(v),\end{aligned}$$

and

$$(\eta \otimes \mathrm{id}_V) \circ \tau(v) = (\eta \otimes \mathrm{id}_V)(\rho(\cdot)\, v) = \rho(I_N)(v) = v = \mathrm{id}_V(v),$$

so V is a well-defined $\mathscr{P}$-comodule. One obtains this way an equivalence of categories between polynomial representations of $\mathrm{GL}(N,\mathbb{C})$, and $\mathscr{P}$-comodules.

Now, fix a $\mathscr{P}$-comodule V of finite dimension, and for each $v \in V$, write

$$\tau(v) = \sum_{n=0}^{\infty} r_n(v) \otimes v_n,$$

where $r_n(v) \in \mathscr{P}_n$. One has

$$\sum_{n=0}^{\infty} r_n(v) \otimes \tau(v_n) = \sum_{n=0}^{\infty} \Delta(r_n(v)) \otimes v_n.$$

Notice that $\Delta(\mathscr{P}_n) \subset \mathscr{P}_n \otimes \mathscr{P}_n$: since $(gh)_{ij} = \sum_{k=1}^{N} g_{ik}\, h_{kj}$, $\Delta(\mathscr{P}_1) \subset \mathscr{P}_1 \otimes \mathscr{P}_1$, and it suffices then to use the property of morphism of algebras for Δ. Therefore, the right-hand side of the previous equation lies in $\left(\sum_{n=0}^{\infty} \mathscr{P}_n \otimes \mathscr{P}_n\right) \otimes V$, and by comparison, each $\tau(v_n)$ is in $\mathscr{P}_n \otimes V$. Notice now that

$$v = (\eta \otimes \mathrm{id}_V) \circ \tau(v) = \sum_{n=0}^{\infty} \eta(r_n(v))\, v_n.$$

Combining this with the fact that each $\tau(v_n)$ is in $\mathscr{P}_n \otimes V$, we have obtained an expansion of V as a direct sum of $\mathscr{P}_n$-comodules V_n, the projection from V to V_n being the map $v \mapsto \eta(r_n(v))\, v_n$. This is equivalent to a decomposition $V = \bigoplus_{n=0}^{\infty} V_n$ of V into homogeneous polynomial representations of $\mathrm{GL}(N,\mathbb{C})$ of degree n, $n \in \mathbb{N}$. The remainder of the theorem is then trivial. □

Actually, Theorem 2.40 leads to a classification of all **rational representations** of the group $\mathrm{GL}(N,\mathbb{C})$ in the sense of algebraic geometry. As an algebraic variety, $\mathrm{GL}(N,\mathbb{C})$ has for ring of regular functions

$$\mathcal{O} = \mathcal{O}(\mathrm{GL}(N,\mathbb{C})) = \mathbb{C}[g_{11},\ldots,g_{NN},(\det g)^{-1}].$$

Call rational a representation V of $\mathrm{GL}(N,\mathbb{C})$ whose coefficient space is included into $\mathcal{O}$. By tensoring by the determinant representation $\det : \mathrm{GL}(N,\mathbb{C}) \to \mathbb{C}^*$ a certain number of times, one gets a polynomial representation $W = V \otimes (\det)^{\otimes k}$. Therefore, any rational representation of $\mathrm{GL}(N,\mathbb{C})$ writes as $W \otimes (\det)^{\otimes k}$, where W is polynomial and $k \in \mathbb{Z}$. Notice that the character of det is the symmetric function $e_N(x_1,\ldots,x_N)$. On the other hand, we have seen that

$$\mathrm{R}_{\mathrm{pol}}(\mathrm{GL}(N,\mathbb{C})) = \mathrm{Sym}_{\mathbb{Z}}^{(N)} = \mathbb{Z}[e_1,\ldots,e_N].$$

So:

Corollary 2.41. *The ring of rational representations of the group* $\mathrm{GL}(N,\mathbb{C})$ *is isomorphic to* $\mathbb{Z}[e_1,e_2,\ldots,e_N,e_N^{-1}]$*, the isomorphism being* $V \mapsto \mathrm{ch}^V$.

Notes and references

The paragraph on conjugacy classes of permutations and their decompositions in transpositions is fairly standard, and the combinatorics exposed here lead to the Coxeter presentation of $\mathfrak{S}(n)$; see the beginning of our Chapter 4, and [Bou68, GP00]. For our presentation of symmetric functions, we followed of course [Mac95, Chapter 1], and we also tried to insist on the Hopf algebra structure, which is only treated as an exercise in loc. cit., and is the central notion in [Zel81]. The full structure of Hopf algebra eases a bit the proof of certain results, e.g., of the Frobenius–Schur isomorphism. The theory of alphabets, and formal sums and differences of them, can be found in many different places, and was explained to us by A. Lascoux; it is presented in a way very similar to ours in [Las99]. During some of the proofs, we used certain non-trivial properties of the rings of polynomials $k[X_1,\ldots,X_N]$, e.g., the fact that these rings are factorial; we refer to [Lan02, Chapter IV] for these results.

For the arguments of Galois theory used in order to prepare the proof of Proposition 2.26, see [Lan94, Chapter IV]. We could have bypassed these arguments by working with sesquilinear forms on Sym and $R_{\mathbb{C}}(\mathfrak{S}(n))$, but this is not very convenient in the setting of symmetric functions. For instance, in Chapter 5, we shall need to work with $\mathbb{C}(q) \otimes_{\mathbb{C}} \mathrm{Sym}$, and on this space, it does not really makes sense to extend a sesquilinear form, whereas a $\mathbb{C}(q)$-bilinear map will be easy to construct. We shall make the same choice of bilinearity instead of sesquilinearity later, when working with quasi-symmetric and non-commutative symmetric functions; see Chapter 6.

We deliberately separated the theory of symmetric functions and its combinatorial counterpart, the theory of partitions and tableaux (cf. Chapter 3). Thus, except for Lemma 2.10, there was no use of Young diagrams in this chapter. This might seem a little awkward, but it enabled in our opinion a clearer presentation, and in theory the reader can skip the next chapter and still understand a substantial part of this book. An important aspect of this approach is that we don't need at first complicated combinatorial notions such as tabloids and polytabloids; on the other hand, we only get an implicit description of the Specht module S^{λ}, which the reader might regret (this will be solved during the next chapter).

Similarly, our treatment of Schur–Weyl duality, for which we refer to [GW09, Chapters 4 and 9], yields an implicit description of all the irreducible polynomial representations of $\mathrm{GL}(N,\mathbb{C})$ (and their characters), but we do not get the explicit matrix form of these representations. This explicit construction is only possible through the highest weight theory (see again [GW09], and also [Var84, Kna02]), and its geometric counterpart, the Borel–Weil construction (see [CG97] and the references therein). More precisely, consider the flag variety G/B, where $G = \mathrm{GL}(N,\mathbb{C})$ and B is the subgroup of upper-triangular matrices; it parameterizes complete flags

$$V_{\bullet} = (\{0\} = V_0 \subset V_1 \subset V_2 \subset \cdots \subset V_N = \mathbb{C}^N)$$

with $\dim V_i = i$ for all i. We associate to any integer partition λ of length smaller than N the line bundle L^{λ} on G/B with fiber at $V_{\bullet}$ given by

$$(V_1/V_0)^{\otimes \lambda_N} \otimes (V_2/V_1)^{\otimes \lambda_{N-1}} \otimes \cdots \otimes (V_N/V_{N-1})^{\otimes \lambda_1}.$$

This line bundle is G-equivariant for the action of G on G/B, therefore, the space of regular sections $\mathrm{H}^0(G/B, L^{\lambda})$ is a representation of G. It turns out that it is irreducible, with character $s_{\lambda}(x_1,\ldots,x_N)$, hence equal to the representation V^{λ} constructed in Theorem 2.36. Moreover, the highest weight of V^{λ} is λ, that is to say that there is a one-dimensional subspace U of V^{λ} such that the torus $T = (\mathbb{C}^*)^n$ acts on U by

$$\mathrm{diag}(t_1,\ldots,t_N) \cdot v = (t_1)^{\lambda_1}(t_2)^{\lambda_2} \cdots (t_N)^{\lambda_N}\, v.$$

Finally, we refer to [Gre07] for a more detailed study of polynomial representations and Schur algebras, and to [Gre76] for our proof of Theorem 2.40, which actually holds for any comodule over a graded coalgebra.

3

Combinatorics of partitions and tableaux

In Chapter 2, we introduced the notion of integer partition, and we saw that integer partitions labeled the bases of the algebra Sym, and the irreducible representations of the symmetric groups. Our goal is now to use these combinatorial objects in order to compute quantities such as the dimensions of the modules S^λ, or their characters on specific conjugacy classes. In this framework, a central role is played by *chains of partitions* $\lambda^{(1)} \subset \lambda^{(2)} \subset \cdots \subset \lambda^{(r)}$. These chains are also called tableaux, and they can be interpreted as numberings of the cells of the Young diagram of the final partition $\lambda^{(r)}$, with some condition of growth of the numbers along the rows and the columns of the Young diagram.

In Section 3.1, we detail this notion of tableau, and we explain how to use it in order to compute the dimensions and the characters of the irreducible representations of $\mathfrak{S}(n)$. These combinatorial formulas are related in several ways to a construction on words known as the Robinson–Schensted–Knuth algorithm, and which will be pervasive in the fourth part of this book. We devote Section 3.2 to the presentation of this algorithm, and of a variant of it that was invented by Schützenberger, and which we use as a tool to prove properties of the RSK map.

In Section 3.3, we use the combinatorics of tableaux to build *concretely* the irreducible representation S^λ of $\mathfrak{S}(n)$ that is sent by the characteristic map to the Schur function s_λ. This is mainly to be complete, and so that the reader gets an idea of this other (more classical) approach to the representation theory of symmetric groups. Later, the description of S^λ with tabloids will only be used in Section 10.2. Finally, in Section 3.4, we discuss the famous hook-length formula, which is a formula for $\dim S^\lambda$ that is entirely explicit in terms of the parts of λ.

3.1 Pieri rules and Murnaghan–Nakayama formula

In this section, we give combinatorial rules in order to compute:

- the expansion of Schur functions in monomial symmetric functions;
- the product of Schur functions with power sums, homogeneous and elementary symmetric functions.

By the Frobenius–Schur isomorphism, these rules turn into results regarding representations of symmetric groups. In particular, they will lead to a combinatorial formula for the irreducible character values of the symmetric groups, see Proposition 3.10.

▷ *Schur functions and semistandard tableaux.*

Let λ be an integer partition. We recall that its Young diagram is the array of boxes (or **cells**) with λ_1 boxes on the first row, λ_2 boxes on the second row, etc.

Definition 3.1. *A* ***tableau*** *of shape λ is a numbering of the boxes of the Young diagram of λ by integers in $\mathbb{N}^*$.*

For instance,

$$\begin{array}{|c|c|c|c|} \hline 5 \\ \cline{1-1} 2 & 4 & 4 \\ \cline{1-3} 1 & 1 & 2 & 3 \\ \hline \end{array}$$

is a tableau of shape $(4,3,1)$. A tableau is called **semistandard** if the numbering is weakly increasing along the rows, and strictly increasing along the columns. For instance, the tableau above is semistandard. The semistandard tableaux occur naturally in the theory of Schur functions, and they yield their expansion in monomial symmetric functions. Actually, one can write easily in terms of semistandard tableaux the expansion of any *skew* Schur function in monomial symmetric functions. To this purpose, it is convenient to introduce the notions of skew Young diagrams and of skew tableaux. Recall that a skew partition is a pair of integer partitions $\mu \subset \lambda$. We will also denote it $\lambda \setminus \mu$, and the Young diagram associated to $\lambda \setminus \mu$ is the set of cells of the Young diagram of λ that are not in the Young diagram of μ. Thus,

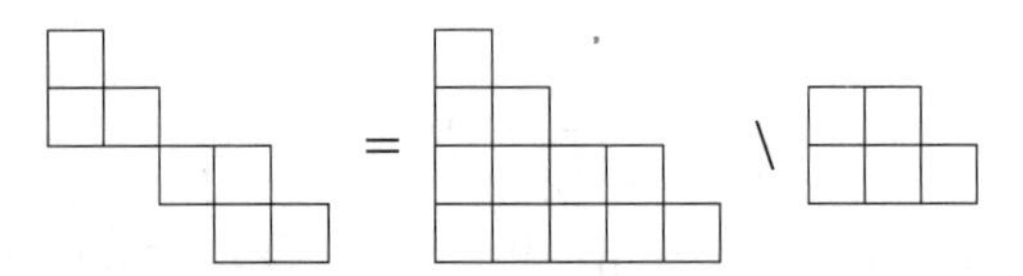

is the Young diagram of the skew partition $(5,4,2,1) \setminus (3,2)$. In this setting, a **skew tableau** of shape $\lambda \setminus \mu$ is a numbering of the cells of the Young diagram of $\lambda \setminus \mu$, and it is called semistandard if again it is weakly increasing along the rows and strictly increasing along the columns. For example,

$$\begin{array}{ccccc} 2 & & & & \\ 1 & 5 & & & \\ & & 3 & 3 & \\ & & & 1 & 4 \end{array}$$

is a semistandard skew tableau of shape $(5,4,2,1) \setminus (3,2)$.

To a tableau or a skew tableau T, we associate the monomial

$$x^T = \prod_{\square \in T} x_{T(\square)},$$

which is the product over the cells of the variables $x_{\text{numbering of the cell}}$. With the previous example of skew tableau, we obtain for instance $x^T = x_1^2 x_2 x_3^2 x_4 x_5$.

Theorem 3.2. *Let $\lambda \setminus \mu$ be a skew partition. The skew Schur polynomial $s_{\lambda\setminus\mu}$ writes as*

$$s_{\lambda\setminus\mu}(x_1, \ldots, x_N) = \sum_{T \in \mathrm{SST}(N, \lambda\setminus\mu)} x^T,$$

where $\mathrm{SST}(N, \lambda \setminus \mu)$ is the set of semistandard tableaux of shape $\lambda \setminus \mu$ and with entries in $[\![1, N]\!]$.

In order to prove Theorem 3.2 and some subsequent results, we shall need the notions of strip and of ribbon. A skew partition $\lambda \setminus \mu$ is called a **ribbon** if its Young diagram is connected and contains no square of cells of size 2×2. For instance,

$$(5,4,2,1) \setminus (3,1) =$$

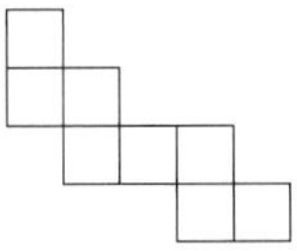

is a ribbon, whereas

$$(5,4,2,1) \setminus (3,2) = \quad ; \quad (5,4,2,1) \setminus (2,1) =$$

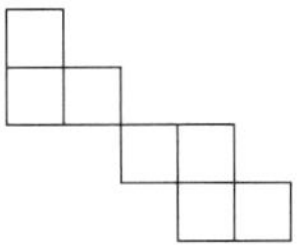

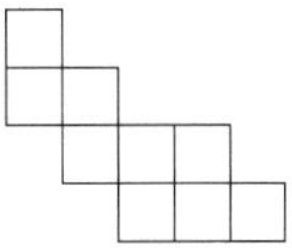

are not ribbons (the first skew partition is not connected, and the second skew partition contains a block of size 2×2). On the other hand, a skew partition $\lambda \setminus \mu$ is called a **horizontal strip** if its Young diagram contains at most one cell by column; and a **vertical strip** if its Young diagram contains at most one cell by row. In terms of parts of partitions, a skew partition $\lambda \setminus \mu$ is a vertical strip if and only if $\lambda_i - \mu_i \leq 1$ for all $i \geq 1$; and it is a horizontal strip if $\lambda_i' - \mu_i' \leq 1$ for all $i \geq 1$.

Lemma 3.3. *For any skew partition $\lambda \setminus \mu$, the specialization in one variable $s_{\lambda\setminus\mu}(x)$ is equal to*

$$\begin{cases} x^{|\lambda\setminus\mu|} = x^{|\lambda|-|\mu|} & \text{if } \lambda \setminus \mu \text{ is a horizontal strip,} \\ 0 & \text{otherwise.} \end{cases}$$

Proof. We use the second Jacobi–Trudi identity $s_{\lambda\setminus\mu} = \det(e_{\lambda_i'-\mu_j'-i+j})_{1\leq i,j\leq M}$. Notice that the specializations in one variable of elementary symmetric functions are:

$$e_0(x) = 1 \quad ; \quad e_1(x) = x \quad ; \quad e_{k\geq 2}(x) = 0.$$

Suppose that $\lambda'_i - \mu'_i > 1$ for some index i. Then, the matrix $(e_{\lambda'_i-\mu'_j-i+j})_{1\leq i,j\leq M}$ has a block of zeros in its top right corner, starting from the diagonal coefficient (i,i). Hence, $s_{\lambda\setminus\mu}(x) = 0$ if $\lambda\setminus\mu$ is not a horizontal strip. In this last situation, the matrix previously considered is block diagonal, each block being triangular with x's on the diagonal for each i such that $\lambda'_i - \mu'_i = 1$, and 1's everywhere else on the diagonal. Thus, $s_{\lambda\setminus\mu}(x) = x^{|\lambda|-|\mu|}$ if $\lambda\setminus\mu$ is a horizontal strip. □

Proof of Theorem 3.2. A semistandard tableau of shape $\lambda\setminus\mu$ and with entries in $[\![1,N]\!]$ is equivalent to a sequence of partitions $\mu = \mu^{(0)} \subset \mu^{(1)} \subset \cdots \subset \mu^{(N)} = \lambda$, such that each $\mu^{(i)} \setminus \mu^{(i-1)}$ is a horizontal strip and corresponds to the cells numbered by i. On the other hand, by the duplication rule 2.23 for skew Schur functions,

$$s_{\lambda\setminus\mu}(x_1,\ldots,x_N) = \sum_{\mu=\mu^{(0)}\subset\mu^{(1)}\subset\cdots\subset\mu^{(N)}=\lambda} s_{\mu^{(1)}\setminus\mu^{(0)}}(x_1)\cdots s_{\mu^{(N)}\setminus\mu^{(N-1)}}(x_N).$$

By the previous lemma, one can restrict the sum to sequences of horizontal strips, and then,

$$\begin{aligned} &s_{\mu^{(1)}\setminus\mu^{(0)}}(x_1)\, s_{\mu^{(2)}\setminus\mu^{(1)}}(x_2)\cdots s_{\mu^{(N)}\setminus\mu^{(N-1)}}(x_N) \\ &= (x_1)^{|\mu^{(1)}|-|\mu^{(0)}|}\cdots(x_N)^{|\mu^{(N)}|-|\mu^{(N-1)}|} = x^T, \end{aligned}$$

where T is the skew semistandard tableau corresponding to the sequence of partitions. □

Example. Take $N = 3$ and consider the integer partition $(3,2)$. The semistandard tableaux with this shape are listed in Figure 3.1:

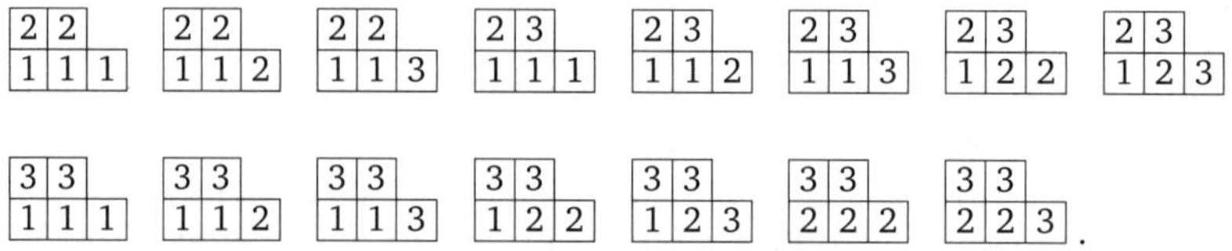

$\begin{smallmatrix}2&2\\1&1&1\end{smallmatrix}$ $\begin{smallmatrix}2&2\\1&1&2\end{smallmatrix}$ $\begin{smallmatrix}2&2\\1&1&3\end{smallmatrix}$ $\begin{smallmatrix}2&3\\1&1&1\end{smallmatrix}$ $\begin{smallmatrix}2&3\\1&1&2\end{smallmatrix}$ $\begin{smallmatrix}2&3\\1&1&3\end{smallmatrix}$ $\begin{smallmatrix}2&3\\1&2&2\end{smallmatrix}$ $\begin{smallmatrix}2&3\\1&2&3\end{smallmatrix}$

$\begin{smallmatrix}3&3\\1&1&1\end{smallmatrix}$ $\begin{smallmatrix}3&3\\1&1&2\end{smallmatrix}$ $\begin{smallmatrix}3&3\\1&1&3\end{smallmatrix}$ $\begin{smallmatrix}3&3\\1&2&2\end{smallmatrix}$ $\begin{smallmatrix}3&3\\1&2&3\end{smallmatrix}$ $\begin{smallmatrix}3&3\\2&2&2\end{smallmatrix}$ $\begin{smallmatrix}3&3\\2&2&3\end{smallmatrix}$.

Figure 3.1
Semistandard tableaux of shape $(3,2)$.

Therefore,

$$s_{3,2}(x_1,x_2,x_3) = m_{3,2}(x_1,x_2,x_3) + m_{3,1,1}(x_1,x_2,x_3) + 2\,m_{2,2,1}(x_1,x_2,x_3).$$

The previous example is generalized by the following proposition. Given a (semistandard) tableau T of shape $\lambda\setminus\mu$, we call **weight** of T the sequence $(\nu_1(T),\nu_2(T),\ldots)$, where $\nu_i(T)$ is the number of entries i in T.

Proposition 3.4. *In the expansion of skew Schur functions in monomial symmetric functions*

$$s_{\lambda\setminus\mu}(X) = \sum_{\nu} K_{\lambda\setminus\mu,\,\nu}\, m_\nu(X),$$

the coefficient $K_{\lambda\setminus\mu,\,\nu}$ is the number of semistandard tableaux of shape $\lambda \setminus \mu$ and weight ν.

In particular, any (skew) Schur function is a positive linear combination with integer coefficients of monomial symmetric functions. The numbers $K_{\lambda\setminus\mu,\,\nu}$ are called the **Kostka numbers**; they will appear later in several places.

Proof. It suffices to show the identity with a finite alphabet of size $N = |\lambda \setminus \mu|$. Then, the coefficient of $m_\nu(X)$ is the one of the monomial $(x_1)^{\nu_1} \cdots (x_N)^{\nu_N}$, and by Theorem 3.2, this is the number of semistandard tableaux of shape $\lambda \setminus \mu$ with ν_1 entries equal to 1, ν_2 entries equal to 2, etc. □

On the other hand, Theorem 3.2 leads to a combinatorial formula for the dimensions of the polynomial representations of $\mathrm{GL}(N,\mathbb{C})$. Thus:

Proposition 3.5. *Let λ be an integer partition of length $\ell(\lambda) \leq N$, and V^λ the corresponding irreducible representation of $\mathrm{GL}(N,\mathbb{C})$, with character $s_\lambda(x_1,\ldots,x_N)$. The complex dimension of V^λ is the number of semistandard tableaux of shape λ and entries in $[\![1,N]\!]$.*

Proof. Since $\dim V^\lambda = \mathrm{ch}^{V^\lambda}(I_N) = s_\lambda(1,\ldots,1)$, this follows from Theorem 3.2. □

Remark. There is also a formula for $\dim V^\lambda$ that is explicit in terms of the parts of the partition λ, namely,

$$\dim V^\lambda = \prod_{1\leq i<j\leq N} \frac{\lambda_i - \lambda_j + j - i}{j-i}.$$

It is an immediate consequence of Weyl's character formula; see Theorem A.14 in the Appendix. We shall never need this formula, but we shall prove an analogue formula for $\dim S^\lambda$ in Section 3.4.

▷ *Pieri rules and branching rules.*

Fix an integer $k \geq 1$. In this paragraph, we compute $h_k(X)s_\lambda(X)$ and $e_k(X)s_\lambda(X)$ for an arbitrary partition λ; and we give an interpretation of these **Pieri rules** in terms of representations of symmetric groups.

Theorem 3.6 (Pieri). *For any integer partition λ,*

$$h_k(X)\,s_\lambda(X) = \sum_{\Lambda} s_\Lambda(X),$$

where the sum runs over integer partitions Λ such that $\Lambda \setminus \lambda$ is a horizontal strip of size k. Similarly,

$$e_k(X)\,s_\lambda(X) = \sum_{\Lambda} s_\Lambda(X),$$

where the sum runs over integer partitions Λ such that $\Lambda \setminus \lambda$ is a vertical strip of size k.

Proof. Recall that Schur functions form an orthonormal basis of Sym with respect to Hall's scalar product. Therefore, in the expansion in Schur functions of $h_k(X)\,s_\lambda(X)$, the coefficient of $s_\Lambda(X)$ is given by

$$\langle h_k\,s_\lambda \,|\, s_\Lambda\rangle_{\mathrm{Sym}} = \left\langle h_k \,\middle|\, s_{\Lambda\setminus\lambda}\right\rangle_{\mathrm{Sym}} = \text{coefficient of } m_k \text{ in } s_{\Lambda\setminus\lambda}$$

since $(h_\lambda)_{\lambda\in\mathfrak{Y}}$ and $(m_\lambda)_{\lambda\in\mathfrak{Y}}$ are dual bases of Sym. By Proposition 3.4, this coefficient is the number of semistandard tableaux of shape $\Lambda \setminus \lambda$ with k entries equal to 1; hence, 1 if $\Lambda \setminus \lambda$ is a horizontal strip of size k, and 0 otherwise. The first identity is thus shown, and the second identity follows by applying the antipode ω of Sym, since $\omega(s_\lambda) = (-1)^{|\lambda|}\,s_{\lambda'}$ and $\omega(h_k) = (-1)^k\,e_k$. □

Example. Starting from $s_{3,2}(X)$, one has:

$$h_2(X)\,s_{3,2}(X) = s_{3,2,2}(X) + s_{3,3,1}(X) + s_{4,2,1}(X) + s_{4,3}(X) + s_{5,2}(X),$$

because the five partitions Λ obtained by adding a horizontal strip of size 2 to $(3,2)$ are

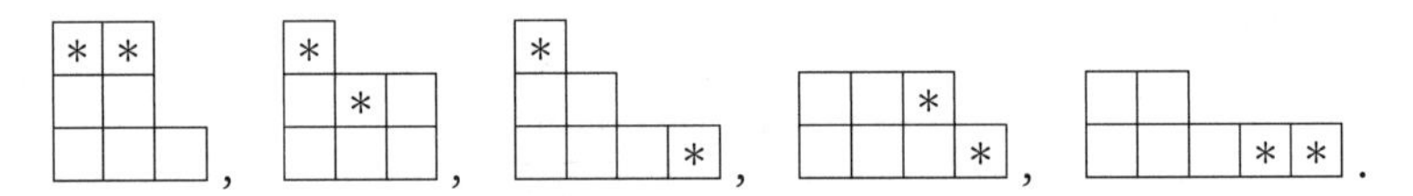

Similarly,

$$e_2(X)\,s_{3,2}(X) = s_{3,2,1,1}(X) + s_{3,3,1}(X) + s_{4,2,1}(X) + s_{4,3}(X),$$

this expansion corresponding to the four following vertical strips:

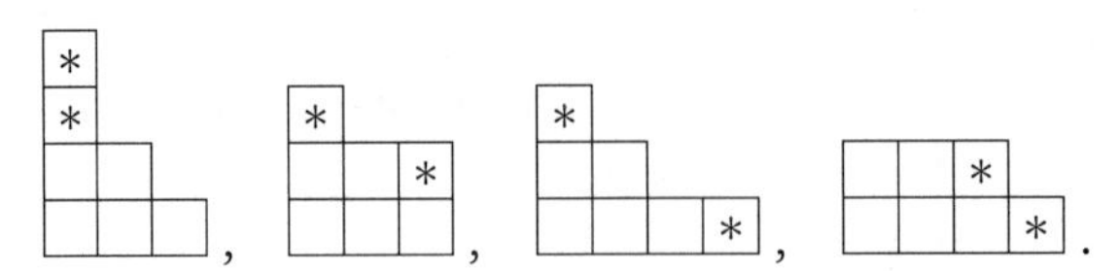

Notice that $h_k(X) = s_k(X)$ and $e_k(X) = s_{1^k}(X)$ by the Jacobi–Trudi formulas. Therefore, the homogeneous and elementary symmetric functions correspond to representations of symmetric groups via the Frobenius–Schur isomorphism. More precisely, we saw in the proof of Theorem 2.31 that the partition (k) corresponds to the character $\mathrm{ch}^{(k)}(\mu) = 1$, which is the character of the trivial representation of $\mathfrak{S}(k)$. Hence, $\Psi^{-1}(h_k)$ is the trivial representation of $\mathfrak{S}(k)$, and by applying the antipodes of Sym and of $R_{\mathbb{C}}(\mathfrak{S})$, $\Psi^{-1}(e_k)$ is the signature representation of $\mathfrak{S}(k)$. From this one deduces:

Corollary 3.7. *Consider the irreducible representation S^λ of $\mathfrak{S}(n)$, where λ is an arbitrary integer partition of size n. One has the decomposition in irreducibles*

$$\mathrm{Ind}_{\mathfrak{S}(n)\times\mathfrak{S}(k)}^{\mathfrak{S}(n+k)}(S^\lambda \boxtimes 1_k) = \bigoplus_{\Lambda} S^\Lambda,$$

where the sum runs over partitions Λ of size $n+k$, such that $\Lambda \setminus \lambda$ is a horizontal strip of size k. Similarly,

$$\mathrm{Ind}_{\mathfrak{S}(n)\times\mathfrak{S}(k)}^{\mathfrak{S}(n+k)}(S^\lambda \boxtimes \varepsilon_k) = \bigoplus_{\Lambda} S^\Lambda,$$

where the sum runs over partitions Λ of size $n+k$, such that $\Lambda \setminus \lambda$ is a vertical strip of size k.

Proof. It suffices to combine the Frobenius–Schur isomorphism with Theorem 3.6. □

As a particular case of this corollary, we get the **branching rules** for representations of symmetric groups: for any $\lambda \in \mathfrak{Y}(n)$,

$$\mathrm{Ind}_{\mathfrak{S}(n)}^{\mathfrak{S}(n+1)}(S^\lambda) = \bigoplus_{\lambda \nearrow \Lambda} S^\Lambda \quad ; \quad \mathrm{Res}_{\mathfrak{S}(n)}^{\mathfrak{S}(n+1)}(S^\Lambda) = \bigoplus_{\lambda \nearrow \Lambda} S^\lambda$$

where $\lambda \nearrow \Lambda$ means that Λ is obtained from λ by adding exactly one cell. Indeed, the first rule comes from the identification

$$\mathrm{Ind}_{\mathfrak{S}(n)}^{\mathfrak{S}(n+1)}(S^\lambda) = \mathrm{Ind}_{\mathfrak{S}(n)\times\mathfrak{S}(1)}^{\mathfrak{S}(n+1)}(S^\lambda \boxtimes 1_1),$$

and the second formula follows by Frobenius' reciprocity (Proposition 1.12). As a consequence, we get a combinatorial formula for the dimension of an irreducible representation S^λ of $\mathfrak{S}(n)$. Call **standard** a tableau whose rows and columns are strictly increasing, and such that if n is the number of boxes of the tableau, then each integer between 1 and n appears exactly once in the tableau. For instance,

$$\begin{array}{|c|c|c|c|}\cline{1-1} 9 \\ \cline{1-1} 7 \\ \cline{1-3} 2 & 5 & 6 \\ \hline 1 & 3 & 4 & 8 \\ \hline\end{array}$$

is a standard tableau of shape $(4, 3, 1, 1)$.

Proposition 3.8. *The complex dimension of S^λ is the number of standard tableaux of shape λ.*

Proof. To give a standard tableau of shape λ amounts to giving a sequence of integer partitions $\emptyset = \mu^{(0)} \nearrow \mu^{(1)} \nearrow \mu^{(2)} \nearrow \cdots \nearrow \mu^{(n)} = \lambda$, where $\mu^{(i)} \setminus \mu^{(i-1)}$ corresponds to the cell labeled i in the tableau. By applying inductively the branching rule,

$$\begin{aligned}\dim S^\lambda = \dim \mathrm{Res}^{\mathfrak{S}(n)}_{\mathfrak{S}(n-1)}(S^\lambda) &= \sum_{\mu^{(n-1)} \nearrow \lambda} \dim S^{\mu^{(n-1)}} \\ &= \sum_{\emptyset = \mu^{(0)} \nearrow \mu^{(1)} \nearrow \mu^{(2)} \nearrow \cdots \nearrow \lambda} \dim S^{\emptyset} = \mathrm{card}\, \mathrm{ST}(\lambda)\end{aligned}$$

since the only irreducible representation of $\mathfrak{S}(0)$ is the trivial one, with dimension 1. □

Example. The dimension of $S^{(3,2)}$ is 5, since there are five standard tableaux with this shape (Figure 3.2):

$$\begin{array}{|c|c|}\hline 4 & 5 \\ \hline \end{array}\!\!\begin{array}{l}\\\end{array} \qquad \begin{array}{|c|c|}\hline 3 & 5 \\ \hline \end{array} \qquad \begin{array}{|c|c|}\hline 3 & 4 \\ \hline \end{array} \qquad \begin{array}{|c|c|}\hline 2 & 4 \\ \hline \end{array} \qquad \begin{array}{|c|c|}\hline 2 & 5 \\ \hline \end{array}$$
$$\begin{array}{|c|c|c|}\hline 1 & 2 & 3 \\ \hline \end{array} \quad \begin{array}{|c|c|c|}\hline 1 & 2 & 4 \\ \hline \end{array} \quad \begin{array}{|c|c|c|}\hline 1 & 2 & 5 \\ \hline \end{array} \quad \begin{array}{|c|c|c|}\hline 1 & 3 & 5 \\ \hline \end{array} \quad \begin{array}{|c|c|c|}\hline 1 & 3 & 4 \\ \hline \end{array}\,.$$

Figure 3.2
Standard tableaux of shape $(3,2)$.

▷ *Murnaghan–Nakayama formula.*

There exists a rule similar to Theorem 3.6 for the computation of $p_k(X)s_\lambda(X)$, which involves ribbons instead of strips. Call **height** of a ribbon the number of rows that it occupies *minus one*. For instance, the ribbon

$$(5,4,2,1) \setminus (3,1) \;=\; \begin{array}{ccccc} \square & & & & \\ \square & \square & & & \\ & \square & \square & \square & \\ & & & \square & \square \end{array}$$

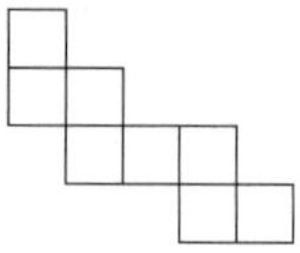

has height $\mathrm{ht}((5,4,2,1) \setminus (3,1)) = 3$.

Proposition 3.9. *The product $p_k(X)s_\lambda(X)$ is equal to*

$$\sum_{\Lambda} (-1)^{\mathrm{ht}(\Lambda\setminus\lambda)}\, s_\Lambda(X),$$

where the sum runs over partitions Λ such that $\Lambda \setminus \lambda$ is a ribbon of size k.

Proof. We prove the identity with a finite alphabet $(x_1,\ldots,x_N)$ of size sufficiently large (bigger than $k+|\lambda|$). Consider the antisymmetric polynomial

$$p_k(x_1,\ldots,x_N)\,a_{\lambda+\rho}(x_1,\ldots,x_N)=\sum_{j=1}^{N} a_{\lambda+\rho+ke_j}(x_1,\ldots,x_N),$$

where $e_j=(0,\ldots,0,1_j,0,\ldots,0)$. If $\lambda+\rho+ke_j$ contains two identical coordinates, then the antisymmetric function $a_{\lambda+\rho+ke_j}(x_1,\ldots,x_N)$ vanishes, so one can restrict the sum to indices j such that there exists $i\leq j$ with

$$\lambda_{i-1}+N-i+1>\lambda_j+N-j+k>\lambda_i+N-i.$$

However, in this case, the decreasing reordering of $\lambda+\rho+ke_j$ is $\Lambda+\rho$, where Λ is the integer partition

$$\lambda_1,\ldots,\lambda_{i-1},\lambda_j+i-j+k,\lambda_i+1,\ldots,\lambda_{j-1}+1,\lambda_{j+1},\ldots,\lambda_N.$$

This new partition is obtained from λ by "gluing" a ribbon of size k to the right of the Young diagram λ, starting from the row j and going down. For example, if $\lambda=(6,3,1,1)$, $j=3$ and $k=5$, then one obtains for Λ:

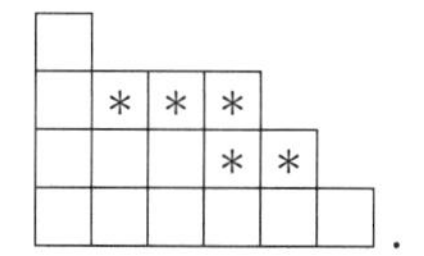

Moreover, the reordering is provided by a cycle of length $j-i+1$, hence of signature $(-1)^{j-i}=(-1)^{\mathrm{ht}(\Lambda\setminus\lambda)}$. We conclude that

$$p_k(x_1,\ldots,x_N)\,a_{\lambda+\rho}(x_1,\ldots,x_N)=\sum_{\Lambda\setminus\lambda\text{ ribbon of size }k}(-1)^{\mathrm{ht}(\Lambda\setminus\lambda)}\,a_{\Lambda+\rho}(x_1,\ldots,x_N),$$

whence the formula by dividing by $a_\rho(x_1,\ldots,x_N)$. □

Example. We get the product $p_2(X)s_{3,2}(X)=-s_{3,2,1,1}(X)+s_{3,2,2}(X)+s_{5,2}(X)$, since the three ribbons of size 2 that one can add to $(3,2)$ are

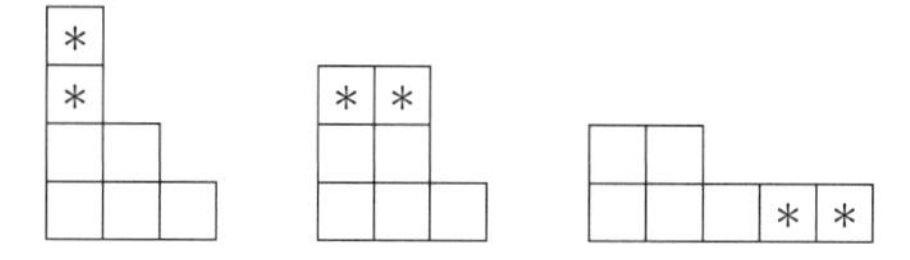

with respective heights 1, 0 and 0.

The previous proposition yields a combinatorial formula for the values of the irreducible characters of the symmetric groups, the so-called **Murnaghan–Nakayama formula**. Call **ribbon tableau** of shape λ an increasing sequence of partitions $\emptyset=\mu^{(0)}\subset\mu^{(1)}\subset\cdots\subset\mu^{(\ell)}=\lambda$ such that each $\mu^{(i)}\setminus\mu^{(i-1)}$ is a ribbon.

The weight of the ribbon tableau is $(|\mu^{(1)} \setminus \mu^{(0)}|, |\mu^{(2)} \setminus \mu^{(1)}|, \ldots, |\mu^{(\ell)} \setminus \mu^{(\ell-1)}|)$, and the height of the ribbon tableau is the sum

$$\mathrm{ht}(T) = \sum_{i=1}^{\ell} \mathrm{ht}(\mu^{(i)} \setminus \mu^{(i-1)}).$$

Theorem 3.10 (Murnaghan–Nakayama). *For any partitions λ, μ of same size n,*

$$\mathrm{ch}^{\lambda}(\mu) = \sum_{T} (-1)^{\mathrm{ht}(T)},$$

where the sum runs over ribbon tableaux T of shape λ and weight μ.

In particular, we recover the fact that the **character table** of $\mathfrak{S}(n)$, which is the matrix $(\mathrm{ch}^{\lambda}(\mu))_{\lambda,\mu \in \mathfrak{Y}(n)}$, has integer coefficients.

Proof. By Frobenius formula 2.32, $p_\mu = \sum_\lambda \mathrm{ch}^{\lambda}(\mu) s_\lambda$, that is to say that $\mathrm{ch}^{\lambda}(\mu)$ is the coefficient of s_λ in $p_\mu = p_{\mu_1} \cdots p_{\mu_\ell}$. By applying ℓ times Proposition 3.9, this is indeed the number of ribbon tableaux of shape λ and weight μ, each counted with coefficient $(-1)^{\mathrm{ht}(T)}$. □

Example. The ribbon tableaux of shape $(3, 1)$ are listed in Figure 3.3:

1	1	2	1	1
1 1 1	1 1 2	1 1 1	1 2 2	1 2 3

3	2	4	3	2
1 1 2	1 1 3	1 2 3	1 2 4	1 3 4 .

Figure 3.3
Ribbon tableaux of shape $(3, 1)$.

The heights of these tableaux are respectively 1, 1, 0, 1, 1, 0, 0, 0, 0 and 0. Consequently, the irreducible character $\mathrm{ch}^{(3,1)}$ of $\mathfrak{S}(4)$ has values

$$\mathrm{ch}^{(3,1)}(4) = -1 \quad ; \quad \mathrm{ch}^{(3,1)}(3,1) = 0 \quad ; \quad \mathrm{ch}^{(3,1)}(2,2) = -1$$
$$\mathrm{ch}^{(3,1)}(2,1,1) = 1 \quad ; \quad \mathrm{ch}^{(3,1)}(1^4) = 3.$$

3.2 The Robinson–Schensted–Knuth algorithm

The tableau combinatorics of the representations of symmetric groups lead to some surprising identities. Consider for instance the $(\mathrm{GL}(N,\mathbb{C}), \mathfrak{S}(n))$-bimodule

$(\mathbb{C}^N)^{\otimes n}$. By Theorem 2.33 and its proof, it admits for decomposition

$$(\mathbb{C}^N)^{\otimes n} = \bigoplus_{|\lambda|=n,\ \ell(\lambda)\leq N} V^\lambda \otimes S^\lambda,$$

and Propositions 3.5 and 3.8 give the dimensions of these irreducible components. Hence,

$$N^n = \sum_{\lambda\in\mathfrak{Y}(n)} \operatorname{card}\mathrm{SST}(N,\lambda) \times \operatorname{card}\mathrm{ST}(\lambda).$$

Similarly, consider the Schur algebra $\mathscr{S}_n(N,\mathbb{C})$, which is of dimension $\binom{N^2+n-1}{n}$. It is semisimple, and its expansion as a sum of matrix algebras yields the identity

$$\binom{N^2+n-1}{n} = \sum_{\lambda\in\mathfrak{Y}(n)} (\operatorname{card}\mathrm{SST}(N,\lambda))^2.$$

Finally, the group algebra of the symmetric group $\mathbb{C}\mathfrak{S}(n)$ is also semisimple, and its expansion as a sum of matrix algebras yields the identity

$$n! = \sum_{\lambda\in\mathfrak{Y}(n)} (\operatorname{card}\mathrm{ST}(\lambda))^2.$$

The goal of this section is to give a combinatorial explanation of these identities. In the process, we shall obtain a new proof of the Cauchy identity; and a way to study the longest increasing subsequences in words, which will prove very useful in the last part of the book when examining these subsequences for *random* words (cf. Theorem 13.10).

▷ *Robinson–Schensted–Knuth algorithm.*

Let N_1, N_2 and n be positive integers.

Definition 3.11. *A **two-line array** of length n and entries in* $[\![1,N_1]\!] \times [\![1,N_2]\!]$ *is a pair of sequences* $\binom{a_1,\ldots,a_n}{b_1,\ldots,b_n}$ *such that:*

1. *The* a_i*'s are ordered and smaller than* N_1*:* $1 \leq a_1 \leq a_2 \leq \cdots \leq a_n \leq N_1$.
2. *Each* b_i *is in* $[\![1,N_2]\!]$*, and if* $a_i = a_{i+1}$*, then* $b_i \leq b_{i+1}$.

Example. The pair of sequences

$$\binom{11112333}{12443223}$$

is a two-line array of length 8 and entries in $[\![1,3]\!] \times [\![1,4]\!]$.

The two-line arrays are generalizations of words and permutations. More precisely, suppose that $(a_1,a_2,\ldots,a_n) = (1,2,\ldots,n)$. Then, a two-line array $\binom{1,\ldots,n}{b_1,\ldots,b_n}$

with lower entries in $[\![1,N]\!]$ is just a **word** of length n with entries in $[\![1,N]\!]$, that is to say a sequence $(b_1,\ldots,b_n)$ in $[\![1,N]\!]^n$. If one supposes moreover that $N=n$ and $\{b_1,\ldots,b_n\}=\{1,2,\ldots,n\}$, then the word $b_1b_2\ldots b_n$ is the notation in line for the permutation $i\mapsto b_i$ in $\mathfrak{S}(n)$.

Denote $\mathfrak{A}(n;N_1,N_2)$ the finite set of all two-line arrays of length n and with entries in $[\![1,N_1]\!]\times[\![1,N_2]\!]$.

Proposition 3.12. *The cardinality of* $\mathfrak{A}(n;N_1,N_2)$ *is* $\binom{N_1N_2+n-1}{n}$.

Proof. The number of non-decreasing sequences of length n and with entries in $[\![1,N]\!]$ is $\binom{N+n-1}{n}$. Indeed, such a sequence is entirely determined by a list

$$\cdot\cdot|\cdot||\cdot\cdot|\cdot$$

in $\{\cdot,|\}^{N+n-1}$ with $N-1$ separators $|$. From this list, one gets back a non-decreasing sequence by replacing the dots $\cdot$ between the i-th and the $(i+1)$-th separators by entries $i+1$ (in the previous example, one obtains 112445).

Consider then a non-decreasing sequence of length n and with entries in $[\![1,N_1]\!]\times[\![1,N_2]\!]$, this set being endowed with the lexicographic order. For example, with $n=8$, $N_1=3$ and $N_2=4$, one can consider

$$(1,1)(1,2)(1,4)(1,4)(2,3)(3,2)(3,2)(3,2)(3,3).$$

This sequence corresponds bijectively to a two-line array in $\mathfrak{A}(n;N_1,N_2)$ (in our example, one gets back $\binom{111123333}{124443223}$). Since $\mathrm{card}([\![1,N_1]\!]\times[\![1,N_2]\!])=N_1N_2$, the proof is completed. $\square$

The **Robinson–Schensted–Knuth algorithm** is a combinatorial bijection between the set of two-line arrays $\mathfrak{A}(n;N_1,N_2)$, and the set of pairs of semistandard tableaux (P,Q) with the same shape $\lambda\in\mathfrak{Y}(n)$, and with entries respectively in $[\![1,N_2]\!]$ and $[\![1,N_1]\!]$. It relies on the following insertion procedure. Fix a two-line array

$$\sigma=\begin{pmatrix}a_1,\ldots,a_n\\ b_1,\ldots,b_n\end{pmatrix}$$

in $\mathfrak{A}(n;N_1,N_2)$, and let us construct by recursion on n two semistandard tableaux $P(\sigma)$ and $Q(\sigma)$. We denote

$$\sigma_{|(n-1)}=\begin{pmatrix}a_1,\ldots,a_{n-1}\\ b_1,\ldots,b_{n-1}\end{pmatrix};$$

this is a two-line array in $\mathfrak{A}(n-1;N_1,N_2)$. Suppose that the two tableaux $P(\sigma_{|n-1})$ and $Q(\sigma_{|n-1})$ are already constructed. The **Schensted insertion** of b_n in $P(\sigma_{|n-1})$ is defined as follows:

1. If b_n is larger than all the entries of the first (bottom) row of $P(\sigma_{|n-1})$, we put b_n in a new cell at the end of this row to get $P(\sigma)$.

2. Otherwise, consider the first entry b strictly larger than b_n when one reads the first row from left to right. One replaces b by b_n, and one inserts the "bumped" entry b into the second row of the tableau, following the same procedure (thus, possibly bumping other entries of the tableau to higher rows).

At the end of the insertion of b_n, one obtains a new tableau $P(\sigma)$ whose shape Λ has one more cell than λ, that is to say that $\lambda \nearrow \Lambda$. We define $Q(\sigma)$ to be the tableau of shape λ such that the cells in λ are labeled as in $Q(\sigma_{|n-1})$, and the remaining cell is labeled by a_n. One says that $P(\sigma)$ is the **insertion tableau** and that $Q(\sigma)$ is the **recording tableau** of the RSK algorithm.

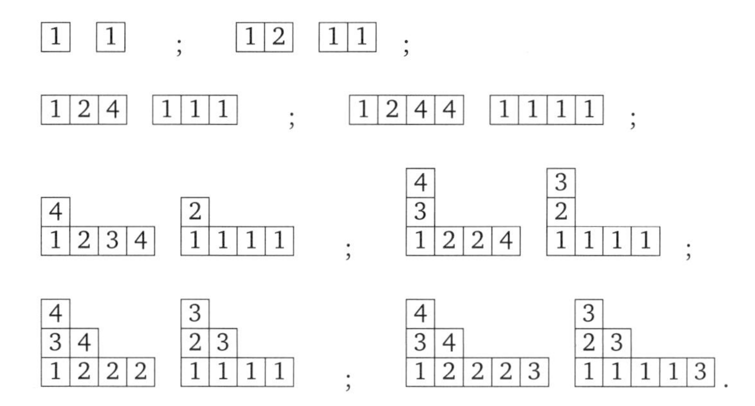

Figure 3.4
The RSK algorithm applied to the two-line array $\binom{11112333}{12443223}$.

Example. Consider the two-line array $\sigma = \binom{11112333}{12443223}$. The sequence of pairs of tableaux obtained by the procedure described previously is given by Figure 3.4. The last pair is the result $(P(\sigma), Q(\sigma))$ of the RSK algorithm.

It is clear from the definition of the algorithm that RSK produces for any two-line array a pair of tableaux with entries $\{b_1, \ldots, b_n\}$ and $\{a_1, \ldots, a_n\}$, and with the same shape in $\mathfrak{Y}(n)$. It is much less clear that $P(\sigma)$ and $Q(\sigma)$ are always *semistandard* tableaux. If a is a number and T is a semistandard tableau, denote $T \leftarrow a$ the new tableau with a inserted into T.

Lemma 3.13. *If T is semistandard, then $T \leftarrow a$ is semistandard.*

Proof. If a is inserted into the first row of T, then the result is trivial, because this means that a is larger than all the entries of the first row of T. One then creates on the right of this row a new cell with a, and there are no cells to compare with on top or below this cell:

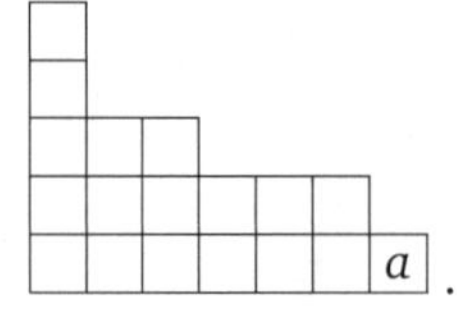

Suppose now that a bumps a value $b > a$ to the next row. This means that T has one of the following forms:

1. either

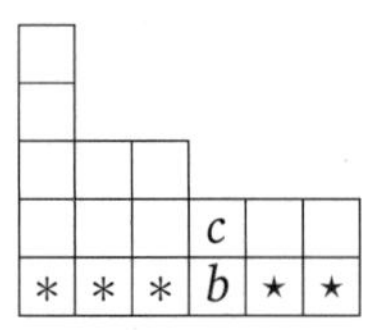

   with all the entries $*$ strictly smaller than a, all the entries $\star$ larger than b, and $c > b > a$;

2. or,

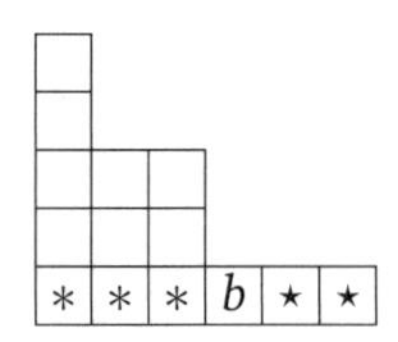

   with all the entries $*$ strictly smaller than a, all the entries $\star$ larger than b, and no cell on top of the cell labeled by b.

In both cases, by replacing b by a in the first row, one still obtains a non-decreasing first row

$$\boxed{*}\boxed{*}\boxed{*}\boxed{a}\boxed{\star}\boxed{\star}\,.$$

One then has to check that after the insertion of b in the second row, the cell $\square$ on top of a will be strictly larger. In the first case, this cell on top of a contains either c, or b if b bumps c, and we have seen that b and c are strictly larger than a. In the second case, if there is a new cell on top of a, then it contains b, which is strictly larger than a. $\square$

Proposition 3.14. *For any two-line array σ, the two tableaux $P(\sigma)$ and $Q(\sigma)$ are semistandard tableaux.*

Proof. For the first tableau, this is an immediate consequence of Lemma 3.13. Consider now the second tableau, and let us prove by induction on n that it is also semistandard. By construction, to obtain $Q(\sigma)$ from $Q(\sigma_{|n-1})$, one inserts on the right border of the semistandard tableau $Q(\sigma_{|n-1})$ a cell $\square$ with an entry a_n larger than all the entries of $Q(\sigma_{|n-1})$. In particular, a_n is automatically larger than the entry in the cell directly to the left of □, so the only thing to check is that a_n is *strictly* larger than the entry in the cell directly below □.

To this purpose, it is convenient to introduce the notion of **bumping route** of the Schensted insertion of a number a in a semistandard tableau T. This is the set of all cells of $T \leftarrow a$ changed during the insertion of a into T, including the last cell created. For instance, if one inserts $a = 2$ inside

$$T = \begin{array}{|c|c|c|c|} \hline 3 & 6 \\ \cline{1-3} 2 & 3 & 4 \\ \hline 1 & 1 & 3 & 5 \\ \hline \end{array}$$

then the bumping route of $T \leftarrow a$ is

$$\begin{array}{|c|c|c|c|} \hline 6 \\ \cline{1-2} 3 & 4 \\ \cline{1-3} 2 & 3 & 3 \\ \hline 1 & 1 & 2 & 5 \\ \hline \end{array}\;.$$

It follows from the proof of Lemma 3.13 that a bumping route contains exactly one cell per row that it intersects, and that it always goes north or west. Consider now two consecutive insertions $T \leftarrow a$ and $(T \leftarrow a) \leftarrow b$. We claim that if $a \leq b$, then the bumping route of the insertion $T \leftarrow a$ is strictly to the left of the bumping route of $(T \leftarrow a) \leftarrow b$. Indeed, by induction, one sees that on each row, the number inserted during the second procedure is bigger than the number inserted during the first procedure. Moreover, when $a \leq b$, the bumping route of a stops at a row above the end row of the bumping route of b, and therefore, b is in a column strictly left to the column of a. For example, if one inserts 3 in the previous tableau, then one gets

$$(T \leftarrow a) \leftarrow b = \begin{array}{|c|c|c|c|} \hline 6 \\ \cline{1-2} 3 & 4 \\ \hline 2 & 3 & 3 & 5 \\ \hline 1 & 1 & 2 & 3 \\ \hline \end{array}$$

and the second bumping route is indeed to the left of the first one, and stops before this first bumping route.

Let us now go back to our problem, and assume that a_n is not strictly larger than the entry in the cell directly below its cell □. Then, this other cell contains some $a_i = a_n$, and since σ is a two-line array,

$$a_i = a_{i+1} = a_{i+2} = \cdots = a_n \qquad ; \qquad b_i \leq b_{i+1} \leq b_{i+2} \leq \cdots \leq b_n.$$

By the previous discussion, the cells created by the insertions of $b_i, \ldots, b_n$ are in different columns, which is a contradiction since a_i is in the same column as a_n. □

Theorem 3.15 (Robinson–Schensted–Knuth). *The RSK map is a bijection between two-line arrays in $\mathfrak{A}(n;N_1,N_2)$ and pairs of semistandard tableaux (P,Q) with the same shape $\lambda \in \mathfrak{Y}(n)$, with $P \in \mathrm{SST}(N_2,\lambda)$ and $Q \in \mathrm{SST}(N_1,\lambda)$.*

Proof. The theorem amounts to the possibility to invert the RSK algorithm. Notice first that given a tableau $T \leftarrow b$, if one knows the cell of $T \leftarrow b$ which was added to T (that is to say, the end of the bumping route), then one can recover the tableau T, by reversing the Schensted insertion procedure. This is better explained on an example. Consider the tableau

$$\begin{array}{|c|c|c|c|} \hline 6 \\ \cline{1-1} 3 & 4 \\ \cline{1-2} 2 & 3 & 3 \\ \cline{1-3} 1 & 1 & 2 & 5 \\ \hline \end{array}$$

where one knows that the cell containing 6 is the new cell. The entry 6 was bumped to the fourth row by the insertion of the largest number of the third row strictly smaller than 6, namely, 4; this entry 4 was bumped by the insertion of the largest number of the second row strictly smaller than 4, namely, 3; and this entry 4 was bumped by 2 on the first row. Thus, one gets back the whole bumping route, and by deleting 2 and shifting the other entries of the bumping route to lower rows, one gets back the original tableau

$$\begin{array}{|c|c|c|c|} \hline 3 & 6 \\ \cline{1-2} 2 & 3 & 4 \\ \cline{1-3} 1 & 1 & 3 & 5 \\ \hline \end{array}$$

together with the information $b = 2$.

Fix now two semistandard tableaux $P \in \mathrm{SST}(N_2,\lambda)$ and $Q \in \mathrm{SST}(N_1,\lambda)$. The discussion with bumping routes of Proposition 3.14 ensures that if P and Q comes from the RSK map, then a_n is the rightmost largest entry of Q. Thus, consider the cell $\square$ of λ that contains the largest entry a_n of Q, and that is the rightmost among cells containing a_n in Q. We define Q_{n-1} as the tableau obtained from Q by removing this cell $\square$, and P_{n-1} as the tableau obtained from P by reversing the Schensted insertion starting from $\square$. We denote b_n the entry obtained as the result of the reversed Schensted insertion. By induction, P_{n-1} and Q_{n-1} correspond to a unique two-line array $\sigma_{|n-1} = \binom{a_1,\ldots,a_{n-1}}{b_1,\ldots,b_{n-1}}$ in $\mathfrak{A}(n-1;N_1,N_2)$, and then P and Q are the RSK tableaux of

$$\sigma = \begin{pmatrix} a_1,\ldots,a_{n-1},a_n \\ b_1,\ldots,b_{n-1},b_n \end{pmatrix}.$$

$\square$

Corollary 3.16. *The RSK map yields by restriction a bijection between:*

- *words in $[\![1,N]\!]^n$, and pairs of tableaux (P,Q) of same shape $\lambda \in \mathfrak{Y}(n)$, with P semistandard tableau with entries in $[\![1,N]\!]$, and Q standard tableau;*
- *permutations in $\mathfrak{S}(n)$, and pairs of standard tableaux (P,Q) of same shape $\lambda \in \mathfrak{Y}(n)$.*

Proof. A two-line array is a word if and only if its first row is $12\dots n$. In this case, Q is semistandard and contains all numbers from 1 to n; hence, it is a standard tableau. For a permutation, the first tableau P is also semistandard and contains all numbers from 1 to n; hence, it is also a standard tableau. □

In particular, one obtains a combinatorial proof of the identities stated at the beginning of this section, and the more general identity

$$\binom{N_1N_2+n-1}{n} = \sum_{\lambda \in \mathfrak{Y}(n)} \operatorname{card} \mathrm{SST}(N_1,\lambda) \times \operatorname{card} \mathrm{SST}(N_2,\lambda).$$

Another important consequence of Theorem 3.15 is a new easy proof of the Cauchy identity of Theorem 2.18 for Schur functions:

$$\prod_{i,j} \frac{1}{1-x_i y_j} = \sum_{\lambda \in \mathfrak{Y}} s_\lambda(X)\, s_\lambda(Y).$$

Indeed, define the monomial m^σ associated to a two-line array $\sigma = \binom{a_1,\dots,a_n}{b_1,\dots,b_n}$ as the product $x_{a_1}\cdots x_{a_n} y_{b_1}\cdots y_{b_n}$. Notice that given a two-line array σ, if one knows for any fixed pair $\binom{a}{b}$ the number of indices i such that $\binom{a_i}{b_i}$ is equal to $\binom{a}{b}$, then σ is entirely determined, because the pairs are then ordered lexicographically in σ. As a consequence, the series of monomials m^σ over all possible two-line arrays is given by

$$\sum m^\sigma = \prod_{i,j}\left(\sum_{k_{ij}=0}^{\infty} (x_i y_j)^{k_{ij}}\right) = \prod_{i,j} \frac{1}{1-x_i y_j}.$$

If one splits the series according to the shape λ of the tableaux $P(\sigma)$ and $Q(\sigma)$, one gets

$$\sum_{\lambda \in \mathfrak{Y}} \left(\sum_{T \in \mathrm{SST}(\lambda)} x^T\right)\left(\sum_{T \in \mathrm{SST}(\lambda)} y^T\right)$$

according to Theorem 3.15. It suffices then to use the expansion of Schur functions over semistandard tableaux (cf. Theorem 3.2).

▷ *The plactic monoid.*

In the remainder of this section, we shall consider the restriction of the RSK map to words, and look for an interpretation of the shape $\lambda(w)$ of the two tableaux $P(w)$ and $Q(w)$ associated to a word. To this purpose, we shall need another combinatorial algorithm due to Schützenberger and which produces the insertion tableau $P(w)$. To start with, let us introduce the so-called **plactic monoid**. Let $\mathscr{W} = (\mathbb{N}^*)^{(\mathbb{N})} = \bigsqcup_{n=0}^{\infty} (\mathbb{N}^*)^n$ be the set of all words with integer letters and arbitrary length. This set is a monoid for the operation of concatenation of words

$$(a_1a_2\dots a_r)\cdot(b_1b_2\dots b_s) = a_1\dots a_r b_1\dots b_s.$$

An elementary **Knuth transformation** on a word is a transformation of 3 consecutive letters according to the following rules:

$$yxz \longleftrightarrow yzx \quad \text{if } x < y \leq z;$$
$$zxy \longleftrightarrow xzy \quad \text{if } x \leq y < z.$$

Two words are said to be **Knuth equivalent** if they differ by a finite number of elementary Knuth transformations (notation: $w_1 \equiv w_2$). If two pairs of words are Knuth equivalent, then so are their concatenates, so the set of equivalence classes of words is a quotient monoid of $\mathscr{W}$. We shall denote $[w]$ the Knuth class of a word, and $\mathscr{P}$ the monoid of all classes, also called plactic monoid.

We shall prove in a moment that two words w and w' are Knuth equivalent if and only if they have the same insertion tableau $P(w) = P(w')$. Thus, the classes of the plactic monoid shall be labeled by semistandard tableaux. The best formulation of this result is as an isomorphism of monoids. If T is a skew tableau of shape $\lambda \setminus \mu$, we allow the addition of a special cell labeled by $\bullet$, and placed either on the lower left side of the tableau, or on the upper right side of the tableau. Thus, the semistandard skew tableau of shape $(5,4,2,1) \setminus (3,2)$

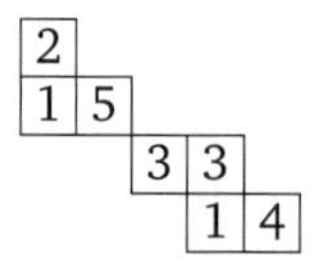

is identified with the tableaux

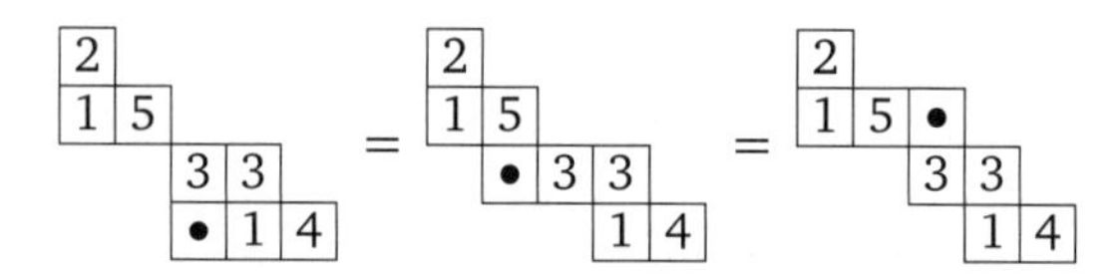

and with several other tableaux, corresponding to the other possibilities for a special cell on the upper right side. More generally, we call **generalized tableau** a skew tableau that possibly contains one cell with the special label $\bullet$ (not necessarily on the border of the tableau). This allows one to deal with skew tableaux containing one hole (the cell labeled by $\bullet$).

We call semistandard a generalized tableau whose rows and columns are respectively non-decreasing and strictly increasing (not taking into account the special label $\bullet$). We denote $\mathscr{T}^{\bullet}$ (respectively, $\mathscr{T}$) the set of all semistandard generalized tableaux (respectively, the set of all semistandard skew tableaux). The **reading word** of a semistandard generalized tableau is the sequence of its entries, read from left to right and from top to bottom (again, omitting the special label $\bullet$ if it occurs). For instance, the reading word of the previous generalized tableaux is 2153314. On the other hand, an elementary **Schützenberger transformation** on a semistandard generalized tableau is one of the following slidings:

$$\begin{array}{|c|c|}\hline y & z \\ \hline \bullet & x \\ \hline \end{array} \longleftrightarrow \begin{array}{|c|c|}\hline y & z \\ \hline x & \bullet \\ \hline \end{array} \qquad \text{if } x < y \leq z;$$

$$\begin{array}{|c|c|}\hline x & z \\ \hline \bullet & y \\ \hline \end{array} \longleftrightarrow \begin{array}{|c|c|}\hline \bullet & z \\ \hline x & y \\ \hline \end{array} \qquad \text{if } x \leq y < z.$$

One also authorizes slidings when the box of z is an empty cell; so for instance, if $x \leq y$, then

$$\begin{array}{|c|c}\hline x & \\ \hline \bullet & \multicolumn{1}{c|}{y} \\ \hline \end{array} \longleftrightarrow \begin{array}{|c|c}\hline \bullet & \\ \hline x & \multicolumn{1}{c|}{y} \\ \hline \end{array}$$

is a valid Schützenberger transformation, and if $x < y$, then

$$\begin{array}{|c|c}\hline y & \\ \hline \bullet & \multicolumn{1}{c|}{x} \\ \hline \end{array} \longleftrightarrow \begin{array}{|c|c}\hline y & \\ \hline x & \multicolumn{1}{c|}{\bullet} \\ \hline \end{array}$$

is also a valid Schützenberger transformation. Two generalized tableaux in $\mathscr{T}^{\bullet}$ are said to be **Schützenberger equivalent** if they differ by a finite number of Schützenberger transformations. In the sequel, we shall consider the restriction of this equivalence relation to $\mathscr{T}$.

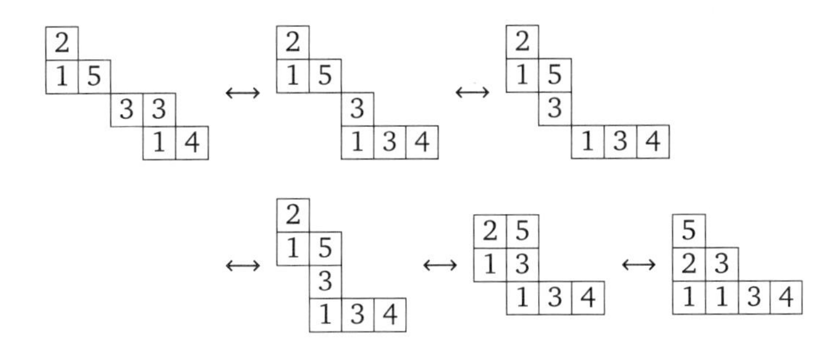

Figure 3.5
A sequence of Schützenberger transformations.

Example. The slidings drawn in Figure 3.5 are compositions of Schützenberger transformations. Each time, one took an empty cell in the left bottom corner, and one slid it to the right top corner. Notice that the last tableau is a semistandard tableau, and that it is exactly the insertion tableau of the reading word 2153314 of the initial semistandard skew tableau.

We make $\mathscr{T}$ into a monoid with the following construction. The height (respectively, the width) of a skew tableau is the label of its highest non-empty row (respectively, of its rightmost non-empty column). Given two semistandard skew tableaux T_1 and T_2 with $\text{height}(T_i) = h_i$ and $\text{width}(T_i) = w_i$, we set:

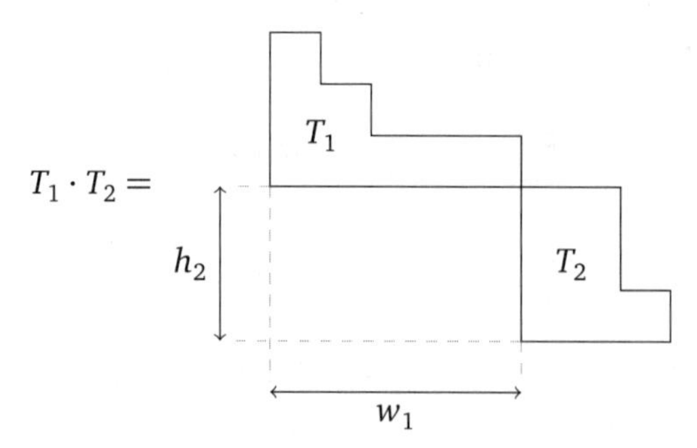

This product is obviously compatible with Schützenberger equivalence, so one can consider the monoid $\mathscr{P}'$ of classes of semistandard skew tableaux. As before, we denote $[T]$ the equivalence class of a semistandard skew tableau T.

Given a generalized semistandard tableau T, we denote $W(T)$ its reading word. On the other hand, given a word $w = a_1 a_2 \dots a_n$, we denote $R(w)$ the unique ribbon semistandard tableau whose reading word is w (and whose Young diagram touches the two borders of the quadrant $\mathbb{N}^* \times \mathbb{N}^*$). For example, if $w = 2153314$, then its ribbon tableau $R(w)$ is

<table>
<tr><td>2</td><td></td><td></td><td></td></tr>
<tr><td>1</td><td>5</td><td></td><td></td></tr>
<tr><td></td><td>3</td><td>3</td><td></td></tr>
<tr><td></td><td></td><td>1</td><td>4</td></tr>
</table>

.

Theorem 3.17 (Knuth, Lascoux–Schützenberger). *The maps*

$$\begin{aligned}\phi : \mathscr{P} &\to \mathscr{P}' \\ [w] &\mapsto [R(w)]\end{aligned} \quad ; \quad \begin{aligned}\phi' : \mathscr{P}' &\to \mathscr{P} \\ [T] &\mapsto [W(T)]\end{aligned}$$

are well defined, and they are isomorphisms of monoids, with $\phi' \circ \phi = \mathrm{id}_{\mathscr{P}}$. Moreover, for any word w, the insertion tableau $P(w)$ belongs to the class of $[R(w)]$.

Lemma 3.18. *For any semistandard tableau T and any number a, $T \leftarrow a$ is Schützenberger equivalent to $T \cdot \boxed{a}$.*

Proof. For any skew semistandard tableau, notice that if

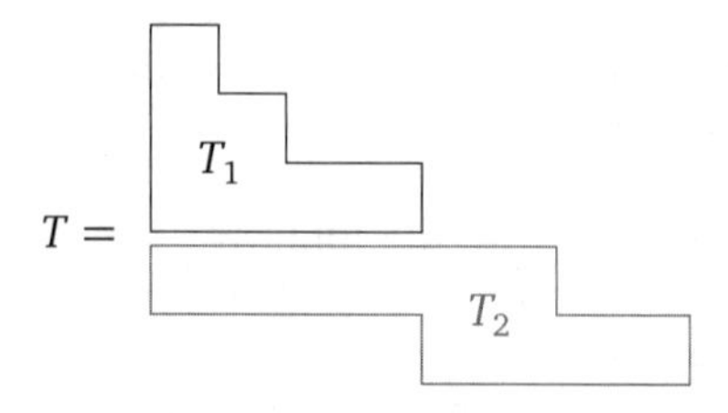

then $T \equiv T_1 \cdot T_2$; indeed, one can use Schützenberger transformations in order to slide entirely T_2 to the right. As a consequence, it suffices to show that if $T = \boxed{a_1}\cdots\boxed{a_r}$, and if i is the smallest index of an entry $a_i > a$, then

$$\begin{array}{ccccc} a_1 \cdots a_i \cdots a_r & \\ & a \end{array} \equiv \begin{array}{cc} a_i & \\ & a_1 \cdots a \cdots a_r \end{array}.$$

This will imply that Schützenberger transformations enable us to realize row insertions, and to bump entries to higher rows, as in the definition of the RSK algorithm. However, this last assertion is easy to prove:

$$\begin{array}{cc} a_1 \cdots a_i \cdots a_r & \\ & a \end{array} \equiv \begin{array}{cc} a_1 \cdots a_i \cdots a_r \\ a \end{array} \equiv \begin{array}{cc} a_1 \cdots a_i \cdots & \\ & a\; a_r \end{array}$$

$$\equiv \begin{array}{cc} a_1 \cdots a_i & \\ & a \cdots a_r \end{array} \equiv \begin{array}{cc} a_1\, a_i & \\ & \cdots\; a \cdots a_r \end{array}$$

$$\equiv \begin{array}{c} a_i \\ a_1 \cdots a \cdots a_r \end{array} \equiv \begin{array}{cc} a_i & \\ & a_1 \cdots a \cdots a_r \end{array}. \qquad \square$$

Lemma 3.19. *For any semistandard tableau T and any number a, $W(T \leftarrow a)$ is Knuth equivalent to $W(T) \cdot a$.*

Proof. The sequence of elementary Schützenberger transformations used in the previous lemma corresponds to a sequence of elementary Knuth transformations for the reading words:

$$\begin{aligned} a_1 \ldots a_i \ldots a_{r-1} a_r a &\equiv a_1 \ldots a_i \ldots a_{r-1} a a_r \quad (a < a_{r-1} \leq a_r) \\ &\equiv \;\vdots \\ &\equiv a_1 \ldots a_i a a_{i+1} \ldots a_r \quad (a < a_i \leq a_{i+1}) \\ &\equiv a_1 \ldots a_{i-2} a_i a_{i-1} a a_{i+1} \ldots a_r \quad (a_{i-1} \leq a < a_i) \\ &\equiv \;\vdots \\ &\equiv a_1 a_i a_2 \ldots a_{i-1} a a_{i+1} \ldots a_r \quad (a_2 \leq a_3 < a_i) \\ &\equiv a_i a_1 \ldots a_{i-1} a a_{i+1} \ldots a_r \quad (a_1 \leq a_2 < a_i). \end{aligned}$$

On the other hand, the same trick of "separation of rows" of tableaux holds for reading words, so the proof is completed. $\square$

Lemma 3.20. *If two words are Knuth equivalent, then their ribbon tableaux are Schützenberger equivalent. If two semistandard skew tableaux are Schützenberger equivalent, then their reading words are Knuth equivalent.*

Proof. For the first part of the lemma, notice that for any word $w = w_1 \cdot w_2$, $R(w)$ is Schützenberger equivalent to $R(w_1) \cdot R(w_2)$. Indeed, assuming for instance that the last letter of w_1 is smaller or equal to the first letter of w_2, one has

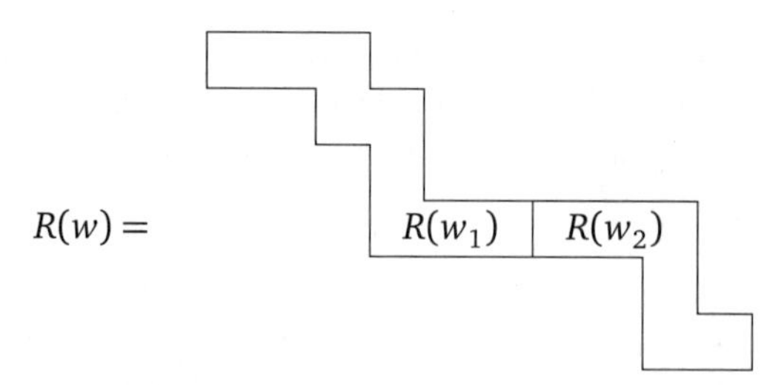

and then, Schützenberger transformations allow one to slide $R(w_1)$ one step up, thus getting $R(w_1) \cdot R(w_2)$. As a consequence, given $x < y \leq z$, in order to show that the ribbons of two Knuth equivalent words $w_1(yxz)w_2$ and $w_1(yzx)w_2$ are Schützenberger equivalent, it suffices to show it for the ribbons of yxz and yzx. This last statement is obvious:

$$R(yxz) = \begin{array}{|c|c}\hline y & \\ \hline x & \multicolumn{1}{c|}{z} \\ \hline\end{array} \equiv \begin{array}{|c|c|}\hline y & z \\ \hline x & \\ \cline{1-1}\end{array} \equiv \begin{array}{c|c|}\hline \multicolumn{1}{|c|}{y} & z \\ \hline & x \\ \cline{2-2}\end{array} = R(yzx).$$

Similarly, if $x \leq y < z$, then

$$R(zxy) = \begin{array}{|c|c}\hline z & \\ \hline x & \multicolumn{1}{c|}{y} \\ \hline\end{array} \equiv \begin{array}{c|c|}\cline{2-2} & z \\ \hline \multicolumn{1}{|c|}{x} & y \\ \hline\end{array} \equiv \begin{array}{c|c|}\hline \multicolumn{1}{|c|}{x} & z \\ \hline & y \\ \cline{2-2}\end{array} = R(xzy).$$

Since these transformations generate Knuth's equivalence relation, we have shown that ϕ is well defined, and also that it is compatible with the product, hence a morphism of monoids.

The second part of the lemma is a little more difficult. Using the same trick of "separation of rows" as before, it suffices to show the result for generalized semistandard tableaux that occupy two rows of same size and differ by an elementary Schützenberger transformation. Let us then treat the two cases of elementary transformations:

1. Consider two generalized semistandard tableaux

$$\begin{array}{|c|c|c|c|c|c|c|c|}\hline * & * & * & y & z & \star & \star & \star \\ \hline \times & \times & \times & \bullet & x & \circ & \circ & \circ \\ \hline\end{array} \equiv \begin{array}{|c|c|c|c|c|c|c|c|}\hline * & * & * & y & z & \star & \star & \star \\ \hline \times & \times & \times & x & \bullet & \circ & \circ & \circ \\ \hline\end{array}$$

   with $x < y \leq z$. In this case, the tableaux have the same reading word, so there is nothing to prove.

2. Consider two generalized semistandard tableaux

$$\begin{array}{|c|c|c|c|c|c|c|c|}\hline * & * & * & x & z & \star & \star & \star \\ \hline \times & \times & \times & \bullet & y & \circ & \circ & \circ \\ \hline\end{array} \equiv \begin{array}{|c|c|c|c|c|c|c|c|}\hline * & * & * & \bullet & z & \star & \star & \star \\ \hline \times & \times & \times & x & y & \circ & \circ & \circ \\ \hline\end{array}$$

   with $x \leq y < z$. We give names a_i, b_i, c_i and d_i to the cells around x, y and z, so that the two tableaux are

$$\begin{array}{|c|c|c|c|c|c|c|c|}\hline a_1 & \cdots & a_r & x & z & b_1 & \cdots & b_s \\ \hline c_1 & \cdots & c_r & \bullet & y & d_1 & \cdots & d_s \\ \hline\end{array} \equiv \begin{array}{|c|c|c|c|c|c|c|c|}\hline a_1 & \cdots & a_r & \bullet & z & b_1 & \cdots & b_s \\ \hline c_1 & \cdots & c_r & x & y & d_1 & \cdots & d_s \\ \hline\end{array}$$

   and we then have to show that $axzbcyd$ is Knuth equivalent to $azbxyd$, where $a = a_1 \ldots a_r$, $b = b_1 \ldots b_s$, $c = c_1 \ldots c_r$ and $d = d_1 \ldots d_s$.

(a) Suppose first that $r = 0$. Then, using several times Lemma 3.19,

$$
\begin{aligned}
xzb_1\dots b_s y d_1\dots d_s &\equiv W\left(\begin{array}{|c|c|c|c|c|}\hline x & z & b_1 & \cdots & b_s \\ \hline\end{array} \leftarrow y\right)\cdot d_1\dots d_s \\
&\equiv W\left(\begin{array}{|c|c|c|c|c|}\cline{1-1} z \\ \hline x & y & b_1 & \cdots & b_s \\ \hline\end{array}\right)\cdot d_1\dots d_s \\
&\equiv W\left(\begin{array}{|c|c|c|c|c|}\cline{1-1} z \\ \hline x & y & b_1 & \cdots & b_s \\ \hline\end{array} \leftarrow d_1\right)\cdot d_2\dots d_s \\
&\equiv W\left(\begin{array}{|c|c|c|c|c|}\cline{1-2} z & b_1 \\ \hline x & y & d_1 & \cdots & b_s \\ \hline\end{array}\right)\cdot d_2\dots d_s \\
&\equiv \quad \vdots \\
&\equiv W\left(\begin{array}{|c|c|c|c|c|}\cline{1-4} z & b_1 & \cdots & b_s \\ \hline x & y & d_1 & \cdots & d_s \\ \hline\end{array}\right) = zb_1\dots b_s xyd_1\dots d_s,
\end{aligned}
$$

hence the result in this case.

(b) Suppose the result to be true up to order $r-1$. We set $a' = a_2\dots a_r$ and $c' = c_2\dots c_r$. Then, with $2\times r$ cells on the right of the square of the Schützenberger transformation, one has:

$$
\begin{aligned}
axzbcyd &\equiv W\left(\begin{array}{|c|c|c|c|}\hline a & x & z & b \\ \hline\end{array} \leftarrow c_1\right)\cdot c'yd \equiv W\left(\begin{array}{|c|c|c|c|c|}\cline{1-1} a_1 \\ \hline c_1 & a' & x & z & b \\ \hline\end{array}\right)\cdot c'yd \\
&\equiv a_1c_1\cdot W\left(\begin{array}{|c|c|c|c|}\hline a' & x & z & b \\ \hline c' & \bullet & y & d \\ \hline\end{array}\right) \equiv a_1c_1\cdot W\left(\begin{array}{|c|c|c|c|}\hline a' & \bullet & z & b \\ \hline c' & x & y & d \\ \hline\end{array}\right) \\
&\equiv a_1c_1a'zbc'xyd = W\left(\begin{array}{|c|c|c|c|}\cline{1-1} a_1 \\ \hline c_1 & a' & z & b \\ \hline\end{array}\right)\cdot c'xyd \\
&\equiv W\left(\begin{array}{|c|c|c|}\hline a & z & b \\ \hline\end{array} \leftarrow c_1\right)\cdot c'xyd \equiv azbcxyd
\end{aligned}
$$

by using the induction hypothesis on the second line.

This ends the proof of the lemma, which ensures that ϕ and ϕ' are well defined, and that ϕ is a morphism of monoids. The compatibility of ϕ' with the product is even more obvious. $\square$

Proof of Theorem 3.17. We saw in the previous lemma that ϕ and ϕ' are well-defined morphisms of monoids. Moreover, for any word w, $W(R(w)) = w$, so $\phi'\circ\phi = \mathrm{id}_{\mathscr{P}}$. If $w = w_1\dots w_r$ is a word, then Lemma 3.18 ensures that

$$
P(w) = (\emptyset \leftarrow w_1 \leftarrow w_2 \leftarrow \cdots \leftarrow w_r) \equiv \begin{array}{|c|c|c|c|}\cline{1-1} w_1 & \multicolumn{3}{c}{} \\ \cline{1-2} \multicolumn{1}{c|}{} & w_2 & \multicolumn{2}{c}{} \\ \cline{2-3} \multicolumn{2}{c|}{} & \ddots & \multicolumn{1}{c}{} \\ \cline{3-4} \multicolumn{3}{c|}{} & w_r \\ \cline{4-4}\end{array} \equiv R(w)
$$

so $P(w)$ is in the equivalence class of the ribbon tableau of w. $\square$

▷ *Greene invariants.*

Theorem 3.17 can be completed by the following statement: $P(w)$ is the *unique* semistandard tableau in the class of $R(w)$, so each Schützenberger class contains a unique semistandard tableau, and the plactic classes are labeled by these tableaux. The proof of this unicity relies on the notion of **increasing subsequences** in words. Let $w = w_1 w_2 \dots w_n$ be an arbitrary word. A subword, or subsequence of w is a word $w' = w_{i_1} w_{i_2} \dots w_{i_r}$ with $i_1 < i_2 < \cdots < i_r$; in other words, one has deleted some letters in w. The subword w' is said to be weakly increasing if $w_{i_1} \leq w_{i_2} \leq \cdots \leq w_{i_r}$. In the following we shall only deal with weakly increasing words, hence drop the adjective "weakly."

Example. In the word $w = 2\mathbf{1}5\mathbf{33}1\mathbf{4}$, $w' = 1334$ is a subword that is increasing.

A family of subwords of w is said to be disjoint if the subwords correspond to disjoint subsets of letters of w.

Definition 3.21. *The k-th* **Greene invariant** *of a word w is the integer*

$$L_k(w) = \max(\ell(w^{(1)}) + \cdots + \ell(w^{(k)})),$$

where the maximum is taken over families $(w^{(1)}, \dots, w^{(k)})$ of disjoint increasing subwords of w.

If $\ell(w) = n$, then $L_k(w) \geq \min(n, k)$ for any k, since one can take k different letters of w to get a family of disjoint increasing subwords. Obviously,

$$L_1(w) \leq L_2(w) \leq \cdots \leq L_k(w) \leq \cdots \leq n.$$

Example. It can be checked that with $w = 2153314$,

$$\begin{aligned} L_1(w) &= \ell(1334) = 4; \\ L_2(w) &= \ell(1334) + \ell(25) = 6; \\ L_3(w) &= \ell(1334) + \ell(25) + \ell(1) = 7 \end{aligned}$$

and $L_{k \geq 3}(w) = \ell(w) = 7$. Beware that if $(w^{(1)}, \dots, w^{(k)})$ is a maximizer for the k-th Greene invariant of w, then it is not true in general that a maximizer for the $(k-1)$-th Greene invariant can be obtained by removing one word in the family $(w^{(1)}, \dots, w^{(k)})$. For instance, $(133, 25, 14)$ is another maximal family for $L_3(2153314) = 7$, and removing a word cannot give a family of total length $L_2(2153314) = 6$.

Proposition 3.22. *Let w be the reading word of a semistandard tableau T of shape $\lambda = (\lambda_1, \dots, \lambda_r)$. Then,*

$$\forall k \in [\![1, r]\!],\ L_k(w) = \lambda_1 + \lambda_2 + \cdots + \lambda_k.$$

Proof. If w' is a subword of w, then it corresponds to a sequence of cells in T (going in general from top to bottom and from left to right, row by row). Suppose now w' increasing, and consider two consecutive letters w_i and w_j of w'. The corresponding cells $\square_i$ and $\square_j$ are either on the same row, with $\square_j$ strictly to the right of $\square_i$; or, $\square_j$ is in a row strictly below $\square_i$, but then it must also be strictly to the right, because otherwise one would have $w_i > w_j$. Hence, increasing subwords of w correspond to sequences of cells in T that go downwards and occupy at most one cell by column. For instance, in the tableau

$$\begin{array}{|c|c|c|c|} \hline 3 & 6 & \multicolumn{2}{c}{} \\ \cline{1-3} 2 & 3 & 4 & \multicolumn{1}{c}{} \\ \hline 1 & 1 & 3 & 5 \\ \hline \end{array}$$

with reading word 362341135, the increasing subword 3345 corresponds to the sequence of cells

$$\begin{array}{|c|c|c|c|} \hline \colorbox{lightgray}{3} & 6 & \multicolumn{2}{c}{} \\ \cline{1-3} 2 & \colorbox{lightgray}{3} & \colorbox{lightgray}{4} & \multicolumn{1}{c}{} \\ \hline 1 & 1 & 3 & \colorbox{lightgray}{5} \\ \hline \end{array}\ .$$

The result follows immediately: such a sequence of cells has for maximal length the number of columns λ_1, so $L_1(w) = \lambda_1$ by taking for increasing subword of maximal length the word of the bottom row; and more generally, a family of k disjoint sequences of cells with this property has always total length smaller than $\lambda_1 + \cdots + \lambda_k$, the equality being obtained with the family of the k bottom rows. $\square$

Proposition 3.23. *Let w and w' be two Knuth equivalent words. For any k, $L_k(w) = L_k(w')$.*

Proof. We treat the case when $w = a \cdot yxz \cdot b$ and $w' = a \cdot yzx \cdot b$, with $x < y \leq z$; the other elementary Knuth transformation is very similar.

1. If $(w^{(1)}, \ldots, w^{(k)})$ is a disjoint family of increasing subwords of w', then none of these words contains zx since $z > x$, so $(w^{(1)}, \ldots, w^{(k)})$ is also a disjoint family of increasing subwords of w. Therefore, $L_k(w) \geq L_k(w')$.

2. Conversely, fix $(w^{(1)}, \ldots, w^{(k)})$ a disjoint family of increasing subwords of w such that $L_k(w) = \ell(w^{(1)}) + \cdots + \ell(w^{(k)})$. If no subword $w^{(i)}$ contains xz, then $(w^{(1)}, \ldots, w^{(k)})$ is also a disjoint family of increasing subwords of w', so $L_k(w') \geq L_k(w)$. Suppose now that one subword $w^{(i)}$ contains xz, and thus writes as $w^{(i)} = a^{(i)} \cdot xz \cdot b^{(i)}$, where $a^{(i)}$ and $b^{(i)}$ are (increasing) subwords of a and b. We distinguish two cases:

   (a) There is one word $w^{(j)} = a^{(j)} \cdot y \cdot b^{(j)}$ that contains y. We then set

$$\begin{aligned} \widetilde{w}^{(i)} &= a^{(i)} \cdot x \cdot b^{(j)} \\ \widetilde{w}^{(j)} &= a^{(j)} \cdot yz \cdot b^{(i)} \\ \widetilde{w}^{(k \neq i,j)} &= w^{(k)} \end{aligned}$$

and get a sequence of increasing subwords of w' with the same total length, so $L_k(w') \geq L_k(w)$.

(b) There is no word $w^{(j)}$ containing y. Then, we set $\widetilde{w}^{(i)} = a^{(i)} \cdot yz \cdot b^{(i)}$ and get a sequence of increasing subwords of w' with the same total length, so again $L_k(w') \geq L_k(w)$.

In all cases, by double inequality, $L_k(w) = L_k(w')$. □

Corollary 3.24. *For any word w, the sequence $(L_k(w) - L_{k-1}(w))_{k\geq 1}$ is an integer partition.*

Proof. By Theorem 3.17, the word w is Knuth equivalent to the reading word w' of the semistandard tableau $P(w)$, and if λ is the shape of $P(w)$, then its Greene invariants verify $\lambda_k = L_k(w') - L_{k-1}(w')$. Then, by Proposition 3.23,

$$L_k(w') - L_{k-1}(w') = L_k(w) - L_{k-1}(w).$$ □

We can finally state:

Theorem 3.25 (Knuth, Lascoux–Schützenberger). *For any word w, the insertion tableau $P(w)$ is the unique semistandard tableau in $[R(w)]$. Therefore, classes of Knuth equivalent words and classes of Schützenberger equivalent tableaux are labeled by semistandard tableaux. Moreover, $w \equiv w'$ if and only if $P(w) = P(w')$.*

Lemma 3.26. *Let w and w' be two Knuth equivalent words, and $w_i = w'_j$ the largest rightmost letter of w and w'. We set $\widetilde{w} = w \setminus w_i$ and $\widetilde{w}' = w' \setminus w'_j$. Then, $\widetilde{w}$ and $\widetilde{w}'$ are Knuth equivalent.*

Proof. It suffices to show the result when w and w' differ by an elementary Knuth transformation. Moreover, one can assume without loss of generality that this transformation involves w_i and w'_j, since otherwise the result is trivial. Thus, suppose that $w = a \cdot yxz \cdot b$ and $w' = a \cdot yzx \cdot b$, with $z = w_i = w'_j$. Then, $\widetilde{w} = a \cdot yx \cdot b = \widetilde{w}'$. The other elementary transformation is treated similarly. □

Proof of Theorem 3.25. Let T and T' be two Schützenberger equivalent semistandard tableaux; we want to show that $T = T'$. We reason by induction on the size n of these tableaux. Set $w = W(T)$ and $w' = W(T')$; these words are Knuth equivalent, so they have the same Greene invariants. By Propositions 3.22 and 3.23, these Greene invariants dictate the shape λ of T and T':

$$\lambda_k = L_k(w) - L_{k-1}(w) = L_k(w') - L_{k-1}(w').$$

Thus, T and T' have the same shape. We denote $\widetilde{w}$ and $\widetilde{w}'$ the words obtained from w and w' by deleting the largest and rightmost letter z; and $\widetilde{T}$ and $\widetilde{T}'$ the tableaux obtained from T and T' by deleting the largest and rightmost cell labeled by z.

Since $\widetilde{w} = W(\widetilde{T})$ and $\widetilde{w}' = W(\widetilde{T}')$ are Knuth equivalent by the previous lemma, $\widetilde{T}$ and $\widetilde{T}'$ are Schützenberger equivalent, so by the induction hypothesis they are equal. Then, T is obtained from $\widetilde{T}$ by adding on the right side a cell labeled z, and similarly with T' and $\widetilde{T}'$. Since T and T' have the same shape, the position of this cell is fixed, so $T = T'$. The last part of the theorem follows then from the fact that $P(w) \in \phi([w])$ according to Theorem 3.17. □

Remark. As an application of the two Theorems 3.17 and 3.25, one can actually recover the Pieri rules of Section 3.1. Indeed, consider the plactic monoid $\mathscr{P} = \mathscr{P}'$: it has a basis labeled by semistandard tableaux, with the rule

$$\begin{aligned}&[T_1] * [T_2]\\ &= [\text{unique semistandard tableau } T \text{ Schützenberger equivalent to } T_1 \cdot T_2].\end{aligned}$$

We consider in the following the monoid algebra $\mathbb{C}\mathscr{P}$ built upon the plactic monoid, and the linear map

$$\begin{aligned}\Xi : \mathbb{C}\mathscr{P} &\to \mathbb{C}[x_1, x_2, \ldots]\\ [T] &\mapsto x^T.\end{aligned}$$

This map Ξ is a morphism of algebras, and by Theorem 3.2,

$$\Xi\left(\sum_{T \in \mathrm{SST}(\lambda)} [T]\right) = s_\lambda(X).$$

If one inserts a non-decreasing sequence $(a_1, \ldots, a_k)$ of length k into a semistandard tableau of shape λ, by the discussion of Proposition 3.14, the cells created during this procedure occupy different columns, hence, the shape Λ of the end tableau differs from λ by a horizontal strip of size k. It follows that in $\mathbb{C}\mathscr{P}$,

$$\left(\sum_{T \in \mathrm{SST}(\lambda)} [T]\right) * \left(\sum_{R \in \mathrm{SST}(k)} [R]\right) = \sum_{\Lambda \setminus \lambda \text{ horizontal strip of size } k} \left(\sum_{U \in \mathrm{SST}(\Lambda)} [U]\right)$$

since the product of tableaux in the plactic monoid can be realized by Schensted insertions. By applying the morphism Ξ, one recovers the first Pieri rule for the product $s_\lambda(X)\, h_k(X)$.

▷ *Symmetry of the RSK algorithm.*

To conclude our presentation of the RSK algorithm, let us explain how Schützenberger slidings provide a new algorithm in order to compute the two tableaux $P(\sigma)$ and $Q(\sigma)$ associated to a *permutation* σ. If w is any word, starting from the ribbon tableau $R(w)$, one can slide inner empty cells (on the left of the skew tableau) to the right of the skew tableau, cell by cell; at the end one gets a semistandard tableau which is Schützenberger equivalent to $R(w)$, hence equal to $P(w)$. Now, if $w = \sigma$ is a permutation, one can also apply these transformations to $w' = \sigma^{-1}$.

Example. Consider the permutation $\sigma = 927513648$. It has for inverse $\sigma^{-1} = 526847391$. The Schützenberger slidings applied to the ribbons of these words yield the two tableaux $P(\sigma)$ and $Q(\sigma)$; see Figure 3.6.

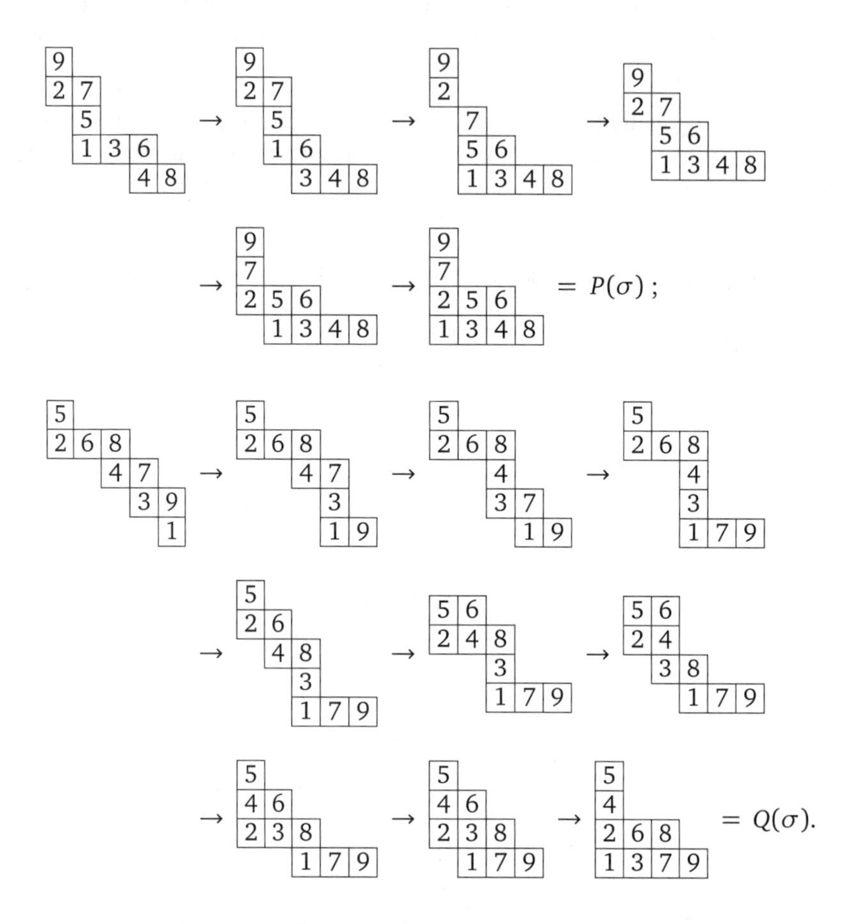

Figure 3.6
The Schützenberger transformations applied to $\sigma = 927513648$ and its inverse $\sigma^{-1} = 526847391$.

This phenomenon is general, so one can also use Schützenberger transformations in order to compute the recording tableau $Q(\sigma)$ of a permutation. Indeed, one has the following **symmetry theorem**:

Theorem 3.27. *For any permutation $\sigma \in \mathfrak{S}(n)$, $Q(\sigma) = P(\sigma^{-1})$.*

Remark. In fact, there is a symmetry theorem for the RSK algorithm on two-line arrays. If $\sigma = \binom{a_1,\ldots,a_n}{b_1,\ldots,b_n}$ is a two-line array, denote $\overline{\sigma}$ the two-line array obtained by lexicographic reordering of $\binom{b_1,\ldots,b_n}{a_1,\ldots,a_n}$. For instance,

$$\sigma = \begin{pmatrix} 11112333 \\ 12443223 \end{pmatrix} \quad \Rightarrow \quad \overline{\sigma} = \begin{pmatrix} 12223344 \\ 11332311 \end{pmatrix}.$$

Then, it can be shown that $P(\overline{\sigma}) = Q(\sigma)$, which generalizes the previous result. The usual proof rclies on Viennot's geometric interpretation of the RSK algorithm; see the references at the end of the chapter. Since we shall mostly deal with permutations in this book, we shall only present the version of the symmetry theorem that is valid for permutations.

The known proofs of Theorem 3.27 are all surprisingly difficult. We present here the first proof due to Schützenberger, and which relies on the notion of shifting; it is not too hard to understand, except the last case in Lemma 3.29, which admittedly will be a bit difficult to follow.

Let $\sigma \in \mathfrak{S}(n)$ be a permutation, A be a subset of $[\![1,n]\!]$, and a and b be two elements of A. We set $A_a = A \cap [\![1,a]\!]$. On the other hand, we denote $P(\sigma,A)$ the tableau obtained by Schensted insertion of the subword of $\sigma(1)\sigma(2)\ldots\sigma(n)$ corresponding to A. For instance, if $\sigma = 927513648$ and $A = \{1,3,5,6,8\}$, then $\sigma_{|A} = 97134$, and

$$P(\sigma,A) = \begin{array}{|c|cc} \hline 9 \\ \hline 7 \\ \hline 1 & \multicolumn{1}{|c|}{3} & \multicolumn{1}{c|}{4} \\ \hline \end{array}$$

is the restricted insertion tableau. Finally, given a tableau T and an entry x of this tableau, we denote $T(x)$ the coordinates (i,j) of the cell that contains x (with i corresponding to the row and j to the column).

Definition 3.28. *The **shifting*** $\mathrm{Shift}(a,b,\sigma,A)$ *of* $\sigma(b)$ *by* $\sigma(a)$ *in* $P(\sigma,A)$ *is the quantity*

$$\begin{cases} 0 & \text{if } a < b; \\ ((0,\infty),(i,j)) & \text{if } a = b, \text{ with } (i,j) = P(\sigma,A_a)(\sigma(a)); \\ 0 & \text{if } b < a \text{ and } P(\sigma,A_{a-1})(\sigma(b)) = P(\sigma,A_a)(\sigma(b)); \\ ((i',j'),(i,j)) & \text{otherwise, with } (i',j') = P(\sigma,A_{a-1})(\sigma(b)) \\ & \qquad\qquad\qquad (i,j) = P(\sigma,A_a)(\sigma(b)). \end{cases}$$

To say that the shifting is non-zero amounts to saying that $\sigma(b)$ moves when one performs the Schensted insertion of $\sigma(a)$ into $P(\sigma,A_{a-1})$. Moreover, if the shifting is non-zero, then $i = i'+1$ and $j \leq j'$ by usual properties of bumping routes.

Example. Consider as before $\sigma = 927513648$, $A = \{1,3,5,6,8\}$, $a = 5$ and $b = 3$. We then have:

$$P(\sigma,A_{a-1}) = P(97) = \begin{array}{|c|} \hline 9 \\ \hline 7 \\ \hline \end{array} \quad ; \quad P(\sigma,A_a) = P(971) = \begin{array}{|c|} \hline 9 \\ \hline 7 \\ \hline 1 \\ \hline \end{array}$$

and the shifting is $((1,1),(2,1))$.

Notice that if the shifting is non-zero, then there exists $a' \leq a$ such that $\sigma(a')$ bumps $\sigma(b)$, and therefore such that

$$\sigma(a) \leq \sigma(a') < \sigma(b) \qquad ; \qquad \mathrm{Shift}(a,a',\sigma,A) = ((i'',j''),(i',j')).$$

From this we deduce the following important restriction property for shiftings:

$$\mathrm{Shift}(a,b,\sigma,A) = \mathrm{Shift}(a,b,\sigma,\{a' \in A \mid a' \leq a \text{ and } \sigma(a') < \sigma(b)\}).$$

On the other hand, notice that the shape of the tableau $P(\sigma,A)$ is entirely determined by the family of sets of non-zero shiftings

$$\Big\{\{\mathrm{Shift}(a,b,\sigma,A) \neq 0,\ b \in A\},\ a \in A\Big\}.$$

Indeed, such a family corresponds to a unique family of growing partitions $\emptyset \nearrow \lambda^{(1)} \nearrow \cdots \nearrow \lambda^{(r)}$, associated to the construction of the insertion tableau $P(\sigma,A)$. Hence, the family determines the shape of $P(\sigma,A)$ (and in fact, the whole standard tableau $Q(\sigma,A)$).

Lemma 3.29. *For any $a,b \in A \subset [\![1,n]\!]$,*

$$\mathrm{Shift}(a,b,\sigma,A) = \mathrm{Shift}(\sigma(b),\sigma(a),\sigma^{-1},\sigma(A)).$$

Moreover, $P(\sigma,A)$ and $P(\sigma^{-1},\sigma(A))$ have the same shape.

Proof. We reason by induction on the size of A. Suppose the result to be true up to size $r-1$, and take A of size r. Notice that by the previous remark, it suffices to prove the first part of the lemma. In the sequel, we set $C = \sigma(A)$, and

$$a^* = \max A \qquad ; \qquad c^+ = \sigma(a^*)$$
$$c^* = \max C \qquad ; \qquad a^+ = \sigma^{-1}(c^*).$$

If $a,b \neq a^+$, then by the restriction property of shiftings,

$$\mathrm{Shift}(a,b,\sigma,A) = \mathrm{Shift}(a,b,\sigma,A \setminus \{a^+\});$$
$$\mathrm{Shift}(\sigma(b),\sigma(a),\sigma^{-1},C) = \mathrm{Shift}(\sigma(b),\sigma(a),\sigma^{-1},C \setminus \{c^*\}),$$

and the two right-hand sides are equal by induction. One has by symmetry the same conclusions if $\sigma(a),\sigma(b) \neq c^+$, which is equivalent to $a,b \neq a^*$. Therefore, it remains to treat the case when $\{a,b\} = \{a^+,a^*\}$. This case can be split into three situations:

1. Suppose $a = b = a^+ = a^*$. Then, $\mathrm{Shift}(a,a,\sigma,A) = ((0,\infty),(1,j))$, where j is the index of the column containing a after its insertion in $P(\sigma,A_{a-1})$, and is equal to

$$j = 1 + \text{ size of the first row of } P(\sigma,A_{a-1}).$$

Since $\sigma(b) = \sigma(a) = c^+ = c^*$,

$$\mathrm{Shift}(\sigma(a), \sigma(a), \sigma^{-1}, \sigma(A)) = ((0, \infty), (1, k))$$

for the same reasons as before, with

$$k = 1 + \text{ size of the first row of } P(\sigma^{-1}, C_{\sigma(a)-1}).$$

By hypothesis, $P(\sigma, A_{a-1})$ and $P(\sigma^{-1}, C_{\sigma(a)-1})$ have the same shape, so $j = k$ and this case is treated.

2. Suppose now $a = a^+$ and $b = a^*$, with $a^+ \neq a^*$. Then, $a < b$ and $\mathrm{Shift}(a, b, \sigma, A) = 0$. On the other hand,

$$\mathrm{Shift}(\sigma(b), \sigma(a), \sigma^{-1}, C) = \mathrm{Shift}(c^+, c^*, \sigma^{-1}, C),$$

and $c^+ < c^*$, so this second shifting is also zero.

3. Suppose finally $a = a^*$ and $b = a^+$, again with $a^+ \neq a^*$. It is sufficient to show that if $\mathrm{Shift}(a, b, \sigma, A) = ((i, j), (i + 1, \bar{j}))$, then the shifting $\mathrm{Shift}(\sigma(b), \sigma(a), \sigma^{-1}, C)$ is also equal to $((i, j), (i + 1, \bar{j}))$. Indeed, by symmetry, this will imply that if $\mathrm{Shift}(a, b, \sigma, A) = 0$, then the shifting $\mathrm{Shift}(\sigma(b), \sigma(a), \sigma^{-1}, C)$ also vanishes.

Thus, set $\mathrm{Shift}(a, b, \sigma, A) = ((i, j), (i + 1, \bar{j}))$: this implies that

$$P(\sigma, A_{a-1})(\sigma(b)) = P(\sigma, A_{a^*-1})(c^*) = (i, j).$$

We claim that one also has

$$P(\sigma^{-1}, C_{c^*-1})(a^*) = (i, j).$$

Indeed, there exists $y < a$ such that $\sigma(a) \leq \sigma(y) < \sigma(b) = c^*$ and $\mathrm{Shift}(a, y, \sigma, A) = ((i', j'), (i, j))$. Then, $\sigma(y)$ is the largest element in C_{c^*-1} that is shifted when inserting $\sigma(a^*)$ in $P(\sigma, A_{a^*-1})$. Applying the restriction property and the induction hypothesis, one has therefore:

$$\begin{aligned}
&\sigma(y)\\
&= \max\left\{c \in C_{c^*-1},\ c = \sigma(a') \text{ and } \mathrm{Shift}(a^*, a', \sigma, A) \neq 0\right\}\\
&= \max\left\{c \in C_{c^*-1},\ c = \sigma(a') \text{ and } \mathrm{Shift}(a^*, a', \sigma, A \setminus \{a^+\}) \neq 0\right\}\\
&= \max\left\{c \in C_{c^*-1},\ \mathrm{Shift}(c, c^+, \sigma^{-1}, C_{c^*-1}) \neq 0\right\}\\
&= \text{last element such that } y \text{ shifts } a^* \text{ when constructing } P(\sigma^{-1}, C_{c^*-1}).
\end{aligned}$$

Then, by the induction hypothesis applied to σ^{-1} and C_{c^*-1},

$$\mathrm{Shift}(\sigma(y), \sigma(a^*), \sigma^{-1}, C_{c^*-1}) = \mathrm{Shift}(a^*, y, \sigma, A) = ((i', j'), (i, j)),$$

hence the previous claim.

Now, notice that $\bar{j}$ is entirely determined by the position (i,j) of c^* in $P(\sigma,A_{a^*-1})$: indeed, since $c^* = \max C$, it is bumped *at the end of* the next row, so $\bar{j}$ depends only on i and on the shape of $P(\sigma,A_{a^*-1})$. Actually, if one knows i, then the shape of $P(\sigma,A\setminus\{a^+,a^*\})$ determines the shape of $P(\sigma,A_{a^*-1})$, because c^* is placed at the end of the i-th row of $P(\sigma,A\setminus\{a^+,a^*\})$). So,

$$\bar{j} \text{ depends only on } i \text{ and on the shape of } P(\sigma,A\setminus\{a^+,a^*\}).$$

We want to compute $\mathrm{Shift}(\sigma(b),\sigma(a),\sigma^{-1},C) = \mathrm{Shift}(c^*,c^+,\sigma^{-1},C)$. As we saw that $P(\sigma^{-1},C_{c^*-1})(\sigma(c^+)) = (i,j)$, it suffices now to show that $\sigma^{-1}(c^*) = a^+$ shifts $\sigma^{-1}(c) = a^*$ (to the next row, with position $(i,\tilde{j})$). Indeed, we shall then know that

$$\tilde{j} \text{ depends only on } i \text{ and on the shape of } P(\sigma^{-1},C\setminus\{c^+,c^*\}),$$

and by induction $P(\sigma,A\setminus\{a^+,a^*\})$ and $P(\sigma^{-1},C\setminus\{c^+,c^*\})$ have the same shape. As for this last statement, since $P(\sigma,A_{a^*-1})(c^*) = (i,j)$, there exists $x < a^*$ such that $\mathrm{Shift}(x,b,\sigma,A_{a^*-1}) = ((i'',j''),(i,j))$. By the induction hypothesis,

$$\mathrm{Shift}(c^*,\sigma(x),\sigma^{-1},C\setminus\{c^+\}) = ((i'',j''),(i,j)),$$

that is to say that $\sigma^{-1}(c^*)$ shifts some entry x to the row (i,j). Since $x < a^*$, this x bumps a^* to the next row; this is what we wanted to prove.

□

Proof of Theorem 3.27. We shall prove more generally that for any bijection σ between two finite sets of integers A and C, one has

$$Q(\sigma^{-1},C) = P(\sigma,A),$$

with obvious notations (the insertion tableau of $\sigma : A \to C$ has entries in C, and the reading tableau has entries in A). Suppose the result to be true for sets of size up to $n-1$, and take $\sigma : A \to C$, with $|A| = |C| = n$. One has

$$P(\sigma,A) = P(\sigma,A\setminus\{\sigma^{-1}(c^*)\}) + \boxed{c^*}_{(i,j)},$$

where (i,j) are the coordinates of the difference cell beween $P(\sigma,A\setminus\{a^+\})$ and $P(\sigma,A)$. By Lemma 3.29, this is also the difference cell between $P(\sigma^{-1},C\setminus\{c^*\})$ and $P(\sigma^{-1},C)$, because it is determined by the same family of non-zero shiftings. Hence, using the induction hypothesis,

$$P(\sigma,A) = Q(\sigma^{-1},C\setminus\{c^*\}) + \boxed{c^*}_{(i,j)} = Q(\sigma^{-1},C).$$

□

Example. We saw before that $n! = \sum_{\lambda\in\mathfrak{Y}(n)} |\mathrm{ST}(\lambda)|^2$. Using the symmetry theorem for the RSK algorithm, we also obtain a formula for $\sum_{\lambda\in\mathfrak{Y}(n)} |\mathrm{ST}(\lambda)|$. Call **involution** a permutation $\sigma \in \mathfrak{S}(n)$ such that $\sigma^2 = \mathrm{id}_{[\![1,n]\!]}$, or, equivalently, such that $\sigma = \sigma^{-1}$. An easy consequence of Theorem 3.27 is that σ is an involution if and only if $P(\sigma) = Q(\sigma)$. Indeed,

$$\begin{aligned} \sigma = \sigma^{-1} \quad &\Longleftrightarrow \quad (P(\sigma), Q(\sigma)) = (P(\sigma^{-1}), Q(\sigma^{-1})) \\ &\Longleftrightarrow \quad (P(\sigma), Q(\sigma)) = (Q(\sigma), P(\sigma)). \end{aligned}$$

Therefore, $\sum_{\lambda \in \mathfrak{Y}(n)} |\mathrm{ST}(\lambda)|$ is the number of involutions of size n.

In terms of cycle decomposition, a permutation is an involution if and only if it writes as a product of disjoint transpositions, plus a certain number of fixed points. It follows that the number of involutions of size n is

$$\begin{aligned} \sum_{\lambda \in \mathfrak{Y}(n)} \operatorname{card} \mathrm{ST}(\lambda) &= \sum_{k=0}^{\lfloor \frac{n}{2} \rfloor} \binom{n}{2k} \frac{1}{k!} \binom{2k}{2} \binom{2k-2}{2} \cdots \binom{2}{2} \\ &= \sum_{k=0}^{\lfloor \frac{n}{2} \rfloor} \frac{n!}{2^k\, k!\, (n-2k)!} \end{aligned}$$

by counting according to the number k of transpositions in the cycle decomposition. For instance, if $n = 9$, one gets

$$\frac{9!}{0!\,9!} + \frac{1}{2}\,\frac{9!}{1!\,7!} + \frac{1}{4}\,\frac{9!}{2!\,5!} + \frac{1}{8}\,\frac{9!}{3!\,3!} + \frac{1}{16}\,\frac{9!}{4!\,1!} = 2620$$

involutions, and the same number of standard tableaux with this size.

3.3 Construction of the irreducible representations

In Chapter 2, we saw that the irreducible representations S^λ of the symmetric group of size n are labeled by integer partitions of size n; and at the beginning of this chapter, that the dimension of S^λ is the number of standard tableaux of shape λ. Therefore, there should be an action of the symmetric group on (standard) tableaux that allows the construction of the morphism $\rho^\lambda : \mathfrak{S}(n) \to \mathrm{GL}(S^\lambda)$. The goal of this section is to prove the following theorem, which indeed provides such a construction. If T is a tableau (not necessarily standard), denote

$$\Delta_T(x_1, \ldots, x_n) = \prod_{\text{columns } C \text{ of } T} \left(\prod_{i \text{ under } j \text{ in } C} x_i - x_j \right).$$

For instance,

$$T = \begin{array}{|c|c|c|c|} \hline 9 \\ \cline{1-1} 7 \\ \cline{1-3} 2 & 5 & 6 \\ \hline 1 & 3 & 4 & 8 \\ \hline \end{array} \quad \Rightarrow \quad \Delta_T(x_1, \ldots, x_9) = \begin{array}{c} (x_1-x_2)\times(x_1-x_7)\times(x_1-x_9)\times(x_2-x_7) \\ \times(x_2-x_9)\times(x_7-x_9)\times(x_3-x_5)\times(x_4-x_6) \end{array}.$$

Given a partition λ, we associate to it the vector subspace U^λ of $\mathbb{C}[x_1,\ldots,x_n]$ spanned by all the polynomials Δ_T, where T runs over the set of standard tableaux of shape λ. It is a vector space of homogeneous polynomials of degree

$$n(\lambda)=\sum_{j=1}^{\lambda_1}\binom{\lambda'_j}{2}=\sum_{i=1}^{\ell(\lambda)}(i-1)\,\lambda_i,$$

the identity between the two sums coming from the fact that both count the number of choices of two cells in the same row of λ.

Theorem 3.30 (Frobenius, Specht, Young). *Consider the left action of $\mathfrak{S}(n)$ on $\mathbb{C}[x_1,\ldots,x_n]$ by permutation of the variables (cf. the beginning of Section 2.2). Then, U^λ is stable by this action, and it is isomorphic as a representation of $\mathfrak{S}(n)$ to S^λ, the irreducible representation corresponding to the Schur function s_λ by the characteristic map. In particular, the character of $S^\lambda = U^\lambda$ is given by Theorem 3.10. Moreover,*

$$\{\Delta_T,\ T\in \mathrm{ST}(\lambda)\}$$

is a linear basis of this irreducible representation.

The beauty of this formulation of the classification of irreducible representations of symmetric groups is that it only involves *polynomials*, and it does not require the introduction of *tabloids* and *polytabloids*. Then, it is difficult to give a proof of Theorem 3.30 without using these notions; however, the combinatorics of symmetric functions will still allow us to skip an important part of the analysis of these objects.

▷ *Tableaux and permutation modules.*

The first step of the proof of Theorem 3.30 consists in building for every partition $\lambda\in\mathfrak{Y}(n)$ a (reducible) representation M^λ of $\mathfrak{S}(n)$ such that

$$\Psi(M^\lambda)=h_\lambda(X).$$

Consider the set $\mathrm{N}(\lambda)$ of all numberings of λ by labels in $[\![1,n]\!]$, each label appearing exactly once. The set of standard tableaux $\mathrm{ST}(\lambda)$ is a subset of $\mathrm{N}(\lambda)$, and the cardinality of $\mathrm{N}(\lambda)$ is obviously $n!$, the symmetric group acting freely transitively on $\mathrm{N}(\lambda)$ by permutation of the cells. We denote N^λ the complex vector space of dimension $n!$ that has for basis the numberings $T\in\mathrm{N}(\lambda)$. The symmetric group $\mathfrak{S}(n)$ acts on the left of N^λ by permutation of the cells. The **row subgroup** of a numbering T of λ is the subgroup $\mathfrak{R}(T)$ of $\mathfrak{S}(n)$ that is isomorphic to $\mathfrak{S}(\lambda_1)\times\mathfrak{S}(\lambda_2)\times\cdots\times\mathfrak{S}(\lambda_r)$, and consists in all the permutations of T that stabilize the rows. Similarly, the **column subgroup** of a numbering T of λ is the subgroup $\mathfrak{C}(T)$ of $\mathfrak{S}(n)$ that is isomorphic to $\mathfrak{S}(\lambda'_1)\times\mathfrak{S}(\lambda'_2)\times\cdots\times\mathfrak{S}(\lambda'_s)$, and consists in all the permutations of T that stabilize the columns.

Example. With

$$T = \begin{array}{|c|c|c} \cline{1-2} 3 & 4 & \multicolumn{1}{c}{} \\ \hline 1 & 2 & \multicolumn{1}{c|}{5} \\ \hline \end{array}$$

one has $\mathfrak{R}(T) = \mathfrak{S}(\{1,2,5\}) \times \mathfrak{S}(\{3,4\})$, and $\mathfrak{C}(T) = \mathfrak{S}(\{1,3\}) \times \mathfrak{S}(\{2,4\}) \times \mathfrak{S}(\{5\})$.

Two numberings T and T' are said **row equivalent** if T' can be obtained from T by a permutation in $\mathfrak{R}(T)$. This is an equivalence relation on numberings of shape λ, and each class of numberings, called **tabloid**, contains the same number of terms, namely,

$$\prod_{i=1}^{\ell(\lambda)} (\lambda_i)!\,.$$

The **permutation module** M^λ is the subspace of N^λ spanned by these classes of row equivalent numberings.

Example. If $\lambda = (2,1)$, then there are 3 tabloids with this shape (Figure 3.7):

$$\begin{array}{|c|c} \cline{1-1} 3 & \multicolumn{1}{c}{} \\ \hline 1 & \multicolumn{1}{c|}{2} \\ \hline \end{array} + \begin{array}{|c|c} \cline{1-1} 3 & \multicolumn{1}{c}{} \\ \hline 2 & \multicolumn{1}{c|}{1} \\ \hline \end{array} \quad ; \quad \begin{array}{|c|c} \cline{1-1} 2 & \multicolumn{1}{c}{} \\ \hline 1 & \multicolumn{1}{c|}{3} \\ \hline \end{array} + \begin{array}{|c|c} \cline{1-1} 2 & \multicolumn{1}{c}{} \\ \hline 3 & \multicolumn{1}{c|}{1} \\ \hline \end{array} \quad ; \quad \begin{array}{|c|c} \cline{1-1} 1 & \multicolumn{1}{c}{} \\ \hline 2 & \multicolumn{1}{c|}{3} \\ \hline \end{array} + \begin{array}{|c|c} \cline{1-1} 1 & \multicolumn{1}{c}{} \\ \hline 3 & \multicolumn{1}{c|}{2} \\ \hline \end{array}\,.$$

Figure 3.7
Tabloids with shape $(2,1)$.

It will be convenient to have a simple graphical representation of these tabloids; thus, we denote

$$\begin{array}{c} \hline 3 \\ \hline 1\ 2 \\ \hline \end{array} \quad ; \quad \begin{array}{c} \hline 2 \\ \hline 1\ 3 \\ \hline \end{array} \quad ; \quad \begin{array}{c} \hline 1 \\ \hline 2\ 3 \\ \hline \end{array}\,.$$

the tabloids corresponding to the previous formal sums of tableaux. With these notations, it is understood that

$$\begin{array}{c} \hline 3 \\ \hline 1\ 2 \\ \hline \end{array} = \begin{array}{c} \hline 3 \\ \hline 2\ 1 \\ \hline \end{array}\,,$$

and similarly for any permutation of the rows of a tableau in a given class.

Proposition 3.31. *The permutation module M^λ is a subrepresentation of N^λ, and it is isomorphic to*

$$\mathrm{Ind}_{\mathfrak{S}(\lambda)}^{\mathfrak{S}(n)}(1_{\mathfrak{S}(\lambda)}),$$

where $\mathfrak{S}(\lambda) = \mathfrak{S}(\lambda_1) \times \cdots \times \mathfrak{S}(\lambda_r)$, and $1_{\mathfrak{S}(\lambda)}$ is the trivial representation of dimension 1 of this group.

Proof. The map which sends a numbering to its reading word is an isomorphism of $\mathfrak{S}(n)$-representations between $\mathrm{N}(\lambda)$, and $\mathbb{C}\mathfrak{S}(n)$ viewed as a left module. In this setting, M^λ corresponds to the space of cosets $\mathbb{C}[\mathfrak{S}(n)/\mathfrak{S}(\lambda)]$, which is also the induced representation $\mathrm{Ind}_{\mathfrak{S}(\lambda)}^{\mathfrak{S}(n)}(1_{\mathfrak{S}(\lambda)})$. $\square$

By the Frobenius–Schur theorem 2.31, since $1_{\mathfrak{S}(\lambda_i)}$ corresponds to h_{λ_i},

$$M^\lambda = \mathrm{Ind}_{\mathfrak{S}(\lambda_1)\times\cdots\times\mathfrak{S}(\lambda_r)}^{\mathfrak{S}(n)}(1_{\mathfrak{S}(\lambda_1)} \boxtimes \cdots \boxtimes 1_{\mathfrak{S}(\lambda_r)})$$

corresponds to $\prod_{i=1}^r h_{\lambda_i} = h_\lambda$.

▷ *Polytabloids and Specht modules.*

If T is a numbering, we denote

$$[T] = \mathfrak{R}(T)\cdot T = \sum_{\sigma\in\mathfrak{R}(T)} \sigma\cdot T$$

the tabloid that it generates. Notice then that the action of $\mathfrak{S}(n)$ on the permutation module M^λ is given by $\sigma\cdot[T] = [\sigma\cdot T]$, because $\mathfrak{R}(\sigma\cdot T) = \sigma\,\mathfrak{R}(T)\,\sigma^{-1}$. The **polytabloid** associated to T is defined as

$$e_T = \sum_{\sigma\in\mathfrak{C}(T)} \varepsilon(\sigma)\,[\sigma\cdot T].$$

For instance,

$$T = \begin{array}{|c|c|c}\hline 4 & 5 & \multicolumn{1}{|c}{} \\ \hline 1 & 2 & \multicolumn{1}{|c|}{3} \\ \hline\end{array} \quad\Rightarrow\quad e_T = \begin{array}{c}\hline 4\ 5 \\ \hline 1\ 2\ 3 \\ \hline\end{array} - \begin{array}{c}\hline 1\ 5 \\ \hline 2\ 3\ 4 \\ \hline\end{array} - \begin{array}{c}\hline 2\ 4 \\ \hline 1\ 3\ 5 \\ \hline\end{array} + \begin{array}{c}\hline 1\ 2 \\ \hline 3\ 4\ 5 \\ \hline\end{array}.$$

Lemma 3.32. *For any numbering T and any permutation $\sigma\in\mathfrak{S}(n)$, $\sigma\cdot e_T = e_{\sigma\cdot T}$ in the permutation module.*

Proof. Notice that $\mathfrak{C}(\sigma\cdot T) = \sigma\,\mathfrak{C}(T)\,\sigma^{-1}$. Therefore,

$$\begin{aligned}\sigma\cdot e_T &= \sum_{\rho\in\mathfrak{C}(T)} \varepsilon(\rho)\,\sigma\cdot[\rho\cdot T] = \sum_{\rho\in\mathfrak{C}(T)} \varepsilon(\rho)\,[\sigma\rho\cdot T] \\ &= \sum_{\rho'\in\mathfrak{C}(\sigma\cdot T)} \varepsilon(\rho')\,[\rho'\sigma\cdot T] = e_{\sigma\cdot T}.\end{aligned}$$ □

An immediate consequence of the previous lemma is:

Definition 3.33. *The subspace T^λ of M^λ spanned by all polytabloids e_T with $T\in$ $\mathrm{N}(\lambda)$ is a subrepresentation for $\mathfrak{S}(n)$, called* **Specht module** *of label λ.*

The main result of this paragraph is **James' submodule theorem**:

Theorem 3.34 (James). *The Specht modules T^λ are non-isomorphic irreducible representations of $\mathfrak{S}(n)$. Moreover, if T^λ appears with positive multiplicity in the expansion in irreducibles of M^μ, then $\lambda \succeq \mu$ for the dominance order on integer partitions.*

We split the proof of Theorem 3.34 into several combinatorial lemmas. For any numbering $T \in \mathrm{N}(\lambda)$, we denote

$$\mathfrak{C}(T)^{\varepsilon} = \sum_{\sigma \in \mathfrak{C}(T)} \varepsilon(\sigma)\, \sigma,$$

viewed as an element of the group algebra $\mathbb{C}\mathfrak{S}(n)$. By definition, $e_T = \mathfrak{C}(T)^{\varepsilon} \cdot [T]$.

Lemma 3.35. *Let T and U be two numberings of shape λ and μ. If $\mathfrak{C}(T)^{\varepsilon} \cdot [U] \neq 0$, then $\lambda \succeq \mu$. Moreover, if $\lambda = \mu$, then $\mathfrak{C}(T)^{\varepsilon} \cdot [U] = \pm e_T$.*

Proof. The lemma relies on the following remark: if a, b are two elements in the same row of U, then they cannot be in the same column of T. Indeed, otherwise, the transposition (a, b) belongs to $\mathfrak{C}(T)$, and by looking at cosets in $\mathfrak{C}(T)/(\mathrm{id}, (a, b))$, this leads to a factorization

$$\mathfrak{C}(T)^{\varepsilon} = D\,(\mathrm{id} - (a, b))$$

in the symmetric group algebra. Then,

$$\mathfrak{C}(T)^{\varepsilon} \cdot [U] = D\,(\mathrm{id} - (a, b)) \cdot [U] = D \cdot ([U] - [U]) = 0.$$

We claim that this property implies that $\lambda \succeq \mu$. Indeed, consider the μ_1 entries in the first row of U. They must be placed in different columns of T, so T has at least μ_1 columns, and $\lambda_1 \geq \mu_1$. Then, consider the μ_2 entries in the second row of U. They must also be placed in different columns of T, and the resulting diagram (after placement of the entries of the two first rows of U) has now at most 2 entries by column. Therefore, $\lambda_1 + \lambda_2 \geq \mu_1 + \mu_2$ (slide the filled cells of T to the bottom of λ to get the inequality). By induction, one gets $\lambda_1 + \cdots + \lambda_i \geq \mu_1 + \cdots + \mu_i$ for all i, that is to say that $\lambda \succeq \mu$.

Suppose now that $\lambda = \mu$. Then, by using the same argument as before, one can choose $\sigma \in \mathfrak{R}(U)$ to rearrange the entries of U such that x is in the i-th column of $\sigma \cdot U$ if and only if x is in the i-th column of T. This implies that there exists $\rho \in \mathfrak{C}(T)$ such that

$$[U] = [\rho \cdot T].$$

Then, one computes easily

$$\mathfrak{C}(T)^{\varepsilon} \cdot [U] = \sum_{\sigma \in \mathfrak{C}(T)} \varepsilon(\sigma)\,[\sigma\rho \cdot T] = \varepsilon(\rho)\,\mathfrak{C}(T)^{\varepsilon} \cdot [T] = \varepsilon(\rho)\, e_T. \qquad \Box$$

We endow M^{λ} with a scalar product such that the tabloids form an orthonormal basis. Notice then that the operators $\mathfrak{C}(T)^{\varepsilon}$ are self-adjoint:

$$\begin{aligned}\langle \mathfrak{C}(T)^{\varepsilon} \cdot [U] \,|\, [V] \rangle &= \sum_{\sigma \in \mathfrak{C}(T)^{\varepsilon}} \varepsilon(\sigma)\, \langle [\sigma \cdot U] \,|\, [V] \rangle \\ &= \sum_{\sigma \in \mathfrak{C}(T)^{\varepsilon}} \varepsilon(\sigma^{-1})\, \left\langle [U] \,\middle|\, [\sigma^{-1} \cdot V] \right\rangle = \langle [U] \,|\, \mathfrak{C}(T)^{\varepsilon} \cdot [V] \rangle.\end{aligned}$$

Lemma 3.36. *If P is a submodule of M^λ, then either $T^\lambda \subset P$, or $P \subset (T^\lambda)^\perp$.*

Proof. Let $x \in M^\lambda$, which we decompose as a sum of tabloids $x = \sum_i x_i\,[U_i]$. By the second part of Lemma 3.35, for any fixed tableau T of shape λ,

$$\mathfrak{C}(T)^\varepsilon \cdot x = \sum_i x_i\, \mathfrak{C}(T)^\varepsilon \cdot [U_i] = \left(\sum_i \pm_i x_i\right) e_T$$

is proportional to e_T. We can then distinguish two cases:

1. Suppose that there exists some $x \in P$ such that $\mathfrak{C}(T)^\varepsilon \cdot x = \alpha e_T$ with $\alpha \neq 0$. Then, since P is a submodule, $e_{\sigma\cdot T} = \sigma \cdot e_T$ belongs to P for every σ, so $T^\lambda \subset P$.

2. On the opposite, suppose that $\mathfrak{C}(T)^\varepsilon \cdot x = 0$ for every x in P. Then, for every permutation σ and every $x \in P$,

$$\begin{aligned}\left\langle x \,\middle|\, e_{\sigma\cdot T}\right\rangle &= \langle x \mid \sigma \cdot e_T\rangle = \left\langle \sigma^{-1}\cdot x \,\middle|\, e_T\right\rangle \\ &= \left\langle \mathfrak{C}(T)^\varepsilon \cdot \sigma^{-1} \cdot x \,\middle|\, [T]\right\rangle = \langle 0 \mid [T]\rangle = 0\end{aligned}$$

   since $\sigma^{-1} \cdot x \in P$. Hence, $P \subset (T^\lambda)^\perp$. □

Proof of Theorem 3.34. The previous lemma implies readily that T^λ is irreducible: if $P \subset T^\lambda$ is a submodule, then either $T^\lambda \subset P$ and $T^\lambda = P$, or $P \subset (T^\lambda)^\perp$ and $P = 0$. Suppose now that T^λ appears as a component of M^μ. This is equivalent to the existence of a non-zero morphism of representations $\phi : T^\lambda \to M^\mu$. Since $M^\lambda = T^\lambda \oplus (T^\lambda)^\perp$, one can extend ϕ in a non-zero morphism $M^\lambda \to M^\mu$, by setting $\phi((T^\lambda)^\perp) = 0$. Then, given a numbering T such that $\phi(e_T) \neq 0$,

$$\phi(e_T) = \phi(\mathfrak{C}(T)^\varepsilon \cdot [T]) = \mathfrak{C}(T)^\varepsilon \cdot \phi([T]) = \mathfrak{C}(T)^\varepsilon \cdot \left(\sum_i x_i\,[U_i]\right),$$

where the U_i's are numberings of shape μ. By the first part of Lemma 3.35, since this expression is non-zero, $\lambda \succeq \mu$.

Finally, suppose that T^λ and T^μ are isomorphic. Then, T^λ appears as a component of M^μ, and T^μ appears as a component of M^λ, so $\lambda \succeq \mu$ and $\mu \succeq \lambda$; hence, $\lambda = \mu$. Since we get the right number of irreducible representations, $(T^\lambda)_{\lambda \in \mathfrak{Y}(n)}$ is a complete collection of irreducible representations of $\mathfrak{S}(n)$. □

▷ *Characters and bases of Specht modules.*

We can now relate the construction of the irreducible representation of $\mathfrak{S}(n)$ with tabloids and polytabloids to the implicit construction of Chapter 2:

Theorem 3.37. *The Specht module T^λ is isomorphic to the irreducible module S^λ constructed in Chapter 2.*

Lemma 3.38. *The multiplicity of T^λ as a component of M^λ is 1.*

Proof. Let ϕ be a morphism of representations from T^λ to M^λ. Pursuing the computation performed during the proof of Theorem 3.34, we see that

$$\phi(e_T) = \mathfrak{C}(T)^\varepsilon \cdot \left(\sum_i x_i\,[U_i]\right),$$

where the U_i's are numberings of shape λ. By the second part of Lemma 3.35, this is equal to

$$\left(\sum_i \pm x_i\right) e_T = \alpha e_T,$$

so ϕ is the multiplication by a scalar, and $\dim \mathrm{Hom}_{\mathfrak{S}(n)}(T^\lambda, M^\lambda) = 1$. □

Proof of Theorem 3.37. Using Theorem 3.2 and the fact that $(m_\mu)_{\mu\in\mathfrak{Y}}$ and $(h_\mu)_{\mu\in\mathfrak{Y}}$ are dual bases of Sym, one sees that $K_{\lambda\mu}$, the number of semistandard tableaux of shape λ and weight μ, is equal to $\langle s_\lambda \mid h_\mu \rangle$, and therefore, that

$$h_\mu(X) = \sum_\lambda K_{\lambda\mu}\, s_\lambda(X).$$

In this decomposition, $K_{\lambda\mu} \neq 0$ if $\lambda \succeq \mu$. Indeed, given a semistandard tableau T of shape λ and weight μ, one can apply to it the discussion of Lemma 3.35 with U tableau of shape μ containing μ_1 entries 1 on its first row, μ_2 entries 2 on its second row, etc. Notice moreover that $K_{\mu\mu} = 1$, the unique semistandard tableau of shape and weight μ being precisely given by this construction.

Then, consider a linear extension $\geqslant$ of the dominance order $\succeq$ on $\mathfrak{Y}(n)$, that is to say, a total order $\geqslant$ such that $\lambda \succeq \mu \Rightarrow \lambda \geqslant \mu$. One can always construct such a total order by adding relations to the dominance order. In this setting, the previous argument shows that with respect to this total order and to the Hall scalar product, $(s_\mu)_{\mu\in\mathfrak{Y}(n)}$ is the Gram–Schmidt orthonormal basis obtained from the basis $(h_\mu)_{\mu\in\mathfrak{Y}(n)}$. However, combining the previous lemma with Theorem 3.34, one sees that in $\mathrm{R}_\mathbb{R}(\mathfrak{S}(n)) = \mathbb{R} \otimes_\mathbb{Z} \mathrm{R}_0(\mathfrak{S}(n))$, $(T^\mu)_{\mu\in\mathfrak{Y}(n)}$ is the Gram–Schmidt orthonormal basis obtained from the basis $(M^\mu)_{\mu\in\mathfrak{Y}(n)}$. Since the restriction of the characteristic map

$$\Psi_n : \mathrm{R}_\mathbb{R}(\mathfrak{S}(n)) \to \mathrm{Span}_\mathbb{R}(\{h_\mu\}_{\mu\in\mathfrak{Y}(n)})$$

is an isometry with $\Psi_n(S^\mu) = s_\mu(X)$, and since $\Psi_n(M^\mu) = h_\mu(X)$, by unicity of the Gram–Schmidt orthonormalization, $\Psi_n(T^\mu) = s_\mu(X)$, hence, $S^\mu = T^\mu$. □

From now on, we shall denote S^λ instead of T^λ the Specht module of label λ. By the previous theorem and Theorem 3.10, we know how to compute its character. In order to relate S^λ to the space of polynomials U^λ of Theorem 3.34, consider the following construction. To an arbitrary numbering

$$T = \begin{array}{|c|c|c|c|c|} \hline n_{3,1} & n_{3,\lambda_3} \\ \cline{1-3} n_{2,1} & n_{2,2} & n_{2,\lambda_2} \\ \hline n_{1,1} & n_{1,2} & \cdots & \cdots & n_{1,\lambda_1} \\ \hline \end{array}$$

of shape $\lambda = (\lambda_1, \ldots, \lambda_r)$, we associate the monomial

$$X_T = (x_{n_{1,1}} x_{n_{1,2}} \cdots x_{n_{1,\lambda_1}})^0 \, (x_{n_{2,1}} x_{n_{2,2}} \cdots x_{n_{2,\lambda_2}})^1 \cdots (x_{n_{r,1}} x_{n_{r,2}} \cdots x_{n_{r,\lambda_r}})^{r-1}.$$

Notice that this monomial has total degree $n(\lambda) = \sum_{i=1}^{\ell} (i-1)\lambda_i$. Obviously, $X_{\sigma \cdot T} = \sigma \cdot X_T$, so X extends to a morphism of representations

$$X : N^\lambda \to \mathbb{C}[x_1, \ldots, x_n],$$

with values in the space of homogeneous polynomials of degree $n(\lambda)$. By construction, if T is row equivalent to U, then $X_T = X_U$. Therefore, X factors through M^λ, and gives a morphism of representations

$$[X] : M^\lambda \to \mathbb{C}[x_1, \ldots, x_n],$$

with $[X]([T]) = \left(\prod_{i=1}^{\ell(\lambda)} (\lambda_i)!\right) X_T = \lambda! \, X_T$ for any numbering T. Then, notice that

$$X\,(\mathfrak{C}(T)^\varepsilon \cdot T) = \Delta_T(x_1, \ldots, x_N),$$

because X factorizes over the columns of the tableau, and the alternating sums give rise to Vandermonde determinants. As a consequence,

$$[X](e_T) = X(e_T) = \mathfrak{C}(T)^\varepsilon \cdot X([T]) = \lambda! \; \mathfrak{C}(T)^\varepsilon \cdot X_T = \lambda! \; \Delta_T.$$

Thus, if U^λ is the subspace of $\mathbb{C}[x_1, \ldots, x_N]$ spanned linearly by polynomials $\Delta_T(x_1, \ldots, x_N)$, where T runs over numberings of shape λ, then U^λ is the nonzero image by the morphism of representations $[X]$ of the irreducible submodule $S^\lambda \subset M^\lambda$, so U^λ is isomorphic to S^λ.

To end the proof of Theorem 3.30, it remains to see that one can restrict oneself to standard tableaux T, that is to say that:

Proposition 3.39. *For any $\lambda \in \mathfrak{Y}(n)$, $(e_T)_{T \in \mathrm{ST}(\lambda)}$ is a linear basis of S^λ.*

By Proposition 3.8, since the family $(e_T)_{T \in \mathrm{ST}(\lambda)}$ has the right cardinality, it suffices to show that it is independent. To this purpose, we introduce the dominance order on tabloids. If $[T]$ is a tabloid of shape $(\lambda_1, \ldots, \lambda_r)$, and if $i \in [\![1, n]\!]$, denote $[T]_i$ the sequence

$$\big(\mathrm{card}\,(\mathrm{row}_1(T) \cap [\![1,i]\!]), \mathrm{card}\,(\mathrm{row}_2(T) \cap [\![1,i]\!]), \ldots, \mathrm{card}\,(\mathrm{row}_r(T) \cap [\![1,i]\!])\big).$$

For instance, with the tabloid

$$\begin{array}{c} \hline 2\;3 \\ \hline 1\;4 \\ \hline \end{array}\,,$$

one has $[T]_1 = (1,0)$, $[T]_2 = (1,1)$, $[T]_3 = (1,2)$ and $[T]_4 = (2,2)$. Though these sequences are no partitions, the dominance order still makes sense for them, and we shall say that a tabloid $[T]$ dominates $[U]$ if $[T]_i \succeq [U]_i$ for all $i \geq 1$. As an example,

$$\begin{array}{c} \hline 2\;3 \\ \hline 1\;4 \\ \hline \end{array} \succeq \begin{array}{c} \hline 1\;3 \\ \hline 2\;4 \\ \hline \end{array}\,.$$

Given two tabloids, if $[T]$ dominates $[U]$, then $[T]_n$ dominates $[U]_n$, so this implies dominance for the shapes of the tabloids.

Lemma 3.40. *If $[T]$ is the tabloid of a standard tableau, and if $[U]$ appears in e_T, then $[T] \succeq [U]$.*

Proof. The proof relies on a remark analogous to the remark of Lemma 3.35: if $a < b$ and a is in a row above the row of b in a tabloid $[U]$, then $(a,b)\cdot[U] \succeq [U]$. Indeed, if a is in row k and b is in row $j < k$, then the transposition (a,b) only modifies the j-th and the k-th coordinates of $[U]_a, [U]_{a+1}, \ldots, [U]_{b-1}$: it adds one to the j-th coordinate and substract one to the k-th coordinate. Clearly this makes $(a,b)\cdot[U]_i$ dominate $[U]_i$ for all i, so $(a,b)\cdot[U] \succeq [U]$.

Now, let $U = \sigma \cdot T$ be a numbering with $\sigma \in \mathfrak{C}(T)$, such that $[U]$ appears in e_T. To prove that $[T] \succeq [U]$, we reason by induction on the number of column inversions of U, that is to say the number of pairs (a,b) with $a < b$ appearing in the same column, and a in a row above the row of b. If U has no column inversion, then $U = T$ and the proof is done. Otherwise, let (a,b) be a column inversion. Then, $(a,b) \in \mathfrak{C}(T)$, so $(a,b)\cdot[U]$ appears in e_T, and by the previous discussion and the induction hypothesis,

$$[T] \succeq (a,b)\cdot[U] \succeq [U]. \qquad \square$$

Proof of Proposition 3.39. Consider a linear combination $\sum_{T\in\mathrm{ST}(\lambda)} c_T\, e_T = 0$, and assume that some c_T are non-zero. We choose a standard tableau T such that $c_T \neq 0$, and such that $[T]$ is maximal among

$$\{[U],\ U \in \mathrm{ST}(\lambda),\ c_U \neq 0\}.$$

By the previous Lemma, $[T]$ only appears in $c_T\, e_T$, so $c_T = 0$, hence a contradiction. $\square$

Thus, we have obtained a construction of the Specht module S^λ as a homogeneous submodule of the natural graded representation of $\mathfrak{S}(n)$ over $\mathbb{C}[x_1,\ldots,x_N]$, together with a distinguished basis labeled by standard tableaux of shape λ. For other explicit "polynomial" constructions of the Specht modules, we refer to the notes at the end of this chapter.

3.4 The hook-length formula

If $\lambda \in \mathfrak{Y}(n)$, we saw that the dimension of the irreducible representation S^λ is the number of standard tableaux of shape λ. In this section, we explain how to compute this number in terms of $\lambda = (\lambda_1, \ldots, \lambda_r)$. We shall actually give *two* explicit formulas for $\dim \lambda = \dim S^\lambda$.

▷ *Hook lengths and the probabilistic proof of the Frame–Robinson–Thrall formula.*

If $\square$ is a cell in a Young diagram, its **hook length** $h(\square)$ is equal to the number of cells to its right, plus the number of cells above it, plus one. For instance, the Young diagram $(4, 3, 1, 1)$ has its hook lengths calculated in Figure 3.8.

1			
2			
5	2	1	
7	4	3	1

Figure 3.8
Hook lengths of the cells of the Young diagram $(4, 3, 1, 1)$.

The **hook length formula** states that:

Theorem 3.41 (Frame–Robinson–Thrall). *For any integer partition $\lambda \in \mathfrak{Y}(n)$,*

$$\dim \lambda = \frac{n!}{\prod_{\square \in \lambda} h(\square)}.$$

Example. The hook lengths of the partition $(3, 2)$ are (from top to bottom and left to right) $2, 1, 4, 3, 1$. Therefore,

$$\dim \lambda = \frac{120}{2 \times 1 \times 4 \times 3 \times 1} = 5,$$

which agrees with the enumeration of standard tableaux of shape $(3, 2)$ on page 106.

Denote $f(\lambda)$ the ratio involving hook lengths in Theorem 3.41. By the branching rules for representations of symmetric groups, it suffices to show that for any partition Λ,

$$f(\Lambda) = \sum_{\lambda \nearrow \Lambda} f(\lambda).$$

There is a clever probabilistic proof of this fact due to Greene, Nijenhuis and Wilf. Fix Λ of size n, and consider a random cell $\square_1$ of Λ that is chosen with uniform probability $\frac{1}{n}$. If $\square_1$ is a cell in the top right corner of Λ, meaning that $h(\square_1) = 1$, then removing this cell yields a random partition λ with $\lambda \nearrow \Lambda$. Otherwise, we choose a random cell $\square_2$ of Λ in the hook of $\square_1$ and with uniform probability $\frac{1}{h(\square_1)-1}$. If $\square_2$ is in the top right corner, we remove this cell to obtain a new partition λ with $\lambda \nearrow \Lambda$, and otherwise, we continue the process and choose $\square_3$ in the hook of $\square_2$, then $\square_4$ in the hook of $\square_3$, etc., until one obtains a cell that one can remove. Since the hook length of the random cells decreases at each step, the algorithm always terminates, and we denote $p(\lambda)$ the probability to obtain in the end the integer partition $\lambda \nearrow \Lambda$. For instance, with $\Lambda = (4,3,1,1)$, a possible succession of cells is

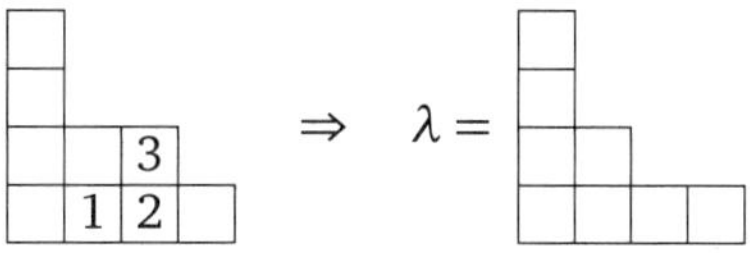

$$\begin{array}{|c|c|c|c|} \cline{1-1} \ & \multicolumn{3}{c}{} \\ \cline{1-1} \ & \multicolumn{3}{c}{} \\ \cline{1-3} \ & \ & 3 & \multicolumn{1}{c}{} \\ \hline \ & 1 & 2 & \ \\ \hline \end{array} \quad \Rightarrow \quad \lambda = \begin{array}{|c|c|c|c|} \cline{1-1} \ & \multicolumn{3}{c}{} \\ \cline{1-1} \ & \multicolumn{3}{c}{} \\ \cline{1-2} \ & \ & \multicolumn{2}{c}{} \\ \hline \ & \ & \ & \ \\ \hline \end{array} .$$

Proposition 3.42. *The probability $p(\lambda)$ is equal to $\frac{f(\lambda)}{f(\Lambda)}$.*

This immediately implies the hook-length formula, since then

$$1 = \sum_{\lambda \nearrow \Lambda} p(\lambda) = \sum_{\lambda \nearrow \Lambda} \frac{f(\lambda)}{f(\Lambda)}.$$

In the proof we shall need the notions of **leg length** and **arm length** of a cell in a Young diagram: they are respectively the number of cells above and to the right of the cell, so that

$$h(\square) = l(\square) + a(\square) + 1.$$

Proof. Let (y,z) be the coordinates of a cell in the top right corner of Λ. We have

$$\frac{f(\lambda)}{f(\Lambda)} = \frac{1}{n} \frac{\prod_{\square \in \Lambda} h(\square, \Lambda)}{\prod_{\square \in \lambda} h(\square, \lambda)},$$

and the only cells in Λ that have a different hook length than in λ are those in the y-th row or in the z-th column. Thus,

$$\begin{aligned} \frac{f(\lambda)}{f(\Lambda)} &= \frac{1}{n} \left(\prod_{i=1}^{y-1} \frac{h(i,z)}{h(i,z)-1} \right) \left(\prod_{j=1}^{z-1} \frac{h(y,j)}{h(y,j)-1} \right) \\ &= \frac{1}{n} \left(\prod_{i=1}^{y-1} 1 + \frac{1}{h(i,z)-1} \right) \left(\prod_{j=1}^{z-1} 1 + \frac{1}{h(y,j)-1} \right) \\ &= \frac{1}{n} \sum_{\substack{I \subset [\![1,y-1]\!] \\ J \subset [\![1,z-1]\!]}} \left(\prod_{i \in I} \frac{1}{h(i,z)-1} \right) \left(\prod_{j \in J} \frac{1}{h(y,j)-1} \right). \end{aligned}$$

where the hook lengths are denoted with respect to Λ.

On the other hand, consider a sequence of cells

$$(c,d)=(c_1,d_1)\to(c_2,d_2)\to\cdots\to(c_{r-1},d_{r-1})\to(c_r,d_r)=(y,z),$$

with each (c_k,d_k) in the hook of (c_{k-1},d_{k-1}), and that has been obtained by the previously described random process. The probability of this sequence is

$$p(c_1,\ldots,c_r;d_1,\ldots,d_r)=\frac{1}{n}\,\frac{1}{h(c_1,d_1)-1}\,\frac{1}{h(c_2,d_2)-1}\cdots\frac{1}{h(c_{r-1},d_{r-1})-1}.$$

Let $I=\{c_1,\ldots,c_{r-1}\}$ and $J=\{c_1,\ldots,c_{r-1}\}$. Beware that there are usually repetitions in the sequences (c_k) and (d_k), so I and J can have cardinality smaller than r, and different cardinalities. We claim that the probability $p(I,J;y,z)$ that a sequence of cells starts with (c,d), ends with (y,z) and gives transversal sets I and J is

$$p(I,J;y,z)=\frac{1}{n}\left(\prod_{i\in I}\frac{1}{h(i,z)-1}\right)\left(\prod_{j\in J}\frac{1}{h(y,j)-1}\right).$$

This will end the proof by summing over sets $I\subset[\![1,y-1]\!]$ and $J\subset[\![1,z-1]\!]$, with $c=\min(I\cup\{y\})$ and $d=\min(J\cup\{z\})$. We reason by induction on $s=|I|+|J|\leq 2r$. If $s=0$, then this is trivial since

$$p(\emptyset,\emptyset;y,z)=\frac{1}{n}.$$

Suppose the result to be true up to order $s-1$. Then, by induction, with $c=\min I$ and $d=\min J$, one has

$$p(I\setminus\{c\},J;y,z)=\frac{1}{n}\left(\prod_{i\in I\setminus\{c\}}\frac{1}{h(i,z)-1}\right)\left(\prod_{j\in J}\frac{1}{h(y,j)-1}\right)=(h(c,z)-1)Q$$

$$p(I,J\setminus\{d\};y,z)=\frac{1}{n}\left(\prod_{i\in I}\frac{1}{h(i,z)-1}\right)\left(\prod_{j\in J\setminus\{d\}}\frac{1}{h(y,j)-1}\right)=(h(y,d)-1)Q,$$

where Q is the quantity expected for the value of $p(I,J;y,z)$. However,

$$\begin{aligned}p(I,J;y,z)&=\frac{1}{h(c,d)-1}\,(p(I\setminus\{c\},J;y,z)+p(I,J\setminus\{d\};y,z))\\&=\frac{h(c,z)+h(y,d)-2}{h(c,d)-1}\,Q,\end{aligned}$$

and

$$\begin{aligned}h(c,z)+h(y,d)-2&=l(c,z)+a(c,z)+l(y,d)+a(y,d)\\&=(l(c,z)+l(y,d))+(a(c,z)+a(y,d))\\&=l(c,d)+a(c,d)=h(c,d)-1.\end{aligned}$$

Hence, by induction, the formula for $p(I,J;y,z)$ is always true, and the proof is done. □

▷ *Determinantal formula for the dimensions.*

There is another formula for $\dim \lambda$ that is similar to the hook length formula, and involves a Vandermonde determinant. Set $\lambda = (\lambda_1, \lambda_2, \ldots, \lambda_n)$, and $\mu = \lambda + \rho = (\lambda_i + n - i)_{i \in [\![1,n]\!]}$. We denote $\mu! = \prod_{i=1}^{n} (\mu_i)!$, and $\Delta(\mu) = \prod_{1 \leq i < j \leq n} (\mu_i - \mu_j)$.

Proposition 3.43. *For any partition $\lambda \in \mathfrak{Y}(n)$, if $\mu = \lambda + \rho$, then*

$$\dim \lambda = \frac{n!}{\mu!}\, \Delta(\mu).$$

Example. If $\lambda = (3,2)$, then $\mu = (7,5,2,1,0)$, and again

$$\begin{aligned}\frac{n!}{\mu!}\, \Delta(\mu) &= \frac{5!}{7!\,5!\,2!\,1\,0!}\,(7-5)(7-2)(7-1)(7-0) \\ &\quad \times (5-2)(5-1)(5-0)(2-1)(2-0)(1-0) \\ &= 5.\end{aligned}$$

Proof. By the Frobenius formula 2.32,

$$(p_1(X))^n = \sum_{\lambda \in \mathfrak{Y}(n)} \mathrm{ch}^{\lambda}(1^n)\, s_\lambda(X) = \sum_{\lambda \in \mathfrak{Y}(n)} (\dim \lambda)\, s_\lambda(X),$$

so $\dim \lambda$ is the coefficient of $s_\lambda(X)$ in $(p_1(X))^n$, or, by taking an alphabet of size n, the coefficient of $a_{\lambda+\rho}(x_1, \ldots, x_n)$ in the antisymmetric polynomial $(p_1(x_1, \ldots, x_n))^n\, a_\rho(x_1, \ldots, x_n)$. If one looks at monomials, then this is also the coefficient of $x^{\lambda+\rho} = x^\mu$. Hence,

$$\dim \lambda = [x^\mu]\left(\left(\sum_{i=1}^{n} x_i\right)^n \sum_{\sigma \in \mathfrak{S}(n)} \varepsilon(\sigma)\, x^{\rho \cdot \sigma}\right).$$

Fix a permutation σ. One has

$$[x^\mu]\left(x^{\rho\cdot\sigma}\left(\sum_{i=1}^{n} x_i\right)^n\right) = \frac{n!}{\prod_{i=1}^{n} (\mu_i - n + \sigma(i))!},$$

so

$$\begin{aligned}\dim \lambda &= n! \sum_{\sigma \in \mathfrak{S}(n)} \frac{\varepsilon(\sigma)}{\prod_{i=1}^{n} (\mu_i - n + \sigma(i))!} = n!\, \det\left(\frac{1}{(\mu_i - n + j)!}\right)_{1 \leq i,j \leq n} \\ &= \frac{n!}{\mu!}\, \det(\mu_i(\mu_i - 1) \cdots (\mu_i - n + j - 1))_{1 \leq i,j \leq n} \\ &= \frac{n!}{\mu!}\, \det\left(\mu_i^{n-j}\right)_{1 \leq i,j \leq n} = \frac{n!}{\mu!}\, \Delta(\mu).\end{aligned}$$

□

Remark. It is not entirely evident that the hook length formula and the determinantal formula are equivalent. This can be proven by clever manipulations of the matrix $(\frac{1}{\mu_i - n + j})_{i,j} = (\frac{1}{\lambda_i - i + j})_{i,j}$; see the references hereafter.

Notes and references

Following again [Mac95, Chapter 1], we proved all the combinatorial results of Section 3.1 by using only symmetric functions. The use of symmetric functions also allows one to shorten a lot the study of Specht modules, by using the argument of Gram–Schmidt orthonormal bases in the proof of Theorem 3.30. In particular, one does not need to introduce the straightening algorithm (see [Sag01, §2.6]), or to prove Young's rule for the coefficients of the expansion of the permutation module M^μ in Specht modules S^λ (loc. cit., §2.9-2.10).

In fact, there exists a way to construct Specht modules with polynomials, but this time without ever using tabloids or polytabloids, see [Las04, Chapter 3]. Let us state this construction without proof. We call coinvariant space the quotient of the ring of polynomials $\mathbb{C}[x_1,\ldots,x_n]$ by the ideal generated by symmetric polynomials in $\mathrm{Sym}^{(n)}$ without constant term:

$$\mathrm{Coinv}^{(n)} = \mathbb{C}[x_1,\ldots,x_n]/(\mathrm{Sym}^{(n),*}),$$

with $\mathrm{Sym}^{(n),*} = \{P \in \mathrm{Sym}^{(n)},\ P(0,\ldots,0) = 0\}$. It can be shown that $\mathrm{Coinv}^{(n)}$ has its dimension equal to $n!$ (see Corollary 4.73 in the next chapter), and that it is isomorphic to $\mathbb{C}\mathfrak{S}(n)$ as a $\mathfrak{S}(n)$-module. Set

$$e(\lambda) = (0^{\lambda_1}, (\lambda_1)^{\lambda_2}, (\lambda_1+\lambda_2)^{\lambda_3}, \ldots, (\lambda_1+\cdots+\lambda_{r-1})^{\lambda_r}),$$

where exponents denote multiplicities. Then, the orbit of the monomial $x^{e(\lambda)}$ in the ring of coinvariants spans linearly a space that is isomorphic to the Specht module S^λ.

A use of symmetric functions that is not presented in this chapter is the Littlewood–Richardson rule for the structure coefficients $c^\nu_{\lambda\mu}$ in the product of Schur functions

$$s_\lambda(X)\, s_\mu(X) = \sum_{|\nu|=|\lambda|+|\mu|} c^\nu_{\lambda\mu}\, s_\nu(X).$$

Notice that the Pieri rules yield these coefficients in certain special cases. In general, the Littlewood–Richardson coefficients describe the tensor products of irreducible polynomial representations of $\mathrm{GL}(N,\mathbb{C})$, and the induction of irreducible representations from $\mathfrak{S}(m) \times \mathfrak{S}(n)$ to $\mathfrak{S}(m+n)$. We refer to [Ful97, Chapter 5] for a combinatorial interpretation of the Littlewood–Richardson coefficients in terms of skew tableaux whose reading words are reverse lattice words. Later in this book, we shall provide a way to compute $c^\nu_{\lambda\mu}$, in the setting of the Littelmann path model for the weights of the representations of $\mathrm{GL}(N,\mathbb{C})$.

Our presentation of the RSK algorithm is extremely similar to the one of [Ful97], though not done in the same order. Indeed, there is essentially one way to prove the combinatorial properties of RSK and of the Schützenberger slidings. We also refer to [LLT02], which contains in particular another accessible proof of the symmetry theorem for the RSK algorithm. Our proof of the symmetry theorem for permutations comes from the original paper by Schützenberger [Sch63]. For the general symmetry theorem for RSK on two-line arrays, and a geometric version of the RSK algorithm, we refer to the classical paper by Viennot [Vie77]. One thing that we did not discuss and prove for RSK is that the *columns* of the shape $\lambda(w)$ of the two tableaux $P(w)$ and $Q(w)$ associated to a word w correspond to the longest strictly decreasing subsequences in w. Hence, $\lambda_1'(w)$ is the length of a longest decreasing subsequence, and more generally, $\lambda_1'(w) + \cdots + \lambda_k'(w)$ is the maximal total length of k disjoint strictly decreasing subwords of w. The proof is the same as for the Greene invariants, computing the quantities for reading words of semistandard tableaux, and then showing that they are invariant by Knuth equivalence (see again [Ful97]).

Finally, for the equivalence between the hook length formula and the determinantal formula for $\dim \lambda$, we refer to the original proof due to Frame, Robinson and Thrall [FRT54]. The probabilistic proof comes from [GNW79].

Part II

Hecke algebras and their representations

4

Hecke algebras and the Brauer–Cartan theory

Let k be a field, and G be an algebra over k. A **deformation** of G is given by a commutative ring A, an algebra H over A, and a morphism of rings $\theta : A \to k$ such that

$$k \otimes_A H = G,$$

where the structure of A-modules on k is given by $a \cdot x = \theta(a)x$. In particular, a one-parameter deformation of G is a deformation defined over $A = k[q]$. In this case, notice that every morphism of algebra $\theta : k[q] \to k$ is a specialization of the parameter q:

$$\theta(P(q)) = P(z) \quad \text{for some } z \in k.$$

We denote θ_z the specialization of $k[q]$ given by a value $z \in k$. Thus, a one-parameter deformation of a k-algebra G is an algebra H over $k[q]$, such that for some parameter $z \in k$, the tensor product $H_z = k \otimes_A H$ associated to the morphism is equal to G. One then has a whole family of algebras over k, namely, all the other tensor products H_y associated to values $y \in k$. In this framework, it is natural to study how the representation theory of H_y varies with y. More precisely, one can expect that generically, H_y and $H_z = G$ have the same representation theory: for instance, if H_z is semisimple and isomorphic to a direct sum of matrix algebras $\bigoplus_{\lambda \in \widehat{G}} \mathrm{M}(d_\lambda, k)$, then for y "close to z," one can expect that H_y is also a semisimple algebra, isomorphic to a sum of matrix algebras with the same numerical invariants d_λ.

In this chapter, we introduce the Hecke algebra of the symmetric group $\mathfrak{S}(n)$, which is a one-parameter deformation $\mathfrak{H}(n, \mathbb{C}[q])$ of the group algebra $\mathbb{C}\mathfrak{S}(n)$:

$$(\mathfrak{H}(n, \mathbb{C}[q]))_1 = \mathbb{C}\mathfrak{S}(n).$$

We construct this algebra in Section 4.1, by introducing the formal parameter q in the Coxeter presentation of $\mathfrak{S}(n)$. The study of the family of algebras $((\mathfrak{H}(n, \mathbb{C}[q]))_z)_{z \in \mathbb{C}}$ then requires some prerequisites of representation theory of general algebras. We devote Section 4.2 to this topic, and explain there how to go beyond the framework of semisimple algebras of Section 1.4. We then present in Section 4.3 the Brauer–Cartan deformation theory, which relates the representation theory of

$$\text{the family of algebras } ((\mathfrak{H}(n, \mathbb{C}[q]))_z)_{z \in \mathbb{C}},$$

and the representation theory of

$$\text{the generic Hecke algebra } \mathbb{C}(q) \otimes_{\mathbb{C}[q]} \mathfrak{H}(n, \mathbb{C}[q]),$$

that is defined over the field of rational functions $\mathbb{C}(q)$. This can serve as an introduction to the so-called *modular* representation theory. In Section 4.4, we apply this theory to the Hecke algebras, and we show that the generic Hecke algebra and most of the specialized Hecke algebras $(\mathfrak{H}(n, \mathbb{C}[q]))_z$ are semisimple, and have the same representation theory as the symmetric group $\mathfrak{S}(n)$.

In Section 4.5, we close the chapter with an explicit "polynomial" description of the simple modules over the Hecke algebras which are semisimple. The construction that is proposed is due to A. Lascoux, and its main interest is that it is a direct extension of Theorem 3.30. Unfortunately, the proof of the fact that one indeed obtains in this way a complete family of representatives of the simple modules of $\mathfrak{H}(n)$ relies on *another* construction due to Murphy of the q-Specht modules $S^{\lambda,\mathbb{C}(q)}$, and on considerations on q-Jucys–Murphy elements (see Chapter 8 for the theory of Jucys–Murphy elements). As a consequence, we chose to only describe Lascoux' construction, and to omit the proof of its validity.

The motivations for the study of the Hecke algebras are given in Chapter 5: there, we shall see that $\mathfrak{H}(n, \mathbb{C}[q])$ connects the representation theories of:

- the symmetric groups $\mathfrak{S}(n)$;
- the quantum groups $U_q(\mathfrak{gl}(N))$, that are quantizations of the complex general linear groups $\mathrm{GL}(N, \mathbb{C})$;
- and the finite general linear groups $\mathrm{GL}(n, \mathbb{F}_q)$.

We shall also explain in Chapters 5 and 6 the character theory of the Hecke algebras, in a fashion similar to the treatment of Chapter 2 for characters of symmetric groups. The computation of the characters of the Hecke algebras relies either on the theory of quantum groups (generic case, see Chapter 5), or on the theory of non-commutative symmetric functions and quasi-symmetric functions (case $q = 0$, cf. Chapter 6); this is why we have to devote two independent chapters to these computations.

For this chapter, a certain familiarity with commutative algebra will prove useful for the reader. In particular, we shall deal with: exact sequences of modules, quotient modules; ring and field extensions, integral and algebraic closures; local rings and valuation rings. We shall also admit a few results from commutative algebra, namely:

1. Jordan–Hölder and Krull–Schmidt theorems for the existence and unicity of a composition series, and of a decomposition in indecomposables of a module;
2. a few elementary facts from algebraic number theory, regarding Dedekind domains and their extensions.

We will give at the end of the chapter the references for these results.

4.1 Coxeter presentation of symmetric groups

As explained in the introduction of this chapter, the Hecke algebra $\mathfrak{H}(n,\mathbb{C}[q])$ of the symmetric group $\mathfrak{S}(n)$ is a one-parameter deformation of $\mathbb{C}\mathfrak{S}(n)$ whose specialization with $q=1$ gives back the group algebra $\mathbb{C}\mathfrak{S}(n)$. A way to produce such a deformation of the group algebra is to introduce the parameter q in a **presentation** of the symmetric group. Recall that if G is a group, then a presentation $\langle S \mid R\rangle$ of G is given by:

- a set $S=\{s_1,s_2,\ldots,s_r\}$ of **generators** of G: the smallest subgroup of G containing S is G itself.
- a subset R of the free group F_S (the **relations**), such that the smallest normal subgroup $N(R)$ of F_S that contains R is the kernel of the natural (surjective) morphism $\mathrm{F}_S \to G$.

Then, by definition, G is isomorphic to the quotient $\mathrm{F}_S/N(R)$. The set of relations corresponds to a set of identities $s_{i_1}s_{i_2}\cdots s_{i_l}=e_G$ in the group G. More generally, one can include identities $s_{i_1}s_{i_2}\cdots s_{i_l}=s_{j_1}s_{j_2}\cdots s_{j_k}$ in R, being understood that this identity corresponds to the element $s_{i_1}s_{i_2}\cdots s_{i_l}(s_{j_k})^{-1}(s_{j_{k-1}})^{-1}\cdots(s_{j_1})^{-1}$ of the free group F_S, which one sets equal to e_G in G.

The main goal of this section is to prove the following presentation of $\mathfrak{S}(n)$:

Theorem 4.1. *Denote s_i the transposition $(i,i+1)$ in $\mathfrak{S}(n)$, for $i\in[\![1,n-1]\!]$. A presentation of the symmetric group $\mathfrak{S}(n)$ is given by:*

$$\begin{aligned}
&\textit{generators:}\quad s_1,s_2,\ldots,s_{n-1};\\
&\textit{relations:}\quad (s_i)^2=1 \qquad &&\forall i\in[\![1,n-1]\!],\\
&\qquad s_is_{i+1}s_i=s_{i+1}s_is_{i+1} &&\forall i\in[\![1,n-2]\!],\\
&\qquad s_is_j=s_js_i &&\forall |i-j|\geq 2.
\end{aligned}$$

By replacing the quadratic relations $(s_i)^2=1 \iff (s_i+1)(s_i-1)=0$ by quadratic relations $(T_i+1)(T_i-q)=0$, we shall obtain the Hecke algebras of symmetric groups. A consequence of Theorem 4.1 is that symmetric groups belong to the class of **Coxeter groups**; see the notes at the end of the chapter for references on this notion.

▷ *Matsumoto's theorem.*

Theorem 4.1 mainly relies on an abstract result known as **Matsumoto's theorem**. To begin with, notice that by Lemma 2.3, the elementary transpositions $(s_i)_{i\in[\![1,n-1]\!]}$ indeed generate $\mathfrak{S}(n)$. Moreover, they obviously satisfy the **quadratic relations**

$$(s_i)^2=1$$

and the **commutation relations**

$$s_i s_j = s_j s_i \quad \text{if } \{i,i+1\} \cap \{j,j+1\} = \emptyset \iff |i-j| \geq 2.$$

As for the remaining **braid relations**, they also hold since

$$s_i s_{i+1} s_i = (i,i+1)(i+1,i+2)(i,i+1) = (i,i+2);$$
$$s_{i+1} s_i s_{i+1} = (i+1,i+2)(i,i+1)(i+1,i+2) = (i,i+2).$$

Therefore, if $G = \mathrm{F}_S/N(R)$ is the group with presentation given by Theorem 4.1, then there is a surjective morphism of groups $\phi : G \to \mathfrak{S}(n)$, and it suffices to show that this is an isomorphism.

If $\sigma \in \mathfrak{S}(n)$, we denote $\ell = \ell(\sigma)$ the minimal number required to write

$$\sigma = s_{i_1} s_{i_2} \cdots s_{i_\ell}$$

as a product of elementary transpositions. Beware that it differs from the quantity $n(\sigma) = \ell(t(\sigma))$, which is the number of cycles of σ. By the discussion of Section 2.1, the **length** $\ell(\sigma)$ of the permutation σ is also equal to the number of inversions of σ. A decomposition of σ as a product of $\ell(\sigma)$ elementary transpositions is called a **minimal** or **reduced decomposition**. For example, $s_3 s_1 s_2 s_3$ is a reduced decomposition of the permutation $(1,2,4)$ in $\mathfrak{S}(4)$.

Lemma 4.2. *Let σ be a permutation, and s be an elementary transposition in $\mathfrak{S}(n)$. Then, $\ell(s\sigma) = \ell(\sigma)+1$ or $\ell(\sigma)-1$. Moreover, if $\ell(s\sigma) = \ell(\sigma)-1$, then there exists a reduced decomposition of σ that starts with s.*

Proof. If $s_{i_1} \cdots s_{i_r}$ is a reduced expression for σ, then $s s_{i_1} \cdots s_{i_r}$ is a possibly non-reduced expression for $s\sigma$, so $\ell(s\sigma) \leq \ell(\sigma)+1$. By symmetry, $\ell(\sigma) \leq \ell(s\sigma)+1$, so

$$\ell(\sigma)-1 \leq \ell(s\sigma) \leq \ell(\sigma)+1.$$

Since $(-1)^{\ell(s\sigma)} = \varepsilon(s\sigma) = \varepsilon(s)\,\varepsilon(\sigma) = -\varepsilon(\sigma)$, the value $\ell(\sigma)$ is not allowed for $\ell(s\sigma)$, so the proof of the first part of the lemma is done. For the second part, if $\ell(s\sigma) = \ell(\sigma)-1$, then given a reduced decomposition $s_{i_1} \cdots s_{i_r}$ for $s\sigma$,

$$\sigma = s(s\sigma) = s s_{i_1} \cdots s_{i_r}$$

is a reduced expression of σ that starts with s. □

Lemma 4.3. *In the same setting, if $s = (i,i+1)$, then $\ell(s\sigma) = \ell(\sigma)-1$ if and only if $\sigma^{-1}(i) > \sigma^{-1}(i+1)$ (that is to say that i is a descent of σ^{-1}).*

Proof. In the discussion of Section 2.1, we saw that $N(\sigma s) = N(\sigma)-1$ if and only if $\sigma(i) > \sigma(i+1)$. However, this number of inversions is also the length, which is invariant by the involution $\sigma \mapsto \sigma^{-1}$. Hence,

$$\ell(s\sigma) = \ell(\sigma^{-1}s) = \ell(\sigma^{-1})-1 = \ell(\sigma)-1$$

if and only if $\sigma^{-1}(i) > \sigma^{-1}(i+1)$. One then says that i is a **backstep**, or **recoil** of σ. □

Remark. The backsteps of a permutation σ are easily computed on its word: they are the values i such that, when one enumerates the letters $1,2,3,\dots,n$ of the word, reading them from left to right, one has to go back to the beginning of the word of σ when reading $i+1$. For instance, consider the permutation $\sigma = 5612743$. One can read from left to right $1,2,3$, and then one has to go back to the beginning of σ to read 4; hence, 3 is a backstep. Similarly, one can read 4, and one goes back to the beginning to read 5, so 4 is a backstep. One can finally read from left to right $5,6,7$, so the only backsteps are 3 and 4.

Example. Consider the permutation $\sigma = 5612743$. A reduced decomposition of it is

$$s_3s_4s_5s_6s_3s_4s_5s_2s_1s_3s_2,$$

and on the other hand, we have just seen that its backsteps are 3 and 4. Consider then $s_4\sigma$. It has indeed length $11-1=10$, since

$$\begin{aligned} s_4(s_3s_4s_5s_6s_3s_4s_5s_2s_1s_3s_2) &= s_3s_4s_3s_5s_6s_3s_4s_5s_2s_1s_3s_2 \\ &= s_3s_4s_5s_6(s_3)^2s_4s_5s_2s_1s_3s_2 = s_3s_4s_5s_6s_4s_5s_2s_1s_3s_2 \end{aligned}$$

by using the braid and commutation relations.

Lemma 4.4. *Let $s \neq t$ be two elementary transpositions, and $\sigma \in \mathfrak{S}(n)$. If $\ell(s\sigma) < \ell(\sigma)$ and $\ell(t\sigma) < \ell(\sigma)$, then there exists $\tau \in \mathfrak{S}(n)$ such that:*

1. *If s and t are contiguous ($s = (i,i+1)$ and $t = (i+1,i+2)$), then $\sigma = sts\tau$ and $\ell(\sigma) = 3 + \ell(\tau)$.*

2. *If s and t are not contiguous, then $\sigma = st\tau$ and $\ell(\sigma) = 2 + \ell(\tau)$.*

Proof. In the first case, i and $i+1$ are both backsteps of σ, that is to say that

$$\sigma^{-1}(i) > \sigma^{-1}(i+1) > \sigma^{-1}(i+2).$$

Multiplying σ^{-1} on the right by $sts = (i,i+2)$ exchanges the values of σ^{-1} on i and $i+2$, hence, decreases by 3 the number of inversions of σ^{-1}. Indeed, the inversions deleted are $(i,i+1)$, $(i+1,i+2)$ and $(i,i+2)$, and the other ones are kept, up to the replacement of i by $i+2$ and conversely. So, if $\tau = sts\sigma$, then $\sigma = sts\tau$ and

$$\ell(\tau) = \ell(\tau^{-1}) = \ell(\sigma^{-1}) - 3 = \ell(\sigma) - 3.$$

The second case is entirely similar: if $s = (i,i+1)$ and $t = (j,j+1)$, then i and j are both backsteps of σ, and $\sigma^{-1}ts$ has two less inversions than σ^{-1}. □

Recall that a **monoid** is a set M endowed with an associative product $\cdot$, and that admits a neutral element e_M for this product. The difference with the notion of group is that one does not ask for the existence of inverses.

Theorem 4.5 (Matsumoto). *Set $S = \{s_1, s_2, \dots, s_{n-1}\}$. Let M be a monoid, and $\pi : S \to M$ be a map such that*

$$\pi(s_i)\,\pi(s_{i+1})\,\pi(s_i) = \pi(s_{i+1})\,\pi(s_i)\,\pi(s_{i+1});$$
$$\pi(s_i)\,\pi(s_j) = \pi(s_j)\,\pi(s_i) \quad \text{if } |i-j| \geq 2.$$

Then, there is a unique map $\pi : \mathfrak{S}(n) \to M$ that extends π and such that for any reduced decomposition $\sigma = s_{i_1} s_{i_2} \cdots s_{i_\ell}$, one has

$$\pi(\sigma) = \pi(s_{i_1})\,\pi(s_{i_2}) \cdots \pi(s_{i_\ell}).$$

Proof. We show the existence of π by induction on the length of σ; the unicity is then obvious. Suppose that for any permutation σ of length less than $k-1 \geq 1$, $\pi(\sigma)$ is well defined and satisfies the property stated above. One then has to show that if σ has length k and admits two reduced decompositions $\sigma = s_{i_1} s_{i_2} \cdots s_{i_k} = s_{j_1} s_{j_2} \cdots s_{j_k}$, then

$$\pi(s_{i_1})\,\pi(s_{i_2}) \cdots \pi(s_{i_k}) = \pi(s_{j_1})\,\pi(s_{j_2}) \cdots \pi(s_{j_k}).$$

1. Suppose first that $s_{i_1} = s_{j_1}$. Then, $\pi(s_{i_2}) \cdots \pi(s_{i_k}) = \pi(s_{j_2}) \cdots \pi(s_{j_k})$ since $s_{i_2} \cdots s_{i_k}$ and $s_{j_2} \cdots s_{j_k}$ are two reduced decompositions of length $k-1$ of the same element of $\mathfrak{S}(n)$. The result follows by multiplying by $\pi(s_{i_1}) = \pi(s_{j_1})$.

2. Suppose now that $s = s_{i_1} \neq s_{j_1} = t$. Then, the previous lemma applies and there exists τ such that

$$\sigma = st s\tau \quad ; \quad \ell(\sigma) = 3 + \ell(\tau)$$

or

$$\sigma = st\tau \quad ; \quad \ell(\sigma) = 2 + \ell(\tau).$$

In the first case, $s\sigma$ and $t\sigma$ have length $k-1$, so the induction hypothesis applies and

$$\pi(s\sigma) = \pi(s_{i_2}) \cdots \pi(s_{i_k}) = \pi(t)\,\pi(s)\,\pi(\tau);$$
$$\pi(t\sigma) = \pi(s_{j_2}) \cdots \pi(s_{j_k}) = \pi(s)\,\pi(t)\,\pi(\tau).$$

Then, since $\pi(s)\,\pi(t)\,\pi(s) = \pi(t)\,\pi(s)\,\pi(t)$,

$$\begin{aligned} \pi(s_{i_1})\,\pi(s_{i_2}) \cdots \pi(s_{i_k}) &= \pi(s)\,\pi(t)\,\pi(s)\,\pi(\tau) \\ &= \pi(t)\,\pi(s)\,\pi(t)\,\pi(\tau) \\ &= \pi(s_{j_1})\,\pi(s_{j_2}) \cdots \pi(s_{j_k}). \end{aligned}$$

The second case is entirely similar and left to the reader.

□

Proof of Theorem 4.1. If G is the group with presentation given by Theorem 4.1, then we have a morphism of groups $\phi : G \to \mathfrak{S}(n)$ such that $\phi_{|S} = \mathrm{id}_S$. On the other hand, since the elements of S satisfy the braid and commutation relations in G, we have a map $\pi : S \to G$ that satisfies the condition of Matsumoto's theorem 4.5, hence, a unique map $\pi : \mathfrak{S}(n) \to G$ that factorizes on reduced decompositions. Let us show that π is a morphism of groups. It suffices to prove that for every permutation σ and every elementary transposition s, $\pi(s\sigma) = \pi(s)\,\pi(\sigma)$. If $\ell(s\sigma) > \ell(\sigma)$, then given a reduced decomposition $s_{i_1} \cdots s_{i_r}$ of σ, $ss_{i_1} \cdots s_{i_r}$ is a reduced decomposition of $s\sigma$, so

$$\pi(s\sigma) = \pi(s)\,\pi(s_{i_1}) \cdots \pi(s_{i_r}) = \pi(s)\,\pi(\sigma).$$

On the other hand, if $\ell(s\sigma) < \ell(\sigma)$, then there exists a reduced decomposition of σ that starts with s: $\sigma = ss_{i_1} \cdots s_{i_r}$. Then, since $s^2 = 1$,

$$\pi(s\sigma) = \pi(s_{i_1} \cdots s_{i_r}) = \pi(s)\,\pi(s)\,\pi(s_{i_1}) \cdots \pi(s_{i_r}) = \pi(s)\,\pi(\sigma).$$

Thus, we have a morphism of groups $\pi : \mathfrak{S}(n) \to G$, and such that $\pi_{|S} = \mathrm{id}_S$. Since π and ϕ both yield the identity map on the generating set S, $\phi \circ \pi = \mathrm{id}_{\mathfrak{S}(n)}$, and $\mathfrak{S}(n)$ is isomorphic to G. □

If A is an algebra over $\mathbb{C}$, then there is a notion of **presentation** of A similar to the notion of presentation of groups. Hence, one says that A has presentation $\langle S \mid R \rangle$ if:

- $S \subset A$ and the smallest $\mathbb{C}$-subalgebra of A that contains S is A itself.
- R is a subset of the free associative algebra $\mathbb{C}\langle S \rangle$ with generators the elements of S, and the bilateral ideal (R) generated by R in $\mathbb{C}\langle S \rangle$ is the kernel of the natural surjective morphism of algebras $\mathbb{C}\langle S \rangle \to A$.

Then, A is isomorphic to $\mathbb{C}\langle S \rangle/(R)$. It should be noticed that given a *group* G, there is in general no way to deduce a presentation of the group algebra $\mathbb{C}G$ from a presentation of group $G = \langle S \mid R \rangle$. For instance, $G = \mathbb{Z}$ has presentation $\mathbb{Z} = \langle x \mid \emptyset \rangle$, but the algebra with presentation $\langle x \mid \emptyset \rangle$ is $\mathbb{C}[x]$, whereas the group algebra $\mathbb{C}\mathbb{Z} = \mathbb{C}[x, x^{-1}]$ has presentation $\langle x, y \mid xy = 1 \rangle$. In fact, the presentation of algebras is functorially related to the presentation of *monoids*. Let S be a set, and S^* be the set of words with arbitrary length and letters in S. A monoid M has presentation $\langle S \mid R \rangle$ if:

- $S \subset M$ and the smallest submonoid of M that contains S is M itself.
- $R \subset S^* \times S^*$, and if $\equiv_R$ is the equivalence relation on S^* which is the reflexive, transitive and symmetric closure of

  $$w_1 \equiv_R w_2 \quad \Longleftrightarrow \quad w_1 = usv,\ w_2 = utv,\ (s, t) \in R,$$

  then the natural map $S^* \to M$ induces an isomorphism of monoids between the quotient monoid $S^*/\equiv_R$ and M.

With these definitions, it is easily seen that if M has presentation $\langle S \mid R\rangle$, then the monoid algebra $\mathbb{C}M$ has also presentation $\langle S \mid R\rangle$, so the functor $\mathfrak{Monoids} \to \mathbb{C}$-$\mathfrak{Algebras}$ is compatible with presentations.

Remark. Through the book, given an algebra A defined over a field k and a subset $R \subset A$, we shall need to consider two kinds of generated subsets:

- the (bilateral, or left, or right) *ideal* generated by R, denoted (R), $A \cdot R$ or AR.
- the k-subalgebra of A generated by R, which we shall always denote between brackets: $\langle R\rangle$, or $k\langle R\rangle$ if one needs to specify the base field k.

The two different notations will allow us to avoid any possible confusion. In the special case when A is a commutative algebra, we shall usually denote $k\langle R\rangle = k[R]$, thereby copying the notation for polynomial rings.

Proposition 4.6. *The group algebra $\mathbb{C}\mathfrak{S}(n)$ has presentation:*

$$\begin{array}{lll} \textit{generators:} & T_1, T_2, \ldots, T_{n-1}; & \\ \textit{relations:} & (T_i)^2 = 1 & \forall i \in [\![1, n-1]\!], \\ & T_i T_{i+1} T_i = T_{i+1} T_i T_{i+1} & \forall i \in [\![1, n-2]\!], \\ & T_i T_j = T_j T_i & \forall |i-j| \geq 2. \end{array}$$

Proof. It suffices to show that this is the presentation of $\mathfrak{S}(n)$ as a monoid (instead of a group presentation). However, Matsumoto's theorem was stated with monoids, so one can follow the proof of Theorem 4.1 with monoids instead of groups. □

▷ *Bruhat order on the symmetric groups.*

In a moment we shall construct the Hecke algebra of the symmetric group by introducing a parameter q in the presentation of Proposition 4.6. Before that, it is useful to introduce a partial order on $\mathfrak{S}(n)$ whose existence is closely related to Matsumoto's theorem.

Definition 4.7 (Bruhat–Chevalley). *Given two permutations σ and τ, denote $\sigma \leq \tau$ if there exists a reduced decomposition $\tau = s_{i_1} \cdots s_{i_l}$ such that σ corresponds to a subword of this writing:*

$$\sigma = s_{i_{j_1}} s_{i_{j_2}} \cdots s_{i_{j_k}}$$

for some sequence $1 \leq j_1 < \cdots < j_k \leq l$. We call $\leq$ the **Bruhat order** *on $\mathfrak{S}(n)$.*

Example. In $\mathfrak{S}(3)$, the Bruhat order has the Hasse diagram drawn in Figure 4.1 (in this diagram, the smaller elements are on the bottom and the larger elements are on top).

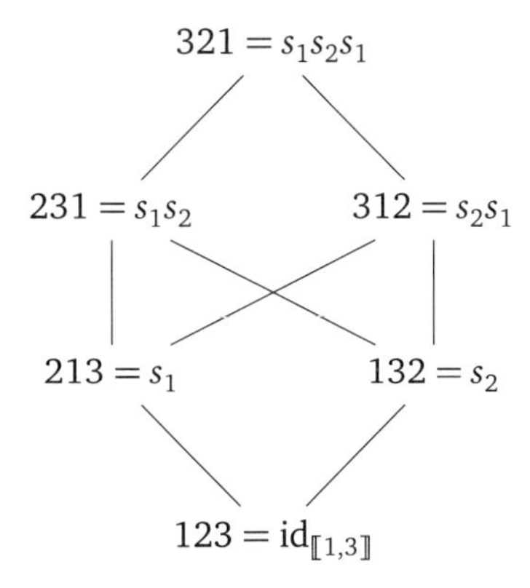

Figure 4.1
Hasse diagram of the Bruhat order on $\mathfrak{S}(3)$.

Theorem 4.8. *The Bruhat order is a partial order on $\mathfrak{S}(n)$. Moreover, if $\sigma \leq \tau$, then for any reduced decomposition $\tau = s_{i_1} \cdots s_{i_l}$, σ corresponds to a subword of this decomposition.*

Proof. Consider the monoid $M = \mathfrak{P}(\mathfrak{S}(n))$ whose elements are the subsets of $\mathfrak{S}(n)$, and whose product is $A \cdot B = \{ab \mid a \in A \text{ and } b \in B\}$. The map $\pi : S \to M$ which associates to s the subset $\{\mathrm{id}, s\}$ satisfies the hypotheses of Matsumoto's theorem, hence, it extends to a map $\pi : \mathfrak{S}(n) \to M$, such that

$$\pi(\sigma = s_{i_1} \cdots s_{i_l}) = \prod_{j=1}^{l} \{\mathrm{id}, s_{i_j}\} = \{s_{i_{j_1}} \cdots s_{i_{j_k}} \mid 1 \leq j_1 < \cdots < j_k \leq l\}$$

for *any* reduced expression $s_{i_1} \cdots s_{i_l}$ of σ. This shows that the definition of the Bruhat order does not depend on the choice of a reduced expression, and then, $\sigma \leq \tau$ if and only if $\pi(\sigma) \subset \pi(\tau)$, so $\leq$ is indeed a partial order. □

There are many non-trivial combinatorial properties of reduced expressions for permutations that are closely related to the existence of the Bruhat order; see the references at the end of the chapter, as well as some of the results of Chapter 6 (see, e.g., Lemma 6.26). For the moment, we shall only need the following result:

Proposition 4.9. *If $\sigma = s_{i_1}s_{i_2} \cdots s_{i_\ell} = s_{j_1}s_{j_2} \cdots s_{j_\ell}$ are two different reduced decompositions of σ, then there is a sequence of braid transformations $s_is_{i+1}s_i \longleftrightarrow s_{i+1}s_is_{i+1}$ and of commutation transformations $s_is_j \longleftrightarrow s_js_i$ on subwords of the reduced writings that allow us to go from one reduced writing to another reduced writing.*

Proof. We reason by induction on ℓ, the case $\ell = 1$ being trivial. If $s_{i_1} = s_{j_1}$, then it suffices to apply the induction hypothesis on $s_{i_2}\cdots s_{i_\ell} = s_{j_2}\cdots s_{j_\ell}$. Otherwise, the conclusions of Lemma 4.4 hold. Suppose for instance that $s_{i_1} = s$ and $s_{j_1} = t$ are contiguous elementary transpositions, and write $\sigma = sts\,\tau = tst\,\tau$. Fix a reduced writing $s_{k_4}\cdots s_{k_\ell}$ of τ. By induction, there exists a sequence of braid and commutation transformations that relates the two reduced decompositions

$$s\sigma = ts\,s_{k_4}\cdots s_{k_\ell} = s_{i_2}\cdots s_{i_\ell},$$

and similarly for

$$t\sigma = st\,s_{k_4}\cdots s_{k_\ell} = s_{j_2}\cdots s_{j_\ell}.$$

Therefore, since a braid transformation relates sts to tst, there exists indeed a sequence of braid and commutation transformations that relates the following reduced decompositions:

$$\sigma = s_{i_1}\cdots s_{i_\ell} = sts\,s_{k_4}\cdots s_{k_\ell} = tst\,s_{k_4}\cdots s_{k_\ell} = s_{j_1}\cdots s_{j_\ell}.$$

Again, the other case where s and t are not contiguous is extremely similar. □

▷ *Hecke algebras of symmetric groups.*

The **Hecke algebra** of $\mathfrak{S}(n)$ is obtained by deformation of the Coxeter presentation of $\mathbb{C}\mathfrak{S}(n)$:

Definition 4.10. *The Hecke algebra of $\mathfrak{S}(n)$ is the $\mathbb{C}[q]$-algebra with generators $T_1,\ldots,T_{n-1}$, and relations:*

$$\begin{aligned}
&(T_i - q)(T_i + 1) = 0 && \forall i \in [\![1, n-1]\!],\\
&T_i T_{i+1} T_i = T_{i+1} T_i T_{i+1} && \forall i \in [\![1, n-2]\!],\\
&T_i T_j = T_j T_i && \forall |i-j| \geq 2.
\end{aligned}$$

The Hecke algebra will be denoted $\mathfrak{H}(n,\mathbb{C}[q])$. A specialization of $\mathfrak{H}(n,\mathbb{C}[q])$ with q fixed to a value $z \in \mathbb{C}$ will be denoted $\mathfrak{H}_z(n)$; by Proposition 4.6, $\mathfrak{H}_1(n) = \mathbb{C}\mathfrak{S}(n)$. On the other hand, if A is a commutative algebra over $\mathbb{C}[q]$, we shall denote

$$\mathfrak{H}(n,A) = A \otimes_{\mathbb{C}[q]} \mathfrak{H}(n,\mathbb{C}[q]).$$

The **specialized Hecke algebra** $\mathfrak{H}_z(n)$ are examples of such tensor products: $\mathfrak{H}_z(n) = \mathfrak{H}(n,\mathbb{C})$, with the structure of $\mathbb{C}[q]$-algebra on $\mathbb{C}$ given by the morphism $\theta_z : \mathbb{C}[q] \to \mathbb{C}$. We shall also call **generic Hecke algebra** the tensor product

$$\mathfrak{H}(n,\mathbb{C}(q)) = \mathbb{C}(q) \otimes_{\mathbb{C}[q]} \mathfrak{H}(n,\mathbb{C}[q]),$$

and denote it simply $\mathfrak{H}(n)$.

The goal of this chapter is to study the structure and the representations of the generic Hecke algebra and of its various specializations. In particular, in Section 4.4, we shall prove that $\mathfrak{H}(n) = \mathfrak{H}(n,\mathbb{C}(q))$ is a semisimple algebra, and that $\mathfrak{H}_z(n)$

is also semisimple when z is not equal to zero or to a non-trivial root of unity. It is for the moment unclear why Definition 4.10 is a pertinent deformation of the symmetric group algebra. The motivations for this construction will be given in Chapter 5, where $\mathfrak{H}(n)$ and some of its specializations shall appear as the Schur–Weyl duals of quantum groups and of general linear groups over finite fields. In this section, we shall only give a linear basis of the Hecke algebra, as well as some elementary computation rules.

Theorem 4.11 (Bourbaki). *If $\sigma \in \mathfrak{S}(n)$ has minimal decomposition $\sigma = s_{i_1}s_{i_2}\cdots s_{i_\ell}$, denote*

$$T_\sigma = T_{i_1}T_{i_2}\cdots T_{i_\ell}.$$

1. *The element $T_\sigma \in \mathfrak{H}(n,\mathbb{C}[q])$ does not depend on the chosen minimal decomposition.*

2. *For any commutative algebra A over $\mathbb{C}[q]$, the Hecke algebra $\mathfrak{H}(n,A)$ is free as an A-module, and $(T_\sigma)_{\sigma\in\mathfrak{S}(n)}$ is a linear basis of it.*

A first step in the proof of Theorem 4.11 consists in giving a new definition and presentation of $\mathfrak{H}(n,\mathbb{C}[q])$:

Proposition 4.12. *The elements $(T_\sigma)_{\sigma\in\mathfrak{S}(n)}$ are well defined, and a new presentation of the algebra $\mathfrak{H}(n,\mathbb{C}[q])$ over $\mathbb{C}[q]$ is given by:*

$$\textit{generators:}\quad T_\sigma,\ \sigma\in\mathfrak{S}(n);$$

$$\textit{relations:}\quad T_sT_\sigma = \begin{cases} T_{s\sigma} & \text{if } \ell(s\sigma)>\ell(\sigma),\\ q\,T_{s\sigma}+(q-1)\,T_\sigma & \text{if } \ell(s\sigma)<\ell(\sigma),\end{cases}$$

where for the relations σ runs over $\mathfrak{S}(n)$ and s runs over the set of elementary transpositions.

Proof. If $s_{i_1}\cdots s_{i_\ell} = s_{j_1}\cdots s_{j_\ell}$ are two reduced writings of the same permutation σ, then by Proposition 4.9, they differ by a sequence of braid and commutation transformations. These transformations are valid for the corresponding elements of $\mathfrak{H}(n,\mathbb{C}[q])$, so

$$T_{i_1}\cdots T_{i_\ell} = T_{j_1}\cdots T_{j_\ell}$$

and T_σ is well defined. For the second part of the proposition, to show that two presentations $\langle S_1\,|\,R_1\rangle$ and $\langle S_2\,|\,R_2\rangle$ yield the same algebra amounts to showing that:

- S_1 spans S_2 and conversely;
- the relations of R_1 imply those of R_2 and conversely.

Set $S_1 = \{T_i,\ i\in[\![1,n-1]\!]\}$ and $S_2 = \{T_\sigma,\ \sigma\in\mathfrak{S}(n)\}$. One has $S_1\subset S_2$, so S_2 generates S_1, and conversely, by definition, each element of S_2 is a product of elements of S_1, so S_1 spans S_2. Now, let us see the equivalence between the set of relations R_1 of Definition 4.10 and the set of relations R_2 of Proposition 4.12. Assume that the set of identities R_1 holds, and let us compute $T_s T_\sigma$. If $\ell(s\sigma) = \ell(\sigma)+1$, then given a minimal decomposition $\sigma = s_{i_1} \cdots s_{i_\ell}$, $s\, s_{i_1} \cdots s_{i_\ell}$ is a reduced decomposition of $s\sigma$, so

$$T_s T_\sigma = T_s\, T_{i_1} \cdots T_{i_\ell} = T_{s\sigma}.$$

If $\ell(s\sigma) = \ell(\sigma) - 1$, then by Lemma 4.2, there exists a reduced decomposition of σ that starts with s:

$$\sigma = s\, s_{i_2} \cdots s_{i_\ell}.$$

Then, since $(T_s)^2 = (q-1)\, T_s + q$ is a quadratic relation in R_1, one has

$$T_s T_\sigma = (T_s)^2\, T_{i_2} \cdots T_{i_\ell} = (q-1)\, T_s T_{i_2} \cdots T_{i_\ell} + q\, T_{i_2} \cdots T_{i_\ell} = (q-1)\, T_\sigma + q\, T_{s\sigma}.$$

Hence, the relations in R_2 can be deduced from the relations in R_1. Conversely, it is easily seen that the braid and commutation relations in R_1 are particular cases of the first alternative in R_2, and that the quadratic relations in R_1 are particular cases of the second alternative in R_2. So, the two presentations are equivalent. □

Proof of Theorem 4.11. An immediate consequence of the second presentation of $\mathfrak{H}(n, \mathbb{C}[q])$ is that $(T_\sigma)_{\sigma \in \mathfrak{S}(n)}$ spans linearly $\mathfrak{H}(n, A)$ over A. Hence, it remains to see that this family is free. It is sufficient to treat the case of $\mathfrak{H}(n, \mathbb{C}[q])$, because if $\mathfrak{H}(n, \mathbb{C}[q]) = \bigoplus_{\sigma \in \mathfrak{S}(n)} \mathbb{C}[q]\, T_\sigma$, then for any commutative algebra A over $\mathbb{C}[q]$, one has

$$\mathfrak{H}(n, A) = A \otimes_{\mathbb{C}[q]} \mathfrak{H}(n, \mathbb{C}[q]) = \bigoplus_{\sigma \in \mathfrak{S}(n)} A \otimes_{\mathbb{C}[q]} \mathbb{C}[q]\, T_\sigma = \bigoplus_{\sigma \in \mathfrak{S}(n)} A\, T_\sigma.$$

Suppose that the T_σ's are not linearly independent over $\mathbb{C}[q]$ in $\mathfrak{H}(n, \mathbb{C}[q])$; then, there exist polynomials $P_\sigma(q)$, not all equal to zero, and such that

$$\sum_{\sigma \in \mathfrak{S}(n)} P_\sigma(q)\, T_\sigma = 0.$$

Without loss of generality, one can assume that the polynomials $P_\sigma(q)$ have for greatest common divisor 1 in the factorial ring $\mathbb{C}[q]$. However, if one specializes this identity by $\theta_1 : P(q) \in \mathbb{C}[q] \mapsto P(1) \in \mathbb{C}$, one gets in $\mathbb{C} \otimes_{\mathbb{C}[q]} \mathfrak{H}(n, \mathbb{C}[q]) = \mathfrak{H}_1(n) = \mathbb{C}\mathfrak{S}(n)$:

$$\sum_{\sigma \in \mathfrak{S}(n)} P_\sigma(1)\, \sigma = 0.$$

Since $(\sigma)_{\sigma \in \mathfrak{S}(n)}$ is a basis of $\mathfrak{S}(n)$, this forces $P_\sigma(1) = 0$ for all σ, and therefore, $(q-1) \mid P_\sigma(q)$ for all σ. This contradicts the previous assumption that the P_σ's have no non-trivial common divisor.

So, for any commutative algebra A over the ring $\mathbb{C}[q]$, $\mathfrak{H}(n, A)$ has for A-basis $(T_\sigma)_{\sigma \in \mathfrak{S}(n)}$, and in particular, the specialized Hecke algebra $\mathfrak{H}_z(n)$ is always of dimension $n!$ over $\mathbb{C}$, and the generic Hecke algebra $\mathfrak{H}(n)$ is a $\mathbb{C}(q)$-algebra also of dimension $n!$. □

4.2 Representation theory of algebras

Since $\mathfrak{H}(n)$ was built in Section 4.1 as a deformation of the group algebra $\mathbb{C}\mathfrak{S}(n)$, it is natural to compare the representation theory of this Hecke algebra with the representation theory of $\mathfrak{S}(n)$ detailed in Chapters 2 and 3. Intuitively, since $\mathfrak{H}_1(n) = \mathbb{C}\mathfrak{S}(n)$, one can expect that for z "close to" 1, the specialized Hecke algebra $\mathfrak{H}_z(n)$ has the same representation theory as $\mathfrak{S}(n)$, i.e., it is semisimple and has simple modules S_z^λ labeled by integer partitions in $\mathfrak{Y}(n)$, with $\dim S_z^\lambda = \mathrm{card}\,\mathrm{ST}(\lambda)$. In fact, this result is true for almost all $z \in \mathbb{C}$, and also for the generic Hecke algebra $\mathfrak{H}(n)$. We shall prove this by combining two kind of arguments:

- **Brauer–Cartan deformation theory** (Section 4.3), which relates the representation theory of algebras obtained from one another by specializations or by extensions of scalars;
- the theory of so-called **symmetric algebras** (Section 4.4), which is a class of algebras that share certain representation-theoretic properties with group algebras.

One of the main difficulties is that we do not know a priori that the Hecke algebras are semisimple. Actually, some specializations $\mathfrak{H}_z(n)$ are not semisimple $\mathbb{C}$-algebras, so the theory of Section 1.4 does not apply. For instance, if $z = 0$, then we shall see in Section 6.3 that $\mathfrak{H}_0(n \geq 3)$ is not semisimple, though many things can be said about its modules. As a consequence, and even though we want to prove that many Hecke algebras are indeed semisimple, we need to introduce a bit of the representation theory of general (possibly not semisimple) algebras. This section is devoted to this topic.

Here and in the remainder of the chapter, every k-algebra A will be assumed finite-dimensional over k, as well as every A-module M.

▷ *Composition series and the Grothendieck group* $\mathrm{R}_0(A)$.

If A is a k-algebra and M is an A-module, then usually one cannot split M as a direct sum of simple A-modules: $M = \bigoplus_\lambda m_\lambda M^\lambda$. However, there are two notions that play a similar role and that can be used without the assumption of semisimplicity: composition series, and decompositions in direct sums of indecomposable modules. Let us start with the notion of **composition series**:

Definition 4.13. *Let A be a k-algebra, and M be an A-module. A composition series of M is a sequence of submodules*

$$0 = M_0 \subset M_1 \subset M_2 \subset \cdots \subset M_r = M$$

such that each quotient M_i/M_{i-1} is a simple A-module. Every A-module that is finite-dimensional over k admits a composition series. For any module M, the multiset of *composition factors $\{M_i/M_{i-1},\ i \in [\![1,r]\!]\}$ does not depend on the choice of the composition series (up to isomorphisms of A-modules).*

The unicity of the multiset of composition factors is the Jordan–Hölder theorem for modules, which already appeared in Section 1.4. It allows one to introduce the Grothendieck group of a general algebra (possibly not semisimple). Thus, the **Grothendieck group of finite-dimensional modules** over A is the group $\mathrm{R}_0(A)$ with presentation:

generators: $[M]$, with M module over A;

relations: $[P]=[M]+[N]$ if there is an exact sequence

$$0 \longrightarrow M \longrightarrow P \longrightarrow N \longrightarrow 0.$$

Notice then that for any modules M and N, $[M\oplus N]=[M]+[N]$ since there is an exact sequence

$$0 \longrightarrow M \longrightarrow M\oplus N \longrightarrow N \longrightarrow 0.$$

Proposition 4.14. *The Grothendieck group $\mathrm{R}_0(A)$ is a free abelian group, with basis over $\mathbb{Z}$ given by the classes of isomorphism of simple A-modules. Moreover, given two modules M and N, one has $[M]=[N]$ in the Grothendieck group if and only if M and N have the same composition factors (counting multiplicities, up to isomorphisms and permutations).*

Proof. Consider a module M with composition series $0=M_0 \subset M_1 \subset \cdots \subset M_r = M$. By induction on r, one has $[M] = \sum_{i=1}^r [M_i/M_{i-1}]$, since there is an exact sequence

$$0 \longrightarrow M_{r-1} \longrightarrow M_r \longrightarrow M_r/M_{r-1} \longrightarrow 0$$

which implies $[M] = [M_r] = [M_r/M_{r-1}]+[M_{r-1}]$. Let $\{V_s\}_{s\in S}$ be a complete family of representatives of the isomorphism classes of simple modules over A (we shall see later that S is a finite set). Every module M is Grothendieck equivalent to a sum $\bigoplus_{s\in S} m_s V_s$, and by the Jordan–Hölder theorem, the coefficients $m_s = m_s(M)$ are uniquely determined by M. Moreover, suppose that one has an exact sequence

$$0 \longrightarrow M \overset{i}{\longrightarrow} P \overset{s}{\longrightarrow} N \longrightarrow 0,$$

so that $[P] = [M]+[N]$. If $0 = M_0 \subset M_1 \subset \cdots \subset M_q = M$ and $0 = N_0 \subset N_1 \subset \cdots \subset N_r = N$ are two composition series of M and N, then setting $P_i = s^{-1}(N_i)$ for $i \in [\![0,r]\!]$, the quotient $P_i/P_{i-1} = (P_i/M)/(P_{i-1}/M) = N_i/N_{i-1}$ is a simple A-module for every i, and

$$0=M_0 \subset M_1 \subset \cdots \subset M_q = P_0 \subset P_1 \subset \cdots \subset P_r = P$$

is a composition series for P whose composition factors are those of $M\oplus N$. As a consequence, the map

$$\begin{aligned}\psi : \mathrm{R}_0(A) &\to \bigoplus_{s\in S} \mathbb{Z}V_s \\ M &\mapsto \sum_{s\in S} m_s(M)\, V_s\end{aligned}$$

is a well-defined map that is a surjective morphism of groups. Its inverse is the map

$$\begin{aligned}\bigoplus_{s\in S} \mathbb{Z}V_s &\to \mathrm{R}_0(A) \\ \sum_{s\in S} m_s V_s &\mapsto \sum_{s\in S} m_s [V_s],\end{aligned}$$

so $\mathrm{R}_0(A)$ is indeed the free abelian group whose basis elements are the classes of simple modules. Finally, if M and N are two A-modules, then

$$\begin{aligned}[M] = [N] \text{ in } \mathrm{R}_0(A) &\iff \psi([M]) = \psi([N]) \\ &\iff m_s(M) = m_s(N) \text{ for every } s \in S \\ &\iff M \text{ and } N \text{ have the same composition factors.} \quad \square\end{aligned}$$

If A is semisimple, then two modules are Grothendieck equivalent in $\mathrm{R}_0(A)$ if and only if they are isomorphic: indeed, in this case, the composition factors of M are the components of the unique decomposition in simple modules $M = \bigoplus_{\lambda \in \widehat{A}} m_\lambda M^\lambda$. Thus, the notion of Grothendieck group presented above generalizes the notion of Grothendieck group of representations of a finite group or a finite-dimensional semisimple algebra (see Chapter 1).

▷ *Indecomposable projective modules and the Grothendieck group* $\mathrm{K}_0(A)$.

Call **indecomposable** a module M over A that cannot be written as a direct sum of two non-zero submodules. This is weaker than the notion of simple module. For instance, the $\mathbb{C}[X]$-module

$$M = \mathbb{C}[X]/(X^2) = \{aX + b,\ a, b \in \mathbb{C}\}$$

is not simple, since it admits as a submodule $X\mathbb{C}[X]/X^2\mathbb{C}[X] = \{aX,\ a \in \mathbb{C}\}$. However, it is indecomposable, since this submodule (and any other non-trivial $\mathbb{C}[X]$-submodule) has no complement that is a $\mathbb{C}[X]$-submodule. If A is a k-algebra and M is a (finite-dimensional) A-module, then for dimension reasons, M can always be written as a direct sum of indecomposable modules. Then, one has the analogue of the Jordan–Hölder theorem:

Theorem 4.15 (Krull–Schmidt). *With the usual assumption of finite dimension, if*

$$M = \bigoplus_{i=1}^{r} M_i = \bigoplus_{j=1}^{s} N_j$$

are two decompositions of M in indecomposable A-modules, then $r = s$, and up to a permutation of the terms, there exists an isomorphism of A-module between M_i and N_i for every $i \in [\![1, r]\!]$.

As this is a bit outside the scope of this book, we shall just admit this result (cf. the notes at the end of the chapter). From there, it might seem adequate to introduce another Grothendieck group of classes of isomorphism of (finite-dimensional) modules over A, with basis over $\mathbb{Z}$ given by the (classes of isomorphism of) indecomposable modules. However, for several reasons, it is better to restrict oneself to the so-called **projective modules**. Let A be a k-algebra, and M be a module over A.

Definition 4.16. *The module M is said to be projective if, for every exact sequence*

$$0 \longrightarrow K \longrightarrow L \xrightarrow{s} M \longrightarrow 0$$

of morphisms of A-modules, there exists a section $r : M \to L$ such that $s \circ r = \mathrm{id}_M$ (one then says that the exact sequence is split by r).

In the setting of Hecke algebras, the notion of projective modules will prove particularly useful, because of the following characterization of semisimplicity:

Proposition 4.17. *Let A be a k-algebra. The following assertions are equivalent:*

1. *The algebra A is semisimple.*
2. *Every finite-dimensional A-module is semisimple.*
3. *Every finite-dimensional A-module is projective.*

Lemma 4.18. *An A-module M is semisimple if and only if every exact sequence*

$$0 \longrightarrow L \longrightarrow M \longrightarrow N \longrightarrow 0$$

splits.

Proof. Suppose that M satisfies the condition with exact sequences. If M is not a simple module, let L be a submodule of M, and consider the exact sequence

$$0 \longrightarrow L \xrightarrow{i} M \xrightarrow{s} M/L \longrightarrow 0$$

where i is the inclusion $L \subset M$, and s is the canonical surjection $M \to M/L$. If $r \in \mathrm{Hom}_A(M/L, M)$ is such that $s \circ r = \mathrm{id}_{M/L}$, then one checks at once that $M = L \oplus r(M/L)$, and L admits a complement in M that is an A-module. By Proposition 1.20, M is semisimple. Conversely, suppose that M is semisimple, and consider an exact sequence

$$0 \longrightarrow L \xrightarrow{i} M \xrightarrow{s} N \longrightarrow 0.$$

By semisimplicity, there exists a complement K of L in M, which is an A-module. Then, $s_{|K}$ yields an isomorphism between K and N, as it is surjective and has kernel reduced to $K \cap L = \{0\}$. A section of the exact sequence is then given by $r = (s_{|K})^{-1}$. □

Proof of Proposition 4.17. The two first statements are equivalent by definition of a semisimple algebra. If every A-module is semisimple, then every exact sequence splits by the previous lemma, so every A-module is projective by Definition 4.16; and the converse is true. □

On the other hand, for a general algebra A, the notion of projective modules is important because of the following characterization:

Theorem 4.19. *A module P over A is projective if and only if:*

1. *It is (isomorphic to) a direct summand of some free module A^n, that is to say that there exists a complement $Q \subset A^n$ which is an A-module such that $P \oplus Q = A^n$.*
2. *It is a direct sum $P = \bigoplus_{i=1}^{r} P_i$ of indecomposable projective modules, with each P_i isomorphic to an indecomposable left ideal of A that is a direct summand of A.*

An indecomposable left ideal of A that is also a direct summand of A is called a **principal indecomposable module** of A. Using the unicity of Krull–Schmidt theorem, one sees that given a decomposition $A = \bigoplus_{i=1}^{r} P_i$ of A in principal indecomposable modules, any other principal indecomposable module Q is isomorphic to a module P_i. So, the notion of principal indecomposable module is independent of the choice of a decomposition of A.

The second part of Theorem 4.19 is the analogue of the second part of Theorem 1.29 for algebras that are not semisimple. If A is semisimple, we know that every (left-)A-module writes uniquely as a sum of copies of simple left ideals of A. Then, Theorem 4.19 states that in the general case, every *projective* A-module writes uniquely as a sum of copies of *principal indecomposable* left ideals of A.

Lemma 4.20. *A module M over A is projective if and only if it is a direct summand of a free A-module.*

Proof. Suppose M is projective. Since it is finite-dimensional, it is finitely generated over A, hence, there exists a surjective morphism of A-modules $A^n \to M$ for some $n \geq 1$, and an exact sequence

$$0 \longrightarrow K \longrightarrow A^n \xrightarrow{\ s\ } M \longrightarrow 0.$$

Let r be a section of this exact sequence; by definition r is injective. We claim that $K \oplus r(M) = A^n$. Indeed, if $x \in K \cap r(M)$, then $x = r(m)$ for some $m \in M$, and

$m = s(x) = 0$ since $x \in K$, so $x = 0$. On the other hand, for any $x \in A^n$, one can write $x = (x - r(s(x))) + r(s(x))$, and this is a decomposition in elements of K and $r(M)$. Hence, if M is projective, then it is isomorphic to a direct summand $r(M)$ of some free module A^n.

Conversely, if $M \oplus N = A^n$ for some $n \geq 1$, let

$$0 \longrightarrow K \longrightarrow L \xrightarrow{s} M \longrightarrow 0$$

be an exact sequence involving M. Setting $s_{|N} = \mathrm{id}_N$, one gets an exact sequence

$$0 \longrightarrow K \longrightarrow L \oplus N \xrightarrow{s} A^n \longrightarrow 0.$$

Let $(e_1, \ldots, e_n)$ be an A-basis of A^n (e.g., the canonical basis). Since s is surjective, there exist elements $a_1, \ldots, a_n \in L \oplus N$ such that $s(a_i) = e_i$ for all i. We then set $r(e_i) = a_i$, and extend this by linearity to get a map $r : A^n \to L \oplus N$ which is a morphism of A-modules. By construction, $s \circ r = \mathrm{id}_{A^n}$, since this is true on an A-basis of A^n. To conclude, we only need to show that $r_{|M}$ takes its values in L. However, if $m \in M$, then $r(m) = l + n$ with $l \in L$ and $n \in N$, and $s(r(m)) = m = s(l) + n$, so $s(l) = m$ and $n = 0$, hence $r(m) = l \in L$. □

Proof of Theorem 4.19. We only have to prove the second part of the theorem, since the first part is contained in Lemma 4.20. If $P = \bigoplus_{i=1}^r P_i$ is a direct sum of principal indecomposable modules, then since each P_i is a direct summand of A, P is a direct summand of A^r, hence projective by Lemma 4.20. Conversely, suppose that P is projective, hence a direct summand of some A^n. Let $P = \bigoplus_{i=1}^r P_i$ be a decomposition of P in indecomposable A-modules, and Q a complement of P in A^n, with decomposition $Q = \bigoplus_{j=1}^s Q_j$. Then,

$$A^n = \left(\bigoplus_{i=1}^r P_i\right) \oplus \left(\bigoplus_{j=1}^s Q_j\right)$$

is a decomposition of A^n in indecomposable modules, and by Theorem 4.15, this decomposition is unique up to isomorphisms. Since A^n contains n copies of each principal indecomposable module, each P_i (and each Q_j) is isomorphic to a principal indecomposable module, and P is a direct sum of principal indecomposables. □

Remark. The second part of Theorem 4.19 ensures that a projective module is indecomposable if and only if it is a principal indecomposable module.

The **Grothendieck group of finite-dimensional projective modules** over A is the group $\mathrm{K}_0(A)$ built by symmetrization of the monoid of classes of isomorphism of projective A-modules. By Theorem 4.19, it admits for basis over $\mathbb{Z}$ the principal indecomposable modules of A. There is a natural map $c : \mathrm{K}_0(A) \to \mathrm{R}_0(A)$ (the so-called **Cartan map**) obtained by sending the class of isomorphism of a projective module M to the class $[M] \in \mathrm{R}_0(A)$; indeed, the relations of $\mathrm{R}_0(A)$ are compatible with isomorphisms. Usually, the Cartan map c is not an isomorphism. However, we shall see in a moment that the two Grothendieck groups $\mathrm{K}_0(A)$ and $\mathrm{R}_0(A)$ are always isomorphic (though not by c).

▷ *Top and projective cover of a module.*

Let A be a finite-dimensional k-algebra, and M be an A-module.

Definition 4.21. *A* ***projective cover*** *of M is a projective module P endowed with a surjective map $s : P \to M$ that is* ***essential****, which means that any restriction of s to a strict submodule of P is not surjective.*

Proposition 4.22. *Every finite-dimensional A-module M admits a projective cover P. Moreover, given two projective covers $s_1 : P_1 \to M$ and $s_2 : P_2 \to M$, there exists an isomorphism of modules $\phi : P_1 \to P_2$ such that $s_1 = s_2 \circ \phi$.*

Lemma 4.23. *Let P be a projective module, and L, M be two other modules. Suppose that we have two surjective maps $s : P \to M$ and $t : L \to M$. Then, there exists a morphism of A-modules $\phi : P \to L$ such that $s = t \circ \phi$.*

Proof. Let N be a module such that $P \oplus N = A^n$ is free over A. One obtains a surjective map $A^n \to M$ by composing the projection $\pi : A^n \to P$ and $s : P \to M$. Let $(e_i)_{i\in[\![1,n]\!]}$ be a basis of A^n. Since t is surjective, there exists $(f_i)_{i\in[\![1,n]\!]}$ in Q such that

$$t(f_i) = s \circ \pi(e_i) \quad \text{for all } i \in [\![1,n]\!].$$

Set then $\phi(e_i) = f_i$: this defines a morphism $\phi : A^n \to Q$, such that $s \circ \pi = t \circ \phi$. By restriction of these maps to P, we get $s = t \circ \phi_{|P}$, hence the existence of a map $\phi : P \to Q$ such that the following diagram is commutative:

□

Remark. One can show that the content of Lemma 4.23 is actually another characterization of projective modules.

Proof of Proposition 4.22. We call **length** of a module its number of composition factors; by Jordan–Hölder theorem, this number is well defined. If M is a finite-dimensional A-module, it is finitely generated, so there exists a surjective morphism $A^n \to M$ for some n. In particular, the set

$$\{s : P \to M,\ P \text{ projective module, } s \text{ surjective morphism}\}$$

is not empty, and one can choose an element of minimal length in it. Let us show that in this case, the map $s : P \to M$ is essential. Suppose that there exists a submodule $Q \subset P$ such that $s_{|Q} : Q \to M$ is surjective; one can suppose that it is of minimal dimension over k. Then, by the previous lemma, there exists a morphism $\phi : P \to Q$ such that $s_{|Q} \circ \phi = s$. It follows from this statement that $P = Q \oplus (\mathrm{Ker}\,\phi)$. Indeed, since $s_{|Q} = s_{|Q} \circ \phi_{|Q}$ is surjective, $s_{|\phi(Q)}$ is surjective, and by minimality of Q, $\phi(Q) = Q$, so $\phi_{|Q}$ is an automorphism of Q (it is a surjective endomorphism of a k-vector space of finite dimension). As a consequence, $Q \cap (\mathrm{Ker}\,\phi) = \{0\}$. On the other hand, if $x \in P$, then there exists $y \in Q$ such that $\phi(y) = \phi(x)$, and then $x = y + (x - y)$ with $y \in Q$ and $(x - y) \in \mathrm{Ker}\,\phi$. Therefore, Q is a direct summand of a projective module, hence projective and endowed with a surjective map $s : Q \to M$: this contradicts the minimality of P. Thus, we have proven that $s : P \to M$ is essential, and that M admits a projective cover.

For the unicity, consider two projective covers $s_1 : P_1 \to M$ and $s_2 : P_2 \to M$. Again by Lemma 4.23, there exists a map $\phi : P_1 \to P_2$ such that the following diagram is commutative:

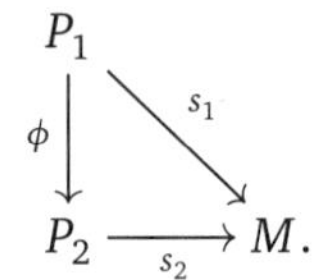

By the same argument as above, $P_1 = (\mathrm{Ker}\,\phi) \oplus \phi^{-1}(P_2)$. Since s_1 is essential, $\mathrm{Ker}\,\phi = \{0\}$, hence ϕ is an isomorphism between P_1 and P_2. □

As a consequence, it makes sense to speak of *the* projective cover $P(M)$ of an A-module M; it is in some sense the "best projective approximation" of M. There is a similar construction which from an arbitrary module M yields a semisimple module $T(M)$, the so-called top of M. To begin with, let us see how to build a semisimple algebra starting from an arbitrary algebra A.

Proposition 4.24. *Let A be a finite-dimensional k-algebra. If $a \in A$, then the following assertions are equivalent:*

(J1) For any simple A-module M, $aM = 0$.

(J2) The element a is in the intersection of all maximal left ideals of A.

(J3) For every $x \in A$, $1 - xa$ is left invertible.

The set of all such elements a is a left ideal of A called the ***Jacobson radical*** *of A, and denoted* $\mathrm{rad}(A)$.

Proof. (J1) ⇒ (J2). Let I be a maximal ideal of A, and $M = A/I$, which is a simple module. Since $aM = 0$, $a \mod I = 0$, and a belongs to I.

(J2) ⇒ (J3). Suppose that $1 - xa$ is not left invertible. Then, 1_A does not belong to $A(1 - xa)$, so one can choose a maximal ideal I such that I contains $A(1 - xa)$. If a belongs also to I, then $1 - xa$ and xa belong to I, hence $1 \in I$, which is impossible for a maximal ideal. So, $a \notin I$.

(J3) $\Rightarrow$ (J1). Let M be a simple A-module, and suppose that there exists $m \in M$ such that $am \neq 0$. Then, by simplicity, $Aam = M$, so there exists $x \in A$ such that $m = xam$. Since $(1 - xa)$ is left invertible, there is some $y \in A$ such that

$$m = (y(1 - xa))m = y0 = 0.$$

This is a contradiction. □

Proposition 4.25. *The Jacobson radical* rad(A) *is also the intersection of all maximal right ideals of A. In particular,* rad(A) *is a two-sided ideal of A, and $A/$*rad(A) *is an algebra.*

Proof. It suffices to find a condition for $a \in \mathrm{rad}(A)$ that is left-right symmetric. We claim that $a \in \mathrm{rad}(A)$ if and only if $1 - xay$ is invertible for every $x, y \in A$. Clearly, this implies the third condition in the previous proposition, so this is a sufficient condition for being in rad(A). Conversely, suppose that $a \in \mathrm{rad}(A)$. For $y \in A$, and any simple module M, $ayM \subset aM = 0$, so $ay \in \mathrm{rad}(A)$ and there exists a left inverse $(1 - b)$ with

$$(1 - b)(1 - axy) = 1.$$

Then, $b = -(1 - b)axy$, and since rad(A) is a left ideal, $b \in \mathrm{rad}(A)$ and $(1 - b)$ has a left inverse, which is necessarily $1 - axy$. Hence, $1 - axy$ is invertible for any $x, y \in A$. □

Theorem 4.26 (Jacobson). *The quotient $A/$*rad(A) *is a semisimple algebra, and any other semisimple quotient A/I with I bilateral ideal satisfies* rad(A) $\subset I$. *An algebra A is semisimple if and only if* rad(A) $= 0$.

Lemma 4.27 (Nakayama). *Let M be a non-zero module over A. Then,* rad(A)$M \neq M$.

Proof. Let $N \subsetneq M$ be a maximal strict submodule of M. The quotient M/N is a simple A-module, hence, for any $a \in \mathrm{rad}(A)$, $a\,M/N = 0$, i.e., $aM \subset N$. Therefore, $\mathrm{rad}(A)\,M \subset N \subsetneq M$. □

Proof of Theorem 4.26. Since rad(A) is the intersection of all maximal left ideals of A, and since A is finite-dimensional, one can choose maximal ideals $I_1, \ldots, I_r$ such that

$$A \supsetneq I_1 \supsetneq I_1 \cap I_2 \supsetneq \cdots \supsetneq I_1 \cap \cdots \cap I_r = \mathrm{rad}(A).$$

In this setting, no I_i contains the intersection $J_i = \bigcap_{j \neq i} I_j$. Therefore, $I_i + J_i = A$ for all i, since I_i is a maximal ideal. Consider then the map

$$\begin{aligned} \psi : A/\mathrm{rad}(A) &\to \prod_{i=1}^{r} A/I_i \\ [a]_{\mathrm{rad}(A)} &\mapsto ([a]_{I_i})_{i \in [\![1,r]\!]}. \end{aligned}$$

If $\psi([a]_{\mathrm{rad}(A)}) = 0$, then $a \in I_i$ for all i, so $a \in \mathrm{rad}(A)$ and $[a]_{\mathrm{rad}(A)} = 0$; hence, ψ is an injective morphism of A-module. For $i \in [\![1,r]\!]$, write $1 = x_i + y_i$ with $x_i \in I_i$ and $y_i \in J_i$. Then,

$$\psi([y_i]_{\mathrm{rad}(A)}) = (0,\dots,0,[1]_{I_i},0,\dots,0),$$

so $\psi(A/\mathrm{rad}(A))$ contains a generating set of $\prod_{i=1}^r A/I_i$, and ψ is surjective. Thus, $A/\mathrm{rad}(A)$ is isomorphic via ψ to a direct sum of simple modules A/I_i, hence is semisimple as an A-module. However, given a two-sided ideal I, the structure of A-module on A/I is the same as the structure of A/I-module, so $A/\mathrm{rad}(A)$ is also semisimple as an algebra.

If $\mathrm{rad}(A) = 0$, then by the previous discussion $A = A/\mathrm{rad}(A)$ is semisimple as an A-module, hence is a semisimple algebra. Conversely, suppose that A is a semisimple algebra, and decompose it as a sum of simple left ideals:

$$A = \bigoplus_{i=1}^r M_i.$$

By Nakayama's lemma, for all i, $\mathrm{rad}(A)\,M_i \subsetneq M_i$, and on the other hand, $\mathrm{rad}(A)\,M_i$ is a left submodule of M_i, because $\mathrm{rad}(A)$ is a right ideal:

$$A(\mathrm{rad}(A)\,M_i) = (A\,\mathrm{rad}(A))\,M_i = \mathrm{rad}(A)\,M_i.$$

By simplicity, $\mathrm{rad}(A)\,M_i = 0$, and therefore, $\mathrm{rad}(A) = \mathrm{rad}(A)A = \bigoplus_{i=1}^r \mathrm{rad}(A)\,M_i = 0$. Finally, let I be a bilateral ideal such that A/I is semisimple. The previous discussion shows that there are maximal left ideals $J_1,\dots,J_r$ of A/I such that $A/I = \bigoplus_{i=1}^r (A/I)/J_i$ is a decomposition in simple modules. These ideals correspond to maximal left ideals I_i of A such that $I_i/I = J_i$. Then, $I = \bigcap_{i=1}^r I_i \supseteq \mathrm{rad}(A)$ since $\mathrm{rad}(A)$ is the intersection of *all* maximal left ideals of A. □

Jacobson's theorem 4.26 ensures that $A/\mathrm{rad}(A)$ is the "largest" quotient of A that is semisimple. Now, if M is an A-module, then the **top** of M is defined as the quotient module $T(M) = M/\mathrm{rad}(A)\,M$.

Proposition 4.28. *If M is an A-module, then its top $T(M)$ is semisimple, and for any other semisimple quotient M/N, there exists a morphism of A-modules $\phi : T(M) \to M/N$ that makes the following diagram commutative:*

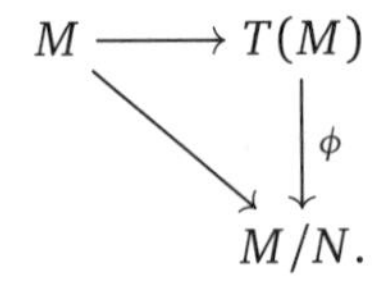

Proof. The structure of $M/\mathrm{rad}(A)\,M = A/\mathrm{rad}(A) \otimes_A M$ as an $A/\mathrm{rad}(A)$-module is the same as its structure as an A-module, and by Jacobson theorem $A/\mathrm{rad}(A)$

is semisimple, so $T(M)$ is a semisimple A-module. Suppose that M/N is another quotient that is a semisimple A-module. Then, $\mathrm{rad}(A)(M/N) = 0$, hence $\mathrm{rad}(A)M \subset N$, and this inclusion of submodules yields the factorization of projection maps by a morphism $\phi : T(M) \to M/N$. □

We can now relate the two notions of projective cover and top:

Proposition 4.29. *For any A-module M,*

$$P(T(M)) = P(M) \quad ; \quad T(P(M)) = T(M).$$

Proof. Consider the quotient map $\pi : M \to T(M)$; this is an essential surjection. Indeed, let N be a submodule of M, such that $\pi(N) = T(M)$. Then, $N + \mathrm{rad}(A)M = M$, and by quotienting by N, $\mathrm{rad}(A)M/N = M/N$. By Nakayama's lemma, $M/N = 0$, hence $M = N$. As a consequence, the composition of maps $P(M) \to M \to T(M)$ is an essential surjection, and since $P(M)$ is projective, by unicity of the projective cover, $P(T(M)) = P(M)$.

On the other hand, let N be the kernel of the essential surjection $s_M : P(M) \to M$. If $Q \subset P(M)$ is a maximal strict submodule of $P(M)$, then $N \subset Q$, because otherwise $Q + N = M$ and $s_M : Q \to M$ would be surjective. So, N is included in the intersection of all maximal strict submodules of $P(M)$. We claim that this intersection is $\mathrm{rad}(A)P(M)$. Indeed, if Q is a maximal strict submodule of a module P, then $\mathrm{rad}(A)(P/Q) = 0$ because P/Q is simple, so $\mathrm{rad}(A)P \subset Q$, and $\mathrm{rad}(A)P$ is included in the intersection of all maximal submodules. Conversely, consider a decomposition of the semisimple module $P/\mathrm{rad}(A)P = \bigoplus_{i\in I} S_i$, where the S_i are simple A-modules. The projection map $P/\mathrm{rad}(A)P \to S_i$ has for kernel a maximal submodule of $P/\mathrm{rad}(A)P$, which is of the form $Q_i/\mathrm{rad}(A)P$ for some maximal submodule Q_i of P. Then, $\mathrm{rad}(A)P$ contains $\bigcap_{i\in I} Q_i$, and a fortiori the intersection of all maximal submodules of P. Going back to our initial problem, we now have $N \subset \mathrm{rad}(A)P(M)$, and on the other hand, since s_M is surjective, it maps $\mathrm{rad}(A)P(M)$ to $\mathrm{rad}(A)M$. It follows that

$$\begin{aligned} T(P(M)) = P(M)/\mathrm{rad}(A)P(M) &= (P(M)/N)/((\mathrm{rad}(A)P(M))/N) \\ &= M/\mathrm{rad}(A)M = T(M). \end{aligned}$$

□

Proposition 4.30. *Let M be a finite-dimensional module over A. If M is simple, then $P(M)$ is a principal indecomposable module. If M is a principal indecomposable module, then $T(M)$ is simple.*

Proof. Suppose M simple. $P(M) = K \oplus L$, then the surjection $s : P(M) \to M$ yields two morphisms $s_K : K \to M$ and $s_L : L \to M$. As M is simple, and $s_K + s_L \neq 0$, one of the two maps s_K and s_L is surjective. As s is essential, $P(M) = K$ or $P(M) = L$, hence $P(M)$ is indecomposable.

Conversely, notice that if N is a module with indecomposable projective cover, then N is itself indecomposable. Indeed, for any modules K and L, $P(K \oplus L) = P(K) \oplus P(L)$, since both sides are projective and the direct sum of essential surjections is an essential surjection. Now, let M be a principal indecomposable module; it is the projective cover of $T(M)$ since $M = P(M) = P(T(M))$. Therefore, $T(M)$ is indecomposable, and since it is also semisimple, it is a simple A-module. □

We can finally state the isomorphism theorem for Grothendieck groups of a finite-dimensional algebra; it is one of the cornerstones of the general representation theory of algebras.

Theorem 4.31. *Let A be a finite-dimensional algebra. The maps P and T yield a bijection between the set $\{S_1, \ldots, S_r\}$ of isomorphism classes of simple modules over A, and the set $\{P_1, \ldots, P_r\}$ of isomorphism classes of principal indecomposable modules over A. In particular, these two finite sets have the same cardinality, so the Grothendieck groups $\mathrm{K}_0(A)$ and $\mathrm{R}_0(A)$ are free abelian groups of same rank r, hence isomorphic.*

Proof. Let S_1 and S_2 be two simple A-modules. If $P_1 = P(S_1)$ and $P_2 = P(S_2)$ are two isomorphic indecomposable A-modules, then $S_1 = T(S_1) = T(P(S_1)) = T(P_1) = T(P_2) = S_2$ are isomorphic, and conversely. So, the maps P and T indeed induce bijections between the classes of isomorphism of simple and indecomposable projective modules. On the other hand, the set of classes of principal indecomposable modules over A is finite, because it is a subset of a decomposition of A as a sum of indecomposable left ideals. □

Hence, in some sense, the classification of all simple modules of A is equivalent to the classification of all principal indecomposable modules, the correspondence being given by projective covers and semisimple tops. In the case of semisimple algebras, the two notions coincide and $\mathrm{K}_0(A) = \mathrm{R}_0(A)$; in the general case, one still has a natural isomorphism between $\mathrm{K}_0(A)$ and $\mathrm{R}_0(A)$. One also has a natural perfect pairing $\mathrm{K}_0(A) \times \mathrm{R}_0(A) \to \mathbb{Z}$ given by

$$\langle P_i \mid S_j \rangle = \delta_{ij},$$

where $S_i = T(P_i)$ and $P_i = P(S_i)$. This pairing will play an important role in the representation theory of the specialized Hecke algebra $\mathfrak{H}_0(n)$ (cf. Section 6.3); it is a generalization of the pairing $\mathrm{R}_0(\mathbb{C}G) \times \mathrm{R}_0(\mathbb{C}G) \to \mathbb{Z}$ introduced in Chapters 1 and 2 for group algebras of finite groups.

Remark. Using the Jacobson radical, one can show that every simple module over an algebra A appears as a composition factor of A viewed as an A-module. This is similar to the statement that for a semisimple algebra, every simple module appears as a component of the decomposition in simple modules of A. To prove this claim, consider a simple module M over A. By Nakayama's lemma, notice that $\mathrm{rad}(A)$ acts by zero on M, since $\mathrm{rad}(A)M \subsetneq M$. Therefore, M can be considered as a simple module for the semisimple algebra $A/\mathrm{rad}(A)$. In this semisimple setting,

M appears necessarily in a composition series (and actually a semisimple decomposition) of $A/\text{rad}(A)$. Such a composition series is a part of a larger composition series for A; whence the claim. This gives another proof of the fact that there is only a finite number of classes of isomorphism of simple A-modules.

Before we go on, let us summarize our presentation of the general representation theory of algebras. If A is an algebra that is not semisimple, then:

1. The notion of decomposition of an A-module M in simple components is replaced by the two notions of composition series, and of decomposition in indecomposable components.

2. These two notions correspond to two Grothendieck groups $\mathrm{R}_0(A)$ and $\mathrm{K}_0(A)$, which are free over $\mathbb{Z}$ and with the same rank. The correspondence between the generators of $\mathrm{R}_0(A)$ (simple modules) and the generators of $\mathrm{K}_0(A)$ (indecomposable projective modules) is given by projective covers and semisimple tops.

4.3 Brauer–Cartan deformation theory

In order to compare the representation theory of generic and of specialized Hecke algebras, we now turn to the so-called *deformation theory of algebras*. A general problem is the effect of a change of base field or of a specialization on the Grothendieck groups of an algebra. Consider the following setting. We fix a field k, a commutative k-algebra A, another field K containing A, and a morphism of k-algebras $\theta : A \to k$. Let H be an algebra over A (here we extend the definition of algebra over a field given at the beginning of Section 1.4, by replacing the base field by a base ring or a base algebra). By using tensor products, one obtains the two algebras $kH = k \otimes_A H$ (over the field k) and $KH = K \otimes_A H$ (over the field K). It is then a natural idea to try to compare the two Grothendieck groups $\mathrm{R}_0(kH)$ and $\mathrm{R}_0(KH)$ (one could also try to compare $\mathrm{K}_0(kH)$ and $\mathrm{K}_0(KH)$).

Example. With $A = \mathbb{C}[q]$, $k = \mathbb{C}$, $K = \mathbb{C}(q)$, $H = \mathfrak{H}(n, \mathbb{C}[q])$, and $\theta_z = \theta : \mathbb{C}[q] \to \mathbb{C}$ given by $\theta(P) = P(z)$ for some $z \in \mathbb{C}$, one obtains the two Hecke algebras $kH = \mathfrak{H}_z(n)$ (specialized) and $KH = \mathfrak{H}(n)$ (generic). It is the purpose of this section and of the next one to compare their two Grothendieck groups.

Example. If one removes the hypothesis that all fields and algebras are defined over the base field k, then the following problem falls in the previously described setting: we take $A = \mathbb{Z}$, $k = \mathbb{Z}/p\mathbb{Z}$, $K = \mathbb{C}$ and $H = \mathbb{Z}\mathfrak{S}(n)$. Then, $kH = (\mathbb{Z}/p\mathbb{Z})\mathfrak{S}(n)$ and $KH = \mathbb{C}\mathfrak{S}(n)$, and one wants to compare the representation theory of the symmetric group in characteristic p, and its representation theory in characteristic 0 over $\mathbb{C}$. This is the purpose of *modular representation theory*, and the discussion of this paragraph will shed light on some tools of this theory.

To get an idea of the kind of results that one can expect, suppose for a moment that $A = k$ and that $H = kH$ is a semisimple k-algebra; then, it is isomorphic as a k-algebra to a direct sum $\bigoplus_{\lambda \in \widehat{H}} \mathrm{M}(d_\lambda, C^\lambda)$, and

$$\mathrm{R}_0(kH) = \bigoplus_{\lambda \in \widehat{H}} \mathbb{Z}\,(C^\lambda)^{d_\lambda}.$$

Suppose to simplify that the algebra is split, i.e., all the division rings C^λ are equal to k. If $K \mid k$ is an extension field, then

$$kH = \bigoplus_{\lambda \in \widehat{H}} \mathrm{M}(d_\lambda, k)$$
$$KH = K \otimes_k kH = \bigoplus_{\lambda \in \widehat{H}} \mathrm{M}(d_\lambda, K)$$

and

$$\mathrm{R}_0(KH) = \bigoplus_{\lambda \in \widehat{H}} \mathbb{Z}\,K^{d_\lambda}.$$

Thus, for a split semisimple algebra and in the case of a change of the base field k by an extension K, the Grothendieck groups $\mathrm{R}_0(kH)$ and $\mathrm{R}_0(KH)$ are isomorphic. We shall actually see that if kH is a split algebra, then the result holds without the hypothesis of semisimplicity, see Theorem 4.37. In the general case of a diagram

$$\begin{array}{ccc} A & \hookrightarrow & K \\ {\scriptstyle\theta}\downarrow & & \\ k & & \end{array},$$

one can still expect relations between the two Grothendieck groups $\mathrm{R}_0(kH)$ and $\mathrm{R}_0(KH)$. These relations have been studied in particular by Brauer, Cartan and Nesbitt, and they lead to Tits' deformation theorem 4.39, which gives sufficient conditions to have an isomorphism of groups between $\mathrm{R}_0(kH)$ and $\mathrm{R}_0(KH)$. In this section, we present this deformation theory, which will then be applied to the Hecke algebras in Section 4.4.

▷ *The Brauer–Cartan setting and the Brauer–Nesbitt map.*

To get sufficiently strong results, we need some assumptions on the morphism of specialization $\theta : A \to k$. Recall that a **valuation ring** $\mathscr{O} \subset K$ is an integral domain with $K = \mathrm{Frac}(\mathscr{O})$, and such that, for every $x \in K$, $x \in \mathscr{O}$ or $x^{-1} \in \mathscr{O}$. Such a ring is always a **local ring**, that is to say a ring with a unique maximal ideal. Actually, a valuation ring has its ideals totally ordered for the inclusion. Indeed, suppose that $\mathfrak{p}$ and $\mathfrak{q}$ are two ideals of $\mathscr{O}$ with $\mathfrak{p} \not\subset \mathfrak{q}$ and $\mathfrak{q} \not\subset \mathfrak{p}$. Then, if $x \in \mathfrak{p} \setminus \mathfrak{q}$ and $y \in \mathfrak{q} \setminus \mathfrak{p}$, x/y or y/x belong to $\mathscr{O}$, and therefore, either $x = (x/y)y \in \mathfrak{q}$ or $y = (y/x)x \in \mathfrak{p}$; hence a contradiction. In the sequel, we fix A, k and K as before, and we assume that there exists a valuation ring $\mathscr{O}$ for K such that:

(BC1) One has $A \subset \mathscr{O} \subset K$, and A is **integrally closed** in K, i.e., the solutions in K of monic polynomial equations $X^n + a_{n-1}X^{n-1} + \cdots + a_0 = 0$ with the a_i's in A also belong to A.

(BC2) If $\mathfrak{m}_{\mathscr{O}}$ is the maximal ideal of $\mathscr{O}$, then θ is the map $A \to \mathscr{O}/\mathfrak{m}_{\mathscr{O}} = k$.

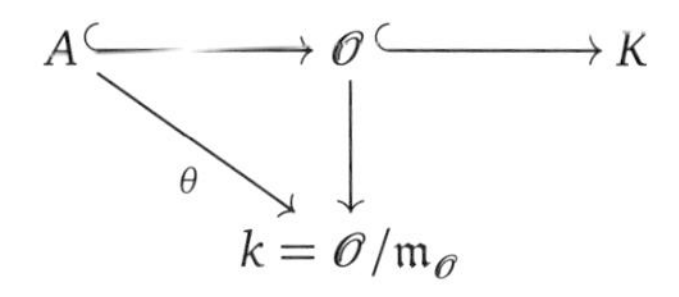

Figure 4.2
The Brauer–Cartan setting: $K = \mathrm{Frac}\,\mathscr{O}$ is the fraction field, and $k = \mathscr{O}/\mathfrak{m}_{\mathscr{O}}$ is the residue field.

We shall refer to this situation (Conditions (BC1) and (BC2)) as the **Brauer–Cartan setting**. Under these conditions, with H finitely generated and free A-algebra, we are going to prove the existence of a morphism of groups $d_\theta : \mathrm{R}_0(KH) \to \mathrm{R}_0(kH)$ called the decomposition map, and which actually does not depend on the choice of a valuation ring $\mathscr{O}$ that realizes the morphism θ. Our theory will apply in particular to the case where $A = \mathbb{C}[q]$, $k = \mathbb{C}$, $K = \mathbb{C}(q)$, $H = \mathfrak{H}(n, \mathbb{C}[q])$ and

$$\mathscr{O} = \mathbb{C}[q]_{(q-z)}.$$

Thus, $\mathscr{O}$ is the localization of $\mathbb{C}[q]$ with respect to the ideal $(q-z)\,\mathbb{C}[q]$, that is to say, the set of rational fractions without factor $(q-z)^{-p}$ with $p \geq 1$ in their decompositions as products of irreducible polynomials. This ring is indeed a valuation ring, since if $R(q) \in \mathbb{C}(q)$, then either $R(q) \in \mathbb{C}[q]_{(q-z)}$, or $R(q) = (q-z)^{-p}\, R_1(q)$ where $R_1(q)$ has no factor $q - z$. In the latter case, $R(q)^{-1} = (q-z)^p\, R_1(q)^{-1}$ belongs to $\mathbb{C}[q]_{(q-z)}$. On the other hand, the projection of $\mathscr{O}$ on its residue field is trivially the evaluation at $q = z$, so $\theta_z = \theta : \mathbb{C}[q] \to \mathbb{C}$ yields the specialized Hecke algebra $kH = \mathfrak{H}_z(n)$, whereas $KH = \mathfrak{H}(n)$ is the generic Hecke algebra.

Let H be a finitely generated free A-algebra; this means that there exists an A-basis $(h_1, \ldots, h_n)$ of H such that every element $h \in H$ writes uniquely as $h = \sum_{i=1}^n c_i h_i$ with the c_i's in A. We set $\mathscr{O}H = \mathscr{O} \otimes_A H$, and $KH = K \otimes_A H = K \otimes_{\mathscr{O}} \mathscr{O}H$. An arbitrary A-basis $(h_1, \ldots, h_n)$ of H is also an $\mathscr{O}$-basis of $\mathscr{O}H$ and a K-basis of KH.

Lemma 4.32. *Let $\mathscr{O}$ be a valuation ring, and U be a finitely generated and torsion-free module over $\mathscr{O}$. Then, U is free over $\mathscr{O}$.*

Proof. Fix a minimal generating set $(u_1, \ldots, u_d)$ of U over $\mathscr{O}$, and let us show that it is free. Suppose that $\sum_{i=1}^d o_i\, u_i = 0$, with some o_i's non-zero. Since ideals of $\mathscr{O}$ are totally ordered for the inclusion, by looking at principal ideals, one sees that elements of $\mathscr{O}$ are totally ordered for the divisibility. Therefore, there is some o_i, say $o_1 \neq 0$ that divides in $\mathscr{O}$ all the other o_j. But then, since the module is torsion-free, this means that

$$u_1 = -\sum_{i=2}^{d} (o_i/o_1)\, u_i,$$

which contradicts the minimality of the generating set $(u_1, \ldots, u_d)$. □

Corollary 4.33. *If V^K is a module over KH, then one can find a K-basis $(v_1, \ldots, v_d)$ of V^K such that if $\rho^{V^K} : KH \to \mathrm{End}_K(V^K)$ is the morphism of algebras underlying the module structure of V^K, then the matrix of $\rho^{V^K}(h)$ with respect to the basis $(v_1, \ldots, v_d)$ belongs to $\mathrm{M}(d, \mathscr{O})$ for any $h \in H \subset KH$.*

Proof. Fix an arbitrary K-basis $(u_1, \ldots, u_d)$ of V^K, and consider the $\mathscr{O}$-submodule $U \subset V^K$ spanned linearly by the elements u_j. By construction, $K \otimes_{\mathscr{O}} U = V^K$. On the other hand, U is a finitely generated $\mathscr{O}$-module, and since it is contained in the KH-module V^K, it is torsion free, hence free by the previous lemma. Let $(v_1, \ldots, v_d)$ be an $\mathscr{O}$-basis of U; it is also a K-basis of KH, and for every $h \in H$ (and even in $\mathscr{O}H$), the action of h on $(v_1, \ldots, v_d)$ is indeed given by a matrix with coefficients in $\mathscr{O}$. □

As a consequence, for every KH module V^K, there exists an $\mathscr{O}H$-module $V^{\mathscr{O}}$ such that

$$V^K = K \otimes_{\mathscr{O}} V^{\mathscr{O}} = KH \otimes_{\mathscr{O}H} V^{\mathscr{O}}.$$

Beware that $V^{\mathscr{O}}$ is not necessarily unique up to isomorphism of $\mathscr{O}H$-modules. Nonetheless, we call **modular reduction** of V^K the kH-module

$$V^k = k \otimes_{\mathscr{O}} V^{\mathscr{O}} = H \otimes_{\mathscr{O}H} V^{\mathscr{O}}.$$

This construction depends a priori on the choice of a realization $V^{\mathscr{O}}$ of the module V^K over the valuation ring $\mathscr{O}$. In Theorem 4.38, we shall see that the class of the modular reduction V^k in the Grothendieck ring $\mathrm{R}_0(kH)$ is usually independent of this choice.

Let H be an A-algebra, $K \supset A$ and V^K be a finitely generated KH-module. The **Brauer–Nesbitt map** p^K sends a module V^K over KH to the map $p^K(V^K)$ in $\mathrm{Maps}(H \to K[X])$ that is defined by the values of characteristic polynomials

$$p^K(V^K)(h) = \det\left(\rho^{V^K}(h) - X\,\mathrm{id}_{V^K}\right).$$

By Corollary 4.33, in the setting $A \subset \mathscr{O} \subset K$ previously described, p^K actually takes its values in $\mathrm{Maps}(H \to \mathscr{O}[X])$, since one can find a basis of V^K such that the representation matrices of elements of H are in $\mathrm{M}(d, \mathscr{O})$.

Lemma 4.34. *Assuming A integrally closed in K, the ring A is the intersection of all valuation rings $\mathscr{O}$ with $A \subset \mathscr{O} \subset K$ and $K = \mathrm{Frac}(\mathscr{O})$. Therefore, $\mathrm{p}^K(V^K)$ belongs to* $\mathrm{Maps}(H \to A[X])$.

Proof. Denote A' the intersection of all the valuation rings $\mathscr{O}$ that contain A; by construction, $A' \supset A$. Conversely, let x be an element that is not in A, and $y = x^{-1}$. The unity 1 is not in the ideal $yA[y]$ of $A[y]$: indeed, otherwise,

$$1 = a_{n-1} + a_{n-2}y + \cdots + a_0 y^{n-1}$$

gives by multiplication by x^n a monic equation satisfied by x, and since A is integrally closed, this would imply that $x \in A$. Let $\mathfrak{m}$ be a maximal ideal of $B = A[y]$ that contains $yA[y]$. We claim that there exists a valuation ring $\mathscr{O}$ that contains B, and such that $B \cap \mathfrak{m}_{\mathscr{O}} = \mathfrak{m}$. Notice that this implies $A = A'$: indeed, one has then $y \in \mathfrak{m} \subset \mathfrak{m}_{\mathscr{O}}$, and therefore, $x = y^{-1} \notin \mathscr{O}$, and $x \notin A'$.

To prove the claim, consider the set $\mathscr{R}$ of all subrings C of K that contain B, and such that $\mathfrak{m}C \neq C$, which is equivalent to $1 \notin \mathfrak{m}C$. By Zorn's lemma, one can choose a maximal element $\mathscr{O}$ for the inclusion in $\mathscr{R}$. Let $\mathfrak{m}_{\mathscr{O}}$ be a maximal ideal of $\mathscr{O}$ that contains the ideal $\mathfrak{m}\mathscr{O}$. Notice that the localization $\mathscr{O}_{\mathfrak{m}_{\mathscr{O}}}$ contains $\mathscr{O}$ and is in $\mathscr{R}$, so by maximality $\mathscr{O} = \mathscr{O}_{\mathfrak{m}_{\mathscr{O}}}$ and $\mathscr{O}$ is a local ring. Let us show that it is actually a valuation ring for K. If $z \in K$ is not in $\mathscr{O}$, then $\mathscr{O}[z]$ is strictly bigger than $\mathscr{O}$, so it cannot be in $\mathscr{R}$. Therefore, $1 \in \mathfrak{m}\mathscr{O}[z]$, that is to say that one has an equation $1 = a_0 + a_1 z + \cdots + a_n z^n$ with the a_i's in $\mathfrak{m}\mathscr{O} \subset \mathfrak{m}_{\mathscr{O}}$. As $1 - a_0 \notin \mathfrak{m}_{\mathscr{O}}$, it is inversible, so it leads to another equation

$$1 = b_1 z + \cdots + b_n z^n$$

with the b_i's in $\mathfrak{m}\mathscr{O}$. Suppose now that z^{-1} also does not belong to $\mathscr{O}$, and write an equation

$$1 = c_1 z^{-1} + \cdots + c_p z^{-p}$$

with the c_i's in $\mathfrak{m}\mathscr{O}$. One can assume that n and p are minimal integers for such equations. If $n \geq p$, then

$$1 - (b_1 z + \cdots + b_{n-1} z^{n-1}) = b_n z^n = b_n (c_1 z^{n-1} + \cdots + c_p z^{n-p}),$$

which gives an equation for z with a degree $n - 1 \leq n$. One gets the same contradiction if $n < p$ by interchanging the roles of z and z^{-1}. Therefore, $\mathscr{O}$ is indeed a valuation ring. Finally, the intersection $\mathfrak{m}_{\mathscr{O}} \cap B$ contains $\mathfrak{m}$, and since $\mathfrak{m}$ is a maximal ideal, it is equal to $\mathfrak{m}$. This ends the proof of the claim. The second part of the lemma is then a trivial consequence of the first part and of Corollary 4.33. □

In the following, given an algebra H, we denote $\mathrm{R}_0^+(H)$ the submonoid of $\mathrm{R}_0(H)$ that is spanned by the classes of (simple) modules.

Lemma 4.35 (Brauer–Nesbitt). *The map*

$$\begin{aligned} p^K : \mathrm{R}_0^+(KH) &\to \mathrm{Maps}(H \to A[X]) \\ [V^K] &\mapsto (h \mapsto \det(\rho^{V^K}(h) - X\,\mathrm{id}_{V^K})) \end{aligned}$$

is well defined, and a morphism of monoids, the set $\mathrm{Maps}(H \to A[X])$ *being endowed with the pointwise product of functions. Moreover, if the characters of the simple* KH*-modules are linearly independent, then this map is injective.*

Proof. Let us first show that p^K is compatible with the relations of the Grothendieck group. If

$$0 \longrightarrow M \longrightarrow P \longrightarrow N \longrightarrow 0$$

is an exact sequence of KH-modules, then one can find an adequate basis of P such that

$$\rho^P(h) = \begin{pmatrix} \rho^M(h) & * \\ 0 & \rho^N(h) \end{pmatrix}$$

for every $h \in KH$. As a consequence,

$$p^K(P) = p^K(M) \times p^K(N) = p^K(M \oplus N),$$

which shows the compatibility of p^K with classes in $\mathrm{R}_0^+(KH)$. Suppose now that the characters of simple KH-modules are independent, and let us show the injectivity of p^K. Since simple modules generate $\mathrm{R}_0^+(KH)$, if one denotes $S_1, \ldots, S_r$ a set of representatives of the classes of isomorphism of simple modules, then the injectivity amounts to

$$p^K\left(\bigoplus_{i=1}^r m_i S_i\right) = p^K\left(\bigoplus_{i=1}^r n_i S_i\right) \quad \Rightarrow \quad \forall i,\ m_i = n_i.$$

One can divide both sides by factors $p^K(S_i)$ if m_i and n_i are both positive, so without loss of generality, one can assume that for every i, only one of the two coefficients m_i and n_i is non-zero. One then has to show that all the m_i's and n_i's vanish in this situation. However, if χ_i denotes the irreducible character of S_i, then

$$\begin{aligned} \sum_{i=1}^r m_i\, \chi_i(h) &= (-1)^{d-1}\,[X]^{d-1}\left(p^K\left(\bigoplus_{i=1}^r m_i S_i\right)(h)\right) \\ &= (-1)^{d-1}[X]^{d-1}\left(p^K\left(\bigoplus_{i=1}^r n_i S_i\right)(h)\right) = \sum_{i=1}^r n_i\, \chi_i(h). \end{aligned}$$

Indeed, the trace of a morphism is up to a sign the second leading coefficient of its characteristic polynomial. Since the χ_i's are supposed linearly independent, in characteristic zero, this implies $m_i = n_i = 0$ for all index i. In positive characteristic ℓ, the identity implies that ℓ divides m_i and n_i for all i. By applying this argument several times, ℓ^p divides m_i and n_i for every i and every $p > 0$, so again $m_i = n_i = 0$ for every i. □

To apply the Brauer–Nesbitt lemma 4.35, we need sufficient conditions to get the linear independence of characters. Let K be an arbitrary field, and KH be an algebra over K. By Schur's lemma from Section 1.4, if M is a simple KH-module, then $\mathrm{End}_{KH}(M)$ is a K-division ring. We say that KH is a **split algebra** if for every simple KH-module, $\mathrm{End}_{KH}(M) = K$. This is for instance the case if K is an algebraically closed field.

Lemma 4.36. *Let KH be a split semisimple finite-dimensional K-algebra. The characters of the simple KH-modules are linearly independent, and they form a linear K-basis of the space of linear functions $\tau : KH \to K$ such that $\tau(ab) = \tau(ba)$. If KH is only assumed split, then the characters of simple modules are still linearly independent, so the conclusion of Lemma 4.35 holds.*

Proof. Suppose KH split semisimple. It is well known that on a matrix space $\mathrm{M}(d,K)$, the only linear functions that satisfy $\tau(ab) = \tau(ba)$ are the scalar multiples of the trace. Then, the lemma is a trivial consequence of the Wedderburn decomposition 1.24

$$KH = \bigoplus_{\lambda \in \widehat{KH}} \mathrm{End}_K(M^\lambda),$$

using the fact that the algebra is split to write each matrix block as a space of matrices over the field K.

Suppose now that KH is only a split algebra. Then, any simple KH-module yields a simple $KH/\mathrm{rad}(KH)$-module, since $\mathrm{rad}(KH)$ acts by 0 by Nakayama's lemma 4.27. It follows that the characters of simple KH-modules correspond to characters of simple $KH/\mathrm{rad}(KH)$-module, and this last algebra is semisimple, so one can use the first part of the lemma. □

The injectivity of the Brauer–Nesbitt map is a powerful tool in order to construct maps between Grothendieck groups. As a first example, let us state a general result on the change of base field for algebras:

Theorem 4.37. *Let kH be a finite-dimensional algebra over k, and $K \mid k$ be a field extension. The map*

$$\begin{aligned} e_K^k : \mathrm{R}_0^+(kH) &\to \mathrm{R}_0^+(KH) \\ [V] &\mapsto [K \otimes_k V] \end{aligned}$$

is well defined, and it induces a morphism of monoids that makes the following diagram commutative:

$$\begin{array}{ccc} \mathrm{R}_0^+(kH) & \xrightarrow{p^k} & \mathrm{Maps}(kH \to k[X]) \\ {\scriptstyle e_K^k}\downarrow & & \downarrow \\ \mathrm{R}_0^+(KH) & \xrightarrow[p^K]{} & \mathrm{Maps}(kH \to K[X]) \end{array}$$

If kH is split, then this map is an isomorphism that sends classes of simple kH-modules towards classes of simple KH-modules.

Proof. The compatibility of the map e_K^k with the relations of Grothendieck groups is evident, since if

$$0 \longrightarrow M \longrightarrow P \longrightarrow N \longrightarrow 0$$

is an exact sequence of kH-modules, then

$$0 \longrightarrow K \otimes_k M \longrightarrow K \otimes_k P \longrightarrow K \otimes_k N \longrightarrow 0$$

is an exact sequence of KH-modules. The commutativity of the diagram is obtained as follows: if $(v_1, \ldots, v_n)$ is a k-basis of a kH-module V, then it is also a K-basis of the KH-module $K \otimes_k V$, and therefore, for any $h \in kH$, the two matrices $\rho^V(h)$ and $\rho^{K \otimes_k V}(h)$ in this basis are the same, so $p^k(V)$ and $p^K(K \otimes_k V)$ are equal; in other words, $p^K \circ e_K^k = p^k$.

If kH is split, consider a simple kH-module V. The natural maps $kH \to kH/\mathrm{rad}(kH)$ and $kH/\mathrm{rad}(kH) \to \mathrm{End}_k(V)$ are surjective morphism of algebras. Therefore, by tensoring by K, $KH \to \mathrm{End}_K(K \otimes_k V)$ is a surjective morphism of algebras, which implies that $K \otimes_k V$ is a split simple KH-module. On the other hand, every simple KH-module appears in a composition series of KH, and such a composition series can be obtained from a composition series of kH by tensoring by K. So, the map e_K^k:

1. sends simple split modules of kH to simple split modules of KH;
2. allows one to construct every simple module of KH from a simple module of kH.

In particular, if kH is split, then KH is also split. It remains to see that if V_1 and V_2 are two non-isomorphic simple kH-modules, then $K \otimes_k V_1$ and $K \otimes_k V_2$ are non-isomorphic. However, this follows easily from Schur's lemma:

$$\begin{aligned}
\dim_K \mathrm{Hom}_{KH}(K \otimes_k V_1, K \otimes_k V_2) &= \dim_K(K \otimes_k \mathrm{Hom}_{kH}(V_1, V_2)) \\
&= \dim_k \mathrm{Hom}_{kH}(V_1, V_2) \\
&= 0 \text{ if } V_1 \text{ and } V_2 \text{ are not isomorphic.}
\end{aligned}$$

Therefore, e_k^K is indeed an isomorphism of monoids that conserves simple (split) modules. □

▷ *Decomposition maps and Tits' theorem.*

Using again the Brauer–Nesbitt map, we can finally relate modules over KH and modules over kH through a **decomposition map** that is a morphism between the Grothendieck groups of these algebras:

Theorem 4.38 (Brauer–Cartan). *We fix as before a ring A, a valuation ring $\mathscr{O} \subset K$ that contains A, and the specialization $\theta : A \to k = \mathscr{O}/\mathfrak{m}_{\mathscr{O}}$. We suppose that kH is a split algebra (for instance with k algebraically closed). Then, the modular reduction induces a morphism of monoids $d_\theta : \mathrm{R}_0^+(KH) \to \mathrm{R}_0^+(kH)$:*

$$d_\theta([V^K]) = [V^k].$$

This decomposition map d makes the following diagram commutative:

$$\begin{array}{ccc} \mathrm{R}_0^+(KH) & \xrightarrow{p^K} & \mathrm{Maps}(H \to A[X]) \\ {\scriptstyle d_\theta}\downarrow & & \downarrow{\scriptstyle \theta} \\ \mathrm{R}_0^+(kH) & \xrightarrow[p^k]{} & \mathrm{Maps}(H \to k[X]) \end{array}$$

and it is uniquely determined by this condition. In particular, it does not depend on the choice of a valuation ring $\mathscr{O}$.

Proof. Let V^K be a KH-module, and V^k be a modular reduction of V^K with respect to a fixed valuation ring $\mathscr{O}$. As in Corollary 4.33, one fixes a basis of V^K such that $\rho^{V^K}(h) \in \mathrm{M}(d, \mathscr{O})$ for every $h \in H$. If $\pi_{\mathscr{O}} : \mathscr{O} \to \mathscr{O}/\mathfrak{m}_{\mathscr{O}}$ is the canonical projection, then $\pi_{\mathscr{O}} \circ \rho^{V^K} = \rho^{V^k}$ as matrices with coefficients in k corresponding to the previous choice of basis, by definition of the modular reduction. As a consequence,

$$p^k(V^k) = \theta(p^K(V^K)).$$

Consider then two $\mathscr{O}H$-modules V_1 and V_2 such that $K \otimes_{\mathscr{O}} V_1 = K \otimes_{\mathscr{O}} V_2 = V^K$, and let us show that the corresponding modular reduction V_1^k and V_2^k are Grothendieck equivalent. By the previous discussion, they have the same image by the Brauer–Nesbitt map:

$$p^k(V_1^k) = \theta(p^K(V^K)) = p^k(V_2^k).$$

Since kH is assumed split, by Brauer–Nesbitt lemma, p^k is injective on $\mathrm{R}_0^+(kH)$, so $[V_1^k] = [V_2^k]$. Thus, the map $d_\theta : V^K \mapsto [V^k]$ is well defined (it does not depend on the choice of realizations over $\mathscr{O}$). Moreover, $d_\theta(V^K)$ only depends on the class $[V^K]$ in $\mathrm{R}_0^+(KH)$, since if $[V_1^K] = [V_2^K]$, then $p^K(V_1^K) = p^K(V_2^K)$, and the corresponding modular reductions V_1^k and V_2^k have then the same image by p^k, hence are again Grothendieck equivalent. This ensures the existence of the map $d_\theta : \mathrm{R}_0^+(KH) \to \mathrm{R}_0^+(kH)$, and the commutativity of the diagram. The unicity follows once more from the injectivity of p^k. □

So, to restate Theorem 4.38, assuming kH split, if M^λ is a simple KH-module, then for every valuation domain $\mathscr{O}$ such that $A \subset \mathscr{O} \subset K$ and $A \to \mathscr{O}/\mathfrak{m}_{\mathscr{O}}$ realizes the specialization $\theta : A \to k$, and every realization $M^{\lambda,\mathscr{O}}$ of M^λ over $\mathscr{O}H$, the composition factors of the modular reduction $M^{\lambda,k} = k \otimes_{\mathscr{O}} M^{\lambda,\mathscr{O}}$ do not depend on the previous choices, so that $M^{\lambda,k}$ is Grothendieck equivalent to a unique direct sum $\bigoplus_\mu c_{\lambda\mu} N^\mu$ of simple kH-modules N^μ. The coefficients $c_{\lambda\mu}$ form the **decomposition matrix** of the algebra H with respect to the diagram

$$\begin{array}{ccccc} A & \hookrightarrow & \mathscr{O} & \hookrightarrow & K. \\ & \theta \searrow & \downarrow \pi_{\mathscr{O}} & & \\ & & k & & \end{array}$$

In general, the decomposition matrix takes a non-trivial form, that is to say that it is not the identity matrix. The following result due to Tits give a sufficient condition for the triviality of the decomposition matrix:

Theorem 4.39 (Tits). *Suppose that kH and KH are split, and that kH is semisimple. Then, KH is also semisimple, and d_θ is an isomorphism of monoids that conserves simple modules. Hence, there is a common labeling $\{M^{\lambda,K}\}_\lambda$ and $\{M^{\lambda,k}\}_\lambda$ of the classes of isomorphism of simple KH- and kH-modules, such that the modular reduction reads as*

$$d_\theta(M^{\lambda,K}) = M^{\lambda,k}.$$

Proof. Let $\{M^\lambda\}_\lambda$ and $\{N^\mu\}_\mu$ be complete families of representatives of the isomorphism classes of simple KH- and kH-modules, and $(c_{\lambda\mu})$ be the decomposition matrix, which has non-negative integer coefficients. Since kH is split semisimple,

$$\dim_k kH = \sum_\mu (\dim_k N^\mu)^2.$$

On the other hand, since $KH/\mathrm{rad}(KH)$ is split semisimple and has for simple modules the M^λ's,

$$\dim_K KH/\mathrm{rad}(KH) = \sum_\lambda (\dim_K M^\lambda)^2.$$

The modular reduction conserves the dimensions, therefore,

$$\begin{aligned} \sum_\mu (\dim_k N^\mu)^2 &= \dim_k kH = \dim_K KH \\ &\geq \dim_K KH/\mathrm{rad}(KH) = \sum_\lambda \left(\sum_\mu c_{\lambda\mu} \dim_k N^\mu \right)^2 \\ &\geq \sum_\mu \left(\sum_\lambda c_{\lambda\mu}^2 \right) (\dim_k N^\mu)^2. \end{aligned}$$

Since the modular reduction of KH is kH, and since every simple module of an algebra A appears as a composition factor of A viewed as an A-module, for every μ, there exists at least one λ such that $c_{\lambda\mu} > 0$. The previous inequality shows that there can be only one such λ, with $c_{\lambda\mu} = 1$. Moreover, $\dim_K \mathrm{rad}(KH) = 0$, i.e., KH is semisimple. Thus, the decomposition matrix contains exactly one non-zero entry by column, equal to 1. However, it also contains at least one non-zero entry by row, since $d_\theta[M^\lambda] \neq 0$ for every λ. Thus, it is a permutation matrix, which ends the proof. □

Call **trace** of a k-algebra A a linear map $\tau : A \to k$ such that $\tau(ab) = \tau(ba)$ for every $a, b \in A$. We have seen before that if A is a split semisimple k-algebra, then the characters of the simple A-modules form a linear basis of the space of traces of A. Consider in the Brauer–Cartan setting a character ψ of a KH-module M^K. Since M^K can be realized over $\mathscr{O}$ for every valuation domain $\mathscr{O}$ containing A, the restriction of ψ to H yields a trace $\psi : H \to A$. By tensoring by k we get a trace $\psi^k = \theta \circ \psi$ of the algebra kH. If kH is split, then the decomposition matrix yields in this setting

$$(\psi^\lambda)^k = \sum_\mu c_{\lambda\mu} \xi^\mu$$

if (ψ^λ) is the family of simple characters of KH, and (ξ^μ) is the family of simple characters of kH. Assuming kH semisimple and kH and KH split, Tits' deformation theorem ensures that there is a labeling of the simple characters such that

$$(\psi^\lambda)^k = \xi^\lambda,$$

that is to say that *the simple characters of kH are just obtained by specialization of the simple characters of KH*. In particular, this will be the case for Hecke algebras, and all the combinatorial formulas given in Chapters 2 and 3 for the characters of $\mathfrak{S}(n)$ will be specializations of formulas for the characters of the generic Hecke algebra $\mathfrak{H}(n)$; see Section 5.5.

4.4 Structure of generic and specialized Hecke algebras

After our presentation of the representation theory of algebras and of the Brauer–Cartan setting for decomposition maps, we can give a clear program to deal with the representation theory of Hecke algebras. We are going to show that:

1. The generic Hecke algebra $\mathfrak{H}(n)$, as well as the specialized Hecke algebra $\mathfrak{H}_z(n)$, are split. For the specialized algebras, there is nothing to show since they are defined over the algebraically closed field $\mathbb{C}$. For the generic Hecke algebra, since $\mathfrak{H}_1(n) = \mathbb{C}\mathfrak{S}(n)$ is semisimple, the result will imply that $\mathfrak{H}(n)$ is also semisimple and has the same representation theory as $\mathbb{C}\mathfrak{S}(n)$: its simple modules are labeled by integer partitions of size n, and these modules have their dimensions given by the numbers of standard tableaux.
2. If z is not zero and is not a non-trivial root of unity, then every $\mathfrak{H}_z(n)$-module

is projective. Therefore, by Proposition 4.17, the specialized Hecke algebra $\mathfrak{H}_z(n)$ is a semisimple $\mathbb{C}$-algebra, and by Tits' deformation theorem 4.39, it has the same representation theory as the generic Hecke algebra $\mathfrak{H}(n)$.

The first item will be related to the notion of splitting field, and both parts shall rely on the theory of symmetric algebras.

▷ *Existence of a splitting field for the generic Hecke algebra.*

In order to apply Tits' deformation theorem to the case

$$k = \mathbb{C},\ K = \mathbb{C}(q),\ A = \mathbb{C}[q],\ H = \mathfrak{H}(n, \mathbb{C}[q]) \text{ and } \theta_1(P(q)) = P(1),$$

one only needs to show that the generic Hecke algebra $\mathfrak{H}(n, \mathbb{C}(q))$ is split. Indeed, $kH = \mathbb{C}\mathfrak{S}(n)$ is the group algebra of the symmetric group, hence is split semisimple. Unfortunately, proving directly that the generic Hecke algebra $\mathfrak{H}(n) = \mathfrak{H}(n, \mathbb{C}(q))$ is split is quite hard unless one constructs by hand all the simple modules of it, in a fashion similar to the constructions of Section 3.3. This is a possible approach (see the notes), but it requires difficult combinatorial constructions that we wanted to avoid by using the Brauer–Cartan theory. Another approach relies on the following argument, which we are going to develop in this paragraph and the next ones: the conditions of Tits' theorem 4.39 can be shown to be satisfied for the generic Hecke algebra *up to a possible finite field extension K' of $\mathbb{C}(q)$*, that is to say for $\mathfrak{H}(n, K')$ instead of $\mathfrak{H}(n, \mathbb{C}(q))$. An ad hoc argument will then prove that one can return from K' to $K = \mathbb{C}(q)$ and keep the same structure and representation theory. To make our argument rigorous, it is convenient to introduce the notion of **splitting field** of a finite-dimensional algebra B over a field k.

Definition 4.40. *Let B be a finite-dimensional algebra over a field k. An extension $K \mid k$ is called a splitting field for B if $KB = K \otimes_k B$ is a split algebra, that is to say that for every simple module M over KB, $\mathrm{End}_{KB}(M) = K$.*

Theorem 4.41. *Every finite-dimensional algebra B over k has a splitting field K that is a finite-dimensional field extension of k.*

Proof. Let $\overline{k}$ be an algebraic closure of k. Since $\overline{k}$ has no non-trivial division ring, every simple module M over $\overline{k}B$ satisfies $\mathrm{End}_{\overline{k}B}(M) = \overline{k}$, so $\overline{k}B$ is a split algebra. We fix a $\overline{k}$-basis $(b_1, \ldots, b_n)$ of $\overline{k}B$, such that $(b_{p+1}, \ldots, b_n)$ is a basis of $\mathrm{rad}(\overline{k}B)$. If $\pi : \overline{k}B \to \overline{k}B/\mathrm{rad}(\overline{k}B)$ is the canonical projection, then one can choose the other $b_{h \leq p}$'s so that $\{\pi(b_1), \ldots, \pi(b_p)\}$ corresponds in the Artin–Wedderburn decomposition

$$\overline{k}B/\mathrm{rad}(\overline{k}B) = \bigoplus_{\lambda} \mathrm{End}_{\overline{k}}(M^\lambda)$$

to the elementary matrices e^λ_{ij}. Rewrite $\{b_1, \ldots, b_p\} = \{b^\lambda_{ij}\}_{\lambda, 1 \leq i,j \leq \dim M^\lambda}$, so that

$\pi(b_{ij}^{\lambda}) = e_{ij}^{\lambda}$. If $b_i b_j = \sum_{h=1}^{n} \gamma^h(b_i, b_j) b_h$ for $i, j \leq n$, then $\pi(b_{ij}^{\lambda} b_{kl}^{\mu}) = e_{ij}^{\lambda} e_{kl}^{\mu} = \delta_{\lambda,\mu} \delta_{j,k} e_{il}^{\lambda}$, so

$$b_{ij}^{\lambda} b_{kl}^{\mu} = \delta_{\lambda,\mu} \delta_{j,k} b_{il}^{\lambda} + \sum_{h=p+1}^{n} \gamma^h(b_{ij}^{\lambda}, b_{kl}^{\mu}) b_h.$$

Let $(a_1, \ldots, a_n)$ be a basis of B over k. Then, it is also a basis of $\overline{k}B$ over $\overline{k}$, so there exist coefficients $c_{ij} \in \overline{k}$ such that

$$b_j = \sum_{i=1}^{n} c_{ij} a_i.$$

We set $K = k[c_{ij}, \gamma^h(b_i, b_j)]$. This is a finite field extension of k, and $KB = K \otimes_k B$ admits for K-basis $b_1, \ldots, b_n$. Let

$$KN = \bigoplus_{h=p+1}^{n} K b_h;$$

we claim that $KN = \mathrm{rad}(KB)$. In $\overline{k}B$, the Jacobson radical is a nilpotent bilateral ideal: indeed, by Nakayama's lemma, $\overline{k}B \supsetneq \mathrm{rad}(\overline{k}B) \supsetneq (\mathrm{rad}(\overline{k}B))^2 \supsetneq \cdots$ until one hits zero. As $(b_{p+1}, \ldots, b_n)$ is a $\overline{k}$-basis of $\mathrm{rad}(\overline{k}B)$, this means that there exists an integer $k \geq 1$ such that $b_{h_1} b_{h_2} \cdots b_{h_k} = 0$ for any choice of indices. This stays true in KB, so KN is also a nilpotent ideal in KB. This implies that $KN \subset \mathrm{rad}(KB)$. Indeed, for any $x \in KN$, and any $b \in KB$, since $bx \in KN$ is nilpotent, $1 - bx$ is invertible, so $x \in \mathrm{rad}(KB)$ by the third characterization of the Jacobson ideal. On the other hand, $KB/KN = \bigoplus_{h=1}^{p} K b_h$ is a direct sum of matrix algebras over K, because

$$\begin{aligned} b_{ij}^{\lambda} b_{kl}^{\mu} &= \delta_{\lambda,\mu} \delta_{j,k} b_{il}^{\lambda} + \sum_{h=p+1}^{n} \gamma^h(b_{ij}^{\lambda}, b_{kl}^{\mu}) b_h \\ &\equiv \delta_{\lambda,\mu} \delta_{j,k} b_{il}^{\lambda} \mod KN \end{aligned}$$

Hence, KB/KN is semisimple, and by Jacobson theorem 4.26, $\mathrm{rad}(KB) \subset KN$, so $KN = \mathrm{rad}(KB)$. Then, KB is a split algebra, since $KB/\mathrm{rad}(KB)$ is a sum of matrix algebras over K. Notice that our proof implies that in a finite-dimensional algebra A over a field, the Jacobson radical, which is the intersection of all maximal ideals of A, is also the largest nilpotent ideal of A. □

As an application of the existence of splitting fields, consider the Brauer–Cartan setting with $A = \mathbb{C}[q]$, $K = \mathbb{C}(q)$, $k = \mathbb{C}$, $\mathscr{O} = \mathbb{C}[q]_{(q-1)}$ and $H = \mathfrak{H}(n, \mathbb{C}[q])$. The specialized Hecke algebra $kH = \mathfrak{H}_1(n)$ is the group algebra $\mathbb{C}\mathfrak{S}(n)$, hence, it is split and semisimple. On the other hand, there exists a finite field extension $K' \mid K$ such that $K'H = \mathfrak{H}(n, K')$ is split. Let A' be the integral closure of A into K', that is to say the set of solutions in K' of monic polynomial equations

$$x^n + a_{n-1} x^{n-1} + \cdots + a_0 = 0$$

with the a_i's in A.

Lemma 4.42. *In the previous setting, if $\theta_1 = \theta : A \to k$ is the specialization morphism, then there exists a valuation ring $\mathscr{O}'$ for K' with:*

1. $A' \subset \mathscr{O}' \subset K'$;
2. *$\mathscr{O}'$ has residue field k, and the restriction of the projection $\pi' : \mathscr{O}' \to \mathscr{O}'/\mathfrak{m}_{\mathscr{O}'} = k$ to A is equal to θ.*

We shall prove this lemma in a moment. An immediate consequence of it and of Theorem 4.39 is:

Theorem 4.43. *If K' is a finite extension field of $\mathbb{C}(q)$ such that $\mathfrak{H}(n,K')$ splits, then $\mathfrak{H}(n,K')$ is a semisimple split algebra, and moreover, the decomposition map*

$$d : \mathrm{R}_0(\mathfrak{H}(n,K')) \to \mathrm{R}_0(\mathbb{C}\mathfrak{S}(n))$$

associated to the modular reduction $\mathbb{C}[q] \hookrightarrow \mathscr{O}' \to \mathbb{C}$ yields an isomorphism that conserves the simple modules. Hence, there is a labeling of the simple modules over $\mathfrak{H}(n,K')$ by integer partitions $\lambda \in \mathfrak{Y}(n)$, such that

$$d(S^{\lambda,K'}) = S^{\lambda}.$$

In particular, $\dim_{K'}(S^{\lambda,K'}) = \dim_{\mathbb{C}}(S^{\lambda}) = \mathrm{card}\,\mathrm{ST}(\lambda)$.

Indeed, one can apply the Brauer–Cartan deformation theory to $A' \subset \mathscr{O}' \subset K'$, and since the map $\theta' : A' \to k$ obtained by composition of the inclusion $A' \hookrightarrow \mathscr{O}'$ and of $\pi' : \mathscr{O}' \to k$ is an extension of θ, $kH = \mathbb{C}\mathfrak{S}(n)$ and has its representation theory covered by the results of Chapters 2 and 3. Hereafter, we shall prove that one can in fact take $K' = K$, that is to say that the previous theorem actually holds for the generic Hecke algebra $\mathfrak{H}(n) = \mathfrak{H}(n,\mathbb{C}(q))$. However, it is necessary to know beforehand that there exists a larger field K' such that $\mathfrak{H}(n,K')$ is split semisimple.

The proof of the technical Lemma 4.42 relies on simple arguments from algebraic number theory (see the notes for references). Notice that $A = \mathbb{C}[q]$ is a **Dedekind ring**, that is to say a ring that is noetherian (every ideal of A is finitely generated), integrally closed in its fraction field $K = \mathbb{C}(q)$, and such that every non-zero prime ideal is maximal. Indeed, it is even a principal ring, which is a stronger property. A Dedekind ring A always has the following properties:

1. Every non-zero ideal I of A factorizes uniquely as a product of prime ideals:

$$I = (\mathfrak{p}_1)^{n_1}\,(\mathfrak{p}_2)^{n_2}\cdots(\mathfrak{p}_r)^{n_r}.$$

2. For every finite-dimensional field extension $K' \mid K$ of the fraction field of A, the integral closure A' of A in K' is again a Dedekind domain.

3. For any prime ideal $\mathfrak{p}$ of A, the localization $A_{\mathfrak{p}}$ is a valuation ring.

Consider then a prime ideal $\mathfrak{p}$ of A, and a finite field extension $K' \mid K = \mathrm{Frac}(A)$. It decomposes in a unique way in A' as a product

$$\mathfrak{p}A' = (\mathfrak{P}_1)^{e_1}\,(\mathfrak{P}_2)^{e_2} \cdots (\mathfrak{P}_r)^{e_r}$$

of prime ideals of A'. The ideals $\mathfrak{P}_i$ that appear in this decomposition are precisely those such that $(A',\mathfrak{P})$ **dominates** $(A,\mathfrak{p})$, that is to say that $A \cap \mathfrak{P} = \mathfrak{p}$. Therefore, for each index i, one has an inclusion of rings $A/\mathfrak{p} \to A'/\mathfrak{P}_i$, and these rings are in fact fields. Indeed, A (respectively A') is a Dedekind domain and $\mathfrak{p}$ (respectively $\mathfrak{P}_i$) is a non-zero prime ideal, hence a maximal ideal; so the quotient is a field. Therefore, if $k = A/\mathfrak{p}$, then each quotient $A'/\mathfrak{P}_i$ is a field extension of k. The multiplicity e_i of $\mathfrak{P}_i$ in $\mathfrak{p}A'$ is called the **ramification index** of $\mathfrak{P}_i$ in $\mathfrak{p}$, and the dimension $f_i = [A'/\mathfrak{P}_i : A/\mathfrak{p}]$ is called the **inertia degree**. One then has, using the Chinese remainder theorem,

$$A'/\mathfrak{p}A' = \bigoplus_{i=1}^{r} A'/(\mathfrak{P}_i)^{e_i};$$

$$[K' : K] = \dim_k(A'/\mathfrak{p}A') = \sum_{i=1}^{r} \dim_k(A'/(\mathfrak{P}_i)^{e_i}) = \sum_{i=1}^{r} e_i f_i.$$

In particular, f_i is finite for each i.

Take now $\mathfrak{p} = (q-1)\mathbb{C}[q]$. The quotient $kA/\mathfrak{p}$ is $\mathbb{C}$, which is algebraically closed, so $f_i = 1$ for every prime ideal $\mathfrak{P}_i$ of A' that lies over $\mathfrak{p}$. We choose such an ideal $\mathfrak{P}$, and set $\mathcal{O}' = A'_{\mathfrak{P}}$; by the previous discussion, $\mathbb{C} = k = A/\mathfrak{p} = A'/\mathfrak{P}$. On the other hand, since A' is a Dedekind ring, its localization $\mathcal{O}'$ is a valuation ring. Its residue field is $\mathcal{O}'/\mathfrak{P}\mathcal{O}' = A'/\mathfrak{P} = \mathbb{C}$, and we then have the commutative diagram

$$\begin{array}{ccccc}
A & \hookrightarrow & A' & \hookrightarrow & \mathcal{O}' \\
\downarrow{\scriptstyle \theta} & & & \searrow^{\theta'} & \downarrow{\scriptstyle \pi'} \\
\mathbb{C} = A/\mathfrak{p} & & = & & \mathbb{C} = A'/\mathfrak{P}.
\end{array}$$

This is exactly what we wanted to prove in Lemma 4.42.

▷ *Symmetric algebras and a criterion of semisimplicity.*

If M' is a module over $K'H = K' \otimes_K KH$, we say that M' can be realized over K if there exists a module M over KH such that $M' = K'H \otimes_{KH} M = K' \otimes_K M$. In order to deduce from Theorem 4.43 that $\mathfrak{H}(n)$ is split semisimple, we will have to prove that all the simple modules $S^{\lambda,K'}$ of $\mathfrak{H}(n,K')$ can be realized over K. Our proof will use the notion of **symmetric algebra**; this also provides a "numerical" criterion of semisimplicity that we shall use to study the specialized Hecke algebras $\mathfrak{H}_z(n)$. In the following, we fix a field k, and a finite-dimensional k-algebra B.

Definition 4.44. *A* ***symmetrizing trace*** *on B is a trace $\tau : B \to k$ that induces a non-degenerate bilinear form*

$$\begin{aligned} B \otimes_k B &\to k \\ b \otimes b' &\mapsto \tau(bb'). \end{aligned}$$

Hence, for any basis $(b_1, \dots, b_n)$ of B over k, there exists a (unique) dual basis $(b'_1, \dots, b'_n)$ of B such that $\tau(b_i b'_j) = \delta_{ij}$. If B admits a symmetrizing trace τ, we call the pair (B, τ) a **symmetric algebra**.

Example. Let G be a finite group, and k be a field. The algebra kG is a symmetric algebra for the regular trace $\tau(g) = \delta_{g,e_G}$. Its canonical basis $(g)_{g\in G}$ admits for dual basis $(g^{-1})_{g\in G}$.

Proposition 4.45. *The generic Hecke algebra $\mathfrak{H}(n)$, and the specialized Hecke algebras $\mathfrak{H}_z(n)$ with $z \neq 0$ are symmetric algebras for the trace*

$$\tau(T_\sigma) = \delta_{\sigma, \mathrm{id}_{[\![1,n]\!]}}.$$

The dual basis of the family $(T_\sigma)_{\sigma\in\mathfrak{S}(n)}$ is $(q^{-\ell(\sigma)} T_{\sigma^{-1}})_{\sigma\in\mathfrak{S}(n)}$ in the generic case, and $(z^{-\ell(\sigma)} T_{\sigma^{-1}})_{\sigma\in\mathfrak{S}(n)}$ in the specialized case.

Proof. We treat the generic case, the specialized case with $z \neq 0$ being identical. We have to show that

$$\tau(T_\sigma T_\rho) = \begin{cases} q^{\ell(\sigma)} & \text{if } \rho = \sigma^{-1}; \\ 0 & \text{otherwise.} \end{cases}$$

We reason by induction on the length of σ. If $\ell(\sigma) = 0$, $\sigma = 1$ and the result is trivial. Suppose now the result is true up to length $\ell - 1$, and take σ with $\ell(\sigma) = \ell$. We choose an elementary permutation s_i such that $\sigma = \sigma' s_i$ with $\ell(\sigma') = \ell(\sigma) - 1$.

- If $\ell(s_i \rho) = 1 + \ell(\rho)$, then $T_\sigma T_\rho = T_{\sigma'} T_i T_\rho = T_{\sigma'} T_{s_i\rho}$, so by the induction hypothesis,

  $$\tau(T_\sigma T_\rho) = q^{\ell(\sigma')} \delta_{(\sigma')^{-1}, s_i\rho} = q^{\ell(\sigma)-1} \delta_{\sigma^{-1},\rho} = 0 = q^{\ell(\sigma)} \delta_{\sigma^{-1},\rho}.$$

  Indeed, σ^{-1} cannot be equal to ρ, since the length of σ^{-1} decreases when multiplying on the left by s_i, whereas the length of ρ grows by hypothesis. So, the result holds in this case.

- Suppose now that $\ell(s_i\rho) = \ell(\rho) - 1$; then, by Lemma 4.2, $\rho = s_i \rho'$ with $\ell(\rho') = \ell(\rho) - 1$. Notice then that $\rho = \sigma^{-1}$ if and only if $\rho' = (\sigma')^{-1}$. By the induction hypothesis, we therefore have:

  $$\begin{aligned} \tau(T_\sigma T_\rho) &= \tau(T_{\sigma'} (T_i)^2 T_{\rho'}) = (q-1)\,\tau(T_{\sigma'} T_\rho) + q\,\tau(T_{\sigma'} T_{\rho'}) \\ &= (q-1)\,\tau(T_{\sigma'} T_\rho) + q^{1+\ell(\sigma')} \delta_{(\sigma')^{-1},\rho'} = (q-1)\,\tau(T_{\sigma'} T_\rho) + q^{\ell(\sigma)} \delta_{\sigma^{-1},\rho}. \end{aligned}$$

However, $\tau(T_{\sigma'} T_\rho) = 0$: $(\sigma')^{-1}$ cannot be equal to ρ, as one of these permutations has its length that grows by multiplication on the left by s_i, and it is not the case of the other permutation. Therefore, $\tau(T_\sigma T_\rho) = q^{\ell(\sigma)} \delta_{\sigma^{-1},\rho}$, hence the result also in this case. □

In the following, we fix a symmetric k-algebra (B, τ), with a basis $(b_1, \ldots, b_n)$ and its dual basis $(b'_1, \ldots, b'_n)$. If M_1 and M_2 are modules over B and $u \in \mathrm{Hom}_k(M_1, M_2)$, we denote

$$I(u)(m) = \sum_{i=1}^{n} b_i \, u(b'_i m),$$

which is a priori a new element of $\mathrm{Hom}_k(M_1, M_2)$.

Proposition 4.46. *For any u, $I(u)$ belongs to $\mathrm{Hom}_B(M_1, M_2)$, and moreover, the definition does not depend on the choice of a basis $(b_1, \ldots, b_n)$ of B.*

Proof. Denote the action of B on itself by:

$$c\, b_j = \sum_i \rho^B_{ij}(c)\, b_i.$$

Notice then that $\tau(b'_i c b_j) = \rho^B(c)_{ij}$. Moreover,

$$b'_i\, c = \sum_j \rho^B_{ij}(c)\, b'_j,$$

because for every j, both sides give the same result when computing $\tau(x b_j)$. As a consequence,

$$\begin{aligned} c\, I(u)(m) &= \sum_{j=1}^{n} c b_j \, u(b'_j m) = \sum_{i,j=1}^{n} \tau(b'_i c b_j)\, b_i \, u(b'_j m) \\ &= \sum_{i=1}^{n} b_i \, u\left(\sum_{j=1}^{n} \tau(b'_i c b_j) b'_j \, m \right) = \sum_{i=1}^{n} b_i \, u(b'_i \, cm) \end{aligned}$$

so $I(u)$ belongs to $\mathrm{Hom}_B(M_1, M_2)$. Now, let $(c_1, \ldots, c_n)$ be another basis of B, and $(a_{ij})_{1 \le i,j \le n}$ be the invertible matrix such that $c_j = \sum_{i=1}^n a_{ij} b_i$. One has then $c'_j = \sum_{i=1}^n \alpha_{ij} b'_i$, where $(\alpha_{ij})_{i,j}$ is the inverse of the transpose of the matrix $(a_{ij})_{i,j}$. Consequently,

$$\sum_{j=1}^{n} c_j \, u(c'_j m) = \sum_{i_1, i_2, j} a_{i_1 j} \alpha_{i_2 j} \, b_{i_1} \, u(b'_{i_2} m) = \sum_{i=1}^{n} b_i \, u(b'_i m),$$

hence the independence of $I(u)$ from the choice of a basis of B. □

Theorem 4.47 (Gaschütz, Higman, Ikeda). *Let (B, τ) be a symmetric algebra over k, and M be a B-module. The module M is projective if and only if there exists $u \in \mathrm{End}_k(M)$ such that $I(u) = \mathrm{id}_M$.*

Proof. Suppose that there exists $u \in \mathrm{End}_k(M)$ such that $I(u) = \mathrm{id}_M$, and consider an exact sequence

$$0 \longrightarrow K \xrightarrow{\ i\ } L \xrightarrow{\ s\ } M \longrightarrow 0$$

of finite-dimensional B-modules. If one considers the sequence above as an exact sequence with k-vector spaces, then it splits: k-vector spaces are free over k, hence projective. So, there exists $r \in \mathrm{Hom}_k(M, L)$ such that $s \circ r = \mathrm{id}_M$. Then,

$$\begin{aligned} u &= s \circ r \circ u \\ \mathrm{id}_M &= I(u) = I(s \circ r \circ u) = s \circ I(r \circ u) \end{aligned}$$

since s is a morphism of B-modules. Therefore, $I(r \circ u)$ yields a section of the exact sequence that is a morphism of B-modules; and M is projective.

Conversely, suppose that M is projective; then, there exists a section r of the morphism of B-modules

$$\begin{aligned} s : B \otimes_k M &\to M \\ \sum_i b_i \otimes m_i &\mapsto \sum_i b_i\, m_i. \end{aligned}$$

We define $t \in \mathrm{End}_k(B \otimes_k M)$ by $t(c \otimes m) = \tau(c)\, 1 \otimes m$. Then,

$$I(t)(c \otimes m) = \sum_{i=1}^{n} b_i\, t(b_i' c \otimes m) = \left(\sum_{i=1}^{n} \tau(b_i' c)\, b_i \right) \otimes m.$$

The last term is simply $c \otimes m$, because for any b_j',

$$\tau\left(b_j' \left(\sum_{i=1}^{n} \tau(b_i' c)\, b_i \right) \right) = \tau(b_j' c).$$

So, $I(t) = \mathrm{id}_B$, and then $I(s \circ t \circ r) = s \circ I(t) \circ r = s \circ r = \mathrm{id}_M$, so there exists $u = s \circ t \circ r \in \mathrm{End}_k(M)$ such that $I(u) = \mathrm{id}_M$. □

Theorem 4.47 can be combined with the notion of **Schur element** of a module M over B to get a criterion of semisimplicity.

Definition 4.48. *Let M be a simple module over B, that is also split; hence, $\mathrm{End}_B(M) = k\,\mathrm{id}_M$. There exists a unique constant c_M, called the Schur element of M, such that for any $u \in \mathrm{End}_k(M)$,*

$$I(u) = \mathrm{tr}(u)\, c_M\, \mathrm{id}_M.$$

Proof. Fix a basis $(m_1,\dots,m_d)$ of M. Since M is a split simple module, for every $u \in \mathrm{End}_k(M)$, $I(u) = c_u\,\mathrm{id}_M$ for some constant c_u. Let e_{ij} be the linear map that sends m_j to m_i and vanishes on the other basis elements $m_{j'}$; we set $c_{ij} = c_{e_{ij}}$. By definition,

$$c_{ij}\,\mathrm{id}_M = I(e_{ij}) = \sum_{k=1}^{n} \rho^M(b_k) \circ e_{ij} \circ \rho^M(b'_k),$$

and by taking matrices with respect to the basis $(m_1,\dots,m_d)$, this leads to

$$\sum_{k=1}^{n} \rho^M_{hi}(b_k)\,\rho^M_{jl}(b'_k) = \delta_{hl}\,c_{ij}.$$

However, we can exchange the roles played by the b_i's and the b'_j's without changing the value of $I(e_{ij})$, so the previous expression is also equal to

$$\sum_{k=1}^{n} \rho^M_{jl}(b_k)\,\rho^M_{hi}(b'_k) = \delta_{ji}\,c_{lh}.$$

Set $c_M = c_{11}$. If $i = j$, then $c_{ii} = \delta_{11}c_{ii} = \delta_{ii}c_{11} = c_M$. On the other hand, if $i \neq j$, then $c_{ij} = \delta_{11}c_{ij} = \delta_{ji}c_{11} = 0$. It follows that for every e_{ij}, $I(e_{ij}) = \mathrm{tr}(e_{ij})\,c_M\,\mathrm{id}_M$. As $(e_{ij})_{1\leq i,j\leq d}$ forms a basis of $\mathrm{End}_k(M)$, the proof is done. □

As a corollary of the computations for the proof of the existence of the Schur element, we get the analogue in the setting of symmetric algebras of Schur's orthogonality relations (Theorem 1.8):

Proposition 4.49. *Let M_1 and M_2 be simple split modules over a symmetric algebra, with representation matrices $(\rho^{M_1}_{i_1 j_1})_{i_1,j_1}$ and $(\rho^{M_2}_{i_2 j_2})_{i_2,j_2}$. One has:*

$$\sum_{k=1}^{n} \rho^{M_1}_{i_1 j_1}(b_k)\,\rho^{M_2}_{i_2 j_2}(b'_k) = \begin{cases} c_M\,\delta_{i_1 j_2}\,\delta_{i_2 j_1} & \text{if } M = M_1 = M_2 \\ 0 & \text{otherwise.} \end{cases}$$

Therefore, for characters of simple split modules,

$$\sum_{k=1}^{n} \mathrm{ch}^{M_1}(b_k)\,\mathrm{ch}^{M_2}(b'_k) = \delta_{M_1,M_2}\,c_{M_1}\,\dim_k M_1.$$

Proof. The previous computations treat the case $M_1 = M_2$, and if $M_1 \neq M_2$, then by Schur's lemma $\mathrm{Hom}_B(M_1,M_2) = 0$, which leads to the vanishing of the sums of matrix coefficients by computing $0 = I(e_{ij})$ for $(e_{ij})_{i,j}$ basis of $\mathrm{Hom}_k(M_1,M_2)$. □

Combining Theorem 4.47 with Definition 4.48, we get:

Theorem 4.50. *Let B be a split symmetric algebra over k. A simple module M over B is projective if and only if $c_M \neq 0$. As a consequence, B is a semisimple split algebra if and only if $c_M \neq 0$ for all simple modules M over B. In this case, one has a decomposition of the symmetrizing trace in terms of the characters of the simple modules:*

$$\tau = \sum_{M^\lambda \in \widehat{B}} \frac{\mathrm{ch}^{M^\lambda}}{c_{M^\lambda}}.$$

Proof. If M is a simple split module over B, then there exists c_M such that for all $u \in \mathrm{End}_k(M)$, $I(u) = \mathrm{tr}(u)\, c_M\, \mathrm{id}_M$. As a consequence, there exists u with $I(u) = \mathrm{id}_M$ if and only if $c_M \neq 0$, and by Theorem 4.47, this is equivalent to the fact that M is a projective B module.

Suppose now that $c_M \neq 0$ for any simple module M, that is to say that all the simple modules of B are projective. If $\{S_1, \ldots, S_r\}$ is a complete family of non-isomorphic simple modules of B, then all these modules are projective and simple, hence projective and indecomposable. But there are as many projective indecomposable modules as there are simple modules (Theorem 4.31), so $\{S_1, \ldots, S_r\}$ is also a complete family of non-isomorphic projective indecomposable modules. As a consequence, if $B = \bigoplus_j P_j$ is a decomposition of B in projective indecomposable modules, then all these modules are simple, so B is semisimple. Conversely, if B is semisimple, then all its simple modules M are projective, hence they have non-zero Schur elements by the previous discussion.

Let us finally prove the decomposition of the symmetrizing trace in the semisimple case. If $B = \bigoplus_{\lambda \in \widehat{B}} \mathrm{End}_k(M^\lambda)$ is a split semisimple algebra, then a symmetrizing trace on it writes uniquely as $\tau = \sum_{\lambda \in \widehat{B}} f_\lambda\, \mathrm{ch}^\lambda$ with $f_\lambda \neq 0$, because the only symmetrizing traces on a matrix space $\mathrm{End}_k(M^\lambda)$ are the non-zero scalar multiples of the usual matrix trace. We choose for basis of $B = \bigoplus_{\lambda \in \widehat{B}} \mathrm{End}_k(M^\lambda)$ the reunion of the bases $(e^\lambda_{ij})_{1 \leq i,j \leq \dim M^\lambda}$ formed by the elementary matrices in each space $\mathrm{End}_k(M^\lambda)$. For the trace $\tau = \sum_{\lambda \in \widehat{B}} f_\lambda\, \mathrm{ch}^\lambda$, the dual element of e^λ_{ij} is $\frac{1}{f_\lambda} e^\lambda_{ji}$. Then, by definition of the Schur element $c_\lambda = c_{M^\lambda}$:

$$I(\mathrm{id}_{M^\lambda}) = c_\lambda\, (\dim M^\lambda)\, \mathrm{id}_{M^\lambda} = \frac{1}{f_\lambda} \sum_{i,j=1}^{\dim M^\lambda} e^\lambda_{ij}\, e^\lambda_{ji} = \frac{1}{f_\lambda} (\dim M^\lambda)\, \mathrm{id}_{M^\lambda}.$$

Thus, $f_\lambda = \frac{1}{c_\lambda}$. □

Example. Given a finite group G, consider the trace on kG defined by $\tau(g) = \delta_{g,e_G}$. We assume that k is algebraically closed, and that its characteristic does not divide $\mathrm{card}\, G$; then, kG is necessarily split, and it is also semisimple (cf. the example at the end of Section 1.4). Therefore, $\tau = \sum_{\lambda \in \widehat{kG}} \frac{\mathrm{ch}^\lambda}{c_\lambda}$. However, the Schur element c_λ of a simple module M^λ over kG is easily computed:

$$I(\mathrm{id}_{M^\lambda}) = c_\lambda\, (\dim M^\lambda)\, \mathrm{id}_{M^\lambda} = \sum_{g \in G} \rho^\lambda(g)\, \rho^\lambda(g^{-1}) = (\mathrm{card}\, G)\, \mathrm{id}_{M^\lambda}.$$

Therefore, $\tau = \sum_{\lambda} \frac{\dim M^{\lambda}}{\operatorname{card} G} \operatorname{ch}^{\lambda}$. This generalizes the discussion on Plancherel measures of the end of Section 1.3.

Corollary 4.51. *Suppose that (B, τ) is a semisimple split symmetric algebra, and for $\lambda \in \widehat{B}$, denote e^{λ} the central idempotent that projects B on its matrix block $\operatorname{End}_k(M^{\lambda})$. Then, for any basis $(b_1, \dots, b_n)$ of B,*

$$e^{\lambda} = \sum_{k=1}^{n} \frac{\operatorname{ch}^{\lambda}(b_k)}{c_{\lambda}} b'_k,$$

where c_{λ} denotes the Schur element of the simple module M^{λ}.

Proof. Theorem 4.50 and its proof imply that the central idempotent e^{λ} is characterized by the following property: for any $b \in B$,

$$\tau(e^{\lambda} b) = \frac{\operatorname{ch}^{\lambda}(b)}{c_{\lambda}}.$$

Therefore, it suffices to show that the right-hand side of the formula r^{λ} satisfies this identity. We compute, using Proposition 4.49:

$$\begin{aligned}
\tau(r^{\lambda} b) &= \sum_{k=1}^{n} \frac{\operatorname{ch}^{\lambda}(b_k)}{c_{\lambda}} \tau(b'_k b) = \sum_{k=1}^{n} \sum_{\mu \in \widehat{B}} \sum_{1 \leq h,i,j \leq \dim M^{\mu}} \frac{\rho^{\lambda}_{hh}(b_k)\, \rho^{\mu}_{ij}(b'_k)\, \rho^{\mu}_{ji}(b)}{c_{\lambda}\, c_{\mu}} \\
&= \sum_{k=1}^{n} \sum_{1 \leq h \leq \dim M^{\lambda}} \frac{\rho^{\lambda}_{hh}(b_k)\, \rho^{\lambda}_{hh}(b'_k)\, \rho^{\lambda}_{hh}(b)}{(c_{\lambda})^2} = \sum_{1 \leq h \leq \dim M^{\lambda}} \frac{\rho^{\lambda}_{hh}(b)}{c_{\lambda}} \\
&= \frac{\operatorname{ch}^{\lambda}(b)}{c_{\lambda}}.
\end{aligned}$$

□

This corollary leads to a criterion of realization of a simple module of a K-symmetric algebra over a smaller field $k \subset K$. Fix a field k, an extension of field $K \mid k$, and a k-symmetric algebra (B, τ). The K-linear extension of τ to $KB = K \otimes_k B$ is a symmetrizing trace, so (KB, τ) is also a symmetric algebra.

Proposition 4.52. *In the previous setting, suppose $KB = K \otimes_k B$ split semisimple, and consider a simple module M over K. Assume that $\operatorname{ch}^M(b) \in k$ for all $b \in B$, and that M appears with multiplicity 1 as a component of some module $KN = KB \otimes_B N$, where N is a B-module. Then, M can be realized over k: there exists a module L over B such that $KL = KB \otimes_B L$ is isomorphic to M.*

Proof. Fix a basis $(b_1, \dots, b_n)$ of B over k; it is also a basis of KB over K, with the same dual basis with respect to τ in B and in KB. By the previous corollary, the central idempotent e^M corresponding to M writes as

$$e^M = \sum_{k=1}^{n} \frac{\operatorname{ch}^M(b_k)}{c_M} b'_k,$$

and the Schur element writes as

$$c_M = \frac{1}{\dim_K M} \sum_{k=1}^{n} \mathrm{ch}^M(b_k)\,\mathrm{ch}^M(b'_k).$$

As the character values belong to k, $c_M \in k$ and $e^M \in B$. Fix N as in the statement of the proposition, and $L = e^M N$, which is a module over B. Then, $KL = e^M KN = M$. □

▷ *Reduction of the splitting field and structure of the generic Hecke algebra.*

In this paragraph, we fix an extension K' of $K = \mathbb{C}(q)$ such that Theorem 4.43 holds for $\mathfrak{H}(n,K')$. In view of Proposition 4.52, to prove that the simple modules of $\mathfrak{H}(n,K')$ are realized over K, we need to prove that:

1. Restricted to $\mathfrak{H}(n)$, the characters of these simple modules take their values in K;

2. Each simple module M over $\mathfrak{H}(n,K')$ appears with multiplicity 1 in some module $\mathfrak{H}(n,K') \otimes_{\mathfrak{H}(n)} N$, with N module over $\mathfrak{H}(n)$.

The proof of these facts relies on the study of the operations of induction and restriction of modules between $\mathfrak{H}(n)$ and its **parabolic subalgebras**. Call **composition** of size n a sequence $c = (c_1, \ldots, c_\ell)$ of positive integers with $|c| = \sum_{i=1}^{\ell} c_i = n$. For instance, $(2,3,2)$ is a composition of size 7. A parabolic subalgebra of the Hecke algebra $\mathfrak{H}(n,\mathbb{C}[q])$ is an algebra

$$\mathfrak{H}(c,\mathbb{C}[q]) = \mathfrak{H}(c_1,\mathbb{C}[q]) \otimes_{\mathbb{C}[q]} \mathfrak{H}(c_2,\mathbb{C}[q]) \otimes_{\mathbb{C}[q]} \cdots \otimes_{\mathbb{C}[q]} \mathfrak{H}(c_\ell,\mathbb{C}[q]),$$

with c in the set $\mathfrak{C}(n)$ of compositions of size n. It can indeed be seen as a subalgebra of $\mathfrak{H}(n,\mathbb{C}[q])$, namely, the subalgebra generated by the elements T_i with

$$i \notin \{c_1,\ c_1 + c_2,\ c_1 + \cdots + c_{\ell-1}\}.$$

A linear basis of the parabolic subalgebra $\mathfrak{H}(c,\mathbb{C}[q])$ consists in the elements T_σ with $\sigma \in \mathfrak{S}(c) = \mathfrak{S}(c_1) \times \mathfrak{S}(c_2) \times \cdots \times \mathfrak{S}(c_r)$, this group being identified with a subgroup of $\mathfrak{S}(n)$ by making act $\mathfrak{S}(c_1)$ on $[\![1, c_1]\!]$, $\mathfrak{S}(c_2)$ on $[\![c_1+1, c_1+c_2]\!]$, etc., $\mathfrak{S}(c_\ell)$ on $[\![c_1 + \cdots + c_\ell - 1, n]\!]$. We can then introduce as before

$$\mathfrak{H}(c) = \mathfrak{H}(c,\mathbb{C}(q)) = \mathbb{C}(q) \otimes_{\mathbb{C}[q]} \mathfrak{H}(n,\mathbb{C}[q]);$$
$$\mathfrak{H}_z(c) = \mathbb{C} \otimes_{\mathbb{C}[q]} \mathfrak{H}(c,\mathbb{C}[q]),$$

where in the second case $\mathbb{C}$ is viewed as a $\mathbb{C}[q]$-module by means of the specialization $\theta_z(P(q)) = P(z)$, with $z \in \mathbb{C}$. These algebras are respectively subalgebras of $\mathfrak{H}(n)$ and of $\mathfrak{H}_z(n)$. More generally, if A is a commutative algebra over $\mathbb{C}[q]$, we set $\mathfrak{H}(c,A) = A \otimes_{\mathbb{C}[q]} \mathfrak{H}(n,\mathbb{C}[q])$.

Definition 4.53. *The* ***index representation*** $1_{n,\mathbb{C}[q]}$ *of* $\mathfrak{H}(n,\mathbb{C}[q])$ *is the* $\mathfrak{H}(n,\mathbb{C}[q])$*-module that is free of rank 1 over* $\mathbb{C}[q]$*, and defined by* $\rho(T_\sigma)=q^{\ell(\sigma)}$*.*

Let us check that this affords an $\mathfrak{H}(n,\mathbb{C}[q])$-module. If s is an elementary transposition and $\sigma\in\mathfrak{S}(n)$, let us compute $\rho(T_sT_\sigma)$. If $\ell(s\sigma)=\ell(\sigma)+1$, then

$$\rho(T_sT_\sigma)=\rho(T_{s\sigma})=q^{\ell(\sigma)+1}=q\,q^{\ell(\sigma)}=\rho(T_s)\,\rho(T_\sigma).$$

On the other hand, if $\ell(s\sigma)=\ell(\sigma)-1$, then there exists σ' with $\sigma=s\sigma'$ and $\ell(\sigma')=\ell(\sigma)-1$, so

$$\begin{aligned}\rho(T_sT_\sigma)&=\rho(q\,T_{s\sigma}+(q-1)\,T_\sigma)=q\,\rho(T_{\sigma'})+(q-1)\,\rho(T_\sigma)\\&=q^{\ell(\sigma)}+(q-1)\,q^{\ell(\sigma)}=q^{\ell(\sigma)+1}=\rho(T_s)\,\rho(T_\sigma).\end{aligned}$$

Thus, we get a representation of $\mathfrak{H}(n,\mathbb{C}[q])$, and by tensoring by $\mathbb{C}(q)$ or by $\mathbb{C}$, it yields a representation of $\mathfrak{H}(n)$ of dimension 1 over $\mathbb{C}(q)$, and a representation of $\mathfrak{H}_z(n)$ of dimension 1 over $\mathbb{C}$. We shall denote these modules respectively $1_{n,\mathbb{C}(q)}$ and $1_{n,z}$. Notice in particular that $1_{n,z=1}$ is the trivial representation of $\mathfrak{S}(n)$, so the index representation is a deformation of this representation that is compatible with the Hecke algebras.

In the following, we fix a field K that is defined over $\mathbb{C}[q]$, and a Hecke algebra $\mathfrak{H}(n,K)$; this includes the cases of generic ($K=\mathbb{C}(q)$) and specialized ($K=\mathbb{C}$) Hecke algebras.

Definition 4.54. *Suppose* $\mathfrak{H}(n,K)$ *semisimple. A simple module* M *over* $\mathfrak{H}(n,K)$ *is called* ***parabolic*** *with respect to a composition* $c\in\mathfrak{C}(n)$ *if it occurs with multiplicity 1 in*

$$\mathrm{Ind}_{\mathfrak{H}(c,K)}^{\mathfrak{H}(n,K)}(1_{c,K}),$$

where $1_{c,K}$ *is the index representation* $1_{c_1,K}\otimes_K\cdots\otimes_K 1_{c_\ell,K}$ *of* $\mathfrak{H}(c,K)=\mathfrak{H}(c_1,K)\otimes_K\cdots\otimes_K\mathfrak{H}(c_\ell,K)$*.*

Proposition 4.55. *Consider* $\mathfrak{H}_1(n)=\mathbb{C}\mathfrak{S}(n)$*. Every irreducible representation* S^λ *of* $\mathfrak{S}(n)$ *is a parabolic module with respect to some composition* $c\in\mathfrak{C}(n)$*.*

Proof. If $c=\lambda$, then $\mathrm{Ind}_{\mathfrak{S}(\lambda)}^{\mathfrak{S}(n)}(1_{\lambda,1})$ is the permutation module M^λ, and according to the lemma just after Theorem 3.37, the multiplicity of the Specht module S^λ in M^λ is 1. □

Actually, the following more general statement is true:

Proposition 4.56. *If* K' *is a field extension of* $\mathbb{C}(q)$ *such that* $\mathfrak{H}(n,K')$ *is split semisimple, then every simple module* M *over* $\mathfrak{H}(n,K')$ *is parabolic.*

Proof. Fix a valuation ring $\mathscr{O}'$ as in Lemma 4.42, and denote $1_{c,\mathscr{O}'}$ the index representation of $\mathfrak{H}(c,\mathscr{O}')$. One has of course $\mathfrak{H}(c,K')\otimes_{\mathfrak{H}(c,\mathscr{O}')} 1_{c,\mathscr{O}'} = 1_{c,K'}$ and $\mathfrak{H}_1(c)\otimes_{\mathfrak{H}(c,\mathscr{O}')} 1_{c,\mathscr{O}'} = 1_{c,1}$. Then, recall that changes of base rings are associative, in the sense that if C is an algebra over B, and B is an algebra over A, then $C\otimes_B (B\otimes_A M) = C\otimes_A M$ for any A-module M. As a consequence,

$$\begin{aligned}\mathrm{Ind}_{\mathfrak{H}(c,K')}^{\mathfrak{H}(n,K')}(1_{c,K'}) &= \mathfrak{H}(n,K')\otimes_{\mathfrak{H}(c,\mathscr{O}')} 1_{c,\mathscr{O}'}\\ &= \mathfrak{H}(n,K')\otimes_{\mathfrak{H}(n,\mathscr{O}')}\left(\mathfrak{H}(n,\mathscr{O}')\otimes_{\mathfrak{H}(c,\mathscr{O}')} 1_{c,\mathscr{O}'}\right);\\ \mathrm{Ind}_{\mathfrak{S}(c)}^{\mathfrak{S}(n)}(1_{c,1}) &= \mathfrak{H}_1(n)\otimes_{\mathfrak{H}(c,\mathscr{O}')} 1_{c,\mathscr{O}'} = \mathfrak{H}_1(n)\otimes_{\mathfrak{H}(n,\mathscr{O}')}\left(\mathfrak{H}(n,\mathscr{O}')\otimes_{\mathfrak{H}(c,\mathscr{O}')} 1_{c,\mathscr{O}'}\right).\end{aligned}$$

It follows that the modular reduction of $\mathrm{Ind}_{\mathfrak{H}(c,K')}^{\mathfrak{H}(n,K')}(1_{c,K'})$ with respect to $\mathbb{C}[q]\hookrightarrow \mathscr{O}'\to\mathbb{C}$ is $\mathrm{Ind}_{\mathfrak{S}(c)}^{\mathfrak{S}(n)}(1_{c,1})$. However, by Theorem 4.43, the modular reduction is an isomorphism of Grothendieck groups that conserves the simple modules, so if

$$\mathrm{Ind}_{\mathfrak{H}(c,K')}^{\mathfrak{H}(n,K')}(1_{c,K'}) = \bigoplus_{\lambda\in\mathfrak{Y}(n)} m_\lambda\, S^{\lambda,K'},$$

then

$$\mathrm{Ind}_{\mathfrak{S}(c)}^{\mathfrak{S}(n)}(1_{c,1}) = \bigoplus_{\lambda\in\mathfrak{Y}(n)} m_\lambda\, S^{\lambda}$$

with the same multiplicities m_λ. Since every irreducible representation of $\mathfrak{S}(n)$ is parabolic, the same is therefore true for every simple module over $\mathfrak{H}(n,K')$. □

Corollary 4.57. *If M is a simple module over $\mathfrak{H}(n,K')$, then it appears with multiplicity 1 in some module $\mathfrak{H}(n,K')\otimes_{\mathfrak{H}(n)} N$ with N module over the generic Hecke algebra $\mathfrak{H}(n)$.*

Proof. This is trivial since every simple module is parabolic, and $\mathrm{Ind}_{\mathfrak{H}(c,K')}^{\mathfrak{H}(n,K')}(1_{c,K'})$ is realized over K, and comes from the module $\mathrm{Ind}_{\mathfrak{H}(c,K)}^{\mathfrak{H}(n,K)}(1_{c,K})$. □

Corollary 4.58. *If M is a simple module over $\mathfrak{H}(n,K')$, and ch^M is its character, then $\mathrm{ch}^M(h)\in K$ for every $h\in\mathfrak{H}(n,K)$.*

Proof. Let λ be an integer partition of size n, and consider the simple module $M = S^{\lambda,K'}$. By the Jacobi–Trudi formula 2.16, one can write the Schur function s_λ as an integer linear combination of homogeneous functions h_μ:

$$s_\lambda(X) = \sum_{\mu\in\mathfrak{Y}(n)} c_{\lambda\mu}\, h_\mu(X).$$

In $\mathrm{R}_0(\mathfrak{S}(n))$, this identity corresponds to:

$$S^\lambda = \bigoplus_{\mu\in\mathfrak{Y}(n)} c_{\lambda\mu}\, M^\mu,$$

where the M^{μ}'s are the permutation modules. Since the modular reduction yields an isomorphism between the two groups $\mathrm{R}_0(\mathfrak{H}(n,K'))$ and $\mathrm{R}_0(\mathfrak{S}(n))$, it follows that in $\mathrm{R}_0(\mathfrak{H}(n,K'))$,

$$S^{\lambda,K'} = \bigoplus_{\mu\in\mathfrak{Y}(n)} c_{\lambda\mu}\,\mathrm{Ind}_{\mathfrak{H}(\mu,K')}^{\mathfrak{H}(n,K')}(1_{\mu,K'}).$$

In particular, at the level of characters,

$$\mathrm{ch}^{\lambda} = \sum_{\mu\in\mathfrak{Y}(n)} c_{\lambda\mu}\,\mathrm{ch}^{\mathrm{Ind}_{\mathfrak{H}(\mu,K')}^{\mathfrak{H}(n,K')}(1_{\mu,K'})}.$$

However, we saw that the modules $\mathrm{Ind}_{\mathfrak{H}(\mu,K')}^{\mathfrak{H}(n,K')}(1_{\mu,K'})$ are defined over K, so their characters restricted to $\mathfrak{H}(n,K)$ take their values in K. □

Theorem 4.59. *The generic Hecke algebra $\mathfrak{H}(n) = \mathfrak{H}(n,\mathbb{C}(q))$ is split semisimple. The modular reduction associated to the diagram*

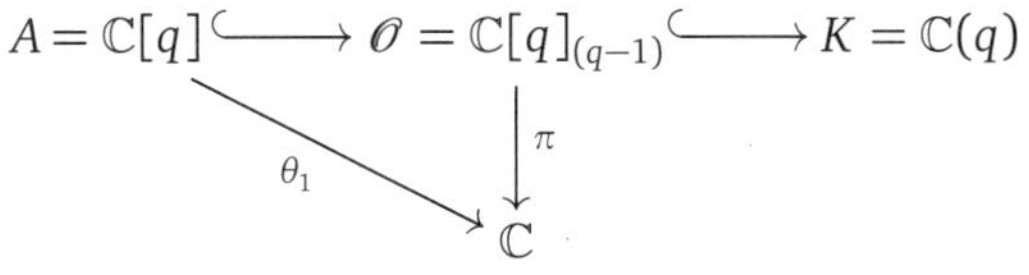

$$\begin{array}{ccccc} A=\mathbb{C}[q] & \hookrightarrow & \mathcal{O}=\mathbb{C}[q]_{(q-1)} & \hookrightarrow & K=\mathbb{C}(q) \\ & \searrow^{\theta_1} & \downarrow \pi & & \\ & & \mathbb{C} & & \end{array}$$

yields a decomposition map $d : \mathrm{R}_0(\mathfrak{H}(n)) \to \mathrm{R}_0(\mathfrak{S}(n))$ that is an isomorphism, and that conserves the simple modules. Hence, there is a labeling of the simple modules of $\mathfrak{H}(n)$ by integer partitions such that

$$\mathfrak{H}(n) = \bigoplus_{\lambda\in\mathfrak{Y}(n)} \mathrm{End}_{\mathbb{C}(q)}(S^{\lambda,\mathbb{C}(q)}) \quad ; \quad d(S^{\lambda,\mathbb{C}(q)}) = S^{\lambda}.$$

If $\mathrm{ch}^{\lambda,q}$ denotes the irreducible character associated to the simple $\mathfrak{H}(n)$-module $S^{\lambda,\mathbb{C}(q)}$, then for every permutation σ, $\mathrm{ch}^{\lambda,q}(T_{\sigma})$ belongs to $\mathbb{C}[q]$ and

$$\theta_1(\mathrm{ch}^{\lambda,q}(T_{\sigma})) = \mathrm{ch}^{\lambda}(\sigma),$$

where the right-hand side is the character of the Specht module S^{λ} for $\mathfrak{S}(n)$.

Proof. Fix an integer partition λ, and consider the simple module $S^{\lambda,K'}$ of $\mathfrak{H}(n,K')$; we denote $\mathrm{ch}^{\lambda,K'}$ its character. By the previous corollaries, $S^{\lambda,K'}$ appears with multiplicity one in some module $\mathfrak{H}(n,K') \otimes_{\mathfrak{H}(n,K)} N$, and the restriction of the character ψ^{λ} to $\mathfrak{H}(n,K)$ takes its values in $K = \mathbb{C}(q)$. Therefore, by Proposition 4.52, $S^{\lambda,K'}$ is realized over K, i.e., there exists a module $S^{\lambda,K}$ over $\mathfrak{H}(n)$ such that $S^{\lambda,K'} = \mathfrak{H}(n,K') \otimes_{\mathfrak{H}(n)} S^{\lambda,K}$. This is necessarily a simple $\mathfrak{H}(n)$-module, since a submodule of $S^{\lambda,K}$ gives by tensorization by $\mathfrak{H}(n,K')$ a submodule of $S^{\lambda,K'}$, which is simple. So, we have a collection

$$\left(S^{\lambda,\mathbb{C}(q)}\right)_{\lambda\in\mathfrak{Y}(n)} \text{ of simple modules over } \mathfrak{H}(n).$$

If $\lambda \neq \mu$, then $S^{\lambda,\mathbb{C}(q)}$ and $S^{\mu,\mathbb{C}(q)}$ are not isomorphic as $\mathfrak{H}(n)$-modules, as otherwise by tensorization by $\mathfrak{H}(n,K')$, one would have $S^{\lambda,K'} = S^{\mu,K'}$. So, the previous family consists in non-isomorphic simple $\mathfrak{H}(n)$-modules.

If $u \in \mathrm{End}_{\mathfrak{H}(n)}(S^{\lambda,K})$, then $\mathrm{id}_{K'} \otimes_K u$ belongs to $\mathrm{End}_{\mathfrak{H}(n,K')}(S^{\lambda,K'})$, and the extended Hecke algebra $\mathfrak{H}(n,K')$ is split, so $\mathrm{id}_{K'} \otimes_K u \in K'$. Thus, there exists $x \in K'$ such that for every $v \in S^{\lambda,K'}$,

$$\mathrm{id}_{K'} \otimes_K u(v) = xv.$$

However, by choosing a basis $(1, x_2, \ldots, x_d)$ of K' over K, one gets inclusions $K \subset K'$ and $S^{\lambda,K} \subset S^{\lambda,K'}$, and since $u(S^{\lambda,K}) \subset S^{\lambda,K}$, this implies that $x \in K$. Therefore, it is shown that $S^{\lambda,K}$ is split for any integer partition λ. As a consequence,

$$\begin{aligned} n! = \dim_K \mathfrak{H}(n) \geq \dim_K \mathfrak{H}(n)/\mathrm{rad}(\mathfrak{H}(n)) &\geq \sum_{\lambda \in \mathfrak{Y}(n)} \dim_K \mathrm{End}_K(S^{\lambda,K}) \\ \geq \sum_{\lambda \in \mathfrak{Y}(n)} (\dim_K S^{\lambda,K})^2 = \sum_{\lambda \in \mathfrak{Y}(n)} (\dim_{K'} S^{\lambda,K'})^2 = n! \end{aligned}$$

since $\mathfrak{H}(n,K')$ is split semisimple, with simple modules the $S^{\lambda,K'}$. So, we have an equality everywhere above, and $\mathfrak{H}(n)$ is split semisimple, with $(S^{\lambda,\mathbb{C}(q)})_{\lambda \in \mathfrak{Y}(n)}$ forming a complete family of representatives of the simple $\mathfrak{H}(n)$-modules. The remainder of the theorem is then an application of Theorem 4.39, exactly as in the proof of Theorem 4.43, but with K instead of K'. □

▷ *Computation of the Schur elements and structure of the specialized Hecke algebras.*

To conclude our study of the representation theory of Hecke algebras, we want to prove that Theorem 4.59 also holds with respect to almost all the other specializations θ_z of $\mathbb{C}[q]$. Then, a formula for the characters of the semisimple split algebra $\mathfrak{H}(n)$ will yield by specialization all the character values of the simple modules of the Hecke algebras $\mathfrak{H}_z(n)$, including those of the symmetric group. To this purpose, the only missing ingredient now is the semisimplicity of the algebra $\mathfrak{H}_z(n)$, and by Theorem 4.50, it suffices to show that for every simple $\mathfrak{H}_z(n)$-module M, the Schur element c_M is non-zero. Actually, one can reduce the computation of Schur elements to the generic case. Fix $z \neq 0$, $\mathscr{O} = \mathbb{C}[q]_{(q-z)}$, and

$$\mathscr{O}H = \mathfrak{H}(n,\mathscr{O}) \quad ; \quad KH = \mathfrak{H}(n) \quad ; \quad kH = \mathfrak{H}_z(n).$$

We denote π_z the projection $\mathscr{O} \to \mathbb{C}$ associated to the specialization $q = z$. Since $\mathbb{C}$ is algebraically closed, $\mathfrak{H}_z(n)$ is split, so by Theorem 4.38 one has a decomposition map $d_\theta : \mathrm{R}_0(\mathfrak{H}(n)) \to \mathrm{R}_0(\mathfrak{H}_z(n))$ associated to the modular reduction. On the other hand, all the algebras above are symmetric algebras for the trace $\tau(T_\sigma) = \delta_{\sigma,\mathrm{id}_{[1,n]}}$, so for any module M over kH, KH or $\mathscr{O}H$, if u is an endomorphism over the base ring of the module M, then

$$I(u)(\cdot) = \sum_{\sigma \in \mathfrak{S}(n)} q^{-\ell(\sigma)}\, T_{\sigma^{-1}}\, u(T_\sigma(\cdot))$$

is a well-defined endomorphism of module over the Hecke algebra (with $q = z$ in the specialized case).

Proposition 4.60. *Let M^K be a simple module over $\mathfrak{H}(n,K)$. Then, c_{M^K} belongs to $\mathscr{O}$, and if $\pi_z(c_{M^K}) \neq 0$, then $d_{\theta_z}(M^K) = [M^k]$ is a (class of isomorphism of) simple module over $\mathfrak{H}_z(n)$. As a consequence, the Hecke algebra $\mathfrak{H}_z(n)$ is semisimple split if and only if, for any integer partition $\lambda \in \mathfrak{Y}(n)$, $\pi_z(c_\lambda) \neq 0$, where c_λ denotes the Schur element of $S^{\lambda,\mathbb{C}(q)}$.*

Proof. Let $M^{\mathscr{O}}$ be a realization of M^K over $\mathscr{O}$, and $M^k = k \otimes_{\mathscr{O}} M^{\mathscr{O}}$ be a modular reduction of M^K. Since M^K is a simple module over the symmetric split algebra $\mathfrak{H}(n,K)$, the Schur element is given by the formula

$$I(\mathrm{id}_{M^K}) = c_{M^K}\,(\dim_K M^K)\,\mathrm{id}_{M^K} = \sum_{\sigma\in\mathfrak{S}(n)} q^{-\ell(\sigma)}\,\rho^{M^K}(T_{\sigma^{-1}})\,\rho^{M^K}(T_\sigma).$$

Restricted to $M^{\mathscr{O}}$, the right-hand side is a well-defined element of $\mathrm{End}_{\mathscr{O}}(M^{\mathscr{O}})$, so one has also

$$c_{M^K}\,(\dim_K M^K)\,\mathrm{id}_{M^{\mathscr{O}}} = \sum_{\sigma\in\mathfrak{S}(n)} q^{-\ell(\sigma)}\,\rho^{M^{\mathscr{O}}}(T_{\sigma^{-1}})\,\rho^{M^{\mathscr{O}}}(T_\sigma),$$

and $c_{M^K} \in \mathscr{O}$. We extend the notation for the projection $\pi_z : \mathscr{O} \to k$ to the projections

$$\begin{aligned} &M^{\mathscr{O}} \to M^k;\\ &\mathrm{End}_{\mathscr{O}}(M^{\mathscr{O}}) \to \mathrm{End}_k(M^k); \end{aligned}$$

etc. Notice then that for any $u \in \mathrm{End}_k(M^k)$, choosing an element $u^{\mathscr{O}} \in \mathrm{End}_{\mathscr{O}}(M^{\mathscr{O}})$ such that $\pi_z(u^{\mathscr{O}}) = u$, one has

$$\begin{aligned} \sum_{\sigma\in\mathfrak{S}(n)} z^{-\ell(\sigma)}\,\rho^{M^k}(T_{\sigma^{-1}})\,u\,\rho^{M^k}(T_\sigma) &= \pi_z\left(\sum_{\sigma\in\mathfrak{S}(n)} q^{-\ell(\sigma)}\,\rho^{M^{\mathscr{O}}}(T_{\sigma^{-1}})\,u^{\mathscr{O}}\,\rho^{M^{\mathscr{O}}}(T_\sigma)\right)\\ &= \pi_z\left(c_{M^K}\,(\mathrm{tr}\,u^{\mathscr{O}})\,\mathrm{id}_{M^{\mathscr{O}}}\right) = \pi_z(c_{M^K})\,(\mathrm{tr}\,u)\,\mathrm{id}_{M^k}. \end{aligned}$$

Therefore, for the first part of the proposition, it is sufficient to prove the following claim: if B is a split symmetric algebra over k and M is a module such that there exists $c \neq 0$ with $I(u) = c\,(\mathrm{tr}\,u)\,\mathrm{id}_M$ for all $u \in \mathrm{End}_k(M)$, then M is a simple module (and $c = c_M$). Suppose that M is not simple, and fix a basis $(m_1, \ldots, m_n)$ of M such that $m_1, \ldots, m_{k<n}$ is a basis of a B-submodule $N \subsetneq M$. Then, if u is the endomorphism defined by

$$u(m_i) = \begin{cases} m_i & \text{if } i > k,\\ 0 & \text{if } i \le k, \end{cases}$$

one has $\mathrm{tr}\,u = n-k \neq 0$, but

$$I(u)(m_1) = \sum_b b\,u(b'(m_1)) = 0 \quad \text{since } b'm_1 \in N \text{ and } u(N) = 0.$$

So, $I(u)$ is not a non-zero multiple of the identity. We have therefore shown that if $\pi_z(c_{M^K}) \neq 0$, then M^k is a simple module over kH. In the Grothendieck group, two simple modules M and N have equivalent classes $[M] = [N]$ if and only if they are isomorphic, so the modular reduction M^k is well defined up to isomorphism, and a simple module by the previous discussion.

With $c_\lambda = c_{S^{\lambda,K}}$, suppose that $\pi_z(c_\lambda) \neq 0$ for all $\lambda \in \mathfrak{Y}(n)$. We denote S_z^λ the modular reduction of $S^{\lambda,K}$, which is then a simple module for all $\lambda \in \mathfrak{Y}(n)$. If $\lambda \neq u$, then for every $u \in \mathrm{Hom}_K(S^{\lambda,K}, S^{\mu,K})$, $I(u) = 0$, because $S^{\lambda,K}$ and $S^{\mu,K}$ are not isomorphic and $\mathrm{Hom}_{KH}(S^{\lambda,K}, S^{\mu,K}) = 0$. By modular reduction, it follows that for every $u \in \mathrm{Hom}_k(S_z^\lambda, S_z^\mu)$, $I(u) = 0$; therefore, S_z^λ and S_z^μ are not isomorphic $\mathfrak{H}_z(n)$-module. For dimension reasons, this implies that $(S_z^\lambda)_{\lambda \in \mathfrak{Y}(n)}$ is a complete family of representatives of the classes of isomorphism of simple modules of the semisimple algebra $\mathfrak{H}_z(n)$. Conversely, if $\pi_z(c_\lambda) = 0$ for some partition λ, then $\mathfrak{H}_z(n)$ is a symmetric algebra that admits a non-zero module M with $I(u) = 0$ for all $u \in \mathrm{End}_k(M)$. By Theorem 4.47, this module is not projective, so $\mathfrak{H}_z(M)$ is not semisimple by Proposition 4.17. □

Our task is now to compute the Schur elements of the simple modules of the *generic* Hecke algebra $\mathfrak{H}(n)$. These calculations are related to a certain specialization of the algebra of symmetric functions Sym, and to the combinatorics of **Pochhammer symbols**. Let q and x be indeterminates, and $k \geq 1$. We denote

$$(x;q)_k = \prod_{i=1}^{k} (1 - xq^{i-1}) = (1-x)(1-qx)\cdots(1-q^{k-1}x);$$

this is the Pochhammer symbol of x and q of rank k. By working in the algebra of power series $\mathbb{C}[[x,q]]$, one can also introduce the infinite symbol

$$(x;q)_\infty = \prod_{i=1}^{\infty} (1 - xq^{i-1}) = (1-x)(1-qx)(1-q^2x)\cdots.$$

One can see $(x;q)_\infty$ as a q-deformation of $(1-x)$: indeed, setting $q=0$ gives back the polynomial $1-x$. In particular, the identity of power series $\frac{1}{1-x} = \sum_{k=0}^\infty x^k$ admits the following generalization:

Proposition 4.61. *In the algebra $\mathbb{C}((x,q))$ of Laurent series in x and q,*

$$\frac{1}{(x;q)_\infty} = \sum_{k=0}^{\infty} \frac{x^k}{(q;q)_k}.$$

Proof. If $\lambda = (\lambda_1, \ldots, \lambda_\ell)$ is an integer partition of size n, we associate to it the monomial $X(\lambda) = x^\ell q^n$. The generating function of integer partitions with respect to this association is then

$$G(x,q) = \sum_{\lambda \in \mathfrak{Y}} X(\lambda) = \sum_{m_1, m_2, \ldots, m_s, \ldots \geq 0} x^{\sum_{i=1}^{\infty} m_i} q^{\sum_{i=1}^{\infty} i m_i}$$

$$= \prod_{i=1}^{\infty} \left(\sum_{m_i=0}^{\infty} (xq^i)^{m_i} \right) = \prod_{i=1}^{\infty} \frac{1}{1 - xq^i} = \frac{1}{(qx;q)_\infty}.$$

Set $F_k(q) = \sum_{\lambda \in \mathfrak{Y},\, \ell(\lambda)=k} q^{|\lambda|}$. The previous series is $G(x,q) = \sum_{k=0}^{\infty} x^k F_k(q)$. On the other hand, if one removes the first column of the Young diagram of a partition of length k, then one obtains the Young diagram of a partition of length smaller than k, so

$$F_k(q) = q^k \left(\sum_{i=0}^{k} F_i(q) \right).$$

This recurrence relation rewrites as

$$F_k(q) = \frac{q^k}{1-q^k} \left(\sum_{i=0}^{k-1} F_i(q) \right) = \frac{q^k}{1-q^k} \left(F_{k-1}(q) + \frac{1-q^{k-1}}{q^{k-1}} F_{k-1}(q) \right)$$

$$= \frac{q}{1-q^k} F_{k-1}(q),$$

from which it follows that $F_k(q) = \frac{q^k}{(q;q)_k}$. Thus, $\frac{1}{(qx;q)_\infty} = \sum_{k=0}^{\infty} \frac{(qx)^k}{(q;q)_k}$, whence the result by replacing qx by x. □

Corollary 4.62. *Let $k \geq 1$, and q be an indeterminate. We denote X_q the infinite alphabet $\{1-q, (1-q)q, (1-q)q^2, \ldots, (1-q)q^n, \ldots\}$. Then,*

$$h_k(X_q) = \frac{1}{[k]_q!} = \prod_{i=1}^{k} \frac{1}{[k]_q},$$

where $[k]_q = \frac{1-q^k}{1-q} = 1 + q + \cdots + q^{k-1}$ denotes the q-analogue of the integer k.

Proof. One has $h_k(X_q) = (1-q)^k h_k(1, q, q^2, \ldots)$, and the specializations of the homogeneous symmetric functions at $(1, q, q^2, \ldots)$ are given by the power series

$$H(x) = \sum_{k=0}^{\infty} h_k(1, q, q^2, \ldots) x^k = \prod_{i=1}^{\infty} \frac{1}{1 - xq^{i-1}}.$$

The formula of the previous proposition enables one to conclude. □

Proposition 4.63. *With the same notations as above, for any integer partition λ,*

$$s_\lambda(X_q) = q^{n(\lambda)} \prod_{\square \in \lambda} \frac{1}{[h(\square)]_q}, \quad \text{with } n(\lambda) = \sum_{i=1}^{\ell(\lambda)} (i-1)\lambda_i.$$

Proof. We are going to compute $s_\lambda(1,q,\ldots,q^{n-1}) = \frac{a_{\lambda+\rho}(1,q,\ldots,q^{n-1})}{a_\rho(1,q,\ldots,q^{n-1})}$, where as usual ρ denotes the staircase partition $(n-1,n-2,\ldots,1,0)$. If $\mu=\lambda+\rho$, then one has the Vandermonde determinant

$$\begin{aligned} a_\mu(1,q,\ldots,q^{n-1}) &= \det(q^{i(\mu_j)})_{1\leq i,j\leq n} = \prod_{1\leq i<j\leq n}(q^{\mu_j}-q^{\mu_i}) \\ &= q^{n(\lambda)+\frac{n(n-1)(n-2)}{6}} \prod_{1\leq i<j\leq n}(1-q^{\lambda_i-\lambda_j-i+j}). \end{aligned}$$

In the proof of Jacobi–Trudi Theorem 2.16, we saw that if $n\geq\lambda_1'$ and $m\geq\lambda_1$, then the two sets $\{\lambda_{n+1-j}+j\}_{j\in[\![1,n]\!]}$ and $\{n+j-\lambda_j'\}_{j\in[\![1,m]\!]}$ are complementary subsets of $[\![1,m+n]\!]$. Therefore,

$$\sum_{j=1}^{n} q^{\lambda_j+n+1-j} + \sum_{j=1}^{m} q^{n+j-\lambda_j'} = \sum_{j=1}^{n+m} q^j = q\,\frac{1-q^{n+m}}{1-q}.$$

If we interchange the roles played by λ and λ', this gives, with $m=\lambda_1$ and $n\geq\lambda_1'$

$$\begin{aligned} \sum_{j=1}^{m} q^{\lambda_j'+\lambda_1-j} + \sum_{j=1}^{n} q^{\lambda_1-\lambda_j+j-1} &= \frac{1-q^{n+\lambda_1}}{1-q} \\ \sum_{j=1}^{\lambda_1} q^{h(1,j)} + \sum_{j=2}^{n} q^{\mu_1-\mu_j} &= \sum_{j=1}^{\mu_1} q^j. \end{aligned}$$

If one applies the result to the partition $(\lambda_i,\ldots,\lambda_n)$, one obtains

$$\sum_{j=1}^{\lambda_i} q^{h(i,j)} + \sum_{j=i+1}^{n} q^{\mu_i-\mu_j} = \sum_{j=1}^{\mu_i} q^j,$$

hence,

$$\sum_{\square\in\lambda} q^{h(\square)} + \sum_{1\leq i<j\leq n} q^{\mu_i-\mu_j} = \sum_{i=1}^{n}\sum_{j=1}^{\mu_i} q^j.$$

As a consequence,

$$\begin{aligned} \prod_{\square\in\lambda}(1-q^{h(\square)}) &= \exp\left(\sum_{\square\in\lambda}\log(1-q^{h(\square)})\right) = \exp\left(-\sum_{k=1}^{\infty}\frac{1}{k}\left(\sum_{\square\in\lambda} q^{k\,h(\square)}\right)\right) \\ &= \exp\left(\sum_{k=1}^{\infty}\frac{1}{k}\left(\sum_{1\leq i<j\leq n} q^{k(\mu_i-\mu_j)} - \sum_{i=1}^{n}\sum_{j=1}^{\mu_i} q^{kj}\right)\right) \\ &= \exp\left(\sum_{i=1}^{n}\sum_{j=1}^{\mu_i}\log(1-q^j) - \sum_{1\leq i<j\leq n}\log(1-q^{\mu_i-\mu_j})\right) \\ &= \frac{\prod_{i=1}^{n}\prod_{j=1}^{\mu_i}(1-q^j)}{\prod_{1\leq i<j\leq n}(1-q^{\lambda_i-\lambda_j-i+j})}. \end{aligned}$$

It follows that

$$s_\lambda(1,q,\ldots,q^{n-1}) = \frac{a_\mu(1,q,\ldots,q^{n-1})}{a_\rho(1,q,\ldots,q^{n-1})} = q^{n(\lambda)} \prod_{1\leq i<j\leq n} \frac{1-q^{\lambda_i-\lambda_j+j-i}}{1-q^{j-i}}$$

$$= q^{n(\lambda)} \prod_{\square\in\lambda} \frac{1}{1-q^{h(\square)}} \left(\frac{\prod_{i=1}^{n}\prod_{j=1}^{\mu_i}(1-q^j)}{\prod_{i=1}^{n}\prod_{j=1}^{n-i}(1-q^j)}\right)$$

$$= q^{n(\lambda)} \prod_{\square\in\lambda} \frac{1}{1-q^{h(\square)}} \left(\prod_{i=1}^{n}\prod_{j=n-i+1}^{n-i+\lambda_i}(1-q^j)\right).$$

In the right-hand side, the term in parentheses rewrites as:

$$\prod_{i=1}^{\ell(\lambda)}\prod_{j=1}^{\lambda_i}(1-q^{n-i+j}) = \prod_{\square\in\lambda}(1-q^{n+c(\square)}),$$

where $c(\square)$ denotes the **content** of a box of a Young diagram, and is defined by $c(i,j) = j-i$. Therefore,

$$s_\lambda(1,q,\ldots,q^{n-1}) = q^{n(\lambda)} \prod_{\square\in\lambda} \frac{1-q^{n+c(\square)}}{1-q^{h(\square)}}.$$

With $|q|<1$, the limit of the left-hand side as n goes to infinity is $(q-1)^{-|\lambda|}\, s_\lambda(X_q)$, whereas the limit of the right-hand side is $q^{n(\lambda)} \prod_{\square\in\lambda} \frac{1}{1-q^{h(\square)}}$. We conclude to the equality of power series in q

$$s_\lambda(X_q) = q^{n(\lambda)} \prod_{\square\in\lambda} \frac{1}{[h(\square)]_q}.$$

□

Let us now state the connection between the specialization X_q of Sym, and the Schur elements of the generic Hecke algebra $\mathfrak{H}(n)$:

Theorem 4.64 (Steinberg). *For any integer partition $\lambda \in \mathfrak{Y}(n)$, if c_λ denotes the Schur element of the simple module $S^{\lambda,\mathbb{C}(q)}$, then $\frac{1}{c_\lambda} = s_\lambda(X_q)$.*

Lemma 4.65. *Let $k \geq 1$. One has*

$$\sum_{\sigma\in\mathfrak{S}(k)} q^{\ell(\sigma)} = [k]_q! = \prod_{i=1}^{k}[i]_q.$$

Proof. We reason by induction on k, and we use the fact that the length $\ell(\sigma)$ is equal to the number of inversions of the permutation. If $\sigma = \sigma(1)\sigma(2)\cdots\sigma(k-1)$ is the word of a permutation of size k, then the insertion of k before the i-th letter of σ in this word creates a permutation of size k with $k-i$ more inversions. Every permutation of size k is obtained in this way, so,

$$\sum_{\tau\in\mathfrak{S}(k)} q^{\ell(\tau)} = \sum_{\sigma\in\mathfrak{S}(k-1)} \sum_{i=1}^{k} q^{k-i+\ell(\sigma)} = [k]_q \sum_{\sigma\in\mathfrak{S}(k-1)} q^{\ell(\sigma)}.$$

By induction on k, this yields the formula announced. □

Lemma 4.66. *Let c be a composition of size n, and M and N be two modules respectively over the semisimple split algebras $\mathfrak{H}(c)$ and $\mathfrak{H}(n)$. One has the Frobenius reciprocity:*

$$\mathrm{Hom}_{\mathfrak{H}(n)}\left(\mathrm{Ind}_{\mathfrak{H}(c)}^{\mathfrak{H}(n)}(M),N\right) = \mathrm{Hom}_{\mathfrak{H}(c)}\left(M,\mathrm{Res}_{\mathfrak{H}(c)}^{\mathfrak{H}(n)}(N)\right).$$

In particular, if M and N are simple modules, then the multiplicity of N in the induced module $\mathrm{Ind}_{\mathfrak{H}(c)}^{\mathfrak{H}(n)}(M)$ is equal to the multiplicity of M in the reduced module $\mathrm{Res}_{\mathfrak{H}(c)}^{\mathfrak{H}(n)}(N)$.

Proof. This generalized version of Frobenius' reciprocity, to be compared with Proposition 1.12, actually holds for any k-algebras $A \subset B$ and any pair (M,N), where M is an A-module and N is a B-module:

$$\mathrm{Hom}_B(\mathrm{Ind}_A^B(M),N) = \mathrm{Hom}_A(M,\mathrm{Res}_A^B(N)).$$

Indeed, let $u : M \to N$ be an endomorphism of A-modules. We define $\psi(u) : B \otimes_A M \to N$ by $\psi(u)(b \otimes m) = b \cdot (u(m))$. This is obviously an endomorphism of B-modules, so we get a map $\psi : \mathrm{Hom}_A(M,\mathrm{Res}_A^B(N)) \to \mathrm{Hom}_B(\mathrm{Ind}_A^B(M),N)$. Its inverse is given by $\psi^{-1}(v)(m) = v(1 \otimes m)$. □

Proof of Theorem 4.64. Recall the definition of the Kostka numbers $K_{\lambda,\mu}$: $K_{\lambda,\mu} = \langle s_\lambda \,|\, h_\mu \rangle_{\mathrm{Sym}}$, so that

$$s_\lambda(X) = \sum_{|\mu|=|\lambda|} K_{\lambda,\mu}\, m_\mu(X) \quad ; \quad h_\mu(X) = \sum_{|\lambda|=|\mu|} K_{\lambda,\mu}\, s_\lambda(X).$$

By the Frobenius–Schur isomorphism and the representation theoretic interpretation of the symmetric functions h_μ and s_λ, the second identity becomes in $\mathrm{R}_0(\mathfrak{S}(n))$:

$$M^\mu = \mathrm{Ind}_{\mathfrak{S}(\mu)}^{\mathfrak{S}(n)}(1_\mu) = \bigoplus_{|\lambda|=|\mu|} K_{\lambda,\mu}\, S^\lambda.$$

Since the decomposition map $d : \mathrm{R}_0(\mathfrak{H}(n)) \to \mathrm{R}_0(\mathfrak{S}(n))$ is an isomorphism of groups that preserves the simple modules, in $\mathrm{R}_0(\mathfrak{H}(n))$, we have therefore

$$\mathrm{Ind}_{\mathfrak{H}(\mu)}^{\mathfrak{H}(n)}(1_{\mu,\mathbb{C}(q)}) = \bigoplus_{|\lambda|=|\mu|} K_{\lambda,\mu}\, S^{\lambda,\mathbb{C}(q)},$$

where $1_{\mu,\mathbb{C}(q)}$ is the index representation of the parabolic subalgebra $\mathfrak{H}(\mu)$ of $\mathfrak{H}(n)$.

In the following, if M is a simple module over $\mathfrak{H}(\mu)$ and N is a simple module over $\mathfrak{H}(n)$, we denote K_{MN} the coefficient

$$K_{MN} = \dim_{\mathbb{C}(q)} \mathrm{Hom}_{\mathfrak{H}(n)}\left(\mathrm{Ind}_{\mathfrak{H}(\mu)}^{\mathfrak{H}(n)}(M), N\right) = \dim_{\mathbb{C}(q)} \mathrm{Hom}_{\mathfrak{H}(\mu)}\left(M, \mathrm{Res}_{\mathfrak{H}(\mu)}^{\mathfrak{H}(n)}(N)\right).$$

One then has:

$$\mathrm{Ind}_{\mathfrak{H}(\mu)}^{\mathfrak{H}(n)}(M) = \bigoplus_N K_{MN}\, N \quad ; \quad \mathrm{Res}_{\mathfrak{H}(\mu)}^{\mathfrak{H}(n)}(N) = \bigoplus_M K_{MN}\, M.$$

On the other hand, notice that the parabolic subalgebra $\mathfrak{H}(\mu)$ is a *symmetric* subalgebra of $\mathfrak{H}(n)$, i.e., the restriction of the trace τ from $\mathfrak{H}(n)$ to $\mathfrak{H}(\mu)$ is again nondegenerate. Indeed, a basis of $\mathfrak{H}(\mu)$ over $\mathbb{C}(q)$ is $(T_\sigma)_{\sigma\in\mathfrak{S}(\mu)}$, and its dual basis with respect to τ is $(q^{-\ell(\sigma)}\, T_{\sigma^{-1}})_{\sigma\in\mathfrak{S}(\mu)}$. Therefore, the one-dimensional representation $1_{\mu,\mathbb{C}(q)}$ of $\mathfrak{H}(\mu)$, which is a split simple module, admits a Schur element, which we denote b_μ. We claim that

$$\frac{1}{b_\mu} = \sum_\lambda K_{\lambda\mu}\, \frac{1}{c_\lambda}.$$

Indeed, using Theorem 4.50, one sees that the Schur elements b_M, where M runs over the classes of simple $\mathfrak{H}(\mu)$-modules, are characterized by the expansion

$$\tau_{|\mathfrak{H}(\mu)} = \sum_M \frac{\mathrm{ch}^M}{b_M}.$$

However,

$$\tau_{|\mathfrak{H}(\mu)} = \left(\sum_N \frac{\mathrm{ch}^N}{c_N}\right)_{|\mathfrak{H}(\mu)} = \sum_M \left(\sum_N \frac{K_{MN}}{c_N}\right) \mathrm{ch}^M,$$

so $\frac{1}{b_M} = \sum_N \frac{K_{MN}}{c_N}$. The Schur element b_μ is now easily computed as

$$\begin{aligned} b_\mu &= \sum_{\sigma\in\mathfrak{S}(\mu)} q^{-\ell(\sigma)} \rho^{1_{\mu,\mathbb{C}(q)}}(T_\sigma)\, \rho^{1_{\mu,\mathbb{C}(q)}}(T_{\sigma^{-1}}) \\ &= \sum_{\sigma\in\mathfrak{S}(\mu)} q^{\ell(\sigma)} = \prod_{i=1}^{\ell(\mu)} \left(\sum_{\sigma\in\mathfrak{S}(\mu_i)} q^{\ell(\sigma)}\right) = \prod_{i=1}^{\ell(\mu)} [\mu_i]_q!. \end{aligned}$$

By Corollary 4.62, $\frac{1}{b_\mu} = h_\mu(X_q)$, and then, the specializations $s_\lambda(X_q)$ of the Schur functions solve the equations

$$\frac{1}{b_\mu} = \sum_\lambda K_{\lambda\mu}\, s_\lambda(X_q).$$

Since the matrix $(K_{\lambda\mu})_{\lambda,\mu}$ is invertible (it is a matrix of change of basis in the algebra of symmetric functions), these equations characterize the Schur elements c_λ, so we have indeed $\frac{1}{c_\lambda} = s_\lambda(X_q)$. Proposition 4.63 then allows us to compute

$$c_\lambda = \frac{1}{q^{n(\lambda)}} \prod_{\square \in \lambda} [h(\square)]_q.$$

Notice that it belongs indeed to $\mathbb{C}[q]_{(q-z)}$ for $z \neq 0$. □

We finally have the analogue of Theorem 4.59 for specialized Hecke algebras:

Theorem 4.67. *Let $z \neq 0$, such that $z^2, z^3, \ldots, z^n \neq 1$. The specialized Hecke algebra $\mathfrak{H}_z(n)$ is split semisimple. The modular reduction associated to the diagram*

$$\begin{array}{ccccc} A = \mathbb{C}[q] & \hookrightarrow & \mathcal{O} = \mathbb{C}[q]_{(q-z)} & \hookrightarrow & K = \mathbb{C}(q) \\ & \searrow^{\theta_z} & \downarrow \pi & & \\ & & \mathbb{C} & & \end{array}$$

yields a decomposition map $d : \mathrm{R}_0(\mathfrak{H}(n)) \to \mathrm{R}_0(\mathfrak{H}_z(n))$ that is an isomorphism, and that preserves the simple modules. Hence, there is a labeling of the simple modules of $\mathfrak{H}_z(n)$ by integer partitions such that

$$\mathfrak{H}_z(n) = \bigoplus_{\lambda \in \mathfrak{Y}(n)} \mathrm{End}_{\mathbb{C}}(S_z^\lambda) \quad ; \quad d(S^{\lambda,\mathbb{C}(q)}) = S_z^\lambda.$$

If ch_z^λ denotes the irreducible character associated to the simple $\mathfrak{H}(n)_z$-module S_z^λ, then for every permutation σ,

$$\theta_z(\mathrm{ch}^{\lambda,q}(T_\sigma)) = \mathrm{ch}_z^\lambda(T_\sigma),$$

where the left-hand side involves the character of the simple module $S^{\lambda,\mathbb{C}(q)}$ for the generic Hecke algebra $\mathfrak{H}(n)$.

Conversely, for $n \geq 2$, if $z = 0$ or if $z^k = 1$ for some $k \in [\![2, n]\!]$, then $\mathfrak{H}_z(n)$ is not semisimple.

Proof. From Proposition 4.60, we know that in the case $z \neq 0$, the semisimplicity is equivalent to the condition $\pi_z(c_\lambda) \neq 0$ for every $\lambda \in \mathfrak{Y}(n)$ (then, one uses Tits deformation theorem exactly as before). However, in the formula for c_λ, the numerator is a product of polynomials $[h]_q = 1+q+\cdots+q^{h-1}$ with h hook-length of a partition, and therefore that belongs to $[\![1, n]\!]$. These polynomials vanish at roots of unity, hence the result. As for the non-semisimplicity of $\mathfrak{H}_0(n)$, we refer to Section 6.3, where it will be shown that $\dim_{\mathbb{C}} \mathfrak{H}_0(n)/\mathrm{rad}(\mathfrak{H}_0(n)) = 2^{n-1} < n!$. □

Hence, it is shown as planned that the group algebra $\mathbb{C}\mathfrak{S}(n)$, the specialized Hecke algebras $\mathfrak{H}_z(n)$ with $z \notin \{0, \text{roots of unity}\}$, and the generic Hecke algebra $\mathfrak{H}(n)$ are semisimple split and have the same representation theory, their modules being related by decomposition maps that are bijections between the isomorphism classes of simple modules.

4.5 Polynomial construction of the q-Specht modules

In Chapter 3, we gave an explicit construction of each irreducible representation S^λ of $\mathfrak{S}(n)$, by means of an action of the symmetric group on polynomials $\Delta_T(x_1,\dots,x_n)$ associated to the standard tableaux T of shape λ. We now wish to do the same for the generic Hecke algebra $\mathfrak{H}(n)$, and to provide an explicit construction of each simple module $S^{\lambda,\mathbb{C}(q)}$. These modules will then give by specialization the simple modules S_z^λ of the specialized Hecke algebras $\mathfrak{H}_z(n)$ for $z \notin \{0, \text{roots of unity}\}$, and in particular we will get back the Specht modules $S_1^\lambda = S^\lambda$ of $\mathfrak{S}(n)$.

▷ *Action of the Hecke algebra on polynomials.*

If $P(x_1,\dots,x_n)$ is a polynomial in n variables and with coefficients in $\mathbb{C}[q]$, denote

$$(s_i \cdot P)(x_1,\dots,x_n) = P(x_1,\dots,x_{i+1},x_i,\dots,x_n);$$
$$(\pi_i \cdot P)(x_1,\dots,x_n) = \frac{x_i\,P(x_1,\dots,x_n) - x_{i+1}\,(s_i \cdot P)(x_1,\dots,x_n)}{x_i - x_{i+1}};$$

and $T_i \cdot P = (q-1)\pi_i \cdot P + s_i \cdot P$. Notice that $\pi_i \cdot P$ is again a polynomial (instead of a rational function), because

$$s_i \cdot (x_i\,P - x_{i+1}\,(s_i \cdot P)) = (x_{i+1}\,(s_i \cdot P) - x_i\,P) = -(x_i\,P - x_{i+1}\,(s_i \cdot P)),$$

and therefore, $x_i - x_{i+1}$ divides $x_i\,P - x_{i+1}\,(s_i \cdot P)$ (this is the same argument as in the proof of Proposition 2.12).

Proposition 4.68. *The previous rule yields a left action of the Hecke algebra $\mathfrak{H}(n,\mathbb{C}[q])$ on $\mathbb{C}[q](x_1,\dots,x_n)$.*

Proof. We just have to check that the action of the T_i's is compatible with the braid, commutation and quadratic relations. The previous computation shows that for every polynomial P,

$$s_i \cdot (\pi_i \cdot P) = \frac{-(x_i\,P - x_{i+1}\,(s_i \cdot P))}{-(x_i - x_{i+1})} = \pi_i \cdot P.$$

On the other hand, if $P = s_i \cdot P$, then $\pi_i \cdot P = P$, so π_i is a projection on the set of polynomials that are invariant by the elementary transposition s_i. In particular, $(\pi_i)^2 = \pi_i$. Therefore,

$$\begin{aligned}(T_i)^2 \cdot P &= (q-1)^2\,((\pi_i)^2 \cdot P) + \pi_i \cdot (q-1)(P + s_i \cdot P) + P\\ &= (q-1)^2\,(\pi_i \cdot P) + (q-1)(P + s_i \cdot P) + P\\ &= (q-1)((q-1)\pi_i \cdot P + s_i \cdot P) + q\,P\\ &= (q-1)\,(T_i \cdot P) + q\,P.\end{aligned}$$

For the commutation relations, if $|j-i|\geq 2$, then

$$\pi_j\cdot(\pi_i\cdot P)=\frac{x_ix_jP-x_{i+1}x_j(s_i\cdot P)-x_ix_{j+1}(s_j\cdot P)+x_{i+1}x_{j+1}(s_is_j\cdot P)}{x_ix_j-x_{i+1}x_j-x_ix_{j+1}+x_{i+1}x_{j+1}},$$

and one obtains the same formula for $\pi_i\cdot(\pi_j\cdot P)$. Hence, π_i and π_j commute, and similarly, π_i and s_j commute, π_j and s_i commute, and of course s_i and s_j commute. It follows that $T_i=(q-1)\pi_i+s_i$ and $T_j=(q-1)\pi_j+s_j$ commute.

The verification for the braid relations is much more subtle. Notice that if $P=QR$ with R symmetric in x_i and x_{i+1}, then

$$\pi_i\cdot P=\frac{x_iP-x_{i+1}(s_i\cdot P)}{x-x_{i+1}}=\frac{x_iQR-x_{i+1}(s_i\cdot Q)R}{x_i-x_{i+1}}=(\pi_i\cdot Q)R,$$

and therefore, $T_i\cdot P=(T_i\cdot Q)R$. As a consequence, it suffices to show that $T_iT_{i+1}T_i$ and $T_{i+1}T_iT_{i+1}$ have the same action on a generating family of the module $\mathbb{C}[q][x_1,x_2,\ldots,x_n]$ over $\mathbb{C}[q][x_1,x_2,\ldots,x_n]^{\mathfrak{S}(3)}$, where in this last ring $\mathfrak{S}(3)$ acts by permutation of the three variables x_i, x_{i+1}, x_{i+2}. This reduces the computations to the case $n=3$. We then claim that $\mathbb{C}[q][x,y,z]$ is freely generated as a $\mathbb{C}[q][x,y,z]^{\mathfrak{S}(3)}$-module by the following polynomials:

$$\begin{aligned}&(x-y)(y-z)(x-z)\quad;\\&(x-y)(x+y-2z)\quad;\quad(y-z)(y+z-2x)\quad;\\&(x-y)\quad;\quad(y-z)\quad;\\&1.\end{aligned}$$

The proof of this claim will be given in the next paragraph. Taking it for granted, we then compute easily:

$$\begin{aligned}T_1T_2T_1\cdot(x-y)(y-z)(x-z)&=T_2T_1T_2\cdot(x-y)(y-z)(x-z)\\&=q^3\,yz^2+(q^3-4q^2+8q-6)\,x^2y-2q(q-1)\,xyz\\&\quad+q(q-2)^2\,(xy^2+x^2z)+q^2(q-2)\,(xz^2+y^2z);\\T_1T_2T_1\cdot(x-y)(x+y-2z)&=T_2T_1T_2\cdot(x-y)(x+y-2z)\\&=q^3\,z^2+q^2(q-2)\,y^2+q(q-1)^2\,x^2+q(2q^2-5q+1)\,xz\\&\quad+q(2q^2-7q+7)\,xy+2q^2(q-1)\,yz;\\T_1T_2T_1\cdot(x-y)&=T_2T_1T_2\cdot(x-y)=q^3\,z+q^2(q-2)\,y+q(q-1)^2\,x;\\T_1T_2T_1\cdot 1&=T_2T_1T_2\cdot 1=q^3,\end{aligned}$$

and similarly for the two remaining elements of the basis of $\mathbb{C}[q][x,y,z]$ viewed as a module over $\mathbb{C}[q][x,y,z]^{\mathfrak{S}(3)}$. □

The next paragraph is devoted to the proof of the claim about the basis of the module $\mathbb{C}[q][x,y,z]$ over $\mathbb{C}[q][x,y,z]^{\mathfrak{S}(3)}$; more generally, we shall describe the structure of $\mathrm{Sym}^{(n)}$-module on $\mathbb{C}[x_1,\ldots,x_n]$. Before that, let us state the main result of this section, which we shall simply admit. For $\lambda \in \mathfrak{Y}(n)$, we fill the boxes of the Young diagram of λ with the integers of $[\![1,n]\!]$ column by column, and we denote $T(\lambda)$ the corresponding standard tableau. For instance, if $\lambda = (4,3,1,1)$, then

$$T(\lambda) = \begin{array}{|c|c|c|c|} \hline 4 \\ \cline{1-1} 3 \\ \cline{1-1} 2 & 6 & 8 \\ \cline{1-3} 1 & 5 & 7 & 9 \\ \hline \end{array} \;.$$

We then set

$$\Delta_{\lambda,\mathbb{C}(q)}(x_1,\ldots,x_n) = \prod_{\text{column } C \text{ of } T(\lambda)} \left(\prod_{i \text{ under } j \text{ in } C} x_i - q x_j \right).$$

This is a q-deformation of the polynomial $\Delta_{T(\lambda)}$ introduced in Section 3.3: setting $q = 1$ in the formula for $\Delta_{\lambda,\mathbb{C}(q)}$, one gets back this product of Vandermonde polynomials. In Section 3.3, we saw that the left $\mathbb{C}\mathfrak{S}(n)$-module

$$\mathbb{C}\mathfrak{S}(n)\,\Delta_{T(\lambda)}$$

is isomorphic to the irreducible representation S^λ of $\mathfrak{S}(n)$, and thus, every irreducible representation of $\mathfrak{S}(n)$ is obtained uniquely by this polynomial construction. The q-analogue of this result is true:

Theorem 4.69 (Lascoux). *Consider the $\mathfrak{H}(n)$-module $\mathfrak{H}(n)\,\Delta_{\lambda,\mathbb{C}(q)}$. It is a simple module over $\mathfrak{H}(n)$, isomorphic to the generic Specht module $S^{\lambda,\mathbb{C}(q)}$ described by Theorem 4.59. Thus, every simple module of $\mathfrak{H}(n)$ is obtained uniquely by this polynomial construction.*

Remark. The difficult part of the theorem is to show that $\mathfrak{H}(n)\,\Delta_{\lambda,\mathbb{C}(q)}$ is a simple $\mathfrak{H}(n)$-module. Then, since $\theta_1(\Delta_{\lambda,\mathbb{C}(q)}) = \Delta_{T(\lambda)}$, the modular reduction of $\mathfrak{H}(n)\,\Delta_{\lambda,\mathbb{C}(q)}$ contains S^λ, which necessarily implies that $\mathfrak{H}(n)\,\Delta_{\lambda,\mathbb{C}(q)} = S^{\lambda,\mathbb{C}(q)}$.

Remark. Beware that given a polynomial $P \in \mathbb{C}[q][x_1,x_2,\ldots,x_N]$, the modular reduction of the module $\mathfrak{H}(n)\,P$ *is not* in general equal to $\mathbb{C}\mathfrak{S}(n)\,\theta_1(P)$; this is only true if $\mathfrak{H}(n)\,P$ is a simple module over $\mathfrak{H}(n)$. For example, consider with $n = 2$ the polynomial $P = x - y$. Then,

$$\mathfrak{H}(n,\mathbb{C}[q])\,P = \mathbb{C}[q]\,(x-y) + \mathbb{C}[q]\,(q-1)\,(x+y),$$

with $T(x-y) = -(x-y) + (q-1)(x+y)$ and $T(x+y) = q(x+y)$. The modular reduction of $\mathfrak{H}(2)\,P$ is therefore given by the matrix

$$\theta_1\begin{pmatrix} -1 & 0 \\ (q-1) & q \end{pmatrix} = \begin{pmatrix} -1 & 0 \\ 0 & 1 \end{pmatrix},$$

so it is the regular representation $S^{(1,1)} \oplus S^{(2)} = \mathbb{C}\mathfrak{S}(2)$; and $\mathfrak{H}(2)\,P = \mathfrak{H}(2)$, although $\mathbb{C}\mathfrak{S}(2)\,P = S^{(1,1)}$.

Example. If $\lambda = (n)$, then $\Delta_{\lambda,\mathbb{C}(q)} = 1$, and the action of the Hecke algebra on the constant polynomial 1 is given by $T_i\,1 = q$. So, $T_\sigma\,1 = q^{\ell(\sigma)}\,1$, and one recovers the index representation $S^{(n),\mathbb{C}(q)} = 1_{n,\mathbb{C}(q)}$.

Example. The only other partition λ that admits a unique standard tableau is $\lambda = (1^n)$, and it corresponds to the unique other representation of $\mathfrak{H}(n)$ that is of dimension 1 over $\mathbb{C}(q)$. Notice then that $\Delta_{(1^n),\mathbb{C}(q)} = \prod_{1\leq i<j\leq n}(x_i - qx_j)$ and

$$\begin{aligned}
&T_i \cdot \Delta_{(1^n),\mathbb{C}(q)} \\
&= \frac{\Delta_{(1^n),\mathbb{C}(q)}}{x_i - qx_{i+1}} \left((q-1)\frac{x_i(x_i - qx_{i+1}) - x_{i+1}(x_{i+1} - qx_i)}{x_i - x_{i+1}} + (x_{i+1} - qx_i) \right) \\
&= \frac{qx_{i+1} - x_i}{x_i - qx_{i+1}}\,\Delta_{(1^n),\mathbb{C}(q)} = -\Delta_{(1^n),\mathbb{C}(q)}.
\end{aligned}$$

Thus, one obtains the signature representation $\rho(T_\sigma) = (-1)^{\ell(\sigma)}$, which extends the classical signature representation of $\mathfrak{S}(n)$ to the Hecke setting.

▷ *Theory of invariants and harmonic polynomials.*

To close this chapter, let us give an explanation of the claim made during the proof of Proposition 4.68. We are going to prove the following important result from the theory of invariants:

Theorem 4.70. *Fix an integer $N \geq 1$, and consider the algebra of polynomials $\mathbb{C}[x_1,\ldots,x_N]$. Viewed as a module over its subalgebra $\mathrm{Sym}^{(N)} = \mathbb{C}[x_1,\ldots,x_N]^{\mathfrak{S}(N)}$, it is free of rank $N!$, and it is spanned linearly by the Vandermonde polynomial*

$$\Delta = \prod_{1\leq i<j\leq N} x_i - x_j.$$

and by all its partial derivatives with respect to $x_1,\ldots,x_N$.

The claim of Proposition 4.68 is an immediate consequence of this result in the case $N = 3$. The proof of Theorem 4.70 relies on the notion of **harmonic polynomial** with respect to the symmetric group $\mathfrak{S}(N)$. If $P(x_1,\ldots,x_N) = \sum_{k=(k_1,\ldots,k_N)} a_k x^k$, we set

$$P(\partial) = \sum_k a_k \frac{\partial^{|k|}}{\partial x^k},$$

which acts on the ring of polynomials $\mathbb{C}[x_1,\ldots,x_N]$. For instance, if $P(x,y) = x^2 - 3xy$, then $P(\partial) = \frac{\partial^2}{\partial x^2} - 3\frac{\partial^2}{\partial x \partial y}$. Let $\mathrm{Sym}^{(N),*}$ be the vector space of symmetric polynomials in the variables $x_1,\ldots,x_N$ without constant term:

$$P(0,\ldots,0) = 0 \quad ; \quad \forall \sigma \in \mathfrak{S}(N),\ \sigma \cdot P = P.$$

A polynomial Q is called harmonic if, for every $P \in \mathrm{Sym}^{(N),*}$, $P(\partial)(Q) = 0$. Since

$$\mathrm{Sym}^{(N)} = \mathbb{C}[e_1(x_1,\ldots,x_N),\ldots,e_N(x_1,\ldots,x_N)]$$

and $P \mapsto P(\partial)$ is a morphism of algebras, this is equivalent to the condition:

$$\forall k \in [\![1,N]\!],\ e_k(\partial)(Q) = 0.$$

We denote $\mathrm{H}^{(N)}$ the space of harmonic polynomials in N variables. Notice that if P is harmonic, then all its homogeneous components are also harmonic, so $\mathrm{H}^{(N)} = \bigoplus_{k=0}^{\infty} \mathrm{H}_k^{(N)}$, where $\mathrm{H}_k^{(N)}$ is the set of harmonic polynomials that are homogeneous of degree k.

Proposition 4.71. *The mutiplication map*

$$\begin{aligned} m : \mathrm{H}^{(N)} \otimes_{\mathbb{C}} \mathrm{Sym}^{(N)} &\to \mathbb{C}[x_1,\ldots,x_N] \\ h \otimes f &\mapsto hf \end{aligned}$$

is an isomorphism of (graded) vector spaces.

Lemma 4.72. *Consider non-zero homogeneous polynomials P_i, and non-zero homogeneous symmetric polynomials Q_i, such that*

$$Q_r \notin \sum_{i=1}^{r-1} Q_i\, \mathrm{Sym}^{(N)} \qquad ; \qquad \sum_{i=1}^{r} P_i\, Q_i = 0.$$

The polynomial P_r belongs to the ideal $\mathbb{C}[x_1,\ldots,x_N]\,\mathrm{Sym}^{(N),*}$.

Proof. Suppose first that $\deg P_r = 0$, that is to say that P_r is a constant. We then claim that $P_r = 0$, which implies that $P_r \in \mathbb{C}[x_1,\ldots,x_N]\,\mathrm{Sym}^{(N),*}$. Indeed, otherwise, one could write

$$Q_r = -\frac{1}{P_r}\left(\sum_{i=1}^{r-1} P_i\, Q_i\right) = -\frac{1}{P_r}\left(\sum_{i=1}^{r-1} \mathscr{S}(P_i)\, Q_i\right),$$

where $\mathscr{S}$ is the symmetrization operator that was introduced in Section 2.2. Since $Q_r \notin \sum_{i=1}^{r-1} Q_i\, \mathrm{Sym}^{(N)}$, this is not possible, so $P_r = 0$ and the result is proven.

We now reason by induction on the degree d of the products $P_i\, Q_i$, the case $d = 0$ following from the previous argument. Suppose the result is true up to degree $d-1$, and consider an equation of degree d. By the previous argument one can assume without loss of generality that $\deg P_r \geq 1$. If $s = (j, j+1)$ is an elementary transposition, notice that

$$0 = \sum_{i=1}^{r} P_i\, Q_i = \sum_{i=1}^{r} (s \cdot P_i)(s \cdot Q_i) = \sum_{i=1}^{r} (s \cdot P_i)\, Q_i,$$

so $0 = \sum_{i=1}^{r} (P_i - s \cdot P_i)\, Q_i$. However, the polynomial $x_j - x_{j+1}$ divides $P_i - s \cdot P_i$, so there exist polynomials O_i such that

$$0 = (x_j - x_{j+1}) \sum_{i=1}^{r} O_i\, Q_i.$$

We then have $\deg(O_i Q_i) = d-1$, so by the induction hypothesis, $O_r \in \mathbb{C}[x_1,\ldots,x_N]\,\mathrm{Sym}^{(N),*}$, and $P_r - s\cdot P_r$ belongs also to this ideal, that is to say that

$$P_r \equiv s\cdot P_r \mod \mathrm{Sym}^{(N),*}$$

in the ring of polynomials $\mathbb{C}[x_1,\ldots,x_N]$. Applying this result several times, we conclude that

$$P_r \equiv \sigma\cdot P_r \mod \mathrm{Sym}^{(N),*}$$

for every permutation $\sigma = s_{j_1}\cdots s_{j_\ell}$ in $\mathfrak{S}(N)$. In particular,

$$P_r \equiv \left(\frac{1}{N!}\sum_{\sigma\in\mathfrak{S}(N)} \sigma\cdot P_r\right) \mod \mathrm{Sym}^{(N),*}.$$

However, the term $\mathscr{S}(P)$ on the right-hand side is in $\mathrm{Sym}^{(N)}$, and since P_r is homogeneous of degree ≥ 1, it is in $\mathrm{Sym}^{(N),*}$. Therefore, $P_r \equiv 0 \mod \mathrm{Sym}^{(N),*}$. □

Proof of Proposition 4.71. For the surjectivity, we prove by induction on k that if P is a homogeneous polynomial of degree k in $\mathbb{C}[x_1,\ldots,x_N]$, then it is attained by m. The initial case $k=0$ is obvious; in the following we suppose the result to be true up to order $k-1$. Consider the bilinear form

$$\langle P \mid Q\rangle = (P(\partial)(Q))(0,\ldots,0).$$

Since $\left\langle x^k \mid x^l\right\rangle = \delta_{k,l}\prod_{i=1}^N (k_i!)$, it is non-degenerate on $\mathbb{C}[x_1,\ldots,x_N]$, and the subspaces $\mathbb{C}[x_1,\ldots,x_N]_k$ of homogenenous polynomials of degree k are mutually orthogonal. Moreover,

$$\mathbb{C}[x_1,\ldots,x_N]_k = \mathrm{H}_k^{(N)} \oplus (\mathbb{C}[x_1,\ldots,x_N]\,\mathrm{Sym}^{(N),*})_k.$$

Indeed, a polynomial P is harmonic if and only if

$$\begin{aligned}
&\forall Q\in\mathrm{Sym}^{(N),*},\ Q(\partial)(P)=0\\
\iff\quad &\forall R\in\mathbb{C}[x_1,\ldots,x_N],\ \forall Q\in\mathrm{Sym}^{(N),*},\ \langle R\mid Q(\partial)(P)\rangle = 0\\
\iff\quad &\forall R\in\mathbb{C}[x_1,\ldots,x_N],\ \forall Q\in\mathrm{Sym}^{(N),*},\ \langle RQ\mid P\rangle = 0
\end{aligned}$$

that is to say that P is in the orthogonal of $\mathbb{C}[x_1,\ldots,x_N]\,\mathrm{Sym}^{(N),*}$. Fix now P homogeneous of degree k, and write

$$P = H + \sum_{i=1}^r R_i\,Q_i,$$

where $H\in\mathrm{H}_k^{(N)}$ and $R_i\otimes Q_i \in (\mathbb{C}[x_1,\ldots,x_N]\otimes\mathrm{Sym}^{(N),*})_k$. By the induction hypothesis, since each R_i has degree smaller than $k-1$, they are in $m(\mathrm{H}^{(N)}\otimes\mathrm{Sym}^{(N)})$. This proves that P is also in $m(\mathrm{H}^{(N)}\otimes\mathrm{Sym}^{(N)})$.

For the injectivity, suppose that one has a relation $m(\sum_{i=1}^{r} P_i \otimes Q_i) = \sum_{i=1}^{r} P_i Q_i = 0$, with the P_i's harmonic and the Q_i's symmetric, all these polynomials being non-zero. One can assume without loss of generality that the P_i's and the Q_i's are homogeneous, that r is minimal in this formula, and that $\deg Q_1 \geq \deg Q_2 \geq \cdots \geq \deg Q_r$. Notice then $Q_r \notin \sum_{i=1}^{r-1} Q_i \,\mathrm{Sym}^{(N)}$, as otherwise one could write a relation with $r-1$ terms. By Lemma 4.72, $P_r \in \mathbb{C}[x_1, \ldots, x_N]\,\mathrm{Sym}^{(N),*}$. However, P_r is also harmonic, and we have just seen that harmonic polynomials are orthogonal to elements of $\mathbb{C}[x_1, \ldots, x_N]\,\mathrm{Sym}^{(N),*}$. Hence, $P_r = 0$, which contradicts the initial hypothesis. □

Corollary 4.73. *The space $\mathrm{H}^{(N)}$ has finite dimension $N!$ over $\mathbb{C}$. More precisely, its Poincaré series*

$$\mathrm{H}^{(N)}(t) = \sum_{k=0}^{\infty} \left(\dim_{\mathbb{C}} \mathrm{H}_k^{(N)}\right) t^k$$

is equal to $\prod_{i=1}^{N}(1 + t + t^2 + \cdots + t^{i-1})$.

Proof. Given a graded vector space $V = \bigoplus_{k=0}^{\infty} V_k$ with $\dim_{\mathbb{C}} V_k < \infty$ for all k, we introduce its **Poincaré series** $V(t) = \sum_{k=0}^{\infty} (\dim_{\mathbb{C}} V_k)\, t^k$, to be considered as a formal power series in $\mathbb{C}[[t]]$. Trivially, if $V = W_1 \otimes W_2$ in the sense of graded tensor product, then $V(t) = W_1(t)\, W_2(t)$. Therefore, Proposition 4.71 implies that

$$\mathrm{H}^{(N)}(t)\,(\mathrm{Sym}^{(N)})(t) = (\mathbb{C}[x_1, \ldots, x_N])(t) = (\mathbb{C}[x](t))^n = \left(\frac{1}{1-t}\right)^n.$$

Indeed, the Poincaré series of $\mathbb{C}[x]$ is $\sum_{k=0}^{\infty} t^k$, since there is one generator x^k in each degree. Now, recall that $\mathrm{Sym}^{(N)} = \mathbb{C}[e_1, \ldots, e_N]$, where each e_i has degree i. Therefore,

$$\mathrm{Sym}^{(N)}(t) = \prod_{i=1}^{N} \mathbb{C}[e_i](t) = \prod_{i=1}^{N} \frac{1}{1-t^i}.$$

We conclude that

$$\mathrm{H}^{(N)}(t) = \prod_{i=1}^{N} \frac{1-t^i}{1-t} = \prod_{i=1}^{N} (1 + t + t^2 + \cdots + t^{i-1}),$$

and this implies in particular that $\dim_{\mathbb{C}} \mathrm{H}^{(N)} = \mathrm{H}^{(N)}(1) = N!$. □

Corollary 4.74. *The ring of polynomials $\mathbb{C}[x_1, \ldots, x_N]$ is free of rank $N!$ over its subring $\mathrm{Sym}^{(N)}$.*

Proof. A basis of $\mathrm{H}^{(N)}$ over $\mathbb{C}$ yields a basis of $\mathbb{C}[x_1, \ldots, x_N]$ over $\mathrm{Sym}^{(N)}$ by Proposition 4.71. We can even choose a basis of homogeneous polynomials, with $[t^k](\prod_{i=1}^{N}(1 + t + \cdots + t^{i-1}))$ terms of degree k for each $k \in [\![1, N]\!]$. □

Proof of Theorem 4.70. It remains to see that a generating family of $\mathrm{H}^{(N)}$ consists in the Vandermonde determinant Δ and its partial derivatives. First, notice that for every permutation σ, and every symmetric polynomial $P \in \mathrm{Sym}^{(N),*}$,

$$\sigma \cdot (P(\partial)(\Delta)) = P(\partial)(\sigma \cdot \Delta) = \varepsilon(\sigma)\, P(\partial)(\Delta),$$

so $P(\partial)(\Delta)$ is an antisymmetric polynomial. By Proposition 2.12, it is therefore divisible by Δ, but also of lower degree, so $P(\partial)(\Delta) = 0$. Moreover, harmonic polynomials are stable by partial derivatives, so any partial derivative of Δ is in $\mathrm{H}^{(N)}$. Thus, if V is the vector space generated by Δ and its partial derivatives, then we have

$$V \subset \mathrm{H}^{(N)}.$$

To prove the converse, it suffices to show that the orthogonal of V with respect to $\langle P \mid Q \rangle = (P(\partial)(Q^*))(0)$ is included in the orthogonal of $\mathrm{H}^{(N)}$, which is the ideal $\mathbb{C}[x_1,\ldots,x_N]\,\mathrm{Sym}^{(N),*}$. Thus, we have to show that

$$\{P \mid P(\partial)(\Delta) = 0\} \subset \mathbb{C}[x_1,\ldots,x_N]\,\mathrm{Sym}^{(N),*},$$

because P is in the orthogonal of all the derivatives of Δ if and only if $P(\partial)(\Delta) = 0$. Fix P with this property; without loss of generality, one can assume P homogeneous of degree k. If k is sufficiently large, then there is nothing to prove, because

$$\mathbb{C}[x_1,\ldots,x_N]_k = \mathrm{H}^{(N)}_k \oplus (\mathbb{C}[x_1,\ldots,x_N]\,\mathrm{Sym}^{(N),*})_k$$

and $\mathrm{H}^{(N)}_k = 0$ for k large (larger than $\binom{N}{2}$ by identification of the Poincaré series). We can then reason by reverse induction. Suppose the result to be true at order $k+1$. As before, we introduce an elementary transposition $s = (j, j+1)$. If $P(\partial)(\Delta) = 0$, then the same is true for $(x_j - x_{j+1})\,P$, and by induction, there exist polynomials Q_i and symmetric polynomials $R_i \in \mathrm{Sym}^{(N),*}$ such that

$$(x_j - x_{j+1})\,P = \sum_{i=1}^{r} Q_i\,R_i.$$

Applying s to this identity yields:

$$(x_j - x_{j+1})(P + s \cdot P) = \sum_{i=1}^{r} (Q_i - s \cdot Q_i)\,R_i.$$

However, $x_j - x_{j+1}$ divides each $Q_i - s \cdot Q_i$, so there exist polynomials $\widetilde{Q}_i$ with $P + s \cdot P = \sum_{i=1}^{r} \widetilde{Q}_i\,R_i$. In terms of congruence, this means that

$$P \equiv -s \cdot P \mod \mathrm{Sym}^{(N),*}.$$

Applying several times this result, we obtain that

$$P \equiv \varepsilon(\sigma)(\sigma \cdot P) \mod \mathrm{Sym}^{(N),*}$$

for any permutation $\sigma \in \mathfrak{S}(N)$. In particular,

$$P \equiv \left(\frac{1}{N!} \sum_{\sigma \in \mathfrak{S}(N)} \varepsilon(\sigma)(\sigma \cdot P) \right) \mod \mathrm{Sym}^{(N),*}.$$

The right-hand side $\mathscr{A}(P)$ is an antisymmetric polynomial, hence writes as $O\,\Delta$ for some $O \in \mathrm{Sym}^{(N)}$. If $O \in \mathrm{Sym}^{(N),*}$, then our proof is done. Otherwise, since P is homogeneous, this implies that O is a constant, so one obtains $P \equiv c\Delta \mod \mathrm{Sym}^{(N),*}$ for some constant c. Then,

$$0 = P(\partial)(\Delta) = c\,\Delta(\partial)(\Delta) = c\,\langle \Delta \mid \Delta \rangle,$$

so $c = 0$ and again $P \in \mathrm{Sym}^{(N),*}$. Our proof is then completed. □

Notes and references

All the results of the first Section 4.1 can be generalized to the setting of Coxeter groups; see [Bou68, Chapter 4]. This requires the introduction of the notion of root system (see Appendix Appendix A), but then one can prove an analogue of Matsumoto's theorem, and construct a Hecke algebra for any Coxeter group; see [GP00, Chapters 1 and 4]. In particular, one can produce a Hecke algebra for the hyperoctahedral group $\mathfrak{B}(n) = (\mathbb{Z}/2\mathbb{Z})^n \rtimes \mathfrak{S}(n)$ of signed permutations, and for its subgroup of even signed permutations $\mathfrak{D}(n) = \mathfrak{B}(n)/\{\pm 1\}$. These groups are the Weyl groups of the simple complex Lie algebras of type B_n and D_n, whereas $\mathfrak{S}(n)$ is the Weyl group of the simple complex Lie algebra of type A_{n-1}. Our terminology and notations for the generic and specialized Hecke algebras differ a bit from what is found usually in the literature: thus, a more generic Hecke algebra would have different elements q_i for each generator T_i; see [GP00, Chapter 8].

Regarding Theorem 4.11 and the freeness of the family $(T_\sigma)_{\sigma \in \mathfrak{S}(n)}$ over $\mathbb{C}[q]$ in $\mathfrak{H}(n, \mathbb{C}[q])$, the original proof is due to Bourbaki, see [Bou68, Chapter 4, §2, ex. 23], and it relies on an interpretation of $\mathfrak{H}(n, A)$ as an algebra of linear operators on $A[\mathfrak{S}(n)]$. This realization can be used to prove the linear independence of the T_σ's. Alternatively, one could use the realization of the elements T_σ as operators on polynomials defined in terms of divided differences, see [Las99, Las13] and our Section 4.5. The proof that we presented is much simpler, and was found in the first chapter of [Mat99].

For the general representation theory of algebras, two excellent references are [CR81] and [Coh89, Coh91]. The restriction to finite-dimensional algebras defined over a field k allowed us to bypass a few complications, that are solved in general with the assumption of noetherian or artinian rings. Thus, most of the results stated in Section 4.2 hold in fact for artinian rings and their finite-length modules. For a diagrammatic proof of the Jordan–Hölder theorem, we refer to the very beginning of [Lan02]; for a proof of the Krull–Schmidt theorem, one can consult [CR81, Volume I, §6].

The Brauer–Cartan deformation theory is treated in detail in [CR81, Volume I, Chapter 2], with a less restrictive setting than what is presented in this book: in general, one only assumes the specialization θ to take values in a subfield k of $\mathscr{O}/\mathfrak{m}_{\mathscr{O}}$. In Sections 4.3 and 4.4, we tried to present in a self-contained and accessible way all the arguments of [GR96] and [GP00, Chapters 7 and 9], being a bit more focused as we only treat the type A case. Therefore, our study is in essence nearly identical to the study of Geck, Pfeiffer and Rouquier; and it relies deeply on the use of valuation rings. However, we simplified the arguments of parabolic induction and of Schur elements by using the theory of symmetric functions of Chapter 2. For the arguments of algebraic number theory used in Lemma 4.42, we refer to [Neu99, Chapter 1, §8]. We also used [Mat86, Theorem 10.4] for the proof of the fact that every integrally closed ring A is the intersection of the valuation rings $\mathscr{O}$ with $A \subset \mathscr{O} \subset \mathrm{Frac}(A)$; and we refer to this book for details on modules over valuation rings. For the calculations related to the specialization X_q of Sym, we followed the exercises of [Mac95, Chapter 1].

As mentioned in Section 4.3, the Brauer–Cartan deformation theory can also be used in order to study the linear representations of $\mathfrak{S}(n)$ over fields different from $\mathbb{C}$; e.g., with $k = \mathbb{Z}/p\mathbb{Z}$, see [JK81, Chapters 6 and 7]. Thus, one gets from Theorem 4.38 the existence of a decomposition map $d : \mathrm{R}_0(\mathbb{C}\mathfrak{S}(n)) = \mathrm{R}_0(\mathbb{Q}\mathfrak{S}(n)) \to \mathrm{R}_0((\mathbb{Z}/p\mathbb{Z})\mathfrak{S}(n))$, and the blocks of the associated decomposition matrix are described by Nakayama's conjecture (loc. cit., Theorem 6.1.21), in connection with the combinatorial notions of p-core and p-quotient of Young diagrams.

For a construction of the q-Specht modules $S^{\lambda,\mathbb{C}(q)}$ based on the combinatorics of tableaux, we refer to [Mur92], as well as [DJ86, DJ87]. These results were used by Lascoux in [Las13] to construct the modules $S^{\lambda,\mathbb{C}(q)}$ by using the action of $\mathfrak{H}(n)$ on polynomials. Another "explicit" construction of the simple modules relies on the so-called Kazhdan-Lusztig polynomials, see the original paper [KL79], and the theory of cellular algebras that was developed from it, cf. [GL96, Mat99]. Though the definition of the Kazhdan-Lusztig basis of $\mathfrak{H}(n)$ is purely combinatorial and could have been used here, the proof of its representation theoretic properties relies on a positivity result that, as far as we know, can only be proven by using the étale cohomology of Schubert varieties. This is the reason why we preferred to present Lascoux' construction.

5

Characters and dualities for Hecke algebras

In the previous chapter, given an integer $n \geq 1$, we constructed a family of complex algebras $(\mathfrak{H}_z(n))_{z\in\mathbb{C}}$ of dimension $n!$, that are deformations of the symmetric group algebras $\mathbb{C}\mathfrak{S}(n) = \mathfrak{H}_1(n)$. Almost all of these algebras have the same representation theory: they are semisimple and write as

$$\mathfrak{H}_z(n) = \bigoplus_{\lambda\in\mathfrak{Y}(n)} \mathrm{End}_{\mathbb{C}}(S_z^\lambda),$$

with $\dim S_z^\lambda = \dim S^\lambda = \mathrm{card}\,\mathrm{ST}(\lambda)$. The main goal of this chapter is to compute the characters of the simple modules S_z^λ. More precisely, by the discussion of Chapter 4, there exists for every partition $\lambda \in \mathfrak{Y}(n)$ and every permutation $\sigma \in \mathfrak{S}(n)$ a polynomial $\mathrm{ch}^{\lambda,q}(T_\sigma)$ in $\mathbb{C}[q]$, such that if $\mathfrak{H}_z(n)$ is semisimple, then the specialization

$$\theta_z(\mathrm{ch}^{\lambda,q}(T_\sigma)) = \mathrm{ch}_z^\lambda(T_\sigma)$$

is the value of the character of the simple module S_z^λ of $\mathfrak{H}_z(n)$ on the basis element T_σ. In this chapter, we shall give explicit combinatorial rules in order to compute these polynomials. These rules will in particular involve a Frobenius formula similar to the one of Theorem 2.32 for the case $z = 1$. Notice that if σ and τ are two permutations with the same cycle type $\mu \in \mathfrak{Y}(n)$, then there is no reason for $\mathrm{ch}^{\lambda,q}(T_\sigma)$ and $\mathrm{ch}^{\lambda,q}(T_\tau)$ to be the same polynomial, and indeed, we shall see that in general they are not the same. However, there is still a notion of *character table* for the generic Hecke algebra $\mathfrak{H}(n)$: its coefficients are the polynomials $\mathrm{ch}^{\lambda,q}(T_{\sigma_\mu})$ with $\lambda, \mu \in \mathfrak{Y}(n)$, and where the permutations σ_μ are certain specific representatives of the conjugacy classes C_μ of $\mathfrak{S}(n)$.

In Section 5.1, we introduce the main tool for the computation of characters of simple modules over $\mathfrak{H}(n)$: it is a deformation $\mathrm{GL}_z(N,\mathbb{C})$ of the linear group $\mathrm{GL}(N,\mathbb{C})$, or more precisely of its universal enveloping algebra $U(\mathfrak{gl}(N))$, which is called the quantum group. Its representation theory is detailed in Section 5.2 (Theorems 5.18, 5.21 and 5.25), and in Section 5.3, we establish a generalization of the Schur–Weyl duality for the commuting actions of the quantum group $\mathrm{GL}_z(N,\mathbb{C})$ and of the specialized Hecke algebra $\mathfrak{H}_z(n)$ on the space of tensors $(\mathbb{C}^N)^{\otimes n}$ (Theorem 5.28). This duality is due to Jimbo, and it is one of the main motivations for the study of the Hecke algebras.

In Section 5.4, we complete Jimbo's duality result by another interpretation of $\mathfrak{H}_z(n)$; namely, when $z = q = p^e$ is a prime power, the Hecke algebra $\mathfrak{H}_z(n)$

is isomorphic to the commutant algebra of the finite linear group $G = \mathrm{GL}(n, \mathbb{F}_q)$ for a certain finite-dimensional bimodule. Thus, in the sense of the bicommutant theory, the Hecke algebras connect the representation theory of the finite general linear groups, of the quantum groups and of the symmetric groups.

Finally, by using the Jimbo–Schur–Weyl duality, we compute in Section 5.5 the character values of the specialized Hecke algebras $\mathfrak{H}_z(n)$, and we make the connection with symmetric functions, in a fashion similar to Chapter 2 (see Theorems 5.49 and 5.50). Hence, the character values of the simple modules of the Hecke algebras are afforded by the matrix of change of basis between Schur functions and a deformation of the power sums, namely, the Hall–Littlewood polynomials. These results are due to A. Ram.

5.1 Quantum groups and their Hopf algebra structure

In Section 2.5, we observed that $\mathrm{GL}(N, \mathbb{C})$ and $\mathfrak{S}(n)$ are in duality for their actions on $V = (\mathbb{C}^N)^{\otimes n}$: their images in $\mathrm{End}_{\mathbb{C}}((\mathbb{C}^N)^{\otimes n})$ generate algebras A and B with $\mathrm{Com}(A, V) = B$ and $\mathrm{Com}(B, V) = A$. There exists an analogous result for an action of the Hecke algebra $\mathfrak{H}_z(n)$ on $(\mathbb{C}^N)^{\otimes n}$, which will be in duality with an adequate deformation of the group $\mathrm{GL}(N, \mathbb{C})$ called the quantum group, and denoted $\mathrm{GL}_z(N, \mathbb{C})$. In this section, we construct this quantum group, which is endowed with a natural structure of Hopf algebra. The duality between $\mathrm{GL}_z(N, \mathbb{C})$ and $\mathfrak{H}_z(n)$ will then be established in Section 5.3, and in Section 5.5, we shall use this duality in order to compute the irreducible characters of $\mathfrak{H}(n)$.

▷ *The quantum group $\mathfrak{U}(\mathfrak{gl}(N))$.*

We refer to our Appendix Appendix A for a concise presentation of the representation theory of $\mathrm{GL}(N, \mathbb{C})$, its Lie algebra $\mathfrak{gl}(N) = \mathfrak{gl}(N, \mathbb{C})$, and the corresponding universal enveloping algebra $U(\mathfrak{gl}(N))$. As a consequence of Serre's theorem A.10, $\mathfrak{gl}(N)$ admits the following presentation of Lie algebra. Denote $(e_{ij})_{1 \leq i,j \leq N}$ the basis of elementary matrices in $\mathfrak{gl}(N)$, and introduce the Chevalley–Serre elements

$$\begin{aligned} e_i &= e_{i(i+1)}, & i &\in [\![1, N-1]\!]; \\ f_i &= e_{(i+1)i}, & i &\in [\![1, N-1]\!]; \\ \varepsilon_i &= e_{ii}, & i &\in [\![1, N]\!]. \end{aligned}$$

We also set $h_i = \varepsilon_i - \varepsilon_{i+1}$ for $i \in [\![1, N-1]\!]$. Let $(c_{ij})_{1 \leq i,j \leq N}$ be the Cartan matrix defined by $c_{ii} = 2$, $c_{i(i+1)} = c_{(i+1)i} = -1$, and $c_{ij} = 0$ if $|j - i| \geq 2$. Then, $U(\mathfrak{gl}(N))$

is the algebra generated by $(\varepsilon_i, e_i, f_i)$ and with relations

$$\varepsilon_i\varepsilon_j = \varepsilon_j\varepsilon_i \quad ; \quad h_i e_j - e_j h_i = c_{ij}\, e_j \quad ; \quad h_i f_j - f_j h_i = -c_{ij}\, f_j \quad ;$$
$$(\varepsilon_1 + \varepsilon_2 + \cdots + \varepsilon_N)x = x(\varepsilon_1 + \varepsilon_2 + \cdots + \varepsilon_N) \quad \forall x \in U(\mathfrak{gl}(N)) \quad ;$$
$$e_i^2 e_{i\pm1} - 2\, e_i e_{i\pm1} e_i + e_{i\pm1} e_i^2 = 0 \quad ;$$
$$f_i^2 f_{i\pm1} - 2\, f_i f_{i\pm1} f_i + f_{i\pm1} f_i^2 = 0 \quad ;$$
$$e_i e_j = e_j e_i \quad \text{if } |j-i| \geq 2 \quad ;$$
$$f_i f_j = f_j f_i \quad \text{if } |j-i| \geq 2 \quad ;$$
$$e_i f_j - f_j e_i = \delta_{ij}\, h_i.$$

A holomorphic representation of $\mathrm{GL}(N, \mathbb{C})$ on a complex vector space V is then entirely determined by elements $\rho(e_i)$, $\rho(f_i)$ and $\rho(\varepsilon_i)$ in $\mathrm{End}_{\mathbb{C}}(V)$ that satisfy the previous relations.

The **(generic) quantum group** $\mathfrak{U}(\mathfrak{gl}(N))$ is a deformation of $U(\mathfrak{gl}(N))$ which is obtained, in a rough sense, by "replacing each generator ε_i by q^{ε_i}." Of course, there are many ways to do this, but there is one specific construction that makes $\mathfrak{U}(\mathfrak{gl}(N))$ into a Hopf algebra; see Theorem 5.3. We give the following definition:

Definition 5.1. *Let $q^{1/2}$ be a variable. The quantum group $\mathfrak{U}(\mathfrak{gl}(N))$ is the algebra over the field of fractions $\mathbb{C}(q^{1/2})$ with generators $(e_i)_{i\in[\![1,N-1]\!]}$, $(f_i)_{i\in[\![1,N-1]\!]}$ and $(q^{\pm\varepsilon_i/2})_{i\in[\![1,N]\!]}$, and relations*

$$q^{\frac{\varepsilon_i}{2}}\, q^{-\frac{\varepsilon_i}{2}} = q^{-\frac{\varepsilon_i}{2}}\, q^{\frac{\varepsilon_i}{2}} = 1 \quad ; \quad q^{\frac{\varepsilon_i}{2}}\, q^{\frac{\varepsilon_j}{2}} = q^{\frac{\varepsilon_j}{2}}\, q^{\frac{\varepsilon_i}{2}} \quad ;$$
$$q^{\frac{\varepsilon_i}{2}}\, e_i\, q^{-\frac{\varepsilon_i}{2}} = q^{\frac{1}{2}}\, e_i \quad ; \quad q^{\frac{\varepsilon_i}{2}}\, e_{i-1}\, q^{-\frac{\varepsilon_i}{2}} = q^{-\frac{1}{2}}\, e_{i-1} \quad ;$$
$$q^{\frac{\varepsilon_i}{2}}\, f_i\, q^{-\frac{\varepsilon_i}{2}} = q^{-\frac{1}{2}}\, f_i \quad ; \quad q^{\frac{\varepsilon_i}{2}}\, f_{i-1}\, q^{-\frac{\varepsilon_i}{2}} = q^{\frac{1}{2}}\, f_{i-1} \quad ;$$
$$q^{\frac{\varepsilon_i}{2}}\, e_{j\neq i,i-1}\, q^{-\frac{\varepsilon_i}{2}} = e_j \quad ; \quad q^{\frac{\varepsilon_i}{2}}\, f_{j\neq i,i-1}\, q^{-\frac{\varepsilon_i}{2}} = f_j \quad ;$$
$$e_i^2 e_{i+1} - (q+1)\, e_i e_{i+1} e_i + q\, e_{i+1} e_i^2 = 0 \quad ;$$
$$q\, e_{i+1}^2 e_i - (q+1)\, e_{i+1} e_i e_{i+1} + e_i e_{i+1}^2 = 0 \quad ;$$
$$q\, f_i^2 f_{i+1} - (q+1)\, f_i f_{i+1} f_i + f_{i+1} f_i^2 = 0 \quad ;$$
$$f_{i+1}^2 f_i - (q+1)\, f_{i+1} f_i f_{i+1} + q\, f_i f_{i+1}^2 = 0 \quad ;$$
$$e_i e_j = e_j e_i \quad \textit{if } |j-i| \geq 2 \quad ;$$
$$f_i f_j = f_j f_i \quad \textit{if } |j-i| \geq 2 \quad ;$$
$$e_i f_j - f_j e_i = \delta_{ij}\, \frac{q^{\varepsilon_i} - q^{\varepsilon_{i+1}}}{q-1}.$$

Notice that these relations imply that $q^{\frac{\varepsilon_1+\varepsilon_2+\cdots+\varepsilon_N}{2}}\, x = x\, q^{\frac{\varepsilon_1+\varepsilon_2+\cdots+\varepsilon_N}{2}}$ for any $x \in \mathfrak{U}(\mathfrak{gl}(N))$. Also, if $q^{\frac{h_i}{2}} = q^{\frac{\varepsilon_i-\varepsilon_{i+1}}{2}}$, then

$$q^{\frac{h_i}{2}}\, e_j\, q^{-\frac{h_i}{2}} = q^{\frac{c_{ij}}{2}}\, e_j \quad ; \quad q^{\frac{h_i}{2}}\, f_j\, q^{-\frac{h_i}{2}} = q^{-\frac{c_{ij}}{2}}\, f_j.$$

Let us now see why Definition 5.1 is pertinent. First, the following proposition ensures that $\mathfrak{U}(\mathfrak{gl}(N))$ is endowed with a natural action on $(\mathbb{C}(q^{1/2}))^N$, that can be seen as a deformation of the geometric representation of $\mathrm{GL}(N,\mathbb{C})$ on $\mathbb{C}^N$.

Proposition 5.2. *Consider the map*

$$\begin{aligned}\rho : \mathfrak{U}(\mathfrak{gl}(N)) &\to \mathrm{End}_{\mathbb{C}(q^{1/2})}((\mathbb{C}(q^{1/2}))^N)\\ e_i &\mapsto e_{i(i+1)};\\ f_i &\mapsto e_{(i+1)i};\\ q^{\pm\frac{\varepsilon_i}{2}} &\mapsto q^{\pm\frac{1}{2}} e_{ii} + \sum_{j\neq i} e_{jj}.\end{aligned}$$

It yields a morphism of $\mathbb{C}(q^{1/2})$-algebras, hence a representation of $\mathfrak{U}(\mathfrak{gl}(N))$ on $(\mathbb{C}(q^{1/2}))^N$.

Proof. It suffices to check that the relations defining $\mathfrak{U}(\mathfrak{gl}(N))$ are satisfied for the images of the generators in $\mathrm{End}_{\mathbb{C}(q^{1/2})}((\mathbb{C}(q^{1/2}))^N)$. One has indeed

$$\begin{aligned}\rho\left(q^{\frac{\varepsilon_i}{2}}\right)\rho(e_j)\rho\left(q^{-\frac{\varepsilon_i}{2}}\right) &= \mathrm{diag}\left(1,\ldots,(q^{\frac{1}{2}})_i,\ldots,1\right) e_{j(j+1)}\,\mathrm{diag}\left(1,\ldots,(q^{-\frac{1}{2}})_i,\ldots,1\right)\\ &= \begin{cases} q^{\frac{1}{2}}\, e_{i(i+1)} & \text{if } j=i,\\ q^{-\frac{1}{2}}\, e_{(i-1)i} & \text{if } j=i-1,\\ e_{j(j+1)} & \text{if } j\neq i,i-1,\end{cases}\\ &= \rho\left(q^{\frac{\varepsilon_i}{2}}\, e_j\, q^{-\frac{\varepsilon_i}{2}}\right);\end{aligned}$$

and

$$\begin{aligned}\rho(e_i)\rho(f_j)-\rho(f_j)\rho(e_i) &= \delta_{ij}\,(e_{ii}-e_{(i+1)(i+1)})\\ &= \delta_{ij}\,\frac{(q-1)\,e_{ii}+\sum_j e_{jj}-(q-1)\,e_{(i+1)(i+1)}-\sum_j e_{jj}}{q-1}\\ &= \delta_{ij}\,\frac{\rho\,(q^{\varepsilon_i})-\rho\,(q^{\varepsilon_{i+1}})}{q-1} = \rho([e_i,f_j]).\end{aligned}$$

The other relations are either trivially satisfied, or similar to the previous calculations. □

If $z\in\mathbb{C}$ is not 0 and not a root of unity, then in particular $z^{1/2}$ and $z-1\neq 0$, and one can define the quantum group $\mathfrak{U}(\mathfrak{gl}(N))$ over the valuation ring $\mathscr{O} = \mathbb{C}[q^{1/2}]_{(q^{1/2}-z^{1/2})}$, instead of $\mathbb{C}(q^{1/2})$. As a consequence, as in Section 4.3, one can work with specializations of $\mathfrak{U}(\mathfrak{gl}(N))$, that are the algebras defined over $\mathbb{C}$ by replacing the variable q by the constant z in the presentation of the quantum group. We shall denote $U_z(\mathfrak{gl}(N))$ the **specialized quantum group**, and it will be convenient later to denote η_i the image of q^{ε_i} by the specialization morphism (thus making the letter q disappear). We shall only work with specializations corresponding to parameters z that are non-zero, and not a root of unity (exactly as in the case of Hecke algebras, but also forbidding the value $z = 1$). Because of the term $q - 1$ in the denominator of $e_i f_i - f_i e_i$, it is a bit more complicated to make $U(\mathfrak{gl}(N))$ appear as the specialization $z = 1$ of the quantum group; we shall explain this construction during the proof of the Lusztig–Rosso correspondence, see Theorem 5.21.

▷ *The Hopf algebra structure on $\mathfrak{U}(\mathfrak{gl}(N))$.*

For $z \in \mathbb{C}$, there is no Lie group G_z such that $U(\mathrm{Lie}(G_z)) = U_z(\mathfrak{gl}(N))$, but the next best thing in this framework is a structure of Hopf algebra on $\mathfrak{U}(\mathfrak{gl}(N))$. Indeed, notice that if G is a complex Lie group with Lie algebra $\mathfrak{g}$, then the algebra $U(\mathfrak{g})$ is a Hopf algebra for the coproduct

$$\Delta(x) = 1 \otimes x + x \otimes 1 \quad \forall x \in \mathfrak{g},$$

the counity

$$\eta(x) = 0 \quad \forall x \in \mathfrak{g},$$

and the antipode

$$\omega(x) = -x \quad \forall x \in \mathfrak{g}.$$

These operations on $\mathfrak{g}$ can be extended in a unique way to yield endomorphisms of algebras of $T(\mathfrak{g})$, and a Hopf algebra structure on $T(\mathfrak{g})$ (see Section 2.3). Then, they are compatible with the relations of $U(\mathfrak{g})$:

$$\begin{aligned}
&\Delta(x \otimes y - y \otimes x - [x,y]) \\
&\quad = (1 \otimes x + x \otimes 1)(1 \otimes y + y \otimes 1) - (1 \otimes y + y \otimes 1)(1 \otimes x + x \otimes 1) \\
&\qquad - (1 \otimes [x,y] + [x,y] \otimes 1) \\
&\quad = 1 \otimes (x \otimes y - y \otimes x - [x,y]) + (x \otimes y - y \otimes x - [x,y]) \otimes 1 \\
&\quad \equiv 0 \text{ in } U(\mathfrak{g}),
\end{aligned}$$

and similarly for the antipode and the counity. Therefore, they descend to well defined operations on $U(\mathfrak{g})$, which is a Hopf algebra.

In the following, we endow $\mathfrak{U}(\mathfrak{gl}(N))$ with the following operations

$$\begin{aligned}
&\Delta\left(q^{\pm\frac{\varepsilon_i}{2}}\right) = q^{\pm\frac{\varepsilon_i}{2}} \otimes q^{\pm\frac{\varepsilon_i}{2}} &&;\ \Delta(e_i) = 1 \otimes e_i + e_i \otimes q^{\varepsilon_i} &&;\ \Delta(f_i) = q^{\varepsilon_{i+1}} \otimes f_i + f_i \otimes 1 \\
&\eta\left(q^{\pm\frac{\varepsilon_i}{2}}\right) = 1 &&;\ \eta(e_i) = 0 &&;\ \eta(f_i) = 0 \\
&\omega\left(q^{\pm\frac{\varepsilon_i}{2}}\right) = q^{\mp\frac{\varepsilon_i}{2}} &&;\ \omega(e_i) = -e_i\, q^{-\varepsilon_i} &&;\ \omega(f_i) = -q^{-\varepsilon_{i+1}} f_i.
\end{aligned}$$

Theorem 5.3. *These operations can be extended to $\mathfrak{U}(\mathfrak{gl}(N))$ so that it becomes a Hopf algebra.*

Lemma 5.4. *In a Hopf algebra H, the antipode ω_H always satisfies $\omega_H(xy) = \omega_H(y)\,\omega_H(x)$, i.e., it is an anti-endomorphism of algebras.*

Proof. In the following, we use Sweedler's notation for coproducts:

$$\Delta(x) = \sum_i x_{i1} \otimes x_{i2}.$$

In a coalgebra C, the counit η_C satisfies

$$x = \sum_i \eta_C(x_{i1})\, x_{i2} = \sum_i \eta_C(x_{i2})\, x_{i1}$$

since $\mathrm{id}_C = (\eta_C \otimes \mathrm{id}_C) \circ \Delta = (\mathrm{id}_C \otimes \eta_C) \circ \Delta$. Given a coalgebra C and an algebra A over a field k, we endow $\mathrm{Hom}_k(C,A)$ with the convolution product:

$$(f * g)(x) = \sum_i f(x_{i1})\, g(x_{i2}).$$

Notice then that in a Hopf algebra H, the neutral element for the convolution product on $\mathrm{Hom}_k(H,H)$ is $1_H \circ \eta_H : H \to H$:

$$(f * (1_H \circ \eta_H))(x) = \sum_i f(x_{i1})\, \eta_H(x_{i2}) = f\left(\sum_i x_{i1}\, \eta_H(x_{i2})\right) = f(x)$$

and similarly for $(1_H \circ \eta_H) * f$. The antipode ω_H is then the inverse of the identity map id_H for the convolution product:

$$(\omega_H * \mathrm{id}_H)(x) = \sum_i \omega_H(x_{i1})\, x_{i2} = \nabla \circ (\omega_H \otimes \mathrm{id}_H) \circ \Delta(x) = 1_H \circ \eta_H(x)$$

and similarly for $\mathrm{id}_H * \omega_H$. In particular, given a bialgebra H, there is at most one antipode ω_H that makes H into a Hopf algebra.

Now, consider the linear maps $u, v, w : H \otimes H \to H$ defined by $u(x \otimes y) = \omega_H(xy)$, $v(x \otimes y) = xy$ and $w(x \otimes y) = \omega_H(y)\omega_H(x)$. One has

$$\begin{aligned}
(u * v)(x \otimes y) &= \sum_i u((x \otimes y)_{i1})\, v((x \otimes y)_{i2}) \\
&= \sum_{j,k} u(x_{j1} \otimes y_{k1})\, v(x_{j2} \otimes y_{k2}) \\
&= \sum_{j,k} \omega_H(x_{j1} y_{k1})\, x_{j2} y_{k2} \\
&= \sum_i \omega_H((xy)_{i1})\, (xy)_{i2} = \nabla \circ (\omega_H \otimes \mathrm{id}) \circ \Delta(xy) = 1_H \circ \eta_H(xy);
\end{aligned}$$

$$\begin{aligned}(v*w)(x\otimes y) &= \sum_i v((x\otimes y)_{i1})\,w(x\otimes y)_{i2}\\ &= \sum_{j,k} x_{j1}y_{k1}\,\omega_H(y_{k2})\,\omega_H(x_{j2})\\ &= \sum_j x_{j1}\,((\mathrm{id}_H*\omega_H)(y))\,\omega_H(x_{j2})\\ &= (1_H\circ\eta_H)(y)\left(\sum_j x_{j1}\,\omega_H(x_{j2})\right)\\ &= (1_H\circ\eta_H)(y)\,(1_H\circ\eta_H)(x) = (1_H\circ\eta_H)(xy)\end{aligned}$$

using on the last lines the commutativity of the base field k. From these computations, one deduces that

$$u = u*v*w = w,$$

which amounts to the fact that ω_H is an antimorphism of algebras. □

Proof of Theorem 5.3. In a Hopf algebra, the coproduct and the counity are morphisms of algebras, whereas the antipode is an antimorphism of algebras by the previous lemma. Hence, the operations written before Theorem 5.3 for generators of $\mathfrak{U}(\mathfrak{gl}(N))$ can be extended to linear maps on $\mathfrak{U}(\mathfrak{gl}(N))$ as long as they are compatible with the relations defining this algebra. We first check this, only treating the non-trivial cases and using the q-commutation relations between the $q^{\varepsilon_i/2}$'s and the e_i's or f_i's:

$$\begin{aligned}\Delta\left(q^{\frac{\varepsilon_i}{2}}\right)\Delta(e_j)\Delta\left(q^{-\frac{\varepsilon_i}{2}}\right) &= \left(q^{\frac{\varepsilon_i}{2}}\otimes q^{\frac{\varepsilon_i}{2}}\right)\left(1\otimes e_j + e_j\otimes q^{\varepsilon_j}\right)\left(q^{-\frac{\varepsilon_i}{2}}\otimes q^{\frac{-\varepsilon_i}{2}}\right)\\ &= \begin{cases} q^{\frac{1}{2}}\,(1\otimes e_i + e_i\otimes q^{\varepsilon_i}) & \text{if } j=i,\\ q^{-\frac{1}{2}}\,(1\otimes e_{i-1} + e_{i-1}\otimes q^{\varepsilon_{i-1}}) & \text{if } j=i-1,\\ 1\otimes e_j + e_j\otimes q^{\varepsilon_j} & \text{if } j\neq i, i-1,\end{cases}\\ &= \Delta\left(q^{\frac{\varepsilon_i}{2}}\,e_j\,q^{-\frac{\varepsilon_i}{2}}\right);\\ \Delta(e_i)\Delta(f_j) - \Delta(f_j)\Delta(e_i) &= (1\otimes e_i + e_i\otimes q^{\varepsilon_i})\left(q^{\varepsilon_{j+1}}\otimes f_j + f_j\otimes 1\right)\\ &\quad - \left(q^{\varepsilon_{j+1}}\otimes f_j + f_j\otimes 1\right)(1\otimes e_i + e_i\otimes q^{\varepsilon_i})\\ &= q^{\varepsilon_{j+1}}\otimes [e_i, f_j] + [e_i, f_j]\otimes q^{\varepsilon_i}\\ &= \delta_{ij}\,\frac{q^{\varepsilon_i}\otimes q^{\varepsilon_i} - q^{\varepsilon_{i+1}}\otimes q^{\varepsilon_{i+1}}}{q-1} = \delta_{ij}\,\Delta\left(\frac{q^{\varepsilon_i} - q^{\varepsilon_{i+1}}}{q-1}\right).\end{aligned}$$

The computations for

$$\Delta(e_i^2 e_{i+1} - (q+1)\,e_i e_{i+1} e_i + q\,e_{i+1} e_i^2) = 0$$

are a bit more cumbersome but of the same kind. As before, the other relations for Δ are either of the same kind (with f_j instead of e_j), or trivial; hence, Δ is a well-defined endomorphism of the algebra $\mathfrak{U}(\mathfrak{gl}(N))$. For the counity η, the compatibility with the relations of $\mathfrak{U}(\mathfrak{gl}(N))$ is evident, and for the antipode ω, one has

$$\omega\left(q^{-\frac{\varepsilon_i}{2}}\right)\omega(e_j)\,\omega\left(q^{\frac{\varepsilon_i}{2}}\right)=-q^{\frac{\varepsilon_i}{2}}\,e_j\,q^{-\frac{\varepsilon_i}{2}-\varepsilon_j}=\omega\left(q^{\frac{\varepsilon_i}{2}}\,e_j\,q^{-\frac{\varepsilon_i}{2}}\right);$$

$$\omega(f_j)\,\omega(e_i)-\omega(e_i)\,\omega(f_j)=q^{-\varepsilon_{j+1}}\,[f_j,e_i]\,q^{-\varepsilon_i}=\delta_{ij}\,q^{-\varepsilon_{i+1}}\,\frac{q^{\varepsilon_{i+1}}-q^{\varepsilon_i}}{q-1}\,q^{-\varepsilon_i}$$
$$=\delta_{ij}\,\frac{q^{-\varepsilon_i}-q^{-\varepsilon_{i+1}}}{q-1}=\delta_{ij}\,\omega\left(\frac{q^{\varepsilon_i}-q^{\varepsilon_{i+1}}}{q-1}\right)$$

so ω extends to a well-defined anti-endomorphism of the algebra $\mathfrak{U}(\mathfrak{gl}(N))$ (again we only wrote the non-trivial calculations).

Now that Δ, η and ω are well defined on the whole of $\mathfrak{U}(\mathfrak{gl}(N))$, we have to check that they define a Hopf algebra structure, that is to say that the diagrams drawn at the beginning of Section 2.3 are commutative for $\mathfrak{U}(\mathfrak{gl}(N))$. It is sufficient to check it on the generators of the algebra:

1. counity:

$$(\eta\otimes\mathrm{id})\circ\Delta\left(q^{\frac{\varepsilon_i}{2}}\right)=(\eta\otimes\mathrm{id})\left(q^{\frac{\varepsilon_i}{2}}\otimes q^{\frac{\varepsilon_i}{2}}\right)=q^{\frac{\varepsilon_i}{2}};$$
$$(\eta\otimes\mathrm{id})\circ\Delta(e_i)=(\eta\otimes\mathrm{id})(1\otimes e_i+e_i\otimes q^{\varepsilon_i})=e_i+0=e_i;$$
$$(\eta\otimes\mathrm{id})\circ\Delta(f_i)=(\eta\otimes\mathrm{id})(q^{\varepsilon_{i+1}}\otimes f_i+f_i\otimes 1)=f_i+0=f_i$$

   and similarly for $(\mathrm{id}\otimes\eta)\circ\Delta$.

2. coassociativity:

$$(\mathrm{id}\otimes\Delta)\circ\Delta\left(q^{\frac{\varepsilon_i}{2}}\right)=q^{\frac{\varepsilon_i}{2}}\otimes q^{\frac{\varepsilon_i}{2}}\otimes q^{\frac{\varepsilon_i}{2}}=(\Delta\otimes\mathrm{id})\circ\Delta\left(q^{\frac{\varepsilon_i}{2}}\right);$$
$$(\mathrm{id}\otimes\Delta)\circ\Delta(e_i)=1\otimes 1\otimes e_i+1\otimes e_i\otimes q^{\varepsilon_i}+e_i\otimes q^{\varepsilon_i}\otimes q^{\varepsilon_i}=(\Delta\otimes\mathrm{id})\circ\Delta(e_i)$$

   and similarly for the calculation with f_i.

3. bialgebra structure: by construction Δ and η are morphisms of algebras.

4. antipode:

$$\nabla\circ(\mathrm{id}\otimes\omega)\circ\Delta\left(q^{\frac{\varepsilon_i}{2}}\right)=q^{\frac{\varepsilon_i}{2}}\,q^{-\frac{\varepsilon_i}{2}}=1=(1\circ\eta)\left(q^{\frac{\varepsilon_i}{2}}\right);$$
$$\nabla\circ(\mathrm{id}\otimes\omega)\circ\Delta(e_i)=-e_i\,q^{-\varepsilon_i}+e_i\,q^{-\varepsilon_i}=0=(1\circ\eta)(e_i)$$

   and similarly for f_i, and for $\nabla\circ(\omega\otimes\mathrm{id})\circ\Delta$.

Thus, $\mathfrak{U}(\mathfrak{gl}(N))$ is indeed a Hopf algebra for the operations Δ, η and ω. □

As an application of the structure of Hopf algebra, we can now define an action of the quantum group $\mathfrak{U}(\mathfrak{gl}(N))$ on tensors in $((\mathbb{C}(q^{1/2}))^N)^{\otimes n\geq 2}$. Indeed, if ρ is the morphism of algebras $\mathfrak{U}(\mathfrak{gl}(N)) \to \mathrm{End}_{\mathbb{C}(q^{1/2})}((\mathbb{C}(q^{1/2}))^N)$ of Proposition 5.2, then one has a morphism of algebras

$$\rho^{\otimes n} : (\mathfrak{U}(\mathfrak{gl}(N)))^{\otimes n} \to \mathrm{End}_{\mathbb{C}(q^{1/2})}\left(((\mathbb{C}(q^{1/2}))^N)^{\otimes n}\right)$$
$$g_1 \otimes \cdots \otimes g_n \mapsto (v_1 \otimes \cdots \otimes v_n \mapsto \rho(g_1)(v_1) \otimes \cdots \rho(g_n)(v_n)).$$

One can compose it with

$$\Delta^{(n)} = (\Delta \otimes \mathrm{id}_{(\mathfrak{U}(\mathfrak{gl}(N)))^{\otimes n-2}}) \circ (\Delta \otimes \mathrm{id}_{(\mathfrak{U}(\mathfrak{gl}(N)))^{\otimes n-3}}) \circ \cdots \circ \Delta,$$

which is a morphism of algebras $\mathfrak{U}(\mathfrak{gl}(N)) \to (\mathfrak{U}(\mathfrak{gl}(N)))^{\otimes n}$, to get a well-defined representation of the quantum group on tensors of order n. Later in this section, we shall identify the commutant of this action, or more precisely the commutant of the specialized action of $U_z(\mathfrak{gl}(N))$ on $(\mathbb{C}^N)^{\otimes n}$ that is obtained by applying the previous construction to the specialized quantum group.

Remark. Denote $\mathrm{GL}_z(N,\mathbb{C})$ the $\mathbb{C}$-subalgebra of $U_z(\mathfrak{gl}(N))$ spanned over $\mathbb{C}$ by the e_i's, the f_i's and the $q^{\pm\varepsilon_i}$ (thus, we remove the square roots $q^{\pm\varepsilon_i/2}$). This is actually a Hopf subalgebra, and later we shall work with this slightly smaller deformation of $\mathrm{GL}(N,\mathbb{C})$, which we call the **reduced specialized quantum group**. The reason why we introduced the non-reduced version of the quantum group with square roots of the generators q^{ε_i} is that it simplifies the analysis of representations. Indeed, one can then find subalgebras $U_{z,i}(\mathfrak{sl}(2)) = \langle e_i', f_i', k_i^{\pm 1/2}\rangle$ in a quotient of $U_z(\mathfrak{gl}(N))$, with a symmetric relation

$$[e_i', f_i'] = \frac{k_i^{1/2} - k_i^{-1/2}}{z^{1/2} - z^{-1/2}}.$$

These subalgebras will allow a weight analysis of the $U_z(\mathfrak{gl}(N))$-modules, that will be similar to Weyl's classical theory.

▷ *A linear basis of the quantum group* $\mathfrak{U}(\mathfrak{gl}(N))$.

The representation theory of the quantum group $\mathfrak{U}(\mathfrak{gl}(N))$ is very similar to the representation theory of $U(\mathfrak{gl}(N))$, and there are analogues of Theorems A.12, A.13 and A.14 in the quantum setting. The objective of Section 5.2 will be to prove this correspondence, due to Lusztig and Rosso (see Theorems 5.18, 5.21 and 5.25).

In the construction of the Verma modules M^λ for $\mathfrak{sl}(N)$ (see Section A.3 of the Appendix), the Poincaré–Birkhoff–Witt theorem and the decomposition $\mathfrak{g} = \mathfrak{n}_+ \oplus \mathfrak{h} \oplus \mathfrak{n}_-$ play an important role, as $M^\lambda = U(\mathfrak{g}) \otimes_{U(\mathfrak{n}_+ \oplus \mathfrak{h})} \mathbb{C}$ becomes isomorphic as a complex vector space to $U(\mathfrak{n}_-)$. We start by proving an analogous result for $\mathfrak{U}(\mathfrak{gl}(N))$ (Theorem 5.6); in the process, we shall identify a linear basis of $\mathfrak{U}(\mathfrak{gl}(N))$. In the following, we denote $\mathfrak{h} = \mathfrak{h}(N)$ the Cartan subalgebra of $\mathfrak{sl}(N)$ spanned by the matrices $h_i = e_{ii} - e_{(i+1)(i+1)}$, $\mathfrak{h}^*$ its complex dual, and $\mathfrak{h}^*_{\mathbb{R}}$ the real part of $\mathfrak{h}^*$ generated by the roots. It is endowed with a natural scalar product (the restriction of the dual of the Killing form of $\mathfrak{sl}(N)$), and a basis of $\mathfrak{h}^*_{\mathbb{R}}$ consists in the simple roots $\alpha_1,\dots,\alpha_{N-1}$ of $\mathfrak{sl}(N)$. We denote $Y = \bigoplus_{i=1}^{N-1} \mathbb{Z}\alpha_i$ the root lattice of $\mathfrak{sl}(N)$, which is a lattice of maximal rank in $\mathfrak{h}^*_{\mathbb{R}}$, and a sublattice of the weight lattice X of $\mathfrak{sl}(N)$. Notice that h_i is the vector of $\mathfrak{h}$ representing $2\frac{\langle \cdot \mid \alpha_i\rangle}{\langle \alpha_i \mid \alpha_i\rangle}$ on $\mathfrak{h}^*_{\mathbb{R}}$.

Lemma 5.5. *The generic quantum group $\mathfrak{U}(\mathfrak{gl}(N))$ is graded by Y, where an element x is said to be homogeneous of degree $\alpha \in Y$ if and only if*

$$q^{\frac{h_i}{2}}\, x\, q^{-\frac{h_i}{2}} = q^{\frac{\alpha(h_i)}{2}}\, x$$

for all $i \in [\![1, N-1]\!]$.

Proof. The quantum group $\mathfrak{U}(\mathfrak{gl}(N))$ is generated as an algebra by the $q^{\pm\varepsilon_i/2}$'s of rank 0, the e_i's of rank α_i, and the f_i's of rank $-\alpha_i$. Notice then that the gradation is compatible with the algebra structure. □

We set $\mathfrak{U}(\mathfrak{n}_+)$, $\mathfrak{U}(\mathfrak{t})$ and $\mathfrak{U}(\mathfrak{n}_-)$ for the subalgebras of $\mathfrak{U}(\mathfrak{gl}(N))$ spanned over $\mathbb{C}(q^{1/2})$ respectively by the e_i's, the $q^{\pm\varepsilon_i/2}$'s and the f_i's. We also write $\mathfrak{U}(\mathfrak{b})$ for the subalgebra generated by the e_i's and the $q^{\pm\varepsilon_i/2}$'s.

Theorem 5.6. *One has the isomorphisms of $\mathbb{C}(q^{1/2})$-vector spaces*

$$\mathfrak{U}(\mathfrak{gl}(N)) = \mathfrak{U}(\mathfrak{n}_-) \otimes_{\mathbb{C}(q^{1/2})} \mathfrak{U}(\mathfrak{t}) \otimes_{\mathbb{C}(q^{1/2})} \mathfrak{U}(\mathfrak{n}_+)$$
$$\mathfrak{U}(\mathfrak{b}) = \mathfrak{U}(\mathfrak{t}) \otimes_{\mathbb{C}(q^{1/2})} \mathfrak{U}(\mathfrak{n}_+).$$

Therefore, $\mathfrak{U}(\mathfrak{gl}(N))$ is a free $\mathfrak{U}(\mathfrak{b})$-module.

Lemma 5.7. *For every $n_1,\dots,n_{N-1} \in \mathbb{N}$, $(e_1)^{n_1}\cdots(e_{N-1})^{n_{N-1}} \neq 0$ in $\mathfrak{U}(\mathfrak{gl}(N))$.*

Proof. We first start by proving that $(e_i)^n \neq 0$ for all $n \geq 0$. This is true if $n = 1$, as e_i is sent to a non-zero endomorphism by the geometric representation of $\mathfrak{U}(\mathfrak{gl}(N))$. For an arbitrary value of n, it suffices to prove that $\Delta^{(n)}((e_i)^n)$ has a non-zero component in $(\mathfrak{U}(\mathfrak{gl}(N)))^{\otimes n}$ of Y^n-degree $(\alpha_i,\alpha_i,\dots,\alpha_i)$. Notice that

$$\Delta^{(n)}(e_i) = \sum_{m=1}^{n} u_m, \quad \text{with } u_m = \underbrace{1 \otimes \cdots \otimes 1}_{m-1 \text{ times}} \otimes e_i \otimes \underbrace{q^{\varepsilon_i} \otimes \cdots \otimes q^{\varepsilon_i}}_{n-m \text{ times}}.$$

The u_j's satisfy the anticommutation relation $u_j u_k = q\, u_k u_j$ for $j < k$. Therefore, by the q-analogue of the multinomial Newton formula,

$$\begin{aligned}\Delta^{(n)}((e_i)^n) &= (u_1 + \cdots + u_n)^n \\ &= \sum_{m_1+\cdots+m_n=n} \frac{(q^{-1};q^{-1})_n}{(q^{-1};q^{-1})_{m_1}\cdots(q^{-1};q^{-1})_{m_n}} (u_1)^{m_1}\cdots(u_n)^{m_n}.\end{aligned}$$

The part of Y^n-rank $(\alpha_i, \dots, \alpha_i)$ of this sum corresponds to the choice $m_1 = \cdots = m_n = 1$, which gives a term:

$$[n]_{q^{-1}}! \left(e_i \otimes q^{\varepsilon_i} e_i \otimes \cdots \otimes q^{(n-1)\varepsilon_i} e_i\right) = [n]_q! \left(e_i \otimes e_i q^{\varepsilon_i} \otimes \cdots \otimes e_i q^{(n-1)\varepsilon_i}\right).$$

Since $1 \otimes q^{\varepsilon_i} \otimes \cdots \otimes q^{(n-1)\varepsilon_i}$ is invertible, the quantity above is non-zero: indeed, $e_i \otimes \cdots \otimes e_i \neq 0$. So, $(e_i)^n \neq 0$.

Finally, fix integers $n_1, \dots, n_{N-1}$, and consider $\Pi = (e_1)^{n_1} \cdots (e_{N-1})^{n_{N-1}}$, and its image by the map $\Delta^{(N-1)}$. The part of Y^{N-1}-degree $(n_1\alpha_1, \dots, n_{N-1}\alpha_{N-1})$ in $\Delta^{(N-1)}(\Pi)$ is

$$(e_1)^{n_1} \otimes q^{n_1\varepsilon_1} (e_2)^{n_2} \otimes \cdots \otimes q^{n_1\varepsilon_1 + \cdots + n_{N-2}\varepsilon_{N-2}} (e_{N-1})^{n_{N-1}},$$

which is non-zero by the previous discussion; therefore, $\Pi \neq 0$. □

Lemma 5.8. *A linear basis of* $\mathfrak{U}(\mathfrak{t})$ *consists in the vectors* $q^{\frac{1}{2}\sum_{i=1}^N k_i\varepsilon_i}$, *where the* k_i*'s run over* $\mathbb{Z}$.

Proof. We only have to prove that the vectors $q^{\frac{1}{2}\sum_{i=1}^N k_i\varepsilon_i}$ are linearly independent. Suppose that we have a linear combination

$$\sum_{j=1}^m a_j q^{\frac{1}{2}K_j} = 0, \quad \text{where the } K_j = \sum_{i=1}^N k_{ij}\varepsilon_i \text{ are all distinct.}$$

Notice that any vector $K = \sum_{i=1}^N k_i\varepsilon_i$ can be rewritten as

$$k_1 h_1 + (k_1 + k_2) h_2 + \cdots + (k_1 + \cdots + k_{N-1}) h_{N-1} + (k_1 + \cdots + k_N)\varepsilon_N.$$

Let $\phi : \mathfrak{U}(\mathfrak{gl}(N)) \to \mathbb{C}(q^{1/2})[t, t^{-1}] \otimes_{\mathbb{C}(q^{1/2})} \mathfrak{U}(\mathfrak{gl}(N))$ be the morphism of algebras defined by $\phi(e_i) = t e_i$, $\phi(f_i) = t f_i$, and $\phi(q^{\pm\varepsilon_i/2}) = t^{\pm 1} q^{\pm\varepsilon_i/2}$. This map is indeed compatible with the relations of the algebra $\mathfrak{U}(\mathfrak{gl}(N))$, and it sends a vector $q^{K/2}$ to $t^{k_1+\cdots+k_n} q^{K/2}$. By applying ϕ to the previous relation, for every fixed sum $l = k_1 + \cdots + k_n$, we obtain by looking at the homogeneous term of degree t^l:

$$\sum_{j \mid k_1 + \cdots + k_N = l} a_j q^{\frac{1}{2}K_j} = 0.$$

By multiplying by $q^{-l\varepsilon_n/2}$, we are reduced to the proof of the following easier fact: the vectors $q^{\sum_{i=1}^N l_i h_i/2}$ with the l_i's in $\mathbb{Z}$ are linearly independent, hence form a linear basis of $\mathfrak{U}(\mathfrak{h})$, the subalgebra of $\mathfrak{U}(\mathfrak{gl}(N))$ spanned by the vectors $q^{h_i/2}$. Thus, in the following, we consider a linear combination

$$\sum_{j=1}^m a_j q^{\frac{1}{2}L_j} = 0, \quad \text{where the } L_j = \sum_{i=1}^{N-1} l_{ij} h_i \text{ are all distinct.}$$

Denote L and R the left and right regular representation of $\mathfrak{U}(\mathfrak{gl}(N))$ on itself, and consider

$$0 = ((L \otimes R) \circ (\mathrm{id} \otimes \omega) \circ \Delta)\left(\sum_{j=1}^{m} a_j\, q^{\frac{1}{2}L_j}\right) = \sum_{j=1}^{m} a_j\, L\left(q^{\frac{1}{2}L_j}\right) \otimes R\left(q^{-\frac{1}{2}L_j}\right).$$

By looking at its action on a monomial $(e_1)^{n_1} \cdots (e_{N-1})^{n_{N-1}}$, one obtains

$$0 = \sum_{j=1}^{m} a_j\, q^{\frac{1}{2}\left(\sum_{i=1}^{N-1} n_i \alpha_i\right)(L_j)},$$

since $(e_1)^{n_1} \cdots (e_{N-1})^{n_{N-1}} \neq 0$ by the previous lemma, and is an element of Y-degree $\sum_{i=1}^{N-1} n_i \alpha_i$. Therefore, for any choice of integers $n_1, \ldots, n_{N-1}$, the coefficients

$$\sum_{j \mid (\sum_{i=1}^{N-1} n_i \alpha_i)(L_j) = \text{constant}} a_j$$

of this polynomial are all equal to 0. We now reason by induction on m, the initial case $m = 1$ being trivial. Suppose that we know that for all integers $m' \leq m-1$, a linear combination of m' terms q^L that vanishes has all its coefficients equal to 0. With a linear combination of m terms, it suffices then to show that there exist integers $n_1, \ldots, n_{N-1}$ such that

$$\left(\sum_{i=1}^{N-1} n_i \alpha_i\right)(L_m) \neq \left(\sum_{i=1}^{N-1} n_i \alpha_i\right)(L_j) \text{ for all } j \neq m.$$

Indeed, we shall then know that $a_m = 0$, which will allow us to reduce the situation to the case of a sum of $m-1$ terms. However, the linear forms $(L_m - L_j)_{j \leq m-1}$ on $\mathfrak{h}^*$ are all non-zero, so the union of the hyperplanes

$$\left\{\alpha \in \mathfrak{h}^*_{\mathbb{R}} = \bigoplus_{i=1}^{N-1} \mathbb{R}\alpha_i \mid \forall j \leq m-1,\ \alpha(L_m - L_j) = 0\right\}$$

cannot contain the whole root lattice Y, which is of maximal rank. Therefore, there indeed exists some element $\sum_{i=1}^{N-1} n_i \alpha_i$ that separates L_m from the other elements L_j. □

Lemma 5.9. *Let $(E_t)_{t \in T}$ be a Y-graded basis of $\mathfrak{U}(\mathfrak{n}_+)$ that consists in monomials in the e_i's. Then,*

$$\left(q^{\frac{1}{2}\sum_{i=1}^{N} k_i \varepsilon_i}\, E_t\right)_{k_i \in \mathbb{Z},\, t \in T}$$

is a linear basis of $\mathfrak{U}(\mathfrak{b})$.

Proof. The defining relations of $\mathfrak{U}(\mathfrak{gl}(N))$ allows one to rewrite every product of terms $q^{\pm\varepsilon_i/2}$ and e_j as a product $q^{K/2} e_{i_1} \cdots e_{i_r}$; therefore, the family

$(q^{\sum_{i=1}^{n} k_i \varepsilon_i/2} E_t)_{k_i \in \mathbb{Z}, t \in T}$ spans $\mathfrak{U}(\mathfrak{b})$ over $\mathbb{C}(q^{1/2})$. Suppose that we have a vanishing linear combination

$$0 = \sum_{j=1}^{m} a_j \, q^{\frac{1}{2}K_j} \, E_{t_j}.$$

By looking at the homogeneous components for the Y-degree, without loss of generality, one can assume all the E_{t_j}'s of Y-degree α. The image by Δ of this linear combination has its component of Y^2-degree $(0, \alpha)$ equal to

$$0 = \sum_{j=1}^{m} a_j \, q^{\frac{1}{2}K_j} \otimes q^{\frac{1}{2}K_j} \, E_{t_j}.$$

By the previous lemma, the different elements $q^{K_j/2}$ are linearly independent, so, for all $K \in \bigoplus_{i=1}^{n} \mathbb{Z}\varepsilon_i$,

$$\sum_{j \,|\, K_j = K} a_j \, q^{\frac{1}{2}K} \, E_{t_j} = 0.$$

Since the E_t's are linearly independent, we conclude that all the a_j's vanish. □

Proof of Theorem 5.6. The second part of the theorem is an immediate consequence of the previous lemma, since $(q^{\sum_{i=1}^{n} k_i \varepsilon_i/2})_{k_i \in \mathbb{Z}}$ is a basis of $\mathfrak{U}(\mathfrak{t})$ and $(E_t)_{t \in T}$ is a basis of $\mathfrak{U}(\mathfrak{n}_+)$. Let $(F_s)_{s \in S}$ be a linear basis of $\mathfrak{U}(\mathfrak{n}_-)$ that consists in monomials in the f_i's. The first part of the theorem amounts to showing that $(F_s \, q^{K/2} \, E_t)_{K \in \bigoplus_{i=1}^{n} \mathbb{Z}\varepsilon_i, s \in S, t \in T}$ is a linear basis of $\mathfrak{U}(\mathfrak{gl}(N))$, and the only difficult part is the linear independence. Suppose that we have a vanishing linear combination

$$0 = \sum_{j=1}^{m} a_j \, F_{s_j} \, q^{\frac{1}{2}K_j} \, E_{t_j},$$

with all the $a_j \neq 0$. Using the Y-gradation on $\mathfrak{U}(\mathfrak{gl}(N))$, one can assume that if F_{s_j} is of degree $-\gamma_j$ and E_{t_j} is of degree δ_j, then $\delta_j - \gamma_j = \alpha$ is a constant that does not depend on j. Notice that δ_j and γ_j always belong to $Y_+ = \bigoplus_{i=1}^{N-1} \mathbb{N}\alpha_i$.

We endow Y with the lexicographic order with respect to the decomposition of an element as a linear combination $\sum_{i=1}^{N-1} k_i \, \alpha_i$ of simple roots, and denote γ the maximal element of $\{\gamma_1, \ldots, \gamma_m\}$, and J the set of indices j such that $\gamma_j = \gamma$. We also set $\delta = \alpha + \gamma$. Then, in $\Delta(\sum_{j=1}^{m} a_j \, F_{s_j} \, q^{K_j/2} \, E_{t_j})$, the component of degree $(-\gamma, \delta)$ is

$$\sum_{j \in J} a_j F_{s_j} q^{\frac{1}{2}K_j} \otimes q^{\frac{1}{2}K_j} E_{t_j} = 0.$$

By the previous discussion, the elements $q^{K_j/2} \, E_{t_j}$ are linearly independent, and the proof of the previous lemma is easily adapted to the algebra $\mathfrak{U}(\mathfrak{b}_-) = \mathbb{C}(q^{1/2})\langle f_i, q^{\pm\varepsilon_i/2} \rangle$ instead of $\mathfrak{U}(\mathfrak{b})$, so the elements $F_{s_j} \, q^{K_j/2}$ are also linearly independent. So, $a_j = 0$ for all $j \in J$. This ensures the linear independence of the vectors $F_s \, q^{K/2} \, E_t$. □

To close this section, notice that the analogue of Theorem 5.6 holds for the specialized quantum groups with $q^{1/2} = z^{1/2}$ not equal to zero, and not a root of unity: indeed, in Lemma 5.7, the rational coefficients that are functions of $q^{1/2} = z^{1/2}$ do not vanish by assumption on z, and in Lemma 5.8, the extraction of the coefficients of the polynomials in $q^{1/2}$ is valid, since what is true for $q^{1/2} = z^{1/2}$ is also true for all positive powers of $z^{1/2}$. Thus,

$$U_z(\mathfrak{gl}(N)) = U_z(\mathfrak{n}_-) \otimes_{\mathbb{C}} U_z(\mathfrak{t}) \otimes_{\mathbb{C}} U_z(\mathfrak{n}_+)$$
$$U_z(\mathfrak{b}) = U_z(\mathfrak{t}) \otimes_{\mathbb{C}} U_z(\mathfrak{n}_+).$$

and $U_z(\mathfrak{gl}(N))$ is a free $U_z(\mathfrak{b})$-module.

5.2 Representation theory of the quantum groups

Recall the Schur–Weyl duality proved in Section 2.5: as a $(\mathrm{GL}(N,\mathbb{C}), \mathfrak{S}(n))$-bimodule, $(\mathbb{C}^N)^{\otimes n}$ can be decomposed as

$$(\mathbb{C}^N)^{\otimes n} = \bigoplus_{\lambda \in \mathfrak{Y}(n) \,|\, \ell(\lambda) \leq N} V^\lambda \otimes_{\mathbb{C}} S^\lambda,$$

where S^λ is the (right) Specht module for $\mathfrak{S}(n)$ of label λ, and V^λ is the irreducible representation of the general linear group $\mathrm{GL}(N,\mathbb{C})$ whose character on diagonal elements is given by the Schur function

$$\mathrm{ch}^\lambda(\mathrm{diag}(x_1, \ldots, x_N)) = s_\lambda(x_1, \ldots, x_N).$$

In Section 5.3, we shall prove that $(\mathbb{C}^N)^{\otimes n}$ is also a $(\mathrm{GL}_z(N,\mathbb{C}), \mathfrak{H}_z(n))$-bimodule, which expands as

$$(\mathbb{C}^N)^{\otimes n} = \bigoplus_{\lambda \in \mathfrak{Y}(n) \,|\, \ell(\lambda) \leq N} V_z^\lambda \otimes_{\mathbb{C}} S_z^\lambda,$$

where the V_z^λ's are certain simple modules over the reduced specialized quantum group $\mathrm{GL}_z(N,\mathbb{C}) \subset U_z(\mathfrak{gl}(N))$.

With this in mind, the goal of the present section is to explain what are the modules V_z^λ, and how they are related to the irreducible representations V^λ of $\mathrm{GL}(N,\mathbb{C})$. Thus, we shall demonstrate the highest weight theorem for $U_z(\mathfrak{gl}(N))$ (Theorem 5.18), with a proof that follows the usual arguments of the classical case. In particular, we shall see that there exists a strong (fonctorial) connection between the representation theory of the specialized quantum group $U_z(\mathfrak{gl}(N))$, and the classical representation theory of the general linear group $\mathrm{GL}(N,\mathbb{C})$; see Theorem 5.21. This correspondence is due to Lusztig and Rosso.

▷ *Highest weights of the modules over the specialized quantum group.*

We start our discussion by recalling the definition of a highest weight. Beware that in the following, we shall consider certain modules that are not a priori finite-dimensional over $\mathbb{C}$. On the other hand, recall that $(\eta_i)^{1/2}$ is the image of $q^{\varepsilon_i/2}$ by the specialization morphism $\mathfrak{U}(\mathfrak{gl}(N)) \to U_z(\mathfrak{gl}(N))$.

Definition 5.10. *Let V be a module over $U_z(\mathfrak{gl}(N))$. A vector $\theta = (\theta_1, \ldots, \theta_N) \in (\mathbb{C}^\times)^N$ is called a weight of V if the space*

$$V_\theta = \{v \in V \mid \forall i \in [\![1, N-1]\!],\ (\eta_i)^{1/2} \cdot v = \theta_i\, v\}$$

has strictly positive dimension. It is called a highest weight if, moreover, $e_i \cdot V_\theta = 0$ for all $i \in [\![1, N-1]\!]$.

Proposition 5.11. *Every finite-dimensional module V over $U_z(\mathfrak{gl}(N))$ admits at least one highest weight.*

Proof. Denote $\rho : U_z(\mathfrak{gl}(N)) \to \mathrm{End}_{\mathbb{C}}(V)$ the defining morphism of a finite-dimensional representation. Since the elements $(\eta_i)^{1/2}$ commute, their images $\rho((\eta_i)^{1/2})$ are simultaneously trigonalizable, so in particular, looking at the first vector of a basis of simultaneous trigonalization, the set of weights of V is non-empty. We denote V' the vector subspace of V generated by the weight spaces V_θ. It is a $U_z(\mathfrak{gl}(N))$-submodule: indeed, if v is a vector of weight $(\theta_1, \ldots, \theta_N)$, then $e_j \cdot v$ is a vector of weight $(\theta_1, \ldots, z^{1/2}\theta_j, z^{-1/2}\theta_{j+1}, \ldots, \theta_N)$, since

$$(\eta_i)^{\frac{1}{2}} \cdot (e_j \cdot v) = \begin{cases} z^{\frac{1}{2}}\, e_j \cdot \left((\eta_i)^{\frac{1}{2}} \cdot v\right) & \text{if } i = j, \\ z^{-\frac{1}{2}}\, e_j \cdot \left((\eta_i)^{\frac{1}{2}} \cdot v\right) & \text{if } i = j+1, \\ e_j \cdot \left((\eta_i)^{\frac{1}{2}} \cdot v\right) & \text{otherwise,} \end{cases}$$
$$= z^{0 \text{ or } \frac{1}{2} \text{ or } -\frac{1}{2}}\, \theta_i\, (e_j \cdot v).$$

Similarly, $f_j \cdot v$ is a vector of weight $(\theta_1, \ldots, z^{-1/2}\theta_j, z^{1/2}\theta_{j+1}, \ldots, \theta_N)$; so, V', the space spanned by the weight vectors, is indeed stable. We continue to denote ρ the morphism $U_z(\mathfrak{gl}(N)) \to \mathrm{End}_{\mathbb{C}}(V')$. We now claim that V' admits a highest weight. It suffices to show that $\bigcap_{i=1}^{N-1} \mathrm{Ker}\, \rho(e_i) \neq \{0\}$: indeed, this subspace is stable by the $\rho((\eta_i)^{1/2})$'s (for the same reasons as above), so if non-zero it contains a weight vector. We shall prove a stronger result: for M large enough, and for every $i_1, \ldots, i_p \in [\![1, N-1]\!]$ with $p \geq M$,

$$\rho(e_{i_1})\rho(e_{i_2}) \cdots \rho(e_{i_p}) = 0 \text{ in } \mathrm{End}_{\mathbb{C}}(V').$$

This indeed proves the result: if M is the minimal integer satisfying the property above, then there exists a non-zero vector in V' that writes as $(e_{i_2} \cdots e_{i_M}) \cdot v$, and this vector falls in the intersection of the kernels $\mathrm{Ker}\, \rho(e_{i_1})$.

The previous statement is a consequence of the finiteness of the number of weights of V' (since V' is finite-dimensional). Indeed, let $\{\mu^{(1)}, \ldots, \mu^{(t)}\}$ be the set of distinct weights of V', and assume that for all $p \geq 0$, there exists a choice of indices such that $\rho(e_{i_{1,p}}) \cdots \rho(e_{i_{p,p}})$ is a non-zero endomorphism of V'. Then, by the pigeon-hole principle, and using the computation performed at the beginning of the proof, one sees that there exists one weight of V', say $\mu = \mu^{(1)}$, such that for $(t_1, \ldots, t_N)$ running over an infinite subset of $\mathbb{Z}^N$,

$$\left(\mu_1 z^{\frac{t_1}{2}}, \mu_2 z^{\frac{t_2}{2}}, \ldots, \mu_N z^{\frac{t_N}{2}}\right)$$

is also a weight of V'. We thus get an infinite number of distinct weights, since $z \neq 0$ and z is not a root of unity; whence a contradiction. □

Proposition 5.12. *Every simple finite-dimensional module V over $U_z(\mathfrak{gl}(N))$ is the direct sum of its weight spaces, and has a unique highest weight θ, which is of multiplicity* $\dim V_\theta = 1$.

Proof. In the proof of the previous proposition, we saw that the sum of all weight spaces is a non-zero $U_z(\mathfrak{gl}(N))$-submodule of V, so if V is simple, then it is equal to the direct sum of its weight spaces:

$$V = \bigoplus_{\mu} V_\mu.$$

Let θ be a highest weight of V, and v be a non-zero weight vector for θ. By simplicity, the vector v spans V as a $U_z(\mathfrak{gl}(N))$-module, and using Theorem 5.6, one sees that it is even a generator as a $U_z(\mathfrak{n}_-)$-module, since $U_z(\mathfrak{b})\, v = \mathbb{C}v$ by hypothesis on the weight θ. If F_s is a monomial in the f_i's with Y-degree $-\gamma = -\sum_{i=1}^{N-1} n_i \alpha_i$ with $\gamma \in Y_+$, then $F_s \cdot v$ is a weight vector of weight

$$\left(\theta_1 z^{-\frac{n_1}{2}}, \theta_2 z^{\frac{n_1-n_2}{2}}, \ldots, \theta_N z^{\frac{n_{N-1}}{2}}\right).$$

As a consequence of this observation and of the existence of a basis of $U_z(\mathfrak{n}_-)$ consisting in monomials F_s, we get that:

1. there exists a basis of V that consists in weight vectors $F_s \cdot v$;
2. among them, the only vector of weight θ is v, so $\dim V_\theta = 1$. □

More generally, call **cyclic module** of highest weight θ a $U_z(\mathfrak{gl}(N))$-module V, not necessarily finite-dimensional, but which is spanned as a module by a vector v which is a weight vector for θ. The same discussion as in the proof of Proposition 5.12 holds; therefore, V is the direct sum of its weight spaces, and for every weight

$$\left(\theta_1 z^{\frac{-n_1}{2}}, \theta_2 z^{\frac{n_1-n_2}{2}}, \ldots, \theta_N z^{\frac{n_{N-1}}{2}}\right),$$

the dimension of the weight space V_μ is finite, since there are only a finite number of monomials F_s in a basis of $U_z(\mathfrak{n}_-)$ that have degree $-\gamma = -\sum_{i=1}^{N-1} n_i \alpha_i$. The result actually holds for all submodules $W \subset V$. Indeed, suppose that W is a submodule, and for $w \in W$, write $w = w_1 + w_2 + \cdots + w_r$, where the w_i's are weight vectors in V; we have to show that every w_i is in W. If this is not the case, one can assume $r \geq 2$ minimal, and then, all w_i's are not in W. We denote $\mu^{(1)}, \ldots, \mu^{(r)}$ the weights associated to the vectors $w_1, \ldots, w_r$, and we see them as functions on $U_z(\mathfrak{t})$, with $t \cdot w_i = \mu^{(i)}(t)\, w_i$ for any $t \in U_z(\mathfrak{t})$. Fix now $t \in U_z(\mathfrak{t})$, such that $\mu^{(1)}(t) \neq \mu^{(2)}(t)$. Then,

$$t \cdot w = \sum_{i=1}^{r} \mu^{(i)}(t)\, w_i;$$

$$(t - \mu^{(1)}(t)\, 1) \cdot w = \sum_{i=2}^{r} (\mu^{(i)}(t) - \mu^{(1)}(t))\, w_i$$

and the second line belongs again to W, and satisfies the same assumptions as before but with $r-1$ terms; hence a contradiction. So, every submodule of a cyclic module is the direct sum of its weight spaces, all of them being finite-dimensional.

On the other hand, notice that any homomorphic image of a cyclic module of highest weight θ is again a cyclic module of highest weight θ, or the zero module. These remarks lead to a precision of the previous proposition:

Proposition 5.13. *Let V be a simple finite-dimensional module over $U_z(\mathfrak{gl}(N))$. Its highest weight θ is unique, and it determines the isomorphism class of V.*

Proof. The uniqueness of the highest weight θ is immediate, since it is the only weight μ of V such that $V = U_z(\mathfrak{n}_-)\, V_\mu$. Suppose now that we have two simple modules V and W with the same highest weight θ; we denote v and w the corresponding highest weight vectors. In $V \oplus W$, we set

$$U = U_z(\mathfrak{gl}(N))(v + w).$$

This is a cyclic module of highest weight θ. Let $p_V : V \oplus W \to V$ and $p_W : V \oplus W \to W$ be the projection maps, which are morphisms of $U_z(\mathfrak{gl}(N))$-modules; we continue to denote p_V and p_W their restrictions to U. Then, $p_V(U)$ and $p_W(U)$ are non-zero submodules of V and W, so

$$V = p_V(U) \quad ; \quad W = p_W(U)$$

by simplicity of V and W. Thus, V and W appear as quotients of the cyclic module U of highest weight θ. However, every cyclic module of highest weight θ is an indecomposable $U_z(\mathfrak{gl}(N))$-module with a unique maximal proper submodule, and therefore a unique simple quotient. Indeed, by the previous discussion, a proper submodule of a cyclic module U of highest weight θ is necessarily strictly included in the direct sum of the weight spaces U_μ, $\mu \neq \theta$. Therefore, the union U_+ of all proper submodules is again a proper submodule, and the maximal proper submodule of U; and U/U_+ is the unique simple quotient of U. Thus, $V = U/U_+ = W$. □

▷ *Lusztig–Rosso correspondence for the highest weights.*

In the previous paragraph, we saw that the simple finite-dimensional modules of $U_z(\mathfrak{gl}(N))$ are classified up to isomorphism by their highest weight vector $\theta = (\theta_1, \ldots, \theta_N) \in (\mathbb{C}^\times)^N$. To complete the study of the finite-dimensional representations of the specialized quantum group, we now have to:

1. give the list of all the possible highest weight vectors of finite-dimensional simple modules over $U_z(\mathfrak{gl}(N))$;

2. decompose any finite-dimensional representation of $U_z(\mathfrak{gl}(N))$ as a sum of simple modules.

A preliminary step consists in the classification of all the one-dimensional representations of $U_z(\mathfrak{gl}(N))$:

Lemma 5.14. *Let W be a representation of $U_z(\mathfrak{gl}(N))$ of dimension 1. There exists a complex number $v \neq 0$, and signs $\phi_1, \ldots, \phi_{N-1} \in \{\pm 1\}$, such that W is of highest weight*

$$(\phi_1 v, \ldots, \phi_{N-1} v, v),$$

and such that the action of $U_z(\mathfrak{gl}(N))$ is given by

$$e_i \cdot w = 0 \quad ; \quad f_i \cdot w = 0 \quad ; \quad (\eta_i)^{1/2} \cdot w = \phi_i v\, w.$$

Proof. If W is of dimension 1, then every non-zero vector w is a highest weight vector, hence vanishes under the action of the elements e_i. Therefore, the defining morphism $\rho : U_z(\mathfrak{gl}(N)) \to \mathrm{End}_{\mathbb{C}}(W)$ satisfies $\rho(e_i) = 0$ for all i. One has also $\rho(f_i) = 0$: otherwise, W has a weight different from the highest weight, and $\dim W \geq 2$, which contradicts the hypothesis $\dim W = 1$. Since

$$[e_i, f_i] = \frac{\eta_i - \eta_{i+1}}{z - 1},$$

this forces $\rho(\eta_i - \eta_{i+1}) = 0$, so $\rho(\eta_i)$ does not depend on i. Write $\rho(\eta_i) = v^2\, \mathrm{id}_W$; then, $v \neq 0$ since η_i is invertible in $U_z(\mathfrak{gl}(N))$, and W has highest weight $(\phi_1 v, \ldots, \phi_{N-1} v, v)$, where the signs ϕ_i are chosen such that $\rho((\eta_i)^{1/2}) = \phi_i v\, \mathrm{id}_W$ for $i \leq N-1$, and $\rho((\eta_N)^{1/2}) = v\, \mathrm{id}_W$. Therefore, the action of the quantum group is indeed the one given in the statement of the lemma. □

The previous lemma is completed by the following result, which ensures that the set of allowed highest weights is stable by the action of the two multiplicative groups $\mathbb{C}^\times$ and $\{\pm 1\}^{N-1}$. Given a complex number $v \neq 0$ and a sequence of signs $\phi = (\phi_1, \ldots, \phi_{N-1}, 1)$, we denote $\mathbb{C}^{v,\phi}$ the one-dimensional representation of highest weight $(\phi_1 v, \ldots, \phi_{N-1} v, v)$. If W^θ is a simple $U_z(\mathfrak{gl}(N))$-module of highest weight $(\theta_1, \theta_2, \ldots, \theta_N)$, we denote $\rho^{v,\phi}$ and ρ^θ the defining morphisms $U_z(\mathfrak{gl}(N)) \to \mathrm{End}_{\mathbb{C}}(\mathbb{C}^{v,\phi})$ and $U_z(\mathfrak{gl}(N)) \to \mathrm{End}_{\mathbb{C}}(W^\theta)$, and we consider the module

$$W' = \mathbb{C}^{v,\phi} \otimes_{\mathbb{C}} W^\theta,$$

endowed with the action of $U_z(\mathfrak{gl}(N))$ given by the morphism $(\rho^{\upsilon,\phi} \otimes_{\mathbb{C}} \rho^{\theta}) \circ \Delta$. We write $W' = \mathbb{C}^{\upsilon,\phi} \times W^{\theta}$ to signify that W' is considered as a module over $U_z(\mathfrak{gl}(N))$.

Lemma 5.15. *For any simple module W^{θ} of highest weight $(\theta_1, \ldots, \theta_N)$, $\mathbb{C}^{\upsilon,\phi} \times W^{\theta}$ is a simple $U_z(\mathfrak{gl}(N))$-module of highest weight $(\upsilon\phi_1\theta_1, \ldots, \upsilon\phi_{N-1}\theta_{N-1}, \upsilon\theta_N)$.*

Proof. We denote w a highest weight vector in W^{θ}, and we identify $\mathbb{C}^{\upsilon,\phi}$ with the field of complex numbers $\mathbb{C}$. Set $w' = 1 \otimes w$. The action of e_i on w' is

$$e_i \cdot w' = (1 \otimes e_i + e_i \otimes \eta_i)(1 \otimes w) = 0.$$

Therefore, w' is a highest weight vector, with associated weight given by

$$(\eta_i)^{\frac{1}{2}} \cdot w' = \left((\eta_i)^{\frac{1}{2}} \otimes (\eta_i)^{\frac{1}{2}}\right)(1 \otimes w) = \upsilon\phi_i\theta_i\, w'.$$

It remains to see that W' is a simple module. However, for any module W,

$$\mathbb{C}^{\upsilon^{-1},\phi^{-1}} \times (\mathbb{C}^{\upsilon,\phi} \times W) = W,$$

because the defining morphism of this module is $(\rho^{\upsilon^{-1},\phi^{-1}} \otimes ((\rho^{\upsilon,\phi} \otimes \rho^{W}) \circ \Delta)) \circ \Delta$, and by expanding the coproducts, one gets

$$\begin{aligned}
\left((\rho^{\upsilon^{-1},\phi^{-1}} \otimes ((\rho^{\upsilon,\phi} \otimes \rho^{W}) \circ \Delta)) \circ \Delta\right)(e_i) &= 1 \otimes 1 \otimes \rho^{W}(e_i);\\
\left((\rho^{\upsilon^{-1},\phi^{-1}} \otimes ((\rho^{\upsilon,\phi} \otimes \rho^{W}) \circ \Delta)) \circ \Delta\right)(f_i) &= \upsilon^{-2} \otimes \upsilon^{2} \otimes \rho^{W}(f_i)\\
&= 1 \otimes 1 \otimes \rho^{W}(f_i);\\
\left((\rho^{\upsilon^{-1},\phi^{-1}} \otimes ((\rho^{\upsilon,\phi} \otimes \rho^{W}) \circ \Delta)) \circ \Delta\right)\left((\eta_i)^{\frac{1}{2}}\right) &= \upsilon^{-1}\phi_i^{-1} \otimes \upsilon\phi_i \otimes \rho^{W}\left((\eta_i)^{\frac{1}{2}}\right)\\
&= 1 \otimes 1 \otimes \rho^{W}\left((\eta_i)^{\frac{1}{2}}\right).
\end{aligned}$$

As a consequence, if there were a decomposition of $\mathbb{C}^{\upsilon,\phi} \times W^{\theta}$ in non-trivial submodules, there would exist a similar decomposition of $\mathbb{C}^{\upsilon^{-1},\phi^{-1}} \times (\mathbb{C}^{\upsilon,\phi} \times W^{\theta}) = W^{\theta}$, which is absurd since W^{θ} is a simple $U_z(\mathfrak{gl}(N))$-module. □

If $(\theta_1, \theta_2, \ldots, \theta_N)$ is the highest weight of some simple finite-dimensional $U_z(\mathfrak{gl}(N))$-module, then by the previous lemma, the same holds for

$$\left(\frac{\theta_1}{(\theta_1\theta_2\cdots\theta_N)^{1/N}}, \frac{\theta_2}{(\theta_1\theta_2\cdots\theta_N)^{1/N}}, \ldots, \frac{\theta_N}{(\theta_1\theta_2\cdots\theta_N)^{1/N}}\right),$$

so one can assume without loss of generality that $\theta_1\theta_2\cdots\theta_N = 1$. Notice then that under this hypothesis, the element $(\eta_1\eta_2\cdots\eta_N)^{1/2}$ acts on W^{θ} by the identity. Indeed, we saw during the proof of Proposition 5.12 that all the weights of W^{θ} wrote as

$$\mu = \left(\theta_1 z^{\frac{-n_1}{2}}, \theta_2 z^{\frac{n_1-n_2}{2}}, \ldots, \theta_N z^{\frac{n_{N-1}}{2}}\right),$$

so they all have the same product as θ. So, $(\eta_1\eta_2\cdots\eta_N)^{1/2} w = w$ with w running over a basis of weight vectors of W^{θ}. Therefore, the action of $U_z(\mathfrak{gl}(N))$ on a module with highest weight $(\theta_1, \theta_2, \ldots, \theta_N)$ such that $\theta_1 \theta_2 \cdots \theta_N = 1$ factorizes through the specialization $(\eta_1 \eta_2 \cdots \eta_N)^{1/2} = 1$. In the following, we denote $\widetilde{U_z}(\mathfrak{gl}(N))$ the quotient of $U_z(\mathfrak{gl}(N))$ by the relation $(\eta_1\, \eta_2 \cdots \eta_N)^{1/2} = 1$, and $\widetilde{e}_i$, $\widetilde{f}_i$ and $(\widetilde{\eta}_i)^{1/2}$ the images of the generators of $U_z(\mathfrak{gl}(N))$ in this quotient. It will also be convenient to set

$$e_i' = \widetilde{e}_i \quad ; \quad f_i' = z^{\frac{1}{2}}\, \widetilde{f}_i\, (\widetilde{\eta}_i \widetilde{\eta}_{i+1})^{-\frac{1}{2}} \quad ; \quad (k_i)^{\frac{1}{2}} = (\widetilde{\eta}_i)^{\frac{1}{2}}\, (\widetilde{\eta}_{i+1})^{-\frac{1}{2}}.$$

Notice then that for any index $i \in [\![1, N-1]\!]$, the subalgebra $U_{z,i}(\mathfrak{sl}(2))$ that is spanned by $(e_i', f_i', k_i^{\pm 1/2})$ admits for presentation over $\mathbb{C}$:

$$\begin{aligned} k^{\frac{1}{2}}\, k^{-\frac{1}{2}} &= k^{-\frac{1}{2}}\, k^{\frac{1}{2}} = 1; \\ k^{\frac{1}{2}}\, e'\, k^{-\frac{1}{2}} &= z\, e'; \\ k^{\frac{1}{2}}\, f'\, k^{-\frac{1}{2}} &= z^{-1}\, f'; \\ e'f' - f'e' &= \frac{k^{\frac{1}{2}} - k^{-\frac{1}{2}}}{z^{\frac{1}{2}} - z^{-\frac{1}{2}}}. \end{aligned}$$

Indeed, one computes for the last relation

$$\begin{aligned} e_i'f_i' - f_i'e_i' &= \frac{\widetilde{\eta}_i - \widetilde{\eta}_{i+1}}{z^{\frac{1}{2}} - z^{-\frac{1}{2}}}\, (\widetilde{\eta}_i \widetilde{\eta}_{i+1})^{-\frac{1}{2}} \\ &= \frac{1}{z^{\frac{1}{2}} - z^{-\frac{1}{2}}} \left((\widetilde{\eta}_i \widetilde{\eta}_{i+1}^{-1})^{\frac{1}{2}} - (\widetilde{\eta}_{i+1} \widetilde{\eta}_i^{-1})^{\frac{1}{2}} \right) = \frac{k_i^{\frac{1}{2}} - k_i^{-\frac{1}{2}}}{z^{\frac{1}{2}} - z^{-\frac{1}{2}}}. \end{aligned}$$

As in the classical case, the identification of all the possible highest weights of the simple $U_z(\mathfrak{gl}(N))$-modules starts then with an ad hoc study of the representation theory of the algebra $U_z(\mathfrak{sl}(2))$. Lemmas 5.16 and 5.17 hereafter summarize this theory. By analogy with Definition 5.10, we call weight of a module V over $U_z(\mathfrak{sl}(2))$ a number $\theta \in \mathbb{C}^\times$ such that the subspace of vectors $v \in V$ with $k^{1/2} \cdot v = \theta\, v$ is not reduced to 0. An highest weight is a weight such that the corresponding weight space is sent to 0 by the element e'. The previous results (Propositions 5.12 and 5.13) on (highest) weights of finite-dimensional modules over $U_z(\mathfrak{gl}(N))$ extend mutatis mutandis to $U_z(\mathfrak{sl}(2))$).

Lemma 5.16. *Let V be a simple finite-dimensional module over $U_z(\mathfrak{sl}(2))$. An highest weight of V writes as $\theta = \pm z^{n/2}$, with $n \in \mathbb{N}$. Conversely, every weight of this form corresponds to a simple module over $U_z(\mathfrak{sl}(2))$, of complex dimension $n+1$.*

Proof. Let v be a highest weight vector of a fixed finite-dimensional representation

V of $U_z(\mathfrak{sl}(2))$, associated to a weight θ. Set $v_p = \frac{1}{p!}(f')^p \cdot v$. Notice that

$$[e', f'] = \frac{k^{\frac{1}{2}} - k^{-\frac{1}{2}}}{z^{\frac{1}{2}} - z^{-\frac{1}{2}}};$$

$$[e', (f')^2] = [e', f']f' + f'[e', f'] = \frac{k^{\frac{1}{2}} - k^{-\frac{1}{2}}}{z^{\frac{1}{2}} - z^{-\frac{1}{2}}} f' + f' \frac{k^{\frac{1}{2}} - k^{-\frac{1}{2}}}{z^{\frac{1}{2}} - z^{-\frac{1}{2}}}$$

$$= f' \frac{z^{-1}k^{\frac{1}{2}} - zk^{-\frac{1}{2}}}{z^{\frac{1}{2}} - z^{-\frac{1}{2}}} + f' \frac{k^{\frac{1}{2}} - k^{-\frac{1}{2}}}{z^{\frac{1}{2}} - z^{-\frac{1}{2}}} = f' \frac{z^{-\frac{1}{2}} + z^{-\frac{1}{2}}}{z^{\frac{1}{2}} - z^{-\frac{1}{2}}} \left(z^{-\frac{1}{2}}k^{\frac{1}{2}} - z^{\frac{1}{2}}k^{-\frac{1}{2}}\right)$$

$$= f' \frac{z - z^{-1}}{\left(z^{\frac{1}{2}} - z^{-\frac{1}{2}}\right)^2} \left(z^{-\frac{1}{2}}k^{\frac{1}{2}} - z^{\frac{1}{2}}k^{-\frac{1}{2}}\right);$$

and more generally,

$$[e', (f')^p] = (f')^{p-1} \frac{z^{\frac{p}{2}} - z^{-\frac{p}{2}}}{\left(z^{\frac{1}{2}} - z^{-\frac{1}{2}}\right)^2} \left(z^{-\frac{p-1}{2}}k^{\frac{1}{2}} - z^{\frac{p-1}{2}}k^{-\frac{1}{2}}\right).$$

Therefore,

$$f' \cdot v_p = (p+1)\, v_{p+1};$$

$$k^{\frac{1}{2}} \cdot v_p = \frac{1}{p!} k^{\frac{1}{2}}(f')^p \cdot v = \frac{z^{-p}}{p!} (f')^p k^{\frac{1}{2}} \cdot v = \frac{\theta\, z^{-p}}{p!} (f')^p \cdot v = \theta\, z^{-p}\, v_p;$$

$$e' \cdot v_p = \frac{1}{p!} e'(f')^p \cdot v = \frac{1}{p!}[e', (f')^p] \cdot v = \frac{z^{\frac{p}{2}} - z^{-\frac{p}{2}}}{\left(z^{\frac{1}{2}} - z^{-\frac{1}{2}}\right)^2} \left(z^{-\frac{p-1}{2}}\theta - z^{\frac{p-1}{2}}\theta^{-1}\right) \frac{v_{p-1}}{p}.$$

Let n be the largest integer such that $v_n \neq 0$. Then, $(v_0, \ldots, v_n)$ is a linear basis of $U_z(\mathfrak{sl}(2))\, v$ that consists in weight vectors, and since $e' \cdot v_{n+1} = e' \cdot 0 = 0$, this forces

$$z^{-\frac{n}{2}}\theta - z^{\frac{n}{2}}\theta^{-1} = 0 \iff \theta^2 = z^n \iff \theta = \pm z^{\frac{n}{2}}.$$

Conversely, given a weight $\theta = \pm z^{n/2}$, the previous formulas for the v_p's and their images by e', f' and $k^{\pm 1/2}$ yield indeed a simple $U_z(\mathfrak{sl}(2))$-module of dimension $n+1$. □

Lemma 5.17. *Let V be a finite-dimensional module over $U_z(\mathfrak{sl}(2))$. Then, V is a direct sum of finite-dimensional simple modules over $U_z(\mathfrak{sl}(2))$.*

Proof. Let V' be a proper submodule of V; we have to construct a complement submodule V'' such that $V = V' \oplus V''$. Without loss of generality, one can assume that V' is a simple module, with highest weight $\theta = \sigma' z^{n/2}$ with $\sigma' \in \{\pm 1\}$. Fix a sign $\sigma \in \{\pm 1\}$, and set

$$C = \frac{z^{\frac{1}{2}}k^{\frac{1}{2}} + z^{-\frac{1}{2}}k^{-\frac{1}{2}} - 2\sigma}{\left(z^{\frac{1}{2}} - z^{-\frac{1}{2}}\right)^2} + f'e';$$

$$D = C - \sigma\,\frac{z^{\frac{1}{2}} + z^{-\frac{1}{2}} - 2}{\left(z^{\frac{1}{2}} - z^{-\frac{1}{2}}\right)^2}.$$

One sees readily that C and D commute with $k^{1/2}$, e' and f'; therefore, these elements belong to the center $Z(U_z(\mathfrak{sl}(2)))$ of the quantum group $U_z(\mathfrak{sl}(2))$. Their actions on any simple finite-dimensional module U are consequently in $Z(\mathfrak{gl}(U))$, so, they act by multiplication by a scalar, which can be computed by taking a highest weight vector. Suppose U simple, of highest weight $\sigma\, z^{m/2}$. Then, C acts on U by

$$\sigma\,\frac{z^{\frac{m+1}{2}} + z^{-\frac{m+1}{2}} - 2}{\left(z^{\frac{1}{2}} - z^{-\frac{1}{2}}\right)^2} = \sigma\left(\frac{z^{\frac{m+1}{4}} - z^{-\frac{m+1}{4}}}{z^{\frac{1}{2}} - z^{-\frac{1}{2}}}\right)^2,$$

which is non-zero since z is not a root of unity. Similarly, D acts by

$$\frac{\sigma}{\left(z^{\frac{1}{2}} - z^{-\frac{1}{2}}\right)^2}\left(\left(z^{\frac{m+1}{2}} + z^{-\frac{m+1}{2}}\right) - \left(z^{\frac{1}{2}} + z^{-\frac{1}{2}}\right)\right) = \frac{\sigma}{\left(z^{\frac{1}{2}} - z^{-\frac{1}{2}}\right)^2}\left(z^{\frac{m}{2}} - 1\right)\left(z^{\frac{1}{2}} - z^{-\frac{m+1}{2}}\right)$$

which is non-zero unless $m = 0$, and U is one-dimensional.

Assume first that V' is of codimension 1, and set $U = V/V'$: it is a simple module of weight $\sigma \in \{\pm 1\}$.

1. If $\dim V' \geq 2$, then D acts by a non-zero scalar on V', and it acts by zero on $U = V/V'$. Therefore, if $V'' = \operatorname{Ker} D$ is the kernel of the action of D on V, then it is one-dimensional, and a complement submodule of V': $V = V' \oplus \operatorname{Ker} D$.

2. If $\dim V' = 1$, then $\dim V = 2$ and V' and $U = V/V'$ are one-dimensional modules of highest weight σ' and σ in $\{\pm 1\}$. So, there exists a basis of V such that $k^{\frac{1}{2}}$, e' and f' act by the matrices

$$\begin{pmatrix} \sigma' & a \\ 0 & \sigma \end{pmatrix} \quad ; \quad \begin{pmatrix} 0 & b \\ 0 & 0 \end{pmatrix} \quad ; \quad \begin{pmatrix} 0 & c \\ 0 & 0 \end{pmatrix}.$$

Since $k^{\frac{1}{2}}e'k^{-\frac{1}{2}} = ze'$, $\sigma'\sigma^{-1}b = z\,b$, hence $b = 0$ since z is not a root of unity. Similarly, $c = 0$, so e' and f' both act by zero on V. Now, if $\sigma \neq \sigma'$, then the matrix of the action of $k^{1/2}$ can be diagonalized, so there is another basis of V such that $k^{1/2}$ acts by the diagonal matrix $\left(\begin{smallmatrix} \sigma' & 0 \\ 0 & \sigma \end{smallmatrix}\right)$, and V' admits a complement submodule, namely, the vector space spanned by the second vector of the diagonalization basis. Suppose finally that $\sigma = \sigma'$. Then, since $[e', f'] = \frac{k^{1/2} - k^{-1/2}}{z^{1/2} - z^{-1/2}}$ acts by 0,

$$0 = \begin{pmatrix} \sigma & a \\ 0 & \sigma \end{pmatrix} - \begin{pmatrix} \sigma & -a \\ 0 & \sigma \end{pmatrix} = \begin{pmatrix} 0 & 2a \\ 0 & 0 \end{pmatrix},$$

and $a = 0$. This leads to the same conclusion as before.

So, we have just shown that if V' is a simple submodule of a module V with codimension 1, then V' admits a complement submodule. Suppose now that V' is of arbitrary codimension, and set

$$W = \{u \in \mathrm{Hom}_{\mathbb{C}}(V, V') \mid u_{|V'} = t\,\mathrm{id}_{V'} \text{ with } t \in \mathbb{C}\};$$
$$W' = \{u \in \mathrm{Hom}_{\mathbb{C}}(V, V') \mid u_{|V'} = 0\}.$$

We consider W and W' as submodules of $\mathrm{Hom}_{\mathbb{C}}(V, V') = V' \otimes_{\mathbb{C}} V^*$, where $U_z(\mathfrak{sl}(2))$ acts on V^* by the contragredient representation of V:

$$\forall \varphi \in V^*,\ \forall v \in V,\ \forall x \in U_z(\mathfrak{sl}(2)),\ (x \cdot \varphi)(v) = \varphi(\omega(x) \cdot v),$$

where ω is the antipode of the quantum group $U_z(\mathfrak{sl}(2))$, given by

$$\omega(e') = -e'k^{-1} \quad ; \quad \omega(f') = -kf' \quad ; \quad \omega\left(k^{\frac{1}{2}}\right) = k^{-\frac{1}{2}}.$$

If ρ and ρ^* are the defining morphisms of V and V^*, then the action of $U_z(\mathfrak{sl}(2))$ on $\mathrm{Hom}_{\mathbb{C}}(V, V')$ is given by $(\rho \otimes \rho^*) \circ \Delta$, where Δ is the coproduct that makes $U_z(\mathfrak{sl}(2))$ into a Hopf algebra:

$$\Delta(e') = 1 \otimes e' + e' \otimes k \quad ; \quad \Delta(f') = k^{-1} \otimes f' + f' \otimes 1 \quad ; \quad \Delta\left(k^{\frac{1}{2}}\right) = k^{\frac{1}{2}} \otimes k^{\frac{1}{2}}.$$

Let us check that W is stable: if $u \in W$ and $u_{|V'} = t\,\mathrm{id}_{V'}$, then for any $x \in U_z(\mathfrak{sl}(2))$,

$$(x \cdot u)_{|V'} = \sum_i \rho(x_{i1}) \circ u \circ \rho(\omega(x_{i2})) = t\,\rho\left(\sum_i x_{i1}\,\omega(x_{i2})\right) = t\,\eta(x)\,\mathrm{id}_{V'},$$

where η is the counity of $U_z(\mathfrak{sl}(2))$:

$$\eta(e') = \eta(f') = 0 \quad ; \quad \eta\left(k^{\frac{1}{2}}\right) = 1.$$

Thus, W is indeed a $U_z(\mathfrak{sl}(2))$-module, and the same computation shows that W' is a codimension 1 submodule of W. By the previous discussion, there is a complement submodule W'' of W' in W: $W = W' \oplus W''$. Take $u \neq 0$ in W'': it acts by a non-zero multiple of the identity on V', therefore, it has a non-zero kernel $\mathrm{Ker}\,u$, with $V = V' \oplus \mathrm{Ker}\,u$. On the other hand, since W'' is one-dimensional, for any $x \in U_z(\mathfrak{sl}(2))$, $x \cdot u = \sigma\,\eta(x)\,u$ for some sign σ. Therefore:

$$k^{\frac{1}{2}} \cdot u = \rho\left(k^{\frac{1}{2}}\right) \circ u \circ \rho\left(k^{-\frac{1}{2}}\right) = \sigma\,u;$$
$$e' \cdot u = \rho(e') \circ u \circ \rho(k^{-1}) - u \circ \rho(e'k^{-1}) = 0;$$
$$f' \cdot u = \rho(f') \circ u - \rho(k^{-1}) \circ u \circ \rho(kf') = 0.$$

Since u commutes with $\rho(k)$, we can rewrite the two last lines as $\rho(e') \circ u = u \circ \rho(e')$ and $\rho(f') \circ u = u \circ \rho(f')$. As a consequence, $\mathrm{Ker}\,u$ is stable by the action of $k^{1/2}$, of e' and of f'. So, $\mathrm{Ker}\,u$ is indeed a complement submodule of V' in V. □

We now give a list of necessary conditions on $(\theta_1,\theta_2,\ldots,\theta_N)$ to be the highest weight of a simple $U_z(\mathfrak{gl}(N))$-module. This classification involves the weight lattice of $\mathrm{SL}(N,\mathbb{C})$, and we need to explain how to embed it in the dual of the torus of $\mathrm{GL}(N,\mathbb{C})$. Let $\mathfrak{t}_{\mathbb{R}}$ be the real vector space spanned by vectors $\varepsilon_1,\varepsilon_2,\ldots,\varepsilon_N$, and $\mathfrak{t}_{\mathbb{R}}^*$ be its dual. We endow the weight lattice $X=X(\mathfrak{sl}(N))$ of $\mathfrak{sl}(N)$ into $\mathfrak{t}_{\mathbb{R}}^*$ as follows: if $\omega_1,\ldots,\omega_{N-1}$ are the fundamental weights of $X(\mathfrak{sl}(N))$, we set

$$\omega_i(\varepsilon_j)=\begin{cases}1-\frac{i}{N} & \text{if } j\leq i,\\ -\frac{i}{N} & \text{if } j>i.\end{cases}$$

This is the same convention as in Appendix Appendix A, and we have $\omega_i(h_j)=\delta_{ij}$ for any $h_j=\varepsilon_j-\varepsilon_{j+1}$, as in the case of the Lie algebra $\mathfrak{sl}(N)$. We denote in the following X_+ the set of dominant weights, that is to say linear combinations $\sum_{j=1}^{N-1} m_j\,\omega_j$ with all the m_j's in $\mathbb{N}$.

Theorem 5.18 (Lusztig, Rosso). *If $\theta=(\theta_1,\ldots,\theta_N)$ is the highest weight of a finite-dimensional simple $U_z(\mathfrak{gl}(N))$-module W, then there exists a unique complex number $\upsilon\in\mathbb{C}^\times$, a unique sequence of signs $\phi_1,\ldots,\phi_{N-1}\in\{\pm 1\}$, and a unique dominant weight $\omega\in X_+$ such that*

$$\theta_i=\upsilon\,\phi_i\,z^{\frac{\omega(\varepsilon_i)}{2}}$$

for every $i\in[\![1,N]\!]$.

Proof. Without loss of generality, one can assume that $\theta_1\theta_2\cdots\theta_N=1$, up to multiplication of $\theta_1,\ldots,\theta_N$ by a common factor $\upsilon\in\mathbb{C}^\times$. In other words, one can assume that W is a simple finite-dimensional $\widetilde{U_z}(\mathfrak{gl}(N))$-module. In this setting, if w is a highest weight vector for $U_z(\mathfrak{gl}(N))$, then it is a highest weight vector for the action of each subalgebra $U_{z,i}(\mathfrak{sl}(2))$ generated by $(e_i',f_i',k_i^{\pm 1/2})$:

$$(k_i)^{\frac{1}{2}}\cdot w=(\widetilde{\eta}_i)^{\frac{1}{2}}(\widetilde{\eta}_{i+1})^{-\frac{1}{2}}\cdot w=(\eta_i)^{\frac{1}{2}}(\eta_{i+1})^{-\frac{1}{2}}\cdot w=\left(\frac{\theta_i}{\theta_{i+1}}\right)w.$$

Therefore, by using Lemma 5.16, we get that for every $i\in[\![1,N-1]\!]$,

$$\left(\frac{\theta_i}{\theta_{i+1}}\right)=\sigma_i\,z^{\frac{m_i}{2}}$$

for some signs σ_i and some non-negative integers m_i. We conclude that there exist signs $\phi_i\in\{+1,-1\}$ and a sequence of integers $n_1\geq n_2\geq\cdots\geq n_{N-1}\geq 0$ such that

$$\theta_i=\phi_i\,z^{\frac{n_i}{2}}\,\theta_N$$

for all $i\in[\![1,N-1]\!]$. As we can modify the signs of $\theta_1,\ldots,\theta_{N-1}$, we can assume in the following $\phi_1=\phi_2=\cdots=\phi_{N-1}=1$. Thus, up to the choice of signs and of some complex number $\upsilon\in\mathbb{C}^\times$, we are reduced to the case where

$$\theta_1\theta_2\cdots\theta_N=1\qquad;\qquad\forall i\in[\![1,N-1]\!],\ \theta_i=z^{\frac{n_i}{2}}\,\theta_N$$

where $n_1 \geq n_2 \geq \cdots \geq n_{N-1} \geq 0$. The combination of these conditions yields:

$$\theta_N = z^{-\frac{n_1+n_2+\cdots+n_{N-1}}{2N}},$$

and since $n_1 + \cdots + n_{N-1} = m_1 + 2m_2 \cdots + (N-1)m_{N-1}$,

$$\theta_i = z^{\frac{1}{2}\left(m_i+\cdots+m_{N-1}-\frac{1}{N}(m_1+2m_2+\cdots+(N-1)m_{N-1})\right)} = z^{\frac{1}{2}\sum_{i=1}^{N-1} m_j \omega_j(\varepsilon_i)} = z^{\frac{\omega(\varepsilon_i)}{2}},$$

where $\omega = \sum_{j=1}^{N-1} m_j \, \omega_j$ belongs to the set of dominant weights X_+. The discussion on one-dimensional representations of $U_z(\mathfrak{gl}(N))$ shows then that the general form of a highest weight of a finite-dimensional simple module over $U_z(\mathfrak{gl}(N))$ is as predicted

$$\theta = v \, \phi \, z^{\frac{\omega}{2}}$$

with $v \in \mathbb{C}^\times$, $\phi \in \{\pm 1\}^{N-1}$ and $\omega \in X_+$. □

Theorem 5.18 is completed by the following important converse statement:

Proposition 5.19. *Conversely, every element of* $(\mathbb{C}^\times)^N$ *that writes as* $\theta = v \, \phi \, z^{\frac{\omega}{2}}$ *is the highest weight of a (unique) finite-dimensional simple* $U_z(\mathfrak{gl}(N))$*-module.*

Proof. Again, without loss of generality, we can assume $\phi_1 = \phi_2 = \cdots = \phi_{N-1} = 1$ and $v = 1$, so it suffices to perform the construction when $\theta = z^{\frac{\omega}{2}}$, with $\omega = \sum_{j=1}^{N-1} m_j \omega_j$ in X_+. We shall use the same arguments as for the classical Verma modules, see Section A.3. Consider the one-dimensional $U_z(\mathfrak{b})$-module $\mathbb{C} = \mathbb{C}^\theta$, where $(\eta_i)^{1/2}$ acts on $\mathbb{C}^\theta$ by

$$(\eta_i)^{\frac{1}{2}} \cdot w = \theta_i \, w = z^{\frac{\omega(\varepsilon_i)}{2}} \, w,$$

and e_i acts on $\mathbb{C}^\theta$ by the zero map. We set $M_z^\theta = U_z(\mathfrak{gl}(N)) \otimes_{U_z(\mathfrak{b})} \mathbb{C}^\theta$, where $\mathbb{C}^\theta$ is endowed with the previously described structure of left $U_z(\mathfrak{b})$-module. This is a cyclic $U_z(\mathfrak{gl}(N))$-module with highest weight θ, and highest weight vector $w = 1 \otimes 1$; therefore, it admits a unique simple quotient $W_z^\theta = M_z^\theta / N_z^\theta$, where N_z^θ is the unique maximal proper $U_z(\mathfrak{gl}(N))$-submodule of M_z^θ (cf. the proof of Proposition 5.13). To end the proof, we have to show that W_z^θ is finite-dimensional.

Let w be a highest weight vector in W_z^θ. By the same arguments as in Lemma 5.16, $(f_i')^{m_i+1} \cdot w = 0$, and the space $U_{z,i}(\mathfrak{sl}(2))(w) = \langle e_i', f_i', k_i^{\pm 1/2} \rangle(w)$ is a finite-dimensional subspace of W_z^θ, of dimension larger than 1. So, W_z^θ contains for each i a non-zero finite-dimensional $U_{z,i}(\mathfrak{sl}(2))$-module. Let W_i be the sum of all finite-dimensional $U_{z,i}(\mathfrak{sl}(2))$-modules included in W_z^θ, where i is fixed in $[\![1, N-1]\!]$. It is a non-zero subspace of W_z^θ; we claim that it is a $U_z(\mathfrak{gl}(N))$-submodule, and therefore, by simplicity, that $W_i = W_z^\theta$. If $i \neq j$, we set

$$e_{i,j}' = e_i' e_j' - e_j' e_i' \quad ; \quad f_{i,j}' = f_i' f_j' - f_j' f_i'.$$

The relations between the e_i's and the f_i's in $\mathfrak{U}(\mathfrak{gl}(N))$ become

$$\begin{aligned}
&(e'_i)^2 e'_{i+1} - (z+1)\, e'_i e'_{i+1} e'_i + z\, e'_{i+1} (e'_i)^2 = 0;\\
&z\,(e'_{i+1})^2 e'_i - (z+1)\, e'_{i+1} e'_i e'_{i+1} + e'_i (e'_{i+1})^2 = 0;\\
&(f'_i)^2 f'_{i+1} - (z+1)\, f'_i f'_{i+1} f'_i + z\, f'_{i+1} (f'_i)^2 = 0;\\
&z\,(f'_{i+1})^2 f'_i - (z+1)\, f'_{i+1} f'_i f'_{i+1} + f'_i (f'_{i+1})^2 = 0
\end{aligned}$$

in the quotient $\widetilde{U_z}(\mathfrak{gl}(N))$. Notice that if $|i-j| \geq 2$, then

$$e'_i\, e'_j = e'_j\, e'_i \quad ; \quad f'_i\, e'_j = e'_j\, f'_i \quad ; \quad (k_i)^{\frac{1}{2}}\, e'_j = e'_j\, (k_i)^{\frac{1}{2}},$$

so if W is a stable $U_{z,i}(\mathfrak{sl}(2))$-submodule of W^θ_z, then $e'_j W$ is also a stable $U_{z,i}(\mathfrak{sl}(2))$-submodule of W^θ_z. Suppose now that $j = i+1$. Then,

$$U_{z,i}(\mathfrak{sl}(2))(e'_{i+1}\, W) \subset e'_{i+1}\, W + e'_{i,i+1}\, W,$$

and since

$$\begin{aligned}
e'_i\, e'_{i,i+1} &= z\, e'_{i,i+1}\, e'_i;\\
f'_i\, e'_{i,i+1} &= z^{-\frac{1}{2}}\, e'_{i,i+1}\, f'_i + e'_{i+1}\, (k_i)^{\frac{1}{2}};\\
(k_i)^{\frac{1}{2}}\, e'_{i,i+1} &= z^{1/2}\, e'_{i,i+1}\, (k_i)^{\frac{1}{2}},
\end{aligned}$$

$e'_{i+1}\, W + e'_{i,i+1}\, W$ is a stable $U_{z,i}(\mathfrak{sl}(2))$-submodule of W^θ_z. By using similar arguments when $j = i-1$ and for the elements f'_j and $(\widetilde{\eta}_j)^{\pm 1/2}$, we conclude that if W is a finite-dimensional $U_{z,i}(\mathfrak{sl}(2))$-submodule of W^θ_z, then the span of the spaces

$$e'_j\, W,\ f'_j\, W,\ (\widetilde{\eta}_j)^{\pm\frac{1}{2}}\, W,\ e'_{i,j}\, W,\ f'_{i,j}\, W$$

is again a finite-dimensional subspace of $U_{z,i}(\mathfrak{sl}(2))$-submodule of W^θ_z, which contains $e'_j\, W$, $f'_j\, W$ and $(\widetilde{\eta}_j)^{\pm 1/2}\, W$. Consequently, W_i is stable by the action of $\widetilde{U_z}(\mathfrak{gl}(N))$; which is what we wanted to prove.

The fact that W^θ_z is the sum of its finite-dimensional $U_{z,i}(\mathfrak{sl}(2))$-submodules (Lemma 5.17) will allow us to construct an action of the Weyl group $\mathfrak{S}(N)$ on the set of weights of W^θ_z. Let μ be a weight of W^θ_z, and w be a weight vector for it. If $\theta = z^{\omega/2}$, then

$$\mu = z^{\frac{1}{2}\left(\omega - \sum_{i=1}^{N-1} k_i \alpha_i\right)},$$

where the α_i's are the simple roots of $\mathfrak{sl}(N)$, and the k_i's are some non-negative integers. Therefore, $\mu = z^{\omega'/2}$, where $\omega' = \sum_{j=1}^{N-1} m'_j \omega_j$ is some weight in $X = \bigoplus_{j=1}^{N-1} \mathbb{Z}\omega_j$ (not necessarily dominant). Consider then the subspace $W' = \bigoplus_{k\in\mathbb{Z}} (W^\theta_z)_{\mu z^{k\alpha_i/2}}$ of W^θ_z. It is invariant under $U_{z,i}(\mathfrak{sl}(2))$, so by the previous discussion, there is a finite-dimensional subspace $W'' \subset W'$ that contains w and is a

$U_{z,i}(\mathfrak{sl}(2))$-module. However, it follows from the classification of simple modules over $U_z(\mathfrak{sl}(2))$ (Lemmas 5.16 and 5.17) that if $\mu_i = z^{m_i/2}$ is a weight of a (simple) finite-dimensional $U_{z,i}(\mathfrak{sl}(2))$-module, then $z^{-m_i/2}$ is also a weight of this module. Indeed, the $n+1$ weights of the unique simple module over $U_z(\mathfrak{sl}(2))$ with highest weight $z^{n/2}$ are:

$$z^{\frac{n}{2}},\ z^{\frac{n-2}{2}},\ z^{\frac{n-4}{2}},\ldots,z^{-\frac{n-2}{2}},\ z^{-\frac{n}{2}}$$

the operators e' and f' allowing to raise or decrease the power of z in a weight. On the other hand, the weight space $(W_z^\theta)_\mu$ is a space of weight $z^{\omega'(h_i)/2} = z^{m_i'/2}$ for the action of $U_{z,i}(\mathfrak{sl}(2))$, and similarly, the weight space $(W_z^\theta)_{\mu z^{k\alpha_i/2}}$ is a space of weight $z^{(m_i'+2k)/2}$ for the action of $U_{z,i}(\mathfrak{sl}(2))$. So, there exists $k \in \mathbb{Z}$ such that $(W_z^\theta)_{\mu z^{k\alpha_i/2}} \neq \{0\}$, and such that $m_i' + 2k = -m_i'$, i.e., $k = -m_i'$. We have therefore shown:

$$(W_z^\theta)_{\mu = z^{\omega'/2}} \neq \{0\} \quad \Rightarrow \quad (W_z^\theta)_{z^{s_i(\omega')/2}} \neq \{0\},$$

where s_i is the reflection with respect to the simple root α_i. Thus, the set of weights of W_z^θ is stable by the action of the symmetric group $\mathfrak{S}(N)$.

It follows immediately that the set of weights of W_z^θ is finite. Indeed, if μ is a weight of W_z^θ, then it is conjugated by the action of the symmetric group to a dominant weight, which must fall in the Weyl chamber of $\mathfrak{sl}(N)$, but also be smaller than the highest weight θ. However, the set of dominant weights that are smaller than a given dominant weight θ is always of finite cardinality C: if $\varphi = \omega - \sum_{i=1}^{N-1} k_i \alpha_i$ is dominant, then

$$\forall i \in [\![1,N-1]\!],\ \varphi(h_i) \geq 0 \quad \iff \quad \forall i \in [\![1,N-1]\!],\ \omega(h_i) \geq 2k_i - k_{i-1} - k_{i+1},$$

which imposes a finite bound on each k_i. Therefore, the set of weights of W_z^θ is also of finite cardinality, smaller than $C\,N!$. As the weight spaces of a cyclic module are finite-dimensional, we have finally shown that W_z^θ is a finite-dimensional simple module over $U_z(\mathfrak{gl}(N))$. □

Thus, up to the choice of parameters $v \in \mathbb{C}^\times$ and $\phi_1, \ldots, \phi_{N-1} \in \pm 1$, which determine a one-dimensional representation by which one can twist the action of the specialized quantum group, the simple modules over $U_z(\mathfrak{gl}(N))$ have the same classification as the irreducible representations of the classical Lie algebra $\mathfrak{sl}(N)$. In a moment, we shall complete this result by showing that this correspondence also holds for the weights multiplicities $\dim(W_z^\theta)_\mu$. Since in the end we are only interested in the *reduced* (specialized) quantum group $\mathrm{GL}_z(N,\mathbb{C})$ (without the square roots $q^{\varepsilon_i/2}$), we shall switch to this setting, which will allow us to get rid of the signs $\phi_1, \ldots, \phi_{N-1}$. If V is a simple module over $\mathrm{GL}_z(N,\mathbb{C}) = \langle e_i, f_i, \eta_i^{\pm 1} \rangle$, we call highest weight of V the (unique) element $\mu = (\mu_1, \ldots, \mu_N) \in (\mathbb{C}^\times)^N$ such that there exists $v \in V$ with $e_i \cdot v = 0$ for any $i \in [\![1,N-1]\!]$, and $\eta_i \cdot v = \mu_i v$ for any $i \in [\![1,N]\!]$. More generally, a weight of a $\mathrm{GL}_z(N,\mathbb{C})$-module V is a N-uple $(\mu_1, \ldots, \mu_N) \in (\mathbb{C}^\times)^N$ such that η_i (instead of $(\eta_i)^{1/2}$) acts by multiplication by μ_i on a non-zero vector subspace of V. As for $U_z(\mathfrak{gl}(N))$, the finite-dimensional simple modules over $\mathrm{GL}_z(N,\mathbb{C})$ are classified by their highest weights.

Corollary 5.20. *If V is a finite-dimensional simple module over $\mathrm{GL}_z(N,\mathbb{C})$ of highest weight μ, then there exists a dominant weight $\omega \in X_+$, and a complex number $\nu \in \mathbb{C}^\times$, such that*

$$\mu_i = \nu z^{\omega(\varepsilon_i)}$$

for any $i \in [\![1,N]\!]$. Conversely, every such pair (ν,ω) corresponds to a finite-dimensional simple module over the reduced specialized quantum group $\mathrm{GL}_z(N,\mathbb{C})$.

Proof. If $\theta = \upsilon\,\phi\, z^{\omega/2}$ is the highest weight of a simple module over $U_z(\mathfrak{gl}(N))$, then $V = W_z^\theta$ is by restriction a module over $\mathrm{GL}_z(N,\mathbb{C})$. It is actually a simple module over $\mathrm{GL}_z(N,\mathbb{C})$, with highest weight

$$\mu = \theta^2 = \upsilon^2 z^\omega.$$

Indeed, let v be an element of V such that $e_i \cdot v = 0$ for every $i \in [\![1,N-1]\!]$. Then, v is a highest weight vector for V viewed as a $U_z(\mathfrak{gl}(N))$-module, so it falls necessarily in the one-dimensional vector space $(W_z^\theta)_\theta$. Therefore, the space of highest weight vectors of V viewed as a $\mathrm{GL}_z(N,\mathbb{C})$-module is one-dimensional, and V is simple over the algebra $\mathrm{GL}_z(N,\mathbb{C})$. Moreover, the highest weight of V viewed as a $\mathrm{GL}_z(N,\mathbb{C})$-module is indeed

$$\mu_i = \frac{\eta_i \cdot v}{v} = (\theta_i)^2 = \upsilon^2 z^{\omega(\varepsilon_i)}.$$

We thus have constructed for every vector $\mu = \nu z^\omega$ a simple $\mathrm{GL}_z(N,\mathbb{C})$-module with highest weight μ.

Conversely, consider a finite-dimensional and simple $\mathrm{GL}_z(N,\mathbb{C})$-module V, with highest weight $\mu \in (\mathbb{C}^\times)^N$. We denote $v \in V$ a highest weight vector for the action of $\mathrm{GL}_z(N,\mathbb{C})$, and set $W = U_z(\mathfrak{gl}(N)) \otimes_{\mathrm{GL}_z(N,\mathbb{C})} V$, which is a $U_z(\mathfrak{gl}(N))$-module. A basis of $U_z(\mathfrak{gl}(N))$ over $\mathrm{GL}_z(N,\mathbb{C})$ consists in the vectors

$$\prod_{i=1}^N (\eta_i)^{\frac{\delta_i}{2}},$$

where $(\delta_i)_{i\in[\![1,N]\!]}$ belongs to $\{0,1\}^N$. Therefore, if $\dim_\mathbb{C} V = D$, then $\dim_\mathbb{C} W = 2^N D$, and moreover, the space $H \subset W$ of vectors $h \in W$ such that $e_i \cdot h = 0$ for all i has dimension 2^N. Choose a highest weight vector $h \in H$, with weight $\theta = \upsilon\,\phi\, z^{\omega/2}$ for the quantum group $U_z(\mathfrak{gl}(N))$. Then, h is a linear combination of the tensors

$$\left(\prod_{i=1}^N (\eta_i)^{\frac{\delta_i}{2}}\right) \otimes v$$

and the action of η_i on each of these tensors is given by μ_i, so $\mu_i = (\theta_i)^2$ for all $i \in [\![1,N]\!]$. Therefore, $\mu = \nu z^\omega$ for some $\omega \in X_+$ and some $\nu \in \mathbb{C}^\times$. □

We can finally explain what are the modules V_z^λ that will appear in the generalization of Schur–Weyl duality. Recall from Section A.3 that elements of the Weyl chamber $\bigoplus_{i=1}^{N-1} \mathbb{R}_+\omega_i$ of the Lie algebra $\mathfrak{sl}(N)$ can be identified with linear forms

$$n_1 d_1^* + n_2 d_2^* + \cdots + n_N d_n^*$$

with $n_1 \geq n_2 \geq \cdots \geq n_N$ and $\sum_{i=1}^N n_i = 0$, and where $(d_1^*, \ldots, d_N^*)$ is the dual basis of $(\varepsilon_1, \ldots, \varepsilon_N)$. This correspondence is obtained by setting as before

$$\omega_i = (d_1^* + d_2^* + \cdots + d_i^*) - \frac{i}{N}(d_1^* + d_2^* + \cdots + d_N^*).$$

In this Weyl chamber, the dominant weights correspond to the linear forms such that $n_i - n_{i+1} = m_i \in \mathbb{N}$ for every $i \in [\![1, N-1]\!]$. Suppose that V is a finite-dimensional simple module over $\mathrm{GL}_z(N,\mathbb{C})$ with parameters $\nu \in \mathbb{C}^\times$ and $\omega \in X_+$, such that $\nu = z^t$ with $t \in \mathbb{R}$, and such that ω corresponds to a linear form $n_1 d_1^* + \cdots + n_N d_N^*$. Then, the highest weight μ of V can be rewritten as

$$\mu_i = z^{t+\omega(\varepsilon_i)} = z^{t+n_i}.$$

The sequence $\lambda = (\lambda_1, \lambda_2, \ldots, \lambda_N) = (n_1 + t, n_2 + t, \ldots, n_N + t)$ is an arbitrary non-increasing sequence of real numbers. We define V_z^λ as the simple module over $\mathrm{GL}_z(N,\mathbb{C})$ with highest weight $\mu_i = z^{\lambda_i}$. In particular, the definition makes sense for every integer partition $\lambda \in \mathfrak{Y}^{(N)}$, and the V_z^λ with $|\lambda| = n$ and $\ell(\lambda) \leq N$ are precisely the modules that will appear in the expansion of the space of tensors $(\mathbb{C}^N)^{\otimes n}$, viewed as $\mathrm{GL}_z(N,\mathbb{C})$-module.

▷ *Lusztig–Rosso correspondence for the weight multiplicities.*

Fix a dominant weight $\omega \in X_+$, and consider a simple $\mathrm{GL}_z(N,\mathbb{C})$-module V^μ of highest weight $\mu = \nu z^\omega$, where ν is some parameter in $\mathbb{C}^\times$. We want to compute the weight multiplicities of V_z^μ, and in particular the sum of all these multiplicities, that is to say the complex dimension of V_z^μ. By tensoring by the one-dimensional representation $\mathbb{C}^{\nu^{-1}}$, where $\mathrm{GL}_z(N,\mathbb{C})$ acts on $\mathbb{C}^\nu = \mathbb{C}$ by

$$e_i \cdot 1 = 0 \quad ; \quad f_i \cdot 1 = 0 \quad ; \quad \eta_i \cdot 1 = \nu,$$

one can assume without loss of generality that $\nu = 1$. Indeed, if V_z^μ has a weight $\vartheta = \nu z^\varpi$ with multiplicity m_ϑ, then $\mathbb{C}^{\nu^{-1}} \times V_z^\mu$ has weight $\vartheta' = z^\varpi$ with the same multiplicity m_ϑ, since

$$\eta_i \cdot (1 \otimes w) = \nu^{-1} \otimes \nu z^{\varpi(\varepsilon_i)} w = z^{\varpi(\varepsilon_i)}(1 \otimes w).$$

if w is a weight vector for the weight ϑ. Thus, in the following, we fix a dominant weight $\mu = z^\omega$ of a simple $\mathrm{GL}_z(N,\mathbb{C})$-module, with $\omega \in X_+$. The main result regarding the weight multiplicities is then:

Theorem 5.21 (Lusztig). *Fix a dominant weight $\omega \in X_+$, and denote $\mu = z^{\omega}$, and V_z^{μ} the finite-dimensional simple $\mathrm{GL}_z(N,\mathbb{C})$-module with highest weight μ. The weights of V_z^{μ} all write as z^{π}, where π is one of the weights of the irreducible representation V^{ω} of highest weight ω of the Lie algebra $\mathfrak{sl}(N)$. Moreover, for any π weight of V^{ω},*

$$\dim_{\mathbb{C}}\left(V_z^{z^{\omega}}\right)_{z^{\pi}} = \dim_{\mathbb{C}}(V^{\omega})_{\pi}.$$

In particular, $\dim_{\mathbb{C}}(V_z^{z^{\omega}}) = \dim_{\mathbb{C}}(V^{\omega})$.

Thus, the Lusztig–Rosso correspondence also holds for the weight multiplicities. In particular, fix an integer partition $\lambda = (\lambda_1, \ldots, \lambda_N)$ of length less than N, and consider the character

$$X_z^{\lambda} = \mathrm{tr}\, \rho_z^{\lambda}$$

of the $\mathrm{GL}_z(N,\mathbb{C})$-module V_z^{λ}.

Corollary 5.22. *The value of X_z^{λ} on a basis element $z^{\sum_{i=1}^N k_i \varepsilon_i} = \prod_{i=1}^N (\eta_i)^{k_i}$ of $U_z(\mathfrak{t})$ is given by the Schur function s_{λ} :*

$$X_z^{\lambda}\left(z^{\sum_{i=1}^N k_i \varepsilon_i}\right) = s_{\lambda}(z^{k_1}, z^{k_2}, \ldots, z^{k_N}).$$

Proof. The integer partition $\lambda = (\lambda_1, \ldots, \lambda_N)$ corresponds to the highest weight $\mu = z^{t + \sum_{i=1}^{N-1}(\lambda_i - \lambda_{i+1})\omega_i}$, where $t = \frac{|\lambda|}{N}$. Therefore,

$$X_z^{\lambda}\left(z^{\sum_{i=1}^N k_i \varepsilon_i}\right) = \sum_{z^{t+\pi} \text{ weight of } V_z^{\lambda}} (\dim (V_z^{\lambda})_{z^{\pi}})\, z^{\sum_{i=1}^N t k_i}\, z^{\pi(\sum_{i=1}^N k_i \varepsilon_i)}.$$

By using the correspondence for weight multiplicities, this can be rewritten as

$$X_z^{\lambda}\left(z^{\sum_{i=1}^N k_i \varepsilon_i}\right) = z^{\sum_{i=1}^N t k_i} \sum_{\pi \text{ weight of } V^{\omega}} (\dim (V^{\omega})_{\pi})\, z^{\pi(\sum_{i=1}^N k_i \varepsilon_i)}.$$

By Weyl's formula (cf. Theorem A.14),

$$\sum_{\pi \text{ weight of } V^{\omega}} (\dim (V^{\omega})_{\pi})\, z^{\pi(\sum_{i=1}^N k_i \varepsilon_i)} = \frac{\sum_{\sigma \in \mathfrak{S}(N)} \varepsilon(\sigma)\, z^{(\sigma(\omega+\rho))(\sum_{i=1}^N k_i \varepsilon_i)}}{\sum_{\sigma \in \mathfrak{S}(N)} \varepsilon(\sigma)\, z^{(\sigma(\rho))(\sum_{i=1}^N k_i \varepsilon_i)}}$$

where ρ is the half-sum of positive roots of $\mathfrak{sl}(N)$, or equivalently the sum $\sum_{i=1}^N \omega_i$ of all fundamental weights. For any weight $\omega = \sum_{i=1}^N m_i \omega_i$, we have

$$\begin{aligned}\omega\left(\sum_{j=1}^N k_j \varepsilon_j\right) &= \sum_{i=1}^N m_i \left(\left(1 - \frac{i}{N}\right)\sum_{j \le i} k_j - \frac{i}{N}\sum_{j > i} k_j\right) \\ &= \left(\sum_{j=1}^N \left(k_j \sum_{i=j}^N m_i\right)\right) - \left(\frac{1}{N}\sum_{i=1}^N i m_i\right)\left(\sum_{j=1}^N k_j\right).\end{aligned}$$

With $m_i = \lambda_i - \lambda_{i+1} + 1$ for all i, we obtain

$$\omega\left(\sum_{j=1}^{N} k_j \varepsilon_j\right) = \left(\sum_{j=1}^{N} k_j(\lambda_j + N + 1 - j)\right) - \left(\frac{|\lambda|}{N} + \frac{N+1}{2}\right)\left(\sum_{j=1}^{N} k_j\right),$$

whereas with $m_i = 1$ for all i, we obtain

$$\omega\left(\sum_{j=1}^{N} k_j \varepsilon_j\right) = \left(\sum_{j=1}^{N} k_j(N + 1 - j)\right) - \left(\frac{N+1}{2}\right)\left(\sum_{j=1}^{N} k_j\right).$$

We conclude that

$$X_z^{\lambda}\left(z^{\sum_{i=1}^{N} k_i \varepsilon_i}\right) = \frac{\sum_{\sigma \in \mathfrak{S}(N)} \varepsilon(\sigma)\, z^{k_{\sigma(j)}(\lambda_j + N + 1 - j)}}{\sum_{\sigma \in \mathfrak{S}(N)} \varepsilon(\sigma)\, z^{k_{\sigma(j)}(N+1-j)}},$$

which is a quotient of two antisymmetric functions in the variables $z^{k_1}, \ldots, z^{k_N}$, and is equal to $s_\lambda(z^{k_1}, z^{k_2}, \ldots, z^{k_N})$. □

In order to prove Theorem 5.21, we need to construct a "modular" map that relates the quantum group $\mathfrak{U}(\mathfrak{gl}(N))$ and the universal enveloping algebra $U(\mathfrak{sl}(N))$; and the simple modules $V_z^{z^\omega}$ and V^ω. We start by a few remarks. In the following, we work with the **reduced generic quantum group** $\mathfrak{U}'(\mathfrak{gl}(N))$, which is the $\mathbb{C}(q)$-subalgebra of $\mathfrak{U}(\mathfrak{gl}(N))$ that is spanned by the e_i's, the f_i's and the $q^{\pm\varepsilon_i}$'s. Its specialization $q = z$ is the reduced specialized quantum group $\mathrm{GL}_z(N, \mathbb{C})$ introduced before. More generally, the notation $\mathfrak{U}'$ will stand for the analogue of $\mathfrak{U}$ but defined over $\mathbb{C}(q)$ (instead of $\mathbb{C}(q^{1/2})$), and without the square roots $q^{\pm\varepsilon_i/2}$, so for instance $\mathfrak{U}'(\mathfrak{t}) = \mathbb{C}(q)[q^{\pm\varepsilon_1}, \ldots, q^{\pm\varepsilon_N}]$.

Define a weight of a module V over $\mathfrak{U}'(\mathfrak{gl}(N))$ as a vector $\Theta = (\Theta_1, \ldots, \Theta_N)$ of $(\mathbb{C}(q)^\times)^N$ which determines the action of the elements q^{ε_i} on a non-zero $\mathbb{C}(q)$-vector subspace V_Θ of V:

$$\forall v \in V_\Theta,\ q^{\varepsilon_i} \cdot v = \Theta_i\, v.$$

Then, everything stated before stays true in the generic case, but the existence of a highest weight for any finite-dimensional module: $\mathbb{C}(q)$ is not algebraically closed, and the q^{ε_i}'s cannot necessarily be simultaneously trigonalized. For any $\Theta \in (\mathbb{C}(q)^\times)^N$, this does not prevent us from defining the generic Verma module

$$\mathfrak{M}^\Theta = \mathfrak{U}'(\mathfrak{gl}(N)) \otimes_{\mathfrak{U}'(\mathfrak{b})} \mathbb{C}(q),$$

where $\mathfrak{U}'(\mathfrak{n}_+)$ acts on $\mathbb{C}(q)$ by $\mathfrak{U}'(\mathfrak{n}_+) \cdot \mathbb{C}(q) = 0$, and $\mathfrak{U}'(\mathfrak{t})$ acts in a way prescribed by the weight Θ. This is a cyclic module of highest weight Θ, equal to the sum of its finite $\mathscr{O}$-rank weight spaces $(\mathfrak{M}^\Theta)_\Pi$, and with a unique simple quotient $\mathfrak{V}^\Theta$. In the following we fix a highest weight vector $v \in \mathfrak{V}^\Theta$; thus, $\mathfrak{U}'(\mathfrak{n}_-)\, v = \mathfrak{V}^\Theta$.

We now consider an analogue of $\mathfrak{U}'(\mathfrak{gl}(N))$ where one can indeed specialize q to z to get back $\mathrm{GL}_z(N, \mathbb{C})$, and where one can also specialize q to 1 to get back $U(\mathfrak{sl}(N))$. Let $\mathscr{O}$ be the ring $\mathbb{C}[q, q^{-1}]$, which is a subring of $\mathbb{C}(q)$.

Remark. Everything below could be done with the valuation ring $\mathscr{O} = \mathbb{C}[q]_{(q-1)}$, thus sticking to the framework of change of base ring developed in Section 4.3. We worked with $\mathbb{C}[q, q^{-1}]$, in order to make appear more clearly the belonging of certain coefficients to the subring $\mathbb{C}[q, q^{-1}] \subset \mathbb{C}(q)$.

For $i \in [\![1, N-1]\!]$ and $l \geq 1$, we set

$$e_i^{(l)} = \left(\prod_{i=1}^{l} \frac{q-1}{q^i - 1}\right)(e_i)^l = \frac{(e_i)^l}{[l]_q!} \quad ; \quad f_i^{(l)} = \left(\prod_{i=1}^{l} \frac{q-1}{q^i - 1}\right)(f_i)^l = \frac{(f_i)^l}{[l]_q!}.$$

It is also convenient to introduce

$$\binom{q^{\varepsilon_i}, q^{\varepsilon_{i+1}}; s}{t} = \prod_{r=1}^{t} \frac{q^{\varepsilon_i} q^{1+s-r} - q^{\varepsilon_{i+1}}}{q^r - 1},$$

with $s \in \mathbb{Z}$ and $t \in \mathbb{N}$. These coefficients satisfy the recursion

$$\binom{q^{\varepsilon_i}, q^{\varepsilon_{i+1}}; s+1}{t} = q^t \binom{q^{\varepsilon_i}, q^{\varepsilon_{i+1}}; s}{t} + q^{\varepsilon_{i+1}} \binom{q^{\varepsilon_i}, q^{\varepsilon_{i+1}}; s}{t-1},$$

as well as the commutation relation

$$e_i^{(m)} f_i^{(n)} = \sum_{0 \leq t \leq \min(m,n)} f_i^{(n-t)} \binom{q^{\varepsilon_i}, q^{\varepsilon_{i+1}}; 2t-m-n}{t} e_i^{(m-t)},$$

which can be shown by induction on m and n. We then denote $\mathfrak{U}^{\mathscr{O}}(\mathfrak{gl}(N)) = \mathfrak{U}^{\mathscr{O}}(\mathfrak{g})$, $\mathfrak{U}^{\mathscr{O}}(\mathfrak{n}_+)$, $\mathfrak{U}^{\mathscr{O}}(\mathfrak{t})$ and $\mathfrak{U}^{\mathscr{O}}(\mathfrak{n}_-)$ the unital $\mathscr{O}$-subalgebras of $\mathfrak{U}'(\mathfrak{gl}(N))$ generated by:

$$\begin{aligned} \mathfrak{U}^{\mathscr{O}}(\mathfrak{g}) &= \langle e_i^{(l)}, f_i^{(l)}, q^{\pm\varepsilon_i} \rangle; \\ \mathfrak{U}^{\mathscr{O}}(\mathfrak{n}_+) &= \langle e_i^{(l)} \rangle; \\ \mathfrak{U}^{\mathscr{O}}(\mathfrak{t}) &= \left\langle q^{\pm\varepsilon_i}, \binom{q^{\varepsilon_i}, q^{\varepsilon_{i+1}}; 0}{t} \right\rangle; \\ \mathfrak{U}^{\mathscr{O}}(\mathfrak{n}_-) &= \langle f_i^{(l)} \rangle. \end{aligned}$$

Notice that $\binom{q^{\varepsilon_i}, q^{\varepsilon_{i+1}}; s}{t}$ belongs to $\mathfrak{U}^{\mathscr{O}}(\mathfrak{g})$ for any $s \in \mathbb{Z}$ and any $t \geq 0$. Indeed, suppose the result to be true up to order $t-1$, and let us show it for an integer t. By using the recursion relation, it suffices to show it for one integer s, e.g., $s = 0$. However, $\binom{q^{\varepsilon_i}, q^{\varepsilon_{i+1}}; 0}{t}$ is indeed in $\mathfrak{U}^{\mathscr{O}}(\mathfrak{g})$, because of the commutation relation written with $m = n = t$, and of the induction hypothesis.

Proposition 5.23. *Fix a weight $\Theta = q^{\omega}$ with $\omega \in X_+$, and consider the simple $\mathfrak{U}'(\mathfrak{g})$-module $\mathfrak{V} = \mathfrak{V}^{\Theta} = \mathfrak{M}^{\Theta}/\mathfrak{N}^{\Theta}$ of highest weight Θ. If v is a highest weight vector in $\mathfrak{V}$, we set $V^{\mathscr{O}} = \mathfrak{U}^{\mathscr{O}}(\mathfrak{n}_-)\, v$. Then, $V^{\mathscr{O}}$ is a $\mathfrak{U}^{\mathscr{O}}(\mathfrak{g})$-submodule of $\mathfrak{V}$, and*

$$\mathbb{C}(q) \otimes_{\mathscr{O}} V^{\mathscr{O}} = \mathfrak{V}.$$

The space $V^{\mathscr{O}}$ is the direct sum of the spaces $V^{\mathscr{O}}_{\Pi} = V^{\mathscr{O}} \cap \mathfrak{V}_{\Pi}$, where Π runs over the set of weights of $\mathfrak{V}$. Each of these intersections is a free $\mathscr{O}$-module of finite rank.

Proof. First, there is an analogue of Theorem 5.6 with scalars restricted to $\mathscr{O} = \mathbb{C}[q,q^{-1}]$: every element of $\mathfrak{U}^{\mathscr{O}}(\mathfrak{g})$ writes as a product $x = x_- x_t x_+$, with each element x_-, x_t, x_+ in one of the three aforementioned subalgebras of $\mathfrak{U}^{\mathscr{O}}(\mathfrak{g})$. This follows from the commutation relation previously written with the elements $\binom{q^{\varepsilon_i},q^{\varepsilon_{i+1}};s}{t}$, as well as the other relations

$$\begin{aligned} e_i^{(m)} f_j^{(n)} &= f_j^{(n)} e_i^{(m)} \quad \text{for any } i \neq j; \\ e_j^{(l)} \binom{q^{\varepsilon_i},q^{\varepsilon_{i+1}};s}{t} &= q^{t\alpha_i(\varepsilon_j)} \binom{q^{\varepsilon_i},q^{\varepsilon_{i+1}};s-lc_{ij}}{t} e_j^{(l)}; \\ f_j^{(l)} \binom{q^{\varepsilon_i},q^{\varepsilon_{i+1}};s}{t} &= q^{-t\alpha_i(\varepsilon_j)} \binom{q^{\varepsilon_i},q^{\varepsilon_{i+1}};s+lc_{ij}}{t} f_j^{(l)}. \end{aligned}$$

It follows immediately that $V^{\mathscr{O}}$ is an $\mathfrak{U}^{\mathscr{O}}(\mathfrak{g})$-submodule of $\mathfrak{V}$: indeed, $x_+ \cdot v = 0$ for any $x_+ \in \mathfrak{U}^{\mathscr{O}}(\mathfrak{n}_+)$, and

$$q^{\pm\varepsilon_i} \cdot v = q^{\pm\omega(\varepsilon_i)}\, v$$

$$\binom{q^{\varepsilon_i},q^{\varepsilon_{i+1}};0}{t} \cdot v = \begin{cases} \left(\prod_{r=1}^{t} \frac{q^{\omega(\varepsilon_i)+1-r} - q^{\omega(\varepsilon_{i+1})}}{q^r - 1}\right) v & \text{if } \omega(h_i) \geq t, \\ 0 & \text{otherwise,} \end{cases}$$

so $V^{\mathscr{O}} = \mathfrak{U}^{\mathscr{O}}(\mathfrak{n}_-)\, v = \mathfrak{U}^{\mathscr{O}}(\mathfrak{g})\, v$.

Write $\mathfrak{V} = \mathfrak{V}_\Theta \oplus \mathfrak{W}$, where $\mathfrak{W}$ is the unique complement subspace of the weight space $\mathfrak{V}_\Theta$ that is stable by $\mathfrak{U}'(\mathfrak{t})$ (it is the sum of the other weight spaces). If $\pi : \mathfrak{V} \to \mathfrak{V}_\Theta$ is the $\mathbb{C}(q)$-linear projection corresponding to this decomposition, then we have $\pi(V^{\mathscr{O}}) = \mathscr{O}\, v$: indeed, $V^{\mathscr{O}}$ is spanned over $\mathscr{O}$ by vectors $w = f_{i_1}^{(l_1)} \cdots f_{i_r}^{(l_r)}\, v$, and if $r \geq 1$, then $\pi(w) = 0$, whereas $\pi(v) = v$. Now, consider the canonical map $\mathbb{C}(q) \otimes_{\mathscr{O}} V^{\mathscr{O}} \to \mathfrak{V}$; it is clearly surjective, and we claim that it is also injective. Otherwise, we can find coefficients $a_j \in \mathbb{C}(q)$, and vectors $v_j \in V^{\mathscr{O}}$, such that $\sum_{j=1}^m a_j \otimes v_j \neq 0$ in the tensor product, and $\sum_{j=1}^m a_j v_j = 0$ in $\mathfrak{V}$. One can assume without loss of generality m minimal; then, all the a_j's and v_j's are non-zero. There exist indices $i_1, \ldots, i_r$ such that $\pi(e_{i_1} \cdots e_{i_r} \cdot v_1) \neq 0$: otherwise, $\mathfrak{U}'(\mathfrak{n}_+)\, v_1 \subset \mathfrak{W}$, and

$$\mathfrak{U}'(\mathfrak{gl}(N))\, v_1 \subset \mathfrak{U}'(\mathfrak{n}_-)(\mathfrak{W}) \subset \mathfrak{W},$$

which is absurd since $\mathfrak{V}$ is a simple $\mathfrak{U}'(\mathfrak{gl}(N))$-module. Set $x = e_{i_1} \cdots e_{i_r}$; for any $j \in [\![1,m]\!]$, there exists $b_j \in \mathscr{O}$ such that $\pi(x \cdot v_j) = b_j v$, since $\pi(V^{\mathscr{O}}) = \mathscr{O}\, v$ by a previous remark. Moreover, by construction, $b_1 \neq 0$. Consequently,

$$0 = \pi\left(x \cdot \sum_{j=1}^m a_j v_j\right) = \left(\sum_{j=1}^m a_j b_j\right) v,$$

so $\sum_{j=1}^m a_j b_j = 0$. For $j \geq 2$, we set $w_j = b_1 v_j - b_j v_1$, which belongs to $V^{\mathscr{O}}$ since the v_j's are in $V^{\mathscr{O}}$ and the b_j's are in $\mathscr{O}$. Then,

$$\sum_{j=2}^m a_j w_j = b_1 \left(\sum_{j=2}^m a_j v_j\right) - \left(\sum_{j=2}^m a_j b_j\right) v_1 = -a_1 b_1 v_1 + a_1 b_1 v_1 = 0,$$

so by minimality of m, $w_j = 0$ for all $j \in [\![1,m]\!]$. But then, since $b_1 \neq 0$,

$$0 \neq \sum_{j=1}^{m} b_1 a_j \otimes v_j = \left(\sum_{j=1}^{m} b_j a_j\right) \otimes v_1,$$

which contradicts the fact that $\sum_{j=1}^{m} a_j b_j = 0$. So, $\mathbb{C}(q) \otimes_{\mathscr{O}} V^{\mathscr{O}} = \mathfrak{V}$.

Finally, let us show that if $w = w_1 + w_2 + \cdots + w_m$ is a decomposition of a vector $w \in V^{\mathscr{O}}$ in weight vectors associated to weights $\Pi^{(1)} \neq \Pi^{(2)} \neq \cdots \neq \Pi^{(m)}$, then each w_i is in $V^{\mathscr{O}}$. We write $\Pi^{(i)} = q^{\sum_{j=1}^{N-1} k_{ij}\omega_j}$, where the ω_j's are the fundamental weights of $\mathfrak{sl}(N)$. Let s be an integer larger than $|k_{ij} - k_{1j}|$ for all $i \in [\![1,m]\!]$ and $j \in [\![1,N-1]\!]$, and

$$x = \prod_{j=1}^{N-1} \binom{q^{\varepsilon_j}, q^{\varepsilon_{j+1}}; s - k_{1j}}{s} \binom{q^{\varepsilon_j}, q^{\varepsilon_{j+1}}; -1 - k_{1j}}{s},$$

which belongs to $U^{\mathscr{O}}(\mathfrak{t})$. It acts on a weight vector w_i by multiplication by

$$\prod_{j=1}^{N-1} \prod_{r=1}^{s} \frac{(q^{\omega(\varepsilon_j)+1+s-k_{1j}-r} - q^{\omega(\varepsilon_{j+1})})(q^{\omega(\varepsilon_j)-k_{1j}-r} - q^{\omega(\varepsilon_{j+1})})}{(q^r - 1)^2}$$
$$= \prod_{j=1}^{N-1} \prod_{r=1}^{s} \frac{q^{2\omega(\varepsilon_{j+1})}(q^{k_{ij}-k_{1j}+1+s-r} - 1)(q^{k_{ij}-k_{1j}-r} - 1)}{(q^r - 1)^2}$$

If one of the quantity $k_{ij} - k_{1j}$ is non-zero, then it belongs to $[\![1,s]\!]$ or $[\![-s,-1]\!]$, and therefore the quantity above vanishes. Thus, x acts on a weight vector w_i by zero unless $i = 1$, in which case $x \cdot w_1 = \pm q^t\, w_1$ for some $t \in \mathbb{Z}$. Therefore,

$$w_1 = \pm q^{-t}\, x \cdot w \in V^{\mathscr{O}},$$

and the same reasoning can be used for all the other weight vectors w_i. So, $V^{\mathscr{O}}$ is the direct sum of the restricted weight spaces $V^{\mathscr{O}}_{\Pi} = V^{\mathscr{O}} \cap \mathfrak{V}_{\Pi}$, and these intersection submodules are necessarily free $\mathscr{O}$-modules, since $V^{\mathscr{O}}$ is itself a finitely generated free $\mathscr{O}$-module, with a basis that consists in vectors

$$f_{i_1} \cdots f_{i_r} \cdot v,$$

which form also a basis of $\mathfrak{V}$ over $\mathbb{C}(q)$. □

Now, $\mathbb{C}$ is a $\mathscr{O}$-module thanks to the specialization $q = 1$, so it makes sense to consider $V = \mathbb{C} \otimes_{\mathscr{O}} V^{\mathscr{O}}$ and its subspaces $V_\pi = \mathbb{C} \otimes_{\mathscr{O}} V^{\mathscr{O}}_{q^\pi}$, which are finite-dimensional complex vector spaces with

$$\dim_{\mathbb{C}}(V) = \operatorname{rank}_{\mathscr{O}}(V^{\mathscr{O}}) = \dim_{\mathbb{C}(q)}(\mathfrak{V});$$
$$\dim_{\mathbb{C}}(V_\pi) = \operatorname{rank}_{\mathscr{O}}(V^{\mathscr{O}}_{q^\pi}) = \dim_{\mathbb{C}(q)}(\mathfrak{V}_{q^\pi}).$$

Theorem 5.24. *If $\Theta = q^{\omega}$ is the highest weight of $\mathfrak{V}$, then the $\mathbb{C}$-vector space V constructed above is endowed with a natural structure of $U(\mathfrak{sl}(N))$-module, and it is the simple module of highest weight ω for this structure. Moreover, for any weight π of this $U(\mathfrak{sl}(N))$-module, V_{π} is indeed the weight space associated to π in V.*

Proof. By extension of the scalars, the elements $e_i^{(l)}$, $f_i^{(l)}$, $q^{\pm\varepsilon_i}$ and $\binom{q^{\varepsilon_i}, q^{\varepsilon_{i+1}};0}{t}$ of $\mathfrak{U}^{\mathcal{O}}(\mathfrak{g})$ act on V. Denote $\rho(e_i^{(l)})$, $\rho(f_i^{(l)})$, $\rho(q^{\pm\varepsilon_i})$ and $\rho(\binom{q^{\varepsilon_i}, q^{\varepsilon_{i+1}};0}{t})$ the corresponding $\mathbb{C}$-endomorphisms of V; these linear maps satisfy the same relations as

$$\frac{(e_i)^l}{l!} \quad ; \quad \frac{(f_i)^l}{l!} \quad ; \quad 1 \quad \text{and} \quad \binom{h_i}{t}$$

in $U(\mathfrak{sl}(N))$, where

$$\binom{h_i}{t} = \frac{h_i(h_i-1)\cdots(h_i-t+1)}{t!}.$$

Therefore, one has a *bona fide* structure of $U(\mathfrak{sl}(N))$-module on V, and V admits for highest weight vector $1 \otimes v$, with highest weight ω: indeed, one obtains

$$h_i \cdot v = \omega(h_i)\, v$$

by setting $q = 1$ in the formula $\binom{q^{\varepsilon_i};0}{1} \cdot v = \frac{q^{\omega(\varepsilon_i)} - q^{\omega(\varepsilon_{i+1})}}{q-1}\, v$. So, V is indeed the simple $U(\mathfrak{sl}(N))$-module of highest weight ω, and similar computations show that if $\Pi = z^{\pi}$ is a weight of $\mathfrak{V}$, then $\rho(\binom{h_i;0}{1})$ acts by multiplication by $\pi(h_i)$ on $V_{\pi} = \mathbb{C} \otimes_{\mathcal{O}} V_{\Pi}^{\mathcal{O}}$, so V_{π} is the weight space of V for the weight π. □

Proof of Theorem 5.21. Let z^{π} be a weight of the simple $\mathrm{GL}_z(N,\mathbb{C})$-module $V_z^{z^{\omega}}$. It comes from the weight q^{π} of the generic module $\mathfrak{V}^{q^{\omega}}$ over $\mathfrak{U}'(\mathfrak{gl}(N))$, and we have

$$\dim_{\mathbb{C}} (V_z^{z^{\omega}})_{z^{\pi}} = \dim_{\mathbb{C}(q)} (\mathfrak{V}^{q^{\omega}})_{q^{\pi}}.$$

Indeed, one can lead the same discussion as before when comparing $\mathfrak{U}'(\mathfrak{gl}(N))$ and $U(\mathfrak{sl}(N))$, though with much simpler arguments, since the relations between the elements e_i, f_i and $q^{\pm\varepsilon_i}$ are this time well defined over the ring $\mathbb{C}[q]_{(q-z)}$. Then,

$$\dim_{\mathbb{C}} (V_z^{z^{\omega}})_{z^{\pi}} = \dim_{\mathbb{C}(q)} (\mathfrak{V}^{q^{\omega}})_{q^{\pi}} = \dim_{\mathbb{C}} (V^{\omega})_{\pi}.$$ □

The Lusztig–Rosso correspondence (Theorems 5.18 and 5.21) can be completed by the following theorem, which leads to an understanding of the representation theory of quantum groups that is at the same level as what is presented in Appendix Appendix A for the classical groups:

Theorem 5.25 (Rosso). *If V is a finite-dimensional module over $U_z(\mathfrak{gl}(N))$ or over $\mathrm{GL}_z(N,\mathbb{C})$, then V is completely reducible, that is to say that it can be written as a direct sum of finite-dimensional simple modules.*

Notice that Lemma 5.17 treats the important case $n = 2$ of this theorem. We shall admit the general case, as its proof is quite long and does not facilitate the understanding of the other aspects of the representation theory of the quantum groups (see the notes for a complete bibliography on this semisimplicity result).

Thus, to summarize the results of this section, for any $z \notin \{0, \text{ roots of unity}\}$:

- We constructed for any integer partition $\lambda \in \mathfrak{Y}^{(N)}$ a representation V_z^λ of $\mathrm{GL}_z(N,\mathbb{C})$, which is finite-dimensional and irreducible, and with a unique non-zero vector (up to a scalar multiple) v such that

$$e_i \cdot v = 0 \qquad ; \qquad \eta_i \cdot v = z^{\lambda_i}\, v.$$

- The restriction to $\mathbb{C}[(\eta_1)^{\pm 1}, \ldots, (\eta_N)^{\pm 1}]$ of the irreducible character X_z^λ of the simple module V_z^λ is given by the Schur function s_λ:

$$X_z^\lambda\left(\prod_{i=1}^N (\eta_i)^{k_i}\right) = s_\lambda\left(z^{k_1}, z^{k_2}, \ldots, z^{k_N}\right).$$

Example. Consider the geometric representation of $\mathrm{GL}_z(N,\mathbb{C})$ on $\mathbb{C}^N$, which is obtained by specialization of the construction of Proposition 5.2. It admits for highest weight vector $v = (1, 0, \ldots, 0)$, and the action of η_i on v is

$$\eta_i \cdot v = \begin{cases} z\,v & \text{if } i = 1, \\ v & \text{if } i \neq 1. \end{cases}$$

Therefore, the geometric representation is the simple module V_z^λ associated to the integer partition $\lambda = (1, 0, \ldots, 0)$. This generalizes the analogous result for the geometric representation of $\mathrm{GL}(N,\mathbb{C})$ on $\mathbb{C}$.

5.3 Jimbo–Schur–Weyl duality

In this section, we establish the analogue of Theorem 2.33 with respect to the specialized Hecke algebra $\mathfrak{H}_z(n)$, and to the reduced quantum group $\mathrm{GL}_z(N,\mathbb{C})$ (instead of $\mathfrak{S}(n)$ and $\mathrm{GL}(N,\mathbb{C})$).

▷ *Action of $\mathfrak{H}_z(n)$ on tensors.*

Until the end of this section, $V = (\mathbb{C}^N)^{\otimes n}$ denotes the space of n-tensors. A basis of V consists in words $w = w_1 w_2 \ldots w_n$ of length n and with letters in $[\![1, N]\!]$, and one has a right action of $\mathfrak{S}(n)$ on V by permutation of the letters in these words.

We define an action of the Hecke algebra $\mathfrak{H}_z(n)$ on V by the following rules:

$$w \cdot T_i = \begin{cases} z\,w & \text{if } w_i = w_{i+1}, \\ w \cdot s_i & \text{if } w_i < w_{i+1}, \\ z\,(w \cdot s_i) + (z-1)\,w & \text{if } w_i > w_{i+1}. \end{cases}$$

Theorem 5.26. *The previous relations define a structure of right $\mathfrak{H}_z(n)$-module on $V = (\mathbb{C}^N)^{\otimes n}$.*

Proof. Fix a word w. We have to check that

$$\begin{aligned} ((w \cdot T_i) \cdot T_{i+1}) \cdot T_i &= ((w \cdot T_{i+1}) \cdot T_i) \cdot T_{i+1}; \\ (w \cdot T_i) \cdot T_j &= (w \cdot T_j) \cdot T_i \qquad \text{if } |i-j| \geq 2; \\ (w \cdot T_i) \cdot T_i &= (z-1)(w \cdot T_i) + z\,w. \end{aligned}$$

For the first relation, suppose for instance that $w_i > w_{i+1} > w_{i+2}$. Then,

$$\begin{aligned} ((w \cdot T_i) \cdot T_{i+1}) \cdot T_i &= (((z-1)\,w + z\,(w \cdot s_i)) \cdot T_{i+1}) \cdot T_i \\ &= ((z-1)^2\,w + (z-1)\,z\,(w \cdot s_i + w \cdot s_{i+1}) + z^2\,(w \cdot s_i s_{i+1})) \cdot T_i \\ &= (z-1)(z^2 - z + 1)\,w + (z-1)^2\,z\,(w \cdot s_i + w \cdot s_{i+1}) \\ &\quad + (z-1)\,z^2\,(w \cdot s_{i+1} s_i + w \cdot s_i s_{i+1}) + z^3\,(w \cdot s_i s_{i+1} s_i). \end{aligned}$$

Since $s_i s_{i+1} s_i = s_{i+1} s_i s_{i+1}$, one obtains the exact same result when computing $((w \cdot T_{i+1}) \cdot T_i) \cdot T_{i+1}$, hence the first relation when $w_i > w_{i+1} > w_{i+2}$. The other cases, e.g., when there are some equalities between the letters w_i, w_{i+1}, w_{i+2} lead to similar (and easier) computations.

The commutation relations between T_i and T_j for $|j-i| \geq 2$ are obviously satisfied, since they involve the elementary transpositions s_i and s_j that commute with one another when $|j-i| \geq 2$. Finally, let us verify the quadratic relations. If $w_i = w_{i+1}$, then

$$\begin{aligned} (w \cdot T_i) \cdot T_i &= z\,(w \cdot T_i) = z^2\,w; \\ (z-1)(w \cdot T_i) + z\,w &= z(z-1)\,w + z\,w = z^2\,w. \end{aligned}$$

If $w_i < w_{i+1}$, then

$$\begin{aligned} (w \cdot T_i) \cdot T_i = (w \cdot s_i) \cdot T_i &= (z-1)\,(w \cdot s_i) + z\,w \\ &= (z-1)(w \cdot T_i) + z\,w. \end{aligned}$$

Finally, if $w_i > w_{i+1}$, then

$$\begin{aligned} (w \cdot T_i) \cdot T_i &= (z\,(w \cdot s_i) + (z-1)\,w) \cdot T_i \\ &= (z^2 - z + 1)\,w + z(z-1)\,(w \cdot s_i); \\ (z-1)(w \cdot T_i) + z\,w &= (z-1)(z\,(w \cdot s_i) + (z-1)\,w) + z\,w \\ &= (z^2 - z + 1)\,w + z(z-1)\,(w \cdot s_i). \end{aligned}$$

Hence, in every case, the quadratic relations are compatible with the right action of $\mathfrak{H}_z(n)$ on the space of tensors V. □

Remark. Set $z = 1$. Then, $\mathfrak{H}_z(n)$ specializes to $\mathbb{C}\mathfrak{S}(n)$, and the action above becomes $w \cdot T_i = w \cdot s_i$ in any case. So, the definition above generalizes the standard permutation action of $\mathfrak{S}(n)$.

Proposition 5.27. *The action of $\mathfrak{H}_z(n)$ on the right of V commutes with the action of the reduced quantum group $\mathrm{GL}_z(N)$ on the left of V.*

Proof. Fix a word w, say with $w_j > w_{j+1}$ (we only treat this case, the other cases being extremely similar). The elements e_i, f_i and η_i act on V by

$$\Delta^{(n)}(e_i) = \sum_{m=1}^{n} 1^{\otimes(m-1)} \otimes e_i \otimes (\eta_i)^{\otimes(n-m)};$$
$$\Delta^{(n)}(f_i) = \sum_{m=1}^{n} (\eta_{i+1})^{\otimes(m-1)} \otimes f_i \otimes 1^{\otimes(n-m)};$$
$$\Delta^{(n)}(\eta_i) = (\eta_i)^{\otimes n}.$$

Therefore, for any word w,

$$e_i \cdot w = \sum_{m=1}^{n} 1_{w_m=i+1}\, z^{\sum_{k=m+1}^{n} 1_{w_k=i}}\, (w_1 \dots w_{m-1}\, i\, w_{m+1} \dots w_n);$$
$$f_i \cdot w = \sum_{m=1}^{n} 1_{w_m=i}\, z^{\sum_{k=1}^{m-1} 1_{w_k=i+1}}\, (w_1 \dots w_{m-1}\, (i+1)\, w_{m+1} \dots w_n);$$
$$\eta_i \cdot w = z^{\sum_{k=1}^{n} 1_{w_k=i}}\, w.$$

Suppose first $w_j, w_{j+1} \neq i+1$. Then,

$$\begin{aligned}
A = (e_i \cdot w) \cdot T_j &= \sum_{m=1}^{n} 1_{w_m=i+1}\, z^{\sum_{k=m+1}^{n} 1_{w_k=i}}\, (w_1 \dots w_{m-1}\, i\, w_{m+1} \dots w_n) \cdot T_j \\
&= \sum_{m \neq j,j+1} 1_{w_m=i+1}\, z^{1+\sum_{k=m+1}^{n} 1_{w_k=i}}\, (w_1 \dots w_{m-1}\, i\, w_{m+1} \dots w_n) \cdot s_j \\
&\quad + \sum_{m \neq j,j+1} 1_{w_m=i+1}\, z^{\sum_{k=m+1}^{n} 1_{w_k=i}}\, (z-1)\, (w_1 \dots w_{m-1}\, i\, w_{m+1} \dots w_n);
\end{aligned}$$

$$\begin{aligned}
B = e_i \cdot (w \cdot T_j) &= e_i \cdot \big(z\,(w \cdot s_j) + (z-1)\,w\big) \\
&= \sum_{m \neq j,j+1} 1_{w_m=i+1}\, z^{1+\sum_{k=m+1}^{n} 1_{w_k=i}}\, (w_1 \dots w_{m-1}\, i\, w_{m+1} \dots w_n) \cdot s_j \\
&\quad + \sum_{m \neq j,j+1} 1_{w_m=i+1}\, z^{\sum_{k=m+1}^{n} 1_{w_k=i}}\, (z-1)\, (w_1 \dots w_{m-1}\, i\, w_{m+1} \dots w_n).
\end{aligned}$$

So, $A = B$ in this first case. The other cases are treated as follows:

1. If $w_{j+1} = i+1$, then one adds to A

$$z^{\sum_{k=j+2}^{m} 1_{w_k=i}} \big(z\,(w_1 \dots w_{j-1}\, i\, w_j w_{j+2} \dots w_n) \\ + (z-1)(w_1 \dots w_{j-1}\, w_j\, i\, w_{j+2} \dots w_n)\big),$$

   and one adds to B the same quantity.

2. If $w_j = i+1$ and $w_{j+1} \neq i$, then one adds to A

$$z^{\sum_{k=j+2}^{m} 1_{w_k=i}} \big(z\,(w_1 \dots w_{j-1}\, w_{j+1}\, i\, w_{j+2} \dots w_n) \\ + (z-1)(w_1 \dots w_{j-1}\, i\, w_{j+1}\, w_{j+2} \dots w_n)\big),$$

   and one adds to B the same quantity.

3. Suppose finally that $w_j = i+1$ and $w_{j+1} = i$. One adds to A

$$z^{2+\sum_{k=j+2}^{m} 1_{w_k=i}} (w_1 \dots w_{j-1}\, i\, i\, w_{j+2} \dots w_n),$$

   and one adds to B the same quantity.

Thus, $A = B$ in every case, and e_i commutes with T_j. The commutation of f_i and T_j is similar, and for η_i, since w and $w \cdot s_j$ have the same number of entries i, we have trivially

$$\eta_i \cdot (w \cdot T_j) = z^{\sum_{k=1}^{n} 1_{w_k}=i} (z\,(w \cdot s_j) + (z-1)\,w) = (\eta_i \cdot w) \cdot T_j.$$

We leave the reader to check the two other cases $w_j = w_{j+1}$ and $w_j < w_{j+1}$. □

Denote $\mathscr{S}_{n,z}(N,\mathbb{C})$ the image of $\mathrm{GL}_z(N,\mathbb{C})$ in $\mathrm{End}_{\mathbb{C}}((\mathbb{C}^N)^{\otimes n})$; this is the z-analogue of the Schur algebra $\mathscr{S}_n(N,\mathbb{C})$ studied in Section 2.5. We can finally state our duality result, which was the main motivation for the introduction and study of the quantum groups:

Theorem 5.28 (Jimbo). *Let z be a generic parameter in $\mathbb{C}$ (not zero, and not a root of unity). The two algebras $\mathscr{S}_{n,z}(N,\mathbb{C})$ and $\mathfrak{H}_z(n)$ are in duality for their actions on $(\mathbb{C}^N)^{\otimes n}$.*

Before we prove Theorem 5.28, let us complete this result by the representation theoretic consequence of it. Assume to simplify that $N \geq n$. Then, the morphism of algebras $\mathfrak{H}_z(n) \to \mathrm{End}_{\mathbb{C}}((\mathbb{C}^N)^{\otimes n})$ is injective. Indeed, consider a permutation σ, and a reduced decomposition $\sigma = s_{i_1} s_{i_2} \cdots s_{i_\ell}$ of it. By definition, $T_\sigma = T_{i_1} T_{i_2} \cdots T_{i_\ell}$, and moreover, since $\ell(\sigma)$ is the number of inversions of σ, for every $k \in [\![1,l]\!]$,

$$s_{i_1} \cdots s_{i_k}(i_{k+1}) < s_{i_1} \cdots s_{i_k}(i_{k+1}+1),$$

and one adds an inversion by each elementary transposition s_{i_k}. As a consequence, by induction on ℓ, we get that

$$(e_1 \otimes e_2 \otimes \cdots \otimes e_n) \cdot T_\sigma = e_{\sigma(1)} \otimes e_{\sigma(2)} \otimes \cdots \otimes e_{\sigma(n)}.$$

Therefore, the elements T_σ of $\mathfrak{H}_z(n)$ are sent in $\mathrm{End}_{\mathbb{C}}((\mathbb{C}^N)^{\otimes n})$ to a family of independent endomorphisms, since the vectors $e_{\sigma(1)}\otimes e_{\sigma(2)}\otimes\cdots\otimes e_{\sigma(n)}$ are themselves independent in the space of tensors. Thus, $\mathfrak{H}_z(n)$ can be considered as a subalgebra of $\mathrm{End}_{\mathbb{C}}((\mathbb{C}^N)^{\otimes n})$ if $N \geq n$. Now, since z is not a root of unity, the Hecke algebra $\mathfrak{H}_z(n)$ is a semisimple $\mathbb{C}$-algebra (Theorem 4.67). Therefore, by the double commutant theory (Section 1.5), $\mathscr{S}_{n,z}(N,\mathbb{C})$ is also a semisimple algebra, and there exists a decomposition of the space of tensors

$$(\mathbb{C}^N)^{\otimes n} = \bigoplus_{\lambda\in\mathfrak{Y}(n)} V_z^\lambda \otimes_{\mathbb{C}} S_z^\lambda,$$

where the S_z^λ are the (right) Specht modules of the specialized Hecke algebra $\mathfrak{H}_z(n)$, and the V_z^λ form a complete collection of simple modules over $\mathscr{S}_{n,z}(N,\mathbb{C})$. As $\mathscr{S}_{n,z}(N,\mathbb{C})$ is the image of $\mathrm{GL}_z(N,\mathbb{C})$ by a morphism of algebras, the V_z^λ's are also simple modules over $\mathrm{GL}_z(N,\mathbb{C})$, and the following theorem ensures that our notations are compatible with the discussion of Section 5.2.

Theorem 5.29. *Suppose that $N \geq n$. Then, one has the following expansion for the bimodule $(\mathbb{C}^N)^{\otimes n}$ over the pair $(\mathrm{GL}_z(N,\mathbb{C}),\mathfrak{H}_z(n))$:*

$$(\mathbb{C}^N)^{\otimes n} = \bigoplus_{\lambda\in\mathfrak{Y}(n)} V_z^\lambda \otimes_{\mathbb{C}} S_z^\lambda,$$

where the S_z^λ are the (right) Specht modules of $\mathfrak{H}_z(n)$ (as constructed in Section 4.5), and the V_z^λ are the simple modules of highest weight z^λ for the reduced quantum group $\mathrm{GL}_z(N,\mathbb{C})$.

Remark. The theorem can be extended to the case where $N \leq n$; then, the decomposition of the bimodule is the same, but with the direct sum restricted to integer partitions λ with size n and length smaller than N. As our final goal is to compute the characters ch_z^λ of the irreducible modules S_z^λ of the Hecke algebra, this extension will not be needed.

Remark. As the bicommutant theory guarantees the semisimplicity of the commutant of a semisimple algebra, we shall not need to use Theorem 5.25 in order to expand $(\mathbb{C}^N)^{\otimes n}$ as a direct sum of simple $\mathrm{GL}_z(N,\mathbb{C})$-modules.

The proof of Theorem 5.28 splits in two parts. We first identify the commutant of the Hecke algebra $\mathfrak{H}_z(n)$, by giving a linear basis of it (Proposition 5.32). Then, we show that the elements of the basis indeed belong to $\mathscr{S}_{n,z}(N,\mathbb{C})$, so that $(\mathfrak{H}_z(n))' \subset \mathscr{S}_{n,z}(N,\mathbb{C})$. Since we have already shown that $\mathscr{S}_{n,z}(N,\mathbb{C}) \subset (\mathfrak{H}_z(n))'$, this will prove that $(\mathfrak{H}_z(n))' = \mathscr{S}_{n,z}(N,\mathbb{C})$. A basis of the space of tensors consists in the words

$$I = i_1 i_2 \ldots i_n, \qquad (i_1,\ldots,i_N) \in [\![1,N]\!]^n.$$

As in Section 2.5, if $U \in \mathrm{End}((\mathbb{C}^N)^{\otimes n})$, we denote $(U_{I,J})_{I,J}$ its matrix in this basis. Suppose that U commutes with every element $T_k \in \mathfrak{H}_z(n)$. Then, the coefficients $U_{I,J}$ satisfy:

1. If $j_k = j_{k+1}$, then given a multi-index I, the coefficient of I in $U(J \cdot T_k)$ is $z\, U_{I,J}$.
   (a) Suppose $i_k = i_{k+1}$. Then, $[I](U(J) \cdot T_k) = z\, U_{I,J}$, so there is no condition to verify in this case.
   (b) Suppose $i_k < i_{k+1}$. Then, $[I](U(J) \cdot T_k) = z\, U_{I \cdot s_k, J}$, so $U_{I,J} = U_{I \cdot s_k, J}$.
   (c) Finally, if $i_k > i_{k+1}$, then $[I](U(J) \cdot T_k) = U_{I \cdot s_k, J} + (z-1)\, U_{I,J}$, so again $U_{I,J} = U_{I \cdot s_k, J}$.

   Therefore, by reunion of the three cases, if $j_k = j_{k+1}$, then $U_{I,J} = U_{I \cdot s_k, J}$ for any $I \in [\![1,N]\!]^n$.

2. If $j_k < j_{k+1}$, then given a multi-index I, the coefficient of I in $U(J \cdot T_k)$ is $U_{I,J \cdot s_k}$.
   (a) Suppose $i_k = i_{k+1}$. Then, $[I](U(J) \cdot T_k) = z\, U_{I,J}$, so $U_{I,J \cdot s_k} = z\, U_{I,J}$.
   (b) Suppose $i_k < i_{k+1}$. Then, $[I](U(J) \cdot T_k) = z\, U_{I \cdot s_k, J}$, so $U_{I,J \cdot s_k} = z\, U_{I \cdot s_k, J}$.
   (c) Finally, if $i_k > i_{k+1}$, then $[I](U(J) \cdot T_k) = U_{I \cdot s_k, J} + (z-1)\, U_{I,J} = U_{I,J \cdot s_k}$.

   Exchanging J and $J \cdot s_k$, we conclude that if $j_k > j_{k+1}$, then

$$U_{I,J} = \begin{cases} z\, U_{I \cdot s_k, J \cdot s_k} & \text{if } i_k \leq i_{k+1}, \\ U_{I \cdot s_k, J \cdot s_k} + (z-1)\, U_{I, J \cdot s_k} & \text{if } i_k > i_{k+1}. \end{cases}$$

The case when $j_k < j_{k+1}$ leads to the same condition. Thus:

Proposition 5.30. *An element $U = (U_{I,J})_{I,J}$ in $\mathrm{End}((\mathbb{C}^N)^{\otimes n})$ is in the commutant of the Hecke algebra if and only if its matrix elements satisfy*

$$U_{I,J} = \begin{cases} U_{I \cdot s_k, J} & \text{if } j_k = j_{k+1}, \\ z\, U_{I \cdot s_k, J \cdot s_k} & \text{if } j_k > j_{k+1} \text{ and } i_k \leq i_{k+1}, \\ U_{I \cdot s_k, J \cdot s_k} + (z-1)\, U_{I, J \cdot s_k} & \text{if } j_k > j_{k+1} \text{ and } i_k > i_{k+1}. \end{cases}$$

Recall from Section 3.2 that a **two-line array** of length n with entries in $[\![1,N]\!] \times [\![1,N]\!]$ is a pair of sequences $\binom{J}{I} = \binom{j_1,\ldots,j_n}{i_1,\ldots,i_n}$ with the i_k's and the j_k's in $[\![1,N]\!]$, J being weakly increasing, and I satisfying $i_k \leq i_{k+1}$ if $j_k = j_{k+1}$. As in Section 3.2, we denote $\mathfrak{A}(n;N)$ the set of two-line arrays of length n and with entries in $[\![1,N]\!] \times [\![1,N]\!]$. It has cardinality $\binom{N^2+n-1}{n}$.

Lemma 5.31. *There exist polynomials $p_{(I,J);(K,L)}(z) \in \mathbb{C}[z]$ for every two-line array $\binom{J}{I}$ and every $K, L \in [\![1,N]\!]^n$, such that if $U \in (\mathfrak{H}_z(n))'$, then*

$$U_{K,L} = \sum_{\binom{J}{I} \in \mathfrak{A}(n;N)} p_{(I,J);(K,L)}(z)\, U_{I,J}.$$

Proof. We reason by induction on the number of inversions of the sequence L, that is to say the number of pairs (a,b) with $1 \leq a < b \leq n$ and $l_a > l_b$.

- If there is no inversion in L, then L is weakly increasing, and on each interval $[\![a,b]\!]$ on which $c \mapsto l_c$ is constant, one can use elementary transpositions s_c to modify the sequence K without changing the value of $U_{K,L}$ (this is the first condition of Proposition 5.30). Therefore, if L has no inversion, then $U_{K,L} = U_{I,L}$, where I is the unique permutation of K such that $\binom{L}{I}$ is a two-line array.

- Suppose now that the result holds for any sequence L with less than $p-1$ inversions, and consider a sequence L of indices with $p \geq 1$ inversions. We can find a descent a of L, that is to say an index $a \in [\![1,n-1]\!]$ such that $l_a > l_{a+1}$; then, $L \cdot s_a$ has $p-1$ inversions, so there exist polynomials $p_{(I,J);(K,L\cdot s_a)}(z)$ such that
$$U_{K,L\cdot s_a} = \sum_{\binom{J}{I}\in\mathfrak{A}(n;N)} p_{(I,J);(K,L\cdot s_a)}(z)\, U_{I,J}.$$
If $k_a \leq k_{a+1}$, then
$$U_{K,L} = z\, U_{K\cdot s_a,L\cdot s_a} = \sum_{\binom{J}{I}\in\mathfrak{A}(n;N)} \left(z\, p_{(I,J);(K\cdot s_a,L\cdot s_a)}(z)\right) U_{I,J}.$$
Otherwise, if $k_a < k_{a+1}$, then
$$\begin{aligned} U_{K,L} &= U_{K\cdot s_a,L\cdot s_a} + (z-1)\, U_{K,L\cdot s_a} \\ &= \sum_{\binom{J}{I}\in\mathfrak{A}(n;N)} \left(p_{(I,J);(K\cdot s_a,L\cdot s_a)}(z) + (z-1)\, p_{(I,J);(K,L\cdot s_a)}(z)\right) U_{I,J}. \end{aligned}$$
In both cases, the existence of polynomials $p_{(I,J);(K,L)}(z)$ is thus ensured, hence the result. □

Proposition 5.32. *There exists a linear basis $(M_{I,J})_{\binom{J}{I}\in\mathfrak{A}(n;N)}$ of $(\mathfrak{H}_z(n))'$, with the following characterization: for every non-decreasing sequence $L \in [\![1,N]\!]^n$,*
$$M_{I,J}(L) = \begin{cases} \{I\}_J & \text{if } J = L, \\ 0 & \text{otherwise,} \end{cases}$$
where $\{I\}_J$ is the sum of all distinct permutations of I by $\mathfrak{S}(J)$, the symmetry group of the sequence $(j_1,\ldots,j_n)$, that is to say the Young subgroup
$$\mathfrak{S}(n_1) \times \mathfrak{S}(n_2) \times \cdots \times \mathfrak{S}(n_l) \subset \mathfrak{S}(n)$$
such that $a \mapsto j_a$ is constant on every interval $[\![1,n_1]\!]$, $[\![n_1+1,n_1+n_2]\!]$, etc.

Proof. The previous lemma shows that an element $U \in (\mathfrak{H}_z(n))'$ is entirely determined by the coefficients $U_{I,J}$ with $\binom{J}{I} \in \mathfrak{A}(n;N)$. As one can choose these coefficients arbitrarily, a linear basis of the commutant algebra $(\mathfrak{H}_z(n))'$ consists in elements $(M_{I,J})_{\binom{J}{I}}$ such that
$$(M_{I,J})_{K,L} = \begin{cases} 1 & \text{if } \binom{J}{I} = \binom{L}{K}, \\ 0 & \text{for every other two-line array } \binom{L}{K}. \end{cases}$$

Then, if L is a non-decreasing sequence, we saw before that for any sequence K, $(M_{I,J})_{K,L} = (M_{I,J})_{K\cdot\sigma,L}$, where $K\cdot\sigma$ is the reordering of K that is non-decreasing on each interval $[\![a,b]\!]$ on which $c \mapsto l_c$ is constant. So, if L is non-decreasing, then $(M_{I,J})_{K,L} = 0$ if $J \neq L$, and $(M_{I,J})_{K,L} = 1$ if $J = L$ and K is one of the allowed reordering of I, that is to say if $K = I\cdot\sigma$ with $\sigma \in \mathfrak{S}(J)$. This leads to the formula that characterizes the basis $(M_{I,J})_{\binom{J}{I}\in\mathfrak{A}(n;N)}$. □

Example. Suppose $n = 4$ and $N = 4$. Then, a possible two-line array in $\mathfrak{A}(4;4)$ is $\binom{J}{I} = \binom{1,2,2,4}{3,2,3,1}$, and with $L = (1,2,2,4)$, we obtain

$$M_{I,J}(L) = (3,2,3,1) + (3,3,2,1),$$

whereas any other non-decreasing sequence L vanishes under $M_{I,J}$.

In order to prove Theorem 5.28, it suffices now to exhibit elements $M_{I,J} \in \mathscr{S}_{n,z}(N,\mathbb{C})$ that satisfy the characteristic property of Proposition 5.32. The following lemma reduces further the task to accomplish:

Lemma 5.33. *Suppose that there exist elements $N_{I,J} \in \mathscr{S}_{n,z}(N,\mathbb{C})$ labeled by two-line arrays $\binom{J}{I}$, such that $N_{I,J}(J) = \{I\}_J$. Then, there also exist elements $M_{I,J} \in \mathscr{S}_{n,z}(N,\mathbb{C})$ labeled by two-line arrays $\binom{J}{I}$ and that satisfy the condition of Proposition 5.32.*

Proof. The action of $(\eta_1)^{k_1}(\eta_2)^{k_2}\cdots(\eta_N)^{k_N}$ on a non-decreasing word L is

$$(\eta_1)^{k_1}(\eta_2)^{k_2}\cdots(\eta_N)^{k_N}\,L = z^{\sum_{i=1}^N k_i n_i}\,L$$

where $(n_1,\ldots,n_N)$ is the sequence with $n_i = \mathrm{card}\,\{a \in [\![1,n]\!] \,|\, l_a = i\}$. Let $\mathscr{A}_{n;N}$ be the subalgebra of the algebra of functions on

$$C(n;N) = \{(n_1,\ldots,n_N) \in \mathbb{N}^N \,|\, n_1 + \cdots + n_N = n\}$$

that is generated by the functions $(n_1,\ldots,n_N) \mapsto z^{\sum_{i=1}^N k_i n_i}$, with $k_1,\ldots,k_N \in \mathbb{Z}$. We claim that $\mathscr{A}_{n;N}$ is the full algebra of functions on $C(n;N)$. Indeed, fix a sequence $(m_1,\ldots,m_N) \in C(n;N)$, and let us show that the indicator of $(m_1,\ldots,m_N)$ belongs to $\mathscr{A}_{n,N}$. If $(n_1,\ldots,n_N) \in C(n;N)$, then the allowed values for n_i are in $[\![0,N]\!]$, and since all the numbers $z^0, z^1, \ldots, z^N$ are distinct, there exists for each i a polynomial $P_i(y) = \sum_{k=0}^K b_{i,k}\,y^k$ such that $P_i(z^{m_i}) = 1$, and $P(z^m) = 0$ if $m \in [\![0,N]\!] \setminus \{m_i\}$. Then,

$$\begin{aligned}\sum_{k_1,\ldots,k_N} b_{1,k_1} b_{2,k_2}\cdots b_{N,k_N}\, z^{k_1 n_1 + k_2 n_2 + \cdots + k_N n_N} &= \prod_{j=1}^N P_i(z^{n_i})\\ &= \begin{cases} 1 & \text{if } n_i = m_i \text{ for all } i \in [\![1,N]\!],\\ 0 & \text{otherwise.}\end{cases}\end{aligned}$$

As a consequence, there exists for every non-decreasing word J an element T_J in $\mathscr{S}_{n,z}(N,\mathbb{C})$ (and in fact in the commutative algebra generated by the actions of the elements η_i) such that $T_J(J) = J$, and $T_J(L) = 0$ if $L \neq J$ is another non-decreasing word with entries in $[\![1,N]\!]$. Now, under the hypothesis of the lemma, there also exists $N_{I,J} \in \mathscr{S}_{n,z}(N,\mathbb{C})$ such that $N_{I,J}(J) = \{I\}_J$. Therefore, $M_{I,J} = N_{I,J} \circ T_J$ belongs to $\mathscr{S}_{n,z}(N,\mathbb{C})$, and it satisfies the characteristic property of Proposition 5.32. □

Proof of Theorem 5.28. We now explain how to construct the elements $N_{I,J} \in \mathscr{S}_{n,z}(N,\mathbb{C})$ of the previous lemma. We fix a non-decreasing sequence J; notice then that the element T_J can be chosen as $N_{J,J}$, so the case where $I = J$ is done. To deal with the general case, we shall introduce operators on sequences of integers that allow us to go from J to an arbitrary sequence I such that $\binom{J}{I}$ is a two-line array. As J is fixed, the set of allowed sequences I is in bijection with $C(n_1;N)\times C(n_2;N)\times\cdots\times C(n_N;N)$, where $(n_1,\ldots,n_N)\in C(n;N)$ is the sequence associated to the non-decreasing sequence J: n_1 is the number of 1's in J, n_2 the number of 2's, *etc.* In the following, we denote m_i^k the number of elements of I of indices $a \in [\![n_1+\cdots+n_{k-1}+1, n_1+\cdots+n_k]\!]$ such that $i_a = i$. Thus, the sequence I is equal to

$$\left(1^{m_1^1},\ldots,N^{m_N^1},1^{m_1^2},\ldots,N^{m_N^2},\ \ldots,1^{m_1^N},\ldots,N^{m_N^N}\right)$$

with $m_1^k + m_2^k + \cdots + m_N^k = n_k$ for every k. For $i \in [\![1,N-1]\!]$ and $k \in [\![1,N]\!]$, we then define operators $C(n_k;N) \to C(n_k;N)\cup\{\emptyset\}$ as follows:

$$(m_1^k,\ldots,m_N^k)^{k,i+} = \begin{cases} (m_1^k,\ldots,m_i^k+1,m_{i+1}^k-1,\ldots,m_N^k) & \text{if } m_{i+1}^k \geq 1, \\ \emptyset & \text{if } m_{i+1}^k = 0; \end{cases}$$

$$(m_1^k,\ldots,m_N^k)^{k,i-} = \begin{cases} (m_1^k,\ldots,m_i^k-1,m_{i+1}^k+1,\ldots,m_N^k) & \text{if } m_i^k \geq 1, \\ \emptyset & \text{if } m_i^k = 0. \end{cases}$$

These operators can be extended to $C(n_1;N)\times C(n_2;N)\times\cdots\times C(n_N;N)$, which is the set of allowed sequences I (J being fixed). In this setting, one computes readily for any allowed sequence I

$$e_i \cdot (\{I\}_J) = \sum_{k=1}^{N} [m_i^k+1]_z \, z^{m_i^{k+1}+\cdots+m_i^N} \, \{I^{k,i+}\}_J;$$

$$f_i \cdot (\{I\}_J) = \sum_{k=1}^{N} [m_{i+1}^k+1]_z \, z^{m_{i+1}^1+\cdots+m_{i+1}^{k-1}} \, \{I^{k,i-}\}_J$$

where as in the previous chapter, $[k]_z = 1+z+\cdots+z^{k-1}$ is the z-analogue of the integer k. To explain the apparition of these coefficients, it suffices to treat the case when $N = 2$ and J is the constant sequence (say, $J = 1^n$). Indeed, one can convince oneself that the computations can always be reduced to this case, and then, one has to show for any pair (n_1,n_2) such that $n_1+n_2 = n$,

$$e_1 \cdot \{1^{n_1}2^{n_2}\} = [n_1+1]_z \, \{1^{n_1+1}2^{n_2-1}\}$$

if $n_2 \geq 1$, and similarly,

$$f_1 \cdot \{1^{n_1}2^{n_2}\} = [n_2+1]_z \{1^{n_1-1}2^{n_2+1}\}$$

if $n_1 \geq 1$ (in these formulas, $\{w\}$ denotes the set of all distinct permutations of the word w). Let us prove for instance the first identity. On each side of the formula, the number of terms $z^k\, w$ with $k \in \mathbb{N}$ and w word in $\{1,2\}^n$ is $\frac{n!}{(n_1)!\,(n_2-1)!}$. However, every word with n_1+1 entries 1 and n_2-1 entries 2 can be obtained in n_1+1 ways from a word with n_1 entries 1 and n_2 entries 2 by changing a 2 into a 1. Moreover, each way yields a different factor z^k, where $k \in [\![0,n_1]\!]$ is the number of entries 1 after the position where the switch is made; hence the formula.

Given a non-decreasing sequence $I_k = (1^{m_1^k}, 2^{m_2^k}, \ldots, N^{m_N^k})$ of size n_k, one can always obtain it from (k^{n_k}) by applying operators $(I_k) \mapsto (I_k)^{k,i+}$ or $(I_k) \mapsto (I_k)^{k,i-}$ a certain number of times d_k, with

$$d_k = \sum_{j=1}^{k-1}(k-j)\, m_j^k + \sum_{j=k+1}^{N}(j-k)\, m_j^k.$$

If I is an allowed sequence, we set $d(I) = (d_1, d_2, \ldots, d_N)$, where d_k is the number of operations $I \mapsto I^{k,i+}$ or $I \mapsto I^{k,i-}$ needed to transform (k^{n_k}) into the part I_k of I that is non-decreasing and corresponds to the n_k indices between $n_1+\cdots+n_{k-1}+1$ and $n_1+\cdots+n_k$. We denote $D(L)$ the set of all sequences $d(I)$ with I allowed sequence with respect to J; it is a product of intervals

$$D(L) = \prod_{i=1}^{N} [\![0, D_i]\!],$$

which can be endowed with the lexicographic order. On the other hand, notice that

$$d(I^{k,i+}) = \begin{cases} (d_1,\ldots,d_k+1,\ldots,d_N) & \text{if } k > i, \\ (d_1,\ldots,d_k-1,\ldots,d_N) & \text{if } k \leq i. \end{cases}$$

Similarly,

$$d(I^{k,i-}) = \begin{cases} (d_1,\ldots,d_k-1,\ldots,d_N) & \text{if } k \geq i, \\ (d_1,\ldots,d_k+1,\ldots,d_N) & \text{if } k < i. \end{cases}$$

We now prove by induction on $d(I) \in D(L)$ that for every I allowed sequence, there exists $N_{I,J} \in \mathscr{S}_{n,z}(N,\mathbb{C})$ such that $N_{I,J}(J) = \{I\}_J$. If $d(I) = (0,0,\ldots,0)$, then $I = J$ and we already treated this case. We suppose now that the result is true for any I'' with $d(I'') \leq (d_1,\ldots,d_N)$, and we consider an allowed sequence I' such that $d(I')$ is the direct successor of $(d_1,\ldots,d_N)$ with respect to the lexicographic order on $D(L)$.

1. If $d_N < D_N$, then
$$d(I') = (d_1,\ldots,d_{N-1},d_N+1).$$

Set $I = (I')^{N,(N-1)-}$. We have $d(I) = (d_1, \dots, d_N)$, so by the induction hypothesis, there exists $N_{I,J}$ in $\mathscr{S}_{n,z}(N, \mathbb{C})$ such that $N_{I,J}(J) = \{I\}_J$. Applying e_{N-1} to the sequence I yields

$$e_{N-1} \cdot \{I'\}_J$$
$$= [m^N_{N-1}(I) + 1]_z \, \{I\}_J + \sum_{k=1}^{N-1} [m^k_{N-1}(I) + 1]_z \, z^{m^{k+1}_{N-1}(I) + \dots + m^N_{N-1}(I)} \, \{I^{k,(N-1)+}\}_J.$$

By using the transform rules for the sequences $d(I)$ described before, we see that all the terms $I^{k,(N-1)+}$ with $k \leq N-1$ satisfy $d(I^{k,(N-1)+}) < d(I)$, so there exist elements $N_{I^{k,(N-1)+},J}$ in $\mathscr{S}_{n,z}(N, \mathbb{C})$ such that $N_{I^{k,(N-1)+},J}(J) = \{I^{k,(N-1)+}\}_J$. So, $\{I'\}_J$ is the image of J by

$$\sum_{k=1}^{N-1} \frac{[m^k_{N-1}(I) + 1]_z}{[m^N_{N-1}(I) + 1]_z} \, z^{m^{k+1}_{N-1}(I) + \dots + m^N_{N-1}(I)} \, N_{I^{k,(N-1)+},J} - \frac{1}{[m^N_{N-1}(I) + 1]_z} e_{N-1} N_{I,J},$$

hence the existence of $N_{I',J} \in \mathscr{S}_{n,z}(N, \mathbb{C})$ with the desired property.

2. Suppose now that $d_N = D_N$, and in fact that

$$(d_1, \dots, d_N) = (d_1, \dots, d_k, D_{k+1}, \dots, D_N),$$

with $d_k < D_k$ for some $k \in [\![1, N-1]\!]$. We then have

$$d(I') = (d_1, \dots, d_{k-1}, d_k + 1, 0, \dots, 0),$$

so $I' = (I_1, \dots, I_k, ((k+1)^{n_{k+1}}), \dots, (N^{n_N}))$. We set $I = (I')^{k,k+}$; we have

$$d(I) = (d_1, \dots, d_{k-1}, d_k, 0, \dots, 0) \leq (d_1, \dots, d_N).$$

By the induction hypothesis, there exists an element $N_{I,J} \in \mathscr{S}_{n,z}(N, \mathbb{C})$ with $N_{I,J}(J) = \{I\}_J$. Then,

$$f_k \cdot \{I\}_J = [m^k_{k+1}(I) + 1]_z \, z^{m^1_{k+1}(I) + \dots + m^{k-1}_{k+1}(I)} \, \{I'\}_J$$
$$+ \sum_{l=1}^{k-1} [m^l_{k+1}(I) + 1]_z \, z^{m^1_{k+1}(I) + \dots + m^{l-1}_{k+1}(I)} \, \{I^{l,k-}\}_J$$

and one gets the same conclusion as before. □

We now deal with the representation theoretic counterpart of Theorem 5.28. By the theory of bicommutants, for $N \geq n$, there exist simple modules U^λ_z over $\mathscr{S}_{n,z}(N, \mathbb{C})$, and therefore over $\mathrm{GL}_z(N, \mathbb{C})$, such that

$$(\mathbb{C}^N)^{\otimes n} = \bigoplus_{\lambda \in \mathfrak{Y}(n)} U^\lambda_z \otimes S^\lambda_z$$

as a bimodule for the pair $(\mathrm{GL}_z(N, \mathbb{C}), \mathfrak{H}_z(n))$. By the discussion of Section 5.2, each simple module U^λ_z is entirely determined by its highest weight νz^ω. We want to prove that this highest weight is actually z^λ, so that $U^\lambda_z = V^\lambda_z$ with the notations previously introduced. A partial result in this direction is:

Lemma 5.34. *Suppose that μ is a weight of the $\mathrm{GL}_z(N,\mathbb{C})$-module $V = (\mathbb{C}^N)^{\otimes n}$. Then, $\mu = (z^{m_1},\ldots,z^{m_N})$, where all the m_i's are non-negative integers, and $\sum_{i=1}^N m_i = n$. If μ is a highest weight, then $m_1 \geq m_2 \geq \cdots \geq m_N \geq 0$.*

Proof. Each word w is a weight vector for the action of $\mathrm{GL}_z(N,\mathbb{C})$, with

$$\eta_i \cdot w = z^{n_i(w)}\, w,$$

where $n_i(w)$ is the number of entries i in w. Thus, we have an explicit basis of weight vectors of V, and each of these words gives a weight $\mu = (z^{m_1},\ldots,z^{m_N})$ with the m_i's non-negative integers of sum n. Moreover, if μ is a highest weight, then $\mu = \nu z^{\omega}$ with $\omega \in X_+$, which implies that $m_1 \geq m_2 \geq \cdots \geq m_N \geq 0$. □

As a consequence, the set of allowed highest weights of simple $\mathrm{GL}_z(N,\mathbb{C})$-submodules of $(\mathbb{C}^N)^{\otimes n}$ is $\{z^\lambda,\ \lambda \in \mathfrak{Y}(n)\}$. We conclude that for every z generic, there exists a permutation $\rho_z \in \mathfrak{S}(\mathfrak{Y}(n))$ such that

$$(\mathbb{C}^N)^{\otimes n} = \bigoplus_{\lambda \in \mathfrak{Y}(n)} V_z^{\rho_z(\lambda)} \otimes S_z^\lambda.$$

We postpone to Section 5.5 the proof of the fact that

$$\rho_z(\lambda) = \lambda$$

for any z and any integer partition λ; it will lead immediately to Theorem 5.29.

5.4 Iwahori–Hecke duality

In the previous section, we established the complete duality between $\mathrm{GL}_z(N,\mathbb{C})$ and $\mathfrak{H}_z(n)$ for their actions on the space of tensors $(\mathbb{C}^N)^{\otimes n}$. The specialized Hecke algebras $\mathfrak{H}_z(n)$ appear in another duality result of this kind when $z = q = p^e$ is a positive power of a prime number p. Fix such a prime power q, and consider the general linear group $\mathrm{GL}(n,\mathbb{F}_q)$, which consists in invertible square matrices of size $n \times n$ and with coefficients in the finite field $\mathbb{F}_q$ of cardinality q. This group admits for subgroup

$$\mathrm{B}(n,\mathbb{F}_q) = \{M \in \mathrm{GL}(n,\mathbb{F}_q) \,|\, M \text{ is upper-triangular}\}.$$

We already computed in Chapter 1 the cardinality of $\mathrm{GL}(n,\mathbb{F}_q)$: it is the number of linear basis of the $\mathbb{F}_q$-vector space $(\mathbb{F}_q)^n$, and it is given by the formula

$$\operatorname{card} \mathrm{GL}(n,\mathbb{F}_q) = (q^n-1)(q^n-q)(q^n-q^2)\cdots(q^n-q^{n-1}).$$

On the other hand, the Borel subgroup $\mathrm{B}(n,\mathbb{F}_q)$ has for cardinality

$$\operatorname{card} \mathrm{B}(n,\mathbb{F}_q) = (q-1)^n\, q^{\frac{n(n-1)}{2}}.$$

Indeed, to choose a matrix in $\mathrm{B}(n,\mathbb{F}_q)$, there are $q-1$ possible choices for each of the n diagonal coefficients (all the non-zero elements of $\mathbb{F}_q$), and then the $\frac{n(n-1)}{2}$ remaining upper-triangular coefficients to choose are arbitrary in $\mathbb{F}_q$. As a consequence, the space of cosets $\mathrm{GL}(n,\mathbb{F}_q)/\mathrm{B}(n,\mathbb{F}_q)$ admits for cardinality

$$\mathrm{card}\,\big(\mathrm{GL}(n,\mathbb{F}_q)/\mathrm{B}(n,\mathbb{F}_q)\big) = \prod_{i=1}^{n} \frac{q^i-1}{q-1} = [n]_q!.$$

The main result of this section is:

Theorem 5.35 (Iwahori). *Consider the action of the group* $\mathrm{GL}(n,\mathbb{F}_q)$ *on the left of the module* $\mathbb{C}[\mathrm{GL}(n,\mathbb{F}_q)/\mathrm{B}(n,\mathbb{F}_q)]$, *which is the space of right-*$\mathrm{B}(n,\mathbb{F}_q)$*-invariant functions on* $\mathrm{GL}(n,\mathbb{F}_q)$. *The commutant algebra of the action of* $\mathrm{GL}(n,\mathbb{F}_q)$ *on this module is isomorphic to the specialized Hecke algebra* $\mathfrak{H}_q(n)$.

The proof of Theorem 5.35 consists in two parts. By a discussion of Section 1.5, if $G=\mathrm{GL}(n,\mathbb{F}_q)$ and $B=\mathrm{B}(n,\mathbb{F}_q)$, then the ideal of cosets $\mathbb{C}[G/B]\subset\mathbb{C}[G]$ admits for commutant of the action of $\mathbb{C}[G]$ the algebra of double cosets $\mathbb{C}[B\backslash G/B]$. In this setting:

1. A classical result due to Bruhat will allow us to construct a basis $(T_\sigma)_{\sigma\in\mathfrak{S}(n)}$ of this algebra of double cosets; see Theorem 5.38.

2. Then, a sequence of calculations will show that the T_σ's satisfy the relations of the Hecke algebra $\mathfrak{H}_q(n)$.

▷ *Bruhat decomposition and the flag variety of* $\mathrm{GL}(n,\mathbb{F}_q)$.

To study the $\mathrm{GL}(n,\mathbb{F}_q)$-module $\mathbb{C}[\mathrm{GL}(n,\mathbb{F}_q)/\mathrm{B}(n,\mathbb{F}_q)]$, it will be convenient to have a combinatorial description of the set of cosets $\mathrm{GL}(n,\mathbb{F}_q)/\mathrm{B}(n,\mathbb{F}_q)$. Call **complete flag** of $(\mathbb{F}_q)^n$ a strictly increasing family

$$F = (\{0\} \subsetneq F_1 \subsetneq F_2 \subsetneq \cdots \subsetneq F_n)$$

of vector subspaces of $(\mathbb{F}_q)^n$. For dimension reasons, one has necessarily $\dim F_i = i$ for every i if F is a complete flag.

Lemma 5.36. *The elements of* $\mathrm{GL}(n,\mathbb{F}_q)/\mathrm{B}(n,\mathbb{F}_q)$ *correspond bijectively to complete flags of* $(\mathbb{F}_q)^n$.

Proof. We see an invertible matrix in $\mathrm{GL}(n,\mathbb{F}_q)$ as a family of vectors $(e_1,\ldots,e_n)$ whose coordinates are written in columns. Then, if $(e_1,\ldots,e_n)$ and $(f_1,\ldots,f_n)$ are two linear bases with matrices M and N, if $M=NT$ with $T\in\mathrm{B}(n,\mathbb{F}_q)$, we have

$$(e_1,e_2,\ldots,e_n) = (T_{11}f_1, T_{12}f_1+T_{22}f_2, T_{13}f_1+T_{23}f_2+T_{33}f_3,\ldots,T_{1n}f_1+\cdots+T_{nn}f_n).$$

It follows that the flags

$$\begin{aligned}\mathrm{Flag}(e_1,e_2,\ldots,e_n) &= (\mathrm{Span}(e_1),\mathrm{Span}(e_1,e_2),\ldots,\mathrm{Span}(e_1,e_2,\ldots,e_n))\\ \mathrm{Flag}(f_1,f_2,\ldots,f_n) &= (\mathrm{Span}(f_1),\mathrm{Span}(f_1,f_2),\ldots,\mathrm{Span}(f_1,f_2,\ldots,f_n))\end{aligned}$$

are the same. Conversely, given two bases $(e_1,\ldots,e_n)$ and $(f_1,\ldots,f_n)$ with the same associated complete flag, one constructs readily a matrix T that is upper triangular and such that $M = NT$ if M and N are the two matrices of the two bases. We conclude that a coset $M\,\mathrm{B}(n,\mathbb{F}_q)$ of matrices corresponds to a unique flag F, the correspondence being $M = \mathrm{mat}(e_1,\ldots,e_n) \mapsto F = \mathrm{Flag}(e_1,\ldots,e_n)$. □

In the sequel, we denote $\mathrm{Flag}(n,\mathbb{F}_q) = \mathrm{GL}(n,\mathbb{F}_q)/\mathrm{B}(n,\mathbb{F}_q)$ the set of all complete flags of $(\mathbb{F}_q)^n$, also called the **flag variety**. The module $\mathbb{C}[G/B]$ can then be reinterpreted as the set of all formal linear combinations of flags $F = (F_1,\ldots,F_n)$, the action of an isomorphism $u \in \mathrm{GL}(n,\mathbb{F}_q)$ on a flag being

$$u\cdot(F_1 \subsetneq F_2 \subsetneq \cdots \subsetneq F_n) = (u(F_1) \subsetneq u(F_2) \subsetneq \cdots \subsetneq u(F_n)).$$

The main combinatorial result regarding this action is:

Proposition 5.37. *Let $E = (E_1 \subsetneq E_2 \subsetneq \cdots \subsetneq E_n)$ and $F = (F_1 \subsetneq F_2 \subsetneq \cdots \subsetneq F_n)$ be two complete flags in $\mathrm{Flag}(n,\mathbb{F}_q)$. There exists a basis $(e_1,\ldots,e_n)$ of $(\mathbb{F}_q)^n$, and a unique permutation $\sigma \in \mathfrak{S}(n)$, such that*

$$E = \mathrm{Flag}(e_1,\ldots,e_n) \quad \text{and} \quad F = \mathrm{Flag}(e_{\sigma(1)},\ldots,e_{\sigma(n)}).$$

Proof. We prove the existence by induction on n, the case $n = 1$ being trivial. Let e_1 be a basis vector of the one-dimensional vector space E_1. We define $i = \min\{k \in [\![1,n]\!] \,|\, e_1 \in F_k\}$, and we denote

$$\pi : (\mathbb{F}_q)^n \to ((\mathbb{F}_q)^n)/(\mathbb{F}_q\, e_1)$$

the projection from $(\mathbb{F}_q)^n$ to its quotient by the vector line spanned by e_1. Notice then that

$$(\pi(E_2),\ldots,\pi(E_n))$$

and

$$(\pi(F_1),\ldots,\pi(F_{i-1}),\pi(F_{i+1}),\ldots,\pi(F_n))$$

are complete flags of the $(n-1)$-dimensional space $\pi((\mathbb{F}_q)^n)$. Therefore, there exists a basis $(\widetilde{e}_2,\ldots,\widetilde{e}_n)$, and a bijection $\tau : [\![2,n]\!] \to [\![2,n]\!]$ such that

$$\begin{aligned}\mathrm{Flag}(\widetilde{e}_2,\ldots,\widetilde{e}_n) &= (\pi(E_2),\ldots,\pi(E_n));\\ \mathrm{Flag}(\widetilde{e}_{\tau(2)},\ldots,\widetilde{e}_{\tau(n)}) &= (\pi(F_1),\ldots,\pi(F_{i-1}),\pi(F_{i+1}),\ldots,\pi(F_n)).\end{aligned}$$

Now, if $e_2,\ldots,e_n$ are preimages in $(\mathbb{F}_q)^n$ of $\widetilde{e}_2,\ldots,\widetilde{e}_n$ in $\pi((\mathbb{F}_q)^n)$, then it is immediate that $\mathrm{Flag}(e_1,\ldots,e_n) = (E_1,\ldots,E_n)$. We set

$$\begin{aligned}&\sigma(1) = \tau(2),\ \sigma(2) = \tau(3),\ldots,\ \sigma(i-1) = \tau(i),\ \sigma(i) = 1,\\ &\sigma(i+1) = \tau(i+1),\ldots,\ \sigma(n) = \tau(n).\end{aligned}$$

Then, $\mathrm{Flag}(e_{\sigma(1)},\ldots,e_{\sigma(n)}) = (F_1,\ldots,F_n)$, hence the existence of a linear basis $(e_1,\ldots,e_n)$ and of a compatible permutation $\sigma \in \mathfrak{S}(n)$ for any pair of complete flags (E,F) in $\mathrm{Flag}(n,\mathbb{F}_q)$.

For the unicity of the permutation, suppose that there exist two bases $(e_1,\ldots,e_n)$ and $(f_1,\ldots,f_n)$, and two permutations σ and τ, such that

$$E = \mathrm{Flag}(e_1,\ldots,e_n) = \mathrm{Flag}(f_1,\ldots,f_n);$$
$$F = \mathrm{Flag}(e_{\sigma(1)},\ldots,e_{\sigma(n)}) = \mathrm{Flag}(f_{\tau(1)},\ldots,f_{\tau(n)}).$$

Since $\mathrm{Flag}(n,\mathbb{F}_q) = \mathrm{GL}(n,\mathbb{F}_q)/\mathrm{B}(n,\mathbb{F}_q)$, there exist two upper-triangular matrices S and T such that

$$\mathrm{mat}(e_1,\ldots,e_n) = \mathrm{mat}(f_1,\ldots,f_n) \times S;$$
$$\mathrm{mat}(e_{\tau(1)},\ldots,e_{\tau(n)}) = \mathrm{mat}(f_{\tau(1)},\ldots,f_{\tau(n)}) \times T.$$

Therefore, if P_σ and P_τ are the permutation matrices of σ and τ, then

$$SP_\sigma T^{-1} = P_\tau.$$

Notice now that a permutation matrix $P = P_\tau$ is entirely determined by the ranks of all its upper left submatrices, and that these ranks are invariant by multiplication on the left or on the right by triangular matrices. Therefore, $P_\sigma = P_\tau$. □

A reformulation of the previous proposition is the celebrated:

Theorem 5.38 (Bruhat's decomposition). *Every* $\mathrm{B}(n,\mathbb{F}_q)$*-double coset belonging to* $\mathrm{B}(n,\mathbb{F}_q)\backslash\mathrm{GL}(n,\mathbb{F}_q)/\mathrm{B}(n,\mathbb{F}_q)$ *contains a unique permutation matrix. Therefore,*

$$\mathrm{GL}(n,\mathbb{F}_q) = \bigsqcup_{\sigma\in\mathfrak{S}(n)} \mathrm{B}(n,\mathbb{F}_q)\,\sigma\,\mathrm{B}(n,\mathbb{F}_q).$$

Proof. Let M be an arbitrary invertible matrix in $\mathrm{GL}(n,\mathbb{F}_q)$, which sends the canonical basis $(e_1,\ldots,e_n)$ to a basis $(f_1,\ldots,f_n)$. If E and F are the two flags associated to these bases, then there exists a basis $(g_1,\ldots,g_n)$ and a permutation σ such that

$$E = \mathrm{Flag}(g_1,\ldots,g_n) \quad ; \quad F = \mathrm{Flag}(g_{\sigma(1)},\ldots,g_{\sigma(n)}).$$

As $(e_1,\ldots,e_n)$ and $(g_1,\ldots,g_n)$ correspond to the same flag E, they differ by an upper-triangular matrix S. Similarly, as $(g_{\sigma(1)},\ldots,g_{\sigma(n)})$ and $(f_1,\ldots,f_n)$ correspond to the same flag F, they differ by an upper-triangular matrix T. Then, $M = SP_\sigma T$. So, we have shown that every matrix is in the same double coset as a permutation matrix, and the unicity comes from the same argument as in the proof of the previous proposition. □

Remark. Since we never used the finiteness of the defining field $\mathbb{F}_q$, the Bruhat decomposition holds in fact for any general linear group $\mathrm{GL}(n,k)$.

▷ *Iwahori–Hecke duality and computation of the generic degrees.*

As a consequence of Theorem 5.38, a linear basis of the algebra $\mathbb{C}[B\backslash G/B]$ consists in elements

$$T_\sigma = \frac{1}{\operatorname{card} \mathrm{B}(n,\mathbb{F}_q)} \left(\mathrm{B}(n,\mathbb{F}_q)\,\sigma\,\mathrm{B}(n,\mathbb{F}_q)\right) = \frac{1}{(q-1)^n\, q^{\binom{n}{2}}} \sum_{\substack{M=T_1 P_\sigma T_2 \\ T_1,T_2\in \mathrm{B}(n,\mathbb{F}_q)}} M.$$

In particular, since the algebra of double cosets $\mathbb{C}[B\backslash G/B]$ is the commutant of $\mathbb{C}[G]$ for the module $\mathbb{C}[G/B]=\mathbb{C}[\mathrm{Flag}(n,\mathbb{F}_q)]$, the dimension of this commutant is $n!$. Notice on the other hand that $T_{\mathrm{id}_{[\![1,n]\!]}}$ is the unit element of the algebra of double cosets.

Proposition 5.39. *The elements T_σ satisfy the relations*

$$T_s T_\sigma = \begin{cases} T_{s\sigma} & \text{if } \ell(s\sigma) > \ell(\sigma), \\ q\,T_{s\sigma} + (q-1)\,T_\sigma & \text{if } \ell(s\sigma) < \ell(\sigma) \end{cases}$$

for any elementary transposition s.

Lemma 5.40. *Let $(A_i)_{i\in[\![1,n-1]\!]}$ be a family of subsets of $\mathfrak{S}(n)$ that satisfies the three following assertions:*

1. $\mathrm{id}_{[\![1,n]\!]} \in A_i$ *for any* $i \in [\![1,n-1]\!]$.
2. A_i *and* $s_i A_i$ *are disjoint for any* $i \in [\![1,n-1]\!]$.
3. *for any* i,j *and any* $\sigma \in \mathfrak{S}(n)$, *if* $\sigma \in A_i$ *and* $\sigma s_j \notin A_i$, *then* $s_i\sigma = \sigma s_j$.

Then, $A_i = \{\sigma \in \mathfrak{S}(n) \,|\, \ell(s_i\sigma) > \ell(\sigma)\}$ *for any* i.

Proof. Fix an elementary transposition s_i and a permutation σ, with reduced decomposition $\sigma = s_{j_1}s_{j_2}\cdots s_{j_\ell}$.

- Suppose that $\sigma \notin A_i$. If $\sigma_k = s_{j_1}s_{j_2}\cdots s_{j_k}$, then $\sigma_0 \in A_i$ and $\sigma_\ell = \sigma \notin A_i$, so there exists some index k such that $\sigma_k = \tau \in A_i$ and $\sigma_{k+1} = \tau\, s_{j_{k+1}} \notin A_i$. Therefore, setting $s_j = s_{j_{k+1}}$, τ and τs_j satisfy

$$\tau \in A_i \qquad ; \qquad \tau s_j \notin A_i,$$

so $s_i\tau = \tau s_j$ by assumption. Then,

$$s_i\sigma = s_i\,\tau\, s_j\, s_{j_{k+2}}\cdots s_{j_\ell} = \tau\, s_{j_{k+2}}\cdots s_{j_\ell} = s_{j_1}\cdots s_{j_k}\, s_{j_{k+2}}\cdots s_{j_\ell}$$

so $\ell(s_i\sigma) = \ell - 1 < \ell(\sigma)$.

- Suppose that $\sigma \in A_i$. Then, $s_i\sigma \notin A_i$ since $A_i \cap s_iA_i = \emptyset$, so by the previous discussion applied to $\sigma' = s_i\,\sigma$, $\ell(\sigma) < \ell(s_i\,\sigma)$. ◻

Lemma 5.41. *The product of double cosets $T_s T_\sigma$ is always equal to a linear combination $a_{s\sigma} T_{s\sigma} + a_\sigma T_\sigma$, with $a_{s\sigma}, a_\sigma \in \mathbb{Q}$.*

Proof. Since $(T_\sigma)_{\sigma \in \mathfrak{S}(n)}$ is a linear basis of $\mathbb{C}[B\backslash G/B]$, there exist coefficients $a_\tau \in \mathbb{C}$ such that $T_s T_\sigma = \sum_{\tau \in \mathfrak{S}(n)} a_\tau T_\tau$. Moreover, for any permutation τ,

$$\begin{aligned} a_\tau &= (q-1)^n q^{\binom{n}{2}} [\tau](T_s T_\sigma) \\ &= \frac{1}{(q-1)^n q^{\binom{n}{2}}} \operatorname{card}\{(\tau_1, \tau_2) \in BsB \times B\sigma B \mid \tau_1 \tau_2 = \tau\} \\ &= \frac{1}{(q-1)^n q^{\binom{n}{2}}} \operatorname{card}\{(BsB\tau) \cap (B\sigma B)\}. \end{aligned}$$

This formula leads us to consider sets $sB\tau$, where $s = s_i$ is an elementary transposition and τ is an arbitrary permutation. We claim that

$$s_i B\tau \subset B\tau B \sqcup Bs_i\tau B$$

for any permutation τ; equivalently,

$$s_i B \subset BB' \sqcup Bs_i B',$$

where $B' = \tau B \tau^{-1}$. This claim will imply that $a_\tau = 0$ unless $\tau \in \{s\sigma, \sigma\}$, hence the lemma.

If $(e_1, \ldots, e_n)$ is the canonical basis of $(\mathbb{F}_q)^n$, we denote G_i the subgroup of $\mathrm{GL}(n, \mathbb{F}_q)$ that stabilizes the plane spanned by the vectors e_i and e_{i+1}, and that fixes the other vectors $e_{j \neq i, i+1}$; it is isomorphic to $\mathrm{GL}(2, \mathbb{F}_q)$. We have $G_i B = BG_i$, both sides of the formula corresponding to the set of block upper-triangular matrices with blocks of sizes determined by the sequence

$$(\underbrace{1, 1, \ldots, 1}_{i-1 \text{ terms}}, 2, \underbrace{1, \ldots, 1}_{n-i-1 \text{ terms}}).$$

Therefore, $s_i B \subset G_i B = BG_i$, and it suffices to prove that

$$G_i \subset (B \cap G_i)(B' \cap G_i) \cup (B \cap G_i) s_i (B' \cap G_i).$$

However, $B \cap G_i$ is isomorphic to $\mathrm{B}(2, \mathbb{F}_q)$, and is the set of block diagonal matrices

$$\operatorname{diag}(1^{i-1}, b, 1^{n-i-1})$$

with $b \in \mathrm{B}(2, \mathbb{F}_q)$. Similarly, the subgroup $B' \cap G_i$ is:

- either the same subgroup isomorphic to $\mathrm{B}(2, \mathbb{F}_q)$ if $\tau^{-1}(i) < \tau^{-1}(i+1)$;
- or, if $\tau^{-1}(i) > \tau^{-1}(i+1)$, the group of block diagonal matrices

  $$\operatorname{diag}(1^{i-1}, c, 1^{n-i-1})$$

  with $c \in \mathrm{B}^-(2, \mathbb{F}_q)$, the group of *lower* triangular matrices of size 2.

In the first case, we thus have to prove that

$$\mathrm{GL}(2,\mathbb{F}_q) = \mathrm{B}(2,\mathbb{F}_q) \cup \big(\mathrm{B}(2,\mathbb{F}_q)\, s\, \mathrm{B}(2,\mathbb{F}_q)\big),$$

but this is the case $n = 2$ of the Bruhat decomposition. In the second case, we have to prove that

$$\mathrm{GL}(2,\mathbb{F}_q) = \big(\mathrm{B}(2,\mathbb{F}_q)\mathrm{B}^-(2,\mathbb{F}_q)\big) \cup \big(\mathrm{B}(2,\mathbb{F}_q)\, s\, \mathrm{B}^-(2,\mathbb{F}_q)\big),$$

but since $\mathrm{B}^-(2,\mathbb{F}_q) = s\,\mathrm{B}(2,\mathbb{F}_q)\,s$, this follows actually from the first case by multiplication by $s = \left(\begin{smallmatrix} 0 & 1 \\ 1 & 0 \end{smallmatrix}\right)$. So, the claim is shown. □

Lemma 5.42. *For any* $i \in [\![1,n-1]\!]$, $(T_{s_i})^2 = (q-1)\,T_{s_i} + q\,T_{\mathrm{id}_{[\![1,n]\!]}}$.

Proof. The previous lemma shows that $(T_{s_i})^2$ is a linear combination of T_{s_i} and $T_{\mathrm{id}_{[\![1,n]\!]}}$. Notice that the matrices that appear in $T_{\mathrm{id}_{[\![1,n]\!]}}$ and T_{s_i} are all block upper-triangular matrices, with blocks of sizes determined by the sequence $(1,\ldots,1,2,1,\ldots,1)$. Denote P the subgroup of $\mathrm{GL}(n,\mathbb{F}_q)$ that consists in such block upper-triangular matrices. As a formal sum of matrices,

$$P = (q-1)^n\, q^{\binom{n}{2}}\, (T_{s_i} + T_1);$$

the matrices in $T_{\mathrm{id}_{[\![1,n]\!]}}$ are those that are upper-triangular, and the matrices that appear in T_{s_i} are those that are not upper-triangular, but are block upper-triangular. Since P is a subgroup of $G = \mathrm{GL}(n,\mathbb{F}_q)$,

$$P^2 = (\mathrm{card}\, P)\, P = (q-1)^{n-1}\, q^{\binom{n}{2}}\, (q^2-1)\, P.$$

Therefore, $(T_{s_i} + T_{\mathrm{id}_{[\![1,n]\!]}})^2 = (q+1)(T_{s_i} + T_{\mathrm{id}_{[\![1,n]\!]}})$, and one obtains the desired identity by expansion of the square. □

Proof of Proposition 5.39. In the following, $C(\sigma) = B\sigma B$ is the double coset of a permutation σ, considered as a subset of $\mathrm{GL}(n,\mathbb{F}_q)$. For any $i \in [\![1,n-1]\!]$, denote A_i the set of permutations σ such that $C(s_i)\,C(\sigma) = C(s_i\sigma)$ (this is an identity as subsets, and not as formal sums in the algebra $\mathbb{C}[B\backslash G/B]$). We verify that the family $(A_i)_{i\in[\![1,n-1]\!]}$ satisfies the three conditions of Lemma 5.40.

1. Obviously, $C(s_i)\,C(\mathrm{id}_{[\![1,n]\!]}) = C(s_i)\,B = C(s_i)$, so $\mathrm{id}_{[\![1,n]\!]} \in A_i$ for any i.

2. Suppose that σ belongs to A_i and to s_iA_i. Then,

$$C(s_i)\,C(s_i)\,C(\sigma) = C(s_i)\,C(s_i\sigma) = C(\sigma)$$

but on the other hand $C(s_i)\,C(s_i) = C(s_i) \sqcup C(\mathrm{id}_{[\![1,n]\!]})$, this following from the previous lemma. Therefore,

$$C(\sigma) = C(s_i)\,C(\sigma) \cup C(\mathrm{id}_{[\![1,n]\!]})\,C(\sigma) = C(s_i\sigma) \cup C(\sigma),$$

which is not possible since the double cosets are disjoint (cf. Theorem 5.38).

3. Suppose finally that $\sigma \in A_i$ and $\sigma s_j \notin A_i$. Then, $C(s_i)\,C(\sigma) = C(s_i\sigma)$, and on the other hand, if $\tau = \sigma s_j$, then $C(s_i)\,C(\sigma s_j)$ is a reunion of double cosets

$$C(s_i)\,C(\tau) = \begin{cases} & C(s_i\tau) \\ \text{or} & C(s_i\tau) \sqcup C(\tau). \end{cases}$$

Since $\tau = \sigma s_j \notin A_i$, the first case is excluded, so $C(s_i)\,C(\tau) = C(s_i\tau) \sqcup C(\tau)$, and in particular

$$C(\tau) \subset C(s_i)\,C(\tau).$$

In this identity, the left-hand side $C(\tau)$ is a union of cosets gB, and the right-hand side can be rewritten as $C(s_i)\tau B$, so $C(s_i)\tau$ has a non-empty intersection with $C(\tau)$. However,

$$C(s_i\sigma) = C(s_i)\,C(\sigma) = C(s_i)\,\tau\, s_j B$$

so $C(s_i\sigma)$ has a non-empty intersection with $C(\tau)s_j B \subset C(\tau) \cup C(\sigma)$. As double cosets form a partition of $\mathrm{GL}(n, \mathbb{F}_q)$, we conclude that $C(s_i\sigma)$ is either $C(\tau)$ or $C(\sigma)$, and since $s_i\sigma \neq \sigma$, we have necessarily $C(s_i\sigma) = C(\tau) = C(\sigma s_j)$, hence $s_i\sigma = \sigma s_j$.

By Lemma 5.40, we conclude that $C(s_i)C(\sigma) = C(s_i\sigma)$ if and only if $\ell(s_i\sigma) > \ell(\sigma)$. Therefore, the coefficient a_σ in the expansion $T_s T_\sigma = a_{s\sigma} T_{s\sigma} + a_\sigma T_\sigma$ vanishes if and only if $\ell(s\sigma) > \ell(\sigma)$.

In this situation $\ell(s\sigma) > \ell(\sigma)$, let us count the number of elements of the intersection $BsBs\sigma \cap B\sigma B$. Notice that $BsBs$ is included in the parabolic subgroup P introduced in Lemma 5.42, and therefore it is included in $BsB \cup B$. We then have

$$\begin{aligned}(BsBs\sigma) \cap (B\sigma B) &= ((BsBs)\sigma) \cap (B\sigma B) \\ &\subset (BsB\sigma \cup B\sigma) \cap (B\sigma B) \\ &\subset (C(s)C(\sigma) \cup B\sigma) \cap C(\sigma) = (C(s\sigma) \cup B\sigma) \cap C(\sigma) = B\sigma.\end{aligned}$$

In other words, the only elements of the intersection are those that write as $b\sigma$ with $b \in B$. Therefore, $a_{s\sigma} = \frac{\operatorname{card} \mathrm{B}(n,\mathbb{F}_q)}{\operatorname{card} \mathrm{B}(n,\mathbb{F}_q)} = 1$, and the relation is proven when $\ell(s\sigma) > \ell(\sigma)$.

Suppose finally that $\ell(s\sigma) < \ell(\sigma)$. If $T_s\, T_\sigma = a_{s\sigma} T_{s\sigma} + a_\sigma T_\sigma$, then by using the previous case and the quadratic relation for T_s shown in Lemma 5.42, we get:

$$\begin{aligned}(T_s)^2\, T_\sigma &= a_\sigma a_{s\sigma} T_{s\sigma} + (a_{s\sigma} + a_\sigma^2)\, T_\sigma \\ &= ((q-1)T_s + q T_{\mathrm{id}_{[\![1,n]\!]}})\, T_\sigma = (q-1)\, a_{s\sigma}\, T_{s\sigma} + ((q-1)\, a_\sigma + q) T_\sigma.\end{aligned}$$

Therefore, $a_\sigma = (q-1)$, and $a_{s\sigma} = q$. □

Proof of Theorem 5.35. The elements T_σ in $\mathbb{C}[\mathrm{B}(n,\mathbb{F}_q)\backslash\mathrm{GL}(n,\mathbb{F}_q)/\mathrm{B}(n,\mathbb{F}_q)]$ and T_σ in $\mathfrak{H}_q(n)$ satisfy the same relations, so one has a well-defined morphism of algebras

$$\mathfrak{H}_q(n) \to \mathbb{C}[\mathrm{B}(n,\mathbb{F}_q)\backslash\mathrm{GL}(n,\mathbb{F}_q)/\mathrm{B}(n,\mathbb{F}_q)].$$

It is an isomorphism since the two algebras have the same dimension $n!$. □

By the bicommutant theory, there exist simple modules U_q^λ over $\mathrm{GL}(n,\mathbb{F}_q)$ such that

$$\mathbb{C}[\mathrm{Flag}(n,\mathbb{F}_q)] = \mathbb{C}[\mathrm{GL}(n,\mathbb{F}_q)/\mathrm{B}(n,\mathbb{F}_q)] = \bigoplus_{\lambda\in\mathfrak{Y}(n)} U_q^\lambda \otimes_{\mathbb{C}} S_q^\lambda$$

as a $(\mathrm{GL}(n,\mathbb{F}_q),\mathfrak{H}_q(n))$-bimodule. The U_q^λ's are called the **unipotent modules** of the general linear group $\mathrm{GL}(n,\mathbb{F}_q)$. They do not form a complete collection of simple modules over $\mathrm{GL}(n,\mathbb{F}_q)$, but they are important building blocks in the representation theory of this finite Lie group. The following result gives the complex dimension of U_q^λ:

Proposition 5.43. *The generic degree* $\dim U_q^\lambda$ *is equal to*

$$q^{n(\lambda)}\,\frac{[n]_q!}{\prod_{\square\in\lambda}[h(\square)]_q},$$

where $n(\lambda) = \sum_{i=1}^{\ell(\lambda)}(i-1)\lambda_i$ *as in the previous chapters.*

Proof. Let us compute the character of the action of $\mathfrak{H}_q(n)$ on the right of $\mathbb{C}[G/B]$. If gB is a fixed left coset and T_σ acts on the right of it with $\sigma \neq \mathrm{id}_{[\![1,n]\!]}$, then

$$(gB\cdot T_\sigma) = \frac{1}{\mathrm{card}\,B}\,(gB)(B\sigma B) = g\,B\sigma B$$

does not make appear any element of the coset gB, since B and $B\sigma B$ are disjoint in $\mathrm{GL}(n,\mathbb{F}_q)$. Therefore, the trace of T_σ is zero, unless $\sigma = \mathrm{id}_{[\![1,n]\!]}$, in which case the trace is $\dim\mathbb{C}[G/B] = [n]_q!$. So, the character $\mathrm{ch}^{\mathbb{C}[G/B]}$ is a multiple of the symmetrizing trace of $\mathfrak{H}_q(n)$:

$$\mathrm{ch}^{\mathbb{C}[G/B]} = [n]_q!\,\tau.$$

By Theorem 4.50, the right-hand side of this formula admits for expansion

$$\tau = \sum_{\lambda\in\mathfrak{Y}(n)} \frac{1}{c_q^\lambda}\,\mathrm{ch}_q^\lambda$$

where c_q^λ is the Schur element of the simple module S_q^λ, and is equal to $\frac{1}{s_\lambda(X_q)}$ with the notations of Chapter 4. On the other hand, the bimodule expansion yields

$$\mathrm{ch}^{\mathbb{C}[G/B]} = \sum_{\lambda\in\mathfrak{Y}(n)} (\dim U_q^\lambda)\,\mathrm{ch}_q^\lambda.$$

Since the irreducible characters over the semisimple split algebra $\mathfrak{H}_q(n)$ are linearly independent, by comparison,

$$\dim U_q^\lambda = [n]_q!\, s_\lambda(X_q) = q^{n(\lambda)} \frac{[n]_q!}{\prod_{\square\in\lambda}[h(\square)]_q}. \qquad \square$$

Remark. It should be noticed that when $q = 1$, the formula for the generic degree specializes to the hook-length formula of Theorem 3.41. Therefore, the unipotent modules U_q^λ can be considered as new deformations of the Specht modules S^λ (different from the S_q^λ's, and defined over the groups $\mathrm{GL}(n, \mathbb{F}_q)$).

5.5 Hall–Littlewood polynomials and characters of Hecke algebras

Let z be a generic complex number (not 0, and not a root of unity), and $\sigma \in \mathfrak{S}(n)$. We now have all the tools required to give an explicit formula for the character value $\mathrm{ch}_z^\lambda(T_\sigma)$, where ch_z^λ is the character of the irreducible Specht module S_z^λ of $\mathfrak{H}_z(n)$. By the discussion of Chapter 4, these character values depend polynomially on z, so they all come from a common polynomial $\mathrm{ch}^{\lambda,q}(T_\sigma) \in \mathbb{C}[q]$. Thus, a candidate for a **character table** of the Hecke algebra $\mathfrak{H}(n)$ is the family of polynomials $\mathfrak{W} = (\mathrm{ch}^{\lambda,q}(T_\sigma))_{\lambda\in\mathfrak{Y}(n),\sigma\in\mathfrak{S}(n)}$. This differs a bit from the group case, where the character table is a *square* matrix of size $\mathrm{card}\,\mathfrak{Y}(n) \times \mathrm{card}\,\mathfrak{Y}(n)$. A priori, to know all the character values in $\mathfrak{W}$, it is not sufficient to give the values $\mathrm{ch}^{\lambda,q}(T_\sigma)$ with σ running over a set of representatives of the conjugacy classes of $\mathfrak{S}(n)$. Indeed, if σ and σ' are two conjugated permutations, then in general it is not true that $\mathrm{ch}^{\lambda,q}(T_\sigma) = \mathrm{ch}^{\lambda,q}(T_{\sigma'})$. For instance, if one considers the index representation $T_\sigma \mapsto q^{\ell(\sigma)}$, then it takes different values over a same conjugacy class, because the length function ℓ is not invariant by conjugation. Nonetheless, we shall prove in this section (Theorem 5.46) that there exists indeed a submatrix

$$\mathfrak{X} = \left(\mathrm{ch}^{\lambda,q}(T_{\sigma_\mu})\right)_{\lambda\in\mathfrak{Y}(n),\mu\in\mathfrak{Y}(n)}$$

that allows us to reconstruct $\mathfrak{W}$ entirely. Moreover, we shall express the polynomials $\mathrm{ch}^{\lambda,q}(T_{\sigma_\mu})$ of this smaller character table as the coefficients of change of basis between the Schur functions and certain new symmetric functions in $\mathbb{C}[q]\otimes_{\mathbb{C}}\mathrm{Sym}$, called the Hall–Littlewood polynomials. This generalization of the Frobenius formula (Theorem 5.49) is due to A. Ram, and it leads to a generalization of the Murnaghan–Nakayama rule (Theorem 5.50), which allows an explicit calculation of the characters of the generic Hecke algebra.

▷ *Relations between the character values of the Hecke algebra.*

Let μ be an integer partition, and $\sigma \in C_\mu$ be a permutation with cycle type μ. We say that σ is **minimal in its conjugacy class** if $\ell(\sigma)$ is the minimum of the set of integers $\{\ell(\tau),\ \tau \text{ conjugated to } \sigma\}$.

Proposition 5.44. *Let μ be an integer partition. The minimal allowed length for a permutation $\sigma \in \mathfrak{S}(n)$ of cycle type μ is $|\mu| - \ell(\mu)$. Moreover, if $c = (c_1, \ldots, c_\ell)$ is a composition of size n whose parts are the same as μ, and if*

$$\sigma_c = (1, 2, \ldots, c_1)(c_1 + 1, \ldots, c_1 + c_2) \cdots (c_1 + \cdots + c_{\ell-1} + 1, \ldots, c_1 + \cdots + c_\ell),$$

then σ_c is minimal in the conjugacy class C_μ.

Proof. Recall that the length of a permutation is also its number of inversions. Consider a permutation $\sigma = \sigma_1\sigma_2 \cdots \sigma_\ell$, which is a product of disjoint cycles σ_k with lengths μ_k. We associate to each cycle σ_k a support S_k of size μ_k, such that $[\![1, n]\!] = \bigsqcup_{k=1}^{\ell} S_k$. Set

$$N(\sigma_k, \sigma_l) = \{(i, j) \in [\![1, n]\!] \,|\, i \in S_k,\ j \in S_l,\ i < j \text{ and } \sigma(i) > \sigma(j)\}.$$

Then,

$$N(\sigma) = \sum_{k,l=1}^{\ell} N(\sigma_k, \sigma_l) \geq \sum_{k=1}^{\ell} N(\sigma_k, \sigma_k).$$

The first part of the proposition comes now from the fact that $N(\sigma_k, \sigma_k) \geq \mu_k - 1$. Indeed, a cycle $(a_1, \ldots, a_{\mu_k})$ cannot be written as the product of less than $\mu_k - 1$ transpositions, and a fortiori, it cannot be written as the product of less than $\mu_k - 1$ *elementary* transpositions. As for the second part of the proposition, it comes from the decomposition

$$\sigma_c = (s_1 s_2 \cdots s_{c_1 - 1})(s_{c_1+1} \cdots s_{c_1+c_2-1}) \cdots (s_{c_1+\cdots+c_{\ell-1}+1} \cdots s_{n-1}),$$

which has $|\mu| - \ell(\mu)$ terms. □

Proposition 5.45. *Let c and d be two compositions of size n with the same parts (in a different order). Then, for any integer partition λ, $\mathrm{ch}^{\lambda,q}(T_{\sigma_c}) = \mathrm{ch}^{\lambda,q}(T_{\sigma_d})$.*

Proof. We fix an integer $N \geq n$, and we use the duality between $\mathrm{GL}_z(N, \mathbb{C})$ and $\mathfrak{H}_z(n)$. If $h = h_1 h_2$ belongs to $\langle T_1, \ldots, T_{n_1-1}, T_{n_1+1}, \ldots, T_{n-1}\rangle = \mathfrak{H}_z(n_1) \times \mathfrak{H}_z(n_2) \subset \mathfrak{H}_z(n)$ with $n_1 + n_2 = n$, then the action of h on a word $w = w_1 w_2$ is

$$w \cdot h = (w_1 \cdot h_1)(w_2 \cdot h_2).$$

On the other hand, with the same decomposition $w = w_1 w_2$ of a word in two parts of size n_1 and n_2, one has

$$(\eta_1)^{k_1} \cdots (\eta_N)^{k_N} \cdot w = ((\eta_1)^{k_1} \cdots (\eta_N)^{k_N} \cdot w_1)((\eta_1)^{k_1} \cdots (\eta_N)^{k_N} \cdot w_2).$$

As a consequence, with $h = h_1 h_2$ and $x = (\eta_1)^{k_1}(\eta_2)^{k_2}\cdots(\eta_N)^{k_N}$,

$$\begin{aligned}
\mathrm{btr}(x,h) &= \sum_{w\in[\![1,N]\!]^n} [w](x\cdot w\cdot h) \\
&= \sum_{w_1\in[\![1,N]\!]^{n_1},\, w_2\in[\![1,N]\!]^{n_2}} [w_1w_2](x\cdot w_1\cdot h_1)(x\cdot w_2\cdot h_2) \\
&= \left(\sum_{w_1\in[\![1,N]\!]^{n_1}} [w_1](x\cdot w_1\cdot h_1)\right)\left(\sum_{w_2\in[\![1,N]\!]^{n_2}} [w_1](x\cdot w_2\cdot h_2)\right) \\
&= \mathrm{btr}(x,h_1)\,\mathrm{btr}(x,h_2),
\end{aligned}$$

where the bitraces are computed with respect to the actions on the spaces of tensors. More generally, for any $x = (\eta_1)^{k_1}(\eta_2)^{k_2}\cdots(\eta_N)^{k_N}$, any composition $c = (c_1,\ldots,c_\ell) \in \mathfrak{C}(n)$ and any $h = h_1h_2\cdots h_\ell$ in the subalgebra $\mathfrak{H}_z(c) = \prod_{i=1}^{\ell} \mathfrak{H}_z(c_i)$,

$$\mathrm{btr}(x,h) = \prod_{i=1}^{\ell} \mathrm{btr}(x,h_i).$$

Consequently, $\mathrm{btr}(x, T_{\sigma_c}) = \mathrm{btr}(x, T_{\sigma_d})$, since they involve the same factors. However, by using Theorem 5.28 and its representation theoretic counterpart, we get that for any integers $k_1,\ldots,k_N \in \mathbb{Z}$ and any $h \in \mathfrak{H}_z(n)$:

$$\begin{aligned}
\mathrm{btr}((\eta_1)^{k_1}(\eta_2)^{k_2}\cdots(\eta_N)^{k_N}, h) &= \sum_{\lambda\in\mathfrak{Y}(n)} \chi_z^{\rho_z(\lambda)}((\eta_1)^{k_1}(\eta_2)^{k_2}\cdots(\eta_N)^{k_N})\,\mathrm{ch}_z^{\lambda}(h) \\
&= \sum_{\lambda\in\mathfrak{Y}(n)} s_{\rho_z(\lambda)}(z^{k_1}, z^{k_2},\ldots,z^{k_N})\,\mathrm{ch}_z^{\lambda}(h),
\end{aligned}$$

where ρ_z is some permutation of the integer partitions in $\mathfrak{Y}(n)$. Combining this with the previous observations, we obtain:

$$\sum_{\lambda\in\mathfrak{Y}(n)} s_{\rho_z(\lambda)}(z^{k_1}, z^{k_2},\ldots,z^{k_N})\,\mathrm{ch}_z^{\lambda}(T_{\sigma_c}) = \sum_{\lambda\in\mathfrak{Y}(n)} s_{\rho_z(\lambda)}(z^{k_1}, z^{k_2},\ldots,z^{k_N})\,\mathrm{ch}_z^{\lambda}(T_{\sigma_d}).$$

Both sides are polynomials in $z^{k_1},\ldots,z^{k_N}$, and since these variables can all take an infinite number of values, we have in fact the identity of symmetric polynomials

$$\sum_{\lambda\in\mathfrak{Y}(n)} s_{\rho_z(\lambda)}(x_1,x_2,\ldots,x_N)\,\mathrm{ch}_z^{\lambda}(T_{\sigma_c}) = \sum_{\lambda\in\mathfrak{Y}(n)} s_{\rho_z(\lambda)}(x_1,x_2,\ldots,x_N)\,\mathrm{ch}_z^{\lambda}(T_{\sigma_d})$$

in $\mathrm{Sym}^{(N)}$. As the Schur functions of degree less than N are linearly independent in the ring $\mathbb{C}[x_1,\ldots,x_N]$, we can conclude by identification of the coefficients that

$$\mathrm{ch}_z^{\lambda}(T_{\sigma_c}) = \mathrm{ch}_z^{\lambda}(T_{\sigma_d})$$

for any generic complex number z. Finally, since both sides of the equality are specializations of polynomials, the identity for any z is equivalent to the identity of the polynomials $\mathrm{ch}^{\lambda,q}(T_{\sigma_c})$ and $\mathrm{ch}^{\lambda,q}(T_{\sigma_d})$. □

Using the previous propositions, we can now establish the following important result, which ensures the existence of a character table for $\mathfrak{H}(n)$ that is a square matrix of size $\operatorname{card}\mathfrak{Y}(n) \times \operatorname{card}\mathfrak{Y}(n)$.

Theorem 5.46. *There exist polynomials $f_{\sigma,\mu}(q) \in \mathbb{C}[q]$ labeled by permutations σ and integer partitions μ of size n, such that for any character ch^M of a finite-dimensional module M over $\mathfrak{H}(n)$,*

$$\mathrm{ch}^M(T_\sigma) = \sum_{\mu \in \mathfrak{Y}(n)} f_{\sigma,\mu}(q)\, \mathrm{ch}^M(T_{\sigma_\mu}),$$

where $(\sigma_\mu)_{\mu \in \mathfrak{Y}(n)}$ is the family of representatives of the conjugacy classes C_μ of $\mathfrak{S}(n)$ defined by

$$\sigma_\mu = (1,2,\ldots,\mu_1)(\mu_1+1,\ldots,\mu_1+\mu_2)\cdots(\mu_1+\cdots+\mu_{\ell-1}+1,\ldots,n).$$

Therefore, the character values $\mathrm{ch}^{\lambda,q}(T_\sigma)$ of $\mathfrak{H}(n)$ can all be computed from the square character table

$$\mathfrak{X} = \left(\mathrm{ch}^{\lambda,q}(T_{\sigma_\mu})\right)_{\lambda,\mu \in \mathfrak{Y}(n)}.$$

Proof. If c is a composition of size n and μ is the integer partition with the same parts as c, then $\mathrm{ch}^M(T_{\sigma_c}) = \mathrm{ch}^M(T_{\sigma_\mu})$ by the previous proposition. Therefore, it suffices to show that there exist polynomials $f_{\sigma,c}(q)$ such that

$$\mathrm{ch}^M(T_\sigma) = \sum_{\mu \in \mathfrak{Y}(n)} f_{\sigma,c}(q)\, \mathrm{ch}^M(T_{\sigma_c});$$

we shall then take $f_{\sigma,\mu}(q) = \sum_{c \text{ composition with the same parts as } \mu} f_{\sigma,c}(q)$.

Notice that if σ is not of the form $\sigma = \sigma_c$ with $c \in \mathfrak{C}(n)$, then there exists an index i such that $\sigma(i) > i+1$. Let $i \in [\![1,n]\!]$ be the first index that has this property, and $r = \sigma(i)-(i+1)$; we reason by induction on (i,r). We set $j = \sigma(i)-1$; notice that $\sigma^{-1}(j) \neq i$, and that $\sigma^{-1}(j)$ cannot be smaller than i, as otherwise one would have

$$\sigma(i)-1 = j = \sigma(\sigma^{-1}(j)) \leq \sigma^{-1}(j)+1 \leq i$$

which contradicts the hypothesis $\sigma(i) > i+1$. Therefore, $\sigma^{-1}(j) > i$, and on the other hand, $\sigma^{-1}(j+1) = i$, so j is a backstep of σ and $\ell(s_j\sigma) < \ell(\sigma)$ (Lemma 4.3).

1. If $\ell(s_j\sigma s_j) > \ell(s_j\sigma)$, then

   $$\mathrm{ch}^M(T_\sigma) = \mathrm{ch}^M(T_j T_{s_j\sigma}) = \mathrm{ch}^M(T_{s_j\sigma}\, T_{s_j}) = \mathrm{ch}^M(T_{s_j\sigma s_j}),$$

   and if $\sigma' = s_j\sigma s_j$, then $\sigma'(i) = s_j\sigma(i) = s_j(j+1) = j = \sigma(i)-1$ and $r' = \sigma'(i)-(i+1)$ is smaller than r.

2. If $\ell(s_j \sigma s_j) < \ell(s_j \sigma)$, then

$$\mathrm{ch}^M(T_\sigma) = \mathrm{ch}^M(T_{s_j\sigma}\, T_j) = q\, \mathrm{ch}^M(T_{s_j\sigma s_j}) + (q-1)\, \mathrm{ch}^M(T_{s_j\sigma}).$$

Both permutations $\sigma' = s_j \sigma s_j$ and $\sigma' = s_j \sigma$ satisfy $\sigma'(i) = \sigma(i) - 1$, so they correspond to a pair (i, r') with $r' < r$. By the induction hypothesis, there exist polynomials $f_{\sigma',c}(q)$ for these permutations, hence a family of polynomials $f_{\sigma,c}$. □

Example. Consider the permutation $\sigma = 43251 = (1,4,5)(2,3)$, which admits for reduced decomposition $s_2s_3s_2s_1s_2s_3s_4$. The algorithm presented above shows that, for any finite-dimensional module M over $\mathfrak{H}(5)$,

$$\mathrm{ch}^M(T_\sigma) = (q^2+1)(q-1)\, \mathrm{ch}^M(T_{\sigma_{(5)}}) + q^2\, \mathrm{ch}^M(T_{\sigma_{(3,2)}}) + q(q-1)^2\, \mathrm{ch}^M(T_{\sigma_{(4,1)}}).$$

In particular, $\mathrm{ch}^M(T_\sigma)$ is not equal to $\mathrm{ch}^M(T_{\sigma_{(3,2)}})$, though σ has cycle type $(3,2)$.

Remark. One can show that if σ and σ' are two permutations that are conjugated and that are both minimal in their conjugacy class, then $\mathrm{ch}^M(T_\sigma) = \mathrm{ch}^M(T_{\sigma'})$ for any finite-dimensional module M over $\mathfrak{H}(n)$. Theorem 5.46 only ensures that one can compute any character value of $\mathfrak{H}(n)$ in terms of the character values of *certain* permutations that are minimal in their respective conjugacy classes.

▷ *Computation of the bitrace and Hall–Littlewood polynomials.*

We now focus on the computation of the elements $\mathrm{ch}^{\lambda,q}(T_{\sigma_\mu})$ of the character table $\mathfrak{X}$. Since these characters (or more precisely, their specializations with $q = z$) appear as coefficients of the symmetric polynomial

$$\mathrm{btr}((\eta_1)^{k_1} \cdots (\eta_N)^{k_N}, T_{\sigma_\mu})$$

in the variables $x_1 = z^{k_1}, \ldots, x_N = z^{k_N}$, it suffices to compute these bitraces. Moreover, we saw that they factorize over the parts of the integer partition μ, so we only have to compute $\mathrm{btr}(x, T_{\sigma_{(n)}})$, where $x = (\eta_1)^{k_1} \cdots (\eta_N)^{k_N}$ and the integers k_i are arbitrary.

Proposition 5.47. *Let $x = (\eta_1)^{k_1} \cdots (\eta_N)^{k_N}$, and $n \geq 1$. The bitrace of x and $T_{\sigma_{(n)}} = T_1 T_2 \cdots T_{n-1}$ for the actions of $\mathrm{GL}_z(N,\mathbb{C})$ and $\mathfrak{H}_z(n)$ on $(\mathbb{C}^N)^{\otimes n}$ is*

$$\mathrm{btr}(x, T_1T_2\cdots T_{n-1}) = \sum_{\lambda \in \mathfrak{Y}(n)} z^{n-\ell(\lambda)}\, (z-1)^{\ell(\lambda)-1}\, m_\lambda(x_1, \ldots, x_N),$$

where $x_i = z^{k_i}$.

Proof. We compute by induction on n the expansion of the bitrace $\mathrm{btr}(x, T_{\sigma_{(n)}})$ in monomials $x_{i_1}x_{i_2}\dots x_{i_n}$. To begin with, one can use the tracial property $\mathrm{ch}^M(ab) = \mathrm{ch}^M(ba)$ and the commutation relations $T_iT_j = T_jT_i$ for $|j-i| \geq 2$ to show that

$$\mathrm{ch}^M(T_1T_2\cdots T_{n-1}) = \mathrm{ch}^M(T_{n-1}T_{n-2}\cdots T_1)$$

for any finite-dimensional module over $\mathfrak{H}(n)$. For instance, with $n = 5$, one has indeed

$$\begin{aligned}\mathrm{ch}^M(T_1T_2T_3T_4) &= \mathrm{ch}^M(T_4T_1T_2T_3) = \mathrm{ch}^M(T_1T_2T_4T_3) = \mathrm{ch}^M(T_2T_4T_3T_1)\\ &= \mathrm{ch}^M(T_2T_1T_4T_3) = \mathrm{ch}^M(T_4T_3T_2T_1)\end{aligned}$$

and the same kind of transformations allows one to treat the general case (this is also a particular case of the unproven remark about the character values of an element T_σ with σ minimal in its conjugacy class). As a consequence,

$$\mathrm{btr}(x, T_1T_2\cdots T_{n-1}) = \mathrm{btr}(x, T_{n-1}\cdots T_2T_1),$$

the second form being more adapted to an induction.

Consider now a word $w = w_1w_2\cdots w_n$ of length n and with entries in $[\![1, N]\!]$. We set $\widetilde{w} = w_1w_2\cdots w_{n-1}$.

1. If $w_{n-1} = w_n$, then

$$\begin{aligned}[w](x\cdot w\cdot T_{n-1}\cdots T_2T_1) &= z\,[w](x\cdot w\cdot T_{n-2}\cdots T_2T_1)\\ &= z\,x_{w_n}\,[\widetilde{w}](x\cdot\widetilde{w}\cdot T_{n-2}\cdots T_2T_1).\end{aligned}$$

2. If $w_{n-1} < w_n$, then

$$[w](x\cdot w\cdot T_{n-1}\cdots T_2T_1) = [w](x\cdot(w_1\cdots w_{n-2}w_nw_{n-1})\cdot T_{n-2}\cdots T_2T_1) = 0$$

   since x acts on a word by a scalar, and $T_{n-2}\cdots T_2T_1$ acts only on the $n-1$ first letters of the word $w_1\cdots w_{n-2}w_nw_{n-1}$, hence yields words with the last letter equal to w_{n-1}.

3. Finally, if $w_{n-1} > w_n$, then

$$\begin{aligned}[w](x\cdot w\cdot T_{n-1}\cdots T_2T_1) &= (z-1)\,[w](x\cdot w\cdot T_{n-2}\cdots T_2T_1)\\ &= (z-1)\,x_{w_n}\,[\widetilde{w}](x\cdot\widetilde{w}\cdot T_{n-2}\cdots T_2T_1).\end{aligned}$$

As an immediate consequence, by induction on n,

$$\mathrm{btr}(x, T_{n-1}\cdots T_2T_1) = \sum_{N\geq w_1\geq w_2\geq\cdots\geq w_n\geq 1} z^{e(w)}\,(z-1)^{g(w)}\,x_{w_1}x_{w_2}\cdots x_{w_n}$$

where $e(w)$ is the number of indices i such that $w_i = w_{i+1}$, and $g(w)$ is the number of indices i such that $w_i > w_{i+1}$. If we gather in the sum the monomials that are permutations by some $\sigma \in \mathfrak{S}(N)$ of $x^\lambda = (x_1)^{\lambda_1}(x_2)^{\lambda_2}\cdots(x_N)^{\lambda_N}$, then we obtain

$$z^{\sum_{i=1}^{\ell(\lambda)}(\lambda_i-1)}\,(z-1)^{\ell(\lambda)-1}\,m_\lambda(x_1,\dots,x_N),$$

and this ends the proof. $\square$

Proposition 5.47 leads to the introduction of new symmetric functions known as the Hall–Littlewood polynomials. More precisely, for any $n \geq 1$, set

$$q_n(q;X) = \sum_{\lambda \in \mathfrak{Y}(n)} (1-q)^{\ell(\lambda)}\, m_\lambda(X),$$

which is an element of the tensor product $\mathrm{Sym}[q] = \mathbb{C}[q] \otimes_{\mathbb{C}} \mathrm{Sym}$. We say that $q_n(q;X)$ is the **Hall–Littlewood symmetric function** of degree n, and we have $\mathrm{btr}(x, T_1T_2\cdots T_{n-1}) = \frac{q^n}{(q-1)}\, q_n(q^{-1}; x_1, \ldots, x_N)$ for any n. More generally,

$$\mathrm{btr}(x, T_{\sigma_\mu}) = \prod_{i=1}^{\ell(\mu)} \mathrm{btr}(x, T_{\sigma_{(\mu_i)}}) = \frac{q^{|\mu|}}{(q-1)^{\ell(\mu)}}\, q_\mu(q^{-1}; x_1, \ldots, x_N),$$

where $q_\mu(q;X) = \prod_{i=1}^{\ell(\mu)} q_{\mu_i}(q;X)$.

Proposition 5.48. *Consider the formal alphabet $(1-q)X = X - qX$, defined by the identities*

$$p_k((1-q)X) = p_k(X) - p_k(qX) = (1-q^k)\,p_k(X)$$

for any $k \geq 1$. One has $q_\mu(q;X) = h_\mu((1-q)X)$ for any integer partition $\mu \in \mathfrak{Y}$.

Proof. Since the symmetric functions $q_\mu(q;X)$ and $h_\mu(X)$ factorize over the parts of μ, it suffices to prove the identity $q_n(q;X) = h_n((1-q)X)$ for any $n \geq 1$. Consider the generating series $Q(q;t) = \sum_{n=0}^{\infty} t^n\, q_n(q;X)$. It is equal to

$$\begin{aligned} Q(q;t) &= \sum_{\mu \in \mathfrak{Y}} t^{|\mu|}\,(1-q)^{\ell(\mu)}\, m_\mu(X) = \prod_{i \geq 1} \left(1 + (1-q)(tx_i + (tx_i)^2 + \cdots)\right) \\ &= \prod_{i \geq 1} \left(\frac{1 - qtx_i}{1 - tx_i}\right) = \exp\left(\sum_{k=1}^{\infty} \frac{t^k\, p_k(X) - (qt)^k\, p_k(X)}{k}\right) \\ &= \exp\left(\sum_{k=1}^{\infty} \frac{t^k\, p_k((1-q)X)}{k}\right) = \sum_{n=0}^{\infty} t^n\, h_n((1-q)X), \end{aligned}$$

whence the result. □

After Proposition 5.48, one can expand the symmetric function h_n in power sums, to get

$$q_n(q;X) = \sum_{\mu \in \mathfrak{Y}(n)} \frac{\prod_{i=1}^{\ell(\mu)} (1-q^{\mu_i})}{z_\mu}\, p_\mu(X).$$

As a consequence, for any $n \geq 1$, $\frac{q_n(q;X)}{1-q}$ belongs to $\mathrm{Sym}[q]$, and the specialization $q = 1$ of $\frac{q_n(q;X)}{1-q}$ is $p_n(X)$. So,

$$\frac{q_\mu(q;X)}{(1-q)^{\ell(\mu)}}$$

is for any $\mu \in \mathfrak{Y}$ a symmetric function in $\mathrm{Sym}[q]$ which interpolates between $p_\mu(X)$ (when $q=1$) and $h_\mu(X)$ (when $q=0$). Set

$$\widetilde{q}_\mu(q;X) = \frac{q^{|\mu|}}{(q-1)^{\ell(\mu)}}\, q_\mu(q^{-1};X) = \frac{h_\mu((q-1)X)}{(q-1)^{\ell(\mu)}}.$$

The modified Hall–Littlewood polynomials $\widetilde{q}_\mu(q;X)$ are again in $\mathrm{Sym}[q]$, and their specializations with $q=1$ are the power sums $p_\mu(X)$. Moreover,

$$\mathrm{btr}((\eta_1)^{k_1}\cdots(\eta_N)^{k_N}, T_{\sigma_\mu}) = \widetilde{q}_\mu(q;z^{k_1},\ldots,z^{k_N})$$

for any N and any integer partition μ.

▷ *Generalization of the Frobenius and Murnaghan–Nakayama formulas.*

We can finally compute the characters $\mathrm{ch}^{\lambda,q}(T_{\sigma_\mu})$ of the Hecke algebra. Set $\mathrm{Sym}(q) = \mathbb{C}(q)\otimes_{\mathbb{C}}\mathrm{Sym}$. We endow $\mathrm{Sym}(q)$ with the $\mathbb{C}(q)$-bilinear form that comes from the Hall scalar product on Sym.

Theorem 5.49. *The family $(\widetilde{q}_\mu(q;X))_{\mu\in\mathfrak{Y}}$ is a $\mathbb{C}(q)$-linear basis of* $\mathrm{Sym}(q)$. *Moreover, for any integer partitions λ and μ of the same size,*

$$\mathrm{ch}^{\lambda,q}(T_{\sigma_\mu}) = \left\langle \widetilde{q}_\mu(q;X) \,\middle|\, s_\lambda(X)\right\rangle_{\mathrm{Sym}(q)},$$

or equivalently,

$$\widetilde{q}_\mu(q;X) = \sum_{\lambda\in\mathfrak{Y}(n)} \mathrm{ch}^{\lambda,q}(T_{\sigma_\mu})\, s_\lambda(X).$$

Proof. We start by proving the remaining part of Theorem 5.29, that is to say that the permutation ρ_z is in fact the identity of $\mathfrak{Y}(n)$ for any z. Since the map $z \mapsto \mathrm{ch}^\lambda_z(T_{\sigma_\mu})$ is a polynomial in z, it is continuous. The same holds for the map $z \mapsto \widetilde{q}_\mu(z;x_1,\ldots,x_N)$, where $x_1,\ldots,x_N$ are fixed complex numbers. In the sequel, we set $C = \{z\in\mathbb{C} \,|\, |z|>1\}\cup\{1\}$; it is a connected subset of $\mathbb{C}$. For any $z\in C$, any permutation $\rho\in\mathfrak{S}(\mathfrak{Y}(n))$, and any $x_1,\ldots,x_N\in\mathbb{C}$, we also set

$$F(\rho;z;x_1,\ldots,x_N) = \left(\widetilde{q}_\mu(z;x_1,\ldots,x_N) - \sum_{\lambda\in\mathfrak{Y}(n)} \mathrm{ch}^\lambda_z(T_{\sigma_\mu})\, s_{\rho(\lambda)}(x_1,x_2,\ldots,x_N)\right)_{\mu\in\mathfrak{Y}(n)}.$$

By definition of ρ_z, and by looking at bitraces, we get $F(\rho_z;z;x_1,\ldots,x_N)=0$ for any $z\in C\setminus\{1\}$, and this is also true for $z=1$ if one convenes that $\rho_1 = \mathrm{id}_{\mathfrak{Y}(n)}$. In the latter case, one uses the usual Frobenius formula 2.32, as well as the identity $\widetilde{q}_\mu(1;x_1,\ldots,x_N) = p_\mu(x_1,\ldots,x_N)$.

If $\rho\in\mathfrak{S}(\mathfrak{Y}(n))$, denote

$$C_\rho = \{z\in C \,|\, \forall x_1,\ldots,x_N,\; F(\rho;z;x_1,\ldots,x_N)=0\}.$$

We have just explained that $z \in C_{\rho_z}$ for any z, so $C = \bigcup_{\rho \in \mathfrak{S}(\mathfrak{Y}(n))} C_\rho$. On the other hand, the sets C_ρ are closed, since they are reciprocal images of the vector 0 by a continuous map. Finally, the sets C_ρ are disjoint. Indeed, suppose that $z \in C_\rho \cap C_\tau$ with $\rho \neq \tau$; it implies

$$\sum_{\lambda \in \mathfrak{Y}(n)} \mathrm{ch}_z^\lambda(T_{\sigma_\mu})\, s_{\rho(\lambda)}(x_1,\ldots,x_N) = \sum_{\lambda \in \mathfrak{Y}(n)} \mathrm{ch}_z^\lambda(T_{\sigma_\mu})\, s_{\tau(\lambda)}(x_1,\ldots,x_N).$$

Therefore, as the Schur functions of degree n of variables $x_1,\ldots,x_{N\geq n}$ are linearly independent, there is a non-trivial permutation $\rho' = \tau^{-1} \circ \rho \neq \mathrm{id}_{\mathfrak{Y}(n)}$ such that, for any μ,

$$\mathrm{ch}_z^{\rho'(\lambda)}(T_{\sigma_\mu}) = \mathrm{ch}_z^\lambda(T_{\sigma_\mu}).$$

Since these character values determine all the other character values, we conclude to the equality of functions $\mathrm{ch}_z^{\rho'(\lambda)} = \mathrm{ch}_z^\lambda$, which is absurd since irreducible characters of a split semisimple algebra (here, $\mathfrak{H}_z(n)$) are linearly independent. So, $C = \bigsqcup_{\rho \in \mathfrak{S}(\mathfrak{Y}(n))} C_\rho$ is a disjoint union of closed subsets, and by connectedness, there is one permutation ρ such that $C = C_\rho$. Since $1 \in C_{\mathrm{id}_{\mathfrak{Y}(n)}}$, we conclude that $C = C_{\mathrm{id}_{\mathfrak{Y}(n)}}$, and therefore that $\rho_z = \mathrm{id}_{\mathfrak{Y}(n)}$ for any z with $|z| > 1$. The same argument applies to parameters z with $|z| < 1$ and $z \neq 0$, and if z is not a root of unity but has modulus 1, then one can add it to the set C and use again the same argument, so for any generic z (not 0 and not a root of unity), $\rho_z = \mathrm{id}_{\mathfrak{Y}(n)}$. Thus, we know now that for any z generic, and any $\mu \in \mathfrak{Y}(n)$,

$$\widetilde{q}_\mu(z; x_1,\ldots,x_N) = \sum_{\lambda \in \mathfrak{Y}(n)} \mathrm{ch}_z^\lambda(T_{\sigma_\mu}) s_\lambda(x_1,\ldots,x_N).$$

Since both sides depend polynomially on the parameter z, the identity also holds with the polynomials $\widetilde{q}_\mu(q; x_1,\ldots,x_N)$ and $\mathrm{ch}^{\lambda,q}(T_{\sigma_\mu})$ in $\mathbb{C}[q; x_1,\ldots,x_N]$ or $\mathbb{C}[q]$. As we are dealing with symmetric polynomials of degree $n \leq N$, there is actually an identity of symmetric functions in $\mathrm{Sym}[q]$:

$$\widetilde{q}_\mu(q; X) = \sum_{\lambda \in \mathfrak{Y}(n)} \mathrm{ch}^{\lambda,q}(T_{\sigma_\mu}) s_\lambda(X).$$

Since $(s_\lambda)_\lambda$ is an orthonormal basis of Sym, this implies the identity

$$\mathfrak{X} = \mathrm{ch}^{\lambda,q}(T_{\sigma_\mu}) = \left\langle \widetilde{q}_\mu(q; X) \,\middle|\, s_\lambda(X) \right\rangle_{\mathrm{Sym}(q)}.$$

Finally, let us explain why $(\widetilde{q}_\mu(q; X))_{\mu \in \mathfrak{Y}}$ is a (graded) linear basis of $\mathrm{Sym}(q)$. The formula for $q_n(X; q)$ in terms of the power sums $p_\mu(X)$ shows that for any integer partition μ,

$$\widetilde{q}_\mu(X; q) = \sum_{\substack{|\nu| = |\mu| \\ \nu \preceq \mu}} a_\mu^\nu(q)\, p_\nu(X),$$

where the $a_\mu^\nu(q)$ are certain polynomials in q, and the sum runs only on smaller integer partitions with respect to the dominance order. Moreover,

$$a_\mu^\mu(q) = \prod_{i=1}^{\ell(\mu)} \frac{[\mu_i]_q}{\mu_i} \neq 0.$$

So, $(\widetilde{q}_\mu(q;X))_{\mu\in\mathfrak{Y}(n)}$ is indeed a linear basis of the space of homogeneous symmetric functions of degree n in $\mathrm{Sym}(q)$, with an upper-triangular matrix of change of basis between it and the basis of power sums $(p_\mu(X))_{\mu\in\mathfrak{Y}(n)}$. □

Example. Let us compute the character tables of the Hecke algebras in size $n=2$, $n=3$ and $n=4$. The first modified Hall–Littlewood polynomials are:

$$\begin{aligned}
\widetilde{q}_1(q;X) &= p_1(X);\\
\widetilde{q}_2(q;X) &= \frac{q+1}{2}\,p_2(X)+\frac{q-1}{2}\,p_{1,1}(X);\\
\widetilde{q}_3(q;X) &= \frac{q^2+q+1}{3}\,p_3(X)+\frac{q^2-1}{2}\,p_{2,1}(X)+\frac{(q-1)^2}{6}\,p_{1,1,1}(X);\\
\widetilde{q}_4(q;X) &= \frac{q^3+q^2+q+1}{4}\,p_4(X)+\frac{q^3-1}{3}\,p_{3,1}(X)+\frac{(q+1)(q^2-1)}{8}\,p_{2,2}(X)\\
&\quad+\frac{(q-1)(q^2-1)}{4}\,p_{2,1,1}(X)+\frac{(q-1)^3}{24}\,p_{1,1,1,1}(X).
\end{aligned}$$

By expanding the power sums in Schur functions, we obtain the following character tables:

- $n=2$:

$\lambda\backslash\mu$	(2)	$(1,1)$
(2)	q	1
$(1,1)$	-1	1

- $n=3$:

$\lambda\backslash\mu$	(3)	$(2,1)$	$(1,1,1)$
(3)	q^2	q	1
$(2,1)$	$-q$	$q-1$	2
$(1,1,1)$	1	-1	1

- $n=4$:

$\lambda\backslash\mu$	(4)	$(3,1)$	$(2,2)$	$(2,1,1)$	$(1,1,1,1)$
(4)	q^3	q^2	q^2	q	1
$(3,1)$	$-q^2$	q^2-q	q^2-2q	$2q-1$	3
$(2,2)$	0	$-q$	q^2+1	$q-1$	2
$(2,1,1)$	q	$1-q$	$1-2q$	$q-2$	3
$(1,1,1,1)$	-1	1	1	-1	1

In each case, the first line (n) corresponds to the index representation $T_\sigma \mapsto q^{\ell(\sigma)}$, and the last line (1^n) corresponds to the signature representation $T_\sigma \mapsto (-1)^{\ell(\sigma)}$.

To close this section, we shall give an analogue of the Murnaghan–Nakayama rule (Theorem 3.10), which amounts to a recursive algorithm in order to compute the character values $\mathrm{ch}^{\lambda,q}(T_{\sigma_\mu})$. Recall that we computed in Chapter 3

$$\mathrm{ch}^{\lambda}(\mu) = \sum_{T} (-1)^{\mathrm{ht}(T)},$$

where the sum runs over ribbon tableaux of shape λ and weight μ. The q-analogue of this formula involves **generalized ribbons**: a generalized ribbon is a skew partition $\lambda \setminus \mu$ that does not contain any square of cells of size 2×2, but that is not necessarily connected (this is the difference with the usual notion of ribbon). For instance,

$$(5,4,2,1) \setminus (3,2) \;=$$

is a generalized ribbon, but it is not a ribbon in the traditional sense (it has two connected components). We define the q-weight of a generalized ribbon by the following formula:

$$\mathrm{wt}^q(\lambda\setminus\mu) = (q-1)^{\text{number of connected components of } \lambda\setminus\mu - 1} \prod_{C \text{ connected component}} q^{\mathrm{w}(C)}\,(-1)^{\mathrm{ht}(C)},$$

where $\mathrm{ht}(C)$ is the number of rows *minus one* occupied by C (the **height** of C), and $\mathrm{w}(C)$ is the number of columns *minus one* occupied by the connected component C (the **width** of C). For instance, the previous generalized ribbon $(5,4,2,1)\setminus(3,2)$ has q-weight $\mathrm{wt}^q(\lambda \setminus \mu) = (q-1)q^3$. Call **generalized ribbon tableau** of shape $\lambda \in \mathfrak{Y}(n)$ and weight $\mu \in \mathfrak{Y}(n)$ a sequence of partitions $\emptyset = \mu^{(0)} \subset \mu^{(1)} \subset \cdots \subset \mu^{(\ell)} = \lambda$ such that each skew partition $\mu^{(i)} \setminus \mu^{(i-1)}$ is a generalized ribbon tableau with $|\mu^{(i)} \setminus \mu^{(i-1)}| = \mu_i$. The q-weight $\mathrm{wt}^q(T)$ of a generalized ribbon tableau is the product $\prod_{i=1}^{\ell} \mathrm{wt}^q(\mu^{(i)} \setminus \mu^{(i-1)})$ of its constituting ribbons.

Theorem 5.50. *For any partitions λ, μ of same size n,*

$$\mathrm{ch}^{\lambda,q}(T_{\sigma_\mu}) = \sum_{T} \mathrm{wt}^q(T),$$

where the sum runs over generalized ribbon tableaux T of shape λ and weight μ.

Example. Let us compute $\mathrm{ch}^{(3,1),q}(T_{\sigma_{(2,2)}})$. There are two generalized ribbon tableaux of shape $(3,1)$ and weight $(2,2)$:

$$\begin{array}{|c|cc}\hline 1 & & \\ \hline 1 & \multicolumn{1}{|c|}{2} & \multicolumn{1}{c|}{2} \\ \hline\end{array} \quad \text{and} \quad \begin{array}{|c|cc}\hline 2 & & \\ \hline 1 & \multicolumn{1}{|c|}{1} & \multicolumn{1}{c|}{2} \\ \hline\end{array}\,.$$

The first ribbon tableau has weight $-q$, and the second tableau has weight $q(q-1)$, so $\mathrm{ch}^{(3,1),q}(T_{\sigma_{(2,2)}}) = -q + q(q-1) = q^2 - 2q$.

Remark. Suppose $q = 1$. Then, the weight of a generalized ribbon $\lambda \setminus \mu$ specializes to 0 unless the $\lambda \setminus \mu$ is connected, in which case it is equal to $(-1)^{\mathrm{ht}(\lambda\setminus\mu)}$. So, Theorem 5.50 generalizes the classical Murnaghan–Nakayama formula.

Lemma 5.51. *For any* $n \geq 1$,

$$\widetilde{q}_n(q;X) = \sum_{m=1}^{n} (-1)^{n-m}\, q^{m-1}\, s_{(m)\sqcup 1^{n-m}}(X).$$

Proof. Recall that the coproduct of $h_n = s_n$ in Sym is $\Delta(h_n) = \sum_{m=0}^{n} h_m \otimes h_{n-m}$, and that the antipode of $h_n(X)$ is $h_n(-X) = \omega(h_n)(X) = (-1)^n\, e_n(X)$. Therefore,

$$\widetilde{q}_n(q;X) = \frac{1}{q-1}\, h_n(qX - X) = \frac{1}{q-1} \sum_{m=0}^{n} h_m(qX)\, h_{n-m}(-X)$$
$$= \frac{1}{q-1} \sum_{m=0}^{n} (-1)^{n-m}\, q^m\, s_m(X)\, e_{n-m}(X).$$

We can then compute the product $s_m(X)\, e_{n-m}(X)$ by using the Pieri rules 3.6: it involves the Schur functions associated to partitions that are obtained by adding to (m) a vertical strip of size $n-m$, so

$$s_m(X)\, e_{n-m}(X) = \begin{cases} s_{(m)\sqcup 1^{n-m}}(X) + s_{(m+1)\sqcup 1^{n-m-1}}(X) & \text{if } m \neq 0, n, \\ s_{1^n}(X) & \text{if } m = 0, \\ s_n(X) & \text{if } m = n. \end{cases}$$

This implies immediately the result, by gathering the coefficients of the Schur functions $s_{(m)\sqcup 1^{n-m}}(X)$. □

Proof of Theorem 5.50. Since $\mathrm{ch}^{\lambda,q}(T_{\sigma_\mu})$ is the coefficient of $s_\lambda(X)$ in the expansion in Schur functions of $\widetilde{q}_\mu(q;X)$, the theorem is equivalent to the following multiplication rule: for any $n \geq 1$, and any integer partition μ,

$$\widetilde{q}_n(q;X)\, s_\mu(X) = \sum \mathrm{wt}^q(\lambda \setminus \mu)\, s_\lambda(X),$$

where the sum runs over generalized ribbons $\lambda \setminus \mu$ of size n. We shall use the previous lemma, or more precisely the expansion

$$\widetilde{q}_n(q;X) = \frac{1}{q-1} \sum_{m=0}^{n} (-1)^{n-m}\, q^m\, h_m(X)\, e_{n-m}(X)$$

obtained during its proof. By the Pieri rules, $h_m(X)\, e_{n-m}(X)\, s_\mu(X)$ is a sum of Schur functions $s_\lambda(X)$, where λ is obtained from μ by adding first a vertical strip of size $n-m$, and then a horizontal strip of size m. Therefore, $\lambda \setminus \mu$ cannot contain a square of size 2×2: indeed, in such a square, the possible horizontal strips are

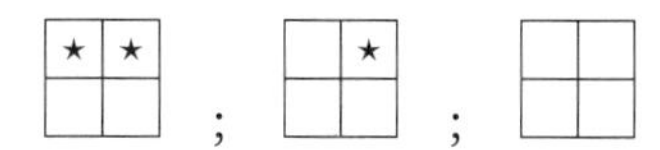

and what remains is never a vertical strip. Therefore,

$$h_m(X)\, e_{n-m}(X)\, s_\mu(X) = \sum_\lambda c_{\lambda,\mu,n,m}\, s_\lambda(X)$$

where the sum runs over partitions λ such that $\lambda \setminus \mu$ is a generalized ribbon of size n, and where $c_{\lambda,\mu,n,m}$ is the number of ways of obtaining λ by adding to μ first a vertical strip of size $n-m$, and then a horizonal strip of size m.

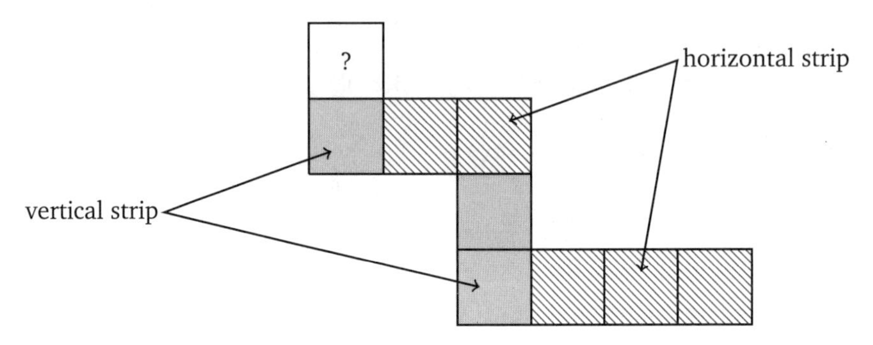

Figure 5.1
Origins of the cells of a ribbon obtained by adding a vertical strip and then a horizontal strip.

Consider a connected component C of such a generalized ribbon. It is a ribbon whose cells have well-determined origins, except the top-left cell which can come from the vertical strip or from the horizontal strip; see Figure 5.1. If one removes the top-left cell of unknown origin, then the number of cells from the horizontal strip is also the width of the ribbon, and the number of cells from the vertical strip is also the height of the ribbon. As a consequence, one can make the previous formula for $h_m(X)\, e_{n-m}(X)\, s_\mu(X)$ more explicit:

$$h_m(X)\, e_{n-m}(X)\, s_\mu(X) = \sum_\lambda \binom{\mathrm{cc}(\lambda \setminus \mu)}{m - \mathrm{w}(\lambda \setminus \mu)} s_\lambda(X)$$

where the sum runs over generalized ribbons $\lambda \setminus \mu$ of size n, and where $\mathrm{cc}(\lambda \setminus \mu)$ is equal to the number of connected components of $\lambda \setminus \mu$. Indeed, given such a generalized ribbon with $\mathrm{cc}(\lambda \setminus \mu)$ components, one has to choose which connected components will have their top-left cell coming from a horizontal strip, and one has then the three equations

$$\begin{aligned} m &= ?_h + \mathrm{w}(\lambda \setminus \mu); \\ n - m &= ?_v + \mathrm{ht}(\lambda \setminus \mu); \\ \mathrm{cc}(\lambda \setminus \mu) &= ?_h + ?_v \end{aligned}$$

where $?_h$ (respectively, $?_v$) is the number of connected components of the generalized ribbon where the top-left cell ? comes from a horizontal strip (respectively, from a vertical strip). Thus, the choice of the origins of the top-left cells of the connected components explains the apparition of the binomial coefficient $\binom{\mathrm{cc}(\lambda\setminus\mu)}{?_h}$.

We can now compute the product $\widetilde{q}_n(q;X)s_\mu(X)$. The coefficient of $s_\lambda(X)$ in this product is equal to 0 unless $\lambda\setminus\mu$ is a generalized ribbon of weight n. In this case, it is equal to

$$\frac{1}{q-1}\sum\binom{\mathrm{cc}(\lambda\setminus\mu)}{m-\mathrm{w}(\lambda\setminus\mu)}(-1)^{n-m}\,q^m,$$

where the sum runs over integers m such that $n-\mathrm{ht}(\lambda\setminus\mu)\geq m\geq \mathrm{w}(\lambda\setminus\mu)$ (we need the numbers $?_h$ and $?_v$ to be non-negative). Thus, one obtains

$$\begin{aligned}[s_\lambda(X)]\left(\widetilde{q}_n(q;X)s_\mu(X)\right)&=\frac{1}{q-1}\sum_{m=\mathrm{w}(\lambda\setminus\mu)}^{n-\mathrm{ht}(\lambda\setminus\mu)}\binom{\mathrm{cc}(\lambda\setminus\mu)}{m-\mathrm{w}(\lambda\setminus\mu)}(-1)^{n-m}\,q^m\\&=\frac{q^{\mathrm{w}(\lambda\setminus\mu)}\,(-1)^{\mathrm{ht}(\lambda\setminus\mu)}}{q-1}\sum_{k=0}^{\mathrm{cc}(\lambda\setminus\mu)}\binom{\mathrm{cc}(\lambda\setminus\mu)}{k}(-1)^{\mathrm{cc}(\lambda\setminus\mu)-k}\,q^k\\&=q^{\mathrm{w}(\lambda\setminus\mu)}\,(-1)^{\mathrm{ht}(\lambda\setminus\mu)}\,(q-1)^{\mathrm{cc}(\lambda\setminus\mu)-1}\end{aligned}$$

and this ends the proof. □

Notes and references

Our definition of the quantum groups is due to Jimbo; see [Jim85, Jim86]. However, there exist many variants of the presentation of $\mathfrak{U}(\mathfrak{gl}(N))$, and the one that we used comes from the paper [KT99]. Our treatment of the representation theory of $U_z(\mathfrak{gl}(N))$ is inspired from [Lus88, Ros88], but we rewrote all the arguments for the particular presentations of $U_z(\mathfrak{gl}(N))$ and its subalgebra $\mathrm{GL}_z(N,\mathbb{C})$ that we chose. We refer to [Ros90, Part C] for the proof of Theorem 5.25. Notice that most of the discussion of Sections 5.1 and 5.2 can be adapted to the case of an arbitrary simple or reductive complex Lie algebra $\mathfrak{g}$.

The definition of the action of $\mathfrak{H}_z(n)$ on tensors is due again to Jimbo, and the duality between $\mathfrak{H}_z(n)$ and $\mathrm{GL}_z(N,\mathbb{C})$ is stated without proof in [Jim86]. As far as we know, the only papers where the duality is actually proven are [KT99] and [LZ00]; we followed the first paper. The use of the Jimbo–Schur–Weyl duality to compute the characters $\mathrm{ch}^{\lambda,q}(T_{\sigma_\mu})$ is then due to A. Ram, cf. [Ram91]. However, the action of $U_z(\mathfrak{gl}(N))$ on tensors that is defined in loc. cit. is not compatible with the presentation of the quantum group that is given in this paper; this is why we prefered the presentation of [KT99]. The Hall–Littlewood polynomials $q_n(q;X)$ that we introduced in Section 5.5 are studied in more detail in [Mac95,

Chapter 3]. On the other hand, the q-Murnaghan–Nakayama rule (Theorem 5.50) is shown in [Ram91, Section 6], and it is also discussed in detail in [RR97].

The Iwahori–Hecke duality (Theorem 5.35) has been proven by Iwahori in [Iwa64]; we also followed [GP00, Section 8.4] and [Bou68, Chapter IV, §2], in particular for the exchange lemma 5.40. It should be noticed that there is an important generalization of this duality result to modules obtained by parabolic or Harish–Chandra induction of a cuspidal character ρ on a Levi subgroup L^F of a finite Lie group G^F. Thus, the commutant of the action of G^F on such an induced module $\mathrm{R}_{L^F}^{G^F}(\rho)$ is always a Hecke algebra of some Coxeter group $W(\rho)$, possibly with distinct parameters q_i for the generators s_i of the group. This result is due to Geck, Howlett and Lehrer; see [HL80, Gec93]. As a consequence, the Iwahori–Hecke duality can be seen as a first step in the understanding of the representation theory of finite Lie groups G^F such as $\mathrm{GL}(n, \mathbb{F}_q)$. In this specific case of the general linear groups, there is in fact a complete classification of the irreducible representations of the group due to J. A. Green; see [Gre55]. We refer also to [Mac95, Chapter IV] and [Zel81] for a modern treatment of this topic, and for a $\mathrm{GL}(n, \mathbb{F}_q)$-analogue of the Frobenius–Schur isomorphism theorem.

6

Representations of the Hecke algebras specialized at $q = 0$

Though the Hecke algebras $\mathfrak{H}_z(n)$ are in general isomorphic to $\mathfrak{H}_1(n) = \mathbb{C}\mathfrak{S}(n)$, the isomorphism fails for certain special values of the parameter $z \in \mathbb{C}$, e.g., if $z = 0$. In this chapter, we focus on these specialized Hecke algebras $\mathfrak{H}_0(n)$ and their representation theory. The results from Chapter 4 do not say anything about this case, and indeed, we shall see that $\mathfrak{H}_0(n)$ is not a semisimple algebra as soon as $n \geq 3$. In this setting, the pertinent objects that capture the representation theory of $\mathfrak{H}_0(n)$ are the two Grothendieck groups $\mathrm{R}_0(\mathfrak{H}_0(n))$ and $\mathrm{K}_0(\mathfrak{H}_0(n))$, as well as their perfect pairing

$$\left\langle P_i \,\middle|\, S_j \right\rangle = \delta_{ij}.$$

Here, $(P_i)_{i\in I}$ is a complete family of non-isomorphic principal indecomposable modules over $\mathfrak{H}_0(n)$, $(S_i)_{i\in I}$ is a complete family of non-isomorphic simple modules over $\mathfrak{H}_0(n)$, and the labeling is chosen so that S_i is the top of P_i, and P_i is the projective cover of S_i (cf. Section 4.2).

In the case of the symmetric groups, one has a product

$$\mathrm{Ind}_{\mathfrak{S}(m)\times\mathfrak{S}(n)}^{\mathfrak{S}(m+n)} : \mathrm{R}_0(\mathfrak{S}(m)) \otimes_{\mathbb{Z}} \mathrm{R}_0(\mathfrak{S}(n)) \to \mathrm{R}_0(\mathfrak{S}(m+n))$$

given by the induction functor. The same kind of functor can be used for the Grothendieck groups of the 0–Hecke algebras: thus, the functor $\mathrm{Ind}_{\mathfrak{H}_0(m)\times\mathfrak{H}_0(n)}^{\mathfrak{H}_0(m+n)}$ yields morphisms of groups

$$\mathrm{R}_0(\mathfrak{H}_0(m)) \otimes_{\mathbb{Z}} \mathrm{R}_0(\mathfrak{H}_0(n)) \to \mathrm{R}_0(\mathfrak{H}_0(m+n))$$

and

$$\mathrm{K}_0(\mathfrak{H}_0(m)) \otimes_{\mathbb{Z}} \mathrm{K}_0(\mathfrak{H}_0(n)) \to \mathrm{K}_0(\mathfrak{H}_0(m+n)).$$

Indeed:

1. For the Grothendieck groups of finite-dimensional modules, recall that a presentation of $\mathrm{R}_0(\mathfrak{H}_0(m)) \otimes_{\mathbb{Z}} \mathrm{R}_0(\mathfrak{H}_0(n)) = \mathrm{R}_0(\mathfrak{H}_0(m) \times \mathfrak{H}_0(n))$ is:

   generators: $[M]$, with M module over $\mathfrak{H}_0(m) \times \mathfrak{H}_0(n)$;
   relations: $[P] = [M] + [N]$ if there is an exact sequence
   $$0 \longrightarrow M \longrightarrow P \longrightarrow N \longrightarrow 0.$$

However, tensoring by $B = \mathfrak{H}_0(m+n)$ an exact sequence of modules over $A = \mathfrak{H}_0(m) \times \mathfrak{H}_0(n)$ yields again an exact sequence, because we are dealing with finite-dimensional $\mathbb{C}$-vector spaces. Therefore, if $[P] = [M] + [N]$ in $\mathrm{R}_0(\mathfrak{H}_0(m) \times \mathfrak{H}_0(n))$, then the sequence

$$0 \longrightarrow \mathrm{Ind}_A^B(M) \longrightarrow \mathrm{Ind}_A^B(P) \longrightarrow \mathrm{Ind}_A^B(N) \longrightarrow 0$$

is also an exact sequence of $\mathfrak{H}_0(m+n)$-modules, so

$$\left[\mathrm{Ind}_{\mathfrak{H}_0(m)\times\mathfrak{H}_0(n)}^{\mathfrak{H}_0(m+n)}(P)\right] = \left[\mathrm{Ind}_{\mathfrak{H}_0(m)\times\mathfrak{H}_0(n)}^{\mathfrak{H}_0(m+n)}(M)\right] + \left[\mathrm{Ind}_{\mathfrak{H}_0(m)\times\mathfrak{H}_0(n)}^{\mathfrak{H}_0(m+n)}(N)\right]$$

in $\mathrm{R}_0(\mathfrak{H}_0(m+n))$, and the induction functor is compatible with the relations of the Grothendieck groups R_0. Therefore, it gives a well-defined map $\mathrm{R}_0(\mathfrak{H}_0(m) \times \mathfrak{H}_0(n)) \to \mathrm{R}_0(\mathfrak{H}_0(m+n))$.

2. For the Grothendieck groups of projective modules, suppose that M and N are direct summands of free $\mathfrak{H}_0(m)$- and $\mathfrak{H}_0(n)$-modules:

$$M \oplus M' = (\mathfrak{H}_0(m))^a \quad ; \quad N \oplus N' = (\mathfrak{H}_0(n))^b.$$

Then, $M \boxtimes N$ is also a direct summand of a free $(\mathfrak{H}_0(m) \times \mathfrak{H}_0(n))$-module:

$$\begin{aligned}(M \oplus M') \boxtimes (N \oplus N') &= (M \boxtimes N) \oplus (M' \boxtimes N) \oplus (M \boxtimes N') \oplus (M' \boxtimes N') \\ &= (\mathfrak{H}_0(m))^a \boxtimes (\mathfrak{H}_0(n))^b = (\mathfrak{H}_0(m) \times \mathfrak{H}_0(n))^{ab}.\end{aligned}$$

As a consequence, $\mathrm{Ind}_{\mathfrak{H}_0(m)\times\mathfrak{H}_0(n)}^{\mathfrak{H}_0(m+n)}(M \boxtimes N)$ is also a direct summand of a free $\mathfrak{H}_0(m+n)$-module, namely, $(\mathfrak{H}_0(m+n))^{ab}$. Therefore, the induction functor $\mathrm{Ind}_{\mathfrak{H}_0(m)\times\mathfrak{H}_0(n)}^{\mathfrak{H}_0(m+n)}$ sends products of projective modules towards projective modules over $\mathfrak{H}_0(m+n)$, and it corresponds to a well-defined map $\mathrm{K}_0(\mathfrak{H}_0(m)) \otimes_{\mathbb{Z}} \mathrm{K}_0(\mathfrak{H}_0(n)) \to \mathrm{K}_0(\mathfrak{H}_0(m+n))$.

Following Chapter 2, it is then natural to consider the graded rings

$$\mathrm{R}_{\mathbb{C}}(\mathfrak{H}_0) = \bigoplus_{n=0}^{\infty} (\mathbb{C} \otimes_{\mathbb{Z}} \mathrm{R}_0(\mathfrak{H}_0(n))) \quad ; \quad \mathrm{K}_{\mathbb{C}}(\mathfrak{H}_0) = \bigoplus_{n=0}^{\infty} (\mathbb{C} \otimes_{\mathbb{Z}} \mathrm{K}_0(\mathfrak{H}_0(n))),$$

whose products $\cdot \times \cdot$ are given by the induction functors $\mathrm{Ind}_{\mathfrak{H}_0(m)\times\mathfrak{H}_0(n)}^{\mathfrak{H}_0(m+n)}(\cdot \boxtimes \cdot)$. There is a non-degenerate bilinear form

$$\langle \cdot \mid \cdot \rangle : \mathrm{K}_{\mathbb{C}}(\mathfrak{H}_0) \times \mathrm{R}_{\mathbb{C}}(\mathfrak{H}_0) \to \mathbb{C},$$

which is the orthogonal sum of the perfect pairings $\mathrm{K}_0(\mathfrak{H}_0(n)) \times \mathrm{R}_0(\mathfrak{H}_0(n)) \to \mathbb{Z}$ defined by $\langle P \mid S \rangle = \delta_{S=T(P)} = \delta_{P=P(S)}$ if S is a simple module and P is a principal indecomposable module. In this setting, a good understanding of the representation theory of the 0–Hecke algebras amounts to the description of the rings $\mathrm{R}_{\mathbb{C}}(\mathfrak{H}_0)$ and $\mathrm{K}_{\mathbb{C}}(\mathfrak{H}_0)$, as well as their pairing $\langle \cdot \mid \cdot \rangle$. Notice that the two Grothendieck rings

$R_{\mathbb{C}}(\mathfrak{H}_0)$ and $K_{\mathbb{C}}(\mathfrak{H}_0)$ have no reason to be isomorphic, in opposition to the case of the symmetric groups, where

$$R_{\mathbb{C}}(\mathfrak{S}) = K_{\mathbb{C}}(\mathfrak{S}) = \mathrm{Sym}$$

by the Frobenius–Schur isomorphism 2.31.

In Sections 6.1 and 6.2, we construct two graded Hopf algebras of functions NCSym and QSym, which are extensions of the algebra of symmetric functions Sym, and which are dual of one another: the algebra of non-commutative symmetric functions, and the algebra of quasi-symmetric functions. We shall then see in Section 6.3 that there are isomorphisms of $\mathbb{C}$-algebras

$$R_{\mathbb{C}}(\mathfrak{H}_0) = \mathrm{QSym} \qquad ; \qquad K_{\mathbb{C}}(\mathfrak{H}_0) = \mathrm{NCSym},$$

and that these isomorphisms are compatible with the pairings of $(K_{\mathbb{C}}(\mathfrak{H}_0), R_{\mathbb{C}}(\mathfrak{H}_0))$ and of $(\mathrm{NCSym}, \mathrm{QSym})$; cf. Theorem 6.18. We shall thus obtain an analogue of the Frobenius–Schur isomorphism in the setting of the 0–Hecke algebras and their two Grothendieck groups of representations.

As in Chapter 2, the two combinatorial Hopf algebras NCSym and QSym have interest in themselves, and they shall appear in later parts of the book:

- the algebra of quasi-symmetric functions QSym will be closely related to an algebra of generalized observables of Young diagrams; see Chapter 10;
- the algebra of *free* quasi-symmetric functions FQSym, which is a common extension of NCSym and of QSym, will allow a combinatorial study of the central measures on partitions; see Chapter 12.

On the other hand, the notion of composition introduced in Section 4.4 will be the main combinatorial tool in the proofs of the results of this chapter; thus, the compositions $c \in \mathfrak{C}(n)$ shall replace the integer partitions $\lambda \in \mathfrak{Y}(n)$ in the representation theory of $\mathfrak{H}_0(n)$ (instead of $\mathfrak{S}(n)$).

6.1 Non-commutative symmetric functions

▷ *Non-commutative elementary and homogeneous functions.*

Given an infinite alphabet $X = \{x_1, x_2, \ldots, x_n, \ldots\}$ of commutative variables, we defined in Chapter 2 the elementary symmetric functions

$$e_k(X) = \sum_{i_1 > i_2 > \cdots > i_k} x_{i_1} x_{i_2} \cdots x_{i_k},$$

as well as the homogeneous symmetric functions

$$h_k(X) = \sum_{i_1 \leq i_2 \leq \cdots \leq i_k} x_{i_1} x_{i_2} \cdots x_{i_k}.$$

The relations between these functions are encoded in the formula

$$E(t) = \frac{1}{H(-t)},$$

where $E(t) = 1 + \sum_{k=1}^{\infty} e_k(X)\, t^k$ and $H(t) = 1 + \sum_{k=1}^{\infty} h_k(X)\, t^k$. Suppose now that $x_1, x_2, \ldots$ do not commute anymore; thus, we place ourselves in $\mathbb{C}\langle X\rangle = \mathbb{C}\langle x_1, x_2, \ldots\rangle$, which is the projective limit in the category of graded algebras of the free associative algebras

$$\mathbb{C}\langle x_1, \ldots, x_n\rangle$$

over n generators. We can still consider in $\mathbb{C}\langle X\rangle$ the sums

$$\Lambda_k(X) = \sum_{i_1 > i_2 > \cdots > i_k} x_{i_1} x_{i_2} \cdots x_{i_k};$$

$$S_k(X) = \sum_{i_1 \leq i_2 \leq \cdots \leq i_k} x_{i_1} x_{i_2} \cdots x_{i_k}.$$

Here, we use new notations Λ_k and S_k instead of e_k and h_k in order to make clear that we are dealing with non-commutative variables. The associated generating series $\Lambda(t) = \sum_{k=0}^{\infty} \Lambda_k(X)\, t^k$ and $S(t) = \sum_{k=0}^{\infty} S_k(X)\, t^k$, which live in $\varprojlim_{n\to\infty} (\mathbb{C}\langle x_1, \ldots, x_n\rangle)[[t]]$, are obtained by expansion of the *ordered* infinite products

$$\Lambda(t) = \overleftarrow{\prod_{i \geq 1}} (1 + t x_i) = \cdots (1 + t x_n) \cdots (1 + t x_2)(1 + t x_1);$$

$$S(t) = \overrightarrow{\prod_{i \geq 1}} \frac{1}{1 - t x_i} = \frac{1}{1 - t x_1} \frac{1}{1 - t x_2} \cdots \frac{1}{1 - t x_n} \cdots.$$

One has again $\Lambda(t) S(-t) = 1$, which implies that any (non-commutative) polynomial in the elements $\Lambda_k(X)$ can be written in terms of the elements $S_k(X)$, and vice versa. For instance, one can compute

$$S_3(X) = \Lambda_3(X) - \Lambda_2(X)\Lambda_1(X) - \Lambda_1(X)\Lambda_2(X) + (\Lambda_1(X))^3.$$

Definition 6.1. *A* ***non-commutative symmetric function*** *is an element of the graded subalgebra*

$$\mathrm{NCSym} = \mathbb{C}\langle \Lambda_1, \Lambda_2, \ldots\rangle = \mathbb{C}\langle S_1, S_2, \ldots\rangle$$

of $\mathbb{C}\langle X\rangle$. We call ***non-commutative elementary functions*** *and* ***non-commutative homogeneous functions*** *the series $\Lambda_{k\geq 1}(X)$ and $S_{k\geq 1}(X)$.*

If $c = (c_1, \dots, c_\ell)$ is a composition of size n, we denote

$$\Lambda_c(X) = \Lambda_{c_1}(X)\,\Lambda_{c_2}(X) \cdots \Lambda_{c_\ell}(X);$$
$$S_c(X) = S_{c_1}(X)\,S_{c_2}(X) \cdots S_{c_\ell}(X).$$

We also set $\mathfrak{C} = \bigsqcup_{n \in \mathbb{N}} \mathfrak{C}(n)$. Notice that if $\deg(\cdot)$ is the restriction to NCSym of the gradation $\deg(x_{i_1} x_{i_2} \dots x_{i_k}) = k$ on $\mathbb{C}\langle X \rangle$, then

$$\deg \Lambda_c = \deg S_c = |c| = \sum_{i=1}^{\ell(c)} c_i$$

for any composition c.

Proposition 6.2. *The two families* $(\Lambda_c)_{c \in \mathfrak{C}}$ *and* $(S_c)_{c \in \mathfrak{C}}$ *are graded linear bases of* NCSym.

Proof. In the sequel, we shall use many times the notion of **descent** of a composition: if $c = (c_1, c_2, \dots, c_\ell) \in \mathfrak{C}(n)$, then a descent of c is an element of the set

$$D(c) = (c_1, c_1 + c_2, \dots, c_1 + c_2 + \dots + c_{\ell-1}).$$

By definition of the algebra NCSym, the products Λ_c and S_c of elements Λ_k or S_k span linearly NCSym, so the only thing to prove is the linear independence. We shall only treat the case of the functions Λ_c, the case of the functions S_c being extremely similar. Suppose that one has a non-trivial linear combination $\sum_c a_c \Lambda_c(X)$ that vanishes, with $a_c \neq 0$ for any composition c involved in the sum. By using the gradation of NCSym, one can assume that the sum is over a set of compositions c that all have the same size n. Notice then that elements of $\mathbb{C}\langle X \rangle_n$ can be considered as formal linear combinations of words of size n, with entries in $\mathbb{N}^*$. With this point of view,

$$\Lambda_c(X) = \sum_{w \text{ word whose rises belong to } D(c)} w,$$

where we convene that a rise in a word is an index i such that $w_i \leq w_{i+1}$. Let $c_* = (c_1, c_2, \dots, c_\ell)$ be a composition involved in the sum that is maximal with respect to the inclusion of descent sets. Then, any word whose rises are exactly the descents of c_* appears in the sum with coefficient a_{c_*}, as $D(c_*)$ is not included in any other descent set $D(c)$. Since words are linearly independent in $\mathbb{C}\langle X \rangle$, $a_{c_*} = 0$, which contradicts the previous hypothesis that $a_c \neq 0$ for any composition c involved in the sum. Thus, the elements Λ_c are indeed linearly independent. □

If $c \in \mathfrak{C}(n)$, then it is entirely determined by its descent set, which is an arbitrary subset of $[\![1, n-1]\!]$. Therefore, the number of compositions of size n, and the dimension of the space of homogeneous degree n non-commutative symmetric functions is

$$\dim_{\mathbb{C}}(\mathrm{NCSym}_n) = \operatorname{card} \mathfrak{C}(n) = 2^{n-1}.$$

On the other hand, there is a natural morphism of algebras $\Phi : \mathrm{NCSym} \to \mathrm{Sym}$, which associates to a function of the non-commutative alphabet X the same function but with a commutative alphabet X. Then, by construction, $\Phi(\Lambda_k) = e_k$ and $\Phi(S_k) = h_k$ for any $k \geq 1$.

$\triangleright$ *The Hopf algebra structure of* NCSym.

When we introduced in Section 2.3 the Hopf algebra of symmetric functions, the coproduct Δ came from the operation of sum of alphabets of commutative variables:

$$X + Y = (x_1, x_2, \ldots, y_1, y_2, \ldots) \quad \text{if } X = (x_1, x_2, \ldots) \text{ and } Y = (y_1, y_2, \ldots).$$

We keep the same definition with alphabets of non-commutative variables. Beware though that when dealing with non-commutative variables and non-commutative symmetric functions of them, the labeling or ordering $x_1, x_2, \ldots$ of the variables is important, as

$$\Lambda_k(x_1, x_2, \ldots) \neq \Lambda_k(x_{\sigma(1)}, x_{\sigma(2)}, \ldots)$$

if $\sigma \in \mathfrak{S}(\infty)$. Therefore, the sum $X + Y$ defined above has to be considered as a sum of **ordered alphabets** of non-commutative variables. To highlight this difference, we shall denote ordered alphabets with parentheses ( ), and non-ordered alphabets with brackets { }; and we shall use the symbol $\oplus$ instead of $+$ for a sum of ordered alphabets.

Let $f(X \oplus Y)$ be a non-commutative symmetric function in the sum of ordered alphabets $X \oplus Y$. It is easily seen that

$$\Lambda_k(X \oplus Y) = \sum_{l=0}^{k} \Lambda_l(Y)\, \Lambda_{k-l}(X);$$

$$S_k(X \oplus Y) = \sum_{l=0}^{k} S_l(X)\, S_{k-l}(Y).$$

As a consequence, if one makes the two ordered alphabets X and Y commute with one another ($x_i y_j = y_j x_i$ for any i, j), then for any function $f \in \mathrm{NCSym}$, $f(X \oplus Y)$ belongs to the tensor product $\mathrm{NCSym}(X) \otimes_{\mathbb{C}} \mathrm{NCSym}(Y)$. This allows one to define the coproduct $\Delta : \mathrm{NCSym} \to \mathrm{NCSym} \otimes_{\mathbb{C}} \mathrm{NCSym}$: it is the unique morphism of algebras such that

$$\Delta(\Lambda_k) = \sum_{l=0}^{k} \Lambda_l \otimes \Lambda_{k-l} \quad ; \quad \Delta(S_k) = \sum_{l=0}^{k} S_l \otimes S_{k-l},$$

with by convention $\Lambda_0 = S_0 = 1$. The coassociativity is trivial from these formulas. On the other hand, since $\Delta(e_k)$ and $\Delta(h_k)$ satisfy the same formulas in Sym, the projection $\Phi : \mathrm{NCSym} \to \mathrm{Sym}$ is then a morphism of bialgebras, the counity η

of NCSym being the projection on the one-dimensional space NCSym_0. Finally, notice that NCSym is cocommutative: $\Delta = \tau \circ \Delta$, where $\tau(f \otimes g) = g \otimes f$ for any $f, g \in \mathrm{NCSym}$. Indeed, since Δ and τ are morphisms of algebras, it suffices to show this on a generating family of NCSym, say the non-commutative elementary functions Λ_k. Then,

$$\tau \circ \Delta(\Lambda_k) = \sum_{l=0}^{k} \tau(\Lambda_l \otimes \Lambda_{k-l}) = \sum_{l=0}^{k} \Lambda_{k-l} \otimes \Lambda_l = \Delta(\Lambda_k).$$

We also introduce an antipode ω for NCSym: it is the unique antimorphism of algebras such that

$$\omega(\Lambda_k) = (-1)^k S_k.$$

Proposition 6.3. *Endowed with its product ∇, its unity $\varepsilon : \mathbb{C} \to \mathrm{NCSym}$, its coproduct Δ, its counity $\eta : \mathrm{NCSym} \to \mathbb{C}$ and its antipode ω, NCSym is a non-commutative and cocommutative graded Hopf algebra, and the map $\Phi : \mathrm{NCSym} \to \mathrm{Sym}$ is a morphism of graded Hopf algebras.*

Proof. To show that NCSym is a Hopf algebra, it remains to check the relation

$$\nabla \circ (\mathrm{id} \otimes \omega) \circ \Delta = \varepsilon \circ \eta$$

(the symmetric relation with $\omega \otimes \mathrm{id}$ is shown by the same argument). For any $k \geq 1$, the left-hand side of the formula evaluated on Λ_k yields

$$(\nabla \circ (\mathrm{id} \otimes \omega) \circ \Delta)(\Lambda_k) = \sum_{l=0}^{k} (\nabla \circ (\mathrm{id} \otimes \omega))(\Lambda_l \otimes \Lambda_{k-l}) = \sum_{l=0}^{k} (-1)^{k-l} \Lambda_l \, S_{k-l}.$$

The vanishing of the right-hand side is equivalent to the formula $\Lambda(t)\,S(-t) = 1$, which we know to be true. More generally, given a composition $c = (c_1, \ldots, c_\ell)$ of size $|c| \geq 1$, we have:

$$\begin{aligned}(\nabla \circ (\mathrm{id} \otimes \omega) \circ \Delta)(\Lambda_c) &= \sum_{\substack{d=(d_1,\ldots,d_\ell)\\ 0 \leq d_i \leq c_i}} (\nabla \circ (\mathrm{id} \otimes \omega))(\Lambda_d \otimes \Lambda_{c-d}) \\ &= \sum_{\substack{d=(d_1,\ldots,d_\ell)\\ 0 \leq d_i \leq c_i}} (-1)^{|c|-|d|} \Lambda_d \, S_{\widetilde{(c-d)}}\end{aligned}$$

where $\widetilde{c} = (c_\ell, \ldots, c_1)$ if $c = (c_1, \ldots, c_\ell)$. The vanishing of the right-hand side is now equivalent to the formula

$$\Lambda(t_1)\,\Lambda(t_2) \cdots \Lambda(t_\ell)\,S(t_\ell) \cdots S(t_2)\,S(t_1) = 1.$$

So, for any $c \neq 0$, $(\nabla \circ (\mathrm{id} \otimes \omega) \circ \Delta)(\Lambda_c) = 0 = \varepsilon \circ \eta(\Lambda_c)$, and the relation is true on a linear basis of NCSym. The fact that Φ is a morphism of Hopf algebras is then trivial. □

Proposition 6.4. *The antipode ω of* NCSym *is involutive, so $\omega(S_k) = (-1)^k \Lambda_k$ for any $k \geq 1$.*

Proof. Since $S_1 = \Lambda_1$, the formula $\omega(S_k) = (-1)^k \Lambda_k$ is true for $k = 1$. Suppose that it is true up to order $k-1 \geq 1$, and consider the case of S_k. One has

$$0 = \varepsilon \circ \eta(S_k) = (\nabla \circ (\omega \otimes \mathrm{id}) \circ \Delta)(S_k) = \sum_{k=0}^{l} \omega(S_l) S_{k-l} = \omega(S_k) + \sum_{k=0}^{l-1} (-1)^l \Lambda_l S_{k-l}.$$

However, we also know that $0 = \sum_{k=0}^{l} (-1)^l \Lambda_l S_{k-l}$, so $\omega(S_k) = (-1)^k \Lambda_k$, and by induction the formula is true for any k. Hence, ω is involutive. □

Remark. In fact, in any commutative or cocommutative Hopf algebra H, the antipode ω is involutive. Indeed, suppose for instance that H is cocommutative, and let us prove that $\omega \circ \omega = \mathrm{id}_H$. Recall from Lemma 5.4 that ω is the inverse of id_H with respect to the convolution product on linear functions in $\mathrm{End}(H)$. Therefore, it suffices to prove that $\omega^2 * \omega = 1_H \circ \eta_H$: this will imply that $\omega^2 = \mathrm{id}_H$. However, using the fact that ω is an anti-endomorphism of algebras, we get for any $h \in H$:

$$\begin{aligned}(\omega^2 * \omega)(h) = \nabla \circ (\omega^2 \otimes \omega) \circ \Delta(h) &= \nabla \circ (\omega^2 \otimes \omega)\left(\sum_i h_{i1} \otimes h_{i2}\right)\\ &= \sum_i \omega^2(h_{i1})\,\omega(h_{i_2}) = \omega\left(\sum_i h_{i2}\,\omega(h_{i1})\right)\\ &= \omega \circ \nabla \circ (\mathrm{id}_H \otimes \omega)\left(\sum_i h_{i2} \otimes h_{i1}\right).\end{aligned}$$

As H is supposed cocommutative, $\sum_i h_{i2} \otimes h_{i1} = \sum_i h_{i1} \otimes h_{i2} = \Delta(h)$, so,

$$(\omega^2 * \omega)(h) = \omega \circ \nabla \circ (\mathrm{id}_H \otimes \omega) \circ \Delta(h) = \omega \circ (1_H \circ \eta_H)(h).$$

Finally, in H, $1 = (1_H \circ \eta_H)(1) = (\nabla \circ (\omega \otimes \mathrm{id}_H) \circ \Delta)(1) = \omega(1)1 = \omega(1)$, so $\omega \circ 1_H = 1_H$, and finally $\omega^2 * \omega = 1_H \circ \eta_H$. The proof is analogue when H is supposed commutative.

▷ *Ribbon Schur functions.*

We now have a graded involutive non-commutative Hopf algebra NCSym, which will turn out to be isomorphic to $\mathrm{K}_{\mathbb{C}}(\mathfrak{H}_0)$ at the end of the chapter. To establish this isomorphism, we shall need to manipulate functions in NCSym that correspond to the principal indecomposable modules over $\mathfrak{H}_0(n)$. These functions will be the **non-commutative ribbon Schur functions**, and with respect to the isomorphism $\Psi_{\mathrm{K}} : \mathrm{K}_{\mathbb{C}}(\mathfrak{H}_0) \to \mathrm{NCSym}$, they will play a role that is similar to the role of Schur functions in the Frobenius–Schur isomorphism $\Psi : \mathrm{R}_{\mathbb{C}}(\mathfrak{S}) \to \mathrm{Sym}$.

Definition 6.5. *Let $c = (c_1, \dots, c_\ell)$ be a composition of size n. We set*

$$R_c = \sum_{d \preceq c} (-1)^{\ell(c)-\ell(d)} S_d,$$

where the sum runs over compositions d such that $D(d) \subset D(c)$ (notation: $d \preceq c$). We then say that R_c is the non-commutative ribbon Schur function of label c.

The inverse matrix of the change of basis matrix from $(R_c)_{c\in\mathfrak{C}(n)}$ to $(S_c)_{c\in\mathfrak{C}(n)}$ is easily computed to be:

$$S_c = \sum_{d \preceq c} R_d.$$

Indeed, one then has

$$S_c = \sum_{d,e \,|\, e \preceq d \preceq c} (-1)^{\ell(d)-\ell(e)} S_e,$$

and if $c \neq e$, then $\sum_{d \,|\, e \preceq d \preceq c} (-1)^{\ell(d)-\ell(e)} = 0$: this sum is the sum over all subsets D of $D(c) \setminus D(e)$ of $(-1)^{\mathrm{card}\, D}$, and

$$\sum_{D \subset D(c)\setminus D(e)} (-1)^{\mathrm{card}\, D} = \prod_{i \in D(c)\setminus D(e)} (1-1) = 0.$$

As a consequence, $(R_c)_{c\in\mathfrak{C}}$ is another graded linear basis of NCSym. The remainder of the paragraph is devoted to the description of the Hopf algebra operations of NCSym on this basis.

Proposition 6.6. *For any compositions $c = (c_1, \dots, c_l)$ and $d = (d_1, \dots, d_m)$,*

$$R_c R_d = R_{c \bowtie d} + R_{c \cdot d},$$

where $c \bowtie d = (c_1, \dots, c_{l-1}, c_l + d_1, d_2, \dots, d_m)$ and $c \cdot d = (c_1, \dots, c_l, d_1, \dots, d_m)$.

Proof. We expand R_c and R_d over the basis of non-commutative homogeneous functions:

$$R_c R_d = \sum_{\substack{a \preceq c \\ b \preceq d}} (-1)^{\ell(c\cdot d)-\ell(a\cdot b)} S_{a\cdot b}.$$

On the other hand, when one expands $R_{c\cdot d}$, one gets terms S_e with the composition $e \preceq c \cdot d$ that can be of two types:

- either $|c|$ is a descent of e, in which case $e = a \cdot b$ with $a \preceq c$ and $b \preceq d$.
- or, $|c|$ is not a descent of e, in which case $e = a \bowtie b$ with $a \preceq c$ and $b \preceq d$.

Therefore,

$$R_{c\cdot d} = \sum_{\substack{a \preceq c \\ b \preceq d}} (-1)^{\ell(c\cdot d)-\ell(a\cdot b)} S_{a\cdot b} + \sum_{e \preceq c \bowtie d} (-1)^{\ell(c\cdot d)-\ell(e)} S_e = R_c R_d - R_{c\bowtie d},$$

the sign minus in the right-hand side coming from $\ell(c \cdot d) = \ell(c \bowtie d) + 1$. □

For the next computations, we need to associate a Young diagram to any composition $c \in \mathfrak{C}$. If $c = (c_1, \ldots, c_\ell) \in \mathfrak{C}$, the **ribbon diagram** of c, or simply **ribbon** of c, is the unique ribbon Young diagram with c_1 cells on its first line, c_2 cells on the second line, etc. For instance, the ribbon of the composition $(4, 1, 3, 2)$ is

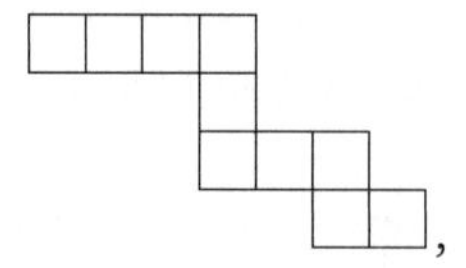

and the descents of the composition correspond to the places where its ribbon descends. This graphical representation allows one to define the conjugate c' of a composition: it is the unique composition whose ribbon is obtained by symmetrizing the ribbon of c with respect to the first diagonal. For instance, $(4, 1, 3, 2)' = (1, 2, 1, 3, 1, 1, 1)$.

Remark. There are several skew partitions $\lambda \setminus \mu$ that correspond to the same graphical ribbon, as one can add rows or columns of cells below or to the right of λ and μ without changing the shape of $\lambda \setminus \mu$. In the sequel, when we speak of the skew partition $\lambda \setminus \mu$ associated to a ribbon and to a composition c, we shall always mean the one for which $|\mu|$ and $|\lambda|$ are minimal. In other words, the top-left cell of the ribbon touches the y-axis, and the bottom-right cell of the ribbon touches the x-axis. Actually, all of the results hereafter are independent of this choice, and it is only here to ensure that there is no ambiguity of notation.

Proposition 6.7. *For any composition c, $\omega(R_c) = (-1)^{|c|} R_{c'}$.*

Lemma 6.8. *For any $k \geq 1$, $R_k = S_k$ and $R_{1^k} = \Lambda_k$.*

Proof. The identity $R_k = S_k$ is trivial, since (k) is the only composition smaller than (k) for the partial order $\preceq$. On the other hand, by definition,

$$R_{1^k} = \sum_{c \in \mathfrak{C}_k} (-1)^{k-\ell(c)} S_c,$$

and we have to prove that the right-hand side is Λ_k. Let $w = x_{i_1} x_{i_2} \cdots x_{i_k}$ be a word of length k. It appears as a component of S_c if and only if

$$1, \ldots, c_1 - 1, c_1 + 1, \ldots, c_1 + c_2 - 1, \ldots, c_1 + \cdots + c_{\ell-1} + 1, \ldots, c_1 + \cdots + c_\ell - 1$$

are rises of w. As a consequence, w appears in S_c if and only if the descents of w belong to the set $D(c)$ of descents of c. It follows that

$$[w](R_{1^k}) = \sum_{c \mid D(c) \supset D(w)} (-1)^{k-1-\operatorname{card} D(c)}.$$

This alternate sum vanishes unless $D(w) = [\![1, k-1]\!]$, in which case it is equal to 1. So, we conclude that

$$R_{1^k} = \sum_{w \mid D(w) = [\![1,k-1]\!]} w = \Lambda_k. \qquad \square$$

Proof of Proposition 6.7. We reason by induction on the length of c. If $c = (k)$ has length 1, then $R_{(k)} = S_k$, and $\omega(R_k) = (-1)^k \Lambda_k = (-1)^k R_{1^k}$. So, the result is true in this case, as $(k)' = 1^k$. Suppose now the result is true for compositions of length $\ell - 1 \geq 1$, and consider a composition c of length ℓ. By Proposition 6.6,

$$R_c = R_{(c_1,\ldots,c_{\ell-1})} R_{c_\ell} - R_{(c_1,\ldots,c_{\ell-2},c_{\ell-1}+c_\ell)}.$$

Since ω is an anti-isomorphism, this implies

$$\begin{aligned}\omega(R_c) &= (-1)^{|c|} \left(R_{1^{c_\ell}} R_{(c_1,\ldots,c_{\ell-1})'} - R_{(c_1,\ldots,c_{\ell-2},c_{\ell-1}+c_\ell)'}\right) \\ &= (-1)^{|c|} \left(R_{(1^{c_\ell}) \bowtie (c_1,\ldots,c_{\ell-1})'} + R_{(1^{c_\ell})\cdot(c_1,\ldots,c_{\ell-1})'} - R_{(c_1,\ldots,c_{\ell-2},c_{\ell-1}+c_\ell)'}\right)\end{aligned}$$

by using the induction hypothesis on the first line, and Proposition 6.6 on the second line. The result follows then from the identities of ribbons

$$(c_1,\ldots,c_\ell)' = (1^{c_\ell}) \bowtie (c_1,\ldots,c_{\ell-1})'$$

and

$$(1^{c_\ell}) \cdot (c_1,\ldots,c_{\ell-1})' = (c_1,\ldots,c_{\ell-2},c_{\ell-1}+c_\ell)'.$$

These identities are particular cases of the more general identity $(a \cdot b)' = b' \bowtie a'$, which is geometrically obvious. □

If $\lambda \setminus \mu$ is a ribbon Young diagram, then for any integer partition ν such that $\mu \subset \nu \subset \lambda$, the two skew Young diagrams $\nu \setminus \mu$ and $\lambda \setminus \nu$ are both disjoint unions of ribbons. We say that $(\nu \setminus \mu, \lambda \setminus \nu)$ is a **decomposition of the ribbon** $\lambda \setminus \mu$. In the following, if $\lambda \setminus \mu = c^{(1)} \sqcup c^{(2)} \sqcup \cdots \sqcup c^{(m)}$ is a disjoint union of ribbons, we set

$$R_{\lambda\setminus\mu} = \prod_{j=1}^{m} R_{c^{(j)}}.$$

By Proposition 6.6, it is a sum of (distinct) ribbon Schur functions.

Proposition 6.9. *The coproduct of a ribbon Schur function R_c is*

$$\Delta(R_c) = \sum_{\mu\subset\nu\subset\lambda} R_{\nu\setminus\mu} \otimes R_{\lambda\setminus\nu},$$

where $\lambda \setminus \mu$ is the ribbon of the composition c, and the sum runs over decompositions $(\nu \setminus \mu, \lambda \setminus \nu)$ of this ribbon.

Lemma 6.10. *For any composition $c \in \mathfrak{C}(n)$, $R_c(X)$ is the sum of all words $w = x_{i_1} x_{i_2} \ldots x_{i_n}$ of length n and whose descents are exactly those of the composition c.*

Proof. Notice that this result generalizes the previous Lemma 6.8. By definition,

$R_c(X) = \sum_{d \preceq c}(-1)^{\ell(c)-\ell(d)} S_d(X)$, and on the other hand, $S_d(X)$ is the sum of words whose descents belong to $D(d)$. Therefore,

$$R_c(X) = \sum_{w \,|\, D(w) \subset D(c)} \left(\sum_{d \,|\, D(w) \subset D(d) \subset D(c)} (-1)^{\ell(c)-\ell(d)} \right) w.$$

By the same argument as in Lemma 6.8, the sum in parentheses vanishes unless the set of descents $D(w)$ of the word w is $D(c)$. So, $R_c(X) = \sum_{w \,|\, D(w)=D(c)} w$. □

Proof of Proposition 6.9. We compute $R_c(X \oplus Y)$ and expand it as a sum of product of functions of X and of Y. Let w be a word of length n that appears in $R_c(X \oplus Y)$; it is a succession of $\ell = \ell(c)$ increasing words $w_1, \ldots, w_\ell$ with entries in $X \oplus Y$, with the last letter of w_i that is strictly bigger than the first letter of w_{i+1} for any $i \in [\![1, \ell-1]\!]$. Denote each subword $w_i = u_i v_i$, where u_i is the part with entries in X, and v_i is the part with entries in Y. We associate to this decomposition a coloring of the cells of the ribbon $\lambda \setminus \mu$ of the composition c: on each row i of length c_i, we mark the $|u_i|$ first cells. The marked cells correspond then to a decomposition $(\nu \setminus \mu, \lambda \setminus \nu)$ of the ribbon $\lambda \setminus \mu$. Indeed, we only have to check that if the last cell of the i-th row is marked, then the first cell of the $(i+1)$-th row is also marked (this is equivalent to the fact that the marked cells correspond to a skew partition $\nu \setminus \mu$). However, if the last cell of the i-th row is marked, then w_i finishes with an entry in X, and as the first letter of w_{i+1} is smaller, it is also in X, so the first cell of the $(i+1)$-th row is also marked. The result follows immediately by gathering the words w of the series $R_c(X \oplus Y)$ according to the decompositions $(\nu \setminus \mu, \lambda \setminus \nu)$ that are associated to them. □

Example. The decompositions of the ribbon associated to the composition $(2,2)$ are drawn below:

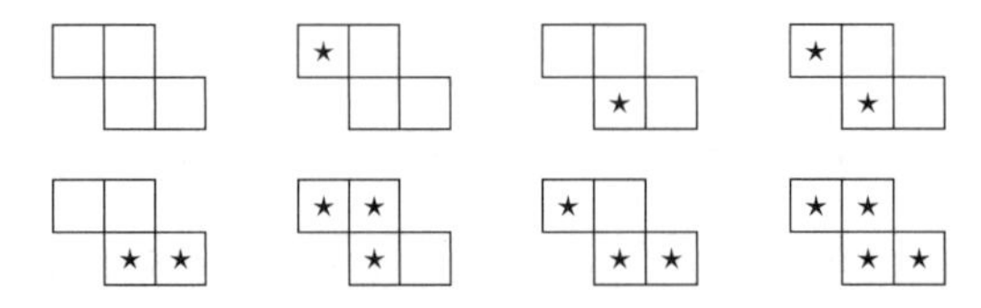

Therefore,

$$\begin{aligned}
&\Delta(R_{(2,2)}) \\
&= 1 \otimes R_{(2,2)} + R_1 \otimes R_{(1,2)} + R_1 \otimes (R_2 R_1) + (R_1)^2 \otimes (R_1)^2 \\
&\quad + R_2 \otimes R_2 + R_{(2,1)} \otimes R_1 + (R_1 R_2) \otimes R_1 + R_{(2,2)} \otimes 1 \\
&= 1 \otimes R_{(2,2)} + R_1 \otimes R_{(1,2)} + R_1 \otimes R_{(2,1)} + R_1 \otimes R_3 + R_{(1,1)} \otimes R_{(1,1)} + R_2 \otimes R_{(1,1)} \\
&\quad + R_{(1,1)} \otimes R_2 + 2R_2 \otimes R_2 + R_{(2,1)} \otimes R_1 + R_{(1,2)} \otimes R_1 + R_3 \otimes R_1 + R_{(2,2)} \otimes 1.
\end{aligned}$$

To conclude our study of the non-commutative ribbon Schur functions, let us compute their images by the morphism $\Phi : \mathrm{NCSym} \to \mathrm{Sym}$. The result is exactly what one can expect:

Proposition 6.11. *Let c be a composition with associated ribbon $\lambda \setminus \mu$. The symmetric function $\Phi(R_c)$ is the skew Schur function $s_{\lambda\setminus\mu}$.*

Proof. A reformulation of Lemma 6.10 is as follows: $R_c(X)$ is the sum of all reading words $W(T)$ of skew semistandard tableaux T of shape $\lambda\setminus\mu$. Making the variables commutative transforms the reading word $W(T)$ into the monomial x^T, so

$$\Phi(R_c)(X) = \sum_{T\in\mathrm{SST}(\lambda\setminus\mu)} x^T.$$

By Theorem 3.2, the right-hand side of the formula is $s_{\lambda\setminus\mu}(X)$. □

Remark. Since $(R_c)_{c\in\mathfrak{C}(n)}$ is a linear basis of NCSym_n, and since the projection $\Phi : \mathrm{NCSym} \to \mathrm{Sym}$ is surjective, we conclude that the *commutative* ribbon Schur functions $s_{\lambda\setminus\mu}$ with $\lambda \setminus \mu$ ribbon of size n span linearly Sym_n; this was not entirely obvious.

6.2 Quasi-symmetric functions

Since NCSym is a non-commutative but cocommutative Hopf algebra, there is no hope to endow it with a scalar product for which it will be a self-dual Hopf algebra. Instead, one can try to construct *another* Hopf algebra which will be commutative but non-cocommutative, and which will be the dual of NCSym with respect to a non-degenerate bilinear form. The solution to this problem is the algebra of **quasi-symmetric functions**; its construction is due to I. Gessel.

▷ *Monomial and fundamental quasi-symmetric functions.*

In the sequel, we fix an ordered alphabet $X = (x_1, x_2, \ldots)$, but this time with commutative variables x_i. A series $f(X)$ in $\mathbb{C}[X] = \varprojlim_{n\to\infty} \mathbb{C}[x_1, x_2, \ldots, x_n]$ is a symmetric function in Sym if and only if, for any families of distinct indices $i_1 \neq i_2 \neq \cdots \neq i_k$ and $j_1 \neq j_2 \neq \cdots \neq j_k$, and any family of positive exponents $\alpha_1, \ldots, \alpha_k$,

$$[x_{i_1}^{\alpha_1} x_{i_2}^{\alpha_2} \cdots x_{i_k}^{\alpha_k}](f) = [x_{j_1}^{\alpha_1} x_{j_2}^{\alpha_2} \cdots x_{j_k}^{\alpha_k}](f).$$

The notion of *quasi*-symmetric function is given by the slightly weaker following condition:

Definition 6.12. *Consider a series $f(X)$ in $\mathbb{C}[X]$, the projective limit in the category of graded algebras of the rings of polynomials $\mathbb{C}[x_1, x_2, \ldots, x_n]$. One says that $f(X)$ is a quasi-symmetric function if, for any increasing families of indices $i_1 < i_2 < \cdots < i_k$ and $j_1 < j_2 < \cdots < j_k$, and any family of positive exponents $\alpha_1, \ldots, \alpha_k$,*

$$[x_{i_1}^{\alpha_1} x_{i_2}^{\alpha_2} \cdots x_{i_k}^{\alpha_k}](f) = [x_{j_1}^{\alpha_1} x_{j_2}^{\alpha_2} \cdots x_{j_k}^{\alpha_k}](f).$$

We denote QSym the algebra of quasi-symmetric functions; it is a graded subalgebra of $\mathbb{C}[X]$, and it contains Sym.

Example. The series

$$M_{(2,1)}(X) = \sum_{i<j} (x_i)^2 x_j$$

is a quasi-symmetric function, but it is not symmetric: the coefficient of $(x_1)^2 x_2$ in $M_{(2,1)}(X)$ is 1, whereas the coefficient of $x_1(x_2)^2$ is 0. More generally, if $c = (c_1, \ldots, c_\ell)$ is a composition of size n, denote

$$M_c(X) = \sum_{i_1 < i_2 < \cdots < i_\ell} (x_{i_1})^{c_1} (x_{i_2})^{c_2} \cdots (x_{i_\ell})^{c_\ell};$$

it is a quasi-symmetric function of degree $|c|$, and actually, $(M_c)_{c \in \mathfrak{C}}$ is a graded linear basis of QSym. Indeed, if $f(X) \in$ QSym, then for any composition c, the coefficients in $f(X)$ of all the monomials $(x_{i_1})^{c_1}(x_{i_2})^{c_2} \cdots (x_{i_\ell})^{c_\ell}$ with $i_1 < i_2 < \cdots < i_\ell$ are the same, so,

$$f(X) = \sum_{c \in \mathfrak{C}} [x^c](f) \, M_c(X)$$

where $x^c = (x_1)^{c_1} \cdots (x_\ell)^{c_\ell}$. This shows that the M_c's span QSym, and since they involve disjoint sets of monomials x^k, they are linearly independent. We call the functions $M_c(X)$ **monomial quasi-symmetric functions**. The relation between them and the monomial symmetric functions is the following: for any integer partition λ,

$$m_\lambda(X) = \sum_{c \text{ permutation of } \lambda} M_c(X).$$

Thus, for instance, $m_{(2,1)}(X) = M_{(2,1)}(X) + M_{(1,2)}(X)$.

As for non-commutative symmetric functions, the coproduct of the algebra QSym is related to the operation of sum of (ordered) alphabets. More precisely, if X and Y are ordered alphabets of commutative variables, and if $X \oplus Y = (x_1, x_2, \ldots, y_1, y_2, \ldots)$ is their sum, then for any $c = (c_1, \ldots, c_\ell) \in \mathfrak{C}$,

$$M_c(X \oplus Y) = \sum_{k=0}^{\ell(c)} M_{(c_1,\ldots,c_k)}(X) \, M_{(c_{k+1},\ldots,c_\ell)}(Y)$$

belongs to $\mathrm{QSym}(X) \otimes \mathrm{QSym}(Y)$. Therefore, one has a well-defined morphism of algebras $\Delta : \mathrm{QSym} \to \mathrm{QSym} \otimes_{\mathbb{C}} \mathrm{QSym}$, given by

$$\Delta(M_c) = \sum_{k=0}^{\ell(c)} M_{(c_1,\ldots,c_k)} \otimes M_{(c_{k+1},\ldots,c_\ell)}.$$

The coassociativity of Δ is trivial from this formula, hence, QSym has a structure of bialgebra.

In order to define the antipode of QSym, it is convenient to introduce another graded linear basis of this algebra. The **fundamental quasi-symmetric functions** are defined by

$$L_c(X) = \sum_{d \succeq c} M_d(X),$$

the sum running over compositions d with the same size as c, and with a larger descent set. For instance,

$$L_{(2,2)}(X) = M_{(2,2)}(X) + M_{(1,1,2)}(X) + M_{(2,1,1)}(X) + M_{(1,1,1,1)}(X).$$

Then, $(L_c)_{c \in \mathfrak{C}}$ is a graded linear basis of QSym, and one has the inverse relation

$$M_c(X) = \sum_{d \succeq c} (-1)^{\ell(d)-\ell(c)} L_d(X).$$

Proposition 6.13. *For any composition c of size n,*

$$\Delta(L_c) = \sum_{k=0}^{n} L_{d_k} \otimes L_{e_k}$$

where d_k (respectively, e_k) is the composition of size k (respectively, $n-k$) with descent set $D(d_k) = D(c) \cap [\![1, k-1]\!]$ (resp., $D(e_k) = (D(c) \cap [\![k+1, n-1]\!]) - k$).

Example. One has

$$\Delta(L_{(2,2)}) = 1 \otimes L_{(2,2)} + L_{(1)} \otimes L_{(1,2)} + L_{(2)} \otimes L_{(2)} + L_{(2,1)} \otimes L_{(1)} + L_{(2,2)} \otimes 1.$$

Remark. The rule for the coproduct of a fundamental quasi-symmetric function is better explained in terms of ribbon diagrams: $\Delta(L_c)$ is the sum of all tensors $L_d \otimes L_e$, where d is a "beginning" of the ribbon of c, and e is the remainder $c \setminus d$. Thus, the previous computation rewrites as

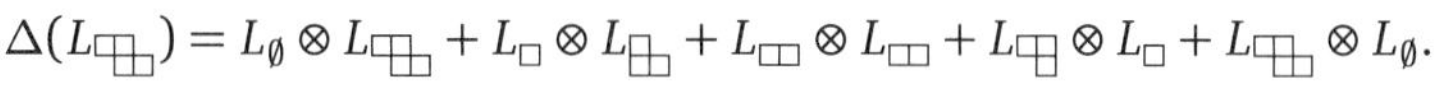

$$\Delta(L_{\substack{\square\square\\ \square\square}}) = L_{\emptyset} \otimes L_{\substack{\square\square\\ \square\square}} + L_{\square} \otimes L_{\substack{\square\\ \square\square}} + L_{\square\square} \otimes L_{\square\square} + L_{\substack{\square\square\\ \square}} \otimes L_{\square} + L_{\substack{\square\square\\ \square\square}} \otimes L_{\emptyset}.$$

Proof. Let $(x_{i_1})^{\alpha_1}(x_{i_2})^{\alpha_2} \cdots (x_{i_l})^{\alpha_l} (y_{j_1})^{\beta_1}(y_{j_2})^{\beta_2} \cdots (y_{i_m})^{\beta_m}$ be a monomial that appears in the series $L_c(X \oplus Y)$, with $i_1 < i_2 < \cdots < i_l$ and $j_1 < j_2 < \cdots < j_m$. The composition $(\alpha_1, \ldots, \alpha_l, \beta_1, \ldots, \beta_m)$ has more descents than c, because $L_c(X) = \sum_{C \succeq c} M_C(X)$. Now, if $k = \alpha_1 + \alpha_2 + \cdots + \alpha_l$, then $D_k = (\alpha_1, \ldots, \alpha_l)$ is a composition of size k, which has more descents than the composition d_k defined in the statement of the proposition; and $E_k = (\beta_1, \ldots, \beta_m)$ is a composition of size $n-k$, which has more descents than the composition e_k. Consequently, by gathering the monomials of $L_c(X \oplus Y)$ according to the associated pair of compositions,

$$L_c(X \oplus Y) = \sum_{k=0}^{n} \sum_{\substack{D_k \succeq d_k \\ E_k \succeq e_k}} M_{D_k}(X)\, M_{E_k}(Y) = \sum_{k=0}^{n} L_{d_k}(X)\, L_{e_k}(Y). \qquad \square$$

▷ *The Hopf algebra structure of* QSym.

We now define the antipode of QSym by:

$$\omega(L_c) = (-1)^{|c|} L_{c'},$$

where as before c' is the composition whose ribbon is the conjugate of the ribbon of c. Notice that by construction, ω is an involutive linear map on QSym.

Proposition 6.14. *Endowed with its product ∇, its unity $\varepsilon : \mathbb{C} \to$ QSym, its coproduct Δ, its counity η : QSym $\to$ QSym$_0 = \mathbb{C}$ and its antipode ω, QSym is a commutative and non-cocommutative graded Hopf algebra.*

The proof of this proposition relies on formulas for the product and the antipode of monomial quasi-symmetric functions. Let $c = (c_1, \ldots, c_l)$ and $d = (d_1, \ldots, d_m)$ be two compositions. If $k \in [\![\max(l,m), l+m]\!]$, we call compatible map $s : [\![1,l]\!] \sqcup [\![1',m']\!] \to [\![1,k]\!]$ a map that is surjective, and strictly increasing on $[\![1,l]\!]$ and separately on $[\![1',m']\!]$. Then, each $a \in [\![1,k]\!]$ is attained at least by one element in $[\![1,l]\!] \sqcup [\![1',m']\!]$, and at most by two elements, with one preimage in $[\![1,l]\!]$ and one preimage in $[\![1',m']\!]$. To such a compatible map, we can associate a new composition e of length k and size $|c| + |d|$:

$$e_a = \sum_{x \in [\![1,l]\!]} 1_{s(x)=a}\, c_x + \sum_{y \in [\![1,m]\!]} 1_{s(y')=a}\, d_y.$$

Thus, each part of e is either some c_x, or some d_y, or a sum $c_x + d_y$. We say that $e = (e_1, \ldots, e_k)$ is a **mixing** of the compositions c and d if e is the composition associated to a map s compatible with the compositions c and d. For instance, the list of all mixings of $(2,1)$ and (3) is:

$$(2,1,3),\ (2,3,1),\ (3,2,1),\ (2,4),\ (5,1).$$

Lemma 6.15. *For any compositions $c = (c_1, \ldots, c_l)$ and $d = (d_1, \ldots, d_m)$,*

$$M_c(X)\, M_d(X) = \sum M_e(X),$$

where the sum runs over mixings e of the compositions c and d.

Proof. Consider a monomial $(x_{i_1})^{c_1} \cdots (x_{i_l})^{c_l}$ in $M_c(X)$, with $i_1 < \cdots < i_l$; and a monomial $(x_{j_1})^{d_1} \cdots (x_{j_m})^{d_m}$ that appears in $M_d(X)$, with $j_1 < \cdots < j_m$. The nature of the product of these monomials depends on the possible equalities of indices $i_a = j_b$. We shall encode these identities by a compatible map. Let $K = \{i_x, j_y\}$, $k = \operatorname{card} K$, and s be the map from $[\![1,l]\!] \sqcup [\![1',m']\!]$ to $[\![1,k]\!]$ defined by

$$s(x \in [\![1,l]\!]) = \operatorname{card}\,(K \cap [\![1,i_x]\!]) \qquad ; \qquad s(y' \in [\![1',m']\!]) = \operatorname{card}\left(K \cap [\![1,j_y]\!]\right).$$

In other words, s associates to an index i_x or j_y its position in K. The map s

is therefore surjective, and it is strictly increasing on the interval $[\![1,l]\!]$ and on the interval $[\![1',m']\!]$. So, it is a compatible map, and by construction, if e is the corresponding mixing of c and d, then the type of the product

$$\left((x_{i_1})^{c_1}\cdots(x_{i_l})^{c_l}\right)\left((x_{j_1})^{d_1}\cdots(x_{j_m})^{d_m}\right)=(x_{p_1})^{e_1}\cdots(x_{p_k})^{e_k}$$

is e. The result follows by gathering the product of monomials according to the associated compositions e. □

Lemma 6.16. *For any composition $c=(c_1,\ldots,c_\ell)$, if $\widetilde{c}=(c_\ell,\ldots,c_1)$, then*

$$(\omega(M_c))(X)=(-1)^{\ell(c)}\sum_{a\preceq\widetilde{c}}M_a(X).$$

Example. If $c=(2,3,1,1)$, then $\widetilde{c}=(1,1,3,2)$, and

$$\omega(M_{2,3,1,1})=M_{(1,1,3,2)}+M_{(2,3,2)}+M_{(1,4,2)}+M_{(1,1,5)}+M_{(5,2)}+M_{(2,5)}+M_{(1,6)}+M_{(7)}.$$

Proof. We compute

$$\begin{aligned}\omega(M_c)&=\omega\left(\sum_{d\succeq c}(-1)^{\ell(d)-\ell(c)}L_d\right)=(-1)^{|c|}\sum_{d\succeq c}(-1)^{\ell(d)-\ell(c)}L_{d'}\\&=(-1)^{|c|}\sum_{b\preceq c'}(-1)^{\ell(b)-\ell(c')}L_b=(-1)^{|c|}\sum_{a,b\,|\,b\preceq c'\wedge a}(-1)^{\ell(b)-\ell(c')}M_a,\end{aligned}$$

where in the last sum $c'\wedge a$ is the composition of size $n=|c|$ with descent set $D(c')\cap D(a)$. For the same reasons as in many previous computations, the sum of signs $\sum_{b\,|\,b\preceq c'\wedge a}(-1)^{\ell(b)}$ vanishes unless it is trivial, that is to say that $c'\wedge a=(n)$ has no descent. As a consequence, and using the identity $\ell(c')+\ell(c)=|c|+1$, we obtain

$$\omega(M_c)=(-1)^{\ell(c)}\sum_{a\,|\,c'\wedge a=(n)}M_a.$$

To end the proof, we have to check that $c'\wedge a=(n)$ if and only if $a\preceq\widetilde{c}$. This is because $\widetilde{c}$ is the unique composition whose descent set is the complement of the descent set of c' (this is easy to see on ribbon diagrams). Therefore,

$$c'\wedge a=(n)\iff D(a)\cap D(c')=\emptyset\iff D(a)\subset D(\widetilde{c})\iff a\preceq\widetilde{c}.$$ □

Proof of Proposition 6.14. Again, the only non-trivial thing to check is the relation

$$\nabla\circ(\mathrm{id}\otimes\omega)\circ\Delta=\varepsilon\circ\eta.$$

Let $c=(c_1,\ldots,c_l)$ be a composition of size $|c|\geq 1$. We have:

$$\begin{aligned}(\nabla\circ(\mathrm{id}\otimes\omega)\circ\Delta)(M_c)&=\sum_{k=0}^{l}(\nabla\circ(\mathrm{id}\otimes\omega))(M_{(c_1,\ldots,c_k)}\otimes M_{(c_{k+1},\ldots,c_l)})\\&=\sum_{k=0}^{l}(-1)^{l-k}\sum_{a\preceq(c_l,\ldots,c_{k+1})}M_{(c_1,\ldots,c_k)}M_a.\end{aligned}$$

To explain the vanishing of this alternate sum, it will be useful to look at an example. Thus, we consider the case of $M_{(1,3,2)}$, and we write below the different terms of the sum:

$$\begin{aligned}
&- M_{(1,3,2)} \\
&+ M_{(1,3,2)} + M_{(1,2,3)} + M_{(1,5)} + M_{(2,1,3)} + M_{(3,3)} \\
&- M_{(1,2,3)} - M_{(1,5)} - M_{(2,1,3)} - M_{(3,3)} - M_{(2,3,1)} - M_{(2,4)} - M_{(5,1)} - M_{(6)} \\
&+ M_{(2,3,1)} + M_{(2,4)} + M_{(5,1)} + M_{(6)}.
\end{aligned}$$

Notice that every term on the k-th line is cancelled by a term on the $(k-1)$-th line or by a term on the $(k+1)$-th line. This phenomenon is general, as shown by the following argument. Let e be a composition such that M_e appears as a component of the k-th term $\sum_{a \preceq (c_l,\ldots,c_{k+1})} M_{(c_1,\ldots,c_k)} M_a$ (beware that M_e can appear with a multiplicity > 1). The composition $e = (e_1, \ldots, e_m)$ is a mixing of $(c_1, \ldots, c_k)$ and of some composition $a \preceq (c_l, \ldots, c_{k+1})$. As a consequence, e can be described as follows. If $m \in [\![k, l]\!]$, let $s : [\![1, l]\!] \to [\![1, m]\!]$ be a map such that:

- s is surjective;
- s is strictly increasing on $[\![1, k]\!]$;
- s is weakly decreasing on $[\![k+1, l]\!]$.

We define from s a composition $e = e(s) = (e_1, \ldots, e_m)$ of length m, with

$$e_a = \sum_{x=1}^{l} 1_{s(x)=a}\, c_x.$$

Then, any composition that is obtained by mixing of $(c_1, \ldots, c_k)$ and of some composition $a \preceq (c_l, \ldots, c_{k+1})$ corresponds to some map s satisfying the three conditions above, and conversely. Thus,

$$\sum_{a \preceq (c_l,\ldots,c_{k+1})} M_{(c_1,\ldots,c_k)} M_a = \sum_{s \text{ map satisfying the three conditions}} M_{e(s)}.$$

We then have two cases to distinguish:

1. If $s(k) < \ell(e) = m$, that is to say that the last part of e does not contain c_k, then the same composition e also appears in the $(k+1)$-th sum. Indeed, the last part of e must be attained by integers in $[\![k+1, l]\!]$, and as s is weakly decreasing on this interval, one has necessarily $s(k+1) = m$. Then, s is a surjective map that is strictly increasing on $[\![1, k+1]\!]$ and weakly decreasing on $[\![k+2, l]\!]$, so it is involved in the $(k+1)$-th sum.

2. Otherwise, if $s(k) = m$, then s is a surjective map that is strictly increasing on $[\![1, k-1]\!]$ and weakly decreasing on $[\![k, l]\!]$, so it is involved in the $(k-1)$-th sum.

As the sums of index k have alternate signs, we have thus shown that they cancel one another. So, for any $|c| \geq 1$, $(\nabla \circ (\mathrm{id} \otimes \omega) \circ \Delta)(M_c) = 0 = (\varepsilon \circ \eta)(M_c)$, and we have proved that QSym is a Hopf algebra, which contains Sym as a sub–Hopf algebra. □

▷ *The pairing between* NCSym *and* QSym.

To conclude our presentation of the algebras of non-commutative symmetric functions and quasi-symmetric functions, we shall prove that they are dual of one another. We define a bilinear map $\langle \cdot \mid \cdot \rangle : \mathrm{NCSym} \times \mathrm{QSym} \to \mathbb{C}$ by linear extension of the rule

$$\forall c, d \in \mathfrak{C},\ \langle S_c \mid M_d \rangle = \begin{cases} 1 & \text{if } c = d, \\ 0 & \text{otherwise,} \end{cases}$$

where S_c is a product of non-commutative homogeneous functions, and M_d is a monomial quasi-symmetric function. Since $(S_c)_{c \in \mathfrak{C}}$ and $(M_c)_{c \in \mathfrak{C}}$ are graded linear bases of NCSym and of QSym, this is a perfect pairing which realizes an isomorphism of vector spaces $\psi : \mathrm{QSym} \to \mathrm{NCSym}^* = \mathrm{Hom}_{\mathbb{C}}(\mathrm{NCSym}, \mathbb{C})$. The following theorem ensures that ψ is in fact an isomorphism of Hopf algebras.

Theorem 6.17. *With respect to the pairing between* NCSym *and* QSym*,* Δ_{QSym} *is the dual linear operator of* ∇_{NCSym}*,* Δ_{NCSym} *is the dual of* ∇_{QSym}*, and the antipodes* ω_{QSym} *and* ω_{NCSym} *are dual of one another. Moreover,*

$$\forall c, d \in \mathfrak{C},\ \langle R_c \mid L_d \rangle = \begin{cases} 1 & \text{if } c = d, \\ 0 & \text{otherwise.} \end{cases}$$

Proof. If $f, g \in \mathrm{NCSym}$ and $h \in \mathrm{QSym}$, we have to prove that

$$\langle f \otimes g \mid \Delta(h) \rangle = \langle f g \mid h \rangle .$$

It suffices to do so with $f = S_c$, $g = S_d$ and $h = M_e$, and indeed,

$$\langle S_c \otimes S_d \mid \Delta(M_e) \rangle = 1_{c \cdot d = e} = \langle S_c S_d \mid M_e \rangle .$$

Similarly, for the relation

$$\langle \Delta(f) \mid g \otimes h \rangle = \langle f \mid gh \rangle$$

with $f \in \mathrm{NCSym}$ and $g, h \in \mathrm{QSym}$, it suffices to prove it with $f = S_c$, $g = M_d$ and $h = M_e$, and then,

$$\langle \Delta(S_c) \mid M_d \otimes M_e \rangle = 1_{c \text{ is a mixing of } d \text{ and } e} = \langle S_c \mid M_d M_e \rangle .$$

The orthogonality relation between the non-commutative ribbon Schur functions and the fundamental quasi-symmetric functions comes by expansion on the bases of non-commutative homogeneous functions and monomial quasi-symmetric functions:

$$\begin{aligned}\langle R_c \mid L_d\rangle &= \sum_{\substack{a\preceq c\\ b\succeq d}} (-1)^{\ell(c)-\ell(a)}\,\langle S_a \mid M_b\rangle \\ &= \sum_{a\,|\,d\preceq a\preceq c} (-1)^{\ell(c)-\ell(a)} = \begin{cases} 1 & \text{if } c=d,\\ 0 & \text{otherwise.}\end{cases}\end{aligned}$$

Finally, we can use this second orthogonality relation to show that ω_{NCSym} and ω_{QSym} are dual:

$$\begin{aligned}\langle \omega(R_c) \mid L_d\rangle &= (-1)^{|c|}\,\langle R_{c'} \mid L_d\rangle = (-1)^{|c|}\,\delta_{c',d} = (-1)^{|d|}\,\delta_{c,d'} \\ &= (-1)^{|d|}\,\langle R_c \mid L_{d'}\rangle = \langle R_c \mid \omega(L_d)\rangle\,.\end{aligned}$$

This ends the proof of the duality between NCSym and QSym. □

6.3 The Hecke–Frobenius–Schur isomorphisms

In this last section of the chapter, we relate the two Hopf algebras NCSym and QSym to the two Grothendieck rings $\mathrm{K}_{\mathbb{C}}(\mathfrak{H}_0)$ and $\mathrm{R}_{\mathbb{C}}(\mathfrak{H}_0)$ of the Hecke algebras specialized at $q=0$. The specialized Hecke algebra $\mathfrak{H}_0(n)$ is the $\mathbb{C}$-algebra of dimension $n!$ with generators $T_1,\ldots,T_{n-1}$, and relations:

$$\begin{aligned}(T_i)^2 &= -T_i;\\ T_iT_{i+1}T_i &= T_{i+1}T_iT_{i+1};\\ T_iT_j &= T_jT_i \quad \text{if } |j-i|\geq 2.\end{aligned}$$

Our main objective is to give a proof of the following result, to be compared with Theorem 2.31:

Theorem 6.18. *The number of projective indecomposable modules or of simple modules over $\mathfrak{H}_0(n)$ is 2^{n-1}. There exists a labeling of these modules by compositions $c\in\mathfrak{C}(n)$, such that if $(P^c)_{c\in\mathfrak{C}(n)}$ (respectively, $(S^c)_{c\in\mathfrak{C}(n)}$) is a complete family of representatives of the isomorphism classes of principal indecomposable modules (respectively, simple modules) over $\mathfrak{H}_0(n)$ with $T(P^c)=S^c$ and $P(S^c)=P^c$, then the maps*

$$\begin{aligned}\Psi_{\mathrm{R}} : \mathrm{R}_{\mathbb{C}}(\mathfrak{H}_0) &\to \mathrm{QSym}\\ S^c &\mapsto L_c(X)\end{aligned}$$

and

$$\begin{aligned}\Psi_{\mathrm{K}} : \mathrm{K}_{\mathbb{C}}(\mathfrak{H}_0) &\to \mathrm{NCSym}\\ P^c &\mapsto R_c(X)\end{aligned}$$

are isomorphisms of graded algebras, and are also isometries with respect to the pairings between $\mathrm{R}_{\mathbb{C}}(\mathfrak{H}_0)$ *and* $\mathrm{K}_{\mathbb{C}}(\mathfrak{H}_0)$*, and between* QSym *and* NCSym.

▷ *Simple modules over* $\mathfrak{H}_0(n)$.

The first part of the proof of Theorem 6.18 consists in the identification of the simple modules over $\mathfrak{H}_0(n)$. By the discussion on Jacobson radicals in Section 4.2, a simple module M over $\mathfrak{H}_0(n)$ corresponds to a simple module over the semisimple algebra $\mathfrak{H}_0(n)/\mathrm{rad}(\mathfrak{H}_0(n))$, where $\mathrm{rad}(\mathfrak{H}_0(n))$ is the Jacobson radical.

Proposition 6.19. *The Jacobson radical of* $\mathfrak{H}_0(n)$ *is the bilateral ideal* I *spanned by the commutators* $[T_i, T_j] = T_iT_j - T_jT_i$. *The quotient* $\mathfrak{H}_0(n)/\mathrm{rad}(\mathfrak{H}_0(n))$ *is the commutative algebra of dimension* 2^{n-1} *with generators* $\widetilde{T}_1, \ldots, \widetilde{T}_{n-1}$ *and relations:*

$$(\widetilde{T}_i)^2 = -\widetilde{T}_i \quad ; \quad \widetilde{T}_i\widetilde{T}_j = \widetilde{T}_j\widetilde{T}_i.$$

In order to prove Proposition 6.19, we need a new characterization of the Jacobson radical of a finite-dimensional algebra A over a field k. Call **nilpotent ideal** of A a bilateral ideal I such that $I^k = 0$ for some $k \geq 1$, that is to say that the product of k elements of I is always 0. On the other hand, call **nil ideal** of A a bilateral ideal I whose elements are nilpotent: $\forall a \in I,\ \exists k \geq 1,\ a^k = 0$. Of course, a nilpotent ideal is a nil ideal, with a common integer k such that $a^k = 0$ for all $a \in I$.

Lemma 6.20. *In a finite-dimensional k-algebra, a bilateral ideal* I *is nilpotent if and only if it is nil.*

Proof. Notice first that the Jacobson radical $\mathrm{rad}(A)$ is a nilpotent ideal. Indeed, consider a composition series for A:

$$0 = I_0 \subset I_1 \subset \cdots \subset I_k = A$$

where all the I_i are left ideals, and with I_i/I_{i-1} simple A-module for any $i \in [\![1,k]\!]$. Since $\mathrm{rad}(A) \cdot (I_i/I_{i-1}) = 0$ for any i, $(\mathrm{rad}(A)) \cdot I_i \subset I_{i-1}$, hence $(\mathrm{rad}(A))^k = (\mathrm{rad}(A))^k \cdot I_k \subset I_0 = 0$ and $\mathrm{rad}(A)$ is nilpotent. Now, on the other hand, any nil ideal is contained in $\mathrm{rad}(A)$. Indeed, if I is nil and $a \in I$, then xa is nilpotent for any $x \in A$, hence, $1 - xa$ is invertible, with inverse $1 + (xa) + (xa)^2 + \cdots + (xa)^k$ for some k large enough. By the characterization (J3) of the Jacobson radical, $a \in \mathrm{rad}(A)$, so $I \subset \mathrm{rad}(A)$. As a consequence, a nil ideal is always included in a nilpotent ideal, hence is nilpotent. □

Corollary 6.21. *In a finite-dimensional k-algebra A,* $\mathrm{rad}(A)$ *is the largest nilpotent ideal for the inclusion.*

Proof. If $I \subset A$ is a nilpotent ideal, then it is nil, and the proof of the previous lemma shows that $I \subset \mathrm{rad}(A)$. So, $\mathrm{rad}(A)$ contains all the nilpotent ideals, and it is itself nilpotent by the argument of composition series. □

Lemma 6.22. *The dimension of the Jacobson radical of $\mathfrak{H}_0(n)$ is $n!-2^{n-1}$. In particular, $\mathfrak{H}_0(n)$ is not semisimple ($\mathrm{rad}(\mathfrak{H}_0(n)) \neq 0$) as soon as $n \geq 3$.*

Proof. We construct a composition series for $\mathfrak{H}_0(n)$ as follows: we choose a total order on $\mathfrak{S}(n)$ such that

$$\sigma \leq \tau \Rightarrow \ell(\sigma) \leq \ell(\tau),$$

and if $\sigma_1 > \sigma_2 > \cdots > \sigma_{n!}$ is the sequence of permutations with respect to this order, we set $A_k = \mathrm{Span}(\sigma_1, \sigma_2, \ldots, \sigma_k)$ for all $k \in [\![1, n!]\!]$. We claim that the A_k's are bilateral ideals of $\mathfrak{H}_0(n)$. Recall that the relations of $\mathfrak{H}_0(n)$ can be rewritten as:

$$T_i\, T_\sigma = \begin{cases} T_{s_i\sigma} & \text{if } \ell(s_i\sigma) > \ell(\sigma), \\ -T_\sigma & \text{if } \ell(s_i\sigma) < \ell(\sigma). \end{cases}$$

Therefore, $T_i T_\sigma$ is always a linear combination of elements $T_{\sigma'}$ with $\ell(\sigma') \geq \ell(\sigma)$, which proves that A_k is a left ideal; and the proof for the right side is identical. So, we have a composition series

$$0 \subset A_1 \subset A_2 \subset \cdots \subset A_{n!} = \mathfrak{H}_0(n),$$

where each quotient A_k/A_{k-1} is of dimension 1, and spanned by the class of T_{σ_k} modulo A_{k-1}. Remark now that

$$(T_{\sigma_k})^2 = 0 \text{ or } (-1)^{\ell(\sigma_k)}\, T_{\sigma_k} \mod A_{k-1}$$

for any k. Indeed, if $\sigma_k = s_{i_1} s_{i_2} \cdots s_{i_\ell}$ is a reduced decomposition, then when computing $(T_{\sigma_k})^2 = T_{i_1} T_{i_2} \cdots T_{i_\ell}\, T_{\sigma_k}$:

- either one has $\ell(s_{i_j}(s_{i_{j+1}} \cdots s_{i_\ell}\sigma_k)) < \ell(s_{i_{j+1}} \cdots s_{i_\ell}\sigma_k)$ for any $j \in [\![1, \ell]\!]$, in which case $(T_{\sigma_k})^2 = (-1)^\ell\, T_{\sigma_k}$;
- or, there is some index j such that $\ell(s_{i_j}(s_{i_{j+1}} \cdots s_{i_\ell}\sigma_k)) > \ell(s_{i_{j+1}} \cdots s_{i_\ell}\sigma_k)$, and then $(T_{\sigma_k})^2 = \pm T_{\sigma_{k'}}$ with $\ell(\sigma_{k'}) > \ell(\sigma_k)$, and therefore $k' < k$.

As a consequence, the element $(-1)^{\ell(\sigma_k)}\, T_{\sigma_k}$ is always either a nilpotent element of A_k/A_{k-1}, or an idempotent element. In other words, there are two types of quotients in the composition series of $\mathfrak{H}_0(n)$:

- quotients A_k/A_{k-1} generated by a nilpotent element;
- and quotients A_k/A_{k-1} generated by an idempotent element.

Consider now the sequence of inclusions $0 \subset \mathrm{rad}(\mathfrak{H}_0(n)) \subset \mathfrak{H}_0(n)$. It can be refined in a composition series

$$0 \subset B_1 \subset B_2 \subset \cdots \subset B_d = \mathrm{rad}(\mathfrak{H}_0(n)) \subset \cdots \subset B_{n!} = \mathfrak{H}_0(n),$$

for the left-module $\mathfrak{H}_0(n)$, and by the Jordan–Hölder theorem, the isomorphism classes of the quotients B_k/B_{k-1} of this composition series are the same as the isomorphism classes of the quotients A_k/A_{k-1}, up to some permutation of these quotients (in particular, $\dim B_k/B_{k-1} = 1$ for any k). Therefore, any B_k/B_{k-1} is either generated by a nilpotent element, or by an idempotent element. However, $\mathrm{rad}(\mathfrak{H}_0(n))$ is a nilpotent ideal, so all the quotients B_k/B_{k-1} with $k \leq d$ are nilpotent. On the other hand, $\mathfrak{H}_0(n)/\mathrm{rad}(\mathfrak{H}_0(n))$ is a semisimple algebra with a composition series whose quotients are all of dimension 1, so it is a commutative algebra isomorphic to $\mathbb{C}^{n!-d}$, and all the quotients of its composition series are generated by idempotents. Therefore, to prove the lemma, it suffices now to show that the number of elements $\sigma \in \mathfrak{S}(n)$ such that $(T_\sigma)^2 = (-1)^{\ell(\sigma)} T_\sigma$ is $n! - d = 2^{n-1}$.

Fix $\sigma = s_{i_1} s_{i_2} \cdots s_{i_\ell}$ such that $(T_\sigma)^2 = (-1)^{\ell(\sigma)} T_\sigma$. If $I = \{i_j,\ j \in [\![1,\ell]\!]\}$, then each $i \in I$ is a backstep of σ, and this condition is equivalent to the equation $(T_\sigma)^2 = (-1)^{\ell(\sigma)} T_\sigma$. Denote c the composition associated to I, and notice that $\sigma \in \mathfrak{S}(c)$. If $\mathfrak{S}(c) = \prod_{i=1}^t \mathfrak{S}(c_i)$ and σ corresponds to permutations $\sigma_1, \ldots, \sigma_t$ on the subintervals $[\![1, c_1]\!]$, $[\![c_1+1, c_1+c_2]\!]$, etc., then each σ_i is an element of $\mathfrak{S}(c_i)$ that has all the possible backsteps, hence is equal to the permutation

$$(c_1 + \cdots + c_{i-1} + 1, \ldots, c_1 + \cdots + c_i) \to (c_1 + \cdots + c_i, \ldots, c_1 + \cdots + c_{i-1} + 1)$$

which is the longest element in $\mathfrak{S}(c_i)$. Therefore, σ is equal to

$$c_1(c_1-1)\ldots 1\ (c_1+c_2)(c_1+c_2-1)\ldots(c_1+1)\ \ldots\ (c_1+\cdots+c_t)\ldots(c_1+\cdots+c_{t-1}+1),$$

which is the longest element of $\mathfrak{S}(c) = \prod_{i=1}^t \mathfrak{S}(c_i) \subset \mathfrak{S}(n)$, of length $\sum_{i=1}^t \binom{c_i}{2}$. So, the number of idempotent factors in a composition series of $\mathfrak{H}_0(n)$ is the number of compositions c of size n, that is to say 2^{n-1}. □

Proof of Proposition 6.19. We are going to construct a semisimple quotient $A = \mathfrak{H}_0(n)/I$ of $\mathfrak{H}_0(n)$ that has dimension 2^{n-1}; since $\mathfrak{H}_0(n)/\mathrm{rad}(\mathfrak{H}_0(n))$ is the largest semisimple quotient, for dimension reasons, this will imply that $I = \mathrm{rad}(\mathfrak{H}_0(n))$. The relations of $\mathfrak{H}_0(n)$ imply

$$\begin{aligned}
([T_i, T_{i+1}])^2 &= (T_i T_{i+1} - T_{i+1} T_i)^2 \\
&= T_i T_{i+1} T_i T_{i+1} + T_{i+1} T_i T_{i+1} T_i - T_i (T_{i+1}^2) T_i - T_{i+1} (T_i)^2 T_{i+1} \\
&= (T_i)^2 T_{i+1} T_i + (T_{i+1})^2 T_i T_{i+1} - T_i (T_{i+1}^2) T_i - T_{i+1} (T_i)^2 T_{i+1} \\
&= -T_i T_{i+1} T_i - T_{i+1} T_i T_{i+1} + T_i T_{i+1} T_i + T_{i+1} T_i T_{i+1} = 0.
\end{aligned}$$

Thus, $[T_i, T_{i+1}]$ is nilpotent for any i, and the same holds also for $[T_i, T_j]$ if $|j-i| \neq 1$, since in this case $[T_i, T_j] = 0$. In the sequel, we denote I the bilateral ideal spanned by these commutators $[T_i, T_j]$.

Since the relations $(\widetilde{T}_i)^2 = -\widetilde{T}_i$ and $[\widetilde{T}_i, \widetilde{T}_j] = 0$ imply the commutation relations and the braid relations of $\mathfrak{H}_0(n)$, a presentation of $A = \mathfrak{H}_0(n)/I$ is

$$(\widetilde{T}_i)^2 = -\widetilde{T}_i \quad ; \quad \widetilde{T}_i \widetilde{T}_j = \widetilde{T}_j \widetilde{T}_i,$$

where $\widetilde{T}_i$ is the image of T_i by the projection $\mathfrak{H}_0(n) \to \mathfrak{H}_0(n)/I$. If $D \subset [\![1,n-1]\!]$, set

$$\widetilde{T}_D = \prod_{d \in D} \widetilde{T}_d.$$

By using the presentation of A, one shows readily that the family $(\widetilde{T}_D)_{D\subset[\![1,n-1]\!]}$ spans linearly the algebra A. To prove that it is a linear basis, we reason by induction on n. Suppose that $(\widetilde{T}_D)_{D\subset[\![1,n-2]\!]}$ is known to be a family of linearly independent vectors in A, and consider a linear combination

$$\sum_{D\subset[\![1,n-1]\!]} a_D\, \widetilde{T}_D = 0$$

that vanishes. If A' is the subalgebra of A with generators $\widetilde{T}_1,\ldots,\widetilde{T}_{n-2}$, then there is a morphism of algebras $\pi_0 : A \to A'$ defined by $\pi_0(\widetilde{T}_{i\leq n-2}) = \widetilde{T}_i$ and $\pi_0(\widetilde{T}_{n-1}) = 0$; indeed, π_0 is compatible with the presentation of A. Applying this morphism to the vanishing linear combination, we obtain

$$\sum_{D\subset[\![1,n-2]\!]} a_D\, \widetilde{T}_D = 0,$$

so by the induction hypothesis, all the coefficients a_D with $(n-1) \notin D$ are equal to 0. Consider then the other morphism $\pi_{-1} : A \to A'$ defined by $\pi_{-1}(\widetilde{T}_{i\leq n-2}) = \widetilde{T}_i$ and $\pi_{-1}(\widetilde{T}_{n-1}) = -1$; if one applies it to the linear combination, one gets

$$-\sum_{D\subset[\![1,n-1]\!]\,|\,(n-1)\in D} a_D\, \widetilde{T}_{D\setminus\{n-1\}} = 0,$$

so all the other coefficients a_D with $(n-1) \in D$ are also equal to 0. So, the result is also true in size n.

Thus, $A = \mathfrak{H}_0(n)/I$ is a commutative algebra of dimension 2^{n-1}, with basis $(\widetilde{T}_D)_{D\subset[\![1,n-1]\!]}$. Now, for any $C \subset [\![1,n-1]\!]$, the linear map

$$\begin{aligned} \chi^C &: A \to \mathbb{C} \\ \widetilde{T}_D &\mapsto 1_{D\subset C}\,(-1)^{\mathrm{card}\, D} \end{aligned}$$

is a morphism of algebras, because it satisfies

$$\begin{aligned} \chi^C((\widetilde{T}_i)^2) &= \chi^C(-\widetilde{T}_i) = 1_{i\in C} = (-1_{i\in C})^2 = (\chi^C(\widetilde{T}_i))^2; \\ \chi^C(\widetilde{T}_i\widetilde{T}_j) &= \chi^C(\widetilde{T}_{\{i,j\}}) = 1_{\{i,j\}\subset C} = (-1_{i\in C})(-1_{j\in C}) = \chi^C(\widetilde{T}_i)\,\chi^C(\widetilde{T}_j). \end{aligned}$$

Therefore, we have 2^{n-1} distinct one-dimensional representations of A, which can be aggregated into an isomorphism of algebras

$$\chi = \bigoplus_{C\subset[\![1,n-1]\!]} \chi^C : A \to \bigoplus_{C\subset[\![1,n-1]\!]} \mathbb{C}.$$

To see that χ is an isomorphism, it suffices to prove the injectivity, since both algebras have the same dimension 2^{n-1}. Suppose thus that $\chi(a) = 0$, that is to say that $\chi^C(a) = 0$ for any subset C. If $a = \sum_D a_D T_D$ is not zero, consider D subset that is minimal for the inclusion among the subsets such that $a_D \neq 0$.Then, $\chi^D(a) = a_D$, which is a contradiction. Hence, $a = 0$ and χ is an isomorphism. We conclude that A is a *semisimple* commutative algebra of dimension 2^{n-1}, and since $\mathfrak{H}_0(n)/\mathrm{rad}(\mathfrak{H}_0(n))$ is the largest semisimple quotient of $\mathfrak{H}_0(n)$ (Theorem 4.26), $\mathrm{rad}(\mathfrak{H}_0(n)) = I$. □

Theorem 6.23. *The number of simple modules of $\mathfrak{H}_0(n)$ is 2^{n-1}, and all of them have dimension 1. More precisely, if $c \in \mathfrak{C}(n)$, then one can associate to c the representation S^c defined by*

$$\rho^c : \mathfrak{H}_0(n) \to \mathbb{C}$$
$$T_i \mapsto \begin{cases} -1 & \text{if } i \in D(c) \\ 0 & \text{otherwise,} \end{cases}$$

and $(S_c)_{c \in \mathfrak{C}(n)}$ is then a complete family of representatives of the isomorphism classes of simple modules over $\mathfrak{H}_0(n)$, that is to say a basis of $\mathrm{R}_0(\mathfrak{H}_0(n))$.

Proof. If (M, ρ) is a simple module over $\mathfrak{H}_0(n)$, then by the remark at the very end of Section 4.2, ρ factorizes via the projection $\mathfrak{H}_0(n) \to \mathfrak{H}_0(n)/\mathrm{rad}(\mathfrak{H}_0(n))$, and M is a simple module over this semisimple quotient. The result follows immediately from the discussion of the proof of Proposition 6.19. □

▷ *The Grothendieck ring of finite-dimensional modules.*

If $[M] = \sum_{c \in \mathfrak{C}} a_c [S^c]$ is an element of the Grothendieck ring $\mathrm{R}_{\mathbb{C}}(\mathfrak{H}_0) = \bigoplus_{n \in \mathbb{N}} \mathrm{R}_{\mathbb{C}}(\mathfrak{H}_0(n))$, we define its characteristic by

$$\Psi_{\mathrm{R}}([M]) = \sum_{c \in \mathfrak{C}} a_c L_c,$$

which is an element of QSym. Thus, any finite-dimensional module M over a Hecke algebra $\mathfrak{H}_0(n)$ is sent by the **characteristic map** Ψ_{R} to a quasi-symmetric function $\Psi_{\mathrm{R}}(M) \in \mathrm{QSym}_n$.

Theorem 6.24. *The characteristic map Ψ_{R} is an isomorphism of graded algebras.*

Since $([S^c])_{c \in \mathfrak{C}}$ is a graded basis of $\mathrm{R}_{\mathbb{C}}(\mathfrak{H}_0)$ and $(L_c)_{c \in \mathfrak{C}}$ is a graded basis of QSym, we already know that Ψ_{R} is an isomorphism of graded vector spaces. Therefore, the only problem that arises is the computation of the product

$$S^c \times S^d = \mathrm{Ind}_{\mathfrak{H}_0(m) \times \mathfrak{H}_0(n)}^{\mathfrak{H}_0(m+n)} (S^c \boxtimes S^d)$$

of two simple modules S^c and S^d over $\mathfrak{H}_0(m)$ and $\mathfrak{H}_0(n)$. We start with the corresponding computation in the algebra QSym, that is to say the product of two fundamental quasi-symmetric functions. If σ and τ are two permutations of size m and n, we define their **shuffle product** as the set $\sigma \sqcup\!\sqcup \tau$ of permutations in $\mathfrak{S}(m+n)$ whose words are obtained by interlacing the letters i of the word of σ with the translations $j+m$ of the letters j of the word of τ, in any possible way (there are $\binom{m+n}{m} = \frac{(m+n)!}{m!\,n!}$ possibilities).

Example. If $\sigma = 312$ and $\tau = 21$, then the translation of the word of τ is 54, and the shuffle product is therefore

$$\sigma \sqcup\!\sqcup \tau$$
$$= \{31254, 31524, 35124, 53124, 31542, 35142, 53142, 35412, 53412, 54312\}.$$

Proposition 6.25. *Let c and d be two compositions, and σ and τ be arbitrary permutations such that $D(\sigma) = c$ and $D(\tau) = d$, where $D(\sigma \in \mathfrak{S}(n))$ denotes the set of descents of the permutations σ. Then,*

$$L_c(X)\, L_d(X) = \sum_{\rho \in \sigma \sqcup\!\sqcup \tau} L_{c(\rho)}(X),$$

where $c(\rho)$ is the unique composition such that $D(c(\rho)) = D(\rho)$.

We postpone the proof of this proposition to Section 12.1, where we shall build a Hopf algebra FQSym of permutations, endowed with the shuffle product. Then, the formula above will come from a formula in FQSym by means of a surjective morphism of algebras FQSym $\to$ QSym (cf. the proof of Theorem 12.7).

Example. With $c = (1,2)$ and $d = (1,1)$, we can take $\sigma = 312$ and $\tau = 21$, and by looking at the descents of the permutations in $\sigma \sqcup\!\sqcup \tau$, we get:

$$\begin{aligned} L_{(1,2)}\, L_{(1,1)} = {} & L_{(1,1,1,2)} + L_{(1,1,2,1)} + L_{(1,2,1,1)} + L_{(1,1,3)} + L_{(1,3,1)} \\ & + L_{(2,2,1)} + L_{(2,1,2)} + 2\, L_{(1,2,2)} + L_{(2,3)}. \end{aligned}$$

It turns out that the shuffle product is also the right object to describe the right cosets of $(\mathfrak{S}(m) \times \mathfrak{S}(n)) \backslash \mathfrak{S}(m+n)$. Thus, for any permutations $\sigma \in \mathfrak{S}(m)$ and $\tau \in \mathfrak{S}(n)$, $\sigma \sqcup\!\sqcup \tau$ is a set of representatives of the right cosets:

$$\mathfrak{S}(m+n) = \bigsqcup_{\rho \in \sigma \sqcup\!\sqcup \tau} (\mathfrak{S}(m) \times \mathfrak{S}(n))\, \rho.$$

Indeed, notice that $\sigma \sqcup\!\sqcup \tau = (\sigma, \tau)(\mathrm{id}_{[\![1,m]\!]} \sqcup\!\sqcup \mathrm{id}_{[\![1,n]\!]})$, so this decomposition is equivalent to

$$\mathfrak{S}(m+n) = \bigsqcup_{\rho \in (12\ldots m) \sqcup\!\sqcup (12\ldots n)} (\mathfrak{S}(m) \times \mathfrak{S}(n))\, \rho.$$

However, given a permutation word $w = w_1 w_2 \ldots w_{m+n}$ in $\mathfrak{S}(m+n)$, one can use permutations of $\mathfrak{S}(m) \times \mathfrak{S}(n)$ on the left of w to rearrange the letters $1, 2, \ldots, m$ in increasing order, and the letters $m+1, \ldots, m+n$ also in increasing order. Therefore, $w \in (\mathfrak{S}(m) \times \mathfrak{S}(n))\, \widetilde{w}$, where $\widetilde{w}$ is a permutation word where the letters $1,2,\dots,m$ appear in this order, and similarly for the letters $m+1,\dots,m+n$. This means that $\widetilde{w} \in (12\dots m) \sqcup (12\dots n)$, so we have shown that $\mathfrak{S}(m+n) \subset \bigcup_{\rho \in (12\dots m)\sqcup(12\dots n)} (\mathfrak{S}(m)\times\mathfrak{S}(n))\,\rho$. The union is then a disjoint union for cardinality reasons.

Remark. We work in the following with *right* modules over Hecke algebras, so in particular, $M \times N = (M \boxtimes N) \otimes_{\mathfrak{H}_0(m)\times\mathfrak{H}_0(n)} \mathfrak{H}_0(m+n)$ if M is a right module over $\mathfrak{H}_0(m)$ and N is a right module over $\mathfrak{H}_0(n)$. This convention does not modify the results (or, if the reader prefers it, he can consider Theorem 6.18 as a result on right modules).

Lemma 6.26. *Let $\sigma = s_{i_1}s_{i_2}\cdots s_{i_m}$ be a decomposition of a permutation σ that is not a reduced decomposition ($m > \ell(\sigma)$). There exist two indices $j < k$ such that*

$$\sigma = s_{i_1}\cdots s_{i_{j-1}}s_{i_{j+1}}\cdots s_{i_{k-1}}s_{i_{k+1}}\cdots s_{i_m}.$$

This result is the ***cancellation law*** *for non-reduced decompositions.*

Proof. Let $k \geq 2$ be an index such that $\ell(s_{i_1}\cdots s_{i_{k-1}}) = k-1$ and $\ell(s_{i_1}\cdots s_{i_k}) < k$; there is necessarily one such index, and one has then $\ell(s_{i_1}\cdots s_{i_k}) = k-2$. If $\tau = s_{i_1}\cdots s_{i_{k-1}}$, then $\tau s_{i_k} = s_{i'_1}\cdots s_{i'_{k-2}}$ for some indices $i'_1,\dots,i'_{k-2}$, so

$$\tau = s_{i'_1}\cdots s_{i'_{k-2}}s_{i_k}$$

is a reduced decomposition of τ, with $\tau' = s_{i'_1}\cdots s_{i'_{k-2}}$ lower than τ with respect to the Bruhat order. However, the Bruhat order is independent of the choice of a reduced decomposition, so $\tau' = s_{i_1}\cdots s_{i_k}$ has a reduced decomposition that is obtained from the decomposition $\tau = s_{i_1}\cdots s_{i_{k-1}}$ by deleting one term s_{i_j}. This proves the result. □

Proposition 6.27. *Let c and d be two compositions of size m and n, and $M = S^c \times S^d$, where $S^c = \mathbb{C}$ and $S^d = \mathbb{C}$ are considered as right modules. A linear basis of M consists in the elements $e \cdot T_\sigma$, where $e = 1 \otimes 1$, and σ runs over $\mathrm{id}_{[\![1,m]\!]} \sqcup \mathrm{id}_{[\![1,n]\!]}$.*

Proof. We are going to prove that for any $\rho \in \mathfrak{S}(m+n)$, there is a factorization $\rho = \tau\sigma$, where $\tau \in \mathfrak{S}(m)\times\mathfrak{S}(n)$, $\sigma \in \mathrm{id}_{[\![1,m]\!]} \sqcup \mathrm{id}_{[\![1,n]\!]}$, and $\ell(\rho) = \ell(\tau) + \ell(\sigma)$. Then, M will be spanned by the vectors

$$(1\otimes 1)\cdot T_\rho = \underbrace{((1\otimes 1)\cdot T_\tau)}_{\in S^c\boxtimes S^d}\cdot T_\sigma = f(\tau)(1\otimes 1)\cdot T_\sigma,$$

where $f(\tau) \in \{0,1,-1\}$. Thus, the module M will indeed be spanned by the vectors $e \cdot T_\sigma$ with $\sigma \in \mathrm{id}_{[\![1,m]\!]} \sqcup \mathrm{id}_{[\![1,n]\!]}$, and for dimension reasons this will be a linear basis.

Fix $\rho \in \mathfrak{S}(m+n)$, and consider some permutation σ of minimal length in the right coset $(\mathfrak{S}(m)\times\mathfrak{S}(n))\,\rho$. We set $\rho = \tau\sigma$ with $\tau \in \mathfrak{S}(m)\times\mathfrak{S}(n)$. We claim that

$\ell(\rho) = \ell(\tau) + \ell(\sigma)$. Indeed, otherwise, $\ell(\rho) < \ell(\tau) + \ell(\sigma)$, so by the cancellation law, there exist permutations τ' and σ' smaller than τ and σ for the Bruhat order, such that:

1. $\rho = \tau'\sigma'$;

2. τ' is strictly smaller than τ, or σ' is strictly smaller than σ for the Bruhat order.

However, $\mathfrak{S}(m) \times \mathfrak{S}(n)$ is a closed interval of the Bruhat order in $\mathfrak{S}(m+n)$, so τ' belongs to $\mathfrak{S}(m) \times \mathfrak{S}(n)$, and then, by minimality of σ, $\sigma' = \sigma$, which implies that $\tau' = \tau$, hence a contradiction. So, $\ell(\rho) = \ell(\tau) + \ell(\sigma)$. In this situation, we have $\sigma \in \mathrm{id}_{[\![1,m]\!]} \sqcup \mathrm{id}_{[\![1,n]\!]}$. Indeed, for any $j \neq m$, $\ell(s_j\sigma) > \ell(\sigma)$, as otherwise, one would create another decomposition $\rho = (\tau s_j)(s_j\sigma)$ with $\tau s_j \in \mathfrak{S}(m) \times \mathfrak{S}(n)$, and $s_j\sigma$ of length strictly smaller than the length of σ. However, $\mathrm{id}_{[\![1,m]\!]} \sqcup \mathrm{id}_{[\![1,n]\!]}$ is exactly the set of permutations of size $m+n$ for any $j \neq m$, $\ell(s_j\sigma) > \ell(\sigma)$ (in other words, the only allowed backstep is m). So, we have shown the claim made at the beginning of the proof. $\square$

Remark. The description of right coset representatives, and the arguments that we use here can be generalized to any Young subgroup $\mathfrak{S}(c)$ associated to a composition $c \in \mathfrak{C}(n)$.

Proposition 6.28. *Let c and d be two compositions of size n, and τ_1 and τ_2 be two permutations with $D(\tau_1) = c$ and $D(\tau_2) = d$. The action of an element T_i with $i \in [\![1, m+n-1]\!]$ on the basis $(e \cdot T_\sigma)_{\sigma \in \mathrm{id}_{[\![1,m]\!]} \sqcup \mathrm{id}_{[\![1,n]\!]}}$ of $M = S^c \times S^d$ writes as follows:*

1. *If $\sigma s_i = \sigma'$ is again in $\mathrm{id}_{[\![1,m]\!]} \sqcup \mathrm{id}_{[\![1,n]\!]}$ and $\ell(\sigma s_i) > \ell(\sigma)$, then $(e \cdot T_\sigma) \cdot T_i = e \cdot T_{\sigma'}$.*

2. *Otherwise, $(e \cdot T_\sigma) \cdot T_i = \chi^{D((\tau_1,\tau_2)\sigma)}(T_i)(e \cdot T_\sigma)$.*

Lemma 6.29. *Let $\sigma \in \mathfrak{S}(n)$, and s and t be two elementary transpositions. If $\ell(s\sigma t) = \ell(\sigma)$ and $\ell(s\sigma) = \ell(\sigma t)$, then $s\sigma = \sigma t$.*

Proof. Suppose first that $\ell(\sigma t) > \ell(\sigma)$. Then, $\ell(s\sigma t) < \ell(\sigma t)$, so there exists a reduced decomposition $s\, s_{i_1} \cdots s_{i_\ell}$ that starts with s. Then,

$$\sigma = (\sigma t)t = s\, s_{i_1} \cdots s_{i_\ell}\, t$$

is not a reduced decomposition, so there exists by the cancellation law 6.26 a reduced decomposition of σ that is obtained by deleting two terms in the previous formula. Now, $\ell(\sigma) < \ell(\sigma t)$ and $\ell(\sigma) < \ell(s\sigma)$, so this reduced decomposition cannot start with s or end with t. Hence, it is necessarily

$$\sigma = s_{i_1} \cdots s_{i_\ell},$$

which implies immediately $s\sigma = s\, s_{i_1} \cdots s_{i_\ell} = (s\, s_{i_1} \cdots s_{i_\ell}\, t)t = \sigma t$.

Suppose now $\ell(\sigma t) < \ell(\sigma)$, and set $\sigma' = \sigma t$. Then, $\ell(s\sigma' t) = \ell(s\sigma) = \ell(\sigma t) = \ell(\sigma')$, and $\ell(s\sigma') = \ell(\sigma t) = \ell(\sigma' t)$, so σ' satisfies the hypotheses of the previous case. Hence, $s\sigma' = \sigma' t$, and by remultiplying by t, one gets $s\sigma = \sigma t$ also in this case. □

Lemma 6.30 (Deodhar). *Fix $\sigma \in \mathrm{id}_{[\![1,m]\!]} \sqcup\!\sqcup\, \mathrm{id}_{[\![1,n]\!]}$, and an elementary transposition s_i.*

1. *If $\ell(\sigma s_i) < \ell(\sigma)$, that is to say that σs_i is a left prefix of a reduced decomposition of σ, then $\sigma s_i \in \mathrm{id}_{[\![1,m]\!]} \sqcup\!\sqcup\, \mathrm{id}_{[\![1,n]\!]}$.*
2. *If $\ell(\sigma s_i) > \ell(\sigma)$, then either σs_i belongs to $\mathrm{id}_{[\![1,m]\!]} \sqcup\!\sqcup\, \mathrm{id}_{[\![1,n]\!]}$, or there exists an elementary transposition s_j with $j \neq m$ such that $\sigma s_i = s_j \sigma$.*

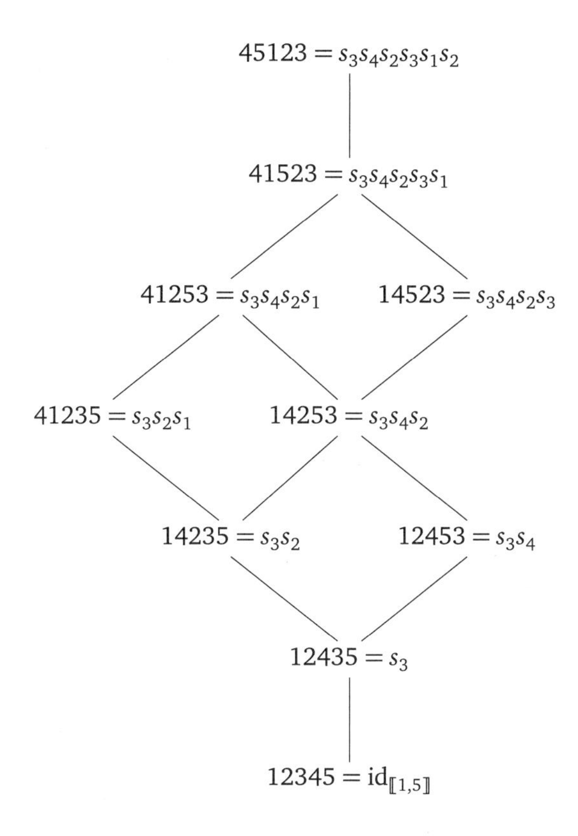

Figure 6.1
The interval $123 \sqcup\!\sqcup\, 45$ of the permutohedron of $\mathfrak{S}(5)$.

Example. Figure 6.1 shows the reduced decompositions of the elements of the shuffle product $123 \sqcup\!\sqcup 45$. These 10 permutations form a closed interval of the (left) **permutohedron** of $\mathfrak{S}(5)$, which is the Hasse diagram of the partial order

$$\sigma \leq \tau \iff \sigma \text{ is a left prefix of a reduced decomposition of } \tau.$$

Proof. Suppose that σs_i is not in $\mathrm{id}_{[\![1,m]\!]} \sqcup\!\sqcup \mathrm{id}_{[\![1,n]\!]}$. Then, since

$$\mathrm{id}_{[\![1,m]\!]} \sqcup\!\sqcup \mathrm{id}_{[\![1,n]\!]} = \{\sigma \mid \forall j \neq m,\ \ell(s_j\sigma) > \ell(\sigma)\},$$

there exists $s_{j\neq m}$ with $\ell(s_j\sigma s_i) < \ell(\sigma s_i)$. On the other hand, we saw that the decomposition $\mathfrak{S}(m+n) = (\mathfrak{S}(m) \times \mathfrak{S}(n)) \times (\mathrm{id}_{[\![1,m]\!]} \sqcup\!\sqcup \mathrm{id}_{[\![1,n]\!]})$ is compatible with the lengths, so $\ell(s_j\sigma) > \ell(\sigma)$. Therefore, $\ell(s_j\sigma s_i) \geq \ell(s_j\sigma) - 1 = \ell(\sigma)$. On the other hand, $\ell(s_j\sigma s_i) < \ell(\sigma s_i) \leq \ell(\sigma) + 1$, so the only possibility is

$$\ell(s_j\sigma s_i) = \ell(\sigma) \quad ; \quad \ell(s_j\sigma) = \ell(\sigma s_i) = \ell(\sigma) + 1.$$

By the previous lemma, $\sigma s_i = s_j\sigma$, and this case can only happen if $\ell(\sigma s_i) > \ell(\sigma)$. So, if $\ell(\sigma s_i) < \ell(\sigma)$, then this forces $\sigma s_i \in \mathrm{id}_{[\![1,m]\!]} \sqcup\!\sqcup \mathrm{id}_{[\![1,n]\!]}$. □

Proof of Proposition 6.28. The first case of the proposition is obvious, so it suffices to treat the second case.

1. Suppose that $\ell(\sigma s_i) < \ell(\sigma)$. Then, $T_\sigma T_i = -T_\sigma$, and on the other hand, i is in the descent set of σ, and also of $(\tau_1, \tau_2)\sigma$ for any $(\tau_1, \tau_2) \in \mathfrak{S}(m) \times \mathfrak{S}(n)$. Indeed, by Deodhar's lemma 6.30, σs_i is in $\mathrm{id}_{[\![1,m]\!]} \sqcup\!\sqcup \mathrm{id}_{[\![1,n]\!]}$, therefore, for any pair (τ_1, τ_2) in $\mathfrak{S}(m) \times \mathfrak{S}(n)$,

   $$\ell((\tau_1,\tau_2)\sigma s_i) = \ell(\tau_1) + \ell(\tau_2) + \ell(\sigma s_i) < \ell(\tau_1) + \ell(\tau_2) + \ell(\sigma) = \ell((\tau_1,\tau_2)\sigma).$$

   So,
   $$(e \cdot T_\sigma) \cdot T_i = -(e \cdot T_\sigma) = \chi^{D((\tau_1,\tau_2)\sigma)}(T_i)(e \cdot T_\sigma).$$

2. Suppose now that $\ell(\sigma s_i) > \ell(\sigma)$, but σs_i is not in the shuffle product $\mathrm{id}_{[\![1,m]\!]} \sqcup\!\sqcup \mathrm{id}_{[\![1,n]\!]}$. By Deodhar's lemma, $\sigma s_i = s_j\sigma$ for some elementary transposition $s_j \in \mathfrak{S}(m) \times \mathfrak{S}(n)$. Therefore,

   $$(e \cdot T_\sigma) \cdot T_i = e \cdot T_{\sigma s_i} = e \cdot T_{s_j\sigma} = (e \cdot T_j) \cdot T_\sigma = \chi^{D((\tau_1,\tau_2))}(T_j)(e \cdot T_\sigma).$$

   Since

   $$\begin{aligned} \ell((\tau_1,\tau_2)s_j) < \ell((\tau_1,\tau_2)) &\iff \ell((\tau_1,\tau_2)s_j\sigma) < \ell((\tau_1,\tau_2)\sigma) \\ &\iff \ell((\tau_1,\tau_2)\sigma s_i) < \ell((\tau_1,\tau_2)\sigma) \end{aligned}$$

   the coefficient is also $\chi^{D((\tau_1,\tau_2))\sigma}(T_i)$, which ends the proof. □

Example. We take $c = (1,2)$, $d = (1,1)$, $\tau_1 = 312$ and $\tau_2 = 21$. The linear basis $(e \cdot T_\sigma)_{\sigma \in \mathrm{id}_{[\![1,m]\!]} \sqcup\!\sqcup \mathrm{id}_{[\![1,n]\!]}}$ of $S^c \times S^d$, as well as the action of the elements T_1, T_2, T_3, T_4 on this basis, is drawn in Figure 6.2.

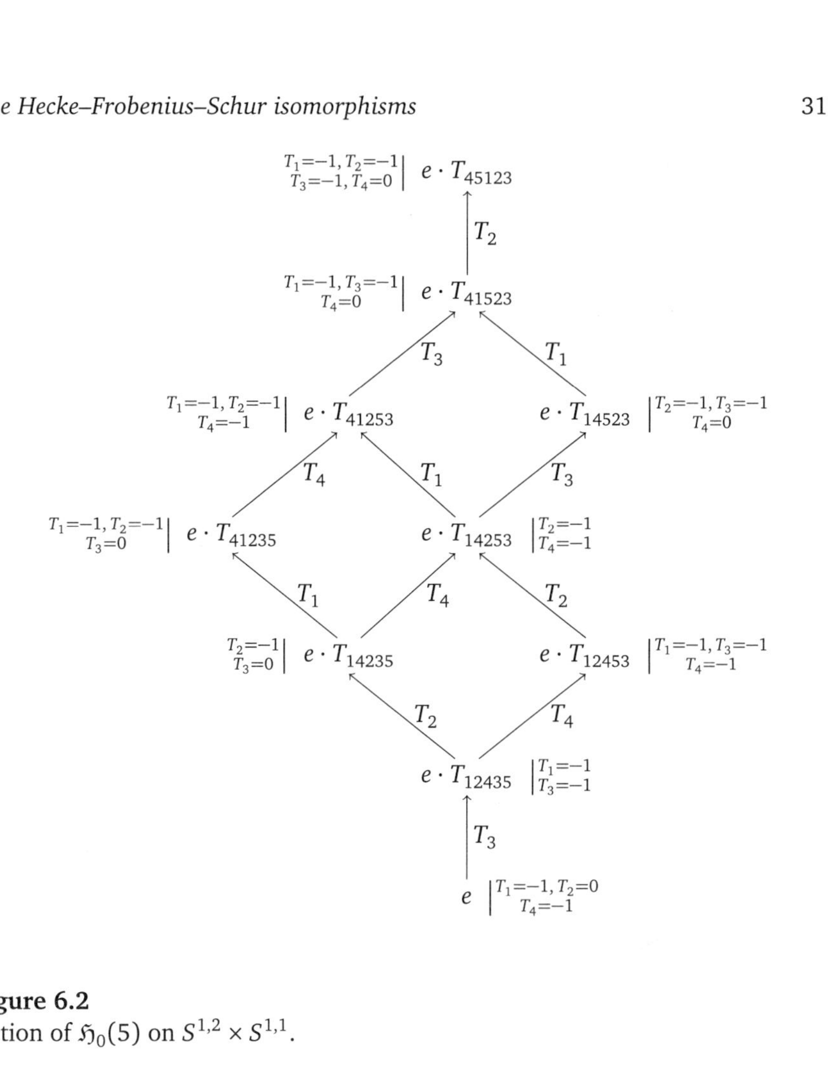

Figure 6.2
Action of $\mathfrak{H}_0(5)$ on $S^{1,2} \times S^{1,1}$.

Proposition 6.31. *If $d \in \mathfrak{C}(n)$, denote T_{σ_d} the basis element of $\mathfrak{H}_0(n)$ corresponding to the permutation*

$$\sigma_d = (1,2,\ldots,d_1)(d_1+1,\ldots,d_1+d_2)\cdots(d_1+\cdots+d_{\ell-1}+1,\ldots,d_1+\cdots+d_\ell).$$

The character of S^c on T_{σ_d} is

$$\mathrm{ch}^c(T_{\sigma_d}) = (-1)^{|d|-\ell(d)} \langle \Lambda_d \mid L_c \rangle .$$

Proof. The value of ch^c on T_{σ_d} is 0 unless all the elements s_i that appear in a reduced decomposition of σ_d satisfy $i \in D(c)$. Thus,

$$\begin{aligned}\mathrm{ch}^c(T_{\sigma_d}) &= \varepsilon(\sigma_d)\, 1_{\{1,2,\ldots,d_1-1,d_1+1,\ldots,d_1+d_2-1,\ldots,d_1+\cdots+d_{\ell-1}+1,\ldots,n-1\}\subset D(c)} \\ &= \varepsilon(\sigma_d)\, 1_{[\![1,n-1]\!]\setminus D(d)\subset D(c)} = \varepsilon(\sigma_d)\, 1_{[\![1,n-1]\!]\setminus D(\widetilde{c})\subset D(\widetilde{d})}.\end{aligned}$$

We already noticed that the composition whose descent set is the complementary of $D(\widetilde{c})$ is the conjugate c'. So, $\mathrm{ch}^c(T_{\sigma_d}) = \varepsilon(\sigma_d)\,1_{c' \preceq \widetilde{d}}$. On the other hand, since ω_{NCSym} and ω_{QSym} are dual involutive isometries of NCSym and QSym,

$$\langle \Lambda_d \mid L_c \rangle = \langle \omega(\Lambda_d) \mid \omega(L_c) \rangle = \langle S_{\widetilde{d}} \mid L_{c'} \rangle = \sum_{e \succeq c'} \langle S_{\widetilde{d}} \mid M_e \rangle = 1_{\widetilde{d} \succeq c'}.$$

Thus, the character formula is proven. Notice that it allows one to compute the character of any element T_σ with $\sigma \in \mathfrak{S}(n)$. Indeed, the reduction algorithm presented during the proof of Theorem 5.46 can also be used when $q=0$, so there exist integers $a_{\sigma,c}$ such that, for any finite-dimensional module M over $\mathfrak{H}_0(n)$,

$$\mathrm{ch}^M(T_\sigma) = \sum_{c \in \mathfrak{C}_n} a_{\sigma,c}\, \mathrm{ch}^M(T_{\sigma_c}).$$

The difference with the generic case is that $\mathrm{ch}^M(T_{\sigma_c})$ now depends on the order of the parts of c. Thus, we need $2^{n-1} = \mathrm{card}\,\mathfrak{C}(n)$ values to determine the character of a module over $\mathfrak{H}_0(n)$, instead of $\mathrm{card}\,\mathfrak{Y}(n)$ values in the generic case. ☐

Corollary 6.32. *The character ch^M of a finite-dimensional module over $\mathfrak{H}_0(n)$ entirely determines its Grothendieck equivalence class in $\mathrm{R}_0(\mathfrak{H}_0(n))$.*

Proof. If $[M] = \sum_{c \in \mathfrak{C}(n)} a_c\,[S^c]$ in $\mathrm{R}_0(\mathfrak{H}_0(n))$, then $\mathrm{ch}^M = \sum_{c \in \mathfrak{C}(n)} a_c\,\mathrm{ch}^c$, therefore,

$$\mathrm{ch}^M(T_{\sigma_d}) = (-1)^{|d|-\ell(d)} \left\langle \Lambda_d \,\middle|\, \sum_{c \in \mathfrak{C}(n)} a_c\, L_c \right\rangle = (-1)^{|d|-\ell(d)} \,\langle \Lambda_d \mid \Psi_{\mathrm{R}}(M) \rangle\,.$$

However, the pairing between NCSym and QSym is perfect, so one can reconstruct $\Psi_{\mathrm{R}}(M)$ from the character values of ch^M. We use then the fact that Ψ_{R} is an isomorphism of vector spaces. ☐

Proof of Theorem 6.24. We endow $\mathrm{id}_{[\![1,m]\!]} \sqcup\!\sqcup\, \mathrm{id}_{[\![1,n]\!]}$ with a total order such that $\sigma \geq \tau \Rightarrow \ell(\sigma) \geq \ell(\tau)$, and we denote $\sigma_1 > \sigma_2 > \cdots > \sigma_{\binom{m+n}{m}}$ the list of elements of the shuffle product. If $M_i = \mathrm{Span}(e \cdot T_{\sigma_j},\ j \leq i)$, then

$$0 = M_0 \subset M_1 \subset M_2 \subset \cdots \subset M_{\binom{m+n}{m}} = S^c \times S^d$$

is a composition series of $S^c \times S^d$, with each quotient M_i/M_{i-1} spanned by $e \cdot T_{\sigma_i}$, hence of dimension one and simple over $\mathfrak{H}_0(m+n)$. Moreover, the action of T_j on M_i/M_{i-1} is always given by the character $\mathrm{ch}^{D((\tau_1,\tau_2)\sigma_i)}(T_j)$, by disjunction of the two cases of Proposition 6.28. Therefore, the character of $M = S^c \times S^d$ is

$$\mathrm{ch}^M = \sum_{i=1}^{\binom{m+n}{m}} \mathrm{ch}^{M_i/M_{i-1}} = \sum_{\sigma \in \mathrm{id}_{[\![1,m]\!]} \sqcup\!\sqcup\, \mathrm{id}_{[\![1,n]\!]}} \mathrm{ch}^{D((\tau_1,\tau_2)\sigma)} = \sum_{\sigma \in \tau_1 \sqcup\!\sqcup\, \tau_2} \mathrm{ch}^{D(\sigma)}.$$

As a consequence,

$$\Psi_{\mathrm{R}}(S^c \times S^d) = \sum_{\sigma \in \tau_1 \sqcup\!\sqcup \tau_2} S^{D(\sigma)},$$

and by using Proposition 6.25, we conclude that Ψ_{R} is a morphism of algebras. □

▷ *The Grothendieck ring of projective modules.*

We now have the first half of Theorem 6.18, and the second part regarding NCSym and $\mathrm{K}_{\mathbb{C}}(\mathfrak{H}_0)$ will follow from Frobenius' reciprocity for modules over algebras (cf. the proof of Lemma 4.66). To begin with, notice that the pairing between principal indecomposable and simple modules over a finite-dimensional complex algebra A can be rewritten as:

$$\langle P \mid S \rangle = \dim \mathrm{Hom}_A(P, S).$$

Indeed, if there exists a non-trivial morphism $P \to S$ with P indecomposable projective module and S simple module, then by Proposition 4.28 it factorizes through a morphism $T(P) \to S$, so necessarily $S = T(P)$. The same proposition shows that $\dim \mathrm{Hom}_A(P, T(P)) = \dim \mathrm{Hom}_A(T(P), T(P))$, which is equal to 1 by Schur's lemma.

We can now end the proof of the Hecke–Frobenius–Schur isomorphism:

Proof of Theorem 6.18. Since any finite-dimensional algebra A has the same number of simple modules and of principal indecomposable modules, $\mathrm{K}_0(\mathfrak{H}_0(n))$ has rank 2^{n-1}. For $c \in \mathfrak{C}(n)$, denote P^c the unique indecomposable projective module over $\mathfrak{H}_0(n)$ whose simple top is S^c. For dimension reasons, we already know that the linear map Ψ_{K} which associates to P^c the non-commutative symmetric function R_c is an isomorphism of graded vector spaces between $\mathrm{K}_{\mathbb{C}}(\mathfrak{H}_0)$ and NCSym. By Proposition 6.6, it is an isomorphism of algebras if and only if one has the identity

$$P^c \times P^d = P^{c \cdot d} \oplus P^{c \bowtie d}.$$

This identity is equivalent to

$$\begin{aligned} &\dim \mathrm{Hom}_{\mathfrak{H}_0(m+n)}(P^c \times P^d, S^e) = \delta_{e,c\cdot d} + \delta_{e,c\bowtie d} \\ \iff &\dim \mathrm{Hom}_{\mathfrak{H}_0(m)\times\mathfrak{H}_0(n)}(P^c \boxtimes P^d, \mathrm{Res}^{\mathfrak{H}_0(m+n)}_{\mathfrak{H}_0(m)\times\mathfrak{H}_0(n)}(S^e)) = \delta_{e,c\cdot d} + \delta_{e,c\bowtie d}. \end{aligned}$$

However, the restriction of the one-dimensional simple module S^e to the algebra $\mathfrak{H}_0(m) \times \mathfrak{H}_0(n)$ is trivial to compute: it is $S^{e^{(1)}} \boxtimes S^{e^{(2)}}$, where $e^{(1)}$ is the composition of size m whose descent set is $D(e) \cap [\![1, m-1]\!]$, and $e^{(2)}$ is the composition of size n whose descent set is $(D(e) \cap [\![m+1, m+n-1]\!]) - m$. Therefore, it is equal to $S^c \boxtimes S^d$ if and only if $e = c \cdot d$ or $e = c \bowtie d$. Thus, the identity above is true, and we have indeed shown that Ψ_{K} is an isomorphism of graded algebras. □

By using the isomorphisms Ψ_{R} and Ψ_{K}, one can transport the Hopf algebra structures of QSym and NCSym to $\mathrm{R}_{\mathbb{C}}(\mathfrak{H}_0)$ and $\mathrm{K}_{\mathbb{C}}(\mathfrak{H}_0)$; then, Ψ_{R} and Ψ_{K} become morphisms of graded Hopf algebras for these structures. The same argument as above shows then that the coproduct of the two Grothendieck rings is given by

$$\begin{aligned}\Delta : \mathrm{R}_0(\mathfrak{H}_0(n)) &\to \bigoplus_{m=0}^{n} \mathrm{R}_0(\mathfrak{H}_0(m)) \otimes \mathrm{R}_0(\mathfrak{H}_0(n-m))\\ [M] &\mapsto \sum_{m=0}^{n} \left[\mathrm{Res}^{\mathfrak{H}_0(n)}_{\mathfrak{H}_0(m)\times\mathfrak{H}_0(n-m)}(M)\right]\end{aligned}$$

and similarly for projective modules. As for the antipode, for any $n \geq 1$, there is an involutive isomorphism of algebras $\theta_n : \mathfrak{H}_0(n) \to \mathfrak{H}_0(n)$ given by

$$T_i \mapsto (-1 - T_{n-i}).$$

The composition on the right by θ_n yields involutive isomorphisms $\phi_{n,\mathrm{R}} : \mathrm{R}_0(\mathfrak{H}_0(n)) \to \mathrm{R}_0(\mathfrak{H}_0(n))$ and $\phi_{n,\mathrm{K}} : \mathrm{K}_0(\mathfrak{H}_0(n)) \to \mathrm{K}_0(\mathfrak{H}_0(n))$. Actually, $\phi_{n,\mathrm{R}}(S^c) = S^{c'}$. Indeed,

$$\begin{aligned}\mathrm{ch}^c \circ \theta_n(T_i) &= -1 - \begin{cases} -1 & \text{if } n-i \in D(c) \\ 0 & \text{if } n-i \notin D(c) \end{cases}\\ &= \begin{cases} 0 & \text{if } i \in D(\widetilde{c}) \\ -1 & \text{if } i \notin D(\widetilde{c}) \end{cases}\\ &= \mathrm{ch}^{c'}(T_i)\end{aligned}$$

since the complementary subset of $D(c')$ is $D(\widetilde{c})$. By duality between simple modules and principal indecomposable modules, one has also $\phi_{n,\mathrm{K}}(P^c) = P^{c'}$. Then, since $\omega_{\mathrm{NCSym}}(R_c) = (-1)^{|c|} R_{c'}$ and $\omega_{\mathrm{QSym}}(L_c) = (-1)^{|c|} L_{c'}$, the antipodes of the Grothendieck groups are given by the formula

$$\omega = \sum_{n=0}^{\infty} (-1)^n \phi_n.$$

If A is a finite-dimensional algebra, recall that there is a natural map

$$M \in \mathrm{K}_0(A) \mapsto [M] \in \mathrm{R}_0(A)$$

called the Cartan map (cf. Section 4.2). In the setting of 0–Hecke algebras, one has the following description of this map:

Theorem 6.33. *If* $c : \mathrm{K}_{\mathbb{C}}(\mathfrak{H}_0) \to \mathrm{R}_{\mathbb{C}}(\mathfrak{H}_0)$ *is the direct sum of the Cartan maps*

$$\begin{aligned}\mathrm{K}_0(\mathfrak{H}_0(n)) &\to \mathrm{R}_0(\mathfrak{H}_0(n))\\ P &\mapsto [P]\end{aligned}$$

then c is conjugated by the isomorphisms Ψ_{R} and Ψ_{K} to the commutative projection

$$\Phi : \mathrm{NCSym} \to \mathrm{Sym} \subset \mathrm{QSym}.$$

Hence, a finite-dimensional module M over $\mathfrak{H}_0(n)$ is Grothendieck equivalent to a projective module if and only if its characteristic $\Psi_{\mathrm{R}}(M) \in \mathrm{QSym}$ is a symmetric function in Sym.

Proof. Denote $\Phi' = \Psi_{\mathrm{R}} \circ c \circ \Psi_{\mathrm{K}}^{-1}$; it is a morphism of graded (Hopf) algebras between NCSym and QSym. To prove that $\Phi' = \Phi$, since both maps are morphisms of graded algebras, it suffices to show that $\Phi'(S_n) = h_n$ for any n; indeed, $\mathrm{NCSym} = \mathbb{C}\langle S_1, S_2, \ldots\rangle$. Consider the one-dimensional vector subspace V of $\mathfrak{H}_0(n)$ that is spanned by the vector $v_n = \frac{1}{n!}\sum_{\sigma \in \mathfrak{S}(n)} T_\sigma$. It is an ideal of $\mathfrak{H}_0(n)$, because for any i,

$$\begin{aligned} n!\, T_i \cdot v_n &= \sum_{\sigma \in \mathfrak{S}(n)\,|\,\ell(s_i\sigma) > \ell(\sigma)} T_{s_i\sigma} - \sum_{\sigma \in \mathfrak{S}(n)\,|\,\ell(s_i\sigma) < \ell(\sigma)} T_\sigma \\ &= \sum_{\sigma \in \mathfrak{S}(n)\,|\,\ell(s_i\sigma) < \ell(\sigma)} T_\sigma - \sum_{\sigma \in \mathfrak{S}(n)\,|\,\ell(s_i\sigma) < \ell(\sigma)} T_\sigma = 0. \end{aligned}$$

The previous formula shows that V is the simple module of $\mathfrak{H}_0(n)$ of label $c = (n)$: $V = S^{(n)}$. However, V is also a projective module, because it admits for complement submodule $\mathrm{Span}(T_\sigma,\ \sigma \neq \mathrm{id}_{[\![1,n]\!]})$. Since V is one-dimensional, it is a principal indecomposable module, and moreover, it admits a non-zero morphism towards $S^{(n)}$, namely, the identity of V. So,

$$V = P^{(n)} = S^{(n)},$$

which shows that for every n,

$$\Phi'(S_n) = \Phi'(R_n) = \Psi_{\mathrm{R}} \circ c(P^{(n)}) = \Psi_{\mathrm{R}}(S^{(n)}) = M_n = h_n = \Phi(S_n). \qquad \square$$

Corollary 6.34. *The dimension of the principal indecomposable module P^c over $\mathfrak{H}_0(n)$ is the number of permutations $\sigma \in \mathfrak{S}(n)$ such that $D(\sigma) = c$. More precisely, for any compositions c and d, the number of terms S^d in a composition series of P^c is*

$$\mathrm{card}\,\{\sigma \in \mathfrak{S}(n)\,|\, D(\sigma) = D(c) \text{ and } D(\sigma^{-1}) = D(d)\}.$$

Proof. The commutative image of R_c is the ribbon Schur function $s_{\lambda\setminus\mu}$, where $\lambda \setminus \mu$ is the skew Young diagram associated to the composition c. Therefore, the corollary is equivalent to the decomposition

$$s_{\lambda\setminus\mu}(X) = \sum_{d \in \mathfrak{C}(n)} C_{c,d}\, L_d(X)$$

in QSym, where $C_{c,d}$ is the number of permutations with descent composition c

and backstep composition d. Equivalently,

$$s_{\lambda\backslash\mu}(X) = \sum_{d \in \mathfrak{C}(n)} B_{c,d}\, M_d(X),$$

where

$$B_{c,d} = \mathrm{card}\,\{\sigma \in \mathfrak{S}(n) \,|\, D(\sigma) = D(c) \text{ and } D(\sigma^{-1}) \subset D(d)\}.$$

Recall that we have the expansion of skew Schur functions

$$s_{\lambda\backslash\mu}(X) = \sum_{T \in \mathrm{SST}(\lambda\backslash\mu)} x^T.$$

The coefficient of $M_d(X)$ in this expansion is the number of skew semistandard tableaux with shape the ribbon of c, and with weight d, that is to say d_1 entries 1, d_2 entries 2, etc. To such a skew semistandard tableau T, we can associate its reading word $w = R(T)$, which is **packed**: if $i+1$ appears in the word, then i also appears. For instance,

$$\begin{array}{|c|c|c|c|c|}
\cline{1-1} 4 & \multicolumn{4}{c}{} \\
\cline{1-3} 1 & 1 & 3 & \multicolumn{2}{c}{} \\
\cline{1-5} \multicolumn{2}{c|}{} & 1 & 2 & 5 \\
\cline{3-5} \multicolumn{4}{c|}{} & 4 \\
\cline{5-5} \multicolumn{4}{c|}{} & 1 \\
\cline{5-5}
\end{array}$$

corresponds to the packed word 411312541, with monomial $(x_1)^4 x_2 x_3 (x_4)^2 x_5$. We call **standardization** of a word w of length n the permutation in $\mathfrak{S}(n)$ obtained by:

- replacing the d_1 entries 1 in w by the numbers $1, 2, \ldots, d_1$, going from left to right;
- then, replacing the d_2 entries 2 in w by the numbers $d_1+1, \ldots, d_1+d_2$, going from left to right;
- etc.

For instance, the standardization of $w = 411312541$ is $\mathrm{Std}(w) = 712635984$. By construction, if $\sigma = \mathrm{Std}(w)$, then the set of descents of $\mathrm{Std}(w)$ is the same as the set of descents of w, that is to say $D(c)$ if w is the reading word of $T \in \mathrm{SST}(\lambda \backslash \mu)$. Suppose on the other hand that w has weight d. Then, the standardization algorithm ensures that the set of backsteps of σ is included into $D(d)$. So, to any skew semistandard tableau T of shape c and weight d, we can associate a permutation with $D(\sigma) = D(c)$ and $D(\sigma^{-1}) \subset D(d)$. This construction can be reversed, so we have indeed

$$s_{\lambda\backslash\mu}(X) = \sum_{d} \left(\mathrm{card}\,\{\sigma \in \mathfrak{S}(n) \,|\, D(\sigma) = D(c),\, D(\sigma^{-1}) \subset D(d)\}\right) M_d(X). \qquad \square$$

Example. The principal indecomposable module $P^{(1,2)}$ of $\mathfrak{H}_0(3)$ has dimension 2, since there are two permutations of size 3 with descent set $D(\sigma) = \{1\}$, namely, 213 and 312. The first permutation has backstep composition $(1,2)$, and the second permutation has backstep composition $(2,1)$, so a composition series for $P^{(1,2)}$ has factors $S^{(1,2)}$ (the top of the module) and $S^{(2,1)}$, each with multiplicity 1.

It is quite remarkable that one can compute the Cartan matrix of the 0–Hecke algebra $\mathfrak{H}_0(n)$ without ever describing explicitly its projective indecomposable modules. A similar phenomenon occurred in Chapter 2, where we were able to compute the irreducible characters of $\mathfrak{S}(n)$ before we described explicitly the irreducible representations.

Notes and references

The algebra of non-commutative symmetric functions appeared first in [GKL+95], and its theory was then developed in the series of papers [KLT97, DKKT97, KT97, KT99, DHT02]. An important result which we did not discuss is the isomorphism between NCSym endowed with an internal product $\mathrm{NCSym}_n \times \mathrm{NCSym}_n \to \mathrm{NCSym}_n$, and the direct sum of the descent algebras $D(\mathfrak{S}(n))$, that are the subalgebras of the group algebras $\mathbb{C}\mathfrak{S}(n)$ that are spanned by the sums of permutations with fixed descent set; see [Sol76, MR95].

We only introduced the elements of NCSym that are related to the representation theory of the 0–Hecke algebras, and we refer to the aforementioned papers for the study of other basis elements, e.g., the non-commutative power sums, which can for instance be defined by the identity

$$\sum_{k=1}^{\infty} \frac{\Phi_k(X)\, t^k}{k} = \log\left(\sum_{k=0}^{\infty} S_k(X)\, t^k\right).$$

There are actually several families of non-commutative power sums, which come from the various non-commutative generalizations of the relation between commutative power sums and homogeneous functions in the algebra Sym.

On the other hand, the algebra of quasi-symmetric functions has its construction due to Gessel; see [Ges84], and also [Sta99a, §7.19]. The link between the two algebras QSym and NCSym and the Grothendieck rings of 0–Hecke algebras is due to Krob and Thibon; see [KT97]. This chapter can be considered as a complete development of the arguments of Section 5 in loc. cit.

The classification of simple and principal indecomposable modules over $\mathfrak{H}_0(n)$ is due to Norton, cf. [Nor79], and see also [Car86]. Our classification of the simple modules follows closely the arguments of Norton. On the other hand, as mentioned at the end of Section 6.3, the Hopf algebraic approach allows us to prove numerous properties of the principal indecomposable modules without constructing them explicitly (this explicit construction is possible; see again the paper by Norton). In particular, though the result appeared already in [Nor79], our approach to the computation of the decomposition matrix (Corollary 6.34) is to our knowledge new.

Finally, the arguments involved in the proof of Theorem 6.24 and related to the decomposition

$$\mathfrak{S}(m+n) = (\mathfrak{S}(m) \times \mathfrak{S}(n))\,(\mathrm{id}_{[\![1,m]\!]} \sqcup\!\sqcup\, \mathrm{id}_{[\![1,n]\!]})$$

come from the second chapter of [GP00], and they can in fact be applied to any parabolic right coset $\mathfrak{S}(c)\backslash\mathfrak{S}(n)$. Hence, for any composition $c \in \mathfrak{C}(n)$, one has a decomposition

$$\mathfrak{S}(n) = \mathfrak{S}(c) \times X_c,$$

where X_c is the set of permutations whose backsteps belong to $D(c)$, and is equal to the shuffle product $\mathrm{id}_{[\![1,c_1]\!]} \sqcup\!\sqcup \mathrm{id}_{[\![1,c_2]\!]} \sqcup\!\sqcup \cdots \sqcup\!\sqcup \mathrm{id}_{[\![1,c_\ell]\!]}$. Moreover, this decomposition is compatible with the length of permutations, so if $\rho = \sigma\tau$ with $\sigma \in \mathfrak{S}(c)$ and $\tau \in X_c$, then $\ell(\rho) = \ell(\sigma) + \ell(\tau)$.

Part III

Observables of partitions

7

The Ivanov–Kerov algebra of observables

In this chapter, we present the foundations of the so-called *dual combinatorics* of the characters of the symmetric groups. This theory relies on the following fundamental idea. In the first part of this book (in particular, Chapters 2 and 3), we detailed the representation theory of the symmetric group $\mathfrak{S}(n)$, and we constructed the characters ch^λ, which are functions from $\mathfrak{S}(n)$ to $\mathbb{C}$, or more precisely, from the set $\mathfrak{Y}(n)$ of conjugacy classes of $\mathfrak{S}(n)$ to $\mathbb{C}$ (and in fact to $\mathbb{Z}$):

$$\mathrm{ch}^\lambda : \mu \in \mathfrak{Y}(n) \mapsto \mathrm{ch}^\lambda(\mu) \in \mathbb{C}.$$

It turns out that a powerful tool of study of the representations of $\mathfrak{S}(n)$ comes from the dual point of view, which consists in looking at the functions

$$\lambda \in \mathfrak{Y}(n) \mapsto \mathrm{ch}^\lambda(\mu) \in \mathbb{C}.$$

These objects are functions of the irreducible representations S^λ of $\mathfrak{S}(n)$, and one can use them to construct an *algebra of observables* of the irreducible representations of all the symmetric groups $\mathfrak{S}(n)$ with $n \in \mathbb{N}$. A basis of this algebra will consist in renormalized versions of the maps $S^\lambda \mapsto \mathrm{ch}^\lambda(\mu)$. The definition of this algebra $\mathscr{O}$ is related to the construction of a projective limit of the group algebras $\mathbb{C}\mathfrak{S}(n)$, which is not possible in the strict sense, but is allowed if one introduces the notion of *partial permutation*; see Section 7.1.

The *algebraic observables* built as combinations of renormalized versions of the character values happen to be related to certain *geometric observables* of Young diagrams. Thus, the algebra $\mathscr{O}$ admits several other bases, which consist in functions $\lambda \mapsto f(\lambda)$ that capture the shape and geometry of the Young diagrams λ (cf. Section 7.2). The formulas of change of basis in $\mathscr{O}$ are detailed in Section 7.3; these formulas are usually quite involved, but they become simple if one looks only at the first terms of them with respect to certain gradations of $\mathscr{O}$. Finally, in Section 7.4, we extend the range of the observables in $\mathscr{O}$ to the space $\mathscr{Y}$ of so-called *continuous Young diagrams*. This will prove very useful in the fourth part of the book, when studying large random partitions stemming from the representation theory of the groups $\mathfrak{S}(n)$. We detail in particular the weak topology on $\mathscr{Y}$ associated to the algebra of functions $\mathscr{O}$.

7.1 The algebra of partial permutations

▷ *Partial permutations of size n.*

If $\sigma \in \mathfrak{S}(n)$ is a permutation, we call **admissible support** of σ a part $A \subset [\![1,n]\!]$, such that σ only moves elements from A :

$$\forall k \in [\![1,n]\!] \setminus A,\ \ \sigma(k)=k.$$

In terms of the cycles of σ, a part $A \subset [\![1,n]\!]$ is an admissible support if and only if it contains all the supports of the cycles of σ of length greater than 2.

Example. If $n=8$ and $\sigma=(2,5,3)(6,8)$, then the set $\{1,2,3,5,6,8\}$ is an admissible support, and so is $\{2,3,5,6,8\}$, but $\{1,2,3,4,5,6\}$ is not an admissible support.

Definition 7.1. *A* ***partial permutation*** *of size n is a pair (σ,A), where $\sigma \in \mathfrak{S}(n)$ and A is an admissible support for σ.*

We denote $\mathfrak{PS}(n)$ the set of partial permutations of size n. If A is a fixed part of cardinality k in $[\![1,n]\!]$, then there exist $k!$ permutations σ that admit A as a support, namely, the permutations in $\mathfrak{S}(A) \subset \mathfrak{S}(n)$. Therefore,

$$\operatorname{card}\mathfrak{PS}(n)=\sum_{k=0}^{n}\binom{n}{k}k!=\sum_{k=0}^{n}\frac{n!}{(n-k)!}.$$

In the sequel, we shall use several times the notation $n^{\downarrow k}$ for the falling factorial

$$\frac{n!}{(n-k)!}=n\times(n-1)\times\cdots\times(n-k+1);$$

we agree that $n^{\downarrow k}=0$ if $k>n$. Thus, we have just shown that $\operatorname{card}\mathfrak{PS}(n)=\sum_{k=0}^{n} n^{\downarrow k}$. The first values of $\operatorname{card}\mathfrak{PS}(n)$ are $|\mathfrak{PS}(0)|=1$, $|\mathfrak{PS}(1)|=2$, $|\mathfrak{PS}(2)|=5$, $|\mathfrak{PS}(3)|=16$ and $|\mathfrak{PS}(4)|=65$.

An important feature of partial permutations is the possibility to multiply them. Let (σ,A) and (τ,B) be two partial permutations in $\mathfrak{PS}(n)$. Since σ leaves invariant the elements not in A, and τ leaves invariant the elements not in B, the composition $\sigma\tau$ leaves invariant the elements that do not belong to $A\cup B$, hence, $(\sigma\tau,A\cup B)$ is a partial permutation. We define it to be the product of (σ,A) by (τ,B):

$$(\sigma,A)\times(\tau,B)=(\sigma\tau,A\cup B).$$

Endowed with this operation, $\mathfrak{PS}(n)$ becomes a non-commutative monoid, with neutral element $(\mathrm{id}_\emptyset, \emptyset)$. Moreover, one has a natural surjective morphism of monoids

$$\pi_n : \mathfrak{PS}(n) \to \mathfrak{S}(n)$$
$$(\sigma, A) \mapsto \sigma.$$

We also denote π_n the corresponding morphism of algebras between the monoid algebras $\mathbb{C}\mathfrak{PS}(n)$ and $\mathbb{C}\mathfrak{S}(n)$. These maps π_n will serve the following objective. Suppose that one wants to make generic calculations in symmetric groups and symmetric group algebras, that is to say calculations that do not depend on the size n of the symmetric group, or more generally that do depend on n in a prescribed and simple way. For instance, one could be interested in the computation of a product of conjugacy classes, say,

$$(C_{(2,1,\ldots,1)})^2 = \left(\sum_{1 \leq i < j \leq n} (i,j) \right)^2$$

in $\mathbb{C}\mathfrak{S}(n)$. It turns out that the result of this calculation depends on n in a very simple way: for any $n \geq 4$,

$$(C_{(2,1,\ldots,1)})^2 = 3\, C_{(3,1,\ldots,1)} + 2\, C_{(2,2,1,\ldots,1)} + \binom{n}{2} C_{(1,1,\ldots,1)}.$$

Indeed, if one expands the square of the conjugacy class of transpositions, then each term $(i,j)(k,l)$ can be:

- a product of two disjoint transpositions, if $\{i,j\} \cap \{k,l\} = 0$; each such product is obtained twice, by $(i,j) \times (k,l)$ and $(k,l) \times (i,j)$.
- a 3-cycle if $\{i,j\} \cap \{k,l\} = 1$. Each 3-cycle (a,b,c) is obtained three times, namely, by $(a,b) \times (b,c)$, by $(b,c) \times (c,a)$ and by $(c,a) \times (a,b)$.
- the identity if $\{i,j\} = \{k,l\}$, which happens $\binom{n}{2}$ times.

In a moment, we shall construct a projective limit $\mathbb{C}\mathfrak{PS}(\infty)$ of the algebras $\mathbb{C}\mathfrak{PS}(n)$, endowed with maps $\phi_n : \mathbb{C}\mathfrak{PS}(\infty) \to \mathbb{C}\mathfrak{PS}(n)$, and which contains generic elements Σ_μ such that

$$\pi_n \circ \phi_n(\Sigma_\mu) = c(n,\mu)\, C_{\mu \sqcup 1^{n-|\mu|}}.$$

In this formula, $c(n,\mu)$ is a polynomial in n, and $\mu \sqcup 1^{n-|\mu|}$ denotes the integer partition of size n and with parts $(\mu_1, \mu_2, \ldots, \mu_{\ell(\mu)}, 1, 1, \ldots, 1)$. Then, a relation

$$\Sigma_\mu \times \Sigma_\nu = \sum_\lambda a_{\mu\nu}^\lambda\, \Sigma_\lambda$$

in $\mathbb{C}\mathfrak{PS}(\infty)$ will lead by projection by the maps π_n and ϕ_n to an identity

$$C_{\mu \sqcup 1^{n-|\mu|}}\, C_{\nu \sqcup 1^{n-|\nu|}} = \sum_\lambda a_{\mu\nu}^\lambda \frac{c(n,\lambda)}{c(n,\mu)\, c(n,\nu)}\, C_{\lambda \sqcup 1^{n-|\lambda|}},$$

hence a generic identity for the product of the conjugacy classes $C_{\mu\sqcup 1^{n-|\mu|}}$ and $C_{\nu\sqcup 1^{n-|\nu|}}$. Let us remark that it is not possible to do so by constructing directly a projective limit of the group algebras $\mathbb{C}\mathfrak{S}(n)$. Indeed, it is a well-known fact that for $n \geq 5$, the only normal subgroups of $\mathfrak{S}(n)$ are $\{1\}$, $\mathfrak{S}(n)$ and the alternating group $\mathfrak{A}(n) = \{\sigma \in \mathfrak{S}(n) \,|\, \varepsilon(\sigma) = 1\}$. Therefore, for $n \geq 5$, there is no surjective morphism of groups $\mathfrak{S}(n) \to \mathfrak{S}(n-1)$: otherwise, the kernel would have cardinality $\frac{\operatorname{card}\mathfrak{S}(n)}{\operatorname{card}\mathfrak{S}(n-1)} = n$ and be a normal subgroup of $\mathfrak{S}(n)$, which is not possible. This prohibits the construction of a projective limit $\varprojlim_{n\to\infty} \mathfrak{S}(n)$ in the category of groups. The trick is then to construct a projective limit $\varprojlim_{n\to\infty} \mathfrak{P}\mathfrak{S}(n)$, and to use the maps π_n to deduce from it results on the groups $\mathfrak{S}(n)$.

Proposition 7.2. *Consider the linear map*

$$\begin{aligned}\psi : \mathbb{C}\mathfrak{P}\mathfrak{S}(n) &\to \bigoplus_{A\subset [\![1,n]\!]} \mathbb{C}\mathfrak{S}(A)\\ (\sigma,B) &\mapsto \sum_{A\supset B} \sigma\end{aligned}$$

This map ψ is an isomorphism of complex algebras.

Proof. The two algebras $\mathbb{C}\mathfrak{P}\mathfrak{S}(n)$ and $\bigoplus_{A\subset[\![1,n]\!]} \mathbb{C}\mathfrak{S}(A)$ have the same dimension $|\mathfrak{P}\mathfrak{S}(n)| = \sum_{k=0}^{n} n^{\downarrow k}$, so it suffices to show that ψ is an injective morphism of algebras. The linearity of ψ is a part of its definition; let us check that ψ is compatible with the product. If (σ,B) and (τ,C) are two partial permutations of size n, then

$$\begin{aligned}\psi\left((\sigma,B)(\tau,C)\right) = \psi((\sigma\tau, B\cup C)) &= \sum_{A\supset B\cup C} \sigma\tau = \sum_{A\,|\,A\supset B \text{ and } A\supset C} \sigma\tau\\ &= \left(\sum_{A\supset B} \sigma\right) \times \left(\sum_{A\supset C} \tau\right) = \psi((\sigma,B)) \times \psi((\tau,C)).\end{aligned}$$

Thus, ψ is a morphism of algebras. Let us now prove the injectivity. If

$$x = \sum_{(\sigma,B)\in\mathfrak{P}\mathfrak{S}(n)} c_{(\sigma,B)}\,(\sigma,B)$$

is a non-zero element of $\mathbb{C}\mathfrak{P}\mathfrak{S}(n)$, let A be a part of $[\![1,n]\!]$ such that there exists $\sigma \in \mathfrak{S}(A)$ with $c_{(\sigma,A)} \neq 0$. Without loss of generality, we can assume that A is minimal with respect to inclusion in the set of subsets with this property. If $\theta_A : \bigoplus_{A'\subset[\![1,n]\!]} \mathbb{C}\mathfrak{S}(A') \to \mathbb{C}\mathfrak{S}(A)$ is the projection on the block $\mathbb{C}\mathfrak{S}(A)$, then

$$\begin{aligned}\theta_A(\psi(x)) &= \sum_{(\sigma,B)\in\mathfrak{P}\mathfrak{S}(n)} c_{(\sigma,B)}\,\theta_A(\psi(\sigma,B)) = \sum_{(\sigma,B)\in\mathfrak{P}\mathfrak{S}(n)} c_{(\sigma,B)}\,\theta_A\left(\sum_{B\subset A'} \sigma\right)\\ &= \sum_{(\sigma,B)\in\mathfrak{P}\mathfrak{S}(n)\,|\,B\subset A} c_{(\sigma,B)}\,\sigma.\end{aligned}$$

Since A is minimal for the inclusion, the last term is just $\sum_{\sigma \in \mathfrak{S}(A)} c_{(\sigma,A)}\, \sigma$, which is non-zero by assumption on A. So, $\theta_A(\psi(x)) \neq 0$, and $\psi(x) \neq 0$. □

In particular, the complex algebras $\mathbb{C}\mathfrak{P}\mathfrak{S}(n)$ are all semisimple, since they are isomorphic to direct sums of group algebras.

▷ *The projective limit in the category of graded algebras.*

We generalize Definition 7.1 as follows: if A is a finite set of positive integers, we define $\mathfrak{P}\mathfrak{S}(A)$ to be the set of partial permutations (σ, B) with $B \subset A$. We then connect the various algebras $\mathbb{C}\mathfrak{P}\mathfrak{S}(A)$ by morphisms of algebras

$$\begin{aligned} \phi_{A \to A'} : \mathbb{C}\mathfrak{P}\mathfrak{S}(A) &\to \mathbb{C}\mathfrak{P}\mathfrak{S}(A') \\ (\sigma, B) &\mapsto \begin{cases} (\sigma, B) & \text{if } B \subset A', \\ 0 & \text{otherwise} \end{cases} \end{aligned}$$

for $A \supset A'$. These maps are compatible with one another: if $A \supset A' \supset A''$, then $\phi_{A' \to A''} \circ \phi_{A \to A'} = \phi_{A \to A''}$. Moreover, they are compatible with the gradations of algebras

$$\deg(\sigma, B) = \operatorname{card} B.$$

Hence, if $x \in \mathbb{C}\mathfrak{P}\mathfrak{S}(A)$ and $A \supset A'$, then $\deg \phi_{A \to A'}(x) \leq \deg x$. So, the maps $\phi_{A \to A'}$ are morphisms of graded algebras, a graded algebra G being an algebra endowed with a map $\deg : G \to \mathbb{N}$ such that

$$\begin{aligned} \deg(\lambda x + \mu y) &\leq \max(\deg x, \deg y); \\ \deg(x \times y) &\leq \deg x + \deg y. \end{aligned}$$

We abbreviate $\phi_{[\![1,N]\!] \to [\![1,n]\!]} = \phi_n^N$. One then has an inverse system of graded algebras

$$\cdots \longrightarrow \mathbb{C}\mathfrak{P}\mathfrak{S}(n+2) \xrightarrow{\phi_{n+1}^{n+2}} \mathbb{C}\mathfrak{P}\mathfrak{S}(n+1) \xrightarrow{\phi_n^{n+1}} \mathbb{C}\mathfrak{P}\mathfrak{S}(n) \longrightarrow \cdots,$$

We define $\mathbb{C}\mathfrak{P}\mathfrak{S}(\infty)$ to be the projective limit

$$\mathbb{C}\mathfrak{P}\mathfrak{S}(\infty) = \varprojlim_{n \to \infty} \mathbb{C}\mathfrak{P}\mathfrak{S}(n),$$

this limit being taken in the category of graded $\mathbb{C}$-algebras and with respect to the maps ϕ_n^{n+1}. Let us describe more precisely the elements of $\mathbb{C}\mathfrak{P}\mathfrak{S}(\infty)$. Since we take the projective limit in the category of *graded* algebras, an element $x \in \mathbb{C}\mathfrak{P}\mathfrak{S}(\infty)$ must have projections $\phi_n(x)$ in each $\mathbb{C}\mathfrak{P}\mathfrak{S}(n)$ that have a uniformly bounded degree. As a consequence, any element of $\mathbb{C}\mathfrak{P}\mathfrak{S}(\infty)$ can be written as a possibly infinite linear combination

$$x = \sum_{(\sigma,A)} c_{\sigma,A}\, (\sigma, A)$$

of partial permutations, such that $\max(\deg(\sigma,A)) = \max(\operatorname{card} A) = \deg x < +\infty$. Thus,

$$x = \sum_{1 \leq i < j} \big((i,j),\{i,j\}\big)$$

is an element of $\mathbb{C}\mathfrak{P}\mathfrak{S}(\infty)$ (of degree $\deg x = 2$), but

$$y = \sum_{k=1}^{\infty} \big((1,2,\ldots,k),\{1,2,\ldots,k\}\big)$$

is not an element of $\mathbb{C}\mathfrak{P}\mathfrak{S}(\infty)$, because the partial permutations appearing in the formal sum do not have a bounded degree.

We are now ready to define the main object of interest of this chapter. Recall that $\mathfrak{S}(\infty) = \varinjlim_{n\to\infty} \mathfrak{S}(n)$ is the set of finite permutations, that is to say bijections $\sigma : \mathbb{N}^* \to \mathbb{N}^*$ that move only a finite number of integers. There is a natural action by conjugation of $\mathfrak{S}(\infty)$ on $\mathfrak{P}\mathfrak{S}(\infty) = \bigcup_{n\in\mathbb{N}} \mathfrak{P}\mathfrak{S}(n)$:

$$\sigma \cdot (\tau, B) = (\sigma\tau\sigma^{-1}, \sigma(B)).$$

This action extends to a linear representation of $\mathfrak{S}(\infty)$ on $\mathbb{C}\mathfrak{P}\mathfrak{S}(\infty)$, which is compatible with the degree and with the algebra structure:

$$\forall \sigma \in \mathfrak{S}(\infty),\ \deg(\sigma\cdot x) = \deg x \qquad ; \qquad \sigma\cdot(x\times y) = (\sigma\cdot x)\times(\sigma\cdot y).$$

Definition 7.3. *An* ***observable of partition*** *is an element of the subalgebra of invariants*

$$\mathscr{O} = (\mathbb{C}\mathfrak{P}\mathfrak{S}(\infty))^{\mathfrak{S}(\infty)} = \{x \in \mathbb{C}\mathfrak{P}\mathfrak{S}(\infty)\,|\, \forall \sigma \in \mathfrak{S}(\infty),\ \sigma\cdot x = x\}.$$

The terminology of observables will be fully explained in Section 7.3; for the moment, we have simply defined observables as formal sums of partial permutations, with a property of invariance by conjugation.

Example. The sum $\Sigma_2 = \sum_{i\neq j}((i,j),\{i,j\})$ is an observable in $\mathscr{O}$, since for every permutation $\sigma \in \mathfrak{S}(\infty)$,

$$\sigma\cdot\big((i,j),\{i,j\}\big) = \big((\sigma(i),\sigma(j)),\{\sigma(i),\sigma(j)\}\big).$$

Hence, the conjugation by σ permutes the elements of the formal sum of partial permutations Σ_2. More generally, given an integer partition $\mu = (\mu_1,\ldots,\mu_r)$ of size n, we denote

$$\Sigma_\mu = \sum \big((i_1,\ldots,i_{\mu_1})(i_{\mu_1+1},\ldots,i_{\mu_1+\mu_2})\cdots(i_{\mu_1+\cdots+\mu_{r-1}+1},\ldots,i_n),\{i_1,\ldots,i_n\}\big)$$

where the sum runs over **arrangements** of size n in $\mathbb{N}^*$, that is to say injective maps

$$i : j \in [\![1,n]\!] \mapsto i_j \in \mathbb{N}^*.$$

Each element appearing in the formal sum Σ_μ has the form (σ,A), where A is a set of n integers, and $\sigma \in \mathfrak{S}(A)$ has cycle type μ. For the same reasons as in the case when $\mu = (2)$, Σ_μ is invariant by conjugation by any element $\sigma \in \mathfrak{S}(\infty)$, hence belongs to the algebra of observables $\mathscr{O}$. Moreover, $\deg \Sigma_\mu = |\mu|$.

Theorem 7.4. *The algebra of observables $\mathscr{O}$ admits for graded basis over $\mathbb{C}$:*

$$\Sigma_\mu,\ \mu \in \mathfrak{Y} = \bigsqcup_{n=0}^{\infty} \mathfrak{Y}(n).$$

One has $\deg \Sigma_\mu = |\mu|$. Moreover, $(\Sigma_k = \Sigma_{(k)})_{k\in\mathbb{N}}$ is a transcendance basis of $\mathscr{O}$ over $\mathbb{C}$, and for every partition μ,

$$\prod_{i=1}^{\ell(\mu)} \Sigma_{\mu_i} = \Sigma_\mu + \textit{terms of degree smaller than } |\mu|-1.$$

Proof. Notice that two partial permutations (σ,B) and (τ,C) are conjugated under the action of $\mathfrak{S}(\infty)$ if and only if $|B| = |C|$ and σ and τ have the same cycle type as permutations of B and C respectively. Consequently, the conjugacy classes under the action of $\mathfrak{S}(\infty)$ in $\mathfrak{PS}(\infty)$ are labeled by integer partitions $\mu \in \mathfrak{Y}$, an integer partition μ corresponding to partial permutations (σ,B) with $\operatorname{card} B = |\mu|$ and $t(\sigma) = \mu$. Moreover, if A_μ is the sum of all partial permutations with type μ in $\mathbb{C}\mathfrak{PS}(\infty)$, then $(A_\mu)_{\mu\in\mathfrak{Y}}$ is a linear basis of $\mathscr{O}$. It remains to see that

$$A_\mu = \frac{\Sigma_\mu}{z_\mu}.$$

Indeed, each partial permutation (σ,B) of type μ is attained z_μ times by the map

$$\begin{aligned}&(i_1 \neq i_2 \neq \cdots \neq i_n) \text{ arrangement of size } n\\ &\mapsto \big((i_1,\ldots,i_{\mu_1})(i_{\mu_1+1},\ldots,i_{\mu_1+\mu_2})\cdots(i_{\mu_1+\cdots+\mu_{r-1}+1},\ldots,i_n),\{i_1,\ldots,i_n\}\big) \in \mathfrak{PS}(\infty).\end{aligned}$$

This ends the proof of the first part of the theorem. For the second part, denote $C(\mu,i)$ the partial permutation of type μ associated to an arrangement $i = (i_1,\ldots,i_n)$ by the map above. We then have $\Sigma_\mu = \sum_{i\in\mathfrak{A}(\mathbb{N}^*,n)} C(\mu,i)$, where $\mathfrak{A}(\mathbb{N}^*,n)$ is the set of n-arrangements of positive integers. As a consequence,

$$\Sigma_{\mu_1}\Sigma_{\mu_2}\cdots\Sigma_{\mu_r} = \sum_{\substack{i^{(1)}\in\mathfrak{A}(\mathbb{N}^*,\mu_1)\\ i^{(2)}\in\mathfrak{A}(\mathbb{N}^*,\mu_2)\\ \cdots}} C(\mu_1,i^{(1)})\,C(\mu_2,i^{(2)})\cdots C(\mu_r,i^{(r)}).$$

Now, notice that $\deg((\sigma,A)(\tau,B)) < \deg(\sigma,A) + \deg(\tau,B)$ unless the sets A and B are disjoint. Therefore,

$$\begin{aligned}\Sigma_{\mu_1}\Sigma_{\mu_2}\cdots\Sigma_{\mu_r} = &\sum_{i^{(1)},\ldots,i^{(r)} \text{ disjoint arrangements}} C(\mu_1,i^{(1)})\,C(\mu_2,i^{(2)})\cdots C(\mu_r,i^{(r)})\\ &+ \text{terms of degree smaller than } |\mu|-1.\end{aligned}$$

However, if $i^{(1)},\dots,i^{(r)}$ are disjoint arrangements of integers of respective sizes $\mu_1,\dots,\mu_r$, then $C(\mu_1,i^{(1)})\cdots C(\mu_r,i^{(r)}) = C(\mu,i)$, where i is the arrangement that is the union of the arrangements $i^{(1)},\dots,i^{(r)}$ read in this order. Thus, the sum in the last equality is simply Σ_μ, and this ends the proof of the theorem. Indeed, the "factorization in higher degree" of the symbols Σ_μ implies the algebraic independence of the symbols Σ_k over $\mathbb{C}$ in $\mathscr{O}$. □

▷ *Partial permutations and products of conjugacy classes.*

In the next Sections 7.2 and 7.3, we shall consider the symbols $\Sigma_\mu \in \mathscr{O}$ as functions of the irreducible representations S^λ of the symmetric groups $\mathfrak{S}(n)$, thereby justifying the terminology of "observables" given to the elements of $\mathscr{O}$. Before, we shall explain how to use the elements Σ_μ in order to solve the problem of the generic calculation of products of conjugacy classes in the symmetric group algebras $\mathbb{C}\mathfrak{S}(n)$. We start with an explicit formula for the product of two symbols Σ_μ, which precises the second part of Theorem 7.4. A **partial pairing** between two finite sets A and B is an injective map i from a subset of A to a subset of B. It can always be described as a set of pairs (a,b) with $a \in A$ and $b \in B$, such that if $a = a'$ or $b = b'$, then $(a,b) = (a',b')$. For instance, if $A = [\![1,3]\!]$ and $B = [\![1,2]\!]$, then $\pi_1 = \{(2,1)\}$ and $\pi_2 = \{(3,1),(2,2)\}$ are two partial pairings, represented by the graphs

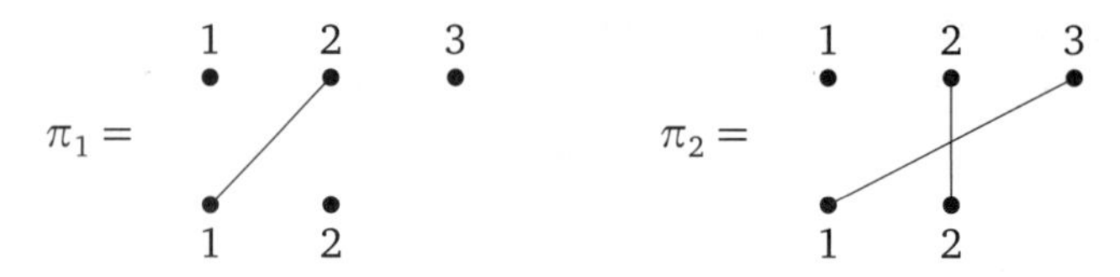

We denote $|\pi|$ the number of pairs in a partial pairing. The number of partial pairings between two sets A and B of respective sizes k and l is

$$\sum_{m=0}^{\min(k,l)} m! \binom{k}{m}\binom{l}{m} = \sum_{m=0}^{\min(k,l)} \frac{k!\,l!}{(k-m)!\,(l-m)!\,m!},$$

as can be seen by counting the partial pairings according to their size $|\pi| = m$.

If i and j are two arrangements in $\mathfrak{A}(\mathbb{N}^*,k)$ and $\mathfrak{A}(\mathbb{N}^*,l)$, we associate to them a pairing $\pi(i,j)$ between the sets $[\![1,k]\!]$ and $[\![1,l]\!]$: it consists in the pairs (m,n) such that $i_m = j_n$. For instance, if $i = (4,5,2)$ and $j = (5,3)$, then the corresponding pairing between $[\![1,3]\!]$ and $[\![1,2]\!]$ is $\pi_1 = \{(2,1)\}$, because $i_2 = j_1$ and this is the only equality.

Lemma 7.5. *Consider two integer partitions μ and ν of respective sizes k and l, and two arrangements i and j also of sizes k and l. The product of partial permutations $C(\mu,i) \times C(\nu,j)$ is a partial permutation whose type λ (with respect to the conjugacy action of $\mathfrak{S}(\infty)$ on $\mathfrak{P}\mathfrak{S}(\infty)$) depends only on μ, ν and the pairing $\pi(i,j)$.*

Proof. We are looking at the product

$$x = \big((i_1,\ldots,i_{\mu_1})(i_{\mu_1+1},\ldots,i_{\mu_1+\mu_2})\cdots(i_{\mu_1+\cdots+\mu_{r-1}+1},\ldots,i_k),\{i_1,\ldots,i_k\}\big)$$
$$\times\big((j_1,\ldots,j_{\nu_1})(j_{\nu_1+1},\ldots,j_{\nu_1+\nu_2})\cdots(j_{\nu_1+\cdots+\nu_{s-1}+1},\ldots,j_l),\{j_1,\ldots,j_l\}\big).$$

Let σ the permutation in $\mathfrak{S}(\infty)$ which sends:

1. i_1 to 1, i_2 to 2, etc., i_k to k;
2. the $l-|\pi(i,j)|$ elements $j_{n_1},\ldots,j_{n_{l-|\pi|}}$ that are not equal to some i_m's to the integers $k+1,\ldots,k+l-|\pi|$.

Notice that this permutation σ depends only on the pairing $\pi(i,j)$. Then, $\sigma\cdot x$ has support $[\![1,k+l-|\pi|]\!]$, and underlying permutation which is the composition of

$$(1,\ldots,\mu_1)(\mu_1+1,\ldots,\mu_1+\mu_2)\cdots(\mu_1+\cdots+\mu_{r-1}+1,\ldots,k)$$

and of a permutation

$$(\sigma(j_1),\ldots,\sigma(j_{\nu_1}))(\sigma(j_{\nu_1+1}),\ldots,\sigma(j_{\nu_1+\nu_2}))\cdots(\sigma(j_{\nu_1+\cdots+\nu_{s-1}+1}),\ldots,\sigma(j_l))$$

that depends only on $\pi(i,j)$. Since x and $\sigma\cdot x$ have the same type in $\mathfrak{PS}(\infty)$, the lemma is shown. □

Given a partial pairing π between $[\![1,|\mu|]\!]$ and $[\![1,|\nu|]\!]$, we denote $\lambda=\lambda(\mu,\nu,\pi)$ the type of a partial permutation $C(\mu,i)\times C(\nu,j)$ such that $\pi(i,j)=\pi$.

Proposition 7.6. *In the algebra of observables $\mathscr{O}$, we have*

$$\Sigma_\mu\,\Sigma_\nu=\sum_\pi \Sigma_{\lambda(\mu,\nu,\pi)},$$

where the sum runs over partial pairings π between $[\![1,|\mu|]\!]$ and $[\![1,|\nu|]\!]$.

Proof. Denote k and l the sizes of μ and ν. There is a bijection between pairs of arrangements $(i,j)\in\mathfrak{A}(\mathbb{N}^*,k)\times\mathfrak{A}(\mathbb{N}^*,l)$, and pairs $(\pi,h)=\psi(i,j)$, where π is a partial pairing between $[\![1,k]\!]$ and $[\![1,l]\!]$, and h is an arrangement in $\mathfrak{A}(\mathbb{N}^*,k+l-|\pi|)$. This bijection ψ is given by $\pi=\pi(i,j)$, and then h obtained by writing first $i_1,\ldots,i_k$, and the j_n's not equal to some i_m's, in the same order as in the arrangement j. According to the previous lemma, $C(\mu,i)\times C(\nu,j)=C(\lambda(\mu,\nu,\pi(i,j)),h')$, where h' is some reordering of h prescribed by μ, ν and $\pi(i,j)$. This reordering comes from the decomposition of the product permutation as a product of cycles of decreasing sizes $\lambda_1\geq\lambda_2\geq\cdots\geq\lambda_t$; this will be made clear in the example hereafter. So,

$$\Sigma_\mu\,\Sigma_\nu=\sum_{\substack{i\in\mathfrak{A}(\mathbb{N}^*,k)\\ j\in\mathfrak{A}(\mathbb{N}^*,l)}} C(\mu,i)\,C(\nu,j)=\sum_\pi\sum_{h\in\mathfrak{A}(\mathbb{N}^*,k+l-|\pi|)} C(\lambda(\mu,\nu,\pi),h')$$
$$=\sum_\pi\sum_{h'\in\mathfrak{A}(\mathbb{N}^*,k+l-|\pi|)} C(\lambda(\mu,\nu,\pi),h')=\sum_\pi \Sigma_{\lambda(\mu,\nu,\pi)},$$

the sum running over partial pairings between $[\![1,k]\!]$ and $[\![1,l]\!]$. □

Example. Let us detail the calculation of $\Sigma_3\,\Sigma_2$. By definition, it is equal to the sum of partial permutations

$$((i_1,i_2,i_3)(j_1,j_2),\{i_1,i_2,i_3\}\cup\{j_1,j_2\}),$$

where i and j run over arrangements of size 3 and 2.

1. If $\{i_1,i_2,i_3\}\cap\{j_1,j_2\}=\emptyset$, then the pairing $\pi(i,j)$ is the empty pairing, and these situations yield a contribution $\Sigma_{(3,2)}$ to the product.

2. If $|\{i_1,i_2,i_3\}\cap\{j_1,j_2\}|=1$, then there are 6 pairings corresponding to this situation, and all of them correspond to a product which is a 4-cycle. For instance, if $i_1=j_1$, then $(i_1,i_2,i_3)(j_1,j_2)=(i_2,i_3,i_1)(i_1,j_2)=(i_2,i_3,i_1,j_2)$. Thus, the 6 pairings of size 1 between $[\![1,3]\!]$ and $[\![1,2]\!]$ yield a contribution $6\,\Sigma_4$ to the product.

3. Finally, if $|\{i_1,i_2,i_3\}\cap\{j_1,j_2\}|=2$, then there are 6 pairings corresponding to this situation, and all of them correspond to a product which has type $(2,1)$. For instance, if $i_1=j_1$ and $i_2=j_2$, then $(i_1,i_2,i_3)(j_1,j_2)=(i_3,i_1,i_2)(i_2,i_1)=(i_3,i_1)$, and on the other hand the support of the partial permutation is $\{i_1,i_2,i_3\}$.

We conclude that $\Sigma_3\,\Sigma_2=\Sigma_{(3,2)}+6\,\Sigma_4+6\,\Sigma_{(2,1)}$.

Let us now compute the image of Σ_μ by the morphisms of algebras $\pi_n\circ\phi_n$, where $\phi_n:\mathbb{C}\mathfrak{PS}(\infty)\to\mathbb{C}\mathfrak{PS}(n)$ is the projection map, defined by linear extension of the rule

$$\phi_n((\sigma,A))=\begin{cases}(\sigma,A) & \text{if } A\subset[\![1,n]\!]\\ 0 & \text{otherwise.}\end{cases}$$

If μ is an integer partition of size k, and $n\geq k$, we denote $\mu\uparrow n$ the completed partition of size n, obtained by adding parts of size 1 at the end of μ. So,

$$\mu\uparrow n=\mu\sqcup 1^{n-|\mu|}\quad\text{if } n\geq|\mu|.$$

The map $\mu\mapsto(\mu\uparrow n)$ yields a bijection between partitions of size $k\leq n$ without parts of size 1, and partitions of size n.

Proposition 7.7. *Let μ be an integer partition. One has*

$$\pi_n\circ\phi_n(\Sigma_\mu)=\begin{cases}z_\mu\binom{n-|\mu|+m_1(\mu)}{m_1(\mu)}\,C_{\mu\uparrow n} & \text{if } n\geq|\mu|,\\ 0 & \text{otherwise,}\end{cases}$$

where $m_1(\mu)$ is the number of parts of size 1 in μ.

Proof. Set $k=|\mu|$. If $n<k$, then every partial permutation in Σ_μ has degree k.

Since there is no subset $A \subset [\![1,n]\!]$ such that $\operatorname{card} A = k$, $\phi_n(\Sigma_\mu) = 0$. Suppose now that $n \geq k$. Then,

$$\Sigma_\mu = \sum_{i \in \mathfrak{A}(\mathbb{N}^*,k)} C(\mu,i)$$

$$\phi_n(\Sigma_\mu) = \sum_{i \in \mathfrak{A}(n,k)} C(\mu,i)$$

where $\mathfrak{A}(n,k)$ denotes the set of k-arrangements in $[\![1,n]\!]$, that is to say sequences of k distinct integers $(i_1, i_2, \ldots, i_k)$ with all the i_j's in $[\![1,n]\!]$. For any such arrangement i, the projection $\pi_n(C(\mu,i))$ is of cycle type $(\mu \uparrow n)$ in $\mathfrak{S}(n)$. On the other hand, since Σ_μ is invariant by conjugation by $\mathfrak{S}(\infty)$, $\phi_n(\Sigma_\mu)$ is invariant by conjugation by $\mathfrak{S}(n)$, therefore, $\pi_n \circ \phi_n(\Sigma_\mu)$ is an element of $\mathbb{C}\mathfrak{S}(n)$ invariant by conjugation, hence, in $Z(\mathbb{C}\mathfrak{S}(n))$. So, it is a linear combination of conjugacy classes, and by the previous argument, there is a scalar c such that

$$\pi_n \circ \phi_n(\Sigma_\mu) = c\, C_{\mu\uparrow n}.$$

The coefficient c is the quotient of the number of partial permutations in $\phi_n(\Sigma_\mu)$, namely, $n^{\downarrow|\mu|}$, by the cardinality of $C_{\mu\uparrow n}$. Finally, a simple computation yields

$$\frac{n^{\downarrow|\mu|}}{\operatorname{card} C_{\mu\uparrow n}} = z_\mu \binom{n-|\mu|+m_1(\mu)}{m_1(\mu)}. \qquad \square$$

An application of Proposition 7.7 and of the algebra structure of $\mathscr{O}$ yields the following deep result:

Theorem 7.8 (Farahat–Higman). *If μ and ν are integer partitions without parts of size 1, then there exists for every integer partition λ of size $|\lambda| \leq |\mu| + |\nu|$ and without part of size 1 a polynomial $c^\lambda_{\mu\nu}(n)$ with rational coefficients, such that for any $n \geq |\mu|, |\nu|$*

$$C_{\mu\uparrow n}\, C_{\nu\uparrow n} = \sum_\lambda c^\lambda_{\mu\nu}(n)\, C_{\lambda\uparrow n}$$

in $Z(\mathbb{C}\mathfrak{S}(n))$.

Proof. Proposition 7.6 can be rewritten as:

$$\Sigma_\mu\, \Sigma_\nu = \sum_{|\lambda| \leq |\mu|+|\nu|} a^\lambda_{\mu\nu}\, \Sigma_\lambda$$

for any integer partitions μ and ν. Suppose that μ and ν do not have parts of size 1. Then, for any $n \geq \max(|\mu|, |\nu|)$, $\pi_n \circ \phi_n(\Sigma_\mu) = z_\mu\, C_{\mu\uparrow n}$ and $\pi_n \circ \phi_n(\Sigma_\nu) = z_\nu\, C_{\nu\uparrow n}$, and since $\pi_n \circ \phi_n$ is a morphism of algebras, we obtain by projection of the previous identity:

$$C_{\mu\uparrow n}\, C_{\nu\uparrow n} = \sum_{|\lambda| \leq |\mu|+\nu} a^\lambda_{\mu\nu} \frac{z_\lambda}{z_\mu z_\nu} \binom{n-|\lambda|+m_1(\lambda)}{m_1(\lambda)} C_{\lambda\uparrow n}.$$

The result follows by gathering integer partitions λ that have the same completion $\lambda \uparrow n$: they all differ by their number of parts 1, and there is among them a unique partition λ without part of size 1. □

Example. The computation of $(C_{2\uparrow n})^2$ performed before comes directly from the identity $(\Sigma_2)^2 = \Sigma_{(2,2)} + 4\,\Sigma_3 + 2\,\Sigma_{(1,1)}$ in $\mathscr{O}$. Similarly, the formula for $\Sigma_3\,\Sigma_2$ yields the identity

$$C_{3\uparrow n}\,C_{2\uparrow n} = C_{(3,2)\uparrow n} + 4\,C_{4\uparrow n} + (2n-4)\,C_{2\uparrow n}.$$

To conclude this section, notice that Proposition 7.7 can also be used to prove that $\mathscr{O}$ is a *commutative* subalgebra of $\mathbb{C}\mathfrak{PS}(\infty)$. Indeed, consider two integer partitions μ and ν, and their products

$$\Sigma_\mu\,\Sigma_\nu = \sum_{|\lambda|\leq|\mu|+|\nu|} a^\lambda_{\mu\nu}\,\Sigma_\lambda \quad ; \quad \Sigma_\nu\,\Sigma_\mu = \sum_{|\lambda|\leq|\mu|+|\nu|} a^\lambda_{\nu\mu}\,\Sigma_\lambda.$$

Suppose that there exists an integer partition λ, such that $a^\lambda_{\mu\nu} \neq a^\lambda_{\nu\mu}$. Without loss of generality, one can assume λ of minimal size $n \leq |\mu| + |\nu|$ among the integer partitions with this property. Then,

$$\begin{aligned}
&\sum_{|\lambda|\leq n} a^\lambda_{\mu\nu}\,z_\lambda \binom{n-|\lambda|+m_1(\lambda)}{m_1(\lambda)} C_{\lambda\uparrow n} \\
&= \pi_n \circ \phi_n\left(\sum_{|\lambda|\leq|\mu|+|\nu|} a^\lambda_{\mu\nu}\,\Sigma_\lambda\right) = \pi_n \circ \phi_n(\Sigma_\mu\,\Sigma_\nu) = C_{\mu\uparrow n}\,C_{\nu\uparrow n} \\
&= C_{\nu\uparrow n}\,C_{\mu\uparrow n} = \pi_n \circ \phi_n(\Sigma_\nu\,\Sigma_\mu) = \pi_n \circ \phi_n\left(\sum_{|\lambda|\leq|\mu|+|\nu|} a^\lambda_{\nu\mu}\,\Sigma_\lambda\right) \\
&= \sum_{|\lambda|\leq n} a^\lambda_{\nu\mu}\,z_\lambda \binom{n-|\lambda|+m_1(\lambda)}{m_1(\lambda)} C_{\lambda\uparrow n}
\end{aligned}$$

since the conjugacy classes $C_{\mu\uparrow n}$ and $C_{\nu\uparrow n}$ commute in $Z(\mathbb{C}\mathfrak{S}(n))$. Since λ is assumed of minimal size, we can remove on both sides of the identity the linear combinations of conjugacy classes of size less than n, which leaves:

$$\sum_{|\lambda|=n} a^\lambda_{\mu\nu}\,z_\lambda\,C_\lambda = \sum_{|\lambda|=n} a^\lambda_{\nu\mu}\,z_\lambda\,C_\lambda.$$

Since the conjugacy classes C_λ form a linear basis of $Z(\mathbb{C}\mathfrak{S}(n))$, this is a contradiction. So, $a^\lambda_{\mu\nu} = a^\lambda_{\nu\mu}$ for every integer partitions λ, μ, ν, and the algebra $\mathscr{O}$ is commutative.

7.2 Coordinates of Young diagrams and their moments

We shall now construct another commutative algebra $\mathscr{O}^{\mathrm{geom}}$, which consists in "polynomial" functions of integer partitions. Here by polynomial function, we do not mean a polynomial in the standard coordinates $\lambda_1, \ldots, \lambda_{\ell(\lambda)}$, but rather a symmetric polynomial of certain variables related to these standard coordinates. In Section 7.3, we shall then prove that the "new" algebra $\mathscr{O}^{\mathrm{geom}}$ is in fact the same as the algebra $\mathscr{O}$ of Section 7.1, thereby establishing a direct link between the geometry of the integer partitions λ, and the values of the irreducible characters ch^{λ}.

▷ *Frobenius coordinates and interlaced coordinates.*

In this whole section, λ is a fixed integer partition of size n. The Frobenius coordinates $A(\lambda)$ and $B(\lambda)$ of λ are defined as follows. We draw the Young diagram of λ, and denote $d \geq 0$ the number of boxes of the diagram that are on the principal diagonal. For instance, if $\lambda = (5,3,2)$, then the Young diagram of λ is

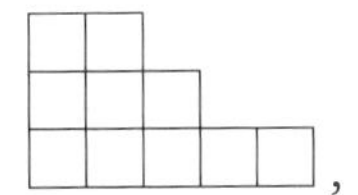

and there are $d = 2$ boxes on the diagonal. We then set $A(\lambda) = (a_1, a_2, \ldots, a_d)$ and $B(\lambda) = (b_1, b_2, \ldots, b_d)$, where $a_i - \frac{1}{2}$ is the number of boxes on the right of the i-th diagonal box, and $b_i - \frac{1}{2}$ is the number of boxes on top of the i-th diagonal box. Equivalently, the a_i's and the b_i's can be described as the lengths of the segments connecting the center of the diagonal boxes to the borders of the Young diagram; see Figure 7.1. With the previous example, we obtain $A = (\frac{9}{2}, \frac{3}{2})$ and $B = (\frac{5}{2}, \frac{3}{2})$.

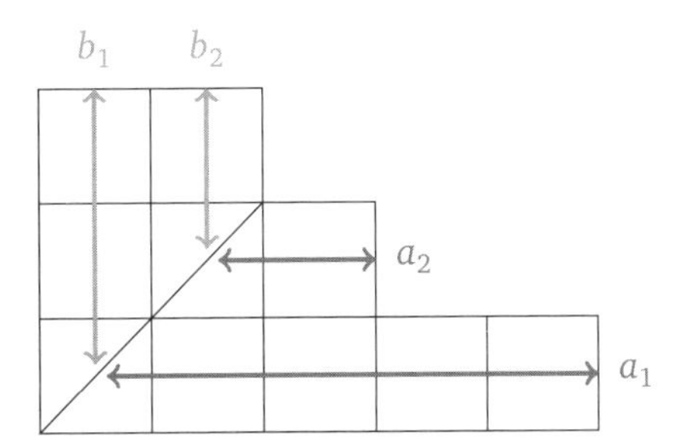

Figure 7.1
Frobenius coordinates of the Young diagram $\lambda = (5,3,2)$.

If instead of lengths of segments, we interpret the a_i's and b_i's as areas of boxes or half boxes, then we see immediately that

$$\sum_{i=1}^{d} a_i + b_i = |\lambda| = n.$$

The pair $(A(\lambda), B(\lambda))$ constitutes the so-called **Frobenius coordinates** of the integer partition λ. They form two strictly decreasing sequences of half-integers in $\mathbb{N} + \frac{1}{2}$, and conversely, given two decreasing sequences of half integers with the same length d, one can reconstruct an integer partition with these sequences as Frobenius coordinates. On the other hand, the operation of conjugation of partition is easily read on Frobenius coordinates: $A(\lambda') = B(\lambda)$ and $B(\lambda') = A(\lambda)$.

An alternative description of the integer partitions is provided by the so-called interlaced coordinates. Given an integer partition λ, we draw its Young diagram rotated by 45 degrees, and so that the boxes have edges of length $\sqrt{2}$ (this is the so-called Russian convention for drawings of Young diagrams, as opposed to the French convention used so far). Thus, the drawing of the partition $(5,3,2)$ becomes (Figure 7.2):

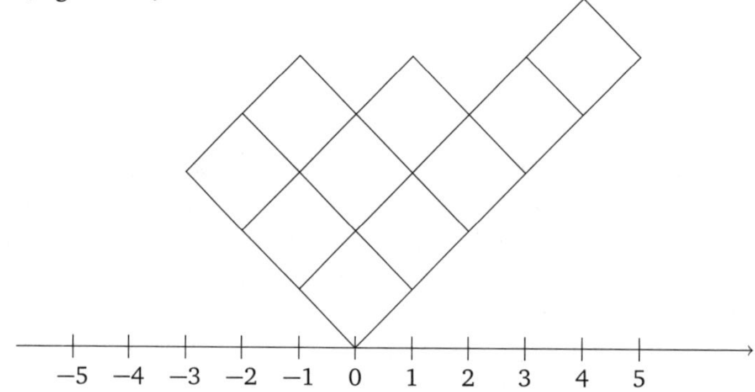

Figure 7.2
"Russian convention" for the drawings of Young diagrams, with $\lambda = (5,3,2)$.

We associate to this drawing its upper boundary, which is a continuous and affine by parts function $s \mapsto \omega_\lambda(s)$, with slopes ± 1, that meets the function $s \mapsto |s|$ at the two points $s = \lambda_1$ and $s = -\lambda'_1 = -\ell(\lambda)$. We extend the definition of ω_λ outside the interval $[-\lambda'_1, \lambda_1]$ by setting $\omega_\lambda(s) = |s|$. So, we obtain a function $\omega_\lambda : \mathbb{R} \to \mathbb{R}_+$; see Figure 7.3. The integer partition λ and the corresponding function ω_λ are entirely determined by the abscissa $x_1 < y_1 < x_2 < y_2 < \cdots < x_{s-1} < y_{s-1} < x_s$ of the local minima (the x's) and local maxima (the y's) of ω_λ. These two sequences $X(\lambda)$ and $Y(\lambda)$ are called the **interlaced coordinates** of λ, and they are interlaced sequences of integers in $\mathbb{Z}$ that satisfy for every partition:

$$\sum_{i=1}^{s} x_i - \sum_{i=1}^{s-1} y_i = 0.$$

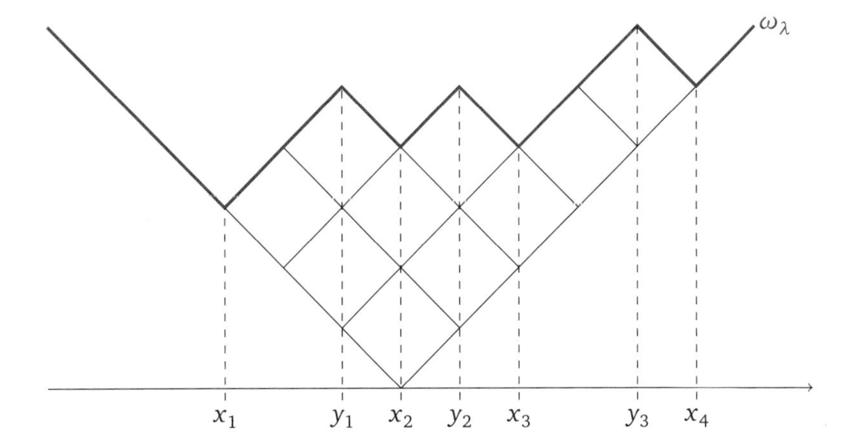

Figure 7.3
Continuous function ω_λ obtained from the Young diagram $\lambda = (5, 3, 2)$.

Indeed, the result is trivial for the empty partition $\emptyset$, which has interlaced sequences $X(\emptyset) = \{0\}$ and $Y(\emptyset) = \emptyset$. It suffices then to remark that one can add a box to the Young diagram of an integer partition only at a local minima x_i.

- If $y_{i-1} < x_i - 1$ and $x_i + 1 < y_i$, this operation replaces in the interlaced sequence x_i by $x_i - 1 < x_i < x_i + 1$, the middle term being a local maxima.
- If $y_{i-1} = x_i - 1$ and $x_i + 1 < y_i$, the operation moves y_{i-1} to $y_{i-1} + 1$ and x_i to $x_i + 1$.
- If $y_{i-1} < x_i - 1$ and $x_i + 1 = y_i$, the operation moves x_i to $x_i - 1$ and y_i to $y_i - 1$.
- Finally, if $y_{i-1} = x_i - 1$ and $y_i = x_i + 1$, then adding a box at x_i removes the local extrema x_i and y_i, and moves y_{i-1} to $y_{i-1} + 1 = x_i$.

In all these cases, the quantity $\sum_{i=1}^{s} x_i - \sum_{i=1}^{s-1} y_i$ is left invariant, hence the result.

Example. The integer partition $\lambda = (5, 4, 2)$ has interlaced coordinates

$$-3_x < -1_y < 0_x < 1_y < 2_x < 4_y < 5_x,$$

and one has indeed $(-3 + 0 + 2 + 5) - (-1 + 1 + 4) = 4 - 4 = 0$.

There is a third way to describe integer partitions, which relates them to systems of interacting particles. We consider the Young diagram of λ rotated by 45 degrees, and we mark by a dot the abscissa of the middles of the decreasing segments $\diagdown$ of the boundary of the Young diagram; see Figure 7.4. We then obtain

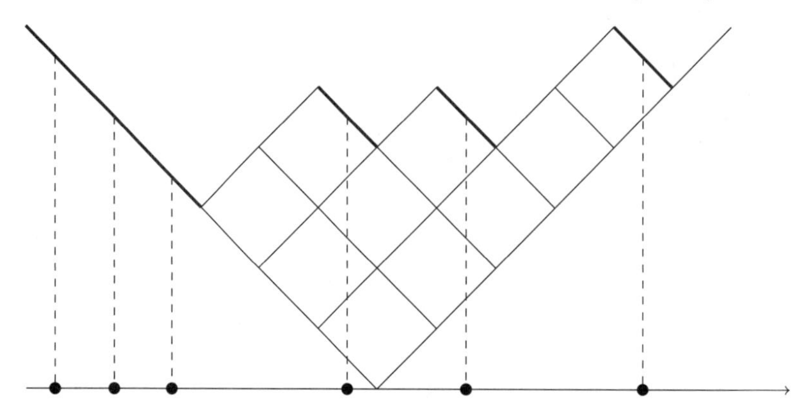

Figure 7.4
System of particles $L(\lambda)$ associated to the Young diagram $\lambda = (5,3,2)$.

a part $L(\lambda)$ of $\mathbb{Z} + \frac{1}{2}$, whose symmetric difference with $-(\mathbb{N} + \frac{1}{2})$ is finite. For instance, if $\lambda = (5,4,2)$, then

$$L(\lambda) = \left(\frac{9}{2}, \frac{3}{2}, -\frac{1}{2}, -\frac{7}{2}, -\frac{9}{2}, -\frac{11}{2}, \dots\right).$$

Notice that adding or removing a box to λ amounts to making one of the particles of $L(\lambda)$ move left or right. On the other hand, if $\lambda = (\lambda_1, \lambda_2, \dots, \lambda_r)$, then

$$L(\lambda) = (\lambda_1 - 1 + \frac{1}{2}, \lambda_2 - 2 + \frac{1}{2}, \dots, \lambda_r - r + \frac{1}{2}, -(r+1) + \frac{1}{2}, -(r+2) + \frac{1}{2}, \dots).$$

The connection between the **system of particles** $L(\lambda)$ (also called system of **shifted coordinates**) and the previous systems of coordinates is given by:

Proposition 7.9. *One has $L(\lambda) \cap (\mathbb{N} + \frac{1}{2}) = A(\lambda)$, and $(-(\mathbb{N} + \frac{1}{2})) \setminus L(\lambda) = -B(\lambda)$.*

Proof. This is geometrically obvious, since decreasing segments ╲ on the positive side of the rotated drawing correspond to boundary boxes on the right of the diagonal of the Young diagram, and increasing segments ╱ on the negative side of the rotated drawing correspond to boundary boxes on top of the diagonal of the Young diagram. □

▷ *Frobenius moments, interlaced moments and their generating functions.*

We define the generating function of the particle system $L(\lambda)$ associated to a Young diagram λ by

$$H_\lambda(z) = \prod_{i=1}^{\infty} \frac{z + i - \frac{1}{2}}{z - \lambda_i + i - \frac{1}{2}} = \frac{\prod_{x \in L(\emptyset)} z - x}{\prod_{x \in L(\lambda)} z - x},$$

which is a rational fraction since the symmetric difference $L(\lambda) \,\Delta\, L(\emptyset)$ is finite. The previous proposition shows that $H_\lambda(z)$ is also a generating function for the Frobenius coordinates of λ:

$$H_\lambda(z) = \prod_{i=1}^{d} \frac{z+b_i}{z-a_i} = \prod_{i=1}^{d} \frac{1+z^{-1}b_i}{1-z^{-1}a_i}.$$

Recall from Section 2.2 that

$$\prod_i \frac{1}{1-ta_i} = \sum_{k=0}^{\infty} h_k(A)\, t^k = \exp\left(\sum_{k=1}^{\infty} \frac{p_k(A)}{k}\, t^k\right)$$

$$\prod_i 1+tb_i = \sum_{k=0}^{\infty} e_k(B)\, t^k = \sum_{k=0}^{\infty} h_k(-\overline{B})\, t^k = \exp\left(\sum_{k=1}^{\infty} \frac{p_k(-\overline{B})}{k}\, t^k\right)$$

where $-B$ is the formal alphabet such that $f(-B) = (\omega(f))(B)$ for any symmetric function f (thus, $p_k(-B) = -p_k(B)$); and $\overline{B} = \{-b_1, \ldots, -b_d\}$ if $B = \{b_1, \ldots, b_d\}$. So,

$$H_\lambda(z) = \sum_{k=0}^{\infty} h_k(A(\lambda) - \overline{B(\lambda)})\, z^{-k} = \exp\left(\sum_{k=1}^{\infty} \frac{p_k(A(\lambda) - \overline{B(\lambda)})}{k}\, z^{-k}\right),$$

and the generating function of a Young diagram λ is the generating function of the symmetric functions of the formal alphabet $A(\lambda) - \overline{B(\lambda)}$.

Definition 7.10. *A* ***geometric observable*** *of a Young diagram is a symmetric function of the alphabet $A(\lambda) - \overline{B(\lambda)}$. The geometric observables of Young diagrams form an algebra $\mathscr{O}^{\mathrm{geom}}$, which has for algebraic basis over $\mathbb{C}$ the* ***Frobenius moments***

$$p_k(\lambda) = p_k(A(\lambda) - \overline{B(\lambda)}) = \sum_{i=1}^{d} (a_i)^k - (-b_i)^k.$$

This definition leads to several remarks. First, viewed as functions of Young diagrams, the Frobenius moments are indeed algebraically independent over $\mathbb{C}$, so that $\mathscr{O}^{\mathrm{geom}} = \mathbb{C}[p_1, p_2, \ldots]$ is isomorphic to the algebra of symmetric functions Sym. Indeed, suppose that there exists a polynomial $f(p_1, \ldots, p_N)$ such that for any integer partition λ,

$$f(p_1(\lambda), p_2(\lambda), \ldots, p_N(\lambda)) = 0.$$

Fix N parameters $x_1 \geq x_2 \geq \cdots \geq x_N \geq 0$, and consider the integer partitions $\lambda(t) = (\lfloor tx_i \rfloor)_{i \in [\![1,N]\!]}$. It is easily seen that

$$p_k(\lambda(t)) = t^k p_k(x_1, \ldots, x_N) + O(t^{k-1}).$$

Denote f_0 the top homogeneous component of f, where the variable p_k is considered of degree k. Then,

$$0 = \frac{1}{t^{\deg f}} f(p_1(\lambda(t)),\ldots,p_N(\lambda(t))) = f_0(p_1(X),\ldots,p_N(X)) + O\left(\frac{1}{t}\right),$$

so $f_0(p_1(x_1,\ldots,x_N),\ldots,p_N(x_1,\ldots,x_N)) = 0$ for any variables $x_1 \geq x_2 \geq \cdots \geq x_N \geq 0$. Since the power sums $p_1,\ldots,p_N$ are algebraically independent in $\mathrm{Sym}^{(N)}$, we conclude that $f_0 = 0$, which implies that $f = 0$, and the claim $\mathscr{O}^{\mathrm{geom}} = \mathbb{C}[p_1, p_2, \ldots]$.

On the other hand, let us justify the terminology of moments. We associate to an integer partition λ the finite measure on $\mathbb{R}$

$$m_\lambda = \sum_{i=1}^{d} a_i\,\delta_{a_i} + b_i\,\delta_{-b_i},$$

where δ_x is the Dirac measure at x. By a previous remark, the total mass of m_λ is $m_\lambda(\mathbb{R}) = \sum_{i=1}^{d} a_i + b_i = |\lambda|$. Then, for any integer $k \geq 1$,

$$\int_{\mathbb{R}} s^{k-1}\, m_\lambda(ds) = \sum_{i=1}^{d} (a_i)^k + (-1)^{k-1}(b_i)^k = p_k(\lambda).$$

So, the moments of the measure m_λ are the Frobenius moments of λ, hence the terminology. In particular, notice that for any integer partition λ, one has

$$p_1(\lambda) = |\lambda|.$$

Let us now relate the geometric observables to the interlaced coordinates of Young diagrams. The link between the two notions is provided by the following relation:

Proposition 7.11. *For any integer partition λ,*

$$H_\lambda\left(z - \frac{1}{2}\right) = \frac{\prod_{i=1}^{s} \Gamma(z - x_i)}{\Gamma(z)\prod_{i=1}^{s-1}\Gamma(z - y_i)},$$

where Γ is Euler's gamma function (the analytic solution of the equation $\Gamma(z+1) = z\,\Gamma(z)$, with $\Gamma(1) = 1$.)

Proof. The relation is true for $\lambda = \emptyset$, since $H_\emptyset(z - \frac{1}{2}) = 1 = \frac{\Gamma(z)}{\Gamma(z)}$. It suffices now to prove that the relation is preserved by the operation of adding a box to the Young diagram at the corner marked by the local minima x_i of ω_λ. Let us first treat the case when $x_i > 0$; then, one raises one of the Frobenius coordinate a_j by one unity, and one leaves all the other Frobenius coordinates invariant. Therefore, if Λ is the new integer partition, then

$$H_\Lambda\left(z - \frac{1}{2}\right) = \frac{z - a_j - \frac{1}{2}}{z - a_j - \frac{3}{2}}\, H_\lambda\left(z - \frac{1}{2}\right) = \frac{z - x_i}{z - x_i - 1}\, H_\lambda\left(z - \frac{1}{2}\right),$$

because $x_i = a_j + \frac{1}{2}$. We now have to distinguish the cases according to the modification of the interlaced coordinates which we previously analyzed:

- If x_i becomes $x_i - 1 < x_i < x_i + 1$, then the ratio $\frac{\prod_{i=1}^{s} \Gamma(z-x_i)}{\Gamma(z)\prod_{i=1}^{s-1} \Gamma(z-y_i)}$ is multiplied by
$$\frac{1}{\Gamma(z-x_i)} \times \frac{\Gamma(z-x_i-1)\,\Gamma(z-x_i+1)}{\Gamma(z-x_i)} = \frac{z-x_i}{z-x_i-1}.$$
- If $y_{i-1} = x_i - 1$ and x_i are replaced by $y_{i-1}+1$ and $x_i + 1$, then the ratio is multiplied by
$$\frac{\Gamma(z-x_i+1)}{\Gamma(z-x_i)} \times \frac{\Gamma(z-x_i-1)}{\Gamma(z-x_i)} = \frac{z-x_i}{z-x_i-1}.$$
- If x_i and $y_i = x_i + 1$ become $x_i - 1$ and $y_i - 1$, then the ratio is multiplied by
$$\frac{\Gamma(z-x_i-1)}{\Gamma(z-x_i)} \times \frac{\Gamma(z-x_i+1)}{\Gamma(z-x_i)} = \frac{z-x_i}{z-x_i-1}.$$
- Finally, if $x_i - 1 = y_{i-1} < x_i < y_i = x_i + 1$ is replaced by x_i (local maxima), then the ratio is multiplied by
$$\frac{\Gamma(z-x_i-1)\,\Gamma(z-x_i+1)}{\Gamma(z-x_i)} \times \frac{1}{\Gamma(z-x_i)} = \frac{z-x_i}{z-x_i-1}.$$

In all cases, we thus multiply the ratio by $\frac{z-x_i}{z-x_i-1}$, so the relation is still true. We leave to the reader the two other cases $x_i < 0$, which corresponds to the raising of b_j by one unity, with $-b_j = x_i - \frac{1}{2}$; and $x_i = 0$, which corresponds to the addition of $(\frac{1}{2}, \frac{1}{2})$ to the pair of Frobenius coordinates $(A(\lambda), B(\lambda))$. □

By using the previous relation, we can expand the generating series of the interlaced coordinates of λ in terms of the Frobenius moments:

$$\begin{aligned}\frac{\prod_{i=1}^{s-1}(z-y_i)}{\prod_{i=1}^{s}(z-x_i)} &= \frac{\prod_{i=1}^{s}\Gamma(z-x_i)\,\prod_{i=1}^{s-1}\Gamma(z-y_i+1)}{\prod_{i=1}^{s-1}\Gamma(z-y_i)\,\prod_{i=1}^{s}\Gamma(z-x_i+1)} = \frac{1}{z}\,\frac{H_\lambda(z-\frac{1}{2})}{H_\lambda(z+\frac{1}{2})} \\ &= \frac{1}{z}\exp\left(\sum_{j=1}^{\infty}\frac{p_j(\lambda)}{k}\left(\left(z-\frac{1}{2}\right)^{-j} - \left(z+\frac{1}{2}\right)^{-j}\right)\right) \\ &= \frac{1}{z}\exp\left(\sum_{k=2}^{\infty}\left(\sum_{j=0}^{\lfloor\frac{k-1}{2}\rfloor}\binom{k}{2j+1}\frac{p_{k-2j-1}(\lambda)}{2^{2j}}\right)\frac{z^{-k}}{k}\right)\end{aligned}$$

by using on the last line the expansions in power series of $(z-\frac{1}{2})^{-j}$ and $(z+\frac{1}{2})^{-j}$. We define the **interlaced moments** of λ by

$$\widetilde{p}_k(\lambda) = p_k(X(\lambda) - Y(\lambda)) = \sum_{i=1}^{s}(x_i)^k - \sum_{i=1}^{s-1}(y_i)^k.$$

We have

$$\frac{\prod_{i=1}^{s-1}(z-y_i)}{\prod_{i=1}^{s}(z-x_i)} = \frac{1}{z}\exp\left(\sum_{k=1}^{\infty}\frac{\widetilde{p}_k(\lambda)}{k}z^{-k}\right),$$

and on the other hand, by a previous remark, $\widetilde{p}_1(\lambda)=0$. The previous computation shows then that:

Theorem 7.12. *The functions $\widetilde{p}_k$ belong to $\mathscr{O}^{\mathrm{geom}}$, and more precisely, $\mathscr{O}^{\mathrm{geom}} = \mathbb{C}[\widetilde{p}_2,\widetilde{p}_3,\ldots]$.*

Proof. By identification of the coefficient of z in $\log\left(\frac{z\prod_{i=1}^{s-1}(z-y_i)}{\prod_{i=1}^{s}(z-x_i)}\right)$, we see that for any $k\geq 2$,

$$\widetilde{p}_k(\lambda)=\sum_{j=0}^{\lfloor\frac{k-1}{2}\rfloor}\binom{k}{2j+1}\frac{p_{k-2j-1}(\lambda)}{2^{2j}} = k\,p_{k-1}(\lambda)+\text{term of degree lower than } k-2.$$

The result follows immediately from these relations and from the fact that $(p_k)_{k\geq 1}$ is a graded algebraic basis of $\mathscr{O}$. The relations between the first Frobenius moments and interlaced moments are:

$$\begin{aligned}
\widetilde{p}_2 &= 2\,p_1 \qquad ; \qquad & p_1 &= \frac{\widetilde{p}_2}{2} \quad ;\\
\widetilde{p}_3 &= 3\,p_2 \qquad ; \qquad & p_2 &= \frac{\widetilde{p}_3}{3} \quad ;\\
\widetilde{p}_4 &= 4\,p_3+p_1 \qquad ; \qquad & p_3 &= \frac{\widetilde{p}_4}{4}-\frac{\widetilde{p}_2}{8}.
\end{aligned}$$

□

If λ is a Young diagram, denote $\sigma_\lambda(s)=\frac{\omega_\lambda(s)-|s|}{2}$. This is a continuous, affine by parts and compactly supported function on the real line. Moreover, the second derivative of σ_λ is (in the sense of distributions)

$$\sigma_\lambda''=\sum_{i=1}^{s}\delta_{x_i}-\sum_{i=1}^{s-1}\delta_{y_i}-\delta_0.$$

Therefore, for any $k\geq 1$,

$$\widetilde{p}_k(\lambda)=\int_{\mathbb{R}} s^k\,\sigma_\lambda''(s)\,ds.$$

Hence, the $\widetilde{p}_k$'s can again be written as moments of a finite measure, this time signed and with total mass 0. Notice that in particular, if one knows all the observables $\widetilde{p}_k(\lambda)$, then one can reconstruct the signed measure σ_λ'', and thereby the Young diagram (this is why the elements of $\mathscr{O}^{\mathrm{geom}}$ have been called "geometric" observables). Indeed, the measure σ_λ'' is compactly supported by $[x_1,x_s]=$

$[-\lambda_1', \lambda_1]$, and this compactness is related to the asymptotic behavior of the moments of the measure:

$$\lim_{k \to \infty} \frac{\log |\widetilde{p}_{2k}(\lambda)|}{2k} = \max\{x_s, -x_1\}.$$

Thus, one can deduce from the knowledge of the interlaced moments that σ''_λ is compactly supported on an interval $[-A, A]$. Then, it is a classical fact that the moments of a finite signed measure compactly supported on $[-A, A]$ determine entirely the measure, so one can indeed reconstruct σ''_λ, and then λ.

7.3 Change of basis in the algebra of observables

In this section, we shall relate the two algebras $\mathcal{O}$ and $\mathcal{O}^{\mathrm{geom}}$, and show that they are actually the same. We shall then detail the formulas of change of basis between the symbols Σ_k, p_k and $\widetilde{p}_k$.

▷ *Renormalized character values and Wassermann's formula.*

If λ is an integer partition of size n, and μ is an arbitrary integer partition, we define the **renormalized character value** $\Sigma_\mu(\lambda)$ by:

$$\Sigma_\mu(\lambda) = \chi^\lambda \circ \pi_n \circ \phi_n(\Sigma_\mu),$$

where the normalized character χ^λ is considered as in Chapter 1 as a coordinate of the Fourier transform of the center $Z(\mathbb{C}\mathfrak{S}(n))$ of the group algebra (see Theorem 1.17). More explicitly, since $\pi_n \circ \phi_n(\Sigma_\mu)$ is either 0 or a known multiple of the conjugacy class $C_{\mu \uparrow n}$, we have

$$\Sigma_\mu(\lambda) = \begin{cases} n^{\downarrow |\mu|}\, \chi^\lambda(\mu \uparrow n) & \text{if } n \geq |\mu|, \\ 0 & \text{otherwise,} \end{cases}$$

where $\chi^\lambda(\mu \uparrow n) = \frac{\mathrm{ch}^\lambda(\sigma_\mu)}{\dim \lambda}$ is the renormalized value of the irreducible character ch^λ on the permutation $\sigma_\mu = (1, 2, \ldots, \mu_1) \cdots (\mu_1 + \cdots + \mu_{r-1} + 1, \ldots, \mu_1 + \cdots + \mu_r)$ if $\mu = (\mu_1, \ldots, \mu_r)$.

A non-trivial fact is the coherence of this definition with the product of partial permutations on $\mathcal{O}$: hence, for any integer partitions μ and ν,

$$(\Sigma_\mu \times_{\mathcal{O}} \Sigma_\nu)(\lambda) = \Sigma_\mu(\lambda)\, \Sigma_\nu(\lambda).$$

This is because the Fourier transform on $Z(\mathbb{C}\mathfrak{S}(n))$ as well as the maps π_n and ϕ_n are morphisms of algebras. Thus, the symbols Σ_μ defined as linear combinations of partial permutations yield well-defined functions on the set $\mathfrak{Y}$ of all Young diagrams; in other words, the Σ_μ are indeed "observables" of Young diagrams, of a rather algebraic nature. The main result of this section is:

Theorem 7.13. *The algebra $\mathscr{O}$ and the algebra $\mathscr{O}^{\mathrm{geom}}$ are the same. Moreover, the symbols Σ_k and p_k have the same top homogeneous component with respect to the degree:*

$$\forall k \geq 1,\ \deg(\Sigma_k - p_k) \leq k-1.$$

Lemma 7.14. *For any $k \geq 2$ and any λ of size $n \geq k$, $\Sigma_k(\lambda)$ is the coefficient of z^{-1} in the expansion at infinity of*

$$-\frac{1}{k}\, z^{\downarrow k} \prod_{i=1}^{n} \frac{z-\lambda_i-n+i-k}{z-\lambda_i-n+i}.$$

Proof. The proof relies on an argument which is very similar to the one used in the proof of the determinantal formula for $\dim\lambda = \dim S^\lambda$ (Proposition 3.43). By the Frobenius formula 2.32, for $k \geq 2$,

$$p_k(X)\,(p_1(X))^{n-k} = \sum_{\lambda \in \mathfrak{Y}(n)} \mathrm{ch}^\lambda(k1^{n-k})\, s_\lambda(X),$$

so, taking an alphabet $X = (x_1, \ldots, x_n)$ and denoting $\rho = (n-1, n-2, \ldots, 1, 0)$, we see by multiplying by $a_\rho(x_1, \ldots, x_n)$ that $\mathrm{ch}^\lambda(k1^{n-k})$ is the coefficient of $x^{\lambda+\rho}$ in

$$p_k(x_1, \ldots, x_n)\,(p_1(x_1, \ldots, x_n))^{n-k}\, a_\rho(x_1, \ldots, x_n).$$

So,

$$\begin{aligned}
\mathrm{ch}^\lambda(k1^{n-k}) &= [x^{\lambda+\rho}]\left(\left(\sum_{i=1}^{n}(x_i)^k\right)\left(\sum_{i=1}^{n}x_i\right)^{n-k} a_\rho(x_1, \ldots, x_n)\right) \\
&= \sum_{i=1}^{n}[x^{\lambda+\rho-ke_i}]\left(\left(\sum_{i=1}^{n}x_i\right)^{n-k} a_\rho(x_1, \ldots, x_n)\right) \\
&= \sum_{i} \frac{(n-k)!}{(\mu_1)!\cdots(\mu_i-k)!\cdots(\mu_n)!}\, \Delta(\mu_1, \ldots, \mu_i-k, \ldots, \mu_n)
\end{aligned}$$

where $\mu = \lambda+\rho$, and where the last sum runs over indices i such that $\mu_i - k \geq 0$. Here we use the computation of Proposition 3.43. If we multiply the last formula by $\frac{n^{\downarrow k}}{\dim\lambda}$, using $\dim\lambda = \frac{n!}{\mu!}\,\Delta(\mu)$, we get:

$$\begin{aligned}
\Sigma_k(\lambda) &= \sum_{i=1}^{n} (\mu_i)^{\downarrow k}\, \frac{\Delta(\mu_1, \ldots, \mu_i-k, \ldots, \mu_n)}{\Delta(\mu_1, \ldots, \mu_n)} \\
&= \sum_{i=1}^{n} (\mu_i)^{\downarrow k} \prod_{j\neq i} \frac{\mu_i-\mu_j-k}{\mu_i-\mu_j} = -\frac{1}{k}\sum_{i=1}^{n}(\mu_i)^{\downarrow k}\, \frac{\phi_\lambda(\mu_i-k)}{\phi_\lambda'(\mu_i)}
\end{aligned}$$

with $\phi_\lambda(z) = \prod_{i=1}^{n}(z-\mu_i)$. If $F(z) = -\frac{1}{k}\, z^{\downarrow k}\, \frac{\phi_\lambda(z-k)}{\phi_\lambda(z)}$, then this rational fraction has poles at the points μ_i, and the quantity above is the sum of the residues at these poles, so,

$$\Sigma_k(\lambda) = \sum_{i=1}^{n} \mathrm{Res}_{z=\mu_i}(F(z)) = -\mathrm{Res}_{z=\infty}(F(z))$$

since the sum of the residues of a non-zero meromorphic function on the Riemann sphere $\mathbb{S} = \mathbb{C} \sqcup \{\infty\}$ is always equal to zero. However, if $F(z)$ has for expansion at infinity $\sum_{j\geq -k} c_j z^{-j}$, then $\mathrm{Res}_{z=\infty}(F(z)) = -c_1$, hence the claim. □

Proof of Theorem 7.13. Fix $k \geq 2$. Since $H_\lambda(z-n+\frac{1}{2}) = \prod_{i=1}^{n} \frac{z-(n-i)}{z-\lambda_i-(n-i)}$, with the same notations as before, we obtain

$$F(z) = -\frac{1}{k}\,(z-n)^{\downarrow k}\,\frac{H_\lambda(z-n+\frac{1}{2})}{H_\lambda(z-n-k+\frac{1}{2})}.$$

However, the residue at $z = +\infty$ is invariant by the transformation $z \mapsto z+n-\frac{1}{2}$, so,

$$\Sigma_k(\lambda) = [z^{-1}]\left(-\frac{1}{k}\left(z-\frac{1}{2}\right)^{\downarrow k}\frac{H_\lambda(z)}{H_\lambda(z-k)}\right).$$

This holds for $|\lambda| = n \geq k$, but also for $n < k$. Indeed, in this case, the function whose residue at infinity is considered is

$$\left(z-\frac{1}{2}\right)^{\downarrow k}\prod_{i=1}^{d}\frac{z+b_i}{z-a_i}\prod_{j=1}^{d}\frac{z-a_j-k}{z+b_j-k}.$$

For any i and j, $-a_i$ is different from $b_j - k$, because $a_i + b_j \leq n < k$. As a consequence, in the denominator of this fraction, we have $2d$ different terms. But each of them cancels with one factor of $(z-\frac{1}{2})^{\downarrow k}$, because the a_i's and the b_j's belong to

$$\left\{\frac{1}{2},\frac{3}{2},\ldots,n-\frac{1}{2}\right\} \subset \frac{1}{2} + [\![0,k-1]\!].$$

So, if $n < k$, then the function considered is in fact a polynomial, and its residue at infinity is 0.

We now make some manipulations on the identity previously shown, using the fact that $H_\lambda(z)$ is the generating function of the Frobenius moments:

$$\begin{aligned}\Sigma_k(\lambda) &= -\frac{1}{k}[z^{-1}]\left(\prod_{i=1}^{k}\left(z+\frac{1}{2}-i\right)\exp\left(\sum_{j=1}^{\infty}\frac{p_j(\lambda)}{j}\left(z^{-j}-(z-k)^{-j}\right)\right)\right)\\ &= -\frac{1}{k}[t^{k+1}]\left(\prod_{i=1}^{k}\left(1-\left(i-\frac{1}{2}\right)t\right)\exp\left(\sum_{j=1}^{\infty}\frac{p_j(\lambda)\,t^j}{j}\left(1-(1-kt)^{-j}\right)\right)\right).\end{aligned}$$

This is **Wassermann's formula**, and it proves that each symbol $\Sigma_{k\geq 2}$ is a polynomial in the Frobenius moments p_k. The result is also true for $k = 1$, since $\Sigma_1(\lambda) = p_1(\lambda) = |\lambda|$ for any $\lambda \in \mathfrak{Y}$. Now, more precisely, since

$$\Sigma_k = -\frac{1}{k}[t^{k+1}]\left(\prod_{i=1}^{k}\left(1-\left(i-\frac{1}{2}\right)t\right)\exp\left(-\sum_{j=1}^{\infty} kp_j\,(t^{j+1}+O(t^{j+2}))\right)\right)$$
$$= -\frac{1}{k}[t^{k+1}]\left(\prod_{i=1}^{k}\left(1-\left(i-\frac{1}{2}\right)t\right)\left(1-\sum_{j=1}^{\infty} kp_j\,t^{j+1}+\text{terms } p_\lambda t^{m\geq|\lambda|+2}\right)\right)$$

we see that for any $k \geq 1$, $\Sigma_k = p_k +$ terms of lower degree. Since $(\Sigma_k)_{k\geq 1}$ and $(p_k)_{k\geq 1}$ are algebraic bases of $\mathscr{O}$ and $\mathscr{O}^{\text{geom}}$, the theorem is proven. □

Example. The first Frobenius moments and renormalized character values are related by the formulas:

$$\Sigma_1 = p_1 \quad ; \quad \Sigma_2 = p_2;$$
$$\Sigma_3 = p_3 - \frac{3}{2}\,p_{(1,1)} + \frac{5}{4}\,p_1;$$
$$\Sigma_4 = p_4 - 4\,p_{(2,1)} + \frac{11}{2}\,p_2$$

and

$$p_1 = \Sigma_1 \quad ; \quad p_2 = \Sigma_2;$$
$$p_3 = \Sigma_3 + \frac{3}{2}\,\Sigma_{(1,1)} + \frac{1}{4}\,\Sigma_1;$$
$$p_4 = \Sigma_4 + 4\,\Sigma_{(2,1)} + \frac{5}{2}\,\Sigma_2.$$

▷ *The weight gradation of the algebra $\mathscr{O}$.*

By Theorem 7.13, the **degree gradation** on $\mathscr{O}$ can be defined by $\deg \Sigma_k = k$ or by $\deg p_k = k$, since the difference $\Sigma_k - p_k$ is for any k a linear combination of terms of degree less than $k-1$. There is another interesting gradation on $\mathscr{O}$, called the **weight gradation** and which actually comes from a gradation on $\mathbb{C}\mathfrak{PS}(\infty)$. If (σ,A) is a partial permutation, we define its weight by

$$\text{wt}(\sigma,A) = \text{card}\,A + \text{card}\,\{a \in A \,|\, \sigma(a) = a\}.$$

Proposition 7.15. *The weight of partial permutations is compatible with their product:*

$$\forall (\sigma,A),(\tau,B),\ \text{wt}((\sigma,A)(\tau,B)) \leq \text{wt}(\sigma,A) + \text{wt}(\tau,B).$$

For any integer partition λ, $\text{wt}(\Sigma_\lambda) = |\lambda| + m_1(\lambda)$.

Proof. For any permutation σ of a set A, denote $\mathrm{Fix}(\sigma,A)$ the set of fixed points of σ in A. We then have by definition

$$\begin{aligned}
&\mathrm{wt}(\sigma,A)+\mathrm{wt}(\tau,B)-\mathrm{wt}((\sigma,A)(\tau,B))\\
&=|A|+|\mathrm{Fix}(\sigma,A)|+|B|+|\mathrm{Fix}(\tau,B)|-|A\cup B|-|\mathrm{Fix}(\sigma\tau,A\cup B)|\\
&=|A\cap B|+|\mathrm{Fix}(\sigma,A)|+|\mathrm{Fix}(\tau,B)|-|\mathrm{Fix}(\sigma\tau,A\cup B)|.
\end{aligned}$$

Let $x\in A\cup B$ be an element fixed by $\sigma\tau$. If x is in B and not in A, then $x=\sigma(\tau(x))=\tau(x)$, so x belongs to $\mathrm{Fix}(\tau,B)$. Similarly, if x is in A and not in B, then $x=\sigma(\tau(x))=\sigma(x)$, so x belongs to $\mathrm{Fix}(\sigma,A)$. Finally, we have otherwise $x\in A\cap B$. So,

$$|\mathrm{Fix}(\sigma\tau,A\cup B)|\leq|\mathrm{Fix}(\tau,B)|+|\mathrm{Fix}(\sigma,A)|+|A\cap B|,$$

hence the fact that $\mathrm{wt}(\cdot)$ is a gradation on $\mathbb{C}\mathfrak{P}\mathfrak{S}(\infty)$, and by restriction on $\mathscr{O}$. □

In Chapter 9, we shall identify the top weight component of Σ_k in terms of the geometric observables $\widetilde{p}_k$. For the moment, we only prove that:

Proposition 7.16. *For any $k\geq 2$, the weight of $\widetilde{p}_k$ is k, so the weight gradation on the algebra $\mathscr{O}=\mathbb{C}[\widetilde{p}_2,\widetilde{p}_3,\dots]$ can be defined alternatively by the formula*

$$\mathrm{wt}(\widetilde{p}_k)=k.$$

Proof. Notice that

$$\frac{p_j\,t^j}{j}\left(1-(1-kt)^{-j}\right)=-k\,p_j\,t^{j+1}(1+O(t))=-k\,\frac{\widetilde{p}_{j+1}\,t^{j+1}}{j+1}(1+O(t)).$$

We can then rework Wassermann's formula:

$$\begin{aligned}
\Sigma_k&=-\frac{1}{k}[t^{k+1}]\left(\prod_{i=1}^{k}\left(1-\left(i-\frac{1}{2}\right)t\right)\exp\left(\sum_{j=1}^{\infty}\frac{p_j\,t^j}{j}\left(1-(1-kt)^{-j}\right)\right)\right)\\
&=-\frac{1}{k}[t^{k+1}]\left(\prod_{i=1}^{k}\left(1-\left(i-\frac{1}{2}\right)t\right)\exp\left(-k\sum_{j=2}^{\infty}\frac{\widetilde{p}_j\,t^j}{j}(1+O(t))\right)\right)
\end{aligned}$$

The term of degree $k+1$ in the variable t in the exponential is

$$-\frac{k\widetilde{p}_{k+1}}{k+1}+\text{polynomial of total weight } k+1 \text{ in the variables } \widetilde{p}_2,\dots,\widetilde{p}_k.$$

Therefore,

$$\Sigma_k=\frac{\widetilde{p}_{k+1}}{k+1}+\text{polynomial of total weight } k+1 \text{ in the variables } \widetilde{p}_2,\dots,\widetilde{p}_k.$$

The result follows by an immediate recurrence on $k\geq 2$. □

Example. The first interlaced moments can be written in terms of the first renormalized character values as follows:

$$\widetilde{p}_2 = 2\,\Sigma_1;$$
$$\widetilde{p}_3 = 3\,\Sigma_2;$$
$$\widetilde{p}_4 = 4\,\Sigma_3 + 6\,\Sigma_{(1,1)} + 2\,\Sigma_1.$$

To conclude this section, we give a formula for the top weight component of $\widetilde{p}_k$ in terms of the symbols Σ_k:

Proposition 7.17. *For any $k \geq 2$,*

$$\widetilde{p}_k = \sum_{|\mu|+\ell(\mu)=k} \frac{k^{\downarrow \ell(\mu)}}{\prod_{i\geq 1}(m_i(\mu))!} \prod_{i\geq 1} (\Sigma_i)^{m_i(\mu)} + \text{ terms of weight smaller than } k-1.$$

The proof of Proposition 7.17 relies on **Lagrange inversion** and the Wassermann formula.

Lemma 7.18 (Lagrange inversion formula). *Given commutative variables $a_{k\geq 2}$ and $b_{k\geq 2}$, consider the two formal power series $A(t) = 1 + \sum_{k=2}^{\infty} a_k\, t^k$ and $B(u) = 1 + \sum_{k=2}^{\infty} b_k\, u^k$, and assume that the maps $x \mapsto x\,A(x)$ and $x \mapsto \frac{x}{B(x)}$ are inverses of one another. Then, for any $k \geq 2$,*

$$b_k = -\frac{1}{k-1}\,[t^k](A^{-(k-1)}(t))$$
$$a_k = \frac{1}{k+1}\,[u^k](B^{k+1}(u))$$
$$\widetilde{a}_k = [u^k](B^k(u))$$

where the $\widetilde{a}_k$'s are defined by the formal power series $\widetilde{A}(t) = \log A(t) = \sum_{k=2}^{\infty} \frac{\widetilde{a}_k\, t^k}{k}$.

Proof. Set $t = \frac{u}{B(u)}$, so that by hypothesis $u = t\,A(t) = \sum_{k\geq 2} a_k\, t^{k+1}$. By differentiation,

$$du = \sum_{j=2}^{\infty} (j+1)\,a_j t^j\, dt$$
$$\frac{du\, B^{k+1}(u)}{u^{k+1}} = \frac{du}{t^{k+1}} = \sum_{j=2}^{\infty} (j+1)\,a_j t^{j-k-1}\, dt$$

so by taking the residues at infinity, we obtain

$$[u^k](B^{k+1}(u)) = [u^{-1}]\left(\frac{B^{k+1}(u)}{u^{k+1}}\right) = (k+1)\,a_k.$$

Similarly, since $\frac{B(u)}{u} = \frac{1}{t}$, by differentiation,

$$-\frac{du}{u^2} + \sum_{j=2}^{\infty} (j-1)\, b_j u^{j-2}\, du = -\frac{dt}{t^2}$$

$$-\frac{du}{u^{k+1}} + \sum_{j=2}^{\infty} (j-1)\, b_j u^{j-k-1}\, du = -\frac{dt}{u^{k-1}t^2} = -\frac{dt}{A^{k-1}(t)\, t^{k+1}}$$

so taking the residues at infinity, we get

$$[t^k](-A^{-(k-1)}(t)) = [t^{-1}]\left(-\frac{A^{-(k-1)}(t)}{t^{k+1}}\right) = (k-1)\, b_k.$$

Finally, $u = t\ \exp\widetilde{A}(t)$, so $\widetilde{A}(t) = \log \frac{u}{t}$. By differentiation,

$$\sum_{j=2}^{\infty} \widetilde{a}_j\, t^{j-1}\, dt = \frac{du}{u} - \frac{dt}{t}$$

$$\sum_{j=2}^{\infty} \widetilde{a}_j\, t^{j-k-1}\, dt = \frac{du}{u\, t^k} - \frac{dt}{t^{k+1}} = \frac{du\, B^k(u)}{u^{k+1}} - \frac{dt}{t^{k+1}}$$

hence

$$\widetilde{a}_k = [u^{-1}]\left(\frac{B^k(u)}{u^{k+1}}\right) = [u^k](B^k(u)). \qquad \square$$

Proof of Proposition 7.17. In the previous lemma, notice that conversely, if any of the formulas for a_k, $\widetilde{a}_k$ or b_k is satisfied, then one can go backwards in the computations, and prove that $x \mapsto xA(x)$ and $x \mapsto \frac{x}{B(x)}$ are formal inverses of one another. Consequently, the three formulas are equivalent. Set $\widetilde{a}_k = \widetilde{p}_k$, and $b_k = \varSigma_{k-1}$. We saw before (Wassermann's formula) that

$$b_k = -\frac{1}{k-1}\, [t^k]\big(\exp(-(k-1)\widetilde{A}(t))\big)$$
$$+ \text{ terms of lower weight with respect to the grading } \mathrm{wt}(a_j) = j.$$

By using a graded version of the previous lemma, we conclude that

$$\widetilde{a}_k = [u^k](B^k(u)) + \text{term of lower weight with respect to the grading } \mathrm{wt}(b_j) = j.$$

Thus, the top weight component of $\widetilde{p}_k = \widetilde{a}_k$ is

$$[u^k]\left(\left(1 + \sum_{j=2}^{\infty} \varSigma_{j-1}\, u^j\right)^k\right) = [u^k]\left(\left(1 + \sum_{j=1}^{\infty} \varSigma_j\, u^{j+1}\right)^k\right),$$

and the result follows by expansion of the k-th power of the generating series of the symbols $\varSigma_j$. $\square$

7.4 Observables and topology of Young diagrams

If $\lambda \in \mathfrak{Y}$ is a Young diagram, we defined above some observables $f(\lambda)$ of it, with f belonging to an algebra $\mathscr{O}$ which is as large as the algebra of symmetric functions (and in fact isomorphic to it). In the last part of the book, we shall look at sequences of (random) Young diagrams $(\lambda_n)_{n\in\mathbb{N}}$, such that renormalized versions of the observables $f(\lambda_n)$ have a limit. For instance, we shall prove in Chapter 13 that if λ_n is a random Young diagram taken under the Plancherel measure Pl_n (cf. the end of Section 1.3), then for any $f \in \mathscr{O}$,

$$n^{-\frac{\mathrm{wt}(f)}{2}} f(\lambda_n)$$

admits a deterministic limit. One can then ask what this implies for the asymptotic geometry of the Young diagrams λ_n. In this last section of the chapter, we give a partial answer to this question, by extending the range of the observables to so-called *continuous* Young diagrams, and by showing that the observables of $\mathscr{O}$ control a certain topology on these objects.

▷ *Transition measure of a Young diagram.*

If λ is a Young diagram, set $G_\lambda(z) = \frac{\prod_{i=1}^{s-1} z-y_i}{\prod_{i=1}^{s} z-x_i}$, which is the generating function of the interlaced sequences $(X(\lambda), Y(\lambda))$ associated to λ. We say that G_λ is the **generating function** of the diagram λ. We showed in Section 7.2 that:

$$G_\lambda(z) = \frac{1}{z}\exp\left(\sum_{k=1}^{\infty} \frac{\widetilde{p}_k(\lambda)}{k} z^{-k}\right) = \frac{1}{z}\frac{H_\lambda(z-\frac{1}{2})}{H_\lambda(z+\frac{1}{2})}.$$

The expansion of the rational fraction $G_\lambda(z)$ in simple elements involves a probability measure on $X(\lambda) \subset \mathbb{R}$:

$$G_\lambda(z) = \sum_{i=1}^{s} \frac{1}{z-x_i} \frac{\prod_{j=1}^{s-1} x_i - y_j}{\prod_{j\neq i} x_i - x_j} = \int_{\mathbb{R}} \frac{1}{z-s}\,\mu_\lambda(ds),$$

where

$$\mu_\lambda = \sum_{i=1}^{s} \frac{\prod_{j=1}^{s-1} x_i - y_j}{\prod_{j\neq i} x_i - x_j}\,\delta_{x_i}.$$

We call μ_λ the **transition measure** of the Young diagram λ. The terminology is justified by:

Proposition 7.19. *For any Young diagram λ, μ_λ is a probability measure, and if Λ is the Young diagram obtained by adding a box to λ at the corner marked by the local minima x_i of ω_λ, then*

$$\mu_\lambda(x_i) = \frac{\dim S^\Lambda}{(n+1)\dim S^\lambda},$$

with $n = |\lambda|$.

Proof. The identity $\frac{\prod_{j=1}^{s-1} x_i - y_j}{\prod_{j\neq i} x_i - x_j} = \frac{\dim S^\Lambda}{(n+1)\dim S^\lambda}$ is a consequence of the hook-length formula 3.41. Indeed,

$$\frac{\dim S^\Lambda}{(n+1)\dim S^\lambda} = \frac{\prod_{\square\in\lambda} h(\square)}{\prod_{\square\in\Lambda} h(\square)},$$

and the boxes $\square$ whose hook-lengths are modified by the transformation $\lambda \mapsto \Lambda$ have:

- in λ, their hook-lengths equal to

$$\begin{aligned}&x_i-(x_1+1), x_i-(x_1+2), \ldots, x_i-y_1,\\ &x_i-(x_2+1), x_i-(x_2+2), \ldots, x_i-y_2,\\ &\ldots,\\ &x_i-(x_{i-1}+1), x_i-(x_{i-1}+2), \ldots, x_i-y_{i-1},\\ &\ldots\end{aligned}$$

- in Λ, their hook-lengths equal to

$$\begin{aligned}&x_i-x_1, x_i-(x_1+1), \ldots, x_i-(y_1+1),\\ &x_i-x_2, x_i-(x_2+1), \ldots, x_i-(y_2+1),\\ &\ldots,\\ &x_i-x_{i-1}, x_i-(x_{i-1}+1), \ldots, x_i-(y_{i-1}+1);\end{aligned}$$

see Figure 7.5. Therefore, the quotient of hook-lengths is indeed $\frac{\prod_{j=1}^{s-1} x_i - y_j}{\prod_{j\neq i} x_i - x_j}$. The fact that μ_λ is a probability measure comes then from the branching rules (Corollary 3.7), which give

$$\sum_{\lambda\nearrow\Lambda} \dim S^\Lambda = \dim\left(\mathrm{Ind}_{\mathfrak{S}(n)}^{\mathfrak{S}(n+1)} S^\lambda\right) = (n+1)\dim S^\lambda. \qquad \square$$

Thus, the generating function $G_\lambda(z)$ of the interlaced coordinates writes as the **Cauchy transform** $\int_{\mathbb{R}} \frac{1}{z-s}\,\mu_\lambda(ds)$ of a probability measure μ_λ which encodes the branching rules verified by the irreducible representation S^λ of $\mathfrak{S}(n)$.

Remark. Since $\widetilde{p}_1(\lambda) = 0$, the expansion in power series of z^{-1} of $G_\lambda(z)$ has no constant term: $G_\lambda(z) = \frac{1}{z} + \frac{\widetilde{p}_2(\lambda)}{2z^3} + \cdots$. As a consequence, the first moment $\int_{\mathbb{R}} s\,\mu(ds)$ of the transition measure μ_λ of a Young diagram is always equal to zero, since we also have the expansion

$$G_\lambda(z) = \frac{1}{z}\sum_{k=0}^{\infty}\left(\int_{\mathbb{R}}\left(\frac{s}{z}\right)^k \mu_\lambda(ds)\right).$$

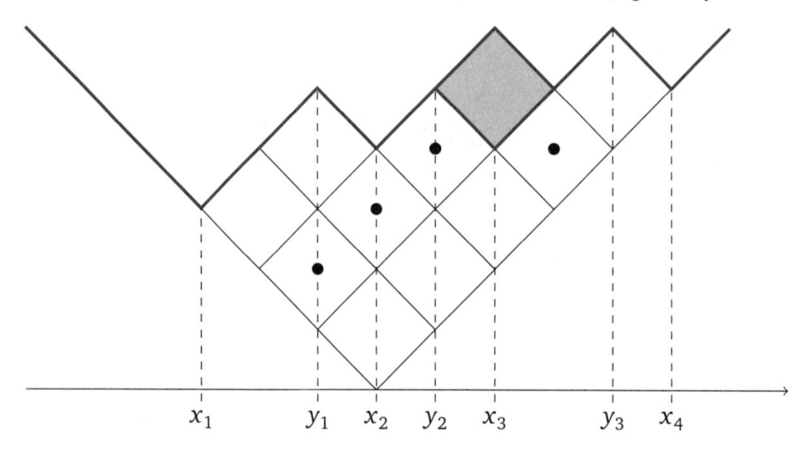

$$h(\boxed{\bullet}, \lambda) = \{x_3 - (y_1 + 1), x_3 - y_1, x_3 - y_2, y_3 - x_3\}$$
$$h(\boxed{\bullet}, \Lambda) = \{x_3 - x_1, x_3 - (x_1 + 1), x_3 - x_2, x_4 - x_3\}$$

Figure 7.5
Modification of the hook-lengths of $\lambda = (5,3,2)$ by addition of a cell at $x_i = x_3$.

▷ *Continuous Young diagrams.*

We now want to extend the domain of the observables to more general objects than Young diagrams of integer partitions. The starting point of this extension is the identity:

$$\begin{aligned} G_\lambda(z) &= \frac{1}{z} \exp\left(\sum_{k=1}^{\infty} \frac{\widetilde{p}_k(\lambda)}{k} z^{-k} \right) = \frac{1}{z} \exp\left(\sum_{k=1}^{\infty} \frac{1}{k} \int_{\mathbb{R}} \sigma''_\lambda(s) \left(\frac{s}{z}\right)^k ds \right) \\ &= \frac{1}{z} \exp\left(-\frac{1}{z} \sum_{k=1}^{\infty} \int_{\mathbb{R}} \sigma'_\lambda(s) \left(\frac{s}{z}\right)^{k-1} ds \right) = \frac{1}{z} \exp\left(- \int_{\mathbb{R}} \frac{\sigma'_\lambda(s)}{z-s} ds \right). \end{aligned}$$

In this identity, one could replace $\sigma'_\lambda(s) = (\frac{\omega_\lambda(s)-|s|}{2})'$ by any measurable bounded function with compact support. This leads to the following definition:

Definition 7.20. *We call* ***continuous Young diagram*** *a function* $\omega : \mathbb{R} \to \mathbb{R}$ *which is Lipschitz with constant* 1*:*

$$\forall s,t,\ \ |\omega(s) - \omega(t)| \leq |s-t|$$

and which is equal to $|s|$ *for* $|s|$ *large enough.*

We denote $\mathscr{Y}$ the set of all continuous Young diagrams. An example of continuous Young diagram is drawn in Figure 7.6. The definition of continuous Young

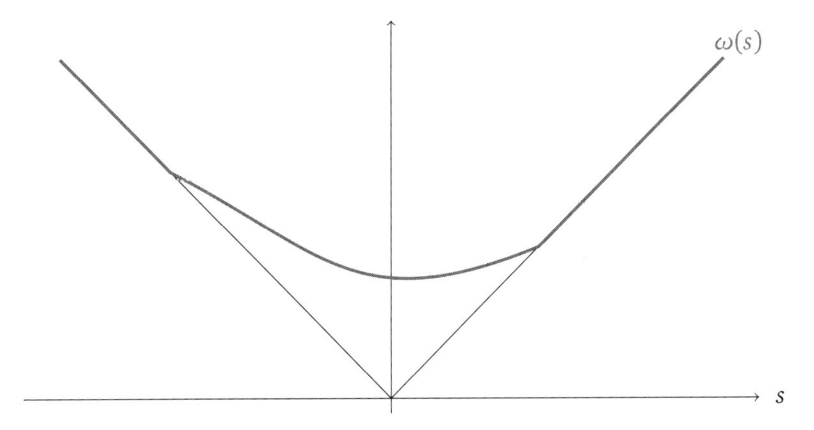

Figure 7.6
A continuous Young diagram ω.

diagrams implies that $\omega(s) \geq |s|$ for any s: indeed, assuming for instance $s \geq 0$ and taking $t \geq s$ large enough so that $\omega(t) = t$, we then have

$$\omega(s) \geq \omega(t) - |\omega(t) - \omega(s)| \geq t - |t - s| = s.$$

For any integer partition λ, ω_λ is a continuous Young diagram, and the same holds for renormalizations of ω_λ:

$$s \mapsto u\,\omega_\lambda\left(\frac{s}{u}\right)$$

which are obtained by multiplying both coordinates of the graph of ω_λ by u.

We define the interlaced moments of a continuous Young diagram ω by:

$$\widetilde{p}_k(\omega) = \int_{\mathbb{R}} \sigma''_\omega(s)\, s^k\, ds = -k \int_{\mathbb{R}} \sigma'_\omega(s)\, s^{k-1}\, ds$$

where $\sigma_\omega(s) = \frac{\omega(s)-|s|}{2}$ as in the case of regular Young diagrams. In the first integral, the derivative σ''_ω is defined in the sense of distributions; this is possible, since σ_ω is a continuous function with compact support. In the second integral, σ'_ω is a well-defined Lebesgue measurable function with values in $[-1, 1]$, since σ_ω is Lipschitz with constant 1. Notice that as in the case of integer partitions, one always has

$$\widetilde{p}_1(\omega) = -\int_{\mathbb{R}} \sigma'_\omega(s)\, ds = 0$$

since σ_ω is compactly supported. As $\mathscr{O} = \mathbb{C}[\widetilde{p}_2, \widetilde{p}_3, \ldots]$, the previous definition allows one to consider $f(\omega)$ for any observable f and any continuous Young diagram ω: it suffices to write f as a polynomial in the $\widetilde{p}_k$'s, and to use the previous definition of the $\widetilde{p}_k(\omega)$'s. In this setting, the weight grading $\mathrm{wt}(\cdot)$ on $\mathscr{O}$ can be related to the operation of scaling of continuous Young diagrams. For any $\omega \in \mathscr{Y}$ and any $u \in \mathbb{R}_+^*$, set

$$\omega^{(u)}(s) = u\,\omega\left(\frac{s}{u}\right);$$

as explained before in the case of regular Young diagrams, this amounts to multiplying the coordinates of the graph of ω (abscissa and ordinate) by u, and $\omega^{(u)}$ belongs again to $\mathscr{Y}$.

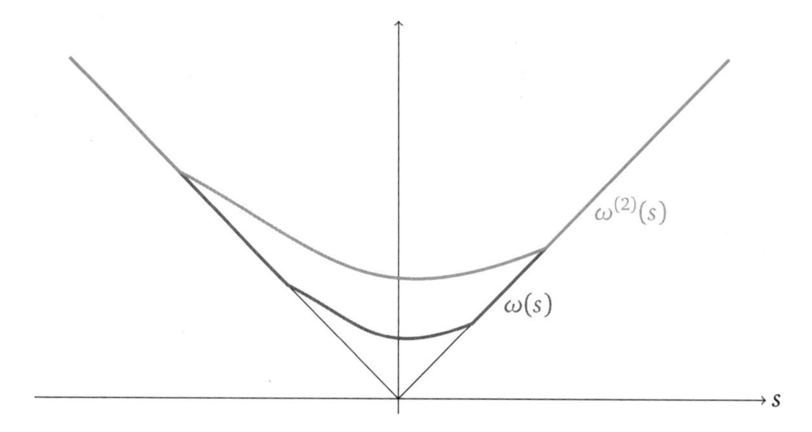

Figure 7.7
Dilation $\omega^{(u)}$ of a continuous Young diagram ω, here by the factor $u = 2$.

Proposition 7.21. *For any continuous Young diagram ω and any observable f,*

$$f(\omega^{(u)}) = u^{\mathrm{wt}(f)}\, f(\omega) + O(u^{\mathrm{wt}(f)-1}).$$

Proof. Since the weight gradation on $\mathscr{O}$ can be defined by $\mathrm{wt}(\widetilde{p}_k) = k$, an observable f of weight K writes as

$$f = P(\widetilde{p}_2, \ldots, \widetilde{p}_K) + Q(\widetilde{p}_2, \ldots, \widetilde{p}_{K-1}),$$

where P is a homogeneous polynomial of total weight K, and Q is of total weight less than $K-1$. Notice then that for any $k \geq 2$,

$$\widetilde{p}_k(\omega^{(u)}) = -k \int_{\mathbb{R}} \sigma'_{\omega^{(u)}}(s)\, s^{k-1}\, ds = -ku^k \int_{\mathbb{R}} \sigma'_{\omega}(t)\, t^{k-1}\, dt = u^k\, \widetilde{p}_k(\omega).$$

Hence, $P(\widetilde{p}_2(\omega^{(u)}), \ldots, \widetilde{p}_K(\omega^{(u)})) = u^K\, P(\widetilde{p}_2(\omega), \ldots, \widetilde{p}_K(\omega))$, and the result follows by an induction on K, which ensures that $Q(\widetilde{p}_2(\omega^{(u)}), \ldots, \widetilde{p}_{K-1}(\omega^{(u)})) = O(u^{K-1})$. $\square$

Given a continuous Young diagram ω, we define its generating function G_ω by

$$G_\omega(z) = \frac{1}{z}\exp\left(-\int_{\mathbb{R}} \frac{\sigma'_\omega(s)}{z-s}\,ds\right) = \frac{1}{z}\exp\left(\sum_{k=1}^{\infty} \frac{\widetilde{p}_k(\omega)}{k}\,z^{-k}\right).$$

It is a well-defined holomorphic function on the **Poincaré half-plane**

$$\mathbb{H} = \{z \in \mathbb{C} \mid \mathrm{Im}(z) > 0\},$$

and it has a negative imaginary part for any $z \in \mathbb{H}$. Moreover, by integration by parts, one can rewrite $G_\omega(z)$ as

$$G_\omega(z) = \frac{1}{z}\exp\left(\left(\int_{\mathbb{R}} \sigma''_\omega(s)\,\log\left(\frac{1}{z-s}\right)ds\right)\right) = \exp\left(\frac{1}{2}\int_{\mathbb{R}} \omega''(s)\,\log\left(\frac{1}{z-s}\right)ds\right).$$

▷ *Markov–Krein correspondence.*

We now want to define the transition measure μ_ω of a continuous Young diagram ω. We want it to be the unique probability measure whose Fourier transform is $G_\omega(z)$:

$$G_\omega(z) = \int_{\mathbb{R}} \frac{1}{z-s}\,\mu_\omega(ds).$$

This raises the question whether the generating function of $\omega \in \mathscr{Y}$ admits such an integral representation. It is actually easier to go in the converse direction, and to try to understand which continuous functions ω can be obtained from Cauchy transforms of probability measures on the real line. In the process, we shall need to extend a bit the space of functions $\mathscr{Y}$, and to define *generalized* continuous Young diagrams. Denote $\mathscr{M}^1 = \mathscr{M}^1(\mathbb{R})$ the set of probability measures on the real line. It is endowed with the topology of weak convergence: a sequence of probability measures $(\mu_n)_{n\in\mathbb{N}}$ converges to a probability measure μ if, for any bounded continuous function f on $\mathbb{R}$,

$$\lim_{n\to\infty} \mu_n(f) = \lim_{n\to\infty} \int_{\mathbb{R}} f(s)\,\mu_n(ds) = \mu(f).$$

It is known that this topology is metrizable by a distance that makes $\mathscr{M}^1(\mathbb{R})$ a complete metric space; see the notes at the end of the chapter. For any $\mu \in \mathscr{M}^1$, we define its Cauchy transform as the function

$$C_\mu(z) = \int_{\mathbb{R}} \frac{1}{z-s}\,\mu(ds),$$

which is well defined and holomorphic on $\mathbb{H}$. For any $z \in \mathbb{H}$, one has

$$\mathrm{Im}(C_\mu(z = x + \mathrm{i}y)) = -\int_{\mathbb{R}} \frac{y}{(x-s)^2 + y^2}\,\mu(ds) < 0,$$

hence, $C_\mu(z)$ has **negative imaginary part** on the whole domain $\mathbb{H}$. Moreover, one has by dominated convergence

$$\lim_{y\to+\infty} \mathrm{i}y\, C_\mu(\mathrm{i}y) = \int_{\mathbb{R}} \mu(ds) = 1.$$

We denote $\mathscr{N}^1 = \mathscr{N}^1(\mathbb{H})$ the set of holomorphic functions N on $\mathbb{H}$ which take values with negative imaginary parts, and such that $\lim_{y\to+\infty} \mathrm{i}y\, N(\mathrm{i}y) = 1$. By the previous discussion, the transformation $\mu \in \mathscr{M}^1 \mapsto C_\mu$ takes its values in $\mathscr{N}^1$. On the other hand, we can endow $\mathscr{N}^1$ with the **Montel topology** of uniform convergence on all compact subsets of $\mathbb{H}$ (local uniform convergence). If $(N_n)_{n\in\mathbb{N}}$ is a sequence of functions of $\mathscr{N}^1$ that converge locally uniformly to N, then the limiting function N is holomorphic on $\mathbb{H}$ and with negative imaginary parts, but it might fail to satisfy the condition $\lim_{y\to\infty} \mathrm{i}y\, N(\mathrm{i}y) = 1$. Therefore, one has to strengthen a bit Montel's topology on $\mathscr{N}^1$ in order to get something interesting (in the end, a metrizable complete space). Thus, we say that $f_n \in \mathscr{N}^1$ **converges properly** towards $f \in \mathscr{N}^1$ if f_n converges locally uniformly on $\mathbb{H}$ to f, and if

$$\lim_{r\to\infty} \sup_{n\in\mathbb{N},\, y\geq r} |\mathrm{i}y\, f_n(\mathrm{i}y) - 1| = 0.$$

Let us check that this new notion of convergence is metrizable. If f and g belong to $\mathscr{N}^1$, set

$$\begin{aligned} d(f,g) = &\sum_{r=1}^{\infty} \frac{1}{2^r} \min\left(1, \sup_{|z|\leq r,\, \mathrm{Im}(z)\geq \frac{1}{r}} |f(z)-g(z)|\right) \\ &+ \sum_{r=1}^{\infty} \frac{1}{2^r} \min\left(1, \sup_{r\leq y} |\mathrm{i}y\,(f(\mathrm{i}y)-g(\mathrm{i}y))|\right). \end{aligned}$$

This distance is well defined and smaller than $2\sum_{r=1}^{\infty} \frac{1}{2^r} = 2$.

Lemma 7.22. *Let $(f_n)_{n\in\mathbb{N}}$ be a sequence of functions in $\mathscr{N}^1$. Then, f_n converges properly towards $f \in \mathscr{N}^1$ if and only if $d(f_n, f)$ converges to 0. Moreover, $(\mathscr{N}^1, d)$ is a complete metric space.*

Proof. Suppose that $f_n \to f$ properly, with the f_n's and f in $\mathscr{N}^1$. For any $r \geq 1$, the set $K_r = \{z : |z| \leq r \text{ and } \mathrm{Im}(z) \geq \frac{1}{r}\}$ is compact, therefore,

$$\sup_{z\in K_r} |f(z) - g(z)| \to 0.$$

As a consequence, the first series $\sum_{r\geq 1} \frac{1}{2^r} \min(1, \sup_{z\in K_r} |f(z)-g(z)|)$ in the definition of $d(f_n, f)$ has limit 0. As for the second series, notice that for any $R \geq 1$

$$\begin{aligned}&\sum_{r=1}^{\infty} \frac{1}{2^r} \min\left(1, \sup_{r\leq y} |\mathrm{i}y\,(f_n(\mathrm{i}y) - f(\mathrm{i}y))|\right)\\&\leq \left(R\sum_{r=1}^{R}\frac{1}{2^r}\right) \sup_{r\leq y\leq R} |f_n(\mathrm{i}y) - f(\mathrm{i}y)|\\&\quad + \left(\sum_{r=1}^{\infty}\frac{1}{2^r}\right)\left(\sup_{y\geq R}|\mathrm{i}y f_n(\mathrm{i}y) - 1| + \sup_{y\geq R}|\mathrm{i}y f(\mathrm{i}y) - 1|\right)\\&\leq R \sup_{1\leq y\leq R+1} |f_n(\mathrm{i}y) - f(\mathrm{i}y)| + \sup_{y\geq R}|\mathrm{i}y f_n(\mathrm{i}y) - 1| + \sup_{y\geq R}|\mathrm{i}y f(\mathrm{i}y) - 1|.\end{aligned}$$

On the last line, the additional condition associated to the proper convergence ensures that the two last terms can be made arbitrarily small for R large enough. Then, R being fixed, the first term goes to zero by local uniform convergence of f_n towards f. So, if f_n converges properly to f, then $\lim_{n\to\infty} d(f_n, f) = 0$.

Conversely, if $d(f_n, f)$ converges to 0, then the convergence of the first series ensures the local uniform convergence, because the sets K_r form an increasing sequence of compact sets such that $\bigcup_{r=1}^{\infty} K_r = \mathbb{H}$. Therefore, if K is a given compact subset of $\mathbb{H}$, then $K \subset K_r$ for some r large enough, and

$$\sup_{z\in K} |f(z) - g(z)| \leq \sup_{z\in K_r} |f(z) - g(z)| \to 0$$

since the first series goes to 0. Then, the properness of the convergence is guaranteed by the convergence to zero of the second series involved in the definition of $d(f_n, f)$.

Finally, let us check that d makes $\mathscr{N}^1$ into a complete metric space. If $(f_n)_{n\in\mathbb{N}}$ is a Cauchy sequence in $\mathscr{N}^1$, then the restrictions $f_{n,|K_r}$ form Cauchy sequences in the Banach algebras $\mathscr{C}^0(K_r)$ of continuous functions; therefore, $(f_n)_{n\in\mathbb{N}}$ admits a local uniform limit f, which is holomorphic on $\mathbb{H}$ by Cauchy's formula, and of negative imaginary type. It remains to see that f belongs to $\mathscr{N}^1$, that is to say that $\lim_{y\to\infty} \mathrm{i}y f(\mathrm{i}y) = 1$. Fix $\varepsilon > 0$; since $(f_n)_{n\in\mathbb{N}}$ is a Cauchy sequence for d, for N large enough and $n, m \geq N$,

$$\sup_{y\geq 1} |\mathrm{i}y\,(f_n(\mathrm{i}y) - f_m(\mathrm{i}y))| \leq \varepsilon$$

by looking at the first coordinate of the second series involved in $d(f_n, f_m)$. Since $f_N \in \mathscr{N}^1$, for r large enough, $\sup_{y\geq r} |\mathrm{i}y\, f_N(\mathrm{i}y) - 1| \leq \varepsilon$, so, for any $n \geq N$,

$$\sup_{y\geq r} |\mathrm{i}y\, f_n(\mathrm{i}y) - 1| \leq \sup_{y\geq 1} |\mathrm{i}y\,(f_n(\mathrm{i}y) - f_N(\mathrm{i}y))| + \sup_{y\geq r} |\mathrm{i}y\, f_N(\mathrm{i}y) - 1| \leq 2\varepsilon.$$

By increasing the value of r, we can assume that this inequality is also satisfied for the functions f_n with $n < N$, since there is only a finite number of them, and they all belong to $\mathscr{N}^1$. Hence, $\sup_{n\in\mathbb{N},y\geq r} |\mathrm{i}y\, f_n(\mathrm{i}y)-1| \leq 2\varepsilon$ for r large enough, i.e.,

$$\lim_{r\to\infty} \sup_{n\in\mathbb{N},\, y\geq r} |\mathrm{i}y\, f_n(\mathrm{i}y)-1| = 0.$$

In the inequality $\sup_{n\in\mathbb{N},y\geq r} |\mathrm{i}y\, f_n(\mathrm{i}y)-1| \leq 2\varepsilon$, one can take the limit in n, since f is the pointwise limit of the f_n's: $\sup_{y\geq r} |\mathrm{i}y\, f(\mathrm{i}y)-1| \leq 2\varepsilon$ for r large enough (the same $r = r(\varepsilon)$ as the for the functions f_n). Therefore,

$$\lim_{r\to\infty} \sup_{y\geq r} |\mathrm{i}y\, f(\mathrm{i}y)-1| = 0,$$

which ends the proof. □

Now that the topology of $\mathscr{N}^1$ is correctly specified, we can state:

Theorem 7.23. *The map $\mu \mapsto C_\mu$ is a homeomorphism between the spaces $\mathscr{M}^1$ and $\mathscr{N}^1$.*

Proof. We shall use without proof the following well-known result from harmonic analysis. Call **harmonic** a real-valued function on an open set $U \subset \mathbb{C}$, such that if the closed disk $\overline{D}_{(x,\varepsilon)} = \{y \in \mathbb{C} \,|\, |y-x| \leq \varepsilon\}$ is included in U, then

$$u(x) = \frac{1}{2\pi\varepsilon} \int_{C_{(x,\varepsilon)}} u(y)\,dy,$$

where dy is the 1-dimensional Lebesgue measure on the circle $C_{(x,\varepsilon)} = \{y \in \mathbb{C} \,|\, |y-x| = \varepsilon\}$. If $u(z)$ is a real continuous function on the disk $\overline{D}_{(0,1)}$ which is harmonic on the interior $D_{(0,1)}$ of the disk, then it admits a **Poisson–Stieltjes representation**:

$$\begin{aligned} u(z) &= \int_{\theta=0}^{2\pi} u(\mathrm{e}^{\mathrm{i}\theta})\,\mathrm{Re}\left(\frac{\mathrm{e}^{\mathrm{i}\theta}+z}{\mathrm{e}^{\mathrm{i}\theta}-z}\right)\frac{d\theta}{2\pi} \\ &= \int_{\theta=0}^{2\pi} u(\mathrm{e}^{\mathrm{i}\theta})\,\frac{1-r^2}{1+r^2-2r\,\cos(\varphi-\theta)}\,\frac{d\theta}{2\pi} \qquad \text{if } z = r\mathrm{e}^{\mathrm{i}\varphi}. \end{aligned}$$

Using this integral representation, one sees that u is positive on the disk $\overline{D}_{(0,1)}$ if and only if u is positive on the circle $C_{(0,1)}$. More generally, suppose that u is a positive continuous harmonic on the open disk $D_{(0,1)}$, possibly with singularities at the boundary of this open domain. Then, there is a unique finite positive measure ρ on the circle such that

$$u(z) = \int_0^{2\pi} \mathrm{Re}\left(\frac{\mathrm{e}^{\mathrm{i}\theta}+z}{\mathrm{e}^{\mathrm{i}\theta}-z}\right) d\rho(\theta)$$

for any z in the open disk. Moreover, the mass of ρ is equal to $u(0)$.

Let us now relate this result to the content of our theorem. Recall that any holomorphic function has its real part and imaginary parts that are real harmonic functions. Moreover, if u is a real harmonic function on $D_{(0,1)}$, then there is up to an additive constant a unique **conjugate harmonic function** v on $D_{(0,1)}$, such that $f(z) = u(z) + \mathrm{i}v(z)$ is holomorphic on the open disk. The function v is given by

$$v(z) = v(0) + \int_{0 \to z} \frac{\partial u}{\partial x}\, dy - \frac{\partial u}{\partial y}\, dx,$$

where the integral of the differential form is taken over any path from 0 to z. The result does not depend on the path because the differential form is closed, as a consequence of the harmonicity of u: $\frac{\partial^2 u}{\partial x^2} + \frac{\partial^2 u}{\partial y^2} = 0$. In the case when u is given by a Poisson–Stieltjes integral, the formula reads simply as:

$$v(z) = v(0) + \int_0^{2\pi} \mathrm{Im}\left(\frac{e^{\mathrm{i}\theta} + z}{e^{\mathrm{i}\theta} - z}\right) d\rho(\theta).$$

Consider now a holomorphic function f on $D_{(0,1)}$, such that $u = \mathrm{Re}(f)$ takes positive values. Then, u admits a Poisson representation with respect to a finite positive measure ρ on $C_{(0,1)}$, and

$$f(z) = u(z) + \mathrm{i}v(z) = \mathrm{i}\alpha + \int_0^{2\pi} \frac{e^{\mathrm{i}\theta} + z}{e^{\mathrm{i}\theta} - z}\, d\rho(\theta).$$

This representation of holomorphic functions is known as the **Riesz–Herglotz theorem**. If one assumes instead that $\mathrm{Im}(f)$ takes negative values, then by multiplying the previous formula by $-\mathrm{i}$, we obtain the representation

$$f(z) = \alpha + \int_0^{2\pi} \mathrm{i}\,\frac{z + e^{\mathrm{i}\theta}}{z - e^{\mathrm{i}\theta}}\, d\rho(\theta),$$

where α is an arbitrary real number. We can then transport this result to the Poincaré half-plane, by means of the Möbius transformation

$$\psi : z \in D_{(0,1)} \mapsto \mathrm{i}\,\frac{1+z}{1-z} \in \mathbb{H},$$

which extends to the circle $C_{(0,1)}$ by sending $e^{\mathrm{i}\theta} \neq 1$ to $-\mathrm{cotan}(\frac{\theta}{2})$, and 1 to ∞. Thus, if N is a holomorphic function on $\mathbb{H}$ of negative imaginary type, setting $z = \psi^{-1}(h) = \frac{h-\mathrm{i}}{h+\mathrm{i}}$ and $f = N \circ \psi$, one obtains

$$\begin{aligned} N(h \in \mathbb{H}) = f(z \in D_{(0,1)}) &= \alpha + \mathrm{i} \int_0^{2\pi} \frac{z + e^{\mathrm{i}\theta}}{z - e^{\mathrm{i}\theta}}\, d\rho(\theta) \\ &= \alpha + \int_0^{2\pi} \frac{h \cos(\frac{\theta}{2}) - \sin(\frac{\theta}{2})}{-h \sin(\frac{\theta}{2}) - \cos(\frac{\theta}{2})}\, d\rho(\theta) \end{aligned}$$

$$= \alpha - \beta h + \int_{\mathbb{R}} \frac{hs+1}{h-s}\,\frac{1}{1+s^2}\,d\mu(s)$$
$$= \alpha - \beta h + \int_{\mathbb{R}} \left(\frac{1}{h-s} + \frac{s}{1+s^2}\right) d\mu(s)$$

where $s = -\mathrm{cotan}(\frac{\theta}{2}) = \psi(\mathrm{e}^{\mathrm{i}\theta})$, $\beta = \rho(\{\theta = 0\}) \geq 0$, and μ is a positive measure on $\mathbb{R}$ such that $\int_{\mathbb{R}} \frac{d\mu(s)}{1+s^2} < +\infty$, which is deduced from ρ by means of the transformation ψ. The term $-\beta h$ comes from the singularity of the map ψ at $z = 1$; moreover,

$$-\beta = \lim_{y \to +\infty} \frac{\mathrm{Im}(N(\mathrm{i}y))}{y},$$

and the **Nevanlinna integral representation** above is unique. Suppose finally that $N \in \mathscr{N}^1$, that is to say that $\lim_{y\to\infty} \mathrm{i}y\, N(\mathrm{i}y) = 1$. This additional condition implies that

$$\beta = 0 \quad ; \quad \alpha = -\int_{\mathbb{R}} \frac{s}{1+s^2}\,\mu(ds)$$

and that μ is a probability measure on $\mathbb{R}$. Therefore, any function N in $\mathscr{N}^1$ writes uniquely as

$$N(h) = \int_{\mathbb{R}} \frac{1}{h-s}\,\mu(ds)$$

with $\mu \in \mathscr{M}^1(\mathbb{R})$. The bijective correspondence is therefore established.

Let us now prove the homeomorphic character of $\mu \mapsto C_\mu$. If $(\mu_n)_{n\in\mathbb{N}}$ is a sequence of probability measures that converges weakly towards μ, then for any $z \in \mathbb{H}$, the map $s \mapsto \frac{1}{z-s}$ is bounded continuous, hence, $C_{\mu_n}(z) \to C_\mu(z)$ by definition of the weak convergence of probability. It is easily seen by a domination argument that this convergence is locally uniform on $\mathbb{H}$. Let us then check that the convergence is proper, i.e.,

$$\lim_{r\to\infty} \sup_{n\in\mathbb{N},\, y\geq r} |\mathrm{i}y\, C_{\mu_n}(\mathrm{i}y) - 1| = 0.$$

Fix $\varepsilon > 0$. Since $(\mu_n)_{n\in\mathbb{N}}$ converges weakly to μ, it is a **tight sequence**, that is to say that there exists a compact interval $[-K_\varepsilon, K_\varepsilon]$ such that

$$\sup_{n\in\mathbb{N}} \mu_n(\mathbb{R} \setminus [-K_\varepsilon, K_\varepsilon]) \leq \varepsilon.$$

This is a general result on probability measures on **polish spaces** (separable and metrizable complete spaces), which is trivial for $\mathbb{R}$: indeed, the convergence $\mu_n \to \mu$ is equivalent by Portmanteau's theorem to the pointwise convergence of the cumulative distribution functions $F_{\mu_n}(t) \to F_\mu(t)$ at any continuity point t of $F_\mu(t)$. Fixing $K = K_\varepsilon$ as above, we then have

$$\mathrm{i}y\, C_{\mu_n}(\mathrm{i}y) = \int_{-K}^{K} \frac{1}{1-s/(\mathrm{i}y)}\,\mu_n(ds) + \int_{\mathbb{R}\setminus[-K,K]} \frac{\mathrm{i}y}{\mathrm{i}y-s}\,\mu_n(ds) = A + B$$

where

$$A = \mu_n([-K,K]) + O\left(\frac{K}{y}\right) = 1 - \mu_n(\mathbb{R} \setminus [-K,K]) + O\left(\frac{K}{y}\right)$$
$$|B| \leq \mu_n(\mathbb{R} \setminus [-K,K]).$$

As a consequence, $\sup_{n\in\mathbb{N},\, y\geq r} |\mathrm{i}y\, C_{\mu_n}(\mathrm{i}y) - 1| \leq 2\varepsilon + O(\frac{K}{r})$, which proves the proper convergence.

Conversely, assume that $(C_{\mu_n})_{n\in\mathbb{N}}$ converges properly to C_μ in $\mathscr{N}^1$. One can then go backwards in the previous argument and show that the sequence $(\mu_n)_{n\in\mathbb{N}}$ is tight. However, a sequence of probability measures on $\mathbb{R}$ is tight if and only if it is relatively compact for the weak topology (this is **Prohorov's theorem**). So, $(\mu_n)_{n\in\mathbb{N}}$ is relatively compact, and by a classical topological argument, it suffices now to show that the only possible limit of a convergent subsequence of $(\mu_n)_{n\in\mathbb{N}}$ is the probability measure μ. Thus, suppose that ν is a probability measure that is a weak limit of a subsequence $(\mu_{n_k})_{k\in\mathbb{N}}$. Since the map $\mu \mapsto C_\mu$ is continuous, $C_{\mu_{n_k}}$ converges locally uniformly towards C_ν, hence, $C_\nu = C_\mu$. Since the map $\mu \mapsto C_\mu$ is bijective, we then have $\nu = \mu$, which ends the proof of the theorem. □

Remark. The inverse of the correspondence $\mu \mapsto C_\mu$ is given by the **Perron–Stieltjes inversion formula**: if μ is a probability measure on $\mathbb{R}$, then

$$\int_a^b \mu(dx) = -\frac{1}{\pi} \lim_{y\to 0}\left(\int_a^b \mathrm{Im}(C_\mu(x+\mathrm{i}y))\,dx\right)$$

for any a and b that is not an atom of μ. This allows us to reconstruct the cumulative distribution function of μ, and thereby μ from its Cauchy transform. To prove the formula, notice that if $P_y(s) = \frac{1}{\pi}\frac{y}{s^2+y^2}$ is the Poisson kernel on the real line, then

$$-\frac{1}{\pi}\mathrm{Im}(C_\mu(x+\mathrm{i}y)) = (\mu * P_y)(x),$$

where $*$ denotes the convolution of probability measures on $\mathbb{R}$. However, P_y is an approximation of the unity, so $\mu * P_y$ converges weakly towards μ as y goes to 0, which implies immediately the aforementioned inversion formula.

The correspondence $\mathscr{M}^1 \leftrightarrow \mathscr{N}^1$ is the first part of the so-called **Markov–Krein correspondence**, which associates to any probability measure μ on $\mathbb{R}$ a generalized continuous diagram ω whose transition measure is $\mu_\omega = \mu$. To build the second part of this correspondence, we shall use an exponential analogue of the integral representation of negative imaginary type functions provided by Theorem 7.23. We call **Rayleigh function** on $\mathbb{R}$ a measurable function $R : \mathbb{R} \to [0,1]$, such that

$$\int_{-\infty}^0 \frac{R(s)}{1-s}\,ds < \infty \quad ; \quad \int_0^\infty \frac{1-R(s)}{1+s}\,ds < \infty.$$

We denote $\mathscr{R}^1$ the space of Rayleigh functions, two Rayleigh functions R_1 and R_2 being identified if they differ on a set of Lebesgue measure equal to 0. Given a Rayleigh function R, we can associate to it a finite positive measure on $\mathbb{R}$ which is absolutely continuous with respect to the Lebesgue measure:

$$m_R = 1_{(-\infty,0]}(s)\,\frac{R(s)}{1-s}\,ds + 1_{[0,+\infty)}(s)\,\frac{1-R(s)}{1+s}\,ds.$$

We endow $\mathscr{R}^1$ with the topology associated to the weak convergence of the finite measures m_R: $(R_n)_{n\in\mathbb{N}}$ is said to converge to R in $\mathscr{R}^1$ if and only if m_{R_n} converges weakly to m_R, which amounts to

$$\forall x \leq 0,\ \lim_{n\to\infty} \int_{-\infty}^{x} \frac{R_n(s)}{1-s}\,ds = \int_{-\infty}^{x} \frac{R(s)}{1-s}\,ds;$$
$$\forall x \geq 0,\ \lim_{n\to\infty} \int_{x}^{\infty} \frac{1-R_n(s)}{1+s}\,ds = \int_{x}^{\infty} \frac{1-R(s)}{1+s}\,ds.$$

Given a Rayleigh function R, we associate to it a holomorphic function $G_R(z)$ on $\mathbb{H}$, defined by the equation

$$G_R(z) = \frac{1}{z}\exp\left(-\int_{-\infty}^{0} \frac{R(s)}{z-s}\,ds + \int_{0}^{\infty} \frac{1-R(s)}{z-s}\,ds\right).$$

Theorem 7.24. *The map $R \mapsto G_R$ is a homeomorphism between the spaces $\mathscr{R}^1$ and $\mathscr{N}^1$.*

Proof. To prove this second integral representation theorem of functions in $\mathscr{N}^1$, notice first that if N is a function with negative imaginary type, then $N(z) = r(z)\,e^{-i\theta(z)}$ with $\theta(z) \in (0,\pi)$ for any $z \in \mathbb{H}$. Therefore,

$$M(z) + i\pi = \log N(z) + i\pi = \log r(z) + i(\pi - \theta(z))$$

is of positive imaginary type. As a consequence, there exists $\alpha \in \mathbb{R}$, $\beta \in \mathbb{R}_+$ and a positive measure μ on $\mathbb{R}$ such that $\int_{\mathbb{R}} \frac{\mu(s)}{1+s^2} < \infty$, and

$$M(z) + i\pi = \alpha + \beta z - \int_{\mathbb{R}} \left(\frac{1}{z-s} + \frac{s}{1+s^2}\right)\mu(ds).$$

Since $\mathrm{Im}(M(z)+i\pi) \in (0,\pi)$, $\beta = \lim_{y\to\infty} \frac{\mathrm{Im}(M(iy)+i\pi)}{y} = 0$. On the other hand, a particular case of the integral formulas above is the representation of the function $\log z$:

$$\log z = i\pi + \int_{0}^{\infty} \left(\frac{1}{z-s} + \frac{s}{1+s^2}\right) ds.$$

Therefore,

$$\begin{aligned}&\log(z\,N(z))\\&=\log z+M(z)\\&=\alpha-\int_{-\infty}^{0}\left(\frac{1}{z-s}+\frac{s}{1+s^2}\right)\mu(ds)+\int_{0}^{\infty}\left(\frac{1}{z-s}+\frac{s}{1+s^2}\right)(ds-\mu(ds)).\end{aligned}$$

However, $-M(z)=-\log r(z)+\mathrm{i}\theta(z)$ is also an analytic function of positive imaginary type, so there exists another integral representation

$$-M(z)=\alpha'+\beta' z-\int_{\mathbb{R}}\left(\frac{1}{z-s}+\frac{s}{1+s^2}\right)\mu'(ds),$$

again with $\alpha'\in\mathbb{R}$, $\beta'\in\mathbb{R}$ and μ' positive measure on $\mathbb{R}$ that integrates $\frac{1}{1+s^2}$. The same argument as before yields $\beta'=0$, hence

$$M(z)=\log N(z)=-\alpha'+\int_{\mathbb{R}}\left(\frac{1}{z-s}+\frac{s}{1+s^2}\right)\mu'(ds).$$

Combining this with

$$\log z=-\int_{-\infty}^{0}\left(\frac{1}{z-s}+\frac{s}{1+s^2}\right)ds,$$

we conclude that

$$\begin{aligned}&\log(z\,N(z))\\&=-\alpha'-\int_{-\infty}^{0}\left(\frac{1}{z-s}+\frac{s}{1+s^2}\right)(ds-\mu'(ds))+\int_{0}^{\infty}\left(\frac{1}{z-s}+\frac{s}{1+s^2}\right)\mu'(ds).\end{aligned}$$

By unicity of the Nevanlinna representation, we conclude that $ds-\mu'(ds)=\mu(ds)$, hence, μ is absolutely continuous with respect to the Lebesgue measure, with density $R(s)\in[0,1]$. On the other hand, if one assumes that $N\in\mathscr{N}^1$, then setting $z=\mathrm{i}y$ in the representation of $\log(z\,N(z))$ in terms of μ, and letting y go to infinity, we obtain

$$\alpha-\int_{-\infty}^{0}\frac{s}{1+s^2}\,\mu(ds)+\int_{0}^{\infty}\frac{s}{1+s^2}\,(ds-\mu(ds))=0.$$

Thus, if $N\in\mathscr{N}^1$, then we can rewrite

$$\log(z\,N(z))=-\int_{-\infty}^{0}\frac{R(s)\,ds}{z-s}+\int_{0}^{\infty}\frac{(1-R(s))\,ds}{z-s},$$

and since these integrals are well defined, R must satisfy the integrability conditions of Rayleigh functions, hence, $R\in\mathscr{R}^1$. So, we have shown that for any

$N \in \mathscr{N}^1$, there exists a Rayleigh function $R \in \mathscr{R}^1$ such that

$$N(z) = \frac{1}{z}\,\exp\left(-\int_{-\infty}^{0} \frac{R(s)}{z-s}\,ds + \int_{0}^{\infty} \frac{1-R(s)}{z-s}\,ds\right) = G_R(z),$$

and by the unicity of Nevanlinna representation, R is unique. Moreover, the calculations can be performed backwards to show that if R is a Rayleigh function, then its transform $G_R(z)$ is an element of $\mathscr{N}^1$, hence, the map $R \mapsto G_R$ is indeed a bijection between $\mathscr{R}^1$ and $\mathscr{N}^1$. Finally, the proof of the homeomorphic character is entirely similar to the proof in Theorem 7.23. □

We can finally relate any probability measure $\mu \in \mathscr{M}^1$ to a continuous Young diagram ω, up to a small modification of Definition 7.20. We call **generalized continuous Young diagram** a function $\omega : \mathbb{R} \to \mathbb{R}$ which is Lipschitz with constant 1, and such that the following integrals converge:

$$\int_{-\infty}^{0} \frac{(1+\omega'(s))}{1-s}\,ds < \infty \quad ; \quad \int_{0}^{\infty} \frac{(1-\omega'(s))}{1+s}\,ds < \infty.$$

Thus, in a sense of convergence of integrals, $\omega'(s)$ is close to -1 when s goes to $-\infty$, and close to $+1$ when s goes to $+\infty$. We identify two generalized continuous Young diagrams if they differ by a constant, that is to say if they have the same derivative. We then denote $\mathscr{Y}^1$ the set of (equivalence classes of) continuous Young diagrams. The set of continuous Young diagrams $\mathscr{Y}$ is the subset of $\mathscr{Y}^1$ that consists in functions such that $\omega'(s) - \mathrm{sgn}(s)$ has compact support. On the other hand, ω belongs to $\mathscr{Y}^1$ if and only if $R_\omega(s) = \frac{\omega'(s)+1}{2}$ is a Rayleigh function. Thus, we have a natural bijection between $\mathscr{Y}^1$ and $\mathscr{R}^1$, which is a homeomorphism if one endows $\mathscr{Y}^1$ with an adequate topology. We say that $\omega_n \to \omega$ in $\mathscr{Y}^1$ if:

- there exist representatives $\omega_n(s)$ and $\omega(s)$ of the classes of functions ω_n and ω such that for any compact interval $[-K,K]$, $\lim_{n\to\infty} \sup_{s\in[-K,K]} |\omega_n(s) - \omega(s)| = 0$.
- the integrals associated to the ω_n's are uniformly bounded:

$$\sup_{n\in\mathbb{N}}\left(\int_{-\infty}^{0} \frac{(1+\omega_n'(s))}{1-s}\,ds\right) < \infty;$$
$$\sup_{n\in\mathbb{N}}\left(\int_{0}^{\infty} \frac{(1-\omega_n'(s))}{1+s}\,ds\right) < \infty.$$

By Fatou's lemma, the second condition above and the local uniform convergence of the diagrams $\omega_n \in \mathscr{Y}^1$ towards a function ω guarantee that ω' also satifies the conditions of integrability, hence belongs to $\mathscr{Y}^1$. Moreover, it is then easily seen that the map $\omega \mapsto R_\omega$ is a homeomorphism, with reciprocal

$$R \mapsto \omega_R(s) = \text{constant} + \int_0^s (2R(s)-1)\,ds.$$

We can then restate Theorems 7.23 and 7.24 as follows:

Theorem 7.25. *There is a sequence of homeomorphisms*

$$\omega \in \mathscr{Y}^1 \longleftrightarrow R_\omega \in \mathscr{R}^1 \longleftrightarrow G_\omega \in \mathscr{N}^1 \longleftrightarrow \mu_\omega \in \mathscr{M}^1,$$

which are characterized by the following identities:

$$\begin{aligned} G_\omega(z) &= \int_{\mathbb{R}} \frac{1}{z-s}\,\mu_\omega(ds) = \frac{1}{z}\exp\left(-\int_{-\infty}^{0} \frac{R_\omega(s)}{z-s}\,ds + \int_0^{\infty} \frac{1-R_\omega(s)}{z-s}\,ds\right) \\ &= \frac{1}{z}\exp\left(-\int_{\mathbb{R}} \frac{\sigma'_\omega(s)}{z-s}\,ds\right), \end{aligned}$$

where as usual $\sigma_\omega(s) = \frac{\omega(s)-|s|}{2}$. In this setting, we say that μ_ω (respectively, G_ω) is the transition measure (respectively, generating function) associated to the generalized continuous Young diagram ω. Moreover, these correspondences restrict to a bijection between

Young diagrams $\lambda \in \mathfrak{Y}$	$\longleftrightarrow$	*generating functions G_λ of interlaced integer sequences (X,Y) with $p_1(X+Y)=0$.*

Example. Consider the **Cauchy distribution** on $\mathbb{R}$:

$$\mu_{\mathrm{Cauchy}}(ds) = \frac{1}{\pi(1+s^2)}\,ds.$$

Its Cauchy transform is

$$C_{\mathrm{Cauchy}}(z) = \frac{1}{\pi}\int_{\mathbb{R}} \frac{1}{(z-s)(1+s^2)}\,ds = \frac{1}{z+\mathrm{i}}.$$

Indeed, $-\frac{1}{\pi}\lim_{y\to 0}\mathrm{Im}(\frac{1}{s+\mathrm{i}y+\mathrm{i}}) = \lim_{y\to 0}\frac{1}{\pi}\frac{1+y}{(1+y)^2+s^2} = \frac{1}{\pi(1+s^2)}$, so the identity follows by the Perron–Stieltjes inversion formula. However, one has also

$$\frac{1}{z+\mathrm{i}} = \exp\left(\int_{\mathbb{R}} \log\left(\frac{1}{z-s}\right)\mu_{\mathrm{Cauchy}}(ds)\right) = \frac{1}{z}\exp\left(-\int_{\mathbb{R}} \frac{\sigma'_{\omega_{\mathrm{Cauchy}}}(s)}{z-s}\,ds\right),$$

where $(\omega_{\mathrm{Cauchy}})'(s) = \frac{2}{\pi}\arctan(s)$. Therefore, the generalized continuous Young diagram

$$\omega_{\mathrm{Cauchy}}(s) = \frac{2}{\pi}\left(s\arctan s - \frac{1}{2}\log(1+s^2)\right)$$

corresponds via the Markov–Krein bijections to the Cauchy probability measure on $\mathbb{R}$. In this particular case, one has also

$$\frac{1}{2}\,\omega''_{\mathrm{Cauchy}} = \mu_{\mathrm{Cauchy}}.$$

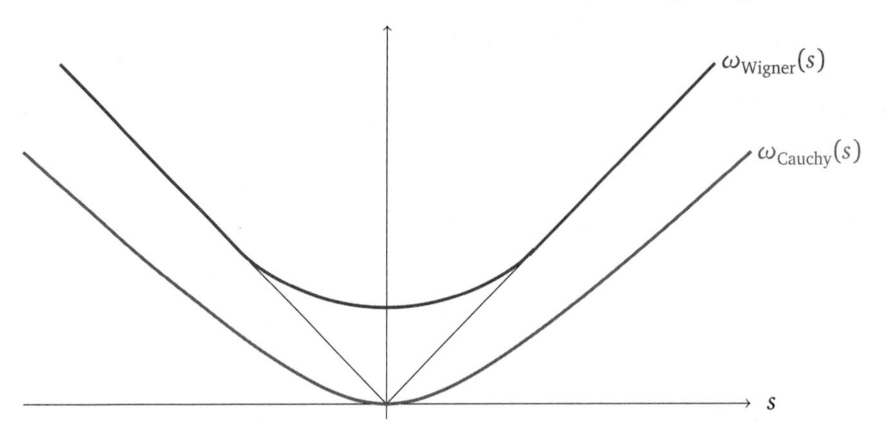

Figure 7.8
The (generalized) continuous Young diagrams ω_{Cauchy} and ω_{Wigner}.

We refer to Figure 7.8 for a drawing of this generalized continuous Young diagram. Notice that because of the term $-\log(1+s^2)$, there is no asymptote to the graph of ω_{Cauchy}.

Example. Consider now the **Wigner distribution** on $\mathbb{R}$:

$$\mu_{\mathrm{Wigner}}(ds) = \frac{\sqrt{4-s^2}}{2\pi}\, 1_{s\in[-2,2]}\, ds.$$

This is a probability measure compactly supported on $[-2,2]$, which is the limit of empirical spectral measures of various models of random matrices. Its Cauchy transform is

$$C_{\mathrm{Wigner}}(z) = \frac{1}{2\pi}\int_{-2}^{2} \frac{\sqrt{4-s^2}}{z-s}\, ds = \frac{2}{z+\sqrt{z^2-4}},$$

where the branch of the square root is chosen so that $C_{\mathrm{Wigner}}(z)$ belongs to $\mathscr{N}^1$. Again, this is immediate by the Perron–Stieltjes formula. Then, one can also write

$$\frac{2}{z+\sqrt{z^2-4}} = \exp\left(\int_{-2}^{2} \log\left(\frac{1}{z-s}\right) \frac{ds}{\pi\sqrt{4-s^2}}\right) = \frac{1}{z}\exp\left(-\int_{\mathbb{R}} \frac{\sigma'_{\omega_{\mathrm{Wigner}}}(s)}{z-s}\, ds\right),$$

where $(\omega_{\mathrm{Wigner}})'(s) = \frac{2}{\pi}\arcsin\frac{s}{2}\, 1_{s\in[-2,2]}$. This second identity, which again is easily shown with Perron–Stieltjes formula, involves the arcsine law $\mu_{\mathrm{arcsine}}(ds) = \frac{1}{\pi\sqrt{4-s^2}}\, 1_{s\in[-2,2]}\, ds$, which satisfies

$$\frac{1}{2}\,\omega''_{\mathrm{Wigner}} = \mu_{\mathrm{arcsine}}.$$

Thus, the continuous Young diagram (which is in $\mathscr{Y}$)

$$\omega_{\mathrm{Wigner}}(s) = \begin{cases} \frac{2}{\pi}\left(s \arcsin\frac{s}{2} + \sqrt{4-s^2}\right) & \text{if } |s| \leq 2, \\ |s| & \text{if } |s| > 2 \end{cases}$$

is the one associated to the Wigner law by the Markov–Krein correspondence. This particular continuous Young diagram will play a prominent role in the asymptotic study of the Plancherel measures of the symmetric groups; see Chapter 13. We refer again to Figure 7.8 for a drawing of ω_{Wigner}.

Remark. In the two previous examples, the Rayleigh function R_ω involved in the exponential integral representation of the Cauchy transform C_ω of the transition probability measure is of bounded variation. In this situation, R_ω can always be written as the cumulative distribution function of a signed measure. Thus, we define the **Rayleigh measure** associated to a Rayleigh function of bounded variation by the equation:

$$R_\omega(s) = \int_{-\infty}^{s} \tau_\omega(dt).$$

Equivalently, the generating function $G_\omega(z)$ is equal to $\exp\left(\int_{\mathbb{R}} \log\left(\frac{1}{z-s}\right) \tau_\omega(ds)\right)$, so τ_ω is just the second derivative of $\frac{\omega}{2}$. One then says that μ_ω is the **Markov transform** of the signed measure τ_ω. This construction yields a bijection between certain signed measures with an interlacing property of their positive and negative Jordan parts, and certain probability measures on the real line. In the previous examples, τ_ω was actually a positive probability measure on the real line, and we have in particular shown that

$$\text{Markov transform}(\mu_{\mathrm{Cauchy}}) = \mu_{\mathrm{Cauchy}};$$
$$\text{Markov transform}(\mu_{\mathrm{arcsine}}) = \mu_{\mathrm{Wigner}}.$$

When $\omega = \omega_\lambda$ is the function attached to a Young diagram $\lambda \in \mathfrak{Y}$, the Rayleigh measure τ_λ is simply the signed sum of Diracs at the interlaced coordinates: $\tau_\lambda = \sum_{i=1}^{s} \delta_{x_i} - \sum_{i=1}^{s-1} \delta_{y_i}$.

▷ *Topology of observables.*

To conclude this section, we relate the Markov–Krein correspondence to the topology controlled by the observables of Young diagrams. We defined previously for any $\omega \in \mathfrak{Y}$ the interlaced moments

$$\widetilde{p}_k(\omega) = -k \int_{\mathbb{R}} s^{k-1} \sigma'_\omega(s)\, ds;$$

when σ'_ω is of bounded variation, these are the moments of the Rayleigh measure τ_ω associated to ω. Notice that we cannot extend this definition to *generalized* continuous Young diagrams. Indeed, the integrals above are convergent for $\omega \in \mathscr{Y}$, because σ_ω is then compactly supported; however, they are usually not convergent when ω belongs to the larger space $\mathscr{Y}^1$. Still, one can ask whether the convergence of all the observables of continuous diagrams ω_n implies the convergence of ω_n in $\mathscr{Y}^1$. Since $\mathscr{Y}$ is not closed in $\mathscr{Y}^1$, we have to suppose that the possible limit is in $\mathscr{Y}$ in order to state a correct result. Thus, we have:

Proposition 7.26. *Let $(\omega_n)_{n\in\mathbb{N}}$ be a sequence of continuous Young diagrams, such that there exists another continuous Young diagram $\omega \in \mathscr{Y}$ with*

$$\forall f \in \mathscr{O}, \ \lim_{n\to\infty} f(\omega_n) = f(\omega).$$

Then, ω_n converges uniformly on $\mathbb{R}$ towards ω.

Proof. To begin with, notice that if ω is a continuous Young diagram with σ_ω compactly supported on $[a,b]$, then the transition measure μ_ω is also compactly supported on $[a,b]$. Indeed,

$$\frac{d\mu_\omega}{dt}(t) = -\frac{1}{\pi}\lim_{y\to 0} \mathrm{Im}\left(\frac{1}{t+\mathrm{i}y}\exp\left(-\int_a^b \frac{\sigma'_\omega(s)}{\mathrm{i}y+t-s}\,ds\right)\right) = 0 \quad \text{if } t \notin [a,b].$$

Moreover, the moments of the transition measure belong to the algebra of observables $\mathscr{O}$: if $\widetilde{h}_k(\omega) = \int_{\mathbb{R}} s^k\,\mu_\omega(ds)$, which is well defined since μ_ω is compactly supported, then

$$\sum_{k=0}^{\infty} \widetilde{h}_k(\omega)\,z^{-k} = z\int_{\mathbb{R}} \frac{\mu_\omega(ds)}{z-s} = \exp\left(\sum_{k=1}^{\infty} \frac{\widetilde{p}_k(\omega)}{k}\,z^{-k}\right),$$

so the $\widetilde{h}_k$'s and the $\widetilde{p}_k$'s can be considered as specializations of the homogeneous symmetric functions and of the power sums, and in particular, they are homogeneous polynomials in one another. Hence, the graded algebra $(\mathscr{O},\mathrm{wt})$ can be defined as

$$\mathscr{O} = \mathbb{C}\left[\widetilde{h}_2, \widetilde{h}_3, \ldots\right] \quad \text{with } \mathrm{wt}(\widetilde{h}_k) = k.$$

Now, the hypotheses of the proposition are equivalent to the following statement: the moments of the transition measures μ_{ω_n} all converge towards those of μ. Since μ is compactly supported, it is entirely determined by its moments, and moreover, the convergence of moments implies the convergence in law $\mu_n \to \mu$ in $\mathscr{M}^1(\mathbb{R})$ (see the notes for a reference for this classical fact from probability theory). By the Markov–Krein correspondence, we conclude that ω_n converges locally uniformly towards ω. This convergence is actually a global uniform convergence, because of the inequalities $\omega_n(s) \geq |s|$ for all $s \in \mathbb{R}$. Indeed, fix $\varepsilon > 0$, and a segment $[a,b]$ which contains the support of the transition measure of the limiting continuous Young diagram ω. By the previous discussion, for n large enough,

$$\sup_{s\in[a,b]} |\omega_n(s) - \omega(s)| \leq \varepsilon.$$

Bu then, for $s \geq b$,

$$\omega(s) = |s| \leq \omega_n(s) \leq \omega_n(b) + |s-b| \leq \omega(b) + \varepsilon + |s-b| = \omega(s) + \varepsilon,$$

and similarly, for $s \leq a$. Therefore, one has in fact

$$\sup_{s\in\mathbb{R}} |\omega_n(s) - \omega(s)| \leq \varepsilon,$$

hence the global uniform convergence. □

So, to summarize, the convergence of all observables controls the global uniform convergence of continuous Young diagrams if one knows beforehand that the possible limit is also compactly supported (i.e., a continuous Young diagram in the non-generalized sense). A probabilistic version of this result will be used several times in Chapter 13.

Notes and references

The observables studied in this chapter correspond to the "polynomial functions on Young diagrams" introduced by Kerov and Olshanski in [KO94]. Their interpretation as invariants of the algebra of partial permutations comes from [IK99]; it is the starting point of many combinatorial arguments of the third and fourth parts of our book. The construction of algebras that project themselves to a family of group algebras or finite-dimensional algebras has also been performed:

- for the centers of the Iwahori–Hecke algebras $\mathfrak{H}(n)$; see [Mél10];
- and for the algebras of double cosets $\mathbb{C}[\mathfrak{B}(n)\backslash\mathfrak{S}(2n)/\mathfrak{B}(n)]$, where $\mathfrak{B}(n)$ denotes the hyperoctahedral group; see [Tou14].

The result of polynomiality of the structure coefficients of the centers of the symmetric group algebras (Theorem 7.8) is due to Farahat and Higman, cf. [FH59]. For the impossibility to build a projective limit $\varprojlim_{n\to\infty} \mathbb{C}\mathfrak{S}(n)$, due to the simplicity of the alternate group $\mathfrak{A}(n)$ for $n \geq 5$, we refer to [Lan02, Chapter 1, Theorem 5.5].

A natural question is whether the algebraic observables Σ_μ admit quantum analogues related to the Hecke algebras $\mathfrak{H}_z(n)$. In the special case of the 0–Hecke algebras, one can construct at once a projective limit $\varprojlim_{n\to\infty} \mathfrak{H}_0(n)$, because the map

$$\begin{aligned} \mathfrak{H}_0(n+1) &\to \mathfrak{H}_0(n) \\ T_i &\mapsto \begin{cases} T_i & \text{if } i < n, \\ 0 & \text{if } i = n \end{cases} \end{aligned}$$

is compatible with the relations that define these algebras. Thus, the infinite Hecke algebra $\mathfrak{H}_0(\infty)$ and some subalgebras of it are natural candidates for an algebra of 0-observables. Unfortunately, because the simple modules over $\mathfrak{H}_0(n)$ are one-dimensional, the theory that comes from this construction is not very interesting. Another reason for this is because the Cartan and decomposition maps of $\mathfrak{H}_0(n)$ are encoded by the algebras NCSym and QSym. As a consequence, a theory of 0-observables of compositions amounts to the combinatorics of descents and back-steps of permutations. However, one can treat these problems by direct counting arguments, and without using representation theory. On the other hand, for the z–Hecke algebras with z generic, the Frobenius–Ram formula 5.49 implies that the Hecke character values $\mathrm{ch}_z^\lambda(T_{\sigma_\mu})$ can be expressed in terms of the standard character values $\mathrm{ch}^\lambda(\mu)$: it suffices to expand the Hall–Littlewood polynomial in the basis of power sums, and then to use the standard Frobenius formula 2.32; see [Ram91, Theorem 5.4] and [FM12, Proposition 10]. This implies that the renormalized character values of the Hecke algebras $\mathfrak{H}_z(n)$ belong to the same algebra $\mathscr{O}$ as the renormalized character values of the symmetric groups; therefore, one does not extend the theory of observables by looking at generic Hecke algebras.

The link between *algebraic* and *geometric* observables is detailed in [IO02], and Theorem 7.13 relies on Wassermann's formula, which can be found in [Was81]. For a study of the gradations of $\mathscr{O}$, we refer to [IK99, Section 10]. The use of the algebra $\mathscr{O}$ to control the topology of (continuous) Young diagrams is the subject of the paper [Ker98], which contains a complete presentation of the Markov–Krein correspondence. We tried to detail as much as possible the topological content of this correspondence, and to make the space of generalized continuous Young diagrams into a *polish space*, that is to say a topological space that is metrizable, complete and separable. This topological setting is the most convenient in order to speak of convergence of random Young diagrams; see the fourth part of the book. We refer to [Bil69] for details on the topology of weak convergence on $\mathscr{M}^1(\mathbb{R})$. We used in particular:

- Prohorov's theorem ([Bil69, Chapter 1, Section 5]), which characterizes the relatively compact subsets of $\mathscr{M}^1(\mathbb{R})$ thanks to the notion of tight sequences of probability measures;
- Portmanteau's theorem ([Bil69, Chapter 1, Section 2]), which gives a list of equivalent definitions of weakly convergent sequences of probability measures.

We refer to [Rud87] for Montel's topology on holomorphic functions, and to [Lan93, Chapter VIII, §3] for the Poisson–Stieltjes integral representation. The two papers [AD56, AD64] detail the various integral representations of holomorphic functions on the upper half-plane, including the Nevanlinna integral representation; and the Markov moment problem is presented in [KN77]. Finally, we used at the end of Section 7.4 the method of moments to prove a convergence of probability measures: if $\mu_n(x^k) \to \mu(x^k)$ for any $k \geq 1$ and if μ is a probability measure determined by its moments, then μ_n converges towards μ in the Skohorod topology of weak convergence. We refer to [Bil95, Chapter 30] for this result.

8

The Jucys–Murphy elements

In Chapter 7, we constructed an algebra $\mathscr{O}$ of observables of Young diagrams, which contained:

- renormalized versions Σ_μ of the character values;
- symmetric functions in the formal alphabet $A - \overline{B}$ of Frobenius coordinates;
- and symmetric functions in the formal alphabet $X - Y$ of interlaced coordinates.

The next chapters introduce new observables in $\mathscr{O}$ or in a larger algebra of functions, which allow one to get a good understanding of the representation theory of the symmetric groups. In this chapter, we shall in particular see that the elements of $\mathscr{O}$ can be written as symmetric functions of the *contents* of the cells of a Young diagram. In representation theoretic terms, this result is related to the so-called *Jucys–Murphy elements* $J_1, J_2, \dots, J_n$, which generate in $\mathbb{C}\mathfrak{S}(n)$ a maximal commutative subalgebra. These elements are presented in Section 8.1, and we prove in particular that the symmetric functions of these elements span the center $\mathrm{Z}(n) = Z(\mathbb{C}\mathfrak{S}(n))$ of the group algebra.

If S^λ is an irreducible representation of $\mathfrak{S}(n)$, we saw (Proposition 3.8) that one can label a complex basis of S^λ by the standard tableaux $T \in \mathrm{ST}(\lambda)$. In Section 8.2, we prove that there exists a canonical basis $(e_T)_{T \in \mathrm{ST}(\lambda)}$ of S^λ called the Gelfand–Tsetlin basis, and such that the elements $J_1, \dots, J_n$ act diagonally on each e_T, with eigenvalues corresponding to the contents of the cells of the tableau. This property is then used in Section 8.3 to relate the algebra $\mathscr{O}$ to symmetric functions of the contents of the cells of Young diagrams.

8.1 The Gelfand–Tsetlin subalgebra of the symmetric group algebra

From now on, we abbreviate the center of the group algebra $\mathbb{C}\mathfrak{S}(n)$ by the notation $\mathrm{Z}(n) = Z(\mathbb{C}\mathfrak{S}(n))$. On the other hand, we recall the branching rules for irreducible representations of symmetric groups (Corollary 3.7): if $\lambda \in \mathfrak{Y}(n)$ and

$\Lambda \in \mathfrak{Y}(n+1)$, then the multiplicity of S^λ in $\mathrm{Res}^{\mathfrak{S}(n+1)}_{\mathfrak{S}(n)}(S^\Lambda)$ or of S^Λ in $\mathrm{Ind}^{\mathfrak{S}(n+1)}_{\mathfrak{S}(n)}(S^\lambda)$ is equal to 1 if $\lambda \nearrow \Lambda$, and to 0 otherwise.

▷ *The Gelfand–Tsetlin basis and the Gelfand–Tsetlin subalgebra.*

Fix an integer partition $\lambda \in \mathfrak{Y}(n)$. To a standard tableau $T \in \mathrm{ST}(\lambda)$, we can as in the proof of Proposition 3.8 associate a unique increasing sequence of partitions $\emptyset = \lambda^{(0)} \nearrow \lambda^{(1)} \nearrow \cdots \nearrow \lambda^{(n)} = \lambda$. This sequence of integer partitions determines a unique sequence of subspaces

$$S^\lambda = V_T^{(n)} \supset V_T^{(n-1)} \supset V_T^{(n-2)} \supset \cdots \supset V_T^{(0)} = \mathbb{C},$$

such that each $V_T^{(i)}$ is an irreducible representation of $\mathfrak{S}(i)$ of type $\lambda^{(i)}$: each time, we take $V^{(i-1)}$ as the component of type $\lambda^{(i-1)}$ in $\mathrm{Res}^{\mathfrak{S}(i)}_{\mathfrak{S}(i-1)}(V^{(i)})$. This sequence of subspaces of S^λ satisfies the property:

$$\forall i \leq j,\ (\mathbb{C}\mathfrak{S}(j))(V_T^{(i)}) = V_T^{(j)}.$$

We call **$[\![i,n]\!]$-stem** of a standard tableau a numbering $t : [\![i,n]\!] \to \{\text{cells of } \lambda\}$ that can be completed to get a full standard tableau of shape λ, that is to say a bijective numbering $T : [\![1,n]\!] \to \{\text{cells of } \lambda\}$ that is increasing along rows and columns. For instance,

$$\begin{array}{|c|cccc}\hline 6 & \multicolumn{3}{c}{} \\ \hline & \multicolumn{1}{|c|}{4} & \multicolumn{1}{c|}{8} & \\ \hline & \multicolumn{1}{|c|}{} & \multicolumn{1}{c|}{5} & \multicolumn{1}{c|}{7} \\ \hline \end{array}$$

is a $[\![4,8]\!]$-stem of a standard tableau of shape $\lambda = (4,3,1)$. Notice then that $V_T^{(i)}$ only depends on the $[\![i,n]\!]$-stem t of T. We can therefore denote without ambiguity $V_T^{(i)} = V_t^{(i)}$.

Lemma 8.1. *For every $i \leq n$,*

$$S^\lambda = \bigoplus_{\substack{t\ [\![i,n]\!]\text{-stem of a standard} \\ \text{tableau of shape } \lambda}} V_t^{(i)}$$

is the decomposition of S^λ in $\mathfrak{S}(i)$-irreducible representations.

Proof. If t is a stem of a standard tableau, denote $\lambda(t)$ the shape of the integer partition $\lambda(t) \subset \lambda$ that consists in unlabeled cells. On the other hand, denote $t \nearrow t'$ if the $[\![i,n]\!]$-stem t is obtained from the $[\![i+1,n]\!]$-stem t' by adding the label i to a cell. By construction, we know that

$$V_t^{(i)} = S^{\lambda(t)} \quad \text{as a } \mathbb{C}\mathfrak{S}(i)\text{-module},$$

and that $V_{t'}^{(i+1)} = \bigoplus_{t \nearrow t'} V_t^{(i)}$. The result follows immediately, by decreasing induction on i. □

Consider in particular the case $i = 0$ of the previous lemma. It yields a decomposition of S^λ in a direct sum of complex lines $V_T^{(0)}$ labeled by standard tableaux (this argument was already used in the proof of Proposition 3.8, but keeping only track of the dimensions). Denote $V_T^{(0)} = \mathbb{C}e_T$, where each vector e_T is uniquely determined up to a multiplicative factor. We have:

Proposition 8.2. *The family $(e_T)_{T\in \mathrm{ST}(\lambda)}$ is a basis of S^λ, entirely determined up to action of a diagonal matrix by the identities*

$$(\mathbb{C}\mathfrak{S}(i))(e_T) = S^{\lambda^{(i)}(T)} \quad \text{as a } \mathbb{C}\mathfrak{S}(i)\text{-module},$$

*where $\lambda^{(i)}(T)$ denotes the integer partition of size i containing the cells of T labeled by $1, 2\ldots, i$. We call $(e_T)_{T\in \mathrm{ST}(\lambda)}$ the **Gelfand–Tsetlin basis** of S^λ. These Gelfand–Tsetlin bases have the following induction property: for every $i < n$, the family of vectors $(e_{T'})_{T'}$ with T' running over standard tableaux with the same $[\![i,n]\!]$-stem as e_T forms a Gelfand–Tsetlin basis of $\mathbb{C}\mathfrak{S}(i)(e_T)$.*

Example. We know from Theorem 3.30 that each Specht module S^λ can be described as a space of polynomials, on which $\mathfrak{S}(n)$ acts by permutation of the variables. This gives another basis $(\Delta_T)_{T\in S^\lambda}$ of S^λ. Let us compute the expansion of the vectors of the new Gelfand–Tsetlin basis in the basis $(\Delta_T)_{T\in S^\lambda}$, when $\lambda = (2,1)$. The two standard tableaux of shape $(2,1)$ are

$$T = \begin{array}{|c|c}\hline 3 & \\ \hline 1 & \multicolumn{1}{c|}{2} \\ \hline\end{array} \quad \text{and} \quad U = \begin{array}{|c|c}\hline 2 & \\ \hline 1 & \multicolumn{1}{c|}{3} \\ \hline\end{array}$$

and the two corresponding polynomials are $\Delta_T = x_1 - x_3$ and $\Delta_U = x_1 - x_2$. Viewed as a space of polynomials, $S^{(2,1)} = \{ax_1 + bx_2 + cx_3 \,|\, a + b + c = 0\}$. The trivial $\mathfrak{S}(2)$-subrepresentation of $S^{(2,1)}$ is $S^{(2)} = \{ax_1 + ax_2 - 2ax_3,\ a \in \mathbb{C}\}$, whereas the $\mathfrak{S}(2)$-subrepresentation of type $(1,1)$ is $S^{(1,1)} = \{ax_1 - ax_2,\ a \in \mathbb{C}\}$. From this we deduce that the Gelfand–Tsetlin basis of $S^{(2,1)}$ is

$$\begin{aligned} e_T &= x_1 + x_2 - 2x_3 = 2\Delta_T - \Delta_U; \\ e_U &= x_1 - x_2 = \Delta_U. \end{aligned}$$

We now introduce a particular subalgebra of the group algebra $\mathbb{C}\mathfrak{S}(n)$, which is closely related to the Gelfand–Tsetlin bases of the irreducible Specht modules S^λ:

Definition 8.3. *In the group algebra $\mathbb{C}\mathfrak{S}(n)$, the **Gelfand–Tsetlin algebra** of order n is the subalgebra $\mathrm{GZ}(n)$ which is generated by the elements of the centers $\mathrm{Z}(1), \mathrm{Z}(2), \ldots, \mathrm{Z}(n)$:*

$$\mathrm{GZ}(n) = \langle \mathrm{Z}(1), \mathrm{Z}(2), \ldots, \mathrm{Z}(n)\rangle.$$

Theorem 8.4. *The Gelfand–Tsetlin algebra consists in the elements $x \in \mathbb{C}\mathfrak{S}(n)$ such that, for any vector e_T of a Gelfand–Tsetlin basis of an irreducible module S^λ,*

$$x \cdot e_T = c_T(x)\, e_T,$$

where $c_T(x)$ is some scalar. Thus, the elements of $\mathrm{GZ}(n)$ *are those that act diagonally on the Gelfand–Tsetlin bases. Moreover,* $\mathrm{GZ}(n)$ *is a maximal commutative subalgebra of* $\mathbb{C}\mathfrak{S}(n)$.

Proof. If $x \in \mathrm{Z}(i)$ and $y \in \mathrm{Z}(j)$, then assuming for instance $i \leq j$, we have $xy = yx$ since $x \in \mathbb{C}\mathfrak{S}(j)$ and $y \in \mathrm{Z}(j)$. As an immediate consequence, $\mathrm{GZ}(n)$ is a commutative algebra, and moreover, any element of $x \in \mathrm{GZ}(n)$ writes as a product

$$x = x_1 x_2 \cdots x_n \quad \text{with } x_i \in \mathrm{Z}(i).$$

Denote A the set of elements $a \in \mathbb{C}\mathfrak{S}(n)$ that act diagonally on the Gelfand–Tsetlin bases $(e_T)_{T\in \mathrm{ST}(\lambda)}$ of the Specht modules. It is obviously a subalgebra of $\mathbb{C}\mathfrak{S}(n)$. We are going to show by double inclusion that $A = \mathrm{GZ}(n)$. If $x_i \in \mathrm{Z}(i)$, then for any λ and any $T \in \mathrm{ST}(\lambda)$, e_T belongs to $(\mathbb{C}\mathfrak{S}(i))(e_T) = S^{\lambda^{(i)}(T)}$, and on the other hand, the center of the group algebra $\mathrm{Z}(i)$ acts by multiplication by the Fourier transform

$$\chi^\mu(x_i) = \sum_{\sigma \in \mathfrak{S}(i)} [\sigma](x_i)\, \chi^\mu(\sigma),$$

cf. the paragraph in Section 1.3 on centers of group algebras. So,

$$x_i \cdot e_T = \chi^{\lambda^{(i)}(T)}(x_i)\, e_T$$

which shows that $\mathrm{Z}(i)$ is included in A. Since A is a subalgebra, we conclude that $\mathrm{GZ}(n) \subset A$.

Conversely, consider a standard tableau T, associated to an increasing sequence of partitions $\lambda^{(0)} \nearrow \lambda^{(1)} \nearrow \cdots \nearrow \lambda^{(n)}$. For any $i \in [\![1,n]\!]$, there exists an element $x_i \in \mathrm{Z}(i)$, whose image by the Fourier transform $\mathbb{C}\mathfrak{S}(i) \to \widehat{\mathbb{C}\mathfrak{S}(i)} = \bigoplus_{\mu \in \mathfrak{Y}(i)} \mathrm{End}(S^\mu)$ is the identity $\mathrm{id}_{S^{\lambda^{(i)}}}$. Then, if $x_T = x_1 x_2 \cdots x_n$, we have

$$x_T \cdot e_T = e_T \qquad ; \qquad x_T \cdot e_{U \neq T} = 0.$$

As a consequence, the elements x_T form a linear basis of A. However, they all belong to $\mathrm{GZ}(n)$, so $\mathrm{GZ}(n) = A$. Finally, this description proves that $\mathrm{GZ}(n)$ is a maximal commutative subalgebra of $\mathbb{C}\mathfrak{S}(n)$. Indeed, it is well known that the space of diagonal matrices $\mathbb{C}^d$ is a maximal commutative subalgebra of $\mathrm{M}(d,\mathbb{C})$ for any d. Therefore, the direct sum of the spaces $\mathbb{C}^{\dim S^\lambda}$ is a maximal commutative subalgebra of $\bigoplus_{\lambda \in \mathfrak{Y}(n)} \mathrm{End}(S^\lambda) = \widehat{\mathbb{C}\mathfrak{S}(n)}$, and it is equal to $\bigoplus_{\lambda, T \in \mathrm{ST}(\lambda)} \mathbb{C}x_T = A$. □

Remark. The proof of Theorem 8.4 adapts readily to any tower of groups

$$\{1\} = G_0 \subset G_1 \subset G_2 \subset \cdots \subset G_n \subset \cdots$$

such that each pair (G_{n-1}, G_n) is a **strong Gelfand pair**, that is to say that the restriction of any irreducible representation of G_n to G_{n-1} is multiplicity free.

▷ *The Young–Jucys–Muphy description of* GZ(n).

We now want to find an adequate generating family of the Gelfand–Tsetlin algebra GZ(n). The solution to this problem is due to Young, Jucys and Murphy.

Definition 8.5. *For any $i \geq 1$, the* **Jucys–Murphy element** J_i *is defined by*

$$J_i = \sum_{h<i} (h,i) = (1,i) + (2,i) + \cdots + (i-1,i) \in \mathbb{C}\mathfrak{S}(i).$$

By convention, $J_1 = 0$.

Theorem 8.6. *For any n,* $\mathrm{GZ}(n) = \mathbb{C}[J_1, J_2, \ldots, J_n]$.

To prove Theorem 8.6, we introduce the **centralizer** of $\mathbb{C}\mathfrak{S}(n-1)$ in $\mathbb{C}\mathfrak{S}(n)$, which is the subalgebra $\mathrm{Z}(n-1,n) \subset \mathbb{C}\mathfrak{S}(n)$ that consists in elements $x \in \mathbb{C}\mathfrak{S}(n)$ such that $xy = yx$ for all $y \in \mathbb{C}\mathfrak{S}(n-1)$. An important result is:

Proposition 8.7. *The centralizer* $\mathrm{Z}(n-1,n)$ *is a commutative subalgebra.*

This commutativity is closely related to the theory of Gelfand pairs, and we propose here a proof which sheds light on this connection.

Lemma 8.8. *Let $H \subset G$ be two finite groups. We consider the representation of $G \times H$ on $\mathbb{C}G$ defined by $(g,h) \cdot g' = g g' h^{-1}$. For any modules V and W over G and H, one has:*

$$\mathrm{Hom}_{G\times H}(V \otimes W, \mathbb{C}G) = \mathrm{Hom}_H(W, \mathrm{Res}_H^G(V')),$$

where V' is the adjoint representation of the G-representation V.

Proof. Let $\phi : V \otimes W \to \mathbb{C}G$ be a morphism of $G \times H$-representations. This means that for any $v \in V$, $w \in W$, $g \in G$ and $h \in H$, we have:

$$\phi(g \cdot v \otimes h \cdot w) = g\phi(v \otimes w)h^{-1}.$$

We want to define a corresponding map $\widetilde{\phi} : W \to \mathrm{Hom}(V, \mathbb{C})$. Since $\phi(v \otimes w)$ belongs to $\mathbb{C}G$, we can set:

$$\big(\widetilde{\phi}(w)\big)(v) = [e_G](\phi(v \otimes w)),$$

where e_G is the neutral element of the group, and $[e_G](x)$ is the coefficient of e_G in an element $x \in \mathbb{C}G$. Then, if $h \in H$, we have

$$\begin{aligned}\big(\widetilde{\phi}(h \cdot w)\big)(v) &= [e_G](\phi(v \otimes w)h^{-1}) = [h](\phi(v \otimes w)) \\ &= [e_G](\phi(h^{-1} \cdot v \otimes w)) = \big(\widetilde{\phi}(w)\big)(h^{-1} \cdot v) = \big(h \cdot \widetilde{\phi}(w)\big)(v)\end{aligned}$$

so $\widetilde{\phi}$ indeed belongs to $\mathrm{Hom}_H(W, \mathrm{Res}^G_H(V'))$. Conversely, given such a map $\widetilde{\phi}$, by setting

$$\phi(v \otimes w) = \sum_{g \in G} \big(\widetilde{\phi}(w)\big)(g^{-1} \cdot v)\, g,$$

we reconstruct $\phi \in \mathrm{Hom}_{G \times H}(V \otimes W, \mathbb{C}G)$. □

Lemma 8.9. *In the same situation, we have the isomorphism of $\mathbb{C}$-algebras:*

$$\mathrm{Hom}_{G \times H}(\mathbb{C}G, \mathbb{C}G)^{\mathrm{opp}} = \mathrm{Z}(H, G),$$

where the second part denotes the subalgebra of $\mathbb{C}G$ that consists in elements that commute with H.

Proof. Let ϕ be an endomorphism of the $\mathbb{C}[G \times H]$-module $\mathbb{C}G$. We associate to it an element of $\mathbb{C}[G]$:

$$\Phi = \phi(e_G).$$

If $\phi_{k,l}$ is the coefficient of k in $\phi(l)$, then $\Phi = \sum_g \phi_{g,e_G}\, g$. Notice that since ϕ is a morphism of $G \times H$ representations, we have $\phi_{k,glh^{-1}} = \phi_{g^{-1}kh,l}$ for any $g, k, l \in G$ and $h \in H$. If ψ is another element of $\mathrm{Hom}_{G \times H}(\mathbb{C}G, \mathbb{C}G)$, then we have $(\psi \circ \phi)_{k,l} = \sum_{r \in G} \psi_{k,r}\, \phi_{r,l}$. Consequently,

$$\begin{aligned}\Phi \times \Psi &= \sum_{k,l \in G} \phi_{k,e_G}\, \psi_{l,e_G}\, kl = \sum_{g \in G} \left(\sum_{k \in G} \psi_{k^{-1}g,e_G}\, \phi_{k,e_G} \right) g \\ &= \sum_{g \in G} \left(\sum_{k \in G} \psi_{g,k}\, \phi_{k,e_G} \right) g = \sum_{g \in G} (\psi \circ \phi)_{g,e_G}\, g,\end{aligned}$$

so the map $\phi \mapsto \Phi$ is a morphism of algebras from $\mathrm{End}_{G \times H}(\mathbb{C}G)^{\mathrm{opp}}$ to $\mathbb{C}[G]$. Its image commutes with H, as

$$h\Phi = h\,\phi(e_G) = \phi(h) = \sum_{g \in G} \phi_{g,h}\, g = \left(\sum_{g \in G} \phi_{gh^{-1},e_G}\, gh^{-1} \right) h = \Phi h.$$

Conversely, from $\Phi \in \mathrm{Z}(H, G)$, we reconstruct ϕ by setting $\phi(g) = g\,\phi(e_G) = g\,\Phi$. □

Proof of Proposition 8.7. If $G = \mathfrak{S}(n)$ and $H = \mathfrak{S}(n-1)$, then for any irreducible representations V and W of these groups, we have by Lemma 8.8

$$\dim \mathrm{Hom}_{G \times H}(V \otimes W, \mathbb{C}G) = \dim \mathrm{Hom}_H(W, \mathrm{Res}^G_H(V')) = 1 \text{ or } 0$$

since the irreducible representations of $\mathfrak{S}(n)$ are multiplicity free when restricted to $\mathfrak{S}(n-1)$. Therefore, as a $\mathfrak{S}(n) \times \mathfrak{S}(n-1)$-module, $\mathbb{C}\mathfrak{S}(n)$ is a multiplicity free sum of irreducible representations. By Schur's lemma, this implies that $\mathrm{Hom}_{G \times H}(\mathbb{C}G, \mathbb{C}G)$ is a direct sum of fields $\mathbb{C}$, hence is commutative. Therefore, the commutant $\mathrm{Z}(H, G) = \mathrm{Z}(n-1, n)$ is commutative. □

The proof of the commutativity of $\mathrm{Z}(n-1,n)$ provided above uses the representation theory of the symmetric groups. One can give a more direct and combinatorial proof, by means of the following description of the centralizer:

Proposition 8.10. *The centralizer* $\mathrm{Z}(n-1,n)$ *is generated as an algebra by* $\mathrm{Z}(n-1)$ *and* J_n.

Proof. It is clear that $\mathrm{Z}(n-1)$ is included in $\mathrm{Z}(n-1,n)$. On the other hand,

$$J_n = C_{(2)\uparrow n} - C_{(2)\uparrow n-1},$$

where $C_{(2)\uparrow n} = \sum_{1\leq i<j\leq n}(i,j)$ is the sum of all transpositions in $\mathfrak{S}(n)$. Since $C_{(2)\uparrow n} \in \mathrm{Z}(n)$ and $C_{(2)\uparrow n-1} \in \mathrm{Z}(n-1)$, we conclude that J_n commutes with every element of $\mathbb{C}\mathfrak{S}(n-1)$. So, we have the inclusion $\langle \mathrm{Z}(n-1), J_n\rangle \subset \mathrm{Z}(n-1,n)$.

To show the converse inclusion, we first describe in a more precise way the elements of the centralizer. They are linear combinations of $\mathfrak{S}(n-1)$-conjugacy classes of permutations in $\mathfrak{S}(n)$, and if σ and ρ are two permutations of size n that are conjugated by an element of $\mathfrak{S}(n-1)$, then they have the same cycle type, and moreover, n is in a cycle of same length in σ and in ρ. Therefore, one can label the $\mathfrak{S}(n-1)$-conjugacy classes of $\mathfrak{S}(n)$ by integer partitions

$$\lambda \in \mathfrak{Y}_{\leq n-1} = \bigsqcup_{k=0}^{n-1} \mathfrak{Y}(k),$$

where $\lambda \in \mathfrak{Y}(k)$ labels the permutations in $\mathfrak{S}(n)$ such that n belongs to a cycle of length $n-k$, and the other cycles have lengths given by the parts of λ. We denote $\widetilde{C}_\lambda$ the $\mathfrak{S}(n-1)$-conjugacy class corresponding to $\lambda \in \mathfrak{Y}_{\leq n-1}$. In particular, $\widetilde{C}_{(1)^{n-2}} = J_n$. On the other hand, if $\lambda \in \mathfrak{Y}(n-1)$, then $\widetilde{C}_\lambda$ is just the conjugacy class C_λ in $\mathbb{C}\mathfrak{S}(n-1)$. Denote $\mathrm{Z}(n-1,n)_h$ the vector subspace of $\mathrm{Z}(n-1,n)$ that consists in classes of permutations with at least $n-h$ fixed points. We have the sequence of inclusions

$$\mathbb{C} = \mathrm{Z}(n-1,n)_0 \subset \mathrm{Z}(n-1,n)_1 \subset \mathrm{Z}(n-1,n)_2 \subset \cdots \subset \mathrm{Z}(n-1,n)_n = \mathrm{Z}(n-1,n),$$

and $\mathrm{Z}(n-1,n)_h$ is spanned linearly by the classes $\widetilde{C}_\lambda$ with $m_1(\lambda)+1_{|\lambda|=n-1} \geq n-h$. In particular, for $h \geq 2$, $\widetilde{C}_{(1)^{n-h}}$ belongs to $\mathrm{Z}(n-1,n)_h$, but not to $\mathrm{Z}(n-1,n)_{h-1}$. The crucial remark is now:

$$(J_n)^k = \widetilde{C}_{(1)^{n-1-k}} + \text{term in } \mathrm{Z}(n-1,n)_k.$$

Indeed, we have the star-factorization of the cycle

$$(n,a_1,a_2,\ldots,a_k) = (a_k,n)(a_{k-1},n)\cdots(a_2,n)(a_1,n),$$

which is proven at once by induction on k. The sum of these terms over arrangements $(a_1,\ldots,a_k)$ in $\mathfrak{A}(n-1,k)$ is equal to $\widetilde{C}_{(1)^{n-1-k}}$, and in the product

$$(J_n)^k = \sum_{a_1,\ldots,a_k\in[\![1,n-1]\!]} (a_k,n)(a_{k-1},n)\cdots(a_2,n)(a_1,n),$$

the other terms move less than k terms and correspond to an element of the subspace $\mathrm{Z}(n-1,n)_k$.

More generally, if $h \geq 2$ and $\mu = (\mu_1, \ldots, \mu_r)$ is a partition without parts of size 1 and such that $|\mu| \leq h-1$, then $\widetilde{C}_{\mu \sqcup 1^{n-h}}$ is in $\mathrm{Z}(n-1,n)_h$ but not in $\mathrm{Z}(n-1,n)_{h-1}$, and

$$\left(\prod_{i=1}^{r} \widetilde{C}_{\mu_i \sqcup 1^{n-1-\mu_i}}\right) (J_n)^{h-|\mu|-1} = c(\mu)\, \widetilde{C}_{\mu \sqcup 1^{n-h}} + \text{term in } \mathrm{Z}(n-1,n)_{h-1},$$

the product involved in the left-hand side being in $\mathrm{Z}(n-1)$, and the coefficient $c(\mu)$ being positive. To prove this result, it is convenient to place oneself in the algebra of partial permutations $\mathbb{C}\mathfrak{P}\mathfrak{S}(n)$. For $k \geq 2$, set

$$\Sigma_{k,n-1} = \sum_{i \in \mathfrak{A}(n-1,k)} C(k,i) = \phi_{n-1}(\Sigma_k);$$

it is sent by $\pi_n : \mathbb{C}\mathfrak{P}\mathfrak{S}(n) \to \mathbb{C}\mathfrak{S}(n)$ to a multiple of the $\mathfrak{S}(n-1)$-conjugacy class $\widetilde{C}_{k1^{n-1-k}} = C_{k1^{n-1-k}}$. We also denote

$$X_n = \sum_{i=1}^{n-1} \big((i,n), \{i,n\}\big);$$

its image by the map $\pi_n : \mathbb{C}\mathfrak{P}\mathfrak{S}(n) \to \mathbb{C}\mathfrak{S}(n)$ is the Jucys–Murphy element J_n, and the previous formula for $(J_n)^k$ comes by projection of the identity

$$(X_n)^k = \sum_{a \in \mathfrak{A}(n-1,k)} \big((n,a_1,\ldots,a_k), \{n,a_1,\ldots,a_k\}\big) + \text{term of degree less than } k$$

in $\mathbb{C}\mathfrak{P}\mathfrak{S}(n)$. Indeed, a partial permutation of degree k in $\mathfrak{P}\mathfrak{S}(n)$ has at least $n-k$ fixed points. Now, by using the second part of Theorem 7.4, we obtain:

$$\begin{aligned}
&\left(\prod_{i=1}^{r} \Sigma_{\mu_i,n-1}\right) (X_n)^{h-|\mu|-1} \\
&= \big(\Sigma_{\mu,n-1}\big) \sum_{a \in \mathfrak{A}(n-1,h-|\mu|-1)} \big((n,a_1,\ldots,a_{h-|\mu|-1}), \{n,a_1,\ldots,a_{h-|\mu|-1}\}\big) \\
&\qquad + \text{term of degree less than } h-1.
\end{aligned}$$

In the product on the right-hand side, the only terms of degree h come from products of partial permutations with disjoint supports. Therefore, these terms of highest degree are

$$\sum_{a \in \mathfrak{A}(n-1,h-1)} C(\mu,(a_1\ldots,a_{|\mu|}))\, C(h-|\mu|,(a_{|\mu|+1},\ldots,a_{h-1},n)),$$

and they are sent by π_n to a multiple of the class $\widetilde{C}_{\mu \sqcup 1^{n-h}}$, whereas the other terms are sent to elements of $\mathrm{Z}(n-1,n)_{h-1}$. This ends the proof of the formula. We can now do an induction on h to prove that $\mathrm{Z}(n-1,n)_h \subset \langle \mathrm{Z}(n-1), J_n\rangle$. Suppose the result to be true up to order $h-1$. Then, the classes $\widetilde{C}_{\mu\sqcup 1^{n-|\mu|-1}}$ form a basis of a complement subspace of $\mathrm{Z}(n-1,n)_h$ in $\mathrm{Z}(n-1,n)_{h-1}$. The previous formulas and the induction hypothesis show precisely that they belong to $\langle \mathrm{Z}(n-1), J_n\rangle$; therefore, $\mathrm{Z}(n-1,n)_h \subset \langle \mathrm{Z}(n-1), J_n\rangle$. In particular, with $h=n$, we conclude that $\mathrm{Z}(n-1,n) = \mathrm{Z}(n-1,n)_n \subset \langle \mathrm{Z}(n-1), J_n\rangle$. This ends the proof of the proposition, and notice that it implies immediately the commutativity of $\mathrm{Z}(n-1,n)$. □

We can finally prove the theorem stated at the beginning of the paragraph:

Proof of Theorem 8.6. We reason by induction on n, the first cases $n=1,2$ being trivial. Suppose the result to be true up to order $n-1$. Then,

$$\begin{aligned}\mathrm{GZ}(n) &= \langle \mathrm{GZ}(n-1), \mathrm{Z}(n)\rangle \\ &\subset \langle \mathrm{GZ}(n-1), \mathrm{Z}(n-1,n)\rangle = \langle \mathrm{GZ}(n-1), \mathrm{Z}(n-1), J_n\rangle = \langle \mathrm{GZ}(n-1), J_n\rangle\end{aligned}$$

since $\mathrm{Z}(n) \subset \mathrm{Z}(n-1,n)$. The inclusion is in fact an equality, because $J_n = C_{(2)\uparrow n} - C_{(2)\uparrow n-1}$, and therefore

$$\langle \mathrm{GZ}(n-1), J_n\rangle \subset \langle \mathrm{GZ}(n-1), \mathrm{Z}(n-1), \mathrm{Z}(n)\rangle = \langle \mathrm{GZ}(n-1), \mathrm{Z}(n)\rangle.$$

Since $\mathrm{GZ}(n-1) = \langle J_1, \ldots, J_{n-1}\rangle$ by the induction hypothesis, we conclude that $\mathrm{GZ}(n) = \langle J_1, \ldots, J_n\rangle$. □

Example. If $n=3$, let us compute the representation matrices of the Jucys–Murphy element. We identify the (non-commutative) Fourier transform of an element $x \in \mathbb{C}\mathfrak{S}(3)$ with a block-diagonal matrix

$$\begin{pmatrix} a & & & \\ & b & c & \\ & d & e & \\ & & & f \end{pmatrix}$$

where the first block corresponds to the action on $S^{(3)}$, the second block to the action on $S^{(2,1)}$ endowed with its Gelfand–Tsetlin basis, and the third block to the action on $S^{(1,1,1)}$. The actions on the two one-dimensional spaces are the trivial representation and the signature representation, hence readily computed. On the other hand, we saw that the Gelfand–Tsetlin basis of $S^{(2,1)}$ can be given by the two polynomials $e_T = x_1 + x_2 - 2x_3$ and $e_U = x_1 - x_2$, on which $\mathfrak{S}(3)$ acts by transposition. We conclude that $\widehat{(1,2)} = \mathrm{diag}(1,1,-1,-1)$ and

$$\widehat{(1,3)} = \begin{pmatrix} 1 & & & \\ & -\frac{1}{2} & -\frac{1}{2} & \\ & -\frac{3}{2} & \frac{1}{2} & \\ & & & -1 \end{pmatrix} \quad ; \quad \widehat{(2,3)} = \begin{pmatrix} 1 & & & \\ & -\frac{1}{2} & \frac{1}{2} & \\ & \frac{3}{2} & \frac{1}{2} & \\ & & & -1 \end{pmatrix}.$$

Therefore, $\widehat{J_2} = \mathrm{diag}(1,1,-1,-1)$ and $\widehat{J_3} = \mathrm{diag}(2,-1,0,-2)$, which leads to:

$$e_1 = \mathrm{diag}(1,0,0,0) = \frac{2\,(J_3)^2 + 3\,J_3 - J_3J_2}{12},$$

and to similar formulas for the other elementary diagonal matrices

$$e_2 = \mathrm{diag}(0,1,0,0),\ e_3 = \mathrm{diag}(0,0,1,0),\ e_4 = \mathrm{diag}(0,0,0,1).$$

Thus, we indeed have $\mathrm{GZ}(3) = \mathbb{C}[J_1, J_2, J_3] = \mathbb{C}[J_2, J_3]$. Notice that the writing of each elementary diagonal matrix in terms of the J_k's is not at all unique (we will see in the next paragraph that there exists a representation of e_1 as a *symmetric* polynomial in J_2 and J_3).

The idea of using partial permutation analogues X_n of the Jucys–Murphy elements J_n will be used again several times in this chapter; they provide easy proofs of certain combinatorial identities in $\mathrm{GZ}(n)$. We call the X_n's the **generic Jucys–Murphy elements.**

▷ *The Young–Jucys–Muphy description of* $\mathrm{Z}(n)$.

The Jucys–Murphy elements also allow one to describe the center $\mathrm{Z}(n)$ of the group algebra $\mathbb{C}\mathfrak{S}(n)$. Thus, the analogue of Theorem 8.6 for $\mathrm{Z}(n)$ is:

Theorem 8.11. *For any n,* $\mathrm{Z}(n)$ *corresponds to the space of symmetric polynomials in the Jucys–Murphy elements:*

$$\mathrm{Z}(n) = \mathrm{Sym}^{(n)}[J_1, \ldots, J_n].$$

Notice that we can indeed specialize symmetric functions in the alphabet of Jucys–Murphy elements, since they belong to a commutative algebra. The inclusion $\supset$ is the easy part of the theorem, and it comes from:

Proposition 8.12 (Jucys). *For any* $k \in [\![1, n]\!]$,

$$e_k(J_1, \ldots, J_n) = \sum_{\lambda \in \mathfrak{Y}(k) \,|\, |\lambda| + \ell(\lambda) \leq n} C_{\lambda \to n},$$

where $\lambda \to n = (\lambda_1 + 1, \lambda_2 + 1, \ldots, \lambda_{\ell(\lambda)} + 1, 1, \ldots, 1)$ *is the unique partition of size n obtained by adding 1 to the* $n - |\lambda|$ *first parts of* λ *(counting if needed parts of size 0).*

Example. We have indeed

$$e_1(J_1, \ldots, J_n) = J_1 + J_2 + \cdots + J_n = C_{(2)\uparrow n} = C_{(1)\to n}$$

and

$$e_2(J_1, \ldots, J_n) = \sum_{\substack{j<l \\ i<j,k<l}} (i,j)(k,l) = C_{(2,2)\uparrow n} + C_{(3)\uparrow n} = C_{(1,1)\to n} + C_{(2)\to n}.$$

Proof. Notice that the sum of conjugacy classes in the statement of the proposition is also $\sum_{\mu \in \mathfrak{Y}(n) \mid \ell(\mu)=n-k} C_\mu$. Therefore, by taking a generating series in $\mathbb{C}[z][\mathfrak{S}(n)]$, the lemma is equivalent to the identity:

$$\prod_{i=1}^{n}(z+J_i) = \sum_{k=0}^{n} z^{n-k}\, e_k(J_1,\ldots,J_n) = \sum_{\sigma \in \mathfrak{S}(n)} z^{\ell(t(\sigma))}\,\sigma = \sum_{\sigma\in\mathfrak{S}(n)} z^{\text{number of cycles of }\sigma}\,\sigma.$$

This identity is related to the following combinatorial fact:

- every permutation $\sigma \in \mathfrak{S}(n)$ factorizes uniquely as
$$\sigma = (i_1,1)(i_2,2)\cdots(i_n,n)$$
where each i_k belongs to $[\![1,k]\!]$.
- in this factorization, one creates a new cycle each time $i_k = k$. Therefore, $\ell(t(\sigma)) = \operatorname{card}\{k \in [\![1,n]\!] \mid i_k = k\}$.

The proof of this factorization is easy by recurrence on n, and on the other hand, it implies immediately the identity of generating series. □

In particular, any elementary function of Jucys–Murphy elements is a sum of conjugacy classes, hence belongs to $\mathrm{Z}(n)$. Since the elementary functions $e_1,\ldots,e_n$ generate the algebra $\mathrm{Sym}^{(n)}$, it follows that $\mathrm{Sym}^{(n)}[J_1,\ldots,J_n] \subset \mathrm{Z}(n)$. To prove that this is an equality, we are going to produce a family of $\dim \mathrm{Z}(n) = \operatorname{card}\mathfrak{Y}(n)$ symmetric polynomials in the J_i's that are linearly independent in the group algebra. Specifically, if $\mu = (\mu_1,\ldots,\mu_r) \in \mathfrak{Y}(n)$, we set

$$J_\mu = m_{\mu_1-1,\ldots,\mu_r-1}(J_1,\ldots,J_n) = \sum_{\text{distinct monomials}} J_{i_1}^{\mu_1-1} J_{i_2}^{\mu_2-1}\cdots J_{i_r}^{\mu_r-1}.$$

It is a monomial function in the Jucys–Murphy elements, hence it belongs to the commutative algebra $\mathrm{Sym}^{(n)}[J_1,\ldots,J_n]$.

Proposition 8.13. *The family $(J_\mu)_{\mu\in\mathfrak{Y}(n)}$ is linearly independent, hence a basis of* $\mathrm{Z}(n)$.

Proof. It is again convenient to manipulate the elements $X_i \in \mathbb{C}\mathfrak{P}\mathfrak{S}(n)$ defined in the proof of Proposition 8.7, and whose images by $\pi_n : \mathbb{C}\mathfrak{P}\mathfrak{S}(n) \to \mathbb{C}\mathfrak{S}(n)$ are the Jucys–Murphy elements J_i. A simple computation shows that these X_i's commute with one another, hence one can consider symmetric polynomials of them without ambiguity. We set

$$X_{\mu,n} = m_{\mu_1-1,\ldots,\mu_r-1}(X_1,\ldots,X_n) = \sum_{\text{distinct monomials}} (X_{i_1})^{\mu_1-1}(X_{i_2})^{\mu_2-1}\cdots(X_{i_r})^{\mu_r-1},$$

with by convention $(X_i)^0 = (\mathrm{id}_\emptyset,\emptyset)$. Let us identify the part of $X_{\mu,n}$ of highest degree. We set $\widetilde{\mu} = (\mu_1,\ldots,\mu_s)$, where μ_s is the last part of μ strictly greater than

1; in other words, we have removed the parts of size 1, which do not contribute to the calculation of $X_{\mu,n}$ and J_μ. Recall then that $\deg(X_{i_j})^{\mu_j-1} = \mu_j$ for $\mu_j \geq 2$, and that the top-degree component of $(X_{i_j})^{\mu_j-1}$ is

$$\sum_{a\in\mathfrak{A}(i_j-1,\mu_j-1)} C(\mu_j,(i_j,a_1,\dots,a_{\mu_j-1})).$$

When taking products of partial permutations, one obtains a maximal degree when the supports of the partial permutations are disjoint. Therefore, the degree of $X_{\mu,n}$ is $n-m_1(\mu)=|\widetilde{\mu}|$, and its top degree component is

$$\sum_a C(\widetilde{\mu},a),$$

where the sum runs over certain arrangements $a \in \mathfrak{A}(n,n-m_1(\mu))$ (not all arrangements, because $X_{\mu,n}$ is the sum over *distinct* monomials). Actually, this restriction is accounted for by a factor $\frac{1}{z_{\widetilde{\mu}}}$, hence,

$$X_{\mu,n} = \frac{\Sigma_{\widetilde{\mu},n}}{z_{\widetilde{\mu}}} + \text{terms of lower degree.}$$

Applying π_n to this identity, we obtain

$$J_\mu = C_\mu + \text{terms with at least } (m_1(\mu)+1) \text{ fixed points.}$$

Thus, the terms with the least number of fixed points in the sum J_μ are the permutations of cycle type μ, and they all appear with multiplicity 1.

Consider then a total order $\leq$ on $\mathfrak{Y}(n)$, such that if $m_1(\mu)<m_1(\lambda)$, then $\mu<\lambda$. Denote $\mu^{(1)}<\mu^{(2)}<\mu^{(3)}<\cdots<\mu^{(\mathrm{card}\,\mathfrak{Y}(n))}$ the ordered list of the partitions of size n with respect to this order. The formula above ensures that the matrix of the family $(J_{\mu^{(1)}},\dots,J_{\mu^{|\mathfrak{Y}(n)|}})$ in the basis $(C_{\mu^{(1)}},\dots,C_{\mu^{(|\mathfrak{Y}(n)|)}})$ of $\mathrm{Z}(n)$ has the form

$$\begin{pmatrix} 1 & 0 & \cdots & 0 \\ * & 1 & \ddots & \vdots \\ \vdots & \ddots & \ddots & 0 \\ * & \cdots & * & 1 \end{pmatrix}.$$

It is invertible, so the family $(J_\mu)_{\mu\in\mathfrak{Y}(n)}$ is another basis of $\mathrm{Z}(n)$. This ends also the proof of Theorem 8.11. □

Example. Pursuing our computations in $\mathbb{C}\mathfrak{S}(3)$, Theorem 8.11 ensures that there exists a *symmetric* polynomial in J_2 and J_3 such that $f(J_2,J_3) = e_1 = \mathrm{diag}(1,0,0,0)$ is the projection on $\mathrm{End}(S^{(3)})$, the one-dimensional endomorphism space of the trivial representation. We find indeed:

$$e_1 = \frac{((J_2)^2+(J_3)^2)+3\,J_2J_3+2\,(J_2+J_3)+1}{18}.$$

8.2 Jucys–Murphy elements acting on the Gelfand–Tsetlin basis

By Theorem 8.4, the elements of $\mathrm{GZ}(n)$ act diagonally on the Gelfand–Tsetlin bases $(e_T)_{T\in \mathrm{ST}(\lambda)}$, and by Theorem 8.6, each Jucys–Murphy element J_i belongs to $\mathrm{GZ}(n)$. Therefore, there exists for any standard tableau T of size n a coefficient $a(J_i,T)$ such that

$$J_i \cdot e_T = a(J_i,T)\, e_T.$$

The objective of this section is to compute the coefficients $a(J_i,T)$, and to prove:

Theorem 8.14. *Recall that the* **content** $c(\square)$ *of a cell of a Young diagram* λ *is* $j-i$, *where* i *and* j *are respectively the number of the row and the number of the column containing the box* $\square$. *For any standard tableau* T *and any* $i \in [\![1,n]\!]$,

$$J_i \cdot e_T = c(i,T)\, e_T,$$

where $c(i,T) = c(\boxed{i})$ *is the content of the cell of label* i *in* T.

To prove this theorem, we shall introduce the two following objects:

- the set of **content vectors** $\mathrm{Cont}(n)$, which is the set of $\mathbb{Z}$-valued vectors of size n

$$\mathrm{Cont}(n) = \{(c(1,T),\ldots,c(n,T)),\ T \in \mathrm{ST}(\lambda),\ \lambda \in \mathfrak{Y}(n)\}.$$

- the set of **spectral vectors** $\mathrm{Spec}(n)$, which is the set of $\mathbb{C}$-valued vectors of size n

$$\mathrm{Spec}(n) = \{(a(J_1,T),\ldots,a(J_n,T)),\ T \in \mathrm{ST}(\lambda),\ \lambda \in \mathfrak{Y}(n)\},$$

  where $a(J_i,T)$ denotes the eigenvalue corresponding to the action of J_i on e_T.

We shall prove that there is a bijection between these two sets (Theorem 8.22), which is compatible with certain equivalence relations on the sets $\mathrm{Cont}(n)$ and $\mathrm{Spec}(n)$. This method, due to Okounkov and Vershik, actually yields an alternative approach to the representation theory of the symmetric groups, though it is easier to understand it if one knows beforehand the branching rules, the dimension formulas and the combinatorics of standard tableaux.

▷ *The set of contents of standard tableaux.*

If T is a standard tableau, denote $c(T)$ its vector of contents. For instance, the vector of contents of

<table>
<tr><td>9</td><td></td><td></td><td></td></tr>
<tr><td>7</td><td></td><td></td><td></td></tr>
<tr><td>2</td><td>5</td><td>6</td><td></td></tr>
<tr><td>1</td><td>3</td><td>4</td><td>8</td></tr>
</table>

is $(0,-1,1,2,0,1,-2,3,-3)$. We are looking for a combinatorial description of all the possible vectors of contents that can be obtained. First, we have:

Proposition 8.15. *The map $T \mapsto c(T)$ is injective.*

Proof. We reconstruct the tableau T from its vector of contents $c = c(T)$ by labeling for every $k \in \mathbb{Z}$ the cells of the diagonal of content k by the indices i such that $c_i = k$. For instance, in the previous example, there are two contents equal to 0, of indices 1 and 5, so we place the labels 1 and 5 on the two first cells of the diagonal of content $k = 0$. By doing so for every k, we indeed reconstruct the standard tableau. □

We now characterize the image of $T \mapsto c(T)$ in $\mathbb{Z}^n$:

Proposition 8.16. *If T is a standard tableau, then its vector of content $c(T) = (c_1, \ldots, c_n)$ satisfies:*

(C1) $c_1 = 0$;

(C2) for any $i \geq 2$, $\{c_i + 1, c_i - 1\} \cap \{c_1, \ldots, c_{i-1}\}$ *is not empty;*

(C3) if $c_i = c_j$ *for* $i < j$, *then* $\{c_j + 1, c_j - 1\} \subset \{c_{i+1}, \ldots, c_{j-1}\}$.

Proof. The first condition is obvious, since the first box of a standard tableau is necessarily on the first row and first column, hence has content $1 - 1 = 0$. For the two other conditions, we can do an induction on n, and therefore it suffices to prove them for the last coordinate c_n. The last box of label n is added on the right of another cell, or on top of another cell, or on the right and on the top of two cells, as shown by Figure 8.1.

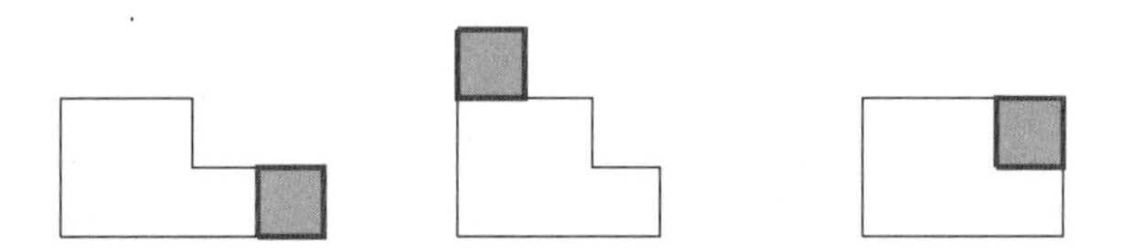

Figure 8.1
Possible positions of the last cell of a standard tableau.

In the first case, $c_n - 1$ belongs to the set of previous contents; in the second case, $c_n + 1$ belongs to the set of previous contents. Finally, in the third case, $c_n = c_i$ for the boxes with label i that are on the same diagonal as the new box of label n, and the two cells contiguous to this new cell have contents $c_n - 1$ and $c_n + 1$, which proves the third condition. □

Theorem 8.17. *The three conditions of Proposition 8.16 characterize the image* *of the map $T \in \bigsqcup_{\lambda \in \mathfrak{Y}(n)} \mathrm{ST}(\lambda) \mapsto c(T) \in \mathbb{Z}^n$. Thus, we have a bijection between standard tableaux of size n, and vectors of integers of size n that satisfy the three conditions (C1), (C2) and (C3).*

Proof. Suppose the result to be true up to order $n-1$, and consider a vector $(c_1, \ldots, c_n)$ which satisfies the three conditions. By induction, $(c_1, \ldots, c_{n-1})$ corresponds to a standard tableau T of size $n-1$.

- If $c_n = c_i$ for some $i \in [\![1, n-1]\!]$, take i as large as possible satisfying this identity. Then, by (C3), the box of label i has a box above it and a box on its right in T, and by construction it has no box after it on its diagonal. Therefore, one can add a cell labeled n on the same diagonal, and keep a standard tableau.
- If $c_n \neq c_i$ for all $i \in [\![1, n-1]\!]$, then (C2) guarantees that one can add the box of label n in a new column or a new row of the standard tableau, according to whether $c_n - 1 \in \{c_1, \ldots, c_{n-1}\}$ or $c_n + 1 \in \{c_1, \ldots, c_{n-1}\}$. □

So, $\mathrm{Cont}(n)$ consists in the sequences in $\mathbb{Z}^n$ that satisfy the three conditions of Proposition 8.16. Let us now give a criterion on such sequences that decide whether they come from standard tableaux with the same shape $\lambda \in \mathfrak{Y}(n)$. We define an equivalence relation on $\mathrm{Cont}(n)$ by:

$$c \sim_{\mathrm{Cont}(n)} d \iff \exists \sigma \in \mathfrak{S}(n),\ c \cdot \sigma = d,$$

where $\mathfrak{S}(n)$ acts on the right of content vectors by permutation of their coordinates. Notice that an arbitrary permutation of a content vector is not necessarily a content vector: for instance, one has $c_1 = 0$ if $c \in \mathrm{Cont}(n)$, which prohibits certain permutations. A permutation $\sigma \in \mathfrak{S}(n)$ will be said **admissible** for a content vector c if $c \cdot \sigma$ is also a content vector.

Proposition 8.18. *Two content vectors c and d are equivalent for $\sim_{\mathrm{Cont}(n)}$ if and only if they come from standard tableaux T and U with the same shape $\lambda \in \mathfrak{Y}(n)$. Moreover, the relation $\sim_{\mathrm{Cont}(n)}$ is the transitive closure of the relation*

$$c \sim d \iff \exists i \in [\![1, n-1]\!],\ s_i \text{ is admissible for } c \text{ and } c \cdot s_i = d$$

where the s_i's are the elementary transpositions $(i, i+1)$.

Proof. If T and U are standard tableaux with the same shape λ, then their vectors of contents have the same underlying multiset, because this multiset depends only on the boxes of λ. Therefore, there is a permutation $\sigma \in \mathfrak{S}(n)$ such that $c(T) \cdot \sigma = c(U)$. Conversely, if T and U do not have the same shape, then the multiset underlying the content vectors $c(T)$ and $c(U)$ are not the same: choosing k such that T and U do not have their k-th diagonals of same size, then there is not the same number of k's in $c(T)$ and in $c(U)$. Therefore, there can be no permutation σ such that $c(T) \cdot \sigma = c(U)$.

Let us now prove that if T and U have the same shape λ, then there is a sequence of admissible elementary transpositions s_i transforming $c(T)$ into $c(U)$. It suffices to do so with T arbitrary and $U = T_\lambda$ equal to a reference tableau, for instance, the canonical standard tableau of shape λ with $1, 2, \ldots, \lambda_1$ on the first row, $\lambda_1 + 1, \ldots, \lambda_2$ on the second row, etc. We reason by induction on n: thus, we suppose that if T' and U' are standard tableaux of size $n-1$ and with the same shape μ, then there exist admissible elementary transpositions $s_{i_1}, \ldots, s_{i_r} \in \mathfrak{S}(n-1)$ such that

$$c(T') \cdot s_{i_1} s_{i_2} \cdots s_{i_r} = c(U').$$

Since content vectors are in bijection with standard tableaux, this is equivalent to the identity of tableaux

$$s_{i_r} \cdots s_{i_2} s_{i_1} \cdot T' = U'$$

(beware that permutations act on the right of content vectors, but on the left of tableaux). Let T be a standard tableau of shape $\lambda \in \mathfrak{Y}(n)$. We denote $U = T_\lambda$, T' the standard tableau of size $n-1$ obtained by removing from T the cell labeled by n, μ the shape of T' and $U' = T_\mu$. By the induction hypothesis, there exists a sequence of admissible transpositions such that $s_{i_r} \cdots s_{i_2} s_{i_1} \cdot T' = U'$. These admissible elementary transpositions are again admissible when applied to T: one does not modify the n-th coordinate of the content vector, and the cell of label n which stays on the boundary of the tableau. Thus,

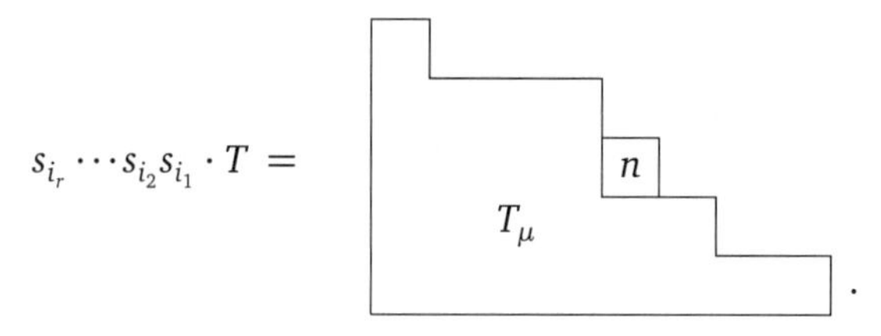

If n was in the top row of T, then we have obtained T_λ. Otherwise, $n-1$ is in the top row of T_μ, and therefore in the top row of $s_{i_r} \cdots s_{i_2} s_{i_1} \cdot T$. In this situation, s_{n-1} is admissible for $s_{i_r} \cdots s_{i_2} s_{i_1} \cdot T$, so

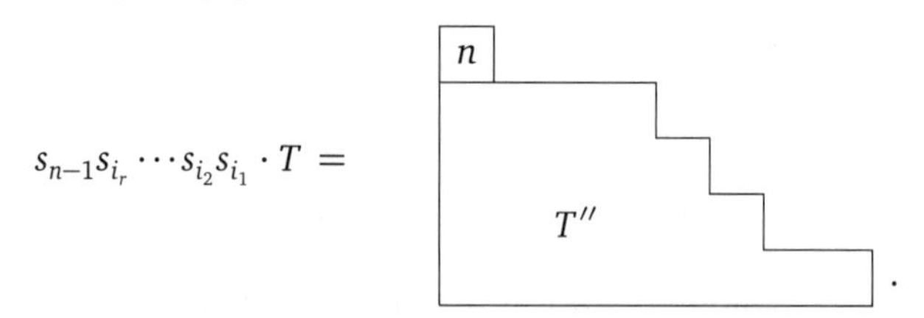

The standard tableau $s_{n-1} s_{i_r} \cdots s_{i_2} s_{i_1} \cdot T$ now has its cell of label n in the top row, so we are reduced to the previous case, and a new sequence of elementary and admissible transpositions yields $s_{j_q} \cdots s_{j_1} s_{n-1} s_{i_r} \cdots s_{i_2} s_{i_1} \cdot T = T_\lambda$. □

To complete this proposition, let us give a simple criterion for an elementary transposition s_i to be admissible with respect to a standard tableau T. One can exchange the cells of label i and $i+1$ in a standard tableau if and only if they are not in the same row or the same column. If they are in the same row or the same column, then they are necessarily contiguous, which implies that $c_i \pm 1 = c_{i+1}$; and the converse is true. So, s_i is admissible for a content vector $c \in \mathrm{Cont}(n)$ if and only if $c_{i+1} \notin \{c_i + 1, c_i - 1\}$.

▷ *Spectra of Jucys–Murphy elements and the degenerate affine Hecke algebra.*

If T is a standard tableau, we denote $a(T)$ the sequence of the eigenvalues of the Jucys–Murphy elements applied to the Gelfand–Tsetlin vector e_T:

$$a(T) = (a(J_1, T), \ldots, a(J_n, T)),$$

and we call it the **spectrum** of the tableau T. We denote $\mathrm{Spec}(n) = \{a(T),\ T \in \mathrm{ST}(\lambda),\ \lambda \in \mathfrak{Y}(n)\}$ the set of spectral vectors. Notice that the map $T \mapsto a(T)$ is injective: indeed, if one knows all the eigenvalues of the Jucys–Murphy elements on e_T, then by Theorem 8.6 one knows the action of $\mathrm{GZ}(n)$ on e_T, so in particular one can reconstruct the projector on $\mathbb{C}e_T$, and the vector e_T.

In a moment we shall prove that $\mathrm{Cont}(n) = \mathrm{Spec}(n)$, and that in fact $a(T) = c(T)$. The main ingredient in the proof of this result is the study of the action of the subalgebras of $\mathbb{C}\mathfrak{S}(n)$

$$H_i = \langle J_i, J_{i+1}, s_i \rangle$$

on the Gelfand–Tsetlin bases of the representations S^λ. The generating elements of the algebra H_i satisfy the relations

$$(s_i)^2 = 1 \quad ; \quad J_i J_{i+1} = J_{i+1} J_i \quad ; \quad s_i J_i + 1 = J_{i+1} s_i.$$

It can be proven that this is actually a presentation of H_i, whose structure as a $\mathbb{C}$-algebra does not depend on i. We call H_i the **degenerate affine Hecke algebra of size** 2. Hereafter, we shall simply use the relations, and we shall not need to know that this set of relations is complete (i.e., H_i is isomorphic to the quotient of the free algebra on symbols J_i, J_{i+1}, s_i by the ideal generated by these relations).

Proposition 8.19. *Let e_T be a Gelfand–Tsetlin vector. If $a(T) = (a_1, \ldots, a_n)$, then:*

1. *If $a_{i+1} = a_i \pm 1$, then $H_i\, e_T = \mathbb{C}e_T$ is one-dimensional, and the action of H_i on e_T is:*

$$J_i\, e_T = a_i\, e_T \quad ; \quad J_{i+1}\, e_T = a_{i+1}\, e_T \quad ; \quad s_i\, e_T = \pm e_T,$$

   where $\pm$ is a certain sign in the last formula.

2. *If $a_{i+1} \neq a_i \pm 1$, then $H_i\, e_T = \mathbb{C}e_T \oplus \mathbb{C}e_U$ has dimension 2, and U is the unique standard tableau such that $a(U) = a(T) \cdot s_i$. Moreover, the action of H_i on $\mathrm{Span}(e_T, e_U)$ is given by the matrices*

$$J_i \mapsto \begin{pmatrix} a_i & 0 \\ 0 & a_{i+1} \end{pmatrix} \quad ; \quad J_{i+1} \mapsto \begin{pmatrix} a_{i+1} & 0 \\ 0 & a_i \end{pmatrix};$$

$$s_i \mapsto \begin{pmatrix} \frac{1}{a_{i+1} - a_i} & 1 - \frac{1}{(a_{i+1} - a_i)^2} \\ 1 & \frac{1}{a_i - a_{i+1}} \end{pmatrix}.$$

In particular, in the second case, $a(T)\cdot s_i$ belongs again to Spec(n).

Lemma 8.20. *Fix $i \in [\![1,n-1]\!]$. For any standard tableau T corresponding to an increasing sequence of partitions $\emptyset = \lambda^{(0)} \nearrow \lambda^{(1)} \nearrow \cdots \nearrow \lambda^{(n)} = T$, $s_i\, e_T$ is a linear combination of vectors $e_{T'}$ with T' of the same shape as T, and corresponding to an increasing sequence of integer partitions*

$$\emptyset = \lambda^{(0)} \nearrow \lambda^{(1)} \nearrow \cdots \nearrow \mu^{(i)} \nearrow \cdots \nearrow \lambda^{(n)} = T$$

that can only differ from the original sequence at the i-th partition $\mu^{(i)}$.

Proof. Since $s_i \in \mathbb{C}\mathfrak{S}(i+1)$, $s_i\, e_T \in \mathbb{C}\mathfrak{S}(i+1)(e_T)$, which is equal to $S^{\lambda^{(i+1)}}$ as a representation of $\mathfrak{S}(i+1)$, and is linearly spanned by the vectors e'_T such that T' has the same $[\![i+1,n]\!]$-stem as T. Hence, if e'_T appears with a non-zero coefficient in $s_i\, e_T$, then T' is associated to a sequence of partitions $\mu^{(0)} \nearrow \mu^{(1)} \nearrow \cdots \nearrow \mu^{(n)}$ with

$$\mu^{(i+1)} = \lambda^{(i+1)} \quad ; \quad \mu^{(i+2)} = \lambda^{(i+2)} \quad ; \quad \mu^{(n)} = \lambda^{(n)}.$$

On the other hand, if $h < i$, then s_i commutes with $\mathbb{C}\mathfrak{S}(h)$, so the map

$$\begin{aligned} \mathbb{C}\mathfrak{S}(h)(e_T) &\to \mathbb{C}\mathfrak{S}(h)(s_i\, e_T) \\ x &\mapsto s_i\, x \end{aligned}$$

is an isomorphism of $\mathfrak{S}(h)$-representations. Therefore, $s_i\, e_T$ belongs to the part of the space $\mathbb{C}\mathfrak{S}(i+1)(e_T)$ that is isomorphic to $S^{\lambda^{(h)}}$, and $\mu^{(h)} = \lambda^{(h)}$. □

Proof of Proposition 8.19. The relations $J_i\, e_T = a_i\, e_T$ and $J_{i+1} = a_{i+1}\, e_T$ are always true by definition of the spectrum $a(T)$. If $s_i\, e_T$ is colinear to e_T, then since $(s_i)^2 = 1$, we have $s_i\, e_T = \pm e_T$ for a certain sign. Then,

$$\pm a_{i+1} e_T = J_{i+1} s_i\, e_T = (s_i J_i + 1)\, e_T = (\pm a_i + 1)\, e_T,$$

so $a_{i+1} = a_i \pm 1$. Conversely, if $a_{i+1} = a_i \pm 1$, then the sequence of identities above forces $s_i\, e_T = \pm e_T$, so the first case of the proposition is shown.

Suppose now that $a_{i+1} \neq a_i \pm 1$. By the previous discussion, $s_i\, e_T$ is not colinear to e_T, but then $H_i\, e_T$ is the two-dimensional space Span($e_T, s_i\, e_T$). Indeed, using the relations satisfied by J_i, J_{i+1} and s_i, one sees that this space is invariant by these three operators. Moreover, the representation matrices in the basis $(e_T, s_i\, e_T)$ of J_i, J_{i+1} and s_i are respectively:

$$\begin{pmatrix} a_i & -1 \\ 0 & a_{i+1} \end{pmatrix} \quad ; \quad \begin{pmatrix} a_{i+1} & 1 \\ 0 & a_i \end{pmatrix} \quad ; \quad \begin{pmatrix} 0 & 1 \\ 1 & 0 \end{pmatrix}.$$

Let us now remark that there exists at most one standard tableau U such that T

and U correspond to increasing sequences of integer partitions that differ only for their i-th coordinate. Indeed, if T and U correspond to

$$\lambda^{(0)} \nearrow \lambda^{(1)} \nearrow \cdots \nearrow \lambda^{(i)} \nearrow \cdots \nearrow \lambda^{(n)}$$
$$\text{and } \lambda^{(0)} \nearrow \lambda^{(1)} \nearrow \cdots \nearrow \mu^{(i)} \nearrow \cdots \nearrow \lambda^{(n)}$$

then one goes from $\lambda^{(i-1)}$ to $\lambda^{(i+1)}$ by adding two boxes $\square = (a,b)$ and $\square' = (c,d)$ to the Young diagram. In this situation, $\lambda^{(i)}$ and $\mu^{(i)}$ differ by the order in which one adds these two cells: $\lambda^{(i)}$ is obtained from $\lambda^{(i-1)}$ by adding $\square$, whereas $\mu^{(i)}$ is obtained by adding $\square'$ to $\lambda^{(i-1)}$. By the previous lemma, we conclude that if $a_{i+1} \neq a_i \pm 1$, then there exists a unique tableau $U \neq T$ of the same shape as T, and such that $\mathrm{Span}(e_T, s_i\, e_T) = \mathrm{Span}(e_T, e_U)$. Moreover,

$$c(U) = c(T) \cdot s_i.$$

Since the Jucys–Murphy elements are diagonalized by the Gelfand–Tsetlin basis, the previous representation matrices $\left(\begin{smallmatrix} a_i & -1 \\ 0 & a_{i+1} \end{smallmatrix}\right)$ and $\left(\begin{smallmatrix} a_{i+1} & 1 \\ 0 & a_i \end{smallmatrix}\right)$ have to be diagonalizable. This implies that $a_i \neq a_{i+1}$, because a matrix $\left(\begin{smallmatrix} a & 1 \\ 0 & a \end{smallmatrix}\right)$ is not diagonalizable (its unique eigenvalue is a, but with a one-dimensional eigenspace). One then calculates the basis of diagonalization:

$$e_T,\; e_U = s_i\, e_T - \frac{1}{a_{i+1} - a_i}\, e_T.$$

In this new basis, J_i and J_{i+1} have representation matrices $\left(\begin{smallmatrix} a_i & 0 \\ 0 & a_{i+1} \end{smallmatrix}\right)$ and $\left(\begin{smallmatrix} a_{i+1} & 0 \\ 0 & a_i \end{smallmatrix}\right)$, and s_i has representation matrix

$$\begin{pmatrix} \frac{1}{a_{i+1}-a_i} & 1 - \frac{1}{(a_{i+1}-a_i)^2} \\ 1 & \frac{1}{a_i - a_{i+1}} \end{pmatrix}.$$

Moreover, the spectrum $a(U)$ of the standard tableau U is now $a(T) \cdot s_i$, so $a(T) \cdot s_i$ belongs to $\mathrm{Spec}(n)$. □

Corollary 8.21. *If $a(T) \in \mathrm{Spec}(n)$, then $a_i \neq a_{i+1}$ for any $i \in [\![1, n-1]\!]$.*

Proof. If $a_{i+1} = a_i \pm 1$, this is obvious, and in the other case, we are in the second situation of Proposition 8.19, and we saw during its proof that $a_{i+1} - a_i$ does not vanish. □

▷ *The bijection between* $\mathrm{Spec}(n)$ *and* $\mathrm{Cont}(n)$.

Theorem 8.22. *For every $n \geq 1$, $\mathrm{Spec}(n) = \mathrm{Cont}(n)$.*

Proof. We reason by induction on $n \geq 1$. If $n = 1$, then the only element of $\mathrm{Cont}(1)$ is $\{(0)\}$, since the only standard tableau of size 1 is $\boxed{1}$. However, $J_1 = 0$, so we also have $\mathrm{Spec}(1) = \{(0)\}$, and the result is true. If $n = 2$, then $\mathrm{Cont}(2) = \{(0,1),(0,-1)\}$, since the two standard tableaux of size 2 are

$$\boxed{1}\boxed{2} \quad \text{and} \quad \begin{array}{|c|}\hline 2 \\ \hline 1 \\ \hline\end{array}.$$

On the other hand, the two irreducible representations of $\mathfrak{S}(2)$ are the trivial representation and the signature representation, on which $J_2 = (1,2)$ acts by 1 and -1, so $\mathrm{Spec}(2) = \{(0,1),(0,-1)\} = \mathrm{Cont}(2)$. Suppose now that the result is true up to order $n-1 \geq 2$, and consider a spectrum vector $a(T) = (a_1,\ldots,a_n)$ with T standard tableau of size n, corresponding to integer partitions $\lambda^{(0)},\ldots,\lambda^{(n)}$. The space $\mathbb{C}\mathfrak{S}(n-1)(e_T)$ is isomorphic as a $\mathfrak{S}(n-1)$-representation to $S^{\lambda^{(n-1)}}$, and e_T is a Gelfand–Tsetlin vector of this space, on which $J_1,\ldots,J_{n-1}$ act by the eigenvalues $(a_1,\ldots,a_{n-1})$. Therefore, $(a_1,\ldots,a_{n-1}) \in \mathrm{Spec}(n-1) = \mathrm{Cont}(n-1)$. It suffices now to prove that a_n satisfies the two last conditions of Proposition 8.16. If $a_n \pm 1 \notin \{a_1,\ldots,a_{n-1}\}$, then in particular $a_n \neq a_{n-1} \pm 1$, so by Proposition 8.19,

$$(a_1,\ldots,a_{n-2},a_{n-1},a_n)\cdot s_{n-1} = (a_1,\ldots,a_{n-2},a_n,a_{n-1})$$

is in $\mathrm{Spec}(n)$, and $(a_1,\ldots,a_{n-2},a_n)$ belongs to $\mathrm{Spec}(n-1)$. But then, by the induction hypothesis, $(a_1,\ldots,a_{n-2},a_n) \in \mathrm{Cont}(n-1)$, so $a_n \pm 1 \in \{a_1,\ldots,a_{n-2}\}$, which is a contradiction. This proves that the second condition is always satisfied.

Suppose now that $a_n = a_i = a$ for some $i \in [\![1,n-1]\!]$. We reason by contradiction, and assume that $a-1 \notin \{a_{i+1},\ldots,a_{n-1}\}$ (the case of $a+1$ is entirely similar). We can take i maximal among the indices that satisfy this assumption; hence, $a \notin \{a_{i+1},\ldots,a_{n-1}\}$.

1. Suppose first that we also have $a+1 \notin \{a_{i+1},\ldots,a_{n-1}\}$. Then, denoting $*$ any quantity different from a, $a-1$ and $a+1$, we can use elementary transpositions to transform
$$a(T) = (a_1,\ldots,a_{i-1},a,*,*,\ldots,*,a)$$
into
$$(a_1,\ldots,a_{i-1},*,\ldots,*,a,a,*,\ldots,*)$$
still keeping a spectral vector by the second part of Proposition 8.19. However, by Corollary 8.21, this is not possible.

2. Therefore, $a+1$ belongs to $\{a_{i+1},\ldots,a_{n-1}\}$. It occurs exactly once: since $(a_1,\ldots,a_{n-1})$ is a content vector by the induction hypothesis, if there were two or more occurrences of $a+1$, then there would be a configuration
$$\begin{array}{ccc} & & \boxed{k} \\ \boxed{i} & \boxed{j} & \end{array}$$
in the tableau corresponding to $(a_1,\ldots,a_{n-1})$, with no cell on the diagonal of i and after it, and this is not possible for a tableau. So,
$$a(T) = (a_1,\ldots,a_{i-1},a,*,\ldots,*,a+1,*,\ldots,*,a),$$

where each symbol $*$ is different from $a, a-1$ and $a+1$. By the second part of Proposition 8.19,

$$a(U) = (a_1, \ldots, a_{i-1}, *, \ldots, *, a, a+1, a, *, \ldots, *)$$

is also a spectral vector. Denote j, $j+1$ and $j+2$ the position of the coordinates a, $a+1$ and a in $a(U)$. By the first part of Proposition 8.19,

$$s_j\, e_U = \varepsilon_j\, e_U \quad ; \quad s_{j+1} = \varepsilon_{j+1}\, e_U$$

for some signs ε_j and ε_{j+1}. Then,

$$\varepsilon_{j+1}\, e_U = s_j s_{j+1} s_j\, e_U = s_{j+1} s_j s_{j+1}\, e_U = \varepsilon_j\, e_U,$$

so $\varepsilon_j = \varepsilon_{j+1}$. On the other hand, $J_{j+1} = s_j J_j s_j + s_j$, so

$$(a+1)\, e_U = J_{j+1}\, e_U = (s_j\, J_j\, s_j + s_j)\, e_U = (a + \varepsilon_j)\, e_U,$$

which forces $\varepsilon_j = \varepsilon_{j+1} = 1$. But $J_{j+2} = s_{j+1} J_{j_1} s_{j+1} + s_{j+1}$, so

$$a\, e_U = J_{j+2}\, e_U = (s_{j+1} J_{j_1} s_{j+1} + s_{j+1})\, e_U = (a + 1 + \varepsilon_{j+1})\, e_U,$$

which forces $\varepsilon_j = \varepsilon_{j+1} = -1$, hence a contradiction.

Thus, $\{a-1, a+1\} \subset \{a_i, \ldots, a_n\}$, and $a(T) = (a_1, \ldots, a_n) \in \mathrm{Cont}(n)$. Since $\mathrm{Spec}(n)$ and $\mathrm{Cont}(n)$ have the same cardinality, we conclude that $\mathrm{Spec}(n) = \mathrm{Cont}(n)$. □

Theorem 8.22 is completed by the following proposition. Introduce the equivalence relation on $\mathrm{Spec}(n) = \mathrm{Cont}(n)$:

$$a(T) \sim_{\mathrm{Spec}(n)} a(U) \iff T \text{ and } U \text{ have the same shape } \lambda \in \mathfrak{Y}(n).$$

Proposition 8.23. *One has $a(T) \sim_{\mathrm{Spec}(n)} a(U)$ if and only if $a(T) \sim_{\mathrm{Cont}(n)} a(U)$.*

Proof. If $a(T) \sim_{\mathrm{Cont}(n)} a(U)$, then there exists a sequence $s_{i_1}, \ldots, s_{i_r}$ of admissible elementary transpositions such that $a(U) = a(T) \cdot (s_{i_1} s_{i_2} \cdots s_{i_r})$, and such that each s_i is applied to a content vector where $a_{i+1} \neq a_i \pm 1$. By the second part of Proposition 8.19, each vector

$$a(T) \cdot s_{i_1},\ a(T) \cdot s_{i_1} s_{i_2}, \ldots,\ a(T) \cdot s_{i_1} s_{i_2} \cdots s_{i_r} = a(U)$$

is a spectral vector corresponding to a tableau with the same shape as T, so T and U have the same shape and $a(T) \sim_{\mathrm{Spec}(n)} a(U)$.

The previous argument shows that the equivalence relation $\sim_{\mathrm{Cont}(n)}$ is finer than the equivalence relation $\sim_{\mathrm{Spec}(n)}$ on $\mathrm{Spec}(n) = \mathrm{Cont}(n)$. However, these relations have the same numbers of equivalence classes, namely, the number of integer partitions $\lambda \in \mathfrak{Y}(n)$, so necessarily $\sim_{\mathrm{Cont}(n)} = \sim_{\mathrm{Spec}(n)}$. □

We denote hereafter $\lambda = [a_1, \ldots, a_n]$ the equivalence class for the relation $\sim_{\mathrm{Spec}(n)}$ of a spectral vector coming from a tableau of shape λ. By the previous proposition, this set of $\mathbb{Z}$-valued vectors is the same as the equivalence class of content vectors $[c_1, \ldots, c_n]$ for the relation $\sim_{\mathrm{Cont}(n)}$. We can finally prove the main theorem of the section, which is equivalent to the statement $c(T) = a(T)$ for any standard tableau of size n.

Proof of Theorem 8.14. We reason again by induction on n, the result being trivial for $n = 1, 2$. Fix a standard tableau T of size n, corresponding to a sequence of integer partitions

$$\emptyset = \lambda^{(0)} \nearrow \lambda^{(1)} \nearrow \cdots \nearrow \lambda^{(n)} = \lambda^{(n)}.$$

We denote $a(T) = (a_1, \ldots, a_n)$ the spectrum of T, and $c(T) = (c_1, \ldots, c_n)$ the content vector of T. In $V^{\lambda^{(n-1)}} = \mathbb{C}\mathfrak{S}(n-1)(e_T)$, e_T is a Gelfand–Tsetlin vector of this representation of $\mathfrak{S}(n-1)$, corresponding to the subtableau T' obtained from T by removing the cell of label n. By the induction hypothesis, the Jucys–Murphy elements $J_1, \ldots, J_{n-1}$ act on it by the eigenvalues

$$(a_1, \ldots, a_{n-1}) = (c(1, T'), \ldots, c(n-1, T')) = (c_1, \ldots, c_{n-1}).$$

The Jucys–Murphy element J_n acts then on e_T by multiplication by an eigenvalue a_n, such that $(a_1, \ldots, a_n)$ is a content vector. Moreover, $[a_1, \ldots, a_n] = \lambda$. However, there is a unique content vector in the equivalence class λ and with first values $c_1, \ldots, c_{n-1}$, namely, the vector $(c_1, \ldots, c_n)$. Therefore, we indeed have $(a_1, \ldots, a_n) = (c_1, \ldots, c_n)$. □

8.3 Observables as symmetric functions of the contents

To conclude this chapter, we relate the results of the two previous sections to the algebra of observables $\mathscr{O}$ constructed in Chapter 7. We shall thereby obtain a new description of the observables as symmetric functions of the contents.

▷ *Observables as symmetric functions of the Jucys–Murphy elements.*

Recall from the proof of Proposition 8.13 that if μ is a partition without parts of size 1, then

$$X_{\mu,n} = \frac{\Sigma_{\mu,n}}{z_\mu} + \text{terms of lower degree},$$

where $X_{\mu,n}$ is a monomial function evaluated on the elements

$$X_i = \sum_{h<i} ((h,i), \{h,i\})$$

with $i \in [\![1,n]\!]$. Since this identity holds for any n, we can consider the projective limit $X_\mu = m_{\mu_1-1,\mu_2-1,\ldots,\mu_r-1}(X_1,X_2,\ldots)$, without restriction on the Jucys–Murphy elements considered. We then have

$$X_\mu = \frac{\Sigma_\mu}{z_\mu} + \text{terms of lower degree.}$$

Lemma 8.24. *The left-hand side X_μ of this formula is an element of $\mathscr{O}$.*

Proof. Since X_μ is a monomial function of the generic Jucys–Murphy elements, it suffices to prove that an algebraic basis of Sym has evaluations on the generic Jucys–Murphy elements $X_1,X_2,\ldots,X_n,\ldots$ that are in $\mathscr{O}$. However, there is an analogue of the identity of generating series given in the proof of Proposition 8.12:

$$\sum_{k=0}^{n} z^{n-k}\, e_k(X_1,X_2,\ldots,X_n) = \prod_{i=1}^{n}(z+X_i) = \sum_{\sigma\in\mathfrak{S}(n)} z^{n-(|t(\sigma)|-\ell(t(\sigma)))}\,(\sigma,\mathrm{Supp}(\sigma)),$$

where $\mathrm{Supp}(\sigma)$ is the essential support of a permutation σ, that is to say the union of its non-trivial cycles; and $t(\sigma)$ is the type of σ viewed as a permutation of its essential support. Taking the coefficient of z^{n-k}, we obtain:

$$e_k(X_1,X_2,\ldots,X_n) = \sum_{\mu \,|\, m_1(\mu)=0 \text{ and } |\mu|-\ell(\mu)=k} A_{\mu,n},$$

where $A_{\mu,n} = \phi_n(A_\mu)$ and $A_\mu = \frac{\Sigma_\mu}{z_\mu}$. So, for every n, $e_k(X_1,X_2,\ldots)$ has its projection by ϕ_n which belongs to $\mathbb{C}\mathfrak{P}\mathfrak{S}(n)^{\mathfrak{S}(n)} = \phi_n(\mathscr{O})$. By definition of a projective limit, $e_k(X_1,X_2,\ldots)$ belongs to $\mathscr{O}$, which ends the proof. □

As a consequence, the lower terms of the formula $X_\mu = \frac{\Sigma_\mu}{z_\mu} + \cdots$ always form a linear combination of other symbols Σ_ν. Now, in order to get arbitrary symbols Σ_μ without the condition $m_1(\mu)=0$, it suffices to multiply the previous identity by a power of Σ_1. Thus, let μ be an arbitrary integer partition, and $\widetilde{\mu}$ be the partition obtained from μ by removing the parts of size 1. Since the products of symbols Σ_μ factorize in top degree, we have

$$(\Sigma_1)^{m_1(\mu)}\, X_{\widetilde{\mu}} = \frac{\Sigma_\mu}{z_{\widetilde{\mu}}} + \text{terms of degree smaller than } |\mu|-1,$$

all the terms belonging to the algebra $\mathscr{O}$. Since $(\Sigma_\mu)_{\mu\in\mathfrak{Y}}$ is a graded linear basis of $\mathscr{O}$, this implies:

Theorem 8.25. *We have the isomorphism of vector spaces*

$$\mathbb{C}[\Sigma_1] \otimes_{\mathbb{C}} \mathrm{Sym}(X_1,X_2,\ldots) = \mathscr{O}.$$

Proof. Consider the two families $(\Sigma_1^n\, m_\nu(X))_{n\in\mathbb{N},\,\nu\in\mathfrak{Y}}$ and $(A_\mu)_{\mu\in\mathfrak{Y}}$. They correspond in top degree if one associates to μ the pair

$$n = m_1(\mu),\ \nu = (\mu_1 - 1, \ldots, \mu_s - 1),$$

where μ_s is the last part of μ greater than 2. This proves that $(\Sigma_1^n\, m_\nu(X))_{n\in\mathbb{N},\,\nu\in\mathfrak{Y}}$ is a basis of $\mathscr{O}$, and the isomorphism of vector spaces. □

▷ *Symmetric functions of the contents with polynomial coefficients.*

Given a symmetric function $f \in \mathrm{Sym}$, we can now consider the observable of Young diagrams $f(X) = f(X_1, X_2, \ldots)$, since we have just shown that it always belongs to $\mathscr{O}$. Let λ be a Young diagram of size n. By definition,

$$\begin{aligned} f(X)(\lambda) &= \chi^\lambda(\pi_n \circ \phi_n(f(X))) \\ &= \chi^\lambda(f(J_1, J_2, \ldots, J_n)) = \frac{\mathrm{ch}^\lambda(f(J_1, J_2, \ldots, J_n))}{\dim S^\lambda} \\ &= \frac{1}{\dim S^\lambda} \sum_{T\in\mathrm{ST}(\lambda)} [e_T](f(J_1, J_2, \ldots, J_n)\cdot e_T) \\ &= \frac{1}{\dim S^\lambda} \sum_{T\in\mathrm{ST}(\lambda)} f(c(1,T), c(2,T), \ldots, c(n,T)) \end{aligned}$$

by Theorem 8.14. Since f is symmetric, in the last sum, $f(c(1,T), \ldots, c(n,T))$ only depends on λ: indeed, the contents of the standard tableaux with the same shape λ only differ by a permutation. Therefore, if $c(\lambda) = \{c(\square),\ \square \in \lambda\}$, which can be viewed as finite alphabet of variables, then the formula of evaluation rewrites simply as:

$$f(X)(\lambda) = f(c(\lambda))$$

By using the isomorphism $\mathbb{C}[\Sigma_1] \otimes_{\mathbb{C}} \mathrm{Sym}(X) = \mathscr{O}$, we obtain a new simple description of the elements of $\mathscr{O}$:

Theorem 8.26. *A function $\lambda \mapsto u(\lambda)$ on the set $\mathfrak{Y}$ of Young diagrams is an observable if and only if there exists a symmetric function $F \in \mathbb{C}[n]\otimes_{\mathbb{C}}\mathrm{Sym}$ with coefficients in the rings of polynomials in one variable, such that:*

$$u(\lambda) = F(c(\lambda))_{|n=|\lambda|}.$$

Proof. This is now immediate, since $\Sigma_1(\lambda) = |\lambda|$. □

So, the observables of Young diagrams can now be described as symmetric functions of the contents, with coefficients that are polynomials in the size n of the Young diagram. There is actually another simple proof of this result, which relies on the interpretation of $\mathscr{O}$ as an algebra of *geometric* observables. We consider a Young diagram λ as a compact domain of the plane $\mathbb{R}^2$, endowed with the multiple of the Lebesgue measure which gives area 1 to every box $\square \in \lambda$. If p is a point inside λ, we define its content by

$c(p) = B - A$, with

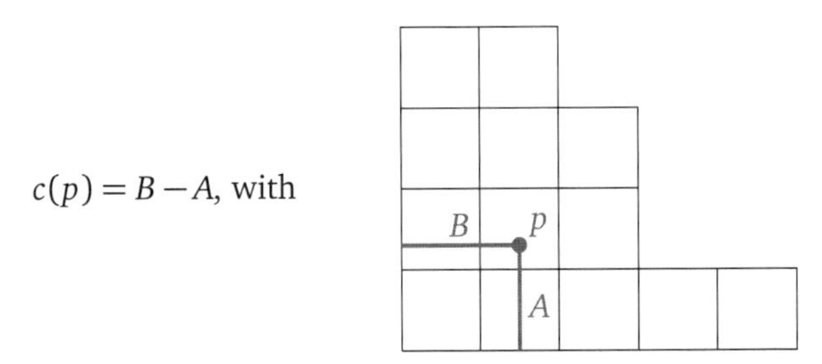

and where boxes in λ are considered to have edges of length 1. In particular, the content of a cell $\square$ is the content of the center of the cell. With our definitions, an integral

$$\iint_{\lambda} f(c(p))\,dp$$

does not depend on the convention of drawing of λ (French or Russian), since we take each time the same normalization of the lengths and areas.

Proposition 8.27. *For any $k \geq 2$, we have:*

$$\widetilde{p}_k(\lambda) = k(k-1) \iint_{\lambda} (c(p))^{k-2}\,dp.$$

Proof. If the diagram is drawn with the Russian convention, then the content of a point $p \in \lambda$ of abscissa s is just s. Therefore, by gathering the points of λ according to their content, we get

$$\iint_{\lambda} (c(p))^{k-2}\,dp = \int_{\mathbb{R}} \sigma_{\omega}(s)\, s^{k-2}\,ds.$$

However, we defined $\widetilde{p}_k(\omega)$ by the integral $\widetilde{p}_k(\omega) = -k \int_{\mathbb{R}} \sigma'_{\omega}(s)\, s^{k-1}\,ds$. The formula follows immediately by an integration by parts. $\square$

We now split the double integral according to the cells of the Young diagram. Notice that if $\square$ has content c, then the points of $\square$ have their content given by $c + y - x$, where y and x belong to $[-\frac{1}{2}, \frac{1}{2}]$. Therefore,

$$\begin{aligned}\frac{\widetilde{p}_k(\lambda)}{k(k-1)} &= \sum_{\square \in \lambda} \iint_{x,y=-\frac{1}{2}}^{\frac{1}{2}} (c(\square) + y - x)^{k-2}\,dx\,dy \\ &= \sum_{r=1}^{k-2} \binom{k-2}{r} p_r(c(\lambda)) \left(\iint_{x,y=-\frac{1}{2}}^{\frac{1}{2}} (y-x)^{k-2-r}\,dx\,dy \right) \\ &\quad + n \left(\iint_{x,y=-\frac{1}{2}}^{\frac{1}{2}} (y-x)^{k-2}\,dx\,dy \right),\end{aligned}$$

where $n = |\lambda|$ is the number of cells $\square$. Since $n = \Sigma_1(\lambda)$, we deduce from this by induction on r that all the power sums $p_r(c(\lambda))$ in the contents are in $\mathscr{O}$. More generally, since $(\widetilde{p}_k)_{k\geq 2}$ is an algebraic basis of $\mathscr{O}$, the formula above can be used to reprove that $\mathbb{C}[n] \otimes_{\mathbb{C}} \mathrm{Sym}(\text{contents}) = \mathscr{O}$.

Notes and references

In many papers, our Gelfand–Tsetlin basis $(e_T)_{T\in \mathrm{ST}(\lambda)}$ of the Specht module S^λ is called the Young basis. The main theorems 8.6 and 8.11 of Section 8.1 are due to Jucys and Murphy; see [Juc74, Mur81]. Their proof is clarified by the use of the algebra of partial permutations. This is to our knowledge is a new approach, the main idea (the use of generic Jucys–Murphy elements) coming from the beautiful paper [Fér12b]. In [CSST10, Chapter 3], a similar argument is developed with *marked* permutations, which are another way to encode additional informations on permutations.

As mentioned at the beginning of Section 8.2, the Jucys–Murphy elements J_i and their relations with the Coxeter elements s_i can be used in an alternative approach to the representation theory of the symmetric groups $\mathfrak{S}(n)$, and small improvements of the arguments of this chapter would allow us to reprove the branching rules. This method was proposed by Okounkov and Vershik in [OV04]. With more details, the same approach is found in [CSST10, Chapter 3], which we followed for most of Section 8.2; our argument differs mainly when we use the already known branching rules (Corollary 3.7) of the symmetric groups. Finally, the interpretation of observables of Young diagrams as symmetric functions of the contents already appeared in the first paper on observables [KO94], and the link with geometric observables is mentioned in [DFŚ10].

An important open problem after Theorem 8.11 is the computation of the expansion of a symmetric function in Jucys–Murphy elements in conjugacy classes C_λ. For elementary symmetric functions e_k, the solution is provided by our Proposition 8.12. For power sums, we refer to [LT01], and for a partial answer for homogeneous symmetric functions h_k, to [MN13, ZJ10, Fér12a].

9

Symmetric groups and free probability

In Chapter 7, we showed that the symbols $(\Sigma_k)_{k\geq 1}$ and $(p_k)_{k\geq 1}$ form two algebraic bases of the algebra of observables $\mathscr{O}$, and that the restriction on $\mathscr{O}$ of the natural gradation $\deg(\cdot)$ of the algebra of partial permutations $\mathbb{C}\mathfrak{PS}(\infty)$ is such that

$$\deg \Sigma_k = \deg p_k = k \quad \text{and} \quad \deg(\Sigma_k - p_k) \leq k-1;$$

thus, the observables Σ_k and p_k have the same top degree component. On the other hand, another algebraic basis of $\mathscr{O}$ is $(\widetilde{p}_k)_{k\geq 2}$, these functions being "smooth" functionals of the shapes ω_λ of the Young diagrams λ. We then know by Proposition 7.17 that another gradation on $\mathscr{O}$ can be defined by the equivalent identities

$$\forall k \geq 1,\ \mathrm{wt}(\Sigma_k) = k+1$$

or

$$\forall k \geq 2,\ \mathrm{wt}(\widetilde{p}_k) = k.$$

The objective of this chapter is to identify the expansion of the top weight component of a symbol Σ_k, or more generally Σ_μ in terms of the smooth functionals $\widetilde{p}_k$. Quite surprisingly, the result (Theorem 9.20) relies on the technology of a quite recent part of probability theory known as *free probability theory*. We give an introduction to this non-commutative probability theory in Section 9.1, describing in particular the operation of free convolution of measures.

In Section 9.2, we give the combinatorial counterpart of free probability theory in terms of the so-called non-crossing partitions. We then show that the top weight component of $\Sigma_k(\lambda)$ is the $(k+1)$-th free cumulant $R_{k+1}(\lambda)$ of the transition measure μ_λ of λ, thereby answering the problem stated above. This use of free probability in the setting of the representation theory of symmetric groups is developed in Section 9.3, where we use an interpretation of $\mathbb{C}\mathfrak{S}(n)$ as a *non-commutative probability space* in order to show that the transition measure μ_λ of a Young diagram can be seen as the distribution of a Jucys–Murphy element. This interpretation of μ_λ can also be proven by introducing an algebra $\mathscr{P}$ of set partitions, endowed with a morphism $\Sigma : \mathscr{P} \to \mathscr{O}$ which allows one to project some identities and to obtain new formulas in the algebra of observables. Section 9.4 of the chapter is devoted to this topic, and we compute in particular the expansion of the observables $\widetilde{h}_k$ in symbols Σ_μ (cf. Theorem 9.33, which can be seen as an improvement of Proposition 7.17).

The identity $\Sigma_k = R_{k+1} +$ term of lower weight has numerous consequences for the asymptotic behavior of a character value

$$\chi^\lambda(\sigma),$$

where $\sigma \in \mathfrak{S}(\infty)$ is a fixed finite permutation and λ is an integer partition of size $n \to +\infty$; cf. Corollary 9.21. It will prove very useful in the fourth part of the book, when dealing with large *random* integer partitions stemming from representations of symmetric groups.

9.1 Introduction to free probability

In this first section, $(\Omega, \mathscr{F}, \mathbb{P})$ is an arbitrary **probability space**. Recall that it means that:

- Ω is a non-empty set;
- $\mathscr{F}$ is a σ**-field** on Ω, that is to say a collection of **measurable** subsets of Ω that contains $\emptyset$ and Ω and that is stable by complementation and by denumerable union (or intersection).
- $\mathbb{P}$ is a **probability measure** on $(\Omega, \mathscr{F})$, that is to say a function $\mathbb{P} : \mathscr{F} \to [0,1]$ which sends $\emptyset$ to 0, Ω to 1, and which is compatible with denumerable disjoint unions of measurable subsets:

$$\forall (U_i)_{i\in I} \text{ denumerable family of disjoint subsets, } \mathbb{P}\left[\bigsqcup_{i\in I} U_i\right] = \sum_{i\in I} \mathbb{P}[U_i].$$

A probability space being fixed, one then says that a property (P) holds almost surely (or $\mathbb{P}$-almost surely if one wants to be precise) if it holds on a measurable subset $A \subset \Omega$ such that $\mathbb{P}[A] = 1$.

▷ *From probability spaces to von Neumann algebras.*

The basic idea of any non-commutative analogue of a mathematical theory is to reinterpret it in terms of commutative algebras. If $(\Omega, \mathscr{F}, \mathbb{P})$ is a probability space, then one can associate to it the algebra

$$\begin{aligned} A &= \mathrm{L}^\infty(\Omega, \mathscr{F}, \mathbb{P}) \\ &= \{\text{measurable functions } f : \Omega \to \mathbb{C} \text{ that are bounded } \mathbb{P}\text{-almost surely}\}, \end{aligned}$$

which is a commutative Banach algebra for the **essential supremum norm**

$$\|f\|_\infty = \operatorname*{ess\,sup}_{\omega\in\Omega} |f(\omega)| = \inf_{\substack{A\in\mathscr{F} \\ \mathbb{P}[A]=1}} \sup_{\omega\in A} |f(\omega)|,$$

two functions being identified if they are equal almost surely. This algebra of (bounded) **random variables** acts by pointwise multiplication on the Hilbert space

$$\begin{aligned} H &= \mathrm{L}^2(\Omega, \mathscr{F}, \mathbb{P}) \\ &= \left\{ \text{measurable functions } f : \Omega \to \mathbb{C} \,\middle|\, \int_\Omega |f(\omega)|^2 \, \mathbb{P}[d\omega] < +\infty \right\}, \end{aligned}$$

and the norm $\|\cdot\|_\infty$ on $A = \mathrm{L}^\infty(\Omega, \mathscr{F}, \mathbb{P})$ appears then as the operator norm restricted to a subalgebra A of the algebra of bounded operators $\mathscr{B}(H)$. Moreover, in this description, the adjoint of the multiplication by f is the multiplication by $\overline{f}$, so A is a $*$-subalgebra of $\mathscr{B}(H)$. As a Banach algebra, it is closed for the operator norm, but as we shall see in a moment, it is also closed for the **weak operator topology**. This topology on $\mathscr{B}(H)$ is defined as the weakest topology (with the smallest number of open sets) that makes the maps $T \mapsto \langle x \mid Ty \rangle$ continuous. A classical result is:

Theorem 9.1 (Von Neumann). *Let $A \subset \mathscr{B}(H)$ be a $*$-subalgebra of the algebra of bounded linear operators on H, and that contains the identity id_H. Then, A is closed for the weak topology if and only if it is equal to its bicommutant A'', where $A' = \{T \in \mathscr{B}(H) \,|\, \forall U \in A,\ TU = UT\}$.*

Proof. If A is a non-empty subset of $\mathscr{B}(H)$, let T be a bounded operator that is not in the commutant of A: there exists $U \in A$, such that $UT - TU \neq 0$. Let $x, y \in H$ such that $\langle x \mid (UT - TU)y \rangle = \langle U^* x \mid Ty \rangle - \langle x \mid TUy \rangle \neq 0$. Since the map

$$f : T \mapsto \langle U^* x \mid Ty \rangle - \langle x \mid TUy \rangle$$

is continuous for the weak operator topology, there exists a neighborhood of T in this topology on which f does not vanish. Then, for any T' in this neighborhood, $UT' - T'U \neq 0$, so the complementary of a commutant is always open in the weak operator topology. Therefore, a commutant A' is always closed, and if $A = A''$, then A is closed for the weak operator topology.

Conversely, let us show that if A is a unital $*$-subalgebra, then A'' is included in the weak operator topology closure of A. If $h \in H$, consider the closure $\overline{Ah}$ of Ah with respect to the norm of the Hilbert space H. It is a closed subspace of a Hilbert space, so it admits an orthogonal projection which we shall denote P_h. We claim that $P = P_h$ belongs to A'. Indeed, if $x \in H$, then $P(x)$ belongs to $\overline{Ah}$, so there is a sequence of operator T_n in A such that $\lim_{n\to\infty} T_n(h) = P(x)$. Taking $T \in A$, we have $TT_n(h) \in Ah$ for every n, and taking the limit as n goes to infinity, $TP(x) \in \overline{Ah}$. Therefore,

$$PTP(x) = TP(x),$$

and this is true for every x, so $PTP = TP$ for any $T \in A$. However, A is a $*$-subalgebra, so,

$$\langle x \mid TP(y) \rangle = \langle x \mid PTP(y) \rangle = \langle PT^*P(x) \mid y \rangle = \langle T^*P(x) \mid y \rangle = \langle x \mid PT(y) \rangle$$

for any x, y in H. This implies that $TP = PT$, hence $P = P_h \in A'$ as claimed.

We reason now by contradiction, and take T that is in the bicommutant A'', but that is not in the weak operator closure of A. The second condition means that there exist elements $x_1, \ldots, x_k, y_1, \ldots, y_k$ in H and a positive real number ε such that

$$T \in \bigcap_{i=1}^{k} U(T; x_i, y_i, \varepsilon) \subset \mathscr{B}(H) \setminus A,$$

where

$$U(T; x, y, \varepsilon) = \left\{T' \in \mathscr{B}(H) \mid \left|\langle x \mid T(y)\rangle - \left\langle x \mid T'(y)\right\rangle\right| < \varepsilon\right\}.$$

Indeed, a subset of $\mathscr{B}(H)$ is a neighborhood of T in the weak operator topology if and only if it contains a finite intersection of subsets $U(T; x_i, y_i, \varepsilon_i)$. We associate to T the diagonal operator

$$\begin{aligned} \operatorname{diag}(T, k) : H^k &\to H^k \\ (h_1, \ldots, h_k) &\mapsto (T(h_1), \ldots, T(h_k)). \end{aligned}$$

It is easy to see that if $T \in A''$, then $\operatorname{diag}(T, k) \in (\operatorname{diag}(A, k))''$. We now use the previous claim with respect to the subalgebra $\operatorname{diag}(A, k) \subset \mathscr{B}(H^k)$ and to the vector $h = (y_1, \ldots, y_k)$. Since the projector $P_{(y_1,\ldots,y_k)}$ belongs to $\operatorname{diag}(A, k)'$, we have

$$\begin{aligned} \operatorname{diag}(T, k)(y_1, \ldots, y_k) &= \operatorname{diag}(T, k) \circ P_{(y_1,\ldots,y_k)}(y_1, \ldots, y_k) \\ &= P_{(y_1,\ldots,y_k)} \circ \operatorname{diag}(T, k)(y_1, \ldots, y_k) \\ &= P_{(y_1,\ldots,y_k)}(T(y_1), \ldots, T(y_k)). \end{aligned}$$

Therefore, there exists a sequence of operators $(T_n)_{n\in\mathbb{N}}$ in A such that

$$\lim_{n\to\infty} \operatorname{diag}(T_n, k)(y_1, \ldots, y_k) = \operatorname{diag}(T, k)(y_1, \ldots, y_k).$$

Then, $\langle x_i \mid T_n(y_i)\rangle \to \langle x_i \mid T(y_i)\rangle$ for any $i \in [\![1, k]\!]$, so in particular, $T_n \in \bigcap_{i=1}^{k} U(T; x_i, y_i, \varepsilon)$ for n large enough. This contradicts the previous assumption, hence A'' is indeed in the weak operator closure of A. □

Corollary 9.2. *The algebra $A = \mathrm{L}^\infty(\Omega, \mathscr{F}, \mathbb{P})$ is a commutative, unital and $*$-stable subalgebra of $\mathscr{B}(H = \mathrm{L}^2(\Omega, \mathscr{F}, \mathbb{P}))$ that is closed for the weak operator topology.*

Proof. We only have to show that A is its own bicommutant. Let $T \in A''$, and set $f = T(1)$, where 1 is the constant function on Ω equal to 1. We claim that $T = m_f$ is the multiplication by f. Let $C = \|T\|_{\mathrm{L}^2\to\mathrm{L}^2}$ be the operator norm of T; we shall first prove that f belongs to L^∞ and is essentially bounded by C. We can assume without loss of generality that $C > 0$, the case $C = 0$ being obvious. If

f is not essentially bounded by C, let $A \subset \Omega$ be a measurable subset of non-zero $\mathbb{P}$-measure, on which $f(\omega) > C$. We define

$$g(\omega) = \frac{1_A(\omega)}{f(\omega)},$$

where 1_A is the indicator of the set A. Since $f(\omega) > C > 0$ on A, g belongs to $\mathrm{L}^\infty \subset \mathrm{L}^2$. Then,

$$T(g) = T \circ m_g(1) = m_g \circ T(1) = m_g(f) = gf = 1_A,$$

so $\mathbb{P}[A] = \|1_A\|_{\mathrm{L}^2} \leq \|T\|_{\mathrm{L}^2 \to \mathrm{L}^2} \|g\|_{\mathrm{L}^2} < \mathbb{P}[A]$, hence a contradiction. So, $f \in \mathrm{L}^\infty$, and we then have for any $g \in \mathrm{L}^\infty$:

$$T(g) = T \circ m_g(1) = m_g \circ T(1) = m_g(f) = gf = m_f(g).$$

Since L^∞ is dense in L^2, $T = m_f$ as claimed. □

If H is a Hilbert space, a unital $*$-subalgebra of $\mathscr{B}(H)$ that is closed for the weak operator topology is called a **von Neumann algebra**. We have just shown that von Neumann subalgebras of $\mathscr{B}(H)$ are those that are equal to their bicommutant, and that for any probability space, $A = \mathrm{L}^\infty(\Omega, \mathscr{F}, \mathbb{P})$ is a *commutative* von Neumann algebra associated to the Hilbert space $H = \mathrm{L}^2(\Omega, \mathscr{F}, \mathbb{P})$.

▷ *Traces on von Neumann algebras and non-commutative probability spaces.*

In a von Neumann algebra A, a self-adjoint element x is called **non-negative** if it writes as $x = a^* a$ with $a \in A$.

Definition 9.3. *Let A be a von Neumann algebra. A (normalized)* ***state*** *on A is a linear form $\tau : A \to \mathbb{C}$ such that:*

1. *for any $a \in A$, $\tau(a^* a) \geq 0$.*
2. $\tau(1_A) = 1$.

If $A \subset \mathscr{B}(H)$, then a vector $h \in H$ of norm 1 yields a state $\tau_h : a \mapsto \langle h \,|\, a(h) \rangle$. In the case when $H = \mathrm{L}^2(\Omega, \mathscr{F}, \mathbb{P})$ and $A = \mathrm{L}^\infty(\Omega, \mathscr{F}, \mathbb{P})$, taking h equal to the constant function 1 on Ω, the corresponding state is simply the expectation:

$$\tau_1(f) = \int_\Omega f(\omega)\, \mathbb{P}[d\omega] = \mathbb{E}[f].$$

It can be shown that the knowledge of the state τ_h of the von Neumann algebra A allows one to reconstruct the action of A on H, and the cyclic vector h; this is the Gelfand–Naimark–Segal construction; see Theorem 11.26. On the other hand, when $H = \mathrm{L}^2(\Omega, \mathscr{F}, \mathbb{P})$ and $A = \mathrm{L}^\infty(\Omega, \mathscr{F}, \mathbb{P})$, the canonical state $\tau = \mathbb{E}$ satisfies also:

$$\forall a, b \in A, \;\; \tau(ab) = \mathbb{E}[ab] = \mathbb{E}[ba] = \tau(ba).$$

These observations lead to the definition of a **non-commutative probability space**:

Definition 9.4. *A non-commutative probability space is a pair (A,τ), where A is a (possibly non-commutative) von Neumann algebra, and τ is a state on A that is a* ***trace****, i.e.,*

$$\forall a,b \in A,\ \ \tau(ab)=\tau(ba).$$

The self-adjoint elements $a \in A$ are called ***non-commutative random variables****.*

By the previous discussion, any standard probability space $(\Omega,\mathscr{F},\mathbb{P})$ yields the non-commutative probability space $(\mathrm{L}^{\infty}(\Omega,\mathscr{F},\mathbb{P}),\mathbb{E})$, so this definition extends the usual setting of probability theory. Let us now give new examples which satisfy our definition:

Example. For $n \geq 1$, consider the matrix space $\mathrm{M}(n,\mathbb{C})$, endowed with the usual adjunction of complex matrices and with the normalized trace

$$\tau(M)=\frac{1}{n}\,\mathrm{tr}(M).$$

For any matrices M and N, $\tau(MN)=\tau(NM)$; $\tau(MM^*)=\frac{1}{n}\sum_{i,j=1}^{n}|M_{ij}|^2$ is indeed non-negative; and $\tau(I_n)=1$. On the other hand, $\mathrm{M}(n,\mathbb{C})$ is a finite-dimensional von Neumann algebra, since it acts faithfully on the Hilbert space $\mathbb{C}^n$. Hence, $(\mathrm{M}(n,\mathbb{C}),\frac{\mathrm{tr}(\cdot)}{n})$ is a non-commutative probability space. Moreover, it is well known that $\tau=\mathrm{tr}(\cdot)/n$ is in fact the unique trace (in the sense of Definition 9.4) on $\mathrm{M}(n,\mathbb{C})$, so it yields the unique non-commutative probability space structure on the space of matrices $\mathrm{M}(n,\mathbb{C})$.

Example. The previous example can be generalized to any semisimple complex algebra. If A is a finite-dimensional semisimple complex algebra, by Wedderburn's theorem 1.24, it is isomorphic to a direct sum of matrix spaces:

$$A=\bigoplus_{\lambda\in\widehat{A}}\mathrm{M}(d_\lambda,\mathbb{C}).$$

Suppose that τ is a trace on A. Then, its restriction to each block $\mathrm{M}(d_\lambda,\mathbb{C})$ is a linear form that satisfies $\tau(MN)=\tau(NM)$, hence a multiple of the normalized trace

$$\tau^\lambda : M\in\mathrm{M}(d_\lambda,\mathbb{C})\mapsto\frac{\mathrm{tr}(M)}{d_\lambda}.$$

Since τ is non-negative on self-adjoint non-negative elements, and with $\tau(1_A)=1$, we conclude that any trace on A writes uniquely as a barycenter

$$\tau=\sum_{\lambda\in\widehat{A}}p_\lambda\,\tau^\lambda,$$

where the weights p_λ are non-negative and with sum $\sum_{\lambda\in\widehat{A}}p_\lambda=\tau(1_A)=1$. We say that the collection $(p_\lambda)_{\lambda\in\widehat{A}}$ is the **spectral measure** of the trace τ; it is a probabilty measure on the set $\widehat{A}$ of (isomorphism classes of) simple modules over A. Thus, we have a bijection between:

$$\begin{aligned}&\text{structures of non-commutative probability space on } A\\ &\longleftrightarrow \text{probability measures on } \widehat{A}.\end{aligned}$$

Example. A particular case of the previous bijection is when $A = \mathbb{C}G$ is a group algebra. Notice then that the adjunction $*$ on $\mathbb{C}G$ is the operation

$$\sum_{g\in G} c_g\, g \mapsto \sum_{g\in G} \overline{c_g}\, g^{-1},$$

since if $(V^\lambda, \rho^\lambda)$ is an irreducible representation of G, then assuming that the matrices $\rho^\lambda(g)$ are unitary, we have $(\rho^\lambda(g))^* = (\rho^\lambda(g))^{-1} = \rho^\lambda(g^{-1})$. By the previous discussion, any probability measure $\mathbb{P}$ on the set $\widehat{G}$ of irreducible representations on G yields a trace τ on $\mathbb{C}G$:

$$\tau = \sum_{\lambda\in\widehat{G}} \mathbb{P}[\lambda]\, \tau^\lambda \circ \rho^\lambda = \sum_{\lambda\in\widehat{G}} \mathbb{P}[\lambda]\, \chi^\lambda$$

and thereby a structure of non-commutative probability space. Conversely, consider a representation $\rho : G \to \mathrm{GL}(V)$, which has for decomposition in irreducible representations

$$V = \bigoplus_{\lambda\in\widehat{G}} m_\lambda\, V^\lambda.$$

The normalized character $\chi^V = \frac{\mathrm{ch}^V}{\dim V}$ is then a trace on $\mathbb{C}G$, with expansion in normalized irreducible characters

$$\chi^V = \sum_{\lambda\in\widehat{G}} \frac{m_\lambda \dim V^\lambda}{\dim V}\, \chi^\lambda.$$

Thus, the spectral measure associated to the representation V is

$$\mathbb{P}_V[\lambda] = \frac{m_\lambda \dim V^\lambda}{\dim V}.$$

In the fourth part of the book, we shall study this kind of probability measure in the case when $G = \mathfrak{S}(n)$ and τ is the normalized character of a "natural" representation of $\mathfrak{S}(n)$, or more generally the restriction of a trace on the infinite symmetric group $\mathfrak{S}(\infty)$. Notice that if τ is a trace on $\mathbb{C}G$, then its interpretation as a non-commutative expectation is natural with respect to the spectral measure, since we can write

$$\tau(a) = \mathbb{E}[a] = \sum_{\lambda\in\widehat{G}} \mathbb{P}[\lambda]\, \chi^\lambda(a),$$

so $\mathbb{E}[a]$ is the expectation in the traditional sense of the random variable $\chi^\lambda(a)$ taken under the spectral measure.

To pursue our construction of a non-commutative probability theory, let us develop the analogue of the notion of conditional expectation. If $(\Omega, \mathscr{F}, \mathbb{P})$ is a standard probability space and $\mathscr{G} \subset \mathscr{F}$ is a sub-σ-field, then $\mathbb{E}[\cdot|\mathscr{G}]$ yields a linear map

$$A = \mathrm{L}^{\infty}(\Omega, \mathscr{F}, \mathbb{P}) \to A' = \mathrm{L}^{\infty}(\Omega, \mathscr{G}, \mathbb{P})$$

such that if $f \in A$ and $g \in A'$, then

$$\mathbb{E}[f\,g|\mathscr{G}] = \mathbb{E}[f|\mathscr{G}]\,g.$$

Moreover, this map is non-negative in the sense that if $f \geq 0$ almost surely, then $\mathbb{E}[f|\mathscr{G}] \geq 0$ almost surely (for $\mathbb{P}$). After these observations, the following definition is not very surprising:

Definition 9.5. *Let $A' \subset A$ be von Neumann algebras. A* ***non-commutative conditional expectation*** *from A to A' is a linear map $\phi : A \to A'$ such that:*

1. *$\phi(1_A) = 1_{A'}$.*
2. *if $a \in A$ and $b, c \in A'$, then $\phi(bac) = b\,\phi(a)\,c$.*
3. *for any $a \in A$, $\phi(a^*a)$ is a non-negative element of A', that is to say that it writes as b^*b with $b \in A'$.*

Example. Consider the two von Neumann algebras $A = \mathbb{C}\mathfrak{S}(n+1)$ and $A' = \mathbb{C}\mathfrak{S}(n)$. We define a linear map:

$$\phi : \mathbb{C}\mathfrak{S}(n+1) \to \mathbb{C}\mathfrak{S}(n)$$
$$\sigma \mapsto \begin{cases} \sigma_{|[\![1,n]\!]} & \text{if } \sigma(n+1) = n+1, \\ 0 & \text{otherwise.} \end{cases}$$

The two first conditions of the previous definition hold trivially. For the third condition, notice that in a matrix algebra, a matrix M writes as $M = N^*N$ (non-negative element) if and only if it is Hermitian and with non-negative spectrum (we can then take $N = M^{1/2}$ in the sense of functional calculus). As a consequence, in a group algebra $\mathbb{C}G$, an element $x \in \mathbb{C}G$ is non-negative if and only if $x = x^*$ and, for any unitary representation (ρ, V) of G and any $v \in V$, $\langle v \,|\, \rho(x)(v)\rangle_V \geq 0$. Fix a representation (V, ρ^V) of $\mathfrak{S}(n)$, and consider $x = a^*a$ non-negative element of the algebra $\mathbb{C}\mathfrak{S}(n+1)$. We write $a = \sum_{\sigma \in \mathfrak{S}(n+1)} c_\sigma\,\sigma$, and we compute

$$\phi(x) = \sum_{\sigma,\tau \in \mathfrak{S}(n+1)} \overline{c_\sigma} c_\tau\,\phi(\sigma^{-1}\tau) = \sum_{k=1}^{n+1} \sum_{\substack{\sigma,\tau \in \mathfrak{S}(n+1) \\ \sigma(n+1)=\tau(n+1)=k}} \overline{c_\sigma} c_\tau\,\sigma^{-1}\tau.$$

This expression shows readily that $\phi(x)$ is self-adjoint in $\mathbb{C}\mathfrak{S}(n)$. On the other hand, notice that the set of transpositions $\{t_k = (k, n+1),\ 1 \leq k \leq n+1\}$ is a set of representatives of the classes in $\mathfrak{S}(n+1)/\mathfrak{S}(n)$, with by convention $t_{n+1} = (n+1, n+1) = \mathrm{id}_{[\![1,n+1]\!]}$. Denote $V' = \mathrm{Ind}_{\mathfrak{S}(n)}^{\mathfrak{S}(n+1)}(V) = \mathbb{C}\mathfrak{S}(n+1) \otimes_{\mathbb{C}\mathfrak{S}(n)} V$. If $(v_r)_{r\in[\![1,\dim V]\!]}$ is an orthonormal basis of V, then

$$\left(t_k \otimes v_r\right)_{k\in[\![1,n+1]\!],\, r\in[\![1,\dim V]\!]}$$

is an orthonormal basis of V', and the action $\rho^{V'}$ of $\mathfrak{S}(n+1)$ on this basis is given by

$$\rho^{V'}(\sigma)(t_k \otimes v_r) = t_{\sigma(k)} \otimes \rho^V(t_{\sigma(k)} \circ \sigma \circ t_k)(v_r).$$

Consequently, if v_r is a basis vector of V, then

$$\begin{aligned}
&\left\langle t_{n+1} \otimes v_r \,\middle|\, \rho^{V'}(x)(t_{n+1} \otimes v_r)\right\rangle_{V'} \\
&= \sum_{\sigma,\tau\in\mathfrak{S}(n+1)} \overline{c_\sigma} c_\tau \left\langle \rho^{V'}(\sigma)(t_{n+1} \otimes v_r) \,\middle|\, \rho^{V'}(\tau)(t_{n+1} \otimes v_r)\right\rangle_{V'} \\
&= \sum_{\sigma,\tau\in\mathfrak{S}(n+1)} \overline{c_\sigma} c_\tau \left\langle t_{\sigma(n+1)} \otimes \rho^V(t_{\sigma(n+1)}\sigma)(v_r) \,\middle|\, t_{\rho(n+1)} \otimes \rho^V(t_{\tau(n+1)\tau})(v_r)\right\rangle_{V'} \\
&= \sum_{k=1}^{n+1} \sum_{\substack{\sigma,\tau\in\mathfrak{S}(n+1) \\ \sigma(n+1)=\tau(n+1)=k}} \overline{c_\sigma} c_\tau \left\langle \rho^V(t_k\sigma)(v_r) \,\middle|\, \rho^V(t_k\tau)(v_r)\right\rangle_V = \left\langle v_r \,\middle|\, \rho^V(\phi(x))(v_r)\right\rangle_V,
\end{aligned}$$

by using on the last line the identity

$$\left\langle \rho^V(t_k\sigma)(v_r) \,\middle|\, \rho^V(t_k\tau)(v_r)\right\rangle_V = \left\langle v_r \,\middle|\, \rho^V(\sigma^{-1}\tau)(v_r)\right\rangle_V.$$

Since $x \in \mathbb{C}\mathfrak{S}(n+1)$ is non-negative, the left-hand side of this sequence of identities is non-negative, so the right-hand side is also non-negative. It is now shown that $\phi(x) \geq 0$ in $\mathbb{C}\mathfrak{S}(n)$; thus, ϕ is a conditional expectation.

▷ *Distribution of a self-adjoint random variable.*

From now on, we fix a non-commutative probability space (A, τ). Our next task is to define the analogue of the notion of the law of a random variable. If $A = \mathrm{L}^\infty(\Omega, \mathscr{F}, \mathbb{P})$ and $X \in A$ is real-valued, then its law is the probability measure μ_X on $\mathbb{R}$ such that, for any continuous function $f : \mathbb{R} \to \mathbb{R}$,

$$\mathbb{E}[f(X)] = \int_\Omega f(X(\omega))\,\mathbb{P}[d\omega] = \int_\mathbb{R} f(x)\,\mu_X(dx).$$

In the non-commutative setting, we shall replace the expectation $\mathbb{E}$ by the trace τ, and the hypothesis of real-valued random variable by self-adjointness in A. The main problem comes then from the notion of continuous functions $f(a)$ of a self-adjoint element a. This problem can only be solved by using the functional calculus of $*$-algebras. Call **spectrum** of a self-adjoint operator $T \in \mathscr{B}(H)$ the set

$$\sigma(T) = \{z \in \mathbb{C} \,|\, z\,\mathrm{id}_H - T \text{ is not invertible}\}.$$

This spectrum is a compact subset included in the segment $[-\|T\|_{\mathscr{B}(H)}, \|T\|_{\mathscr{B}(H)}]$, and we have the classical:

Theorem 9.6 (Spectral theorem). *Let T be a self-adjoint operator in $\mathscr{B}(H)$, where H is a Hilbert space. There is a morphism of algebras*

$$\begin{aligned}\mathscr{C}^0(\sigma(T),\mathbb{C}) &\to \mathscr{B}(H)\\ f &\mapsto f(T)\end{aligned}$$

which agrees with the algebraic definition on polynomials, and which satisfies for any continuous function f

$$\|f(T)\|_{\mathscr{B}(H)} = \|f\|_\infty = \sup_{t\in\sigma(T)} |f(t)|.$$

Its image is the smallest norm-closed subalgebra of $\mathscr{B}(H)$ containing 1 and T, and moreover, it sends non-negative functions to non-negative elements.

We refer to the end of the chapter for details on this result. In the special case where H is a finite-dimensional Hilbert space and (A,τ) is a finite-dimensional non-commutative probability space, it is an immediate consequence of the classical spectral theorem for Hermitian matrices, and this will be sufficient for our purpose, since most of the non-commutative probability spaces hereafter will be group algebras of finite symmetric groups.

Definition 9.7. *We say that a probability measure μ_a on the real line $\mathbb{R}$ is the **law**, or **distribution** of a self-adjoint element $a \in A$ of a non-commutative probability space (A,τ) if, for any polynomial $P \in \mathbb{C}[X]$,*

$$\tau(P(a)) = \int_{\mathbb{R}} P(x)\,\mu_a(dx).$$

Theorem 9.8. *The law μ_a of a self-adjoint element $a \in A$ of a non-commutative probability space exists and is unique. It is compactly supported on a subset of $[-\|a\|, \|a\|]$.*

Proof. Notice that for any $a \in A \subset \mathscr{B}(H)$, $|\tau(a)| \leq \|a\|_{\mathscr{B}(H)}$. Indeed, the map $x,y \mapsto \tau(x^*y)$ is a non-negative sesquilinear form on A, hence it satisfies the Cauchy–Schwarz inequality $|\tau(x^*y)|^2 \leq \tau(x^*x)\,\tau(y^*y)$. Setting $x = 1_A$, we obtain in particular $|\tau(y)|^2 \leq \tau(y^*y)$. Then, the function $t \mapsto \|y\|^2 - t$ is non-negative on $\sigma(y^*y)$, since

$$\sigma(y^*y) \subset [-\|y^*y\|, \|y^*y\|] \subset [-\|y\|^2, \|y\|^2].$$

By the spectral theorem 9.6, $\|y\|^2 - y^*y$ is a non-negative element, so its trace $\tau(\|y\|^2) - \tau(y^*y)$ is non-negative, and

$$|\tau(y)|^2 \leq \tau(y^*y) \leq \tau(\|y\|^2) = \|y\|^2\,\tau(1_A) = \|y\|^2,$$

which ends the proof of the claim. Thus, any state on a von Neumann algebra is automatically continuous.

With a self-adjoint in A, we now consider the map

$$\begin{aligned}\tau_a : \mathscr{C}^0(\sigma(T), \mathbb{R}) &\to \mathbb{R}\\ f &\mapsto \tau(f(a)).\end{aligned}$$

By the previous discussion and the spectral theorem, $|\tau_a(f)| \leq \|f(a)\|_{\mathscr{B}(H)} = \|f\|_\infty$. So, τ_a is a bounded linear form on $\mathscr{C}^0(\sigma(T), \mathbb{R})$, which sends non-negative functions to non-negative real numbers. Since $\sigma(T)$ is a compact set, by **Riesz' representation theorem** of positive functionals on spaces of continuous functions, there exists a unique Borelian positive measure μ_a on $\sigma(T)$ such that

$$\tau_a(f) = \int_{\sigma(T)} f(s)\, \mu_a(ds).$$

Moreover, $1 = \tau_a(1) = \int_{\sigma(T)} \mu_a(ds)$ and μ_a is a probability measure. Finally, the integral representation holds in particular for any polynomial $P \in \mathbb{R}[X]$, and by $\mathbb{C}$-linearity, for any polynomial $P \in \mathbb{C}[X]$. The existence of the law of a is thus established, and the unicity comes from the unicity in Riesz' representation theorem, and from the density of $\mathbb{R}[X]$ in the space of continuous functions $\mathscr{C}^0(\sigma(T), \mathbb{R})$ (**Stone–Weierstrass theorem**). □

Example. Let A be a semisimple complex algebra, endowed with a trace $\tau = \sum_{\lambda \in \widehat{A}} p_\lambda\, \tau^\lambda$, where τ^λ is the normalized trace on $\mathrm{End}(V^\lambda)$. If $a \in A$ is a self-adjoint element, then each matrix $\rho^\lambda(a) \in \mathrm{End}(V^\lambda)$ is Hermitian, with real eigenvalues $s_1^\lambda(a), s_2^\lambda(a), \ldots, s_{d_\lambda}^\lambda(a)$. We then have

$$\mu_a = \sum_{\lambda \in \widehat{A}} \frac{p_\lambda}{d_\lambda} \sum_{i=1}^{d_\lambda} \delta_{s_i^\lambda(a)}.$$

Indeed, if P is a polynomial in $\mathbb{C}[X]$, then the operator $\rho^\lambda(P(a))$ has eigenvalues $P(s_1^\lambda(a)), \ldots, P(s_{d_\lambda}^\lambda(a))$, so

$$\tau^\lambda(P(a)) = \frac{1}{d_\lambda} \sum_{i=1}^{d_\lambda} P(s_i^\lambda(a)) = \frac{1}{d_\lambda} \int_{\mathbb{R}} P(x) \left(\sum_{i=1}^{d_\lambda} \delta_{s_i^\lambda(a)} \right)(dx).$$

The result follows then from the decomposition of the trace as a barycenter of the τ^λ's.

Example. If a is a self-adjoint element of a non-commutative probability space (A, τ), consider the power series

$$C_a(z) = \sum_{k=0}^{\infty} \frac{\tau(a^k)}{z^{k+1}}.$$

This series is absolutely convergent for $|z| > \|a\|_{\mathscr{B}(H)}$, since $|\tau(a^k)| \leq \|a\|^k$ for any k. Moreover,

$$C_a(z) = \lim_{K\to\infty}\left(\sum_{k=0}^{K-1}\frac{\tau(a^k)}{z^{k+1}}\right) = \lim_{K\to\infty}\left(\frac{1}{z}\sum_{k=0}^{K-1}\int_{\mathbb{R}}\left(\frac{s}{z}\right)^k \mu_a(ds)\right)$$
$$= \lim_{K\to\infty}\left(\int_{\mathbb{R}}\frac{1-(\frac{s}{z})^K}{z-s}\,\mu_a(ds)\right) = \int_{\mathbb{R}}\frac{1}{z-s}\,\mu_a(ds)$$

for any $|z| > \|a\|$, so $C_a(z)$ is the Cauchy transform of the distribution of a. It has a unique analytic extension to $\mathbb{C}\setminus[-\|a\|,\|a\|]$. In particular, it is well defined on the whole Poincaré half-plane $\mathbb{H}$, and it belongs to the space $\mathscr{N}^1$ introduced in Section 7.4. Finally, if $|z| > \|a\|_{\mathscr{B}(\mathscr{H})}$, then $z-a = z\,1_A - a$ is invertible in A, with inverse given by the norm convergent power series $\sum_{k=0}^{\infty}\frac{a^k}{z^{k+1}}$. We have in this case $C_a(z) = \tau((z-a)^{-1})$, since τ is continuous.

Remark. Consider conversely a compactly supported probability measure μ on $\mathbb{R}$. Then, there exists a non-commutative probability space (A,τ) and $a \in A$ self-adjoint element such that $\mu = \mu_a$ is the distribution of a. Indeed, set $H = \mathrm{L}^2(\mathbb{R},\mathscr{F}_{\mathbb{R}},\mu)$ where $\mathscr{F}_{\mathbb{R}}$ is the set of Borelian subsets of $\mathbb{R}$. The algebra $A = \mathrm{L}^\infty(\mathbb{R},\mathscr{F}_{\mathbb{R}},\mu)$ is a von Neumann algebra acting on H by the discussion at the beginning of this section, and it is endowed with the trace $\tau(f) = \int_{\mathbb{R}} f(x)\,\mu(dx)$. Set $a = (x \mapsto x)$. This function is essentially bounded, because μ has compact support; so $a \in A$. Moreover, obviously, for any polynomial P, $\tau(P(a)) = \int_{\mathbb{R}} P(x)\,\mu(dx)$, so $\mu = \mu_a$.

▷ *The R-transform and the free convolution of probability measures.*

With the previous definitions, one can extend many notions of classical probability theory to the non-commutative setting: moments of a non-commutative random variable, convergence in distribution of a sequence of non-commutative random variables, etc. The final step in our introduction to non-commutative probability theory is the definition of a non-commutative analogue of the notion of independent random variables. Though one could transpose directly the definition of independence to the non-commutative setting, it turns out that this is not the most interesting thing to do. The notion of **free random variables** is more complex, but leads to a richer theory:

Definition 9.9. *Let $a_1,\ldots,a_n$ be random variables in a non-commutative probability space (A,τ). We say that these variables are free if, for any polynomials $P_1,\ldots,P_r$ and any indices $i_1,\ldots,i_r$ in $[\![1,n]\!]$ such that*

$$i_1 \neq i_2,\ i_2 \neq i_3,\ldots,i_{r-1}\neq i_r,$$

we have

$$\tau\big((P_1(a_{i_1})-\tau(P_1(a_{i_1})))(P_2(a_{i_2})-\tau(P_2(a_{i_2})))\cdots(P_r(a_{i_r})-\tau(P_r(a_{i_r})))\big) = 0.$$

Notice that the notion is stable by translation of the a_i's by constants $\lambda_i\,1_A$.

More generally, if $a_1,\dots,a_n$ are free random variables, then for any polynomials $P_1,\dots,P_n$, the random variables $P_1(a_1),\dots,P_n(a_n)$ are again free random variables.

Remark. In a commutative probability space $\mathrm{L}^\infty(\Omega,\mathscr{F},\mathbb{P})$, the independence of bounded random variables $X_1,\dots,X_n$ is defined by similar conditions: for any polynomials $P_1,\dots,P_n$,

$$\mathbb{E}[(P_1(X_1)-\mathbb{E}[P_1(X_1)])\cdots(P_n(X_n)-\mathbb{E}[P_n(X_n)])]=0.$$

However, the notion of freeness is truly different from the notion of independence. Consider for instance two free random variables a and b with expectation 0: $\tau(a)=\tau(b)=0$. Then, by definition of freeness, $\tau(abab)=0$. This is in strong contrast with the notion of independence of commuting random variables, which would imply $\tau(abab)=\tau(a^2)\,\tau(b^2)$.

Example. Let a and b be two free random variables in a non-commutative probability space (A,τ). We abbreviate $a'=a-\tau(a)$ and $b'=b-\tau(b)$. Let us compute the first joint moments of these variables. We have

$$0=\tau(a'b')=\tau((a-\tau(a))(b-\tau(b))=\tau(ab)-\tau(a)\tau(b),$$

hence $\tau(ab)=\tau(a)\tau(b)$. Since polynomials of a and b are again free, we also have $\tau(ab^2)=\tau(a)\tau(b^2)$ and $\tau(a^2b)=\tau(a^2)\tau(b)$. Then,

$$\begin{aligned}\tau(aba)&=\tau(ab'a)+\tau(a^2)\,\tau(b)=\tau(a'b'a)+\tau(a^2)\,\tau(b)\\&=\tau(a'b'a')+\tau(a^2)\,\tau(b)=\tau(a^2)\,\tau(b).\end{aligned}$$

The first difference between these computations and the case of independent random variables occurs for $\tau(abab)$. Setting $c=ba'b$, we have $\tau(c)=0$ by the previous computation, and

$$\begin{aligned}\tau(abab)&=\tau(a)\,\tau(ab^2)+\tau(ac)=\tau(a)^2\,\tau(b^2)+\tau(a'c)\\&=\tau(a)^2\,\tau(b^2)+\tau(a'ba'b)=\tau(a)^2\,\tau(b^2)+\tau(b)\tau((a')^2b)\end{aligned}$$

by using the previous cases and the vanishing of $\tau(a'b'a'b')$. We conclude that:

$$\tau(abab)=\tau(a)^2\,\tau(b^2)+\tau(a^2)\,\tau(b)^2-(\tau(a)\,\tau(b))^2,$$

which is quite different from the case of independent random variables.

The previous example is generalized by the following result:

Proposition 9.10. *Let $a_1,\dots,a_n$ be free random variables in (A,τ). For any non-commutative polynomial $P\in\mathbb{C}\langle X_1,\dots,X_n\rangle$, $\tau(P(a_1,\dots,a_n))$ is a polynomial in the individual moments $\tau((a_i)^j)$.*

Proof. It suffices to treat the case of monomials, so we can look at the trace $\tau\left((a_{i_1})^{k_1}(a_{i_2})^{k_2}\cdots(a_{i_n})^{k_n}\right)$, where the consecutive a_{i_j}'s are different. Set $(a_i^k)' = a_i^k - \tau(a_i^k)$. Then, we can expand the trace as

$$\tau\left(((a_{i_1}^{k_1})' + \tau(a_{i_1}^{k_1}))((a_{i_2}^{k_2})' + \tau(a_{i_2}^{k_2}))\cdots((a_{i_n}^{k_n})' + \tau(a_{i_n}^{k_n}))\right)$$
$$= \sum_{J\subset[\![1,n]\!]} \left(\prod_{j\in[\![1,n]\!]\setminus J} \tau(a_{i_j}^{k_j})\right) \tau\left(\prod_{j\in J}(a_{i_j}^{k_j})'\right).$$

In this expansion, the term $\tau((a_{i_1}^{k_1})'\cdots(a_{i_n}^{k_n})')$ corresponding to $J = [\![1,n]\!]$ vanishes by freeness, and by an induction on n, the other terms are polynomials in the individual moments, whence the result. □

Let a and b be two free random variables in (A,τ). For any $k \geq 1$, $\tau((a+b)^k)$ is a polynomial in the moments $\tau(a^i)$ and $\tau(b^j)$, which depends only on k. Therefore, under the assumption of freeness, the distribution μ_{a+b} of the random variable $a+b$ only depends on μ_a and μ_b, and not on the way a and b are embedded in the algebra (A,τ).

Definition 9.11. *Let μ and ν be two compactly supported probability measures on $\mathbb{R}$. We denote $\mu \boxplus \nu$ the distribution of a random variable $a+b$, where a and b are non-commutative random variables in a space (A,τ) such that a has distribution μ and b has distribution ν. We call $\mu \boxplus \nu$ the* ***free convolution*** *of μ and ν.*

By the previous discussion, the distribution of $a+b$ with a and b free indeed depends only on μ_a and μ_b, so the previous definition makes sense. On the other hand, one can show that if μ and ν are two compactly probability measures on $\mathbb{R}$, then there exists indeed a non-commutative probability space (C,τ), and a, b self-adjoint elements in C, such that $\mu_a = \mu$, $\mu_b = \nu$, and a and b are free. Indeed, by a previous remark, there exist two non-commutative probability spaces (A,τ_A) and (B,τ_B) with $a \in A$ of law μ, and $b \in B$ of law ν. Set then $H = A * B$, the **free amalgamated product** of A and B, which is the orthogonal direct sum of the spaces

$$A_0 \otimes B_0 \otimes A_0 \otimes B_0 \otimes \cdots \qquad \text{and} \qquad B_0 \otimes A_0 \otimes B_0 \otimes A_0 \otimes \cdots$$

where $A_0 = \{x \in A,\ \tau_A(x) = 0\}$ and $B_0 = \{y \in B,\ \tau_B(y) = 0\}$. The element a acts on H by

$$a\cdot(a_1 \otimes b_1 \otimes \cdots \otimes z_r) = (aa_1 - \tau_A(aa_1)) \otimes b_1 \otimes \cdots \otimes z_r + \tau_A(aa_1)\, b_1 \otimes a_2 \otimes \cdots \otimes z_r$$

and

$$a\cdot(b_1 \otimes a_1 \otimes \cdots \otimes z_r) = (a - \tau_A(a)) \otimes b_1 \otimes a_1 \otimes \cdots \otimes z_r + \tau_A(a)\, b_1 \otimes a_1 \otimes \cdots \otimes z_r.$$

We define similarly the action of b on H, and we obtain self-adjoint elements a and b in $\mathscr{B}(H)$, this space being endowed with the trace $\tau(T) = \langle 1 \mid T(1)\rangle_H$, where 1 is the empty tensor product in H. Let C be the smallest von Neumann algebra containing a and b in $\mathscr{B}(H)$. We verify trivially:

$$\tau((a^{k_1})'(b^{l_1})'(a^{k_2})'(b^{l_2})'\cdots) = \langle 1 \,|\, (a^{k_1})' \otimes (b^{l_1})' \otimes (a^{k_2})' \otimes (b^{l_2})' \otimes \cdots \rangle_H = 0,$$

so a and b are free in the non-commutative probability space (C, τ). Therefore, $\mu \boxplus \nu$ is a well-defined probability measure for *any* pair of compactly supported probability measures μ and ν on $\mathbb{R}$.

In classical probability theory, if μ and ν are the laws of two independent random variables X and Y, then a way to compute their classical convolution $\mu * \nu$ is by means of the Fourier series:

$$\widehat{(\mu * \nu)}(\xi) = \widehat{\mu}(\xi)\,\widehat{\nu}(\xi),$$

where $\widehat{\mu}(\xi) = \int_{\mathbb{R}} \mathrm{e}^{\mathrm{i}\xi s}\,\mu(ds)$. A similar tool exists in non-commutative probability for the computation of the free convolution $\mu \boxplus \nu$ of two probability measures μ and ν. If μ is a compactly supported probability measure μ on the real line, its Cauchy transform $C_\mu(z)$ satisfies

$$C_\mu(z) = \frac{1}{z} + \frac{m_1}{z^2} + \frac{m_2}{z^3} + \cdots = \frac{1}{z\,(1 + o(1))}$$

when $z \to \infty$, and where $m_k = m_k(\mu) = \int_{\mathbb{R}} x^k\,\mu(dx)$ is the k-th moment of μ. Set $w = C_\mu(z)$; since $z \mapsto \frac{1}{z}$ maps bijectively a neighborhood of $+\infty$ to a neighborhood of 0 in the Riemann sphere, for $w \neq 0$ small enough, there exists a unique $z = K_\mu(w)$ such that $w = C_\mu(K_\mu(w))$. The map $w \mapsto K_\mu(w)$ has an analytic expansion around 0:

$$K_\mu(w) = \frac{1}{w} + R_1 + R_2\,w + R_3\,w^2 + \cdots$$

We set $R_\mu(w) = K_\mu(w) - \frac{1}{w} = \sum_{n=1}^{\infty} R_n(\mu)\,w^{n-1}$. This is the **R-transform** of the distribution μ, and the coefficients $R_n(\mu)$ are called the **free cumulants** of the distribution μ. We shall explain in Section 9.2 how to compute these invariants of μ in terms of the moments.

Example. Let μ be the Wigner distribution on $\mathbb{R}$, also called the **semi-circle law**:

$$\mu(dx) = \frac{\sqrt{4 - x^2}}{2\pi}\,1_{x \in [-2,2]}\,dx.$$

It is compactly supported on $[-2, 2]$. The Cauchy transform of μ was computed in Section 7.4: it is

$$C_\mu(z) = \frac{2}{z + \sqrt{z^2 - 4}} = \frac{z - \sqrt{z^2 - 4}}{2}.$$

Its Taylor expansion yields the moments of μ:

$$C_\mu(z) = \sum_{n=0}^{\infty} \frac{C_n}{z^{2n+1}}, \quad \text{with } C_n = \int_{\mathbb{R}} x^{2n}\,\mu(dx) = \frac{1}{n+1}\binom{2n}{n}.$$

One recognizes the well-known **Catalan numbers** C_n, with $C_0 = 1$, $C_1 = 1$, $C_2 = 2$, $C_3 = 5$, $C_4 = 14$ and $C_5 = 42$. Now, the inverse of C_μ at infinity is given by:

$$w = \frac{z - \sqrt{z^2-4}}{2} \quad ; \quad w^2 = \frac{z^2 - 2 - z\sqrt{z^2-4}}{2} = wz - 1 \quad ; \quad z = \frac{1}{w} + w.$$

Therefore, $R_\mu(w) = w$ and the free cumulants of the Wigner distribution are:

$$R_2(\mu) = 1 \quad ; \quad R_{k\neq 2}(\mu) = 0.$$

Theorem 9.12. *Let μ and ν be two compactly supported probability measures on $\mathbb{R}$. The free convolution $\mu \boxplus \nu$ is the unique probability measure such that*

$$R_n(\mu \boxplus \nu) = R_n(\mu) + R_n(\nu)$$

for any $n \geq 1$. In other words, $R_{\mu\boxplus\nu}(w) = R_\mu(w) + R_\nu(w)$.

Proof. We introduce two free random variables a and b of distribution μ and ν, and compute the inverse of the Cauchy transform of a and b. With the same notations as before, we have for any w small enough

$$w = C_\mu(K_\mu(w)) = \tau((K_\mu(w) - a)^{-1}),$$

so there exists an element $\widetilde{a}(w) \in A$ such that $\tau(\widetilde{a}(w)) = 0$ and

$$\begin{aligned} (K_\mu(w) - a)^{-1} &= w(1_A - \widetilde{a}(w)); \\ K_\mu(w) - a &= w^{-1}\,(1_A - \widetilde{a}(w))^{-1}. \end{aligned}$$

In these equalities, $\widetilde{a}(w)$ varies continuously with w, with $\widetilde{a}(0) = 0$. We have the same identities for ν and b, hence,

$$K_\mu(w) + K_\nu(w) - w^{-1} - a - b = w^{-1}\left((1_A - \widetilde{a}(w))^{-1} + (1_A - \widetilde{b}(w))^{-1} - 1\right).$$

If t is the term in parentheses on the right-hand side, then one checks readily that

$$(1_A - \widetilde{a}(w))\, t\, (1_A - \widetilde{b}(w)) = 1_A - \widetilde{a}(w)\widetilde{b}(w).$$

Hence,

$$\begin{aligned} &K_\mu(w) + K_\nu(w) - w^{-1} - a - b \\ &= w^{-1}\left((1_A - \widetilde{a}(w))^{-1}(1_A - \widetilde{a}(w)\widetilde{b}(w))(1_A - \widetilde{b}(w))^{-1}\right); \end{aligned}$$

and

$$\begin{aligned}&(K_\mu(w)+K_\nu(w)-w^{-1}-a-b)^{-1}\\&=w\left((1_A-\widetilde{b}(w))(1_A-\widetilde{a}(w)\widetilde{b}(w))^{-1}(1_A-\widetilde{a}(w))\right).\end{aligned}$$

On the last line, we used the fact that if w is small enough, then $\widetilde{a}(w)\widetilde{b}(w)$ is also small in norm in A by continuity of $\widetilde{a}$ and $\widetilde{b}$, so $1_A-\widetilde{a}(w)\widetilde{b}(w)$ is invertible in A.

For any w, $\widetilde{a}(w)$ can be written as a convergent series of powers of a, and similarly for $\widetilde{b}(w)$ in terms of b. Since a and b are free, the same holds for $\widetilde{a}(w)$ and $\widetilde{b}(w)$ (this is true for any polynomials in a and b, and by extension for any convergent series since the trace τ is continuous on A). In the last formula, we expand $(1-\widetilde{a}(w)\widetilde{b}(w))^{-1}$ in series:

$$\begin{aligned}&(1_A-\widetilde{b}(w))(1_A-\widetilde{a}(w)\widetilde{b}(w))^{-1}(1_A-\widetilde{a}(w))\\&=(1_A-\widetilde{b}(w))\left(\sum_{k=0}^{\infty}(\widetilde{a}(w)\widetilde{b}(w))^k\right)(1_A-\widetilde{a}(w)).\end{aligned}$$

We can take the trace of the last line, and all the terms vanish by freeness of the pair $(\widetilde{a}(w),\widetilde{b}(w))$, but the trivial product 1_A. So:

$$\tau((K_\mu(w)+K_\nu(w)-w^{-1}-a-b)^{-1})=w.$$

This is the identity characterizing $K_{\mu\boxplus\nu}(w)$, so

$$K_{\mu\boxplus\nu}(w)=K_\mu(w)+K_\nu(w)-w^{-1}\quad;\quad R_{\mu\boxplus\nu}(w)=R_\mu(w)+R_\nu(w).\qquad\square$$

Example. If $\mu=\nu$ is the Wigner law, then the free convolution $\mu\boxplus\mu$ is the unique law with R-transform $R_{\mu\boxplus\mu}(w)=w+w=2w$. Hence, $K_{\mu\boxplus\mu}(w)=\frac{1}{w}+2w$, which admits for inverse

$$C_{\mu\boxplus\mu}(z)=w=\frac{z-\sqrt{z^2-16}}{4}=C_\mu\left(\frac{z}{2}\right).$$

Hence, $\mu\boxplus\mu$ is the law of $2a$, where a follows the semi-circle distribution. In other words:

$$(\mu\boxplus\mu)(dx)=\frac{\sqrt{16-x^2}}{8\pi}\,dx.$$

This property of invariance makes the Wigner distribution play a role in free probability theory which is very similar to the role of the Gaussian distribution in classical probability theory. In particular, there exists a free central limit theorem, which shows that the scaled mean of free centered random variables with the same distribution always converges towards the Wigner distribution.

Example. If $\mu = \nu = \frac{1}{2}(\delta_1 + \delta_{-1})$, then the free convolution $\mu \boxplus \mu$ is the unique probability law with R-transform

$$R_{\mu\boxplus\mu}(w) = \frac{\sqrt{1+4w^2}-1}{w}.$$

Indeed, $C_\mu(z) = \frac{z}{z^2-1}$, so $K_\mu(w) = \frac{1+\sqrt{1+4w^2}}{2w}$ and $R_\mu(w) = \frac{\sqrt{1+4w^2}-1}{2w}$. From this we deduce $C_{\mu\boxplus\mu}(z) = \frac{1}{\sqrt{z^2-4}}$, and by the Perron–Stieltjes inversion formula, $\mu \boxplus \mu$ is the law with density

$$\frac{1}{\pi\sqrt{4-x^2}}\, 1_{x\in[-2,2]}\, dx,$$

that is to say the **arcsine law**.

9.2 Free cumulants of Young diagrams

In Section 7.4, we introduced the generating function $G_\omega(z)$ of any (continuous, generalized) Young diagram: it is the Cauchy transform $C_{\mu_\omega}(z)$ of its transition measure μ_ω (the probability measure on $\mathbb{R}$ that is associated to ω by the Markov–Krein correspondence). If $\omega \in \mathscr{Y}$ is a non-generalized continuous Young diagram, then we remarked in Chapter 7 that the transition measure μ_ω was also compactly supported. So, the previous theory applies, and we can define the R-transform $R_{\mu_\omega}(z) = \sum_{k=1}^{\infty} R_k(\mu_\omega)\, z^{k-1}$ of μ_ω (not to be confused with the Rayleigh function $R_\omega(s)$ associated to $\omega(s)$), and the **free cumulants** of the Young diagram

$$R_k(\omega) = R_k(\mu_\omega).$$

The goal of this section is to prove that these free cumulants form a new algebraic basis of $\mathscr{O}$, and that they correspond to the top weight components of the symbols Σ_k.

▷ *The combinatorics of non-crossing partitions.*

A **set partition** of a (finite) set X is a family $\pi = (\pi_1, \ldots, \pi_\ell)$ of non-empty disjoint subsets of X such that $X = \bigsqcup_{i=1}^{\ell} \pi_i$. We denote $\mathfrak{Q}(X)$ the set of set partitions of X; for instance, if $X = [\![1,3]\!]$, then $\mathfrak{Q}(X) = \mathfrak{Q}(3)$ has 5 elements, namely,

$$\{1\} \sqcup \{2\} \sqcup \{3\},\ \{1\} \sqcup \{2,3\},\ \{2\} \sqcup \{1,3\},\ \{3\} \sqcup \{1,2\},\ \{1,2,3\}.$$

If π is a set partition, we usually denote $\ell = \ell(\pi)$ its number of parts. We also abbreviate $\mathfrak{Q}([\![1,n]\!]) = \mathfrak{Q}(n)$. The cardinality of $\mathfrak{Q}(n)$ is the **Bell number** B_n, and we have

$$B_1 = 1 \quad ; \quad B_2 = 2 \quad ; \quad B_3 = 5 \quad ; \quad B_4 = 15.$$

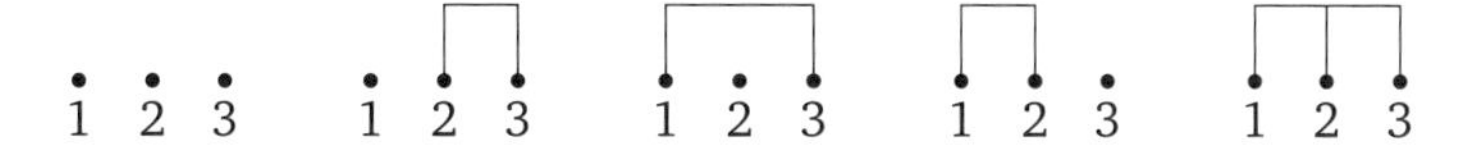

Figure 9.1
The set partitions in $\mathfrak{Q}(3)$, represented as graphs.

It is convenient to represent set partitions of X by graphs that connect elements of X if they are in the same part. Thus, the list of set partitions of $[\![1,3]\!]$ corresponds to the list of graphs in Figure 9.1.

Definition 9.13. *A* ***non-crossing partition*** *of size n is a set partition π in $\mathfrak{Q}(n)$ that does not contain two disjoint parts π_i and π_j with $a < b < c < d$ and*

$$a, c \in \pi_i \qquad ; \qquad b, d \in \pi_j.$$

In other words, the graph associated to π does not contain the configuration

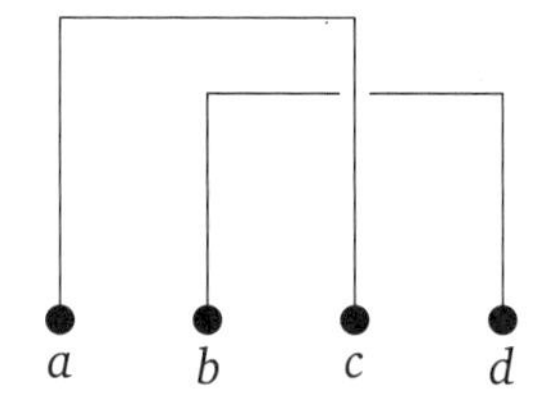

We denote $\mathfrak{NC}(n)$ the set of non-crossing partitions of size n. For instance, $\mathfrak{NC}(4)$ consists of 14 elements, namely, all the set partitions in $\mathfrak{Q}(4)$ but the set partition $\{1,3\} \sqcup \{2,4\}$.

Proposition 9.14. *The cardinality of $\mathfrak{NC}(n)$ is the Catalan number $C_n = \frac{1}{n+1}\binom{2n}{n}$.*

Proof. Set $D_n = \operatorname{card} \mathfrak{NC}(n)$. We are going to show that D_n satisfies the recurrence relation $D_{n+1} = \sum_{k=0}^{n} D_k D_{n-k}$. If $\pi \in \mathfrak{NC}(n+1)$, let $i \in [\![1, n+1]\!]$ be the smallest element in the same part as $n+1$ in i. Then, the graph of π looks as follows:

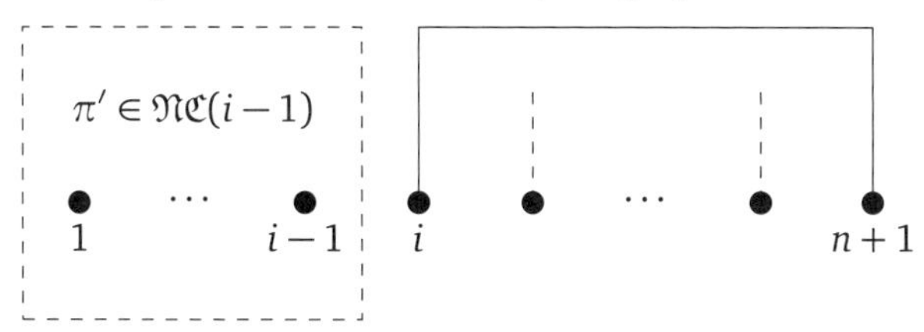

The elements in $[\![1, i-1]\!]$ are in parts that form a non-crossing partition $\pi' \in \mathfrak{NC}(i-1)$. On the other hand, the elements in $[\![i, n+1]\!]$ might be in the same part as $n+1$, or in other parts. If one removes $n+1$ from its part, then what remains is a non-crossing partition on the set $\mathfrak{NC}([\![i,n]\!]) = \mathfrak{NC}(n-i+1)$. Therefore,

$$D_{n+1} = \sum_{i=1}^{n+1} D_{i-1} D_{n-i+1} = \sum_{k=0}^{n} D_k\, D_{n-k}.$$

If $D(z) = \sum_{n=0}^{\infty} D_n\, z^n$, then the previous recurrence yields

$$D(z) - 1 = zD(z)^2 \qquad ; \qquad D(z) = \frac{1-\sqrt{1-4z}}{2z}.$$

We saw before that this is the generating series of the Catalan numbers (and the Cauchy transform of the Wigner distribution). In particular, the formula $C_n = \frac{1}{n+1}\binom{2n}{n}$ is obtained by a Taylor expansion of the function $D(z)$. □

The set $\mathfrak{NC}(n)$ of non-crossing partitions is endowed with the partial order $\leq$ of refinement: a non-crossing partition π is said to be finer than a non-crossing partition ρ if every part of π is included in a part of ρ. The Hasse diagram of this partial order is drawn in Figure 9.2 when $n=4$. As is clear in this figure, the ordered set $(\mathfrak{NC}(n), \leq)$ enjoys the following property:

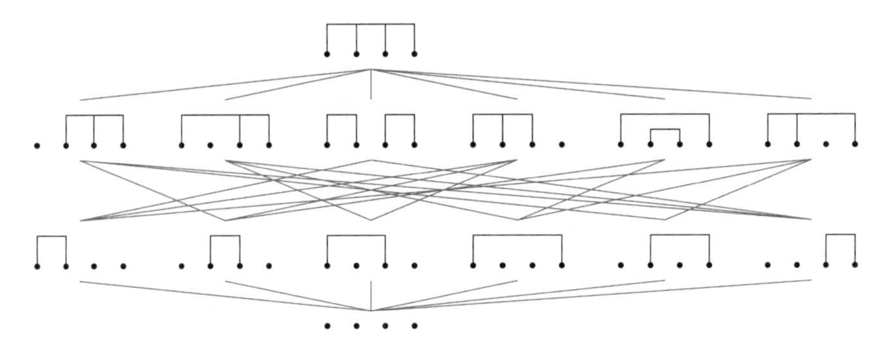

Figure 9.2
The order of refinement on the lattice $\mathfrak{NC}(4)$.

Proposition 9.15. *The ordered set $(\mathfrak{NC}(n), \leq)$ is a* ***lattice****, i.e., for any non-crossing partitions $\pi^{(1)}$ and $\pi^{(2)}$, there exists a non-crossing partition* $\min(\pi^{(1)}, \pi^{(2)})$ *and a non-crossing partition* $\max(\pi^{(1)}, \pi^{(2)})$.

Proof. We treat for instance the case of $\min(\pi^{(1)}, \pi^{(2)})$. Let π be the set partition on $[\![1,n]\!]$ whose parts are defined as follows: a and b are in the same part of π if and only if they are in the same part of $\pi^{(1)}$ and of $\pi^{(2)}$. We claim that π is non-crossing. Indeed, let π_1 and π_2 be two parts of π containing, respectively, elements a and c and elements b and d such that $a<b<c<d$. Since $\pi \leq \pi^{(1)}$,

a and c are in the same part of $\pi^{(1)}$, and b and d are in the same part of $\pi^{(1)}$. Since $\pi^{(1)}$ is non-crossing, a, b, c, d belong to the same part of $\pi^{(1)}$, and also to the same part of $\pi^{(2)}$. By construction of π, they belong then to the same part of π, hence $\pi_1 = \pi_2$ and π is non-crossing. It is then immediate by construction that π is the largest non-crossing partition smaller than $\pi^{(1)}$ and $\pi^{(2)}$. □

Remark. The proof shows that $\mathfrak{NC}(n)$ is a sublattice of the whole lattice of set partitions $\mathfrak{Q}(n)$.

Another important property of $\mathfrak{NC}(n)$ is a symmetry that also appears in Figure 9.2. We embed $[\![1,n]\!]$ into $[\![1,2n]\!]$ by rewriting the elements of $[\![1,2n]\!]$ as

$$1 \leq 1' \leq 2 \leq 2' \leq \ldots \leq n \leq n'.$$

A non-crossing partition in $\mathfrak{NC}(n)$ becomes then a non-crossing partition in $\mathfrak{NC}(2n)$, which has all the elements i' in parts of size 1. The **Kreweras complement** of $\pi \in \mathfrak{NC}(n)$ is defined as the largest non-crossing set partition π' on $[\![1',n']\!]$ such that $\pi \sqcup \pi'$ is a non-crossing partition in $\mathfrak{NC}(2n)$. We leave the reader to check that in Figure 9.2, the non-crossing partitions that are drawn on the second line (starting from the bottom) are all placed just below their Kreweras complements, which are drawn on the third line.

▷ *Möbius inversion and formal free cumulants.*

If $f : \mathfrak{NC}(n) \to \mathbb{C}$ is a function, denote $g = F(f)$ the new function on $\mathfrak{NC}(n)$ defined by

$$g(\pi) = \sum_{\rho \in \mathfrak{NC}(n),\, \rho \leq \pi} f(\rho).$$

A general property of **posets** (partially ordered sets) ensures the existence of an inversion formula for the transformation F:

Theorem 9.16. *There exists a unique function $\mu : \mathfrak{NC}(n) \times \mathfrak{NC}(n) \to \mathbb{Z}$, called the* **Möbius function** *of the lattice $\mathfrak{NC}(n)$, which vanishes on pairs (ρ, π) with $\rho \not\leq \pi$, and such that if $g = F(f)$, then*

$$f(\pi) = \sum_{\rho \in \mathfrak{NC}(n),\, \rho \leq \pi} \mu(\rho, \pi)\, g(\rho).$$

Proof. This is a general property of finite posets, which can be shown by induction on the number of elements. Indeed, we have

$$f(\pi) = g(\pi) - \sum_{\rho < \pi} f(\rho),$$

so if the Möbius function is defined on any elements $\rho_1, \rho_2 < \pi$, then

$$f(\pi) = g(\pi) - \sum_{\rho < \pi} \sum_{\theta \leq \rho} \mu(\theta, \rho)\, g(\theta),$$

which gives a formula for $\mu(\rho,\pi)$ in terms of lower terms:

$$\mu(\pi,\pi)=1 \quad ; \quad \mu(\theta,\pi)=-\sum_{\theta\leq\rho<\pi}\mu(\theta,\rho).$$

This proves the existence of the Möbius function, and the unicity is obvious since $f \mapsto F(f) = g$ is a linear transformation of $\mathbb{C}^{\mathfrak{NC}(n)}$, hence admits at most one inverse. $\square$

Let $(m_1, m_2, \ldots)$ be an infinite sequence of commuting variables. We define the associated sequence of **formal free cumulants** $(k_1, k_2, \ldots)$ by the formulas:

$$\forall n \geq 1,\ m_n = \sum_{\pi\in\mathfrak{NC}(n)} k_\pi,$$

where $k_\pi = k_{|\pi_1|}k_{|\pi_2|}\cdots k_{|\pi_\ell|}$, $|\pi_i|$ being the number of elements in a part π_i. By the previous discussion, these formulas are equivalent to

$$\forall n \geq 1,\ k_n = \sum_{\pi\in\mathfrak{NC}(n)} \mu(\pi)\, m_\pi,$$

where $\mu(\pi) = \mu(\pi, [\![1,n]\!])$, $[\![1,n]\!]$ being the maximal element of $\mathfrak{NC}(n)$ (the set partition with one part of size n). For instance, we have

$$m_1 = k_1 \quad ; \quad m_2 = k_2 + (k_1)^2 \quad ; \quad m_3 = k_3 + 3\,k_2k_1 + (k_1)^3;$$
$$m_4 = k_4 + 4\,k_3k_1 + 2\,(k_2)^2 + 6\,k_2(k_1)^2 + (k_1)^4,$$

and these relations can be inversed to give

$$k_1 = m_1 \quad ; \quad k_2 = m_2 - (m_1)^2 \quad ; \quad k_3 = m_3 - 3\,m_2m_1 + 2\,(m_1)^3;$$
$$k_4 = m_4 - 4\,m_3m_1 + 10\,m_2(m_1)^2 - 2\,(m_2)^2 - 5\,(m_1)^4.$$

We convene that $m_0 = k_0 = 1$.

Proposition 9.17. *The following are equivalent:*

(i) *The sequence $(k_n)_{n\in\mathbb{N}}$ is the sequence of formal free cumulants associated to $(m_n)_{n\in\mathbb{N}}$.*

(ii) *If $M(z) = \sum_{n=0}^{\infty} m_n\, z^n$ and $K(z) = \sum_{n=0}^{\infty} k_n z^n$, then*

$$K(z\,M(z)) = M(z).$$

Proof. Fix a sequence $(m_n)_{n\in\mathbb{N}}$ of variables, and denote $(k_n)_{n\in\mathbb{N}}$ the corresponding sequence of formal free cumulants. We use an argument similar to the one that allowed us to compute the Catalan numbers. If $\pi \in \mathfrak{NC}(n)$, we denote $\pi_1 = \{a_1 < a_2 < \cdots < a_r = n\}$ the part that contains n. Then, setting $a_0 = 0$, the remaining parts of π form non-crossing partitions on the intervals $[\![a_0+1, a_1-1]\!]$,

$[\![a_1+1, a_2-1]\!]$, etc. We denote $\pi^{(1)}, \dots, \pi^{(r)}$ these non-crossing partitions. We then have

$$k_\pi = k_{\pi_1} \prod_{i=1}^{r} k_{\pi^{(i)}}.$$

This leads to the following recurrence equation, for $n \geq 1$:

$$\begin{aligned}
m_n &= \sum_{\pi \in \mathfrak{NC}(n)} k_\pi = \sum_{r=1}^{n} k_r \left(\sum_{a_1 < a_2 < \cdots < a_r = n} \sum_{\pi^{(1)}, \dots, \pi^{(r)}} \prod_{i=1}^{r} k_{\pi^{(i)}} \right) \\
&= \sum_{r=1}^{n} k_r \left(\sum_{a_1 < a_2 < \cdots < a_r = n} m_{a_1 - a_0 - 1} m_{a_2 - a_1 - 1} \cdots m_{a_r - a_{r-1} - 1} \right) \\
&= \sum_{r=1}^{n} k_r \left(\sum_{n_1 + n_2 + \cdots + n_r = n - r} m_{n_1} m_{n_2} \cdots m_{n_r} \right)
\end{aligned}$$

where on the last line we made the change of variables $n_i = a_i - a_{i-1} - 1$. Since this equation determines the m_n's in terms of the k_n's, it is equivalent to the fact that $(k_n)_{n \in \mathbb{N}}$ is the sequence of formal free cumulants associated to $(m_n)_{n \in \mathbb{N}}$. We now take the generating series of the recurrence relation:

$$\begin{aligned}
M(z) = \sum_{n=0}^{\infty} m_n z^n &= 1 + \sum_{n=1}^{\infty} \sum_{r=1}^{n} k_r z^r \left(\sum_{n_1 + n_2 + \cdots + n_r = n - r} m_{n_1} m_{n_2} \cdots m_{n_r} z^{n-r} \right) \\
&= 1 + \sum_{r=1}^{\infty} k_r z^r \sum_{n-r \geq 0} z^{n-r} \, [z^{n-r}](M^r(z)) \\
&= 1 + \sum_{r=1}^{\infty} k_r (z\, M(z))^r = K(z\, M(z)).
\end{aligned}$$

□

▷ *The top-weight component of Σ_k.*

We now relate these formal identities to the relations satisfied by the moments and free cumulants of a compactly supported probability measure.

Theorem 9.18. *The free cumulants $R_n(\mu)$ of a compactly supported probability measure on $\mathbb{R}$ are given by the following formula:*

$$R_n(\mu) = \sum_{\pi \in \mathfrak{NC}(n)} \mu(\pi)\, M_\pi(\mu),$$

where $M_\pi(\mu) = \prod_{i=1}^{\ell(\pi)} M_{|\pi_i|}(\mu) = \prod_{i=1}^{\ell(\pi)} \int_{\mathbb{R}} x^{|\pi_i|} \mu(dx)$.

Proof. Set $M_n(\mu) = m_n$, and $R_n(\mu) = k_n$. The Cauchy transform $C_\mu(z)$ and the

R-transform $R_\mu(z)$ satisfy $C_\mu(\frac{1}{z}+R_\mu(z)) = C_\mu(K_\mu(z)) = z$, and on the other hand,

$$C_\mu(z) = \frac{1}{z}\, M\left(\frac{1}{z}\right) \quad ; \quad R_\mu(z) = \frac{K(z)-1}{z};$$
$$K(z) = 1 + z\, R_\mu(z) = z(K_\mu(z)).$$

Then, $K(z\,M(z)) = z\,M(z)\,K_\mu(z\,M(z)) = z\,M(z)\,K_\mu(C_\mu(\frac{1}{z})) = \frac{z\,M(z)}{z} = M(z)$, hence the result by the previous proposition. □

An immediate consequence of the relation between moments and free cumulants is:

Corollary 9.19. *For any $k \geq 2$, $\lambda \mapsto R_k(\mu_\lambda) = R_k(\lambda)$ is an observable of diagrams of weight k, and $(R_k)_{k\geq 2}$ is an algebraic basis of $\mathscr{O}$.*

Proof. We saw in Section 7.4 that $\widetilde{h}_k(\lambda) = \int_{\mathbb{R}} s^k\, \mu_\lambda(ds)$ is an observable of Young diagrams for any $k \geq 1$, the $\widetilde{h}_k$'s being related to the $\widetilde{p}_k$'s in the same way as the homogeneous symmetric functions and the power sums in Sym. Moreover, one has $\widetilde{h}_1(\lambda) = \widetilde{p}_1(\lambda) = 0$ for any Young diagram, and the others $\widetilde{p}_k$ or $\widetilde{h}_k$ form an algebraic basis of $\mathscr{O}$. Since

$$R_k(\lambda) = \sum_{\pi \in \mathfrak{NC}(k)} \widetilde{h}_\pi(\lambda),$$

we conclude readily that R_k is an observable. As $R_1 = \widetilde{h}_1 = 0$ and

$$R_{k\geq 2} = \widetilde{h}_k + \text{polynomial in } \widetilde{h}_2, \ldots, \widetilde{h}_{k-1} \text{ of total homogeneous degree } k,$$

we then see that $(R_k)_{k\geq 2}$ is a graded basis of $(\mathscr{O}, \mathrm{wt})$, since the same holds for $(\widetilde{h}_k)_{k\geq 2}$. □

Example. We gave before a formula for the first free cumulants in terms of the corresponding moments. Specializing $\widetilde{h}_1$ to 0, we obtain:

$$R_1 = \widetilde{h}_1 = \widetilde{p}_1 = 0;$$
$$R_2 = \widetilde{h}_2 = \frac{\widetilde{p}_2}{2};$$
$$R_3 = \widetilde{h}_3 = \frac{\widetilde{p}_3}{3};$$
$$R_4 = \widetilde{h}_4 - 2(\widetilde{h}_2)^2 = \frac{\widetilde{p}_4}{4} - \frac{3(\widetilde{p}_2)^2}{8}.$$

If one uses the expression of the $\widetilde{p}_k$'s in terms of the symbols Σ_k, one obtains an expansion of the first Σ_k's in terms of the free cumulants:

$$\begin{aligned}
\Sigma_1 &= R_2; \\
\Sigma_2 &= R_3; \\
\Sigma_3 &= R_4 + R_2; \\
\Sigma_4 &= R_5 + 5R_3; \\
\Sigma_5 &= R_6 + 15R_4 + 5(R_2)^2 + 8R_2.
\end{aligned}$$

It is a remarkable fact that we obtain each time a polynomial with positive coefficients in the R_k's, and most of Chapter 10 will be devoted to the study of this phenomenon. An easier fact regarding the expansion of an observable Σ_k in terms of the free cumulants is:

Theorem 9.20. *Call homogeneous of weight $k \geq 2$ an observable $f \in \mathscr{O}$ that is a homogeneous polynomial of total degree k in the $\widetilde{p}_k$'s, or the $\widetilde{h}_k$'s, or the R_k's. The top-weight homogeneous component of Σ_k is the free cumulant R_{k+1}.*

Proof. We can use Lagrange inversion formula to give another formula for the R_k's in terms of the $\widetilde{h}_k$'s:

$$\forall k \geq 2,\ R_k = -\frac{1}{k-1}[t^k](H^{-(k-1)}(t)),$$

where $H(t) = 1 + \sum_{k\geq 2} \widetilde{h}_k\, t^k = \exp\left(\sum_{k\geq 2} \frac{\widetilde{p}_k}{k}\, t^k\right)$. Indeed, the formula $z = C_\mu(K_\mu(z))$ is equivalent to the conditions needed to apply Lemma 7.18. So,

$$R_{k+1} = -\frac{1}{k}[t^{k+1}]\left(\exp\left(-k\sum_{j=2}^{\infty} \frac{\widetilde{p}_j}{j}\, t^j\right)\right).$$

This is the top weight component in Wassermann's formula for Σ_k, whence the result. □

Since $\mathrm{wt}(\cdot)$ is a gradation of algebra, we have more generally

$$\Sigma_\mu = \prod_{i=1}^{\ell(\mu)} R_{\mu_i+1} + \text{term of lower weight}.$$

If ω is a continuous Young diagram, recall that $\omega^{(u)}$ is its dilation by u in both directions (abscissa and ordinate). We have the following important corollary of Theorem 9.20:

Corollary 9.21. *For any continuous Young diagram ω,*

$$\lim_{u\to\infty} \frac{\Sigma_k(\omega^{(u)})}{u^{\frac{k+1}{2}}} = R_{k+1}(\omega).$$

Proof. We know from Theorem 9.20 and Proposition 7.21 that

$$\begin{aligned}\Sigma_k(\omega^{(u)}) &= R_{k+1}(\omega^{(u)}) + \text{observable of weight smaller than } k \text{ evaluated on } \omega^{(u)} \\ &= R_{k+1}(\omega^{(u)}) + O(u^{\frac{k}{2}}).\end{aligned}$$

We also saw during the proof of Proposition 7.21 that if f is a homogeneous observable of weight $k+1$, then $f(\omega^{(u)}) = u^{\frac{k+1}{2}} f(\omega)$. This implies the result. □

9.3 Transition measures and Jucys–Murphy elements

It might come as a bit of a surprise that the formalism of free probability is involved in the combinatorics of the algebra of observables $\mathscr{O}$, and thus in the representation theory of symmetric groups. In this section, we describe another occurrence of non-commutative probability theory in the theory of observables of Young diagrams. Namely, if λ is a Young diagram in $\mathfrak{Y}$, then there is an interpretation of μ_λ as the distribution of a non-commutative random variable in a symmetric group algebra (Theorem 9.23), this finite-dimensional complex algebra being endowed with a certain structure of non-commutative probability space. In this section, we establish this connection by using representation theoretic arguments.

▷ *The symmetric group algebra as a non-commutative probability space.*

Fix $\lambda \in \mathfrak{Y}(n)$. Recall from Section 9.1 that there is a bijection between structures of non-commutative probability spaces on the group algebra $\mathbb{C}\mathfrak{S}(n+1)$, and probability measures on the set of irreducible representations of $\mathfrak{S}(n+1)$, that is to say the set of Young diagrams $\mathfrak{Y}(n+1)$. On the other hand, by Proposition 7.19, μ_λ can be considered as a probability measure on $\mathfrak{Y}(n+1)$, which gives a weight $\frac{\dim S^\Lambda}{(n+1)\dim S^\lambda}$ to $\Lambda \in \mathfrak{Y}(n+1)$ if Λ can be obtained by adding one box to λ. Thus, we can restate the aforementioned proposition as:

Proposition 9.22. *Viewed as a probability measure on $\mathfrak{Y}(n+1)$, the transition measure μ_λ corresponds to the spectral measure on $\mathfrak{Y}(n+1)$ associated to the representation $\mathrm{Ind}_{\mathfrak{S}(n)}^{\mathfrak{S}(n+1)}(S^\lambda)$.*

Another interpretation of μ_λ is obtained by looking at the Jucys–Murphy element $J_{n+1} \in \mathbb{C}\mathfrak{S}(n+1)$, which is self-adjoint since $(i,j)^* = (i,j)^{-1} = (i,j)$ for any transposition (i,j). It belongs to the non-commutative probability space $(A = \mathbb{C}\mathfrak{S}(n+1), \tau)$, where $\tau = \chi^\lambda \circ \phi$, ϕ being the non-commutative conditional expectation

$$\begin{aligned}\mathbb{C}\mathfrak{S}(n+1) &\to \mathbb{C}\mathfrak{S}(n)\\ \sigma &\mapsto \begin{cases}\sigma & \text{if } \sigma(n+1)=n+1,\\ 0 & \text{otherwise.}\end{cases}\end{aligned}$$

studied before. Beware that τ is only a *state* on A, and not a *trace*; however, the tracial property $\tau(ab)=\tau(ba)$ has not been used so far in our presentation of non-commutative probability theory, so this is not a problem. The main result of this section is:

Theorem 9.23. *The transition measure μ_λ is the distribution of the non-commutative random variable J_{n+1} in the space $A=\mathbb{C}\mathfrak{S}(n+1)$ endowed with the state $\tau=\chi^\lambda\circ\phi$.*

As a preliminary step to the proof of Theorem 9.23, let us introduce another non-commutative probability space, which will have the advantage to be tracial. We set $A'=\mathrm{M}(n+1,\mathbb{C})\otimes_{\mathbb{C}}\mathrm{End}(S^\lambda)=\mathrm{M}(n+1,\mathrm{End}(S^\lambda))$, endowed with the trace $\tau'=\frac{\mathrm{tr}(\cdot)}{n+1}\otimes\frac{\mathrm{tr}(\cdot)}{\dim S^\lambda}$. We also denote

$$K_{n+1}=\begin{pmatrix} 0 & \rho^\lambda(2,1) & \rho^\lambda(3,1) & \cdots & \rho^\lambda(n,1) & 1\\ \rho^\lambda(1,2) & 0 & \rho^\lambda(3,2) & \cdots & \rho^\lambda(n,2) & 1\\ \rho^\lambda(1,3) & \rho^\lambda(2,3) & 0 & \ddots & \vdots & 1\\ \vdots & \vdots & \ddots & \ddots & \rho^\lambda(n,n-1) & \vdots\\ \rho^\lambda(1,n) & \rho^\lambda(2,n) & \cdots & \rho^\lambda(n-1,n) & 0 & 1\\ 1 & 1 & 1 & \cdots & 1 & 0\end{pmatrix},$$

which belongs to A' and is Hermitian.

Proposition 9.24. *The distribution of K_{n+1} in (A',τ') is μ_λ.*

Proof. Consider the space $M=\mathbb{C}\mathfrak{S}(n+1)$, on which $\mathbb{C}\mathfrak{S}(n)$ and $\mathbb{C}\mathfrak{S}(n+1)$ act respectively on the left and on the right by multiplication. As a $(\mathbb{C}\mathfrak{S}(n+1),\mathbb{C}\mathfrak{S}(n+1))$-bimodule, we have the decomposition

$$M=\bigoplus_{\Lambda\in\mathfrak{Y}(n+1)} S^\Lambda\otimes_{\mathbb{C}} S^\Lambda$$

by the double commutant theory; see Section 1.5. By restriction and the branching rules, as a $(\mathbb{C}\mathfrak{S}(n),\mathbb{C}\mathfrak{S}(n+1))$-bimodule,

$$M=\bigoplus_{\substack{\lambda\in\mathfrak{Y}(n),\,\Lambda\in\mathfrak{Y}(n+1)\\ \lambda\nearrow\Lambda}} S^\lambda\otimes_{\mathbb{C}} S^\Lambda.$$

The action of J_{n+1} by multiplication on the left on $S^\Lambda\otimes_{\mathbb{C}} S^\Lambda$ is diagonalizable, with eigenvalues the contents of the cells of label $n+1$ in the standard tableaux of shape Λ, each occurring $d_\Lambda = \dim S^\Lambda$ times. In $S^\lambda \otimes_{\mathbb{C}} S^\Lambda \subset S^\Lambda \otimes_{\mathbb{C}} S^\Lambda$, by Theorem 8.14 regarding the action of Jucys–Murphy elements on the Gelfand–Tsetlin basis, the action of J_{n+1} is again diagonalizable, with a unique eigenvalue which is the content of the cell of $\Lambda \setminus \lambda$. Hence, if $\lambda \nearrow \Lambda$ are integer partitions of size n and $n+1$, then J_{n+1} acts on $S^\lambda \otimes_{\mathbb{C}} S^\Lambda$ by multiplication by the content $c(\Lambda \setminus \lambda)$.

Let e^λ be the central idempotent of $\mathbb{C}\mathfrak{S}(n)$ associated to the irreducible representation S^λ; it is an element of $Z(n)$, and if $H = e^\lambda M$, then

$$H = \bigoplus_{\Lambda \in \mathfrak{Y}(n+1) \mid \lambda \nearrow \Lambda} S^\lambda \otimes_{\mathbb{C}} S^\Lambda.$$

Denote T_H the multiplication on the left by J_{n+1} on H. Since $(i,n+1)^{-1} = (i,n+1)$ for any i, and since the adjoint of the multiplication by σ in $\mathbb{C}\mathfrak{S}(n+1)$ is the multiplication by σ^{-1}, the multiplication by J_{n+1} is a self-adjoint operator on M, and by restriction on the Hilbert space H. Moreover, the spectrum of T_H is $\sigma(T_H) = \{c(\Lambda \setminus \lambda) \mid \lambda \nearrow \Lambda\}$, each eigenvalue having multiplicity $d_\lambda\, d_\Lambda$. We endow $\mathrm{End}(H)$ with the standard normalized trace

$$\tau_{\mathrm{End}(H)}(U) = \frac{\mathrm{tr}(U)}{\dim H},$$

and $\mathrm{End}(M)$ with the trace

$$\tau_{\mathrm{End}(M)}(U) = \tau_{\mathrm{End}(H)}(\pi^\lambda U \pi^\lambda),$$

where π^λ is the projection $M \to H$, that is to say the multiplication on the left by the $\mathbb{C}\mathfrak{S}(n)$-central idempotent e^λ. By the previous discussion, for any power k,

$$\begin{aligned}\tau_{\mathrm{End}(H)}((T_H)^k) &= \frac{1}{\dim H} \sum_{\lambda \nearrow \Lambda} d_\lambda\, d_\Lambda\, (c(\Lambda \setminus \lambda))^k \\ &= \sum_{\lambda \nearrow \Lambda} \frac{d_\Lambda}{(n+1)\, d_\lambda} (c(\Lambda \setminus \lambda))^k = \int_{\mathbb{R}} x^k \,\mu_\lambda(dx)\end{aligned}$$

by using the identity $\dim H = d_\lambda \sum_{\lambda \nearrow \Lambda} d_\Lambda = (n+1)\,(d_\lambda)^2$, and Proposition 7.19 for the values of the weights of the probability measure μ_λ. So, the distribution of the self-adjoint element T_H in the non-commutative probability space $(\mathrm{End}(H), \tau_{\mathrm{End}(H)})$ is μ_λ. Similarly, if T_M is the multiplication on the left by J_{n+1} on M, we have $\pi^\lambda T_M \pi^\lambda = T_H$, so, the distribution of the self-adjoint element T_M in the non-commutative probability space $(\mathrm{End}(M), \tau_{\mathrm{End}(M)})$ is again μ_λ.

We now relate the operators T_M and T_H to the matrix K_{n+1}. We identify the group $\mathfrak{S}(n+1)$ with $\mathfrak{S}(n) \times [\![1, n+1]\!]$ by the map $\sigma, k \mapsto \sigma \circ (k, n+1)$, and denote $\pi_k : \mathbb{C}\mathfrak{S}(n+1) \to \mathbb{C}\mathfrak{S}(n)$ the linear map

$$\pi_k(\rho) = \begin{cases} \sigma & \text{if } \rho = \sigma \circ (k, n+1), \\ 0 & \text{otherwise.} \end{cases}$$

We then have an isomorphism of vector spaces

$$\psi = (\pi_1, \ldots, \pi_{n+1}) : \mathbb{C}\mathfrak{S}(n+1) \to \mathbb{C}^{n+1} \otimes_{\mathbb{C}} \mathbb{C}\mathfrak{S}(n) = (\mathbb{C}\mathfrak{S}(n))^{n+1},$$

which yields an isomorphism of algebras

$$\begin{aligned}\Psi : \mathrm{End}(M) &\to \mathrm{End}(\mathbb{C}^{n+1} \otimes_{\mathbb{C}} \mathbb{C}\mathfrak{S}(n)) = \mathrm{M}(n+1, \mathbb{C}) \otimes_{\mathbb{C}} \mathrm{End}(\mathbb{C}\mathfrak{S}(n)) \\ T &\mapsto \psi \circ T \circ \psi^{-1}.\end{aligned}$$

More precisely, if $R_{(k,n+1)} : \mathbb{C}\mathfrak{S}(n) \to \mathbb{C}\mathfrak{S}(n+1)$ is the multiplication on the right by $(k, n+1)$, then, $R_{(k,n+1)} \circ \pi_k$ is a projection of the vector space $\mathbb{C}\mathfrak{S}(n+1)$, which is the identity on permutations of the form $\sigma \circ (k, n+1)$. With these notations, an element $T \in \mathrm{End}(\mathbb{C}\mathfrak{S}(n+1))$ can be represented by a matrix

$$\Psi(T) \in \mathrm{M}(n+1, \mathbb{C}) \otimes_{\mathbb{C}} \mathrm{End}(\mathbb{C}\mathfrak{S}(n)) = \mathrm{M}(n+1, \mathrm{End}(\mathbb{C}\mathfrak{S}(n))),$$

where the (i, j)-th element of the matrix $\Psi(T)$ is

$$(\Psi(T))_{i,j} = \pi_i \circ T \circ R_{(j,n+1)}.$$

Let us then compute $\Psi(T_M)$, where T_M is the multiplication by J_{n+1} on the left of the module $M = \mathbb{C}\mathfrak{S}(n+1)$. Using the fact that $J_{n+1} \in \mathrm{Z}(n, n+1)$, we obtain

$$\begin{aligned}T_M(\sigma(j, n+1)) &= \sum_{i=1}^{n} (j, n+1)\, \sigma\, (i, n+1) = \sigma \sum_{i=1}^{n} (j, n+1)(i, n+1) \\ &= \sigma + \sum_{i \neq j} \sigma\, (i, j)(i, n+1)\end{aligned}$$

for $j \neq n+1$, and $T_M(\sigma) = \sum_{i=1}^{n} \sigma\, (i, n+1)$. As a consequence, if $i \neq j \neq n+1$, then $(\Psi(T_M))_{i,j} = R_{(i,j)}$ is the multiplication on the right by (i, j), whereas $(\Psi(T_M))_{i,i}$ with $i \neq n+1$ vanishes. Similarly, we obtain $(\Psi(T_M))_{i,n+1} = (\Psi(T_M))_{n+1,i} = \mathrm{id}_{\mathbb{C}\mathfrak{S}(n)}$ if $i \neq n+1$. Hence, as an element of $\mathrm{M}(n+1, \mathrm{End}(\mathbb{C}\mathfrak{S}(n)))$, $\Psi(T_M)$ acts on row vectors in $(\mathbb{C}\mathfrak{S}(n))^{n+1}$ by multiplication on the right by the matrix

$$L_{n+1} = \begin{pmatrix} 0 & (2,1) & (3,1) & \cdots & (n,1) & 1 \\ (1,2) & 0 & (3,2) & \cdots & (n,2) & 1 \\ (1,3) & (2,3) & 0 & \ddots & \vdots & 1 \\ \vdots & \vdots & \ddots & \ddots & (n, n-1) & \vdots \\ (1,n) & (2,n) & \cdots & (n-1,n) & 0 & 1 \\ 1 & 1 & 1 & \cdots & 1 & 0 \end{pmatrix}.$$

In other words, for any $x \in \mathbb{C}\mathfrak{S}(n+1)$, $\psi(T_M(x)) = \psi(J_{n+1}\, x) = \psi(x)\, L_{n+1}$. We now restrict ourselves to the case when $x \in H = e^\lambda(M)$. Denote as before π^λ the multiplication on the left by e^λ, either in $\mathbb{C}\mathfrak{S}(n+1)$ or in a subspace. Notice that $(\mathrm{id}_{\mathbb{C}^{n+1}} \otimes \pi^\lambda) \circ \psi = \psi \circ \pi^\lambda$, because $e^\lambda \in \mathrm{Z}(n)$. Hence,

$$\begin{aligned}\psi(H) = \psi(e^\lambda M) &= (\mathrm{id}_{\mathbb{C}^{n+1}} \otimes \pi^\lambda) \circ \psi(M) \\ &= (\mathrm{id}_{\mathbb{C}^{n+1}} \otimes \pi^\lambda)(\mathbb{C}^{n+1} \otimes_{\mathbb{C}} \mathbb{C}\mathfrak{S}(n)) = \mathbb{C}^{n+1} \otimes_{\mathbb{C}} \mathrm{End}(S^\lambda).\end{aligned}$$

From this, one deduces that if $x \in H$, then $\psi(T_H(x)) = \psi(J_{n+1}\,x) = \psi(x)K_{n+1}$, both elements being viewed as elements of $\mathbb{C}^{n+1} \otimes_{\mathbb{C}} \mathrm{End}(S^\lambda)$. Finally, if K_{n+1} has for eigenvalues $(t_1,\ldots,t_{(n+1)d_\lambda})$ as an element of $\mathrm{M}(n+1,\mathbb{C}) \otimes_{\mathbb{C}} \mathrm{End}(S^\lambda) = \mathrm{End}(\mathbb{C}^{n+1} \otimes_{\mathbb{C}} S^\lambda) = \mathrm{End}(X)$ that acts on X, then the operator $R_{K_{n+1}}$ of multiplication by K_{n+1} on the right of $\mathbb{C}^{n+1} \otimes_{\mathbb{C}} \mathrm{End}(S^\lambda) = X \otimes_{\mathbb{C}} S^\lambda$ has spectrum

$$(\underbrace{t_1,\ldots,t_1}_{d_\lambda \text{ occurrences}},\ldots,\underbrace{t_{(n+1)d_\lambda},\ldots,t_{(n+1)d_\lambda}}_{d_\lambda \text{ occurrences}}).$$

So, K_{n+1} and $R_{K_{n+1}}$ have the same normalized spectral measures in their respective endomorphism spaces, and

$$\mu_\lambda = \text{distribution of } T_H = \text{distribution of } R_{K_{n+1}} = \text{distribution of } K_{n+1},$$

which ends the proof. □

Proof of Theorem 9.23. We saw above that

$$\tau_{\mathrm{End}(H)}((T_H)^k) = \tau_{\mathrm{End}(M)}((T_M)^k) = \int_{\mathbb{R}} x^k\,\mu_\lambda(dx),$$

where T_M (respectively, T_H) is the multiplication on the left by J_{n+1} on $M = \mathbb{C}\mathfrak{S}(n+1)$ (respectively, on $H = e^\lambda M$). We decompose $\mathbb{C}\mathfrak{S}(n+1)$ as the direct sum

$$M = \mathbb{C}\mathfrak{S}(n+1) = \bigoplus_{j=1}^{n+1} R_{(j,n+1)}(\mathbb{C}\mathfrak{S}(n)).$$

Applying π^λ to this expansion, we get:

$$H = \bigoplus_{j=1}^{n+1} e^\lambda\,\mathbb{C}\mathfrak{S}(n)(j,n+1).$$

Notice that with the notations of the proof of Proposition 9.24, $\pi_{n+1} = \phi$ is the conditional expectation from $\mathbb{C}\mathfrak{S}(n+1)$ to $\mathbb{C}\mathfrak{S}(n)$. In particular, we can write

$$(J_{n+1})^k = \phi((J_{n+1})^k) + \sum_{j=1}^{n} \pi_j((J_{n+1})^k)(j,n+1).$$

When acting on H by multiplication on the left, a permutation in $\mathbb{C}\mathfrak{S}(n)$ acts block-diagonally with respect to the previous decomposition of H. Indeed, since e^λ is central in $\mathbb{C}\mathfrak{S}(n)$,

$$\sigma\, e^\lambda\,\mathbb{C}\mathfrak{S}(n)(j,n+1) = e^\lambda\,\sigma\,\mathbb{C}\mathfrak{S}(n)(j,n+1) \subset e^\lambda\,\mathbb{C}\mathfrak{S}(n)(j,n+1).$$

Moreover, the normalized trace of this action is $\chi^\lambda(\sigma)$. Therefore, $\phi((J_{n+1})^k)$ acts on H by a block-diagonal matrix with trace $\tau((J_{n+1})^k)$. On the other hand, if $\rho = \sigma(i,n+1)$ with $\sigma \in \mathfrak{S}(n)$ and $i \in [\![1,n]\!]$, then the matrix of the multiplication on the left by ρ has its diagonal blocks that vanish. Indeed, $\pi_{n+1}(\rho\, e^{\lambda}) = 0$, because if $\sigma \in \mathfrak{S}(n)$, then $\rho\,\sigma(n+1) \neq n+1$. Therefore, $\rho\, e^{\lambda} = \sum_{h=1}^{n} \sigma_h\,(h, n+1)$ for some permutations $\sigma_h \in \mathfrak{S}(n)$. Then, if $\sigma \in \mathfrak{S}(n)$,

$$\rho\, e^{\lambda}\,\sigma\,(j,n+1) = \sum_{h=1}^{n} \sigma_h\,(h,n+1)\,\sigma\,(j,n+1)$$

$$= \sum_{h=1}^{n} \sigma_h\,\sigma\,(\sigma^{-1}(h),n+1)\,(j,n+1) = \sum_{g=1}^{n} \sigma_h\,\sigma\,(g,n+1)\,(j,n+1)$$

$$= \left(\sigma_h \sigma + \sum_{g \in [\![1,n]\!],\ g \neq j} (\sigma_h \sigma(g,j))\,(g,n+1) \right) \in \bigoplus_{g \in [\![1,n+1]\!],\, g\neq j} (\mathbb{C}\mathfrak{S}(n))(g,n+1)$$

so the diagonal blocks of ρ indeed vanish. We conclude that $\tau_{\mathrm{End}(H)}((T_H)^k) = \tau((J_{n+1})^k)$, which ends the proof. □

9.4 The algebra of admissible set partitions

The remainder of this chapter is devoted to a more conceptual proof of the identity

$$\forall k \geq 1,\ \tau'((K_{n+1})^k) = \tau((J_{n+1})^k).$$

More precisely, this identity is related to the combinatorics of set partitions, and to a morphism $\mathscr{P} \to \mathscr{O}$ from the algebra of so-called admissible set partitions to the algebra of observables. This construction provides a new combinatorial proof of Theorem 9.23, which sheds light on certain relations in the algebra of observables; in particular, we shall obtain a formula for the expansion of $\widetilde{h}_k$ in symbols Σ_μ (Theorem 9.33). Later, in Chapter 13, we shall need the arguments of this section and the theory of set partitions in order to prove a difficult result on the weight of certain observables; cf. Theorem 13.12.

▷ *Conditional moments of Jucys–Murphy elements as observables.*

For any $k \geq 1$, we have the expansion

$$\tau'((K_{n+1})^k) = \sum_{\substack{i_1,\ldots,i_k \in [\![1,n+1]\!] \\ \forall j \in [\![1,k]\!],\ i_j \neq i_{j+1}}} \frac{1}{n+1}\, \chi^{\lambda}((i_1,i_2)'(i_2,i_3)'\cdots(i_{k-1},i_k)'(i_k,i_1)'),$$

where we convene that $(i,j)' = (i,j)$ if i and j are in $[\![1,n]\!]$, and $(i,j)' = \mathrm{id}$ if one of the indices i,j is equal to $n+1$. For every sequence $(i_1,\ldots,i_k)$, let J be the set of indices $j \in [\![1,k]\!]$ such that $i_j = n+1$, and π' the set partition of $[\![1,k]\!] \setminus J$ associated to the equalities $i_j = i_l$: two elements j and l are in the same part of π' if and only if $i_j = i_l$. Then, the cycle type $t(J,\pi')$ in $\mathfrak{S}(n)$ of the product $(i_1,i_2)'(i_2,i_3)'\cdots(i_{k-1},i_k)'(i_k,i_1)'$ only depends on J and the set partition π', and on the other hand, knowing J and π', the number of possible choices of sequences $i_1,\ldots,i_k$ is $n^{\downarrow \ell(\pi')}$. Finally, the allowed sets J (respectively, the allowed set partitions π') are those without consecutive integers (respectively, without consecutive integers in a same part), which we shall call admissible. Thus,

$$\tau'((K_{n+1})^k) = \frac{1}{n+1} \sum'_{J\subset[\![1,k]\!]} \sum'_{\pi'\in\mathfrak{Q}([\![1,k]\!]\setminus J)} n^{\downarrow\ell(\pi)}\, \chi^\lambda(t(J,\pi')),$$

where the symbol $\sum'$ indicates that the sum is restricted to admissible sets or admissible set partitions. In this formula, we recognize observables Σ_μ of the Young diagram λ. Indeed, if a sequence $i_1,\ldots,i_k$ corresponds to a pair (J,π') and to a permutation σ of type $t(J,\pi')$, then σ moves at most $\ell(\pi')$ points, namely, the points of the set $\{i_j,\ j\notin J\}$. So, one can associate to it a support of size $\ell(\pi')$, which is a subset of $[\![1,n]\!]$; and the type $t(J,\pi')$ of σ can be seen as an integer partition of size $|\ell(\pi')|$. With this point of view,

$$\tau'((K_{n+1})^k) = \frac{1}{n+1} \sum'_{J\subset[\![1,k]\!]} \sum'_{\pi'\in\mathfrak{Q}([\![1,k]\!]\setminus J)} \Sigma_{t(J,\pi')}(\lambda).$$

Up to the factor $\frac{1}{n+1}$, we therefore have an expression of $\tau'((K_{n+1})^k)$ as an n-independent sum of observables of λ. Since this quantity is also equal to $\tau((J_{n+1})^k)$, it is then natural to try to see the elements $\phi((J_{n+1})^k)$ as observables in $(\mathbb{C}\mathfrak{P}\mathfrak{S}(n))^{\mathfrak{S}(n)} = \phi_n(\mathscr{O})$. Thus, we are going to construct elements $M_1^{\mathrm{JM}}, M_2^{\mathrm{JM}},\ldots,M_k^{\mathrm{JM}},\ldots \in \mathscr{O}\subset \mathbb{C}\mathfrak{P}\mathfrak{S}(\infty)$ such that

$$\pi_n\circ\phi_n(M_k^{\mathrm{JM}}) = \phi((J_{n+1})^k)$$

for any n and any k. To this purpose, we introduce a special symbol $*$ not in $\mathbb{N}$; then, the conditional expectation ϕ can be rewritten as

$$\phi : \mathbb{C}\mathfrak{S}([\![1,n]\!]\sqcup\{*\}) \to \mathbb{C}\mathfrak{S}([\![1,n]\!])$$
$$\sigma \mapsto \begin{cases} \sigma & \text{if } \sigma(*)=*,\\ 0 & \text{otherwise.}\end{cases}$$

This map comes from the linear map

$$\Phi : \mathbb{C}\mathfrak{P}\mathfrak{S}([\![1,n]\!]\sqcup\{*\}) \to \mathbb{C}\mathfrak{P}\mathfrak{S}([\![1,n]\!])$$
$$(\sigma,A) \mapsto \begin{cases} (\sigma, A\setminus\{*\}) & \text{if } \sigma(*)=*,\\ 0 & \text{otherwise.}\end{cases}$$

Abbreviating the set $[\![1,n]\!]\sqcup *$ by $n+*$, we have for every n: $\phi\circ\pi_{n+*} = \pi_n\circ\Phi$, where π_{n+*} and π_n are the projection morphisms $\mathbb{C}\mathfrak{P}\mathfrak{S}(n+*)\to\mathbb{C}\mathfrak{S}(n+*)$ and $\mathbb{C}\mathfrak{P}\mathfrak{S}(n)\to\mathbb{C}\mathfrak{S}(n)$. Moreover, $\Phi\circ\phi_{n+*}^{(n+1)+*} = \phi_n^{(n+1)}\circ\Phi$, which allows one to consider Φ as an operator

$$\mathbb{C}\mathfrak{P}\mathfrak{S}(\infty+*)\to\mathbb{C}\mathfrak{P}\mathfrak{S}(\infty).$$

Definition 9.25. *Set $X_* = \sum_{n=1}^{\infty}((n,*),\{n,*\})$, which is an element of $\mathbb{C}\mathfrak{P}\mathfrak{S}(\infty+*)$. The k-th* ***Jucys–Murphy conditional moment*** *is defined as:*

$$M_k^{\mathrm{JM}} = \Phi(X_*^k) = \sum_{n_1,\ldots,n_k=1}^{\infty} \Phi\big((n_1,*)(n_2,*)\cdots(n_k,*),\{n_1,\ldots,n_k,*\}\big).$$

Proposition 9.26. *For any n, $\pi_n \circ \phi_n(M_k^{\mathrm{JM}}) = \phi((J_{n+1})^k)$, and on the other hand, M_k^{JM} belongs to $\mathscr{O}$.*

Proof. Using the commutation rules between the various linear maps $\pi_n, \phi_n, \phi, \Phi$, we obtain

$$\begin{aligned}
\phi_n(M_k^{\mathrm{JM}}) &= \sum_{n_1,\ldots,n_k=1}^{\infty} \phi_n \circ \Phi\big((n_1,*)(n_2,*)\cdots(n_k,*),\{n_1,\ldots,n_k,*\}\big)\\
&= \sum_{n_1,\ldots,n_k=1}^{\infty} \Phi \circ \phi_{n+*}\big((n_1,*)(n_2,*)\cdots(n_k,*),\{n_1,\ldots,n_k,*\}\big)\\
&= \sum_{n_1,\ldots,n_k=1}^{n} \Phi\big((n_1,*)(n_2,*)\cdots(n_k,*),\{n_1,\ldots,n_k,*\}\big);\\
\pi_n \circ \phi_n(M_k^{\mathrm{JM}}) &= \sum_{n_1,\ldots,n_k=1}^{n} \pi_n \circ \Phi\big((n_1,*)(n_2,*)\cdots(n_k,*),\{n_1,\ldots,n_k,*\}\big)\\
&= \sum_{n_1,\ldots,n_k=1}^{n} \phi \circ \pi_{n+*}\big((n_1,*)(n_2,*)\cdots(n_k,*),\{n_1,\ldots,n_k,*\}\big)\\
&= \sum_{n_1,\ldots,n_k=1}^{n} \phi((n_1,*)(n_2,*)\cdots(n_k,*)) = \phi((J_{n+1})^k).
\end{aligned}$$

On the other hand, if $\sigma \in \mathfrak{S}(\infty)$, then the action of σ on M_k^{JM} is a permutation of the indices $n_1,\ldots,n_k$ in the infinite sum, so $\sigma \cdot M_k^{\mathrm{JM}} = M_k^{\mathrm{JM}}$ and $M_k^{\mathrm{JM}} \in \mathscr{O}$. □

Corollary 9.27. *For any $k \geq 1$, we have $M_k^{\mathrm{JM}} = \widetilde{h}_k$ in the algebra $\mathscr{O}$.*

Proof. Since M_k^{JM} is an observable, there is an expansion $M_k^{\mathrm{JM}} = \sum_\mu c_\mu \Sigma_\mu$ (we shall compute this expansion in a moment; see Theorem 9.33). Then, for any integer partition λ,

$$\sum_\mu c_\mu \Sigma_\mu(\lambda) = \chi^\lambda \circ \pi_n \circ \phi_n(M_k^{\mathrm{JM}}) = \chi^\lambda \circ \phi((J_{n+1})^k) = \tau((J_{n+1})^k).$$

By Theorem 9.23, the last term is $\int_{\mathbb{R}} x^k \mu_\lambda(dx) = \widetilde{h}_k(\lambda)$. □

▷ *Set partitions and the map $\pi \mapsto \Sigma(\pi)$.*

In the sequel, we continue to use the notation M_k^{JM} when considering this observable as a linear combination of partial permutations. We then want to compute the expansion of M_k^{JM} in symbols Σ_μ. To this purpose, we shall introduce a set $\mathscr{P}$ of admissible set partitions, and a projection map $\Sigma : \mathscr{P} \to \mathscr{O}$, which will be a morphism with respect to a certain multiplicative structure on $\mathscr{P}$, and which will be involved in an expansion

$$M_k^{\mathrm{JM}} = \sum_{\pi \in \mathscr{P},\, |\pi| = k} \Sigma(\pi)$$

of the conditional moments of Jucys–Murphy elements. If $\pi \in \mathfrak{Q}(n)$ is a set partition, we shall represent it as before by a graph, but with the points $1, 2, \ldots, n$ now placed on a circle. For instance, $\pi = \{1,3,5\} \sqcup \{4,6,8\} \sqcup \{2,7\}$ is represented by

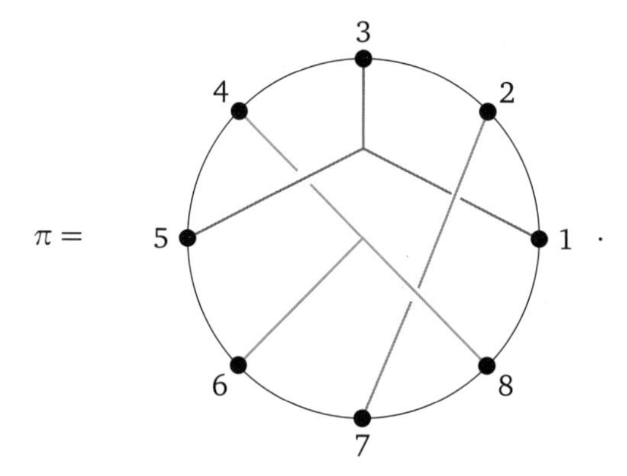

Place integers $1', 2', \ldots, n'$ between 1 and 2, 2 and 3, etc. If $\pi \in \mathfrak{Q}(n)$, we introduce the **fat set partition** π_{fat} of $[\![1,n]\!] \sqcup [\![1',n']\!]$, obtained by "inflating" the edges of the previous graph, and looking at the borders of the fat edges, which connect elements of $[\![1,n]\!]$ to elements of $[\![1',n']\!]$. This is better understood in a picture; see Figure 9.3.

For any set partition π, π_{fat} is a pair set partition of size $2n$, that is to say a set partition whose parts are all of size 2. Moreover, it is non-crossing if and only if π is non-crossing. In this case, if c is the cyclic shift $n' \to n \to (n-1)' \to (n-1) \to \cdots$ on $[\![1,n]\!] \sqcup [\![1',n']\!]$, then $c(\pi_{\mathrm{fat}}) = (\overline{\pi})_{\mathrm{fat}}$ for a unique non-crossing partition $\overline{\pi}$ on $[\![1,n]\!]$, and it is easily seen that $\overline{\pi}$ is the Kreweras complement of π. In particular, one obtains the identity $\overline{\overline{\pi}} = c^2(\pi)$, which was not obvious from the first definition. Since every part of π_{fat} contains an element of $[\![1,n]\!]$ and an element of $[\![1',n']\!]$, it yields a unique bijection $[\![1',n']\!] \to [\![1,n]\!]$, which we continue to denote π_{fat}. Then, $\pi_{\mathrm{fat}} \circ c$ is a permutation of $[\![1,n]\!]$, which we denote $\tau = \tau(\pi)$; see Figure 9.4.

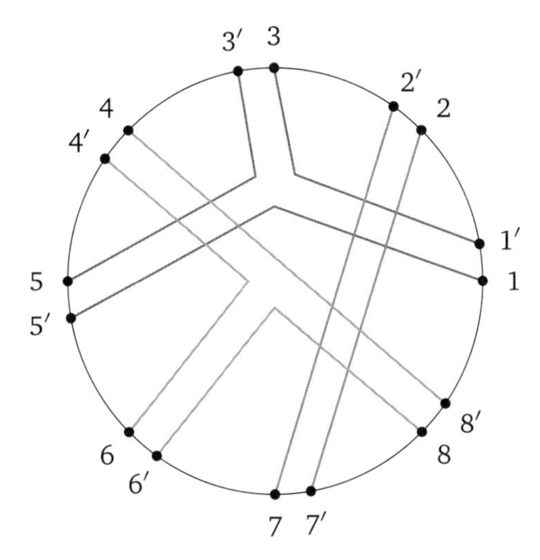

Figure 9.3
Set partition π_{fat} calculated from $\pi = \{1,3,5\} \sqcup \{4,6,8\} \sqcup \{2,7\}$.

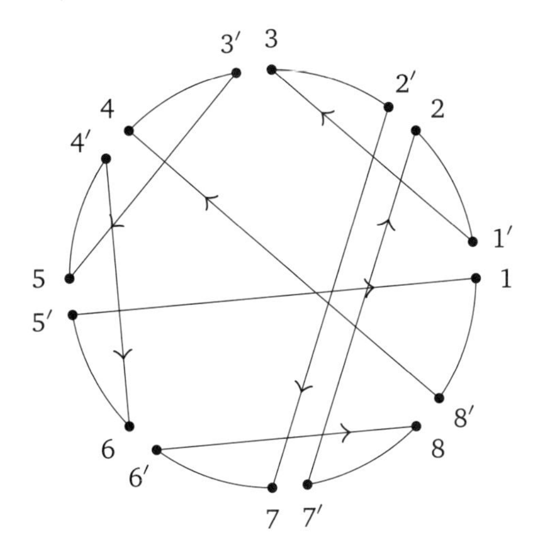

Figure 9.4
The permutation $\tau(\pi)$ associated to $\pi = \{1,3,5\} \sqcup \{4,6,8\} \sqcup \{2,7\}$. The symbol $\Sigma(\pi)$ is in this case $\Sigma_{1,1}$.

For each cycle c_i of the permutation $\tau(\pi)$, we consider the number

$$\begin{aligned} k_i &= \text{number of integers } k \text{ in the cycle} - \text{number of clockwise winds} \\ &= \text{number of counterclockwise winds.} \end{aligned}$$

In the previous example, the two 4-cycles each make 3 clockwise turns, so $k_1 =$

$k_2 = 1$. We reorder the numbers k_i in decreasing order, and set

$$\Sigma(\pi) = \Sigma_{k_1,\ldots,k_r},$$

with the convention that if one of the k_i's vanishes, then $\Sigma_{k_1,\ldots,k_r} = 0$. Obviously, $\Sigma(\pi)$ is invariant by cyclic shift of the set partition π.

▷ *Admissible partitions and admissible sequences.*

We denote $\mathfrak{Q}(\infty)$ the set of pairs (X, π), where X is a finite set of integers, and $\pi \in \mathfrak{Q}(X)$ is a set partition of it. If $(X, \pi) \in \mathfrak{Q}(\infty)$, we set $\Sigma(X, \pi) = \Sigma(\pi')$, where π' is the set partition of $[\![1, n]\!]$ obtained from (X, π) by relabeling the $n = \operatorname{card} X$ elements in increasing order. We then extend by linearity the correspondence $\Sigma : \mathfrak{Q}(\infty) \to \mathscr{O}$ to $\mathbb{C}\mathfrak{Q}(\infty)$.

The map Σ turns out to vanish on a large number of set partitions, so it will be convenient to restrict it to a subspace $\mathscr{P}$ of $\mathbb{C}\mathfrak{Q}(\infty)$. This leads us naturally to the combinatorial notion of **admissible set partitions**. A set partition π on $X = [\![1, n]\!]$ will be said admissible if for any $i \in [\![1, n]\!]$, i and $i + 1$ are not in the same part of π. In this definition, it is understood that $n + 1 = 1$, and on the other hand, if $X \neq [\![1, n]\!]$ is a finite set of integers that is not an interval, then we extend the previous definition by relabeling the elements of X in increasing order. For instance, the set partition $\{1, 3, 5\} \sqcup \{2, 7\} \sqcup \{4, 6, 8\}$ previously drawn is admissible.

Proposition 9.28. *One has $\Sigma(\pi) = 0$ if and only if π is not admissible.*

Proof. If π is not admissible, two consecutive elements i and $i+1$ are in the same part of π, and then the construction $\pi \mapsto \tau(\pi)$ yields a permutation that fixes the integer i. The corresponding cycle has one element and turns clockwise once, hence is associated to the integer $k = 0$. By convention, we then have $\Sigma(\pi) = \Sigma_{k_1,\ldots,0,\ldots,k_r} = 0$. Conversely, if π is admissible, then all the cycles c_i of $\tau(\pi)$ have length greater than 2, and such cycles have at least one descent, hence yield a number $k_i \neq 0$; so, $\Sigma(\pi) \neq 0$. □

We denote $\mathscr{P}$ the set of admissible set partitions, which is a subset of $\mathfrak{Q}(\infty)$. We also use the same notation for the complex vector space that it spans. We then have a linear map $\Sigma : \mathscr{P} \to \mathscr{O}$, which does not vanish on the combinatorial basis of $\mathscr{P}$. The notion of **admissible sequence** will provide us with another interpretation of Σ (Theorem 9.32). An admissible sequence of length l is a sequence $(p_1, \ldots, p_l)$ of elements of $[\![1, +\infty]\!] \sqcup \{*\}$ such that $p_i \neq p_{i+1}$ for any $i \in [\![1, l]\!]$, with $p_1 = p_{l+1} = *$. If $(p_1, \ldots, p_l)$ is a sequence of integers (or the special symbol $*$), we associate to it a set partition $\pi \in \mathfrak{Q}(l)$, defined as follows: i and j are in the same part of π if and only if $p_i = p_j$. Then, a sequence is an admissible sequence if and only if the associated set partition π is admissible and $p_1 = *$. Given an admissible sequence $(p_1, \ldots, p_l)$, we associate to it a sequence of partial permutations $S_1, \ldots, S_{l+1}$ in $\mathfrak{PS}(\infty + *)$ as follows:

1. We set $S_{l+1} = (\mathrm{id}, \{*\})$.
2. Given S_m, we set $n_m = \sigma_m(p_{m-1})$, and $S_{m-1} = \big((n_m,*), \{n_m,*\}\big) \times S_m$.

Recursively, we thus obtain $S_1 = \big((n_1,*)(n_2,*)\cdots(n_l,*), \{n_1,\ldots,n_l,*\}\big)$. We denote

$$\Theta(p_1,\ldots,p_l) = \phi(S_1).$$

Proposition 9.29. *In the previous construction, if $(p_1,\ldots,p_l)$ is admissible, then all the n_m's are in $[\![1,+\infty]\!]$, and moreover, $\Theta(p_1,\ldots,p_l) \neq 0$ (in other words, the permutation σ_1 underlying S_1 sends $*$ to $*$).*

Proof. We write $S_m = (\sigma_m, A_m)$. By descending induction on m, for every $m \in [\![1,l]\!]$, $(\sigma_m)^{-1}(*) = p_m$, so in particular $(\sigma_1)^{-1}(*) = p_1 = *$ since the sequence is admissible. Hence, σ_1 sends $*$ to $*$, and $\Theta(p_1,\ldots,p_l) \neq 0$. Moreover, since $\sigma_m(p_m) = *$, $n_m = \sigma_m(p_{m-1}) \neq *$, as $p_{m-1} \neq p_m$ by admissibility. So, all the n_m's are in $[\![1,+\infty]\!]$. □

Proposition 9.30. *With the same hypothesis of admissibility, the support of $\Theta(p_1,\ldots,p_l)$ is $X = \{p_1,\ldots,p_l\} \setminus \{*\} = \{n_1,\ldots,n_l\}$.*

Proof. Let us compute recursively the supports A_m of the partial permutations S_m. Since $\sigma_{l+1} = (\mathrm{id}, \{*\})$, $A_{l+1} = \{*\}$. Then,

$$S_{m-1} = \big((\sigma_m(p_{m-1}),*), \{\sigma_m(p_{m-1}),*\}\big) \times S_m,$$

so the support of S_{m-1} is:

- either A_m if $p_{m-1} \in A_m$,
- or, $\{p_{m-1}\} \cup A_m$ if $p_{m-1} \notin A_m$.

We conclude that $A_m = \{p_m, p_{m+1},\ldots,p_l,*\}$ for any $m \in [\![1,l+1]\!]$. In particular, $A_1 = X \sqcup \{*\}$, and since $\Theta(p_1,\ldots,p_l) = \phi(S_1)$, its support is X. □

Proposition 9.31. *Conversely, a permutation $\sigma = (n_1,*)(n_2,*)\cdots(n_l,*)$ with $\sigma(*) = *$ and the n_m's in $[\![1,+\infty]\!]$ comes from a unique admissible sequence $(p_1,\ldots,p_l)$.*

Proof. We reconstruct the unique admissible sequence $(p_1,\ldots,p_l)$ such that $\Theta(p_1,\ldots,p_l) = (\sigma, X)$ by setting

$$p_m = ((n_k,*)\cdots(n_m,*))(*).$$

Since $\sigma(*) = *$, $p_1 = *$. On the other hand,

$$\begin{aligned} p_m &= ((n_k,*)\cdots(n_m,*))(*) = ((n_k,*)\cdots(n_{m+1},*))(n_m) \\ &\neq ((n_r,*)\cdots(n_{m+1},*))(*) = p_{m+1}, \end{aligned}$$

so the sequence is admissible. It is then easily seen that this construction is the inverse of the map Θ. $\square$

The previous proposition yields a bijection between the terms

$$\big((n_1,*)(n_2,*)\cdots(n_l,*),\{n_1,\dots,n_l\}\big)$$

of M_l^{JM}, and the set of admissible sequences $(p_1,\dots,p_l)$ of length l. The following result describes the relations between the maps

Θ : {admissible sequences} $\to$ {partial permutations};

Σ : {admissible set partitions} $\to$ {conjugacy classes of partial permutations}.

Theorem 9.32. *Let $(p_1,\dots,p_l)$ be an admissible sequence of length l, and $\pi = \pi(p_1,\dots,p_l)$ the associated admissible set partition in $\mathfrak{Q}(l)$. The type of the partial permutation $\Theta(p_1,\dots,p_l)$ is the integer partition $(k_1,\dots,k_r)$ such that*

$$\Sigma(\pi) = \Sigma_{k_1,\dots,k_r}.$$

Proof. The proof of this theorem relies on another construction of the partial permutation $\Theta(p_1,\dots,p_l)$, on which the cycle type and the connection with π is easily seen. Consider the permutation $\tau(\pi) = \pi_{\mathrm{fat}} \circ c$, which belongs to $\mathfrak{S}(l)$. It writes as a product of disjoint cycles:

$$\tau(\pi) = c_1 \circ c_2 \circ \cdots \circ c_r \quad \text{with } c_i = (b_{i,1}, b_{i,2}, \dots, b_{i,l_i}).$$

In each cycle $c_i = (b_{i,1}, b_{i,2}, \dots, b_{i,l_i})$, let us keep the integers $b_{i,j}$ such that $p_{b_{i,j}} \neq *$, and $b_{i,j}$ is the smallest index in $[\![1,l]\!]$ which yields the value $p_{b_{i,j}}$. For instance, if the admissible sequence considered is $(*,1,*,2,*,2,1,2)$, then the associated set partition is the one studied previously:

$$\pi = \{1,3,5\} \sqcup \{2,7\} \sqcup \{4,6,8\}$$

and the permutation $\tau(\pi)$ was computed to be $(1,4,5,6)(2,3,7,8)$; we then keep

$$(4)(2).$$

By the map $b \mapsto p_b$, this can be considered as a permutation of $X = \{p_1,\dots,p_l\} \setminus \{*\}$, namely,

$$\sigma = \sigma(\pi) = (p_4)(p_2) = (1)(2).$$

We claim that $\Theta(p_1,\dots,p_l) = (\sigma, X)$, and on the other hand that $t(\sigma,X) = (k_1,\dots,k_r)$. The second fact comes from a graphical interpretation of the construction of the permutation $\sigma(\pi)$. On the graph of the permutation $\tau(\pi)$, we add a decoration $p_{b_{i,j}} \neq *$ on each index $b_{i,j}$ such that $b_{i,j}$ is the smallest index in $[\![1,l]\!]$ which yields the value $p_{b_{i,j}}$; see Figure 9.5. Then, the graphical construction of $\tau(\pi) = \pi_{\mathrm{fat}} \circ c$ implies that an index $b_{i,j}$ is decorated in the cycle c_i if and only

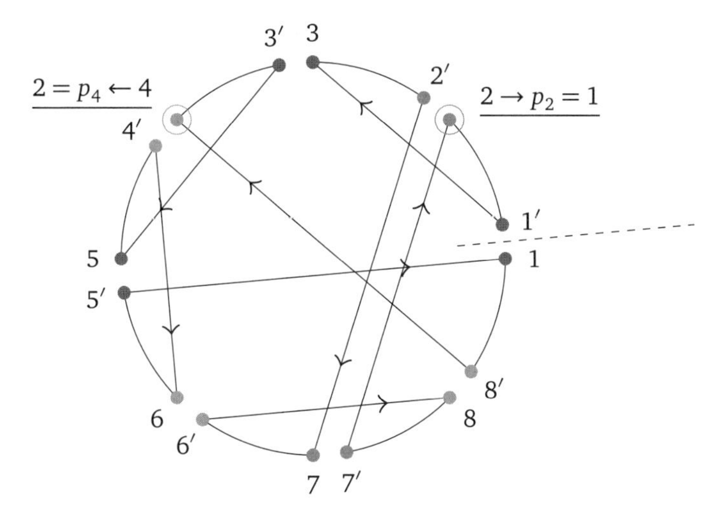

Figure 9.5
The permutation $\sigma(\pi)$ is obtained from the graph of $\tau(\pi)$ by looking only at the decorated vertices.

if, reading the indices counterclockwise, one crosses the line between 1 and 1′ when going from $b_{i,j-1}$ to $b_{i,j}$. On the other hand, by definition, $\sigma(\pi)$ is obtained from the cycles of $\tau(\pi)$ by keeping only the decorated indices. Therefore, the cycles $d_1, \ldots, d_r$ of $\sigma(\pi)$ have for lengths the numbers of counterclockwise winds of the cycles $c_1, \ldots, c_r$ of $\tau(\pi)$, that is to say $k_1, \ldots, k_r$.

It remains to prove that $\Theta(p_1, \ldots, p_l) = (\sigma(\pi), X)$. We prove by descending induction on $m \in [\![1, l+1]\!]$ the following graphical construction of the permutation σ_m of X. If $x \in X \sqcup \{*\}$, then:

1. If $x = p_m$, then $\sigma_m(x) = *$.
2. If $x \notin \{p_m, \ldots, p_l, *\}$, then $\sigma_m(x) = x$.
3. Otherwise, let n be the smallest index in $[\![m+1, l+1]\!]$ such that $x = p_n$ (with by convention $p_{l+1} = p_1 = *$). Then, $\sigma_m(x) = y$ is the first decoration $y = p_k$ seen when following the arrows of the graph of $\tau(\pi)$, starting from n, and not counting p_n if n is decorated.

The case $m = l+1$ is evident. Assume that the result is true for σ_m, and consider $\sigma_{m-1} = (\sigma_m(p_{m-1}), *) \circ \sigma_m$.

1. If $x = p_{m-1}$, then $\sigma_{m-1}(p_{m-1}) = (\sigma_m(p_{m-1}), *)(\sigma_m(p_{m-1})) = *$.
2. If $x \notin \{p_{m-1}, \ldots, p_l, *\}$, then $x \notin A_{m-1}$, the support of S_{m-1}. Therefore, $\sigma_{m-1}(x) = x$.

3. Set in the last case $x = p_n \neq p_{m-1}$, with $n \in [\![m, l+1]\!]$ minimal. If $p_n \neq p_m$, then $\sigma_m(x) = y$ is obtained by the graphical construction, and moreover, it differs from $\sigma_m(p_{m-1})$ and from $*$, so it is also $\sigma_{m-1}(x)$. Therefore, we now only have to treat the case where $x = p_m$ and $m = n$. In this case we have to show that $\sigma_{m-1}(x) = \sigma_{m-1}(p_m) = \sigma_m(p_{m-1})$ is obtained by the graphical construction, starting from m.
   (a) If $p_{m-1} \in A_m = \{p_m, \ldots, p_l, *\}$, let r be the smallest index in $[\![m+1, l+1]\!]$ such that $p_{m-1} = p_r$. Notice that r is not decorated, since $m-1 < r$ are in the same part of π. Then, following the graph of $\tau(\pi)$ starting from m, we see $(m-1)'$, then r, and by the induction hypothesis, the next decoration is $\sigma_m(p_r) = \sigma_m(p_{m-1})$, as wanted.
   (b) Finally, if $p_{m-1} \notin A_m$, then $\sigma_{m-1}(x) = p_{m-1}$, and on the other hand, $m-1$ is the largest element of its part in π. Then, following the graph of $\tau(\pi)$ starting from m, we see $(m-1)'$, and then r, where r is the smallest index such that $p_r = p_{m-1}$, that is to say that r is decorated with label p_{m-1}. Hence, the graphical construction yields again the value of $\sigma_{m-1}(x)$.

In particular, with $m = 1$, we are always in the third case of the alternative above, and $\Theta(p_1, \ldots, p_l) = (\sigma_1, X)$ is obtained by the graphical construction, hence corresponds to $(\sigma(\pi), X)$. □

Now, we can use the previous constructions to get a new proof of Theorem 9.23. In the sequel we denote $\mathscr{P}(k)$ the set of admissible set partitions of $[\![1, k]\!]$. Consider the observable M_k^{JM}: by a previous observation, it is equal to

$$M_k^{\mathrm{JM}} = \sum_{n_1,\ldots,n_k=1}^{\infty} \big((n_1, *)\cdots(n_k, *), \{n_1, \ldots, n_k\}\big)$$
$$= \sum_{(p_1,\ldots,p_k) \text{ admissible sequence}} \Theta(p_1, \ldots, p_k).$$

We gather the admissible sequences according to their associated admissible set partitions $\pi \in \mathfrak{Q}(k)$. If one knows π, then the corresponding admissible sequences are obtained by choosing different integers for the elements of the support of $\Theta(p_1, \ldots, p_k)$, therefore,

$$\sum_{\substack{(p_1,\ldots,p_k) \text{ admissible sequence} \\ \text{with set partition } \pi}} \Theta(p_1, \ldots, p_k) = \Sigma(\pi).$$

So, we have finally computed the expansion of M_k^{JM} in symbols Σ_μ:

Theorem 9.33. *For any* $k \geq 1$,

$$\widetilde{h}_k = M_k^{\mathrm{JM}} = \sum_{\pi \in \mathscr{P}(k)} \Sigma(\pi).$$

Combining this expansion with Proposition 9.26, we obtain for every k:

$$\begin{aligned}\tau((J_{n+1})^k) &= \chi^\lambda \circ \phi((J_{n+1})^k) \\ &= \chi^\lambda \circ \pi_n \circ \phi_n\left(\sum_{\pi \in \mathscr{P}(k)} \Sigma(\pi)\right) = \sum_{\pi \in \mathscr{P}(k)} (\Sigma(\pi))(\lambda).\end{aligned}$$

We rewrite this as a sum over admissible sequences $(p_1, \ldots, p_k)$, with $p_1 = *$ and such that each $p_i \in [\![1, n]\!] \sqcup \{*\}$:

$$\tau((J_{n+1})^k) = \sum_{\substack{(p_1,\ldots,p_k) \text{ admissible sequence} \\ \forall i \in [\![1,k]\!],\ p_i \in [\![1,n]\!] \sqcup \{*\}}} \chi^\lambda(\kappa(\pi(p_1, \ldots, p_k)) \uparrow n),$$

where $\kappa(\pi(p_1, \ldots, p_k)) = (k_1, \ldots, k_r)$ is the type of $\Theta(p_1, \ldots, p_k)$, or, by Theorem 9.32, the sequence of numbers of counterclockwise winds of $\tau(\pi)$, where $\pi = \pi(p_1, \ldots, p_k)$ is the set partition defined by the identities in the sequence $(p_1, \ldots, p_k)$. Now, this definition makes sense if one allows p_1 to be different from $*$, but if one still looks at the numbers $k_1, \ldots, k_r$ of counterclockwise winds of the cycles of the permutation $\tau(\pi)$. Removing the condition $p_1 = *$ multiplies the sum by $n + 1$, hence, we get

$$\tau((J_{n+1})^k) = \frac{1}{n+1} \sum_{\substack{p_1,\ldots,p_k \in [\![1,n]\!] \sqcup \{*\} \\ \forall j \in [\![1,k]\!],\ p_j \neq p_{j+1}}} \chi^\lambda(\kappa(\pi(p_1, \ldots, p_k)) \uparrow n).$$

We saw on the other hand that

$$\tau'((K_{n+1})^k) = \frac{1}{n+1} \sum_{\substack{p_1,\ldots,p_k \in [\![1,n]\!] \sqcup \{*\} \\ \forall j \in [\![1,k]\!],\ p_j \neq p_{j+1}}} \chi^\lambda((p_1, p_2)'(p_2, p_3)' \cdots (p_k, p_1)'),$$

where $*$ plays now the role of $n+1$, and $(a, b)' = \mathrm{id}$ if a or b is the special symbol $*$. Therefore, Theorem 9.23 follows from:

Proposition 9.34. *Given a sequence $(p_1, \ldots, p_k)$ of elements of $[\![1, n]\!] \sqcup \{*\}$ without consecutive terms that are equal, the cycle type of $(p_1, p_2)'(p_2, p_3)' \cdots (p_k, p_1)'$ in $\mathfrak{S}(n)$ is always $\kappa(\pi(p_1, \ldots, p_k)) \uparrow n$.*

Proof. Actually, we have for any admissible sequence $(p_1, \ldots, p_k)$ (so, with $p_1 = *$):

$$\sigma(p_1, \ldots, p_k) = \big((p_1, p_2)'(p_2, p_3)' \cdots (p_k, p_1)'\big)^{-1},$$

where $\sigma(p_1, \ldots, p_k)$ is the permutation of the partial permutation $\Theta(p_1, \ldots, p_k)$. Indeed, the sequence of permutations $(\sigma_m)_{m \in [\![1,k]\!]}$ satisfies the recurrence relation

$$\sigma_{m-1} = (\sigma_m(p_{m-1}), *) \circ \sigma_m = \sigma_m \circ (p_{m-1}, \sigma_m^{-1}(*)) = \sigma_m\,(p_{m-1}, p_m).$$

Therefore, $\sigma(p_1,\ldots,p_k)=\sigma_1=(p_1,p_k)(p_k,p_{k-1})\cdots(p_2,p_1)$. Let us then show by induction on k that this product is inversed by $(p_1,p_2)'(p_2,p_3)'\cdots(p_k,p_1)'$ (this is not trivial, as many transpositions $(p_j,p_{j+1})'$ might be equal to id). The case $k=2$ is obvious:

$$(p_1,p_2)(p_2,p_1)=(p_1,p_2)'(p_2,p_1)'=\mathrm{id}.$$

Assume that the result is true up to order $k\geq 2$, and consider a sequence $P=(p_1,\ldots,p_{k+1})$ of length $k+1$, which is admissible. If $p_k\neq *$, then

$$(p_1,\ldots,p_k)\quad\text{are}\quad(p_1,\ldots,p_{k+1})$$

are both admissible sequences, so by the induction hypothesis,

$$(p_1,p_k)(p_k,p_{k-1})\cdots(p_2,p_1))\circ\big((p_1,p_2)'\cdots(p_{k-1},p_k)'(p_k,p_1)'\big)=\mathrm{id}$$

and

$$\begin{aligned}
&\sigma(p_1,\ldots,p_{k+1})\circ\big((p_1,p_2)'\cdots(p_k,p_{k+1})'(p_{k+1},p_1)'\big)\\
&=(p_1,p_{k+1})(p_{k+1},p_k)(p_1,p_k)\circ\sigma(p_1,\ldots,p_k)\\
&\quad\circ\big((p_1,p_2)'\cdots(p_{k-1},p_k)'(p_k,p_1)'\big)\circ(p_k,p_1)'(p_k,p_{k+1})'(p_{k+1},p_1)'\\
&=(p_1,p_{k+1})(p_{k+1},p_k)(p_1,p_k)(p_k,p_{k+1})\\
&=(p_{k+1},p_k)(p_k,p_{k+1})=\mathrm{id}.
\end{aligned}$$

On the other hand, if $p_k=*$, then $(p_1,\ldots,p_{k-1})$ is admissible, and then $\sigma(p_1,\ldots,p_{k+1})=\sigma(p_1,\ldots,p_{k-1})$ and the result follows trivially from the case $k-1$. So, if $p_1=*$, then the cycle type of $(p_1,p_2)'(p_2,p_3)'\cdots(p_k,p_1)'$ in $\mathfrak{S}(n)$ is indeed the same as the cycle type $\kappa(\pi(p_1,\ldots,p_k))\uparrow n$ of $\sigma(p_1,\ldots,p_k)$. Notice now that the cycle types are invariant by cyclic shift of the indices, so the result also holds if one of the p_i's (not necessarily p_1) is equal to $*$. It remains to treat the case where all the p_i's are different from $*$; then,

$$(p_1,p_2)'(p_2,p_3)'\cdots(p_k,p_1)'=(p_1,p_2)(p_2,p_3)\cdots(p_k,p_1).$$

In this case, if one wants to compute the numbers of counterclockwise winds of the cycles of $\tau(\pi(p_1,\ldots,p_k))$, then one can decide that the special symbol is $*=p_1$, and perform the computations as in the case of admissible sequences. In other words, we are conjugating the permutations by $(*,p_1)$. Then, the permutation written above is obviously the inverse of $(p_1,p_k)(p_k,p_{k-1})\cdots(p_2,p_1)$, which has cycle type $\kappa(\pi(p_1,\ldots,p_k))\uparrow n$. So, the result is true in any case, and this ends the proof. □

▷ *Multiplication of set partitions and the morphism $\mathscr{P} \to \mathscr{O}$.*

An additional interesting property of the map $\pi \mapsto \Sigma(\pi)$ is that it is a morphism for an appropriate product on (admissible) set partitions. The product of elements of $\mathfrak{Q}(\infty)$ will not be defined in general, but it will be defined if $(X^{(1)},\pi^{(1)}),\ldots,(X^{(t)},\pi^{(t)})$ are set partitions of disjoint intervals $X^{(i)}$, or more generally if $X^{(1)} \sqcup X^{(2)} \sqcup \cdots \sqcup X^{(t)}$ is a non-crossing partition.

Definition 9.35. *Let $\rho = (X^{(1)},X^{(2)},\ldots,X^{(t)})$ be a non-crossing partition of a set of integers X, and for each i, $\pi^{(i)}$ a set partition on $X^{(i)}$. Given set partitions on X, we denote $\rho \leq \pi$ if every part of ρ is included in a part of π (this extends the definition previously given for non-crossing partitions). On the other hand, if $\pi \in \mathfrak{NC}(X)$, we set*

$$\widetilde{\pi} = \text{inverse Kreweras complement of } \pi,$$

that is to say that $\overline{\widetilde{\pi}} = \pi$, or equivalently that $\widetilde{\pi} = c^{-2}(\overline{\pi})$. In this situation, we define the product

$$\prod_{i=1}^{t}(X^{(i)},\pi^{(i)}) = \sum_{\pi}(X,\pi)$$

as an element of the vector space $\mathbb{C}\mathfrak{Q}(X) \subset \mathbb{C}\mathfrak{Q}(\infty)$, where the sum runs over set partitions $\pi \in \mathfrak{Q}(X)$ such that:

1. *$\pi \geq \widetilde{\rho}$;*
2. *if $a, b \in X^{(i)}$, then a and b are in the same part of π if and only if they are in the same part of $\pi^{(i)}$.*

Example. In the case of two set partitions of sets $X^{(1)}$ and $X^{(2)}$ forming a non-crossing partition, up to a cyclic shift and relabeling, one can assume that these sets are two consecutive intervals $[\![1,m]\!]$ and $[\![m+1,m+n]\!]$. Then, $\rho = [\![1,m]\!] \sqcup [\![m+1,m+n]\!]$ and

$$\widetilde{\rho} = \{1,m+1\} \sqcup \bigsqcup_{k \neq 1,m+1} \{k\}.$$

Let $\pi^{(1)}$ and $\pi^{(2)}$ be two set partitions respectively on $[\![1,m]\!]$ and on $[\![m+1,m+n]\!]$, such that $1 \in \pi_1^{(1)}$ and $m+1 \in \pi_1^{(2)}$. Then, the product $\pi^{(1)} \cdot \pi^{(2)}$ is the sum of set partitions that are obtained from

$$(\pi_1^{(1)} \sqcup \pi_1^{(2)}), \pi_2^{(1)},\ldots,\pi_r^{(1)}, \pi_2^{(2)},\ldots,\pi_s^{(2)}$$

by joining possibly other parts $\pi_p^{(1)}$ and $\pi_q^{(2)}$. For instance, if

$\pi^{(1)} =$ 1 2 3 4 , $\pi^{(2)} =$ 5 6 7 8 9 10

then the product of these two partitions is the formal sum of the following set partitions:

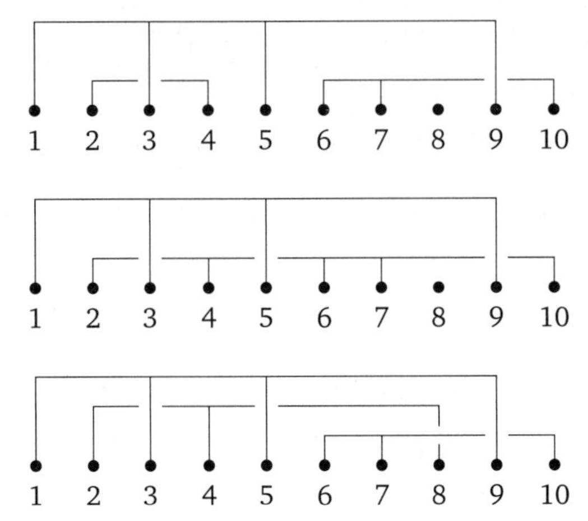

Proposition 9.36. *Suppose that* $(X^{(1)}, \pi^{(1)}), (X^{(2)}, \pi^{(2)}), \ldots, (X^{(r)}, \pi^{(r)})$ *is a family of set partitions such that* $\rho = (X^{(1)}, X^{(2)}, \ldots, X^{(r)})$ *is a non-crossing partition of* X.

1. *The product* $\prod_{i=1}^{r} (X^{(i)}, \pi^{(i)})$ *is associative, in the following sense: if* $\rho \leq \rho'$ *are non-crossing partitions of* X, *with*

$$\rho' = (Y^{(1)}, \ldots, Y^{(q)}) \quad ; \quad Y^{(k)} = \bigsqcup_{j=1}^{r_k} X^{(i_j)}$$

and if one takes the ρ'*-product of the products of partitions* $\prod_{j=1}^{r_k} (X^{(i_j)}, \pi^{(i_j)})$, *then one obtains the same result as the* ρ*-product* $\prod_{i=1}^{r} (X^{(i)}, \pi^{(i)})$.

2. *If all the set partitions* $\pi^{(i)}$ *are admissible, then their product is also a linear combination of admissible set partitions.*

Proof. Set $A = \prod_{k=1}^{q} \left(\prod_{j=1}^{r_k} (X^{(i_j)}, \pi^{(i_j)}) \right)$ and $B = \prod_{i=1}^{r} (X^{(i)}, \pi^{(i)})$. The products A and B are sums of set partitions (X, π), where each set partition π in A or in B has the property $(*)$ that if $x, y \in X^{(i)}$ for some index i, then x and y are in the same part of π if and only if they are in the same part of $\pi^{(i)}$. Therefore, the components (π, X) of A or B are all obtained by joining some parts $\pi_k^{(i)}$ and $\pi_l^{(j)}$ with $i \neq j$. More precisely, by definition of the product of set partitions, the components (X, π) of B are those such that $(*)$ holds for π and $\pi \geq \widetilde{\rho}$. On the other hand, for each $k \in [\![1, q]\!]$, the components $(Y^{(k)}, \theta^{(k)})$ of $\prod_{j=1}^{r_k} (X^{(i_j)}, \pi^{(i_j)})$ are those with the property $(*)$, and such that

$$\theta^{(k)} \geq \widetilde{\rho^{(k)}},$$

the set partition $\rho^{(k)}$ being the restriction of the set partition ρ to the set $Y^{(k)}$.

As a consequence, the components (X, π) of A are those with the property $*$, and such that

$$\pi \geq \max\left(\widetilde{\rho'}, \bigsqcup_{k=1}^{q} \widetilde{\rho^{(k)}}\right),$$

where the maximum is taken in the lattice $\mathfrak{NC}(X)$. So, it suffices to show that

$$\widetilde{\rho} = \max\left(\widetilde{\rho'}, \bigsqcup_{k=1}^{q} \widetilde{\rho^{(k)}}\right).$$

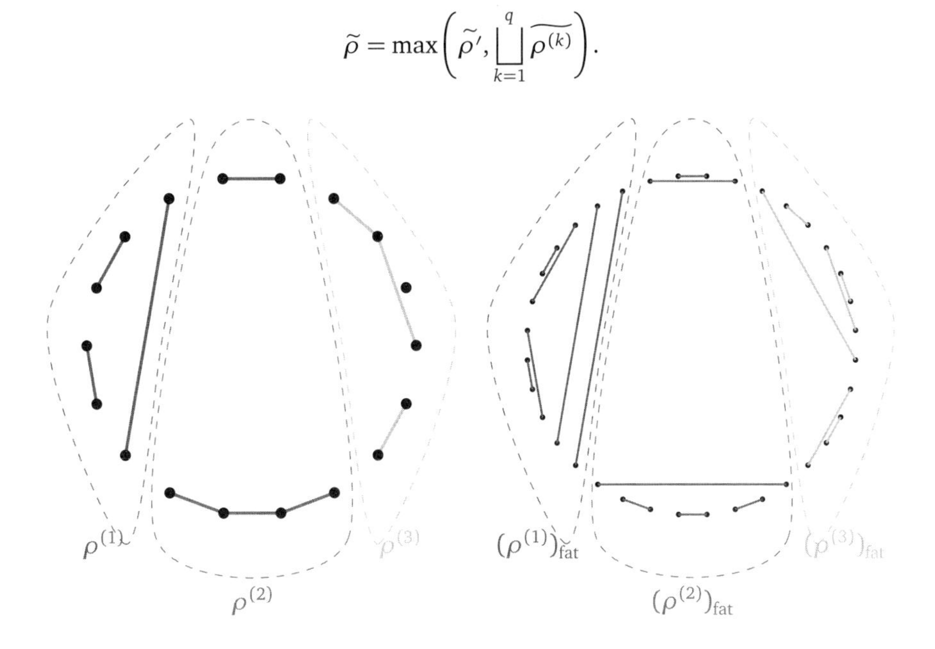

Figure 9.6
The identity $\rho_{\text{fat}} = \bigsqcup_{k=1}^{q} (\rho^{(k)})_{\text{fat}}$.

This computation is again better understood in a picture. An inflated set partition ρ_{fat} is drawn in Figure 9.6; obviously, if $\rho = \bigsqcup_{k=1}^{q} \rho^{(k)}$, then $\rho_{\text{fat}} = \bigsqcup_{k=1}^{q} (\rho^{(k)})_{\text{fat}}$.

Now, when applying the map c^{-1} on a component $(\rho^{(k)})_{\text{fat}}$, one obtains the fat set partition associated to $\widetilde{\rho^{(k)}} \in \mathfrak{Q}(Y^{(k)})$, and one can reconstruct this fat set partition from $c^{-1}(\rho_{\text{fat}}) \cap (Y^{(k)} \sqcup Y^{(k)'})$ by:

- keeping the links *with both ends* in $Y^{(k)} \sqcup Y^{(k)'}$;
- *contracting* the sequences of links

  $$a \in Y^{(k)'} \longleftrightarrow b_1 \in Y^{(l_1)},\ b_1' \in Y^{(l_1)'} \longleftrightarrow b_2 \in Y^{(l_2)}, \ldots, b_s' \in Y^{(l_s)'} \longleftrightarrow d \in Y^{(k)}$$

  with $k \notin \{l_1, l_2, \ldots l_s\}$ into $a \longleftrightarrow d$.

Therefore, $\widetilde{\rho} \geq \bigsqcup_{k=1}^{q} \widetilde{\rho^{(k)}}$; see Figure 9.8. What is missing in $\bigsqcup_{k=1}^{q} \widetilde{\rho^{(k)}}$ in comparison to $\widetilde{\rho}$ is obtained by adding the links of $\widetilde{\rho'}$ (see Figure 9.7); indeed, these links in $\widetilde{\rho}$ come from the links of $c^{-1}((\rho')_{\mathrm{fat}})$ that have ends in different sets $Y^{(k)}$ and $Y^{(l)}$, and they are exactly those forgotten in the contractions previously described. This ends the proof of the identity $\widetilde{\rho} = \max\left(\widetilde{\rho'}, \bigsqcup_{k=1}^{q} \widetilde{\rho^{(k)}}\right)$, and of the associativity of the product of set partitions.

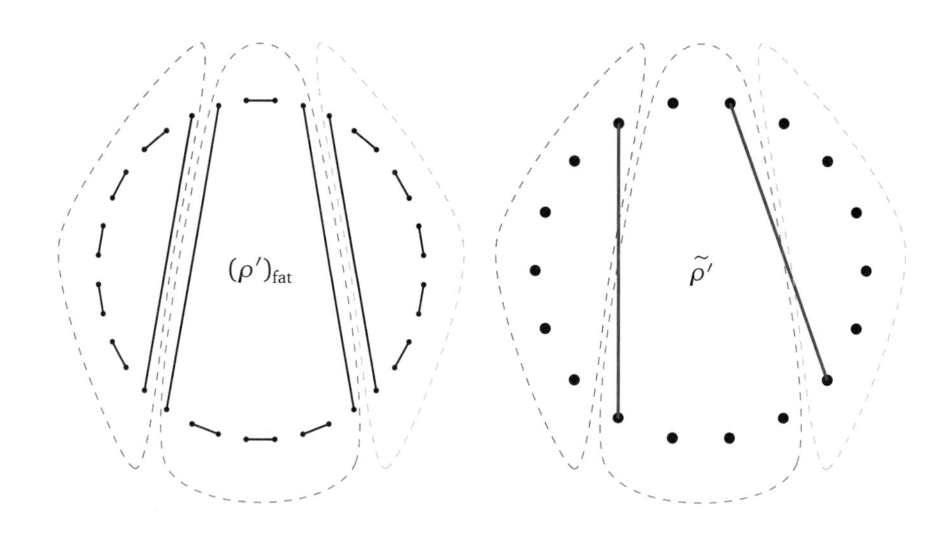

Figure 9.7
The missing links come from $\widetilde{\rho'}$.

Let us now show that the admissibility is conserved by product. Since this product has just been shown to be associative, by induction on the number of parts of the non-crossing partition ρ, it suffices to treat the case of two admissible set partitions $\pi^{(1)}$ and $\pi^{(2)}$, and up to a cyclic shift, one can assume that they are set partitions of consecutive intervals $[\![1,m]\!]$ and $[\![m+1,m+n]\!]$. Notice that we have necessarily $m \geq 2$ and $n \geq 2$: indeed, in $[\![1,1]\!] = \{1\}$, 1 is cyclically consecutive to itself, so the set partition $\{1\}$ is not admissible. Since $\widetilde{\rho} = \{1,m+1\} \sqcup \bigsqcup_{k\neq 1,m+1} \{k\}$ in this case, it is then obvious from the description of the set partitions in $\pi^{(1)} \cdot \pi^{(2)}$ that one cannot create a set partition π in the product with a part containing two consecutive integers. Indeed, these two consecutive integers would necessarily be 1 and $m+n$ or m and $m+1$, and come from the reunion of the corresponding parts in $\pi^{(1)}$ and $\pi^{(2)}$. But this is not possible, since π must contain the reunion of the part containing 1 with the part containing $m+1$, and then this part cannot contain m or $m+n$, since 1 and m are in distinct parts of $\pi^{(1)}$, and $m+1$ and $m+n$ are in distinct parts of $\pi^{(2)}$. □

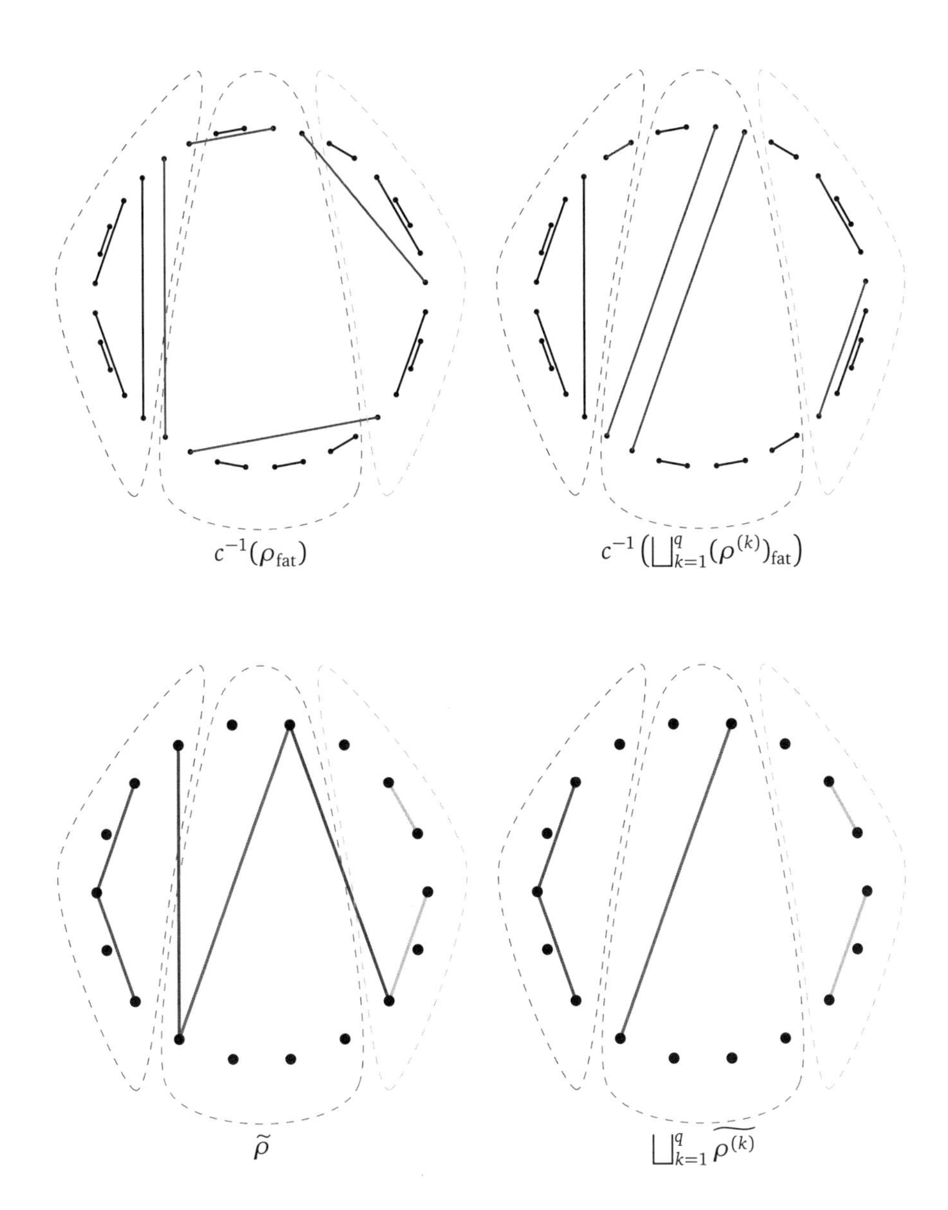

Figure 9.8
The union of set partitions $\widetilde{\rho^{(k)}}$ is finer that the set partition $\widetilde{\rho}$.

Theorem 9.37. *For any pairs $(X^{(1)},\pi^{(1)}),(X^{(2)},\pi^{(2)}),\dots,(X^{(t)},\pi^{(t)})$ such that $\rho=(X^{(1)},\dots,X^{(t)})$ is a non-crossing set-partition of X, one has*

$$\Sigma\left(\prod_{i=1}^{t}(X^{(i)},\pi^{(i)})\right)=\prod_{i=1}^{t}\Sigma(\pi^{(i)}),$$

where the product on the left-hand side is in $\mathbb{C}\mathfrak{Q}(\infty)$, and the product on the right-hand side in $\mathscr{O}$.

Proof. By the previous proposition, it suffices to treat the case of two set partitions $(X^{(1)},\pi^{(1)})$ and $(X^{(2)},\pi^{(2)})$, and by a cyclic shift of indices, one can assume that $X^{(1)}=[\![1,m]\!]$ and $X^{(2)}=[\![m+1,m+n]\!]$ are two consecutive intervals of integers. Notice then that

$$\begin{aligned}&\sum_{\substack{\pi(p_1,\dots,p_m)=\pi^{(1)}\\ \pi(q_1,\dots,q_n)=\pi^{(2)}}}(p_1,\dots,p_m,q_1,\dots,q_n)\\ &=\sum_{(X,\pi)\in(X^{(1)},\pi^{(1)})\cdot(X^{(2)},\pi^{(2)})}\left(\sum_{\pi(r_1,\dots,r_{m+n})=\pi}(r_1,\dots,r_{m+n})\right),\end{aligned}$$

where these formal sums run over admissible sequences. Indeed, given two admissible sequences $(p_1,\dots,p_m)$ and $(q_1,\dots,q_n)$ with associated set partitions $\pi^{(1)}$ and $\pi^{(2)}$, the set partition associated to the concatenation $(p_1,\dots,p_m,q_1,\dots,q_n)$ is obtained by:

- joining the part of $\pi^{(1)}$ containing 1 with the part of $\pi^{(2)}$ containing $m+1$, since $p_1=q_1=*$;
- possibly joining other parts of $\pi^{(1)}$ with parts of $\pi^{(2)}$ according to the other equalities $p_i=q_j$.

This is exactly the description of the components (X,π) of the product $(X^{(1)},\pi^{(1)})\cdot(X^{(2)},\pi^{(2)})$. By applying the map Θ to the previous identity, we get

$$\Sigma\big((X^{(1)},\pi^{(1)})\cdot(X^{(2)},\pi^{(2)})\big)=\sum_{\substack{\pi(p_1,\dots,p_m)=\pi^{(1)}\\ \pi(q_1,\dots,q_n)=\pi^{(2)}}}\Theta(p_1,\dots,p_m,q_1,\dots,q_n).$$

However, given two admissible sequences P and Q, we have for their concatenation $P\cdot Q$ the obvious identity $\sigma_{m+1}(P\cdot Q)=\sigma_1(Q)$ and then, setting $\rho=\sigma_1(Q)$,

$$\sigma_m(P\cdot Q)=(\rho(p_m),*)\circ\rho=(\rho(p_m),\rho(*))\circ\rho=\rho\circ(p_m,*)=\rho\circ(\sigma_m(P))$$

and by descending induction on $l\leq m$,

$$\begin{aligned}\sigma_{l-1}(P\cdot Q)&=((\rho\circ\sigma_l(P))(p_{l-1}),*)\circ\rho\circ\sigma_l(P)\\ &=\rho\circ((\sigma_l(P))(p_{l-1}),*)\circ\sigma_l(P)=\rho\circ(\sigma_{l-1}(P))\end{aligned}$$

that is to say that $\sigma_l(P \cdot Q) = \sigma_1(Q) \circ \sigma_l(P)$ for any $l \in [\![1,m]\!]$. In particular, $\sigma_1(P \cdot Q) = \sigma_1(Q) \circ \sigma_1(P)$, and

$$\Theta(p_1,\dots,p_m,q_1,\dots,q_n) = \Theta(q_1,\dots,q_m)\,\Theta(p_1,\dots,p_m).$$

Therefore, we can rewrite the previous identity as

$$\begin{aligned}
&\Sigma\big((X^{(1)},\pi^{(1)})\cdot(X^{(2)},\pi^{(2)})\big)\\
&= \left(\sum_{\pi(p_1,\dots,p_m)=\pi^{(1)}} \Theta(q_1,\dots,q_n)\right)\left(\sum_{\pi(q_1,\dots,q_n)=\pi^{(2)}} \Theta(p_1,\dots,p_m)\right)\\
&= \Sigma(X^{(1)},\pi^{(1)})\,\Sigma(X^{(2)},\pi^{(2)}),
\end{aligned}$$

which ends the proof. □

Remark. There is a graphical interpretation of the product of admissible set partitions in terms of constructions of Riemann surfaces, and this interpretation makes the notion a bit easier to understand; see Section 13.3 and the proof of Theorem 13.12.

Notes and references

The first part of this chapter relies on a minimal familiarity with some concepts of functional analysis, in particular the theory of bounded linear operators on Hilbert spaces. We refer to [Rud91, Chapter XII] and [Lan93, Chapter XVIII] for the basic results, for instance the spectral theorem 9.6 which we admitted. A more advanced treatment is proposed in the two first chapters of [Tak79], which we followed for the theory of weak operator topologies and of von Neumann algebras (in particular, the proof of the bicommutant theorem). Also, we use for the first time in this book the language of probability theory, for which as a general rule we refer to [Bil69, Bil95] (a more elementary approach to probability theory can be found in [Fel68]).

Our introduction to free probability theory is based on [NS06]. For the results of functional analysis (Riesz' representation theorem, Stone–Weierstrass theorem) used in the proof of the existence of the distribution of a self-adjoint element in a non-commutative probability space, we refer again to [Lan93]. Actually, we only needed the notion of free cumulant, which is why we did not spend too much time studying free random variables and their properties (e.g., the free central limit theorem, or the results due to Voiculescu on random matrices and their asymptotic freeness). The main Theorem 9.20 of this chapter is due to Biane; see [Bia98]. The proof presented here was found in [IO02, Section 10].

As shown in Section 9.3, the technology of free probability is pervasive in the representation theory of symmetric groups. The proof of the main result of this section (Theorem 9.23) was communicated to the author by V. Féray. The other combinatorial proof (Section 9.4) is due to P. Śniady (cf. [Śni06a]), though his article *defines* the transition measure μ_λ as the distribution of J_{n+1}, and only alludes to the fact that this definition is equivalent to the one that we gave in Chapter 7. In particular, our second proof of Theorem 9.23 gives all the details omitted in §6.1 of loc. cit.

We refer again to [Bia98] for other applications of free probability in the study of the characters and representations of large symmetric groups. In particular, one can show that if λ and μ are integer partitions of large size and with

$$\lambda_1 = O\left(\sqrt{|\lambda|}\right), \ \ell(\lambda) = O\left(\sqrt{|\lambda|}\right)$$

and similary for μ, then:

- The spectral measure of $S^\lambda \times S^\nu = \mathrm{Ind}_{\mathfrak{S}(|\lambda|)\times\mathfrak{S}(|\nu|)}^{\mathfrak{S}(|\lambda|+|\mu|)}(S^\lambda \boxtimes S^\nu)$ is concentrated on Young diagrams whose transition measures are close to the free convolution $\mu_\lambda \boxplus \mu_\nu$.

- The spectral measure of $\Delta(S^\lambda) = \mathrm{Res}_{\mathfrak{S}(p)\times\mathfrak{S}(q)}^{\mathfrak{S}(|\lambda|)}(S^\lambda)$ is concentrated on pairs of Young diagrams whose transition measures ν_1 and ν_2 are close to the free compressions
$$\pi_{\frac{p+1}{|\lambda|+1}}(\mu_\lambda) \quad \text{and} \quad \pi_{\frac{q+1}{|\lambda|+1}}(\mu_\lambda),$$
where $\pi_t(\mu)$ is the unique compactly supported probability measure on the real line with $R_k(\pi_t(\mu)) = t^k R_k(\mu)$ for any $k \geq 1$.

Here, "close to" means after some rescaling and with respect to the topology of weak convergence on $\mathscr{M}^1(\mathbb{R})$, and concentration can be understood as a convergence in probability. We shall encounter numerous other results of this kind (concentration of spectral measures) in Chapters 12 and 13.

10

The Stanley–Féray formula for characters and Kerov polynomials

In Chapter 9, we have shown that the top-weight component of the observable Σ_k is the free cumulant R_{k+1} of the transition measure of the Young diagram. More generally, since $(\Sigma_k)_{k\geq 1}$ and $(R_k)_{k\geq 2}$ are two algebraic bases of $\mathscr{O}$, there exists for every k a polynomial K_k such that

$$\Sigma_k = K_k(R_2, R_3, \ldots, R_{k+1}),$$

the dominant term of this polynomial being R_{k+1}. We call K_k the k-th **Kerov polynomial**, and we calculated in Section 9.2 the five first Kerov polynomials. A remarkable fact is that each polynomial K_k has *non-negative integer coefficients*, and the main goal of this chapter will be to give a combinatorial interpretation of these coefficients (and in particular a way to compute them).

The known proofs of this positivity result (Theorem 10.20) all require that we manipulate functions on Young diagrams which are more general than observables $f \in \mathscr{O}$. Thus, in Section 10.1, we shall introduce an extension $\mathscr{Q}$ of $\mathscr{O}$, which consists in the so-called Stanley polynomials, and whose elements will be used throughout this chapter. In Section 10.2, we present and prove the Stanley–Féray formula for characters of symmetric groups, which expresses the observables Σ_μ as Stanley polynomials. This expansion is related to the combinatorial links between permutations and bicolored maps, and its proof also relies on the explicit construction of the Specht modules S^λ given in Section 3.3. Finally, in Section 10.3, we prove the positivity of Kerov polynomials, by using the arguments of the previous sections and by developing a differential calculus on the algebras $\mathscr{O}$ and $\mathscr{Q}$. As an application of the general combinatorial interpretation of the coefficients of K_k, we give a simple formula for the linear terms of the Kerov polynomials.

10.1 New observables of Young diagrams

In this section, we construct an extension $\mathscr{Q}$ of the algebra $\mathscr{O}$ of observables of Young diagrams, which will correspond via some morphisms to the extension

QSym of the algebra of symmetric functions Sym (cf. Section 6.2). This set of "super" observables will be obtained by evaluation of two families of polynomials: the stable polynomials in interlaced coordinates, and the Stanley polynomials in multi-rectangular coordinates. The main result of the section is Theorem 10.7, which states that the same algebra is obtained by both constructions, and which yields a surjective morphism QSym $\to \mathscr{Q}$, with an explicit kernel. We shall also describe a large family of elements of $\mathscr{Q}$, labeled by bipartite graphs, and that will play a crucial role in Sections 10.2 and 10.3.

▷ *Interlaced coordinates and quasi-symmetric functions.*

In Section 7.2, we associated to a Young diagram $\lambda \in \mathfrak{Y}$ two interlaced sequences of coordinates $X(\lambda) = (x_1, \ldots, x_s)$ and $Y(\lambda) = (y_1, \ldots, y_{s-1})$. The observables of Young diagrams can then be seen as symmetric functions of the formal alphabet $X - Y$, with $p_1(X - Y) = 0$ for any λ. Thus,

$$\mathscr{O} = \mathrm{Sym}/(p_1).$$

In this description, notice that if one modifies the alphabets X and Y by inserting in them the same coordinate $x_i = y_i$, then one does not change the value of $p_k(X - Y)$. Thus, observables can be seen as polynomial functions on pairs of interlaced coordinates

$$x_1 \leq y_1 \leq x_2 \leq y_2 \leq \cdots \leq x_{s-1} \leq y_{s-1} \leq x_s$$

with the property that:

$$f(x_1, \ldots, y_{i-1}, x_i = y_i, x_{i+1}, \ldots, x_s) = f(x_1, \ldots, y_{i-1}, x_{i+1}, \ldots, x_s).$$

We rewrite the variables $(x_1, y_1, x_2, y_2, \ldots)$ as $(a_1, a_2, \ldots)$. In the following, we denote $Q(a)$ the set of families of polynomials $(P_s)_{s\geq 1}$ such that:

- $P_s \in \mathbb{C}[a_1, a_2, \ldots, a_s]$;
- $\max_{s\geq 1}(\deg P_s) < \infty$, and the polynomials P_s are compatible with one another:
$$P_{s+1}(a_1, \ldots, a_s, 0) = P_s(a_1, \ldots, a_s);$$
- for any $s \geq 2$ and any $i \in [\![1, s-1]\!]$,
$$P_s(a_1, a_2, \ldots, a_s)_{|a_i = a_{i+1}} = P_{s-2}(a_1, \ldots, a_{i-1}, a_{i+2}, \ldots, a_s).$$

We say that an element of $Q(a)$ is a **stable polynomial** in interlaced coordinates. The second condition ensures that a family of stable polynomials $(P_s)_{s\geq 1}$ corresponds to an element P of

$$\mathbb{C}[a_1, a_2, \ldots] = \varprojlim_{s\to\infty} \mathbb{C}[a_1, a_2, \ldots, a_s],$$

the projective limit being taken in the category of graded rings (thus, elements of the ring $\mathbb{C}[a_1, a_2, \ldots, a_s]$ are possibly infinite linear combinations of monomials a^k, with $|k| = \sum_{i=1}^{\infty} k_i$ bounded). By the previous discussion, if f is an observable of Young diagrams, then it yields an element of $Q(a)$, which in addition is symmetric in the variables x_i, and in the variables y_i separately. Indeed, it suffices to prove this with the interlaced moments $\widetilde{p}_k$, in which case this is trivial.

The algebra $Q(a)$ of stable polynomials is actually much bigger than the algebra of observables. More precisely, we have:

Theorem 10.1. *The algebra $Q(a)$ of stable polynomials is isomorphic to the algebra* QSym *of quasi-symmetric functions.*

The proof of this theorem relies on the use of *formal* ordered alphabets. In Section 6.2, we defined the algebra QSym of quasi-symmetric functions in an infinite ordered family of commutative variables $(x_1, x_2, x_3, \ldots)$. We call **formal ordered alphabet** any ideal X of QSym. A formal ordered alphabet yields a morphism $\mathrm{QSym} \to \mathrm{QSym}/X$, and we denote $P(X)$ the image by this morphism of a quasi-symmetric function $P \in \mathrm{QSym}$. Thus, formal ordered alphabets also correspond to surjective morphisms of algebras m_X starting from QSym.

Example. Consider a finite ordered family $X_N = (x_1, x_2, \ldots, x_N)$ of commuting variables. We identify X_N with the morphism

$$\pi_N : \mathrm{QSym} \to \mathrm{QSym}^{(N)}$$

which is the restriction to $\mathrm{QSym} \subset \mathbb{C}[X]$ of the morphism from $\mathbb{C}[X] = \mathbb{C}[x_1, x_2, \ldots]$ to $\mathbb{C}[x_1, \ldots, x_N]$ defined by setting $x_{N+1} = x_{N+2} = \cdots = 0$. The kernel of this morphism is the ideal generated by the symmetric functions $e_{k \geq N+1}$. So, X_N corresponds to this ideal of QSym. The notation $P(X_N) = P(x_1, x_2, \ldots, x_N)$ for $\pi_N(P)$ is justified, since for any complex numbers $z_1, \ldots, z_N$, the evaluation $P(z_1, \ldots, z_N, 0, \ldots, 0, \ldots)$ of P comes from $\pi_N(P)$ by replacing the variables $x_1, \ldots, x_N$ in $\mathbb{C}[x_1, \ldots, x_N] \supset \mathrm{QSym}^{(N)}$ by these complex numbers.

Example. For any summable sequence of complex numbers $(z_n)_{n \in \mathbb{N}}$, the evaluation $P \mapsto P(z_1, \ldots, z_n, \ldots)$ of quasi-symmetric functions is again a morphism of algebras, hence corresponds to an ideal I of QSym such that QSym/I is isomorphic to $\mathbb{C}$. Thus, summable sequences are (formal) ordered alphabets.

As with true ordered alphabets, we can now use the Hopf algebra structure of QSym to define operations on formal ordered alphabets. Thus, if X and Y are two formal ordered alphabets, we define their **ordered sum** $X \oplus Y$ by the identity

$$F(X \oplus Y) = \sum_i G_i(X) \otimes H_i(Y)$$

for any quasi-symmetric function F such that $\Delta(F) = \sum_i G_i \otimes H_i$. Thus, viewed as a morphism, $X \oplus Y$ is the composition of the coproduct Δ and of the morphism $m_X \otimes m_Y$, where m_X and m_Y are the morphisms associated to the formal alphabets X and Y. Beware that this operation is not commutative (see the example below). We also introduce the formal **opposite** $\ominus X$ of a formal ordered alphabet X, defined by the identity

$$F(\ominus X) = (\omega(F))(X)$$

for any $F \in \mathrm{QSym}$, ω being the antipode of QSym. In terms of morphisms, $m_{\ominus X} = m_X \circ \omega$, and in terms of ideals, $\ominus X = \omega(X)$.

Example. If $W = (w_n)_{n\in\mathbb{N}}$ and $Z = (z_n)_{n\in\mathbb{N}}$ are summable sequences, then their ordered sum is the formal alphabet that corresponds to the sequence $(w_1, w_2, \ldots, z_1, z_2, \ldots)$. On the other hand, there is usually no summable sequence $(y_n)_{n\in\mathbb{N}}$ that corresponds to the ordered formal alphabet $\ominus W$.

Remark. The operations $\oplus$ and $\ominus$ on formal ordered alphabets generalize the definitions given in Section 6.2 for true ordered alphabets.

In the following we consider an infinite sequence $(a_1, a_2, \ldots)$ of commuting random variables, and the **interlaced formal alphabets**

$$I_s = \{a_1\} \ominus \{a_2\} \oplus \{a_3\} \ominus \{a_4\} \oplus \cdots \oplus/\ominus \{a_s\}.$$

The ideals associated to these alphabets are decreasing for the inclusion: $I_S \subset I_s$ if $S \geq s$. With these notations, we can restate Theorem 10.1 as follows:

Proposition 10.2. *A family of polynomials $(P_s)_{s\geq 1}$ consists of stable polynomials if and only if there exists $P \in \mathrm{QSym}$ such that, for any s,*

$$P_s(a_1, \ldots, a_s) = P(I_s),$$

where the right-hand side involves $I_s = \{a_1\} \ominus \{a_2\} \oplus \{a_3\} \ominus \{a_4\} \oplus \cdots \oplus/\ominus \{a_s\}$.

Proof. For any variable x, $\{x\} \ominus \{x\} = \ominus\{x\} \oplus \{x\} = 0$, where 0 is the formal ordered alphabet corresponding to the counity morphism $\eta_{\mathrm{QSym}} : \mathrm{QSym} \to \mathbb{C}$ which projects a quasi-symmetric function onto its homogeneous component of degree 0. Indeed, in the Hopf algebra QSym, we have (cf. the diagrams of the Hopf operations in Section 2.3):

$$1_{\mathrm{QSym}} \circ \eta_{\mathrm{QSym}} = \nabla \circ (\mathrm{id}_{\mathrm{QSym}} \otimes \omega) \circ \Delta = \nabla \circ (\omega \otimes \mathrm{id}_{\mathrm{QSym}}) \circ \Delta,$$

and as a consequence, in terms of morphisms,

$$\begin{aligned} m_{\{x\}\ominus\{x\}} &= (m_{\{x\}} \otimes m_{\{x\}}) \circ (\mathrm{id}_{\mathrm{QSym}} \otimes \omega) \circ \Delta \\ &= m_{\{x\}} \circ \nabla \circ (\mathrm{id}_{\mathrm{QSym}} \otimes \omega) \circ \Delta \\ &= m_{\{x\}} \circ 1_{\mathrm{QSym}} \circ \eta_{\mathrm{QSym}} = \eta_{\mathrm{QSym}} \end{aligned}$$

and similarly for $m_{\ominus\{x\}\oplus\{x\}}$. Notice moreover that the formal alphabet 0 is the neutral element for the operation of sum of formal alphabet, since

$$m_{X\oplus 0} = (m_X \otimes \eta_{\mathrm{QSym}}) \circ \Delta = m_X \circ (\mathrm{id}_{\mathrm{QSym}} \otimes \eta_{\mathrm{QSym}}) \circ \Delta = m_X$$

and similarly for $m_{0\oplus X}$. Now, if P is a quasi-symmetric function, then the polynomials $P_s(a_1, \dots, a_s) = P(I_s)$ form a stable family, since, taking for instance i even,

$$\begin{aligned} P_s(a_1,\dots,a_s)|_{a_i=a_{i+1}} &= P(\{a_1\} \ominus \cdots \oplus \{a_{i-1}\} \ominus \{a_i\} \oplus \{a_i\} \ominus \{a_{i+2}\} \oplus \cdots \oplus/\ominus \{a_s\}) \\ &= P(\{a_1\} \ominus \cdots \oplus \{a_{i-1}\} \oplus 0 \ominus \{a_{i+2}\} \oplus \cdots \oplus/\ominus \{a_s\}) \\ &= P(\{a_1\} \ominus \cdots \oplus \{a_{i-1}\} \ominus \{a_{i+2}\} \oplus \cdots \oplus/\ominus \{a_s\}) \\ &= P_{s-2}(a_1, \dots, a_{i-1}, a_{i+2}, \dots, a_s). \end{aligned}$$

Hence, any quasi-symmetric polynomial yields a stable family of polynomials.

To prove the converse statement, we shall compute the dimension of the space of families $(P_s)_{s\geq 1}$ of stable polynomials that are homogeneous of degree n. We endow the ring $\mathbb{C}[a_1, a_2, \dots]$ with the gradation $d(a^k) = \sum_{i=1}^{\infty} i\, k_i$, and on the other hand, we call **packed** a monomial a^k such that $k = (k_1, \dots, k_r, 0, \dots, 0)$ has all its non-zero terms in its first entries. The packed monomials of degree n are in bijection with the compositions $c \in \mathfrak{C}(n)$. We write P for the element in $\mathbb{C}[a_1, a_2, \dots]$ that projects itself on the polynomials P_s. We now claim that if $(P_s)_{s\geq 1}$ is a family of stable polynomials homogeneous of degree n, then it is entirely determined by the coefficients of the packed monomials a^c in P. Indeed, consider a non-packed monomial a^k that occurs with coefficient γ_k in P. There exists an index i such that $k_i = 0$ and $k_{i+1} \neq 0$. We take s large enough and consider the identity

$$P_s(a_1, \dots, a_i, a_{i+1}, \dots, a_s)_{|a_i=a_{i+1}=a} = P_{s-2}(a_1, \dots, a_{i-1}, a_{i+2}, \dots, a_s).$$

The coefficient of the monomial $(a_1)^{k_1} \cdots (a_{i-1})^{k_{i-1}}\, a^{k_{i+1}}\, (a_{i+2})^{k_{i+2}} \cdots (a_s)^{k_s}$ in the left-hand side is 0, since the right-hand side does not depend on a. Thus, we have a linear relation that expresses γ_k in terms of coefficients $\gamma_{k'}$ with $d(a^k) > d(a^{k'})$. Therefore, P is entirely determined by the coefficients γ_c of the packed monomials a^c. Since $\operatorname{card} \mathfrak{C}(n) = 2^{n-1}$, we conclude that the dimension of the subspace $Q_n(a)$ of $Q(a)$ that consists in stable families of polynomials that are homogeneous of degree n is smaller than 2^{n-1}.

Conversely, given a composition $c \in \mathfrak{C}(n)$, consider the families of stable polynomials $(M_c(I_s))_{s\geq 1}$, where M_c is the monomial quasi-symmetric function labeled by the composition c. In QSym, the functions M_c are linearly independent, and setting $a_2 = a_4 = \cdots = a_{2n} = 0$, we obtain

$$M_c(I_{2n})|_{a_2=a_4=\cdots=a_{2n}=0} = M_c(\{a_1, a_3, \dots, a_{2n-1}\}),$$

the second formal alphabet being the specialization $X_n : \mathrm{QSym} \to \mathrm{QSym}^{(n)} \subset \mathbb{C}[x_1, \dots, x_n]$ presented before as an example. This specialization is injective when restricted to functions of degree less than n, so the families of stable polynomials $(M_c(I_s))_{s\geq 1}$ are linearly independent in $Q(a)$, and the space of stable families of degree n has exactly dimension 2^{n-1}. We then know that the map $f \in \mathrm{QSym} \mapsto (f(I_s))_{s\geq 1} \in Q(a)$ is an isomorphism of graded algebras, so Theorem 10.1 is also shown. □

If $f \in \mathrm{QSym}$ and $\lambda \in \mathfrak{Y}$, we associate to them the function

$$\Pi(f)(\lambda) = f(\{x_1\} \ominus \{y_1\} \oplus \cdots \ominus \{y_{s-1}\} \oplus \{x_s\}),$$

where $(x_1, \dots, x_s)$ and $(y_1, \dots, y_{s-1})$ are the interlaced coordinates of λ. We denote $\mathscr{Q}$ the image of QSym by Π; it is a subalgebra of the algebra of complex functions on $\mathfrak{Y}$, which contains $\mathscr{O}$ since $\Pi(p_k) = \widetilde{p}_k$ for any $k \geq 1$. Restricted to Sym, the morphism Π has kernel $p_1\,\mathrm{Sym}$ and factors through the isomorphism $\mathrm{Sym}/(p_1) \to \mathscr{O}$. Similarly:

Proposition 10.3. *The kernel of the morphism $\Pi : \mathrm{QSym} \to \mathscr{Q}$ is the ideal (M_1), whence the isomorphism*

$$\mathscr{Q} = \mathrm{QSym}/(M_1).$$

Proof. We already know that $\Pi(M_1)(\lambda) = \Pi(p_1)(\lambda) = \widetilde{p}_1(\lambda) = 0$ for any Young diagram λ, so the kernel of Π contains the ideal generated by M_1. Conversely, let $f \in \mathrm{Ker}\,\Pi$; we have to prove that M_1 divides f. If $(P_s)_{s\geq 1}$ is a family of stable polynomials in $Q(a)$, consider the polynomials

$$\widetilde{P}_s(a_1, \dots, a_s) = P_s\left(\sum_{i=1}^{s}(-1)^i a_i, a_2, \dots, a_s\right).$$

They form again a stable family. Indeed, suppose first $j \geq 2$ and $a_j = a_{j+1}$; then

$$\begin{aligned}\widetilde{P}_s(a_1, \dots, a_s)|_{a_j=a_{j+1}} &= P_s\left(\sum_{i=1}^{j-1}(-1)^i a_i + \sum_{i=j+2}^{s}(-1)^i a_i, a_2, \dots, a_s\right) \\ &= P_{s-2}\left(\sum_{i\neq j,j+1}(-1)^i a_i, a_2, \dots, a_{j-1}, a_{j+2}, \dots, a_s\right) \\ &= \widetilde{P}_{s-2}(a_1, \dots, a_{j-1}, a_{j+2}, \dots, a_s).\end{aligned}$$

When $j = 1$ and $a_j = a_{j+1}$, we have

$$\begin{aligned}\widetilde{P}_s(a_1, \dots, a_s)|_{a_1=a_2} &= P_s\left(\sum_{i=3}^{s}(-1)^i a_i, a_2, a_3, \dots, a_s\right) \\ &= P_s\left(\sum_{i=3}^{s}(-1)^i a_i, a_1, a_3, \dots, a_s\right).\end{aligned}$$

This expression does not depend on $a_1 = a_2$. In particular, we can take $a_1 = a_2 = a_3$, and then

$$\begin{aligned}\widetilde{P}_s(a_1,\ldots,a_s)|_{a_1=a_2} &= P_s\left(\sum_{i=3}^{s}(-1)^i a_i, a_3, a_3, \ldots, a_s\right)\\ &= P_{s-2}\left(\sum_{i=3}^{s}(-1)^i a_i, a_4, \ldots, a_s\right)\\ &= \widetilde{P}_{s-2}(a_3,\ldots,a_s).\end{aligned}$$

Thus, we have an involution $P \mapsto \widetilde{P}$ of the space $Q(a)$ of families of stable polynomials. With $f \in \operatorname{Ker}\Pi$, consider an increasing sequence $a_2 < a_3 < \cdots < a_{2s+1}$ of positive integers, and set $a_1 = \sum_{i=2}^{2s+1}(-1)^i a_i$. The sequence $(a_1,\ldots,a_{2s+1})$ is the sequence of interlaced coordinates of a Young diagram, hence, if P is the family of stable polynomials associated to f by Theorem 10.1, then

$$P_{2s+1}\left(\sum_{i=2}^{2s+1}(-1)^i a_i, a_2, \ldots, a_{2s+1}\right) = 0.$$

Using the involution previously constructed, we obtain

$$\widetilde{P}_{2s+1}(0, a_2, \ldots, a_{2s+1}) = 0$$

for any increasing sequence of positive integers, hence also in the ring of polynomials $\mathbb{C}[a_1,\ldots,a_{2s+1}]$. Hence, a_1 divides $\widetilde{P}_{2s+1}$ for any $s \geq 1$, and applying again the involution, we see that $\widetilde{a_1} = \sum_{i=1}^{2s+1}(-1)^i a_i$ divides P_{2s+1} for any $s \geq 1$. In the projective limit, this amounts to the fact that M_1 divides f in QSym. □

▷ *Multi-rectangular coordinates and Stanley polynomials.*

We defined in the previous paragraph an algebra $\mathscr{Q}$ of functions of Young diagrams, which have the property to be polynomials that behave well with the insertion of a "false" corner $x_i = y_i$ in the sequence of interlaced coordinates. In the remainder of this section, we shall relate these functions to another system of coordinates for Young diagrams, namely, the so-called **multi-rectangular Stanley coordinates**. Let $(p,q) = ((p_1,\ldots,p_m),(q_1,\ldots,q_m))$ be two sequences of non-negative integers with the same length m.

Definition 10.4. *We say that the pair (p,q) is a system of Stanley coordinates for $\lambda \in \mathfrak{Y}$ if the Young diagram of λ is*

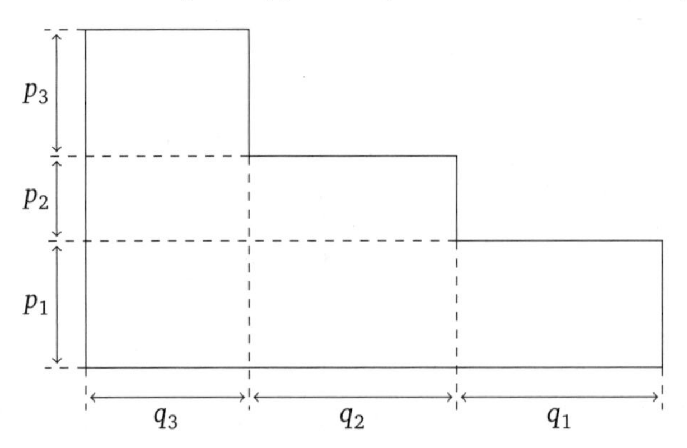

where each label p_i or q_j denotes a number of rows of columns.

Example. The integer partition $\lambda = (6,6,5,2,2)$ drawn below admits for Stanley coordinates $p = (2,1,2)$ and $q = (1,3,2)$.

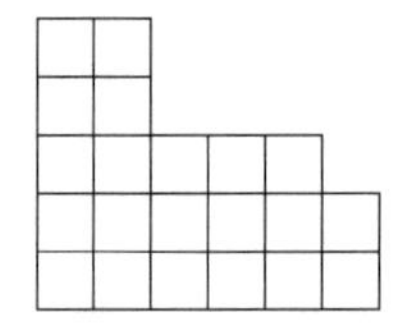

There is unicity of the system of Stanley coordinates if one requires them to be positive. Otherwise, one can construct other systems of Stanley coordinates, e.g., $p = (2,1,0,2)$ and $q = (1,2,1,2)$ for the previous partition. Besides, the unique system of Stanley coordinates of a Young diagram with all p_i and q_j positive has length $m = s-1$ and is given by

$$p_i = x_{s-i+1} - y_{s-i} \quad ; \quad q_i = y_{s-i} - x_{s-i},$$

where the x_i's and the y_i's are the interlaced coordinates of λ.

In the sequel, it will be convenient to denote the Stanley coordinates of a Young diagram in two rows:

$$(p,q) = \begin{pmatrix} p_1,\ldots,p_m \\ q_1,\ldots,q_m \end{pmatrix}.$$

We introduce the polynomial rings $\mathbb{C}[p,q]^{(m)} = \mathbb{C}[p_1,\ldots,p_m,q_1,\ldots,q_m]$, and their projective limit $\mathbb{C}[p,q] = \varprojlim_{m\to\infty} \mathbb{C}[p,q]^{(m)}$ in the category of graded rings and with respect to the morphisms

$$\begin{aligned} \mathbb{C}[p,q]^{(m+1)} &\to \mathbb{C}[p,q]^{(m)} \\ P\begin{pmatrix} p_1,\ldots,p_{m+1} \\ q_1,\ldots,q_{m+1} \end{pmatrix} &\mapsto P\begin{pmatrix} p_1,\ldots,p_m,0 \\ q_1,\ldots,q_m,0 \end{pmatrix}. \end{aligned}$$

The absence of unicity of Stanley coordinates leads to the following definition, which is analogous to the definition of stable families of polynomials:

Definition 10.5. *A **Stanley polynomial** is an element $F \in \mathbb{C}[p,q]$, corresponding to a family of polynomials $(F_m)_{m\geq 1}$ in the rings $\mathbb{C}[p_1,\dots,p_m,q_1,\dots,q_m]$, such that for any $m \geq 1$,*

$$F_{m+1}\begin{pmatrix} p_1,\dots,p_{m+1} \\ q_1,\dots,q_{m+1}\end{pmatrix}\Big|_{p_i=0} = F_m\begin{pmatrix} p_1,\dots,\ p_{i-1}\ ,p_{i+1},\dots,p_{m+1} \\ q_1,\dots,q_{i-1}+q_i,q_{i+1},\dots,q_{m+1}\end{pmatrix};$$
$$F_{m+1}\begin{pmatrix} p_1,\dots,p_{m+1} \\ q_1,\dots,q_{m+1}\end{pmatrix}\Big|_{q_i=0} = F_m\begin{pmatrix} p_1,\dots,p_{i-1},p_i+p_{i+1},\dots,p_{m+1} \\ q_1,\dots,q_{i-1},\ q_{i+1}\ ,\dots,q_{m+1}\end{pmatrix}.$$

We denote $S(p,q)$ the subalgebra of $\mathbb{C}[p,q]$ that consists in Stanley polynomials. The link with Stanley coordinates is provided by:

Proposition 10.6. *If $F=(F_m)_{m\geq 1}$ is a Stanley polynomial, then its evaluation $F(\lambda)$ on the Stanley coordinates of a Young diagram λ does not depend on the choice of the system of Stanley coordinates.*

Proof. Suppose that (p,q) is a system of Stanley coordinates for λ of length $m+1$, that is not **reduced**: some p_i or q_i vanishes. If $p_i=0$, then another system of Stanley coordinates for λ is:

$$\begin{pmatrix} p_1,\dots,\ p_{i-1}\ ,p_{i+1},\dots,p_{m+1} \\ q_1,\dots,q_{i-1}+q_i,q_{i+1},\dots,q_{m+1}\end{pmatrix}$$

as can be seen for instance in the previous example where $\lambda=(6,6,5,2,2)$ and $(p,q)=\binom{2,1,0,2}{1,2,1,2}$. On the other hand, if $q_i=0$, then another system of Stanley coordinates for λ is:

$$\begin{pmatrix} p_1,\dots,p_{i-1},p_i+p_{i+1},\dots,p_{m+1} \\ q_1,\dots,q_{i-1},\ q_{i+1}\ ,\dots,q_{m+1}\end{pmatrix};$$

this reduction happens for instance when $\lambda=(6,6,5,2,2)$ and $(p,q)=\binom{1,1,1,2}{0,1,3,2}$. Thus, we have a standard algorithm of reduction of non-reduced systems of Stanley coordinates, and by definition, if F is a Stanley polynomial, then its evaluations are the same on a system (p,q) and on its reduction $(\widetilde{p},\widetilde{q})$. In particular, if (p^*,q^*) is the unique reduced system of Stanley coordinates, then $F(p,q)=F(p^*,q^*)$ and the evaluation $F(\lambda)$ is indeed independent of the choice of (p,q). □

Theorem 10.7. *The evaluation of Stanley polynomials on Young diagrams yields an isomorphism between $\mathscr{Q}$ and $S(p,q)$. Thus, in the sense of isomorphism of graded complex algebras, we have:*

$$\mathscr{Q} = S(p,q) = Q(a)/\left(\sum_{i\geq 1}(-1)^i a_i\right) = \mathrm{QSym}/(M_1).$$

Proof. Let $P=(P_s)_{s\geq 1}$ be a stable polynomial. We associate to it an element $F=$

$\Psi(P)$ of $\mathbb{C}[p,q]$ by setting:

$$F_m\begin{pmatrix} p_1,\dots,p_m \\ q_1,\dots,q_m \end{pmatrix} = P_{2m+1}(a_1,\dots,a_{2m+1}),$$

where

$$a_{2i+1} = (q_{m+1-i}+\dots+q_m)-(p_1+\dots+p_{m-i});$$
$$a_{2i} = (q_{m+1-i}+\dots+q_m)-(p_1+\dots+p_{m+1-i});$$

these are the inverse formulas of

$$p_i = a_{2m+3-2i}-a_{2m+2-2i} \quad \text{and} \quad q_i = a_{2m+2-2i}-a_{2m+1-2i},$$

with the additional condition $\sum_{i=1}^{2m+1}(-1)^i a_i = 0$. We see straightforwardly that the F_m's are compatible with one another and indeed correspond to a polynomial F in the projective limit $\mathbb{C}[p,q]$:

$$\begin{aligned} F_{m+1}\begin{pmatrix} p_1,\dots,p_m,0 \\ q_1,\dots,q_m,0 \end{pmatrix} &= P_{2m+3}(a_1,\dots,a_{2m+1},a_{2m+1},a_{2m+1}) \\ &= P_{2m+1}(a_1,\dots,a_{2m+1}) = F_m\begin{pmatrix} p_1,\dots,p_m \\ q_1,\dots,q_m \end{pmatrix}. \end{aligned}$$

Let us then check that F is a Stanley polynomial. In a set of Stanley coordinates of length $m+1$, assume for instance $q_i=0$; then, $a_{2m+2-2i}=a_{2m+1-2i}$, so

$$\begin{aligned} &F_{m+1}\begin{pmatrix} p_1,\dots,p_i,\dots,p_{m+1} \\ q_1,\dots,0,\dots,q_{m+1} \end{pmatrix} \\ &= P_{2m+3}(a_1,\dots,a_{2m-2i},a_{2m+1-2i},a_{2m+1-2i},a_{2m+3-2i},\dots,a_{2m+3}) \\ &= P_{2m+1}(a_1,\dots,a_{2m-2i},a_{2m+3-2i},\dots,a_{2m+3}) \\ &= F_m\begin{pmatrix} p_1,\dots,p_{i-1},p_i+p_{i+1},\dots,p_{m+1} \\ q_1,\dots,q_{i-1},\ q_{i+1}\ ,\dots,q_{m+1} \end{pmatrix}. \end{aligned}$$

The case when $p_i=0$ is analogous. Therefore, we have a morphism of algebras $\Psi : Q(a) \to S(p,q)$. Conversely, if F is a Stanley polynomial, we can associate to it an element $P=\Theta(F)$ in $\mathbb{C}[a_1,a_2,\dots]$ by setting

$$P_{2m+1}(a_1,\dots,a_{2m+1}) = F_m\begin{pmatrix} \widetilde{p}_1,\dots,\widetilde{p}_m \\ \widetilde{q}_1,\dots,\widetilde{q}_m \end{pmatrix},$$

where $\widetilde{p}_i = a_{2m+3-2i}-a_{2m+2-2i}$ and $\widetilde{q}_i = a_{2m+2-2i}-a_{2m+1-2i}$. We define the even specializations P_{2m} by using the maps $\mathbb{C}[a_1,\dots,a_{2m+1}] \to \mathbb{C}[a_1,\dots,a_{2m}]$; then, we obtain a compatible family $(P_s)_{s\geq 1}$ corresponding to an element $P \in \mathbb{C}[a_1,a_2,\dots]$, and it is easily seen that it is a stable family. Beware that in this second construction, we do not have $\sum_{i=1}^{2m+1}(-1)^i a_i = 0$. We claim that $\Psi \circ \Theta = \mathrm{id}_{S(p,q)}$. Indeed, if $F \in S(p,q)$, then

$$(\Theta(F))_{2m+1}(a_1,\dots,a_{2m+1}) = F_m\begin{pmatrix} a_{2m+1}-a_{2m},\dots,a_3-a_2 \\ a_{2m}-a_{2m-1},\dots,a_2-a_1 \end{pmatrix} = F_m\begin{pmatrix} \widetilde{p}_1,\dots,\widetilde{p}_m \\ \widetilde{q}_1,\dots,\widetilde{q}_m \end{pmatrix},$$

and then

$$(\Psi \circ \Theta(F))_m \binom{p_1, \ldots, p_m}{q_1, \ldots, q_m} = (\Theta(F))_{2m+1}(a_1, \ldots, a_{2m+1})$$

where

$$a_{2i+1} = (q_{m+1-i} + \cdots + q_m) - (p_1 + \cdots + p_{m-i});$$
$$a_{2i} = (q_{m+1-i} + \cdots + q_m) - (p_1 + \cdots + p_{m+1-i}).$$

We compute $\widetilde{p}_i$:

$$\begin{aligned}\widetilde{p}_i &= a_{2m+3-2i} - a_{2m+2-2i} \\ &= ((q_i + \cdots + q_m) - (p_1 + \cdots + p_{i-1})) - ((q_i + \cdots + q_m) - (p_1 + \cdots + p_i)) = p_i\end{aligned}$$

and similarly, $\widetilde{q}_i = q_i$, so

$$(\Psi \circ \Theta(F))_m \binom{p_1, \ldots, p_m}{q_1, \ldots, q_m} = F_m \binom{p_1, \ldots, p_m}{q_1, \ldots, q_m},$$

$\Psi \circ \Theta(F) = F$. In the other direction, we do not have $\Theta \circ \Psi = \mathrm{id}_{Q(a)}$, as Ψ has a non-zero kernel. Indeed, if $P_s(a_1, \ldots, a_s) = M_1(I_s)$, then

$$(\Psi(P))_m \binom{p_1, \ldots, p_m}{q_1, \ldots, q_m} = P_{2m+1}(a_1, \ldots, a_{2m+1}) = \sum_{i=1}^{2m+1} (-1)^i \, a_i,$$

where a_{2i+1} and a_{2i} are defined in terms of the variables p_i and q_j as before. However, we then have $a_{2i+2} - a_{2i+1} = q_{m-i}$, hence,

$$\sum_{i=1}^{2m+1} (-1)^i \, a_i = -a_{2m+1} + \sum_{i=1}^{m} q_i = 0$$

by definition of $a_{2m+1} = (q_1 + \cdots + q_m) - 0$. Hence, $P = \sum_{i \geq 1} (-1)^i a_i$ belongs to the kernel of Ψ.

Let us now prove the theorem, writing Π' for the morphism from $S(p,q)$ to the algebra $\mathbb{C}^{\mathfrak{Y}}$ of functions on Young diagrams. We have the commutative diagram:

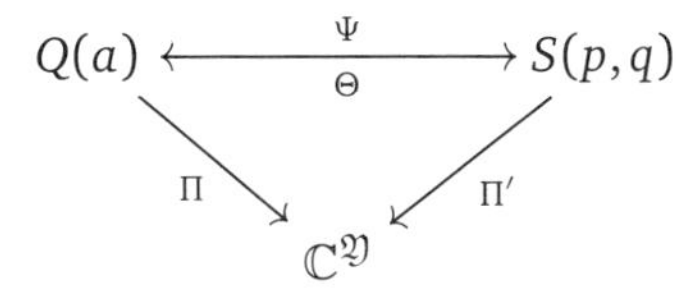

since the evaluation of stable or Stanley polynomials only depends on the Young diagram λ, and not on the use of interlaced or Stanley coordinates to perform these evaluations. Thus, $\Pi' \circ \Psi = \Pi$, and since Ψ is surjective (left factor of $\mathrm{id}_{S(p,q)}$), Π' has the same image as Π, namely, the algebra $\mathscr{Q}$. Let us then show that Π' is injective. If $\Pi'(F) = 0$, then $\Pi \circ \Theta(F) = \Pi'(F) = 0$, so, by Proposition 10.3, $\Theta(F) \in (M_1)$. Then, $F = \Psi \circ \Theta(F) = 0$, since we have shown that $\Psi(M_1) = 0$. This ends the proof of the theorem. We also obtain with the same reasoning $\mathrm{Ker}\,\Psi = (M_1)$ in $Q(a) = \mathrm{QSym}$. $\square$

Example. Consider the observable $\widetilde{p}_2 \in \mathscr{O} \subset \mathscr{Q}$. By the previous theorem, it writes as a Stanley polynomial. Indeed, for any integer partition λ,

$$\widetilde{p}_2(\lambda) = 2\sum_{i=1}^{m} q_i(p_1 + \cdots + p_i) = 2\sum_{i \leq j} p_i q_j,$$

as can be seen by replacing each q_i by the quantity $y_{m+1-i} - x_{m+1-i}$, and each p_i by $x_{m+2-i} - y_{m+1-i}$.

▷ *Stanley polynomials and bipartite graphs.*

To close this section, let us introduce a large class of functions N^G in $\mathscr{Q}$, that are labeled by bipartite graphs. Recall that a (finite, undirected, without loop) graph $G = (V, E)$ is given by a finite set of vertices V, together with a multiset E of pairs $\{v, w\}$ with $v \neq w$ in V. For instance,

$$G = ([\![1,5]\!], \{\{1,4\},\{1,5\},\{2,4\},\{2,5\},\{3,4\}\})$$

is the finite graph represented by

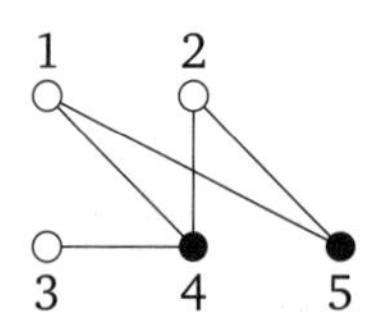

We authorise multiple edges $\{v, w\}$, though this won't change the definition of the function N^G hereafter. A graph is called **bipartite** if there exists a partition $V = V_1 \sqcup V_2$ of its vertex set such that each edge $e \in E$ has one extremity in V_1 and one extremity in V_2. The previous graph is bipartite, with $V_1 = \{1,2,3\}$ and $V_2 = \{4,5\}$. In the following, we consider bipartite graphs as graphs together with a partition $V_1 \sqcup V_2$ for which the aforementioned condition is satified (there might be several such partitions if G is not connected). We then define the function $N^G \in \mathbb{C}[p,q]$ by:

$$N^G(p,q) = \sum_{r:V \to \mathbb{N}^*} \left(\prod_{v_1 \in V_1} p_{r(v_1)}\right)\left(\prod_{v_2 \in V_2} q_{r(v_2)}\right),$$

where the sum runs over functions $r : v \to \mathbb{N}^*$ such that if $e = \{v_1, v_2\} \in E$ with $v_1 \in V_1$ and $v_2 \in V_2$, then $r(v_1) \leq r(v_2)$. One thus obtains a homogeneous polynomial in $\mathbb{C}[p,q]$ of degree card V.

Example. Considering again the previous graph, we have

$$N^{G}(p,q) = \sum_{\substack{d \geq \max(a,b,c) \\ e \geq \max(a,b)}} p_a p_b p_c q_d q_e.$$

Proposition 10.8. *For any bipartite graph G without isolated vertex, the polynomial N^G is a Stanley polynomial in $S(p,q)$.*

Proof. Let us compute $N^G\binom{p_1,\dots,p_{m+1}}{q_1,\dots,q_{m+1}}|_{q_i=0}$. We assume first $i < m+1$. By definition, it is the sum of the monomials

$$m(r) = \prod_{v_1 \in V_1} p_{r(v_1)} \prod_{v_2 \in V_2} q_{r(v_2)}$$

over functions $r : V \to [\![1, m+1]\!]$ that satisfy the condition $r(v_1) \leq r(v_2)$ if $\{v_1, v_2\} \in E$. Since $q_i = 0$, we can restrict the sum to functions r that do not take the value i on V_2:

$$N^G\binom{p_1,\dots,p_{m+1}}{q_1,\dots,q_{m+1}}\Big|_{q_i=0} = \sum_{r:V\to[\![1,m+1]\!],\, i\notin r(V_2)} m(r).$$

To any function $r : V \to [\![1, m+1]\!]$, we associate another function $r' = \Phi(r) : V \to [\![1,m]\!]$:

$$r'(v) = \begin{cases} r(v) & \text{if } r(v) \leq i, \\ r(v) - 1 & \text{if } r(v) > i. \end{cases}$$

If r satisfies the order condition $r(v_1) \leq r(v_2)$ and does not attain i on V_2, then $r' = \Phi(r)$ satisfies the order condition. Moreover, the preimage of a function $r' : V \to [\![1,m]\!]$ by the map Φ is the set of functions r such that

$$\begin{aligned} &r(v) = r'(v) \text{ if } r'(v) < i; \\ &r(v) \in \{i, i+1\} \text{ if } r'(v) = i; \\ &r(v) = r'(v) + 1 \text{ if } r'(v) > i. \end{aligned}$$

Among them, one sees readily that the functions r that avoid i on V_2 satisfy the order condition. We then have:

$$\sum_{r\in\Phi^{-1}(r'),\, i\notin r(V_2)} m(r) = \prod_{v_1\in V_1} \widetilde{p}_{r'(v_1)} \prod_{v_2\in V_2} \widetilde{q}_{r'(v_2)},$$

where the sum on the left-hand side runs over i-avoiding functions r in the preimage of r', and where

$$\widetilde{p}_j = \begin{cases} p_j & \text{if } j < i; \\ p_i + p_{i+1} & \text{if } j = i; \\ p_{j+1} & \text{if } j > i; \end{cases}$$

$$\widetilde{q}_j = \begin{cases} q_j & \text{if } j < i; \\ q_{i+1} & \text{if } j = i; \\ q_{j+1} & \text{if } j > i. \end{cases}$$

Therefore, gathering in the sum defining $N^G\binom{p_1,\ldots,p_{m+1}}{q_1,\ldots,q_{m+1}}|_{q_i=0}$ the functions r according to their image $\Phi(r)$, we get:

$$\begin{aligned}N^G\begin{pmatrix}p_1,\ldots,p_{m+1}\\q_1,\ldots,q_{m+1}\end{pmatrix}\Big|_{q_i=0} &= \sum_{r':V\to[\![1,m]\!]}\left(\prod_{v_1\in V_1}\widetilde{p}_{r'(v_1)}\prod_{v_2\in V_2}\widetilde{q}_{r'(v_2)}\right)\\&= N^G\begin{pmatrix}\widetilde{p}_1,\ldots,\widetilde{p}_m\\\widetilde{q}_1,\ldots,\widetilde{q}_m\end{pmatrix}\\&= N^G\begin{pmatrix}p_1,\ldots,p_{i-1},p_i+p_{i+1},p_{i+2},\ldots,p_{m+1}\\q_1,\ldots,q_{i-1},\ q_{i+1}\ ,q_{i+2},\ldots,q_{m+1}\end{pmatrix}.\end{aligned}$$

On the other hand, if $i = m+1$, then a function r that avoids $m+1$ on V_2 and satisfies the order condition cannot take the value $m+1$ on V_1; indeed, since any vertex of G is assumed to have a neighbor, one would then have $r(v_2) \geq r(v_1) = m+1$ for some vertices $v_1 \in V_1$ and $v_2 \in V_2$, hence $r(v_2) = m+1$, which is excluded. Hence, in this case, the sum is restricted to functions $r : V \to [\![1,m]\!]$, which means that

$$N^G\begin{pmatrix}p_1,\ldots,p_{m+1}\\q_1,\ldots,q_{m+1}\end{pmatrix}\Big|_{q_{m+1}=0} = N^G\begin{pmatrix}p_1,\ldots,p_m\\q_1,\ldots,q_m\end{pmatrix}$$

as wanted. The equations satisfied by the functions N^G with specializations $p_i = 0$ are analogous, and left to the reader. □

10.2 The Stanley–Féray formula for characters of symmetric groups

If μ is an integer partition, then the observable $\Sigma_\mu \in \mathscr{O}$ is a fortiori in the algebra $\mathscr{Q}$, hence, it is a Stanley polynomial in the multi-rectangular coordinates. The objective of this section is to obtain an explicit expression of this polynomial. More precisely, we shall establish an expansion of Σ_μ in functions N^G associated to certain bipartite graphs. This expansion (Theorem 10.11) can be considered as an alternative to the Murnaghan–Nakayama formula 3.10 for the calculation of the characters of the symmetric groups. It plays an important role in the dual combinatorics of their representation theory, which we develop in this third part of the book.

▷ *Permutations, maps and bipartite graphs.*

In the following, we call **bicolored** a bipartite graph without isolated vertex: any vertex of V_1 is connected to at least one vertex of V_2, and conversely. It will be convenient to draw the vertices of V_1 in white and the vertices of V_2 in black, and thus we shall speak of white and black vertices of a bicolored graph. Fix an integer $k \geq 1$, and consider two permutations σ and τ in $\mathfrak{S}(k)$. We are going to associate to them a (bicolored, labeled) **map**, which is a certain kind of graph with additional information on each vertex. If $G = (V, E)$ is a finite graph, a **cyclic order** on a vertex $v \in V$ is a cyclic permutation $(v_1, \ldots, v_d)$ of the neighbors of v. Two such sequences of the neighbors are identified if they differ cyclically, so for instance (v_1, v_2, v_3) is the same cyclic order as (v_2, v_3, v_1) and (v_3, v_1, v_2).

Definition 10.9. *A map with underlying graph $G = (V, E)$ is given by the graph G and by a cyclic order on each of its vertices.*

A **labeled map** is a map endowed with a bijection $E \to [\![1, \operatorname{card} E]\!]$, which gives a number to every edge. Then, every cyclic order on a vertex v can be considered as a cycle on integers, namely, the integers that label the edges stemming from v.

Example. Figure 10.1 represents a labeled map with 10 edges, which is also bipartite as a graph. The cyclic orders of its white vertices are $(1, 2, 3)$, $(4, 5, 6)$ and

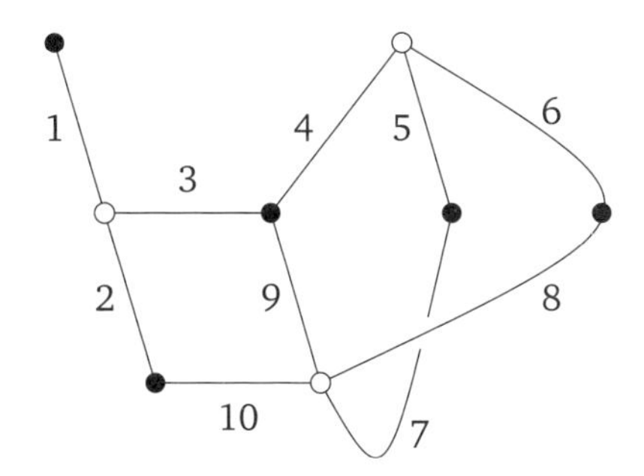

Figure 10.1
A bicolored labeled map with 10 edges.

$(7, 8, 9, 10)$; and the cyclic orders of its black vertices are (1), $(2, 10)$, $(3, 9, 4)$, $(5, 7)$ and $(6, 8)$. Because the graph is bipartite, each integer $i \in [\![1, 10]\!]$ appears in a unique cyclic order of a white vertex, and in a unique cyclic order of a black vertex. Therefore, one can associate to a bicolored labeled map a unique pair of permutations (σ, τ) in $\mathfrak{S}(k) \times \mathfrak{S}(k)$, where σ is the product of the cycles corresponding to the white vertices, and τ is the product of the cycles corresponding to the black vertices. In our example,

$$\sigma = (1, 2, 3)(4, 5, 6)(7, 8, 9, 10) \quad ; \quad \tau = (1)(2, 10)(5, 7)(6, 8)(3, 9, 4).$$

Conversely, given two permutations in $\mathfrak{S}(k)$, we can construct a map corresponding to it, by considering the cycles of σ as white vertices, the cycles of τ as black vertices, and the integers $i \in c \cap d$ with c cycle of σ and d cycle of τ as edges, these edges being ordered cyclically around each vertex according to σ or τ. We thus get a bijection between bicolored labeled maps with k edges, and pairs of permutations of size k.

Call **face** of a map a sequence of vertices $(v_1, v_2, \dots, v_p)$, such that for each i, the edge $\{v_i, v_{i+1}\}$ is the successor of the edge $\{v_i, v_{i-1}\}$ in the cyclic order of the vertex v_i (with by convention $v_0 = v_p$ and $v_{p+1} = v_1$). Alternatively, a face F can be considered as the sequence of oriented edges $(v_1, v_2), (v_2, v_3), \dots, (v_p, v_1)$, with the same condition as before. Each oriented edge (v, w) of a map is contained in exactly one face. For instance, one has in Figure 10.2 the two faces of the map of Figure 10.1. The **word** of a face is the cyclic sequence of its edges (w, b) with

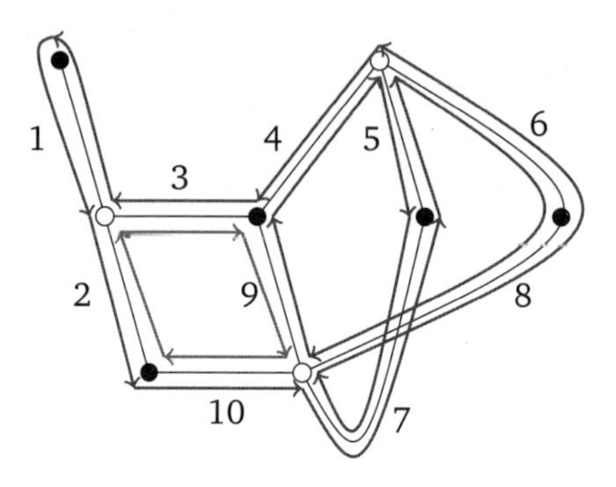

Figure 10.2
Two faces of a map, with corresponding words $(3, 10)$ and $(1, 2, 7, 6, 9, 5, 8, 4)$.

w white vertex and b black vertex (thus, one counts only half of the edges of a face in its word). In the previous example, one obtains the words $(3, 10)$ and $(1, 2, 7, 6, 9, 5, 8, 4)$ for the two faces. Each label appears in the word of exactly one face, and one can consider the permutation ρ whose cycles are the words of the faces of the map. Then:

Proposition 10.10. *Let σ and τ be two permutations of size k, and ρ the permutation coming from the faces of the map associated to (σ, τ). One has $\rho = \sigma\tau$.*

Proof. Let i be an integer in $[\![1, k]\!]$, which we consider as the oriented edge of label i and going from the white extremity $w(i)$ to the black extremity $b(i)$. Its image by τ is the successor of i in the cyclic order of $b(i)$, say, j, with $b(i) = b(j)$. Then, the image of j by σ is the successor of j in the cyclic order of $w(j)$, say, k with $w(j) = w(k)$. Then, k is the next white-to-black oriented edge in the face containing $(w(i), b(i))$ — we skip in this face the black-to-white oriented edge $(b(j), w(j))$. Thus, by definition, $k = \rho(i)$. □

Example. We leave the reader to check that we have indeed

$$\begin{aligned}&\big((1,2,3)(4,5,6)(7,8,9,10)\big)\circ\big((1)(2,10)(5,7)(6,8)(3,9,4)\big)\\&=(3,10)(1,2,7,6,9,5,8,4)\end{aligned}$$

in $\mathfrak{S}(10)$.

Given two permutations σ and τ in $\mathfrak{S}(k)$, we define $N^{\sigma,\tau}$ to be the Stanley polynomial N^G associated to the bicolored graph G underlying the bicolored labeled map associated to σ and τ. By Proposition 10.8, $N^{\sigma,\tau}$ is a Stanley polynomial since the graph G is bicolored, and on the other hand, since we forget labels in the construction, we have $N^{\theta\sigma\theta^{-1},\theta\tau\theta^{-1}}=N^{\sigma,\tau}$ for any $\theta\in\mathfrak{S}(k)$. Finally, $N^{\sigma,\tau}$ is homogeneous of degree $n(\sigma)+n(\tau)$ for any pair of permutations, where $n(\sigma)=\operatorname{card} C(\sigma)=\ell(t(\sigma))$ is the number of cycles of σ.

▷ *The symbols Σ_μ as Stanley polynomials.*

We can now state the main result of this section, which is the expansion of Σ_μ as a Stanley polynomial:

Theorem 10.11 (Féray, Śniady). *Let μ be an integer partition of size k, and ρ_μ be a permutation with cycle type μ in $\mathfrak{S}(k)$. We have*

$$\Sigma_\mu=\sum_{\rho_\mu=\sigma\tau}\varepsilon(\tau)\,N^{\sigma,\tau},$$

where the sum runs over factorizations $\rho_\mu=\sigma\tau$ in $\mathfrak{S}(k)$ (by a previous remark, the terms of the sum do not depend on the choice of ρ_μ in its conjugacy class C_μ).

Before we start the proof of Theorem 10.11, let us reformulate the definition of $N^{\sigma,\tau}(\lambda)$ for an integer partition λ of Stanley coordinates $\binom{p_1,\ldots,p_m}{q_1,\ldots,q_m}$. We have

$$N^{\sigma,\tau}(\lambda)=\sum_{r:C(\sigma)\sqcup C(\tau)\to[\![1,m]\!]}\left(\prod_{c\in C(\sigma)}p_{r(c)}\right)\left(\prod_{d\in C(\tau)}q_{r(d)}\right)$$

where $C(\sigma)$ and $C(\tau)$ are the sets of cycles of σ and τ, and the sum runs over functions such that if $c\in C(\sigma)$ and $d\in C(\tau)$ have a non-empty intersection, then $r(c)\leq r(d)$. We call **λ-coloring** of the cycles of σ and τ a map $s:C(\sigma)\sqcup C(\tau)\to\mathbb{N}^*$ which:

- associates to any cycle of σ the number of a row of λ, in $[\![1,\ell(\lambda)]\!]$;
- associates to any cycle of τ the number of a column of λ, in $[\![1,\lambda_1]\!]$;
- is such that if $c\cap d\neq\emptyset$ with $c\in C(\sigma)$ and $d\in C(\tau)$, then $(s(c),s(d))$ is a box of λ, that is to say that
$$1\leq s(d)\leq\lambda_{s(c)}.$$

Proposition 10.12. *The value of $N^{\sigma,\tau}(\lambda)$ is the number of λ-colorings of the cycles of σ and τ.*

Proof. To a coloring s of the cycles of σ and τ, we can associate a map $r : C(\sigma) \sqcup C(\tau) \to [\![1,m]\!]$ as follows: if $c \in C(\sigma)$ has color $s(c)$ (respectively, if $d \in C(\tau)$ has color $s(d)$) belonging to the rows of λ counted by p_i (respectively, to the columns of λ counted by q_j), then we set $r(c) = i$ (respectively, $r(d) = j$). This is better understood in an example. Suppose that $\lambda = (6,6,5,2,2)$ and (σ,τ) is the pair of permutations in $\mathfrak{S}(10)$ previously considered. Then, a possible λ-coloring is

$$\sigma = (1,2,3)(4,5,6)(7,8,9,10);$$
$$s(1,2,3) = 3,\ s(4,5,6) = 1,\ s(7,8,9,10) = 2;$$
$$\tau = (1)(2,10)(5,7)(6,8)(3,9,4);$$
$$s(1) = 4,\ s(2,10) = 2,\ s(5,7) = 5,\ s(6,8) = 6,\ s(3,9,4) = 3.$$

Indeed, we can represent this coloring by placing each integer $a \in [\![1,k]\!]$ on the row and the column given by the colors of the cycles of σ and τ that contain a; see Figure 10.3. In this situation, the cycles of σ are sent by r to

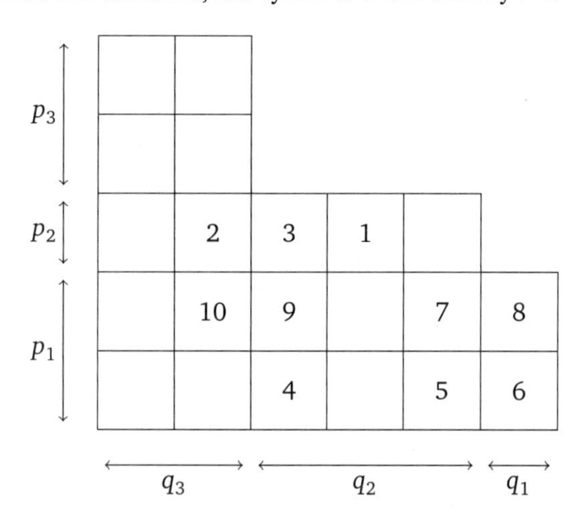

Figure 10.3
A (proper) λ-coloring of the cycles of σ and τ, represented as a function from $[\![1,k]\!]$ to the set of cells of λ.

$$r(1,2,3) = 2,\ r(4,5,6) = r(7,8,9,10) = 1,$$

and the cycles of τ are sent by r to

$$r(6,8) = 1,\ r(1) = r(5,7) = r(3,9,4) = 2,\ r(2,10) = 3.$$

if one takes for a system of Stanley coordinates the reduced system $\binom{2,1,2}{1,3,2}$.

Suppose that c and d have a non-empty intersection. Then, $(s(c), s(d))$ is a box of λ, hence falls in an intersection of p_i rows with q_j columns, where $i = r(c)$ and $j = r(d)$. However, the Stanley coordinates are such that if a box of the Young diagram falls in such an intersection, then $r(c) = i \leq j = r(d)$, so r indeed satisfies the order condition on the cycles of σ and τ. Moreover, given a function $r : C(\sigma) \sqcup C(\tau) \to [\![1, m]\!]$ that satisfies the order condition, the number of colorings c which give the function r is easily seen to be the monomial

$$\left(\prod_{c \in C(\sigma)} p_{r(c)}\right)\left(\prod_{d \in C(\tau)} q_{r(d)}\right).$$

Indeed, knowing r, one has $p_{r(c)}$ possibilities for the color $s(c)$ of any cycle $c \in C(\sigma)$, and $q_{r(d)}$ possibilities for the color $s(d)$ of any cycle $d \in C(\tau)$. Thus, the identity follows by gatherings the λ-colorings according to the functions r that are associated to them. □

We now start the proof of Theorem 10.11. The main ingredient will be a probabilistic formula for the renormalized character value $\chi^\lambda(\sigma)$ (Proposition 10.15), that in turn relies on the theory of Young idempotents of the symmetric group $\mathfrak{S}(k)$. Recall from Section 3.3 that if T is a numbering of the cells of a Young diagram of size k, then $\mathfrak{R}(T)$ and $\mathfrak{C}^\varepsilon(T)$ are respectively the sum of the elements of the subgroup of $\mathfrak{S}(k)$ stabilizing the rows of T, and the alternate sum of the elements of the subgroup of $\mathfrak{S}(k)$ stabilizing the columns of T. With $\lambda \in \mathfrak{Y}(k)$ fixed, we denote $R^\lambda = \mathfrak{R}(T)$ and $C^{\lambda,\varepsilon} = \mathfrak{C}^\varepsilon(T)$ the elements of $\mathbb{C}\mathfrak{S}(k)$ that correspond to the standard numbering T of λ, that is to say with $1, 2, \ldots, \lambda_1$ on the first row, $\lambda_1 + 1, \ldots, \lambda_1 + \lambda_2$ on the second row, etc. (see Figure 10.4). The choice of

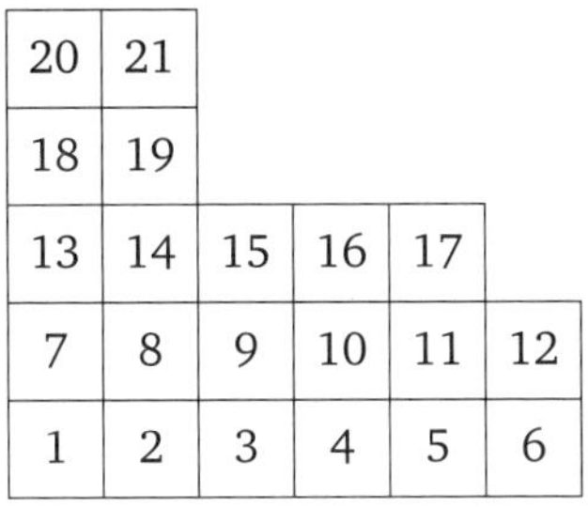

20	21				
18	19				
13	14	15	16	17	
7	8	9	10	11	12
1	2	3	4	5	6

Figure 10.4
The standard numbering T of a Young diagram λ.

the standard numbering is arbitrary, and it does not play an important role in the following; it will amount to the choice of a certain copy of S^λ among the simple left submodules of $\mathbb{C}\mathfrak{S}(k)$ with isomorphism class $\lambda \in \mathfrak{Y}(k)$.

Lemma 10.13. *If $\sigma \in \mathfrak{S}(k)$, consider the numbering $T' = \sigma \cdot T$, the permutations acting on the left of numberings of the cells of the Young diagram λ. The permutation*

σ belongs to the product $\mathfrak{R}(T)\mathfrak{C}(T)$ if and only if, for every pair of integers (i,j) in the same row of T, the integers i and j are not in the same column of T'. Moreover, in this case, if $\sigma = rc$, then this factorization is unique, and if

$$\mathrm{inv}(\sigma) = \text{number of pairs of integers } (i,j) \text{ in the same column of } T', \text{ with } i \text{ in a row under the row of } j \text{ in } T, \text{ and } i \text{ above } j \text{ in } T'$$

then $(-1)^{\mathrm{inv}(\sigma)} = \varepsilon(c)$.

Proof. Since $\mathfrak{R}(T)\cap\mathfrak{C}(T) = \{\mathrm{id}_{[\![1,k]\!]}\}$, a factorization $\sigma = rc$, if it exists, is unique. On the other hand, it is geometrically obvious that permutations in $\mathfrak{R}(T)$ and $\mathfrak{C}(T)$ cannot send two integers in the same row of T to the same column. Conversely, fix a permutation σ that has the property of sending the elements in the same row to different columns. Since the integers $1,2,\ldots,\lambda_1$ are in different columns of $T' = T_0 = \sigma \cdot T$, one can use a permutation c_1 in the column group $\mathfrak{C}(T_0)$ to send them to the first row. Now, the integers $\lambda_1+1,\ldots,\lambda_1+\lambda_2$ are again in different columns of $T_1 = c_1 \cdot T_0$, so one can use a permutation c_2 in the column group $\mathfrak{C}(T_1)$ to send them to the second row. One can also assume that c_2 does not modify the first row of the tableau. By induction on the length ℓ of the shape λ of T, one can find permutations $c_{i\in[\![1,\ell]\!]}$, such that if $T_i = c_i c_{i-1}\cdots c_2 c_1 \cdot T_0$, then $c_{i+1} \in \mathfrak{C}(T_i)$, and T_i has its i first rows that contain, respectively,

$$\{1,2,\ldots,\lambda_1\}, \{\lambda_1+1,\lambda_1+2,\ldots,\lambda_1+\lambda_2\},\ldots,\{\lambda_1+\cdots+\lambda_{i-1}+1,\ldots,\lambda_1+\cdots+\lambda_i\}.$$

We have $c_1 \in \mathfrak{C}(T_0) = \sigma\,\mathfrak{C}(T)\sigma^{-1}$, so $c_1 = \sigma d_1 \sigma^{-1}$ with $d_1 \in \mathfrak{C}(T)$. Similarly,

$$c_i = c_{i-1}\cdots c_1 \sigma d_i \sigma^{-1} c_1 \cdots c_{i-1}$$

with $d_i \in \mathfrak{C}(T)$. Then,

$$c_\ell c_{\ell-1}\cdots c_1 \sigma = \sigma d_1 d_2 \cdots d_\ell.$$

Now, T_ℓ only differs from T by a permutation of its rows, so there exists $r \in \mathfrak{R}(T_\ell)$ such that $r \cdot T_\ell = T$. This permutation r writes as $(\sigma d_1 d_2 \cdots d_\ell)\, e\, (\sigma d_1 d_2 \cdots d_\ell)^{-1}$ with $e \in \mathfrak{R}(T)$. We conclude that

$$\sigma d_1 d_2 \cdots d_\ell\, e = \mathrm{id}_{[\![1,n]\!]},$$

so $\sigma = e^{-1} d_\ell^{-1} \cdots d_1^{-1}$ is indeed in $\mathfrak{R}(T)\mathfrak{C}(T)$. Finally, the identity $(-1)^{\mathrm{inv}(\sigma)} = \varepsilon(c)$ follows from the following observations:

1. The number of inversions $\mathrm{inv}(c)$ with $c \in \mathfrak{C}(T)$ is its number of inversions $N(c)$ as defined in Section 2.1, so $(-1)^{\mathrm{inv}(c)} = \varepsilon(c)$ by definition of the signature.

2. Let us then make some row permutation $r \in \mathfrak{R}(T)$ act on $c \cdot T$, in order to get $\sigma \cdot T$ with $\sigma = rc$. This second action does not change the number of inversions: for instance, if

$$T = \begin{array}{|c|c|c|}\hline 6 \\ \cline{1-1} 4 & 5 \\ \cline{1-2} 1 & 2 & 3 \\ \hline\end{array} \quad\text{and}\quad c \cdot T = \begin{array}{|c|c|c|}\hline 4 \\ \cline{1-1} 1 & 2 \\ \cline{1-2} 6 & 5 & 3 \\ \hline\end{array},$$

then there are 3 inversions in $c \cdot T$, namely, $(1,6)$, $(4,6)$ and $(2,5)$. This is not changed when some row permutation acts, say $r = (1,3)(4,5)$: one still has 3 inversions in

$$rc \cdot T = \begin{array}{|c|c|c|} \hline 5 & & \\ \hline 3 & 2 & \\ \hline 6 & 4 & 1 \\ \hline \end{array}\ ,$$

namely, $(3,6)$, $(5,6)$ and $(2,4)$, that is to say the images by r of the previous inversions. This is a general fact, and thus,

$$(-1)^{\mathrm{inv}(\sigma)} = (-1)^{\mathrm{inv}(c)} = \varepsilon(c). \qquad \square$$

Lemma 10.14. *If $f^\lambda = R^\lambda C^{\lambda,\varepsilon}$, then $(f^\lambda)^2 = \frac{k!}{d_\lambda} f^\lambda$, and moreover, as a left $\mathbb{C}\mathfrak{S}(k)$-module, $\mathbb{C}\mathfrak{S}(k) f^\lambda$ is isomorphic to S^λ.*

Proof. We shall use the following characterization of f^λ: it is up to multiplication by a scalar the unique element $f \in \mathbb{C}\mathfrak{S}(k)$ such that $rfc = \varepsilon(c)f$ for every $r \in \mathfrak{R}(T)$, and for every $c \in \mathfrak{C}(T)$. It is clear that f^λ has this property. Conversely, if $f = \sum_{\sigma \in \mathfrak{S}(k)} \alpha_\sigma \sigma$ has this property, then $\alpha_{r\sigma c} = \varepsilon(c)\,\alpha_\sigma$ for every $r \in \mathfrak{R}(T)$, $c \in \mathfrak{C}(T)$ and $\sigma \in \mathfrak{S}(k)$. In particular, $\alpha_{rc} = \varepsilon(c)\,\alpha_{\mathrm{id}}$ for any r, c, so the sum f contains $\alpha_{\mathrm{id}} f^\lambda$. It suffices now to show that $\alpha_\sigma = 0$ if $\sigma \notin \mathfrak{R}(T)\mathfrak{C}(T)$. Set $T' = \sigma \cdot T$, with $\sigma \notin \mathfrak{R}(T)\mathfrak{C}(T)$. By the previous lemma, there exists a pair of integers (i,j) that appear in the same row of T, and in the same column of T'; we set $\tau = (i,j) \in \mathfrak{S}(k)$. It belongs to the row subgroup $\mathfrak{R}(T)$, and since $\mathfrak{C}(T') = \mathfrak{C}(\sigma \cdot T) = \sigma\,\mathfrak{C}(T)\,\sigma^{-1}$, $\sigma^{-1}\tau\sigma$ belongs to the column subgroup $\mathfrak{C}(T)$. Then, with $r = \tau$ and $c = \sigma^{-1}\tau\sigma$, we get $\alpha_\sigma = \alpha_{r\sigma c} = \varepsilon(\sigma^{-1}\tau\sigma)\,\alpha_\sigma = -\alpha_\sigma$ so $\alpha_\sigma = 0$. This ends the proof of the characterization of f^λ.

If $r \in \mathfrak{R}(T)$ and $c \in \mathfrak{C}(T)$, we now compute

$$r\,(f^\lambda)^2\,c = (r\,f^\lambda)(f^\lambda\,c) = f^\lambda\,(\varepsilon(c)\,f^\lambda) = \varepsilon(c)\,(f^\lambda)^2,$$

so $(f^\lambda)^2$ must be a scalar multiple $n^\lambda f^\lambda$ of f^λ. Actually, we have more generally and for the same reasons:

$$\forall x \in \mathbb{C}\mathfrak{S}(k),\ f^\lambda\,x\,f^\lambda = c(x)\,f^\lambda, \quad \text{with } c(x) \in \mathbb{C}.$$

In particular, if $U^\lambda = \mathbb{C}\mathfrak{S}(k)\,f^\lambda$, which is a non-zero left-module for $\mathbb{C}\mathfrak{S}(k)$, then $f^\lambda U^\lambda \subset \mathbb{C} f^\lambda$. Suppose that $Q \subset U^\lambda$ is a submodule. Then, $f^\lambda Q = 0$ or $f^\lambda Q = \mathbb{C} f^\lambda$.

1. In the first case, $Q^2 \subset U^\lambda Q = \mathbb{C}\mathfrak{S}(k)\,f^\lambda Q = 0$, and this implies that $Q = 0$ (this statement is true for any left-module $Q \subset A$ of a semisimple complex algebra A, as is readily seen on the matrix representation of the elements of A by using the non-commutative Fourier transform).

2. Otherwise, $f^\lambda Q = \mathbb{C} f^\lambda$, and $Q \supset U^\lambda Q = \mathbb{C}\mathfrak{S}(k) f^\lambda Q = \mathbb{C}\mathfrak{S}(k) f^\lambda = U^\lambda$, so $Q = U^\lambda$.

It is therefore shown that U^λ is a simple left $\mathbb{C}\mathfrak{S}(k)$-module, and also that $f^\lambda U^\lambda = \mathbb{C} f^\lambda$, which amounts to the fact that $(f^\lambda)^2 \neq 0$ and $n^\lambda \neq 0$. We now relate U^λ to the Specht module S^λ, which was defined in Section 3.3 as the irreducible submodule of the permutation module M^λ spanned by the polytabloid $e_T = C^{\lambda,\varepsilon} \cdot [T]$. Let

$$\begin{aligned} \phi : \mathbb{C}\mathfrak{S}(k) &\to M^\lambda \\ \sigma &\mapsto \sigma \cdot [T]; \end{aligned}$$

it is a morphism of left $\mathbb{C}\mathfrak{S}(k)$-modules, and it sends $U^\lambda = \mathbb{C}\mathfrak{S}(k) f^\lambda$ to

$$\mathbb{C}\mathfrak{S}(k) R^\lambda C^{\lambda,\varepsilon} \cdot [T] = \mathbb{C}\mathfrak{S}(k) R^\lambda e_T \subset S^\lambda.$$

To prove that ϕ restricts to an isomorphism of modules between U^λ and S^λ, since both modules are simple, it suffices to prove that $\phi(U^\lambda) \neq 0$, or equivalently that $\phi(f^\lambda) \neq 0$. Let us compute the coefficient of $[T]$ in $\phi(f^\lambda)$. If $rc \cdot [T] = [T]$, then $rc \in \mathfrak{R}(T)$, hence $c = \mathrm{id}_{[\![1,n]\!]}$ by unicity of the factorization rc. So,

$$[T](\phi(f^\lambda)) = \mathrm{card}\,\mathfrak{R}(T) \neq 0,$$

and ϕ realizes an isomorphism of modules between U^λ and S^λ.

Finally, consider the right multiplication by f^λ on $\mathbb{C}\mathfrak{S}(k)$. It is the scalar multiplication by n^λ on U^λ, and 0 on the other left submodules of $\mathbb{C}\mathfrak{S}(k)$, because $(f^\lambda)^2 = n^\lambda f^\lambda$. Therefore, its trace is $n^\lambda d_\lambda$. However, for any $\sigma \in \mathfrak{S}(k)$, the coefficient of σ in σf^λ is 1. Indeed, f^λ is a multiplicity free (signed) sum of permutations, and in particular, $\mathrm{id}_{[\![1,k]\!]}$ appears with multiplicity 1 in f^λ. Hence, the trace of the right multiplication by f^λ is $k!$, hence the identity $n^\lambda = \frac{k!}{d_\lambda}$. □

Proposition 10.15. *Let μ be a cycle type in $\mathfrak{Y}(k)$, and σ_μ be a random element of the conjugacy class C_μ in $\mathfrak{S}(k)$, taken under the uniform probability measure on this class. We have*

$$\chi^\lambda(\mu) = \mathbb{E}[(-1)^{\mathrm{inv}(\sigma_\mu)}],$$

where we convene that $(-1)^{\mathrm{inv}(\sigma)} = 0$ if σ is not in $\mathfrak{R}(T)\mathfrak{C}(T)$.

Proof. If $\rho \in \mathfrak{S}(k)$, then the renormalized character $\chi^\lambda(\rho)$ is given by $\frac{1}{d_\lambda}$ times the trace of the operator L_ρ of multiplication on the left by ρ on the module $U^\lambda = \mathbb{C}\mathfrak{S}(k) f^\lambda$. In general, the trace of an operator $F : \mathbb{C}\mathfrak{S}(k) \to \mathbb{C}\mathfrak{S}(k)$ is given by

$$\mathrm{tr}(T) = \sum_{\theta \in \mathfrak{S}(k)} \langle \theta \mid F(\theta) \rangle,$$

where $\langle \theta \mid \rho \rangle = \delta_{\theta\rho}$ for any permutations θ and ρ. If F has values in U^λ, then this trace is also $\frac{1}{n^\lambda} \sum_{\theta \in \mathfrak{S}(k)} \left\langle \theta \,\middle|\, F(\theta) f^\lambda \right\rangle$, so,

$$\chi^\lambda(\mu) = \frac{1}{n^\lambda d_\lambda} \sum_{\theta \in \mathfrak{S}(k)} \left\langle \theta \,\middle|\, \rho\theta f^\lambda \right\rangle = \frac{1}{k!} \sum_{\theta \in \mathfrak{S}(k)} \left\langle \theta^{-1}\rho^{-1}\theta \,\middle|\, f^\lambda \right\rangle$$

for any permutation ρ of cycle type μ. This mean over $\mathfrak{S}(k)$ can be rewritten as a mean over the conjugacy class C_μ:

$$\chi^\lambda(\mu) = \frac{1}{\operatorname{card} C_\mu} \sum_{\rho_\mu \in C_\mu} \left\langle \rho_\mu \,\middle|\, f^\lambda \right\rangle.$$

By Lemma 10.13, each permutation $\rho = rc$ of $\mathfrak{R}(T)\mathfrak{C}(T)$ appears in f^λ with coefficient $\varepsilon(c) = (-1)^{\mathrm{inv}(\sigma)}$. Therefore,

$$\chi^\lambda(\mu) = \frac{1}{\operatorname{card} C_\mu} \sum_{\rho_\mu \in C_\mu} (-1)^{\mathrm{inv}(\sigma_\mu)}. \qquad \square$$

In the proof of Proposition 10.12, we reinterpreted the quantity $N^{\sigma,\tau}(\lambda)$ as a number of colorings of the cycles of σ and τ, and a λ-coloring was represented by placing the integers $1, 2, \ldots, k$ in the cells of the Young diagram λ, possibly with several integers in the same cell. We call **proper** a λ-coloring of the cycles of σ and τ such that each integer $i \in [\![1, k]\!]$ falls in a distinct cell of λ; thus, the example studied during the proof of Proposition 10.12 is proper. We denote $\widetilde{N}^{\sigma,\tau}(\lambda)$ the number of proper λ-colorings of the cycles of σ and τ; it is always smaller than $N^{\sigma,\tau}(\lambda)$.

Proposition 10.16. *We have* $\Sigma_\mu(\lambda) = \sum_{\rho_\mu = \sigma\tau} \varepsilon(\tau) \widetilde{N}^{\sigma,\tau}(\lambda)$, *where the sum runs over factorizations in* $\mathfrak{S}(k)$, $k = |\mu|$.

Proof. Suppose first that λ and μ have the same size k. We proved that

$$\Sigma_\mu(\lambda) = \sum_{\theta \in \mathfrak{S}(k)} \left\langle \theta^{-1}\rho_\mu\theta \,\middle|\, f^\lambda \right\rangle = \sum_{\theta \in \mathfrak{S}(k)} \sum_{r \in \mathfrak{R}(T)} \sum_{c \in \mathfrak{C}(T)} \varepsilon(c)\, 1_{\rho_\mu = \theta r c \theta^{-1}}$$

for any permutation ρ_μ of cycle type μ. Set $\sigma = \theta r \theta^{-1}$ and $\tau = \theta c \theta^{-1}$; they belong respectively to $\mathfrak{R}(\theta \cdot T)$ and $\mathfrak{C}(\theta \cdot T)$, hence, their cycles can be placed respectively on rows of λ and on columns of λ, according to the numbering $\theta \cdot T$. Thus, the numbering $\theta \cdot T$ is a proper λ-coloring of the cycles of σ and τ. Since we are taking the sum over all possible numberings $\theta \cdot T$ of λ, we conclude that:

$$\Sigma_\mu(\lambda) = \sum_{\rho_\mu = \sigma\tau} \varepsilon(\tau) \widetilde{N}^{\sigma,\tau}(\lambda).$$

Suppose now $|\mu| = k < n = |\lambda|$. We can take for representative of the conjugacy class $C_{\mu\uparrow n}$ a permutation ρ_μ with support $[\![1,k]\!]$, and then, we have:

$$\Sigma_\mu(\lambda) = \frac{1}{(n-k)!} \sum_{\theta\in\mathfrak{S}(n)} \sum_{r\in\mathfrak{R}(T)} \sum_{c\in\mathfrak{C}(T)} \varepsilon(c)\, 1_{\rho_\mu=\theta r c \theta^{-1}}$$

by the same argument as before. We claim that if r and c do contribute to this sum, then they do not move points that do not belong to the set $\theta^{-1}([\![1,k]\!])$. Indeed, if $\rho_\mu = \theta r c \theta^{-1}$ and $i \notin \theta^{-1}([\![1,k]\!])$, then

$$rc(i) = \theta^{-1}\rho_\mu\theta(i) = \theta^{-1}\theta(i) = i.$$

This is only possible if $r(i) = i$ and $c(i) = i$: if i is moved along a column with c, then one cannot bring it back to its position with a row permutation r. So, $\sigma = \theta r \theta^{-1}$ and $\tau = \theta c \theta^{-1}$ are in $\mathfrak{S}(k)$, and we obtain:

$$\Sigma_\mu(\lambda) = \frac{1}{(n-k)!} \sum_{\rho_\mu=\sigma\tau,\ \sigma\in\mathfrak{S}(k),\ \tau\in\mathfrak{S}(k)} \varepsilon(\tau)\ \mathrm{card}\left\{ \begin{array}{c} \theta\in\mathfrak{S}(n)\,|\,\theta\cdot T \text{ is a numbering that extends} \\ \text{a proper } \lambda\text{-coloring of the cycles of } \sigma \text{ and } \tau \end{array} \right\}.$$

For each proper λ-coloring of the cycles of σ and τ, there are $n-k$ cells of λ that remain to be numbered to obtain a numbering, hence, $(n-k)!$ possible extensions. We conclude that the formula is again true if $k < n$. Finally, if $k > n$, then $\Sigma_\mu(\lambda) = 0$, and on the other hand, there is no proper coloring of the cycles of permutations of size k in a Young diagram of strictly smaller size, so this case is also trivially true. □

Proof of Theorem 10.11. Consider the signed sum of possibly non-proper λ-colorings $S = \sum_{\rho_\mu=\sigma\tau} \varepsilon(\tau) N^{\sigma,\tau}(\lambda)$. If $s : C(\sigma)\sqcup C(\tau) \to \mathbb{N}^*$ is a λ-coloring, we consider it as before as a function from $[\![1,k]\!]$ to the cells of the Young diagram λ. Suppose that the function s appears in the sum and is not proper, i.e., there exist indices $i < j$ in $[\![1,k]\!]$ that are sent to the same cell. Then, the set of factorizations $\rho_\mu = \sigma\tau$ of which s is a λ-coloring admits for involution

$$(\sigma,\tau) \mapsto (\sigma\circ(i,j), (i,j)\circ\tau).$$

Since this involution changes $\varepsilon(\tau)$ in $-\varepsilon(\tau)$, the contribution of s to the sum vanishes. As a consequence,

$$\sum_{\rho_\mu=\sigma\tau} \varepsilon(\tau) N^{\sigma,\tau}(\lambda) = \sum_{\rho_\mu=\sigma\tau} \varepsilon(\tau) \widetilde{N}^{\sigma,\tau}(\lambda),$$

that is to say that the signed sum can be restricted to proper λ-colorings s. □

Example. Consider the cycle type (2). The possible factorizations of $\rho_{(2)} = (1,2)$ are $\sigma = (1,2), \tau = \mathrm{id}$ and $\sigma = \mathrm{id}, \tau = (1,2)$. For the first factorization, when computing $N^{\sigma,\tau}(\lambda)$, we have to place the cycle $(1,2)$ on a row of λ, and there are $\sum_{i=1}^{\ell(\lambda)} (\lambda_i)^2$ possibilities. For the second factorization, counted with sign -1, we have to place the cycle $(1,2)$ on a column of λ, and there are $\sum_{i=1}^{\lambda_1}(\lambda_i')^2$ possibilities. We conclude that

$$\Sigma_2(\lambda) = \sum_{i\geq 1}(\lambda_i)^2 - \sum_{i\geq 1}(\lambda_i')^2.$$

In terms of Stanley coordinates, the first sum is $\sum_{i\geq 1} p_i(q_i + \cdots + q_m)^2$, and the second sum is $\sum_{i\geq 1} q_i(p_1 + \cdots + p_i)^2$.

Example. Consider more generally a rectangular partition $\lambda = (q,q,\ldots,q) = (q^p)$. Then, $N^{\sigma,\tau}(\lambda) = p^{n(\sigma)}\, q^{n(\tau)}$, as there is a unique function $r : C(\sigma) \sqcup C(\tau) \to \{1\}$. Therefore, we obtain the remarkably simple formula:

$$\Sigma_\mu((q^p)) = \sum_{\rho_\mu = \sigma\tau} p^{n(\sigma)}\,(-q)^{n(\tau)}$$

which was first established by R. Stanley.

To close this section, let us state and prove the analogue of Theorem 10.11 for the observables R_k:

Corollary 10.17. *For any $k \geq 1$,*

$$R_{k+1} = \sum_{\substack{(1,2,\ldots,k)=\sigma\mu \text{ in } \mathfrak{S}(k) \\ n(\sigma)+n(\mu)=k+1}} \varepsilon(\tau) N^{\sigma,\tau}.$$

Proof. We know from Theorem 9.20 that R_{k+1} is the top weight component of Σ_k. However, the weight on observables is the restriction of the weight on Stanley polynomials or stable polynomials coming from the standard gradation of QSym (or more precisely, QSym$/(M_1)$). As $N^{\sigma,\tau}$ has degree $n(\sigma) + n(\tau)$, the result follows immediately. □

In the previous formula for R_{k+1}, there is a simple description of the factorizations $\sigma\tau = (1,2,\ldots,k)$ such that $n(\sigma) + n(\tau) = k + 1$; they are in bijection with the non-crossing partitions of size k. This bijection is better explained as an isomorphism of posets. To a non-crossing partition $\pi \in \mathfrak{NC}(k)$, we associate the permutation σ whose cycles are the parts of π, with a part $\pi_i = \{a_1 < a_2 < \cdots < a_r\}$ sent to the cycle $(a_1, a_2, \ldots, a_r)$. We also denote τ the permutation associated to the Kreweras complement $\overline{\pi}$ of π.

Lemma 10.18. *In the previous setting, $\sigma\tau = (1,2,\ldots,k)$, and $n(\sigma)+n(\tau) = k+1$.*

Proof. Let i be an integer in $[\![1,k]\!]$. If $i + 1$ is in the same part as i for π, then i is an isolated point of $\overline{\pi}$, and then $\tau(i) = i$, $\sigma(i) = i + 1$, and $\sigma\tau(i) = i + 1$. Otherwise, $\tau(i) = j$ is the predecessor of $i + 1$ in the cycle of σ containing $i + 1$; see Figure 10.5. Therefore, we have again $\sigma\tau(i) = i + 1$ in this case (with by convention $k + 1 = 1$). Thus, $\sigma\tau$ is indeed the long cycle $(1,2,\ldots,k)$.

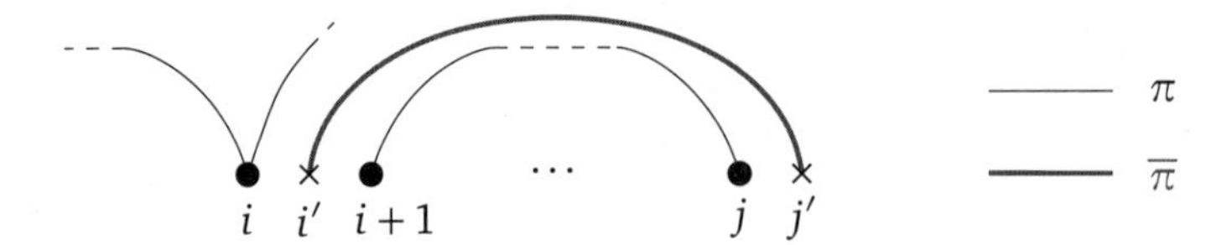

Figure 10.5
If i and $i+1$ are in different parts of π, then $\tau(i) = \sigma^{-1}(i+1)$.

It remains to see that for any non-crossing partition π of size k, $\ell(\pi)+\ell(\overline{\pi}) = k+1$. Suppose the result to be true in size $k-1$, and denote $\pi_{|k-1}$ the restriction of a set partition π of size k to the integers $1, 2, \dots, k-1$ (that is to say that integers i, j smaller than $k-1$ are in the same part of $\pi_{|k-1}$ if and only if they are in the same part of π). If k is an isolated point of π, then π has one more part than $\pi_{|k-1}$, whereas $\overline{\pi}$ is obtained from $\overline{\pi_{|k-1}}$ by putting k in the same part as $k-1$, so

$$\ell(\pi)+\ell(\overline{\pi}) = \big(\ell(\pi_{|k-1})+1\big)+\ell(\overline{\pi_{k-1}}) = k+1.$$

If k is not an isolated point of π, then $\ell(\pi) = \ell(\pi_{|k-1})$, but in this case $\overline{\pi}$ is obtained from $\overline{\pi_{|k-1}}$ by cutting in two the part of $\overline{\pi_{|k-1}}$ that contains $k-1$, one of them being then connected to k. Thus, we have also in this case

$$\ell(\pi)+\ell(\overline{\pi}) = \ell(\pi_{|k-1})+\big(\ell(\overline{\pi_{|k-1}})+1\big) = k+1. \qquad \square$$

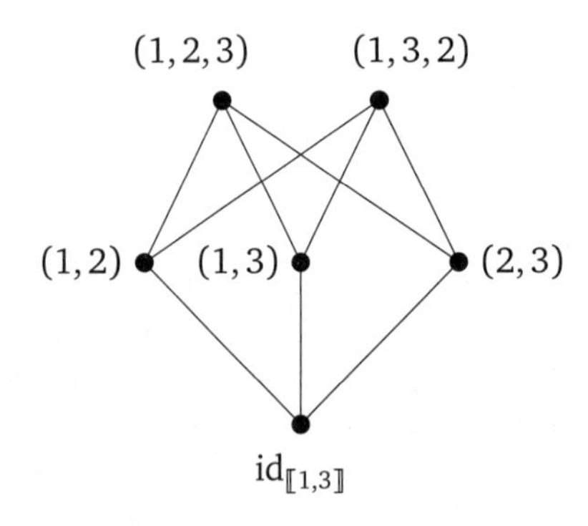

Figure 10.6
The Cayley graph of $\mathfrak{S}(3)$.

We now consider the **Cayley graph** of the symmetric group $\mathfrak{S}(k)$ with respect to the set of generators $S = \{(i,j),\ 1 \leq i < j \leq k\}$. This is the graph with vertices $\sigma \in \mathfrak{S}(k)$, and with an edge between σ and τ if $\sigma\tau^{-1} \in S$. Since S is conjugacy invariant, the edges of the graph are also defined by the equation $\tau^{-1}\sigma \in S$. The

Cayley graph is regular, with $\binom{k}{2}$ edges stemming from each vertex. The Cayley graph of $\mathfrak{S}(3)$ is drawn in Figure 10.6. A **geodesic** of this graph is a path $\sigma = \sigma_0, \sigma_1, \ldots, \sigma_k = \tau$ which is of minimal length k between the permutations σ and τ. An **interval** $[\sigma, \tau]$ of the Cayley graph is the set of permutations ρ that appear in a geodesic from σ to τ. For instance, the interval $[\mathrm{id}_{[\![1,3]\!]}, (1,2,3)]$ is drawn in Figure 10.7. In the following, we endow the interval $[\mathrm{id}_{[\![1,k]\!]}, (1,2,\ldots,k)]$ with the partial order $\sigma \leq \sigma' \iff [\mathrm{id}, \sigma] \subset [\mathrm{id}, \sigma']$.

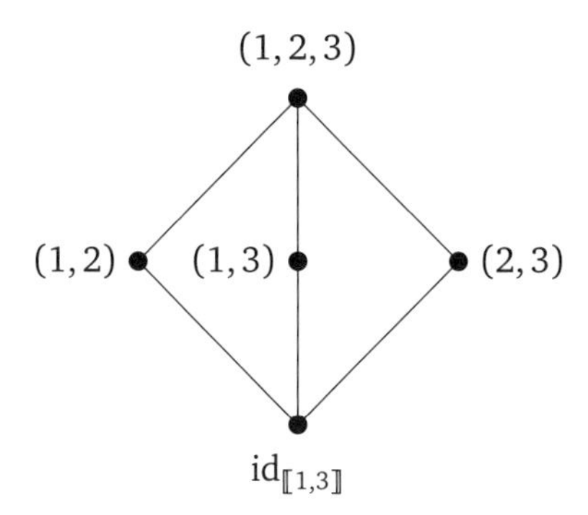

Figure 10.7
The interval $[\mathrm{id}_{[\![1,3]\!]}, (1,2,3)]$ in the Cayley graph of $\mathfrak{S}(3)$.

Theorem 10.19 (Biane). *Consider the map $\Psi : \pi \in \mathfrak{NC}(k) \mapsto \sigma \in \mathfrak{S}(k)$ that associates to a non-crossing partition π the permutation σ whose cycles are the parts of π. It is an isomorphism of partially ordered sets between $\mathfrak{NC}(k)$ endowed with the order of refinement, and the interval $[\mathrm{id}_{[\![1,k]\!]}, (1,2,\ldots,k)]$ endowed with the order coming from the Cayley graph of $\mathfrak{S}(k)$. Moreover, the image of Ψ is the set of permutations σ that appear in a **minimal factorization** $\sigma\tau$ of the long cycle $(1,2,\ldots,k)$, that is to say a factorization with $n(\sigma) + n(\tau) = k+1$.*

Proof. Notice that in the Cayley graph of $\mathfrak{S}(k)$, the distance from $\mathrm{id}_{[\![1,k]\!]}$ to any permutation σ is $d(\mathrm{id}_{[\![1,k]\!]}, \sigma) = k - n(\sigma)$. In particular, $d(\mathrm{id}_{[\![1,k]\!]}, (1,2,\ldots,k)) = k-1$, and a permutation σ lies on a geodesic from $\mathrm{id}_{[\![1,k]\!]}$ to $(1,2,\ldots,k)$ if and only if

$$k-1 = d(\mathrm{id}_{[\![1,k]\!]}, \sigma) + d(\sigma, (1,2,\ldots,k)) = d(\mathrm{id}_{[\![1,k]\!]}, \sigma) + d(\mathrm{id}, \sigma^{-1}(1,2,\ldots,k))$$
$$k+1 = n(\sigma) + n(\sigma^{-1}(1,2,\ldots,k)).$$

By the previous lemma, if $\sigma = \Psi(\pi)$, then $\tau = \sigma^{-1}(1,2,\ldots,k)$ satisfies $n(\sigma) + n(\tau) = k+1$, hence, the image of Ψ is contained in the interval $[\mathrm{id}_{[\![1,k]\!]}, (1,2,\ldots,k)]$. Conversely, take a permutation σ in this interval of the Cayley graph. We prove by induction on the number of cycles $n(\sigma)$ that it is in the image of Ψ. If $n(\sigma) = 1$, then $\sigma = (1,2,\ldots,k)$ is the image of the non-crossing partition with one part $\{1,2,\ldots,k\}$. Suppose the result to be true for permutations with $r-1$ cycles, and take σ on a geodesic from $\mathrm{id}_{[\![1,k]\!]}$ to $(1,2,\ldots,k)$, and with $n(\sigma)=r\geq 2$. Since

$$n(\sigma)=k-d(\sigma,\mathrm{id}_{[\![1,k]\!]})=d(\sigma,(1,2,\ldots,k))+1,$$

there exist permutations $\tau_1,\ldots,\tau_{r-1}$ such that $\sigma=(1,2,\ldots,k)\tau_1\cdots\tau_{r-1}$. Set $\sigma'=\sigma\tau_{r-1}$. It lies at distance $r-2$ from the long cycle $(1,2\ldots,k)$, and since it is on a geodesic from $\mathrm{id}_{[\![1,k]\!]}$ to $(1,2,\ldots,k)$,

$$n(\sigma')=k-d(\sigma',\mathrm{id}_{[\![1,k]\!]})=d(\sigma',(1,2,\ldots,k))+1=r-1=n(\sigma)-1.$$

This is only possible if $\tau_{r-1}=(i,j)$ connects two distinct cycles of σ:

$$\sigma=c\,(a_1,a_2,\ldots,a_s=i)(b_1,b_2,\ldots,b_t=j);$$
$$\sigma'=c\,(a_1,a_2,\ldots,a_s,b_1,b_2,\ldots,b_t),$$

where c has disjoint support from $\{a_1,\ldots,a_s,b_1,\ldots,b_t\}$. Let π' be the non-crossing set partition such that $\Psi(\pi')=\sigma'$ (it exists by the induction hypothesis). The integers i and j are in the same part

$$\{a_1<a_2<\cdots<a_s=i<b_1<b_2<\cdots<b_t=j\}$$

of π', the order $<$ being understood cyclically in $\mathfrak{S}(k)$. Then, the non-crossing set partition π corresponding to σ is obtained from π' by cutting this part of π in two parts $\{a_1<a_2<\cdots<a_s=i\}$ and $\{b_1<b_2<\cdots<b_t=j\}$, so again σ is in the image of Ψ. Thus, Ψ is a bijection between the set of non-crossing partitions $\mathfrak{NC}(k)$, and the interval of the Cayley graph $[\mathrm{id}_{[\![1,k]\!]},(1,2,\ldots,k)]$. Moreover, the previous argument shows that if π' *covers* π in $\mathfrak{NC}(k)$ (i.e., $\pi'>\pi$ and there is no non-crossing set partition ρ such that $\pi'>\rho>\pi$), then $\Psi(\pi')$ covers $\Psi(\pi)$ in the interval of the Cayley graph, that is to say that there is a transposition τ such that $\Psi(\pi')=\Psi(\pi)\tau$, with $d(\Psi(\pi'),\mathrm{id}_{[\![1,k]\!]})=d(\Psi(\pi),\mathrm{id}_{[\![1,k]\!]})+1$. As an immediate consequence, Ψ is compatible with the partial orders on both sets $\mathfrak{NC}(k)$ and $[\mathrm{id}_{[\![1,k]\!]},(1,2,\ldots,k)]$. □

So, there is a natural labeling of the terms of the formula of Corollary 10.17 by non-crossing partitions of size k. Beware that this expansion should not be confused with the expansion of R_{k+1} as the sum over non-crossing partitions of size $k+1$ that relates moments and free cumulants.

Remark. The terminology of *minimal* factorization comes from the geometry of the map associated to a factorization $(1,2,\ldots,k)=\sigma\tau$ with $n(\sigma)+n(\tau)=k+1$. In Section 10.3, we shall explain how to associate to such a factorization a compact orientable surface X, which is then a torus $\mathbb{T}_g$ with genus given by the formula

$$2-2g=n(\sigma)+n(\tau)-k+1.$$

The case of minimal genus $g=0$ corresponds then to the minimal factorizations of the long cycle.

Remark. As a consequence of the previous theorem, one sees that *any* interval of the Cayley graph of the symmetric group is isomorphic to a product of Hasse diagrams of lattices of non-crossing partitions. Indeed, if $\sigma \in \mathfrak{S}(n)$ has cycle type $\mu = (\mu_1, \ldots, \mu_r)$, then it is easily seen from the previous argument that $[\mathrm{id}_{[\![1,n]\!]}, \sigma]$ is isomorphic to the product of graphs $\mathfrak{NC}(\mu_1) \times \mathfrak{NC}(\mu_2) \times \cdots \times \mathfrak{NC}(\mu_r)$, and then, given two permutations σ and τ, the interval $[\tau, \sigma]$ is isomorphic to $[\mathrm{id}_{[\![1,n]\!]}, \sigma\tau^{-1}]$.

10.3 Combinatorics of the Kerov polynomials

Recall from the introduction of this chapter that the k-th Kerov polynomial is the polynomial in free cumulants $R_2, R_3, \ldots, R_{k+1}$ such that, in the algebra of observables $\mathscr{O}$, one has

$$\Sigma_k = K_k(R_2, R_3, \ldots, R_{k+1}).$$

The goal of this last section of the chapter is to obtain a combinatorial interpretation of the coefficient of a monomial $(R_2)^{s_2}(R_3)^{s_3}\cdots$ in the polynomial K_k. Fix a sequence $s = (s_2, s_3, \ldots)$ such that $\mathrm{wt}(s) = 2s_2 + 3s_3 + \cdots \leq k+1$, and write $[R^s](K_k)$ for the coefficient of $(R_2)^{s_2}(R_3)^{s_3}\cdots$ in the polynomial K_k. In the Stanley–Féray formula

$$\Sigma_k = \sum_{\sigma\tau = (1,2,\ldots,k)} \varepsilon(\tau) N^{\sigma,\tau},$$

the term of weight $\mathrm{wt}(s)$ is

$$\sum_{\substack{\sigma\tau = (1,2,\ldots,k) \\ n(\sigma)+n(\tau)=\mathrm{wt}(s)}} \varepsilon(\tau) N^{\sigma,\tau}.$$

This observation leads one to try to write the coefficient $[R^s](K_k)$ as a number of certain factorizations $\sigma\tau = (1, 2, \ldots, k)$ of the long cycle with $n(\sigma) + n(\tau) = \mathrm{wt}(s)$. Thus, we shall prove in this section the following:

Theorem 10.20. *The coefficient $[R^s](K_k)$ is the number of triples (σ, τ, q) where:*

(K1) σ and τ belong to $\mathfrak{S}(k)$, $\sigma\tau = (1, 2, \ldots, k)$ is the long cycle, and

$$\ell(s) = s_2 + s_3 + \cdots = n(\tau);$$
$$\mathrm{wt}(s) = 2s_2 + 3s_3 + \cdots = n(\sigma) + n(\tau);$$

(K2) $q : C(\tau) \to \{2, 3, \ldots\}$ is a coloring of the cycles of τ, with $\mathrm{card}\, q^{-1}(k) = s_k$ for each integer k;

(K3) if $C \subset C(\tau)$ is non-trivial (i.e., $C \neq \emptyset$ and $C \neq C(\tau)$), then there are strictly more than $\sum_{c \in C}(q(c) - 1)$ cycles of σ that intersect $\bigcup_{c \in C} c$.

In particular, $[R^s](K_k) \geq 0$.

In order to prove Theorem 10.20, we shall:

1. develop a differential calculus on $\mathscr{Q}$, and relate the derivatives with respect to free cumulants of an observable $f \in \mathscr{O}$ to its coefficients as a Stanley polynomial (see in particular Theorem 10.24).

2. apply some transformations to the Stanley–Féray formula, by using several vanishing results which are all related to the computation of the Euler characteristic of the simplicial complex given by a family of subsets (Theorem 10.27).

▷ *Applications and computation of the first Kerov polynomials.*

Before we prove Theorem 10.20, let us detail several consequences of it. We can use the theorem to calculate the linear terms $[R_j](K_k)$. Thus:

Proposition 10.21. *If $j \leq k+1$, then the linear term R_j has for coefficient in K_k the number of factorizations $\sigma\tau = (1,2,\ldots,k)$ such that $n(\tau) = 1$ and $n(\sigma) = j-1$.*

Proof. Given a factorization $\sigma\tau = (1,2,\ldots,k)$, Condition (K1) means that $n(\tau) = 1$ and $n(\sigma) = j-1$. Then, there is a unique coloring $C(\tau) \to \{j\}$, and Condition (K3) has no influence: there is no non-trivial subset $C \subset C(\tau)$, as $\operatorname{card} C(\tau) = n(\tau) = 1$. □

In a moment, we shall use this result in order to prove the expansions of the five first Kerov polynomials that were claimed in the previous chapter. To simplify these calculations, we shall also use the following parity property:

Proposition 10.22. *For any k, the homogeneous component of weight j of Σ_k is zero if j and k have the same parity.*

Proof. By the Féray–Stanley formula, it suffices to see that there is no factorization of the long cycle $\sigma\tau = (1,2,\ldots,k)$ with $n(\sigma)+n(\tau) = j$ if j and k have the same parity. This is obvious as

$$(-1)^{2k-j} = \varepsilon(\sigma\tau) = \varepsilon(1,2,\ldots,k) = (-1)^{k-1}. \qquad \square$$

Remark. It seems very difficult to give a proof of this simple statement without using the Stanley–Féray formula. In particular, we need to look at a larger algebra $\mathscr{Q}$ in order to prove this result which regards observables of $\mathscr{O}$.

Remark. The parity result can also be explained by looking at the map associated to a factorization $\sigma\tau = (1,2,\ldots,k)$ with $n(\sigma) + n(\tau) = j$, which is a bicolored map with j vertices, k edges and 1 face. Consider more generally a bicolored labeled map $G = (V,E)$, with $|V|$ vertices, $|E|$ edges and $|F|$ faces. It can always be traced on a compact orientable surface, which is a torus $\mathbb{T}_g$ with g holes. Indeed, for each face $f = (v_1, v_2, \ldots, v_{2p})$ of the map, consider a polygon whose edges are labeled $(v_1, v_2), (v_2, v_3), \ldots, (v_{2p}, v_1)$. Notice that since the graph is bipartite, every face f has an even number of edges. We glue together the collection of polygons according to these edges: we then obtain a surface which is compact, orientable, and with the map drawn on it. For instance, the map considered in Figures 10.1 and 10.2 is obtained by gluing the two polygons

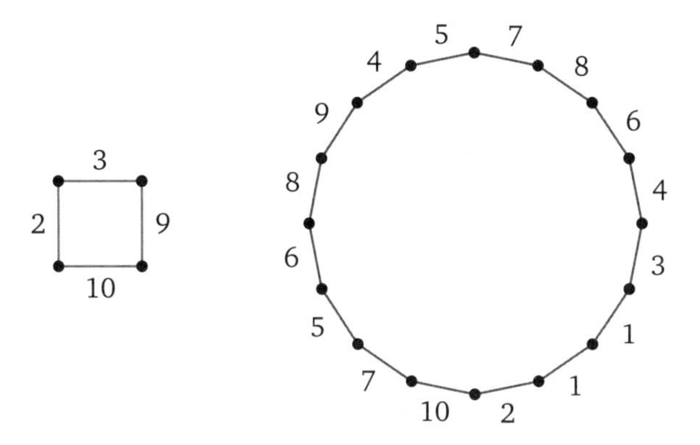

thereby obtaining a torus of genus 1; see Figure 10.8.

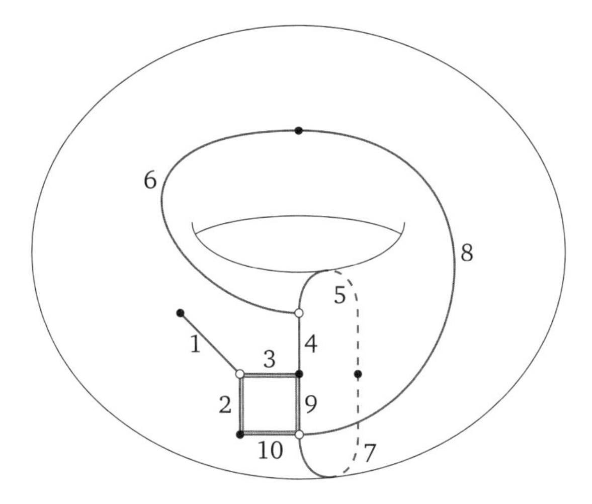

Figure 10.8
The map of Figure 10.1 drawn on a torus.

It is then a well-known fact from algebraic topology that the Euler characteristic of the torus is given by the formula

$$\chi(\mathbb{T}_g) = 2 - 2g = |V| - |E| + |F|.$$

Therefore, in the case of the map associated to a factorization of the long cycle, $j-k+1$ is even, and j and k do not have the same parity. The same argument shows that if μ is an integer partition, then the homogeneous component of weight j in Σ_μ is zero if the quantity $j-(|\mu|-\ell(\mu))$ is not even.

Example. There are $(k-1)!$ long cycles in $\mathfrak{S}(k)$, that is to say permutations τ with $n(\tau)=1$. Therefore, the number of linear terms in Σ_k is equal to $(k-1)!$. In particular,

$$\begin{aligned}\Sigma_1 &= R_2 + \cdots \\ \Sigma_2 &= R_3 + \cdots \\ \Sigma_3 &= R_4 + R_2 + \cdots \\ \Sigma_4 &= R_5 + 5R_3 + \cdots\end{aligned}$$

where the dots $\cdots$ indicate terms that are not linear. Indeed, looking for instance at Σ_4, we already know that $[R_5](\Sigma_4)=1$, and the 5 long cycles τ different from c_4 yield a product $\sigma = c_4\,\tau^{-1}$ which is a 3-cycle or a product of two transpositions, hence with $n(\sigma)=2=3-1$.

Actually, there are no other terms for Σ_1, Σ_2, Σ_3 and Σ_4. Indeed, using the parity result and the bound $\mathrm{wt}(\Sigma_k)=k+1$, we see that the only missing terms might be $a\,(R_2)^2$ for Σ_3 and $b\,R_2R_3$ for Σ_4. However, the homogeneous component of weight $k+1$ of Σ_k is R_{k+1}, so $[(R_2)^2](K_3)$ and $[R_2R_3](K_4)$ vanish and there are no non-linear terms in $\Sigma_{k\leq 4}$.

Let us also compute the linear terms in K_5. When multiplying two long cycles τ and $(1,2,3,4,5)$, the product $\tau(5,4,3,2,1)$ has signature $+1$, hence is of cycle type (5), $(2,2,1)$, $(1,1,1,1,1)$ or $(3,1,1)$. Therefore, the linear terms in K_5 are $R_6 + aR_4 + bR_2$, the coefficient b being the number of long cycles τ such that $\tau(5,4,3,2,1)$ is also a long cycle. This number is 8, the corresponding long cycles τ being

$$\begin{gathered}(1,3,5,4,2),\ (1,3,5,2,4),\ (1,3,2,5,4),\ (1,4,3,5,2),\\ (1,4,2,5,3),\ (1,5,4,3,2),\ (1,5,2,4,3),\ (1,5,3,2,4).\end{gathered}$$

Since $1+a+b=(5-1)!=24$, $a=15$, so $\Sigma_5 = R_6 + 15R_4 + 8R_2 + \cdots$, the dots $\cdots$ standing for non-linear terms.

With a bit more work, we can also determine the quadratic terms R_iR_j of the Kerov polynomials. As an example, let us compute the remaining terms of the fifth Kerov polynomial K_5. By using the parity result and the bound on the weight, we see that we now only need to compute the coefficient $N=[(R_2)^2](K_5)$. By Theorem 10.20, N is the number of pairs of permutations (σ,τ) in $\mathfrak{S}(5)$ such that $n(\sigma)=n(\tau)=2$, $\sigma\tau=(1,2,3,4,5)$, and every cycle of τ intersects with both cycles of σ. The possible cycle types for τ are $(3,2)$ and $(4,1)$, and the second possibility is excluded as the fixed point of τ cannot intersect with both cycles of

σ. For the same reason, the only possibility for the cycle type of σ is $(3,2)$. It is then easy to find with a computer the five solutions to this problem:

$$\begin{array}{lcl} \sigma = (3,5)(1,4,2) & ; & \tau = (1,3)(2,5,4) \; ; \\ \sigma = (2,5)(1,4,3) & ; & \tau = (3,5)(1,4,2) \; ; \\ \sigma = (2,4)(1,5,3) & ; & \tau = (2,5)(1,4,3) \; ; \\ \sigma = (1,3)(2,5,4) & ; & \tau = (1,4)(2,5,3) \; ; \\ \sigma = (1,4)(2,5,3) & ; & \tau = (2,4)(1,5,3). \end{array}$$

So, as claimed in Section 9.2, $\Sigma_5 = R_6 + 15R_4 + 8R_2 + 5(R_2)^2$.

▷ *Differential calculus on $\mathscr{O}$ and $\mathscr{Q}$.*

Let $f \in \mathscr{O}$ be an observable of Young diagrams, viewed as a Stanley polynomial in the coordinates (p,q). An important ingredient of the proof of Theorem 10.20 is a relation between the partial derivative

$$\left.\frac{\partial^l}{\partial R_{k_1} \partial R_{k_2} \cdots \partial R_{k_l}}(f)\right|_{R_2=R_3=\cdots=0},$$

which is up to a combinatorial factor the coefficient of $R_{k_1}R_{k_2}\cdots R_{k_l}$ in f, and the coefficients

$$[p_1 q_1^{a_1}\, p_2 q_2^{a_2} \cdots p_m q_m^{a_m}](f).$$

In this setting, it is convenient to modify the Stanley coordinates and to set

$$q_i' = \sum_{j \geq i} q_j.$$

Thus, a Young diagram λ has modified Stanley coordinates $p = (p_1, \ldots, p_m)$ and $q' = (q_1', \ldots, q_m')$ if

$$\lambda = (\underbrace{q_1', \ldots, q_1'}_{p_1 \text{ times}}, \ldots, \underbrace{q_m', \ldots, q_m'}_{p_m \text{ times}}).$$

Obviously, a polynomial in standard Stanley coordinates is a polynomial in modified Stanley coordinates and conversely. This modification is useful, because the coefficients of an observable f viewed as a polynomial in the $\widetilde{p}_{k\geq 2}$'s are directly related to its coefficients as a polynomial in modified Stanley coordinates:

Proposition 10.23. *For any $k_1, \ldots, k_l \geq 2$ and any $f \in \mathscr{O}$,*

$$\left.\frac{\partial^l}{\partial \widetilde{p}_{k_1} \partial \widetilde{p}_{k_2} \cdots \partial \widetilde{p}_{k_l}}(f)\right|_{\widetilde{p}_2=\widetilde{p}_3=\cdots=0} = \frac{1}{k_1 k_2 \cdots k_l}\,[p_1 (q_1')^{k_1-1}\, p_2 (q_2')^{k_2-1} \cdots p_l (q_l')^{k_l-1}](f).$$

Proof. We have the following expression for

$$\widetilde{p}_k(\lambda) = k(k-1) \int_{\mathbb{R}} \frac{\omega_\lambda(x) - |x|}{2}\, x^{k-2}\, dx.$$

This expression stays true if instead of a Young diagram $\lambda \in \mathfrak{Y}$, we have a sequence of *real* interlaced coordinates $(x_1 < y_1 < \cdots < y_{s-1} < x_s)$ which determines a continuous Young diagram ω which is affine by parts, with slope 1 between x_i and y_i and slope -1 between y_i and x_{i+1}. We associate to such a sequence the Stanley coordinates $p_i = x_{s-i+1} - y_{s-i}$ and $q_i = y_{s-i} - x_{s-i}$. These are positive real numbers, and if one modifies the coordinate q'_i, then it amounts to sliding a part of the boundary ω; see Figure 10.9.

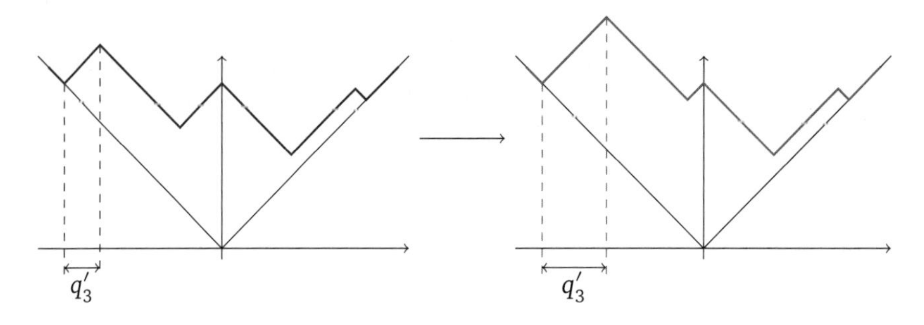

Figure 10.9
Modification of a continuous Young diagram ω along a Stanley coordinate q'_i.

If one raises q'_i by δ, then $\omega(x)$ raises by twice this quantity for

$$q'_i - (p_1 + \cdots + p_i) < x < q'_i - (p_1 + \cdots + p_{i-1}).$$

Consequently,

$$\frac{\partial \widetilde{p}_k}{\partial q'_i} = k(k-1) \int_{q'_i - (p_1 + \cdots + p_i)}^{q'_i - (p_1 + \cdots + p_{i-1})} x^{k-2}\, dx,$$

and by the chain rule,

$$\frac{\partial f}{\partial q'_i} = \sum_{k \geq 2} k(k-1) \int_{q'_i - (p_1 + \cdots + p_i)}^{q'_i - (p_1 + \cdots + p_{i-1})} \frac{\partial f}{\partial \widetilde{p}_k}\, x^{k-2}\, dx.$$

Applying l times this identity, we get:

$$\frac{\partial^l f}{\partial q'_1 \partial q'_2 \cdots \partial q'_l} = \sum_{k_1, \ldots, k_l \geq 2} \left(k_1(k_1-1) \cdots k_l(k_l-1) \frac{\partial^l f}{\partial \widetilde{p}_{k_1} \cdots \partial \widetilde{p}_{k_l}} \right.$$
$$\left. \times \int_{q'_1 - p_1}^{q'_1} \cdots \int_{q'_l - (p_1 + \cdots + p_l)}^{q'_l - (p_1 + \cdots + p_{l-1})} x_1^{k_1 - 2} \cdots x_l^{k_l - 2}\, dx. \right)$$

Denote $g = \frac{\partial^l f}{\partial q'_1 \partial q'_2 \cdots \partial q'_l}$. If we fix $q'_1, \ldots, q'_l$, then g can be considered as a polynomial in $p_1, \ldots, p_l$, and similarly for the right-hand side of the previous formula. We then have, as $p_1, \ldots, p_l$ go to zero:

$$g(p_1,\ldots,p_l) = \sum_{k_1,\ldots,k_l \geq 2} \left(k_1(k_1-1)\cdots k_l(k_l-1) \frac{\partial^l f}{\partial \widetilde{p}_{k_1} \cdots \partial \widetilde{p}_{k_l}} \times \left((q'_1)^{k_1-2} \cdots (q'_l)^{k_l-2} \right) p_1 p_2 \cdots p_l \,(1+o(1)) \right).$$

As a consequence,

$$\begin{aligned}&[p_1(q'_1)^{k_1-2}\, p_2(q'_2)^{k_2-2} \cdots p_l(q'_l)^{k_l-2}](g)\\ &= k_1(k_1-1)\cdots k_l(k_l-1)\, \left.\frac{\partial^l f}{\partial \widetilde{p}_{k_1} \cdots \partial \widetilde{p}_{k_l}}\right|_{p_1=p_2=\cdots=p_l=0}.\end{aligned}$$

For an observable of (continuous) Young diagram f, setting the Stanley coordinates $p_1 = p_2 = \cdots = 0$ is the same as setting $\widetilde{p}_2 = \widetilde{p}_3 = \cdots = 0$, so the right-hand side of the formula is just the coefficient of the monomial $\widetilde{p}_{k_1} \cdots \widetilde{p}_{k_l}$ in f. Thus, the coefficient that we were looking for is indeed

$$\begin{aligned}&\left.\frac{\partial^l}{\partial \widetilde{p}_{k_1} \cdots \partial \widetilde{p}_{k_l}}(f)\right|_{\widetilde{p}_2=\widetilde{p}_3=\cdots=0}\\ &= \frac{1}{k_1(k_1-1)\cdots k_l(k_l-1)} [p_1(q'_1)^{k_1-2} \cdots p_l(q'_l)^{k_l-2}](g)\\ &= \frac{1}{k_1 \cdots k_l} [p_1(q'_1)^{k_1-1} \cdots p_l(q'_l)^{k_l-1}](f),\end{aligned}$$

the second equality coming from $g = \frac{\partial^l f}{\partial q'_1 \cdots \partial q'_l}$. □

Given a polynomial f in modified Stanley coordinates (p, q') and integers $a_1, \ldots, a_m \geq 2$ and $b_1, \ldots, b_m \geq 1$, we set

$$\begin{aligned} f_{(a_1,b_1),\ldots,(a_m,b_m)} &= \left(\prod_{i=1}^{m} (-1)^{b_i-1} (a_i-1)^{\downarrow b_i-1} \right) [p_1(q'_1)^{a_1-1} \cdots p_m(q'_m)^{a_m-1}](f)\\ &= (-1)^l \left(\prod_{i=1}^{m} (-1)^{b_i} (a_i)^{\downarrow b_i} \right) \left.\frac{\partial^m}{\partial \widetilde{p}_{a_1} \cdots \partial \widetilde{p}_{a_m}}(f)\right|_{\widetilde{p}_2=\widetilde{p}_3=\cdots=0} \quad \text{if } f \in \mathscr{O}. \end{aligned}$$

Notice that because of the second formula, when $f \in \mathscr{O}$ is an observable, $f_{(a_1,b_1),\ldots,(a_m,b_m)}$ does not depend on the order of the pairs (a_i, b_i), so it makes sense to write $f_{\{(a,b)\in P\}}$ for some set P of pairs (a, b). These quantities appear in the calculation of the coefficient of a monomial $R_{k_1} R_{k_2} \cdots R_{k_l}$ in f:

Theorem 10.24. *For any $k_1, \ldots, k_l \geq 2$ and any $f \in \mathscr{O}$,*

$$\left.\frac{\partial^l}{\partial R_{k_1} \partial R_{k_2} \cdots \partial R_{k_l}}(f)\right|_{R_2=R_3=\cdots=0} = \sum_{\pi \in \mathfrak{Q}(l)} (-1)^{l-\ell(\pi)} f_{\{(\sum_{j\in\pi_i} k_j,\, |\pi_i|),\, i \in [\![1,\ell(\pi)]\!]\}}.$$

Proof. We start by expressing $\widetilde{p}_j$ in terms of the R_k's, in order to apply the chain rule. Let $P(z)$ be the generating series of the observables $\widetilde{p}_k$:

$$P_\lambda(z) = \sum_{j\geq 2} \frac{\widetilde{p}_j(\lambda)}{j\,z^j} = \log\left(1 + \sum_{j\geq 2} \frac{\widetilde{h}_j(\lambda)}{z^j}\right) = \log(z\,G_\lambda(z)),$$

where $G_\lambda(z)$ is the Cauchy transform of the transition measure. Recall that $G_\lambda(R_\lambda(z) + \frac{1}{z}) = z$, where $R_\lambda(z) = \sum_{k\geq 2} R_k(\lambda)\,z^{k-1}$ is the R-transform. In the following, we consider three power series

$$P(z) = \sum_{j=1}^{\infty} \frac{P_j}{j\,z^j}$$

$$G(z) = \frac{1}{z} + \sum_{j=1}^{\infty} \frac{M_j}{z^{j+1}}$$

$$R(z) = \sum_{j=1}^{\infty} R_j\,z^{j-1}$$

such that $P(z) = \log(zG(z))$ and $G(R(z) + \frac{1}{z}) = z$. These functions all belong to the ring of formal Laurent series $(\mathbb{C}[R_1, R_2, \ldots])[[z]]$, so in particular it makes sense to take their derivatives with respect to their argument z or with respect to one of the free cumulant R_k. Set $t = R(z) + \frac{1}{z}$. One gets by derivation:

$$0 = \frac{\partial z}{\partial R_k} = \frac{\partial G}{\partial R_k}(t) + \frac{\partial G}{\partial t}\,z^{k-1} = \frac{\partial G}{\partial R_k}(t) + \frac{\partial G}{\partial t}\,(G(t))^{k-1},$$

so $\frac{\partial G(t)}{\partial R_k} = -G'(t)\,(G(t))^{k-1}$. As a consequence, for $k \geq 2$,

$$\frac{\partial P(z)}{\partial R_k} = \frac{\partial(\log(z\,G(z)))}{\partial R_k} = -G'(z)\,(G(z))^{k-2} = \frac{\partial G(z)}{\partial R_{k-1}}.$$

More generally, if $k_1, \ldots, k_{l-1} \geq 1$ and $k_l \geq 2$, then

$$\frac{\partial^l P}{\partial R_{k_1} \cdots \partial R_{k_{l-1}} \partial R_{k_l}} = \frac{\partial^l G}{\partial R_{k_1} \cdots \partial R_{k_{l-1}} \partial R_{k_l - 1}};$$

$$\frac{\partial^l P_j}{\partial R_{k_1} \cdots \partial R_{k_{l-1}} \partial R_{k_l}} = j\,\frac{\partial^l M_{j-1}}{\partial R_{k_1} \cdots \partial R_{k_{l-1}} \partial R_{k_l - 1}},$$

the second formula being obtained by taking the coefficient of $\frac{1}{z^j}$. To compute the coefficients of M_{j-1} viewed as a polynomial in the R_k's, let us remark that

$$\begin{aligned}
\frac{\partial^2 G}{\partial R_k \partial R_l} &= \frac{\partial}{\partial R_l}(-G'\,G^{k-1}) = -\left(\frac{\partial G}{\partial R_l}\right)'\,G^{k-1} - (k-1)\,G'\,G^{k-2}\,\frac{\partial G}{\partial R_l} \\
&= (G'\,G^{l-1})'\,G^{k-1} + (k-1)(G')^2\,G^{k+l-3} \\
&= G''\,G^{k+l-2} + (k+l-2)\,(G')^2\,G^{k+l-3} = \frac{1}{k+l-1}\,(G^{k+l-1})''.
\end{aligned}$$

Hence, a multiple derivative of G with respect to free cumulants $R_{k_1}, R_{k_2}, \ldots, R_{k_l}$ only depends on the sum $k_1 + k_2 + \cdots + k_l$:

$$\frac{\partial^l G}{\partial R_{k_1} \cdots \partial R_{k_l}} = \frac{\partial^l G}{(\partial R_1)^{l-1}\, \partial R_{k_1+\cdots+k_l-(l-1)}}.$$

Now, $(R_k)_{k\geq 1}$ is the sequence of formal free cumulants associated to $(M_j)_{j\geq 1}$, so

$$M_j = \sum_{\pi \in \mathfrak{NC}(j)} R_\pi.$$

Therefore, if $j = k_1 + \cdots + k_l$, then the coefficient of $(R_1)^{l-1} R_{k_1+\cdots+k_l-(l-1)}$ in M_j is the number of non-crossing partitions of size j, with an ordering on its blocks, and with $l-1$ blocks of size 1 and one block of size $k_1 + \cdots + k_l - (l-1)$. Such a non-crossing partition with an ordering on blocks is entirely determined by choosing the blocks of size 1, and there are $j^{\downarrow l-1}$ possibilities for this. Therefore, for any $k_1, \ldots, k_{l-1} \geq 1$ and $k_l \geq 2$,

$$\left.\frac{\partial^l M_j}{\partial R_{k_1} \cdots \partial R_{k_l}}\right|_{R_1=R_2=\cdots=0} = \begin{cases} j^{\downarrow l-1} & \text{if } k_1 + \cdots + k_l = j, \\ 0 & \text{otherwise;} \end{cases}$$

$$\left.\frac{\partial^l P_j}{\partial R_{k_1} \cdots \partial R_{k_l}}\right|_{R_1=R_2=\cdots=0} = \begin{cases} j^{\downarrow l} & \text{if } k_1 + \cdots + k_l = j, \\ 0 & \text{otherwise.} \end{cases}$$

By invariance of the result by permutation of the partial derivatives, the result is also true when $(k_1, \ldots, k_l)$ is a vector of positive integers with at least one coordinate $k_i \geq 2$. If $k_1 = k_2 = \cdots = k_l = 1$, the formula also holds: we can then suppose $R_2 = R_3 = \cdots = 0$, in which case $R(z) = R_1$, $G(z) = \frac{1}{z-R_1}$, $P(z) = -\log(1 - \frac{R_1}{z})$ and $M_j = P_j = (R_1)^j$.

We can now send P_j to $\widetilde{p}_j$ and M_j to $\widetilde{h}_j$, which amounts to the specialization $R_1 = 0$ of the previous identities. The previous formulas now hold for the observables $\widetilde{p}_{j\geq 2}$ in the algebra $\mathscr{O}$, taking the derivatives at $R_2 = R_3 = \cdots = 0$. Finally, by the multi-dimensional Faà-di-Bruno formula, for any observable $f \in \mathscr{O}$,

$$\frac{\partial^l f}{\partial R_{k_1} \cdots \partial R_{k_l}} = \sum_{\pi \in \mathfrak{Q}(l)} \sum_{\phi : [\![1,\ell(\pi)]\!] \to [\![2,+\infty]\!]} \frac{\partial^{\ell(\pi)} f}{\prod_{i=1}^{\ell(\pi)} \partial \widetilde{p}_{\phi(i)}} \prod_{i=1}^{\ell(\pi)} \left(\frac{\partial^{|\pi_i|} \widetilde{p}_{\phi(i)}}{\prod_{j\in\pi_i} \partial R_{k_j}} \right)$$

where the second sum runs over colorings ϕ of the blocks of π by integers. When $R_2 = R_3 = \cdots = 0$, all the terms vanish but the ones for which the coloring is $\phi(i) = \sum_{j\in\pi_i} k_j$. Then,

$$\left.\frac{\partial^l f}{\partial R_{k_1} \cdots \partial R_{k_l}}\right|_{R_2=R_3=\cdots=0} = \sum_{\pi \in \mathfrak{Q}(l)} \prod_{i=1}^{\ell(\pi)} \left(\sum_{j\in\pi_i} k_j \right)^{\downarrow|\pi_i|} \left.\frac{\partial^{\ell(\pi)} f}{\prod_{i=1}^{\ell(\pi)} \partial \widetilde{p}_{\sum_{j\in\pi_i} k_j}}\right|_{R_2=R_3=\cdots=0},$$

which ends the proof by distributing the signs $(-1)^{|\pi_i|}$ to make appear the coefficients $f_{\{(\sum_{j\in\pi_i} k_j,|\pi_i|),\ i\leq \ell(\pi)\}}$. □

There is yet another identity coming from the differential calculus on $\mathscr{Q}$ and that we shall use later in the proof of Theorem 10.20. This one involves an unsigned sum of coefficients $f_{(a_1,b_1),\ldots,(a_m,b_m)}$:

Proposition 10.25. *For any $k_1,\ldots,k_l \geq 1$ and any $f \in \mathscr{O}$,*

$$[p_1(q_1')^{k_1-1}\cdots p_l(q_l')^{k_l-1}](f) = \sum_{\pi} f_{\{(\sum_{j\in\pi_i} k_j,\ |\pi_i|),\ i\in[\![1,\ell(\pi)]\!]\}},$$

where the sum runs over set partitions $\pi \in \mathfrak{Q}(l)$ whose blocks have the form

$$\pi_i = \{j_{i1} < j_{i2} < \cdots < j_{im}\} \quad \textit{with } k_{j_{i1}} = \cdots = k_{j_{i(m-1)}} = 1 \textit{ and } k_{j_{im}} \geq 2.$$

Proof. In the following we fix $q_1', q_2', \ldots$ and consider f as a polynomial in $p_1, p_2, \ldots$. To make this clear, we denote $[p_{j_1}\cdots p_{j_l}]_p(f)$ the coefficient of f as a polynomial in the coordinates p_j, this coefficient being in general a polynomial in the coordinates q_j', and not to be confused with $[p_{j_1}\cdots p_{j_l}](f) = [p_{j_1}\cdots p_{j_l}]_{p,q}(f)$. On the other hand, notice that the coefficients $f_{\{(\sum_{j\in\pi_i} k_j,\ |\pi_i|),\ i\in[\![1,\ell(\pi)]\!]\}}$ with π set partition satisfying the hypotheses of the proposition are given by

$$f_{\{(\sum_{j\in\pi_i} k_j,\ |\pi_i|),\ i\in[\![1,\ell(\pi)]\!]\}} = (-1)^{l-\ell(\pi)} \prod_{i=1}^{\ell(\pi)} \left((k_{j_{im}})^{\uparrow|\pi_i|-1}\,[p_{j_{i1}}\cdots p_{j_{im}} q^{k_{j_{im}}-1}](f)\right),$$

where $k^{\uparrow l} = k(k+1)\cdots(k+l-1)$. Suppose first that $f = \widetilde{p}_k$ for some $k \geq 2$. With the notations of the end of Chapter 8, we have

$$\widetilde{p}_k(\lambda) = k(k-1)\iint_{\lambda} (x-y)^{k-2}\,dx\,dy,$$

where x and y are the coordinates of the points inside the Young diagram λ, drawn with the French convention. If we expand this with Newton's formula, we get

$$\begin{aligned}
\widetilde{p}_k(\lambda) &= k!\sum_{r=1}^{k-1}(-1)^{r-1}\iint_{\lambda} \frac{x^{k-1-r}}{(k-1-r)!}\,\frac{y^{r-1}}{(r-1)!}\,dx\,dy \\
&= k!\sum_{r=1}^{k-1}(-1)^{r-1}\operatorname{vol}\left\{(x_1,\ldots,x_{k-r},y_1,\ldots,y_r) \;\middle|\; \substack{0<x_1<\cdots<x_{k-r} \\ 0<y_1<\cdots<y_r \\ (x_{k-r},y_r)\in\lambda}\right\},
\end{aligned}$$

where the volumes are taken with respect to the k-dimensional Lebesgue measure. If we use the Stanley coordinates, we can rewrite these volumes as

$$V_r = \sum_{j=1}^{l}\operatorname{vol}\left\{(x_1,\ldots,x_{k-r},y_1,\ldots,y_r) \;\middle|\; \substack{0<x_1<\cdots<x_{k-r}<q_j' \\ 0<y_1<\cdots<y_r<p_1+\cdots+p_j \\ p_1+\cdots+p_{j-1}<y_r}\right\},$$

which are polynomials in $p_1, \ldots, p_l$. If $[p_1 p_2 \cdots p_l]_p(V_r) \neq 0$, we have necessarily $l = r \leq k-1$, and then

$$[p_1 p_2 \cdots p_l]_p(V_l) = \mathrm{vol}\{0 < x_1 < \cdots < x_{k-l} < q_l'\} = \frac{(q_l')^{k-l}}{(k-l)!}.$$

So,

$$[p_1 p_2 \cdots p_l]_p(\widetilde{p}_k) = (-1)^{l-1} k^{\downarrow l} (q_l')^{k-l},$$

being understood that we obtain 0 if $k \leq l$. More generally, for any sequence of indices $j_1 < j_2 < \cdots < j_l$, by using the coordinates $p_{j_1}, p_{j_2}, \ldots, p_{j_l}$ as Stanley coordinates of the Young diagrams (which means that we specialize to 0 the other p_j's with $j_k < j < j_{k+1}$),

$$[p_{j_1} p_{j_2} \cdots p_{j_l}]_p(\widetilde{p}_k) = \begin{cases} (-1)^{l-1} k^{\downarrow l} (q_{j_l}')^{k-l} & \text{if } k > l \\ 0 & \text{otherwise.} \end{cases}$$

Consider now an observable f which is a monomial in the $\widetilde{p}_k$'s : $f = \widetilde{p}_{n_1} \widetilde{p}_{n_2} \cdots \widetilde{p}_{n_r}$. The monomial $p_1 p_2 \cdots p_l$ appears in f by distributing the p_j's to the observables $\widetilde{p}_n$, according to a set partition $\pi \in \mathfrak{Q}(l)$ with at most r parts, and an injective map $\psi : [\![1, \ell(\pi)]\!] \to [\![1, r]\!]$. Such a set partition yields a coefficient proportional to $(q_1')^{k_1-1} \cdots (q_l')^{k_l-1}$ if and only if all its parts $\pi_i = \{j_{i1} < j_{i2} < \cdots < j_{im}\}$ have their right-most element j_{im} such that

$$k_{j_{im}} - 1 = n_{\psi(i)} - |\pi_i| \geq 1,$$

and all the other $k_{j_{il}}$'s with $j_{il} < j_{im}$ are equal to 1. Then, the contribution to the coefficient $[p_1 (q_1')^{k_1-1} \cdots p_l (q_l')^{k_l-1}](f)$ corresponding to (π, ψ) is

$$(-1)^{l-\ell(\pi)} \prod_{i=1}^{\ell(\pi)} (k_{j_{im}})^{\uparrow |\pi_i|-1},$$

and moreover, the value of $n_{\psi(i)}$ is prescribed by the set partition π. Summing over compatible set partitions π and compatible injective maps ψ yields the formula announced for $f = \widetilde{p}_{n_1} \widetilde{p}_{n_2} \cdots \widetilde{p}_{n_r}$. Since the formula is compatible with sums and since the $\widetilde{p}_{n \geq 2}$'s form an algebraic basis of $\mathscr{O}$, we are done. □

▷ *Euler characteristic of families of subsets.*

Applying Theorem 10.24 to the Stanley–Féray formula, we see that $[R^s](K_k)$ can be expressed as a signed sum over factorizations $\sigma\tau = (1, 2, \ldots, k)$ of the long cycle and functions $r : C(\sigma) \sqcup C(\tau) \to \mathbb{N}^*$. In order to transform this expression in an *unsigned* sum, we shall use several vanishing results, which are all related to the same phenomenon. Let X be a finite set and I be a family of subsets of X.

Definition 10.26. *The* ***Euler characteristic*** *of I is the integer*

$$\chi(I) = \sum_{C = (C_1 \subsetneq C_2 \subsetneq \cdots \subsetneq C_d)} (-1)^{d-1},$$

where the sum runs over chains of subsets $C = (C_1 \subsetneq \cdots \subsetneq C_{d \geq 1})$ where all C_i's are in I.

Example. With $X = [\![1,3]\!]$ and $I = \{\emptyset, \{1\}, \{2,3\}, \{1,2\}\}$, we have

$$\begin{aligned}\chi(I) &= |\{\emptyset, \{1\}, \{2,3\}, \{1,2\}\}| - |\{(\emptyset, \{1\}), (\emptyset, \{2,3\}), (\emptyset, \{1,2\}), (\{1\}, \{1,2\})\}| \\ &\quad + |\{(\emptyset, \{1\}, \{1,2\})\}| \\ &= 4 - 4 + 1 = 1.\end{aligned}$$

Remark. The definition above is a particular case of Euler characteristic of an (abstract) **simplicial complex**. The simplicial complex that we consider here has for faces the chains of subsets in I, the dimension of a face C being $(\mathrm{card}\, C - 1)$. With the previous example, we obtain the following simplicial complex:

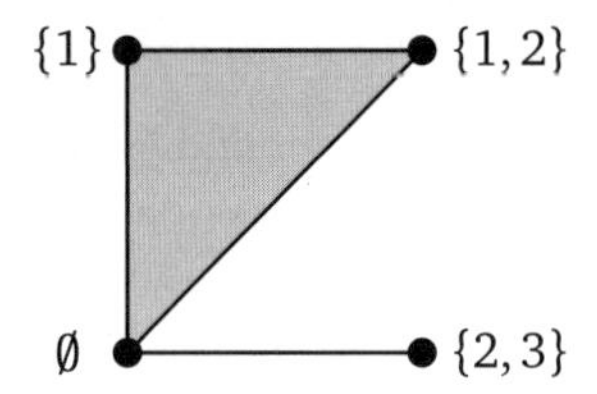

Theorem 10.27. *Suppose that the family of subsets I has the following stability property:*

$$A, B \in I \Rightarrow A \cup B \in I \text{ or } A \cap B \in I.$$

Then, $\chi(I) = 1$.

Proof. Without loss of generality, we can assume that $X = [\![1,n]\!]$. For $k \in [\![0,n]\!]$, set

$$I_k = \{A \cup [\![1,k]\!],\ A \subset I\}.$$

We have $I_0 = I$ and $I_n = \{X\}$, so in particular $\chi(I_n) = 1$. Moreover, each family I_k has the same property of stability as I. We are going to show that $\chi(I_{k-1}) = \chi(I_k)$ for any $k \in [\![1,n]\!]$; this will imply the theorem, as in this case $\chi(I) = \chi(I_0) = \chi(I_n) = 1$. If $C = (C_1 \subsetneq C_2 \subsetneq \cdots \subsetneq C_{d'})$ is a chain in I_{k-1}, denote

$$C + k = (C_1 \cup \{k\} \subset C_2 \cup \{k\} \subset \cdots \subset C_{d'} \cup \{k\}),$$

which gives a chain in I_k either of length d', or of length $d' - 1$ if $C_i \cup \{k\} = C_{i+1}$ for some i. If we gather the chains C of I_{k-1} according to their image D by the map $C \mapsto C + k$, then we get:

$$\begin{aligned}\chi(I_{k-1}) &= \sum_{\text{chain } C \text{ in } I_{k-1}} (-1)^{\ell(C)-1} \\ &= \sum_{\text{chain } D \text{ in } I_k} \left(\sum_{\text{chain } C \text{ in } I_{k-1} \text{ such that } C+k=D} (-1)^{\ell(C)-1} \right).\end{aligned}$$

It suffices now to prove that the sum in parentheses is $(-1)^{\ell(D)-1}$. We write $(D_1 \subsetneq D_2 \subsetneq \cdots \subsetneq D_d)$ for the elements of the chain D, and denote $e \in [\![1,d]\!]$ the largest integer such that $D_e \notin I_{k-1}$, with by convention $e = 0$ if all the elements D_i are in I_{k-1}. If $i \in [\![1,e]\!]$, then $D_i \setminus \{k\} \in I_{k-1}$. Indeed, if $e = 0$, there is nothing to prove, and if $i = e$, then $D_e \notin I_{k-1}$, but $D_e = A_e \cup [\![1,k]\!]$ for some $A_e \in I$, since $D_e \in I_k$. If k were in A_e, then we would have $D_e = A_e \cup [\![1,k-1]\!]$, which is not possible since $D_e \notin I_{k-1}$. Therefore,

$$D_e \setminus \{k\} = (A_e \setminus \{k\}) \cup ([\![1,k]\!] \setminus \{k\}) = A_e \cup [\![1,k-1]\!] \in I_{k-1}.$$

Suppose now $1 \leq i < e$. We can suppose that $D_i \in I_{k-1}$, as otherwise one can use the same argument as above to show that $D_i \setminus \{k\} \in I_{k-1}$. Then, we know that $D_e \setminus \{k\} \in I_{k-1}$ and that $k \in D_i \in I_{k-1}$, hence,

$$D_i \cap (D_e \setminus \{k\}) = D_i \setminus \{k\} \in I_{k-1} \quad \text{or} \quad D_i \cup (D_e \setminus \{k\}) = D_e \in I_{k-1},$$

and the second case is impossible, so $D_i \setminus \{k\} \in I_{k-1}$. As a consequence of this discussion,

$$D_i = \begin{cases} D_i' \sqcup \{k\} & \text{if } i \leq e, \text{ with } D_i' \in I_{k-1} \text{ and } k \notin D_i', \\ D_i \in I_{k-1} & \text{if } i > e. \end{cases}$$

We introduce similarly the largest integer g such that $D_i \setminus \{k\} \in I_{k-1}$ for any $i \in [\![1,g]\!]$, with by convention $g = 0$ if there is no such integer. By the previous discussion, $e \leq g$. Consider now a chain $C = (C_1 \subsetneq C_2 \subsetneq \cdots \subsetneq C_{d'})$ in I_{k-1} such that $C + k = D$, and denote f the smallest integer in $[\![1,d']\!]$ such that $k \in C_f$, with by convention $f = d'+1$ if there is no such integer. We then have $C_i = D_i \setminus \{k\}$ if $i < f$, so in particular $C_{f-1} = D_{f-1} \setminus \{k\} \in I_{k-1}$, and $f - 1 \leq g$. On the other hand:

1. If $d' = d+1$, then $C_f = C_{f-1} \sqcup \{k\}$, and $C_i = D_{i-1}$ for any $i \geq f$. In particular, $D_{f-1} = C_f \in I_{k-1}$, so by definition of e, $f - 1 > e$.

2. If $d' = d$, then $C_i = D_i$ for any $i \geq f$. In particular, $D_f = C_f$, so $f > e$.

Let us then describe the chains $C = (C_1 \subsetneq C_2 \subsetneq \cdots \subsetneq C_{d'})$ such that $C + k = D$. In the previous discussion, if $f - 1$ and d' are known, then the chain C is entirely determined.

1. Set $d' = d+1$, and choose an integer $f - 1 \in [\![e+1,g]\!]$. We have to check that the chain defined by

$$C_i = \begin{cases} D_i \setminus \{k\} & \text{if } i < f \\ D_{i-1} & \text{if } i \geq f \end{cases}$$

is indeed a chain in I_{k-1}. If $i < f$, then $i \leq f - 1 \leq g$, so $D_i \setminus \{k\}$ is in I_{k-1}. If $i \geq f$, then $i - 1 > e$, so D_{i-1} is also in I_{k-1}. So, all the $g - e$ possible choices for $f - 1$ give indeed a chain in I_{k-1}, hence a contribution $(g-e)(-1)^d$ to the sum in parentheses.

2. Set $d' = d$, and choose an integer $f - 1 \in [\![e, g]\!]$. We have to check that the chain defined by

$$C_i = \begin{cases} D_i \setminus \{k\} & \text{if } i < f \\ D_i & \text{if } i \geq f \end{cases}$$

is a chain in I_{k-1}, and this follows from the same calculations as above. Hence, this case gives a contribution $(g - e + 1)(-1)^{d-1}$ to the sum in parentheses.

Notice in particular that there always exists a chain C such that $C + k = D$. Taking the two contributions, we obtain $(-1)^{d-1}$. □

Remark. Theorem 10.27 can be better understood in terms of **contractible** simplicial complexes. Indeed, we used in this proof a sequence $\mathscr{C}(I_1), \mathscr{C}(I_2), \ldots, \mathscr{C}(I_n)$ of contractions of the complex $\mathscr{C}(I)$ associated to I, with $\dim \mathscr{C}(I_k) \leq n - k$. The hypothesis of the theorem ensures that the sets I_k correspond indeed to contractions of $\mathscr{C}(I)$, and that at the end of these contractions one obtains the 0-dimensional complex with just one point. In the case of the example presented just before the statement of Theorem 10.27, the contractions are drawn in Figure 10.10, and the Euler characteristic is conserved by these contractions.

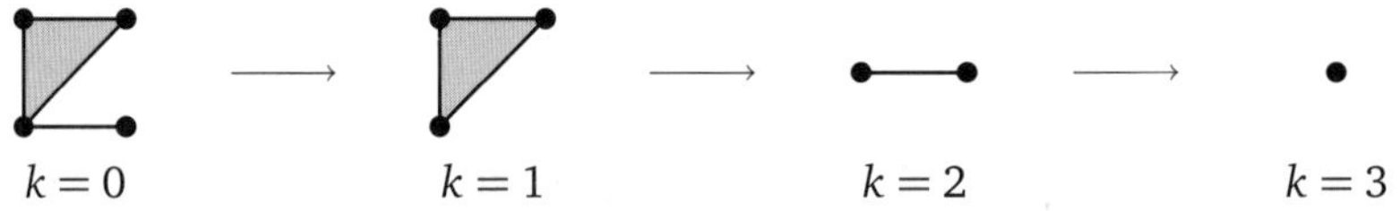

$k = 0$ $k = 1$ $k = 2$ $k = 3$

Figure 10.10
Contractions of the simplicial complex associated to $I = \{\emptyset, \{1\}, \{1, 2\}, \{2, 3\}\}$.

We can now start the proof of Theorem 10.20. It is a bit easier to deal with derivatives of the Kerov polynomials at 0 than with coefficients (the two quantities differ in general by an integer factor, which is a product of factorials). Thus, a statement equivalent to Theorem 10.20 is: for any $k_1, \ldots, k_l \geq 2$,

$$\left. \frac{\partial^l \Sigma_k}{\partial R_{k_1} \cdots \partial R_{k_l}} \right|_{R_2 = R_3 = \cdots = 0}$$

is equal to the number of triples (σ, τ, p), with:

(K1') σ and τ belong to $\mathfrak{S}(k)$, $\sigma\tau = (1, 2, \ldots, k)$ is the long cycle, and

$$n(\tau) = l \quad ; \quad n(\sigma) + n(\tau) = k_1 + k_2 + \cdots + k_l;$$

(K2') $r : C(\tau) \to [\![1, l]\!]$ is a bijection;

(K3') if $C \subset C(\tau)$ is non-trivial, then there are strictly more than $\sum_{c \in C}(k_{r(c)} - 1)$ cycles of σ that intersect $\bigcup_{c \in C} c$.

We shall denote $N(k; k_1, k_2, \ldots, k_l)$ the number of triples that satisfy Conditions (K1'-K3'). The proof of Theorem 10.20 relies now on a sequence of transformations of this quantity, which we present hereafter in a sequence of lemmas.

Lemma 10.28. *We have*

$$N(k; k_1, \ldots, k_l) = \sum_{d \geq 0} \sum_{C = (\emptyset \subsetneq C_1 \subsetneq C_2 \subsetneq \cdots \subsetneq C_d \subsetneq [\![1,l]\!])} (-1)^d \operatorname{Bad}_C,$$

where Bad_C *is the number of "bad" triples for the chain* C*, that is to say triples* (σ, τ, r) *that satisfy Conditions (K1'-K2'), but such that for any* $i \in [\![1, d]\!]$*, there are less than* $\sum_{j \in C_i} (k_j - 1)$ *cycles of* σ *that intersect* $\bigcup_{c \in r^{-1}(C_i)} c$*.*

Proof. Let (σ, τ, r) be a triple that satisfies Conditions (K1'-K2'). We associate to it a coloring $q : C(\tau) \to \{2, 3, \ldots\}$, given by $q(c) = k_{r(c)}$. Let I be the set of subsets $C \subset C(\tau)$ that are non-trivial, and for which the opposite of Condition (K3') holds: there are at most $\sum_{c \in C} (q(c) - 1)$ cycles of σ that intersect $\bigcup_{c \in C} c$. We claim that Theorem 10.27 applies to this set $I = I(\sigma, \tau, r)$. It is convenient to introduce the function on sets of cycles

$$V(C) = \sum_{c \in C} (q(c) - 1) - \operatorname{card} \left\{ b \in C(\sigma),\ b \cap \left(\bigcup_{c \in C} c \right) \neq \emptyset \right\},$$

which characterizes the set I: $I = \{C \mid C \text{ non-trivial and } V(C) \geq 0\}$. If C and D are two sets of cycles, it is easily seen that

$$V(C) + V(D) = V(C \cup D) + V(C \cap D).$$

Indeed, this identity is satisfied by both terms in the definition of V. Therefore, if C and D are in I, then $V(C \cup D) + V(C \cap D) \geq 0$, so at least one of the two sets $C \cup D$ or $C \cap D$ satisfies again the opposite of (K3'). If $C \cup D \neq C(\tau)$ and $C \cap D \neq \emptyset$, this ends the proof of the claim. Otherwise, if one has $C \cup D = C(\tau)$ but $C \cap D \neq \emptyset$, then $V(C(\tau)) = 0$, so $V(C \cap D) = V(C) + V(D) \geq 0$ and $C \cap D$ has the opposite of the property (K3'). The case when $C \cup D \neq C(\tau)$ but $C \cap D = \emptyset$ is similar, as $V(\emptyset) = 0$. Finally, let us show that we cannot have $C \cup D = C(\tau)$ and $C \cap D = \emptyset$. In this case, we have a non-trivial factorization $\tau = \tau_C \tau_D$ as a product of two permutations with disjoint supports $\bigcup_{c \in C} c$ and $\bigcup_{c \in D} c$, and the inequalities on the number of cycles of σ and τ, as well as the identity $\operatorname{card} C(\sigma) = n(\tau) = \sum_{c \in C(\tau)} (q(c) - 1)$ show that we have the same kind of factorization for σ:

$$\sigma = \sigma_C \sigma_D$$

with σ_C supported by $\bigcup_{c \in C} c$, and σ_D supported by $\bigcup_{c \in D} c$. Then, $\sigma\tau = (1, 2, \ldots, k)$ is the product of permutations $\sigma_C \tau_C$ and $\sigma_D \tau_D$ of these disjoint supports, which is absurd. By Theorem 10.27, we therefore have:

$$\sum_{d \geq 0} \sum_{\substack{C = (\emptyset \subsetneq C_1 \subsetneq C_2 \subsetneq \cdots \subsetneq C_d \subsetneq [\![1,l]\!]) \\ \forall i,\ C_i \in I}} (-1)^d = 1 - \chi(I) = 0$$

unless I is empty, in which case this sum is equal to 1. We now sum this formula over all possible triples (σ,τ,r) with properties (K1') and (K2'). On the right-hand side, we obtain the number of triples for which $I=\emptyset$, that is to say such that Condition (K3') is satisfied. On the left-hand side, we get $\sum_{d\geq 0}\sum_{C\,|\,\ell(C)=d}(-1)^d\,\mathrm{Bad}_C$. □

Lemma 10.29. *Fix a chain $C=(\emptyset\subsetneq C_1\subsetneq C_2\subsetneq\cdots\subsetneq C_d\subsetneq[\![1,l]\!])$ as before. We have*

$$\mathrm{Bad}_C=(-1)^{l-1}\sum_{a_1,\dots,a_l}[p_1(q_1')^{a_1-1}\cdots p_l(q_l')^{a_l-1}](\Sigma_k),$$

where the sum runs over families of positive integers $(a_1,\dots,a_l)$ such that $a_1+\cdots+a_l=k_1+\cdots+k_l$, and

$$a_{l+1-|C_j|}+\cdots+a_l\leq\sum_{i\in C_j}k_i$$

for any $j\in[\![1,d]\!]$.

Proof. In terms of the modified Stanley coordinates $(p_1,\dots,p_l)$ and $(q_1',\dots,q_l')$, the functions $N^{\sigma,\tau}$ are given by

$$N^{\sigma,\tau}\begin{pmatrix}p_1,\dots,p_l\\ q_1',\dots,q_l'\end{pmatrix}=\sum_{r:C(\sigma)\to[\![1,l]\!]}\left(\prod_{c\in C(\sigma)}p_{r(c)}\right)\left(\prod_{d\in C(\tau)}q'_{r_{\max}(d)}\right),$$

where $r_{\max}(d)=\max\{r(c)\,|\,c\in C(\sigma),\ c\cap d\neq\emptyset\}$. We rewrite Stanley–Féray formula 10.11 as $\Sigma_k=\sum_{\sigma\tau=(1,2,\dots,k)}\varepsilon(\sigma)N^{\tau,\sigma}$. We then have, if $a_1+\cdots+a_l=k_1+\cdots+k_l$,

$$\begin{aligned}&(-1)^{k+l-(k_1+\cdots+k_l)}[p_1(q_1')^{a_1-1}\cdots p_l(q_l')^{a_l-1}](\Sigma_k)\\ &=\sum_{\substack{\sigma\tau=(1,2,\dots,k)\\ r:C(\tau)\to[\![1,l]\!]}}[p_1(q_1')^{a_1-1}\cdots p_l(q_l')^{a_l-1}]\left(\prod_{c\in C(\tau)}p_{r(c)}\right)\left(\prod_{d\in C(\sigma)}q'_{r_{\max}(d)}\right)\\ &=\sum_{\substack{\sigma\tau=(1,2,\dots,k)\\ n(\sigma)=\sum_{i=1}^l(a_i-1),\ n(\tau)=l\\ r:C(\tau)\to[\![1,l]\!]\text{ bijection}}}[(q_1')^{a_1-1}\cdots(q_l')^{a_l-1}]\left(\prod_{d\in C(\sigma)}q'_{r_{\max}(d)}\right)\\ &=\mathrm{card}\left\{(\sigma,\tau,r)\,\middle|\,\forall i,\ \mathrm{card}\left\{d\in C(\sigma)\,\middle|\,d\cap\bigcup_{\substack{c\in C(\tau)\\ r(c)\geq i}}c\neq\emptyset\right\}=\sum_{j=i}^{l}(a_j-1)\right\}.\end{aligned}$$

Moreover, $(-1)^{k+l-(k_1+\cdots+k_l)}=(-1)^{l-1}$, by computing the signature of $\sigma\tau=(1,2,\dots,k)$. Using now the fact that this result is invariant by permutation by the a_j's, and summing the formula over families of integers $(a_1,\dots,a_l)$ that satisfy the hypotheses of the lemma, we obtain $(-1)^{l-1}\,\mathrm{Bad}_C$. □

We now combine Lemma 10.29 and Proposition 10.25:

$$\mathrm{Bad}_C = (-1)^{l-1} \sum_{a_1,\ldots,a_l}^{\diamond} \sum_{\pi}^{*} (\Sigma_k)_{\{(\sum_{j\in\pi_i} a_j,|\pi_i|),\ i\in[\![1,\ell(\pi)]\!]\}},$$

where the symbol $\diamond$ in the first sum indicates that we are taking the sum over integers $a_1,\ldots,a_l$ that satisfy the condition of Lemma 10.29, and the symbol $*$ in the second sum indicates that we restrict the sum to set partitions $\pi \in \mathfrak{Q}(l)$ that satisfy the condition of Proposition 10.25. If we exchange the order of summation, we get the equivalent expression:

$$\mathrm{Bad}_C = (-1)^{l-1} \sum_{\pi\in\mathfrak{Q}(l)} \sum_{\phi:[\![1,\ell(\pi)]\!]\to\mathbb{N}^*}^{\bullet} (\Sigma_k)_{\{(\phi(i),|\pi_i|),\ i\in[\![1,\ell(\pi)]\!]\}},$$

where the second sum runs over functions $\phi : [\![1,\ell(\pi)]\!] \to \mathbb{N}$ such that:

($\bullet$1) $\sum_{i=1}^{\ell(\pi)} \phi(i) = k_1 + k_2 + \cdots + k_l$;

($\bullet$2) for any block π_i of the set partition, $\phi(i) \geq |\pi_i| + 1$;

($\bullet$3) for any $j \in [\![1,d]\!]$, $\sum_{i\,|\,\pi_i\cap C_j\neq\emptyset} \left(\phi(i) - |\pi_i \setminus C_j|\right) \leq \sum_{i\in C_j} k_i$.

We indicate these hypotheses with the symbol $\bullet$ over the sum, and denote them respectively ($\bullet$1), ($\bullet$2) and ($\bullet$3). Given a function ϕ with these properties, one can define a sequence $a_1,\ldots,a_l$ by setting

$$a_{j_{il}} = \begin{cases} \phi(i) - |\pi_i| + 1 & \text{if } j_{il} = j_{im} \text{ is the right-most element of the part } \pi_i, \\ 1 & \text{otherwise.} \end{cases}$$

Then, the sequence $(a_1,\ldots,a_l)$ and the set partition π have the properties ($\diamond$) and ($*$), hence the two expressions are the same. We leave the reader to check the details of this last transformation of our combinatorial sums. Using now Lemma 10.28, we obtain:

$$N(k;k_1,\ldots,k_l)$$

$$= \sum_{\pi\in\mathfrak{Q}(l)} \sum_{\phi:[\![1,\ell(\pi)]\!]\to\mathbb{N}^*}^{\bullet 1,\bullet 2} (\Sigma_k)_{\{(\phi(i),|\pi_i|),\ i\in[\![1,\ell(\pi)]\!]\}} \left(\sum_{d\geq 0} \sum_{\substack{C=(\emptyset\subsetneq C_1\subsetneq\cdots\subsetneq C_d\subsetneq[\![1,l]\!]) \\ C \text{ satisfies } (\bullet 3)}} (-1)^{l+d-1} \right).$$

We compute the sum in parentheses with a final lemma:

Lemma 10.30. *Let π be a set partition in $\mathfrak{Q}(l)$, and ϕ a function on the parts of π with the properties ($\bullet$1) and ($\bullet$2). The sum in parentheses above equals*

$$(-1)^{l-\ell(\pi)} \quad \text{if } \phi(i) = \sum_{j\in\pi_i} k_j \text{ for any } i \in [\![1,\ell(\pi)]\!],$$

and 0 otherwise.

Proof. We assume $l \geq 2$, the case $l = 1$ being trivial. Consider the special case when $\pi = \{1\} \sqcup \{2\} \sqcup \cdots \sqcup \{l\}$; it contains all the difficulties of the lemma, and we shall explain at the end of the proof how to deal with the general case. Denote I the set of subsets C of $[\![1, l]\!]$ that are non-trivial, and such that

$$\sum_{i \in C} \phi(i) \leq \sum_{i \in C} k_i.$$

Suppose first that $\phi(i) = k_i$. Then, every set C has the property above, so

$$I = \mathfrak{P}([\![1, l]\!]) \setminus \{\emptyset, [\![1, l]\!]\}.$$

Then, a chain C in I of length d is in this case determined by a sequence $(D_1, D_2, \ldots, D_{d+1})$ of disjoint subsets whose union is $[\![1, l]\!]$, and such that $C_i = D_1 \sqcup D_2 \sqcup \cdots \sqcup D_i$. If $d \geq 1$ is given, then the number of such sequences is

$$\left\{ {l \atop d+1} \right\} (d+1)!,$$

where $\left\{ {l \atop k} \right\}$ is the **Stirling number** of the second kind, that is to say the number of set partitions of $[\![1, l]\!]$ with k parts. Therefore, the quantity to compute is

$$A_l = (-1)^l \sum_{d=0}^{l-1} (-1)^{d+1} \left\{ {l \atop d+1} \right\} (d+1)!.$$

However, the Stirling numbers of the second kind satisfy:

$$x^l = \sum_{k=1}^{l} x^{\downarrow k} \left\{ {l \atop k} \right\}$$

Indeed, consider a set X with cardinality x. The left-hand side of this formula is the number of functions from a set L of cardinality l to the set X. Such a function is also determined by choosing $k \leq l$, by cutting L into k parts, and by sending each of these parts to a distinct element of X; and there are $x^{\downarrow k}$ possibilities for the images of these parts. Thus, the identity is true for any $x \in \mathbb{N}$, and therefore it is an identity of polynomials. Taking $x = -1$, we get:

$$(-1)^l = \sum_{k=1}^{l} (-1)^k \, k! \left\{ {l \atop k} \right\} = \sum_{d=0}^{l-1} (-1)^{d+1} (d+1)! \left\{ {l \atop d+1} \right\}.$$

So, $A_l = 1 = (-1)^{l-\ell(\pi)}$ if π is the finest set partition and $\phi(i) = k_i$.

Suppose now that $\phi(i) \neq k_i$ for some i; up to permutation we can assume $\phi(1) \neq k_1$, and even $\phi(1) > k_1$. We introduce as in the proof of Lemma 10.28 a function

$$V(C) = \sum_{i \in C} (k_i - \phi(i)),$$

which characterizes the set $I = \{C \subset [\![1,l]\!] \,|\, C \text{ non-trivial and } V(C) \geq 0\}$. Notice that $\{2,\ldots,l\} \in I$ since $V(\{2,\ldots,l\}) = -V(\{1\}) > 0$. Obviously, for any subsets C and D, $V(C)+V(D) = V(C \cup D)+V(C \cap D)$, so by the same argument as in the proof of Lemma 10.28, if C and D are in I and $(C \cap D, C \cup D) \neq (\emptyset, [\![1,l]\!])$, then $C \cap D \in I$ or $C \cup D \in I$.

Unfortunately, we cannot use directly the result of Theorem 10.27, as we truly need the additional hypothesis $(C \cap D, C \cup D) \neq (\emptyset, [\![1,l]\!])$. However, if one uses the same sequence of sets $I = I_0, I_1, I_2, \ldots, I_l = \emptyset$ as in the proof of Theorem 10.27, which yields a contraction of the simplicial complex associated to I, then the sets $I_1, I_2, \ldots, I_l$ have the property required in Theorem 10.27. Indeed, suppose that $C = \widetilde{C} \cup \{1,\ldots,k\}$ and $D = \widetilde{D} \cup \{1,\ldots,k\}$ are in $I_{k\geq 1}$, with $\widetilde{C}$ and $\widetilde{D}$ in $I_0 = I$. If $(\widetilde{C} \cap \widetilde{D}, \widetilde{C} \cup \widetilde{D}) \neq (\emptyset, [\![1,l]\!])$, then we know that $C \cap D$ or $C \cup D$ is in I_k. Suppose now $(\widetilde{C} \cap \widetilde{D}, \widetilde{C} \cup \widetilde{D}) = (\emptyset, [\![1,l]\!])$. Then, $C \cup D = \{1,2,\ldots,l\} = \{2,\ldots,l\} \cup \{1,2,\ldots,k\}$, and $\{2,\ldots,l\} \in I_0$, so $C \cup D \in I_k$. Therefore, one can use the same arguments as in the proof of Theorem 10.27 to show that $\chi(I_1) = \chi(I_2) = \cdots = \chi(I_l) = 1$. It remains to see that one has also $\chi(I_0) = \chi(I_1)$ (which is not the case when $\phi(i) = k_i$ for any i). We start from the identity:

$$\chi(I_0) = \sum_{D \text{ chain in } I_1} \left(\sum_{C \text{ chain in } I_0 \text{ such that } C+1=D} (-1)^{\ell(C)-1} \right)$$

With the same notations as in the proof of Theorem 10.27, provided that $D_1 \neq \{1\}$ and therefore $D_1 \setminus \{1\} \neq \emptyset$, we obtain the same description of the chains C such that $C+1 = D$ as in Theorem 10.27, and in this case,

$$\sum_{C \text{ chain in } I_0 \text{ such that } C+1=D} (-1)^{\ell(C)-1} = (-1)^{\ell(D)-1}.$$

So, $\chi(I_0)$ equals

$$\sum_{D \text{ chain in } I_1,\, D_1 \neq \{1\}} (-1)^{\ell(D)-1} + \sum_{D \text{ chain in } I_1,\, D_1 = \{1\}} \left(\sum_{C \text{ chain in } I_0 \text{ such that } C+1=D} (-1)^{\ell(C)-1} \right).$$

However, if D is a non-empty chain in I_1 with $D_1 = \{1\}$, then $D_1 = C_1 \cup \{1\}$ with C_1 non-empty in I_0, so $C_1 = \{1\}$, which is absurd since by hypothesis $V(\{1\}) < 0$. Hence, there are no terms in the second sum, and one has indeed $\chi(I_0) = \chi(I_1)$. As the quantity to compute is $1 - \chi(I)$, the proof is done in the case $\pi = \{1\} \sqcup \{2\} \sqcup \cdots \sqcup \{l\}$. The case of a general set partition π is entirely similar, replacing the previous function V by

$$\begin{aligned} V(C) &= \sum_{j \in C} k_j - \left(|C| + \sum_{i \,|\, \pi_i \cap C \neq \emptyset} (\phi(i) - |\pi_i|) \right) \\ &= \sum_{j \in C} k_j - \left(\sum_{i \,|\, \pi_i \cap C \neq \emptyset} (\phi(i) - |\pi_i \setminus C|) \right). \end{aligned}$$

□

We can finally end the proof of Theorem 10.20. By the previous lemma and the formula that precedes it,

$$N(k;k_1,\ldots,k_l)=\sum_{\pi\in\mathfrak{Q}(l)}(-1)^{l-\ell(\pi)}(\Sigma_k)_{\{(\sum_{j\in\pi_i}k_j,|\pi_i|),\,i\in[\![1,\ell(\pi)]\!]\}},$$

as there is for each set partition π only one function ϕ that contributes to the sum, namely, $\phi(i)=\sum_{j\in\pi_i}k_j$. By Theorem 10.24, the right-hand side is the partial derivative of Σ_k with respect to the free cumulants $R_{k_1},\ldots,R_{k_l}$, which ends the proof.

Notes and references

Section 10.1 is inspired by [AFNT15], and the only modification to their argument is the order of the interlaced coordinates, which is the inverse of our conventions in the aforementioned paper. An important result from this theory is the fact that the functions N^G with G bicolored graph generate the algebra $\mathscr{Q}$. However, they do not form a basis, and there are actually many relations between these functions, that can be described in terms of an operation of cyclic inclusion-exclusion; see [Fér09, Fér14]. There are also generalizations of this theory to polynomials in non-commutative variables; see again [AFNT15, Fér14].

The combinatorics of maps on surfaces and the relation with permutations is explained in the first chapter of [LZ04]. In Section 10.2, we also followed [Fér09] and [FŚ11]; the proof that we propose of the Stanley–Féray formula comes from this second paper; for the discussion of Lemma 10.13, we refer to [FH91, Chapter 4]. The formula was initially conjectured by Stanley in [Sta06], following the paper [Sta03] which treated the case of rectangular partitions. Its first proof can be found in [Fér10], and it relies on the use of a combinatorial formula due to Okounkov for shifted Schur functions. For the discussion on minimal factorizations of the long cycle, we refer to [Bia97].

Theorem 10.20 is the second main result of the chapter after the Stanley–Féray formula 10.11, and it comes from [DFŚ10]. We tried to detail its proof as much as possible (in particular, we corrected a small gap in the proof of the result 10.27 on the Euler characteristic; see Lemma 6.1 in loc. cit. where the index $g\geq e$ does not seem to be correctly defined). We refer to the same paper for an explicit formula for the quadratic coefficients of the Kerov polynomials. For the use of the multi-dimensional Faà-di-Bruno in the proof of Theorem 10.24; see [Ma09] and the references therein.

Part IV

Models of random Young diagrams

11

Representations of the infinite symmetric group

In the third part of the book, we developed a theory of observables of Young diagrams, which allowed us to make computations with characters of symmetric groups that were somewhat generic, in the sense that the dependence in n of these computations was well understood. We now want to use this theory in order to find the typical properties of large representations of symmetric groups. To make this program more precise, consider a symmetric group $\mathfrak{S}(n)$ with n large, and a "natural" representation (V, ρ^V) of $\mathfrak{S}(n)$, e.g., the regular representation of $\mathfrak{S}(n)$ on $\mathbb{C}\mathfrak{S}(n)$, or the representation by permutation of the letters of words in the tensor product $(\mathbb{C}^N)^{\otimes n}$. This representation V is in general reducible, and its normalized trace $\chi^V(\cdot) = \frac{\operatorname{tr} \rho^V(\cdot)}{\dim V}$ decomposes as a barycenter of the irreducible characters

$$\chi^V = \sum_{\lambda \in \mathfrak{Y}(n)} \mathbb{P}_V[\lambda]\, \chi^\lambda,$$

where $\mathbb{P}_V$ is the spectral measure of the trace χ^V introduced in Chapter 9. In this setting, one observes frequently a phenomenon of *concentration of measure*: the spectral measure $\mathbb{P}_V$ has most of its mass supported by Young diagrams "with the same shape", in a sense to be precised in Chapters 12 and 13. By using the RSK algorithm (Section 3.2), this probabilistic result will have important combinatorial consequences for the distribution of the lengths of the longest increasing subsequences in random permutations or random words.

Of course, this kind of result cannot be true for *any* representation V of $\mathfrak{S}(n)$, and one has to find an adequate class of large representations for which one can expect such a concentration. The problem is better stated in terms of normalized traces, and then, a natural class to consider is the class of restrictions of traces on the infinite symmetric group $\mathfrak{S}(\infty)$. Thus, consider a normalized trace on $\mathfrak{S}(\infty)$, that is to say a function $\tau : \mathfrak{S}(\infty) \to \mathbb{C}$ such that:

1. $\tau(\mathrm{id}_{\mathbb{N}^*}) = 1$;
2. $\tau(\sigma\rho) = \tau(\rho\sigma)$ for any permutations σ, ρ;
3. $(\tau(\sigma_i \sigma_j^{-1}))_{1 \leq i,j \leq n}$ is a Hermitian non-negative definite matrix for any finite family of permutations $(\sigma_1, \ldots, \sigma_n)$ in $\mathfrak{S}(\infty)$.

If we extend τ by linearity to $\mathbb{C}\mathfrak{S}(\infty)$, and then by continuity to the weak operator topology closure of $\mathbb{C}\mathfrak{S}(\infty)$ in $\mathscr{B}(\ell^2(\mathfrak{S}(\infty)))$, then we recover Definition 9.4 of a trace of a von Neumann algebra. Now, the restriction τ_n of a trace τ of $\mathbb{C}\mathfrak{S}(\infty)$ to the finite group algebra $\mathbb{C}\mathfrak{S}(n)$ is again a trace, so it yields a spectral probability measure $\mathbb{P}_{\tau,n}$ on $\mathfrak{Y}(n)$. We say that $(\tau_n)_{n\in\mathbb{N}}$ is a **coherent family** of traces on the symmetric groups $\mathfrak{S}(n)$. The concentration results that we shall prove in this last part of the book will concern the spectral measures associated to certain coherent families of traces.

In this setting, a preliminary question is the classification of all coherent families of traces, that is to say of all traces of $\mathfrak{S}(\infty)$. In Section 11.1, we shall relate this problem to the harmonic analysis on the Young graph, and to the classification of non-negative specializations of the algebra Sym of symmetric functions. This discussion will allow us to focus on so-called *extremal* characters, which are the traces of $\mathfrak{S}(\infty)$ that generate in the sense of the Krein–Milman theorem the convex set of all traces.

To classify these extremal characters of the group $\mathfrak{S}(\infty)$, it will be convenient to deal with more general objects, namely, the so-called *admissible* representations of the Gelfand pair $(\mathfrak{S}(\infty)\times\mathfrak{S}(\infty),\mathfrak{S}(\infty))$. We introduce them in Section 11.2, and we present the main tool for their classification, which is an infinite semigroup Γ that contains the bi-infinite symmetric group $\mathfrak{S}(\infty)\times\mathfrak{S}(\infty)$, whose construction is due to Olshanski. We then give in Section 11.3 a near complete classification of the admissible representations; it involves Young diagrams and some discrete measures on $[-1,1]$. As a particular case, the extremal characters correspond to the so-called *spherical* representations of the Gelfand pair, and they will be classified by some discrete measures on $[-1,1]$. We thus obtain in Section 11.4 a bijection between the set of extremal characters of $\mathfrak{S}(\infty)$, and a bi-infinite dimensional simplex called the Thoma simplex. It will turn out that every parameter in this Thoma simplex yields a coherent family of traces that has important concentration properties. Thus, to every parameter of the Thoma simplex, we shall be able to associate a model of random integer partitions, that satisfies a law of large numbers and a central limit theorem; cf. Chapters 12 and 13.

11.1 Harmonic analysis on the Young graph and extremal characters

▷ *Extremal characters.*

Given a discrete group G, we denote $\mathscr{T}^1(G)$ the set of normalized traces on G, in the sense of the definition given in the introduction of this chapter. Notice that it is a convex subset of the space of functions from G to $\mathbb{C}$, and that it is closed for the

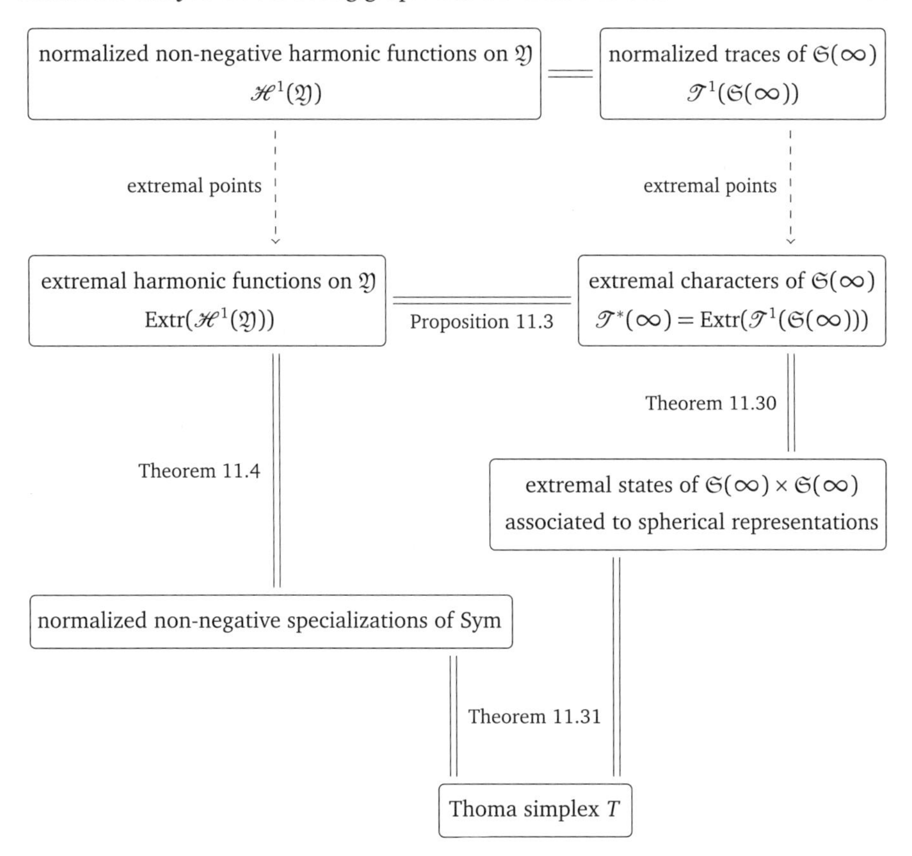

Figure 11.1
The bijections underlying the representation theory of $\mathfrak{S}(\infty)$.

topology of pointwise convergence of functions. By the discussion in Section 9.1, if G is a finite group, then $\mathscr{T}^1(G)$ is in bijection with the set $\mathscr{M}^1(\widehat{G})$ of probability measures on the set of irreducible representations of G, the bijection being

$$p = (p_\lambda)_{\lambda \in \widehat{G}} \quad \mapsto \quad \tau = \sum_{\lambda \in \widehat{G}} p_\lambda \, \chi^\lambda.$$

In particular, we have in this case an inclusion $\widehat{G} \subset \mathscr{T}^1(G)$, each normalized irreducible character being a normalized trace on the group. As a subset of $\mathscr{T}^1(G)$, $\widehat{G}$ is exactly the set of **extremal points**, that is to say traces τ that cannot be written as

$$\tau = x\,\tau_1 + (1-x)\,\tau_2$$

with $0 < x < 1$ and $\tau_1 \neq \tau_2 \in \mathscr{T}^1(G)$.

In the more general case of a discrete (finite or infinite) group, we call **ex-**

tremal character a normalized trace $\tau \in \mathscr{T}^1(G)$ that is not a non-trivial barycenter $\tau = x\,\tau_1 + (1-x)\,\tau_2$ of other normalized traces. We denote $\mathrm{Extr}(\mathscr{T}^1(G))$ the set of these extremal characters. By the previous discussion, if G is finite, then $\mathrm{Extr}(\mathscr{T}^1(G)) = \widehat{G}$, and if G is an infinite group, then $\mathrm{Extr}(\mathscr{T}^1(G))$ is an analogue of the set of irreducible representations. The main goal of this chapter will be to find every extremal character of the infinite symmetric group $\mathfrak{S}(\infty)$. Let us first see why this is a reduction of the problem of finding every normalized trace:

Proposition 11.1. *Let G be a finite or countable group. A function $\tau : G \to \mathbb{C}$ is a normalized trace on G if and only if it lies in the closed convex hull of the extremal characters of G, that is to say that it is a pointwise limit of barycenters*

$$\tau_n = \sum_{i=1}^{N_n} p_{i,n}\,\chi_{i,n}, \quad \text{with} \sum_{i=1}^{N_n} p_{i,n} = 1$$

of extremal characters $\chi_{i,n} \in \mathrm{Extr}(\mathscr{T}^1(G))$.

Proof. We shall use the well-known Krein–Milman theorem (see the references at the end of the chapter), which states that if C is a compact convex set in a locally convex and Hausdorff topological vector space V, then it is the closed convex hull of its extremal points. Here, we take for V the space $\mathbb{C}^G$ of functions from G to $\mathbb{C}$, endowed with the topology of pointwise convergence. It is indeed locally convex and Hausdorff: any function f admits for basis of open neighborhoods the convex sets

$$V_{f,(g_1,\varepsilon_1),\ldots,(g_n,\varepsilon_n)} = \left\{h \in \mathbb{C}^G \mid \forall i \in [\![1,n]\!],\ |f(g_i) - h(g_i)| < \varepsilon_i\right\},$$

and if $f \neq h$, then one can take two such open neighborhoods in order to separate the two functions. Let $C = \mathscr{T}^1(G)$; by a previous remark, it is convex, so it remains to see that it is compact. If $g \in G$ and $\tau \in C$, then the matrix

$$M_g = \begin{pmatrix} 1 & \tau(g) \\ \tau(g^{-1}) & 1 \end{pmatrix}$$

is Hermitian non-negative definite by hypothesis, so $\tau(g^{-1}) = \overline{\tau(g)}$ for any g, and then $\det M_g = 1 - |\tau(g)|^2 \geq 0$, hence $|\tau(g)| \leq 1$. Then, a sequence $(\tau_n)_{n\in\mathbb{N}}$ of traces in C is bounded on any $g \in G$, so, since G is countable, by diagonal extraction, it admits a convergent subsequence. Therefore, C is sequentially compact, and on the other hand, it is easily seen that the topology of V is metrizable, so C is topologically compact. We can therefore apply the Krein–Milman theorem. □

The remainder of the chapter is devoted to an explicit description of the elements of the compact set $\mathrm{Extr}(\mathscr{T}^1(\mathfrak{S}(\infty)))$, which we shall simply denote $\mathscr{T}^*(\infty)$. In this first section, we shall start by establishing the following important result:

Theorem 11.2. *Let τ be an extremal character of $\mathfrak{S}(\infty)$. There exists a unique* *morphism of algebras* $f \in \mathrm{Sym} \mapsto f(X_\tau) \in \mathbb{C}$ *(specialization), such that for any finite permutation* $\sigma = \sigma_\mu \in \mathfrak{S}(n)$ *which has cycle type* $\mu \in \mathfrak{Y}(n)$,

$$\tau(\sigma_\mu) = p_\mu(X_\tau).$$

This correspondence yields a bijection between $\mathscr{T}^*(\infty)$*, and the set of specializations* X *of* Sym *that satisfy the two following conditions:*

1. $s_1(X) = 1$;
2. *for any* $\lambda \in \mathfrak{Y}$, $s_\lambda(X) \geq 0$.

In other words, extremal characters of $\mathfrak{S}(\infty)$ *correspond to non-negative normalized specializations of* Sym *with respect to the basis of Schur functions.*

▷ *Harmonic functions on the Young graph.*

The first step in the proof of Theorem 11.2 is a combinatorial description of the traces on $\mathfrak{S}(\infty)$. Fix $\tau \in \mathscr{T}^1(\mathfrak{S}(\infty))$, and consider the two restrictions $\tau_n = \tau_{|\mathfrak{S}(n)}$ and $\tau_{n+1} = \tau_{|\mathfrak{S}(n+1)}$, associated to the spectral measures $\mathbb{P}_{\tau,n}$ and $\mathbb{P}_{\tau,n+1}$. If $\sigma \in \mathfrak{S}(n)$, denoting $\dim \lambda = \dim S^\lambda$, one has

$$\begin{aligned}
\tau_n(\sigma) &= \sum_{\lambda \in \mathfrak{Y}(n)} \mathbb{P}_{\tau,n}[\lambda]\, \chi^\lambda(\sigma) = \sum_{\lambda \in \mathfrak{Y}(n)} \frac{\mathbb{P}_{\tau,n}[\lambda]}{\dim \lambda} \mathrm{ch}^\lambda(\sigma) \\
&= \tau_{n+1}(\sigma) = \sum_{\Lambda \in \mathfrak{Y}(n+1)} \frac{\mathbb{P}_{\tau,n+1}[\Lambda]}{\dim \Lambda} \mathrm{ch}^\Lambda(\sigma) \\
&= \sum_{\lambda \in \mathfrak{Y}(n)} \left(\sum_{\lambda \nearrow \Lambda} \frac{\mathbb{P}_{\tau,n+1}[\Lambda]}{\dim \Lambda} \right) \mathrm{ch}^\lambda(\sigma)
\end{aligned}$$

by using the restriction rules for irreducible representations of $\mathfrak{S}(n+1)$ (see Corollary 3.7). Therefore, if m_τ is the function on $\mathfrak{Y} = \bigsqcup_{n \in \mathbb{N}} \mathfrak{Y}(n)$ defined by $m_\tau(\lambda) = \frac{\mathbb{P}_{\tau,|\lambda|}[\lambda]}{\dim \lambda}$, then m_τ has the following property:

$$\forall \lambda \in \mathfrak{Y},\; m_\tau(\lambda) = \sum_{\lambda \nearrow \Lambda} m_\tau(\Lambda).$$

Moreover, $m_\tau(\emptyset) = m_\tau(\square) = 1$ since τ is normalized. In the following, we denote $\mathscr{H}^1(\mathfrak{Y})$ the space of non-negative normalized **harmonic functions** on the **Young graph**, that is to say functions $m : \mathfrak{Y} \to \mathbb{R}_+$ such that $m(\emptyset) = 1$ and $m(\lambda) = \sum_{\lambda \nearrow \Lambda} m(\Lambda)$ for any partition λ. By the previous discussion, any trace on $\mathfrak{S}(\infty)$ yields a non-negative normalized harmonic function. Conversely, such a function m on $\mathfrak{Y}$ yields a sequence $(\tau_n)_{n \in \mathbb{N}}$ of traces defined by $\tau_n = \sum_{\lambda \in \mathfrak{Y}(n)} m(\lambda)\, \mathrm{ch}^\lambda$, and which is a coherent family, hence comes from a trace of $\mathfrak{S}(\infty)$. Thus:

Proposition 11.3. *There is a natural bijection between traces in* $\mathscr{T}^1(\mathfrak{S}(\infty))$*, and harmonic functions in* $\mathscr{H}^1(\mathfrak{Y})$.

This bijection being compatible with convex combinations, the extremal characters of $\mathfrak{S}(\infty)$ are in bijection with the extremal points of the compact convex set of non-negative normalized harmonic functions on $\mathfrak{Y}$. There is a nice characterization of these extremal points, which relates this problem of harmonic analysis on the Young graph to the theory of symmetric functions. Recall that a **specialization**, or **formal alphabet** for the algebra Sym is an ideal X of Sym, which yields a surjective morphism of algebras $\mathrm{Sym} \to \mathrm{Sym}/X$, denoted $f \mapsto f(X)$. This is the same notion as in Section 10.1, but with respect to Sym instead of QSym. In the sequel, we only consider numerical specializations, that is to say surjective morphisms $\mathrm{Sym} \to \mathbb{C}$. Among these numerical specializations, there are the summable sequences of complex numbers $X = \{x_1, x_2, \ldots\}$, but they are not the only numerical specializations. For instance, one can consider the so-called **exponential alphabet** E, which is the specialization of Sym defined by $p_1(E) = 1$ and $p_{k\geq 2}(E) = 0$; then, one has a unique extension of these rules that yields a morphism of algebras. This exponential alphabet sends the Schur functions to renormalized dimensions of irreducible representations:

$$s_\lambda(E) = \sum_{\mu \in \mathfrak{Y}(n)} \mathrm{ch}^\lambda(\mu) \frac{p_\mu(E)}{z_\mu} = \mathrm{ch}^\lambda(1^{|\lambda|}) \frac{1}{|\lambda|!} = \frac{\dim S^\lambda}{|\lambda|!}.$$

A (numerical) specialization X of Sym will be called non-negative if it sends any Schur function to a non-negative real number, and normalized if moreover $s_1(X) = 1$. Notice that any numerical specialization such that $s_1(X) = q \neq 0$ can be normalized, by modifying the specialization of the symmetric functions homogeneous of degree n by a factor q^{-n}.

If X is a non-negative normalized specialization of Sym, then it yields a harmonic function on $\mathfrak{Y}$, defined by $m(\lambda) = s_\lambda(X)$. Indeed, by the Pieri rule,

$$m(\lambda) = s_\lambda(X) = s_\lambda(X)\, p_1(X) = \sum_{\lambda \nearrow \Lambda} s_\Lambda(X) = \sum_{\lambda \nearrow \Lambda} m(\Lambda).$$

Theorem 11.4 (Thoma). *A non-negative normalized harmonic function on $\mathfrak{Y}$ is extremal if and only if it comes from a non-negative normalized specialization of the algebra of symmetric functions* Sym.

In order to prove Thoma's theorem 11.4, we shall relate it to a slightly more general result due to Kerov and Vershik, and known as the **ring theorem**:

Theorem 11.5 (Kerov–Vershik). *Let A be an algebra over $\mathbb{R}$, and K be a convex cone of A, that is to say a part of A that contains 0 and that is stable by multiplication by positive real numbers and by convex combinations. We denote $\leq$ the partial order on A associated to this cone: $a \leq b$ if and only if $b - a \in K$. We also assume that:*

1. *The cone K spans linearly the algebra A, and is stable by multiplication: $K \cdot K \subseteq K$.*

2. *The cone K is generated by a countable set.*

3. *The unity 1_A of the algebra has the following property: for any $a \in K$, there exists $\varepsilon > 0$ such that $1_A \geq \varepsilon a$.*

Let P be the convex set of linear forms ϕ on A that take non-negative values on K, and such that $\phi(1_A) = 1$. Then, ϕ is an extremal point of P if and only if it is multiplicative, that is to say that $\phi(ab) = \phi(a)\,\phi(b)$ for any $a, b \in A$.

Proof. Suppose that ϕ is an extremal point of P, and consider two elements a and b of A. We want to prove that $\phi(ab) = \phi(a)\,\phi(b)$. Since K spans linearly A, one can assume that b is in the cone K, and up to multiplication of b by some $\varepsilon > 0$, one can also assume that $b \leq 1$, i.e., $1 - b$ is also in K. Consider then the identity:

$$\phi(a) = \phi(ab) + \phi(a(1-b)).$$

Suppose first that $\phi(b)$ and $\phi(1-b)$ are non-zero. The identity can be rewritten as

$$\phi(a) = \phi(b)\,\phi_b(a) + \phi(1-b)\,\phi_{1-b}(a),$$

where $\phi_b(a) = \frac{\phi(ab)}{\phi(b)}$ and $\phi_{1-b}(a) = \frac{\phi(a(1-b))}{\phi(1-b)}$. These linear forms ϕ_b and ϕ_{1-b} are in P, so, since ϕ is extremal, $\phi = \phi_b = \phi_{1-b}$. Then,

$$\phi(a) = \phi_b(a) = \frac{\phi(ab)}{\phi(b)},$$

which proves the multiplicativity in this case. Now, if $\phi(b) = 0$, then for $a \in K$, $ab \in K$ and is smaller than some multiple of b: $a \leq \frac{1_A}{\varepsilon}$ for some $\varepsilon > 0$, so $ab \leq \frac{b}{\varepsilon}$. Then,

$$0 \leq \phi(ab) \leq \phi\left(\frac{b}{\varepsilon}\right) = 0,$$

so $\phi(ab) = 0 = \phi(a)\phi(b)$. The case when $\phi(1-b) = 0$ is analogous. Therefore, extremal points of P are multiplicative.

Conversely, let ϕ be a multiplicative linear form in P. Using the second property of the theorem, we can find a sequence (finite or infinite) of elements $b_1, b_2, \ldots$ of K that form a linear basis of A. Moreover, without loss of generality, we can assume that each of these elements b_i is smaller than 1_A. Then, any element $\phi \in P$ satisfies $\phi(b_i) \in [0,1]$, so, we get an injective map

$$\begin{aligned} P &\to [0,1]^{\dim A} \\ \phi &\mapsto (\phi(b_i))_{i\geq 1}. \end{aligned}$$

We endow A^* with the weak-$*$ topology. Then, P is a compact convex subset of A^*, since each $\phi(b_i)$ with $\phi \in P$ has its values restricted to a compact interval $[0,1]$. By Choquet's theorem, one can represent every point of P as an integral convex combination of extremal points, that is to say that

$$\phi = \int_{\mathrm{Extr}(P)} \phi_x\,\sigma(dx)$$

for some probability measure σ on $\mathrm{Extr}(P)$, which in general is not unique. In the following we fix such a measure σ.

For any i, denote $X_i = \phi_x(b_i)$, where $\phi_x \in \mathrm{Extr}(P)$ is taken at random according to the probability measure σ. This is a random variable with values in $[0,1]$, and the previous identity shows that

$$\mathbb{E}[X_i] = \int_{\mathrm{Extr}(P)} \phi_x(b_i)\,\sigma(dx) = \phi(b_i).$$

However, since all the linear forms ϕ and ϕ_x written are multiplicative, we also have:

$$\begin{aligned}\mathbb{E}[(X_i)^2] &= \int_{\mathrm{Extr}(P)} (\phi_x(b_i))^2\,\sigma(dx)\\ &= \int_{\mathrm{Extr}(P)} \phi_x((b_i)^2)\,\sigma(dx) = \phi((b_i)^2) = (\phi(b_i))^2 = (\mathbb{E}[X_i])^2.\end{aligned}$$

So, the variance of X_i is zero, which means that every X_i is constant σ-almost surely. Since there is a countable family of such variables, the vector

$$(X_1, X_2, \ldots)$$

is a constant vector σ-almost surely. However, this vector is the image by the injective map $P \to [0,1]^{\dim A}$ of a random linear form ϕ_x under the probability measure σ. So σ is concentrated on a single point $\phi_x \in \mathrm{Extr}(P)$, and $\phi = \phi_x$ is an extremal point. □

Proof of Theorem 11.4. Let $A = \mathrm{Sym}/(p_1 - 1)$ be the quotient of the algebra of symmetric functions by the ideal generated by the function $p_1 - 1$. To a harmonic function $m \in \mathscr{H}^1(\mathfrak{Y})$, we associate a linear form ϕ on A, defined by the formula:

$$\phi(s_\lambda) = m(\lambda).$$

This definition is compatible with the relations of the quotient: if $g = (p_1 - 1)h$ in Sym, then $\phi(g) = 0$: one can assume without loss of generality that $h = s_\lambda$ is a Schur function, and then the relation $\phi(g) = 0$ is an immediate consequence of the harmonicity of m and of the Pieri formula. We therefore have a bijection between non-negative normalized harmonic functions on $\mathfrak{Y}$, and linear forms on A that are normalized and non-negative on the images of the Schur functions by the canonical surjection $\mathrm{Sym} \to A$.

Let J be the cone in Sym which consists in the non-negative linear combinations of Schur functions, and K be its image in A by the projection map $\pi_A : \mathrm{Sym} \to A$. It is clear that J is a convex cone, and moreover, products of Schur functions are non-negative integral combinations of Schur functions (this is obvious if one uses the Frobenius–Schur isomorphism, as it is then a statement on the tensor product of irreducible representations of symmetric groups). Therefore, $J \cdot J \subseteq J$, and since Schur functions form a basis of Sym, J generates linearly Sym. Applying the morphism of algebras π_A, we get similar properties for K in A: K is a convex cone, countably spanned by the images of the Schur functions, that generates linearly A and that is stable by multiplication. Finally, in A, the unity 1_A has also the third property of Theorem 11.5. Indeed, let $a \in K$ be the image by π_A of a non-negative linear combination of Schur functions: $a = \phi(b)$ with $b = \sum_\lambda c_\lambda s_\lambda$, $c_\lambda \geq 0$. We split this sum in homogeneous components:

$$b = \sum_{n=0}^{N} \sum_{\lambda \in \mathfrak{Y}(n)} c_\lambda s_\lambda \leq C \sum_{n=0}^{N} \sum_{\lambda \in \mathfrak{Y}(n)} (\dim \lambda) s_\lambda$$

for some constant $C \geq 0$. The right-hand side of this formula is $C \sum_{n=0}^{N} (p_1)^n$, and applying the map π_A which is compatible with the partial orders on (Sym, J) and (A, K), we get

$$a = \pi_A(b) \leq C \sum_{n=0}^{N} \pi_A((p_1)^n) = CN\, 1_A.$$

Thus, by the Kerov–Vershik ring theorem, the linear form ϕ is extremal if and only if it is multiplicative, that is to say if and only if it is a non-negative normalized specialization of Sym. □

Proof of Theorem 11.2. Let $\tau \in \mathscr{T}^*(\infty)$, which is associated to a unique non-negative normalized and extremal harmonic function on $\mathfrak{Y}$. By Thoma's theorem 11.4, there exists a unique non-negative normalized specialization X_τ such that the weights of $\tau_n = \tau_{|\mathfrak{S}(n)}$ in its expansion in normalized characters are given by the formula

$$\tau_n = \sum_{\lambda \in \mathfrak{Y}(n)} \mathbb{P}_{\tau,n}[\lambda]\, \chi^\lambda, \quad \text{with} \quad \frac{\mathbb{P}_{\tau,n}[\lambda]}{\dim \lambda} = s_\lambda(X_\tau).$$

We then have, by Frobenius formula 2.32,

$$\tau(\sigma_\mu) = \sum_{\lambda \in \mathfrak{Y}(n)} s_\lambda(X_\tau)\, \mathrm{ch}^\lambda(\sigma_\mu) = p_\mu(X_\tau).$$ □

Thus, the classification of extremal characters of $\mathfrak{S}(\infty)$ is equivalent to the classification of non-negative specializations of the algebra Sym. This result allows one to build a candidate for the class of extremal characters. Indeed, if $A = \{a_1 \geq a_2 \geq \cdots \geq 0\}$ is a finite or countable alphabet of non-negative real numbers that are summable, then it yields a non-negative (not necessarily normalized) specialization $f \mapsto f(A)$ of Sym, since Schur functions are non-negative combinations of monomials labeled by semistandard tableaux (see Theorem 3.2).

On the other hand, if A is a non-negative specialization of Sym, then the specialization obtained by linear extension of the rule

$$f \mapsto (-1)^{\deg f}\, \omega(f)(A)$$

for symmetric functions that are homogeneous in degree is also non-negative, since $\omega(s_\lambda) = (-1)^{|\lambda|} s_{\lambda'}$. Therefore, given $B = \{b_1 \geq b_2 \geq \cdots \geq 0\}$ summable alphabet of non-negative real numbers, the specialization

$$p_k(\overline{B}) = (-1)^k \omega(p_k)(B) = (-1)^{k-1} p_k(B)$$

yields again a non-negative specialization. Finally, if C and D are non-negative specializations, then their sum $C + D$ associated to the morphism

$$f \mapsto \Delta(f)(C, D)$$

is again non-negative: indeed, the coproduct $\Delta(s_\lambda)$ of a Schur function is a non-negative linear combination of tensor products $s_\mu \otimes s_\rho$ (this statement is again immediate if one uses the Frobenius–Schur isomorphism 2.31, since it is then a result on the restriction of an irreducible representation of $\mathfrak{S}(p+q)$ to $\mathfrak{S}(p) \times \mathfrak{S}(q)$). As a consequence of this discussion, and of the simple observation that the exponential alphabet E is also a non-negative specialization, we get the following:

Proposition 11.6. *Let T be the set of pairs (α, β), where $\alpha = (\alpha_1 \geq \alpha_2 \geq \cdots \geq \alpha_i \geq \cdots \geq 0)$ and $\beta = (\beta_1 \geq \beta_2 \geq \cdots \geq \beta_i \geq \cdots \geq 0)$ are two non-increasing sequences of non-negative real numbers such that*

$$\sum_{i=1}^{\infty} (\alpha_i + \beta_i) = 1 - \gamma \leq 1.$$

We call T the ***Thoma simplex****. To any $(\alpha, \beta) \in T$, we associate a specialization*

$$A + \overline{B} + \gamma E$$

of Sym, *where A is the specialization associated to the non-negative sequence α, $\overline{B}$ is obtained from the specialization B associated to the non-negative sequence β by using the antipode of* Sym, *and γE is the exponential specialization defined by $p_1(\gamma E) = \gamma$ and $p_{k\geq 2}(\gamma E) = 0$. The specialization $A + \overline{B} + \gamma E$ is a non-negative normalized specialization of* Sym, *corresponding to the extremal character τ of $\mathfrak{S}(\infty)$ defined by*

$$\tau(\sigma) = \prod_{c \text{ cycle of } \sigma \text{ of length } l \geq 2} \left(\sum_{i=1}^{\infty} (\alpha_i)^l + (-1)^{l-1} (\beta_i)^l \right).$$

Proof. The fact that $A + \overline{B} + \gamma E$ is a non-negative specialization is the consequence of the previous discussion. Since $p_1(A + \overline{B} + \gamma E) = p_1(A) + p_1(\overline{B}) + \gamma = p_1(A) + p_1(B) + \gamma = 1$ by definition of γ, it is also normalized. The formula for the associated extremal character τ follows then from Theorem 11.2. ☐

Notice that two pairs (α, β) and (α', β') in T yield the same specialization if and only if they are equal. Indeed, we can consider the generating function of the specialization

$$\exp\left(\sum_{k=1}^{\infty} \frac{p_k(A+\overline{B}+\gamma E)\, z^k}{k}\right),$$

and it is given by the formula

$$\mathrm{e}^{\gamma z} \prod_{i \geq 1} \frac{1+\beta_i z}{1-\alpha_i z}.$$

The $-\frac{1}{\beta_i}$'s are the zeroes of this function, and the $\frac{1}{\alpha_i}$'s are the poles, so one can recover (α, β) from the associated specialization.

The goal of this chapter is to prove that T is in fact the set of *all* the normalized non-negative specializations of Sym, and therefore that $\mathscr{T}^*(\infty) = T$; see Theorem 11.31.

11.2 The bi-infinite symmetric group and the Olshanski semigroup

To prove that the Thoma simplex provides a complete parameterization of the extremal characters of $\mathfrak{S}(\infty)$, we shall see these extremal characters as particular examples of *admissible* representations of the bi-infinite symmetric group $\mathfrak{S}(\infty) \times \mathfrak{S}(\infty)$ (or to be more precise, the spherical functions associated to certain such admissible representations). This point of view, which is due to Olshanski and Okounkov, leads to a representation theoretic proof of Thoma's theorem 11.31, which relies on the theory of infinite Gelfand pairs. In this section, we introduce the notions of Gelfand pairs and admissible representations, and we prove that a representation of the bi-infinite symmetric group is admissible if and only if it extends to a certain semigroup Γ that contains $\mathfrak{S}(\infty) \times \mathfrak{S}(\infty)$; see Theorem 11.12.

▷ *The Gelfand pair* $(\mathfrak{S}(\infty) \times \mathfrak{S}(\infty), \mathfrak{S}(\infty))$.

At the end of the proof of Theorem 8.4, we briefly introduced the notion of a strong Gelfand pair. A weaker concept is the notion of **Gelfand pair**, which we first present for finite groups:

Definition 11.7. *A finite Gelfand pair is a pair of finite groups (G, K) such that $K \subset G$, and such that the induced representation $\mathrm{Ind}_K^G(\mathbb{C})$ from the trivial representation of K has a multiplicity free expansion in irreducible $\mathbb{C}G$-modules.*

A *strong* Gelfand pair is defined similarly, with the multiplicity free condition that holds for *any* induced representation from an irreducible representation of K. For instance, by the branching rules of Corollary 3.7, for any $n \geq 1$, $(\mathfrak{S}(n), \mathfrak{S}(n-1))$ is a strong Gelfand pair.

Proposition 11.8. *Let (G,K) be a pair of finite groups with $K \subset G$. The following conditions are equivalent:*

1. *The pair (G,K) is a Gelfand pair.*
2. *The algebra of double cosets $\mathbb{C}[K\backslash G/K]$ is commutative.*
3. *For any irreducible representation V of G, the space of K-invariant vectors V^K is at most one-dimensional.*
4. *For any finite-dimensional (unitary) representation (V,ρ) of G, if P_K is the orthogonal projection from V to V^K, then $P_K\,\rho(\mathbb{C}G)\,P_K$ is a commutative subalgebra of* $\mathrm{End}_{\mathbb{C}}(V^K)$.

Proof. The induced representation $\mathrm{Ind}_K^G(\mathbb{C}) = \mathbb{C}G \otimes_{\mathbb{C}K} \mathbb{C}$ is simply the space $\mathbb{C}[G/K]$ generated by the left cosets, on the left of which G acts naturally. Then,

$$\mathrm{End}_{\mathbb{C}G}(\mathbb{C}[G/K]) = \mathbb{C}[K\backslash G/K]$$

as explained at the end of Section 1.5 (cf. Section 5.4 for the case when $G = \mathrm{GL}(n,\mathbb{F}_q)$ and $K = \mathrm{B}(n,\mathbb{F}_q)$). On the other hand, if $V = \bigoplus_{\lambda\in\widehat{G}} m_\lambda\, V^\lambda$ is a representation of G, then

$$\mathrm{End}_{\mathbb{C}G}(V) = \bigoplus_{\lambda\in\widehat{G}} \mathrm{M}(m_\lambda,\mathbb{C})$$

so V is multiplicity free if and only if $\mathrm{End}_{\mathbb{C}G}(V)$ is commutative. Combining these two facts yields the equivalence between the first two statements.

The third statement comes then from the fact that if (V,ρ) is an irreducible representation of G, then V^K is an irreducible representation of $\mathbb{C}[K\backslash G/K]$. Indeed, we have a simple formula for the orthogonal projection P_K:

$$P_K = \frac{1}{\mathrm{card}\,K}\sum_{k\in K}\rho(k) = \rho(m_K), \quad \text{with } m_K = \frac{1}{\mathrm{card}\,K}\sum_{k\in K} k.$$

An element x of $\mathbb{C}G$ is in the algebra of double cosets if and only if $x = m_K x m_K$. If (V,ρ) is irreducible, $x \in \mathbb{C}[K\backslash G/K]$ and $v \in V^K$, then

$$x\cdot v = m_K x m_K \cdot v = m_K(x\cdot v) \in V^K,$$

so V^K is indeed a representation of $\mathbb{C}[K\backslash G/K]$. Suppose that V is irreducible, and that one has a decomposition $V^K = W \bigoplus Z$ of V^K in two non-trivial $\mathbb{C}[K\backslash G/K]$-representations. We have $V = \mathbb{C}G\cdot V^K$ by irreducibility of V, so $V = (\mathbb{C}G\cdot W)+$

$(\mathbb{C}G \cdot Z)$, which leads then to $V = \mathbb{C}G \cdot W = \mathbb{C}G \cdot Z$. In particular, if $w \in W$, there exists $x \in \mathbb{C}G$ and $z \in Z$ such that $w = x \cdot z$. In this case,

$$w = m_K \cdot w = m_K x m_K \cdot z,$$

so $w \in \mathbb{C}[K\backslash G/K] \cdot Z = Z$, which is a contradiction if $W \neq 0$ and $Z \neq 0$.

Suppose that $\mathbb{C}[K\backslash G/K]$ is a commutative algebra. Then, for any irreducible representation V of G, V^K is an irreducible representation of $\mathbb{C}[K\backslash G/K]$, and on the other hand, $\mathbb{C}[K\backslash G/K]$ is semisimple, since it is the commutant algebra of the semisimple algebra $\mathbb{C}G$ for the bimodule $\mathbb{C}[G/K]$. Therefore, as an irreducible representation of a commutative semisimple complex algebra, V^K is of dimension 0 or 1, which is the third statement of the proposition. This third statement implies the fourth: if $V = \bigoplus_{\lambda \in \widehat{G}} m_\lambda V^\lambda$ is a representation of G, then it is easily seen that $V^K = \bigoplus_{\lambda \in \widehat{G}} m_\lambda V^{\lambda,K}$ and that $P_K\,\rho(\mathbb{C}G)\,P_K$ stabilizes each 0 or 1-dimensional subspace of V^K, hence is a subalgebra of

$$\bigoplus_{\lambda \in \widehat{G}} m_\lambda \operatorname{End}_{\mathbb{C}}(V^{\lambda,K}),$$

which is commutative since each endomorphism space $\operatorname{End}_{\mathbb{C}}(V^{\lambda,K})$ is 0 or $\mathbb{C}$. Finally, the fourth statement implies the second statement, by taking V equal to the regular representation of G. □

The last part of the previous proposition is easily adapted to the case of infinite groups. Given a discrete group G (finite or infinite), we call **unitary representation** of G a pair (H, ρ), where H is a Hilbert space, $\mathscr{U}(H)$ is the group of unitary isomorphisms of H, and ρ is a morphism of groups $G \to \mathscr{U}(H)$. If G is finite, any representation of G (finite or infinite-dimensional) is conjugated to a representation by unitary morphisms (Lemma 1.4), so up to completion of the space into a Hilbert space, to a unitary representation. The theory of unitary representations is thus a natural extension to infinite-dimensional objects of the theory of finite-dimensional representations presented in this book. Here, we shall only use a few facts from this theory.

Definition 11.9. *An* ***infinite Gelfand pair*** *is a pair (G, K) with G infinite discrete group and $K \subset G$, such that for any unitary representation (H, ρ) of G, if P_K is the orthogonal projection from H to the closed subspace $H^K = \{h \in H \mid \forall k \in K,\ k \cdot h = h\}$, then the morphisms $P_K\,\rho(g)\,P_K$ with $g \in G$ generate a commutative algebra of endomorphisms of H^K.*

This definition of Gelfand pairs is the most convenient when dealing with inductive limits of finite Gelfand pairs. Let $(G(n))_{n \in \mathbb{N}}$ be an increasing sequence of finite groups, $G = G(\infty) = \bigcup_{n \in \mathbb{N}} G(n)$. If $K \subset G$, we denote $K(n) = G(n) \cap K$.

Proposition 11.10. *Suppose that $(G(n), K(n))$ is a finite Gelfand pair for any $n \in \mathbb{N}$. Then, (G, K) is a Gelfand pair.*

Proof. If G is finite, then $(G,K)=(G(n),K(n))$ for n large enough and there is nothing to prove. Suppose now that G infinite, and fix (H,ρ) unitary representation of G. Notice that $H^K=\bigcap_{n\in\mathbb{N}} H^{K(n)}$. We claim that $\lim_{n\to\infty} P_{K(n)}(h)=P_K(h)$ for any $h\in H$. First, notice that the limit $h'=\lim_{N\to\infty} P_{K(N)}(h)$ exists, because it can be written as

$$P_{K(0)}(h)+\lim_{N\to\infty}\sum_{n=1}^{N} P_{K(n)}(h)-P_{K(n-1)}(h)=\lim_{N\to\infty} P_{K(0)}(h)-\sum_{n=1}^{N} P_{K(n)^{\perp}\cap K(n-1)}(h),$$

and the series of orthogonal projections on the spaces $(H^{K(n)})^{\perp}\cap H^{K(n-1)}$ is a series bounded in norm. Now, if $n\leq N$, then $P_{K(n)}\circ P_{K(N)}=P_{K(N)}\circ P_{K(n)}=P_{K(N)}$, so for any n,

$$P_{K(n)}(h')=\lim_{N\to\infty}(P_{K(n)}\circ P_{K(N)})(h)=\lim_{N\to\infty} P_{K(N)}(h)=h'.$$

As a consequence, h' is in $H^{K(n)}$ for any n, so $h'\in H^K$ and $P_K(h')=h'$. Since the limit of operators $\lim_{n\to\infty} P_{K(n)}$ is an idempotent self-adjoint operator, this implies that it is an orthogonal projection on a subspace of K, and conversely, if $h\in K$, then $\lim_{n\to\infty} P_{K(n)}(h)=\lim_{n\to\infty} h=h$, so this orthogonal projection is indeed P_K. Now, if g_1 and g_2 are in G, then for n large enough they are in $G(n)$, and on the other hand,

$$P_K\rho(g_1)P_K\times P_K\rho(g_2)P_K=\lim_{n\to\infty} P_{K(n)}\rho(g_1)P_{K(n)}\times P_{K(n)}\rho(g_2)P_{K(n)}$$

in the sense of pointwise convergence, since all the projection operators written are bounded in norm by 1. Take some n large enough, and $h\in H$. The space $\mathbb{C}G(n)\cdot h$ is a finite-dimensional representation of $G(n)$, hence, since $(G(n),K(n))$ is a Gelfand pair,

$$\big(P_{K(n)}\rho(g_1)P_{K(n)}\times P_{K(n)}\rho(g_2)P_{K(n)}\big)(h)=\big(P_{K(n)}\rho(g_2)P_{K(n)}\times P_{K(n)}\rho(g_1)P_{K(n)}\big)(h).$$

We can take the limit as n goes to infinity, and this ends the proof. □

Corollary 11.11. *The pair $(\mathfrak{S}(\infty)\times\mathfrak{S}(\infty),\mathfrak{S}(\infty))$, where the group $K=\mathfrak{S}(\infty)$ is embedded diagonally in $G=\mathfrak{S}(\infty)\times\mathfrak{S}(\infty)$, is a Gelfand pair.*

Proof. By the previous proposition, it suffices to show that $(\mathfrak{S}(n)\times\mathfrak{S}(n),\mathfrak{S}(n))$ is a finite Gelfand pair for any $n\in\mathbb{N}$. However, for any finite group G, $\mathbb{C}[G\backslash G\times G/G]$ is the same algebra as the center $Z(\mathbb{C}G)$ of the group algebra, the isomorphism being

$$\sum_{g,h\in G} c_{g,h}\,(g,h)\mapsto\sum_{g,h\in G} c_{g,h}\,gh^{-1}.$$

This center is of course commutative, so $(G\times G,G)$ is a Gelfand pair for any finite group. □

▷ *Brauer graphs and the Olshanski semigroup.*

In the following, we denote $\mathfrak{S}_n(\infty)$ the subgroup of $\mathfrak{S}(\infty)$ that consists in finite permutations that fix the n first integers; it commutes with $\mathfrak{S}(n)$, and $\bigcap_{n\in\mathbb{N}} \mathfrak{S}_n(\infty) = \{\mathrm{id}_{\mathbb{N}^*}\}$. We shall sometimes replace the notations of symmetric and bi-symmetric groups by the letters K and G: thus,

$$\begin{aligned} G &= \mathfrak{S}(\infty) \times \mathfrak{S}(\infty) \quad ; \quad G(n) = \mathfrak{S}(n) \times \mathfrak{S}(n) \quad ; \\ K &= \mathfrak{S}(\infty) \quad ; \quad K(n) = \mathfrak{S}(n) \quad ; \\ K_n &= \mathfrak{S}_n(\infty) \quad ; \quad K_n(N) = \mathfrak{S}_n(\infty) \cap \mathfrak{S}(N). \end{aligned}$$

We also convene that the bi-infinite symmetric group $\mathfrak{S}(\infty) \times \mathfrak{S}(\infty)$ acts on $-\mathbb{N}^* \sqcup \mathbb{N}^*$, the first factor acting on negative integers and the second factor on positive integers. On the other hand, given the Gelfand pair (G,K) with $G = \mathfrak{S}(\infty) \times \mathfrak{S}(\infty)$ and $K = \mathfrak{S}(\infty)$, we denote simply k an element (k,k) of G that is diagonal; there is really no possible ambiguity coming from this notation.

A unitary representation H of the group $G = \mathfrak{S}(\infty) \times \mathfrak{S}(\infty)$ will be called **admissible** if

$$H = \overline{\bigcup_{n\in\mathbb{N}} H^{\mathfrak{S}_n(\infty)}},$$

that is to say that any vector $h \in H$ can be approximated in norm by $\mathfrak{S}_n(\infty)$-invariant vectors. We then also say that H, viewed as a unitary representation of $K = \mathfrak{S}(\infty)$, is **tame**. The main goal of this section is to determine all the admissible representations of G. To this purpose, we shall prove that these representations extend naturally to some semigroup $\Gamma(\infty)$, which can be defined abstractly as the limit of the sets of double cosets $K_n\backslash G/K_n$, and which has also a beautiful combinatorial description.

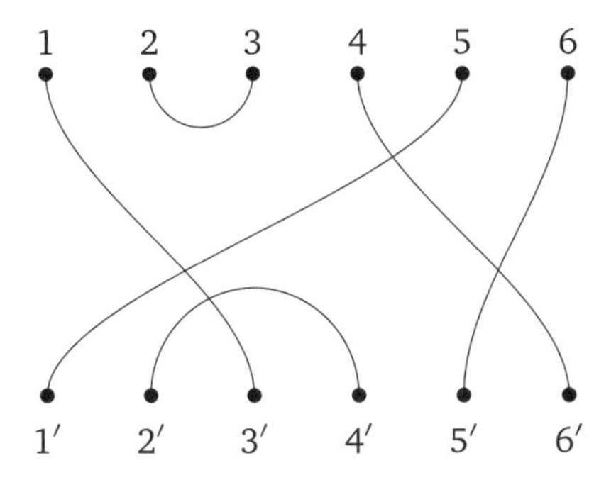

Figure 11.2
A Brauer graph of size 6.

The combinatorial construction of $\Gamma(\infty)$ is a generalization of a classical construction due to Brauer. The **Brauer semigroup** $\mathfrak{B}(n)$ is the set of simple graphs with vertex set $V = [\![1,n]\!] \sqcup [\![1',n']\!]$, and where each vertex has exactly one neighbor. We refer to Figure 11.2 for an example in size 6. For any $n \in \mathbb{N}$, the set $\mathfrak{B}(n)$ has cardinality $(2n-1)!! = (2n-1)(2n-3)\cdots 3\,1$, and it is endowed with a product which is defined by gluing together two Brauer graphs and forgetting the loops thus formed. More precisely, let π and π' be two Brauer graphs, with the vertices of π denoted $1,2,\ldots,n,1',\ldots,n'$, and the vertices of π' denoted $1',\ldots,n',1'',\ldots,n''$. Then, the links of the product of graphs $\pi \times \pi'$ connect i and j in $[\![1,n]\!] \sqcup [\![1'',n'']\!]$ if there exists a sequence $(i_0,i_1),(i_1,i_2),\ldots,(i_{k-1},i_k)$ of links of π or π' such that $i_0 = i$ and $i_k = j$. We refer to Figure 11.3 for an example of calculation of a product.

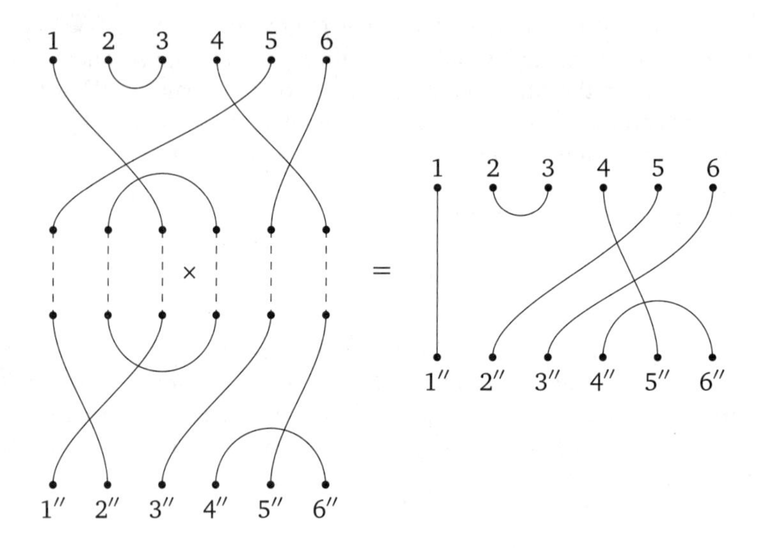

Figure 11.3
The product of two Brauer graphs of size 6.

Notice that one has a canonical embedding of $\mathfrak{S}(n)$ into $\mathfrak{B}(n)$: one associates to a permutation σ the graph which connects i to $(\sigma(i))'$ for any $i \in [\![1,n]\!]$. Then, the restriction of the product of $\mathfrak{B}(n)$ to $\mathfrak{S}(n)$ is *the opposite of* the composition product of permutations. In the remainder of this section, we shall therefore reverse the product of the symmetric groups, and denote $\sigma\tau = \tau \circ \sigma$ for two permutations σ and τ; the reader can check that this modification is purely notational and does not impact the end result (Theorem 11.24, which is the classification of admissible representations of the pair (G,K)).

A **generalized Brauer graph** is a Brauer graph with a non-negative real number associated to any edge of the graph, plus a finite collection of other non-negative real numbers, which we see as numbered loops not connected to any vertex $i \in [\![1,n]\!]$. The product of generalized Brauer graphs is then defined similarly as above, the labels being summed when connecting two links of generalized

Brauer graphs π and π'. With this new product, one does not forget the loops created when computing the product, and one keeps them as numbered loops. For instance, if one takes the two generalized Brauer graphs

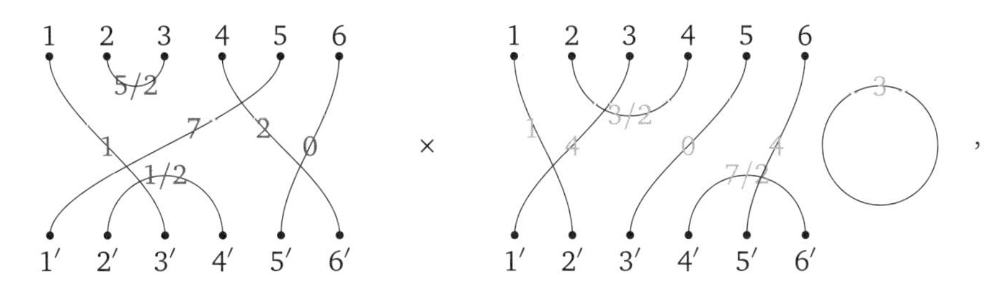

then there is a new loop of label $\frac{1}{2}+\frac{3}{2}=2$ in their product, which is

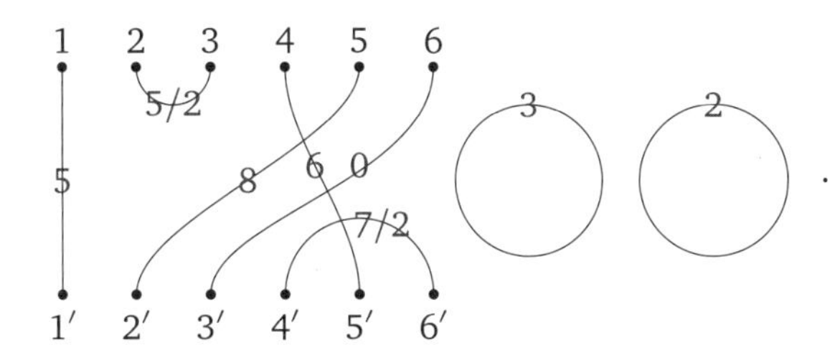

Denote $\widetilde{\mathfrak{B}}(n)$ the semigroup that consists in generalized Brauer graphs π of size n and such that

1. the horizontal edges of π (those that connect i to j with $i,j \in [\![1,n]\!]$ or $i',j' \in [\![1',n']\!]$) have their labels in $\mathbb{N}+\frac{1}{2}=\{\frac{1}{2},\frac{3}{2},\ldots\}$;

2. the vertical edges of π (those that connect $i \in [\![1,n]\!]$ to $j \in [\![1',n']\!]$) have their labels in $\mathbb{N}$;

3. the loops of π have their labels in $\mathbb{N}^* = \{1,2,3,\ldots\}$.

It is easily seen that $\widetilde{\mathfrak{B}}(n)$ is stable by product, and this semigroup admits a projection morphism towards the Brauer semigroup $\mathfrak{B}(n)$, which consists in forgetting the labels and the loops. In the sequel, we shall work with $\widetilde{\mathfrak{B}}(n,n)$, which is the sub-semigroup of

$$\widetilde{\mathfrak{B}}(2n) = \widetilde{\mathfrak{B}}([\![-n,-1]\!] \sqcup [\![1,n]\!])$$

that consists in generalized Brauer graphs whose edges are between $[\![-n',-1']\!] \sqcup [\![1,n]\!]$ and $[\![1',n']\!] \sqcup [\![-n,-1]\!]$. In other words, if $\pi \in \widetilde{\mathfrak{B}}(n,n)$, then its horizontal edges are between elements of $[\![1,n]\!]$ and elements of $[\![-n,-1]\!]$, and between elements of $[\![1',n']\!]$ and elements of $[\![-n',-1']\!]$; and its vertical edges do not connect positive elements to negative elements. It will also be convenient to forget the loops of label 1, that is to say that we introduce the quotient $\Gamma(n)$ of $\widetilde{\mathfrak{B}}(n,n)$ by the relation $C_1 = 1$, with

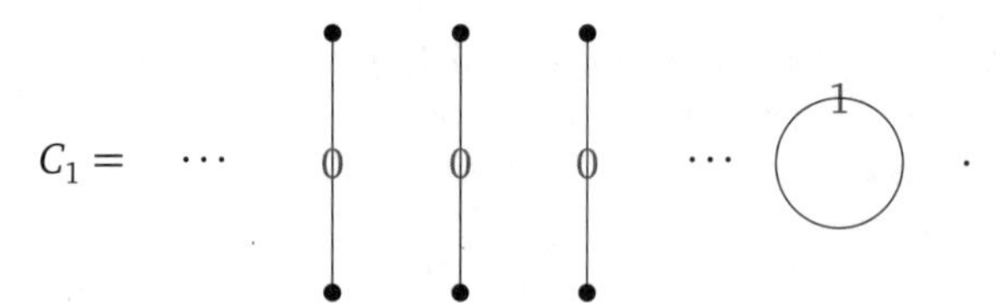

The semigroup $\widetilde{\mathfrak{B}}(n,n)$ admits the following generators:

1. the diagrams of permutations, which have all their edges between $[\![1,n]\!]$ and $[\![1',n']\!]$ and between $[\![-1,-n]\!]$ and $[\![-1',-n']\!]$, and of label 0; they form a group isomorphic to $\mathfrak{S}(n) \times \mathfrak{S}(n)$.

2. the diagrams

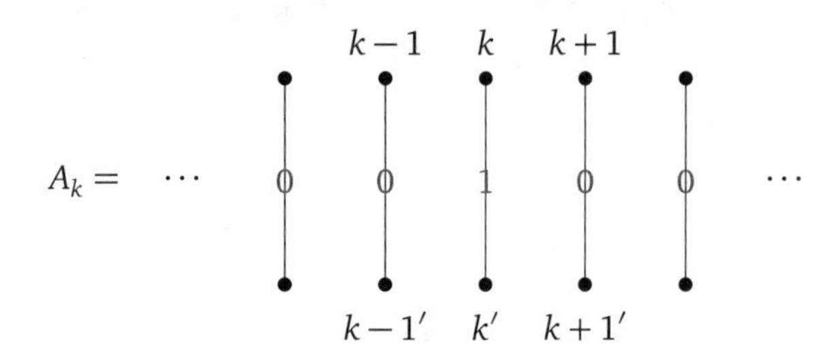

   with $k \in [\![1,n]\!]$ or $k \in [\![-n,-1]\!]$. Actually, A_k can be obtained from A_1 by mutiplying by permutation diagrams, so we only need the generators A_1 and A_{-1}.

3. the loop diagrams

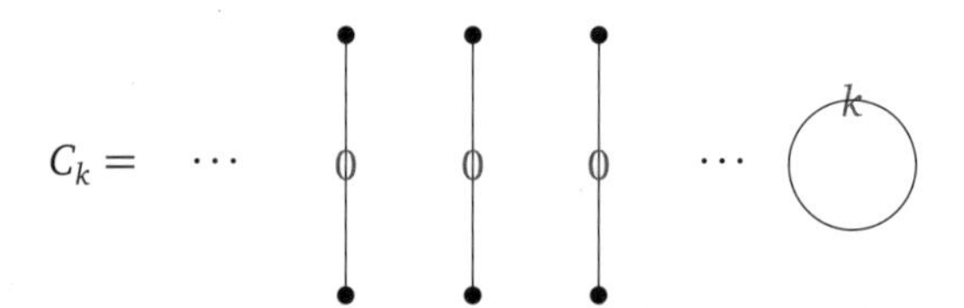

   with $k \in \mathbb{N}^*$.

4. and finally, the horizontal diagrams

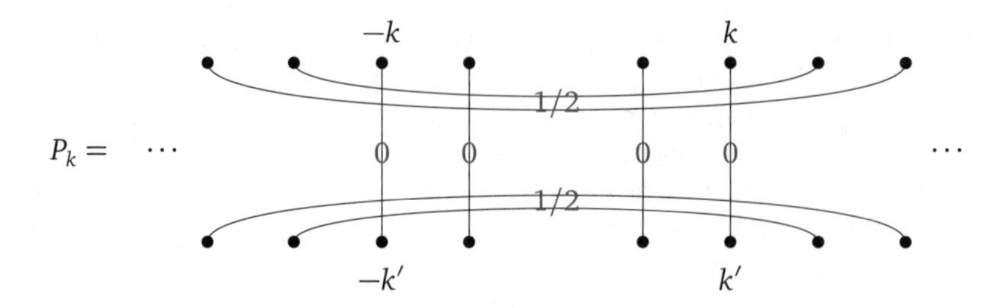

with $k \geq 0$.

Indeed, one can use permutations in $\mathfrak{S}(n) \times \mathfrak{S}(n)$ to transform a graph P_k into a graph in $\widetilde{\mathfrak{B}}(n,n)$ with $2k$ vertical edges, which are prescribed as well as the horizontal edges. Multiplying then by diagrams A_k allows one to put the correct labels on the edges, and multiplying by diagrams C_k allows one to create the correctly labeled loops. So, $\widetilde{\mathfrak{B}}(n,n) = \langle \mathfrak{S}(n) \times \mathfrak{S}(n), A_1, A_{-1}, C_{k\geq 1}, P_{k\geq 0}\rangle$, and

$$\Gamma(n) = \langle \mathfrak{S}(n) \times \mathfrak{S}(n), A_1, A_{-1}, C_{k\geq 1}, P_{k\in [\![0,n-1]\!]}\rangle / (C_1 = 1).$$

We construct $\Gamma(\infty)$ similarly, by introducing the same generators and setting

$$\Gamma(\infty) = \langle \mathfrak{S}(\infty) \times \mathfrak{S}(\infty), A_1, A_{-1}, C_{k\geq 1}, P_{k\geq 0}\rangle / (C_1 = 1).$$

In this definition, the diagrams $P_{k\geq 0}$ have now an infinity of horizontal edges, and $2k$ vertical edges. The multiplication is well defined in $\Gamma(\infty)$, so we get a semigroup which extends $G = \mathfrak{S}(\infty) \times \mathfrak{S}(\infty)$. The semigroup $\Gamma(\infty)$ is endowed with an involution $*$, which consists in reversing horizontally the diagrams (we send i to i' and i' to i). In terms of the generators,

$$\sigma^* = \sigma^{-1} \quad ; \quad (A_k)^* = A_k \quad ; \quad (C_k)^* = C_k \quad ; \quad (P_k)^* = P_k.$$

▷ *The extension theorem for admissible representations.*

We call $\Gamma = \Gamma(\infty)$ the **Olshanski semigroup**. Its interest in the classification of admissible representations of the pair (G,K) lies in the following:

Theorem 11.12 (Olshanski). *Let (H,ρ) be an admissible representation of (G,K). There exists an extension of ρ to Γ, which is a $*$-representation of this semigroup by contractions (operators of norm smaller than 1), and which sends P_n to the orthogonal projection on H^{K_n}.*

In order to prove Theorem 11.12, we shall define directly $\rho(A_k)$, $\rho(C_k)$ and $\rho(P_k)$ for any admissible representation of (G,K), and show that these definitions satisfy sufficiently many relations so that their extension to the whole semigroup Γ is not ambiguous. We start with the second part, which amounts to giving a presentation of the semigroup Γ. If $\gamma \in \Gamma$, we denote γ_0 the underlying graph with vertex set $\mathbb{N}^* \sqcup (\mathbb{N}^*)' \sqcup (-\mathbb{N}^*) \sqcup (-\mathbb{N}^*)'$. It is obtained by forgetting the labels and the loops of γ.

Lemma 11.13. *We convene in the following that P_∞ is the diagram of the pair of permutations $(\mathrm{id}_{\mathbb{N}^*}, \mathrm{id}_{\mathbb{N}^*})$ (the diagram with only vertical edges connecting i to i' for any i). For any graph γ_0 underlying a generalized Brauer graph $\gamma \in \Gamma$, there exists a triple $(\sigma, \tau, n) \in G \times G \times (\mathbb{N} \sqcup \{\infty\})$ such that*

$$\gamma_0 = (\sigma P_n \tau)_0.$$

In this expression, n is entirely determined by γ_0, and then, if (σ,τ) and (σ',τ') are *two solutions of the previous equation, there exists $u \in \mathfrak{S}(n) \times \mathfrak{S}(n)$ and $s_n, t_n \in \mathfrak{S}_n(\infty)$ such that $\sigma' = \sigma s_n u$ and $\tau' = u^{-1} t_n \tau$.*

Proof. If γ belongs to Γ, then:

- either γ does not involve any term $P_{k\geq 0}$ in its expression as a product of generators. Then, it is just a permutation $\sigma \in G$ decorated by labels and loops, and $\gamma_0 = (\sigma)_0$.
- or, γ involves some terms $P_{k\geq 0}$. When one multiplies P_k by any other generator of Γ, one obtains a generalized Brauer graph whose underlying graph γ_0 has an infinity of horizontal links $(i,-i)$ and $(i',-i')$ for i large enough. In particular, for all $\gamma \in \Gamma$, if γ_0 is not the diagram of a permutation, then it has a finite number $2n$ of vertical edges. We can then realize such a diagram by mutiplying the diagram P_n by permutations, hence the existence of an expression $\gamma_0 = (\sigma P_n \tau)_0$.

As an example of the second case, consider the diagram γ_0

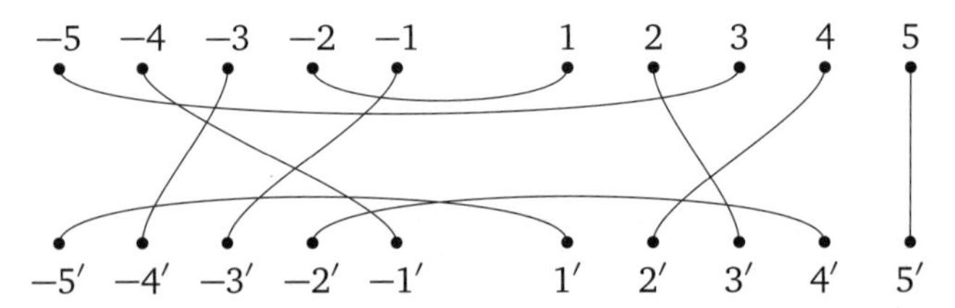

completed by horizontal edges $(i,-i)$ and $(i',-i')$ for $|i| > 5$. It has $6 = 2 \times 3$ vertical edges, hence, there exists a factorization $\gamma_0 = (\sigma\, P_3\, \tau)_0$ with σ and τ permutations. One can check that indeed, γ_0 is the diagram underlying the product

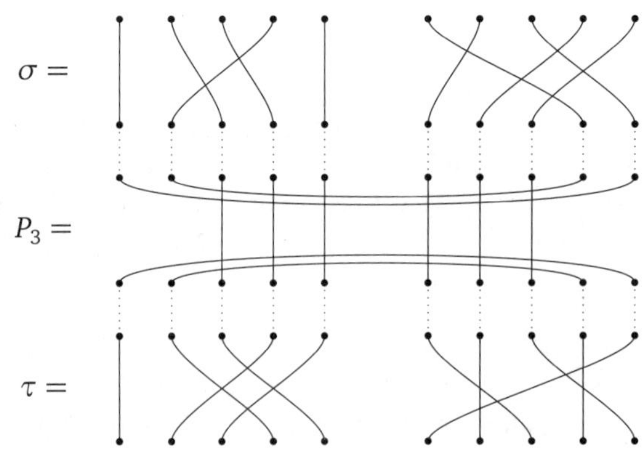

where one omits to draw the horizontal edges $(i,-i)$ or $(i',-i')$ for $|i| > 5$.

Suppose now that $\sigma P_n \tau = \sigma' P_n \tau'$. To solve this equation, one can multiply on both sides by permutations to reduce the problem to $P_n = sP_n t$ with $s, t \in G$. Let us analyze this formula:

1. Suppose $|s(i)| > n$. Then, in the product $sP_n t$, i is connected successively to $s(i)'$, $(-s(i))'$, and finally $s^{-1}(-s(i))$. So, $(i, s^{-1}(-s(i)))$ is a horizontal edge of P_n, hence $|i| > n$ and $s^{-1}(-s(i)) = -i$, or $-s(i) = s(-i)$. It follows that s sends every i with $|i| > n$ to an image $s(i)$ with $|s(i)| > n$, and in such a way that $-s(i) = s(-i)$. Thus,
$$s = s_n s',$$
where $s' \in \mathfrak{S}(n) \times \mathfrak{S}(n)$ and $s_n \in \mathfrak{S}_n(\infty)$.

2. For the same reasons, one has a factorization $t = t' t_n$ with $t' \in \mathfrak{S}(n) \times \mathfrak{S}(n)$ and $t_n \in \mathfrak{S}_n(\infty)$.

3. If $|i| \leq n$, then i is connected by P_n to i', and by $sP_n t$ to $(s't'(i))'$, so $s't' = \mathrm{id}_{\mathbb{N}^*}$.

Conversely, any pair $(s = s_n u, t = u^{-1} t_n)$ with $u \in \mathfrak{S}(n) \times \mathfrak{S}(n)$ and $s_n, t_n \in \mathfrak{S}_n(\infty)$ stabilizes P_n, so the equation is solved. This proves the second part of the lemma when $n \neq \infty$, and this last case is trivial. □

In the following, given a pair (i, j) of positive or of negative integers, we denote $t_{i,j}$ the element of G that exchanges i and j in $-\mathbb{N}^* \sqcup \mathbb{N}^*$. Beware that it is not the same element as $(i, j) \in K \hookrightarrow G$: $(i, j) = t_{-i,-j}\, t_{i,j}$ with our convention of writing k for a diagonal element (k, k) of $K \subset G$.

Proposition 11.14. *Let $\gamma \in \Gamma$. There exists a canonical form:*
$$\gamma = \prod_{k \geq 2} (C_k)^{d_k} \times \prod_{k \in \mathbb{N}^* \sqcup -\mathbb{N}^*} (A_k)^{e_k} \times (\sigma \times P_n \times \tau) \times \prod_{k \in \mathbb{N}^* \sqcup -\mathbb{N}^*} (A_k)^{f_k}$$
with the exponents d_k, e_k, f_k non-negative integers, almost all of them being 0; $n \in \mathbb{N} \sqcup \{\infty\}$; and $\sigma, \tau \in G$. One can assume in this expression that $e_k = 0$ whenever k is not the right-most vertex of an upper horizontal edge of the diagram $\sigma P_n \tau$, and that $f_k = 0$ whenever $k' < 0$ is the left-most vertex of a lower horizontal edge of $\sigma P_n \tau$. This canonical form can then be obtained from any product of generators of Γ by using only the following list of relations:
$$\begin{aligned}
&\forall k \geq 2,\ \forall \gamma \in \Gamma,\ C_k \gamma = \gamma C_k;\\
&\forall i, j,\ A_i A_j = A_j A_i;\\
&\forall i,\ \forall \sigma \in G,\ \sigma A_{\sigma(i)} = A_i \sigma;\\
&\forall |i| \leq n,\ P_n A_i = A_i P_n;\\
&\forall |i| > n,\ P_n A_i = P_n A_{-i} \text{ and } A_i P_n = A_{-i} P_n;\\
&\forall m, n,\ P_m P_n = P_{\min(m,n)};\\
&\forall \sigma \in \mathfrak{S}(n) \times \mathfrak{S}(n),\ P_n \sigma = \sigma P_n;\\
&\forall i, j \text{ with } |i| \leq n < |j|,\ P_n t_{i,j} P_n = P_n A_i;
\end{aligned}$$

$$\forall i,j \text{ with } |i|,|j|>n,\ P_n\, t_{i,j}\, P_n = P_n\, C_2;$$

$$\forall i_1 \neq \cdots \neq i_r \text{ with } |i_j|>n,\ \forall k_1,\ldots,k_r,\ P_n A_{i_1}^{k_1} A_{i_2}^{k_2} \cdots A_{i_r}^{k_r}\, P_n = P_n \prod_{j=1}^{r} C_{k_j+1}.$$

Moreover, this canonical form is unique, up to the transformations $(\sigma,\tau)\mapsto(\sigma',\tau')$ *that are described in Lemma 11.13.*

Proof. The relations stated are easily shown graphically, and we leave this verification to the reader. On the other hand, the existence of a canonical form is clear: one first creates the graph γ_0 underlying γ by taking a product $\sigma P_n \tau$; then one multiplies on both sides by factors A_k to distribute the labels to the edges; and finally one adds loops C_k. In the distribution of labels, one can always add a label to a horizontal edge $(-j,i)$ or $(-j',i')$ by multiplying by A_i (and not A_{-j}) on the left or on the right, whence the possible restriction stated for the integers e_k and f_k. The unicity of the canonical form is then obvious, since $\sigma P_n \tau$ is determined by γ_0, and the parameters d_k, e_k, f_k are determined by the labels and loops of γ. It remains to see that one can obtain the canonical form by using only relations from the list. Consider a product of permutations and elements A_k, C_k and P_k. Since the loops C_k commute with every element of Γ, one can indeed extract them and put them in front of the product. What remains is a sequence of terms

$$P_{n_1} X_1 P_{n_2} X_2 \cdots P_{n_l} X_l P_{n_{l+1}},$$

where each X_i can be written as $A_{i_1}^{k_1} A_{i_2}^{k_2} \cdots A_{i_r}^{k_r} \sigma$ with the i_j's distinct, and $\sigma \in G$ (here we use the commutation relations between the A_i's and the permutations). Actually, we can assume that each X_l is either a product of A_i's, or a permutation. Indeed, if $\sigma \in \mathfrak{S}(m)\times\mathfrak{S}(m)$, then

$$\sigma P_n = \sigma P_{\max(m,n)}\, P_n = P_{\max(m,n)} \sigma P_n,$$

so one can insert terms P_k to reduce the problem to this case. For the same reason, since σ writes as a product of permutations $t_{i,j}$, we can assume that each permutation that appears between two projections P_{n_j} and $P_{n_{j+1}}$ is a transposition t. We now use an induction on l, and transform the last factor $\Pi = P_{n_l} X_l P_{n_{l+1}}$, which we rewrite for simplicity $P_m A_{i_1}^{k_1} A_{i_2}^{k_2} \cdots A_{i_r}^{k_r}\, P_n$ or $P_m\, t\, P_n$.

1. Suppose $m<n$ and $\Pi = P_m A_{i_1}^{k_1} A_{i_2}^{k_2} \cdots A_{i_r}^{k_r}\, P_n$. We can assume that the indices i_j are ordered so that $|i_j|\leq m$ if $j\leq p$, $m<|i_j|\leq n$ if $p+1\leq j\leq q$, and $|i_j|>n$ if $j\geq q+1$, with $p\leq q\leq r$. Then,

$$\begin{aligned}\Pi &= A_{i_1}^{k_1} \cdots A_{i_p}^{k_p}\, P_m\, P_n A_{i_{p+1}}^{k_{p+1}} \cdots A_{i_r}^{k_r}\, P_n \\ &= \left(\prod_{j=1}^{p} A_{i_j}^{k_j}\right) P_m\, P_n A_{i_{q+1}}^{k_{q+1}} \cdots A_{i_r}^{k_r}\, P_n \left(\prod_{j=p+1}^{q} A_{i_j}^{k_j}\right)\end{aligned}$$

$$= \left(\prod_{j=q+1}^{r} C_{k_j+1} \right) \left(\prod_{j=1}^{p} A_{i_j}^{k_j} \right) P_m \left(\prod_{j=p+1}^{q} A_{i_j}^{k_j} \right).$$

We can apply the same kind of transformations if $m \geq n$, as can be seen by using for instance the involution $*$ of Γ. By the induction hypothesis, the remainder of the product $\Pi_l = P_{n_1} X_1 P_{n_2} X_2 \cdots P_{n_l}$ can be rewritten as a canonical form. Thus, the relations of the list can be used to obtain the form

$$\gamma = \prod_{k \geq 2} (C_k)^{d_k} \times \prod_{k \in \mathbb{N}^* \sqcup -\mathbb{N}^*} (A_k)^{e_k} \times (\sigma \times P_l \times \tau) \times \prod_{k \in \mathbb{N}^* \sqcup -\mathbb{N}^*} (A_k)^{f_k} \times P_m \times \prod_{k \in \mathbb{N}^* \sqcup -\mathbb{N}^*} (A_k)^{g_k}$$

for some coefficients d_k, e_k, f_k and g_k. We shall explain in a moment how to transform the middle term $P_l \times \tau \times \prod_{k \in \mathbb{N}^* \sqcup -\mathbb{N}^*} (A_k)^{f_k} \times P_m$ into a product with just one projection P_n.

2. On the other hand, if $\Pi = P_m \, t \, P_n$ with $t = t_{i,j}$ and say $m < n$, then one can use the relations of the list to transform the product Π into $P_m C_2$, $P_m A_i$ or $P_m t_{i,j}$, according to the order of $|i|$, $|j|$, m and n. The two first cases lead to an expression of γ of the same kind as the one of the first item. In the third case, we get a similar expression times $t_{i,j}$, and this expression of γ will also be reduced to a canonical form by the last item of our list of transformations. The case $m \geq n$ is entirely similar.

3. We finally have to reduce a product

$$\Theta = P_l \times \tau \times \prod_{k \in \mathbb{N}^* \sqcup -\mathbb{N}^*} (A_k)^{f_k} \times P_m$$

into a product with just one projection, and permutations σ and terms A_k on the left and on the right of this projection. However, this is exactly the same problem as in the first two items, so one can indeed obtain a canonical form for γ by using only the relations of the list.

Thus, there is an algorithm that yields a canonical form for γ, but possibly without the restriction on the indices e_k and f_k. This restriction can in turn be obtained by using the relations of the list that are satisfied by the A_k's and the P_k's (in particular, $A_i P_n = A_{-i} P_n$ if $|i| > n$). □

Fix now (H, ρ) admissible representation of the pair (G, K). For $k \geq 1$, we define

$$\rho(A_k) = \lim_{n \to \infty} \rho(t_{k,n}),$$

where the limit exists with respect to the weak operator topology, that is to say that $\langle h_1 \mid t_{k,n} \cdot h_2 \rangle$ converges for any $h_1, h_2 \in H$. To prove the existence of these limits, suppose first that h_1, h_2 belong to H^{K_m} for some m. Then, for $|n_1|, |n_2| \geq m$, $(n_1, n_2) = t_{n_1,n_2} t_{-n_1,-n_2} \in K_m$, and on the other hand, $t_{k,n_2} = (n_1, n_2) t_{k,n_1} (n_1, n_2)$, so

$$\begin{aligned} \langle h_1 \mid t_{k,n_2} \cdot h_2 \rangle &= \langle h_1 \mid (n_1, n_2) t_{k,n_1} (n_1, n_2) \cdot h_2 \rangle \\ &= \langle (n_1, n_2) \cdot h_1 \mid t_{k,n_1} (n_1, n_2) \cdot h_2 \rangle = \langle h_1 \mid t_{k,n_1} \cdot h_2 \rangle. \end{aligned}$$

Therefore, the sequence is stationary and its limit exists. Suppose now that h_1 and h_2 are general elements of H, and fix $\varepsilon > 0$. By admissibility, there exists $\widetilde{h_1}$ and $\widetilde{h_2}$ in some space H^{K_m} such that $\|h_1 - \widetilde{h_1}\| \leq \varepsilon$ and $\|h_2 - \widetilde{h_2}\| \leq \varepsilon$. Then, since all the operators $\rho(t_{k,n})$ are unitary symmetries, they all have an operator norm smaller than 1, hence,

$$\begin{aligned}\left|\left\langle h_1 \,\middle|\, t_{k,n} \cdot h_2\right\rangle - \left\langle \widetilde{h_1} \,\middle|\, t_{k,n} \cdot \widetilde{h_2}\right\rangle\right| &\leq \|h_1\| \left\|h_2 - \widetilde{h_2}\right\| + \left\|\widetilde{h_2}\right\| \left\|h_1 - \widetilde{h_1}\right\| \\ &\leq \varepsilon(\|h_1\| + \|h_2\| + \varepsilon).\end{aligned}$$

Since the sequence $(\langle \widetilde{h_1} \mid t_{k,n} \cdot \widetilde{h_2} \rangle)_{n\in\mathbb{N}}$ is convergent, for $n_1, n_2 \geq N$,

$$\left|\left\langle \widetilde{h_1} \,\middle|\, t_{k,n} \cdot \widetilde{h_2}\right\rangle - \left\langle \widetilde{h_1} \,\middle|\, t_{k,n} \cdot \widetilde{h_2}\right\rangle\right| \leq \varepsilon,$$

and combining this with the previous inequality yields

$$\left|\left\langle h_1 \mid t_{k,n_1} \cdot h_2\right\rangle - \left\langle h_1 \mid t_{k,n_2} \cdot h_2\right\rangle\right| \leq \varepsilon(2\|h_1\| + 2\|h_2\| + 2\varepsilon + 1).$$

Therefore, $(\langle h_1 \mid t_{k,n} \cdot h_2 \rangle)_{n\in\mathbb{N}}$ is a Cauchy sequence and is again convergent.

We define similarly $\rho(A_{-k}) = \lim_{n\to\infty} \rho(t_{-k,-n})$, and

$$\rho(C_k) = \lim_{n_1 \neq n_2 \neq \cdots \neq n_k \to \infty} \rho(\mathrm{id}_{\mathbb{N}^*}, (n_1, n_2, \ldots, n_k)),$$

all the limits being taken in the weak operator topology. A simple use of the density of $\bigcup_{n\in\mathbb{N}^*} H^{K_n}$ in H shows that we also have

$$\rho(C_k) = \lim_{n_1 \neq n_2 \neq \cdots \neq n_k \to \infty} \rho((n_1, n_2, \ldots, n_k), \mathrm{id}_{\mathbb{N}^*}).$$

On the other hand, since all the operators considered are of operator norm smaller than 1, we have $\|\rho(A_k)\|_{\mathscr{B}(H)} \leq 1$ and $\|\rho(C_k)\|_{\mathscr{B}(H)} \leq 1$ for any k, and moreover, $\rho(A_k)^* = \rho(A_k)$ and $\rho(C_k)^* = \rho(C_k)$. We finally define $\rho(P_m)$ as the orthogonal projection on the space H^{K_m} for any $m \geq 0$; again, $\|\rho(P_k)\|_{\mathscr{B}(H)} \leq 1$, and $\rho(P_k)^* = \rho(P_k)$ for any k.

Lemma 11.15. *The orthogonal projection $\rho(P_m)$ is also the limit in the weak operator topology*

$$\lim_{n\to\infty} \frac{1}{(n-m)!} \sum_{\sigma \in K_m \cap K(n)} \rho(\sigma).$$

Proof. Denote $K_m(n) = K_m \cap K(n) = \{\sigma \in \mathfrak{S}(n) \mid \forall i \leq m,\ \sigma(i) = i\}$, and

$$T_{m,n} = \frac{1}{(n-m)!} \sum_{\sigma \in K_m(n)} \rho(\sigma),$$

which is a self-adjoint operator on H of operator norm smaller than 1, since it is a barycenter of such operators. We check that $(T_{m,n})^2 = T_{m,n}$:

$$(T_{m,n})^2 = \frac{1}{((n-m)!)^2} \sum_{\sigma,\tau \in K_m(n)} \rho(\sigma\tau) = \frac{1}{((n-m)!)^2} \sum_{\phi,\tau \in K_m(n)} \rho(\phi)$$
$$= \frac{1}{(n-m)!} \sum_{\phi \in K_m(n)} \rho(\phi) = T_{m,n}.$$

Therefore, $T_{m,n}$ is an orthogonal projection, and it is in fact the orthogonal projection on $H^{K_m(n)}$. Since $H^{K_m} = \bigcap_{n\in\mathbb{N}} H^{K_m(n)}$, the result follows immediately, by the same argument as in the proof of Proposition 11.10. □

Proof of Theorem 11.12. If $\gamma \in \Gamma$, we can define $\rho(\gamma)$ by taking its canonical form as in Proposition 11.14, and setting

$$\rho(\gamma) = \prod_{k\geq 2} (\rho(C_k))^{d_k} \prod_{k\in\mathbb{N}^*\sqcup -\mathbb{N}^*} (\rho(A_k))^{e_k} \, (\rho(\sigma)\rho(P_n)\rho(\tau)) \prod_{k\in\mathbb{N}^*\sqcup -\mathbb{N}^*} (\rho(A_k))^{f_k},$$

all the terms of the product being now well defined, and contractions. Let us first see why this definition does not depend on the choice of the pair (σ,τ) such that $(\sigma\, P_n\, \tau)_0$ is the diagram γ_0 underlying γ. If $\sigma\, P_n\, \tau = \sigma'\, P_n\, \tau'$, we know from a previous lemma that there exists $u \in \mathfrak{S}(n) \times \mathfrak{S}(n)$ and $s_n, t_n \in \mathfrak{S}_n(\infty)$ with $\sigma' = \sigma s_n u$ and $\tau' = u^{-1} t_n \tau$. Then,

$$\rho(\sigma')\rho(P_n)\rho(\tau') = \rho(\sigma) \left(\lim_{N\to\infty} \frac{1}{(N-n)!} \sum_{\phi\in K_n(N)} \rho(s_n u \phi u^{-1} t_n) \right) \rho(\tau).$$

If N is large enough so that $s_n, t_n \in K_n(N)$, then $\phi \mapsto s_n u \phi u^{-1} t_n$ is a bijection from $K_n(N)$ to $K_n(N)$, so

$$\rho(\sigma')\rho(P_n)\rho(\tau') = \rho(\sigma) \left(\lim_{N\to\infty} \frac{1}{(N-n)!} \sum_{\phi\in K_n(N)} \rho(\phi) \right) \rho(\tau) = \rho(\sigma)\rho(P_n)\rho(\tau)$$

and our claim is proven.

It remains to see that the extension of ρ to Γ is compatible with the product, and this is equivalent to the compatibility of ρ with the relations of the list given in Proposition 11.14. All these relations are consequences of the definition of ρ as a limit in the weak operator topology. In the sequel, we shall examine in detail the most difficult relation in the list (the last one); we leave the details for the other relations to the reader. Before that, let us remark that in general, the multiplication of operators is not continuous with respect to the weak operator topology. Hence, it is not entirely obvious that one can exchange weak limits and products of permutations in our definition of ρ. However, one can here check at once that there is no problem coming from this phenomenon, so for instance one has

$$\prod_{j=1}^{r} \rho(C_{k_j}) = \lim_{n_{11}<n_{12}<\cdots<n_{rk_r}\to\infty} \rho((n_{11},\ldots,n_{1k_1})\cdots(n_{r1},\ldots,n_{rk_r}))$$

in the weak operator topology.

Consider the relation

$$\forall i_1,\ldots,i_r \text{ with } |i_j|>n,\ \forall k_1,\ldots,k_r,\ P_n A_{i_1}^{k_1} A_{i_2}^{k_2}\cdots A_{i_r}^{k_r} P_n = P_n \prod_{j=1}^{r} C_{k_j+1}.$$

We shall suppose that all the i_j's are positive, this being only in order to simplify a bit the notations. We then introduce a sequence of positive integers

$$N = N_{11} < \cdots < N_{1k_1} < N_{21} < \cdots < N_{2k_2} < \cdots < N_{r1} < \cdots < N_{rk_r},$$

and the group algebra element

$$\Pi_N = \frac{1}{((N-n)!)^2} \sum_{\sigma,\tau\in\mathfrak{S}_n(N)} \sigma\, t_{i_1,N_{11}}\cdots t_{i_r,N_{rk_r}}\, \tau,$$

which will be sent by ρ to an operator that approximates

$$\rho(P_n)\,\rho(A_{i_1})^{k_1}\,\rho(A_{i_2})^{k_2}\cdots\rho(A_{i_r})^{k_r}\,\rho(P_n)$$

as $N_{11},\ldots,N_{rk_r}$ go to $+\infty$ (in the sense of convergence of operators in the weak operator topology). Notice that

$$t_{i_j,N_{j1}}\,t_{i_j,N_{j2}}\cdots t_{i_j,N_{jk_j}} = (\mathrm{id}_{\mathbb{N}^*}, (i_j, N_{j1}, N_{j2}, \ldots, N_{jk_j})),$$

as we have on the the left-hand side a star factorization of a cycle of length k_j+1 (recall that we use the convention that $\sigma\tau = \tau\circ\sigma$ in this section). Therefore,

$$\begin{aligned}
\Pi_N &= \frac{1}{((N-n)!)^2} \sum_{\sigma,\tau\in\mathfrak{S}_n(N)} \sigma(\mathrm{id}_{\mathbb{N}^*}, (i_1, N_{11},\ldots,N_{1k_1})\cdots(i_r,N_{r1},\ldots,N_{rk_r}))\,\tau \\
&= \frac{1}{((N-n)!)^2} \sum_{\sigma,\tau\in\mathfrak{S}_n(N)} (\sigma\tau, \sigma\tau(\tau(i_1), N_{11},\ldots,N_{1k_1})\cdots(\tau(i_r),N_{r1},\ldots,N_{rk_r})) \\
&= \left(\frac{1}{(N-n)!}\sum_{\phi\in\mathfrak{S}_n(N)} \phi\right) \\
&\quad\times\left(\frac{1}{(N-n)!}\sum_{\tau\in\mathfrak{S}_n(N)} (\mathrm{id}_{\mathbb{N}^*}, (\tau(i_1),N_{11},\ldots,N_{1k_1})\cdots(\tau(i_r),N_{r1},\ldots,N_{rk_r}))\right).
\end{aligned}$$

The first term in the product is sent by ρ to an approximation of $\rho(P_n)$, whereas in the second sum, *most of the permutations* $\tau \in \mathfrak{S}_n(N)$ send $i_1, i_2, \ldots, i_r$ to large positive integers, so that $\rho(\mathrm{id}_{\mathbb{N}^*}, (\tau(i_1), N_{11}, \ldots, N_{1k_1}) \cdots (\tau(i_r), N_{r1}, \ldots, N_{rk_r}))$ is close to $\rho(C_{k_1+1} \cdots C_{k_r+1})$ for most permutations τ. Let us precise this argument. Fix $\varepsilon > 0$ and $h_1, h_2 \in H$, and an index $N' \leq N$ such that if $\tau(i_1), \ldots, \tau(i_r) \geq N'$, then

$$\left|\left\langle h_1 \,\middle|\, (\rho(\mathrm{id}_{\mathbb{N}^*}, (\tau(i_1), N_{11}, \ldots, N_{1k_1}) \cdots) - \rho(C_{k_1+1} \cdots C_{k_r+1}))(h_2) \right\rangle\right| \leq \varepsilon.$$

Such an index N' exists by definition of $\rho(C_{k_1+1} \cdots C_{k_r+1})$ as a limit in the weak operator topology. Now, the number of permutations τ in $\mathfrak{S}_n(N)$ which do not satisfy all the inequalities $\tau(i_1) \geq N', \ldots, \tau(i_r) \geq N'$ is smaller than

$$r(N-n-1)!\,N'.$$

Therefore, the proportion of such permutations in $\mathfrak{S}_n(N)$ is smaller than

$$\frac{r(N-n-1)!\,N'}{(N-n)!} = r\frac{N'}{N-n} = O\left(\frac{N'}{N}\right).$$

For N large enough, this proportion is smaller than $\frac{\varepsilon}{\|h_1\|\,\|h_2\|}$. Then,

$$\left\langle h_1 \,\middle|\, \frac{1}{(N-n)!} \sum_{\tau \in \mathfrak{S}_n(N)} (\rho(\mathrm{id}_{\mathbb{N}^*}, (\tau(i_1), \ldots, N_{1k_1}) \cdots) - \rho(C_{k_1+1} \cdots C_{k_r+1}))(h_2) \right\rangle$$

has absolute value smaller than 3ε. Indeed, one ε is coming from the subset of permutations $\tau \in \mathfrak{S}_n(N)$ that yield operators $(\rho(\mathrm{id}_{\mathbb{N}^*}, (\tau(i_1), N_{11}, \ldots, N_{1k_1}) \cdots))$ that are ε-close to $\rho(C_{k_1+1} \cdots C_{k_r+1})$; and the remaining 2ε are coming from a proportion at most $\frac{\varepsilon}{\|h_1\|\,\|h_2\|}$ of permutations of $\mathfrak{S}_n(N)$, to which one applies the trivial inequality

$$\left|\left\langle h_1 \,\middle|\, (\rho(\mathrm{id}_{\mathbb{N}^*}, (\tau(i_1), \ldots, N_{1k_1}) \cdots) - \rho(C_{k_1+1} \cdots C_{k_r+1}))(h_2) \right\rangle\right| \leq 2\,\|h_1\|\,\|h_2\|$$

since we have contractions. We conclude that the limit in the weak operator topology of $\rho(\Pi_N)$ is $\rho(P_n \prod_{j=1}^{r} C_{k_j+1})$, which amounts to the compatibility of $\rho : \Gamma \to \mathscr{B}(H)$ with one of the relations of the list of Proposition 11.14. The other relations are checked similarly. □

11.3 Classification of the admissible representations

We now have the tools required in order to classify the irreducible admissible representations of the pair $(\mathfrak{S}(\infty) \times \mathfrak{S}(\infty), \mathfrak{S}(\infty))$. In Section 11.4, we shall then explain where the extremal characters sit in this classification, and their own classification will be a particular case of Theorem 11.24, which is the main result of this section.

▷ *Irreducible admissible representations and Young distributions.*

We call **irreducible representation** of a discrete group G a unitary representation (H,ρ) of G that does not admit a *closed* subspace $H' \subset H$ which is non-trivial and stable by G. We use the same definition for a $*$-representation of a semigroup Γ with involution by contractions on a Hilbert space H. If (H,ρ) is an admissible irreducible representation of $G = \mathfrak{S}(\infty) \times \mathfrak{S}(\infty)$, then its extension to the Olshanski semigroup $\Gamma = \Gamma(\infty)$ is a fortiori irreducible, since a subspace stable by Γ is stable by G. On the other hand, if (H,ρ) is an admissible representation of G, we call **depth** of (H,ρ) the smallest integer $d(H,\rho) = d \geq 0$ such that $H^{K_d} \neq \{0\}$. Since H is the closure of $\bigcup_{n\in\mathbb{N}} H^{K_n}$, there exists indeed such an integer, unless H is the zero representation.

In the following, we fix $d \geq 0$ and an irreducible admissible representation (H,ρ) of G of depth d. Notice that for any n, the set $P_n\,\Gamma(\infty)\,P_n$ can be identified with $\Gamma(n)$, by sending an infinite generalized Brauer diagram $P_n\gamma P_n$ to the finite generalized Brauer diagram over $[\![-n,-1]\!] \sqcup [\![1,n]\!]$ (or more precisely the class of this diagram modulo cycles C_1 of label 1) which is obtained by forgetting in γ the horizontal edges $(i,-i)$ and $(i',-i')$ for $|i| > n$. Thus, we have (compatible) projection morphisms

$$\phi_n : \gamma \in \Gamma(\infty) \mapsto P_n\gamma P_n \in \Gamma(n)$$

for any $n \in \mathbb{N}$. If (H,ρ) is an admissible representation of G, it yields a representation of Γ, and a representation of $\Gamma(n)$ on H^{K_n} for any n:

$$\begin{aligned} \rho_n : \Gamma(n) &\to \mathscr{B}(H^{K_n}) \\ P_n\gamma P_n &\mapsto \rho(P_n\gamma P_n)_{|H^{K_n}}. \end{aligned}$$

We denote this representation (H_n,ρ_n). The **root** of the representation (H,ρ) is the representation (H_d,ρ_d) of $\Gamma(d)$, where d is the depth of the admissible representation (H,ρ).

Proposition 11.16. *Let (H,ρ) be an admissible representation of G of depth d. If (H,ρ) is irreducible, then its root (H_d,ρ_d) is irreducible (as a representation of $\Gamma(d)$). Moreover, under the same assumption of irreducibility, (H_d,ρ_d) entirely determines (H,ρ) (up to isomorphisms of unitary representations).*

Partial proof. For the moment, we only prove the irreducibility of the root; the fact that the root (H_d,ρ_d) determines (H,ρ) will be a consequence of the Gelfand–Naimark construction; see Section 11.4. Let $V \subset H_d$ be a non-zero vector subspace that is closed and stable by $\Gamma(d)$. We fix $v \neq 0 \in V$ and $w \in H_d$. Since H is irreducible as a representation of G, there exists a sequence of elements $(x_n)_{n\in\mathbb{N}}$ in the group algebra $\mathbb{C}G$ such that $x_n \cdot v \to w$, because $\overline{\mathbb{C}G \cdot v} = H$. Since v and w belong to H_d, $P_d x_n P_d \cdot v = P_d x_n \cdot v \to P_d \cdot w = w$, so w belongs to the closure of the subspace generated by $\Gamma(d) \cdot v$. As V is $\Gamma(d)$-stable and closed, $w \in V$ and $V = H_d$. □

Let (H,ρ) be an irreducible admissible representation of G of depth d. By definition of the depth, $H_{d-1} = \{0\}$, so $\rho(P_{d-1}) = 0$ in $\mathscr{B}(H)$. In particular, the root representation (H_d, ρ_d) sends P_{d-1} to 0, so it corresponds to an irreducible representation of the semigroup

$$\Gamma^{\times}(d) = \Gamma(d) \setminus \Gamma(d)P_{d-1}\Gamma(d) = \langle \sigma \in \mathfrak{S}(d) \times \mathfrak{S}(d), A_{k\in[\![-d,-1]\!]\sqcup[\![1,d]\!]}, C_{k\geq 2}\rangle.$$

This semigroup is much simpler than the previous Brauer and Olshanski semigroups considered, as it writes as a product

$$\Gamma^{\times}(d) = (G(d) \rtimes \mathbb{N}^{2d}) \times \mathbb{N}^{(\infty)}$$

with $G(d) = \mathfrak{S}(d) \times \mathfrak{S}(d)$. The semigroup $\mathbb{N}^{2d}$ corresponds to the powers of the generators A_k, $1 \leq |k| \leq d$, and the semigroup $\mathbb{N}^{(\infty)}$ of infinite sequences of integers with almost all terms equal to 0 corresponds to the powers of the generators $C_{k\geq 2}$. The product in the semigroup $G(d) \rtimes \mathbb{N}^{2d}$ is

$$\begin{aligned}
&(\sigma,(m_{-d},\ldots,m_{-1},m_1,\ldots,m_d))(\tau,(n_{-d},\ldots,n_{-1},n_1,\ldots,n_d))\\
&= \big(\sigma\tau,(m_{\tau^{-1}(-d)}+n_{-d},\ldots,m_{\tau^{-1}(-1)}+n_{-1},m_{\tau^{-1}(1)}+n_1,\ldots,m_{\tau^{-1}(d)}+n_d)\big)\\
&= (\sigma\tau,(m\cdot\tau)+n),
\end{aligned}$$

where τ acts on the right of vectors by permutation of their entries by τ^{-1} (recall that we are working here with the opposite of the usual product of permutations). In the following, we set $\mathfrak{R}(d) = \mathfrak{S}(d) \rtimes \mathbb{N}^d$, so that $G(d) \rtimes \mathbb{N}^{2d} = (\mathfrak{S}(d) \rtimes \mathbb{N}^d) \times (\mathfrak{S}(d) \rtimes \mathbb{N}^d) = \mathfrak{R}(d) \times \mathfrak{R}(d)$.

Definition 11.17. *A **Young distribution** of weight d is a collection of Young diagrams $\Lambda = (\lambda(x))_{x\in[-1,1]}$, such that almost all real numbers in $[-1,1]$ yield the Young diagram $\lambda(x) = \emptyset$, and with*

$$\sum_{\substack{x\in[-1,1]\\ \lambda(x)\neq\emptyset}} |\lambda(x)| = d.$$

We denote $\mathfrak{YD}(d)$ the infinite set of all Young distributions of weight d.

Example. We represent Young distributions by drawing the Young diagrams $\lambda(x)$ over any $x \in [-1,1]$ such that $\lambda(x) \neq 0$. Thus, the representation of the Young distribution

$$\Lambda = \left\{\lambda(0) = (2,1),\ \lambda\left(\frac{1}{3}\right) = (3,2,2),\ \lambda\left(-\frac{1}{2}\right) = (5,4,1)\right\}$$

is drawn in Figure 11.4.

Theorem 11.18. *The irreducible representations of $\mathfrak{R}(d)$ are labeled by Young distributions of weight d.*

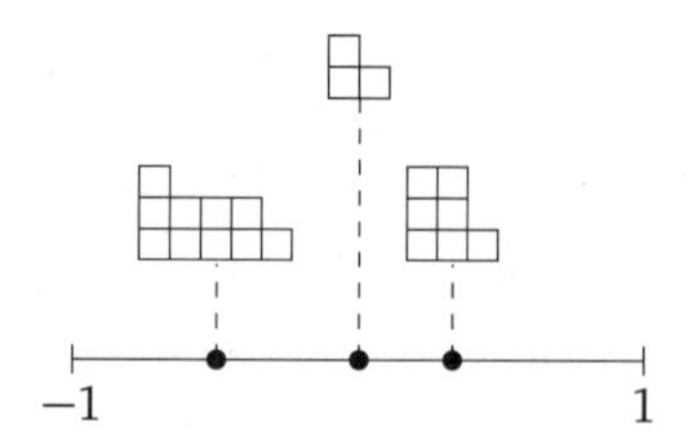

Figure 11.4
A Young distribution of weight $3+7+10=20$.

Lemma 11.19. *An irreducible representation by self-adjoint contractions of the semigroup* $\mathbb{N}^d$ *has dimension 1, and it is entirely determined by a vector* $(x_1,\ldots,x_d)$ *with* $x_i \in [-1,1]$*:*

$$(n_1,\ldots,n_d) \mapsto x_1^{n_1}\, x_2^{n_2} \cdots x_d^{n_d} \in \mathbb{C}.$$

Proof. Let (H,ρ) be an irreducible representation of $\mathbb{N}^d$ by self-adjoint contractions, and $T_i = \rho(0,\ldots,0,1_i,0,\ldots,0)$. It is a bounded self-adjoint operator, to which one can apply the spectral theorem 9.6. We shall use it in the following form: there exists a unitary operator $U_i : H \to \mathrm{L}^2(X_i,\mathscr{F}_i,\mu_i)$ with $(X_i,\mathscr{F}_i,\mu_i)$ measured space, that conjugates T_i to the operator of multiplication by a bounded measurable function $f_i \in \mathrm{L}^\infty(X_i,\mathscr{F}_i,\mu_i)$:

$$\forall g \in \mathrm{L}^2(X_i,\mathscr{F}_i,\mu_i),\ \ U_i T_i U_i^{-1}(g) = f_i g.$$

Moreover, since the operators $T_{i\leq d}$ commute with one another, one can diagonalize them simultaneously: there exists a Hilbert space $\mathrm{L}^2(X,\mathscr{F},\mu)$, a unitary operator $U : H \to \mathrm{L}^2(X,\mathscr{F},\mu)$, and functions $f_i \in \mathrm{L}^\infty(X,\mathscr{F},\mu)$, such that

$$\forall g \in \mathrm{L}^2(X,\mathscr{F},\mu),\ \forall i \in [\![1,d]\!],\ \ UT_iU^{-1}(g) = f_i g.$$

We give references for this multi-operator spectral theorem at the end of this chapter. As a consequence, the representation H of $\mathbb{N}^d$ is unitarily equivalent to the representation on $\mathrm{L}^2(X,\mathscr{F},\mu)$ given by

$$(n_1,\ldots,n_d)\cdot g = (f_1)^{n_1}\cdots(f_d)^{n_d}\cdot g,$$

where the f_i's are in $\mathrm{L}^\infty(X,\mathscr{F},\mu)$, with $|f_i| \leq 1$ for any i.

Suppose that there exists a measurable subset $E \in \mathscr{F}$ with $\mu(E) > 0$ and $\mu(X \setminus E) > 0$. Then, the subspace $\mathrm{L}^2(E,\mathscr{F}_{|E},\mu_{|E}) \subset \mathrm{L}^2(X,\mathscr{F},\mu)$ is a non-trivial closed subrepresentation of $\mathbb{N}^d$, which is impossible by the assumption of irreducibility. Therefore, every measurable subset in $\mathscr{F}$ is either of μ-measure 0, or with a complementary of μ-measure 0. It follows that the functions $f_i \in \mathrm{L}^\infty(X,\mathscr{F},\mu)$ are constant (almost everywhere) to some $x_i \in [-1,1]$, and also that $L^2(X,\mathscr{F},\mu)$ is one-dimensional. Then, (H,ρ) is equivalent to the representation

$$(n_1,\ldots,n_d)\mapsto x_1^{n_1}\,x_2^{n_2}\,\cdots\,x_d^{n_d}\in\mathbb{C}.$$

In the sequel we shall denote $\mathbb{C}_X=\mathbb{C}_{x_1,\ldots,x_d}$ this one-dimensional irreducible representation of $\mathbb{N}^d$. □

Proof of Theorem 11.18. The result relies now on the general theory of representations of semidirect products. We detail here this theory in the special case of $\mathfrak{S}(d)\rtimes\mathbb{N}^d$. We fix a vector $X=(x_1,\ldots,x_d)\in[-1,1]^d$, and denote $\mathfrak{S}(d,X)$ the stabilizer of X for the action of $\mathfrak{S}(d)$ by permutation of the coordinates. This group is isomorphic to a product of symmetric groups

$$\mathfrak{S}(\mu)=\mathfrak{S}(\mu_1)\times\mathfrak{S}(\mu_2)\times\cdots\mathfrak{S}(\mu_r),$$

where μ is an integer partition of size d, r is the number of distinct elements x_{i_j} in the vector X, and the parts μ_j of μ are the multiplicities of these elements. For instance, if $X=(-1,0,-1)$, then $\mu=(2,1)$. We extend the one-dimensional representation $\mathbb{C}_X$ of $\mathbb{N}^d$ to $\mathfrak{S}(d,X)\rtimes\mathbb{N}^d$ by setting

$$(\sigma,(n_1,\ldots,n_d))\cdot 1=(x_1)^{n_1}\cdots(x_d)^{n_d}.$$

This is indeed compatible with the product in $\mathfrak{S}(d,X)\rtimes\mathbb{N}^d$, because if $\sigma,\tau\in\mathfrak{S}(d,X)$, then

$$((\sigma,m)(\tau,n))\cdot 1=(\sigma\tau,m\cdot\tau+n)\cdot 1=x^{m\cdot\tau+n}=x^{m+n}=(\sigma,m)\cdot((\tau,n)\cdot 1)$$

since $x^{m\cdot\tau}=x^m$ for $\tau\in\mathfrak{S}(d,X)$. Take now an irreducible representation (S,ρ) of $\mathfrak{S}(d,X)$, which is determined by a family of integer partitions $\lambda^{(j)}$ of sizes μ_j:

$$S=\boxtimes_{j=1}^r S^{\lambda^{(j)}},$$

where $\boxtimes$ is the outer tensor product of representations (cf. Section 2.4). Since the parts μ_j of μ are associated to distinct coordinates x_{i_j} of the vector X, we can associate to the pair $(X,(\lambda^{(1)},\ldots,\lambda^{(r)}))$ a Young distribution Λ of weight d, with $\lambda(x)=\lambda^{(j)}$ if $x=x_{i_j}$ is a coordinate of the vector X, and $\lambda(x)=\emptyset$ otherwise. This Young distribution Λ is entirely determined by the orbit $O(X)$ of X under the action of $\mathfrak{S}(d)$ (which amounts to giving the distinct coordinates x_{i_j} and their multiplicities) and by the integer partitions $\lambda^{(1)},\ldots,\lambda^{(r)}$, and conversely. The tensor product $S_X=S\otimes_{\mathbb{C}}\mathbb{C}_X$ is isomorphic as a vector space to S, and it is again a representation of $\mathfrak{S}(d,X)\rtimes\mathbb{N}^d$, given by

$$(\sigma,(n_1,\ldots,n_d))\cdot(v\in S)=(x_1)^{n_1}\cdots(x_d)^{n_d}\,\rho(\sigma)(v).$$

This representation S_X of $\mathfrak{S}(d,X)\rtimes\mathbb{N}^d$ is irreducible, because if $T\subset S^X$ is a subspace stable by $\mathfrak{S}(d,X)\rtimes\mathbb{N}^d$, then in particular it is stable by $\mathfrak{S}(d,X)$, whose action on S_X is irreducible (as a $\mathfrak{S}(d,X)$-module, $S_X=S$). We set

$$S^\Lambda=\mathbb{C}[\mathfrak{S}(d)\rtimes\mathbb{N}^d]\otimes_{\mathbb{C}[\mathfrak{S}(d,X)\rtimes\mathbb{N}^d]}S_X,$$

which is a finite-dimensional $*$-representation of the semigroup $\mathfrak{S}(d) \rtimes \mathbb{N}^d$ by contractions. A concrete description of S^Λ is as follows: it is the set of functions $f : \mathfrak{R}(d) \to S$ such that for any $h \in \mathfrak{S}(d,X) \rtimes \mathbb{N}^d$ and any $g \in \mathfrak{S}(d) \rtimes \mathbb{N}^d$, $f(hg) = h \cdot f(g)$. We claim that this representation is irreducible, and that it depends only on the Young distribution Λ. For $Y = \sigma(X)$ vector in the orbit $O(X)$ of X for the action of the symmetric group, set

$$(S^\Lambda)_Y = \{f : \mathfrak{R}(d) \to S \mid \forall \tau \in \mathfrak{S}(d),\ f(\tau) \neq 0 \Rightarrow \tau(Y) = X\}.$$

The decomposition $\mathfrak{S}(d) = \bigsqcup_\sigma (\mathfrak{S}(d,X)\sigma)$ of $\mathfrak{S}(d)$ in cosets of $\mathfrak{S}(d,X)\backslash\mathfrak{S}(d)$ corresponds to a decomposition in direct sum

$$S^\Lambda = \bigoplus_{Y \in O(Y)} (S^\Lambda)_Y,$$

and on the other hand, if $\sigma(X) = Y$, then it is easily seen that $(S^\Lambda)_Y$ is an irreducible representation of $\mathfrak{S}(d,Y) \rtimes \mathbb{N}^d$, that is isomorphic to $(\sigma(S))_Y$, where $\sigma(S)$ is the irreducible representation of $\mathfrak{S}(d,Y)$ that is obtained from the irreducible representation (S,ρ) of $\mathfrak{S}(d,X)$ by setting

$$\tau \cdot (v \in S) = \rho(\sigma^{-1}\tau\sigma)(v)$$

for $\tau \in \mathfrak{S}(d,Y)$. Moreover, the decomposition $S^\Lambda = \bigoplus_{Y \in O(Y)} (S^\Lambda)_Y$ is a decomposition of S^Λ in $\mathbb{N}^d$-submodules, the action of $(n_1,\ldots,n_d) \in \mathbb{N}^d$ on $(S^\Lambda)_Y$ being given by multiplication by $(y_1)^{n_1}(y_2)^{n_2}\cdots(y_d)^{n_d}$. Consider then a $\mathfrak{R}(d)$-submodule W of S^Λ. It is in particular a finite-dimensional representation of $\mathbb{N}^d$ by self-adjoint contractions, so it is a direct sum of spaces $\mathbb{C}_Y$, and necessarily with $Y \in O(X)$ and with the Y-isotypic component W_Y of W that is included in $(S^\Lambda)_Y$:

$$W = \bigoplus_{Y \in O(X)} W_Y \quad ; \quad S^\Lambda = \bigoplus_{Y \in O(X)} (S^\Lambda)_Y \quad ; \quad W_Y \subset (S^\Lambda)_Y.$$

Since each space $(S^\Lambda)_Y$ is an irreducible representation of $\mathfrak{S}(d,X) \rtimes \mathbb{N}^d$, we have $W_Y = \{0\}$ or $W_Y = (S^\Lambda)_Y$ for any Y. Moreover, since W is a $\mathfrak{R}(d)$-module, we have an isomorphism of vector spaces between W_X and W_Y for any $Y \in O(X)$: if $\sigma(X) = Y$, it is given by restriction of the isomorphism $\sigma : W \to W$ to W_X. Therefore, either $W = \{0\}$, or $W = \bigoplus_{Y \in O(X)} (S^\Lambda)_Y = S^\Lambda$. So, S^Λ is an irreducible $*$-representation of $\mathfrak{R}(d)$ by contractions, and the previous discussion shows that it depends only on (S,ρ) and on the orbit of X under the action of $\mathfrak{S}(d)$, that is to say on the Young distribution Λ. Thus, we can associate to any Young distribution $\Lambda \in \mathfrak{YD}(d)$ an irreducible representation of $\mathfrak{R}(d)$. Moreover, two distinct Young distributions Λ and M of weight d yield non-isomorphic representations S^Λ and S^M of $\mathfrak{R}(d)$: indeed, one can recover from S^Λ the orbit $O(X)$ from the isotypic components of S^Λ for the action of $\mathbb{N}^d$, and then the partitions $\lambda^{(j)}$ from the action of $\mathfrak{S}(d,X)$ on the component $(S^\lambda)_X$.

Conversely, let H be an irreducible $*$-representation of $\mathfrak{R}(d)$ by contractions.

If $h \in H$ is a non-zero vector, then H is the closure of the space generated by the vector $\gamma \cdot h$ with $\gamma \in \mathfrak{R}(d)$. Since $\mathfrak{S}(d)$ is a finite group, it follows that H is the closure of a finitely generated $\mathbb{N}^d$-module. By the previous lemma, H has finite dimension and writes as a direct sum of one-dimensional representations $\mathbb{C}_X$ of $\mathbb{N}^d$. If H_X is the X-isotypic component of H as a representation of $\mathbb{N}^d$ (the direct sum of the components isomorphic to $\mathbb{C}_X$), then for any $\sigma \in \mathfrak{S}(d)$, $\sigma(H_X)$ is the $\sigma(X)$-isotypic component of H as a representation of $\mathbb{N}^d$. Moreover, by irreducibility of H, if X and Y are two vectors in $[-1,1]^d$ that correspond to non-zero isotypic components of H, then there exists $\sigma \in \mathfrak{S}(d)$ such that $Y = \sigma(X)$: indeed, $\bigoplus_{Y \in O(X)} H_Y$ is a $\mathfrak{R}(d)$-stable subspace of H. As a consequence, there exists an orbit $O(X)$ of a vector $X \in [-1,1]^d$ such that $H = \bigoplus_{Y \in O(X)} H_Y$. Moreover, for any $Y \in O(X)$, a permutation $\sigma \in \mathfrak{S}(d)$ such that $\sigma(X) = Y$ yields an isomorphism of vector spaces $\psi_{X \to Y}$ between H_X and H_Y that conjugates the actions of $\mathfrak{S}(d,X) \rtimes \mathbb{N}^d$ on H_X and of $\mathfrak{S}(d,Y) \rtimes \mathbb{N}^d$ on H_Y. If H_X were not an irreducible $\mathfrak{S}(d,X) \rtimes \mathbb{N}^d$-representation, then a decomposition $H_X = K_X \oplus L_X$ would lead to a decomposition $H = K \oplus L$ of H in two non-trivial representations of $\mathfrak{R}(d)$, which is excluded by irreducibility of H. Thus, H_X is an irreducible representation of $\mathfrak{S}(d,X) \rtimes \mathbb{N}^d$, on which $(n_1,\dots,n_d) \in \mathbb{N}^d$ acts by $(x_1)^{n_1} \cdots (x_d)^{n_d}$, and that determines entirely H. It remains to see that an irreducible representation of a group $\mathfrak{S}(d,X) \rtimes \mathbb{N}^d$ that is isotypic of type X for the action of $\mathbb{N}^d$ is of the form S_X with S irreducible representation of $\mathfrak{S}(d,X)$; this last fact is quite obvious. So, one can find a Young distribution Λ such that $H = S^\Lambda$ as a representation of $\mathfrak{R}(d)$. This ends the proof of the theorem. □

Consider now the semigroup

$$\Gamma^\times(d) = \mathfrak{R}(d) \times \mathfrak{R}(d) \times \mathbb{N}^{(\infty)}$$

An irreducible $*$-representation of $\mathfrak{R}(d) \times \mathfrak{R}(d)$ is an outer tensor product $S^\Lambda \boxtimes S^M$ of two representations of $\mathfrak{R}(d)$ associated to Young distributions Λ and M in $\mathfrak{YD}(d)$, and a similar argument as before shows that from such a representation, one can construct an arbitrary irreducible $*$-representation of $\Gamma^\times(d)$ by specifying parameters $c_{k\geq 2} \in [-1,1]$ such that the loops C_k act by multiplication by c_k. Thus:

Proposition 11.20. *The irreducible $*$-representations by contractions of $\Gamma^\times(d)$ are labeled by families $(\Lambda, M, (c_k)_{k\geq 2})$ with $\Lambda, M \in \mathfrak{YD}(d)$, and $c_k \in [-1,1]$ for any $k \geq 2$.*

▷ *Spectral measures and the classification of admissible representations.*

By Proposition 11.16, in order to classify the irreducible admissible representations of the pair (G,K) that are of depth d, it remains to decide whether a representation (H_d, ρ_d) of parameters $(\Lambda, M, (c_k)_{k\geq 2})$ of $\Gamma^\times(d)$ can be the root of an irreducible admissible representation (H,ρ) of depth d. In the following, we fix such an irreducible admissible representation (H,ρ), and a vector ξ of norm 1 in H_d. We denote $(\Lambda, M, c_{k\geq 2})$ the parameters of the root (H_d, ρ_d), and θ the **spectral measure** of the self-adjoint action of the element A_{d+1} on H with respect to

ξ: it is the unique positive measure on $[-1,1]$ such that for any $n \geq 0$,

$$\langle \xi \,|\, (\rho(A_{d+1}))^n(\xi)\rangle_H = \int_{-1}^{1} t^n\,\theta(dt)$$

The existence of such a measure is a consequence of the spectral theorem 9.6 and of Riesz' representation theorem: for any function $f \in \mathscr{C}^0(\sigma(\rho(A_{d+1})),\mathbb{R})$, one can consider the quantity $\phi(f) = \langle \xi \,|\, f(\rho(A_{d+1}))(\xi)\rangle_H$, and this yields a non-negative linear form ϕ on the space $\mathscr{C}^0(\sigma(\rho(A_{d+1})),\mathbb{R})$, which can be represented by a positive measure supported on $\sigma(\rho(A_{d+1}))$. Moreover, this spectrum is included in $[-1,1]$ since $\rho(A_{d+1})$ is a contraction of operator norm smaller than 1. In the sequel, we shall always omit the defining morphism ρ in the notations, and thus identify A_k and $\rho(A_k)$, and as well for the other elements of Γ.

Notice now that since $\xi \in H_d$, $P_d \cdot \xi = \xi$, and therefore,

$$\begin{aligned}\langle \xi \,|\, (A_{d+1})^n \cdot \xi\rangle = \langle P_d \cdot \xi \,|\, (A_{d+1})^n P_d \cdot \xi\rangle &= \langle \xi \,|\, P_d (A_{d+1})^n P_d \cdot \xi\rangle \\ &= \langle \xi \,|\, C_{n+1} \cdot \xi\rangle = c_{n+1}.\end{aligned}$$

So, the coefficients c_k are the moments of θ:

$$\forall k \geq 2,\ c_k = \int_{-1}^{1} t^{k-1}\,\theta(dt).$$

This identity shows in particular that θ does not depend on the choice of the vector ξ in H_d. In the sequel, we shall prove that the admissibility of (H,ρ) implies numerous restrictions on the spectral measure θ. The first result in this direction is:

Proposition 11.21. *The spectral measure θ is a discrete probability measure, and its atoms can only accumulate at* 0.

Proof. Suppose first that $d = 0$. In particular, $P_0 \cdot \xi = \xi$, and in fact $P_k \cdot \xi = \xi$ for any $k \geq 0$. In the sequel, we shall use several times the relations of the list in Proposition 11.14. We compute the spectral measure associated to the infinite family of commuting operators A_k:

$$\begin{aligned}\left\langle \xi \,\middle|\, A_{i_1}^{k_1} A_{i_2}^{k_2} \cdots A_{i_r}^{k_r} \cdot \xi \right\rangle = \left\langle P_0 \cdot \xi \,\middle|\, A_{i_1}^{k_1} A_{i_2}^{k_2} \cdots A_{i_r}^{k_r} P_0 \cdot \xi \right\rangle &= \left\langle \xi \,\middle|\, P_0 A_{i_1}^{k_1} A_{i_2}^{k_2} \cdots A_{i_r}^{k_r} P_0 \cdot \xi \right\rangle \\ = \left\langle \xi \,\middle|\, \left(\prod_{j=1}^{r} C_{k_j+1}\right) \cdot \xi \right\rangle &= \prod_{j=1}^{r} c_{k_j+1}.\end{aligned}$$

Therefore, $\theta^{\otimes\infty}$ is the spectral measure of $(A_k)_{k\geq 1}$. We now fix a Borel subset $B \subset [\varepsilon, 1]$ for some $\varepsilon > 0$. Notice that there is an extension of the continuous functional calculus described in Theorem 9.6 to bounded measurable functions;

therefore, it makes sense to consider $1_B(A_k)$, where 1_B is the indicator of the set B. The operator $1_B(A_k)$ acts by an orthogonal projection on the Hilbert space H, since 1_B is positive and $(1_B)^2 = 1_B$. We have on the other hand:

$$\begin{aligned}\varepsilon\,\theta(B) &\leq \int_{-1}^{1} 1_B(t)\,t\,1_B(t)\,\theta(dt) = \langle \xi \mid 1_B(A_1)\,A_1\,1_B(A_1)\cdot\xi\rangle \\ &\leq \langle \xi \mid 1_B(A_1)\,P_1A_1P_1\,1_B(A_1)\cdot\xi\rangle = \langle \xi \mid 1_B(A_1)\,P_1(1,2)P_1\,1_B(A_1)\cdot\xi\rangle \\ &\leq \langle \xi \mid 1_B(A_1)\,(1,2)\,1_B(A_1)\cdot\xi\rangle\,.\end{aligned}$$

Notice now that

$$\begin{aligned}1_B(A_1)\,(1,2)\,1_B(A_1) &= (1_B(A_1))^2\,(1,2)\,1_B(A_1) = (1_B(A_1))^2\,1_B(A_2)\,(1,2) \\ &= 1_B(A_1)\,1_B(A_2)\,1_B(A_1)\,(1,2) = 1_B(A_1)\,(1_B(A_2))^2 \\ &= 1_B(A_1)\,1_B(A_2).\end{aligned}$$

So,

$$\varepsilon\,\theta(B) \leq \langle \xi \mid 1_B(A_1)\,1_B(A_2)\cdot\xi\rangle = \left(\int_{-1}^{1} 1_B(t)\,\theta(dt)\right) = (\theta(B))^2.$$

The same inequality holds for B Borel subset included in $[-1,-\varepsilon]$, and implies that $\theta(B) = 0$ or $\theta(B) \geq \varepsilon$. As a consequence, the cumulative distribution function of the measure θ can only increase by jumps, so θ is discrete. It is a probability measure since

$$\int_{-1}^{1} \theta(dt) = \langle \xi \mid \xi \rangle = 1.$$

Finally, if x is an atom of θ with $|x| = \varepsilon$, then $\theta(x) \geq \varepsilon = |x|$, so the mass of θ at x is always larger than $|x|$. In particular, there cannot be more than $\frac{1}{\varepsilon}$ atoms of θ with absolute value larger than ε, so the only possible accumulation points of these atoms is 0. The general case with an arbitrary depth d is entirely analogous, by replacing P_0 by P_d and A_1, A_2 by A_{d+1}, A_{d+2} in the previous computations. □

Given a Young distribution Λ of weight d, we call **support** of Λ the set $\mathrm{supp}(\Lambda)$ of elements x such that $\lambda(x) \neq \emptyset$.

Lemma 11.22. *In the previous setting, we have* $\mathrm{supp}(\Lambda) \subset \mathrm{supp}(\theta) \cup \{0\}$, *and similarly for* $\mathrm{supp}(M)$.

Proof. Let x be an element in $\mathrm{supp}(\Lambda)$ that is not 0; the description of the irreducible representations of $\mathfrak{R}(d)$ shows that there exists $\xi \in H_d$ such that A_1 acts on ξ by multiplication by x:

$$A_1 \cdot \xi = x\,\xi.$$

Without loss of generality, we can assume that $\|\xi\| = 1$. On the other hand, we can as before consider the element $1_x(A_1)$, which acts on H. Notice that $1_x(A_1)$ sends ξ to ξ. We now have

$$\begin{aligned}|x| &= |\langle \xi \mid A_1 \cdot \xi\rangle| = |\langle \xi \mid P_d A_1 P_d \cdot \xi\rangle| = |\langle \xi \mid P_d(1,d+1)P_d \cdot \xi\rangle| \\ &= |\langle \xi \mid (1,d+1)\cdot \xi\rangle| = |\langle 1_x(A_1)\cdot \xi \mid (1,d+1)1_x(A_1)\cdot \xi\rangle| \\ &= |\langle 1_x(A_1)\cdot \xi \mid 1_x(A_{d+1})(1,d+1)\cdot \xi\rangle| \\ &= |\langle 1_x(A_{d+1})\cdot \xi \mid 1_x(A_1)(1,d+1)\cdot \xi\rangle| \\ &= |\langle 1_x(A_{d+1})\cdot \xi \mid (1,d+1)\,1_x(A_{d+1})\cdot \xi\rangle| \\ &\leq |\langle 1_x(A_{d+1})\cdot \xi \mid 1_x(A_{d+1})\cdot \xi\rangle| = \left|\int_{-1}^{1} 1_x(t)\,\theta(dt)\right| = \theta(\{x\})\end{aligned}$$

by using on the last line the fact that $(1,d+1)$ acts by a unitary transformation. Thus, if $x \in \mathrm{supp}(\Lambda)$ is not 0, then $\theta(\{x\}) \geq |x| > 0$, so x is an atom of θ. The same argument with the operators A_{-k} shows the analogous result for $x \in \mathrm{supp}(M)$. □

Thus, if (H,ρ) is an admissible representation of depth d, then its parameters can be described as follows: they are given by a discrete spectral measure θ on $[-1,1]$ such that $\theta(\{x\}) \geq |x|$ if $\theta(\{x\}) > 0$, and by two Young distributions Λ and M of weight d and supported by $\mathrm{supp}(\theta) \cup \{0\}$. Another important property of θ is:

Proposition 11.23. *Suppose $\theta(\{x\}) > 0$ with $x \neq 0$. Then, $\frac{\theta(\{x\})}{|x|}$ is a positive integer.*

Proof. Again, it suffices to treat the case $d = 0$, the general case following by a shift of indices in the computations. We fix $x > 0$ and denote $k(x) = \frac{\theta(\{x\})}{|x|}$. For $m \geq 0$, we introduce

$$\xi_m = \left(\prod_{i=1}^{m} 1_x(A_i)\right)\cdot \xi,$$

where the operators $1_x(A_i)$ are as before well defined by the theory of measurable functional calculus of self-adjoint operators. For $\sigma \in \mathfrak{S}(m)$, let us compute $\langle \xi_m \mid \sigma \cdot \xi_m\rangle$. We consider more generally two functions $f(A_1,\ldots,A_m) = \prod_{i=1}^m f_i(A_i)$ and $g(A_1,\ldots,A_m) = \prod_{i=1}^m g_i(A_i)$, and compute the scalar product $\langle f(A_1,\ldots,A_m)\cdot \xi \mid \sigma g(A_1,\ldots,A_m)\cdot \xi\rangle$. If $f(A_1,\ldots,A_m) = (A_1)^{k_1}\cdots(A_m)^{k_m}$ and $g(A_1,\ldots,A_m) = (A_1)^{l_1}\cdots(A_m)^{l_m}$ are polynomials, we get

$$\begin{aligned}&\langle f(A_1,\ldots,A_m)\cdot \xi \mid \sigma g(A_1,\ldots,A_m)\cdot \xi\rangle \\ &= \left\langle (A_1)^{k_1}\cdots(A_m)^{k_m} P_0 \cdot \xi \,\middle|\, \sigma (A_1)^{l_1}\cdots(A_m)^{l_m} P_0\cdot \xi\right\rangle \\ &= \left\langle \xi \,\middle|\, P_0 (A_1)^{k_1}\cdots(A_m)^{k_m} \sigma (A_1)^{l_1}\cdots(A_m)^{l_m} P_0\cdot \xi\right\rangle \\ &= \left\langle \xi \,\middle|\, P_0 (A_1)^{k_1+l_{\sigma(1)}}\cdots(A_m)^{k_m+l_{\sigma(m)}} \sigma P_0\cdot \xi\right\rangle.\end{aligned}$$

One computes in the Olshanski semigroup Γ:

$$P_0(A_1)^{t_1}\cdots(A_m)^{t_m}\sigma P_0=\prod_{c\in C(\sigma)} C_{\ell(c)+\sum_{i\in c}t_i}$$

where the product runs over cycles of $\sigma\in\mathfrak{S}(\infty)\times\mathfrak{S}(\infty)$. Therefore, the scalar product is equal to

$$\begin{aligned}\prod_{c\in C(\sigma)} c_{\ell(c)+\sum_{i\in c}k_i+l_{\sigma(i)}}&=\prod_{c\in C(\sigma)} c_{\ell(c)+\sum_{i\in c}k_i+l_i}\\&=\prod_{c\in C(\sigma)}\left(\int_{-1}^{1} t^{\ell(c)-1}\left(\prod_{i\in c}\overline{f}_i(t)\,g_i(t)\right)\theta(dt)\right).\end{aligned}$$

In particular,

$$\langle\xi_m\,|\,\sigma\cdot\xi_m\rangle=\prod_{c\in C(\sigma)} x^{\ell(c)-1}\theta(\{x\})=x^m\,(k(x))^{n(\sigma)}$$

if σ belongs to $\mathfrak{S}(m)\times\mathfrak{S}(m)$ and has $n(\sigma)$ cycles for its action on $[\![-m,-1]\!]\sqcup[\![1,m]\!]$. As a consequence,

$$\begin{aligned}\left\langle\xi_m\,\middle|\,\frac{1}{m!}\sum_{\sigma\in\mathfrak{S}(m)}\varepsilon(\sigma)(\mathrm{id}_{\mathbb{N}^*},\sigma)\cdot\xi_m\right\rangle&=\frac{x^m}{m!}\sum_{\sigma\in\mathfrak{S}(m)}\varepsilon(\sigma)\,(k(x))^{n(\sigma)}\\&=\frac{x^m}{m!}\,k(x)\,(k(x)-1)\cdots(k(x)-m+1)\\&=\frac{x^m}{m!}\,(k(x))^{\downarrow m}.\end{aligned}$$

However, the operator $\mathscr{A}_m=\frac{1}{m!}\sum_{\sigma\in\mathfrak{S}(m)}\varepsilon(\sigma)(\mathrm{id}_{\mathbb{N}^*},\sigma)$ acts on H by a self-adjoint projection, so the scalar product $\langle\xi_m\,|\,\mathscr{A}_m\cdot\xi_m\rangle$ should be non-negative for any m. This is only possible if $(k(x))^{\downarrow m}\geq 0$ for all m, which in turns happens if and only if $k(x)\in\mathbb{N}$. The case when $x<0$ is entirely similar: indeed, if one replaces (H,ρ) by $(H,\rho)\otimes(\mathbb{C},\varepsilon\otimes\varepsilon)$, then this sends the spectral measure $d\theta(x)$ to $d\theta(-x)$. □

Theorem 11.24 (Olshanski, Okounkov). *Let (H,ρ) be an irreducible admissible representation of the Gelfand pair $(G=\mathfrak{S}(\infty)\times\mathfrak{S}(\infty),\ K=\mathfrak{S}(\infty))$, which is of depth d. Let θ be the spectral measure of H, and Λ, M be the two Young distributions of size d that label the root (H_d,ρ_d) viewed as an irreducible representation of $\mathfrak{R}(d)\times\mathfrak{R}(d)$.*

1. *There exists a unique parameter (α,β) in the Thoma simplex T such that*

$$\theta=\left(\sum_{i=1}^{\infty}\alpha_i\,\delta_{\alpha_i}+\beta_i\,\delta_{-\beta_i}\right)+\gamma\delta_0.$$

2. *The supports of the distributions* Λ *and* M *are included in* $\mathrm{supp}(\theta) \cup \{0\}$.

Proof. If x is in the support of θ and not equal to zero, then by the previous proposition, it appears with weight $\theta(\{x\}) = k|x|$ for some $k \in \mathbb{N}^*$. We associate to such an x a sequence $(x, x, \ldots, x)$ with k terms, and we gather all these sequences in two parameters

$$\alpha = (\alpha_1 \geq \alpha_2 \geq \cdots \geq \alpha_r \geq \cdots \geq 0);$$
$$-\beta = (-\beta_1 \leq -\beta_2 \leq \cdots \leq -\beta_s \leq \cdots \leq 0),$$

which correspond to a unique point (α, β) of the Thoma simplex (we have $|\alpha| + |\beta| = \theta([-1,1]\setminus\{0\}) \leq 1$, since the spectral measure θ is a probability measure). By construction, $\theta = \left(\sum_{i=1}^{\infty} \alpha_i\, \delta_{\alpha_i} + \beta_i\, \delta_{-\beta_i}\right) + \gamma\delta_0$, where $\gamma = 1 - |\alpha| - |\beta|$. This proves the first point, and the second point is contained in Lemma 11.22. □

Remark. With a bit more work, one can in fact characterize the allowed Young distributions Λ and M such that (θ, Λ, M) is the set of parameters of an admissible irreducible representation of depth d of (G, K). Hence, it can be shown that the pair of Young distributions (Λ, M) of weight d is allowed if and only if

$$\forall x \in (0,1],\ \ell(\lambda(x)) + \ell(\mu(x)) \leq \frac{\theta(\{x\})}{x};$$
$$\forall x \in [-1,0),\ \ell(\lambda'(x)) + \ell(\mu'(x)) \leq \frac{\theta(\{x\})}{|x|},$$

see the notes at the end of the chapter. We won't need this more precise statement, as we are mostly interested here in the admissible representations of depth 0, for which the Young distributions are trivial.

11.4 Spherical representations and the Gelfand–Naimark–Segal construction

Let (H, ρ) be an irreducible representation of the group $G = \mathfrak{S}(\infty) \times \mathfrak{S}(\infty)$. The representation is called **spherical** if $H^K \neq \{0\}$. An irreducible representation H is spherical if and only if it is admissible of depth 0. Indeed, fix $h \in H^{K_n}$ and $(\sigma, \tau) \in \mathfrak{S}(\infty) \times \mathfrak{S}(\infty)$. By increasing the value of n, one can assume that $(\sigma, \tau) \in \mathfrak{S}(n) \times \mathfrak{S}(n)$. Then, with $k \in K_n = \mathfrak{S}_n(\infty)$,

$$(k,k) \cdot ((\sigma, \tau) \cdot h) = (\sigma, \tau) \cdot ((k,k) \cdot h) = (\sigma, \tau) \cdot h,$$

so $(\sigma, \tau) \cdot h \in H^{K_n}$. Therefore, if $I = \bigcup_{n \in \mathbb{N}} H^{K_n}$, then I is stable by the action of $G = \mathfrak{S}(\infty) \times \mathfrak{S}(\infty)$. Then, $J = \overline{I}$ is a closed G-stable subspace of H, and it is not reduced to $\{0\}$, as it contains $H^K = H^{K_0} \neq \{0\}$. By irreducibility, $H = J$, and the representation is admissible, of depth 0.

By Theorem 11.24, to any spherical representation (H,ρ) of G, one can associate a unique parameter (α,β) of the Thoma simplex, such that the spectral measure θ of the action of A_1 on H is given by

$$\theta = \sum_{i=1}^{\infty} \alpha_i\, \delta_{\alpha_i} + \sum_{i=1}^{\infty} \beta_i\, \delta_{-\beta_i} + \gamma\, \delta_0.$$

Moreover, this parameter $(\alpha,\beta) \in T$ entirely determines (H,ρ) up to unitary equivalence, because in the case of admissible representations of depth 0, there are no additional combinatorial parameters (Λ,M): the unique Young distribution of weight 0 is the empty one. To conclude our classification of the extremal characters of $\mathfrak{S}(\infty)$, we shall explain in this section the link between extremal characters and spherical representations, as well as the reason why the root of an admissible (spherical) representation determines the representation. Both problems are solved by the Gelfand–Naimark–Segal construction (Theorem 11.26), which relates traces on the C^*-algebra of a discrete group, and certain unitary representations of the group.

▷ *The Gelfand–Naimark–Segal construction.*

In this paragraph, we fix a discrete group G, and a unitary irreducible representation (H,ρ) of G. A **C*-algebra** is a complex subalgebra $\mathscr{A}$ of an algebra of operators $\mathscr{B}(H)$, which is stable by adjunction and which is closed for the topology defined by the operator norm on $\mathscr{B}(H)$. In particular, the C^*-algebra of the group G, denoted $\mathscr{A}(G)$, is the closure in $\mathscr{B}(\ell^2(G))$ of the subalgebra generated by the operators $L_g : \sum_{h\in G} c_h\, h \mapsto \sum_{h\in G} c_h\, gh$. Notice that $\mathscr{A}(G)$ is usually not a von Neumann algebra, but the setting of C^*-algebras will be more adequate for our discussion.

We start by explaining how to extend ρ to a $*$-representation of $\mathscr{A}(G)$, that is to say a morphism of algebras $\mathscr{A}(G) \to \mathscr{B}(H)$ that is compatible with the adjunctions $*$ of both algebras. We can of course define $\rho(\sum_{g\in G} c_g\, g) = \sum_{g\in G} c_g\, \rho(g) \in \mathscr{B}(H)$ for any *finite* sum $c = \sum_{g\in G} c_g\, g$ in $\mathbb{C}G$. The key lemma in order to extend this definition to $\mathscr{A}(G)$ is:

Lemma 11.25. *For any $c = \sum_{g\in G} c_g\, g$ in $\mathbb{C}G$, we have*

$$\|\rho(c)\|_{\mathscr{B}(H)} \leq \|c\|_{\mathscr{B}(\ell^2(G))}.$$

Proof. If ξ is a vector in H of norm 1, one has

$$\begin{aligned}\left\|\sum_{g\in G} c_g\, g \cdot \xi\right\|^2 &= \sum_{g,h} \overline{c_g} c_h\, \langle g\cdot\xi \,|\, h\cdot\xi\rangle \\ &= \sum_{g,h} \overline{c_g}\, c_h\, \tau_\xi(g^{-1}h) = \tau_\xi\left(\left(\sum_{g\in G} c_g\, g\right)^* \left(\sum_{g\in G} c_g\, g\right)\right)\end{aligned}$$

where τ_ξ is the function on the $*$-algebra $\mathbb{C}G$ given by $\tau_\xi(c) = \langle \xi \mid \rho(c)(\xi)\rangle_H$. We thus obtain a state on the $*$-algebra $\mathbb{C}G$ (in Chapter 9, the states were defined for von Neumann algebras, but the definition makes sense for any $*$-subalgebra of an algebra of bounded linear operators, here $\mathbb{C}G \subset \mathscr{B}(\ell^2(G))$). We saw in Chapter 9 that the spectral theorem implies

$$|\tau_\xi(c)| \leq \|c\|_{\mathscr{B}(\ell^2(G))}.$$

Therefore,

$$\|c \cdot \xi\|^2 \leq \tau_\xi(c^* c) \leq \|c\|^2_{\mathscr{B}(\ell^2(G))}.$$

Since this inequality is true for any vector ξ of norm 1, we conclude to the same bound for $\|\rho(c)\|^2_{\mathscr{B}(H)}$. □

As a consequence of this lemma, the map $c \mapsto \rho(c)$ is continuous from $(\mathbb{C}G, \|\cdot\|_{\mathscr{B}(\ell^2(G))})$ to $(\mathscr{B}(H), \|\cdot\|_{\mathscr{B}(H)})$, so it admits a unique continuous linear extension to the closure $\mathscr{A}(G)$ of $\mathbb{C}G$, which is a $*$-representation of C^*-algebras. We now fix an arbitrary C^*-algebra $\mathscr{A}$. Given a $*$-representation $\rho : \mathscr{A} \to \mathscr{B}(H)$ of $\mathscr{A}$, if ξ is a vector of norm 1 in H, then it yields a state on $\mathscr{A}$:

$$\tau_\xi(a) = \langle \xi \mid \rho(a)(\xi)\rangle .$$

We say on the other hand that ξ is a **cyclic vector** if $\mathscr{A} \cdot \xi$ is dense in H. Notice that if $\mathscr{A} = \mathscr{A}(G)$ and if the $*$-representation comes from an irreducible representation of G, then *any* vector ξ of norm 1 is cyclic.

Theorem 11.26 (Gelfand–Naimark–Segal). *Suppose that τ is a state on a C^*-algebra $\mathscr{A}$. There exists a $*$-representation of $\mathscr{A}$ on a Hilbert space H, and a cyclic vector $\xi \in H$ of norm 1, such that $\tau = \tau_\xi$. Moreover, if (H, ξ) and (H', ξ') are two $*$-representations of $\mathscr{A}$ with distinguished cyclic vectors and such that $\tau_\xi = \tau_{\xi'}$ are the same state of $\mathscr{A}$, then there exists a unitary isomorphism $U : H \to H'$ that conjugates the two $*$-representations, and such that $U(\xi) = \xi'$.*

Proof. A state τ on $\mathscr{A}$ defines a non-negative sesquilinear form $\langle a \mid b\rangle_{\mathscr{A}} = \tau(a^* b)$. Let Z be the vector subspace of $\mathscr{A}$ that consists in degenerate vectors z, that is to say vectors such that $\langle a \mid z\rangle_{\mathscr{A}} = 0$ for any $a \in \mathscr{A}$. Notice that

$$\begin{aligned} Z &= \{z \in \mathscr{A} \mid \forall a \in \mathscr{A},\ \tau(a^* z) = 0\} = \{z \in \mathscr{A} \mid \tau(z^* z) = 0\} \\ &= \{z \in \mathscr{A} \mid \forall a \in \mathscr{A},\ \tau(z^* a) = 0\}. \end{aligned}$$

Indeed, this is a consequence of the Cauchy–Schwarz inequality $|\tau(a^* z)|^2 \leq \tau(a^* a)\,\tau(z^* z)$. The space Z is obviously an ideal of $\mathscr{A}$, so one can consider the quotient algebra $\mathscr{A}/Z$. If $[a]$ and $[b]$ are classes modulo Z in $\mathscr{A}$, then the scalar product $\langle [a] \mid [b]\rangle = \langle a \mid b\rangle$ is well defined: indeed, if $a, a' \in [a]$ and $b, b' \in [b]$, then there exists $z_1, z_2 \in Z$ such that

$$\left\langle a' \,\middle|\, b'\right\rangle = \langle a + z_1 \mid b + z_2\rangle = \langle a \mid b\rangle + \langle a + z_1 \mid z_2\rangle + \langle z_1 \mid b\rangle = \langle a \mid b\rangle .$$

Moreover, by construction, the scalar product on $\mathscr{A}/Z$ is positive definite, so it induces a prehilbertian structure. We denote H the Hilbert completion of $\mathscr{A}/Z$ for this structure. The algebra $\mathscr{A}$ acts on $\mathscr{A}/Z$ by multiplication on the left, and by similar arguments as before, this action gives rise to a $*$-representation of $\mathscr{A}$ on H, for which the vector $\xi = [1_A]$ is cyclic. Then, by definition,

$$\tau(a) = \langle 1_A \,|\, a\rangle_{\mathscr{A}} = \langle \xi \,|\, a \cdot \xi\rangle_H = \tau_\xi(a),$$

so every state comes indeed from a $*$-representation of the C^*-algebra.

Suppose now that we have another $*$-representation H' of $\mathscr{A}$, with a cyclic vector ξ' that gives rise to the same state τ. We define a linear map $U : \mathscr{A} \to H'$ by

$$U(a) = a \cdot \xi'.$$

If $z \in Z$, then $0 = \tau(z^*z) = \tau_{\xi'}(z^*z) = \langle \xi' \,|\, z^*z \cdot \xi'\rangle = (\|z \cdot \xi'\|_{H'})^2$, so $U(z) = 0$. Therefore, the linear map U factors through the quotient $\mathscr{A}/Z$. It satisfies then, for any $[a] \in \mathscr{A}/Z$:

$$\begin{aligned}(\|U([a])\|_{H'})^2 = (\|U(a)\|_{H'})^2 = (\|a \cdot \xi'\|_{H'})^2 = \left\langle \xi' \,\middle|\, a^*a \cdot \xi'\right\rangle \\ = \tau_{\xi'}(a^*a) = \tau(a^*a) = (\|a\|_{\mathscr{A}/Z})^2.\end{aligned}$$

So, U is an isometry, which extends to an isometry from the completion H of $\mathscr{A}/Z$ to a closed subspace of H'. Since this closed subspace must contain all the vectors $a \cdot \xi'$, and since ξ' is cyclic in H', $U(H) = H'$, and it is then easily seen that we have obtained an isomorphism of $*$-representations of $\mathscr{A}$. Finally, by construction, $U(\xi) = U([1_A]) = 1_A \cdot \xi' = \xi'$. □

The Gelfand–Naimark–Segal construction allows us to end the proof of Proposition 11.16. Indeed, let (H, ρ) be an irreducible admissible representation of the Gelfand pair (G, K), and (H_d, ρ_d) be its root. We take ξ vector of norm 1 in H_d, and consider the associated state:

$$\tau(g) = \langle \xi \,|\, g \cdot \xi\rangle.$$

By the previous discussion, this function entirely determines the representation (H, ρ), up to unitary equivalence. However, since $P_d \cdot \xi = \xi$, the function τ is equal to

$$\tau(g) = \langle P_d \cdot \xi \,|\, g P_d \cdot \xi\rangle = \langle \xi \,|\, P_d g P_d \cdot \xi\rangle = \tau(P_d g P_d),$$

so if one knows its values on $P_d\, G\, P_d \subset \Gamma(d)$, then one can reconstruct the representation (H, ρ). So, if one knows the action of $\Gamma(d)$ on $H_d = \overline{\Gamma(d) \cdot \xi}$, then one can indeed reconstruct (H, ρ).

To complete Theorem 11.26, let us see when a state τ corresponds to an *irreducible* representation of a C^*-algebra $\mathscr{A}$. Call **extremal** a state that cannot be written as a barycenter $\tau = x\,\tau_1 + (1-x)\,\tau_2$ of two distinct states τ_1 and τ_2.

Proposition 11.27. *A state τ of a C^*-algebra $\mathscr{A}$ is extremal if and only if it corresponds to an irreducible $*$-representation of $\mathscr{A}$.*

Lemma 11.28 (Schur). *A $*$-representation H of a C^*-algebra $\mathscr{A}$ is irreducible if and only if $\mathrm{End}_{\mathscr{A}}(H)$, the set of continuous $\mathscr{A}$-morphisms of the Hilbert space H, is one-dimensional.*

Proof. Suppose that H is not irreducible. Then, the orthogonal projection $\pi_K : H \to K$ onto a $\mathscr{A}$-stable invariant closed subspace K of H is a (continuous) $\mathscr{A}$-morphism, because if $h \in H$, then

$$\pi_K(a \cdot h) = \pi_K(a \cdot \underbrace{(h - \pi_K(h))}_{\in K^\perp}) + \pi_K(a \cdot \underbrace{\pi_K(h)}_{\in K}) = 0 + a \cdot \pi_K(h) = a \cdot \pi_K(h)$$

since K and $K^\perp$ are both $\mathscr{A}$-stable subspaces. Therefore, $\mathrm{End}_{\mathscr{A}}(H)$ contains an element π_K that is not colinear to id_H, and it is not one-dimensional.

Conversely, suppose that there exists an element $T \in \mathrm{End}_{\mathscr{A}}(H)$ that is not colinear to the identity. The two elements $\frac{T+T^*}{2}$ and $\frac{T-T^*}{2\mathrm{i}}$ are self-adjoint, and one of them at least is not colinear to the identity; hence, without loss of generality, one can assume $T = T^*$. Then, the C^*-algebra spanned by T is a subalgebra of $\mathrm{End}_{\mathscr{A}}(H)$, and it is commutative, hence isomorphic to $\mathscr{C}^0(\sigma(T), \mathbb{C})$ by Gelfand's representation theorem (see the references at the end of the chapter for this classical result, which is closely related to the spectral theorem 9.6). Moreover, since $T \neq \lambda \mathrm{id}_H$, $\sigma(T)$ contains at least two points $x \neq y$. Let $f \in \mathscr{C}^0(\sigma(T), \mathbb{R})$ be a continuous function that vanishes in a neighborhood of x, and is equal to 1 in a neighborhood of y. The corresponding element F in the C^*-algebra spanned by T has a non-trivial kernel, because the action of F on H is unitarily conjugated to the multiplication by f. This kernel is a non-trivial closed $\mathscr{A}$-submodule of H, so H is not irreducible. $\square$

Proof of Proposition 11.27. Suppose that τ is not extremal, and writes as

$$\tau = x\,\tau_1 + (1-x)\,\tau_2,$$

with τ_1, τ_2 states and $x \neq 0$. In the following, we denote $\sigma = x\,\tau_1$; it is a non-negative linear form on $\mathscr{A}$, such that $\tau(b) \geq \sigma(b)$ for any $b = a^*a$ non-negative element. We also denote (H, ξ) the $*$-representation associated to τ. For $b \in \mathscr{A}$, we define a linear form $f_b : H \to \mathbb{C}$ by:

$$f_b(a \cdot \xi) = \sigma(b^*a).$$

We have the Cauchy–Schwarz inequality

$$\begin{aligned} |f_b(a \cdot \xi)| = |\sigma(b^*a)| &\leq \sqrt{\sigma(a^*a)}\sqrt{\sigma(b^*b)} \\ &\leq \sqrt{\tau(a^*a)}\,\sqrt{\tau(b^*b)} = \|a \cdot \xi\|_H \,\|b \cdot \xi\|_H. \end{aligned}$$

This inequality shows that f_b depends only on $b \cdot \xi$, and also that for a fixed b, f_b is a continuous linear form on H. Therefore, there exists $c = T_\sigma(b \cdot \xi)$ such that

$$f_b(a \cdot \xi) = \langle c \mid a \cdot \xi \rangle_H = \langle T_\sigma(b \cdot \xi) \mid a \cdot \xi \rangle_H .$$

The Cauchy–Schwarz inequality shows then that the map $T_\sigma : H \to H$ is linear and continuous, of operator norm smaller than 1. Moreover,

$$\begin{aligned}\langle T_\sigma(c \cdot (b \cdot \xi)) \mid a \cdot \xi \rangle_H &= \sigma(b^* c^* a) = \langle T_\sigma(b \cdot \xi) \mid c^* \cdot (a \cdot \xi) \rangle_H \\ &= \langle c \cdot (T_\sigma(b \cdot \xi)) \mid a \cdot \xi \rangle_H ,\end{aligned}$$

so T_σ belongs to the space of $\mathscr{A}$-morphisms of H. Finally, T_σ is not a scalar multiple of the identity: otherwise, one would have $x\,\tau_1(b^*a) = \sigma(b^*a) = \langle \lambda\, b \cdot \xi \mid a \cdot \xi \rangle_H = \lambda\, \tau(b^*a)$, which is excluded since $\tau \neq \tau_1$. So, there exists a $\mathscr{A}$-morphism of H which is not a multiple of the identity, and by Schur's lemma, H is not irreducible.

Conversely, suppose that H is not irreducible, and denote K a closed subspace stable by the action of $\mathscr{A}$. We then have the decomposition

$$\tau(a) = \langle \xi \mid a \cdot \xi \rangle = \langle \xi_K \mid a \cdot \xi_K \rangle + \langle \xi_{K^\perp} \mid a \cdot \xi_{K^\perp} \rangle ,$$

where $\xi = \xi_K + \xi_{K^\perp}$ with $\xi_K \in K$ and $\xi_{K^\perp} \in K^\perp$. It is a decomposition of τ in a sum of non-negative linear forms on $\mathscr{A}$, so τ is not extremal. □

Corollary 11.29. *Let (H, ρ) be a spherical representation of the pair (G, K). The space of K-invariants H^K is one-dimensional.*

Proof. We use here the fact that (G, K) is a Gelfand pair. In $\mathscr{B}(H^K)$, consider the closure $\mathscr{A}$ for the operator norm topology of the algebra generated by the $P_K\, \rho(g)\, P_K$'s, $g \in G$. It is a unital commutative C^*-algebra, and H^K is an irreducible representation of it. Thus, it suffices to prove that an irreducible representation of a commutative C^*-algebra is always one-dimensional. By Proposition 11.27, this is equivalent to the fact that the extremal states of such an algebra correspond to one-dimensional representations. In the sequel, we fix an extremal state τ of a commutative C^*-algebra $\mathscr{A}$. By Gelfand's representation theorem of commutative C^*-algebras, $\mathscr{A}$ is isometric and isomorphic to an algebra of continuous functions $\mathscr{C}^0(X, \mathbb{C})$ on a compact space X. Then, by Riesz' representation theorem, a positive continuous linear form τ on $\mathscr{C}^0(X, \mathbb{R})$ corresponds to a positive measure μ on X:

$$\tau(f) = \int_X f(x)\, \mu(dx).$$

If μ is not concentrated on one point $x \in X$, then μ can be decomposed as the sum of two positive measures, and τ is not extremal. Therefore, $\mu = \delta_x$ and $\tau(f) = f(x)$ for some $x \in X$. It follows that the extremal state τ corresponds to the one-dimensional representation which makes $f \in \mathscr{C}^0(X, \mathbb{R})$ acts on $\mathbb{C}$ by multiplication by $f(x)$. This ends the proof of the corollary. □

▷ *The classification of spherical representations.*

We can finally classify all the extremal characters of the group $K = \mathfrak{S}(\infty)$, by relating them to the spherical representations of the pair (G,K). Fix a spherical representation (H,ρ), and a vector $\xi \in H^K$ of norm 1; it is unique up to multiplication by a phase $e^{i\varphi}$. We denote as before $\tau(\cdot)$ the state associated to the pair (H,ξ), and we set $\chi(\sigma)=\tau(\sigma,\mathrm{id}_{\mathbb{N}^*})$. We claim that χ is an extremal character of $\mathfrak{S}(\infty)$. First, notice that χ is invariant by conjugation, because

$$\begin{aligned}\chi(\phi^{-1}\sigma\phi) &= \left\langle \xi \,\middle|\, (\phi^{-1}\sigma\phi,\mathrm{id}_{\mathbb{N}^*})\cdot\xi\right\rangle = \langle (\phi,\phi)\cdot\xi \,|\, (\sigma,\mathrm{id}_{\mathbb{N}^*})(\phi,\phi)\cdot\xi\rangle \\ &= \langle \xi \,|\, (\sigma,\mathrm{id}_{\mathbb{N}^*})\cdot\xi\rangle = \chi(\sigma).\end{aligned}$$

Then, we can compute χ in terms of the spectral measure θ of (H,ρ). Indeed, for any $k_1,\ldots,k_r \geq 2$, the value of χ on a product σ of disjoint cycles of lengths $k_1,\ldots,k_r$ is

$$\begin{aligned}\chi(\sigma) &= \lim_{n_{11}<n_{12}<\cdots<n_{rk_r}\to\infty} \chi((n_{11},\ldots,n_{1k_1})\cdots(n_{r1},\ldots,n_{rk_r})) \\ &= \lim_{n_{11}<n_{12}<\cdots<n_{rk_r}\to\infty} \left\langle \xi \,\middle|\, \rho((n_{11},\ldots,n_{1k_1})\cdots(n_{r1},\ldots,n_{rk_r}),\mathrm{id}_{\mathbb{N}^*})(\xi)\right\rangle \\ &= \left\langle \xi \,\middle|\, \rho(C_{k_1}\cdots C_{k_r})(\xi)\right\rangle,\end{aligned}$$

where on the last line the representation is extended from G to the Olshanski semigroup Γ. Since $C_{k\geq 2}$ acts on an irreducible admissible representation of G by multiplication by

$$c_k = \int_{-1}^{1} t^{k-1}\theta(dt) = \sum_{i=1}^{\infty}(\alpha_i)^k + (-1)^{k-1}(\beta_i)^k,$$

we conclude that

$$\chi(\sigma) = \prod_{c \text{ cycle of } \sigma \text{ of length } l\geq 2} \left(\sum_{i=1}^{\infty}(\alpha_i)^l + (-1)^{l-1}(\beta_i)^l\right)$$

is the extremal character of $\mathfrak{S}(\infty)$ associated to the parameter (α,β) of the Thoma simplex that corresponds to θ.

Conversely, if χ is an extremal character of $\mathfrak{S}(\infty)$, set $\tau(k_1,k_2)=\chi(k_1k_2^{-1})$. We claim that τ is an extremal state of the C^*-algebra $\mathscr{A}(G)$, corresponding to a spherical representation. The fact that τ is a state is an easy calculation. On the other hand, we denote (H,ξ) a unitary representation of G such that $\tau=\tau_\xi$, with ξ cyclic vector in H. For any $k \in K$, $\tau(k)=\chi(kk^{-1})=\chi(\mathrm{id}_{\mathbb{N}^*})=1=\langle \xi \,|\, k\cdot\xi\rangle$, and this is only possible if $k\cdot\xi=\xi$ for any $k\in K$. Therefore, $\xi \in H^K$ and there exists a cyclic vector of H that is left invariant by K. If τ is not extremal, then H is reducible and $H=H_1\oplus H_2$, which leads to a decomposition $\xi=\xi_1\oplus\xi_2$ with

$\xi_1 \in (H_1)^K$ and $\xi_2 \in (H_2)^K$, and $\xi_1, \xi_2 \neq 0$ because ξ is a cyclic vector. Then, $\tau = x\,\tau_1 + (1-x)\,\tau_2$ with

$$x = \|\xi_1\|^2 \quad ; \quad x\,\tau_1(g) = \langle \xi_1 \,|\, g \cdot \xi_1 \rangle \; ;$$
$$1 - x = \|\xi_2\|^2 \quad ; \quad (1-x)\,\tau_2(g) = \langle \xi_2 \,|\, g \cdot \xi_2 \rangle .$$

We set $\chi_1(k) = \tau_1(k, \mathrm{id}_{\mathbb{N}^*})$ and $\chi_2(k) = \tau_2(k, \mathrm{id}_{\mathbb{N}^*})$, so that

$$\chi(k) = \tau(k, \mathrm{id}_{\mathbb{N}^*}) = x\chi_1(k) + (1-x)\chi_2(k).$$

We claim that χ_1 and χ_2 are normalized traces on $K = \mathfrak{S}(\infty)$. It suffices to treat the case of χ_1:

1. normalization: $\chi_1(\mathrm{id}_{\mathbb{N}^*}) = \tau_1(\mathrm{id}_{\mathbb{N}^*}, \mathrm{id}_{\mathbb{N}^*}) = \frac{\langle \xi_1 | \xi_1 \rangle}{x} = 1$.
2. trace property:

$$\chi_1(\sigma\rho) = \frac{\langle \xi_1 \,|\, (\sigma\rho, \mathrm{id}_{\mathbb{N}^*}) \cdot \xi_1 \rangle}{x} = \frac{\langle (\rho,\rho) \cdot \xi_1 \,|\, (\rho\sigma, \mathrm{id}_{\mathbb{N}^*})(\rho,\rho) \cdot \xi_1 \rangle}{x}$$
$$= \frac{\langle \xi_1 \,|\, (\rho\sigma, \mathrm{id}_{\mathbb{N}^*}) \cdot \xi_1 \rangle}{x} = \chi_1(\rho\sigma).$$

3. non-negativity: trivial from the fact that χ_1 and τ_1 are defined from a scalar product.

Therefore, χ is not an extremal character, which contradicts the initial hypothesis. So, τ is an extremal state, and it corresponds to a spherical representation of G. We have thus established:

Theorem 11.30. *The formulas $\chi(k) = \tau(k, \mathrm{id}_{\mathbb{N}^*})$ and $\tau(k_1, k_2) = \chi(k_1 k_2^{-1})$ yield a bijection between the extremal characters χ of $K = \mathfrak{S}(\infty)$, and the states τ of spherical representations of the pair (G, K).*

The previous discussion shows also that if θ is the spectral measure of a spherical representation, and if (α, β) is the corresponding parameter of the Thoma simplex, then the corresponding extremal character of $\mathfrak{S}(\infty)$ is $\chi(\sigma_\mu) = p_\mu(A + \overline{B} + \gamma E)$ as in Section 11.1. We can finally conclude:

Theorem 11.31 (Thoma). *The extremal characters of $\mathfrak{S}(\infty)$ are parameterized by the Thoma simplex T: any $\chi \in \mathscr{T}^*(\infty)$ writes uniquely as*

$$\chi(\sigma) = \prod_{c \text{ cycle of } \sigma \text{ of length } l \geq 2} \left(\sum_{i=1}^{\infty} (\alpha_i)^l + (-1)^{l-1} (\beta_i)^l \right)$$

where $(\alpha, \beta) \in T$. Thus, $\mathscr{T}^*(\infty) = T$

Notes and references

The first section of the chapter uses without proof several important results regarding the topology of convex sets, namely, the Krein–Milman theorem and the Choquet representation theorem. We refer to [Rud91, Chapter 3] and [Lan93, Chapter IV, Appendix] for an introduction to these results, and to [Phe01] for a more thorough study. On the other hand, the spectral theorem and Gelfand's representation theory of (commutative) C^*-algebras plays an important role in Section 11.4, and as in Chapter 9 the reader can consult [Lan93, Chapter XVIII] and [Tak79] for this spectral theory. We also used a multi-operator version of the spectral theorem during the proof of Lemma 11.19, and this can be found in [Att, Chapter 1, §1.6.5].

For a general introduction to infinite-dimensional representation theory; see [Kir94]. In particular, we refer to Theorem 6.2 in loc. cit. for the Gelfand–Naimark–Segal construction. The theory of admissible representations of pairs (G,K) was first developed in the setting of semisimple Lie groups (see e.g. [Kna01, Chapter VIII]), and in the setting of discrete infinite groups such as $\mathfrak{S}(\infty)$, the main contributions are due to Olshanski and Okounkov; see [Ols90, Oko99]. We refer to these papers for the complete classification of the admissible representations of the Gelfand pair $(\mathfrak{S}(\infty)\times\mathfrak{S}(\infty),\mathfrak{S}(\infty))$ ([Oko99, Theorem 3], as well as related pairs, namely,

$$\varinjlim_{n\to\infty}\,(\mathfrak{S}(2n),\mathfrak{S}(n)\rtimes(\mathbb{Z}/2\mathbb{Z})^n)\quad;\quad \varinjlim_{n\to\infty}\,(\mathfrak{S}(2n+1),\mathfrak{S}(n)\rtimes(\mathbb{Z}/2\mathbb{Z})^n).$$

For an introduction to the theory of *finite* Gelfand pairs, see [CSST08, Chapter IV]. For the representation theory of semidirect products and wreath products, we refer to [Ser77, §8.2] and [CSST09].

Let us now mention a few results related to those of this chapter. The Olshanski semigroup is built upon the Brauer semigroup $\mathfrak{B}(n)$, which plays a role analogous to $\mathfrak{S}(n)$ with respect to Schur–Weyl dualities for the orthogonal groups $\mathrm{O}(N)$ and the symplectic groups $\mathrm{Sp}(2N)$ (instead of $\mathrm{GL}(N)$). A quantized version of the semigroup algebra of $\mathfrak{B}(n)$ appeared in [Bra37], where it was shown that certain specializations of this algebra are the commutant of the aforementioned groups. The reader can also consult [Wen88] for a modern presentation and a proof of the semisimplicity of the Brauer algebra in the generic case. On the other hand, it can be shown that the Olshanski semigroup $\Gamma(\infty)$, of which we have given a combinatorial description, can also be defined by operations on double cosets $K_n\backslash G/K_n$; see [Ols90, §2].

Finally, one might ask whether there exists a classification similar to Thoma's theorem for the non-negative specializations for the other combinatorial Hopf algebras studied in this book; for instance, QSym endowed with the basis of fundamental quasi-symmetric functions. An answer in this particular case is provided by [GO06], where we also found our proof of the Kerov–Vershik ring theorem.

12

Asymptotics of central measures

In the previous chapter, we saw that the extremal characters of $\mathfrak{S}(\infty)$ are parameterized by the Thoma simplex

$$T = \left\{ t = (\alpha_1 \geq \alpha_2 \geq \cdots \geq 0, \, \beta_1 \geq \beta_2 \geq \cdots \geq 0) \;\middle|\; \sum_{i=1}^{\infty} \alpha_i + \beta_i \leq 1 \right\}.$$

To a parameter t of the Thoma simplex, we can associate the spectral probability measures $\mathbb{P}_{t,n}$ of the restrictions of the extremal character τ_t to the finite symmetric groups $\mathfrak{S}(n)$. They are given by the formula

$$(\tau_t)_{|\mathfrak{S}(n)} = \sum_{\lambda \in \mathfrak{Y}(n)} \mathbb{P}_{t,n}[\lambda] \, \chi^\lambda.$$

If $f \in \mathrm{Sym}$ and $t \in T$, we denote $f(t) = f(A + \overline{B} + \gamma E)$ the specialization of f on the formal alphabet $A + \overline{B} + \gamma E$ associated to t. By the discussion of the previous chapter, $s_\lambda(t) \geq 0$ for any Schur function s_λ and any $t \in T$. By combining the Frobenius–Schur formula and the formula for τ_t on a permutation σ_μ of cycle type μ, we obtain

$$\tau_t(\sigma_\mu) = p_\mu(t) = \sum_{\lambda \in \mathfrak{Y}(n)} s_\lambda(t) \, d_\lambda \, \chi^\lambda(\sigma_\mu),$$

so, by identification, $\mathbb{P}_{t,n}[\lambda] = d_\lambda \, s_\lambda(t)$, where $d_\lambda = \dim \lambda = \dim S^\lambda$.

The probability measures $\mathbb{P}_{t,n}$ are called **central measures** on partitions. The objective of this chapter is to describe the combinatorial and asymptotic properties of the random partitions $\lambda = \lambda^{(n)}$ chosen according to the central measures $\mathbb{P}_{t,n}$, with a fixed parameter $t \in T$. From a combinatorial point of view, we shall see that a central measure $\mathbb{P}_{t,n}$ is the image by the RSK algorithm of a certain probability measure $\mathbb{Q}_{t,n}$ on permutations of size n (Theorem 12.17). These probability measures $\mathbb{Q}_{t,n}$ are related to a way to produce random permutations in $\mathfrak{S}(n)$ by means of shuffling, and the combinatorics of this operation are encoded in a Hopf algebra FQSym, which is analogous to the algebras QSym and NCSym introduced in Chapter 6. This algebra of *free quasi-symmetric functions* will be studied in Section 12.1, and this theory will yield a combinatorial interpretation of the central measures.

In Section 12.2, by using the algebra of observables $\mathscr{O}$ constructed in the second part of the book, we shall prove that the first rows $\lambda_1^{(n)}, \lambda_2^{(n)}, \ldots$ of a large random partition $\lambda^{(n)}$ chosen under a central measure have a typical size. Thus, if $t = (\alpha, \beta)$ and $\lambda^{(n)} \in \mathfrak{Y}(n)$ is chosen randomly under $\mathbb{P}_{t,n}$, then when n is large, with probability very close to 1, $\lambda_1^{(n)} \simeq n\alpha_1$, $\lambda_2^{(n)} \simeq n\alpha_2$, etc. Similarly, the first columns $\lambda_1^{(n)\prime}, \lambda_2^{(n)\prime}, \ldots$ are typically of size $n\beta_1, n\beta_2$, etc. So, one has a law of large numbers for the rows and columns of a random partition under $\mathbb{P}_{t,n}$ (cf. Theorem 12.19), and this law of large numbers can be reinterpreted in terms of lengths of increasing and decreasing subsequences in a random permutation under $\mathbb{Q}_{t,n}$ (Corollary 12.20).

A major problem is then to precise the law of large numbers and to show a *central limit theorem* for the random partitions $\lambda^{(n)} \sim \mathbb{P}_{t,n}$. In Section 12.3, we shall look at the observables of these random partitions, and by using a technique of moments and cumulants, we shall establish a central limit theorem for these quantities; it is a truly remarkable application of the properties of the algebra $\mathscr{O}$. It is also possible to prove in certain cases a central limit theorem directly for the rows and columns of random partitions $\lambda^{(n)} \sim \mathbb{P}_{t,n}$, but the limit is not always Gaussian, and these results rely on a difficult argument that connects the RSK algorithm, the representation theory of the Lie algebra $\mathfrak{sl}(N)$, and certain transformations on random walks known as Pitman's transforms. We shall not discuss these other asymptotic results, but we shall give references for them at the end of the chapter.

12.1 Free quasi-symmetric functions

We denote $X = (x_1, x_2, \ldots, x_n, \ldots)$ an infinite ordered alphabet, and as before, $\mathbb{C}\langle X\rangle$ is the free associative algebra, that is the projective limit in the category of graded algebras of the free algebras $\mathbb{C}\langle x_1, \ldots, x_n\rangle$; and $\mathbb{C}[X]$ is the algebra of polynomials, that is the quotient of $\mathbb{C}\langle X\rangle$ by the relations $x_i x_j = x_j x_i$, and the projective limit in the category of *commutative* graded algebras of the algebras of polynomials $\mathbb{C}[x_1, \ldots, x_n]$. As in Chapter 6, we denote $\Phi : \mathbb{C}\langle X\rangle \to \mathbb{C}[X]$ the quotient map which makes the variables commutative. We introduced in Chapter 6 the graded subalgebras $\mathrm{NCSym} \subset \mathbb{C}\langle X\rangle$ and $\mathrm{QSym} \subset \mathbb{C}[X]$, and they were related to $\mathrm{Sym} \subset \mathbb{C}[X]$ by the following diagram:

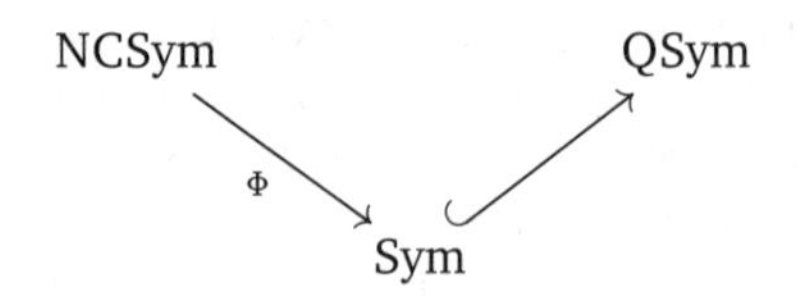

The algebra of free quasi-symmetric functions will be a larger subalgebra FQSym of $\mathbb{C}\langle X\rangle$, which completes the previous diagram:

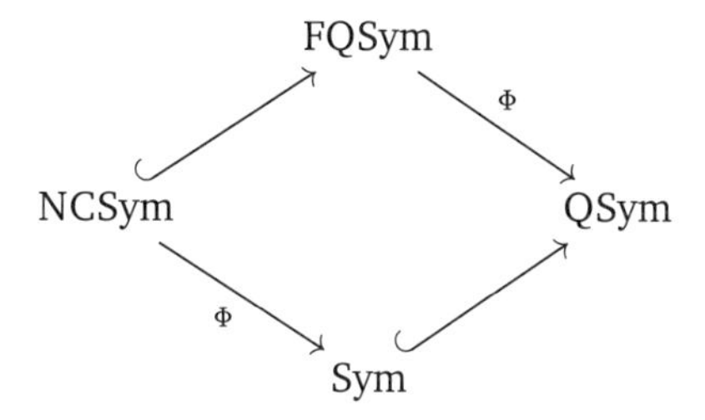

This algebra will provide us with a way to embed in an algebraic structure the combinatorics of the RSK algorithm (Section 3.2); and it will shed a new light on certain relations satisfied in its quotients and subalgebras.

▷ *Free quasi-ribbon functions.*

At the very end of Section 6.3, we introduced the notion of standardization of a word $w \in (\mathbb{N}^*)^{(\mathbb{N})}$: if w has length ℓ and d_1 letters equal to 1, d_2 letters equal to 2, etc., then its standardization is the word $\mathrm{Std}(w)$ with the same length ℓ, and which is obtained by:

- replacing the letters 1 of w by $1, 2, \ldots, d_1$, these numbers appearing in this order in $\mathrm{Std}(w)$;
- replacing the letters 2 of w by $d_1 + 1, \ldots, d_1 + d_2$, these numbers appearing in this order in $\mathrm{Std}(w)$;

and so on. For instance, if $w = 523165331$, then its standardization is $\mathrm{Std}(w) = 734198562$. By construction, the standardization of a word is always the word of a permutation.

Definition 12.1. *Let $\mathfrak{S} = \bigsqcup_{n\in\mathbb{N}} \mathfrak{S}(n)$, and $\sigma \in \mathfrak{S}$. The* ***free quasi-ribbon function*** *$F_\sigma(X)$ is the element of $\mathbb{C}\langle X\rangle$ defined by*

$$F_\sigma(X) = \sum_{w \,|\, \mathrm{Std}(w)=\sigma^{-1}} x^w,$$

where as usual $x^w = x_{w_1} x_{w_2} \cdots x_{w_{\ell(w)}}$ if w is a word in $(\mathbb{N}^*)^{(\mathbb{N})}$. The algebra of ***free quasi-symmetric functions*** *is the graded subspace* FQSym *of $\mathbb{C}\langle X\rangle$ spanned by the functions $F_\sigma(X)$, $\sigma \in \mathfrak{S}$.*

The terminology of algebra is justified by the following result:

Proposition 12.2. *The subspace* FQSym *is a graded subalgebra of $\mathbb{C}\langle X\rangle$, and we have the product rule:*

$$F_\sigma(X)\, F_\tau(X) = \sum_{\rho \in \sigma \sqcup\!\sqcup \tau} F_\rho(X),$$

where ⧢ denotes the shuffle product of permutations introduced in Section 6.3.

Lemma 12.3. *A word w has standardization σ^{-1} if and only if $w \cdot \sigma$ is a non-decreasing word, and σ is a permutation in $\mathfrak{S}(\ell(w))$ which is of minimal length among the permutations with these properties.*

Proof. Suppose that $\mathrm{Std}(w) = \sigma^{-1}$. Then, $w_{\sigma(1)} \leq w_{\sigma(2)} \leq \cdots \leq w_{\sigma(n)}$, so $w \cdot \sigma$ is non-decreasing. Denote $m_1, m_2, \ldots, m_r$ the multiplicities of the entries $1, 2, \ldots, r$ in w. The set of permutations τ such that $w \cdot \tau$ is non-decreasing is then $\sigma\,(\prod_{i=1}^{r} \mathfrak{S}(m_i))$, where $\mathfrak{S}(m_1)$ acts on words by permutation of the m_1 first letters, $\mathfrak{S}(m_2)$ acts by permutation of the m_2 following letters, etc. If σ were not of minimal length, then there would exist some elementary transposition $s = (j, j+1)$ in one of the group $\mathfrak{S}(m_i) \subset \mathfrak{S}(\ell(w))$, such that $\ell(\sigma s) < \ell(\sigma)$, that is to say that $\sigma(j) > \sigma(j+1)$. This is not possible: $w_{\sigma(j)} = w_{\sigma(j+1)}$, so by definition of the standardization, these letters appear in this order in w, that is to say that $\sigma(j) < \sigma(j+1)$. It is easily seen that this discussion can be reversed, so the two conditions are indeed equivalent. ☐

Proof of Proposition 12.2. Denote m and n the sizes of the two permutations σ and τ. Given a word w of length $\ell(w)$, we also denote $NDR(w)$ the subset of $\mathfrak{S}(\ell(w))$ that consists in permutations σ such that $w \cdot \sigma$ is the non-decreasing reordering of w. The lemma before shows that $\mathrm{Std}(w)^{-1}$ is the permutation of minimal length in $NDR(w)$. By definition, $F_\sigma(X)\,F_\tau(X)$ is the sum of all monomials x^w with $w = w_1 w_2$ word such that $\mathrm{Std}(w_1) = \sigma^{-1}$ and $\mathrm{Std}(w_2) = \tau^{-1}$. Since $w_1 \cdot \sigma$ and $w_2 \cdot \tau$ are non-decreasing words, there exists $m_1, m_2, \ldots, m_r \geq 0$ and $n_1, n_2, \ldots, n_r \geq 0$ such that

$$w_1 \cdot \sigma = 1^{m_1} 2^{m_2} \cdots r^{m_r};$$
$$w_2 \cdot \tau = 1^{n_1} 2^{n_2} \cdots r^{n_r}.$$

Let ρ be any permutation in $NDR(w) \subset \mathfrak{S}(m+n)$. Thus,

$$(w_1 w_2) \cdot \rho = 1^{m_1+n_1} 2^{m_2+n_2} \cdots r^{m_r+n_r}.$$

If

$$\phi = 12 \ldots m_1\,(1+m)(2+m) \ldots (n_1+m)$$
$$(m_1+1) \ldots (m_1+m_2)(n_1+1+m) \ldots (n_1+n_2+m) \cdots,$$

we have $(w_1 w_2) \cdot \rho\phi^{-1} = (w_1 \cdot \sigma)(w_2 \cdot \tau) = (w_1 w_2) \cdot (\sigma, \tau)$. Therefore, $\rho\phi^{-1}$ belongs to $NDR(w_1) \times NDR(w_2)$, so there exists $\sigma' \in NDR(w_1)$ and $\tau' \in NDR(w_2)$ such that

$$\rho = (\sigma', \tau')\phi.$$

However, we saw in Section 6.3 that $\mathfrak{S}(m+n) = \bigsqcup_{\phi \in \mathrm{id}_{[\![1,m]\!]} ⧢ \mathrm{id}_{[\![1,n]\!]}} (\mathfrak{S}(m) \times \mathfrak{S}(n))\,\phi$, this decomposition being compatible with lengths. Therefore, in the decomposition above, $\ell(\rho) = \ell(\sigma') + \ell(\tau') + \ell(\phi)$, with ϕ independent of ρ. Then, $\ell(\rho)$ is minimal if and only if $\ell(\sigma')$ and $\ell(\tau')$ are minimal, so we conclude that

$$\rho = (\sigma, \tau)\phi$$

is the inverse of the standardization of $w_1 w_2$. As $(\sigma, \tau)(\mathrm{id}_{[\![1,m]\!]} \sqcup\!\sqcup \mathrm{id}_{[\![1,n]\!]}) = \sigma \sqcup\!\sqcup \tau$, we have thus shown that if x^w is a monomial that appears in $F_\sigma(X)\, F_\tau(X)$, then x^w is some monomial of a free quasi-ribbon function $F_\rho(X)$ with $\rho \in \sigma \sqcup\!\sqcup \tau$.

Conversely, suppose that $x^w = x^{w_1} x^{w_2}$ appears in $F_{\rho(X)}$ with $\rho \in \sigma \sqcup\!\sqcup \tau$. We denote as before $\rho = (\sigma, \tau)\phi$ with $\phi \in \mathrm{id}_{[\![1,m]\!]} \sqcup\!\sqcup \mathrm{id}_{[\![1,n]\!]}$. If $w_1 \cdot \sigma$ or $w_2 \cdot \sigma$ were not non-decreasing, then a shuffle $((w_1 \cdot \sigma)(w_2 \cdot \tau))\phi = w \cdot \rho$ could not be non-decreasing, which contradicts the fact that $\rho = \mathrm{Std}(w)^{-1}$. So, $\sigma \in NDR(w_1)$ and $\tau \in NDR(w_2)$, and as before, $\ell(\rho) = \ell(\sigma) + \ell(\tau) + \ell(\phi)$. Then, as ρ is of minimal length in $NDR(w)$, σ and τ have also to be of minimal length in $NDR(w_1)$ and $NDR(w_2)$, so $\sigma^{-1} = \mathrm{Std}(w_1)$ and $\tau^{-1} = \mathrm{Std}(w_2)$. Therefore, x^w appears as a monomial of $F_\sigma(X)\, F_\tau(X)$. □

Remark. We could have introduced a formal graded $\mathbb{C}$-algebra spanned by the permutations $\sigma \in \mathfrak{S}$, and with product $\sigma \times \tau = \sum_{\rho \in \sigma \sqcup\!\sqcup \tau} \rho$. Actually, this is how the algebra FQSym was originally constructed by Malvenuto and Reutenauer. Proposition 12.2 shows then that one can realize FQSym as a subalgebra of the free associative algebra $\mathbb{C}\langle X \rangle$. This polynomial realization will allow us to relate FQSym and QSym or NCSym.

▷ *The Hopf algebra structure on* FQSym.

The algebra FQSym can in fact be endowed with a structure of self-dual Hopf algebra. Let us define the missing operations for this structure:

- gradation: $\deg(F_\sigma) = |\sigma| = n$ if $\sigma \in \mathfrak{S}(n)$.
- scalar product: as in Chapter 2 for Sym and in Chapter 6 for NCSym and QSym, we work with a scalar product that is a non-degenerate $\mathbb{C}$-bilinear form. We set $\langle F_\sigma \mid F_\tau \rangle_{\mathrm{FQSym}} = 1_{\sigma\tau = \mathrm{id}_{[\![1,n]\!]}} = 1_{\sigma = \tau^{-1}}$, and we extend this definition by linearity. Thus, if $G_\sigma = F_{\sigma^{-1}}$, then $(G_\sigma)_{\sigma \in \mathfrak{S}}$ is the dual basis of the graded basis $(F_\sigma)_{\sigma \in \mathfrak{S}}$.
- coproduct:

$$\Delta(F_\sigma) = \sum_{k=0}^{n} F_{\mathrm{Std}(\sigma(1)\sigma(2)\ldots\sigma(k))}\, F_{\mathrm{Std}(\sigma(k+1)\ldots\sigma(n))}.$$

- unity and counity: the component FQSym_0 of degree 0 in FQSym is isomorphic to $\mathbb{C}$, and the unity 1_{FQSym} and the counity η_{FQSym} realize this isomorphism of vector spaces.

Proposition 12.4. *Endowed with the previous operations,* FQSym *is a graded self-dual bialgebra.*

Proof. The coproduct can be rewritten as $\Delta(F_\sigma) = \sum_{\sigma=\sigma_1\cdot\sigma_2} F_{\mathrm{Std}(\sigma_1)} \otimes F_{\mathrm{Std}(\sigma_2)}$, where the sum runs over all possible deconcatenations (σ_1, σ_2) of the word of σ. Therefore,

$$((\Delta \otimes \mathrm{id}) \circ \Delta)(F_\sigma) = \sum_{\sigma=\sigma_1\cdot\sigma_2\cdot\sigma_3} F_{\mathrm{Std}(\sigma_1)} \otimes F_{\mathrm{Std}(\sigma_2)} \otimes F_{\mathrm{Std}(\sigma_3)},$$

which shows that Δ is coassociative. Let us now show that it is a morphism of algebras. Fix σ and τ two permutations of size k and l, and let ρ be a permutation in $\sigma \sqcup\!\sqcup \tau$, and (ρ_1, ρ_2) be a deconcatenation of ρ: $\rho = \rho_1 \cdot \rho_2$. We denote σ_1 the subword of ρ_1 that consists in letters smaller than k, and $\widetilde{\tau_1}$ the complement, that is to say the subword of ρ_1 that consists in letters in $[\![k+1, k+l]\!]$. This second word $\widetilde{\tau_1}$ is obtained from the word of a permutation $\tau_1 \in \mathfrak{S}(l)$ by translation of all letters by k. We then have by construction $\rho_1 \in \sigma_1 \sqcup\!\sqcup \tau_1$. We define similarly σ_2 and τ_2, such that $\rho_2 \in \sigma_2 \sqcup\!\sqcup \tau_2$. Then, $\sigma = \sigma_1 \cdot \sigma_2$, and $\tau = \tau_1 \cdot \tau_2$: indeed, looking for instance at σ, it is the subword of ρ that consists in letters smaller than k, therefore, it is the concatenation of the subwords σ_1 of ρ_1 and σ_2 of ρ_2 with the same property. Consequently,

$$\begin{aligned}\Delta(F_\sigma F_\tau) &= \sum_{\rho\in\sigma\sqcup\!\sqcup\tau} \sum_{\rho=\rho_1\cdot\rho_2} F_{\rho_1} \otimes F_{\rho_2} \\ &\leq \sum_{\substack{\sigma=\sigma_1\cdot\sigma_2 \\ \tau=\tau_1\cdot\tau_2}} \sum_{\substack{\rho_1\in\sigma_1\sqcup\!\sqcup\tau_1 \\ \rho_2\in\sigma_2\sqcup\!\sqcup\tau_2}} F_{\rho_1} \otimes F_{\rho_2} = \Delta(F_\sigma)\,\Delta(F_\tau),\end{aligned}$$

where the symbol $\leq$ means that there are fewer terms $F_{\rho_1} \otimes F_{\rho_2}$ in the left-hand side than in the right-hand side. Conversely, fix two deconcatenations $\sigma = \sigma_1 \cdot \sigma_2$ and $\tau = \tau_1 \cdot \tau_2$, and $\rho_1 \in \sigma_1 \sqcup\!\sqcup \tau_1$ and $\rho_2 \in \sigma_2 \sqcup\!\sqcup \tau_2$. If $\rho = \rho_1 \cdot \rho_2$, then the letters of ρ smaller than k come by concatenation of σ_1 and σ_2, so they correspond to the subword σ. Similarly, the letters of ρ larger than $k+1$ come by concatenation of τ_1 and τ_2 and translation by k, so they correspond to the subword $\widetilde{\tau}$. Hence, $\rho \in \sigma \sqcup\!\sqcup \tau$. This shows that the previous inequality is in fact an identity in FQSym. We conclude that the multiplication ∇ and the comultiplication Δ are compatible with one another, and the compatibility of the unity and the counity is trivial: hence, FQSym is a (graded) bialgebra.

Let us finally show that FQSym is self-dual with respect to the scalar product $\langle F_\sigma \mid F_\tau \rangle = \delta_{\sigma,\tau^{-1}}$. Given permutations σ, τ_1 and τ_2, we have

$$\left\langle \Delta(F_\sigma) \,\middle|\, F_{\tau_1} \otimes F_{\tau_2} \right\rangle = \begin{cases} 1 & \text{if } \sigma = \sigma_1 \cdot \sigma_2 \text{ with } \mathrm{Std}(\sigma_1) = \tau_1^{-1}, \mathrm{Std}(\sigma_2) = \tau_2^{-1}, \\ 0 & \text{otherwise.} \end{cases}$$

The conditions for this scalar product being equal to 1 can be rewritten as:

$$\sigma = \sigma_1 \cdot \sigma_2,\ \ x^{\sigma_1} \in F_{\tau_1}(X) \text{ and } x^{\sigma_2} \in F_{\tau_2}(X).$$

Therefore,

$$\left\langle \Delta(F_\sigma) \,\middle|\, F_{\tau_1} \otimes F_{\tau_2} \right\rangle = 1_{X^\sigma \in F_{\tau_1}(X) F_{\tau_2}(X)} = 1_{\mathrm{Std}(\sigma)^{-1} \in \tau_1 \sqcup \tau_2} = \left\langle F_\sigma \,\middle|\, F_{\tau_1} F_{\tau_2} \right\rangle.$$

So, the adjoint of Δ with respect to the scalar product on FQSym is the product ∇. □

The construction of the antipode of FQSym is a bit more complicated, and it relies on the following general result:

Proposition 12.5. *Let H be a graded k-bialgebra that is connected, that is to say that its homogeneous component H_0 of degree 0 has dimension 1 and is isomorphic to k by means of the unity and the counity. Then, there exists a (unique) antipode $\omega_H : H \to H$ that makes H into a Hopf algebra.*

Proof. We have to construct an inverse ω_H of id_H with respect to the convolution product on $\mathrm{End}_k(H)$. We noticed in Section 6.1 that $\omega_H \circ 1_H = 1_H$; see the remark just after Proposition 6.4. Therefore, the value of ω_H on H_0 is

$$(\omega_H)_{|H_0} = (\omega_H \circ 1_H \circ \eta_H)_{|H_0} = (1_H \circ \eta_H)_{|H_0} = \mathrm{id}_{|H_0}.$$

Let us construct ω_H on the homogeneous component H_n of degree $n \geq 1$ in H. By induction we can assume that ω_H is already constructed on $H_0, H_1, \ldots, H_{n-1}$. Notice that for any $x \in H_n$,

$$\Delta(x) = x \otimes 1 + 1 \otimes x + R(x),$$

where R belongs to $\bigoplus_{i=1}^{n-1} H_i \otimes H_{n-i}$. Indeed, to find the component (a_n, b_0) of degree $(n, 0)$ in $\Delta(x)$, one can use the identity $(\mathrm{id}_H \otimes \eta_H) \circ \Delta = \mathrm{id}_H$, which applied to x yields:

$$x = \eta_H(b_0)\, a_n \iff a_n \otimes b_0 = x \otimes 1.$$

The same discussion allows one to compute the component $a_0 \otimes b_n = 1 \otimes x$ of degree $(0, n)$ in $\Delta(x)$. Now, in order to have $\omega_H * \mathrm{id}_H = 1_H \circ \eta_H$, we need for $x \in H_{n \geq 1}$

$$\begin{aligned} 0 = (1_H \circ \eta_H)(x) &= (\omega_H * \mathrm{id}_H)(x) \\ &= (\nabla \circ (\omega_H \otimes \mathrm{id}_H))(x \otimes 1 + 1 \otimes x + R(x)) \\ &= \omega_H(x) + x + (\nabla \circ (\omega_H \otimes \mathrm{id}_H))(R(x)). \end{aligned}$$

Since $\omega_H \otimes \mathrm{id}_H$ is already defined on $R(x) \in \bigoplus_{i=1}^{n-1} H_i \otimes H_{n-i}$, this identity shows that there is only one possibility for ω_H on H_n, namely,

$$\omega_H(x) = -x - (\nabla \circ (\omega_H \otimes \mathrm{id}_H))(R(x)).$$

Thus, the antipode ω_H can indeed be constructed by induction on the degree. □

As a consequence, we get for free the existence of an antipode ω_{FQSym} on FQSym, since it is a graded connected bialgebra. Thus, FQSym is a Hopf algebra. Unfortunately, there is no simple explicit formula for the antipode, and in particular, it can be shown that ω_{FQSym} is not involutive, and in fact has infinite order. To conclude our presentation of the Hopf algebra structure of FQSym, let us detail the connection between it and operations on (ordered, non-commutative) alphabets. Recall that if $X=(x_1,x_2,\ldots)$ and $Y=(y_1,y_2,\ldots)$ are two infinite ordered alphabets of non-commutative variables, then their sum $X\oplus Y$ is $(x_1,x_2,\ldots,y_1,y_2,\ldots)$.

Proposition 12.6. *The coproduct Δ of* FQSym *satisfies, for any free quasi-symmetric function $f\in$* FQSym*:*

$$f(X\oplus Y)=(\Delta(f))(X,Y),$$

where X and Y and ordered alphabets that commute with one another.

Proof. It suffices to prove the identity for a free quasi-ribbon function F_σ of degree n. Let w be a word in the non-commutative variables x_i and y_j, such that $\mathrm{Std}(w)=\sigma^{-1}$. This means that

$$w\cdot\sigma=(x_{i_1}\le x_{i_2}\le\cdots\le x_{i_k}\le y_{j_1}\le y_{j_2}\le\cdots\le y_{j_l})$$

and that σ has minimal length among the permutations with these properties. A condition equivalent to the minimality of the length of the permutation σ is: if $w_{\sigma(a)}=w_{\sigma(a+1)}$, then $\sigma(a)<\sigma(a+1)$. Let $\sigma_1\cdot\sigma_2$ be the deconcatenation of σ with σ_1 of size k and σ_2 of size l. Denote also w_1 and w_2 the words in $(\mathbb{N}^*)^{(\mathbb{N})}$ such that $(x,y)^w=x^{w_1}y^{w_2}$ if one makes the variables x commute with the variables y. Then, $w_1\cdot\mathrm{Std}(\sigma_1)=(x_{i_1}\le x_{i_2}\le\cdots\le x_{i_k})$, and $w_2\cdot\mathrm{Std}(\sigma_2)=(y_{j_1}\le y_{j_2}\le\cdots\le y_{j_l})$. Moreover, if

$$(w_1)_{\mathrm{Std}(\sigma_1)(a)}=(w_1)_{\mathrm{Std}(\sigma_1)(a+1)},$$

then $w_{\sigma(a)}=w_{\sigma(a+1)}$, so $\sigma(a)<\sigma(a+1)$, which implies that $\sigma_1(a)<\sigma_1(a+1)$, and a fortiori $\mathrm{Std}(\sigma_1)(a)<\mathrm{Std}(\sigma_1)(a+1)$. Hence, $\mathrm{Std}(\sigma_1)=\mathrm{Std}(w_1)^{-1}$, and similarly, $\mathrm{Std}(\sigma_2)=\mathrm{Std}(w_2)^{-1}$. As the composition $k+l=n$ is arbitrary, we conclude that

$$F_\sigma(X\oplus Y)=\sum_{\sigma=\sigma_1\cdot\sigma_2}\left(\sum_{\substack{w_1\,|\,\mathrm{Std}(w_1)^{-1}=\mathrm{Std}(\sigma_1)\\ w_2\,|\,\mathrm{Std}(w_2)^{-1}=\mathrm{Std}(\sigma_2)}}x^{w_1}y^{w_2}\right)=\sum_{\sigma=\sigma_1\cdot\sigma_2}F_{\mathrm{Std}(\sigma_1)}(X)\,F_{\mathrm{Std}(\sigma_2)}(Y),$$

that is $(\Delta(F_\sigma))(X,Y)$. □

Later, we shall consider specializations FQSym $\to\mathbb{C}$, that is to say surjective morphisms of algebras. As in the case of QSym (cf. Section 10.1), it will be convenient to consider these specializations as formal alphabets, and to extend the notion of sum of alphabets to these settings. Thus, if X and Y are formal alphabets associated to morphisms $m_X : \mathrm{FQSym} \to \mathbb{C}$ and $m_Y : \mathrm{FQSym} \to \mathbb{C}$, then their sum $X \oplus Y$ is defined as the formal alphabet with associated morphism $m_{X\oplus Y}$ given by $m_{X\oplus Y} = (m_X \otimes m_Y) \circ \Delta$.

▷ *The connection between* FQSym *and* QSym.

We now explain the link between FQSym and QSym. It is contained in the following:

Theorem 12.7. *Let $\Phi : \mathbb{C}\langle X\rangle \to \mathbb{C}[X]$ be the morphism of algebras that makes the variables x_i commutative. The restriction of Φ to* FQSym *is a morphism of Hopf algebras from* FQSym *to* QSym*, and for any permutation $\sigma \in \mathfrak{S}$,*

$$\Phi(F_\sigma) = L_{c(\sigma)},$$

where $c(\sigma)$ is the unique composition whose set of descents is the set of descents $D(\sigma)$ of σ, and where the right-hand side of this identity is a fundamental quasi-symmetric function.

Proof. Fix a permutation σ, and let us compute $\Phi(F_\sigma)$. A basis of $\mathbb{C}[X]$ consists in the monomials x^w with w non-decreasing word. By using the characterization of the identity $\mathrm{Std}(w) = \sigma^{-1}$ given by Lemma 12.3, we get

$$\Phi(F_\sigma) = \sum_{\substack{w \,|\, w\cdot\sigma \text{ non-decreasing} \\ \text{and } (\sigma(a)>\sigma(a+1)) \Rightarrow (w_{\sigma(a)}<w_{\sigma(a+1)})}} x^{w\cdot\sigma}.$$

If $w' = w\cdot\sigma$ is a non-decreasing word in this sum, then the second condition says that the descent set of σ is included in the set of rises of $w\cdot\sigma$. So,

$$\Phi(F_\sigma) = \sum_{\substack{w' \,|\, w' \text{ non-decreasing} \\ D(\sigma)\subset R(w')}} x^{w'}.$$

However, for any composition d, $M_d(X) = \sum_{w' \text{ non-decreasing with } D(d)=R(w')} x^{w'}$. Therefore,

$$\Phi(F_\sigma)(X) = \sum_{d \,|\, d \succeq c(\sigma)} M_d(X) = L_{c(\sigma)}(X).$$

As a consequence, $\Phi(\mathrm{FQSym}) = \mathrm{QSym}$. Moreover, Φ is a morphism of coalgebras from FQSym to QSym. Indeed,

$$((\Phi\otimes\Phi)\circ\Delta)(F_\sigma) = \sum_{\sigma=\sigma_1\cdot\sigma_2} L_{c(\mathrm{Std}(\sigma_1))} \otimes L_{c(\mathrm{Std}(\sigma_2))} = \sum_{\sigma=\sigma_1\cdot\sigma_2} L_{c(\sigma_1)} \otimes L_{c(\sigma_2)}$$

because the standardization of words does not modify the sets of descents. By Proposition 6.13, the last term is $\Delta(L_{c(\sigma)}) = (\Delta\circ\Phi)(F_\sigma)$, so Φ is a morphism of coalgebras. It is also compatible with the product, because it is the restriction of a morphism of algebras $\mathbb{C}\langle X\rangle \to \mathbb{C}[X]$; hence, it is a surjective morphism of bialgebras from FQSym to QSym (the compatibility with the unity and counity is trivial). Such a surjective morphism is automatically compatible with the antipodes: $\Phi \circ \omega_{\mathrm{FQSym}} = \omega_{\mathrm{QSym}} \circ \Phi$. Indeed, ω_{QSym} is the unique linear map in $\mathrm{End}_{\mathbb{C}}(\mathrm{QSym})$ such that

$$\nabla \circ (\omega_{\mathrm{QSym}} \otimes \mathrm{id}_{\mathrm{QSym}}) \circ \Delta = 1_{\mathrm{QSym}} \circ \eta_{\mathrm{QSym}}.$$

As Φ is a surjective map, one does not lose any information by composition on the right by Φ, so,

$$1_{\mathrm{QSym}} \circ \eta_{\mathrm{QSym}} \circ \Phi = \nabla \circ (\omega_{\mathrm{QSym}} \otimes \mathrm{id}_{\mathrm{QSym}}) \circ \Delta \circ \Phi = \nabla \circ ((\omega_{\mathrm{QSym}} \circ \Phi) \otimes \Phi) \circ \Delta$$

is an identity that characterizes ω_{QSym} and a fortiori $\omega_{\mathrm{QSym}} \circ \Phi : \mathrm{FQSym} \to \mathrm{QSym}$. However,

$$\begin{aligned} 1_{\mathrm{QSym}} \circ \eta_{\mathrm{QSym}} \circ \Phi = \Phi \circ 1_{\mathrm{FQSym}} \circ \eta_{\mathrm{FQSym}} &= \Phi \circ \nabla \circ (\omega_{\mathrm{FQSym}} \otimes \mathrm{id}_{\mathrm{FQSym}}) \circ \Delta \\ &= \nabla \circ ((\Phi \circ \omega_{\mathrm{FQSym}}) \otimes \Phi) \circ \Delta, \end{aligned}$$

hence, $\Phi \circ \omega_{\mathrm{FQSym}} = \omega_{\mathrm{QSym}} \circ \Phi$. This ends the proof of the theorem, and it implies Proposition 6.25 which was left unproven. Indeed, if c and d are two compositions and if σ and τ are two arbitrary permutations with $D(\sigma) = c$ and $D(\tau) = d$, then

$$F_\sigma(X)\, F_\tau(X) = \sum_{\rho \in \sigma \sqcup\!\sqcup \tau} F_\rho(X)$$

in FQSym, so, by projection by Φ,

$$L_c(X)\, L_d(X) = \sum_{\rho \in \sigma \sqcup\!\sqcup \tau} L_{c(\rho)}(X). \qquad \square$$

Remark. The advantage of working in FQSym instead of QSym is that the product rule for free quasi-ribbon functions is *multiplicity free*, as opposed to the product rule (Proposition 6.25) for fundamental quasi-symmetric functions. Thus, many identities have a simpler explanation in the larger algebra FQSym. For instance, the identity in QSym

$$\begin{aligned} L_{(1,2)}\, L_{(1,1)} = &\; L_{(1,1,1,2)} + L_{(1,1,2,1)} + L_{(1,2,1,1)} + L_{(1,1,3)} + L_{(1,3,1)} \\ &+ L_{(2,2,1)} + L_{(2,1,2)} + 2\, L_{(1,2,2)} + L_{(2,3)} \end{aligned}$$

comes from the multiplicity free identity

$$\begin{aligned} F_{312}\, F_{21} = &\; F_{31254} + F_{31524} + F_{35124} + F_{53124} + F_{31542} \\ &+ F_{35142} + F_{53142} + F_{35412} + F_{53412} + F_{54312} \end{aligned}$$

in FQSym.

▷ *The connection between* FQSym *and* NCSym.

Let us now complete the commutative diagram drawn at the beginning of this section. If $c \in \mathfrak{C}(n)$ is a composition of size n, we consider

$$R'_c(X) = \sum_{c(\sigma)=c} G_\sigma(X) = \sum_{c(\sigma)=c} F_{\sigma^{-1}}(X),$$

where the sum runs over permutations $\sigma \in \mathfrak{S}(n)$ with $D(\sigma) = D(c)$. The notation is reminiscent of ribbon Schur functions, and indeed:

Theorem 12.8. *For any composition $c \in \mathfrak{S}(n)$, $\sum_{c(\sigma)=c} G_\sigma(X) \in \mathbb{C}\langle X\rangle$ is a non-commutative symmetric function in* NCSym*, and it is equal to the ribbon Schur function $R_c(X)$ (cf. Section 6.1). Thus,* NCSym *is a graded Hopf subalgebra of* FQSym.

Proof. By Lemma 6.10, $R_c(X) = \sum_{w \,|\, D(w)=D(c)} x^w$. On the other hand, we have by definition $G_\sigma(X) = \sum_{w \,|\, \mathrm{Std}(w)=\sigma} x^w$, so, if $R'_c(X) = \sum_{\sigma \,|\, c(\sigma)=c} G_\sigma(X)$, then

$$R'_c(X) = \sum_{\sigma \in \mathfrak{S}(n) \,|\, D(\mathrm{Std}(w))=D(c)} x^w.$$

However, the standardization of words does not change the descent set, so

$$R'_c(X) = \sum_{\sigma \in \mathfrak{S}(n) \,|\, D(w)=D(c)} x^w = R_c(X).$$ □

There is a way to be a bit more precise, and to insert an intermediate Hopf subalgebra FSym between NCSym and FQSym. In several proofs above, we used the fact that the descent set of a word w is the same as the descent set of its standardization $\mathrm{Std}(w)$. A more general result is the following: the recording tableau $Q(w)$ in the Robinson–Schensted–Knuth algorithm is the same as the recording tableau $Q(\mathrm{Std}(w))$ of the standardization. Indeed, if $\sigma = \mathrm{Std}(w)$, and if one has the two tableaux $(P(w), Q(w))$, then it is easy to see that one can compute $(P(\sigma), Q(\sigma))$ as follows:

- the recording tableau $Q(\sigma)$ is the same as the recording tableau $Q(w)$;
- the insertion tableau $P(\sigma)$ is obtained from the insertion tableau $P(w)$ by replacing the m_1 entries 1 by $1, 2, \ldots, m_1$, going column by column (there is at most one entry 1 in each column since $P(w)$ is semistandard); then, by replacing the m_2 entries 2 by $m_1 + 1, \ldots, m_1 + m_2$, again going column by column from left to right; etc.

We call **standardization** of (semistandard) Young tableaux the algorithm of the second part.

Example. Consider the word $w = 76511632$, which has standardization $\sigma = 86512743$. Both σ and w have the same recording tableau

$$Q(w) = Q(\sigma) = \begin{array}{|c|cc}\hline 8 \\ \cline{1-1} 4 \\ \cline{1-1} 3 \\ \cline{1-2} 2 & \multicolumn{1}{c|}{7} \\ \hline 1 & \multicolumn{1}{c|}{5} & \multicolumn{1}{c|}{6} \\ \hline \end{array} ,$$

and the two insertion tableaux are

$$P(w) = \begin{array}{|c|cc}\hline 7 \\ \cline{1-1} 6 \\ \cline{1-1} 5 \\ \cline{1-2} 3 & \multicolumn{1}{c|}{6} \\ \hline 1 & \multicolumn{1}{c|}{1} & \multicolumn{1}{c|}{2} \\ \hline \end{array} \quad ; \quad P(\sigma) = \begin{array}{|c|cc}\hline 8 \\ \cline{1-1} 6 \\ \cline{1-1} 5 \\ \cline{1-2} 4 & \multicolumn{1}{c|}{7} \\ \hline 1 & \multicolumn{1}{c|}{2} & \multicolumn{1}{c|}{3} \\ \hline \end{array} .$$

The insertion tableau $P(\sigma)$ is indeed obtained by standardization of the insertion tableau $P(w)$. Thus,

$$P(\mathrm{Std}(w)) = \mathrm{Std}(P(w)) \quad ; \quad Q(\mathrm{Std}(w)) = Q(w).$$

The identity $D(\mathrm{Std}(w)) = D(w)$ follows from the more general identity $Q(\mathrm{Std}(w)) = Q(w)$. Indeed, call **descent** of standard tableau T of size n an integer $i \in [\![1, n-1]\!]$ such that $i+1$ appears in a strictly higher row than i in T. In the previous example, the tableau $T = Q(w)$ has descent set

$$D(T) = \{1, 2, 3, 6, 7\}.$$

This is also the set of descents of $w = 76511632$, and it is a general phenomenon:

Proposition 12.9. *For any word w, $D(w) = D(Q(w))$.*

Proof. Suppose that i is a descent of the word w: $w_i > w_{i+1}$. Denote (P_i, Q_i) the result of the RSK algorithm applied to the i first letters $w_1 w_2 \ldots w_i$ of w. If the new cell $i+1$ in Q_{i+1} created by insertion of w_{i+1} in the insertion tableau P_i were at the end of a row strictly lower than the row of i in Q_i, then at the previous step one could have followed the same bumping route to insert w_i in P_{i-1}, which is a contradiction. It is also not possible to insert w_{i+1} at the end of the row of w_i in P_i, because that would mean that $w_{i+1} \geq w_i$. So, w_{i+1} is inserted in a higher row, and i is a descent of the tableau Q_{i+1}, and therefore of $Q(w)$. Conversely, if $w_i \leq w_{i+1}$, then one can insert w_{i+1} in P_i in the same row as w_i, and therefore $i+1$ appears in a row lower or equal to the row of i in Q_{i+1}. □

If one wants to exploit in FQSym the fact that $Q(w)$ is invariant by standardization, it is natural to introduce the following special free quasi-symmetric functions:

Definition 12.10. *We call **free symmetric function** an element of the vector subspace* FSym ⊂ FQSym *spanned by the **free Schur functions***

$$S_T(X) = \sum_{\sigma \mid P(\sigma)=T} F_\sigma(X) = \sum_{\sigma \mid Q(\sigma)=T} G_\sigma(X),$$

where T runs over the set of all standard tableaux.

If one combines this definition of free Schur functions with the expansion of ribbon Schur functions into free quasi-ribbon functions, one obtains the relation

$$R_c(X) = \sum_{T \mid D(T)=D(c)} S_T(X);$$

therefore, NCSym ⊂ FSym ⊂ FQSym.

Proposition 12.11. *The space* FSym *is a Hopf subalgebra of* FQSym.

Lemma 12.12. *The product rule for the free quasi-symmetric functions $G_\sigma(X)$ in* FQSym *is*

$$G_\sigma(X)\, G_\tau(X) = \sum_{\rho \mid \rho=\rho_1\cdot\rho_2 \text{ with } \mathrm{Std}(\rho_1)=\sigma \text{ and } \mathrm{Std}(\rho_2)=\tau} G_\rho(X).$$

Proof. We use the scalar product of FQSym and compute

$$\langle F_\rho \mid G_\sigma\, G_\tau \rangle = \langle \Delta(F_\rho) \mid G_\sigma \otimes G_\tau \rangle = \sum_{\rho_1,\rho_2 \mid \rho=\rho_1\cdot\rho_2} 1_{\sigma=\mathrm{Std}(\rho_1)}\, 1_{\tau=\mathrm{Std}(\rho_2)}.$$

If the right-hand side is non-zero, then there is only one possible deconcatenation $\rho_1 \cdot \rho_2$ of ρ such that $\sigma = \mathrm{Std}(\rho_1)$ and $\tau = \mathrm{Std}(\rho_2)$, which is imposed by the sizes of the permutations σ and τ. The result follows, since $(F_\rho)_{\rho\in\mathfrak{S}}$ is the dual basis of $(G_\rho)_{\rho\in\mathfrak{S}}$. □

Proof of Proposition 12.11. Fix two standard tableaux T_1 and T_2 of size n_1 and n_2, and consider the product

$$S_{T_1}(X)\, S_{T_2}(X) = \sum_{\substack{P(\sigma_1)=T_1\\ P(\sigma_2)=T_2}} F_{\sigma_1}(X)\, F_{\sigma_2}(X) = \sum_{\substack{P(\sigma_1)=T_1\\ P(\sigma_2)=T_2\\ \rho\in\sigma_1 \sqcup\!\sqcup\, \sigma_2}} F_\rho(X).$$

Suppose that $\rho \in \sigma_1 \sqcup\!\sqcup\, \sigma_2$ is some permutation appearing in this product, and that ρ' is another permutation of size $n = n_1 + n_2$ such that $P(\rho) = P(\rho')$. By the results of Section 3.2, this is equivalent to the fact that $\rho \equiv \rho'$ in the sense of Knuth equivalence and plactic relations. However, if two words w and w' in $(\mathbb{N}^*)^{(\mathbb{N})}$ are Knuth equivalent, then the subwords $\widetilde{w}$ and $\widetilde{w}'$ obtained by erasing certain letters in a part $A \subset \mathbb{N}$ are again Knuth equivalent (the plactic relations stay the same or become equalities by this operation). Therefore, the part τ_1 of ρ' that consists in letters smaller than n_1 is Knuth equivalent to σ_1, and the part τ_2 of ρ' that consists in letters larger than n_1+1 is Knuth equivalent to the translation of σ_2. So, $\rho' \in \tau_1 \sqcup\!\sqcup\, \tau_2$ with $P(\tau_1) = P(\sigma_1) = T_1$ and $P(\tau_2) = P(\sigma_2) = T_2$. Therefore, ρ' is also involved in the product. This reasoning shows that the set of permutations ρ involved in the product $S_{T_1}(X)S_{T_2}(X)$ is a union of classes of permutations for the relation $\rho \equiv \rho' \iff P(\rho) = P(\rho')$. So, there exists a certain set of standard tableaux $T_1 \sqcup\!\sqcup\, T_2$, such that

$$S_{T_1}(X)S_{T_2}(X) = \sum_{T \in T_1 \sqcup\!\sqcup T_2} S_T(X).$$

Therefore, FSym is a graded subalgebra of FQSym.

To prove that $\Delta(\mathrm{FSym}) \subset \mathrm{FSym}\otimes\mathrm{FSym}$, we use the following characterization: $f \in \mathrm{FQSym}$ is a free symmetric function if and only if, whenever σ and τ are two Knuth equivalent permutations, $\langle f \mid G_\sigma - G_\tau\rangle = 0$. Similarly, $F \in \mathrm{FQSym}\otimes\mathrm{FQSym}$ is in the subspace $\mathrm{FSym}\otimes\mathrm{FSym}$ if and only if, whenever $\sigma_1 \equiv \tau_1$ and $\sigma_2 \equiv \tau_2$,

$$\left\langle F \,\middle|\, G_{\sigma_1}\otimes G_{\sigma_2} - G_{\tau_1}\otimes G_{\tau_2}\right\rangle = 0.$$

Fix $f \in \mathrm{FSym}$, $\sigma_1 \equiv \tau_1$ and $\sigma_2 \equiv \tau_2$, and consider

$$\begin{aligned}\left\langle \Delta(f) \,\middle|\, G_{\sigma_1}\otimes G_{\sigma_2} - G_{\tau_1}\otimes G_{\tau_2}\right\rangle &= \left\langle f \,\middle|\, G_{\sigma_1}G_{\sigma_2} - G_{\tau_1}G_{\tau_2}\right\rangle \\ &= \left\langle f \,\middle|\, \sum_{\rho\,\mid\,\rho=\rho_1\cdot\rho_2,\ \mathrm{Std}(\rho_1)=\sigma_1 \text{ and } \mathrm{Std}(\rho_2)=\sigma_2} G_\rho \right\rangle \\ &\quad - \left\langle f \,\middle|\, \sum_{\eta\,\mid\,\eta=\eta_1\cdot\eta_2,\ \mathrm{Std}(\eta_1)=\tau_1 \text{ and } \mathrm{Std}(\eta_2)=\tau_2} G_\eta \right\rangle.\end{aligned}$$

To show that $\Delta(\mathrm{FSym}) \subset \mathrm{FSym}\otimes\mathrm{FSym}$, it suffices to construct a bijection $\Psi : \rho \mapsto \eta = \Psi(\rho)$ between the two sets of summation on the last line, such that $\rho \equiv \Psi(\rho)$ for any ρ. Then, since $f \in \mathrm{FSym}$, we shall have

$$\left\langle \Delta(f) \,\middle|\, G_{\sigma_1}\otimes G_{\sigma_2} - G_{\tau_1}\otimes G_{\tau_2}\right\rangle = \sum_\rho \left\langle f \,\middle|\, G_\rho - G_{\Psi(\rho)}\right\rangle = 0,$$

and the proof will be completed. To this purpose, notice that if σ_1 and σ_2 are of sizes n_1 and n_2, then the set

$$\{\rho \mid \rho = \rho_1\cdot\rho_2,\ \mathrm{Std}(\rho_1) = \sigma_1 \text{ and } \mathrm{Std}(\rho_2) = \sigma_2\}$$

contains $\frac{(n_1+n_2)!}{n_1!\,n_2!}$ permutations. Indeed, if $n = n_1 + n_2$, then a permutation ρ in this set is obtained by choosing a set $X = \{x_1 < x_2 < \cdots < x_{n_1}\} \subset [\![1,n]\!]$, such that the letters of ρ_1 are in X and the letters of ρ_2 are in $[\![1,n]\!]\setminus X$; and there are $\binom{n}{n_1}$ possibilities for X. Suppose now that the two permutations

$$\begin{aligned}\rho &\in \{\rho \mid \rho = \rho_1\cdot\rho_2,\ \mathrm{Std}(\rho_1) = \sigma_1 \text{ and } \mathrm{Std}(\rho_2) = \sigma_2\}\\ \text{and } \eta &\in \{\eta \mid \eta = \eta_1\cdot\eta_2,\ \mathrm{Std}(\eta_1) = \tau_1 \text{ and } \mathrm{Std}(\eta_2) = \tau_2\}\end{aligned}$$

correspond to the same set X; then, the relation $\sigma_1 \equiv \tau_1$ implies $\rho_1 \equiv \eta_1$, because the elementary Knuth transformations that connect σ_1 to τ_1 can be applied to the same places to connect ρ_1 and η_1. Similarly, if ρ and η correspond to the same set X, then $\rho_2 \equiv \eta_2$. Therefore, as the Knuth equivalence is compatible with concatenation, $\rho \equiv \eta$, so if Ψ is the unique bijection between permutations ρ and η that preserves the associated sets $X \subset [\![1,n]\!]$, then $\rho \equiv \Psi(\rho)$ for any permutation ρ in the sum.

Thus, we have shown that FSym is a graded sub-bialgebra of FQSym. It remains to see that ω_{FQSym} leaves FSym stable. However, FSym is a graded connected bialgebra, so by Proposition 12.5 it admits a unique antipode ω_{FSym} : FSym $\to$ FSym which satisfies

$$\nabla \circ (\omega_{\mathrm{FSym}} \otimes \mathrm{id}_{\mathrm{FSym}}) \circ \Delta = 1_{\mathrm{FSym}} \circ \eta_{\mathrm{FSym}}.$$

Since the restriction of ω_{FQSym} to FSym has this property, we conclude that this restriction is ω_{FSym}, hence stabilizes FSym. □

Theorem 12.13. *For any standard tableau T of shape $\lambda \in \mathfrak{Y}(n)$, if Φ is the morphism of algebras that makes the variables commutative, then $\Phi(S_T) = s_\lambda$. As a consequence, $\Phi(\mathrm{FSym}) = \mathrm{Sym}$.*

Proof. The free Schur function is

$$S_T(X) = \sum_{\sigma \,|\, Q(\sigma)=T} G_\sigma(X) = \sum_{w \,\mathrm{word} \,|\, Q(w)=T} x^w.$$

Recall that the RSK algorithm is a bijection between words and pairs (P,Q) of tableaux of same shape λ, with P semistandard and Q standard. Therefore, the last sum can be rewritten as a sum over semistandard tableaux of shape λ:

$$S_T(X) = \sum_{P \in \mathrm{SST}(\lambda)} x^{w(P,T)},$$

where $w(P,T)$ is the unique word with insertion tableau P and recording tableau T. When one makes the variables commutative, $x^{w(P,T)}$ becomes the monomial x^P, so

$$\Phi(S_T(X)) = \sum_{P \in \mathrm{SST}(\lambda)} x^P = s_\lambda(X)$$

by Theorem 3.2. □

So, the commutative diagram of Hopf algebras presented at the beginning of this section can be completed into:

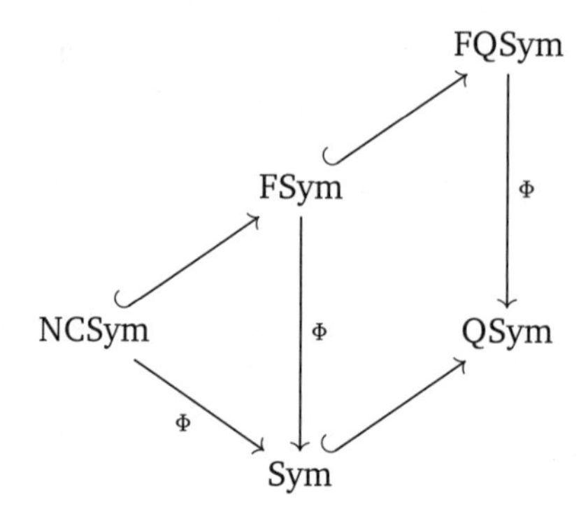

and this diagram encodes many important properties of the RSK algorithm. In the next paragraphs, we shall use these tools in order to study the shape of the Young diagrams associated by the RSK algorithm to *random* words or permutations. To close this section, let us give an application of the construction of FSym:

Corollary 12.14. *Let λ be an integer partition of size n. One has the expansion in fundamental quasi-symmetric functions:*

$$s_\lambda(X) = \sum_{T \in \mathrm{ST}(\lambda)} L_{c(T)}(X),$$

where $c(T)$ is the unique composition such that $D(c(T)) = D(T)$.

Proof. Fix a standard tableau U of shape λ. We have

$$s_\lambda = \Phi(S_U) = \sum_{\sigma \,|\, P(\sigma)=U} \Phi(F_\sigma) = \sum_{\sigma \,|\, P(\sigma)=U} L_{c(\sigma)}.$$

However, by using RSK, one sees that the last sum is also

$$\sum_{T \in \mathrm{ST}(\lambda)} L_{c(\sigma(P,T))},$$

where $\sigma(P,T)$ is the unique permutation with insertion tableau P and recording tableau T. As $c(\sigma(P,T)) = c(T)$, the proof is done. □

12.2 Combinatorics of central measures

In this section, we fix as in the introduction of the chapter a parameter $t = (\alpha, \beta)$ of the Thoma simplex T, and we want to understand the asymptotic behavior of the random partitions $\lambda^{(n)}$ chosen under the central measures $\mathbb{P}_{t,n}$. We have drawn in Figure 12.1 an example of such a random partition, when $n = 200$ and $t = ((\frac{1}{2}, \frac{1}{4}, \frac{1}{8}, \ldots), (0, 0, \ldots))$.

Figure 12.1
A random partition under the probability measure $\mathbb{P}_{t,200}$, where $\alpha = (\frac{1}{2}, \frac{1}{4}, \frac{1}{8}, \ldots)$ and $\beta = 0$.

On this example, the first row has size $101 \simeq 200 \times \alpha_1$, and the second row has size $51 \simeq 200 \times \alpha_2$. The main goal of this section is to prove that this behavior is typical (Theorem 12.19), and that the rows and columns of a large random partition chosen under a central measure satisfy a law of large numbers.

Remark. In the sequel, we shall use freely the language of probability and random variables, and we shall work with sequences of random variables that may converge in several different ways: convergence in probability, convergence in law or distribution, etc. The notation $X \sim \mathbb{P}$ means that the random variable X has its law given by the probability $\mathbb{P}$. On the other hand, $\mathbb{E}[X]$ is the expectation of the random variable X. In most of the chapter, we shall work with random partitions $\lambda^{(n)} \sim \mathbb{P}_{t,n}$, and then, $\mathbb{E}_{t,n}[f] = \mathbb{E}[f(\lambda^{(n)})]$ will be the expectation of a function f of the integer partition $\lambda^{(n)}$, with $\lambda^{(n)}$ chosen randomly according to $\mathbb{P}_{t,n}$.

▷ *Generalized riffle shuffles.*

We shall start by showing that the random partitions $\lambda^{(n)} \sim \mathbb{P}_{t,n}$ can be obtained from certain random permutations $\sigma^{(n)} \in \mathfrak{S}(n)$ by means of the RSK algorithm. This interpretation is not required to prove the law of large numbers, but it will make it quite natural; see the remark after Corollary 12.20.

In the sequel, we shall consider a deck of cards that are numbered from 1 to n. If we enumerate these cards from the top to the bottom of the deck, then we obtain the word $\sigma(1)\sigma(2)\ldots\sigma(n)$ of a permutation $\sigma \in \mathfrak{S}(n)$, so a deck of cards is a convenient way to describe a permutation. Then, a way to produce random permutations is by shuffling randomly these cards. In the following, we shall explain how to associate to any parameter $t \in T$ a procedure of random shuffling, and thus a probability law $\mathbb{Q}_{t,n}$ on $\mathfrak{S}$. We fix a parameter $t = (\alpha, \beta)$ of the Thoma simplex T, and we denote as usual $\gamma = 1 - \sum_{i=1}^{\infty} \alpha_i - \sum_{i=1}^{\infty} \beta_i$. The first step of our algorithm of random shuffling consists in cutting the deck of n cards into several blocks of random sizes, according to a multinomial law of parameters

(α, β, γ). A small difficulty comes from the fact that the two sequences α and β are infinite, but this is merely technical. Let $C = \mathbb{N}^{(\mathbb{N})} \times \mathbb{N}^{(\mathbb{N})} \times \mathbb{N}$ be the set of sequences

$$(a_1, a_2, \ldots), (b_1, b_2, \ldots), c$$

where all the a_i's, all the b_i's and c are non-negative integers, and almost all the a_i's and the b_i's are equal to 0. We denote C_n the subset of C that consists in sequences such that

$$\sum_{i=1}^{\infty} a_i + \sum_{i=1}^{\infty} b_i + c = n,$$

and we endow C_n with the following probability measure:

$$\mathbb{P}[(a_1, a_2, \ldots), (b_1, b_2, \ldots), c] = \frac{n!}{(\prod_{i=1}^{\infty} a_i!)(\prod_{i=1}^{\infty} b_i!)\, c!} \prod_{i=1}^{\infty} (\alpha_i)^{a_i} \prod_{i=1}^{\infty} (\beta_i)^{b_i} \gamma^c.$$

Lemma 12.15. *The previous formula defines indeed a probability measure on C_n:*

$$\sum_{(a,b,c) \in C_n} \mathbb{P}[a, b, c] = 1.$$

Proof. Suppose first that α and β have only a finite number of non-zero terms: $\alpha_i = \beta_i = 0$ for any $i > i_0$. Then, $\mathbb{P}$ vanishes on any sequence (a, b, c) such that $a_i \neq 0$ or $b_i \neq 0$ for some $i > i_0$. The sum of the probabilities is then equal to 1 by Newton's multinomial formula:

$$\begin{aligned}
&\sum_{(a,b,c) \in C_n} \mathbb{P}[a, b, c] \\
&= \sum_{a_1 + \cdots + a_{i_0} + b_1 + \cdots + b_{i_0} + c = n} \binom{n}{a_1, \ldots, a_{i_0}, b_1, \ldots, b_{i_0}, c} \prod_{i=1}^{i_0} (\alpha_i)^{a_i} \prod_{i=1}^{i_0} (\beta_i)^{b_i} \gamma^c \\
&= (\alpha_1 + \cdots + \alpha_{i_0} + \beta_1 + \cdots + \beta_{i_0} + \gamma)^n = 1^n = 1.
\end{aligned}$$

In the general case, denote C_{n,i_0} the subset of C_n that consists in sequences (a, b, c) with $a_i = b_i = 0$ for any $i > i_0$. The set C_n is the increasing union of the sets C_{n,i_0}, therefore,

$$\begin{aligned}
\sum_{(a,b,c) \in C_n} \mathbb{P}[a, b, c] &= \lim_{i_0 \to \infty} \sum_{(a,b,c) \in C_{n,i_0}} \mathbb{P}[a, b, c] = \lim_{i_0 \to \infty} (\alpha_1 + \cdots + \alpha_{i_0} + \cdots + \beta_{i_0} + \gamma)^n \\
&= \left(\sum_{i=1}^{\infty} \alpha_i + \sum_{i=1}^{\infty} \beta_i + \gamma \right)^n = 1^n = 1.
\end{aligned}$$

□

On the other hand, the last step of our algorithm will consist in a **riffle shuffle** of the blocks previously created. Consider two words $v = v_1 v_2 \ldots v_l$ and $w = w_1 w_2 \ldots w_m$, with all the letters of these two words that are distinct and that belong to $[\![1, n]\!]$. These words v and w represent blocks of cards extracted from the deck of cards. A riffle shuffle of these two words is a random element of the shuffle product $v \sqcup\!\sqcup\, w$, which is the set of words of length $l + m$ obtained by interlacing the letters of v and the letters of w; see Figure 12.2. There are $\frac{(l+m)!}{l!\,m!}$ possibilities, and we give to each of them the probability $\frac{l!\,m!}{(l+m)!}$. In the sequel, we denote $v \sqcap\!\sqcap\, w$ the riffle shuffle of two words, which is a random word.

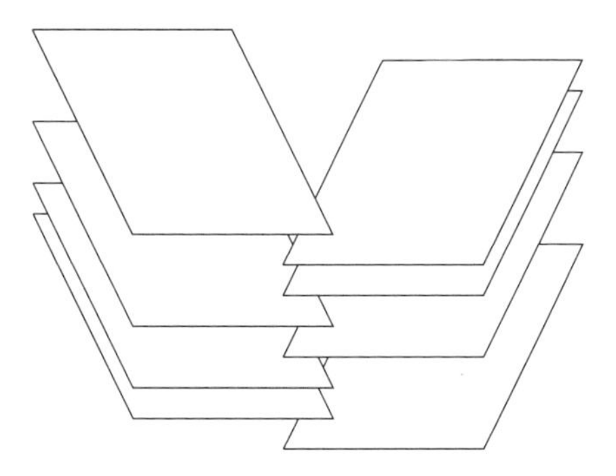

Figure 12.2
The riffle shuffle of two blocks of cards.

The operation of riffle shuffle is associative and commutative, in the following sense. Consider words $v^{(1)}, \ldots, v^{(r)}$ with distinct letters, and an ordered binary tree with leaves $1, 2, \ldots, r$. To this binary tree, we can associate a way to shuffle the words $v^{(1)}, \ldots, v^{(r)}$. For instance, if one considers the binary tree

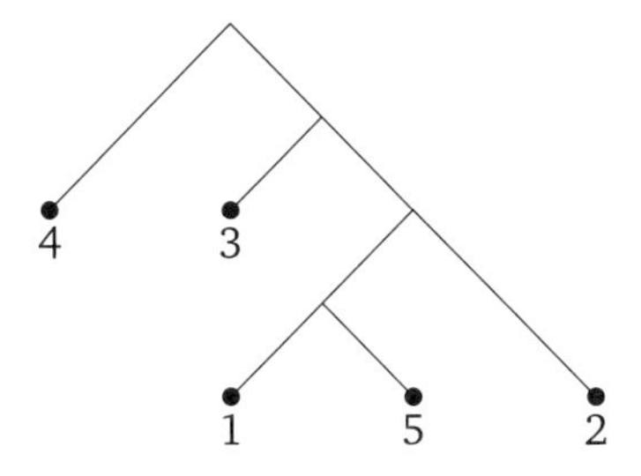

then it is associated to the riffle shuffle

$$v^{(4)} \sqcap\!\sqcap\, (v^{(3)} \sqcap\!\sqcap\, ((v^{(1)} \sqcap\!\sqcap\, v^{(5)}) \sqcap\!\sqcap\, v^{(2)})).$$

Then, the associativity and commutativity of the riffle shuffle amounts to the following statement: the probability distribution on words of length $\ell(v^{(1)})+\ell(v^{(2)})+\cdots+\ell(v^{(r)})$ is independent from the choice of an ordered binary tree. Indeed, this probability measure gives the weight

$$\frac{\ell(v^{(1)})!\,\ell(v^{(2)})!\cdots\ell(v^{(r)})!}{(\ell(v^{(1)})+\ell(v^{(2)})+\cdots+\ell(v^{(r)}))!}$$

to any word w such that, if $w^{(i)}$ is the subword of w with the same letters as $v^{(i)}$, then $w^{(i)} = v^{(i)}$. So, we can denote without ambiguity $v^{(1)} ⧢ v^{(2)} ⧢ \cdots ⧢ v^{(r)}$ the riffle shuffle of any finite sequence of words with distinct letters.

We can now fully describe the algorithm of **generalized riffle shuffling:**

Definition 12.16 (Generalized riffle shuffle). *The generalized riffle shuffle of parameter $t \in T$ is the following procedure to get a random permutation $\sigma \in \mathfrak{S}(n)$:*

0. *We start from an ordered deck of cards* $12\ldots n$.
1. *We cut the deck in blocks of sizes $a_1, a_2, \ldots, b_1, b_2, \ldots, c$, where $(a,b,c) \in C_n$ is chosen randomly according to the probability measure*

$$\mathbb{P}[a,b,c] = \frac{n!}{(\prod_{i=1}^{\infty} a_i!)(\prod_{i=1}^{\infty} b_i!)\,c!} \prod_{i=1}^{\infty} (\alpha_i)^{a_i} \prod_{i=1}^{\infty} (\beta_i)^{b_i}\, \gamma^c.$$

   The blocks of sizes $a_1, a_2, \ldots$ will be called type A, the blocks of sizes $b_1, b_2, \ldots$ will be called type B, and the last block of size c will be called type C.
2. *We reverse each block $(k+1)(k+2)\ldots(k+b_i)$ of type B in order to obtain $(k+b_i)\ldots(k+2)(k+1)$, and we randomize the last block of type C, replacing $(n-c+1)(n-c+2)\ldots n$ by a uniform random permutation of these c letters.*
3. *We shuffle randomly all the blocks, according to the operation ⧢ previously described.*

Example. Suppose that $n = 15$ and $t = ((\frac{1}{6}, \frac{1}{4}, 0, \ldots), (\frac{1}{3}, 0, \ldots))$. We start from the deck of cards

$$123456789\mathrm{ABCDEF}.$$

1. The probability measure on sequences (a,b,c) that is associated to the parameter t is supported by sequences $((a_1,a_2), b_1, c)$, all the other numbers $a_{i\geq 3}$ and $b_{i\geq 2}$ being equal to 0. One possibility is $a_1 = 3$, $a_2 = 3$, $b_1 = 6$, and $c = 3$. We then cut the deck into the four blocks

$$123\,|\,456\,|\,789\mathrm{ABC}\,|\,\mathrm{DEF}.$$

2. For the second step, we reverse the third block which is of type B, and we randomize the last block, one possibility being FDE. Thus, we get

$$123\,|\,456\,|\,\mathrm{CBA}987\,|\,\mathrm{FDE}.$$

3. Finally, we shuffle back together these four blocks, and one possibility is

$$4\mathrm{CB}1\mathrm{A}25\mathrm{F}9387\mathrm{D}6\mathrm{E}.$$

Thus, one possible result of the generalized riffle shuffle of parameter t and with $n = 15$ is the permutation $\sigma = 4CB1A25F9387D6E$.

We denote $\mathbb{Q}_{t,n}$ the probability measure on $\mathfrak{S}(n)$ given by the generalized riffle shuffle of parameter $t \in T$. The next Theorem 12.17 expresses $\mathbb{Q}_{t,n}[\sigma]$ as a certain specialization of the free quasi-symmetric function G_σ. If $\gamma \geq 0$ and $\sigma \in \mathfrak{S}(n)$, we denote

$$F_\sigma(\gamma E) = \frac{\gamma^n}{n!}.$$

If $\sigma \in \mathfrak{S}(m)$ and $\tau \in \mathfrak{S}(n)$, then the product $F_\sigma F_\tau$ involves $\frac{(m+n)!}{m!\,n!}$ terms F_ρ. Consequently, γE is a specialization of FQSym, that is a morphism of algebras FQSym $\to \mathbb{C}$. It is non-negative, in the sense that $F_\sigma(\gamma E) \geq 0$ for any $\sigma \in \mathfrak{S}$. On the other hand, any summable sequence $A = \{\alpha_1, \alpha_2, \ldots\}$ yields a specialization FQSym $\to \mathbb{C}$ by replacing the variables $x_1, x_2, \ldots$ in $\mathbb{C}\langle X \rangle$ by $\alpha_1, \alpha_2, \ldots$ If all the numbers α_i are non-negative, then the specialization $F_\sigma \mapsto F_\sigma(A)$ is also non-negative. We can remark that the two specializations associated to the (formal) alphabets A and γE factorize through the morphism Φ : FQSym $\to$ QSym, and in the setting of QSym, they can be seen as extensions of the corresponding specializations A and γE of Sym defined in Section 11.1. Finally, considering as before a summable sequence $B = \{\beta_1, \beta_2, \ldots\}$ of non-negative numbers, the linear map

$$F_\sigma \mapsto F_\sigma(\overline{B}) = (-1)^{\deg F_\sigma}\,(\omega(F_\sigma))(B)$$

defines again a non-negative specialization. Indeed, by using the fact that B factorizes through Φ, one obtains

$$\begin{aligned}(-1)^{\deg F_\sigma}\,(\omega(F_\sigma))(B) &= (-1)^{\deg F_\sigma}\,\Phi(\omega_{\mathrm{FQSym}}(F_\sigma))(B)\\ &= (-1)^{\deg F_\sigma}\,\omega_{\mathrm{QSym}}(L_{c(\sigma)})(B) = L_{c'(\sigma)}(B),\end{aligned}$$

where $c'(\sigma)$ is the composition conjugated to the composition of descents of σ. This proves the non-negativity, and $\overline{B}$ defines indeed a morphism of algebras, since it is a composition of morphisms.

By using the coproduct of FQSym, we can add formal alphabets (specializations), and define

$$F_\sigma(A \oplus \overline{B} \oplus \gamma E) = \Delta^{(3)}(F_\sigma)(A, \overline{B}, \gamma E).$$

Therefore, to any parameter t of the Thoma simplex, we can associate a non-negative specialization $A \oplus \overline{B} \oplus \gamma E$ of FQSym, which has the following additional properties:

1. It factorizes through the morphism Φ : FQSym $\to$ QSym, and the corresponding specialization $A \oplus \overline{B} \oplus \gamma E$ of QSym is an extension of the non-negative specialization $A + \overline{B} + \gamma E$ of Sym defined in Section 11.1.

2. It is normalized: $F_1(A \oplus \overline{B} \oplus \gamma E) = 1$.

In the sequel, we shall sometimes denote $F_\sigma(A \oplus \overline{B} \oplus \gamma E) = F_\sigma(t)$ for $t \in T$.

Theorem 12.17. *The probability measure induced by the generalized riffle shuffle of parameter t satisfies:*

$$\mathbb{Q}_{t,n}[\sigma] = G_\sigma(t).$$

Proof. We represent a probability measure $\mathbb{P}$ on $\mathfrak{S}(n)$ by the free quasi-symmetric function

$$F_{\mathbb{P}} = \sum_{\sigma \in \mathfrak{S}(n)} \mathbb{P}[\sigma] F_\sigma.$$

Then, the value of $\mathbb{P}[\sigma]$ can be recovered from this free quasi-symmetric function by taking the scalar product $\langle F_{\mathbb{P}} \mid G_\sigma \rangle$. We are going to describe the three steps of the generalized riffle shuffle by using operators on FQSym and on tensor powers of this Hopf algebra. More precisely, if X is a formal alphabet (specialization FQSym $\to \mathbb{C}$), denote t_X the linear map

$$\begin{aligned} t_X : \mathrm{FQSym} &\to \mathrm{FQSym} \\ F_\sigma &\mapsto F_\sigma(X) F_\sigma. \end{aligned}$$

The dual of t_X with respect to the scalar product of FQSym is the operator t_X^* : $F_\sigma \mapsto G_\sigma(X) F_\sigma$. On the other hand, if $X = (x)$ is an alphabet that consists in a single variable, then $F_\sigma(x) = 0$ unless $\sigma = 12 \ldots n$. Therefore,

$$t_{(x)}(F_\sigma) = t_{(x)}^*(F_\sigma) = 1_{\sigma = 12\ldots n}\, x^n\, F_{12\ldots n}.$$

If X is a formal alphabet, we shall also use the linear map

$$\begin{aligned} u_X : \mathrm{FQSym} &\to \mathrm{FQSym} \\ F_\sigma &\mapsto F_\sigma(X) F_{12\ldots n}. \end{aligned}$$

If (x) is a single variable, then $t_{(x)} = u_{(x)}$. Moreover, $u_{\gamma E}^* = R \circ t_{(\gamma)}$, where R : FQSym $\to$ FQSym is the operator of randomization, which is defined by

$$R(F_\sigma) = \frac{1}{n!} \sum_{\tau \in \mathfrak{S}(n)} F_\tau$$

if $\sigma \in \mathfrak{S}(n)$. We assume in the following that $t = (\alpha, \beta)$ has only a finite number of non-zero terms: $\alpha_i = \beta_i = 0$ if $i > i_0$. We shall explain at the end how to remove this assumption.

0. We start from the ordered deck of cards $12 \ldots n$. It is represented by the free quasi-symmetric function $F_{12\ldots n}$. In the following, we abbreviate $F_{12\ldots n} = F_{[n]}$.

1. The result of the first step of the generalized riffle shuffle is represented by the tensor

$$\sum \frac{n!}{a_1! \cdots a_{i_0}!\, b_1! \cdots b_{i_0}!\, c!} \prod_{i=1}^{i_0} (\alpha_i)^{a_i} \prod_{i=1}^{i_0} (\beta_i)^{b_i}\, \gamma^c$$
$$\times F_{[a_1]} \otimes \cdots \otimes F_{[a_{i_0}]} \otimes F_{[b_1]} \otimes \cdots \otimes F_{[b_{i_0}]} \otimes F_{[c]},$$

where the sum runs over sequences (a,b,c) such that $a_1+\cdots+a_{i_0}+b_1+\cdots+b_{i_0}+c=n$. One can rewrite this as $U_1(F_{[n]})$, where

$$U_1=\left(t_{(\alpha_1)}\otimes\cdots\otimes t_{(\alpha_{i_0})}\otimes t_{(\beta_1)}\otimes\cdots\otimes t_{(\beta_{i_0})}\otimes t_{(\gamma)}\right)\circ t_E^{\otimes 2i_0+1}\circ\Delta^{(2i_0+1)}\circ t_E^{-1}.$$

Indeed, the coproduct takes care of all the possible ways of splitting the deck of n cards, the operators t_E take care of the factorials, and the operators $t_{(x)}$ correspond to the powers of α_i, β_i or γ.

2. For the second step, we shall work with the linear map

$$\nu(F_\sigma)=(-1)^{\deg F_\sigma}\,\omega(F_\sigma),$$

where ω is the antipode of FQSym. The value of ω_{FQSym} on F_σ is in general difficult to compute, but it is easy when $F_\sigma=F_{12\ldots n}$:

$$\omega(F_{12\ldots n})=(-1)^n\,F_{n(n-1)\ldots 1}.$$

Indeed, suppose that $\omega(F_{[n]})=\sum_{\tau\in\mathfrak{S}(n)}a_\tau F_\tau$. Since $\Phi:\mathrm{FQSym}\to\mathrm{QSym}$ is a morphism of Hopf algebras,

$$(-1)^n\,L_{1^n}=\omega(L_n)=\Phi(\omega(F_{[n]}))=\sum_{\tau\in\mathfrak{S}(n)}a_\tau\,L_{c(\tau)}.$$

However, $\sigma=n(n-1)\ldots 1$ is the unique permutation such that $c(\sigma)=1^n$, therefore, $a_\sigma=(-1)^n$, and $a_\tau=0$ for all the other permutations. As a consequence, $\nu(F_{12\ldots n})=F_{n(n-1)\ldots 1}$. It follows immediately that the second step of the generalized riffle shuffle can be represented by

$$U_2=(\mathrm{id})^{\otimes i_0}\otimes(\nu)^{\otimes i_0}\otimes R.$$

3. Finally, the last step of the algorithm is represented by the operator

$$U_3=t_E\circ\nabla^{(2i_0+1)}\circ(t_E^{-1})^{\otimes 2i_0+1},$$

where $\nabla^{(2i_0+1)}:\mathrm{FQSym}^{\otimes 2i_0+1}\to\mathrm{FQSym}$ is the product. Indeed, the product of FQSym is the shuffle product (with permutation words that are translated in order to shuffle words with distinct letters), and the operators t_E allow one to introduce the uniform probabilities on shuffles

$$\frac{a_1!\cdots a_{i_0}!\,b_1!\cdots b_{i_0}!\,c!}{n!}.$$

Gathering everything, we conclude that

$$\mathbb{Q}_{t,n}[\sigma]=\left\langle(U_3\circ U_2\circ U_1)(F_{[n]})\,\middle|\,G_\sigma\right\rangle.$$

Now, notice that the operators t_E and t_E^{-1} commute with every degree-preserving operator FQSym $\to$ FQSym. Therefore, we can simplify

$$\begin{aligned}&(U_3 \circ U_2 \circ U_1)(F_{[n]})\\ &= \left(\nabla^{(2i_0+1)} \circ ((\mathrm{id})^{\otimes i_0} \otimes (\nu)^{\otimes i_0} \otimes R) \circ (t_{(\alpha_1)} \otimes \cdots \otimes t_{(\gamma)}) \circ \Delta^{(2i_0+1)}\right)(F_{[n]}).\end{aligned}$$

Then, by using the self-adjointness of the Hopf algebra FQSym, we can rewrite

$$\begin{aligned}&\mathbb{Q}_{t,n}[\sigma]\\ &= \left\langle F_{[n]} \,\middle|\, (\nabla^{(2i_0+1)} \circ (t_{(\alpha_1)} \otimes \cdots \otimes t_{(\gamma)}) \circ ((\mathrm{id})^{\otimes i_0} \otimes (\nu)^{\otimes i_0} \otimes R^*) \circ \Delta^{(2i_0+1)})(G_\sigma)\right\rangle\\ &= \left\langle F_{[n]} \,\middle|\, (\nabla^{(2i_0+1)} \circ (u_{(\alpha_1)} \otimes \cdots \otimes u_{(\beta_{i_0})} \otimes u_{\gamma E}) \circ ((\mathrm{id} \otimes \nu)^{\otimes i_0} \otimes \mathrm{id}) \circ \Delta^{(2i_0+1)})(G_\sigma)\right\rangle.\end{aligned}$$

The quantity

$$Q = ((u_{(\alpha_1)} \otimes \cdots \otimes u_{(\beta_{i_0})} \otimes u_{\gamma E}) \circ ((\mathrm{id})^{\otimes i_0} \otimes (\nu)^{\otimes i_0} \otimes \mathrm{id}) \circ \Delta^{(2i_0+1)})(G_\sigma)$$

is a linear combination of tensor products $F_{[a_1]} \otimes \cdots \otimes F_{[a_{i_0}]} \otimes F_{[b_1]} \otimes \cdots \otimes F_{[b_{i_0}]} \otimes F_{[c]}$, with $a_1 + \cdots + a_{i_0} + b_1 + \cdots + b_{i_0} + c = n$. When one takes the shuffle product of this tensor of free quasi-symmetric functions, only one shuffle will be equal to the identity permutation $F_{[n]}$. Therefore, if

$$Q = \sum_{(a,b,c)} f(a,b,c)\, F_{[a_1]} \otimes \cdots \otimes F_{[a_{i_0}]} \otimes F_{[b_1]} \otimes \cdots \otimes F_{[b_{i_0}]} \otimes F_{[c]},$$

then $\mathbb{Q}_{t,n}[\sigma] = \sum_{(a,b,c)} f(a,b,c)$. Another more algebraic way to say this is: if f : FQSym $\to \mathbb{C}$ is the linear map which sends F_σ to 1 for any permutation σ, then

$$\mathbb{Q}_{t,n}[\lambda] = (f^{\otimes 2i_0+1} \circ (u_{(\alpha_1)} \otimes \cdots \otimes u_{(\beta_{i_0})} \otimes u_{\gamma E}) \circ ((\mathrm{id})^{\otimes i_0} \otimes (\nu)^{\otimes i_0} \otimes \mathrm{id}) \circ \Delta^{(2i_0+1)})(G_\sigma).$$

However, $f \circ u_X = \phi_X$ is the specialization FQSym $\to \mathbb{C}$ associated to the formal alphabet X, so we obtain finally

$$\begin{aligned}\mathbb{Q}_{t,n}[\sigma] &= ((\phi_{(\alpha_1)} \otimes \cdots \otimes \phi_{(\beta_{i_0})} \otimes \phi_{\gamma E}) \circ ((\mathrm{id})^{\otimes i_0} \otimes (\nu)^{\otimes i_0} \otimes \mathrm{id}) \circ \Delta^{(2i_0+1)})(G_\sigma)\\ &= G_\sigma((\alpha_1) \oplus \cdots \oplus (\alpha_{i_0}) \oplus \overline{(\beta_1)} \oplus \cdots \oplus \overline{(\beta_{i_0})} \oplus \gamma E)\\ &= G_\sigma(A \oplus \overline{B} \oplus \gamma E) = G_\sigma(t).\end{aligned}$$

We finally have to explain how to deal with the general case of infinite sequences α and β. If one endows the Thoma simplex T with the topology of pointwise convergence of the coordinates, then the sequences t with only a finite number of non-zero terms are dense in T, and on the other hand, the maps $t \mapsto G_\sigma(t)$ and $t \mapsto \mathbb{Q}_{t,n}[\sigma]$ are continuous. Therefore, the identity on a dense subset implies the identity in the general case. □

Corollary 12.18. *Let σ be a random partition under the probability measure $\mathbb{Q}_{t,n}$, and λ the shape of the two standard tableaux $P(\sigma)$ and $Q(\sigma)$. The random partition λ has law $\mathbb{P}_{t,n}$.*

Proof. Denote RSKsh : $\mathfrak{S}(n) \to \mathfrak{Y}(n)$ the map which sends a permutation σ to the shape λ of the two tableaux $P(\sigma)$ and $Q(\sigma)$. We want to compute the image law

$$(\mathrm{RSKsh}_* \mathbb{Q}_{t,n})[\lambda] = \mathbb{Q}_{t,n}[\{\sigma \in \mathfrak{S}(n) \,|\, \mathrm{RSKsh}(\sigma) = \lambda\}].$$

However, if one gathers the permutations σ according to their recording tableau $Q \in \mathrm{ST}(\lambda)$, one obtains

$$\mathbb{Q}_{t,n}[\{\sigma \in \mathfrak{S}(n) \,|\, \mathrm{RSKsh}(\sigma) = \lambda\}] = \sum_{T \in \mathrm{ST}(\lambda)} \left(\sum_{\sigma \,|\, Q(\sigma) = T} G_\sigma(t) \right) = \sum_{T \in \mathrm{ST}(\lambda)} S_T(t),$$

where S_T is the free Schur function associated to the standard tableau T. Since the non-negative specialization t commutes with the morphism Φ : FQSym $\to$ QSym, $S_T(t) = \Phi(S_T)(t) = s_\lambda(t)$, therefore,

$$(\mathrm{RSKsh}_* \mathbb{Q}_{t,n})[\lambda] = \sum_{T \in \mathrm{ST}(\lambda)} s_\lambda(t) = d_\lambda \, s_\lambda(t) = \mathbb{P}_{t,n}[\lambda]. \qquad \square$$

Thus, any central measure on partitions corresponds via RSK to a model of random permutations obtained by generalized riffle shuffle. As a particular case, consider the parameter $t = (0,0)$, which corresponds to the specialization E : $F_\sigma \mapsto \frac{1}{n!}$, that is to say the uniform probability on permutations $\sigma \in \mathfrak{S}(n)$. Its image by RSK is the Plancherel measure $\mathbb{P}_n[\lambda] = d_\lambda \, s_\lambda(E) = \frac{(d_\lambda)^2}{n!}$.

▷ *The law of large numbers for central measures.*

We now start the asymptotic analysis of random partitions $\lambda^{(n)}$ chosen according to a central measure $\mathbb{P}_{t,n}$, with $t \in T$ fixed. Recall that a sequence of random variables $(X_n)_{n\in\mathbb{N}}$ **converges in probability** towards a constant c if, for every $\varepsilon > 0$,

$$\lim_{n\to\infty} \mathbb{P}[|X_n - c| \geq \varepsilon] = 0.$$

We then denote $X_n \to_{\mathbb{P}} c$, or $X_n \to_{\mathbb{P}_n} c$ if $\mathbb{P}_n$ is the law of X_n. A convenient criterion in order to prove a convergence in probability is the **Bienaymé–Chebyshev inequality**: if $\lim_{n\to\infty} \mathbb{E}[(X_n - c)^2] = 0$, then $X_n \to_{\mathbb{P}} c$. Indeed,

$$\begin{aligned}\mathbb{P}[|X_n - c| \geq \varepsilon] &= \mathbb{P}[(X_n - c)^2 \geq \varepsilon^2] = \mathbb{E}\left[1_{(X_n-c)^2 \geq \varepsilon^2}\right] \\ &\leq \mathbb{E}\left[1_{(X_n-c)^2 \geq \varepsilon^2} \frac{(X_n - c)^2}{\varepsilon^2}\right] \leq \frac{\mathbb{E}[(X_n - c)^2]}{\varepsilon^2} \to_{n\to\infty} 0.\end{aligned}$$

Now, the first main asymptotic result regarding central measures is the following:

Theorem 12.19 (Kerov–Vershik). *Let $\lambda^{(n)}$ be a random partition chosen according to a central measure $\mathbb{P}_{t,n}$, with $t = (\alpha, \beta)$ fixed. For any $i \geq 1$, we have as n goes to infinity*

$$\frac{\lambda_i^{(n)}}{n} \to_{\mathbb{P}_{t,n}} \alpha_i \quad ; \quad \frac{\lambda_i^{(n)\prime}}{n} \to_{\mathbb{P}_{t,n}} \beta_i.$$

Thus, the coordinates α_i and β_i of the parameter t in the Thoma simplex correspond to asymptotic frequencies of the rows and columns of a random partition under $\mathbb{P}_{t,n}$.

Corollary 12.20. *Let $\sigma^{(n)} \in \mathfrak{S}(n)$ be a random permutation of size n obtained by a generalized riffle shuffle of parameter t. The length $\ell^{(n)}$ of a longest increasing subsequence of $\sigma^{(n)}$ satisfies*

$$\frac{\ell^{(n)}}{n} \to_{\mathbb{Q}_{t,n}} \alpha_1.$$

Proof. We know from Section 3.2 that the length of the first row of $\mathrm{RSKsh}(\sigma^{(n)})$ is the first Greene invariant of $\sigma^{(n)}$, that is to say the length of a longest increasing subsequence. The result follows by combining Theorem 12.19 and Theorem 12.17. More generally, for any $k \geq 1$, one has convergence in probability of the k-th Greene invariant

$$\frac{L_k(\sigma^{(n)})}{n} \to_{\mathbb{Q}_{t,n}} \alpha_1 + \alpha_2 + \cdots + \alpha_k. \qquad \square$$

Remark. It is not very hard to see that the k-th Greene invariant $L_k(\sigma^{(n)})$ with $\sigma^{(n)} \sim \mathbb{Q}_{t,n}$ has to be *at least* of size $\simeq n(\alpha_1 + \alpha_2 + \cdots + \alpha_k)$. Indeed, in the generalized riffle shuffle, the k first blocks of cards of type A and of sizes $a_1, \ldots, a_k$ give rise after shuffling to k disjoint increasing subwords in $\sigma^{(n)}$, hence,

$$L_k(\sigma^{(n)}) \geq a_1 + \cdots + a_k.$$

However, the multinomial law of parameters n and (α, β, γ) satisfies

$$\forall i \geq 1, \ \frac{a_i}{n} \to_{\mathbb{P}} \alpha_i.$$

Thus, $L_k(\sigma^{(n)})$ is asymptotically larger than $n(\alpha_1+\cdots+\alpha_k)$. The previous corollary ensures that one does not create much larger increasing subsequences by taking into account the other blocks of cards of type A, and the blocks of type B and C. This could be proven directly by looking more precisely at the positions of the letters of the different blocks, but the algebra of observables $\mathscr{O}$ will provide a much easier proof of the result.

Remark. In the notes at the end of Chapter 3, we mentioned that the first columns $\lambda'_1, \ldots, \lambda'_k$ of the shape of the tableaux associated by RSK to a word w were related to the longest strictly decreasing subwords of w. Thus,

$$\lambda'_1 + \cdots + \lambda'_k = \max(\ell(w^{(1)}) + \cdots + \ell(w^{(k)})),$$

where the maximum runs over families $\{w^{(1)}, \ldots, w^{(k)}\}$ of disjoint decreasing subwords of w. Denote $M_k(w)$ this invariant of words. Combining again Theorems 12.19 and 12.17, we see that if $\sigma^{(n)} \sim \mathbb{Q}_{t,n}$, then

$$\frac{M_k(\sigma^{(n)})}{n} \to_{\mathbb{Q}_{t,n}} \beta_1 + \beta_2 + \cdots + \beta_k.$$

Again, if one looks at the first blocks of type B, then one sees readily that $M_k(\sigma^{(n)})$ is asymptotically at least of size $\simeq n(\beta_1+\cdots+\beta_k)$, and the aforementioned convergence in probability ensures that the generalized riffle shuffle does not create larger decreasing subsequences.

In order to prove Theorem 12.19, we shall use the algebra of observables of partitions $\mathscr{O}$, and first prove the following easier result:

Proposition 12.21. *If $\lambda^{(n)} \sim \mathbb{P}_{t,n}$, then for any integer $k \geq 1$,*

$$\frac{p_k(\lambda^{(n)})}{n^k} \to_{\mathbb{P}_{t,n}} p_k(t),$$

where on the left-hand side, $p_k(\lambda^{(n)})$ is the evaluation of the observable $p_k \in \mathscr{O}$ on the random partition $\lambda^{(n)}$; and on the right-hand side, $p_k(t)$ is the non-negative specialization t of the symmetric function $p_k(X) \in \mathrm{Sym}$.

Proof. For any integer partition μ, recall that the observables Σ_μ and p_μ have the same top homogeneous component with respect to the degree. The observables $(\Sigma_\mu)^2$ and $\Sigma_{\mu\sqcup\mu}$ also have the same top homogeneous component. On the other hand, if σ_μ denotes a permutation with cycle type $\mu \in \mathfrak{Y}(k)$, then

$$\mathbb{E}_{t,n}[\Sigma_\mu] = n^{\downarrow k} \sum_{\lambda \in \mathfrak{Y}(n)} \mathbb{P}_{t,n}[\lambda]\, \chi^\lambda(\sigma_\mu) = n^{\downarrow k}\, \tau_t(\sigma_\mu) = n^{\downarrow k}\, p_\mu(t).$$

Therefore, $\lim_{n\to\infty} \frac{\mathbb{E}_{t,n}[\Sigma_\mu]}{n^{|\mu|}} = p_\mu(t)$ for any $\mu \in \mathfrak{Y} = \bigsqcup_{n\in\mathbb{N}} \mathfrak{Y}(n)$. This actually implies that

$$\frac{\Sigma_\mu(\lambda^{(n)})}{n^{|\mu|}} \to_{\mathbb{P}_{t,n}} p_\mu(t).$$

Indeed,

$$\begin{aligned}
\mathbb{E}_{t,n}\left[\left(\frac{\Sigma_\mu}{n^k} - p_\mu(t)\right)^2\right] &= \frac{\mathbb{E}_{t,n}[(\Sigma_\mu)^2]}{n^{2k}} + \left(1 - \frac{2n^{\downarrow k}}{n^k}\right)(p_\mu(t))^2 \\
&= \frac{\mathbb{E}_{t,n}[f]}{n^{2k}} + \left(1 + \frac{n^{\downarrow 2k}}{n^{2k}} - \frac{2n^{\downarrow k}}{n^k}\right)(p_\mu(t))^2,
\end{aligned}$$

where $f = (\Sigma_\mu)^2 - \Sigma_{\mu\sqcup\mu}$ is an observable of degree at most $2k-1$. By expanding f over the basis $(\Sigma_\nu)_{\nu\in\mathfrak{Y}(n)}$ of Sym, one sees that $\mathbb{E}_{t,n}[f] = O(n^{2k-1})$, therefore,

$$\lim_{n\to\infty} \mathbb{E}_{t,n}\left[\left(\frac{\Sigma_\mu}{n^k} - p_\mu(t)\right)^2\right] = 0.$$

This implies the convergence in probability by Bienaymé–Chebyshev inequality. On the other hand,

$$\mathbb{E}_{t,n}\left[\left(\frac{\Sigma_\mu - p_\mu}{n^k}\right)^2\right] = O\left(\frac{1}{n^2}\right)$$

since $(\Sigma_\mu - p_\mu)^2$ is an observable of degree at most $2k-2$. Therefore,

$$\frac{\Sigma_\mu(\lambda^{(n)}) - p_\mu(\lambda^{(n)})}{n^k} \to_{\mathbb{P}_{t,n}} 0,$$

and since the convergence in probability is compatible with the addition of random variables, this implies

$$\frac{p_\mu(\lambda^{(n)})}{n^{|\mu|}} \to_{\mathbb{P}_{t,n}} p_\mu(t). \qquad \square$$

Proof of Theorem 12.19. To an integer partition λ of size n, we associate a probability measure θ_λ on $[-1,1]$, defined as follows:

$$\theta_\lambda = \sum_{i=1}^{d} \frac{a_i}{n}\,\delta_{\frac{a_i}{n}} + \sum_{i=1}^{d} \frac{b_i}{n}\,\delta_{-\frac{b_i}{n}}$$

where $((a_1,\ldots,a_d),(b_1,\ldots,b_d))$ is the set of Frobenius coordinates of λ. On the other hand, in the previous chapter, we associated to any parameter $t=(\alpha,\beta) \in T$ a probability measure θ_t on $[-1,1]$, defined by

$$\theta_t = \sum_{i=1}^{\infty} \alpha_i\,\delta_{\alpha_i} + \sum_{i=1}^{\infty} \beta_i\,\delta_{-\beta_i} + \gamma\,\delta_0.$$

With these definitions, we have for any $k \geq 1$:

$$\frac{p_k(\lambda)}{n^k} = \int_{-1}^{1} x^{k-1}\,\theta_\lambda(dx) \quad ; \quad p_k(t) = \int_{-1}^{1} x^{k-1}\,\theta_t(dx).$$

Consequently, by Proposition 12.21, the moments of the random probability measure $\theta_{\lambda^{(n)}}$ converge in probability towards those of θ_t. Let f be any continuous function on $[-1,1]$. By Stone–Weierstrass theorem, for any $\varepsilon > 0$, one can approximate f by a polynomial P, such that $\sup_{x\in[-1,1]} |f(x)-P(x)| \leq \varepsilon$. Then, $\theta_{\lambda^{(n)}}(P) \to_{\mathbb{P}_{t,n}} \theta_t(P)$ and

$$|\theta_{\lambda^{(n)}}(f) - \theta_t(f)| \leq |\theta_{\lambda^{(n)}}(P) - \theta_t(P)| + 2\,\|f-P\|_\infty,$$

so $\theta_{\lambda^{(n)}}(f) \to_{\mathbb{P}_{t,n}} \theta_t(f)$.

Let $x \in [-1,1]$ be a point that does not belong to the support $\{\alpha_i,\ i \geq 1\} \sqcup \{\beta_i,\ i \geq 1\} \sqcup \{0\}$ of θ_t; it is a continuity point of the cumulative distribution function of the probability measure θ_t. For any $\varepsilon > 0$, we can find two continuous functions $f^-_{x,\varepsilon} : [-1,1] \to \mathbb{R}_+$ and $f^+_{x,\varepsilon} : [-1,1] \to \mathbb{R}_+$ such that

$$f^-_{x,\varepsilon}(s) = \begin{cases} 1 & \text{if } s \leq x-\varepsilon, \\ 0 & \text{if } s \geq x \end{cases} \quad ; \quad f^+_{x,\varepsilon}(s) = \begin{cases} 1 & \text{if } s \leq x, \\ 0 & \text{if } s \geq x+\varepsilon \end{cases}$$

If ε is small enough, then $[x-\varepsilon, x+\varepsilon]$ does not contain any point of $\mathrm{supp}(\theta_t)$. Then,

$$\theta_{\lambda^{(n)}}(f^-_{x,\varepsilon}) \to_{\mathbb{P}_{t,n}} \theta_t(f^-_{x,\varepsilon}) = \theta_t([-1,x]);$$
$$\theta_{\lambda^{(n)}}(f^+_{x,\varepsilon}) \to_{\mathbb{P}_{t,n}} \theta_t(f^+_{x,\varepsilon}) = \theta_t([-1,x]).$$

Since $f^-_{x,\varepsilon}(s) \leq 1_{s\in[-1,x]} \leq f^+_{x,\varepsilon}(s)$, we conclude that

$$\theta_{\lambda^{(n)}}([-1,x]) \to_{\mathbb{P}_{t,n}} \theta_t([-1,x]),$$

and this is true for any point x that is not in the support of θ_t.

This can only happen if $a_i^{(n)}/n \to_{\mathbb{P}_{t,n}} \alpha_i$ and $b_i^{(n)}/n \to_{\mathbb{P}_{t,n}} \beta_i$ for any $i \geq 1$. Indeed, let us treat for instance the case of $a_1^{(n)}/n$. If $a_1^{(n)}/n$ does not converge in probability towards α_1, then there exists $\varepsilon > 0$ and a subsequence $n_k \to \infty$ such that

$$\forall k \in \mathbb{N},\ \mathbb{P}_{t,n_k}\left[\left|\frac{a_1^{(n_k)}}{n_k} - \alpha_1\right| \geq \varepsilon\right] \geq \varepsilon.$$

In particular, one of the two probabilities

$$\mathbb{P}_{t,n_k}\left[\frac{a_1^{(n_k)}}{n_k} \geq \alpha_1 + \varepsilon\right] \quad ; \quad \mathbb{P}_{t,n_k}\left[\frac{a_1^{(n_k)}}{n_k} \leq \alpha_1 - \varepsilon\right]$$

stays larger than $\frac{\varepsilon}{2}$ for any $k \in \mathbb{N}$. Suppose for instance that this is the first one. Then, with $x = \alpha_1 + \frac{\varepsilon}{2}$, we have

$$\mathbb{P}_{t,n_k}\left[\theta_{\lambda^{(n_k)}}([-1,x]) \leq 1-\alpha_1\right] \geq \frac{\varepsilon}{2}$$

for any $k \in \mathbb{N}$, whereas $\theta_t([-1,x]) = 1$. This contradicts the convergence in law $\theta_{\lambda^{(n)}}([-1,x]) \to_{\mathbb{P}_{t,n}} \theta_t([-1,x])$. So, $a_1^{(n)}/n \to_{\mathbb{P}_{t,n}} \alpha_1$. We can then use an induction on $i \geq 1$ to show that the same result holds for $a_i^{(n)}/n \to_{\mathbb{P}_{t,n}} \alpha_i$ for any i. Indeed, in order to treat for instance the case of $i = 2$, since we already know that the result is true for $i = 1$, we can work with the random measures

$$\theta^{-1}_{\lambda^{(n)}} = \theta_{\lambda^{(n)}} - \frac{a_1^{(n)}}{n}\,\delta_{(a_1^{(n)}/n)},$$

and they have the property that for any x not in the support of θ_t,

$$\theta^{-1}_{\lambda^{(n)}}([-x,1]) \to_{\mathbb{P}_{t,n}} \theta^{-1}_t([-x,1]),$$

where $\theta^{-1}_t = \theta_t - \alpha_1\,\delta_{\alpha_1}$. Thus, by applying the same reasoning as above to these new positive measures on $[-1,1]$, $a_2^{(n)}/n \to_{\mathbb{P}_{t,n}} \alpha_2$. The general case is treated similarly by introducing the random measures

$$\theta^{-i}_{\lambda^{(n)}} = \theta_{\lambda^{(n)}} - \sum_{j<i} \frac{a_j^{(n)}}{n}\,\delta_{(a_j^{(n)}/n)},$$

and their limits in probability $\theta_t^{-i} = \theta_t - \sum_{j<i} \alpha_j \,\delta_{\alpha_j}$. Finally, we can use the same reasoning with the $b_i^{(n)}/n$'s and the β_i's to prove that $b_i^{(n)}/n \to_{\mathbb{P}_{t,n}} \beta_i$ for any $i \geq 1$. The theorem follows immediately, since $\lambda_i^{(n)} = a_i^{(n)} + i - \frac{1}{2}$ and $\lambda_i^{(n)\prime} = b_i^{(n)} + i - \frac{1}{2}$. □

Example. A particular example of central measures is provided by the so-called q-Plancherel measures, which are deformations of the Plancherel measures of the symmetric groups related to the Hecke algebras. Fix a real parameter $q \in (0,1)$, and let τ be the regular trace of the specialized Hecke algebra $\mathfrak{H}_q(n)$: $\tau(T_\sigma) = \delta_{\sigma,\mathrm{id}_{[\![1,n]\!]}}$. We saw at the end of Chapter 4 that

$$\tau = \sum_{\lambda \in \mathfrak{Y}(n)} \frac{1}{c_\lambda}\, \mathrm{ch}_q^\lambda = \sum_{\lambda \in \mathfrak{Y}(n)} d_\lambda\, s_\lambda(X_q)\, \chi_q^\lambda,$$

where ch_q^λ (respectively, χ_q^λ) is the character (resp., the normalized character) of the irreducible $\mathfrak{H}_q(n)$-module S_q^λ, and where X_q is the alphabet

$$(1-q, (1-q)q, (1-q)q^2, \ldots, (1-q)q^k, \ldots).$$

Therefore, the decomposition of τ in normalized irreducible characters of $\mathfrak{H}_q(n)$ yields a spectral measure which is the central measure $\mathbb{P}_{t,n}$, with

$$t = t_q = ((1-q, (1-q)q, (1-q)q^2, \ldots, (1-q)q^k, \ldots), (0, 0, \ldots)).$$

By Kerov–Vershik theorem 12.19, the first rows of a random partition $\lambda^{(n)} \sim \mathbb{P}_{t_q,n}$ satisfy

$$\frac{\lambda_i^{(n)}}{n} \to_{\mathbb{P}} (1-q)\, q^{i-1},$$

whereas the first columns are asymptotically $o(n)$. Figure 12.1 corresponds to the case $q = \frac{1}{2}$ and $n = 200$.

12.3 Gaussian behavior of the observables

If $\lambda^{(n)}$ is a random partition under $\mathbb{P}_{t,n}$, then we have just seen that $\frac{\lambda_i^{(n)}}{n} \to \alpha_i$ in the sense of convergence of probability. If one compares this with the classical law of large numbers for sums of independent identically distributed random variables, one can expect that this result can be precised, and that there is a limit in law for

$$\sqrt{n}\left(\frac{\lambda_i^{(n)}}{n} - \alpha_i\right),$$

and similarly for $\sqrt{n}\left(\frac{\lambda_i^{(n)\prime}}{n} - \beta_i\right)$. In certain special cases, there exists such a limiting result, but the limit distribution can be more complicated than a Gaussian law; see the notes at the end of the chapter. On the other hand, if one looks only at the observables $f(\lambda^{(n)})$ with $f \in \mathscr{O}$, then the law of large numbers given by Proposition 12.21 can *always* be completed by a central limit theorem, that is for any $t \in T$. The goal of this section is to prove this asymptotic Gaussian behavior (see Theorem 12.30), and to this purpose, we shall use a method of cumulants that will also be useful in the study of Plancherel measures (see Chapter 13).

▷ *Cumulants of random variables and their combinatorics.*

We start by recalling facts about the Gaussian distribution and its moments. The **standard Gaussian law** is the probability measure on $\mathbb{R}$, denoted $\mathscr{N}_{\mathbb{R}}(0,1)$, which has density

$$\frac{1}{\sqrt{2\pi}}\,\mathrm{e}^{-\frac{x^2}{2}}\,dx.$$

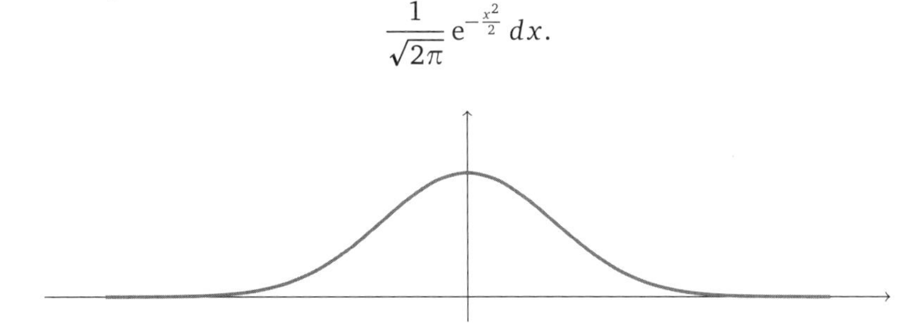

Figure 12.3
Density of the standard Gaussian law $\mathscr{N}_{\mathbb{R}}(0,1)$.

Lemma 12.22. *If $X \sim \mathscr{N}_{\mathbb{R}}(0,1)$, then its Fourier transform is $\mathbb{E}[\mathrm{e}^{\mathrm{i}\xi X}] = \mathrm{e}^{-\frac{\xi^2}{2}}$. Therefore, the moments of X are:*

$$\mathbb{E}[X^{2n}] = (2n-1)!! = (2n-1)(2n-3)\cdots 3\,1,$$

and $\mathbb{E}[X^{2n+1}] = 0$.

Proof. Set $g(\xi) = \mathbb{E}[\mathrm{e}^{\mathrm{i}\xi X}] = \frac{1}{\sqrt{2\pi}}\int_{\mathbb{R}} \mathrm{e}^{\mathrm{i}\xi x - \frac{x^2}{2}}\,dx$. The derivative of g with respect to ξ is

$$g'(\xi) = \frac{1}{\sqrt{2\pi}}\int_{\mathbb{R}} \left(\mathrm{i}\mathrm{e}^{\mathrm{i}\xi x}\right)\left(x\mathrm{e}^{-\frac{x^2}{2}}\right)dx = -\frac{1}{\sqrt{2\pi}}\int_{\mathbb{R}} \left(\mathrm{e}^{\mathrm{i}\xi x}\right)\left(\mathrm{e}^{-\frac{x^2}{2}}\right)dx = -\xi\,g(\xi),$$

by doing an integration by parts. Therefore, g is a solution of a first degree ordinary differential equation, and

$$g(\xi) = K\,\mathrm{e}^{-\frac{\xi^2}{2}}$$

for some constant K. Since $g(0) = \frac{1}{\sqrt{2\pi}} \int_{\mathbb{R}} e^{-\frac{x^2}{2}}\, dx = 1$, $K = 1$ and the value of the Fourier transform is computed. Then,

$$\mathbb{E}[X^k] = (-\mathrm{i})^k \left. \frac{d^k g(\xi)}{d\xi^k} \right|_{\xi=0} = \begin{cases} \frac{2n!}{2^n\, n!} & \text{if } k = 2n \text{ is even,} \\ 0 & \text{otherwise} \end{cases}$$

since $g(\xi) = \sum_{n=0}^{\infty} (-1)^n \frac{\xi^{2n}}{2^n\, n!}$. □

Recall that a sequence of real-valued random variables $(X^{(n)})_{n\in\mathbb{N}}$ is said to **converge in law** towards a random variable X if one of the following equivalent assertions is satisfied:

1. For any bounded and continuous function f on $\mathbb{R}$, $\lim_{n\to\infty} \mathbb{E}[f(X^{(n)})] = \mathbb{E}[f(X)]$.
2. For any closed set $G \subset \mathbb{R}$, $\limsup_{n\to\infty} \mathbb{P}[X^{(n)} \in G] \leq \mathbb{P}[X \in G]$.
3. For any open set $U \subset \mathbb{R}$, $\liminf_{n\to\infty} \mathbb{P}[X^{(n)} \in U] \geq \mathbb{P}[X \in U]$.
4. For any $\xi \in \mathbb{R}$, $\lim_{n\to\infty} \mathbb{E}[e^{\mathrm{i}\xi X^{(n)}}] = \mathbb{E}[e^{\mathrm{i}\xi X}]$.
5. For any $s \in \mathbb{R}$ that is a point of continuity of the cumulative distribution function $F_X(s) = \mathbb{P}[X \leq s]$, one has $\lim_{n\to\infty} F_{X^{(n)}}(s) = F_X(s)$.

We write in this case $X^{(n)} \rightharpoonup X$. When X follows a Gaussian law $\mathcal{N}_{\mathbb{R}}(0,1)$, a sufficient condition for the convergence in law is:

$$\forall k \in \mathbb{N},\ \lim_{n\to\infty} \mathbb{E}[(X^{(n)})^k] = \mathbb{E}[X^k].$$

References for this criterion of convergence are given in the notes at the end of the chapter. Thus, the convergence in law towards a Gaussian distribution amounts to:

$$\lim_{n\to\infty} \mathbb{E}[(X^{(n)})^{2k}] = (2k-1)!! \quad ; \quad \lim_{n\to\infty} \mathbb{E}[(X^{(n)})^{2k+1}] = 0$$

for any $k \geq 0$. In a moment, we shall see that there is a way to rewrite these conditions in terms of other invariants of the random variables $X^{(n)}$, which will allow us to get rid of the double factorials $(2k-1)!!$. Before that, let us generalize the previous discussion to a multi-dimensional setting. Given a positive-definite symmetric matrix A of size $d \times d$, we say that a random vector $\vec{X} = (X_1, \ldots, X_d)$ follows a Gaussian distribution of covariance matrix A (notation: $\vec{X} \sim \mathcal{N}_{\mathbb{R}^d}(0,A)$) if its distribution has density

$$\frac{1}{\sqrt{(2\pi)^d \det A}}\, e^{-\frac{\vec{x}^t A^{-1} \vec{x}}{2}}\, d\vec{x}$$

on $\mathbb{R}^d$. Then, $\mathbb{E}[e^{\mathrm{i}\vec{\xi}^t \vec{X}}] = e^{-\frac{\vec{\xi}^t A \vec{\xi}}{2}}$ for any vector $\vec{\xi} \in \mathbb{R}^d$; this can be shown by using the same argument as in the one-dimensional case, and by diagonalizing A in an orthonormal basis of $\mathbb{R}^d$. In particular, by looking at the first partial derivatives of this formula, one sees that $\mathbb{E}[\vec{X}]=0$, and for any pair $(i,j)\in [\![1,d]\!]^2$,

$$\mathrm{cov}(X_i,X_j)=\mathbb{E}[X_iX_j]-\mathbb{E}[X_i]\,\mathbb{E}[X_j]=A_{ij}.$$

In the sequel, if A is only non-negative definite, we shall still speak of Gaussian distribution $\mathcal{N}_{\mathbb{R}^d}(0,A)$ for a random vector $\vec{X}$ if $\mathbb{E}[\mathrm{e}^{\mathrm{i}\vec{\xi}^t\vec{X}}]=\mathrm{e}^{-\frac{\vec{\xi}^tA\vec{\xi}}{2}}$. Then, everything stated above stays true, but the existence of a density with respect to the d-dimensional Lebesgue measure. In the general case, the distribution of a Gaussian vector $\vec{X}\sim\mathcal{N}_{\mathbb{R}^d}(0,A)$ is supported by the vector subspace $\mathrm{Im}A$.

A sequence of random vectors $(\vec{X}^{(n)})_{n\in\mathbb{N}}$ converges in law towards a random vector $\vec{X}$ if one of the following equivalent assertions is satisfied:

1. For any bounded and continuous function f on $\mathbb{R}^d$, $\lim_{n\to\infty}\mathbb{E}[f(\vec{X}^{(n)})]=\mathbb{E}[f(\vec{X})]$.
2. For any closed set $G\subset\mathbb{R}^d$, $\limsup_{n\to\infty}\mathbb{P}[\vec{X}^{(n)}\in G]\leq\mathbb{P}[\vec{X}\in G]$.
3. For any open set $U\subset\mathbb{R}^d$, $\liminf_{n\to\infty}\mathbb{P}[\vec{X}^{(n)}\in U]\geq\mathbb{P}[\vec{X}\in U]$.
4. For any $\vec{\xi}\in\mathbb{R}^d$, $\lim_{n\to\infty}\mathbb{E}[\mathrm{e}^{\mathrm{i}\vec{\xi}^t\vec{X}^{(n)}}]=\mathbb{E}[\mathrm{e}^{\mathrm{i}\vec{\xi}^t\vec{X}}]$.

Again, if $\vec{X}\sim\mathcal{N}_{\mathbb{R}^d}(0,A)$, then a sufficient condition for the convergence in law $\vec{X}^{(n)}\rightharpoonup\vec{X}$ is:

$$\forall(k_1,\ldots,k_d)\in\mathbb{N}^d,\ \lim_{n\to\infty}\mathbb{E}[(X_1^{(n)})^{k_1}\cdots(X_d^{(n)})^{k_d}]=\mathbb{E}[(X_1)^{k_1}\cdots(X_d)^{k_d}].$$

Then, the joint moments $\mathbb{E}[(X_1)^{k_1}\cdots(X_d)^{k_d}]$ of the coordinates of a Gaussian vector $\vec{X}\sim\mathcal{N}_{\mathbb{R}^d}(0,A)$ can be calculated by using Wick's formula. However, there is a more convenient way to deal with the previous condition, by using the notion of **joint cumulant** of random variables.

Definition 12.23. *Let $X_1,\ldots,X_k$ be random variables that admit joint moments of any order. Their joint cumulant is defined by the formula*

$$\kappa(X_1,\ldots,X_k)=\sum_{\pi\in\mathfrak{Q}(k)}(-1)^{\ell(\pi)-1}(\ell(\pi)-1)!\prod_{j=1}^{\ell(\pi)}\mathbb{E}\left[\prod_{i\in\pi_j}X_i\right],$$

where the sum runs over set partitions $\pi=\pi_1\sqcup\pi_2\sqcup\cdots\sqcup\pi_\ell$ of $[\![1,k]\!]$.

Example. The cumulant of one random variable is its expectation, and the joint cumulant of two random variables is their covariance:

$$\kappa(X)=\mathbb{E}[X]\qquad;\qquad\kappa(X,Y)=\mathrm{cov}(X,Y)=\mathbb{E}[XY]-\mathbb{E}[X]\,\mathbb{E}[Y].$$

The joint cumulant of three random variables is given by the formula

$$\begin{aligned}&\kappa(X,Y,Z)\\&=\mathbb{E}[XYZ]-\mathbb{E}[XY]\,\mathbb{E}[Z]-\mathbb{E}[XZ]\,\mathbb{E}[Y]-\mathbb{E}[YZ]\,\mathbb{E}[X]+2\,\mathbb{E}[X]\,\mathbb{E}[Y]\,\mathbb{E}[Z].\end{aligned}$$

It is clear from the definition that the joint cumulants are invariant by permutation of the random variables. Therefore, given a finite family of random variables $\{X_i,\ i \in I\}$, it makes sense to consider the joint cumulant $\kappa(\{X_i,\ i \in I\})$.

Lemma 12.24. *The joint cumulants of random variables are defined recursively in terms of the joint moments by the relation*

$$\mathbb{E}[X_1 X_2 \cdots X_k] = \sum_{\pi \in \mathfrak{Q}(k)} \prod_{j=1}^{\ell(\pi)} \kappa(\{X_i,\ i \in \pi_j\}).$$

In particular, given a family $\{X_i,\ i \in [\![1,d]\!]\}$ of random variables, the knowledge of all their joint moments $\mathbb{E}[X_{i_1} X_{i_2} \cdots X_{i_k}]$ is equivalent to the knowledge of all their joint cumulants $\kappa(X_{i_1}, X_{i_2}, \ldots, X_{i_k})$.

Proof. The identity is obvious for $k = 1$ or $k = 2$. Suppose that the identity is true for any family of random variables of size smaller than $k-1$, and consider k random variables $X_1, \ldots, X_k$. In the sum S on the right-hand side, given a set partition π on $[\![1,k]\!]$, we can always assume up to a renumbering of its parts that the first part of π is the one that contains k. If we gather the set partitions according to this part $\pi_1 = A \sqcup \{k\}$, we get:

$$\begin{aligned} S &= \sum_{A \subset [\![1,k-1]\!]} \kappa(\{X_i,\ i \in A \sqcup \{k\}\})\ \mathbb{E}\left[\prod_{i \in [\![1,k-1]\!] \setminus A} X_i\right] \\ &= \sum_{A \subset [\![1,k-1]\!]} \sum_{\nu \in \mathfrak{Q}(A \sqcup \{k\})} (-1)^{\ell(\nu)-1} (\ell(\nu)-1)! \left(\prod_{j=1}^{\ell(\nu)} \mathbb{E}\left[\prod_{i \in \nu_j} X_i\right]\right) \mathbb{E}\left[\prod_{i \in [\![1,k-1]\!] \setminus A} X_i\right]. \end{aligned}$$

Here, we have used the induction hypothesis to transform

$$\sum_{\pi \,|\, \pi_1 = A \sqcup \{k\}} \prod_{j=1}^{\ell(\pi)} \kappa(\{X_i,\ i \in \pi_j\}) = \kappa(\{X_i,\ i \in A \sqcup \{k\}\})\ \mathbb{E}\left[\prod_{i \in [\![1,k-1]\!] \setminus A} X_i\right].$$

Now, to a pair (A, ν), we can associate $\pi' = ([\![1,k-1]\!] \setminus A) \sqcup \nu_1 \sqcup \cdots \sqcup \nu_{\ell(\nu)}$, which is a set partition of $[\![1,k]\!]$. This enables the following rewriting:

$$S = \sum_{\pi' \in \mathfrak{Q}(k)} \left(\prod_{j=1}^{\ell(\pi')} \mathbb{E}\left[\prod_{i \in \pi'_j} X_i\right]\right) \left(\sum_{[\![1,k-1]\!] \setminus A = \emptyset \text{ or a part of } \pi'} (-1)^{\ell(\nu)-1} (\ell(\nu)-1)!\right).$$

Let us now distinguish two cases. If $\pi' = \{[\![1,k]\!]\}$ is the trivial partition with one part, then there is only one part A such that $[\![1,k-1]\!] \setminus A$ is $\emptyset$ or a part of π', namely, $A = [\![1,k-1]\!]$. The corresponding alternate sum is equal to 1, so S contains the expectation $\mathbb{E}[X_1 X_2 \cdots X_k]$. On the other hand, if π' is not the trivial partition, then there are $\ell(\pi')$ possibilities for A, namely, $A = [\![1,k-1]\!]$, and on the other hand the complements in $[\![1,k-1]\!]$ of the $\ell(\pi)'-1$ parts of π' that do not contain k. The corresponding alternate sum is then equal to 0. So, in the end, $S = \mathbb{E}[X_1X_2\cdots X_k]$ and the result is true for k random variables. □

Lemma 12.25. *Suppose that the exponential generating function* $\mathbb{E}[\mathrm{e}^{z_1X_1+\cdots+z_kX_k}]$ *is well defined and convergent in a neighborhood of* $(0,0,\ldots,0) \in \mathbb{C}^k$. *Then,*

$$\kappa(X_1,\ldots,X_k) = \left.\frac{\partial^k(\log \mathbb{E}[\mathrm{e}^{z_1X_1+z_2X_2+\cdots+z_kX_k}])}{\partial z_1 \partial z_2 \cdots \partial z_k}\right|_{z_1=z_2=\cdots=z_k=0}.$$

Proof. The exponential generating function is the series

$$\mathbb{E}[\mathrm{e}^{z_1X_1+z_2X_2+\cdots+z_kX_k}] = \sum_{n_1,\ldots,n_k=0}^{\infty} \frac{\mathbb{E}[(X_1)^{n_1}\cdots(X_k)^{n_k}]}{n_1!\cdots n_k!} z_1^{n_1}\cdots z_k^{n_k},$$

and its logarithm is therefore:

$$\log \mathbb{E}[\mathrm{e}^{z_1X_1+z_2X_2+\cdots+z_kX_k}] = \sum_{l=1}^{\infty} \frac{(-1)^{l-1}}{l} \left(\sum_{n_1,\ldots,n_k=0}^{\infty} \frac{\mathbb{E}[(X_1)^{n_1}\cdots(X_k)^{n_k}]}{n_1!\cdots n_k!} z_1^{n_1}\cdots z_k^{n_k} \right)^l.$$

We are looking for the coefficient of $z_1z_2\cdots z_k$ in this power series. If l is fixed, then

$$[z_1z_2\cdots z_k]\left(\sum_{n_1,\ldots,n_k=0}^{\infty} \frac{\mathbb{E}[(X_1)^{n_1}\cdots(X_k)^{n_k}]}{n_1!\cdots n_k!} z_1^{n_1}\cdots z_k^{n_k} \right)^l$$
$$= \sum_{\substack{\pi_1,\ldots,\pi_l \subset [\![1,k]\!] \\ \pi_1 \sqcup \cdots \sqcup \pi_l = [\![1,k]\!]}} \prod_{j=1}^{l} \mathbb{E}\left[\prod_{i\in\pi_j} X_i\right].$$

Indeed, the part π_1 describes which z_i's come from the first factor $\mathbb{E}[\mathrm{e}^{z_1X_1+\cdots+z_kX_k}]$, the part π_2 describes which z_i's come from the second factor $\mathbb{E}[\mathrm{e}^{z_1X_1+\cdots+z_kX_k}]$, etc. If one rewrites this expression as a sum over non-ordered set partitions π with l parts, one obtains a factor $l!$:

$$[z_1z_2\cdots z_k]\left(\sum_{n_1,\ldots,n_k=0}^{\infty} \frac{\mathbb{E}[(X_1)^{n_1}\cdots(X_k)^{n_k}]}{n_1!\cdots n_k!} z_1^{n_1}\cdots z_k^{n_k} \right)^l$$
$$= l! \sum_{\pi\in\mathfrak{Q}(k),\,\ell(\pi)=l} \prod_{j=1}^{l} \mathbb{E}\left[\prod_{i\in\pi_j} X_i\right].$$

Therefore, we have indeed

$$[z_1 z_2 \cdots z_k]\left(\log \mathbb{E}[e^{z_1 X_1+\cdots+z_k X_k}]\right) = \sum_{\pi \in \mathfrak{Q}(k)} (-1)^{\ell(\pi)-1}\,(\ell(\pi)-1)! \prod_{j=1}^{l} \mathbb{E}\left[\prod_{i \in \pi_j} X_i\right]$$
$$= \kappa(X_1,\ldots,X_k). \qquad \square$$

Corollary 12.26. *Let $\vec{X} = (X_1,\ldots,X_d)$ be a random vector with its exponential generating function that is convergent in a neighborhood of $(0,\ldots,0) \in \mathbb{C}^d$. It is a Gaussian vector with covariance matrix A if and only if*

$$\kappa(X_i) = 0 \quad ; \quad \kappa(X_i,X_j) = A_{ij} \qquad \kappa(X_{i_1},\ldots,X_{i_{r\geq 3}}) = 0.$$

Proof. Suppose that $\vec{X} \sim \mathcal{N}_{\mathbb{R}^d}(0,A)$, and let us compute the joint cumulants of its coordinates. Since a joint cumulant $\kappa(X_{i_1},X_{i_2},\ldots,X_{i_r})$ is independent from the order of the indices i_j, it suffices to compute the joint cumulants

$$\kappa\left(\underbrace{X_{i_1},\ldots,X_{i_1}}_{m_1 \text{ terms}},\ldots,\underbrace{X_{i_k},\ldots,X_{i_k}}_{m_k \text{ terms}}\right)$$

with indices $i_1 < i_2 < \cdots < i_k$, and some multiplicities $m_1,\ldots,m_k \geq 1$. Notice that the random vector $(X_{i_1},\ldots,X_{i_k})$ is again Gaussian, with its covariance matrix $B = A_{i_1,\ldots,i_k}$ that is extracted from the covariance matrix A. Therefore,

$$\kappa(\underbrace{X_{i_1},\ldots,X_{i_1}}_{m_1 \text{ terms}},\ldots,\underbrace{X_{i_k},\ldots,X_{i_k}}_{m_k \text{ terms}})$$
$$= \left.\frac{\partial^{m_1+\cdots+m_k} \log \mathbb{E}[e^{(z_{11}+\cdots+z_{1m_1})X_{i_1}+\cdots+(z_{k1}+\cdots+z_{km_k})X_{i_k}}]}{\partial z_{11}\cdots\partial z_{km_k}}\right|_{z=0}$$
$$= \left.\frac{1}{2}\,\frac{\partial^{m_1+\cdots+m_k}}{\partial z_{11}\cdots\partial z_{km_k}}\left(\sum_{i,j=1}^{k} B_{ij}(z_{i1}+\cdots+z_{im_i})(z_{j1}+\cdots+z_{jm_j})\right)\right|_{z=0}.$$

This quantity vanishes unless $m_1+\cdots+m_k = 2$, in which case one gets $\kappa(X_i,X_j) = A_{ij}$. Finally, the Gaussian distributions are characterized by their joint moments, and therefore by their joint cumulants. $\square$

We can finally state a convenient criterion in order to prove that a sequence of random vectors converges in distribution towards a Gaussian vector:

Proposition 12.27. *Let $(\vec{X}^{(n)})_{n\in\mathbb{N}}$ be a sequence of random vectors in $\mathbb{R}^d$ with convergent moment generating series. The sequence $(\vec{X}^{(n)} - \mathbb{E}[\vec{X}^{(n)}])_{n\in\mathbb{N}}$ converges in law towards a Gaussian distribution $\mathcal{N}_{\mathbb{R}^d}(0,A)$ if and only if*

$$\forall i,j \in [\![1,d]\!],\ \lim_{n\to\infty} \kappa(X_i^{(n)},X_j^{(n)}) = A_{ij}$$

and the joint cumulants of order $r \geq 3$ of the coordinates $X_i^{(n)}$ all converge towards 0.

Proof. If X is a random variable, denote $\overline{X} = X - \mathbb{E}[X]$. Since

$$\log \mathbb{E}[\mathrm{e}^{z_1\overline{X}_1+\cdots+z_k\overline{X}_k}] = \log \mathbb{E}[\mathrm{e}^{z_1X_1+\cdots+z_kX_k}] - z_1\mathbb{E}[X_1] - \cdots - z_k\mathbb{E}[X_k],$$

for any $k \geq 2$ and any random variables $X_1, \ldots, X_k$,

$$\kappa(X_1, \ldots, X_k) = \kappa(\overline{X}_1, \ldots, \overline{X}_k).$$

In other words, the only joint cumulants that differ when looking at $\overline{X}$ instead of X are the simple cumulants $\kappa(X) = \mathbb{E}[X]$. As a consequence, the hypotheses of the proposition amount to the convergence of all the joint cumulants of the vector $\vec{X}^{(n)} - \mathbb{E}[\vec{X}^{(n)}]$ towards those of a random vector of law $\mathcal{N}_{\mathbb{R}^d}(0,A)$. By Lemma 12.24, this is equivalent to the convergence of all the joint moments, and when the limit is a Gaussian vector, this convergence of all the joint moments is equivalent to the convergence in law. □

▷ *Cumulants of observables.*

At the end of this section, we shall prove that if $\lambda^{(n)} \sim \mathbb{P}_{t,n}$, then for any integer partitions $\mu^{(1)}, \ldots, \mu^{(r)}$,

$$\kappa(\Sigma_{\mu^{(1)}}(\lambda^{(n)}), \Sigma_{\mu^{(2)}}(\lambda^{(n)}), \ldots, \Sigma_{\mu^{(r)}}(\lambda^{(n)})) = O(n^{|\mu^{(1)}|+\cdots+|\mu^{(r)}|-(r-1)}).$$

This will imply a central limit theorem for an adequate renormalization of the recentered random variables $\Sigma_\mu(\lambda^{(n)}) - \mathbb{E}[\Sigma_\mu(\lambda^{(n)})]$. The proof of this result will rely on a computation of joint cumulants by means of conditioning, which is contained in Proposition 12.28 below. We introduce a new product $\bullet$ for observables of the algebra $\mathscr{O}$. It is defined as follows:

$$\Sigma_\mu \bullet \Sigma_\nu = \Sigma_{\mu \sqcup \nu}.$$

It is the restriction to $\mathscr{O}$ of the disjoint product of partial permutations

$$(\sigma,A) \bullet (\tau,B) = \begin{cases} (\sigma\tau, A \sqcup B) & \text{if } A \cap B = \emptyset, \\ 0 & \text{otherwise.} \end{cases}$$

By using this new operation, we can define three kinds of cumulants for observables of a random partition $\lambda^{(n)} \sim \mathbb{P}_{t,n}$:

1. the usual cumulants:

$$\kappa(f_1, f_2, \ldots, f_k) = \sum_{\pi \in \mathfrak{Q}(k)} (-1)^{\ell(\pi)-1}(\ell(\pi)-1)! \prod_{1 \leq j \leq \ell(\pi)} \mathbb{E}_{t,n}\left[\prod_{i \in \pi_j} f_i)\right].$$

2. the **disjoint cumulants**:

$$\kappa^{\bullet}(f_1,f_2,\ldots,f_k) = \sum_{\pi\in\mathfrak{Q}(k)} (-1)^{\ell(\pi)-1}(\ell(\pi)-1)! \prod_{1\leq j\leq \ell(\pi)} \mathbb{E}_{t,n}\left[\dot{\prod_{i\in\pi_j}} f_i\right].$$

3. the **identity cumulants**:

$$\kappa^{\mathrm{id}}(f_1,f_2,\ldots,f_k) = \sum_{\pi\in\mathfrak{Q}(k)} (-1)^{\ell(\pi)-1}(\ell(\pi)-1)! \dot{\prod_{1\leq j\leq \ell(\pi)}} \left(\prod_{i\in\pi_j} f_i\right).$$

They are related to the commutative diagram of linear maps

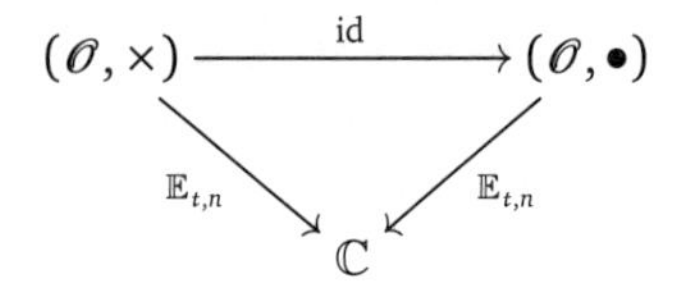

between commutative $\mathbb{C}$-algebras.

Proposition 12.28. *For any observables $f_1,\ldots,f_k$ in $\mathscr{O}$,*

$$\kappa(f_1,f_2,\ldots,f_k) = \sum_{\pi\in\mathfrak{Q}(k)} \kappa^{\bullet}(\kappa^{\mathrm{id}}(\{f_i,\, i\in\pi_1\}),\ldots,\kappa^{\mathrm{id}}(\{f_i,\, i\in\pi_{\ell(\pi)}\})).$$

Proof. If A and B are two commutative $\mathbb{C}$-algebras and $\phi : A \to B$ is a linear map, then one can define more generally the (A,B,ϕ)-cumulant of elements $f_1,\ldots,f_k$ in A:

$$\kappa^{(A,B,\phi)}(f_1,\ldots,f_k) = \left.\frac{\partial^k \log_B \phi(\exp_A(z_1f_1+\cdots+z_kf_k))}{\partial z_1\partial z_2\cdots\partial z_k}\right|_{z=0},$$

where $\exp_A$ is the exponential with products taken in A, $\log_B$ is the logarithm with products taken in B, and the previous identity is in the sense of formal power series in $z_1,\ldots,z_k$. We then have the expansion in power series:

$$\log_B \phi(\exp_A(z_1f_1+\cdots+z_kf_k)) = \sum_{m_1,\ldots,m_k\geq 0} \frac{\kappa^{(A,B,\phi)}(f_1^{\otimes m_1},\ldots,f_k^{\otimes m_k})}{m_1!\cdots m_k!} (z_1)^{m_1}\cdots(z_k)^{m_k},$$

where $f_i^{\otimes m_i}$ means that the element $f_i \in A$ appears m_i times in the joint cumulant. Now, take $A = (\mathscr{O},\times)$, $B = (\mathscr{O},\bullet)$, $C = \mathbb{C}$, $\phi_A^B = \mathrm{id}_{\mathscr{O}}$, and $\phi_A^C = \phi_B^C = \mathbb{E}_{t,n}$. We have

$$\begin{aligned}
&\log_C \circ \phi_A^C \circ \exp_A(z_1f_1+\cdots+z_kf_k)\\
&= \log_C \circ \phi_B^C \circ \exp_B \circ \log_B \circ \phi_A^B \circ \exp_A(z_1f_1+\cdots+z_kf_k)\\
&= \log_C \circ \phi_B^C \circ \exp_B\left(\sum_{m_1,\ldots,m_k\geq 0} \frac{\kappa^{\mathrm{id}}(f_1^{\otimes m_1},\ldots,f_k^{\otimes m_k})}{m_1!\cdots m_k!} (z_1)^{m_1}\cdots(z_k)^{m_k}\right).
\end{aligned}$$

As a consequence,

$$\kappa(f_1,\ldots,f_k) = [z_1\cdots z_k]\left(\log_C \circ \phi_B^C \circ \exp_B\left(\sum_{m_1,\ldots,m_k\geq 0} \frac{\kappa^{\mathrm{id}}(f_1^{\otimes m_1},\ldots,f_k^{\otimes m_k})}{m_1!\cdots m_k!}\,(z_1)^{m_1}\cdots(z_k)^{m_k}\right)\right).$$

Consider the term of order $l \geq 1$ in the expansion of $\log_C \circ \phi_B^C \circ \exp_B$:

$$\frac{(-1)^{l-1}}{l}\left(\mathbb{E}_{t,n}\left[\exp_\bullet\left(\sum_{m_1,\ldots,m_k\geq 0} \frac{\kappa^{\mathrm{id}}(f_1^{\otimes m_1},\ldots,f_k^{\otimes m_k})}{m_1!\cdots m_k!}\,(z_1)^{m_1}\cdots(z_k)^{m_k}\right)\right]\right)^l,$$

where the powers in $\exp_\bullet$ are taken with respect to the disjoint product of observables. The coefficient of $z_1 z_2\cdots z_k$ in this expression is

$$\frac{(-1)^{l-1}}{l}\sum_{\substack{\pi_1,\ldots,\pi_l\subset[\![1,k]\!]\\ \pi_1\sqcup\cdots\sqcup\pi_l=[\![1,k]\!]}} \mathbb{E}_{t,n}\left[\dot{\prod_{1\leq j\leq l}}\kappa^{\mathrm{id}}(\{f_i,\ i\in\pi_j\})\right]$$

$$= (-1)^{l-1}(l-1)!\sum_{\pi\in\mathfrak{Q}(k),\,\ell(\pi)=l} \mathbb{E}_{t,n}\left[\dot{\prod_{1\leq j\leq l}}\kappa^{\mathrm{id}}(\{f_i,\ i\in\pi_j\})\right],$$

for the same reasons as in the proof of Lemma 12.25: one needs to assign the z_i's to different factors of the l-th power, with each z_i that appears exactly once. Therefore,

$$\kappa(f_1,\ldots,f_k) = \sum_{\pi\in\mathfrak{Q}(k)}(-1)^{\ell(\pi)-1}(\ell(\pi)-1)!\ \mathbb{E}_{t,n}\left[\dot{\prod_{1\leq j\leq \ell(\pi)}}\kappa^{\mathrm{id}}(\{f_i,\ i\in\pi_j\})\right]$$

$$= \sum_{\pi\in\mathfrak{Q}(k)}\kappa^\bullet(\kappa^{\mathrm{id}}(\{f_i,\, i\in\pi_1\}),\ldots,\kappa^{\mathrm{id}}(\{f_i,\, i\in\pi_{\ell(\pi)}\})). \qquad \square$$

Proposition 12.28 leads one to study the identity cumulants of observables, which are certain special combinations of observables. The main result that we shall need is:

Theorem 12.29. *For any observables $f_1,\ldots,f_k \in \mathscr{O}$,*

$$\deg \kappa^{\mathrm{id}}(f_1,f_2,\ldots,f_k) \leq \deg(f_1)+\deg(f_2)+\cdots+\deg(f_k)-(k-1).$$

Proof. It is obvious that the operation $\kappa^{\mathrm{id}} : \mathscr{O}^{\otimes k} \to \mathscr{O}$ is multilinear and symmetric. Therefore, it suffices to prove the result for observables $f_i = \Sigma_{\mu^{(i)}}$. Set

$|\mu^{(i)}| = m_i$. Recall that we have an explicit combinatorial description of the product $\Sigma_{\mu^{(1)}}\Sigma_{\mu^{(2)}}\cdots\Sigma_{\mu^{(k)}}$:

$$\Sigma_{\mu^{(1)}}\Sigma_{\mu^{(2)}}\cdots\Sigma_{\mu^{(k)}} = \sum_{\Pi} \Sigma_{\lambda(\mu^{(1)},\mu^{(2)},\ldots,\mu^{(k)};\Pi)},$$

where:

1. the sum runs over partial pairings Π of $[\![1,m_1]\!],[\![1,m_2]\!],\ldots,[\![1,m_k]\!]$, that is to say graphs
   - with vertex set $V = \{(i,j),\ i \in [\![1,k]\!],\ j \in [\![1,m_i]\!]\}$;
   - with the edge set that is transitive: if $(i_1,j_1) \longleftrightarrow (i_2,j_2)$ and $(i_2,j_2) \longleftrightarrow (i_3,j_3)$, then $(i_1,j_1) \longleftrightarrow (i_3,j_3)$;
   - without edge between (i,j_1) and (i,j_2) for $j_1 \neq j_2$.
2. the partition $\lambda(\mu^{(1)},\mu^{(2)},\ldots,\mu^{(k)};\pi)$ is the cycle type of the product of partial permutations $C(\mu^{(1)},a^{(1)})\,C(\mu^{(2)},a^{(2)})\cdots C(\mu^{(k)},a^{(k)})$, where

   $$C(\mu,a) = \big((a_1,\ldots,a_{\mu_1})\cdots(a_{\mu_1+\cdots+\mu_{\ell-1}+1},\ldots,a_{\mu_1+\cdots+\mu_\ell}),\{a_i\}\big)$$

   if a is an arrangement of size $|\mu|$, where the $a^{(i)} = (a^{(i)}_1,\ldots,a^{(i)}_{m_i})$ are arrangements of size m_i, and where the equalities $a^{(i_1)}_{j_1} = a^{(i_2)}_{j_2}$ are prescribed by the partial pairing Π (the cycle type is entirely determined by these equalities).

This formula is an immediate generalization of Proposition 7.6. Now, there is way to split this sum in order to make appear identity cumulants. To a partial pairing Π viewed as a graph on $V = \{(i,j),\ i \in [\![1,k]\!],\ j \in [\![1,m_i]\!]\}$, we associate the set partition $\pi = S(\Pi)$ on $[\![1,k]\!]$, such that i_1 is in the same part as i_2 in π if and only if there exists an edge $(i_1,j_1) \longleftrightarrow (i_2,j_2)$ in Π. Thus, $S(\Pi)$ is the contraction of the graph Π with respect to the first coordinate, and by assumption on a partial pairing, one obtains a set partition of size k. We claim that:

$$\sum_{\Pi\,|\,S(\Pi)=\pi} \Sigma_{\lambda(\mu^{(1)},\mu^{(2)},\ldots,\mu^{(k)};\Pi)} = \dot{\prod_{1\leq j\leq \ell(\pi)}} \kappa^{\mathrm{id}}(\{\Sigma_{\mu^{(i)}},\ i \in \pi_j\}).$$

To prove this identity, we reason by induction on k. Suppose the result is true for $k-1$ terms. The identity cumulants are characterized by the relation:

$$\Sigma_{\mu^{(1)}}\Sigma_{\mu^{(2)}}\cdots\Sigma_{\mu^{(k)}} = \sum_{\pi\in\mathfrak{Q}(k)} \dot{\prod_{1\leq j\leq \ell(\pi)}} \kappa^{\mathrm{id}}(\{\Sigma_{\mu^{(i)}},\ i \in \pi_j\});$$

this is for the same reason as in Lemma 12.24. By the induction hypothesis, if $\pi \neq \{[\![1,k]\!]\}$ is not the trivial partition, then for any part π_j,

$$\kappa^{\mathrm{id}}(\{\Sigma_{\mu^{(i)}},\ i \in \pi_j\}) = \sum_{\Pi^{(j)}\,|\,S(\Pi^{(j)})=\{\pi_j\}} \Sigma_{\lambda(\Pi^{(j)},\{\mu^{(i)},\,i\in\pi_j\})}$$

where the sum runs over partial pairings $\Pi^{(j)}$ of $\{[\![1,m_i]\!],\ i\in\pi_j\}$ with associated set partition $\{\pi_j\}$. When taking the disjoint product of the observables $\Sigma_{\lambda(\{\mu^{(i)},\,i\in\pi_j\};\Pi^{(j)})}$, one obtains the observable

$$\Sigma_{\lambda(\mu^{(1)},\ldots,\mu^{(k)};\Pi)},$$

where $\Pi=\bigsqcup_{1\leq j\leq \ell(\pi)}\Pi^{(j)}$ is the disjoint union of the partial pairings $\Pi^{(j)}$, and therefore has associated set partition π. Thus, if $\pi\neq\{[\![1,k]\!]\}$ is not the trivial partition, then one has indeed

$$\sum_{\Pi\,|\,S(\Pi)=\pi}\Sigma_{\lambda(\mu^{(1)},\mu^{(2)},\ldots,\mu^{(k)};\Pi)}=\dot{\prod_{1\leq j\leq \ell(\pi)}}\kappa^{\mathrm{id}}(\{\Sigma_{\mu^{(i)}},\ i\in\pi_j\}).$$

Now, if $\pi=\{[\![1,k]\!]\}$, then

$$\begin{aligned}&\sum_{\Pi\,|\,S(\Pi)=\{[\![1,k]\!]\}}\Sigma_{\lambda(\mu^{(1)},\mu^{(2)},\ldots,\mu^{(k)};\Pi)}\\&=\Sigma_{\mu^{(1)}}\Sigma_{\mu^{(2)}}\cdots\Sigma_{\mu^{(k)}}-\sum_{\Pi\,|\,S(\Pi)\neq\{[\![1,k]\!]\}}\Sigma_{\lambda(\mu^{(1)},\mu^{(2)},\ldots,\mu^{(k)};\Pi)}\\&=\Sigma_{\mu^{(1)}}\Sigma_{\mu^{(2)}}\cdots\Sigma_{\mu^{(k)}}-\sum_{\pi\neq\{[\![1,k]\!]\}}\dot{\prod_{1\leq j\leq\ell(\pi)}}\kappa^{\mathrm{id}}(\{\Sigma_{\mu^{(i)}},\ i\in\pi_j\})\\&=\kappa^{\mathrm{id}}(\Sigma_{\mu^{(1)}},\Sigma_{\mu^{(2)}},\ldots,\Sigma_{\mu^{(k)}}).\end{aligned}$$

So, the result is again true with k observables $\Sigma_{\mu^{(i)}}$.

The formula for $\kappa^{\mathrm{id}}(\Sigma_{\mu^{(1)}},\Sigma_{\mu^{(2)}},\ldots,\Sigma_{\mu^{(k)}})$ implies immediately the theorem: if Π is a pairing with $S(\Pi)=\{[\![1,k]\!]\}$, then Π consists of at least $k-1$ equalities of indices, and this decreases the degree of the product of observables by at least $k-1$. □

▷ *Central limit theorem for the character values.*

We are now ready to prove:

Theorem 12.30. *We fix a parameter $t\in T$, and integer partitions $\mu^{(1)},\ldots,\mu^{(d)}$. Let $\vec{X}^{(n)}$ be the random vector with coordinates*

$$X_i^{(n)}=\sqrt{n}\,(\chi^{\lambda^{(n)}}(\sigma_{\mu^{(i)}})-p_{\mu^{(i)}}(t)),$$

where $\lambda^{(n)}\sim\mathbb{P}_{t,n}$. One has the convergence in law $\vec{X}^{(n)}\rightharpoonup\vec{X}$, where $\vec{X}\sim\mathcal{N}_{\mathbb{R}}(0,A)$ and

$$A_{ij}=\sum_{a\in\mu^{(i)},\ b\in\mu^{(j)}}ab\,p_{\mu^{(i)}\sqcup\mu^{(j)}\setminus\{a,b\}}(t)\,(p_{a+b-1}(t)-p_a(t)\,p_b(t)).$$

Example. In particular, if $d=1$ and if one looks at the random character value of a cycle of length k, and if $p_{2k-1}(t)-(p_k(t))^2 \neq 0$, then

$$\sqrt{n}\,\frac{\chi^{\lambda^{(n)}}(\sigma_k)-p_k(t)}{k\sqrt{p_{2k-1}(t)-(p_k(t))^2}} \rightharpoonup \mathcal{N}_{\mathbb{R}}(0,1).$$

Thus, the random character value $\chi^{\lambda^{(n)}}(\sigma_k)$ converges in probability to $p_k(t) = \tau_t(\sigma_k)$, and its fluctuations are typically of order $O(1/\sqrt{n})$, and asymptotically Gaussian.

Lemma 12.31. *For any observables $f_1,\ldots,f_k$,*

$$\kappa^\bullet(f_1,\ldots,f_k) = O(n^{\deg f_1+\cdots+\deg f_k-(k-1)}).$$

Proof. Again, it suffices to prove the result for observables $f_i = \Sigma_{\mu^{(i)}}$, where the $\mu^{(i)} \in \mathfrak{Y}(m_i)$ are arbitrary integer partitions. However,

$$\begin{aligned}\kappa^\bullet(\Sigma_{\mu^{(1)}},\ldots,\Sigma_{\mu^{(k)}}) &= \sum_{\pi\in\mathfrak{Q}(k)} (-1)^{\ell(\pi)-1}\,(\ell(\pi)-1)!\prod_{j=1}^{\ell(\pi)} \mathbb{E}_{t,n}\left[\dot{\prod_{i\in\pi_j}} \Sigma_{\mu^{(i)}}\right] \\ &= \left(\prod_{i=1}^{k} p_{\mu^{(i)}}(t)\right) \sum_{\pi\in\mathfrak{Q}(k)} (-1)^{\ell(\pi)-1}\,(\ell(\pi)-1)!\prod_{j=1}^{\ell(\pi)} n^{\downarrow\sum_{i\in\pi_j} m_i} \\ &= \left(\prod_{i=1}^{k} p_{\mu^{(i)}}(t)\right) \kappa^\bullet(\Sigma_{1^{m_1}},\ldots,\Sigma_{1^{m_k}}).\end{aligned}$$

Therefore, it suffices to treat the case where $\mu^{(i)} = 1^{m_i}$. We reason by induction on k, the case of $k=1$ being trivial. Suppose the result to be true for $k-1 \geq 1$ terms, and consider the disjoint cumulant $\kappa^\bullet(\Sigma_{1^{m_1}},\ldots,\Sigma_{1^{m_k}})$. By Proposition 12.28,

$$\begin{aligned}&\kappa^\bullet(\Sigma_{1^{m_1}},\ldots,\Sigma_{1^{m_k}}) \\ &= \kappa(\Sigma_{1^{m_1}},\ldots,\Sigma_{1^{m_k}}) - \sum_{\substack{\pi\in\mathfrak{Q}(k)\\ \pi\neq\{[\![1,k]\!]\}}} \kappa^\bullet(\{\kappa^{\mathrm{id}}(\{\Sigma_{1^{m_i}},\ i\in\pi_j\}),\ j\in[\![1,\ell(\pi)]\!]\}).\end{aligned}$$

However, viewed as functions of a random partition $\lambda^{(n)}$, the observables $\Sigma_{1^{m_i}}$ are constant; therefore, for $k \geq 2$,

$$\begin{aligned}\kappa(\Sigma_{1^{m_1}},\ldots,\Sigma_{1^{m_k}}) &= \kappa(\Sigma_{1^{m_1}} - \mathbb{E}_{t,n}[\Sigma_{1^{m_1}})],\ldots,\Sigma_{1^{m_k}} - \mathbb{E}_{t,n}[\Sigma_{1^{m_k}}]) \\ &= \kappa(0,0,\ldots,0) = 0.\end{aligned}$$

So, for $k \geq 2$,

$$\kappa^\bullet(\Sigma_{1^{m_1}},\ldots,\Sigma_{1^{m_k}}) = -\sum_{\substack{\pi\in\mathfrak{Q}(k)\\ \pi\neq\{1\}\sqcup\{2\}\sqcup\cdots\sqcup\{k\}}} \kappa^\bullet(\{\kappa^{\mathrm{id}}(\{\Sigma_{1^{m_i}},\ i\in\pi_j\}),\ j\in[\![1,\ell(\pi)]\!]\}).$$

Fix a set partition $\pi \in \mathfrak{Q}(k)$ that is different from $\{1\} \sqcup \{2\} \sqcup \cdots \sqcup \{k\}$. For any part π_j, the degree of $\kappa^{\mathrm{id}}(\{\Sigma_{1^{m_i}},\ i \in \pi_j\})$ is smaller than $\sum_{i\in\pi_j} m_i - (|\pi_j|-1)$ by Theorem 12.29. On the other hand, since π has less than $k-1$ parts, by the induction hypothesis, given $\ell(\pi) \le k-1$ observables $f_1,\ldots,f_{\ell(\pi)}$, $\kappa^\bullet(f_1,\ldots,f_{\ell(\pi)}) = O(n^{\deg f_1+\cdots+\deg f_{\ell(\pi)}-(\ell(\pi)-1)})$. So,

$$\begin{aligned}\kappa^\bullet(\{\kappa^{\mathrm{id}}(\{\Sigma_{1^{m_i}},\ i\in\pi_j\}),\ j\in[\![1,\ell(\pi)]\!]\}) &= O\left(n^{\left(\sum_{j=1}^{\ell(\pi)}\sum_{i\in\pi_j} m_i-(|\pi_j|-1)\right)-(\ell(\pi)-1)}\right)\\ &= O\left(n^{\left(\sum_{i=1}^{k} m_i\right)-(k-1)}\right).\end{aligned}$$

Hence, the result is also true for k observables, and the proof is completed. □

Lemma 12.32. *For any observables $f_1,\ldots,f_k$,*

$$\kappa(f_1,\ldots,f_k) = O(n^{\deg f_1+\cdots+\deg f_k-(k-1)}).$$

Proof. By Proposition 12.28, it suffices to bound $\kappa^\bullet(\{\kappa^{\mathrm{id}}(\{f_i,\ i\in\pi_j\}),\ j\in[\![1,\ell(\pi)]\!]\})$ for any set partition $\pi\in\mathfrak{Q}(k)$. However,

$$\deg(\kappa^{\mathrm{id}}(\{f_i,\ i\in\pi_j\}) \le \left(\sum_{i\in\pi_j}(\deg f_i-1)\right)+1$$

by Theorem 12.29, so the result follows from the previous lemma. □

Proof of Theorem 12.30. The random variable $X_i^{(n)}$ can be rewritten as

$$\sqrt{n}\ \frac{\Sigma_{\mu^{(i)}}(\lambda^{(n)})-\mathbb{E}_{t,n}[\Sigma_{\mu^{(i)}}]}{n^{\downarrow m_i}}.$$

It is centered ($\mathbb{E}[X_i^{(n)}]=0$), and for any $r\ge 3$ and any indices $i_1,i_2,\ldots,i_r$,

$$\begin{aligned}\kappa\left(X_{i_1}^{(n)},\ldots,X_{i_r}^{(n)}\right) &= \frac{n^{\frac{r}{2}}}{\prod_{j=1}^{r} n^{\downarrow m_i}}\,\kappa\left(\Sigma_{\mu^{(i_1)}},\ldots,\Sigma_{\mu^{(i_r)}}\right)\\ &= O\left(n^{\frac{r}{2}-(r-1)}\right) = O\left(n^{1-\frac{r}{2}}\right)\to 0.\end{aligned}$$

So, by Proposition 12.27, in order to prove the theorem, it suffices now to compute

$$\lim_{n\to\infty}\operatorname{cov}(X_i^{(n)},X_j^{(n)}) = \lim_{n\to\infty}\frac{1}{n^{m_i+m_j-1}}\left(\mathbb{E}_{t,n}[\Sigma_{\mu^{(i)}}\Sigma_{\mu^{(j)}}]-\mathbb{E}_{t,n}[\Sigma_{\mu^{(i)}}]\,\mathbb{E}_{t,n}[\Sigma_{\mu^{(j)}}]\right).$$

The two quantities $\mathbb{E}_{t,n}[\Sigma_{\mu^{(i)}}\Sigma_{\mu^{(j)}}]$ and $\mathbb{E}_{t,n}[\Sigma_{\mu^{(i)}}]\,\mathbb{E}_{t,n}[\Sigma_{\mu^{(j)}}]$ are polynomials in n of degree m_i+m_j, and we need to compute their two first leading terms. Since

$$n^{\downarrow m} = n^m - \binom{m}{2} n^{m-1} + O(n^{m-2}),$$

we get

$$\begin{aligned}
&\mathbb{E}_{t,n}[\Sigma_{\mu^{(i)}}]\,\mathbb{E}_{t,n}[\Sigma_{\mu^{(j)}}] \\
&= n^{\downarrow m_i}\, n^{\downarrow m_j}\, p_{\mu^{(i)}\sqcup\mu^{(j)}}(t) \\
&= \left(n^{m_i+m_j} - \left(\binom{m_i}{2}+\binom{m_j}{2}\right) n^{m_i+m_j-1}\right) p_{\mu^{(i)}\sqcup\mu^{(j)}}(t) + O(n^{m_i+m_j-2}).
\end{aligned}$$

On the other hand, if one looks at the terms of degree $m_i + m_j$ or $m_i + m_j - 1$ in the product of observables $\Sigma_{\mu^{(i)}}\,\Sigma_{\mu^{(j)}}$, they correspond to partial pairings π between $[\![1, m_i]\!]$ and $[\![1, m_j]\!]$ such that $|\pi| = 0$ or 1. Denote $k = (k_1, \dots, k_{m_i})$ and $l = (l_1, \dots, l_{m_j})$ two arbitrary arrangements of sizes m_i and m_j, and $C(\mu^{(i)}, k)$ and $C(\mu^{(j)}, l)$ the two partial permutations of cycle types $\mu^{(i)}$ and $\mu^{(j)}$ that correspond to these arrangements. If $k \cap l = \emptyset$, then the product $C(\mu^{(i)}, k)\,C(\mu^{(j)}, l)$ is a partial permutation of type $\mu^{(i)} \sqcup \mu^{(j)}$. Thus, the empty partial pairing gives a term $\Sigma_{\mu^{(i)}\sqcup\mu^{(j)}}$ in the product $\Sigma_{\mu^{(i)}}\,\Sigma_{\mu^{(j)}}$. Suppose now that one has a unique identity $k_s = l_t$, with k_s in a cycle $(k_{p+1}, \dots, k_{p+a})$ of length a in $C(\mu^{(i)}, k)$ and l_t in a cycle $(l_{q+1}, \dots, l_{q+b})$ of length b in $C(\mu^{(j)}, l)$. This corresponds to a partial pairing of size 1, and

$$\begin{aligned}
&(k_{p+1}, \dots, k_{p+a})(l_{q+1}, \dots, l_{q+b}) \\
&= (k_{s+1}, \dots, k_{p+a}, k_{p+1}, \dots, k_s)(l_t, \dots, l_{q+b}, l_{q+1}, \dots, l_{t-1}) \\
&= (k_{s+1}, \dots, k_{p+a}, k_{p+1}, \dots, k_s = l_t, \dots, l_{q+b}, l_{q+1}, \dots, l_{t-1}).
\end{aligned}$$

Thus, $C(\mu^{(i)}, k)\,C(\mu^{(j)}, l)$ is in this case of type $(\mu^{(i)} \sqcup \mu^{(j)} \setminus \{a, b\}) \sqcup \{a + b - 1\}$. Each choice of cycles of lengths a and b gives ab such terms, so,

$$\begin{aligned}
\Sigma_{\mu^{(i)}}\,\Sigma_{\mu^{(j)}} = \Sigma_{\mu^{(i)}\sqcup\mu^{(j)}} + \sum_{\substack{a\in\mu^{(i)}\\ b\in\mu^{(j)}}} ab\, \Sigma_{(\mu^{(i)}\sqcup\mu^{(j)}\setminus\{a,b\})\sqcup\{a+b-1\}} \\
+ \text{terms of degree smaller than } m_i + m_j - 2.
\end{aligned}$$

Taking the expectation, we obtain:

$$\begin{aligned}
\mathbb{E}_{t,n}[\Sigma_{\mu^{(i)}}\,\Sigma_{\mu^{(j)}}] &= \left(n^{m_i+m_j} - \binom{m_i+m_j}{2} n^{m_i+m_j-1}\right) p_{\mu^{(i)}\sqcup\mu^{(j)}}(t) \\
&\quad + n^{m_i+m_j-1} \left(\sum_{\substack{a\in\mu^{(i)}\\ b\in\mu^{(j)}}} ab\, p_{(\mu^{(i)}\sqcup\mu^{(j)}\setminus\{a,b\})\sqcup\{a+b-1\}}(t) \right) + O(n^{m_i+m_j-2}).
\end{aligned}$$

Since $\binom{m_i+m_j}{2} - \binom{m_i}{2} - \binom{m_j}{2} = m_i m_j = \sum_{a\in\mu^{(i)},\, b\in\mu^{(j)}} ab$, this ends the proof of the theorem. $\square$

Theorem 12.30 is quite a striking result: given any parameter $t \in T$, the family of central measures $(\mathbb{P}_{t,n})_{n\in\mathbb{N}}$ gives rise to a large family of asymptotically Gaussian random variables, so there is a phenomenon of asymptotic normality that is hidden in the representation theory of symmetric groups. To conclude this chapter, let us restate Theorem 12.30 in a way that sheds more light on this point of view. Given an integer partition λ of size n, we can associate to it canonically an element t_λ in the Thoma simplex:

$$t_\lambda = \left(\left(\frac{a_1}{n},\dots,\frac{a_d}{n},0,\dots\right),\left(\frac{b_1}{n},\dots,\frac{b_d}{n},0,\dots\right)\right),$$

where $((a_1,\dots,a_d),(b_1,\dots,b_d))$ are the Frobenius coordinates of λ. Now, given $t \in T$, we denote $t^{(n)} = t_{\lambda^{(n)}}$, where $\lambda^{(n)}$ is chosen randomly according to the central measure $\mathbb{P}_{t,n}$. The random elements $t^{(n)}$ of the Thoma simplex should be considered as random perturbations of t, and the following result ensures that these perturbations are asymptotically Gaussian:

Proposition 12.33. *For any $t \in T$ and any $k \geq 1$,*

$$\sqrt{n}\,\frac{p_k(t^{(n)})-p_k(t)}{k} \rightharpoonup \mathcal{N}_{\mathbb{R}}(0,p_{2k-1}(t)-(p_k(t))^2).$$

Proof. The result can be generalized to random vectors with coordinates

$$\sqrt{n}\,(p_{\mu^{(i)}}(t^{(n)})-p_{\mu^{(i)}}(t)),$$

and this is only to simplify the presentation that we stick with the one-dimensional random variables $X_k^{(n)} = \sqrt{n}\,\frac{p_k(t^{(n)})-p_k(t)}{k}$. Recall that $\deg(\Sigma_k - p_k) \leq k-1$. Therefore,

$$X_k^{(n)} = \frac{\sqrt{n}}{k}\left(\frac{(p_k-\Sigma_k)(\lambda^{(n)})}{n^k} + \Sigma_k(\lambda^{(n)})\left(\frac{1}{n^k}-\frac{1}{n^{\downarrow k}}\right)+\left(\frac{\Sigma_k(\lambda^{(n)})}{n^{\downarrow k}}-p_k(t)\right)\right).$$

We introduce the random variables

$$A = \frac{\sqrt{n}}{k}\,\frac{(p_k-\Sigma_k)(\lambda^{(n)})}{n^k};$$
$$B = \frac{\sqrt{n}}{k}\,\Sigma_k(\lambda^{(n)})\left(\frac{1}{n^k}-\frac{1}{n^{\downarrow k}}\right);$$
$$C = \frac{\sqrt{n}}{k}\left(\frac{\Sigma_k(\lambda^{(n)})}{n^{\downarrow k}}-p_k(t)\right).$$

By Theorem 12.30, C converges in distribution towards a Gaussian variable of mean 0 and variance $p_{2k-1}(t)-(p_k(t))^2$. On the other hand, $\|A\|_\infty = O(n^{-\frac{1}{2}})$ and $\|B\|_\infty = O(n^{-\frac{1}{2}})$, so A and B converge in probability to 0. By Slutsky's lemma (see the references at the end of the chapter), if A and B converge in probability towards some constants and C converges in distribution, then $A+B+C$ converges in distribution, so $X_k^{(n)} \rightharpoonup \mathcal{N}_{\mathbb{R}}(0,p_{2k-1}(t)-(p_k(t))^2)$. $\square$

Thus, the Thoma simplex is a compact space endowed with the following special structure: by using the representation theory of the symmetric groups, one can generate for every parameter $t \in T$ some random perturbations $t^{(n)}$ of t, such that $t^{(n)} \to_{\mathbb{P}} t$, and such that the fluctuations of a large family of observables $t^{(n)}$ are asymptotically Gaussian, with covariances that depend continuously on the parameter t. This phenomenon leads to the notion of *mod–Gaussian moduli space*, and the Thoma simplex was the first example that we encountered (one can find other interesting examples in the theory of random graphs or of random permutations).

Notes and references

Our presentation of the Hopf algebra of free quasi-symmetric functions follows [DHT02, DHNT11]. It should be noticed that FQSym can be considered as a formal algebra of permutations rather than a subalgebra of $\mathbb{C}\langle X\rangle$; this point of view is the one of [MR95], but several proofs are made easier by using the polynomial realization of this algebra. For the explicit computation of the antipode of FQSym, we refer to [AS05]. The existence and unicity of an antipode in a graded connected bialgebra is a classical result; see [GR15, Proposition 1.36]. Our Corollary 12.14 can be extended to *skew* Schur functions, with the exact same statement; see [Sta99a, Theorem 7.19.7].

The riffle shuffle of cards has been studied thoroughly in [DFP92, Section 5], [BD92, DMP95]. Generalizations related to quasi-symmetric functions were considered in [Sta99b], and our definition with three parameters α, β, γ appeared in [Ful02]. Our Theorem 12.17 is essentially contained in loc. cit., but as far as we know the statement with quasi-symmetric functions is new, as well as the use of free quasi-symmetric functions. The special case when $\beta = \gamma = 0$ was treated in [DHT02].

The law of large numbers (Theorem 12.19) appeared first in [KV81], but its original proof was different from ours and relied on a martingale argument. The special case of the q-Plancherel measures corresponding to the decomposition of the regular traces of the Hecke algebras, was treated in [FM12]; see also [Str08]. As for the central limit theorem for observables, it was proven separately in [Buf12] and [Mél12], and other asymptotic results (large deviations, Berry–Esseen estimates) for these random variables were then obtained by using the formalism of mod–Gaussian convergence; see [FMN16, Section 11]. The technique of cumulants of observables was invented by Śniady; see [Śni06a, Śni06b]. In the next chapter, we shall reuse it, but working with the weight gradation on $\mathscr{O}$ instead of the degree.

If the parameter t of the Thoma simplex has a finite number of non-zero coordinates $\alpha_1 > \alpha_2 > \cdots > \alpha_k$ and $\beta_1 > \beta_2 > \cdots > \beta_l$ that are all distinct, then one can complete the central limit theorem 12.30 for observables by a central limit theorem satisfied by the first rows and columns of the random partition $\lambda^{(n)} \sim \mathbb{P}_{t,n}$:

$$\left(\sqrt{n}\left(\frac{\lambda_i^{(n)}}{n} - \alpha_i\right)\right)_{i \in [\![1,d]\!]} \rightharpoonup \mathscr{N}_{\mathbb{R}^d}(0,A)$$

with $A_{ij} = \delta_{ij}\,\alpha_i - \alpha_i\alpha_j$; see again [Buf12, Mél12]. A similar result holds also in this case for the fluctuations of the first columns of $\lambda^{(n)}$. When there are equalities of coordinates $\alpha_{i+1} = \alpha_{i+2} = \cdots = \alpha_{i+m}$, the limit in law of the fluctuations of the corresponding rows of $\lambda^{(n)}$ still exists, but it is not Gaussian. In particular, consider the case when $t = (\alpha, 0)$ with $\alpha = (\frac{1}{d}, \ldots, \frac{1}{d}, 0, 0, \ldots)$. We then know by the law of large numbers that $\lambda_i^{(n)} \simeq n/d$ for any $i \in [\![1,d]\!]$. Set

$$Y_i^{(n)} = \sqrt{nd}\left(\frac{\lambda_i^{(n)}}{n} - \frac{1}{d}\right).$$

It can be shown that the asymptotic distribution of $(Y_1^{(n)}, \ldots, Y_d^{(n)})$ has density

$$1_{y_1 \geq y_2 \geq \cdots \geq y_d}\,\frac{\Delta(y_1, \ldots, y_d)^2}{1!\,2!\,\cdots\,(d-1)!\,(2\pi)^{\frac{d-1}{2}}}\,\mathrm{e}^{-\frac{\|\vec{y}\|^2}{2}}\,d\vec{y},$$

where $\Delta(\vec{y})$ is the Vandermonde determinant $\prod_{1 \leq i < j \leq d}(y_j - y_i)$; and $d\vec{y}$ is the Lebesgue measure on the $(d-1)$-dimensional hyperplane $H = \{(y_1, \ldots, y_d) \mid y_1 + \cdots + y_d = 0\}$. This is exactly the law of the set of eigenvalues of a Gaussian matrix $M \in \mathfrak{su}(d)$ with covariance matrix

$$\mathbb{E}[\mathrm{tr}((AM)^*(BM))] = \frac{d^2-1}{d}\,\mathrm{tr}(A^*B),$$

see [Dys62].

A general limiting result can be stated when $t = (\alpha, 0)$, with $\sum_{i=1}^{\infty} \alpha_i = 1$ and the α_i's are arbitrary (possibly with equalities $\alpha_{i+1} = \alpha_{i+2} = \cdots = \alpha_{i+m}$). In this case, the asymptotic fluctuations are a sum of "Gaussian fluctuations" coming from different coordinates α_i, and of "eigenvalue fluctuations" coming from equal coordinates α_i; see [Mél12, Theorem 4] for a precise statement. This case can be treated by relating the RSK algorithm to certain operations on (random) walks known as Pitman transforms; cf. [O'C03]. These operations also have an interesting interpretation in the representation theory of complex Lie algebras: starting from a path connecting the origin to a dominant weight in the weight lattice, one can compute all the weights and multiplicities of the corresponding irreducible representation by applying these Pitman operators; see [Lit95], and [BBO05, Bia09] for the probabilistic consequences.

For details regarding the different notions of convergence of random variables (convergence in probability, in law, of the moments), we refer to [Bil69, Chapter 1]. For the method of cumulants and its use to prove asymptotic normality; see [LS59, Jan88, FMN16]. Slutsky's lemma can be found in [Bil69, Chapter 1, Theorem 3.1], and Wick formula for the joint moments of the coordinates of a Gaussian vector appeared in [Wic50]; see also [t'H74] for the extension to Gaussian matrices.

13

Asymptotics of Plancherel and Schur–Weyl measures

In the previous chapter, we showed that for any parameter $t = (\alpha, \beta)$ of the Thoma simplex T, if $\lambda^{(n)}$ is a random partition chosen under the central measure $\mathbb{P}_{t,n}$, then $\lambda_i^{(n)} \simeq n\alpha_i$, and $\lambda_j^{(n)\prime} \simeq n\beta_j$ as n goes to infinity. Consider the particular case when $t = 0 = (0,0)$ is the parameter of the Thoma simplex corresponding to the regular traces of the symmetric group algebras and to the Plancherel measures:

$$\tau_0(\sigma) = 1_{\sigma=\mathrm{id}_{\mathbb{N}^*}} \quad ; \quad \mathbb{P}_{0,n}[\lambda] = d_\lambda \, s_\lambda(E) = \frac{(\dim \lambda)^2}{n!}.$$

By the previous discussion, as n goes to infinity, if λ is chosen randomly according to the Plancherel measure $\mathbb{P}_n = \mathbb{P}_{0,n}$, then

$$\frac{\lambda_i^{(n)}}{n} \to_{\mathbb{P}_n} 0 \quad ; \quad \frac{\lambda_j^{(n)\prime}}{n} \to_{\mathbb{P}_n} 0$$

for any index of row i or index of column j. Thus, the rows and columns are not in this case of typical size $O(n)$, and in this setting, the parameter $t = 0$ appears as a *singular point* of the Thoma simplex. The goal of this chapter is to study the behavior of random partitions corresponding to this singular point (Plancherel measures), or more generally to points in an infinitesimal neighborhood of the singularity (Schur–Weyl measures). We shall see that under adequate hypotheses, these random partitions have rows and columns of order $O(\sqrt{n})$ instead of $O(n)$, and that their Young diagrams have after rescaling a continuous limiting shape.

In Section 13.1, we start from two natural models of random words or random permutations, and we use the RSK algorithm in order to relate them to some spectral measures $\mathbb{P}_{c\geq 0,n}$ of representations of symmetric groups, including the Plancherel measures $\mathbb{P}_{0,n}$. We also prove an a priori bound on the size of the longest weakly increasing subwords in these models (Lemmas 13.2 and 13.3); these bounds shall prove useful later when trying to establish a law of large numbers for these quantities.

In Section 13.2, we study the random integer partitions $\lambda^{(n)} \sim \mathbb{P}_{c,n}$, and we look for an analogue of Theorem 12.19 in this setting. The aforementioned a priori bounds makes one expect $\lambda_1^{(n)}$ to be of size $O(\sqrt{n})$ instead of $O(n)$, and therefore, we shall need a renormalization different than the one for random partitions under central measures (Chapter 12). It will turn out that the *continuous Young diagrams* $\omega_{\lambda^{(n)}}$, after a renormalization by a factor $\sqrt{n}$ in both directions, converge in probability towards limiting shapes $\omega_{c,n}$, the convergence being with respect to the uniform distance in the space $\mathscr{Y}$ of continuous Young diagrams (cf. the discussion at the end of Chapter 7). By using the technology of free cumulants of (continuous) Young diagrams, we shall compute explicitly these limiting shapes and prove a law of large numbers (Theorems 13.6 and 13.7), which is due to Logan, Shepp, Kerov and Vershik in the case of Plancherel measures, and to P. Biane in the case of the Schur–Weyl measures. An important consequence of this law of large numbers is the solution to the so-called *Ulam problem*: in a uniform random permutation $\sigma^{(n)}$ of size n, the length of the longest increasing subsequences is asymptotically of size close to $2\sqrt{n}$ with very large probability.

Finally, in Section 13.3, we complete the law of large numbers by a central limit theorem for the random character values $\chi^{\lambda^{(n)}}(\sigma)$ with $\lambda^{(n)} \sim \mathbb{P}_{c,n}$ (Theorem 13.11). This result takes a form similar to the results of Section 12.3 for central measures, although with a different renormalization of the characters. The proof of this central limit theorem relies on combinatorial and geometric arguments stemming from the theory of admissible set partitions that we developed in Section 9.4. As for central measures, there is a way to deduce from this central limit theorem some information on the fluctuations of the random continuous Young diagrams $\omega_{\lambda^{(n)}}$; we shall not deal with this theory, but as in the previous chapter, we shall give in the notes a detailed account of this geometric counterpart of Theorem 13.11.

13.1 The Plancherel and Schur–Weyl models

In this section, $n \geq 1$ is a fixed integer, and we shall consider the two following models of random words of length n:

1. random uniform permutations: $\mathbb{Q}_{0,n}$ is the uniform probability measure on the permutation group $\mathfrak{S}(n)$:

$$\forall \sigma \in \mathfrak{S}(n),\ \mathbb{Q}_{0,n}[\sigma] = \frac{1}{n!}.$$

2. random uniform words: a parameter $c > 0$ being fixed, we also fix a sequence of integers $(N_n)_{n\in\mathbb{N}}$ such that $\lim_{n\to\infty} \frac{\sqrt{n}}{N_n} = c$. We then denote $\mathbb{Q}_{c,n}$ the uniform probability measure on words in $([\![1,N_n]\!])^n$:

$$\forall w \in ([\![1,N_n]\!])^n,\ \mathbb{Q}_{c,n}[w] = \frac{1}{(N_n)^n}.$$

In the second case, the probability measure $\mathbb{Q}_{c,n}$ depends of course on the choice of the sequence of integers $(N_n)_{n\in\mathbb{N}}$, but we shall see that the asymptotics of the random partitions λ associated to the random words $w \sim \mathbb{Q}_{c,n}$ only depend on c.

$\triangleright$ *Plancherel and Schur–Weyl measures.*

As in the previous chapter, we consider the map $\mathrm{RSKsh} : (\mathbb{N}^*)^n \to \mathfrak{Y}(n)$ which associates to a word w of length n the shape λ of the two RSK tableaux $P(w)$ and $Q(w)$. We then denote

$$\mathbb{P}_{0,n} = \mathrm{RSKsh}_* \mathbb{Q}_{0,n} \quad ; \quad \mathbb{P}_{c,n} = \mathrm{RSKsh}_* \mathbb{Q}_{c,n}$$

the probability measures on $\mathfrak{Y}(n)$ that are the images of $\mathbb{Q}_{0,n}$ and $\mathbb{Q}_{c,n}$; thus, for instance,

$$\mathbb{P}_{0,n}[\lambda] = \mathbb{Q}_{0,n}[\sigma \in \mathfrak{S}(n) \,|\, \mathrm{RSKsh}(\sigma) = \lambda].$$

Notice that $\mathbb{P}_{c,n}[\lambda]$ vanishes if $\ell(\lambda) > N_n$. Indeed, if $\lambda = \mathrm{RSKsh}(w)$, then $\ell(\lambda) = \lambda'_1$ is the length of a longest strictly decreasing subsequence of w (cf. Chapter 3), and since the letters of $w \sim \mathbb{Q}_{c,n}$ belong to $[\![1, N_n]\!]$, this length cannot be larger than N_n.

The following proposition shows that $\mathbb{P}_{0,n}$ and $\mathbb{P}_{c,n}$ are spectral measures of natural representations of $\mathfrak{S}(n)$ (in the case of $\mathbb{P}_{0,n}$, the result was already alluded to in several previous chapters).

Proposition 13.1. *The probability measure $\mathbb{P}_{0,n}$ is the **Plancherel measure** of $\mathfrak{S}(n)$, that is the spectral measure of the regular representation $\mathbb{C}\mathfrak{S}(n)$. On the other hand, the probability measure $\mathbb{P}_{c,n}$ is the spectral measure of the permutation representation of $\mathfrak{S}(n)$ on the space of tensors $(\mathbb{C}^{N_n})^{\otimes n}$. We call $\mathbb{P}_{c,n}$ the **Schur–Weyl measure** of parameters c and n.*

Proof. For $\mathbb{P}_{0,n}$, we have

$$\begin{aligned}\mathbb{P}_{0,n}[\lambda] &= \frac{\mathrm{card}\,\{\sigma \in \mathfrak{S}(n) \,|\, \mathrm{RSKsh}(\sigma) = \lambda\}}{n!} \\ &= \frac{\mathrm{card}\,\{(P,Q) \in \mathrm{ST}(\lambda) \times \mathrm{ST}(\lambda)\}}{n!} = \frac{(d_\lambda)^2}{n!},\end{aligned}$$

and we have already seen that this is the formula for the Plancherel measure (see for instance the end of Section 1.3). Similarly, for $\mathbb{P}_{c,n}$, we have

$$\begin{aligned}\mathbb{P}_{c,n}[\lambda] &= \frac{\mathrm{card}\,\{w \in ([\![1, N_n]\!])^n \,|\, \mathrm{RSKsh}(w) = \lambda\}}{(N_n)^n} \\ &= \frac{\mathrm{card}\,\{(P,Q) \in \mathrm{SST}(N_n, \lambda) \times \mathrm{ST}(\lambda)\}}{(N_n)^n} = \frac{(\dim V^\lambda)(\dim S^\lambda)}{(\dim \mathbb{C}^{N_n})^{\otimes n}}.\end{aligned}$$

On the last line, we have used the fact that the irreducible representation V^λ of

$\mathrm{GL}(N_n)$ of highest weight λ has dimension $\operatorname{card}\mathrm{SST}(N_n,\lambda)$; see Proposition 3.5. We now recognize the spectral measure of the representation $(\mathbb{C}^{N_n})^{\otimes n}$, since by Schur–Weyl duality

$$(\mathbb{C}^{N_n})^{\otimes n} = \bigoplus_{\lambda\in\mathfrak{Y}(n)\,|\,\ell(\lambda)\leq N_n} V^\lambda \otimes S^\lambda. \qquad \square$$

Remark. One can give explicit combinatorial formulas for the Plancherel and Schur–Weyl measures. By the hook-length formula 3.41,

$$\mathbb{P}_{0,n}[\lambda] = \frac{(d_\lambda)^2}{n!} = \frac{n!}{\prod_{\square\in\lambda}(h(\square))^2}.$$

On the other hand, we shall see in a moment that the dimension of the representation V^λ of $\mathrm{GL}(N_n,\mathbb{C})$ is given by the formula

$$\dim V^\lambda = \frac{d_\lambda}{n!}\left(\prod_{\square\in\lambda} N_n + c(\square)\right),$$

where $c(\square)$ is the content of the cell $\square$ in the Young diagram of λ; see the proof of Lemma 13.3. Therefore,

$$\mathbb{P}_{c,n}[\lambda] = \frac{d_\lambda\,(\dim V^\lambda)}{(N_n)^n} = \frac{n!}{\prod_{\square\in\lambda}(h(\square))^2}\left(\prod_{\square\in\lambda} 1+\frac{c(\square)}{N_n}\right).$$

When $c\to 0$, N_n grows and the previous formulas show that $\mathbb{P}_{c,n}[\lambda]$ gets close to $\mathbb{P}_{0,n}[\lambda]$. Actually, most of the results of this chapter tend to prove that the Schur–Weyl measures can be considered as deformations of the Plancherel measures.

Let us now relate these spectral measures to the theory of coherent families of traces and of central measures. For the Plancherel measures, we already mentioned in the introduction to the chapter that $\mathbb{P}_{0,n}$ is the central measure associated to the parameter $t = 0 = ((0,0,\ldots),(0,0,\ldots)) \in T$ and to the non-negative specialization of Sym given by the exponential alphabet E. On the other hand, the Schur–Weyl measures can be written as central measures with varying parameter $t = t_n$, such that $t_n \to 0$. Indeed,

$$\mathbb{P}_{c,n}[\lambda] = \frac{d_\lambda\,(\dim V^\lambda)}{(N_n)^n} = \frac{1}{(N_n)^n}\,d_\lambda\,s_\lambda(\underbrace{1,1,\ldots,1}_{N_n \text{ terms}}) = d_\lambda\,s_\lambda\left(\underbrace{\frac{1}{N_n},\ldots,\frac{1}{N_n}}_{N_n\text{ terms}}\right).$$

Therefore, with the notations of the previous chapter, $\mathbb{P}_{c,n} = \mathbb{P}_{t_n,n}$, with

$$t_n = \left(\left(\underbrace{\frac{1}{N_n},\ldots,\frac{1}{N_n}}_{N_n\text{ terms}},0,\ldots\right),(0,0,\ldots)\right)\in T.$$

Consequently, the Schur–Weyl measures can be considered as infinitesimal deformations of the Plancherel measures, and their study amounts to a kind of differential calculus on the Thoma simplex. We shall prove that the asymptotics of the random partitions $\lambda^{(n)} \sim \mathbb{P}_{c,n}$ vary indeed continuously with the parameter c; see Theorem 13.7.

$\triangleright$ *An upper bound on the lengths of increasing subsequences.*

In the next section, we shall prove that if $\lambda^{(n)} \sim \mathbb{P}_{c,n}$ with $c = 0$ or $c > 0$, then the continuous Young diagram $\omega_{\lambda^{(n)}}$ rescaled by a factor $\sqrt{n}$ converges in probability towards a limit shape $\omega_{c,\infty}$; see Theorems 13.6 and 13.7. This result implies a law of large numbers for the length of a longest increasing subsequence in a random permutation $\sigma^{(n)} \sim \mathbb{P}_{0,n}$ or a random word $w^{(n)} \sim \mathbb{P}_{c,n}$. However, in order to prove this result, it is required to have an a priori bound on $\mathbb{E}_{c,n}[\lambda_1^{(n)}]$. These bounds are provided by the two following lemmas.

Lemma 13.2. *Let $\lambda^{(n)}$ be a random partition under the Plancherel measure $\mathbb{P}_{0,n}$. One has $\mathbb{E}[\lambda_1^{(n)}] \leq 2\sqrt{n}$.*

Proof. We are going to show that for all $n \geq 1$, $\mathbb{E}[\lambda_1^{(n)}] - \mathbb{E}[\lambda_1^{(n-1)}] \leq \frac{1}{\sqrt{n}}$; this will indeed imply

$$\mathbb{E}[\lambda_1^{(n)}] \leq \sum_{k=1}^{n} \frac{1}{\sqrt{k}} \leq 1 + \int_1^n \frac{1}{\sqrt{x}}\,dx = 2\sqrt{n} - 1 \leq 2\sqrt{n}.$$

Let $\sigma^{(n)}$ be a uniform random permutation of size n. Notice that if $\sigma^{(n)}_{|n-1}$ is the subword of $\sigma^{(n)}$ that consists in the $(n-1)$ first letters of $\sigma^{(n)}$, then conditionally to the set of these $(n-1)$ letters, the law of $\sigma^{(n)}_{|n-1}$ is uniform on permutations of these letters. Therefore, if $\lambda^{(n)} \sim \mathbb{P}_{0,n}$ and $\lambda^{(n-1)} \sim \mathbb{P}_{0,n-1}$, then we have the identities in law

$$\lambda^{(n)} = \mathrm{RSKsh}(\sigma^{(n)}) \quad ; \quad \lambda^{(n-1)} = \mathrm{RSKsh}(\sigma^{(n)}_{|n-1}).$$

With this representation, $\lambda^{(n)}$ has one more box in its bottom row than $\lambda^{(n-1)}$ if and only if the recording tableau $Q(\sigma^{(n)})$ contains n on its bottom row. Therefore, $\mathbb{E}[\lambda_1^{(n)}] - \mathbb{E}[\lambda_1^{(n-1)}] = \mathbb{P}[Q(\sigma^{(n)})$ has n on its bottom row$]$. However,

$$\begin{aligned}
&\mathbb{P}[P(\sigma^{(n)}) \text{ and } Q(\sigma^{(n)}) \text{ both have } n \text{ on their bottom rows}] \\
&= \sum_{\lambda \in \mathfrak{Y}(n)} \mathbb{P}_{0,n}[\lambda]\; \mathbb{P}[P(\sigma^{(n)}) \text{ and } Q(\sigma^{(n)}) \text{ have } n \text{ on their bottom rows} \,|\, \lambda^{(n)} = \lambda] \\
&= \sum_{\lambda \in \mathfrak{Y}(n)} \mathbb{P}_{0,n}[\lambda]\; \left(\mathbb{P}[Q(\sigma^{(n)}) \text{ has } n \text{ on its bottom row} \,|\, \mathrm{RSKsh}(\sigma^{(n)}) = \lambda]\right)^2
\end{aligned}$$

$$\geq \left(\sum_{\lambda \in \mathfrak{Y}(n)} \mathbb{P}_{0,n}[\lambda]\, \mathbb{P}[Q(\sigma^{(n)}) \text{ has } n \text{ on its bottom row} \,|\, \mathrm{RSKsh}(\sigma^{(n)}) = \lambda] \right)^2$$
$$\geq \left(\mathbb{P}[Q(\sigma^{(n)}) \text{ has } n \text{ on its bottom row}]\right)^2.$$

If $P(\sigma^{(n)})$ and $Q(\sigma^{(n)})$ both have n on their bottom rows, then this means that n was the last letter inserted during the RSK algorithm, so $\sigma^{(n)}(n) = n$. This happens with probability $\frac{1}{n}$, so the proof is completed. □

Lemma 13.3. *Let $\lambda^{(n)}$ be a random partition under the Schur–Weyl measure $\mathbb{P}_{c,n}$. One has $\mathbb{E}[\lambda_1^{(n)}] \leq \frac{n}{N_n} + 2\sqrt{n} \simeq (c+2)\sqrt{n}$.*

Proof. We fix $N \geq 1$, and we consider a random word $w^{(n)}$ that is chosen uniformly in $([\![1,N]\!])^n$. The law of $w^{(n)}_{|n-1}$ is of course uniform over the set $([\![1,N]\!])^{n-1}$; in other words, we have the equality in law $w^{(n)}_{|n-1} = w^{(n-1)}$. We are going to compute in terms of N and n a bound on

$$\mathbb{E}[(\mathrm{RSKsh}(w^{(n)}))_1] - \mathbb{E}[(\mathrm{RSKsh}(w^{(n-1)}))_1] = \mathbb{P}[Q(w^{(n)}) \text{ has } n \text{ on its bottom row}].$$

For any standard tableau T of shape $\lambda \in \mathfrak{Y}(n)$,

$$\mathbb{P}[Q(w^{(n)}) = T] = \sum_{S \in \mathrm{SST}(N,\lambda)} \mathbb{P}[P(w^{(n)}) = S \text{ and } Q(w^{(n)}) = T] = \sum_{S \in \mathrm{SST}(N,\lambda)} \frac{1}{N^n}$$
$$= \frac{\dim V^\lambda}{N^n} = s_\lambda\left(\underbrace{\frac{1}{N}, \ldots, \frac{1}{N}}_{N \text{ terms}}\right)$$

So, for any $\mu \in \mathfrak{Y}(n-1)$, if μ^+ is the Young diagram obtained from μ by adding one cell to the first row, then

$$\mathbb{P}[Q(w^{(n)}) \text{ has } n \text{ on its bottom row} \,|\, \mathrm{RSKsh}(w^{(n-1)}) = \mu]$$
$$= \frac{1}{\mathbb{P}[\mathrm{RSKsh}(w^{(n-1)}) = \mu]} \sum_{T \in \mathrm{ST}(\mu)} s_{\mu^+}\left(\frac{1}{N}^{\otimes N}\right) = \frac{s_{\mu^+}\left(\frac{1}{N}^{\otimes N}\right)}{s_\mu\left(\frac{1}{N}^{\otimes N}\right)},$$

since $\mathbb{P}[\mathrm{RSKsh}(w^{(n-1)}) = \mu] = d_\mu\, s_\mu(\frac{1}{N}^{\otimes N})$. We can use the Cauchy–Schwarz inequality as in the proof of the previous lemma:

$$(\mathbb{P}[Q(w^{(n)}) \text{ has } n \text{ on its bottom row}])^2$$
$$= \left(\sum_{\mu \in \mathfrak{Y}(n-1)} \mathbb{P}[\mathrm{RSKsh}(w^{(n-1)}) = \mu]\, \frac{s_{\mu^+}(\frac{1}{N}^{\otimes N})}{s_\mu(\frac{1}{N}^{\otimes N})} \right)^2$$

$$\leq \sum_{\mu \in \mathfrak{Y}(n-1)} \mathbb{P}[\mathrm{RSKsh}(w^{(n-1)}) = \mu] \left(\frac{s_{\mu^+}(\frac{1}{N}^{\otimes N})}{s_\mu(\frac{1}{N}^{\otimes N})} \right)^2$$

$$\leq \sum_{\mu \in \mathfrak{Y}(n-1)} d_\mu \, s_{\mu^+}\left(\frac{1}{N}^{\otimes N}\right) \left(\frac{s_{\mu^+}(\frac{1}{N}^{\otimes N})}{s_\mu(\frac{1}{N}^{\otimes N})} \right).$$

Now, recall from Proposition 4.63 that

$$s_\mu(1, q, \ldots, q^{N-1}) = q^{n(\mu)} \prod_{\Box \in \mu} \frac{1 - q^{N+c(\Box)}}{1 - q^{h(\Box)}},$$

where $n(\mu) = \sum_{i=1}^{\ell(\mu)} (i-1)\mu_i$, $c(\Box)$ is the content of the cell $\Box$ and $h(\Box)$ is the hook-length. Specializing q to 1, we obtain:

$$s_\mu(1^{\otimes N}) = \prod_{\Box \in \mu} \frac{N + c(\Box)}{h(\Box)} = \frac{\dim S^\mu}{(n-1)!} \left(\prod_{\Box \in \mu} N + c(\Box) \right)$$

Therefore,

$$\mathbb{P}[Q(w^{(n)}) \text{ has } n \text{ on its bottom row} \,|\, \mathrm{RSKsh}(w^{(n-1)}) = \mu] = \frac{1}{n} \frac{d_{\mu^+}}{d_\mu} \frac{N + \mu_1}{N}$$

and

$$(\mathbb{P}[Q(w^{(n)}) \text{ has } n \text{ on its bottom row}])^2 \leq \frac{1}{n} \sum_{\mu \in \mathfrak{Y}(n-1)} d_{\mu^+} \, s_{\mu^+}\left(\frac{1}{N}^{\otimes N}\right) \frac{N + \mu_1}{N}$$

$$\leq \frac{1}{n} \, \mathbb{E}\left[\frac{N + (\mathrm{RSKsh}(w^{(n)}))_1}{N} \right].$$

Set $\delta_n = \mathbb{E}[(\mathrm{RSKsh}(w^{(n)}))_1] - \mathbb{E}[(\mathrm{RSKsh}(w^{(n-1)}))_1]$. We have just shown that

$$\delta_n \leq \sqrt{\frac{1}{n}\left(1 + \frac{\delta_1 + \cdots + \delta_n}{N}\right)}.$$

From this inequality, we can prove by induction on n that $\delta_n \leq \frac{1}{N} + \frac{1}{\sqrt{n}}$; by summation over n, this implies $\mathbb{E}[(\mathrm{RSKsh}(w^{(n)}))_1] \leq \frac{n}{N} + 2\sqrt{n}$ as claimed. If $n = 1$, then $\delta_1 = 1$ and the inequality is trivial. Suppose the result to be true up to order $n-1$, and consider the quantity δ_n. The map

$$g : \delta \mapsto \delta - \sqrt{\frac{1}{n}\left(1 + \frac{\delta_1 + \cdots + \delta_{n-1} + \delta}{N}\right)}$$

is increasing with respect to δ. Set $\Delta = \frac{1}{N} + \frac{1}{\sqrt{n}}$. We have $g(\delta_n) \leq 0$ by the previous argument, and

$$\begin{aligned}
g(\Delta) &= \frac{1}{N} + \frac{1}{\sqrt{n}} - \sqrt{\frac{1}{n}\left(1 + \frac{\delta_1 + \cdots + \delta_{n-1} + \frac{1}{N} + \frac{1}{\sqrt{n}}}{N}\right)} \\
&\geq \frac{1}{N} + \frac{1}{\sqrt{n}} - \sqrt{\frac{1}{n}\left(1 + \frac{1}{N}\sum_{i=1}^{n}\left(\frac{1}{N} + \frac{1}{\sqrt{i}}\right)\right)} \\
&\geq \frac{1}{N} + \frac{1}{\sqrt{n}} - \sqrt{\frac{1}{nN}\left(N + \frac{n}{N} + 2\sqrt{n}\right)} \\
&\geq \frac{1}{N} + \frac{1}{\sqrt{n}} - \frac{1}{\sqrt{nN}}\left(\sqrt{N} + \sqrt{\frac{n}{N}}\right) = 0
\end{aligned}$$

by using the induction hypothesis on the second line. Therefore, $\delta_n \leq \Delta$ and the proof is completed. □

13.2 Limit shapes of large random Young diagrams

In this section, $\lambda^{(n)}$ is a random partition chosen according to a probability measure $\mathbb{P}_{c,n}$ with $c = 0$ or $c > 0$. We consider the continuous function $\omega_{\lambda^{(n)}}$ associated to $\lambda^{(n)}$ (representation of the Young diagram with the Russian convention), and we denote

$$\omega_{c,n}(s) = \frac{1}{\sqrt{n}}\, \omega_{\lambda^{(n)}}(\sqrt{n}\, s)$$

its renormalization by a factor $\frac{1}{\sqrt{n}}$ in both directions (abscissa and ordinate).

We have drawn in Figures 13.1 and 13.2 some examples of random partitions under the Plancherel and Schur–Weyl measures. We shall prove in this section that there exist limit shapes $\omega_{c,\infty}$ that are continuous Young diagrams in $\mathscr{Y}$, and such that

$$\sup_{s \in \mathbb{R}} |\omega_{c,n}(s) - \omega_{c,\infty}(s)| \to_{\mathbb{P}_{c,n}} 0.$$

Thus, with probability very close to 1, the random curve $\omega_{c,n}$ is uniformly close to the limit shape $\omega_{c,\infty}$ when n is large.

▷ *Convergence in probability of continuous Young diagrams.*

In order to prove the aforementioned convergence of random Young diagrams, we need a probabilistic version of Proposition 7.26. We shall use the following:

Proposition 13.4. *Let $(\omega_n)_{n \in \mathbb{N}}$ be a sequence of random continuous Young diagrams in $\mathscr{Y}$. We suppose that:*

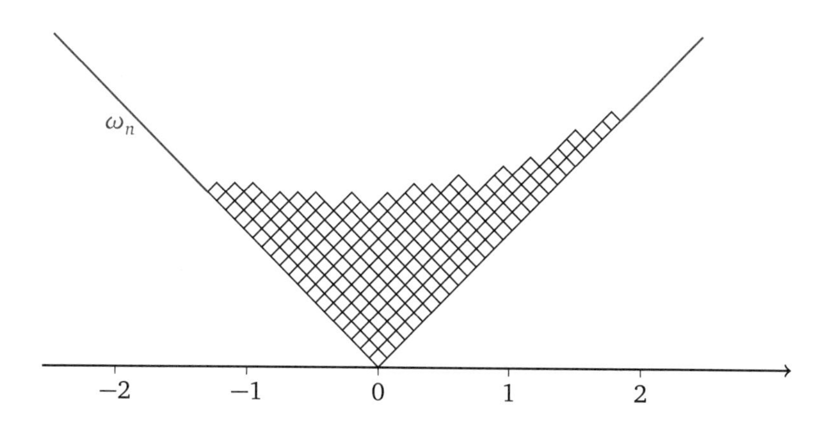

Figure 13.1
A random Young diagram λ under the Plancherel measure $\mathbb{P}_{0,n=200}$.

(C1) There exists a positive real number D, such that

$$\lim_{n\to\infty} \mathbb{P}[\mathrm{supp}(\omega_n(s)-|s|) \subset [-D,D]] = 1.$$

(C2) There exists a continuous Young diagram ω supported on $[-D,D]$ such that, for any observable $f \in \mathscr{O}$,

$$f(\omega_n) \to_{\mathbb{P}} f(\omega).$$

Then, one has the convergence in probability

$$\sup_{s\in\mathbb{R}} |\omega_n(s)-\omega(s)| \to_{\mathbb{P}} 0.$$

Proof. We endow $\mathscr{Y}$ with the topology given by the uniform distance, and we have to prove that with respect to this topology, $(\omega_n)_{n\in\mathbb{N}}$ converges in law towards ω (in a metric space, the convergence in law towards a constant is equivalent to the convergence in distribution). Denote $\mathbb{P}_n$ the law of ω_n, and let us prove that the family of laws $(\mathbb{P}_n)_{n\in\mathbb{N}}$ is **tight**: for every $\varepsilon > 0$, there exists a compact subset $K \subset \mathscr{Y}$ such that $\mathbb{P}_n(K) \geq 1-\varepsilon$ for any $n \in \mathbb{N}$. Notice that since continuous Young diagrams are Lipschitz functions with constant 1, by the Arzelà–Ascoli theorem, for any $C > 0$, the set $\mathscr{Y}_{[-C,C]}$ of continuous Young diagrams ω with

$$\mathrm{supp}(w(s)-|s|) \subset [-C,C]$$

is compact for the topology of uniform convergence. Fix $\varepsilon > 0$. By the hypothesis

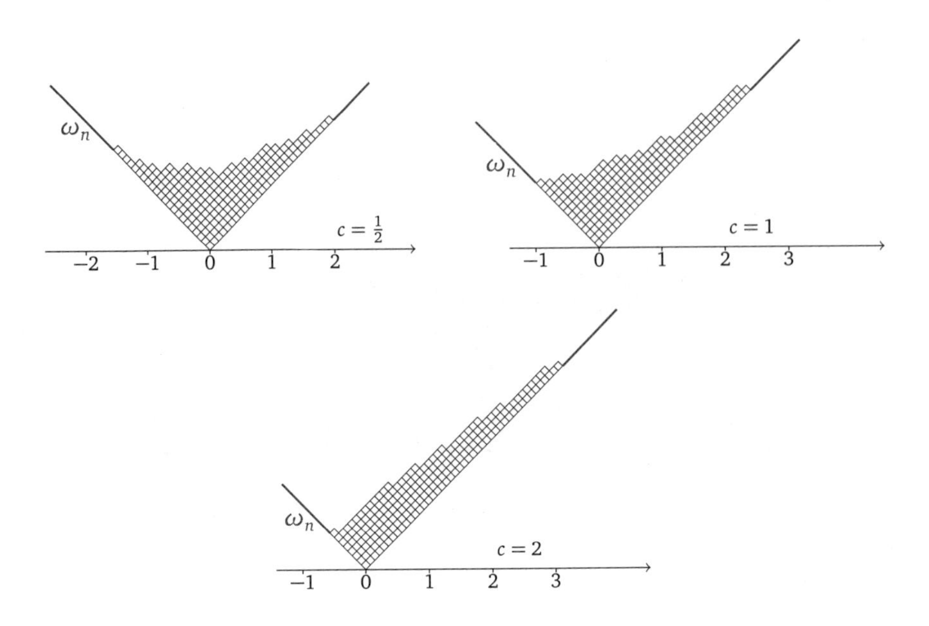

Figure 13.2
Random Young diagrams of size $n = 200$ under the Schur–Weyl measures of parameter $c = \frac{1}{2}$, $c = 1$ and $c = 2$.

(C1), there exists $N \in \mathbb{N}$ such that, for any $n \geq N$, $\mathbb{P}_n[\mathscr{Y}_{[-D,D]}] \geq 1-\varepsilon$. On the other hand, since $\mathscr{Y} = \bigcup_{C>0} \mathscr{Y}_{[-C,C]}$, for every $n < N$, there exists C_n such that $\mathbb{P}_n[\mathscr{Y}_{[-C_n,C_n]}] \geq 1-\varepsilon$. Then, with

$$C = \max(\{C_n,\ n < N\} \cup \{D\})$$

and $K = \mathscr{Y}_{[-C,C]}$, we have $\mathbb{P}_n[K] \geq 1-\varepsilon$ for any $n \in \mathbb{N}$, so the family of probability measures $(\mathbb{P}_n)_{n\in\mathbb{N}}$ is tight. Therefore, it is relatively compact for the topology of convergence in law. It suffices now to prove that the only possible limit in law of a convergent subsequence of $(\mathbb{P}_n)_{n\in\mathbb{N}}$ is the Dirac measure δ_ω. Let $(\mathbb{P}_{n_k})_{k\in\mathbb{N}}$ be such a convergent subsequence, and μ be its limit. The hypothesis (C1) ensures that μ is supported on $\mathscr{Y}_{[-D,D]}$, and also that

$$\mu = \lim_{k\to\infty} \mathbb{P}_{n_k} = \lim_{k\to\infty} \widetilde{\mathbb{P}}_{n_k},$$

where $\widetilde{\mathbb{P}}_{n_k}[A] = \frac{\mathbb{P}_{n_k}[A\cap\mathscr{Y}_{[-D,D]}]}{\mathbb{P}_{n_k}[\mathscr{Y}_{[-D,D]}]}$. Indeed, if G is a closed subset of $\mathscr{Y}$, then

$$\limsup_{k\to\infty} \widetilde{\mathbb{P}}_{n_k}[G] = \limsup_{k\to\infty} \mathbb{P}_{n_k}[G \cap \mathscr{Y}_{[-D,D]}] \leq \mu(G \cap \mathscr{Y}_{[-D,D]}) = \mu(G)$$

since $\mathbb{P}_{n_k} \rightharpoonup \mu$, so by Portmanteau's theorem, $\widetilde{\mathbb{P}}_{n_k} \rightharpoonup \mu$. Now, for any observable

$f \in \mathscr{O}$, $f(\omega_{n_k}) \to_{\mathbb{P}_{n_k}} f(\omega)$, and the same kind of argument as above ensures that we also have

$$f(\widetilde{\omega}_{n_k}) \to_{\widetilde{\mathbb{P}}_{n,k}} f(\omega),$$

where $\widetilde{\omega}_{n_k}$ is a random continuous Young diagram in $\mathscr{Y}_{[-D,D]}$ chosen under $\widetilde{\mathbb{P}}_{n,k}$. Thus, we have reduced the problem to the following situation:

1. we have a sequence of probability measures $(\mu_k = \widetilde{\mathbb{P}}_{n_k})_{k\in\mathbb{N}}$ on $\mathscr{Y}_{[-D,D]}$ such that $\mu_k \rightharpoonup \mu$,
2. for any $f \in \mathscr{O}$, if $\zeta_k \sim \mu_k$ in $\mathscr{Y}_{[-D,D]}$, then $f(\zeta_k) \to_{\mu_k} f(\omega)$.

However, $\mathscr{Y}_{[-D,D]}$ is a complete metric space, and even compact (this was not the case of $\mathscr{Y}$). Therefore, one can use Skorohod's representation theorem, and in the previous setting, one can define random continuous Young diagrams ζ_k and ζ on a common probability space, such that $\zeta_k \sim \mu_k$, $\zeta \sim \mu$, and

$$\zeta_k \to_{\text{almost surely, for the uniform norm}} \zeta.$$

In other words, one can represent the convergence in law by an almost sure convergence (this result holds on any polish space; see the notes at the end of the chapter). Now, the observables $f \in \mathscr{O}$ are continuous on $\mathscr{Y}_{[-D,D]}$, since for any elements $\omega, \zeta \in \mathscr{Y}_{[-D,D]}$ and any $k \geq 2$,

$$|\widetilde{p}_k(\omega) - \widetilde{p}_k(\zeta)| = k(k-1)\left|\int_{-D}^{D} \frac{\omega(s)-\zeta(s)}{2}\, s^{k-2}\, ds\right| \leq k\, D^{k-1}\, \|\omega - \zeta\|_\infty.$$

As a consequence, we also have $f(\zeta_k) \to_{\text{almost surely}} f(\zeta)$ for any $f \in \mathscr{O}$. Since $f(\zeta_k) \to_{\mu_k} f(\omega)$, $f(\zeta) = f(\omega)$ almost surely for any observable $f \in \mathscr{O}$. This implies that $\zeta = \omega$ almost surely, by using for instance Proposition 7.26. Thus, $\mu = \delta_\omega$, and we have shown that the only possible limit of a convergent subsequence of the tight sequence $\mathbb{P}_n$ is δ_ω. Therefore, $\mathbb{P}_n \rightharpoonup \delta_\omega$, and $\omega_n \to_{\mathbb{P}_n} \omega$. □

This criterion of convergence will reduce the asymptotic analysis of Plancherel and Schur–Weyl measures to calculations with observables, because of:

Proposition 13.5. *Let $(\omega_{c,n})_{n\in\mathbb{N}}$ be a sequence of random continuous Young diagrams under the Plancherel ($c = 0$) or Schur–Weyl ($c > 0$) measures. The random sequence satisfies Hypothesis (C1).*

Proof. If $\sigma^{(n)}$ is a random uniform permutation in $\mathfrak{S}(n)$, we are going to prove that for any $\varepsilon > 0$

$$\mathbb{Q}_{0,n}\left[(\mathrm{RSKsh}(\sigma^{(n)}))_1 \leq (\mathrm{e}+\varepsilon)\sqrt{n}\right] \to 1.$$

It implies that

$$\mathbb{P}_{0,n}\left[\lambda_1^{(n)} \leq (\mathrm{e}+\varepsilon)\sqrt{n}\right] = \mathbb{P}_{0,n}[\forall s \geq \mathrm{e}+\varepsilon,\ \omega_n(s) = |s|] \to 1,$$

and since the Plancherel measures are invariant by conjugation of Young diagrams, this gives Hypothesis (C1) for $\mathbb{P}_{0,n}$ with an interval $[-(\mathrm{e}+\varepsilon), \mathrm{e}+\varepsilon]$. To prove this estimate, notice first that

$$\mathbb{P}_{0,n}[\lambda_1^{(n)} = l] \leq \frac{1}{n!}\binom{n}{l}^2 (n-l)! \leq \frac{n^l}{(l!)^2}.$$

Indeed, to construct a permutation of size n with a longest increasing subsequence of size l, one can first choose these l elements and their positions in the permutation ($\binom{n}{l}^2$ possibilities), and then there are at most $(n-l)!$ remaining possibilities for the other elements (in many cases, much less than that, since one does not want larger increasing subsequences). As a consequence,

$$\sum_{n=1}^{\infty} \mathbb{P}_{0,n}\left[\lambda_1^{(n)} \geq D\sqrt{n}\right] \leq \sum_{n=1}^{\infty} \sum_{l=\lceil D\sqrt{n}\rceil}^{n} \frac{n^l}{(l!)^2} = \sum_{l=D}^{\infty} \sum_{n=l}^{\lfloor \frac{l^2}{D^2} \rfloor} \frac{n^l}{(l!)^2} \leq \sum_{l=D}^{\infty} \frac{1}{(l!)^2}\left(\frac{l^2}{D^2}\right)^{l+1}.$$

By using Stirling's estimate $l! \geq (\frac{l}{\mathrm{e}})^l \sqrt{2\pi l}$, one sees that the series is convergent as soon as $D > \mathrm{e}$. In particular, $\mathbb{P}_{0,n}[\lambda_1^{(n)} \geq D\sqrt{n}] \to 0$, hence the result.

Consider now a uniform random word $w^{(n)}$ chosen under the measure $\mathbb{Q}_{c,n}$. Notice first that for any $\varepsilon > 0$

$$\mathbb{Q}_{0,n}\left[(\mathrm{RSKsh}(w^{(n)}))'_1 \leq \left(\frac{1}{c}+\varepsilon\right)\sqrt{n}\right] \to 1.$$

Indeed, a longest strictly decreasing subsequence in $w^{(n)}$ has length always smaller than $N_n \simeq \frac{\sqrt{n}}{c}$, whence the result. For the longest weakly increasing subsequences, we use a reasoning similar to that above. We have

$$\mathbb{P}_{c,n}[\lambda_1^{(n)} = l] \leq \frac{1}{(N_n)^n}\binom{n}{l}\binom{N_n+l-1}{l}(N_n)^{n-l} \leq \frac{n^l}{(l!)^2}\frac{(N_n+l-1)!}{(N_n-1)!\,(N_n)^l},$$

because to construct a word of length n with a longest weakly increasing subsequence of size l, one can first choose this subsequence ($\binom{N_n+l-1}{l}$ possibilities) and its position ($\binom{n}{l}$ possibilities), and then choose the $(n-l)$ other letters. Fix $\varepsilon > 0$ such that $(\frac{1}{c}-\varepsilon)\sqrt{n} \leq N_n \leq (\frac{1}{c}+\varepsilon)\sqrt{n}$ for any $n \geq n_0$, and take

$$D > \max\left(\frac{2\mathrm{e}^2}{\frac{1}{c}-\varepsilon}, \frac{1}{c}+\varepsilon\right).$$

If $l \geq D\sqrt{n} \geq N_n$, then the ratio $\frac{(N_n+l-1)!}{(N_n-1)!\,(N_n)^l}$ is bounded from above by

$$\frac{l+N_n-1}{N_n}\,\frac{l+N_n-2}{N_n}\cdots\frac{N_n+1}{N_n}\,\frac{N_n}{N_n} \leq \left(\frac{2l}{N_n}\right)^l.$$

So,

$$\sum_{n=n_0}^{\infty} \mathbb{P}_{c,n}[\lambda_1^{(n)} \geq D\sqrt{n}]$$

$$\leq \sum_{n=n_0}^{\infty} \sum_{l=\lceil D\sqrt{n}\rceil}^{n} \frac{n^l}{(l!)^2}\left(\frac{2l}{N_n}\right)^l = \sum_{l=\lceil D\sqrt{n_0}\rceil}^{\infty} \frac{(2l)^l}{(l!)^2} \sum_{n=\max(l,n_0)}^{\left\lfloor \frac{l^2}{D^2}\right\rfloor} \left(\frac{n}{N_n}\right)^l$$

$$\leq \sum_{l=\lceil D\sqrt{n_0}\rceil}^{\infty} \frac{(2l)^l}{(l!)^2(\frac{1}{c}-\varepsilon)^l} \sum_{n=\max(l,n_0)}^{\left\lfloor \frac{l^2}{D^2}\right\rfloor} n^{\frac{l}{2}} \leq \sum_{l=\lceil D\sqrt{n_0}\rceil}^{\infty} \frac{(2l)^l}{(l!)^2(\frac{1}{c}-\varepsilon)^l}\left(\frac{l}{D}\right)^{l+2}$$

and the series is convergent as soon as $D > \frac{2e^2}{\frac{1}{c}-\varepsilon}$. We conclude that for any $\varepsilon > 0$,

$$\mathbb{Q}_{0,n}\left[(\mathrm{RSKsh}(w^{(n)}))_1 \leq \left(\max\left(2e^2c, \frac{1}{c}\right)+\varepsilon\right)\sqrt{n}\right] \to 1.$$

Thus, Hypothesis (C1) is satisfied by the Schur–Weyl measures $\mathbb{P}_{c,n}$, with a limiting interval $[-(D+\varepsilon), D+\varepsilon]$ with $D = \max(2e^2c, \frac{1}{c})$. □

▷ *The Logan–Shepp–Kerov–Vershik limiting shape.*

In the sequel, we denote

$$\omega_{0,\infty}(s) = \omega_{\mathrm{Wigner}}(s) = \begin{cases} \frac{2}{\pi}\left(s \arcsin\frac{s}{2} + \sqrt{4-s^2}\right) & \text{if } |s| \leq 2, \\ |s| & \text{if } |s| > 2. \end{cases}$$

It is a continuous Young diagram with support $[-2, 2]$, and its transition measure was computed in Chapter 7: it is the Wigner semicircle law μ_{Wigner} of density

$$\mu_{\mathrm{Wigner}}(ds) = \frac{\sqrt{4-s^2}}{2\pi} 1_{s\in[-2,2]}\, ds.$$

Theorem 13.6 (Logan–Shepp, Kerov–Vershik). *If $\omega_{0,n}$ is the rescaled continuous Young diagram coming from a random partition $\lambda^{(n)}$ chosen under the Plancherel measure $\mathbb{P}_{0,n}$, then*

$$\sup_{s\in\mathbb{R}} |\omega_{0,n}(s) - \omega_{0,\infty}(s)| \to_{\mathbb{P}_{0,n}} 0.$$

Proof. By the discussion of the previous paragraph, it suffices to prove that for any observable $f \in \mathscr{O}$, $f(\omega_{0,n}) \to_{\mathbb{P}_{0,n}} f(\omega_{0,\infty})$ (this is Hypothesis (C2)). We shall prove it on an algebraic basis of $\mathscr{O}$, namely, the basis of free cumulants $(R_k)_{k\geq 2}$. We have

$$R_k(\omega_{0,n}) = \frac{R_k(\lambda^{(n)})}{n^{\frac{k}{2}}},$$

and on the other hand, for any integer partition μ of size $k \leq n$, if τ denotes the normalized regular trace of $\mathbb{C}\mathfrak{S}(n)$, then

$$\mathbb{E}_{0,n}[\Sigma_\mu(\lambda^{(n)})] = n^{\downarrow k}\,\tau(\sigma_\mu) = \begin{cases} n^{\downarrow k} & \text{if } \mu = 1^k, \\ 0 & \text{otherwise.} \end{cases}$$

In particular, for any integer partition μ, $\mathbb{E}_{0,n}[\Sigma_\mu] = O(n^{\frac{|\mu|+\ell(\mu)}{2}})$. By Theorem 9.20, $R_k = \Sigma_{k-1} + \sum_{\mu \,|\, |\mu|+\ell(\mu) \leq k-1} c_\mu\, \Sigma_\mu$ for some coefficients c_μ. Therefore,

$$\mathbb{E}_{0,n}[R_k] = \mathbb{E}_{0,n}[\Sigma_{k-1}] + \sum_{\mu \,|\, |\mu|+\ell(\mu) \leq k-1} c_\mu\, \mathbb{E}_{0,n}[\Sigma_\mu]$$

$$\mathbb{E}[R_k(\omega_{0,n})] = \frac{\mathbb{E}_{0,n}[\Sigma_{k-1}]}{n^{\frac{k}{2}}} + O\left(n^{-\frac{1}{2}}\right) = \begin{cases} 1 + O\left(n^{-\frac{1}{2}}\right) & \text{if } k = 2, \\ O\left(n^{-\frac{1}{2}}\right) & \text{otherwise.} \end{cases}$$

We also have $(R_k)^2 = \Sigma_{k-1,k-1} + \sum_{\mu \,|\, |\mu|+\ell(\mu) \leq 2k-1} d_\mu\, \Sigma_\mu$, so

$$\mathbb{E}_{0,n}[(R_k)^2] = \mathbb{E}_{0,n}[\Sigma_{k-1,k-1}] + \sum_{\mu \,|\, |\mu|+\ell(\mu) \leq 2k-1} d_\mu\, \mathbb{E}_{0,n}[\Sigma_\mu]$$

$$\mathbb{E}[(R_k(\omega_{0,n}))^2] = \frac{\mathbb{E}_{0,n}[\Sigma_{k-1,k-1}]}{n^k} + O\left(n^{-\frac{1}{2}}\right) = \begin{cases} 1 + O\left(n^{-\frac{1}{2}}\right) & \text{if } k = 2, \\ O\left(n^{-\frac{1}{2}}\right) & \text{otherwise.} \end{cases}$$

By using the Bienaymé–Chebyshev inequality as in the previous chapter, we conclude that

$$R_k(\omega_{0,n}) \to_{\mathbb{P}_{0,n}} \begin{cases} 1 & \text{if } k = 2, \\ 0 & \text{otherwise.} \end{cases}$$

We saw in Section 9.1 that the Wigner law μ_{Wigner} is characterized by $R_2(\mu_{\text{Wigner}}) = 1$ and $R_{k\geq 3}(\mu_{\text{Wigner}}) = 0$, so $R_k(\omega_{0,n}) \to_{\mathbb{P}_{0,n}} R_k(\omega_{0,\infty})$ for any $k \geq 2$, and the proof is completed. □

▷ *The limit shapes for Schur–Weyl measures.*

A similar argument allows one to find the limiting shapes of the random Young diagrams under the Schur–Weyl measures $\mathbb{P}_{c,n}$. For $c \in (0,1)$, set

$$\omega_{c,\infty}(s) = \begin{cases} \frac{2}{\pi}\left(s \arcsin\left(\frac{s+c}{2\sqrt{1+sc}}\right) + \frac{1}{c}\arccos\left(\frac{2+sc-c^2}{2\sqrt{1+sc}}\right) + \frac{1}{2}\sqrt{4-(s-c)^2}\right) & \text{if } |s-c| \leq 2, \\ |s| & \text{if } |s-c| > 2. \end{cases}$$

Since $\arccos(1-\varepsilon) = \sqrt{2\varepsilon(1+o(1))}$, $\lim_{c\to 0} \frac{1}{c}\arccos(\frac{2+sc-c^2}{2\sqrt{1+sc}}) = \frac{1}{2}\sqrt{4-s^2}$, and therefore $\lim_{c\to 0} \omega_{c,\infty}(s) = \omega_{0,\infty}(s)$ uniformly in s. On the other hand, when

$c \to 1$ and $s \in [-1, 3]$,

$$\begin{aligned}
\lim_{c\to 1} \omega_{c,\infty}(s) &= \omega_{1,\infty}(s) \\
&= \frac{2}{\pi}\left(s \arcsin\left(\frac{\sqrt{1+s}}{2}\right) + \arccos\left(\frac{\sqrt{1+s}}{2}\right) + \frac{1}{2}\sqrt{4-(s-1)^2}\right) \\
&= 1 + \frac{2}{\pi}\left((s-1) \arcsin\left(\frac{\sqrt{1+s}}{2}\right) + \frac{1}{2}\sqrt{4-(s-1)^2}\right) \\
&= s + \frac{1}{\pi}\left((1-s) \arccos\left(\frac{s-1}{2}\right) + \sqrt{4-(s-1)^2}\right) \\
&= \frac{s+1}{2} + \frac{1}{\pi}\left((s-1) \arcsin\left(\frac{s-1}{2}\right) + \sqrt{4-(s-1)^2}\right).
\end{aligned}$$

by using the relations $\arccos x = \frac{\pi}{2} - \arcsin x$ and $\arcsin x = \frac{1}{2}\arccos(1-2x^2)$. We therefore set

$$\omega_{1,\infty}(s) = \begin{cases} \frac{s+1}{2} + \frac{1}{\pi}\left((s-1)\arcsin\left(\frac{s-1}{2}\right) + \sqrt{4-(s-1)^2}\right) & \text{if } |s-1| \leq 2, \\ |s| & \text{if } |s-1| > 2. \end{cases}$$

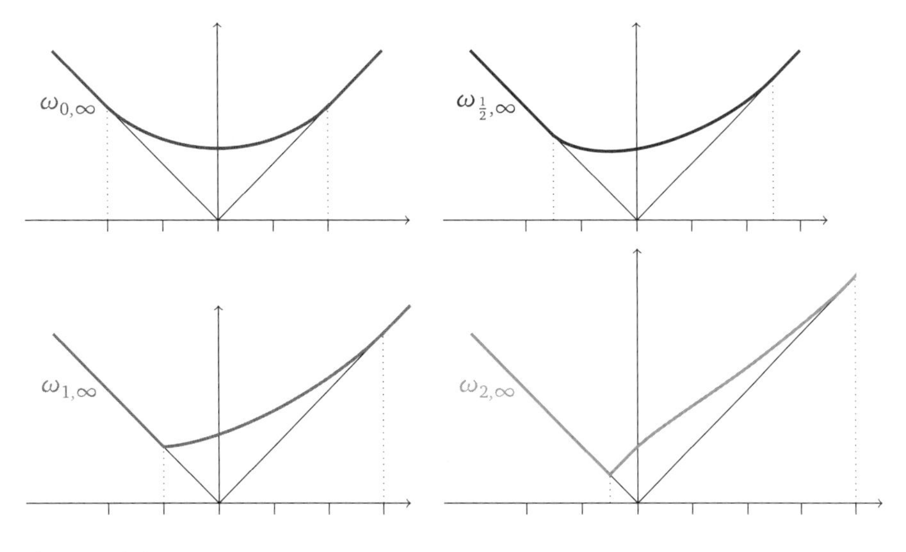

Figure 13.3
The limit shapes $\omega_{c,\infty}$ for $c = 0$, $c = \frac{1}{2}$, $c = 1$ and $c = 2$.

Finally, if $c > 1$, we set

$$\omega_{c,\infty}(s) = \begin{cases} \frac{2}{\pi}\left(s \arcsin\left(\frac{s+c}{2\sqrt{1+sc}}\right) + \frac{1}{c}\arccos\left(\frac{2+sc-c^2}{2\sqrt{1+sc}}\right) + \frac{1}{2}\sqrt{4-(s-c)^2}\right) & \text{if } |s-c| \leq 2, \\ s + \frac{2}{c} & \text{if } s \in [-\frac{1}{c}, c-2], \\ |s| & \text{otherwise.} \end{cases}$$

These continuous Young diagrams are drawn in Figure 13.3.

Theorem 13.7 (Biane). *For any $c > 0$, if $\omega_{c,n}$ is the rescaled continuous Young diagram coming from a random partition $\lambda^{(n)}$ chosen under the Schur–Weyl measure $\mathbb{P}_{c,n}$, then*

$$\sup_{s \in \mathbb{R}} |\omega_{c,n}(s) - \omega_{c,\infty}(s)| \to_{\mathbb{P}_{c,n}} 0.$$

Lemma 13.8. *For any $c > 0$, the generating function $G_{\omega_{c,\infty}}(z)$ is given by the formula*

$$G_{\omega_{c,\infty}}(z) = \frac{2}{z + c + \sqrt{(z-c)^2 - 4}}.$$

Proof. Given a continuous Young diagram ω,

$$G_\omega(z) = \exp\left(\frac{1}{2}\int_{\mathbb{R}} \omega''(s) \log\left(\frac{1}{z-s}\right) ds\right),$$

so

$$-\frac{d}{dz}(\log G_\omega(z)) = \frac{1}{2}\int_{\mathbb{R}} \frac{\omega''(s)}{z-s}\, ds$$

is the Cauchy transform of the Rayleigh measure $\frac{1}{2}\,\omega''(s)\,ds$. Thus, the statement of the lemma is equivalent to

$$\frac{1}{2}\int_{\mathbb{R}} \frac{\omega''_{c,\infty}(s)}{z-s}\, ds = -\frac{d}{dz}\log\left(\frac{2}{z+c+\sqrt{(z-c)^2-4}}\right)$$
$$= \frac{1}{2\sqrt{(z-c)^2-4}}\,\frac{2+cz-c^2+c\sqrt{(z-c)^2-4}}{1+cz}.$$

Using the Perron–Stieltjes inversion formula, we obtain

$$\frac{1}{2}\omega''_{c,\infty}(s)\,ds = \begin{cases} \frac{2+cs-c^2}{2\pi(1+cs)\sqrt{4-(s-c)^2}}\, 1_{s\in[c-2,c+2]}\, ds & \text{if } c \in (0,1), \\ \frac{1}{2\pi\sqrt{4-(s-1)^2}}\, 1_{s\in(-1,3]}\, ds + \frac{1}{2}\,\delta_{-1} & \text{if } c = 1, \\ \frac{2+cs-c^2}{2\pi(1+cs)\sqrt{4-(s-c)^2}}\, 1_{s\in[c-2,c+2]}\, ds + \delta_{-\frac{1}{c}} & \text{if } c > 1. \end{cases}$$

We leave the details of these computations to the reader. These formulas are indeed compatible with the definition of $\omega_{c,\infty}$, because for $s \in [c-2, c+2]$, one computes

$$\begin{aligned}
\frac{\omega_{c,\infty}(s)}{2} &= \frac{1}{\pi}\left(s \arcsin\left(\frac{s+c}{2\sqrt{1+sc}}\right) + \frac{1}{c}\arccos\left(\frac{2+sc-c^2}{2\sqrt{1+sc}}\right) + \frac{1}{2}\sqrt{4-(s-c)^2}\right)\\
\frac{\omega'_{c,\infty}(s)}{2} &= \frac{1}{\pi}\arcsin\left(\frac{s+c}{2\sqrt{1+sc}}\right)\\
\frac{\omega''_{c,\infty}(s)}{2} &= \frac{2+cs-c^2}{2\pi(1+cs)\sqrt{4-(s-c)^2}}
\end{aligned}$$

and one recovers the previous formulas for $\omega''_{c,\infty}$. □

Corollary 13.9. *For any $c > 0$, the free cumulants of the continuous Young diagram $\omega_{c,\infty}$ are $R_k(\omega_{c,\infty}) = c^{k-2}$.*

Proof. The inverse of $G_{\omega_{c,\infty}}(z)$ is $K(w) = \frac{1}{w} + \frac{w}{1-cw} = \frac{1}{w}\left(1 + \sum_{k=2}^{\infty} c^{k-2}\, w^k\right)$, and therefore, the generating series of the free cumulants is

$$\sum_{k=1}^{\infty} R_k(\omega_{c,\infty})\, w^{k-1} = R(w) = K(w) - \frac{1}{w} = \sum_{k=2}^{\infty} c^{k-2}\, w^{k-1}.$$ □

Proof of Theorem 13.7. We have to prove the convergence in probability

$$R_k(\omega_{c,n}) \to_{\mathbb{P}_{c,n}} c^{k-2}$$

for any $k \geq 2$; it will imply Hypothesis (C2), and the convergence $\omega_{c,n} \to_{\mathbb{P}_{c,n}} \omega_{c,\infty}$. We start by computing, for μ integer partition of size $k \leq n$, $\mathbb{E}_{c,n}[\Sigma_\mu]$. We have

$$\mathbb{E}_{c,n}[\Sigma_\mu] = \frac{n^{\downarrow k}}{(N_n)^n} \sum_{\lambda \in \mathfrak{Y}(n)\,|\,\ell(\lambda) \leq N_n} (\dim V^\lambda)\, \mathrm{ch}^\lambda(\sigma_\mu) = \frac{n^{\downarrow k}\, \mathrm{btr}(\mathrm{id}_{\mathbb{C}^N}, \sigma_\mu)}{(N_n)^n},$$

where the bitrace is the one of the $\mathrm{GL}(N_n) \times \mathfrak{S}(n)$-bimodule $(\mathbb{C}^{N_n})^{\otimes n}$. By Proposition 2.38,

$$\mathrm{btr}(\mathrm{id}_{\mathbb{C}^N}, \sigma_\mu) = p_{\mu\uparrow n}(1, 1, \ldots, 1) = (N_n)^{\ell(\mu)+n-k},$$

hence,

$$\mathbb{E}_{c,n}[\Sigma_\mu] = n^{\downarrow k}\, (N_n)^{\ell(\mu)-k} \simeq c^{|\mu|-\ell(\mu)}\, n^{\frac{|\mu|+\ell(\mu)}{2}}.$$

In particular, as in the case of Plancherel measures, $\mathbb{E}_{c,n}[\Sigma_\mu] = O(n^{\frac{|\mu|+\ell(\mu)}{2}})$ for any integer partition μ. The same arguments as in the proof of Theorem 13.6 show then that

$$\begin{aligned}
\mathbb{E}[R_k(\omega_{c,n})] &= \frac{\mathbb{E}_{c,n}[\Sigma_{k-1}]}{n^{\frac{k}{2}}} + O\left(n^{-\frac{1}{2}}\right) = c^{k-2} + o(1);\\
\mathbb{E}[(R_k(\omega_{c,n}))^2] &= \frac{\mathbb{E}_{c,n}[\Sigma_{k-1,k-1}]}{n^k} + O\left(n^{-\frac{1}{2}}\right) = c^{2(k-2)} + o(1);
\end{aligned}$$

hence, using the Bienaymé–Chebyshev inequality, we conclude that $R_k(\omega_{c,n}) \to_{\mathbb{P}_{c,n}} c^{k-2}$. □

Remark. The limit shapes $\omega_{c,\infty}$ for $c = 0$ or $c > 0$ are strongly connected to limiting results in random matrix theory. More precisely, let us compute the transition measures μ_c of these continuous Young diagrams. We have already seen that $\mu_0 = \mu_{\mathrm{Wigner}}$ is the Wigner law of density

$$\frac{\sqrt{4-s^2}}{2\pi} 1_{s\in[-2,2]}\, ds.$$

On the other hand, for $c > 0$, one has

$$\mu_c = \begin{cases} \frac{\sqrt{4-(s-c)^2}}{2\pi(1+sc)} 1_{s\in[c-2,c+2]}\, ds & \text{if } c \in (0,1], \\ \frac{\sqrt{4-(s-c)^2}}{2\pi(1+sc)} 1_{s\in[c-2,c+2]}\, ds + \left(1 - \frac{1}{c^2}\right) \delta_{-\frac{1}{c}} & \text{if } c > 1. \end{cases}$$

This is an immediate calculation with the Perron–Stieltjes formula applied to

$$G_{\omega_{c,\infty}}(z) = \frac{2}{z + c + \sqrt{(z-c)^2 - 4}} = \int_{\mathbb{R}} \frac{\mu_c(ds)}{z - s}.$$

The probability measures μ_c for $c \in \{0, \frac{1}{2}, 1, 2\}$ are drawn in Figure 13.4.

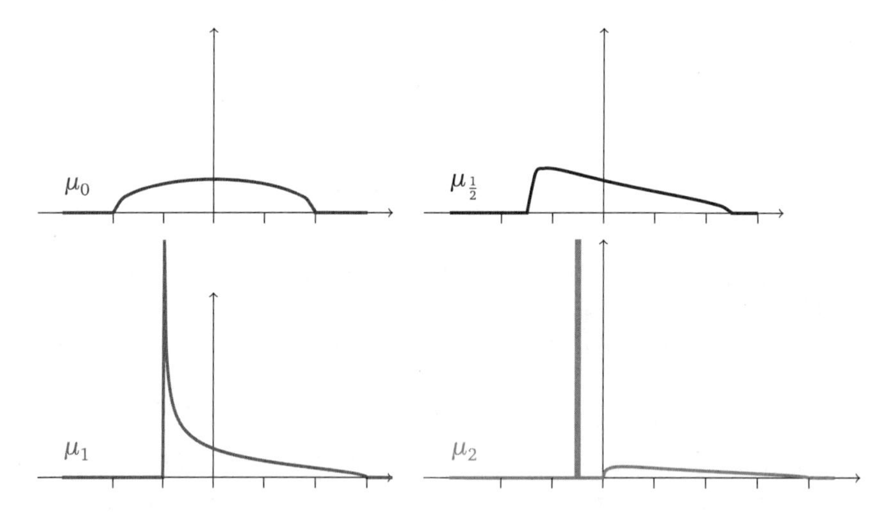

Figure 13.4
The transition measures μ_c for $c = 0$, $c = \frac{1}{2}$, $c = 1$ and $c = 2$.

When $c = 0$, μ_0 is the limit of the spectral measures of large random Hermitian

Gaussian matrices. Hence, consider a random square matrix $M = (M_{ij})_{1\leq i,j\leq N}$ of size $N \times N$ with $M_{ii} \sim \mathscr{N}_{\mathbb{R}}(0,1)$, $M_{ij} = \overline{M_{ji}} = \mathscr{N}_{\mathbb{R}}(0,\frac{1}{2}) + \mathrm{i}\mathscr{N}_{\mathbb{R}}(0,\frac{1}{2})$, and all the coordinates that are independent (except that $M_{ij} = \overline{M_{ji}}$). Then, the eigenvalues $\lambda_1 < \lambda_2 < \cdots < \lambda_N$ of M satisfy:

$$\frac{1}{N}\sum_{i=1}^{N} \delta_{\frac{\lambda_i}{\sqrt{N}}} \to_{N\to\infty} \mu_0,$$

this convergence happening in probability in the (metrizable) space of probability measures on $\mathbb{R}$; see the references at the end of the chapter. This is the starting point of a deep connection between random integer partitions stemming from the representations of the symmetric groups, and random matrices. The transition measures μ_c with $c > 0$ also appear in the theory of random matrices, specifically, large random covariance matrices. Thus, let $X = (X_{ij})_{1\leq i\leq M, 1\leq j\leq N}$ be a random real matrix with independent entries of law $\mathscr{N}_{\mathbb{R}}(0,1)$, and $Y = XX^t$, which is a random non-negative definite symmetric matrix of size $M \times M$. If $M, N \to \infty$ in such a way that $\frac{M}{N} \to c^2$, then the eigenvalues $\lambda_1 < \lambda_2 < \cdots < \lambda_M$ of Y satisfy

$$\frac{1}{M}\sum_{i=1}^{M} \delta_{\frac{\lambda_i}{M}} \to_{M\to\infty} \nu_c,$$

where ν_c is the **Marcenko–Pastur distribution** defined by

$$\nu_c = \begin{cases} \frac{\sqrt{((1+\frac{1}{c})^2-s)(s-(1-\frac{1}{c})^2)}}{2\pi s} 1_{s\in\left[(1-\frac{1}{c})^2,(1+\frac{1}{c})^2\right]}\, ds & \text{if } c \in (0,1], \\ \frac{\sqrt{((1+\frac{1}{c})^2-s)(s-(1-\frac{1}{c})^2)}}{2\pi s} 1_{s\in\left[(1-\frac{1}{c})^2,(1+\frac{1}{c})^2\right]}\, ds + \left(1-\frac{1}{c^2}\right)\delta_0 & \text{if } c > 1. \end{cases}$$

Again, we shall give references for this result at the end of the chapter. Now, let us remark that the image of ν_c by the map $s \mapsto cs - \frac{1}{c}$ is the transition measure μ_c. Therefore, the limiting shapes $\omega_{c>0,\infty}$ of random Young diagrams under Schur–Weyl measures are related to the asymptotics of the spectra of large covariance matrices.

▷ *Asymptotics of the length of the longest increasing subsequences.*

The previous results on the asymptotics of $\omega_{n,c}$ can be restated in terms of the length of a longest weakly increasing subsequence in a random permutation or a random word. Thus:

Theorem 13.10. *Let $\sigma^{(n)}$ be a random permutation chosen under the uniform probability measure $\mathbb{Q}_{0,n}$. The length $\ell^{(n)} = (\mathrm{RSKsh}(\sigma^{(n)}))_1$ of a longest increasing subsequence in $\sigma^{(n)}$ satisfies*

$$\frac{\ell^{(n)}}{\sqrt{n}} \to_{\mathbb{Q}_{0,n}} 2.$$

Similarly, if $w^{(n)} \sim \mathbb{Q}_{c,n}$ *is a random uniform word in* $([\![1,N_n]\!])^n$ *with* $N_n \simeq \frac{\sqrt{n}}{c}$, *then* $\ell^{(n)} = (\mathrm{RSKsh}(w^{(n)}))_1$ *satisfies*

$$\frac{\ell^{(n)}}{\sqrt{n}} \to_{\mathbb{Q}_{0,n}} 2+c.$$

Proof. If $\sigma^{(n)} \sim \mathbb{Q}_{0,n}$, fix $\varepsilon > 0$, and consider the partition $\lambda^{(n)} = \mathrm{RSKsh}(\sigma^{(n)})$. Since $\omega_{0,\infty}(2-\varepsilon) > 2-\varepsilon$, by Theorem 13.6, $\omega_n(2-\varepsilon) = \frac{1}{\sqrt{n}}\,\omega_{\lambda^{(n)}}((2-\varepsilon)\sqrt{n}) > (2-\varepsilon)$ with probability close to 1 for n large. However, for $s > 0$, $\omega_\lambda(s) > s$ if and only if $\lambda_1 > s$, so

$$\lim_{n\to\infty} \mathbb{P}_{0,n}\left[\lambda_1^{(n)} \geq (2-\varepsilon)\sqrt{n}\right] = 1.$$

Suppose now that there exists $\varepsilon' > 0$ such that $\lim_{n\to\infty} \mathbb{P}_{0,n}[\lambda_1^{(n)} \leq (2+\varepsilon')\sqrt{n}] \neq 0$. Then, one can extract a subsequence $n_k \to \infty$ such that

$$\forall k \in \mathbb{N},\ \mathbb{P}_{0,n_k}\left[\lambda_1^{(n_k)} \geq (2+\varepsilon')\sqrt{n_k}\right] \geq \varepsilon'$$

for some $\varepsilon' > 0$. However, since on the other hand

$$\lim_{k\to\infty} \mathbb{P}_{0,n_k}[\lambda_1^{(n_k)} \geq (2-(\varepsilon')^2)\sqrt{n_k}] = 1,$$

this leads to

$$\begin{aligned}
&\liminf_{k\to\infty} \mathbb{E}_{0,n_k}\left[\frac{\lambda_1^{(n_k)}}{\sqrt{n_k}}\right] \\
&\geq \liminf_{k\to\infty}\Big(\big(2-(\varepsilon')^2\big)\,\mathbb{P}_{0,n_k}\left[\lambda_1^{(n_k)} < (2+\varepsilon')\sqrt{n_k}\right] \\
&\qquad\qquad + (2+\varepsilon')\,\mathbb{P}_{0,n_k}\left[\lambda_1^{(n_k)} \geq (2+\varepsilon')\sqrt{n_k}\right]\Big) \\
&\geq (2-(\varepsilon')^2)(1-\varepsilon') + (2+\varepsilon')\varepsilon' = 2+(\varepsilon')^3.
\end{aligned}$$

This contradicts directly Lemma 13.2, so we conclude that for any $\varepsilon' > 0$,

$$\lim_{n\to\infty} \mathbb{P}_{0,n}\left[\lambda_1^{(n)} \leq (2+\varepsilon')\sqrt{n}\right] = 1.$$

Thus, $\frac{\lambda_1^{(n)}}{\sqrt{n}} = \frac{\ell^{(n)}}{\sqrt{n}} \to_{\mathbb{P}_{0,n}} 2$. The case of Schur–Weyl measures is identical, since $\omega_{c,\infty}(2+c-\varepsilon) > (2+c-\varepsilon)$ for $\varepsilon > 0$, and since $\mathbb{E}_{c,n}[\lambda_1^{(n)}] \leq 2\sqrt{n} + \frac{n}{N_n}$ by Lemma 13.3. □

13.3 Kerov's central limit theorem for characters

We now want to complete the laws of large numbers 13.6 and 13.7 by a central limit theorem, which as in Chapter 12 will regard the random character values

$\chi^{\lambda^{(n)}}(\sigma)$ with σ fixed and $\lambda^{(n)} \sim \mathbb{P}_{c,n}$ for $c = 0$ or $c > 0$. We shall concentrate on characters of cycles, and set

$$X_k^{(n)} = n^{\frac{k}{2}} \left(\chi^{\lambda^{(n)}}(\sigma_k) - \mathbb{E}_{c,n}[\chi^{\lambda^{(n)}}(\sigma_k)]\right) = \begin{cases} n^{\frac{k}{2}}\, \chi^{\lambda^{(n)}}(\sigma_k) & \text{if } c = 0, \\ n^{\frac{k}{2}} \left(\chi^{\lambda^{(n)}}(\sigma_k) - (N_n)^{1-k}\right) & \text{if } c > 0, \end{cases}$$

where σ_k is a cycle of length k. Let us remark that the renormalization is not at all the same as in Theorem 12.30, since we have a factor $n^{\frac{k}{2}}$ instead of $\sqrt{n}$. The goal of this section is to establish:

Theorem 13.11 (Kerov, Ivanov–Olshanski, Śniady). *Suppose that $\lambda^{(n)} \sim \mathbb{P}_{c,n}$. If $\vec{X}^{(n)} = (X_1^{(n)}, X_2^{(n)}, \ldots, X_d^{(n)})$ is the vector of the renormalized random character values on the cycles of length $1, 2, \ldots, d$, then we have the convergence in law $\vec{X}^{(n)} \rightharpoonup \vec{X}$, where $\vec{X}$ is the centered Gaussian vector with covariance matrix*

$$\mathrm{cov}(X_i, X_j) = \sum_{r=2}^{\min(i,j)} \binom{i}{r}\binom{j}{r}\, r\, c^{i+j-2r}.$$

In particular, if $c = 0$, then $\vec{X}$ is a vector of independent Gaussian random variables, with $X_k \sim \mathcal{N}_{\mathbb{R}}(0, k)$ for any $k \geq 2$.

As in Chapter 12, in order to establish Theorem 13.11, we shall use the method of cumulants (Proposition 12.27), and prove the asymptotic vanishing of the joint cumulants of order $r \geq 3$ of the random variables $X_k^{(n)}$.

▷ *Weight of identity cumulants and genus of Riemann surfaces.*

The main new tool that we shall use is the following analogue of Theorem 12.29 with respect to the weight gradation on $\mathcal{O}$ (instead of the degree), which we recall to be defined by $\mathrm{wt}(\Sigma_\mu) = |\mu| + \ell(\mu)$.

Theorem 13.12. *For any observables $f_1, \ldots, f_k \in \mathcal{O}$,*

$$\mathrm{wt}\left(\kappa^{\mathrm{id}}(f_1, f_2, \ldots, f_k)\right) \leq \mathrm{wt}(f_1) + \mathrm{wt}(f_2) + \cdots + \mathrm{wt}(f_k) - (2k-2).$$

To prove Theorem 13.12, it suffices again to do it when $f_i = \Sigma_{\mu^{(i)}}$ for arbitrary integer partitions $\mu^{(i)} \in \mathfrak{Y}(m_i)$. During the proof of Theorem 12.29, we proposed a description of the identity cumulant

$$\kappa^{\mathrm{id}}(\Sigma_{\mu^{(1)}}, \Sigma_{\mu^{(2)}}, \ldots, \Sigma_{\mu^{(k)}}) = \sum_{\Pi \,|\, S(\Pi) = [\![1,k]\!]} \Sigma_{\lambda(\mu^{(1)}, \ldots, \mu^{(k)}; \Pi)},$$

where the sum runs over partial pairings Π of $[\![1, m_1]\!], [\![1, m_2]\!], \ldots, [\![1, m_k]\!]$; $S(\Pi)$ denotes the associated set partition of $[\![1, k]\!]$; and $\lambda(\mu^{(1)}, \ldots, \mu^{(k)}; \Pi)$ is the cycle type of a product of partial permutations $\sigma_{\mu^{(1)}}, \ldots, \sigma_{\mu^{(k)}}$ with

$$\sigma_{\mu^{(i)}} = \left(\left(a_1^{(i)}, \ldots, a_{\mu_1^{(i)}}^{(i)}\right) \cdots \left(a_{m_i - \mu_{\ell(\mu^{(i)})}^{(i)} - 1}^{(i)}, \ldots, a_{m_i}^{(i)}\right), \left\{a_j^{(i)}\right\}\right),$$

and with equalities between indices $a_j^{(i)}$ prescribed by the partial pairing Π. On the other hand, in Section 9.4, we introduced symbols $\Sigma(\pi)$ labeled by (admissible) set partitions. In the sequel, we shall prove an analogue of the previous formula with identity cumulants

$$\kappa^{\mathrm{id}}(\Sigma(\pi^{(1)}), \Sigma(\pi^{(2)}), \ldots, \Sigma(\pi^{(k)})),$$

and we shall also relate the weight of $\Sigma(\pi)$ to the geometry of the set partition π. We start with this geometric interpretation of the weight:

Proposition 13.13. *To any admissible set partition π of $[\![1,n]\!]$, one can associate a compact Riemann surface RS_π, such that*

$$\mathrm{wt}(\Sigma(\pi)) = n - 2g(RS_\pi),$$

where $g(RS_\pi)$ is the genus of the Riemann surface RS_π.

Proof. As in Section 9.4, we place the integers $[\![1,n]\!]$ on a circle, and we associate to $\pi \in \mathfrak{Q}([\![1,n]\!])$ a pair set partition π_{fat} in $\mathfrak{Q}([\![1,n]\!] \sqcup [\![1',n']\!])$, cf. Figure 9.3. To construct RS_π, we start from a sphere with one hole, corresponding to the aforementioned circle. For each part $\pi_i = \{a,b,c,\ldots,z\}$ of π, we glue a disk to this hole, the edges of this disk being

- the pairs $(a,a'),(b,b'),\ldots,(z,z')$;
- and the pairs $(a',b),(b',c),\ldots,(z',a)$ that belong to π_{fat}.

For instance, when $n=8$ and $\pi = \{1,3,5\}\sqcup\{4,6,8\}\sqcup\{2,7\}$, we glue 3 disks to the hole, and we obtain the surface with boundary drawn in Figure 13.5. This surface with boundary has a certain number of holes, which correspond to the cycles of the permutation $\tau(\pi) = \pi_{\mathrm{fat}} \circ c$. For each of these remaining holes, we then glue a disk to the boundary of the hole, thereby obtaining a compact Riemann surface RS_π without boundary.

Let us compute the Euler characteristic $2-2g(RS_\pi)$ of RS_π; by Euler's formula, it is equal to $V-E+F$, where V (respectively, E and F) is the number of vertices (respectively, the number of edges and the number of faces) of a map drawn on RS_π. Our construction of RS_π yields a natural map drawn on RS_π, namely, the map whose faces are the disks previously described, whose vertices are the integers in $[\![1,n]\!] \sqcup [\![1',n']\!]$, and whose edges are the pairs $\{i', \pi_{\mathrm{fat}}(i)\}$ with $i \in [\![1,n]\!]$, and the pairs $\{a, c(a)\}$ with $a \in [\![1,n]\!] \sqcup [\![1',n']\!]$.

- The number of vertices of the map is of course $2n$.
- The number of edges of the map is $3n$.
- The number of faces of the map is $1+\ell(\pi)+n(\tau(\pi))$: there is one face coming from the punctured sphere, $\ell(\pi)$ faces coming from the first collection of glued disks, and $n(\tau(\pi))$ faces coming from the second collection of glued disks.

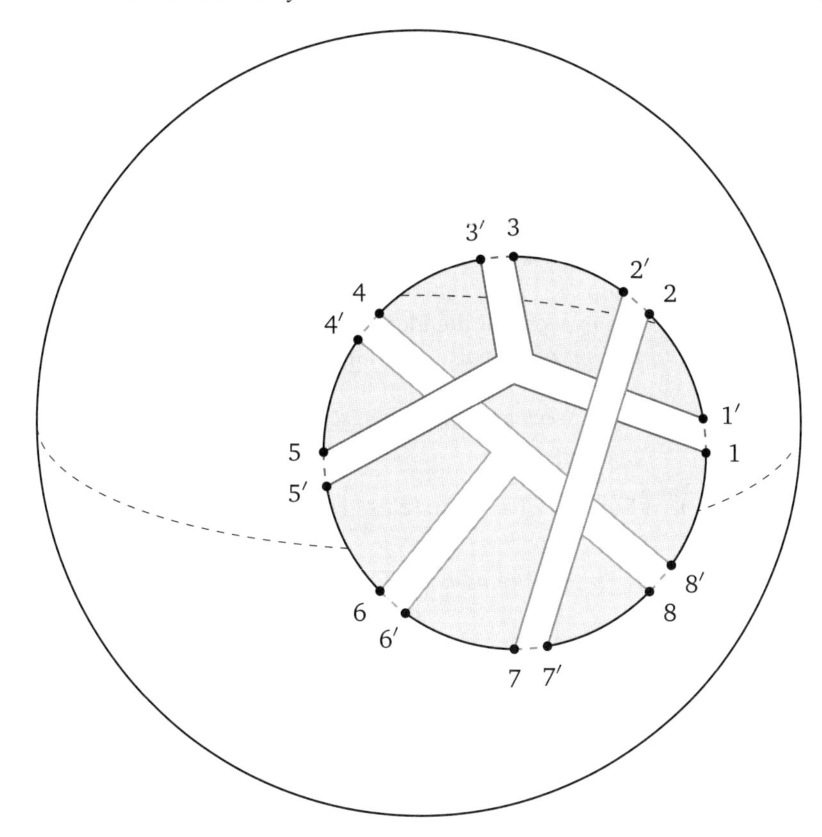

Figure 13.5
Construction of the Riemann surface RS_π.

Therefore, $2-2g(RS_\pi) = 2n - 3n + 1 + \ell(\pi) + n(\tau(\pi))$, and

$$n - 2g(RS_\pi) = (\ell(\pi) - 1) + n(\tau(\pi)),$$

so it suffices to prove that if $\Sigma(\pi) = \Sigma_{(k_1,\ldots,k_r)}$, then $\ell(\pi) - 1 = k_1 + k_2 + \cdots + k_r$ (we have on the other hand $r = n(\tau(\pi))$). To show this identity, notice that $k_1+k_2+\cdots+k_r$ is the total number of counterclockwise winds of the permutation $\pi_{\mathrm{fat}} \circ c$ (by definition of the integers k_i). However, the full cycle c contributes for a clockwise wind, whereas each part of π contributes for a counterclockwise wind of the permutation π_{fat}; hence the identity. □

We now give a formula for the identity cumulant of symbols $\Sigma(\pi^{(i)})$, with the $\pi^{(i)}$'s admissible set partitions of the sets $X^{(i)} = [\![m_1+\cdots+m_{i-1}+1, m_1+\cdots+m_i]\!]$. Recall that in Section 9.4, we defined the product of observables $\Sigma(\pi)$ by the formula

$$\Sigma(\pi^{(1)})\,\Sigma(\pi^{(2)})\cdots\Sigma(\pi^{(k)}) = \sum_{\pi} \Sigma(\pi),$$

where the sum runs over set partitions π of $X = [\![1, m_1 + \cdots + m_k]\!]$ such that

(IC1) The set partition π is coarser than the set partition $A \sqcup \bigsqcup_{j \notin A} \{j\}$, where $A = \{1, m_1 + 1, \ldots, m_1 + \cdots + m_{k-1} + 1\}$.

(IC2) If $a, b \in X^{(i)}$, then a and b are in the same part of π if and only if they are in the same part of $\pi^{(i)}$.

Indeed, this is the description of the ρ-product of set partitions, with ρ equal to the non-crossing partition $X^{(1)} \sqcup X^{(2)} \sqcup \cdots \sqcup X^{(k)}$ of X. Now, from the proof of Theorem 12.29, one can expect that the identity cumulant can be described as a sum over a subset of this set of set partitions, and indeed:

Proposition 13.14. *In the previous setting, with set partitions $\pi^{(i)}$ that are admissible, one has*

$$\kappa^{\mathrm{id}}\left(\Sigma(\pi^{(1)}), \Sigma(\pi^{(2)}), \ldots, \Sigma(\pi^{(k)})\right) = \sum_{\pi} \Sigma(\pi),$$

where the sum runs over set partitions $\pi \in \mathfrak{Q}(X)$ that satisfy the conditions (IC1), (IC2) and:

(IC3) Let π_1 be the part of π that contains the elements of

$$A = \{1, m_1 + 1, \ldots, m_1 + \cdots + m_{k-1} + 1\},$$

and π_- be the set partition obtained from π by replacing the part π_1 by the singletons of this part. Then, $\pi_- \vee (X^{(1)} \sqcup \cdots \sqcup X^{(k)})$ is the trivial set partition:

$$\pi_- \vee (X^{(1)} \sqcup \cdots \sqcup X^{(k)}) = X.$$

Proof. Let $\mathfrak{R}$ be the subset of $\mathfrak{Q}(X)$ that consists in set partitions coarser than the set partition $X^{(1)} \sqcup X^{(2)} \sqcup \cdots \sqcup X^{(k)}$. It is a sublattice of $\mathfrak{Q}(X)$ isomorphic to $\mathfrak{Q}(k)$, and if π is a set partition involved in the product $\Sigma(\pi^{(1)})\, \Sigma(\pi^{(2)}) \cdots \Sigma(\pi^{(k)})$, then $S(\pi) = \pi_- \vee (X^{(1)} \sqcup X^{(2)} \sqcup \cdots \sqcup X^{(k)})$ belongs to $\mathfrak{R}$. In the sequel, we identify an element S from $\mathfrak{R}$ with an element s of $\mathfrak{Q}(k)$, so for instance if $k = 3$, then the set partition with 2 parts $S = (X^{(1)} \cup X^{(2)}) \sqcup X^{(3)}$ corresponds to the element $s = \{1, 2\} \sqcup \{3\}$ in $\mathfrak{Q}(3)$. We denote $s(\pi)$ the element of $\mathfrak{Q}(k)$ that corresponds to $S(\pi) = \pi_- \vee (X^{(1)} \sqcup X^{(2)} \sqcup \cdots \sqcup X^{(k)})$. Now, the identity cumulant is defined by the relation

$$\Sigma(\pi^{(1)})\, \Sigma(\pi^{(2)}) \cdots \Sigma(\pi^{(k)}) = \sum_{\varpi \in \mathfrak{Q}(k)} \dot{\prod_{1 \leq j \leq \ell(\varpi)}} \kappa^{\mathrm{id}}\left(\{\Sigma(\pi^{(i)}), i \in \varpi_j\}\right).$$

Therefore, we can use the exact same induction as in the proof of Theorem 12.29 to prove that for any $\varpi \in \mathfrak{Q}(k)$,

$$\dot{\prod_{1 \leq j \leq \ell(\varpi)}} \kappa^{\mathrm{id}}\left(\{\Sigma(\pi^{(i)}), i \in \varpi_j\}\right) = \sum_{\pi \,|\, s(\pi) = \varpi} \Sigma(\pi)$$

where the sum on the right-hand side runs over set partitions π satisfying the conditions (IC1), (IC2) and $s(\pi) = \varpi$. The content of the proposition corresponds then to the particular case $s(\pi) = [\![1, k]\!]$. $\square$

Proof of Theorem 13.12. Let us estimate the genus of RS_π, where π is a set partition that satisfies the three conditions (IC1), (IC2) and (IC3), hence is involved in the identity cumulant. We consider a sphere punctured by k holes, on which are placed the integers of $X \sqcup X'$, with $X^{(i)} \sqcup X^{(i)\prime}$ placed on the boundary of the i-th hole. We glue to these points of $X^{(i)} \sqcup X^{(i)\prime}$ the disks corresponding to the parts of $\pi^{(i)}$. It will be convenient to consider the following example: $X^{(1)} = [\![1, 4]\!]$, $X^{(2)} = [\![5, 8]\!]$, $\pi^{(1)} = \{1, 3\} \sqcup \{2\} \sqcup \{4\}$, $\pi^{(2)} = \{5, 7\} \sqcup \{6, 8\}$.

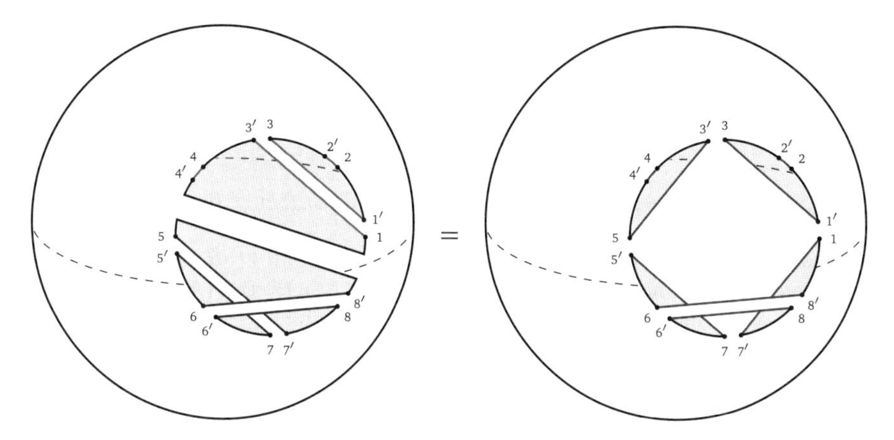

Figure 13.6
Computation of the product of observables $\Sigma(\pi^{(1)})\,\Sigma(\pi^{(2)})$ and of the identity cumulant $\kappa^{\mathrm{id}}(\Sigma(\pi^{(1)}), \Sigma(\pi^{(2)}))$.

The surface with boundary that one obtains is the same as the one coming from the set partition $\alpha \vee (\pi^{(1)} \sqcup \pi^{(2)} \sqcup \cdots \sqcup \pi^{(k)})$ (in our example, $\{1, 3, 5, 7\} \sqcup \{2\} \sqcup \{4\} \sqcup \{6, 8\}$), where $\alpha = A \sqcup \bigsqcup_{j \notin A} \{j\}$ with $A = \{1, m_1 + 1, \ldots, m_1 + \cdots + m_{k-1} + 1\}$. Then, the product of the two set partitions $\pi^{(1)}$ and $\pi^{(2)}$ is given by the set of set partitions π obtained from $\pi_0 = \alpha \vee (\pi^{(1)} \sqcup \pi^{(2)} \sqcup \cdots \pi^{(k)})$ by joining some parts of π_0 without invalidating the hypothesis (IC2). In our example, there are three such set partitions:

$$\pi_0 = \{1, 3, 5, 7\} \sqcup \{2\} \sqcup \{4\} \sqcup \{6, 8\};$$
$$\pi_{\mathrm{a}} = \{1, 3, 5, 7\} \sqcup \{2, 6, 8\} \sqcup \{4\};$$
$$\pi_{\mathrm{b}} = \{1, 3, 5, 7\} \sqcup \{4, 6, 8\} \sqcup \{2\}.$$

Let π be a partition obtained from π_0 by joining some parts, and let

$$\left\{(\pi_{k_1}^{(i_1)}, \pi_{l_1}^{(j_1)}), \ldots, (\pi_{k_r}^{(i_r)}, \pi_{l_r}^{(j_r)})\right\}$$

be an enumeration of the parts of π_0 that are joined in order to obtain π. The order in which these junctions are made is arbitrary, and it will not play any role later. For example, π_a is obtained from π_0 by making a single junction, namely, the junction of the two parts $\{2\} \subset X^{(1)}$ and $\{6,8\} \subset X^{(2)}$. The important point is that each time one joins two parts $\pi_k^{(i)}$ and $\pi_l^{(j)}$ from two previously unconnected sets $X^{(i)}$ and $X^{(j)}$, one adds a handle to the surface with boundary, and after completion by disks to obtain a Riemann surface without boundary, each of these handles raises the genus of the surface by 1. Notice that one can also raise the genus by joining parts $\pi_k^{(i)}$ and $\pi_l^{(j)}$ from already connected sets $X^{(i)}$ and $X^{(j)}$: however, for these secondary junctions, the genus is not always raised (this will be made clear during the proof of Proposition 13.16). Nonetheless, the first junctions always raise the genus, and as a consequence, in the product

$$\Sigma(\pi^{(1)})\,\Sigma(\pi^{(2)})\cdots\Sigma(\pi^{(k)}) = \sum_{\pi} \Sigma(\pi),$$

the genus of a surface RS_π is always larger than

$$g(RS_\pi) \geq g(RS_{\pi^{(1)}}) + g(RS_{\pi^{(2)}}) + \cdots + g(RS_{\pi^{(k)}}) + \text{number of first junctions}.$$

In our example, one can indeed check that $g(RS_{\pi_0}) = 1$ and $g(RS_{\pi_a}) = g(RS_{\pi_b}) = 2$, which agrees with the previous formula since $g(RS_{\pi^{(1)}}) = 0$, $g(RS_{\pi^{(2)}}) = 1$, and the two set partitions π_a and π_b are obtained from π_0 by making one junction.

Given a set partition π involved in the product $\Sigma(\pi^{(1)})\,\Sigma(\pi^{(2)})\cdots\Sigma(\pi^{(k)})$, let $G(\pi)$ be the graph with vertex set $[\![1,k]\!]$, and with one edge between i and j if, in order to obtain π from π_0, one has to join a part in $X^{(i)}$ with a part in $X^{(j)}$. The previous proposition which describes $\kappa^{\mathrm{id}}(\Sigma(\pi^{(1)}), \Sigma(\pi^{(2)}), \ldots, \Sigma(\pi^{(k)}))$ ensures that if $\Sigma(\pi)$ appears in the identity cumulant, then $G(\pi)$ connects all the elements of $[\![1,k]\!]$. As a consequence, the Riemann surface RS_π is obtained from RS_{π_0} by making at least $(k-1)$ first junctions. So, if $\Sigma(\pi)$ appears in the identity cumulant, then

$$g(RS_\pi) \geq g(RS_{\pi^{(1)}}) + \cdots + g(RS_{\pi^{(k)}}) + k - 1.$$

We conclude that

$$\begin{aligned}
&\mathrm{wt}\big(\kappa^{\mathrm{id}}(\Sigma(\pi^{(1)}), \ldots, \Sigma(\pi^{(k)}))\big) \\
&\leq m_1 + \cdots + m_k - 2(g(RS_{\pi^{(1)}}) + \cdots + g(RS_{\pi^{(k)}}) + k - 1) \\
&\leq \mathrm{wt}(\Sigma(\pi^{(1)})) + \cdots + \mathrm{wt}(\Sigma(\pi^{(k)})) - (2k-2),
\end{aligned}$$

which is what we wanted to prove. In our example, Proposition 13.14 shows that

$$\kappa^{\mathrm{id}}(\Sigma(\pi^{(1)}), \Sigma(\pi^{(2)})) = \Sigma(\pi_a) + \Sigma(\pi_b),$$

and the two surfaces RS_{π_a} and RS_{π_b} have indeed genus 2, that is one more than $g(RS_{\pi^{(1)}}) + g(RS_{\pi^{(2)}})$. □

Corollary 13.15. *For any observables $f_1, \ldots, f_k \in \mathscr{O}$,*

$$\kappa^\bullet(f_1, \ldots, f_k) = O\left(n^{\frac{\mathrm{wt}(f_1)+\cdots+\mathrm{wt}(f_k)}{2} - (k-1)}\right);$$

$$\kappa(f_1, \ldots, f_k) = O\left(n^{\frac{\mathrm{wt}(f_1)+\cdots+\mathrm{wt}(f_k)}{2} - (k-1)}\right),$$

where the cumulants are those related to the expectations under $\mathbb{P}_{c,n}$ with $c = 0$ or $c > 0$.

Proof. Recall that $\mathbb{E}_{0,n}[\Sigma_\mu] = n^{\downarrow|\mu|}\, 1_{\mu=1^k}$ and that $\mathbb{E}_{c,n}[\Sigma_\mu] = n^{\downarrow|\mu|}\, (N_n)^{\ell(\mu)-|\mu|}$ for $c > 0$. We start with the disjoint cumulants, and we suppose without loss of generality that $f_i = \Sigma_{\mu^{(i)}}$ with $|\mu^{(i)}| = m_i$. Then,

$$\begin{aligned}
&\kappa^\bullet(\Sigma_{\mu^{(1)}}, \ldots, \Sigma_{\mu^{(k)}}) \\
&= \sum_{\pi \in \mathfrak{Q}(k)} (-1)^{\ell(\pi)-1}\, (\ell(\pi)-1)! \prod_{j=1}^{\ell(\pi)} \mathbb{E}_{c,n}\left[\dot{\prod_{i \in \pi_j}} \Sigma_{\mu^{(i)}}\right] \\
&= \begin{cases} \left(\prod_{i=1}^k 1_{\mu^{(i)}=1^{m_i}}\right) \sum_{\pi\in\mathfrak{Q}(k)} (-1)^{\ell(\pi)-1}\, (\ell(\pi)-1)!\, \prod_{j=1}^{\ell(\pi)} n^{\downarrow(\sum_{i\in\pi_j} m_i)} & \text{if } c = 0, \\ (N_n)^{\sum_{i=1}^k \ell(\mu^{(i)})-|\mu^{(i)}|} \sum_{\pi\in\mathfrak{Q}(k)} (-1)^{\ell(\pi)-1}\, (\ell(\pi)-1)!\, \prod_{j=1}^{\ell(\pi)} n^{\downarrow(\sum_{i\in\pi_j} m_i)} & \text{if } c > 0. \end{cases}
\end{aligned}$$

The alternate sum is the same polynomial in n as the one considered in Lemma 12.31, and we have shown that it was a $O(n^{(\sum_{i=1}^k m_i)-(k-1)})$. Therefore,

$$\begin{aligned}
\kappa^\bullet(\Sigma_{\mu^{(1)}}, \ldots, \Sigma_{\mu^{(k)}}) &= \begin{cases} \left(\prod_{i=1}^k 1_{\mu^{(i)}=1^{m_i}}\right) O\left(n^{(\sum_{i=1}^k m_i)-(k-1)}\right) & \text{if } c = 0, \\ (N_n)^{\sum_{i=1}^k \ell(\mu^{(i)})-|\mu^{(i)}|}\, O\left(n^{(\sum_{i=1}^k m_i)-(k-1)}\right) & \text{if } c > 0, \end{cases} \\
&= O\left(n^{\frac{|\mu^{(1)}|+\ell(\mu^{(1)})+\cdots+|\mu^{(k)}|+\ell(\mu^{(k)})}{2} - (k-1)}\right).
\end{aligned}$$

This ends the proof for disjoint cumulants. For normal cumulants, one can use as in Chapter 12 the formula of Proposition 12.28; therefore, it suffices to prove that for any set partition $\pi \in \mathfrak{Q}(k)$,

$$\kappa^\bullet(\kappa^{\mathrm{id}}(\{f_i,\, i \in \pi_1\}), \ldots, \kappa^{\mathrm{id}}(\{f_i,\, i \in \pi_{\ell(\pi)}\})) = O\left(n^{\frac{\mathrm{wt}(f_1)+\cdots+\mathrm{wt}(f_k)}{2} - (k-1)}\right).$$

However, by Theorem 13.12, the weight of the observable $\kappa^{\mathrm{id}}(\{f_i,\, i \in \pi_j\})$ is smaller than $(\sum_{i\in\pi_j} \mathrm{wt}(f_i)) - 2(|\pi_j| - 1)$. As a consequence,

$$\begin{aligned}
&\kappa^\bullet(\kappa^{\mathrm{id}}(\{f_i,\, i \in \pi_1\}), \ldots, \kappa^{\mathrm{id}}(\{f_i,\, i \in \pi_{\ell(\pi)}\})) \\
&= O\left(n^{\frac{\mathrm{wt}(f_1)+\cdots+\mathrm{wt}(f_k)}{2} - (\sum_{j=1}^{\ell(\pi)} |\pi_j|-1) - (\ell(\pi)-1)}\right) \\
&= O\left(n^{\frac{\mathrm{wt}(f_1)+\cdots+\mathrm{wt}(f_k)}{2} - (k-1)}\right).
\end{aligned}$$

□

▷ *Computation of the limiting covariances.*

The previous corollary ensures that for any $k_1, k_2, \ldots, k_r \geq 1$ with $r \geq 3$,

$$\kappa\left(X^{(n)}_{k_1}, \ldots, X^{(n)}_{k_r}\right) = O\left(n^{-\frac{k_1+\cdots+k_r}{2}}\, \kappa(\Sigma_{k_1}, \ldots, \Sigma_{k_r})\right)$$
$$= O\left(n^{\frac{r}{2}-(r-1)}\right) = O\left(n^{1-\frac{r}{2}}\right) \to 0$$

under $\mathbb{P}_{0,n}$ or under $\mathbb{P}_{c,n}$ with $c > 0$. So, to prove Theorem 13.11, it remains to compute the limits of the covariances

$$\lim_{n\to\infty} \kappa\left(X^{(n)}_i, X^{(n)}_j\right) = \lim_{n\to\infty} n^{-\frac{i+j}{2}}\, \kappa(\Sigma_i, \Sigma_j).$$

Without loss of generality, we can assume $i, j \geq 2$, as otherwise one of the variables is constant and the covariance vanishes. Then, the case $r = 2$ of Proposition 12.28 yields

$$\kappa(\Sigma_i, \Sigma_j) = \kappa^{\bullet}(\Sigma_i, \Sigma_j) + \mathbb{E}_{c,n}[\kappa^{\mathrm{id}}(\Sigma_i, \Sigma_j)],$$

and the first term in the right-hand side is equal to

$$\begin{cases} 0 & \text{if } c = 0, \\ (N_n)^{2-(i+j)}\left(n^{\downarrow(i+j)} - n^{\downarrow i} n^{\downarrow j}\right) & \text{if } c > 0. \end{cases}$$

When $c > 0$, this quantity is equivalent to $-ij\, c^{i+j-2}\, n^{\frac{i+j}{2}}$. Let us now look at the second term $\mathbb{E}_{c,n}[\kappa^{\mathrm{id}}(\Sigma_i, \Sigma_j)]$. We need to compute the leading terms of $\kappa^{\mathrm{id}}(\Sigma_i, \Sigma_j) = \Sigma_i\, \Sigma_j - \Sigma_{(i,j)}$ with respect to the weight gradation.

Proposition 13.16. *For any $i, j \geq 2$,*

$$\kappa^{\mathrm{id}}(\Sigma_i, \Sigma_j) = \sum_{r\geq 1} \sum_{\substack{c_1,\ldots,c_r\geq 1 \\ c_1+\cdots+c_r=i}} \sum_{\substack{d_1,\ldots,d_r\geq 1 \\ d_1+\cdots+d_r=j}} \frac{ij}{r}\, \Sigma_{c_1+d_1-1,\ldots,c_r+d_r-1}$$
$$+ \text{ terms of weight at most } i + j - 2.$$

Proof. If $X = [\![1, i+1]\!]$, then the minimal set partition $\pi_X = \{1\} \sqcup \{2\} \sqcup \cdots \sqcup \{i+1\}$ gives $\Sigma(\pi_X) = \Sigma_i$. Therefore, setting $X^{(1)} = [\![0, i]\!]$ and $X^{(2)} = [\![\overline{0}, \overline{j}]\!]$, we have by Proposition 13.14

$$\kappa^{\mathrm{id}}(\Sigma_i, \Sigma_j) = \sum_{\pi} \Sigma(\pi),$$

where the sum runs over set partitions π of $X = X^{(1)} \sqcup X^{(2)}$ that satisfy the three following conditions:

(IC1) The elements 0 and $\overline{0}$ are in the same part of π.

(IC2) If $0 \leq a < b \leq i$, then a and b are not in the same part of π. Similarly, if $\overline{0} \leq \overline{a} < \overline{b} \leq \overline{j}$, then $\overline{a}$ and $\overline{b}$ are not in the same part of π.

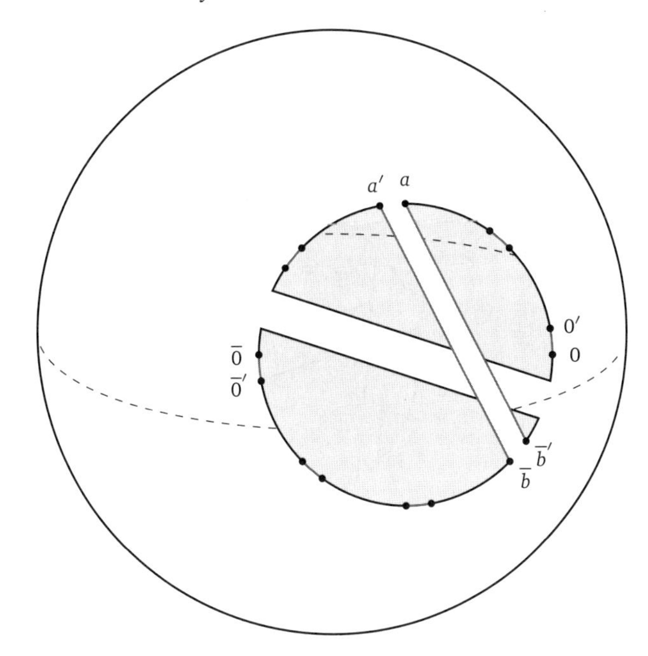

Figure 13.7
A set partition π that contributes to the identity covariance, with $r = 1$.

(IC3) There is at least one part of π that contains an element of $[\![1, i]\!]$ and an element of $[\![\overline{1}, \overline{j}]\!]$.

Clearly, such a set partition π is a partial pairing between $X^{(1)}$ and $X^{(2)}$, and on the other hand, the set partitions which yield elements $\Sigma(\pi)$ of maximal weight $i+j$ are those with genus $g(RS_\pi) = 1$. Therefore, we have to find all partial pairings π between $X^{(1)}$ and $X^{(2)}$ that yield surfaces RS_π of genus 1 (the other partial pairings with genus $g \geq 2$ provide terms with weight $n-2g \leq i+j-2$). In the sequel, we fix such a pairing π, and we denote r the number of pairs $(a, \overline{b}) \in [\![1, i]\!] \times [\![\overline{1}, \overline{j}]\!]$ that form parts of π. If $r = 1$, then the surface RS_π is obtained by filling the holes of the surface with boundary drawn in Figure 13.7, with $a \in [\![1, i]\!]$ and $\overline{b} \in [\![\overline{1}, \overline{j}]\!]$. There are ij possibilities for a and $\overline{b}$, and each pair $(a, \overline{b})$ yields a surface RS_π of genus 1, and a symbol $\Sigma(\pi) = \Sigma_{i+j-1}$. We have thus identified the term $r = 1$ in the sum of the statement of the proposition.

Let us now suppose that $r \geq 2$. Then, most of the partial pairings π yield surfaces RS_π of genus $g(RS_\pi) \geq 2$; see for instance Figure 13.8 for a surface of genus 2 and a term $\Sigma(\pi) = \Sigma_3$ of weight $i + j - 2 = 4$. In this setting, it is not very hard to convince oneself that π yields a surface of genus $g = 1$ if and only if the pairs $(a_1, \overline{b_1}), \ldots, (a_r, \overline{b_r})$ of π with $a_k \in [\![1, i]\!]$ and $\overline{b_k} \in [\![\overline{1}, \overline{j}]\!]$ can be labeled

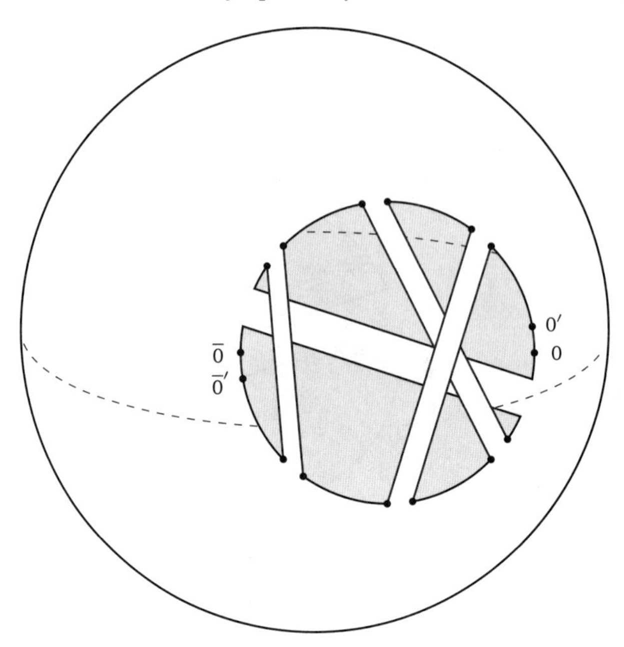

Figure 13.8
A set partition π that contributes to the identity covariance, with $r = 3$ and a genus $g(RS_\pi) = 2$.

in such a way that

$$a_1 \prec a_2 \prec \cdots \prec a_r \quad ; \quad \overline{b_1} \succ \overline{b_2} \succ \cdots \succ \overline{b_r},$$

where $\succ$ denotes the cyclic order on $[\![1, i]\!] = \mathbb{Z}/i\mathbb{Z}$ and on $[\![\overline{1}, \overline{j}]\!] = \mathbb{Z}/j\mathbb{Z}$. Denote

$$c_1 = a_2 - a_1 \bmod i,\ c_2 = a_3 - a_2 \bmod i, \ldots, c_r = a_1 - a_r \bmod i;$$
$$d_1 = b_1 - b_2 \bmod j,\ d_2 = b_2 - b_3 \bmod j, \ldots, d_r = b_r - b_1 \bmod j$$

the distances between the a_i's and the $\overline{b_j}$'s on the circles $\mathbb{Z}/i\mathbb{Z}$ and $\mathbb{Z}/j\mathbb{Z}$. One computes easily under the previous hypothesis:

$$\Sigma(\pi) = \Sigma_{c_1+d_1-1,\ldots,c_r+d_r-1},$$

see Figure 13.9. Moreover, the distances $c_1, \ldots, c_r$ and $d_1, \ldots, d_r$ always satisfy $c_1 + \cdots + c_r = i$ and $d_1 + \cdots + d_r = j$, and if one knows these distances, then there are ij possibilities for a_1 and b_1, and one has then to divide by r in order to take into account the cyclic permutations of the labeling of the pairs $(a_i, \overline{b_i})$. We have therefore identified the terms with $r \geq 2$ in the formula. □

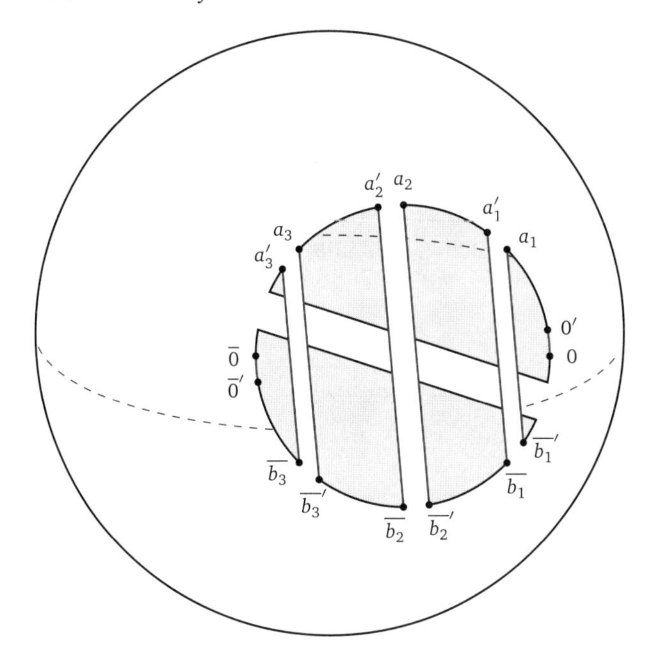

Figure 13.9
A typical set partition that yields a surface of genus $g(RS_\pi) = 1$.

Proof of Theorem 13.11. Suppose first that $c = 0$. Then, $\mathbb{E}_{0,n}[\kappa^{\mathrm{id}}(\Sigma_i, \Sigma_j)]$ vanishes unless one can find compositions $i = c_1 + \cdots + c_r$ and $j = d_1 + \cdots + d_r$ such that $c_k + d_k - 1 = 1$ for any $k \in [\![1, r]\!]$. This is only possible if $i = j = r$ and $c_k = d_k = 1$ for any k, and therefore,

$$\mathbb{E}_{0,n}[\kappa^{\mathrm{id}}(\Sigma_i, \Sigma_j)] = \delta_{i,j}\, i\, n^{\downarrow i};$$

$$\lim_{n\to\infty} \kappa\left(X_i^{(n)}, X_j^{(n)}\right) = \delta_{i,j}\, i.$$

Suppose now that $c > 0$. Then,

$$\begin{aligned}\mathbb{E}_{0,n}[\kappa^{\mathrm{id}}(\Sigma_i, \Sigma_j)] &= \sum_{r=1}^{\min(i,j)} \frac{ij}{r} \sum_{\substack{c_1+\cdots+c_r=i\\ d_1+\cdots+d_r=j}} (N_n)^{2r-i-j}\, n^{\downarrow i+j-r} \\ &= \sum_{r=1}^{\min(i,j)} \binom{i}{r}\binom{j}{r} r\, (N_n)^{2r-i-j}\, n^{\downarrow i+j-r} \\ &= \left(\sum_{r=1}^{\min(i,j)} \binom{i}{r}\binom{j}{r} r\, c^{i+j-2r}\, n^{\frac{i+j}{2}}\right)(1 + o(1)).\end{aligned}$$

If one substracts the term $ij\, c^{i+j-2}\, n^{\frac{i+j}{2}}$ previously computed, this amounts to removing the term $r = 1$ from the sum, hence,

$$\lim_{n\to\infty} \kappa\left(X_i^{(n)}, X_j^{(n)}\right) = \sum_{r=2}^{\min(i,j)} \binom{i}{r}\binom{j}{r} r\, c^{i+j-2r}$$

under the Schur–Weyl measures $\mathbb{P}_{c,n}$. This ends the proof of the theorem, by using Proposition 12.27 that characterizes the convergence to a Gaussian vector in terms of joint cumulants. □

Notes and references

The problem of the longest increasing subsequences in random permutations goes back to the 1960s; see [Ula61]. In 1972, Hammersley proved that there exists a limit in probability $\lambda_1^{(n)}/\sqrt{n} \to C > 0$ under $\mathbb{P}_{0,n}$, cf. [Ham72]. The bound of Proposition 13.5 comes from this paper in the case of Plancherel measures. The value $C = 2$ was computed by Logan and Shepp ([LS77]) and separately by Kerov and Vershik (cf. [KV77]) in 1977; however, to our knowledge, the proof of the inequality $C \leq 2$ was only published in 1985 in [KV85]. The proof of $C \leq 2$ is also alluded to in [KV86], but with a wrong reference to the paper [KV81], which makes finding in the literature the original proof of Lemma 13.2 quite challenging. In the case of Schur–Weyl measures, the inequality $C \leq 2+c$ comes from the much more recent paper [OW16]. We also refer to [AD95] for another approach to the identity $C = 2$ in the case of Plancherel measures.

Our criterion for the convergence in probability of continuous Young diagrams comes from [Ker93b]; see also [IO02, §5]. For the use of the notion of tightness during the proof of this criterion; see [Bil69, Chapter I, Theorem 5.1]. We also used Skorohod's representation theorem, which is Theorem 6.7 in loc. cit., and the Arzelà–Ascoli criterion of compacity in the spaces of continuous functions, which can be found in [Lan93, Chapter III, Theorem 3.1]. The law of large numbers for Plancherel measures (Theorem 13.6) is the main result of the aforementioned papers [LS77, KV77], and its analogue for Schur–Weyl measures (Theorem 13.7) comes from [Bia01]. The connection with random matrix theory is the main topic of the papers [BDJ99, BOO00, Joh01, Oko00], where the following important result is shown. By the law of large numbers of Logan–Shepp–Kerov–Vershik, we know that if $\lambda^{(n)} \sim \mathbb{P}_{0,n}$, then the first rows $\lambda_1^{(n)}, \lambda_2^{(n)}, \ldots$ of this random partition are all of asymptotic size $2\sqrt{n}$. Set

$$X_i^{(n)} = \frac{\lambda_i^{(n)} - 2\sqrt{n}}{n^{1/6}}.$$

On the other hand, let $e_1^{(n)} \geq e_2^{(n)} \geq \cdots \geq e_n^{(n)}$ be the real eigenvalues of a random

Hermitian Gaussian matrix chosen as after the proof of Theorem 13.7, and

$$Y_i^{(n)} = n^{1/6}\left(e_i^{(n)} - 2\sqrt{n}\right).$$

Then, the limiting distribution of $(X_1^{(n)}, X_2^{(n)}, \ldots)$ is the same as the limiting distribution of $(Y_1^{(n)}, Y_2^{(n)}, \ldots)$, and this limiting distribution is given by the Airy kernel and the so-called Tracy–Widom laws (cf. [TW94]). In particular, the second order fluctuations of the length of a longest increasing sequence in a uniform random permutation have the same asymptotics as the second order fluctuations of the largest eigenvalue of a random Hermitian Gaussian matrix. For the law of large numbers satisfied by the spectral measures of large random matrices, we refer to [AGZ10, Theorem 2.2.1] for the Wigner law and the case of the Gaussian unitary ensemble, and to [MP67] for the Marcenko–Pastur distribution and the case of large covariance matrices.

The central limit theorem 13.11 is due to Kerov in the case of Plancherel measures ([Ker93a]), and a complete proof can be found in [IO02]. The more elegant and general approach using cumulants and the theory of admissible set partitions is due to Śniady; see [Śni06a, Śni06b]. It allows us to treat with the same tools the case of Schur–Weyl measures. As in Chapter 12, there is a way to complete Theorem 13.11 by a description of the geometric fluctuations of the random continuous Young diagrams $\omega_{c,n}$. Thus, set

$$\Delta_{c,n}(s) = \sqrt{n}\,(\omega_{c,n}(s) - \omega_{c,\infty}(s)),$$

where $\omega_{c,n}(s) = \frac{1}{\sqrt{n}}\,\omega_{\lambda^{(n)}}(\sqrt{n}\,s)$ with $\lambda^{(n)} \sim \mathbb{P}_{c,n}$. Then, it can be shown that $\Delta_{c,n}$ converges in law towards a random Gaussian distribution, in the following sense: for every polynomial $P(s)$,

$$\int_{\mathbb{R}} P(s)\,\Delta_{c,n}(s)\,ds \rightharpoonup_{n\to\infty} \int_{\mathbb{R}} P(s)\,\Delta(s-c)\,ds,$$

where

$$\Delta(s) = 1_{s\in[-2,2]} \sum_{k\geq 2} \frac{\xi_k}{\sqrt{k}}\,\sin(k\theta) \quad \text{with } s = 2\cos\theta$$

and the variables ξ_k are independent standard Gaussian variables. Thus, the fluctuations of the shapes $\omega_{c,n}$ are asymptotically Gaussian, although the convergence occurs only in the sense of distribution. We refer to [IO02, Sections 7-9] for a proof of this result for Plancherel measures, and to [Mél11] for the case of Schur–Weyl measures.

Appendix

Appendix A

Representation theory of semisimple Lie algebras

In this appendix, we recall the representation theory of semisimple complex Lie algebras, omitting the proofs. This theory is mostly used in this book for the algebras $\mathfrak{gl}(N)$ and $\mathfrak{sl}(N)$. In Chapter 2, it sheds light on the Schur–Weyl duality; see in particular Theorem 2.39 where the Schur functions, that can be defined by Weyl's character formula A.14, appear as the characters of the polynomial representations of the groups $\mathrm{GL}(N,\mathbb{C})$. In Chapter 5, we present the analogue of this theory for the quantum groups $U_q(\mathfrak{gl}(N))$; it leads to the quantum analogue of the Schur–Weyl duality, that involves the Hecke algebras $\mathfrak{H}_q(n)$ instead of the symmetric groups $\mathfrak{S}(n)$.

We refer to [FH91, Kna02, Che04] for a detailed exposition of the results hereafter, and in particular for their proofs. Also, our Section 5.2 contains certain proofs of analogous statements for the quantum groups, which can be adapted quite easily to the corresponding Lie algebras.

A.1 Nilpotent, solvable and semisimple algebras

▷ *Lie algebras.*

Let k be a field of characteristic 0. A vector space $\mathfrak{g}$ over k is called a **Lie algebra** if it is endowed with a bilinear map $[\cdot,\cdot] : \mathfrak{g} \otimes_k \mathfrak{g} \to \mathfrak{g}$ such that

$$[X,Y]+[Y,X]=0 \quad ; \quad [X,[Y,Z]]+[Y,[Z,X]]+[Z,[X,Y]]=0$$

for all $X,Y,Z \in \mathfrak{g}$. The map $[\cdot,\cdot]$ is then called the **bracket** of the Lie algebra, and the second relation satisfied by the bracket is called the **Jacobi identity**. Given two k-Lie algebras $\mathfrak{g}$ and $\mathfrak{h}$, a morphism of Lie algebras between them will be a linear map $\phi : \mathfrak{g} \to \mathfrak{h}$ such that $\phi([X,Y]) = [\phi(X),\phi(Y)]$ for any $X,Y \in \mathfrak{g}$.

Example. Any associative algebra $\mathfrak{g}$ over k admits a structure of Lie algebra, given by the bracket $[X,Y] = XY - YX$. We shall see later that conversely, any Lie algebra over k can be seen as a part of an associative algebra $U(\mathfrak{g})$ endowed with this bracket.

Example. Let G be a real (smooth) Lie group, and $\mathfrak{g} = T_eG$ be the tangent space of G at the neutral element e. One has a map

$$\begin{aligned} \mathrm{Ad} : G &\to \mathrm{GL}(\mathfrak{g}) \\ g &\mapsto \mathrm{Ad}\, g : X \mapsto d_e(c_g)(X) \end{aligned}$$

where $c_g(h) = ghg^{-1}$, and $d_e(c_g)$ is its derivative at the tangent space. The assignation $g \mapsto \mathrm{Ad}\, g$ is smooth, and its derivative

$$\begin{aligned} \mathrm{ad} : \mathfrak{g} &\to \mathrm{End}(\mathfrak{g}) \\ X &\mapsto d_e(\mathrm{Ad})(X) \end{aligned}$$

gives rise to a Lie algebra structure on $\mathfrak{g}$, with bracket $[X,Y] = (\mathrm{ad}X)(Y)$. Conversely, given any finite-dimensional real Lie algebra $\mathfrak{g}$, one can reconstruct a connected and simply connected real Lie group G with $T_eG = \mathfrak{g}$, which is unique up to equivalence (Lie's theorem). Moreover, in this correspondence, a morphism of Lie groups $\phi : G \to H$ corresponds uniquely to a morphism of Lie algebras $d_e\phi : \mathfrak{g} \to \mathfrak{h}$, where $\mathfrak{g}$ and $\mathfrak{h}$ are the Lie algebras associated to G and H. Thus, one has an equivalence of categories between real Lie algebras and simply connected real Lie groups. Similarly, one has an equivalence of categories between complex Lie algebras and simply connected complex (analytic) Lie groups, provided by the same construction.

Example. As a particular case of the previous example, one denotes $\mathfrak{gl}(N)$ and $\mathfrak{sl}(N)$ the Lie algebras associated to the complex Lie groups $\mathrm{GL}(N,\mathbb{C})$ and $\mathrm{SL}(N,\mathbb{C})$. As vector spaces,

$$\begin{aligned} \mathfrak{gl}(N) &= \{M \text{ complex matrix of size } N \times N\} = \mathrm{M}(N,\mathbb{C}); \\ \mathfrak{sl}(N) &= \{M \in \mathfrak{gl}(N) \mid \mathrm{tr}\, M = 0\}. \end{aligned}$$

The bracket on both of these algebras is given by $[X,Y] = XY - YX$.

In the following, if $\mathfrak{g}$ is a Lie algebra and $X \in \mathfrak{g}$, one denotes $\mathrm{ad}X : \mathfrak{g} \to \mathfrak{g}$ the linear map $Y \mapsto [X,Y]$. It is a particular case of a **derivation** of $\mathfrak{g}$:

$$D([Y,Z]) = [D(Y),Z] + [Y,D(Z)].$$

One denotes $\mathrm{Der}(\mathfrak{g})$ the space of derivations of the Lie algebra $\mathfrak{g}$. It is easily seen that it is again a Lie algebra, i.e., if D_1 and D_2 are derivations of $\mathfrak{g}$, then so is $[D_1,D_2] = D_1 \circ D_2 - D_2 \circ D_1$.

▷ *Solvable and nilpotent Lie algebras.*

In the following, every Lie algebra considered will be finite-dimensional over its base field k. An **ideal** of a Lie algebra $\mathfrak{g}$ is a vector subspace $\mathfrak{a}$ such that $[\mathfrak{a},\mathfrak{g}] \subset \mathfrak{a}$. In that case, the quotient space $\mathfrak{g}/\mathfrak{a}$ is naturally endowed with a structure of Lie algebra:

$$[(X)_{\mathfrak{a}},(Y)_{\mathfrak{a}}] = ([X,Y])_{\mathfrak{a}}$$

where $(Z)_{\mathfrak{a}}$ denotes the class modulo $\mathfrak{a}$ of a vector $Z \in \mathfrak{g}$. On the other hand, a Lie algebra is called **abelian** if its bracket is the zero map: $[X,Y]=0$ for all $X,Y \in \mathfrak{g}$.

A Lie algebra is called **solvable** if one of the following equivalent assertions is satisfied:

1. There exists a sequence of Lie subalgebras

$$0=\mathfrak{a}_0 \subset \mathfrak{a}_1 \subset \cdots \subset \mathfrak{a}_r=\mathfrak{g}$$

   such that each $\mathfrak{a}_i$ is an ideal of $\mathfrak{a}_{i+1}$, with $\mathfrak{a}_{i+1}/\mathfrak{a}_i$ abelian Lie algebra.

2. If $\mathfrak{g}_0=\mathfrak{g}$ and $\mathfrak{g}_{i+1}=[\mathfrak{g}_i,\mathfrak{g}_i]$, then for some $r \geq 1$, $\mathfrak{g}_r=0$.

If $\mathfrak{a}$ and $\mathfrak{b}$ are (solvable) ideals of $\mathfrak{g}$, then so is $\mathfrak{a}+\mathfrak{b}$. As a consequence, for any Lie algebra $\mathfrak{g}$, there exists a largest solvable ideal, called the **radical** of $\mathfrak{g}$ and denoted $\operatorname{rad}\mathfrak{g}$. A Lie algebra is called **semisimple** if its radical is equal to zero, which means that there is no non-zero solvable ideal. For any Lie algebra $\mathfrak{g}$, the quotient $\mathfrak{s}=\mathfrak{g}/\operatorname{rad}\mathfrak{g}$ is semisimple, and moreover, one can write $\mathfrak{g}$ as a semidirect product of its radical and its semisimple quotient: there exists a morphism of Lie algebras $\pi : \mathfrak{s} \to \operatorname{Der}(\operatorname{rad}\mathfrak{g})$, such that

$$\mathfrak{g}=\mathfrak{s} \oplus_{\pi} \operatorname{rad}\mathfrak{g},$$

where the Lie structure on the right-hand side is defined for $s_1,s_2 \in \mathfrak{s}$ and $r_1,r_2 \in \operatorname{rad}\mathfrak{g}$ by

$$[s_1+r_1,s_2+r_2]=[s_1,s_2]+\pi(s_1)(r_2)-\pi(s_2)(r_1)+[r_1,r_2].$$

Call **representation** of a k-Lie algebra $\mathfrak{g}$ a morphism of Lie algebras $\rho : \mathfrak{g} \to \mathfrak{gl}(V)$, where V is a finite-dimensional space over k, and $\mathfrak{gl}(V)=\operatorname{End}_k(V)$ is endowed with the bracket $[u,v]=uv-vu$. If k is algebraically closed (e.g., $k=\mathbb{C}$), then the representation theory of solvable Lie algebras is covered by Lie's theorem:

Theorem A.1 (Lie). *Let $\mathfrak{g}$ be a solvable Lie algebra over an algebraically closed field. If V is a representation of $\mathfrak{g}$ of dimension n, then there exists a complete flag of vector subspaces $V_0 \subset V_1 \subset \cdots \subset V_n=V$ with $\dim V_i=i$, and each V_i stable by the action of $\mathfrak{g}$. Thus, if $\rho : \mathfrak{g} \to \mathfrak{gl}(V)$ is the defining morphism of the representation, and if $(e_1,\ldots,e_n)$ is a basis of V such that $V_i=\operatorname{Span}(e_1,\ldots,e_i)$ for all $i \in [\![1,n]\!]$, then the matrix of $\rho(X)$ in this basis is upper triangular for any $X \in \mathfrak{g}$.*

A variant of Lie's theorem A.1 involves **nilpotent** Lie algebras. A Lie algebra $\mathfrak{g}$ is called nilpotent if one of the following assertions is satisfied:

1. For any $X \in \mathfrak{g}$, $\operatorname{ad}X$ is a nilpotent endomorphism of $\mathfrak{g}$: there exists an integer $n(X)$ such that $(\operatorname{ad}X)^{\circ n(X)}=0$.

2. If $\mathfrak{g}_0 = \mathfrak{g}$ and $\mathfrak{g}_{i+1} = [\mathfrak{g}_i, \mathfrak{g}]$, then for some $r \geq 1$, $\mathfrak{g}_r = 0$.

By applying Lie's theorem A.1 to the adjoint representation $\mathrm{ad} : \mathfrak{g} \to \mathfrak{gl}(\mathfrak{g})$, one sees that if $\mathfrak{g}$ is a solvable Lie algebra over an algebraically closed field, then $[\mathfrak{g}, \mathfrak{g}]$ is a nilpotent Lie algebra. On the other hand:

Theorem A.2 (Engel). *Let $\mathfrak{g}$ be a nilpotent Lie algebra over an algebraically closed field. If V is a representation of $\mathfrak{g}$ of dimension n, then there exists a complete flag of vector subspaces $V_0 \subset V_1 \subset \cdots \subset V_n = V$ with $\dim V_i = i$, and $\rho(\mathfrak{g})(V_i) \subset V_{i-1}$ for all i. Thus, if $(e_1, \ldots, e_n)$ is a basis of V such that $V_i = \mathrm{Span}(e_1, \ldots, e_i)$ for all $i \in [\![1, n]\!]$, then the matrix of $\rho(X)$ in this basis is strictly upper triangular for any $X \in \mathfrak{g}$ (upper triangular with 0's on the diagonal).*

▷ *Semisimple Lie algebras and Cartan's criterion.*

Let $\mathfrak{g}$ be a Lie algebra over a field k of characteristic 0. The **Killing form** of $\mathfrak{g}$ is the bilinear symmetric form

$$B(X,Y) = \mathrm{tr}(\mathrm{ad}X \circ \mathrm{ad}Y).$$

It is involved in the following result due to Cartan:

Theorem A.3. *The algebra $\mathfrak{g}$ is semisimple if and only if its Killing form is non-degenerate. It is solvable if and only if $B(\mathfrak{g}, [\mathfrak{g}, \mathfrak{g}]) = 0$.*

When the Killing form B is non-degenerate, one can use it to split $\mathfrak{g}$ into simple blocks. More precisely, if $\mathfrak{g}$ is a semisimple Lie algebra, then for every ideal $\mathfrak{a}$ of $\mathfrak{g}$, its orthogonal supplement $\mathfrak{a}^\perp$ with respect to B is also an ideal, and $\mathfrak{g} = \mathfrak{a} \oplus \mathfrak{a}^\perp$. It follows that $\mathfrak{g}$ can be written in a unique way as a direct sum of **simple** Lie algebras

$$\mathfrak{g} = \mathfrak{a}_1 \oplus \mathfrak{a}_2 \oplus \cdots \oplus \mathfrak{a}_r$$

i.e., Lie algebras without non-trivial ideal.

More generally, call **reductive** a Lie algebra such that for any ideal $\mathfrak{a} \subset \mathfrak{g}$, there exists another ideal $\mathfrak{b} \subset \mathfrak{g}$ with $\mathfrak{g} = \mathfrak{a} \oplus \mathfrak{b}$. By the previous discussion, any semisimple Lie algebra is reductive. In fact, the difference between the two notions is quite small:

Proposition A.4. *Let $\mathfrak{g}$ be a reductive Lie algebra, and $Z(\mathfrak{g})$ be its center, which is the set of elements X such that*

$$\forall Y \in \mathfrak{g},\ \mathrm{ad}X(Y) = [X,Y] = 0.$$

One has the decomposition in direct sum of ideals $\mathfrak{g} = Z(\mathfrak{g}) \oplus [\mathfrak{g}, \mathfrak{g}]$. Moreover, $[\mathfrak{g}, \mathfrak{g}]$ is semisimple. Thus, the reductive Lie algebras are the direct sums of the semisimple and of the abelian Lie algebras.

Example. Consider the Lie algebra $\mathfrak{gl}(N)$. Its center is the set of scalar matrices $\mathbb{C} I_N$, and one has the direct sum $\mathfrak{gl}(N) = \mathbb{C} I_N \oplus \mathfrak{sl}(N)$. The algebra $\mathfrak{sl}(N)$ is semisimple, because one can compute its Killing form

$$B(X,Y) = 2N\operatorname{tr}(XY),$$

which is non-degenerate. Therefore, the Lie algebra $\mathfrak{gl}(N)$ is reductive.

A.2 Root system of a semisimple complex algebra

In this section, $\mathfrak{g}$ denotes a (finite-dimensional) semisimple complex Lie algebra, and we explain the classical Cartan classification of these objects, which is a first important step in the classification of the representations of $\mathfrak{g}$.

▷ *Cartan subalgebras and weights of representations.*

If $\mathfrak{g}$ is an arbitrary Lie algebra over $\mathbb{C}$, a **Cartan subalgebra** of it is a nilpotent subalgebra $\mathfrak{h}$ that is equal to its own normalizer; i.e., for any $X \in \mathfrak{g}$, $[X,\mathfrak{h}] \subset \mathfrak{h}$ if and only if $X \in \mathfrak{h}$. There always exists such a subalgebra, and it is unique in the following sense: if $\mathfrak{h}_1$ and $\mathfrak{h}_2$ are two Cartan subalgebras, then there exists an automorphism ψ of $\mathfrak{g}$ such that:

1. $\psi(\mathfrak{h}_1) = \mathfrak{h}_2$;
2. ψ is in the complex analytic subgroup of $\operatorname{Aut}(\mathfrak{g})$ that corresponds to the Lie algebra $\operatorname{ad}(\mathfrak{g}) \subset \operatorname{Der}(\mathfrak{g}) \subset \mathfrak{gl}(\mathfrak{g})$.

One says that $\mathfrak{h}$ is unique up to **inner automorphisms** of $\mathfrak{g}$. Notice that $\operatorname{Der}(\mathfrak{g})$ is the Lie algebra that corresponds to the group $\operatorname{Aut}(\mathfrak{g})$.

Assume now that $\mathfrak{g}$ is semisimple. Then, a Cartan subalgebra $\mathfrak{h} \subset \mathfrak{g}$ is necessarily abelian, and it can be shown that a Lie subalgebra $\mathfrak{h}$ of $\mathfrak{g}$ is a Cartan subalgebra if and only if it is a maximal subalgebra among those that are abelian and with $\operatorname{ad}_{\mathfrak{g}}(\mathfrak{h})$ consisting in endomorphisms of $\mathfrak{g}$ that are simultaneously diagonalizable. On the other hand, $\operatorname{ad}(\mathfrak{g}) = \operatorname{Der}(\mathfrak{g})$, so a Cartan subalgebra $\mathfrak{h}$ is unique up to automorphisms of the Lie algebra $\mathfrak{g}$. The complex dimension of $\mathfrak{h}$, which is an invariant of $\mathfrak{g}$, is called the **rank** of the Lie algebra.

Example. In the semisimple Lie algebra $\mathfrak{sl}(N)$, the set of diagonal matrices

$$\mathfrak{h}(N) = \{\operatorname{diag}(d_1,\ldots,d_n),\ d_1 + d_2 + \cdots + d_n = 0\}$$

is an abelian subalgebra which is equal to its own normalizer. Indeed, if $[X,D]$ is diagonal for all D, then in terms of matrix coefficients, this means that

$$X_{ij}(d_i - d_j) = 0$$

for all $i \neq j$, and all $d_1, \dots, d_n \in \mathbb{C}$ with sum zero. Obviously, this can only be the case when $X_{ij} = 0$ for all $i \neq j$. Thus, $\mathfrak{h}(N)$ is a Cartan subalgebra of $\mathfrak{sl}(N)$, and the rank of this Lie algebra is $N-1$.

▷ *Root systems and their classification.*

Fix a semisimple complex Lie algebra $\mathfrak{g}$, and a Cartan subalgebra $\mathfrak{h}$. Given a representation $\rho : \mathfrak{g} \to \mathfrak{gl}(V)$ on a finite-dimensional complex vector space V, a **weight** of this representation is a linear form $\omega : \mathfrak{h} \to \mathbb{C}$ such that the **weight space**

$$V_\omega = \{v \in V \mid \forall X \in \mathfrak{h},\ \rho(X)(v) = \omega(X)\,v\}$$

has dimension larger than 1. The dimension of this weight space is then called the multiplicity of the weight ω.

Proposition A.5. *Every representation V of a semisimple complex Lie algebra $\mathfrak{g}$ is the direct sum of its weight spaces: $V = \bigoplus_{\omega \text{ weight of } V} V_\omega$.*

A **root** of $\mathfrak{g}$ is a non-zero weight of the adjoint representation $\mathrm{ad} : \mathfrak{g} \to \mathfrak{gl}(\mathfrak{g})$. The weight space $\mathfrak{g}_0$ corresponding to the weight 0 is $\mathfrak{h}$ by definition of the Cartan subalgebra. Then, by the previous proposition, one can decompose $\mathfrak{g}$ as the direct sum of its root spaces:

$$\mathfrak{g} = \mathfrak{h} \oplus \bigoplus_{\alpha \text{ root}} \mathfrak{g}_\alpha.$$

We denote R the set of roots, $\mathfrak{h}^*$ the dual vector space of $\mathfrak{h}$, and $\mathfrak{h}^*_{\mathbb{R}} = \sum_{\alpha \in R} \mathbb{R}\alpha$ the real vector space generated by the roots.

Proposition A.6. *All roots have multiplicity one, and the roots span $\mathfrak{h}^*$ over $\mathbb{C}$. Two spaces $\mathfrak{g}_\alpha$ and $\mathfrak{g}_\beta$ are orthogonal with respect to the Killing form unless $\alpha + \beta = 0$. In this case, $B_{|\mathfrak{g}_\alpha \times \mathfrak{g}_{-\alpha}}$ is non-degenerate, so in particular, B restricted to $\mathfrak{h} = \mathfrak{g}_0$ is non-degenerate, and can then be written as*

$$B(X,Y) = \sum_{\alpha \in R} \alpha(X)\,\alpha(Y)$$

for any $X, Y \in \mathfrak{h}$.

As a consequence of the non-degeneracy of $B_{|\mathfrak{h} \times \mathfrak{h}}$, B induces a bijection between $\mathfrak{h}$ and $\mathfrak{h}^*$, its complex dual: if $\omega \in \mathfrak{h}^*$, then there exists a unique vector $V_\omega \in \mathfrak{h}$ such that $\omega(\cdot) = B(V_\omega, \cdot)$. We transfer the non-degenerate form B from $\mathfrak{h}$ to $\mathfrak{h}^*$ to obtain a non-degenerate bilinear form $\langle \cdot \mid \cdot \rangle$, defined on $\mathfrak{h}^* \times \mathfrak{h}^*$ by

$$\langle \omega_1 \mid \omega_2 \rangle = B(V_{\omega_1}, V_{\omega_2}) = \omega_1(V_{\omega_2}) = \omega_2(V_{\omega_1}).$$

The set of roots R of a semisimple Lie algebra has certain geometric properties that lead to the notion of **root system**. Let V be a Euclidean real vector space, with scalar product $\langle \cdot \mid \cdot \rangle$. A (reduced) root system in V is a family of non-zero vectors R such that:

1. The roots $\alpha \in R$ span V over $\mathbb{R}$.

2. For any $\alpha \in R$, the only other root that is colinear to α is $-\alpha$, which is required to also belong to R.

3. If $\alpha \in R$, denote $s_\alpha(\omega) = \omega - 2\frac{\langle \alpha \mid \omega \rangle}{\langle \alpha \mid \alpha \rangle}\alpha$ the reflection of X with respect to the hyperplane orthogonal to the line spanned by α. Then, R is stable by s_α for any $\alpha \in R$.

4. For any pair of roots α, β, $2\frac{\langle \alpha \mid \beta \rangle}{\langle \alpha \mid \alpha \rangle} \in \mathbb{Z}$.

Theorem A.7. *The set of roots R of a complex semisimple Lie algebra form a reduced root system in $V = \mathfrak{h}_{\mathbb{R}}^*$, endowed with the restriction of the dual Killing form $\langle \cdot \mid \cdot \rangle$, which is positive-definite. Moreover, if $\mathfrak{g} = \mathfrak{g}_1 \oplus \mathfrak{g}_2 \oplus \cdots \oplus \mathfrak{g}_r$ is the decomposition of $\mathfrak{g}$ in simple Lie algebras, then it corresponds to a decomposition of the Cartan subalgebra $\mathfrak{h} = \mathfrak{h}_1 \oplus \mathfrak{h}_2 \oplus \cdots \oplus \mathfrak{h}_r$, and to a decomposition of the root system $R = R_1 \sqcup R_2 \sqcup \cdots \sqcup R_r$ in mutually orthogonal root systems, that cannot be split further.*

Fix an arbitrary root system R in a Euclidean vector space V of dimension r (not necessarily coming from a semisimple complex Lie algebra). Since R is stable by $\alpha \mapsto -\alpha$, one can always split a root system into two parts R_+ and $R_- = -R_+$, containing **positive** and **negative** roots. In the sequel, we fix such a decomposition $R = R_+ \sqcup R_-$. Then, a positive root $\alpha \in R_+$ is called a **simple** root if it cannot be written as the sum of two positive roots. We denote $S = \{\alpha_1, \alpha_2, \ldots, \alpha_r\}$ the set of simple roots; it forms a linear basis of V over $\mathbb{R}$, and any positive root is a linear combination of the simple roots with non-negative integer coefficients. The choice of the decomposition $R = R_+ \sqcup R_-$, and therefore of the set of simple roots S is unique up to action of the **Weyl group** of the root system, which is the finite subgroup W of $\mathrm{O}(V)$ generated by the reflections s_α, $\alpha \in R$. The **Cartan matrix** of the root system R is the $r \times r$ matrix with integer coefficients

$$c_{ij} = 2\,\frac{\langle \alpha_i \mid \alpha_j \rangle}{\langle \alpha_i \mid \alpha_i \rangle}.$$

The diagonal coefficients of this matrix are all equal to 2, and the other coefficients are non-positive. Since two sets of simple roots S and S' of R are conjugated by an element $w \in W$ which is an isometry, the Cartan matrix of a root system is unique up to conjugation action of the permutation group $\mathfrak{S}(r)$ on its rows and columns. The **Dynkin diagram** associated to the root system and its Cartan matrix is the multigraph with vertices the simple roots $\alpha_i \in S$, and with $c_{ij}\,c_{ji}$ edges between α_i and α_j. These edges are directed towards the shorter root α_j if

$$\langle \alpha_i \mid \alpha_i \rangle > \langle \alpha_j \mid \alpha_j \rangle \iff c_{ij} < c_{ji};$$

they are undirected if the two roots have the same length. If one forgets the labels of the edges, then the Dynkin diagram is uniquely determined by the root system

R, and it is connected if and only if R is irreducible, that is to say that it cannot be split into two mutually orthogonal root systems R' and R''.

Example. Consider the semisimple Lie algebra $\mathfrak{sl}(N)$. We exhibited before a Cartan subalgebra of it, $\mathfrak{h}(N)$, that consists in traceless diagonal matrices. We introduce the following basis of $\mathfrak{sl}(N)$:

$$\{X_{ij} = e_{ij},\ 1 \leq i < j \leq N\} \sqcup \{Y_{ij} = e_{ji},\ 1 \leq i < j \leq N\}$$
$$\sqcup \{H_i = e_{ii} - e_{(i+1)(i+1)},\ 1 \leq i \leq N-1\},$$

where e_{ij} is the elementary matrix with 1 on the i-th row and j-th column, and 0 everywhere else. Notice that $\mathfrak{h} = \mathfrak{h}(N)$ is linearly spanned by the vectors $H_1, \ldots, H_{N-1}$. We compute

$$[\mathrm{diag}(d_1, \ldots, d_N), X_{ij}] = (d_i - d_j)X_{ij} \quad ; \quad [\mathrm{diag}(d_1, \ldots, d_N), Y_{ij}] = (d_j - d_i)Y_{ij}.$$

Therefore, $\mathfrak{sl}(N) = \mathfrak{h} \oplus \bigoplus_{i<j} \mathbb{C}X_{ij} \oplus \bigoplus_{i<j} \mathbb{C}Y_{ij}$ is the decomposition of $\mathfrak{sl}(N)$ in root spaces, and the roots are the linear forms

$$\alpha_{i,j}(\mathrm{diag}(d_1, \ldots, d_N)) = d_i - d_j,\ 1 \leq i \neq j \leq N.$$

For $i < j$, the root vector X_{ij} spans the root space associated to $\alpha_{i,j}$, and the root vector Y_{ij} spans the root space associated to $\alpha_{j,i} = -\alpha_{i,j}$. Set $\alpha_i = \alpha_{i,i+1}$; then,

$$R = \{\alpha_{i,j},\ 1 \leq i \neq j \leq N\};$$
$$R_+ = \{\alpha_{i,j},\ 1 \leq i < j \leq N\};$$
$$S = \{\alpha_i,\ 1 \leq i \leq N-1\}$$

is a possible choice of positive and simple roots for $\mathfrak{sl}(N)$. The scalar product on $\mathfrak{h}^*_{\mathbb{R}}$ obtained by restriction of the dual Killing form on $\mathfrak{h}^*$ is

$$\langle \alpha_i \mid \alpha_j \rangle = \frac{1}{2N} \times \begin{cases} 2 & \text{if } i = j; \\ -1 & \text{if } |i-j| = 1; \\ 0 & \text{if } |i-j| \geq 2. \end{cases}$$

Therefore, the Cartan matrix of $\mathfrak{sl}(N)$ is

$$\begin{pmatrix} 2 & -1 & & & \\ -1 & 2 & -1 & & \\ & -1 & 2 & \ddots & \\ & & \ddots & \ddots & -1 \\ & & & -1 & 2 \end{pmatrix}$$

and the associated Dynkin diagram is

$$\alpha_1 - \alpha_2 - \alpha_3 - \alpha_4 \quad \cdots \quad \alpha_{N-1}.$$

It is connected, which corresponds to the fact that the associated root system is irreducible and that $\mathfrak{sl}(N)$ is a simple Lie algebra.

Theorem A.8. *For any reduced root system, the Weyl group generated by the reflections s_α in $\mathfrak{h}^*_{\mathbb{R}}$ with $\alpha \in \mathbb{R}_+$ is a Coxeter group with presentation:*

$$W = \langle s_\alpha,\ \alpha \in S\rangle,\ \text{with } \forall \alpha \in S,\ (s_\alpha)^2 = 1;$$
$$\forall \alpha \neq \beta \in S,\ (s_\alpha s_\beta)^{m_{\alpha\beta}} = 1$$

where $1 + \cos \frac{2\pi}{m_{\alpha\beta}} = \frac{2(\langle \alpha \mid \beta \rangle)^2}{\langle \alpha \mid \alpha \rangle \langle \beta \mid \beta \rangle}$, or equivalently, $\frac{\pi}{m_{\alpha\beta}}$ is the angle between the hyperplanes orthogonal to the simple roots α and β.

Example. In the case of $\mathfrak{sl}(N)$, one obtains the presentation of W:

$$W = \langle s_i,\ i \in [\![1, N-1]\!]\rangle,\ \text{with } \forall i,\ (s_i)^2 = 1;$$
$$\forall i,\ s_i s_{i+1} s_i = s_{i+1} s_i s_{i+1};$$
$$\forall i, j,\ |i-j| \geq 2 \Rightarrow s_i s_j = s_j s_i.$$

By Theorem 4.1, $W = \mathfrak{S}(N)$, which acts on $\mathfrak{h}^*_{\mathbb{R}}$ by permutation of the coordinates of the diagonal matrices:

$$(\sigma \cdot \alpha)(\mathrm{diag}(d_1, \ldots, d_N)) = \alpha(d_{\sigma(1)}, \ldots, d_{\sigma(N)}).$$

Theorem A.9. *Let R be an irreducible root system. For any pair of distinct simple roots $\alpha_i \neq \alpha_j$, $c_{ij}\, c_{ji} \in \{0, 1, 2, 3\}$. Therefore, assuming without loss of generality that $\|\alpha_i\| \geq \|\alpha_j\|$, the vertices α_i and α_j of the Dynkin diagram of R are either unconnected, or connected as follows:*

The Dynkin diagram of R is one of the following:

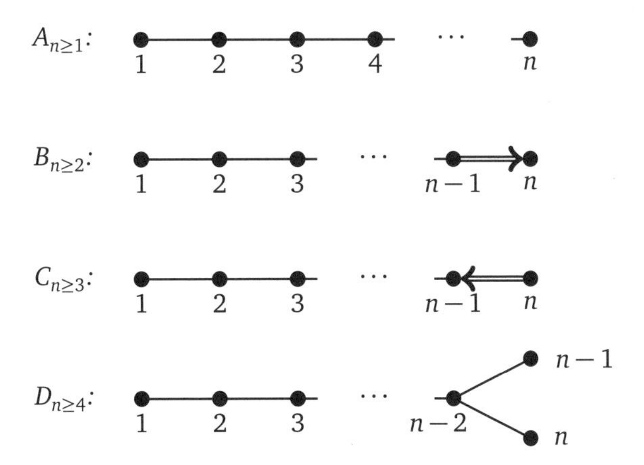

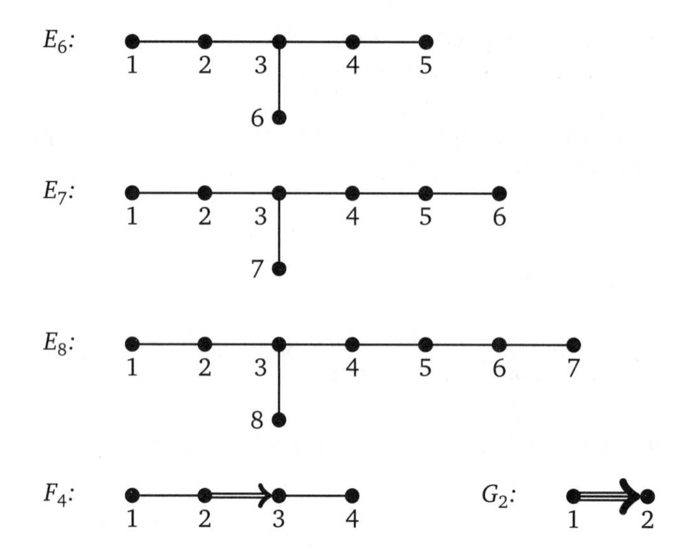

Example. As a consequence of Cartan's classification of the (Dynkin diagrams) of the irreducible root systems, there exist 4 non-isomorphic root systems of rank 2, namely, $A_1 \sqcup A_1$, A_2, B_2 and G_2. The corresponding roots are geometrically arranged as follows in a Euclidean plane:

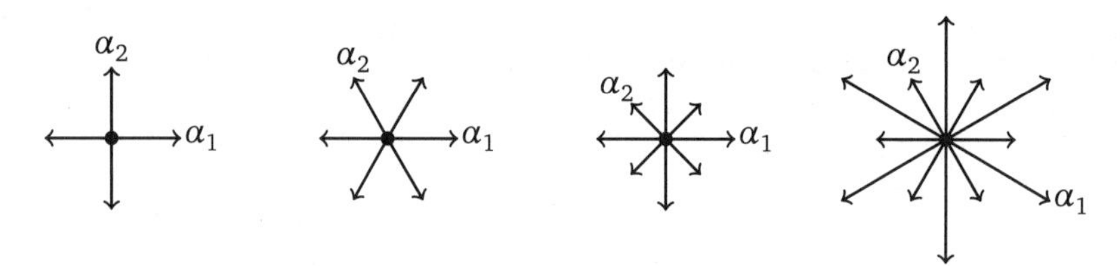

▷ *Serre relations and reconstruction of the Lie algebra from its root system.*

After Cartan's classification of the irreducible root systems, one can ask whether this can be lifted to a classification of the (semi)simple complex Lie algebras. The result is positive, and one can indeed reconstruct (up to isomorphisms) the Lie algebra $\mathfrak{g}$ from its root system R. We fix as before positive and negative roots, and a corresponding set of simple positive roots S. For α positive root, set $H_\alpha = \frac{2V_\alpha}{\langle \alpha \mid \alpha \rangle}$, so that the reflection s_α writes on $\mathfrak{h}^*_{\mathbb{R}}$ as

$$s_\alpha(\omega) = \omega - \omega(H_\alpha)\,\alpha.$$

We also choose vectors $E_\alpha \in \mathfrak{g}_\alpha$ and $F_\alpha \in \mathfrak{g}_{-\alpha}$, such that $B(E_\alpha, F_\alpha) = \frac{2}{\langle \alpha \mid \alpha \rangle}$. Notice then that

$$[H_\alpha, E_\alpha] = 2\,E_\alpha \quad ; \quad [H_\alpha, F_\alpha] = -2\,F_\alpha \quad ; \quad [E_\alpha, F_\alpha] = H_\alpha.$$

Therefore, $\mathrm{Span}(E_\alpha, F_\alpha, H_\alpha)$ is a Lie subalgebra of $\mathfrak{g}$ that is isomorphic to $\mathfrak{sl}(2,\mathbb{C})$, this latter algebra being spanned by the matrices

$$e = \begin{pmatrix} 0 & 1 \\ 0 & 0 \end{pmatrix} \quad ; \quad f = \begin{pmatrix} 0 & 0 \\ 1 & 0 \end{pmatrix} \quad ; \quad h = \begin{pmatrix} 1 & 0 \\ 0 & -1 \end{pmatrix}$$

with relations $[h,e] = 2e$, $[h,f] = 2f$ and $[e,f] = h$. The following theorem, due to Serre, ensures that $\mathfrak{g}$ can be reconstructed by "gluing together" these copies of $\mathfrak{sl}(2,\mathbb{C})$ in a way prescribed by the root system of $\mathfrak{g}$:

Theorem A.10 (Serre). *Let $\mathfrak{g}$ be semisimple complex Lie algebra; we denote $R = R_+ \sqcup R_-$ its root system, $S = \{\alpha_1, \ldots, \alpha_r\}$ the set of simple roots, and $(c_{ij})_{1\leq i,j\leq r}$ the Cartan matrix. For each simple root α_i, set $e_i = E_{\alpha_i}$, $f_i = F_{\alpha_i}$ and $h_i = H_{\alpha_i}$. Then, $\mathfrak{g}$ is generated by $(e_i, f_i, h_i)_{1\leq i\leq r}$, and has for presentation:*

$$\begin{aligned}
&\forall i,j,\ [h_i,h_j] = 0;\\
&\forall i,j,\ [e_i,f_j] = \delta_{ij}\, h_i;\\
&\forall i,j,\ [h_i,e_j] = c_{ij}\, e_j;\\
&\forall i,j,\ [h_i,f_j] = -c_{ij}\, f_j;\\
&\forall i \neq j,\ (\mathrm{ad}\, e_i)^{1-c_{ij}}(e_j) = 0;\\
&\forall i \neq j,\ (\mathrm{ad}\, f_i)^{1-c_{ij}}(f_j) = 0.
\end{aligned}$$

Moreover, any root system of Theorem A.9 corresponds by this construction to a simple complex Lie algebra.

Example. In the Lie algebra $\mathfrak{sl}(N)$, set $e_i = X_{i(i+1)} = e_{i(i+1)}$, $f_i = Y_{i(i+1)} = e_{(i+1)i}$, and $h_i = H_i = e_{ii} - e_{(i+1)(i+1)}$. Then, $\mathfrak{sl}(N)$ is spanned as a Lie algebra by $(e_i, f_i, h_i)_{i\in[\![1,N-1]\!]}$, with relations

$$\begin{aligned}
&\forall i,j,\ [h_i,h_j] = 0 \quad ; \quad [e_i,f_j] = \delta_{ij}\, h_i;\\
&\forall i,\ [h_i,e_i] = 2\,e_i \quad ; \quad [h_i,f_j] = -2\,f_i;\\
&\forall i,\ [h_i,e_{i\pm1}] = -e_{i\pm1} \quad ; \quad [h_i,f_{i\pm1}] = f_{i\pm1};\\
&\forall i,\ [e_i,[e_i,e_{i\pm1}]] = [f_i,[f_i,f_{i\pm1}]] = 0;\\
&\forall |i-j| \geq 2,\ [h_i,e_j] = [h_i,f_j] = [e_i,e_j] = [f_i,f_j] = 0.
\end{aligned}$$

A.3 The highest weight theory

In the previous section, we saw that roots and root systems allowed a classification of the semisimple complex Lie algebras. We now use weights and weight lattices in order to classify their representations; this will lead us to Weyl's highest weight theorem A.12, and to the corresponding character formula A.14.

▷ *Complex Lie groups and universal enveloping algebras.*

Before we attempt to classify the representations of semisimple complex Lie algebras, it is important to relate them to the representations of the corresponding complex Lie groups, and to the modules over the corresponding **universal enveloping algebras**. Let $\mathfrak{g}$ be an arbitrary complex Lie algebra, and G be the unique simply connected complex Lie group with Lie algebra $T_eG = \mathfrak{g}$. If $\phi : G \to \mathrm{GL}(V)$ is a holomorphic morphism of groups with values in the general linear group of a finite-dimensional complex vector space V, then its derivative $d_e\phi : \mathfrak{g} \to \mathfrak{gl}(V)$ is a morphism of Lie algebras, and this correspondence is unique; therefore, the classification of the representations of $\mathfrak{g}$ is equivalent (in the categorical sense) to the classification of the holomorphic representations of G. Suppose now that $\mathfrak{g}$ is semisimple, and let $\mathfrak{h}$ be a Cartan subalgebra of $\mathfrak{g}$, and H be the **Cartan subgroup** of G corresponding to this Cartan subalgebra, that is to say the centralizer of $\mathfrak{h}$ for the adjoint action $\mathrm{Ad} : G \to \mathrm{GL}(\mathfrak{g})$. Then, $H/\exp\mathfrak{h}$ is a discrete group, and the set of elements conjugated to an element $h \in H$ is dense in G. In particular, if ϕ is a representation of G on a complex vector space V, then the restriction of the character $\mathrm{ch}^V = \mathrm{tr}\,\phi$ to H suffices to compute ch^V on the whole group G. Suppose now that $\rho = d_e\phi$, the corresponding morphism of Lie algebras $\mathfrak{g} \to \mathfrak{gl}(V)$, yields a representation of Lie algebras with decomposition in weight spaces equal to

$$V = \sum_{\omega \text{ weight}} V_\omega, \text{ with } \dim V_\omega = d_\omega.$$

The trace of an element $X \in \mathfrak{h}$ acting on V is then given by

$$\mathrm{tr}\,\rho(X) = \sum_{\omega \text{ weight}} d_\omega\, \omega(X);$$

and by taking the exponential, one gets that for every $h = \exp(X)$ in the Cartan subgroup H,

$$\mathrm{ch}^V(h) = \sum_{\omega \text{ weight}} d_\omega\, \mathrm{e}^{\omega(X)}.$$

So, the decomposition in weight spaces of V allows one to compute the character ch^V of the representation $\phi : G \to \mathrm{GL}(V)$. At the end of this section, we shall give a formula that enables the computation of the character of any irreducible representation of a semisimple complex Lie algebra.

We have just seen that a representation of a simply connected complex Lie group G corresponds to a representation of its complex Lie algebra $\mathfrak{g}$. In turn, a representation of a complex Lie algebra $\mathfrak{g}$ corresponds to a module over the **universal enveloping algebra** $U(\mathfrak{g})$ of $\mathfrak{g}$, which is the algebra over $\mathbb{C}$ obtained by quotienting the Fock space $T(\mathfrak{g}) = \bigoplus_{n=0}^{\infty} \mathfrak{g}^{\otimes n}$ by the ideal generated by the elements

$$x \otimes y - y \otimes x - [x, y].$$

If $\psi_\mathfrak{g} : \mathfrak{g} \to U(\mathfrak{g})$ is the composition of the injection $\mathfrak{g} \hookrightarrow T(\mathfrak{g})$ and of the projection map $T(\mathfrak{g}) \to U(\mathfrak{g})$, then for every representation $\phi : \mathfrak{g} \to \mathfrak{gl}(V)$, there exists a morphism of algebras $\phi^U : U(\mathfrak{g}) \to \mathfrak{gl}(V)$ that makes the following diagram commutative:

$$\begin{array}{ccc} \mathfrak{g} & \xrightarrow{\psi_{\mathfrak{g}}} & U(\mathfrak{g}) \\ & \searrow^{\phi} & \downarrow^{\phi^U} \\ & & \mathfrak{gl}(V). \end{array}$$

As a consequence, the category of representations of $\mathfrak{g}$ is equivalent to the category of modules over $U(\mathfrak{g})$. This universal enveloping algebra has a basis described by Poincaré–Birkhoff–Witt theorem: if $(X_i)_{1\leq i\leq n}$ is a basis of $\mathfrak{g}$, then

$$(X_1)^{k_1}(X_2)^{k_2}\cdots(X_n)^{k_n}$$

with all $k_i \geq 0$ is a basis of $U(\mathfrak{g})$.

▷ *The weight lattice of a semisimple Lie algebra.*

As before, $\mathfrak{g}$ is a semisimple complex Lie algebra, $\mathfrak{h}$ is a Cartan subalgebra, and S is a set of simple roots. Recall that a weight ω of a representation V of $\mathfrak{g}$ is a linear form $\omega \in \mathfrak{h}^*$ such that

$$V_\omega = \{v \in V \mid \forall X \in \mathfrak{h},\ X \cdot v = \omega(X)\,v\} \neq \{0\}.$$

Proposition A.11. *For any finite-dimensional representation V of $\mathfrak{g}$, the set of weights of V and their multiplicities are invariant under the action of the Weyl group W. Moreover, these weights belong to $\mathfrak{h}^*_{\mathbb{R}} = \bigoplus_{\alpha\in S}\mathbb{R}\alpha$, and the set of all weights of representations of $\mathfrak{g}$ is the set of linear forms $\omega \in \mathfrak{h}^*_{\mathbb{R}}$ such that*

$$\forall \alpha \in S,\quad \frac{\langle 2\omega \mid \alpha\rangle}{\langle \alpha \mid \alpha\rangle} = \omega(H_\alpha) \in \mathbb{Z}.$$

*Therefore, it is a lattice of maximal rank r in $\mathfrak{h}^*_{\mathbb{R}}$, called the **weight lattice** of $\mathfrak{g}$, and on which the Weyl W group acts.*

If $S = \{\alpha_1,\ldots,\alpha_r\}$, denote $H_{\alpha_j} = \check{\alpha}_j$; it is the **coroot** associated to the simple root α_j. Then, the **fundamental weights** of $\mathfrak{g}$ are the unique elements $\omega_1,\ldots,\omega_r$ of $\mathfrak{h}^*_{\mathbb{R}}$ such that $\omega_i(\check{\alpha}_j) = \delta_{ij}$. The weight lattice of $\mathfrak{g}$ is then $\bigoplus_{i=1}^r \mathbb{Z}\omega_i$, and it is convenient to denote it $X = X(\mathfrak{g})$. A weight is called **dominant** if it is a positive linear combination of the fundamental weights, or, equivalently, if $\langle \omega \mid \alpha\rangle \geq 0$ for all simple roots $\alpha \in S$. The dominant weights are the elements of $\mathfrak{h}^*_{\mathbb{R}}$ that belong to the intersection of the weight lattice X and of the fundamental **Weyl chamber**, which is the cone

$$C = \{\omega \in \mathfrak{h}^*_{\mathbb{R}} \mid \forall \alpha \in S,\ \langle \omega \mid \alpha\rangle \geq 0\}.$$

The images of the interior of the fundamental Weyl chamber by the isometries of the Weyl group $w \in W$ are disjoint, and $\mathfrak{h}^*_{\mathbb{R}} = \bigcup_{w\in W} w(C)$, so the Weyl chambers form a tiling of the dual of the Cartan algebra.

Example. Consider the simple Lie algebra $\mathfrak{sl}(N)$, and denote d_i^* the linear form on $\mathfrak{h}(N)$ defined by $d_i^*(\mathrm{diag}(d_1,\ldots,d_N)) = d_i$. A set of simple coroots for $\mathfrak{sl}(N)$ is $\{\breve{\alpha}_1,\ldots,\breve{\alpha}_{N-1}\}$, where $\breve{\alpha}_i = h_i = e_{ii} - e_{(i+1)(i+1)}$ with our previous notations. The corresponding set of fundamental weights is $\{\omega_1,\ldots,\omega_{N-1}\}$, where

$$\omega_i = (d_1^* + \cdots + d_i^*) - \frac{i}{N}(d_1^* + \cdots + d_N^*).$$

The fundamental Weyl chamber is $\bigoplus_{i=1}^{N-1} \mathbb{R}_+ \omega_i$; thus, it is the set of linear forms

$$\lambda = \lambda_1\, d_1^* + \lambda_2\, d_2^* + \cdots + \lambda_N\, d_N^*$$

with $\lambda_1 \geq \lambda_2 \geq \cdots \geq \lambda_N$ and $\sum_{i=1}^N \lambda_i = 0$. In this fundamental Weyl chamber, the dominant weights are the linear forms λ whose coefficients satisfy $\lambda_i - \lambda_{i+1} \in \mathbb{Z}$ for all $i \in [\![1,N]\!]$. Since $d_1^* + \cdots + d_N^* = 0$ on $\mathfrak{h}_{\mathbb{R}}^*$, replacing

$$\lambda_1\, d_1^* + \lambda_2\, d_2^* + \cdots + \lambda_N\, d_N^*$$

by

$$(\lambda_1 - \lambda_N)\, d_1^* + (\lambda_2 - \lambda_N)\, d_2^* + \cdots + (\lambda_{N-1} - \lambda_N)\, d_{N-1}^*,$$

one can identify the set of dominant weights and the set $\mathfrak{Y}^{(N-1)}$ of integer partitions of length smaller than $N-1$. When $N = 3$, the weight lattice and the root system is drawn in Figure A.1:

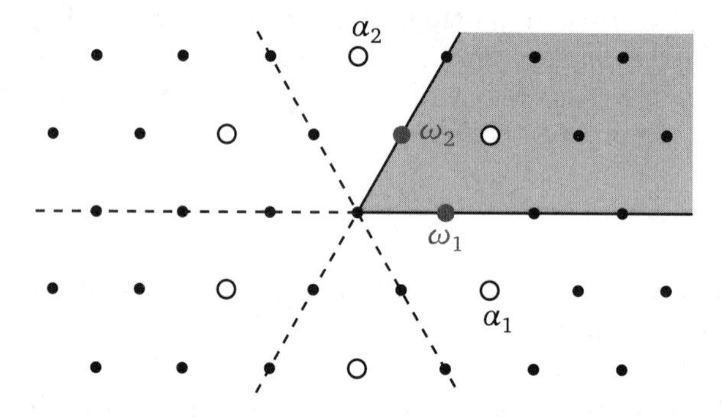

Figure A.1
Roots and weight lattice of $\mathfrak{sl}(3)$.

One has in this case $\alpha_1 = d_1^* - d_2^*$, $\alpha_2 = d_2^* - d_3^*$, $\omega_1 = \frac{2d_1^* - d_2^* - d_3^*}{3}$, and $\omega_2 = \frac{d_1^* + d_2^* - 2d_3^*}{3}$. The gray zone is the fundamental Weyl chamber, and the five other zones are its images by the $6-1$ non-trivial permutations of $W = \mathfrak{S}(3)$.

In the following, we endow the space $\mathfrak{h}_{\mathbb{R}}^*$ with the partial order

$$\lambda \geq \mu \iff \lambda - \mu \in \bigoplus_{\alpha \in S} \mathbb{R}_+ \alpha \iff \forall \omega \text{ fundamental weight}, \langle \omega \,|\, \lambda - \mu \rangle \geq 0.$$

We also denote $X_+ = X \cap C$ the set of dominant weights of $\mathfrak{g}$, and $\mathfrak{n}_+ = \bigoplus_{\alpha \in R_+} \mathfrak{g}_\alpha$, which is a Lie subalgebra of $\mathfrak{g}$.

Theorem A.12. *Every irreducible representation of a semisimple complex Lie algebra $\mathfrak{g}$ admits a unique highest weight λ, which is of multiplicity 1, and a dominant weight. The other weights μ of this representation all satisfy $\lambda - \mu \in \bigoplus_{\alpha \in S} \mathbb{N}\alpha$. Conversely, for every dominant weight $\lambda \in X_+$, there exists a unique irreducible representation V^λ of $\mathfrak{g}$ which has this weight as a highest weight. Moreover, the vectors in the weight space $(V^\lambda)_\lambda$ are characterized by:*

$$\forall X \in \mathfrak{n}_+,\ X \cdot v = 0.$$

Theorem A.13. *Every finite-dimensional representation of a semisimple complex Lie algebra $\mathfrak{g}$ is completely reducible, i.e., can be written (uniquely) as a direct sum*

$$\bigoplus_{\lambda \in X_+} m_\lambda V^\lambda,$$

where the multiplicities m_λ are non-negative integers.

The irreducible representations of $\mathfrak{g}$, which are also the irreducible modules for $U(\mathfrak{g})$, can be constructed as follows. Fix a dominant weight $\lambda \in X_+$. Set $\mathfrak{n}_- = \bigoplus_{\alpha \in R_-} \mathfrak{g}_\alpha$, so that $\mathfrak{g} = \mathfrak{h} \oplus \mathfrak{n}_+ \oplus \mathfrak{n}_-$; and $\mathfrak{b} = \mathfrak{n}_+ \oplus \mathfrak{h}$. Denote M^λ the **Verma module**

$$M^\lambda = U(\mathfrak{g}) \otimes_{U(\mathfrak{b})} \mathbb{C},$$

where $\mathfrak{h}$ acts on $\mathbb{C}$ by

$$X \cdot v = \lambda(X)\, v \quad \forall X \in \mathfrak{h},$$

and $\mathfrak{n}$ acts on $\mathbb{C}$ by

$$X \cdot v = 0 \quad \forall X \in \mathfrak{n}.$$

Since $\mathfrak{g} = \mathfrak{n}_- \oplus \mathfrak{b}$, by Poincaré–Birkhoff–Witt theorem, $U(\mathfrak{g}) = U(\mathfrak{n}_-) \otimes_{\mathbb{C}} U(\mathfrak{b})$, so M^λ is isomorphic as a $\mathbb{C}$-vector space to $U(\mathfrak{n}_-)$. It can be shown that every proper $U_z(\mathfrak{sl}(N))$-submodule of M^λ is included into $\bigoplus_{\omega \neq \lambda} (M^\lambda)_\omega$, and therefore, that there exists a unique maximal proper submodule N^λ of M^λ (the union of all proper submodules). Then, the quotient M^λ / N^λ is finite-dimensional, and the unique (up to isomorphism) irreducible module V^λ for $U(\mathfrak{g})$ of highest weight λ.

Example. For $\mathfrak{sl}(3)$, consider the adjoint representation, whose weights are:

- 0, with multiplicity the rank $r = 2$ of $\mathfrak{sl}(3, \mathbb{C})$.
- and the roots of $\mathfrak{sl}(3, \mathbb{C})$, all with multiplicity 1:

$$\alpha_1, \alpha_2, \alpha_1 + \alpha_2, -\alpha_1, -\alpha_2, -\alpha_1 - \alpha_2.$$

The highest weight of this representation is $\alpha_1 + \alpha_2$, and the weight diagram for $\mathfrak{sl}(3)$ appears in Figure A.2.

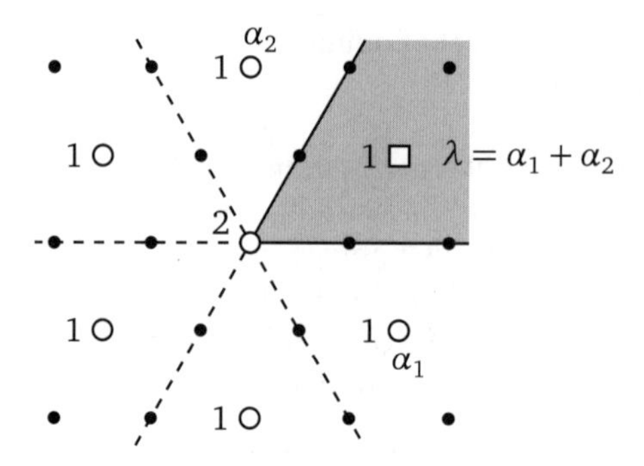

Figure A.2
Weight diagram of the adjoint representation of $\mathfrak{sl}(3)$.

Example. Since $\mathrm{SL}(N,\mathbb{C}) = \{M \in \mathrm{GL}(N,\mathbb{C}) \mid \det M = 1\}$ is the simply connected complex Lie group with Lie algebra $\mathfrak{sl}(N)$, the previous discussion shows that every holomorphic representation of $\mathrm{SL}(N,\mathbb{C})$ is reducible, and that the irreducible representations are in bijection with the integer partitions of length less that $N-1$. Consider now the group $\mathrm{GL}(N,\mathbb{C})$. Its Lie algebra is $\mathfrak{gl}(N) = \mathfrak{sl}(N) \oplus \mathbb{C}$, and the irreducible representations of the abelian Lie algebra $\mathbb{C}$ are all of the form

$$\begin{aligned} W^z : \mathbb{C} &\to \mathbb{C} = \mathrm{End}_{\mathbb{C}}(\mathbb{C}) \\ x &\mapsto zx. \end{aligned}$$

Therefore, the irreducible representations of $\mathfrak{gl}(N)$ are all of the form $V^\lambda \otimes_{\mathbb{C}} W^z$, where $\lambda \in \mathfrak{Y}^{(N-1)}$ and $z \in \mathbb{C}$. However, not all of them correspond to representations of the general linear group. More precisely, consider the exact sequence

$$1 \longrightarrow \mathrm{Ker}\,\psi \longrightarrow \mathrm{SL}(N,\mathbb{C}) \times \mathbb{C} \longrightarrow \mathrm{GL}(N,\mathbb{C}) \longrightarrow 1$$

where $\psi(g,x) = \exp(x)\,g$. The kernel of ψ is the discrete group

$$D = \left\{ \left(\mathrm{e}^{\frac{2ik\pi}{N}} I_N, -\frac{2ik\pi}{N} \right) \mid k \in \mathbb{Z} \right\}.$$

A representation $\phi : \mathrm{SL}(N,\mathbb{C}) \times \mathbb{C} \to \mathrm{GL}(V)$ factors through $\mathrm{GL}(N,\mathbb{C})$ if and only if it is trivial on the kernel D. Notice that D is generated as a group by the pair $(\mathrm{e}^{\frac{2i\pi}{N}} I_N, -\frac{2i\pi}{N})$. One can show that the action of $\mathrm{e}^{\frac{2i\pi}{N}} I_N$ on V^λ is given by multiplication by $\mathrm{e}^{\frac{2i\pi|\lambda|}{N}}$, where $|\lambda| = \sum_{i=1}^{N-1} \lambda_i$. Therefore, the representation $V^\lambda \otimes_{\mathbb{C}} W^z$ is trivial on D if and only if

$$\mathrm{e}^{\frac{2i\pi|\lambda| - 2i\pi z}{N}} = 1 \iff z = |\lambda| + kN$$

for some $k \in \mathbb{Z}$. It follows that every irreducible representation of $\mathrm{GL}(N,\mathbb{C})$ can be written uniquely as the tensor product of the representation coming from $V^\lambda \otimes_{\mathbb{C}}$

$W^{|\lambda|}$, and of a power k of the determinant representation $\det : \mathrm{GL}(N,\mathbb{C}) \to \mathbb{C}^*$. It is convenient to associate to this tensor product $V^\lambda \otimes_{\mathbb{C}} W^{|\lambda|+kN}$ the signed integer partition

$$(\lambda_1 + k, \lambda_2 + k, \ldots, \lambda_{N-1} + k, \lambda_N + k)$$

with $\lambda_N = 0$. Thus, holomorphic irreducible representations of $\mathrm{GL}(N,\mathbb{C})$ are labeled by non-increasing sequences of N integers in $\mathbb{Z}$. Among them, the polynomial representations are those that involve positive powers $k \geq 0$ of the determinant, hence, those that are labeled by true (unsigned) integer partitions of length smaller than N. Moreover, the discussion of Section 2.5 shows that all the holomorphic representations of $\mathrm{GL}(N,\mathbb{C})$ are actually rational algebraic representations.

▷ *Weyl's character formula.*

To conclude this appendix, let us state **Weyl's character formula**. Fix as before a semisimple complex Lie algebra $\mathfrak{g}$, and denote

$$\rho = \frac{1}{2} \sum_{\alpha \in R_+} \alpha$$

the half-sum of positive roots. It is also equal to the sum $\rho = \sum_{i=1}^r \omega_i$ of all fundamental weights. On the other hand, if V^λ is an irreducible representation of $\mathfrak{g}$, we denote $V^\lambda = \bigoplus_\omega (V^\lambda)_\omega$ its decomposition in weight spaces, $d_\omega = \dim (V^\lambda)_\omega$, and

$$\mathrm{ch}^\lambda = \sum_\omega d_\omega \, \mathrm{e}^\omega$$

the character of the representation on $\exp(\mathfrak{h})$. The map

$$\begin{aligned} \mathrm{ch} : \mathrm{R}_0(U(\mathfrak{g})) &\to \mathbb{Z}[X(\mathfrak{g})] \\ M = \bigoplus_{\lambda \in X_+} m_\lambda \, V^\lambda &\mapsto \sum_{\lambda \in X^+} m_\lambda \, \mathrm{ch}^\lambda \end{aligned}$$

takes its values in the invariant ring $\mathbb{Z}[X(\mathfrak{g})]^W$, and is then an isomorphism of groups, and even of rings if $\mathrm{R}_0(U(\mathfrak{g}))$ is endowed with the tensor product of representations.

Theorem A.14. *For any dominant weight $\lambda \in X_+$,*

$$\mathrm{ch}^\lambda = \frac{\sum_{w \in W} \varepsilon(w) \, \mathrm{e}^{w(\lambda+\rho)}}{\sum_{w \in W} \varepsilon(w) \, \mathrm{e}^{w(\rho)}},$$

where $\varepsilon : W \to \{\pm 1\}$ is the sign representation that satisfies $\varepsilon(s_\alpha) = -1$ for every root reflection s_α in the Weyl group. Moreover, the denominator of this formula is equal to $\mathrm{e}^{-\rho} \prod_{\alpha \in R_+} (\mathrm{e}^\alpha - 1)$.

Example. In $\mathfrak{sl}(N)$, the half-sum of positive roots is the linear form

$$\rho = (N-1)d_1^* + (N-2)d_2^* + \cdots + d_{N-1}^*.$$

Therefore, for any integer partition $\lambda \in \mathfrak{Y}^{(N-1)}$ the character of the representation V^λ restricted to the Cartan subgroup H is given by

$$\mathrm{ch}^\lambda(\mathrm{diag}(x_1,\ldots,x_N)) = \frac{\sum_{\sigma\in\mathfrak{S}(N)} \varepsilon(\sigma)\, x^{\sigma(\lambda+\rho)}}{\sum_{\sigma\in\mathfrak{S}(N)} \varepsilon(\sigma)\, x^{\sigma(\lambda+\rho)}},$$

which is the Schur function of label λ. This is actually the value of ch^λ on any matrix of $\mathrm{SL}(N,\mathbb{C})$ with eigenvalues $x_1,\ldots,x_N$. More generally, the character of the irreducible representation of $\mathrm{GL}(N,\mathbb{C})$ labeled by a (signed) integer partition λ of length N is given by the same formula as above, which corresponds to the Schur function s_λ for any true (unsigned) partition $\lambda \in \mathfrak{Y}^{(N)}$.

References

[AD56] N. Aronszajn and W. F. Donoghue. On exponential representations of analytic functions in the upper half-plane with positive imaginary part. *J. Anal. Math.*, 5:321–388, 1956.

[AD64] N. Aronszajn and W. F. Donoghue. A supplement to the paper on exponential representations of analytic functions in the upper half-plane with positive imaginary part. *J. Anal. Math.*, 12:113–127, 1964.

[AD86] D. Aldous and P. Diaconis. Shuffling cards and stopping times. *Amer. Math. Monthly*, 93(5):333–348, 1986.

[AD95] D. Aldous and P. Diaconis. Hammersley's interacting particle process and longest increasing subsequences. *Prob. Th. Rel. Fields*, 103(2):199–213, 1995.

[AFNT15] J.-C. Aval, V. Féray, J.-C. Novelli, and J.-Y. Thibon. Quasi-symmetric functions as polynomial functions on Young diagrams. *Journal of Algebraic Combinatorics*, 41(3):669–706, 2015.

[AGZ10] G. W. Anderson, A. Guionnet, and O. Zeitouni. *An Introduction to Random Matrices*, volume 118 of *Cambridge Studies in Advanced Mathematics*. Cambridge University Press, 2010.

[Alp93] J. L. Alperin. *Local Representation Theory*, volume 11 of *Cambridge Studies in Advanced Mathematics*. Cambridge University Press, 1993.

[AS05] M. Aguiar and F. Sottile. Structure of the Malvenuto–Reutenauer Hopf algebra of permutations. *Adv. Math.*, 191:225–275, 2005.

[Att] S. Attal. Lectures in quantum noise theory. `http://math.univ-lyon1.fr/~attal/chapters.html`.

[BBO05] P. Biane, P. Bougerol, and N. O'Connell. Littelmann paths and Brownian paths. *Duke Math. J.*, 130(1):127–167, 2005.

[BD85] T. Bröcker and T. Dieck. *Representations of Compact Lie Groups*, volume 98 of *Graduate Texts in Mathematics*. Springer–Verlag, 1985.

[BD92] D. Bayer and P. Diaconis. Tailing the dovetail shuffle to its lair. *Ann. Appl. Probab.*, 2(2):294–313, 1992.

[BDJ99] J. Baik, P. Deift, and K. Johansson. On the distribution of the length of the longest increasing subsequence of random permutations. *J. Amer. Math. Soc.*, 12(4):1119–1178, 1999.

[Bia97] P. Biane. Some properties of crossings and partitions. *Discrete Math.*, 175(1):41–53, 1997.

[Bia98] P. Biane. Representations of symmetric groups and free probability. *Adv. Math.*, 138:126–181, 1998.

[Bia01] P. Biane. Approximate factorization and concentration for characters of symmetric groups. *Intern. Math. Res. Notices*, 2001:179–192, 2001.

[Bia09] P. Biane. From Pitman's theorem to crystals. *Advanced Studies in Pure Mathematics*, 55:1–13, 2009.

[Bil69] P. Billingsley. *Convergence of Probability Measures*. John Wiley & Sons, 1969.

[Bil95] P. Billingsley. *Probability and Measure*. John Wiley & Sons, 3rd edition, 1995.

[BOO00] A. Borodin, A. Okounkov, and G. Olshanski. Asymptotics of Plancherel measures for symmetric groups. *J. Amer. Math. Soc.*, 13:491–515, 2000.

[Bou68] N. Bourbaki. *Groupes et algèbres de Lie, 4–6*. Hermann, 1968.

[Bra37] R. Brauer. On algebras which are connected with the semisimple continuous groups. *Ann. Math.*, 38:854–872, 1937.

[Buf12] A. I. Bufetov. A central limit theorem for extremal characters of the infinite symmetric group. *Funct. Anal. Appl.*, 46(2):83–93, 2012.

[Car86] R. W. Carter. Representation theory of the 0–Hecke algebra. *J. Algebra*, 15:89–103, 1986.

[CG97] N. Chriss and V. Ginzburg. *Representation Theory and Complex Geometry*. Modern Birkhäuser Classics. Birkhäuser, 1997.

[Che04] C. Chevalley. *The Classification of Semi-Simple Algebraic Groups*, volume 3 of *Collected Works of Claude Chevalley*. Springer–Verlag, 2004.

[Coh89] P. M. Cohn. *Algebra. Volume 2*. John Wiley & Sons, 2nd edition, 1989.

[Coh91] P. M. Cohn. *Algebra. Volume 3*. John Wiley & Sons, 2nd edition, 1991.

[CR81] C. W. Curtis and I. Reiner. *Methods of Representation Theory*. John Wiley & Sons, 1981.

[CSST08] T. Ceccherini-Silberstein, F. Scarabotti, and F. Tolli. *Harmonic Analysis on Finite Groups*, volume 108 of *Cambridge Studies in Advanced Mathematics*. Cambridge University Press, 2008.

[CSST09] T. Ceccherini-Silberstein, F. Scarabotti, and F. Tolli. Representation theory of wreath products of finite groups. *Journal of Mathematical Sciences*, 156(1):44–55, 2009.

[CSST10] T. Ceccherini-Silberstein, F. Scarabotti, and F. Tolli. *Representation Theory of the Symmetric Groups. The Okounkov–Vershik Approach, Character Formulas, and Partition Algebras*, volume 121 of *Cambridge Studies in Advanced Mathematics*. Cambridge University Press, 2010.

[DFP92] P. Diaconis, J. Fill, and J. Pitman. Analysis of top to random shuffles. *Combinatorics, Probability, and Computing*, 1:135–155, 1992.

[DFŚ10] M. Dołęga, V. Féray, and P. Śniady. Explicit combinatorial interpretation of Kerov character polynomials as numbers of permutation factorizations. *Adv. Math.*, 225(1):81–120, 2010.

[DHNT11] G. Duchamp, F. Hivert, J.-C. Novelli, and J.-Y. Thibon. Noncommutative symmetric functions VII: Free quasi-symmetric functions revisited. *Annals of Combinatorics*, 15:655–673, 2011.

[DHT02] G. Duchamp, F. Hivert, and J.-Y. Thibon. Noncommutative symmetric functions VI: Free quasi-symmetric functions and related algebras. *International Journal of Algebra and Computation*, 12:671–717, 2002.

[Dia86] P. Diaconis. Applications of noncommutative Fourier analysis to probability problems. In *Ecole d'Été de Probabilités de St. Flour, XV–XVII*, volume 1362 of *Lecture Notes in Mathematics*, pages 51–100. Springer–Verlag, 1986.

[DJ86] R. Dipper and G. D. James. Representations of Hecke algebras of general linear groups. *Proc. London Math. Soc.*, 52(3):20–52, 1986.

[DJ87] R. Dipper and G. D. James. Blocks and idempotents of Hecke algebras of general linear groups. *Proc. London Math. Soc.*, 54(3):57–82, 1987.

[DKKT97] G. Duchamp, A. Klyachko, D. Krob, and J.-Y. Thibon. Noncommutative symmetric functions III: Deformations of Cauchy and convolution algebras. *Discrete Mathematics and Theoretical Computer Science*, 1:159–216, 1997.

[DMP95] P. Diaconis, M. McGrath, and J. Pitman. Riffle shuffles, cycles and descents. *Combinatorica*, 15(1):11–29, 1995.

[DS81] P. Diaconis and M. Shahshahani. Generating a random permutation with random transpositions. *Z. Wahr. verw. Gebiete*, 57(2):159–179, 1981.

[Dys62] F. J. Dyson. A brownian-motion model for the eigenvalues of a random matrix. *J. Math. Phys.*, 3:1191–1198, 1962.

[Far08] J. Faraut. *Analysis on Lie Groups: An Introduction*, volume 110 of *Cambridge Studies in Advanced Mathematics*. Cambridge University Press, 2008.

[Fel68] W. Feller. *An Introduction to Probability Theory and Its Applications*. John Wiley & Sons, 3rd edition, 1968.

[Fér09] V. Féray. Combinatorial interpretation and positivity of Kerov's character polynomials. *Journal of Algebraic Combinatorics*, 29(4):473–507, 2009.

[Fér10] V. Féray. Stanley's formula for characters of the symmetric group. *Annals of Combinatorics*, 13(4):453–461, 2010.

[Fér12a] V. Féray. On complete functions in Jucys–Murphy elements. *Annals of Combinatorics*, 16(4):677–707, 2012.

[Fér12b] V. Féray. Partial Jucys–Murphy elements and star factorizations. *European Journal of Combinatorics*, 33:189–198, 2012.

[Fér14] V. Féray. Cyclic inclusion-exclusion. *SIAM J. Discrete Mathematics*, 29(4):2284–2311, 2014.

[FH59] H. K. Farahat and G. Higman. The centres of symmetric group rings. *Proceedings of the Royal Society of London. Series A, Mathematical and Physical Sciences*, 250(1261):212–221, 1959.

[FH91] W. Fulton and J. Harris. *Representation Theory. A First Course*, volume 129 of *Graduate Texts in Mathematics*. Springer–Verlag, 1991.

[FM12] V. Féray and P.-L. Méliot. Asymptotics of q–Plancherel measures. *Prob. Th. Rel. Fields*, 152(3–4):589–624, 2012.

[FMN16] V. Féray, P.-L. Méliot, and A. Nikeghbali. Mod–ϕ convergence: Normality Zones and Precise Deviations. *Springer Briefs in Probability and Mathematical Statistics*, 2016.

[FRT54] J. S. Frame, G. Robinson, and R. M. Thrall. The hook graphs of the symmetric group. *Canad. J. Math.*, 6:316–324, 1954.

[FŚ11] V. Féray and P. Śniady. Asymptotics of characters of symmetric groups related to Stanley character formula. *Ann. Math.*, 173(2):887–906, 2011.

[Ful97] W. Fulton. *Young Tableaux with Applications to Representation Theory and Geometry*, volume 35 of *London Mathematical Society Student Texts*. Cambridge University Press, 1997.

[Ful02] J. Fulman. Applications of symmetric functions to cycle and increasing subsequence structure after shuffles. *Journal of Algebraic Combinatorics*, 16:165–194, 2002.

[Gec93] M. Geck. A note on Harish–Chandra induction. *Manuscripta Math.*, 80:393–401, 1993.

[Ges84] I. Gessel. Multipartite P-partitions and inner products of skew Schur functions. In *Combinatorics and Algebra (Boulder, CO, 1983)*, volume 34 of *Contemporary Mathematics*, pages 289–317. Amer. Math. Soc. (Providence, RI), 1984.

[GKL+95] I. M. Gelfand, D. Krob, A. Lascoux, B. Leclerc, V. S. Retakh, and J.-Y. Thibon. Noncommutative symmetric functions. *Adv. Math.*, 112:218–348, 1995.

[GL96] J. J. Graham and G. I. Lehrer. Cellular algebras. *Invent. Math.*, 123:1–34, 1996.

[GNW79] C. Greene, A. Nijenhuis, and H. Wilf. A probabilistic proof of a formula for the number of Young tableaux with a given shape. *Adv. Math.*, 31:104–109, 1979.

[GO06] A. Gnedin and G. Olshanski. Coherent permutations with descent statistics and the boundary problem for the graph of zigzag diagrams. *Intern. Math. Res. Notices*, 2006:1–39, 2006.

[GP00] M. Geck and G. Pfeiffer. *Characters of Finite Coxeter Groups and Iwahori–Hecke Algebras*, volume 21 of *London Mathematical Society Monographs*. Oxford University Press, 2000.

[GR96] M. Geck and R. Rouquier. Centers and simple modules for Iwahori–Hecke algebras. In M. Cabanes, editor, *Finite Reductive Groups, Related Structures and Representations*, volume 141 of *Progress in Mathematics*, pages 251–272. Birkhäuser, 1996.

[GR15] D. Grinberg and V. Reiner. Hopf algebras in combinatorics. `arXiv: 1409.8356`, 2015.

[Gre55] J. A. Green. The characters of the finite general linear groups. *Trans. Amer. Math. Soc.*, 80:402–447, 1955.

[Gre76] J. A. Green. Locally finite representations. *J. Algebra*, 41:137–171, 1976.

[Gre07] J. A. Green. *Polynomial Representations of* GL_n, volume 830 of *Lecture Notes in Mathematics*. Springer–Verlag, 2nd edition, 2007.

[GS06] P. Gille and T. Szamuely. *Central Simple Algebras and Galois Cohomology*, volume 101 of *Cambridge Studies in Advanced Mathematics*. Cambridge University Press, 2006.

[GW09] R. Goodman and N. R. Wallach. *Symmetry, Representations and Invariants*, volume 255 of *Graduate Texts in Mathematics*. Springer–Verlag, 2009.

[Ham72] J. M. Hammersley. A few seedlings of research. In *Proc. Sixth Berkeley Symp. Math. Statist. and Probability*, volume 1, pages 345–394. University of California Press, 1972.

[HL80] R. B. Howlett and G. I. Lehrer. Induced cuspidal representations and generalised Hecke rings. *Invent. Math*, 58:37–64, 1980.

[IK99] V. Ivanov and S. Kerov. The algebra of conjugacy classes in symmetric groups, and partial permutations. In *Representation Theory, Dynamical Systems, Combinatorial and Algorithmical Methods III*, volume 256 of *Zapiski Nauchnyh Seminarov (POMI)*, pages 95–120, 1999.

[IO02] V. Ivanov and G. Olshanski. Kerov's central limit theorem for the Plancherel measure on Young diagrams. In S. Fomin, editor, *Symmetric Functions 2001: Surveys of Developments and Perspectives*, volume 74 of *NATO Science Series II. Mathematics, Physics and Chemistry*, pages 93–151, 2002.

[Iwa64] N. Iwahori. On the structure of a Hecke ring of a Chevalley group over a finite field. *J. Fac. Sci. Univ. Tokyo*, 10(2):216–236, 1964.

[Jan88] S. Janson. Normal convergence by higher semiinvariants with applications to sums of dependent random variables and random graphs. *Ann. Probab.*, 16(1):305–312, 1988.

[Jim85] M. Jimbo. A q-difference analogue of $U(\mathfrak{g})$ and the Yang–Baxter equation. *Letters in Math. Phys.*, 10:63–69, 1985.

[Jim86] M. Jimbo. A q-analogue of $U(\mathfrak{gl}(n+1))$, Hecke algebra and the Yang–Baxter equation. *Letters in Math. Phys.*, 11:247–252, 1986.

[JK81] G. James and A. Kerber. *The Representation Theory of the Symmetric Group*, volume 16 of *Encyclopedia of Mathematics and Its Applications*. Addison–Wesley, 1981.

[JL93] G. D. James and M. Liebeck. *Representations and Characters of Groups*. Cambridge University Press, 2nd edition, 1993.

[Joh01] K. Johansson. Discrete orthogonal polynomial ensembles and the Plancherel measure. *Ann. Math.*, 153:259–296, 2001.

[Juc74] A. A. A. Jucys. Symmetric polynomials and the center of the symmetric group ring. *Rep. Math. Phys.*, 5(1):107–112, 1974.

[Ker93a] S. V. Kerov. Gaussian limit for the Plancherel measure of the symmetric group. *Comptes Rendus Acad. Sci. Paris Série I*, 316:303–308, 1993.

[Ker93b] S. V. Kerov. Transition probabilities of continual Young diagrams and Markov moment problem. *Funct. Anal. Appl.*, 27:104–117, 1993.

[Ker98] S. V. Kerov. Interlacing measures. *Amer. Math. Soc. Transl.*, 181(2):35–83, 1998.

[Kir94] A. A. Kirillov. Introduction to the theory of representations and noncommutative harmonic analysis. In *Representation Theory and Noncommutative Harmonic Analysis I*, volume 22 of *Encyclopaedia of Mathematical Sciences*. Springer–Verlag, 1994.

[KL79] D. A. Kazhdan and G. Lusztig. Representations of Coxeter groups and Hecke algebras. *Invent. Math.*, 53:165–184, 1979.

[KLT97] D. Krob, B. Leclerc, and J.-Y. Thibon. Noncommutative symmetric functions II: Transformations of alphabets. *International Journal of Algebra and Computation*, 7:181–264, 1997.

[KN77] M. G. Krein and A. A. Nudelman. *The Markov Moment Problem and Extremal Problems*, volume 50 of *Translations of Mathematical Monographs*. American Mathematical Society, 1977.

[Kna01] A. W. Knapp. *Representation Theory of Semisimple Groups: An Overview Based on Examples*. Princeton University Press, 2001.

[Kna02] A. W. Knapp. *Lie Groups Beyond an Introduction*, volume 140 of *Progress in Mathematics*. Birkhäuser, 2nd edition, 2002.

[KO94] S. V. Kerov and G. Olshanski. Polynomial functions on the set of Young diagrams. *Comptes Rendus Acad. Sci. Paris Série I*, 319:121–126, 1994.

[KT97] D. Krob and J.-Y. Thibon. Noncommutative symmetric functions IV: Quantum linear groups and Hecke algebras at $q = 0$. *Journal of Algebraic Combinatorics*, 6:339–376, 1997.

[KT99] D. Krob and J.-Y. Thibon. Noncommutative symmetric functions V: A degenerate version of $U_q(gl_N)$. *Internat. J. Alg. Comput.*, 9:405–430, 1999.

[KV77] S. V. Kerov and A. M. Vershik. Asymptotics of the Plancherel measure of the symmetric group and the limiting form of Young tableaux. *Soviet Mathematics Doklady*, 18:527–531, 1977.

[KV81] S. V. Kerov and A. M. Vershik. Asymptotic theory of characters of the symmetric group. *Funct. Anal. Appl.*, 15(4):246–255, 1981.

[KV85] S. V. Kerov and A. M. Vershik. Asymptotics of maximal and typical dimensions of irreducible representations of the symmetric group. *Funct. Anal. Appl.*, 19(1):25–36, 1985.

[KV86] S. V. Kerov and A. M. Vershik. The characters of the infinite symmetric group and probability properties of the Robinson–Schensted–Knuth algorithm. *Siam J. Alg. Disc. Meth.*, 7(1):116–124, 1986.

[Lan93] S. Lang. *Real and Functional Analysis*, volume 142 of *Graduate Texts in Mathematics*. Springer–Verlag, 3rd edition, 1993.

[Lan94] S. Lang. *Algebraic Number Theory*, volume 110 of *Graduate Texts in Mathematics*. Springer–Verlag, 2nd edition, 1994.

[Lan02] S. Lang. *Algebra*, volume 211 of *Graduate Texts in Mathematics*. Springer–Verlag, 2002.

[Las99] A. Lascoux. *Symmetric Functions and Combinatorial Operators on Polynomials*, volume 99 of *CBMS Regional Conference Series in Mathematics*. American Mathematical Society, 1999.

[Las04] A. Lascoux. Operators on polynomials. In *École d'été ACE*, 2004.

[Las13] A. Lascoux. Polynomial representations of the Hecke algebra of the symmetric group. *Int. J. Algebra Comput.*, 23(4):803–818, 2013.

[Lit95] P. Littelmann. Paths and root operators in representation theory. *Ann. Math.*, 142(3):499–525, 1995.

[LLT02] A. Lascoux, B. Leclerc, and J.-Y. Thibon. The plactic monoid. In *Algebraic Combinatorics on Words*. Cambridge University Press, 2002.

[LS59] V. P. Leonov and A. N. Shiryaev. On a method of calculation of semi-invariants. *Theory Prob. Appl.*, 4:319–329, 1959.

[LS77] B. F. Logan and L. A. Shepp. A variational problem for random Young tableaux. *Adv. Math.*, 26:206–222, 1977.

[LT01] A. Lascoux and J.-Y. Thibon. Vertex operators and the class algebras of symmetric groups. *Zapisky Nauchnyh Seminarov (POMI)*, 283:156–177, 2001.

[Lus88] G. Lusztig. Quantum deformations of certain simple modules over enveloping algebras. *Adv. Math.*, 70:237–249, 1988.

[LZ00] J. Lining and W. Zhengdong. The Schur–Weyl duality between quantum group of type A and Hecke algebra. *Advances in Mathematics China*, 29(5):444–456, 2000.

[LZ04] S. K. Lando and A. K. Zvonkin. *Graphs on Surfaces and Their Applications*, volume 141 of *Encyclopædia of Mathematical Sciences*. Springer–Verlag, 2004.

[Ma09] T.-S. Ma. Higher chain formula proved by combinatorics. *Electronic Journal of Combinatorics*, 16(1):N21, 2009.

[Mac95] I. G. Macdonald. *Hall Polynomials and Symmetric Functions*. Oxford University Press, 2nd edition, 1995.

[Mat86] H. Matsumura. *Commutative Ring Theory*. Cambridge University Press, 1986.

[Mat99] A. Mathas. *Iwahori–Hecke Algebras and Schur Algebras of the Symmetric Group*, volume 15 of *University Lecture Series*. American Mathematical Society, 1999.

[Mél10] P.-L. Méliot. Products of Geck–Rouquier conjugacy classes and the algebra of composed permutations. In *Proceedings of the 22nd International Conference on Formal Power Series and Algebraic Combinatorics (San Francisco, USA)*, pages 789–800, 2010.

[Mél11] P.-L. Méliot. Kerov's central limit theorem for Schur–Weyl and Gelfand measures. In *Proceedings of the 23rd International Conference on Formal Power Series and Algebraic Combinatorics (Reykjavík, Iceland)*, pages 669–680, 2011.

[Mél12] P.-L. Méliot. Fluctuations of central measures on partitions. In *Proceedings of the 24th International Conference on Formal Power Series and Algebraic Combinatorics (Nagoya, Japan)*, pages 387–398, 2012.

[MN13] S. Matsumoto and J. Novak. Jucys–Murphy elements and unitary matrix integrals. *Intern. Math. Res. Notices*, 2013(2):362–397, 2013.

[MP67] V. A. Marcenko and L. A. Pastur. Distribution of eigenvalues for some sets of random matrices. *Math. USSR Sb.*, 1(4):457–483, 1967.

[MR95] C. Malvenuto and C. Reutenauer. Duality between quasi-symmetric functions and the Solomon descent algebra. *J. Algebra*, 177:967–982, 1995.

[Mur81] G. E. Murphy. A new construction of Young's seminormal representation of the symmetric groups. *J. Algebra*, 69(2):287–297, 1981.

[Mur92] G. E. Murphy. On the representation theory of the symmetric groups and associated Hecke algebras. *J. Algebra*, 152:492–513, 1992.

[Neu99] J. Neukirch. *Algebraic Number Theory*, volume 322 of *Grundlehren der mathematischen Wissenschaften*. Springer–Verlag, 1999.

[Nor79] P. N. Norton. 0–Hecke algebras. *J. Austral. Math. Soc. Ser. A*, 27:337–357, 1979.

[NS06] A. Nica and R. Speicher. *Lectures on the Combinatorics of Free Probability*, volume 335 of *London Mathematical Society Lecture Notes Series*. Cambridge University Press, 2006.

[O'C03] N. O'Connell. A path-transformation for random walks and the Robinson–Schensted correspondence. *Trans. American Math. Soc.*, 355:3669–3697, 2003.

[Oko99] A. Okounkov. On representations of the infinite symmetric group. *Journal of Mathematical Sciences*, 96(5):3550–3589, 1999.

[Oko00] A. Okounkov. Random matrices and random permutations. *Intern. Math. Res. Notices*, 2000(20):1043–1095, 2000.

[Ols90] G. Olshanski. Unitary representations of (G, K)-pairs connected with the infinite symmetric group $S(\infty)$. *Leningrad Math. J.*, 1:983–1014, 1990.

[OV04] A. Okounkov and A. M. Vershik. A new approach to the representation theory of the symmetric groups. II. *Zapiski Nauchnyh Seminarov (POMI)*, 307:57–98, 2004.

[OW16] R. O'Donnell and J. Wright. Efficient quantum tomography. In *48th ACM Symposium on the Theory of Computing*, 2016.

[Phe01] R. R. Phelps. *Lectures on Choquet's Theorem*, volume 1727 of *Lecture Notes in Mathematics*. Springer–Verlag, 2nd edition, 2001.

[Ram91] A. Ram. A Frobenius formula for the characters of the Hecke algebras. *Invent. Math.*, 106:461–488, 1991.

[Ros88] M. Rosso. Finite dimensional representations of the quantum analog of the enveloping algebra of a complex simple lie algebra. *Comm. Math. Phys.*, 117:581–593, 1988.

[Ros90] M. Rosso. Analogue de la forme de Killing et du théorème d'Harish–Chandra pour les groupes quantiques. *Annales Scientifiques de l'École Normale Supérieure 4ème série*, 23(3):445–467, 1990.

[RR97] A. Ram and J. B. Remmel. Applications of the Frobenius formulas for the characters of the symmetric group and the Hecke algebra of type A. *Algebraic Combinatorics*, 5:59–87, 1997.

[Rud87] W. Rudin. *Real and Complex Analysis*. McGraw–Hill, 3rd edition, 1987.

[Rud91] W. Rudin. *Functional Analysis*. McGraw–Hill, 2nd edition, 1991.

[Sag01] B. E. Sagan. *The Symmetric Group: Representations, Combinatorial Algorithms and Symmetric Functions*, volume 203 of *Graduate Texts in Mathematics*. Springer–Verlag, 2nd edition, 2001.

[Sch63] M. P. Schützenberger. Quelques remarques sur une construction de Schensted. *Math. Scand.*, 12:117–128, 1963.

[Ser77] J.-P. Serre. *Linear Representations of Finite Groups*, volume 42 of *Graduate Texts in Mathematics*. Springer–Verlag, 1977.

[Śni06a] P. Śniady. Asymptotics of characters of symmetric groups, genus expansion and free probability. *Discrete Math.*, 306(7):624–665, 2006.

[Śni06b] P. Śniady. Gaussian fluctuations of characters of symmetric groups and of Young diagrams. *Prob. Th. Rel. Fields*, 136(2):263–297, 2006.

[Sol76] L. Solomon. A Mackey formula in the group ring of a Coxeter group. *J. Algebra*, 41(2):255–264, 1976.

[Sta99a] R. P. Stanley. *Enumerative Combinatorics, Vol. 2*. Cambridge University Press, 1999.

[Sta99b] R. P. Stanley. Generalized riffle shuffles and quasisymmetric functions. `arXiv:math.CO/9912025`, 1999.

[Sta03] R. P. Stanley. Irreducible symmetric group characters of rectangular shape. *Sém. Lotharingien de Combinatoire*, 50, 2003.

[Sta06] R. P. Stanley. A conjectured combinatorial interpretation of the normalized irreducible character values of the symmetric group. `arXiv:math.CO/0606467`, 2006.

[Str08] E. Strahov. A differential model for the deformation of the Plancherel growth process. *Adv. Math.*, 217(6):2625–2663, 2008.

[Tak79] M. Takesaki. *Theory of Operator Algebras I*, volume 124 of *Encyclopædia of Mathematical Sciences*. Springer–Verlag, 1979.

[t'H74] G. t'Hooft. A planar diagram theory for strong interactions. *Nuclear Physics B*, 72:461–473, 1974.

[Tou14] O. Tout. Structure coefficients of the Hecke algebra of $(\mathscr{S}_{2n}, \mathscr{B}_n)$. *Electronic Journal of Combinatorics*, 21(4):#P4.35, 2014.

[TW94] C. A. Tracy and H. Widom. Level-spacing distributions and the Airy kernel. *Commun. Math. Phys.*, 159:151–174, 1994.

[Ula61] S. M. Ulam. Monte–Carlo calculations in problems of mathematical physics. In *Modern Mathematics for the Engineers*, pages 261–281. McGraw–Hill, 1961.

[Var84] V. S. Varadarajan. *Lie Groups, Lie Algebras, and Their Representations*, volume 102 of *Graduate Texts in Mathematics*. Springer–Verlag, 1984.

[Var89] V. S. Varadarajan. *An Introduction to Harmonic Analysis on Semisimple Lie Groups*, volume 16 of *Cambridge Studies in Advanced Mathematics*. Cambridge University Press, 1989.

[Vie77] G. Viennot. Une forme géométrique de la correspondance de Robinson–Schensted. In *Combinatoire et Représentation du Groupe Symétrique*, volume 579 of *Lecture Notes in Mathematics*, pages 29–58. Springer–Verlag, 1977.

[Was81] A. J. Wassermann. *Automorphic actions of compact groups on operator algebras*. PhD thesis, University of Pennsylvania, 1981.

[Wen88] H. Wenzl. On the structure of Brauer's centralizer algebras. *Ann. Math.*, 128:173–193, 1988.

[Wic50] G. C. Wick. The evaluation of the collision matrix. *Physical Review*, 80(2):268–272, 1950.

[Zel81] A. Zelevinsky. *Representations of Finite Classical Groups. A Hopf Algebra Approach*, volume 869 of *Lecture Notes in Mathematics*. Springer–Verlag, 1981.

[ZJ10] P. Zinn-Justin. Jucys–Murphy elements and Weingarten matrices. *Letters in Math. Physics*, 91(2):119–127, 2010.

Index

C^*-algebra, 539
λ-coloring, 467
    proper, 473
σ-field, 402

admissible sequence, 436
algebra, 27
    deformation, 149
    graded, 28
    Grothendieck group
        of modules, 162
        of projective modules, 166
    Hecke, 44, 158
    Hopf, 69
    ideal, 30
    Jacobson radical, 168
    Lie, 631
    partition of the unity, 35
    presentation, 155
    Schur, 88
    semisimple, 32
    split, 179
    splitting field, 184
    symmetric, 161, 188
    tensor, 71
    trace, 183
    von Neumann, 405
alphabet, 58
    exponential, 506
    formal, 453, 506
    interlaced, 454
    opposite, 73, 454
    ordered, 292
    ordered sum, 453
    product, 73
    sum, 71
antisymmetrization, 64
arrangement, 332

bialgebra, 70
Bienaymé–Chebyshev inequality, 571
bimodule, 40
bipartite graph, 462
    bicolored, 464
bitrace, 45
branching rules, 105
Brauer graph, 516
Brauer semigroup, 515
Brauer–Cartan setting, 161, 175
Brauer–Nesbitt map, 176
Bruhat order, 156

cancellation law, 313
Cartan map, 166
Cartan matrix, 637
Cartan subalgebra, 635
Cartan subgroup, 642
Catalan number, 416
Cauchy product, 73
Cayley graph, 476
    geodesic, 477
    interval, 477
central idempotent, 34
    primitive, 34
central measures, 547
centralizer, 379
character, 13
    extremal, 504
    normalized, 13
    table, 108, 272
characteristic map, 81, 311
coalgebra, 70
    cocommutative, 71

coproduct, 70
county, 70
comodule, 96
composition, 194
descent, 291
mixing, 302
ribbon diagram, 296
content vector, 387
continuous Young diagram, 356
convolution product, 18
Coxeter group, 151
braid relations, 152
commutation relations, 152
quadratic relations, 151
cumulant, 579
disjoint, 584
formal free, 422
free, 415
identity, 584
joint, 579
cyclic order, 465
cyclic vector, 540

decomposition map, 180
degenerate affine Hecke algebra, 391
domination, 187
duality, 43
Schur–Weyl, 45
duplication rule, 78
Dynkin diagram, 637

essential supremum, 402
essential surjection, 167
Euler characteristic, 489
extremal point of a convex set, 503

flag, 264
flag variety, 265
free quasi-symmetric function, 549
free quasi-ribbon, 549
free symmetric function, 559
freeness, 412
function with negative imaginary part, 360
Nevanlinna integral representation, 364

Gelfand pair, 511
infinite, 513
strong, 378
Gelfand–Tsetlin basis, 377
group, 3
algebra, 19
center, 23
commutative, 4
conjugacy class, 23
Coxeter, 151
generator, 151
isomorphism, 4
morphism, 4
non-commutative, 6
presentation, 151
relation, 151
subgroup, 4
symmetric, 5
Weyl, 637

Hall scalar product, 74
harmonic function, 362, 505
conjugate, 363
Poisson–Stieltjes representation, 362
Hecke algebra, 158
generic, 158
index representation, 195
parabolic subalgebra, 194
specialized, 158
hook length formula, 140
Hopf algebra, 69
antipode, 70
involutive, 71
horizontal strip, 101

ideal, 30
nil, 307
nilpotent, 307

Jacobi identity, 631
Jacobi–Trudi determinants, 67
James submodule theorem, 134

Jucys–Murphy element, 379
 generic, 384
 spectrum, 391

Kerov polynomial, 451
Kerov–Vershik ring theorem, 506
Killing form, 634
Knuth equivalence, 116
Knuth transformation, 116
Kostka number, 103

Lagrange inversion, 352
lattice, 420
 Möbius function, 421
Lie algebra, 631
 abelian, 633
 bracket, 631
 coroot, 643
 derivation, 632
 fundamental weight, 643
 ideal, 632
 inner automorphism, 635
 nilpotent, 633
 radical, 633
 rank, 635
 reductive, 634
 representation, 633
 root, 636
  positive, 637
  simple, 637
 semisimple, 633
 simple, 634
 solvable, 633
 universal enveloping algebra, 642
 weight, 636
  dominant, 643
 weight lattice, 643

map, 465
 face, 466
  word, 466
 labeled, 465
Markov–Krein correspondence, 365
Matsumoto's theorem, 151
measurable, 402
modular reduction, 176
module, 29
 commutant, 41
 composition series, 161
 cyclic, 232
 indecomposable, 163
 left, 29
 length, 167
 modular reduction, 176
 outer tensor product, 79
 parabolic, 195
 principal indecomposable, 165
 projective, 164
 projective cover, 167
 right, 29
 semisimple, 31
 simple, 31
 top, 170
 unipotent, 271
monoid, 153
Montel topology, 360
Murnaghan–Nakayama formula, 107

Newton identities, 63
non-commutative Fourier transform, 18
non-commutative probability space, 406
 conditional expectation, 408
 free amalgamated product, 414
 random variable, 406
non-commutative symmetric function, 290
 elementary, 290
 homogeneous, 290
 ribbon Schur, 294

observable, 332
 degree, 350
 Frobenius moment, 343
 geometric, 343
 Jucys–Murphy conditional moment, 433
 renormalized character value, 347

Stanley polynomial, 459
weight, 350
Olshanski semigroup, 519
orbit, 7
orthogonality relations, 46

partial pairing, 334
partition, 49
conjugate, 59
dominance order, 60
Frobenius coordinates, 340
interlaced coordinates, 340
length, 50
observable, 332
size, 49
skew, 77, 100
horizontal strip, 101
ribbon, 101
vertical strip, 101
Young diagram, 59
permutation, 5
admissible, 389
admissible support, 328
backstep, 153
cycle, 6
minimal factorization, 477
cycle type, 50
descent, 54
inversion, 52
involution, 130
length, 152
minimal decomposition, 152
minimal in its conjugacy class, 273
partial, 328
recoil, 153
reduced decomposition, 152
shuffle product, 312
signature, 19
transposition, 7
elementary, 51
permutohedron, 316
Perron–Stieltjes formula, 365
Pieri rules, 103
plactic monoid, 115
Plancherel measure, 25, 597
Pochhammer symbol, 200
Poincaré half-plane, 359
Poincaré series, 213
polish space, 364
polynomial, 54
antisymmetric, 64
harmonic, 210
stable, 452
symmetric, 55
polytabloid, 134
posets, 421
probability measure, 402
arcsine, 418
Cauchy, 369
Cauchy transform, 355
free convolution, 414
Gaussian, 577
Marcenko–Pastur, 613
Markov transform, 371
R-transform, 415
semi-circle, 370, 415
tight sequence, 364, 603
Wigner, 370
probability space, 402
non-commutative, 406
Prohorov theorem, 365
proper convergence, 360

quantum group, 219
reduced, 225, 247
specialized, 220
quasi-symmetric function, 299
fundamental, 301
monomial, 300

ramification index, 187
random variable, 403
convergence in law, 578
convergence in probability, 571
free, 412
free cumulant, 415
joint cumulant, 579
non-commutative, 406
Rayleigh function, 365

Rayleigh measure, 371
real form, 92
representation, 8
    admissible, 515
    character, 13
    coefficient space, 90
    depth, 528
    direct sum, 9
    Grothendieck group, 12
    index, 195
    induced, 17
    irreducible, 9, 528
    morphism, 9
    outer tensor product, 79
    permutation, 8
    polynomial, 90
    rational, 97
    regular, 9
    restricted, 16
    signature, 19, 52
    spherical, 538
    subrepresentation, 9
    tame, 515
    tensor product, 16
    trivial, 8
    unitary, 513
    weight, 636
ribbon, 101
    decomposition, 297
    generalized, 282
    height, 106, 282
    width, 282
Riesz representation theorem, 411
Riesz–Herglotz theorem, 363
riffle shuffle, 564
    generalized, 566
ring, 28
    Dedekind, 186
    division, 32
    integrally closed, 175
    local, 174
    valuation, 174
Robinson–Schensted–Knuth
        algorithm, 110
    insertion tableau, 111
    recording tableau, 111
    symmetry theorem, 126
root system, 636

Schützenberger equivalence, 117
Schützenberger transformation, 116
Schensted insertion, 110
    bumping route, 113
Schur algebra, 88
Schur element, 190
Schur polynomial, 67
Schur–Weyl measure, 597
set partition, 418
    admissible, 436
    fat, 434
    non-crossing, 419
        Kreweras complement, 421
shifting, 127
simplicial complex, 490
    contractible, 492
Specht module, 134
spectral measure, 407, 533
spectral vector, 387
spectrum, 391, 409
standardization, 322
state, 405
    extremal, 541
Stirling number, 496
Stone–Weierstrass theorem, 411
symmetric function, 56
    elementary, 55
    free, 559
    free Schur, 559
    graded basis, 63
    Hall–Littlewood, 278
    homogeneous, 56
    monomial, 59
    non-commutative, 290
    power sum, 56
    Schur, 67
    skew Schur, 76
    specialization, 506
symmetric group, 5
    Grothendieck ring of
        representations, 79

Hecke algebra, 158
symmetrization, 64

tableau, 100
column subgroup, 132
descent, 558
generalized, 116
reading word, 116
ribbon, 107, 282
row equivalence, 133
row subgroup, 132
semistandard, 100
skew, 100
standard, 105
standardization, 557
stem, 376
weight, 102
tabloid, 133
Thoma simplex, 510
trace, 14, 183, 406
coherent family, 502
symmetrizing, 188
two-line array, 109, 257

unitarian trick, 92
universal enveloping algebra, 642

Vandermonde determinant, 64
Verma module, 645
vertical strip, 101
von Neumann algebra, 405
non-negative element, 405
state, 405
trace, 406

Wassermann's formula, 350
weak operator topology, 403
weight lattice, 643
weight space, 636
Weyl chamber, 643
Weyl character formula, 647
Weyl group, 637
word, 110
Greene invariant, 122
increasing subsequence, 122
packed, 322, 455
standardization, 322

Young diagram, 59
cell, 100
arm length, 141
content, 203, 387
hook length, 140
leg length, 141
continuous, 356
free cumulant, 418
Frobenius moment, 343
generating function, 354
interlaced moment, 345
shifted coordinates, 342
Stanley coordinates, 457
reduced, 459
system of particles, 342
transition measure, 354
Young distribution, 529
support, 535
Young graph, 84, 505

◎编辑手记

当人类迈向21世纪之际，数学在当今社会的作用日益受到重视.数学处于所有科学的核心，因为所有的科学分支都需要数学来阐明其真理.在数学之中，群的对称性又是被视为灵魂的部分，表示论是用线性代数来研究群对称的基本工具.群对称遍及数学的各个分支：几何、分析和代数.表示论在当今数学和理论物理的发展上起了十分重要的作用.21世纪著名的数学家I. Gerfand曾经说过："所有的数学都可视作某种表示理论."举个简单的例子：

对称群S_n的互不等价的不可约表示的个数等于S_n的共轭类的个数，也就是n的不同划分的个数.

S_n中的共轭类可以用以下序列表出

$$\mu=(\mu_1,\mu_2,\cdots,\mu_n)$$

满足$\sum_{k=1}^{n}k\mu_k=n$.也就是说这个共轭类中的元素是不相交的μ_1个1－循环，μ_2个2－循环，直至μ_n个n－循环的乘积.当然这里的μ_k可以为0.这个共轭类我们记之为C_μ.

对于λ的一个划分

$$\lambda=(\lambda_1,\lambda_2,\cdots,\lambda_k)$$

满足关系$\lambda_1\geqslant\cdots\geqslant\lambda_k$，我们定义

$$l_1=\lambda_1+k-1,l_2=\lambda_2+k-2,\cdots,l_k=\lambda_k$$

这样就得到了一个严格单调递减序列，我们以$x_1,\cdots,x_k$为变量，定义以下多元多项式

$$P_j(x)=x_1^j+x_2^j+\cdots+x_k^j \quad (j=1,\cdots,n)$$

$$\Delta(x)=\prod_{i<j}(x_i-x_j)$$

如果 $f(x)=f(x_1,\cdots,x_k)$ 为一个多元多项式，我们记 $[f(x)]_{(l_1,\cdots,l_k)}$ 为 $f(x)$ 中 $x_1^{l_1}\cdots x_k^{l_k}$ 这一项的系数.

我们可以得到著名的弗罗比尼乌斯(Frobenius,1849—1917)定理.

定理 S_n 的不可约表示 V_λ 的特征 x_λ 在共轭类 C_μ 上的值可以由下列公式算出

$$x_\lambda(C_\mu)=\left[\Delta(x)\prod_{j=1}^{n}p_j(x)^{\mu_j}\right](l_1,\cdots,l_k)$$

这个定理有一个非常有计算价值的推论.

推论 不可约表示 V_λ 的维数为

$$\dim V_\lambda=\frac{n!}{l_1!\ \cdots l_k!}\prod_{i<j}(l_i-l_j)$$

香港科技大学的黄劲松教授1999年曾以在南开大学为全国数学暑期学校开设的“李群的表示论”的课程的英文讲稿为基础写了一本小册子，并在湖南教育出版社出版.

世界著名数学家E.阿廷(E. Artin,1898—1962)曾说过：

> 研究给定数学结构的对称性的工作，总是能得到大量的结果.

本书就是一部研究对称性的英文版数学专著，中文书名可译为《对称群的表示论》.

本书的作者为皮埃尔—洛伊克·梅利奥特(Pierre-Loïc Méliot)，法国数学家，巴黎第十一大学教授.

要想理解本书，我们首先要了解现代数学中可能是最常见、也最重要的名词——群.

据辽宁师范大学杜瑞芝教授所编的《数学史辞典新编》(山东教育出版社，2017年)中所撰：

> **群(group)** 只具有一种运算的抽象代数结构，是数学中的重要概念之一. 研究群的性质的理论被称为群论，它是抽象代数学的重要组成部分.
>
> 历史上，群的概念是抽象代数系统最初的例子之一. 最先产生的是由 n 个文字的一些置换所构成的置换群. 在18世纪末至19世纪初，代数学的中心问题是五次以上的一元多项式方程是否可以用根式求解. 拉格朗日(Lagrange,1736—1813)、鲁菲尼(Ruffini,1765—1822)、阿贝尔(Abel,1802—1829)、伽罗瓦(Galois,1811—1832)和若尔当(Jordan,1838—1922)先后研究并解决了这个问题，他们引入和发展的置换群的概念是最终产生抽象群论的第一个主要来源. 范德蒙德(Vandermonde,1735—1796)也是最早注意到研究代数方程根的对称函数对解决四次以上方程根式的求解问题的重要性的学者之一，因而在数学史文献中，常常与拉格朗日一起被列为群论的先行者.
>
> 在数论的研究中，早在1761年欧拉(Euler,1707—1783)就明确地使用了同余式和它们分成的剩余类. 这相当于在群论中把一个群分解成若干子集的陪

集.拉格朗日和高斯(Gauss,1777—1855)曾研究过具有同一判别式的二次型类,对于两个型的"复合"乘法,构成了一个交换群.戴德金(Dedekind,1831—1916)在1858年,克罗内克(Kronecker,1823—1891)在1870年给出了有限群和有限交换群的抽象定义,凯莱(Cayley,1821—1895)在1878年发表了关于有限抽象群的研究工作.这些都是抽象群论的第二个主要来源.

群的思想也以独立的方式产生于几何学中.到了19世纪中叶,除了欧几里得几何学,又出现了许多其他的"几何学".而找出它们之间的关系成为一个迫切的问题.以凯莱为代表的英国学派给出了关于几何学的一个系统的分类.在若尔当系统地介绍并发展伽罗瓦理论的专著的影响下,C.F.克莱因(C.F. Klein,1849—1925)于1872年在他著名的"埃朗根纲领"中指出,无限交换群可用来对几何进行分类,并给出这些群连续的含义.C.F.克莱因和庞加莱(Poincaré,1854—1912)在研究自守函数的过程中曾用到其他类型的无限群.1870年前后,S.李(S. Lie,1842—1899)开始研究连续变换群的概念,并用它们阐明微分方程的解,将微分方程进行分类.1874年,S.李建立了连续群的一般理论,这些由研究几何学所派生出来的工作是抽象群论的第三个主要来源.

1854年凯莱就接触过抽象群的概念,1878年他又发表论文研究有限抽象群,指出一个群可以看作一个普遍的概念;德国数学家内托(Netto,1848—1919)在他的专著《置换理论及其对代数的应用》中阐明了群的概念的抽象性,还探讨了同构和同态的概念;C.F.克莱因的学生迪克把前述三个主要来源的工作都纳入了抽象群的概念之中,在1882～1883年的论文中建立了(抽象)群的定义,并明确地运用了群的生成子的概念.在抽象群中,集合元素本身的具体内容无关紧要,关键是联系这些元素的运算法则.到了19世纪80年代,数学家们终于成功地概括出抽象群论的公理体系,大约在1890年得到确认.20世纪初,美国数学家亨廷顿(Huntington,1874—1952)、莫尔(Moore,1862—1932)和迪克森(Dickson,1874—1954)等人都给出过抽象群的种种公理系统,这些公理系统和现代的一致.代数学由于群的概念的引进和发展而获得了新生,它不再仅仅是研究代数方程,而更多的是研究各种抽象的"对象"之间的运算关系,这种抽象的"对象"层出不穷,为20世纪代数结构观念的产生奠定了基础.

抽象群的公理体系建立之后,数学家们建立了抽象群的一些定理,并且发展了抽象群的许多研究方向,如确定具有给定阶的群的全体,确定复合群或可解群与单群,特别是建立有限群的矩阵表示论,这些工作表明有限群论已经形成.在这方面做出重要贡献的有德国数学家弗罗比尼乌斯、英国数学家伯恩塞德(Burnside,1852—1927)和德国数学家舒尔(Schur,1875—1941)等.苏联数学家O. Ю.施密特(O. Ю. Shmidt,1891—1956)在1916年还发表了无限群论的专著《抽象群论》.抽象群论的发展最终导致20世纪30年代抽象代数学的兴起.

到20世纪80年代,群的概念已经普遍地被认为是数学及其许多应用中最基本的概念之一.它不但渗透到诸如几何学、代数拓扑学、函数论、泛函分析及其他许多数学外支中而起着重要的作用，还形成了一些新学科，如拓扑群、李群、代数群、算术群等，它们还具有与群结构相联系的其他结构，如拓扑、解析流形、代数簇等，并在结晶学、理论物理、量子化学以至编码学、自动机理论等方面都有着重要的应用.作为推广“群”的产物，半群、幺半群理论及其近年来对计算机科学和算子理论的应用，也有很大的发展.群论的计算机方法和程序的研究，已经迅速地发展起来.

一般我们稍微专业一点的读者都希望从胡作玄研究员的《近代数学史》(山东教育出版社，2006年)中做更进一步的了解.胡先生指出：

群是数学中的第一个抽象概念，虽然其观念可追溯到很久以前，但直到19世纪末，数学家才对它有了比较清楚的认识.从那时起，几乎所有大数学家无不强调其重要性，把它视为数学的中心概念以及统一数学的基础.S.李说：“群和不变式的概念，每日每时在数学中都占有优越地位，而且越来越趋于支配整个这门科学领域.”外尔(Weyl，1885—1955)说：“没有群就不可能理解现代数学.”

群论是当代数学最重要的分支之一，也是发展得比较完整的一门学科，已成为现代抽象代数学的模式.在伽罗瓦之后的100年间，群论的研究大致有彼此多少有些重叠的五个方面：

① 抽象群的定义及公理化；

② 具体群的构造、刻画及分类；

③ 抽象群的具体表示；

④ 有限群的列举；

⑤ 一般的结构定理.

1　抽象群的定义及公理化

抽象群与具体群的不同之处有三点：一是摆脱掉具体的作为变换对象的集合，群的元素不一定是具体的变换、置换、运动、作用、运算或数，而是抽象的符号，元素的数目不加限制；二是用公理来刻画；三是各种同构的具体群只不过是同一抽象群的不同表示，对此我们不加区分.在19世纪，无用的推广和抽象不受欢迎.因此，抽象群概念的产生过程是极为缓慢的.通常认为，最早的抽象群是凯莱在1854年定义的，实际上，他只是把群看成抽象符号的集合.符号间有二元运算(或复合运算、结合运算)，集合对此二元运算是封闭的.至于结合性、存在幺元等只是事后假定，而不在定义中出现.他认识到运算是结合的，但不一定可交换.克罗内克在1870年的论文中明确表述了交换律及结合律.迪克(Walther Franz Anton von Dyck，1856—1934)在1882年引用凯莱的定义，他认识到同构的群应被当作同一个群，他定义的群不限于有限群，包括有限生成的“离散群”.1883年他还明确要求存在逆元素.同年，H.韦伯(H.Weber，1842—1913)给出了第一个抽象群的定义：任何种类的元素$\theta_1,\theta_2,\cdots,\theta_n$的系统$G$被称为$h$阶群，如果满足下列条件：

① 由系统中的两个元素，可以通过组合或乘法导出同系统中的一个新元素，用符号表示为

$$\theta_r\theta_s=\theta_t$$

② 结合律，即

$$(\theta_r\theta_s)\theta_t=\theta_r(\theta_s\theta_t)=\theta_r\theta_s\theta_t$$

③ 左消去律和右消去律，即由 $\theta\theta_r=\theta\theta_s$ 或 $\theta_r\theta=\theta_s\theta$ 可得出

$$\theta_r=\theta_s$$

由此可推出系统存在幺元，且每个元素存在逆元，这个定义的局限性在于他假设抽象群为有限群．1893 年，他去掉这个限制并写入他的极有影响的著作《代数学教程（第二卷）》（1896 年）中．1889 年，赫尔德（Hölder，1859—1937）也给出了类似的定义，正式的公理定义是 20 世纪初在希尔伯特（Hilbert，1862—1943）《几何学基础》的影响下掀起的公理化热潮中出现的，主要是 1902～1905 年莫尔、亨廷顿及迪克森等人的论文，其后仍不断有许多研究．

从历史上看，从具体群到抽象群的研究经历了很长时期，到 20 世纪初，对有限群和无限群的研究还是分开的．第一部用抽象群统一论述群论的著作是法国数学家德·塞基埃（J. A. de Seguier，1862—1937）在 1904 年出版的《抽象群论原理》，其中谈到“抽象群的观念，不依赖于其元素是什么，而只考虑群本身，来自代数、分析及几何中不同的特殊的群．许多过去属于不同领域的研究组成一个更一般的理论”．这本书预示了结构数学的理论结构，即首先置于 G. 康托尔（G. Cantor，1845—1918）集合论的基础上，然后通过四个公理定义，并通过反例证明其独立性．这本书的重点在有限群上．从无限群开始的抽象群论著作是苏联数学家施密特在 1916 年出版的．

2　群表示论

19 世纪 90 年代，群论最主要的成就是群表示论的出现，它是由德国数学家弗罗比尼乌斯奠定的，后来由他的学生舒尔所发展，成为研究群论所不可缺少的工具．所谓群表示，即把群具体实现为某种结构的自同构群．例如，对于域 F 上的有限维线性空间的线性变换群，通常把群的元素与 F 上的 $n\times n$ 可逆矩阵相对应．英国数学家伯恩塞德的经典著作《有限阶群理论（第 2 版）》（1911 年）中已有这方面的综述及应用．

群表示论来源于群的特征标理论，而阿贝尔群的特征标理论最早来自高斯关于二元二次型的种和特征标理论，他对每个二次型 F 定义一个特征标，并对两个二次型同种给出定义，即如果其行列式及特征标对应相等．高斯的理论由迪利克雷（Dirichlet，1805—1859）在 1839 年表示为有限阿贝尔群及其特征标理论．戴德金的理论在迪利克雷的《数论讲义（第 3 版）》中被进一步推广到代数数域的理想类群上．他在 1886 年引进了群行列式，并向弗罗比尼乌斯提出问题，这直接导致群表示论在 1895 年的诞生．

在弗罗比尼乌斯发表群表示论两年之后，1897 年他又举出一些有趣的例子，但看不出有什么用处，当时人们对群表示论普遍持怀疑的态度，这时，伯恩塞德在他的《有限阶群理论（第 1 版）》的序言中写道：“…… 而我们现在所知道

的、纯粹理论中的许多结果，大多数可以马上从置换群的性质中得出，很难找出一个结果，它更直接从线性变换群的考虑中得到.”这可以说明他在第1版中为什么没有论述群表示论，但他的态度在几年之后就有了明显的转变. 在该书的第2版(1911年)中，他承认他完全忽略群表示论“不再合适”. 他写道：“现在，似乎这么说更正确：为了进一步发展抽象理论，必须更着重把群看成线性代换群的表示.”14年间这种观点的大转变来源于他所说的“群论中最一般的问题”. 什么是一门学科最基本的问题以及对这个问题的看法的转变，是一门学科历史发展中最关键的主题. 在1897年，群论的主要问题是列举问题：对于给定阶 n 的群，列举其所有可能不同的类型. 而这类问题慢慢变成另外一种提法，逐步变成结构问题：什么样的群具有可解性？由于可解群是大家感兴趣的群，是否从阶数就能看出群的可解性结构当然是重要问题. 这个历史上的主题移位(shift)使群表示论大有用武之地，而这个问题的历史发展进程是这样的：

①1872年西罗(Sylow，1832—1918)证明，p^α 阶群一定是可解群，这里 p 为素数，α 为正整数.

②1894年，弗罗比尼乌斯证明，如果 G 的阶 $|G|$ 不被任何素数的平方除尽(即无平方因子)，那么 G 是可解群.

③ 下一个目标当然是 $p^\alpha q^\beta$，其中 p,q 为素数，α,β 为正整数，弗罗比尼乌斯和伯恩塞德等人陆续用各种方法，有的用群表示论，有的不用，得出在特殊条件下，群是可解群.

④1904年，伯恩塞德在确确实实使用群表示论的情况下证明了：每个阶为 $p^\alpha q^\beta$ 的群均为可解群. 这是一个了不起的定理，而且可以说是用群表示论证明的第一个大定理. 这种情况当然出乎人们的意料，甚至也出乎弗罗比尼乌斯本人的意料. 因为就在1902年他证明定理的一个特殊情形时，他说“定理的证明完全用的是纯群论的考察，而没有借助于任何代换群理论，即没有用到任何群的表示”. 顺便说一下，直到20世纪60年代，J. G. 汤普森(J. G. Thompson)才设计了一个纯群论的证明，1970年D. 哥德什密特(D. Goldschnidt)首先发表完全的证明. 不过这个证明极难，伯恩塞德用群表示论得到的证明仍是最好的. 由于这个定理的证明，人们对群表示论的怀疑一扫而空.

1895～1905年弗罗比尼乌斯等人建立了有限群表示论的基础之后，舒尔在1905年做了简化的论述，特别是证明了著名的舒尔引理. 1904～1911年舒尔还引进了有限群的射影表示. 而首先发展一般群表示论的是A. E. 诺特(A. E. Noether，1882—1935)，她在1927～1928年冬季学期的讲课中，首先把超复量及表示论这两大领域联系在一起，把表示论推广到一般基域之上，形成了群与代数完整的抽象理论. 其后群表示论不仅在理论上得到了很大发展(建立模表示论及李群表示论)，而且在物理学上也取得了重要应用.

3 有限单群的分类

20世纪有限群论的中心问题是有限单群的分类. 20世纪初，人们所知道的有限单群，除素数阶循环群及交错群之外，还有许多特殊的线性射影变换群 $\mathrm{PSL}(n,q)$，它们通过行列式为1的 $n\times n$ 矩阵群(元素取在有限域 $\mathrm{GL}(q)$ 中)的商群被构造出来.另外,对于正交矩阵、辛矩阵、酉矩阵也可以造出一批单群来.这些“典型群”从若尔当时候起就已知道,后来美国数学家迪克森、荷兰数学家范·德·瓦尔登(van der Waerden,1903—1996)、法国数学家迪厄多内(Dieudonné,1906—1992)分别对此进行了具体研究.真正的重大突破是1955年C.谢瓦莱(C. Chevalley,1909—1984)在日本的《东京数学杂志》上发表的论文《论某些单群》,他把这些单群通过李代数方法系统地构造出来,被称为谢瓦莱群.这些谢瓦莱群经过美国数学家R.斯坦伯格(R. Steinberg)、韩国数学家李林学(Rim Leon Dock S.)、比利时数学家J.蒂茨(J. Tits)、日本数学家铃木通夫(Suzuki Michio)等人加以扩充,最终得出了全部李型单群的16个系列.除上述18个系列的有限单群之外,还有一些“散在单群”,1861 ~ 1873年马蒂厄(Mathieu,1835—1890)得出的5个马蒂厄群不属于上面任何一个系列,但直到1965年才找到第6个,其后10年间又发现了20个,共26个散在单群.到这时候的问题是:是否所有单群均已被找到,也就是有限单群的分类已经完成了吗?证明这点经历了漫长的时期.最早的突破是一系列群论性质及表示论的成果,其中包括1955年R.布饶尔(R. Brauer,1901—1977)的工作.第二个突破是1963年美国数学家W.费特(W. Feit)和汤普森证明,除循环群之外,奇阶群都是可解群,这篇长达250页的论文包含了极其丰富的信息.20世纪70年代在群的结构研究上有了新的突破,最终导致1981年有限单群分类的彻底完成,不过全文需要15 000页以上,这是各国上百位群论专家通力合作的结果.到了20世纪90年代,有关证明已陆续发表,篇幅也大大地缩减了.

本书的内容正如本书作者在前言中所述:

本书的目的是对对称群的表示论进行现代化的介绍.现在有大量关于有限群的一般表示论的文献,例如经典的J.-P.塞尔(J.-P. Serre)([Ser77])的《有限群的线性表示》;在这些文献中,有几本书专门针对对称群的情况,例如B.萨根(B. Sagan)的《对称群:表示,组合算法和对称函数》(参见[Sag01]).本书的观点和着重点如下:我们将论述对称群的大部分计算都是可以执行的,或者至少可以通过使用一些适当的函数代数来简化.众所周知,自弗罗比尼乌斯和舒尔的著作以来,对称函数的代数可对对称群的大部分特征标理论进行编码.在本书中,我们应把对称函数的代数作为对称群表示论的起始点,然后再通过引入其他有趣的代数来进行深入研究,比如:

① 分解可观测的代数,原名叫“杨(Young)氏图的多项式函数”,其建造归功于克罗夫(Kerov)和奥尔尚斯基(Olshanski).

② 非交换对称函数的H.霍普夫(H. Hopf,1894—1971)代数,拟对称函数和自由拟对称函数,其包含并推广了对称函数的代数.

对称群的表示论的代数方法可能与具有组合性质的更传统的方法相对立,这赋予了著名的杨氏表一个重大的角色.函数代数方法有几个优势:

① 首先如果试图用与之相关的有限维代数(所谓的划分代数,或赫克

(Hecke,1887—1947)代数来代替对称群,那么还可以用对称函数的代数来处理这些代数的特征标理论,并且在这种情况下,大多数与对称群相关的结果都有直接的类似物.在本书中,我们将处理赫克代数的情况,它是对称群理论中这一类扩展的一个很好的例子(比如,切凯里尼—西尔伯斯坦(Ceccherini-Silberstein)、斯卡拉博蒂(Scarabotti)和托利(Tolli)在最近的一本书中讨论了划分代数的情况,参见[CSST10]).

② 另外,代数方法导致了一个对称函数不可约特征的新公式的出现,这要归功于斯坦利(Stanley)和费雷(Féray).这个公式背后的组合学与几个有趣的主题有关,比如自由概率论,或黎曼(Riemann,1826—1866)曲面理论及其映射.

③ 最后,这种代数方法也可以处理渐近表示论,也就是说对称群$\mathfrak{S}(n)$的表示,当n趋于无穷大时.在这种情况下,一个自然出现的问题是:具有大的n的$\mathfrak{S}(n)$表示的典型特征是什么,特别是在不可约分支中这种大的表示的分解是什么?由于$\mathfrak{S}(n)$的不可约表示可以用大小为n的整数划分来标记,这个问题导致了对某些随机划分模型的研究,特别是对被称为普朗谢雷尔(Plancherel,1885—1947)测度的研究.这种情况下,可观测的划分代数将被证明是一个非常强大的工具.

此外,我们的方法使我们能够在一本书中介绍组合霍普夫代数理论,这是当今代数组合学中一个非常活跃的研究领域.

现在,让我们更准确地详细介绍本书的内容.本书共分为四个部分:

第一部分:对称群与对称函数.

本书的第一部分致力于介绍对称群表示的经典理论,这些理论主要归功于弗罗比尼乌斯、舒尔和杨.在第1章中,我们解释了有限群的表示论和有限维半单代数,这些构成了阅读本书的大部分先决条件.

我们将在每一章都尝试做的一件事是获得一个大的"黑箱定理",它总结了大部分结果,并让我们立即重新获得了理论的其余部分.对于有限群的线性表示的一般理论,一个总结陈述的事实是有限群的非交换傅里叶(Fourier,1768—1830)变换是代数、希尔伯特空间和双模的同构(定理1.14).类似的结果适用于有限维半单代数,代数和模的语言比群和表示的语言更灵活.

在第2章中,我们引入了对称函数Sym的霍普夫代数,我们证明了与对称群的不可约表示相对应的舒尔函数,而且,Sym与由这些表示形成的格罗腾迪克(Grothendieck)环同构(定理2.31).这个定理归功于弗罗比尼乌斯和舒尔,可以作为表示的组合学的起始点,这部分内容会在第3章中提出,并且依赖杨氏表,即整数划分的杨氏图的编号.这个深层组合理论的其他两个厦块是舒尔—外尔对偶(2.5节),它将$\mathfrak{S}(n)$的表示与一般线性群GL(N)的表示相关联;还有鲁宾逊—申斯特—克努特(Robinson-Schensted-Knuth)算法(参见第3.2节),它将杨氏表与字或排列联系起来.这两个工具将在整本书中被广泛使用.

第二部分:赫克代数及其表示.

在第二部分中,我们解释了如何将对称群理论扩展到其他相关的组合代数中,换句话说,也就是岩堀—赫克(Iwahori-Hecke)代数.这些代数是群代数 $\mathbf{C}\mathfrak{S}(n)$ 的连续形变 $\mathscr{h}_z(n)$,参数 z 可以取 $\mathbf{C}$ 中的任意值;当 $z=1$ 的时候,我们可以重新得到 $\mathbf{C}\mathfrak{S}(n)$.在第 4 章中,我们展示了对 z 的大部分任意值,$\mathscr{h}_z(n)$ 与 $\mathbf{C}\mathfrak{S}(n)$ 是同构的,且它们有相同的表示论:它的不可约模 S_z^λ 再次用大小为 n 的整数划分来加标号,并且它们与 $\mathfrak{S}(n)$ 的不可约表示 S^λ 具有相同的维度(定理 4.67).本章可以被认为是对模表示论的介绍,其焦点集中在具体的例子上.在第 5 章中,我们使用舒尔—外尔对偶性的扩展来计算一般情况下赫克代数的特征,其中对称群被赫克代数取代,线性群被量子群取代.我们得到了一个公式,它概括了弗罗比尼乌斯—舒尔公式,并涉及了霍尔—利特尔伍德(Hall-Littlewood)对称函数(见定理 5.49).在第 6 章中,我们考虑了 $z=0$ 的情况,这种情况不是一般情况,也不会产生半单代数.在这种设置中,我们还可以用组合的霍普夫代数去描述 $\mathscr{h}_0(n)$ 的表示(见定理 6.18):非交换对称函数 NCSym 的代数和拟对称函数 QSym 的代数是对偶的.因此,将对称群的表示论扩展到赫克代数,可以很自然地将对称函数理论扩展到更一般的函数的情况,这些内容在本书的后面会出现(第 10 章和第 12 章).

第三部分:划分的可观测.

本书的第三部分致力于对称群特征标的对偶组合学的阐释.在本书的第一部分中,对称群的特征标被介绍为函数 $\mathrm{ch}^\lambda:\mathfrak{S}(n)\to\mathbf{C}$ 或 $\mathscr{Y}(n)\to\mathbf{C}$,由大小为 n 的整数划分 λ 标记,并且可以在弗罗比尼乌斯—舒尔公式的帮助下进行计算

$$\mathrm{ch}^\lambda(\mu)=\langle s_\lambda \mid p_\mu\rangle$$

其中 $\langle\cdot|\cdot\rangle$ 是对称函数 Sym 的代数的霍尔标量积.然而,还可以将量 $\mathrm{ch}^\lambda(\mu)$ 视为由共轭类 $\mu\in\mathscr{Y}(n)$ 标记的 λ 的函数.这种观点导致人们考虑对称群的不可约表示的函数,并引入由这些函数组成的代数 $\mathscr{O}$,我们称它为划分的可观测代数.我们的第 7 章展示了这个代数和它的许多基,并且这个代数解释了对称群的特征标 ch^λ 是如何与整数划分 λ 的杨氏图的几何形状产生关系的(参见定理 7.13 和 7.25).在第 8 章和第 9 章中,我们引入了其他划分的可观测,与所谓的朱西—墨菲(Jucys-Murphy)元素或自由概率论有关.我们特别展示了 $\mathscr{O}$ 的一个重要的代数基 $(R_k)_{k\geqslant 2}$,其元素被叫作自由累积量,其组合与集合分区上的构造和曲面上的映射有关.第 10 章探索了自由累积量的基 $(R_k)_{k\geqslant 2}$ 与 $\mathscr{O}$ 中可重正化特征值的基 $(\Sigma_k)_{k\geqslant 1}$ 的相互作用.该研究依靠对称群的特征标的一个新公式(定理 10.11)

$$\mathrm{ch}^\lambda(\mu)=\frac{\dim S^\lambda}{|\lambda|(|\lambda|-1)\cdots(|\lambda|-|\mu|+1)}\sum_{\rho_\mu=\sigma\tau}\varepsilon(\tau)N^{\sigma,\tau}(\lambda)$$

其中总和在循环类型为 μ 的扰动 ρ_μ 的因式分解上进行,并且其中的量 $N^{\sigma,\tau}(\lambda)$ 计算杨氏图 λ 的单元的某些编号.而且,如果代替 Sym,我们可以用组合代数 $\mathscr{O}$ 作为对称群表示论的起始点,然后得到另一个不可约特征标的完全不同的公式,斯坦利—费雷公式位于更大的代数 $\mathscr{Q}\supset\mathscr{O}$ 中.仔细分析该公式会导致符号 R_k 和符号 Σ_k 之间的基本公式发生显式变化;见定理 10.20,解释了如何计算克罗夫多项式的系数.

第四部分:随机杨氏图的模型.

在本书的最后部分,为了描述大的对称群表示的特性,我们使用了前面章节得到的结果.在第11章中,从无限对称群 $\mathfrak{S}(\infty)$ 的极值特征标的分类开始(定理11.31).它们扮演着关于 $\mathfrak{S}(\infty)$ 的类似于有限对称群 $\mathfrak{S}(n)$ 的不可约特征标的角色,并且它们允许我们考虑表示的凝聚族 $(\tau_n)_{n\in\mathbf{N}}$ 或这些有限群的更普遍的迹.该分类涉及被称为托马(Thoma)单纯形的无限维凸紧空间.对这个单纯形中的任意参数 $t\in T$,我们可以考虑对称群 $\mathfrak{S}(n)$ 的迹 $\tau_{t,n}$,在大小为 n 的整数划分的集合 $\mathcal{Y}(n)$ 上,它的不可约特征标中的分解产生概率测度 $\mathbf{P}_{t,n}$.而且,$\mathfrak{S}(\infty)$ 的表示论导致人们去研究划分的随机模型,该研究在第12章和第13章中进行讲解.

在第12章中,我们证明了每个测度族 $(\mathbf{P}_{t,n})_{n\in\mathbf{N}}$(所谓的中心测度)满足大数定律测度(定理12.19)和一个中心极限定理(定理12.30).为了达到这个目的,我们引入了一个新的组合霍普夫代数FQSym,其扩展了NCSym和QSym;和一种随机变量的联合累积量的方法,它与划分的可观测理论很好地混合在一起.在第13章中,我们研究了普朗谢雷尔测度和舒尔—外尔测度的特殊例子,它们与其他测度相比拥有退化渐进式,它们还允许我们去解决一致随机排列或均匀一致随机字中最长递增子序列的问题(定理13.10).

本书的目标读者主要是研究生和研究人员.我们试图使我们的表述尽可能独立,但阅读仍然不可避免地存在某些先决条件.因此,本书的读者应非常熟悉代数(代数结构和相关构造)和组合学(计数理论、双射)的基础知识;在本书最后的部分,我们也应使用概率论的参数.我们认为一个理所当然的能帮助读者理解某些结果的先决条件是经典李代数的表示,因此,附录A对这个理论进行了一个简短的描述(没有证明).诚实地说,我们猜想这些先决条件里面也可能存在一些不一致的地方:比如,本书开始的时候我们回忆了什么是群,但后来我们自由地使用了范畴等价的语言.我们希望本书最后给出的长长的参考书目能够帮助读者顺利地攀过可能遇到的困难高峰.

对于参考书目,每一章后面都跟着叫作"注解和参考文献"的一节,在其中我们解释了为了写这本书所使用的参考资料的来源.所有功劳都归功于这些特殊部分中引用材料的作者们,我们真心地希望可以不忘记任何人,并将每个成果与正确的数学家相对应(这项任务有时可能很难完成).尽管如此,本书中还是有一些(据我们所知)或者是新的,或者是直到现在还没有发表的,或者是在文献中很难找到证据的带有证明的结果,都在注释和参考文献中进行了解释.我们还用这些特殊的部分来详细说明了一些我们没有勇气处理,但我们仍然想让它们出现在书中的结果.

正如作者在前言中所指出的那样,关于群表示论的第一本经典著作应该就是法国著名数学家J.-P.塞尔著的《有限群的线性表示》.这部名著的第一部分最初是为量子化学工作者而写的,它阐述了由弗罗比尼乌斯所建立的有关线性表示与特征标之间的对应关系.这是在数学中以及在量子化学或物理中经常要用到的基本结果.1977年斯普林格出版了英译本,1984年科学出版社出版了中译本,由北京师范大学的郝锛新教授翻译.第一次就印了8 200册,这在今天不可想象.

法国人的数学著作的一贯风格是极简洁,比如《有限群的线性表示》的第一章"线性表示通论":

1 定义

令 V 是复数域 $\mathbf{C}$ 上一个向量空间,$GL(V)$ 是由 V 到自身的一切同构所组成的群.按照定义,$GL(V)$ 的一个元素 a 是 V 到 V 内的一个线性映射,它有一个逆映射 a^{-1};这个逆映射也是线性的.当 V 具有一个 n 元有限基 $(\boldsymbol{e}_i)$ 时,每一个线性映射 $a:V\to V$ 由一个 n 阶方阵 (a_{ij}) 所确定,系数 a_{ij} 都是复数;这些系数是通过将象 $a(\boldsymbol{e}_j)$ 用基 $(\boldsymbol{e}_i)$ 来表示

$$a(\boldsymbol{e}_j)=\sum_i a_{ij}\boldsymbol{e}_i$$

而得到的.

说 a 是一个同构相当于说 a 的行列式 $\det(a)=\det(a_{ij})$ 不等于零.因此,群 $GL(V)$ 可以与一切 n 阶可逆方阵所组成的群等同起来.

现在设 G 是一个有限群,具有单位元 1 和运算 $(s,t)\mapsto st$.群 G 到群 $GL(V)$ 内的一个同态 ρ 叫作 G 在 V 内的一个线性表示.换句话说,对于每一个元素 $s\in G$,令 $GL(V)$ 的一个元素 $\rho(s)$ 与它对应,使得对于 $s,t\in G$,等式

$$\rho(st)=\rho(s)\cdot\rho(t)$$

成立(我们常把 $\rho(s)$ 写作 ρ_s).由上面的等式可以推出

$$\rho(1)=1;\rho(s^{-1})=\rho(s)^{-1}$$

当 ρ 被给定时,我们就说 V 是 G 的一个表示空间(或者为了简单起见,就说 V 是 G 的一个表示).今后我们只限于考虑 V 是有限维向量空间的情形.这并不是一个过分的限制.事实上,对于多数应用来说,人们感兴趣的只是 V 的有限个元素 $\boldsymbol{x}_i$,并且总可以找到 V 的一个有限维子表示,它包含这些 $\boldsymbol{x}_i$:就取由这些 $\boldsymbol{x}_i$ 的象 $\rho_s(\boldsymbol{x}_i)$ 所生成的子空间作为表示空间.

现在设 V 是有限维的,令 n 是它的维数.我们也称 n 是所考虑的表示的级.令 $(\boldsymbol{e}_i)$ 是 V 的一个基,$\boldsymbol{R}_s$ 是 ρ_s 关于这个基的矩阵.我们有

$$\det(\boldsymbol{R}_s)\neq 0,\boldsymbol{R}_{st}=\boldsymbol{R}_s\cdot\boldsymbol{R}_t\quad(s,t\in G)$$

如果令 $r_{ij}(s)$ 表示矩阵 $\boldsymbol{R}_s$ 的系数,那么第二个公式变成

$$r_{ik}(st)=\sum_j r_{ij}(s)r_{jk}(t)$$

反之,给了满足上面等式的可逆矩阵 $\boldsymbol{R}_s=(r_{ij}(s))$,相应地就有 G 在 V 内的一个线性表示 ρ;这就是说,用"矩阵形式"给出一个表示.

设 ρ 和 ρ' 是同一个群 G 分别在向量空间 V 和 V' 内的表示.这两个表示说是等价的(或同构的),如果存在一个线性映射 $\tau:V\to V'$,它把 ρ"变为"ρ',也就是说,以下等式成立

$$\tau\circ\rho(s)=\rho'(s)\circ\tau\quad(\forall s\in G)$$

当 ρ 和 ρ' 是通过矩阵形式分别由 $\boldsymbol{R}_s$ 和 $\boldsymbol{R}'_s$ 给出时,这个等式的意义就是,存

在一个可逆矩阵 $\boldsymbol{T}$ 使得

$$\boldsymbol{T}\cdot\boldsymbol{R}_s=\boldsymbol{R}'_s\cdot\boldsymbol{T}\quad(\forall s\in G)$$

或写成 $\boldsymbol{R}'_s=\boldsymbol{T}\cdot\boldsymbol{R}_s\cdot\boldsymbol{T}^{-1}$. 我们可以把这样的两个表示看作同一个(对于每一个 $\boldsymbol{x}\in V$,令 $\tau'(\boldsymbol{x})\in V'$ 与它对应);特别地,ρ 和 ρ' 有相同的级.

2 基本例子

(1) 群 G 的一个 1 级表示是一个同态 $\rho:G\to C^*$,这里 C^* 表示非零复数乘法群. 因为 G 的每一个元素都有有限阶,所以 ρ 的值 $\rho(s)$ 都是单位根. 特别地,我们有 $|\rho(s)|=1$.

如果对一切 $s\in G$ 都取 $\rho(s)=1$,我们就得到 G 的一个表示,叫作单位(或平凡)表示.

(2) 令 g 是群 G 的阶,V 是一个 g 维向量空间,而 $(\boldsymbol{e}_t)_{t\in G}$ 是 V 的一个基,以 G 的元素 t 为指标. 对于 $s\in G$,令 ρ_s 是 V 到 V 内的线性映射,它将 $\boldsymbol{e}_t$ 变到 $\boldsymbol{e}_{st}$;这样就定义了一个线性表示,叫作 G 的正则表示. 这个表示的级等于 G 的阶. 注意 $\boldsymbol{e}_s=\rho_s(\boldsymbol{e}_1)$,所以 $\boldsymbol{e}_1$ 的象组成 V 的一个基. 反之,令 W 是 G 的一个表示,它含有这样一个向量 $\boldsymbol{w}$,使得 $\rho_s(\boldsymbol{w})$,$s\in G$,组成 W 的一个基,那么 W 与正则表示同构(令 $\tau(\boldsymbol{e}_s)=\rho_s(\boldsymbol{w})$,就定义了一个同构 $\tau:V\to W$).

(3) 更一般地,设 G 作用在一个有限集 X 上. 这就是说,对于每一 $s\in G$,都给出 X 的一个置换 $x\mapsto sx$,满足等式

$$1x=x,s(tx)=(st)x$$

这里 $s,t\in G$,$x\in X$. 令 V 是一个向量空间,它具有一个基 $(\boldsymbol{e}_x)_{x\in X}$,这里取 X 的元素作为指标. 对于 $s\in G$,令 ρ_s 是 V 到 V 内的线性映射,它把 $\boldsymbol{e}_x$ 变到 $\boldsymbol{e}_{sx}$;这样得到的 G 的表示叫作 G 关于 X 的置换表示.

3 子表示

令 $\rho:G\to GL(V)$ 是一个线性表示,W 是 V 的一个子空间. 假设 W 在 G 的作用下是稳定的(或者说是"不变的"),换句话说,假设对于一切 $s\in G$,$\boldsymbol{x}\in W$,都有 $\rho_s\boldsymbol{x}\in W$. 于是 ρ_s 在 W 上的限制 ρ_s^W 是 W 到它自身上的一个同构,并且 $\rho_{st}^W=\rho_s^W\cdot\rho_t^W$. 这样一来,$\rho^W:G\to GL(W)$ 就是 G 在 W 内的一个线性表示. W 叫作 V 的一个子表示.

例 取 V 是 G 的正则表示. 令 W 是由元素 $\boldsymbol{x}=\sum_{s\in G}\boldsymbol{e}_s$ 所生成的 V 的一维子空间. 那么对一切 $s\in G$ 都有 $\rho_s\boldsymbol{x}=\boldsymbol{x}$,从而 W 是 V 的一个子表示. 这个子表示与单位表示同构.

在往下进行之前,让我们先来回顾一下线性代数中的某些概念. 向量空间 V 叫作子空间 W 与 W' 的直和,如果每一个 $\boldsymbol{x}\in V$ 可以唯一地写成 $\boldsymbol{x}=\boldsymbol{w}+\boldsymbol{w}'$ 的形式,这里 $\boldsymbol{w}\in W$,$\boldsymbol{w}'\in W'$;这相当于说 W 与 W' 的交 $W\cap W'$ 是 0,并且 $\dim(V)=\dim(W)+\dim(W')$. 这时就写作 $V=W\oplus W'$,而 W' 叫作 W 在 V 里的一个余子空间. 将每一个 $\boldsymbol{x}\in V$ 变到它在 W 中的分量 $\boldsymbol{w}$ 的映射 p 叫作对于分解 $V=W\oplus W'$ 来说 V 在 W 上的射影,p 的象就是 W,并且对于一切 $\boldsymbol{x}\in W$,$p(\boldsymbol{x})=\boldsymbol{x}$. 反之,如果 p 是 V 到自身内满足上述两个条件的一个线性映射,那么

可以验证,V 是 W 与 p 的核(一切满足条件 $p(\boldsymbol{x})=\boldsymbol{0}$ 的 $\boldsymbol{x}$ 所成的集)的直和.这样一来,V 在 W 上的射影和 W 在 V 中的余子空间之间就建立了一个一一对应.

现在让我们再回到子表示上来.

定理 1 令 $\rho:G\to GL(V)$ 是 G 在 V 内的一个线性表示,而 W 是 V 的一个在 G 之下稳定的子空间.那么存在 W 在 V 中的一个余子空间 W^0,它也在 G 之下稳定.

令 W' 是 W 在 V 中的任意一个余子空间,p 是相应的 V 在 W 上的射影.作 p 在 G 的元素之下的共轭,并且求平均得

$$p^0=\frac{1}{g}\sum_{t\in G}\rho_t\cdot p\cdot\rho_t^{-1}\quad (g\text{ 是 }G\text{ 的阶})$$

因为 p 将 V 映入 W,而 ρ_t 使 W 不变,所以 p^0 将 V 映入 W.如果 $\boldsymbol{x}\in W$,那么 $\rho_t^{-1}\boldsymbol{x}\in W$.因此

$$p\cdot\rho_t^{-1}\boldsymbol{x}=\rho_t^{-1}\boldsymbol{x},\quad \rho_t\cdot p\cdot\rho_t^{-1}\boldsymbol{x}=\boldsymbol{x}$$

从而 $p^0\boldsymbol{x}=\boldsymbol{x}$.这样,$p^0$ 是 V 在 W 上的一个射影,它对应着 W 的一个余子空间 W^0.再者,对一切 $s\in G$,我们有

$$\rho_s\cdot p^0=p^0\cdot\rho_s$$

事实上,计算一下 $\rho_s\cdot p^0\cdot\rho_s^{-1}$,就得到

$$\begin{aligned}\rho_s\cdot p^0\cdot\rho_s^{-1}&=\frac{1}{g}\sum_{t\in G}\rho_s\cdot\rho_t\cdot p\cdot\rho_t^{-1}\cdot\rho_s^{-1}\\&=\frac{1}{g}\sum_{t\in G}\rho_{st}\cdot p\cdot\rho_{st}^{-1}=p^0\end{aligned}$$

现在设 $\boldsymbol{x}\in W^0$,$s\in G$,那么 $p^0\boldsymbol{x}=0$,因此 $p^0\cdot\rho_s\boldsymbol{x}=\rho_s\cdot p^0\boldsymbol{x}=0$,所以 $\rho_s\boldsymbol{x}\in W^0$.这就证明了 W^0 在 G 之下是稳定的.定理证毕.

注记 1 设 V 带有一个内积 $(\boldsymbol{x}\mid\boldsymbol{y})$,满足通常的条件:对 $\boldsymbol{x}$ 是线性的,对 $\boldsymbol{y}$ 是半线性的,并且当 $\boldsymbol{x}\neq\boldsymbol{0}$ 时,$(\boldsymbol{x}\mid\boldsymbol{x})>0$.又假设这个内积在 G 之下不变,即对于一切 $s\in G$,都有 $(\rho_s\boldsymbol{x}\mid\rho_s\boldsymbol{y})=(\boldsymbol{x}\mid\boldsymbol{y})$.我们取 $\sum_{t\in G}(\rho_t\boldsymbol{x}\mid\rho_t\boldsymbol{y})$ 来代替 $(\boldsymbol{x}\mid\boldsymbol{y})$,总可以归结为这一情形.在这样的前提下,$W$ 在 V 中的正交余 W^0 是 W 的一个余子空间,并且在 G 之下稳定.这样就得到定理 1 的另一证明.注意内积 $(\boldsymbol{x}\mid\boldsymbol{y})$ 的不变性意味着,如果 $(\boldsymbol{e}_i)$ 是 V 的一个标准正交基,那么 ρ_s 关于这个基的矩阵是一个酉阵.

保持定理 1 的前提和记法.令 $\boldsymbol{x}\in V$,而 $\boldsymbol{w}$ 和 $\boldsymbol{w}^0$ 是 $\boldsymbol{x}$ 在 W 和 W^0 上的射影.我们有 $\boldsymbol{x}=\boldsymbol{w}+\boldsymbol{w}^0$,同时 $\rho_s\boldsymbol{x}=\rho_s\boldsymbol{w}+\rho_s\boldsymbol{w}^0$,并且由于 W 和 W^0 都在 G 之下稳定,所以 $\rho_s\boldsymbol{w}\in W$,$\rho_s\boldsymbol{w}^0\in W^0$.这样,$\rho_s\boldsymbol{w}$ 和 $\rho_s\boldsymbol{w}^0$ 是 $\rho_s\boldsymbol{x}$ 的射影.因此,表示 W 和 W^0 确定了表示 V.我们就说 V 是 W 和 W^0 的直和,并且记作 $V=W\oplus W^0$.V 的元素可以与元素对 $(\boldsymbol{w},\boldsymbol{w}^0)$ 等同起来,这里 $\boldsymbol{w}\in W$,$\boldsymbol{w}^0\in W^0$.如果 W 和 W^0 由矩阵形式 $\boldsymbol{R}_s$ 和 $\boldsymbol{R}_s^0$ 给出,那么 $W\oplus W^0$ 就由矩阵形式

$$\begin{pmatrix}\boldsymbol{R}_s & \boldsymbol{0}\\ \boldsymbol{0} & \boldsymbol{R}_s^0\end{pmatrix}$$

给出.任意有限多个表示的直和可以类似地定义.

4 不可约表示

令 $\rho: G \to GL(V)$ 是 G 的一个线性表示. 我们说, ρ 是不可约的或单的, 如果 V 不是 0, 并且除 0 和 V 之外, V 没有在 G 之下稳定的子空间. 由定理 1, 第二个条件相当于说, V 不是两个表示的直和(除开显然的分解 $V = 0 \oplus V$). 一级的表示自然都是不可约的. 每一个非交换群至少有一个级大于或等于 2 的不可约表示.

通过作直和, 不可约表示被用来构造其他的表示:

定理 2 每一个表示都是不可约表示的直和.

令 V 是 G 的一个线性表示. 我们对 $\dim(V)$ 作归纳法. 若 $\dim(V) = 0$, 定理是显然的(0 是不可约表示的空集的直和). 假设 $\dim(V) \geqslant 1$. 如果 V 不可约, 那么已经没有什么要证的了; 否则, 由定理 1, V 可以分解为直和 $V' \oplus V''$, 而 $\dim(V') < \dim(V)$, $\dim(V'') < \dim(V)$. 由归纳法的假设, V' 和 V'' 都是一些不可约表示的直和, 因而 V 也是一些不可约表示的直和.

注记 2 令 V 是一个表示, $V = W_1 \oplus \cdots \oplus W_k$ 是 V 的一个不可约表示的直和分解. 我们会问, 这个分解是否唯一. 在一切 ρ_s 都等于 1 的情形, 这个分解一般不是唯一的(这时 W_i 都是直线, 而一个向量空间分成一些直线的直和的分解是很多的). 然而, 与一个给定的不可约表示同构的 W_i 的个数不依赖于分解的选取.

5 两个表示的张量积

同直和运算(这个运算具有通常加法的性质)在一起, 还有一种"乘法": 张量积, 有时也叫作克罗内克积. 它的定义如下:

设 V_1 和 V_2 是两个向量空间. 向量空间 W, 连同一个 $V_1 \times V_2$ 到 W 内的映射 $(\boldsymbol{x}_1, \boldsymbol{x}_2) \mapsto \boldsymbol{x}_1 \cdot \boldsymbol{x}_2$, 叫作 V_1 与 V_2 的张量积, 如果下列条件被满足:

①$\boldsymbol{x}_1 \cdot \boldsymbol{x}_2$ 对于变量 $\boldsymbol{x}_1$ 和 $\boldsymbol{x}_2$ 中每一个都是线性的;

② 若$(\boldsymbol{e}_{i_1})$ 是 V_1 的一个基, $(\boldsymbol{e}_{i_2})$ 是 V_2 的一个基, 则一切乘积 $\boldsymbol{e}_{i_1} \cdot \boldsymbol{e}_{i_2}$ 是 W 的一个基.

容易证明, 这样一个空间是存在且唯一的(在同构的意义下). 记作 $V_1 \otimes V_2$. 条件 ② 表明

$$\dim(V_1 \otimes V_2) = \dim(V_1) \cdot \dim(V_2)$$

现在设 $\rho^1: G \to GL(V_1)$ 和 $\rho^2: G \to GL(V_2)$ 是群 G 的两个表示. 对于 $s \in G$, 由以下条件定义 $GL(V_1 \oplus V_2)$ 的一个元素 ρ_s

$$\rho_s(\boldsymbol{x}_1 \cdot \boldsymbol{x}_2) = \rho_s^1(\boldsymbol{x}_1) \cdot \rho_s^2(\boldsymbol{x}_2) \quad (\boldsymbol{x}_1 \in V_1, \boldsymbol{x}_2 \in V_2)$$

(ρ_s 的存在和唯一性容易由条件 ① 和 ② 得出.) 我们记作

$$\rho_s = \rho_s^1 \otimes \rho_s^2$$

ρ_s 定义了 G 到 $V_1 \otimes V_2$ 内的一个线性表示, 叫作原来两个表示的张量积.

把这个定义转换成矩阵的说法就是: 令$(\boldsymbol{e}_{i_1})$ 是 V_1 的一个基, $(r_{i_1 j_1}(s))$ 是 ρ_s^1 关于这个基的矩阵; 完全类似地定义$(\boldsymbol{e}_{i_2})$ 和$(r_{i_2 j_2}(s))$. 由

$$\rho_s^1(\boldsymbol{e}_{j_1}) = \sum_{i_1} r_{i_1 j_1}(s) \boldsymbol{e}_{i_1}$$

$$\rho_s^2(\boldsymbol{e}_{j_2}) = \sum_{i_2} r_{i_2 j_2}(s) \boldsymbol{e}_{i_2}$$

推出

$$\rho_s(\boldsymbol{e}_{j_1} \cdot \boldsymbol{e}_{j_2}) = \sum_{i_1, i_2} r_{i_1 j_1}(s) \cdot r_{i_2 j_2}(s) \cdot \boldsymbol{e}_{i_1} \cdot \boldsymbol{e}_{i_2}$$

因此 ρ_s 的矩阵就是$(r_{i_1 j_1}(s) \quad r_{i_2 j_2}(s))$;它是 ρ_s^1 的矩阵与 ρ_s^2 的矩阵的张量积.

两个不可约表示的张量积一般说来不是不可约的,它被分解成一些不可约表示的直和,这些不可约表示可以利用特征标的理论来确定.

在量子化学中,张量积常常表现为以下形式:V_1 和 V_2 是两个函数空间,它们都在 G 之下稳定,分别有基$(\boldsymbol{\varphi}_{i_1})$和$(\boldsymbol{\psi}_{i_2})$,而 $V_1 \otimes V_2$ 是由乘积 $\boldsymbol{\varphi}_{i_1} \cdot \boldsymbol{\psi}_{i_2}$ 所生成的向量空间,这些乘积是线性无关的.最后的条件是重要的.这里有两个特殊情形,对于这两个情形来说,上面的条件都被满足:

① 一些函数 φ 只依赖于变量$(x, x', \cdots)$,而另一些函数 ψ 只依赖于与第一组变量无关的变量$(y, y', \cdots)$.

② 空间 V_1(或 V_2)具有一个只由单独一个函数 φ 所组成的基,这个函数在任何区域内都不等于零;于是空间 V_1 是一维的.

6　对称方和交错方

假设表示 V_1 和 V_2 都恒同于同一个表示 V.那么 $V_1 \otimes V_2 = V \otimes V$.如果$(\boldsymbol{e}_i)$是 V 的一个基,令 θ 是 $V \otimes V$ 的一个自同构,使得

$$\theta(\boldsymbol{e}_i \cdot \boldsymbol{e}_j) = \boldsymbol{e}_j \cdot \boldsymbol{e}_i \quad (\text{对一切}(i, j))$$

由此推出,对一切 $x, y \in V$ 都有 $\theta(x \cdot y) = y \cdot x$,因此 θ 不依赖于基$(\boldsymbol{e}_i)$的选取.再者,$\theta^2 = 1$.于是空间 $V \otimes V$ 被分解为直和

$$V \otimes V = \mathrm{Sym}^2(V) \oplus \mathrm{Alt}^2(V)$$

这里 $\mathrm{Sym}^2(V)$ 是 $V \otimes V$ 中一切满足条件 $\theta(z) = z$ 的元素 z 所组成的集合,而 $\mathrm{Alt}^2(V)$ 是 $V \otimes V$ 中一切满足条件 $\theta(z) = -z$ 的元素 z 所组成的集合.元素$(\boldsymbol{e}_i \cdot \boldsymbol{e}_j + \boldsymbol{e}_j \cdot \boldsymbol{e}_i)_{i \leqslant j}$ 构成 $\mathrm{Sym}^2(V)$ 的一个基,而元素$(\boldsymbol{e}_i \cdot \boldsymbol{e}_j - \boldsymbol{e}_j \cdot \boldsymbol{e}_i)_{i < j}$ 构成 $\mathrm{Alt}^2(V)$ 的一个基.如果 $\dim(V) = n$,那么

$$\dim \mathrm{Sym}^2(V) = \frac{n(n+1)}{2}$$

$$\dim \mathrm{Alt}^2(V) = \frac{n(n-1)}{2}$$

子空间 $\mathrm{Sym}^2(V)$ 和 $\mathrm{Alt}^2(V)$ 都在 G 之下稳定.这样定义的表示分别叫作所给的表示的对称方和交错方.

区区几页就把线性表示论概括了,但要想真正读懂线性表示论,恐怕还需要做几倍甚至十几倍的笔记才行.记得在 1987 年暑假,南京师范大学举办过一期群表示论的讲习班,笔者作为青年教师被学校派去参加,所用教材就是塞尔的这本书.南京的酷热和表示论的难懂很快就打败了笔者,于是在游完了大明湖和明孝陵之后便逃回了东北.几十年后又与表示论相遇,只不过是以出版人的视角来看,一时间又觉得很亲切,可能也是对短暂的"青椒"生活的一种怀念吧!

本书的版权编辑李丹女士为我们翻译了本书的目录,如下:

前言
Ⅰ 对称群与对称函数
1 有限群的表示与半单代数
1.1 有限群及其表示
1.2 表示的特征与构造
1.3 非交换傅里叶变换
1.4 半单代数与模
1.5 二次交换子理论
2 对称函数与弗罗比尼乌斯—舒尔同构
2.1 对称群的共轭类
2.2 对称函数代数的五个基
2.3 分次自伴霍普夫代数的结构
2.4 弗罗比尼乌斯—舒尔同构
2.5 舒尔—外尔对偶
3 划分与表的组合
3.1 彼里规则与穆纳汉—中山(Murnaghan-Nakayama)公式
3.2 鲁宾逊—申斯特—克努特算法
3.3 不可约表示的构造
3.4 钩长公式
Ⅱ 赫克代数及其表示
4 赫克代数与布饶尔—嘉当(Brauer-Cartan)定理
4.1 对称群的考克斯特表示
4.2 代数的表示理论
4.3 布饶尔—嘉当形变理论
4.4 一般的与特殊的赫克代数结构
4.5 Q—施佩希特(Specht)模
5 赫克代数的特征与对偶
5.1 量子群及其霍普夫代数结构
5.2 量子群的表示理论
5.3 金博—舒尔—外尔(Jimbo-Schur-Weyl)对偶
5.4 岩堀—赫克对偶
5.5 霍尔—利特尔伍德多项式与赫克代数的特征
6 $q=0$ 时的赫克代数特殊化的表示
6.1 非交换对称函数
6.2 拟对称函数
6.3 赫克—弗罗比尼乌斯—舒尔同构
Ⅲ 划分的可观测
7 可观测的伊万诺夫—克罗夫代数

7.1 局部排列代数
7.2 杨氏图的坐标及其矩
7.3 可观测代数中基的改变
7.4 可观测与杨氏图的拓扑
8 朱西—墨菲元素
8.1 对称群代数的盖尔范德—策特林子(Gelfand-Tsetlin)代数
8.2 作用在盖尔范德—策特林基上的朱西—墨菲元素
8.3 把可观测对象作为内容的对称函数
9 对称群与自由概率
9.1 自由概率的介绍
9.2 杨氏图的自由累积量
9.3 转移测度与朱西—墨菲元素
9.4 容许集划分的代数
10 斯坦利—费雷公式与克罗夫多项式
10.1 杨氏图的新的可观测物
10.2 对称群特征的斯坦利—费雷公式
10.3 克罗夫多项式的组合学
Ⅳ 随机杨氏图的模型
11 无限对称群的表示
11.1 杨氏图与极值特征的调和分析
11.2 双无限对称群与奥尔尚斯基半群
11.3 容许表示的分类
11.4 球形表示和GNS构造
12 中心测度的渐近数
12.1 自由拟对称函数
12.2 中心测度的组合学
12.3 可观测的高斯行为
13 普朗谢雷尔测度和舒尔—外尔测度的渐近
13.1 普朗谢雷尔模型与舒尔—外尔模型
13.2 大随机杨氏图的极限形状
13.3 特征的克罗夫中心极限理论
附录A 半单李代数的表示论
A.1 幂零代数、可解代数与半单代数
A.2 一个半单复代数的根系
A.3 最高权理论
参考文献
索引

笔者业余时间喜欢下围棋，一位棋友告诉笔者，要把自己想象成一个大棋士才行. 此刻笔者就是把自己想象成一个大数学家来做出版决定的，但这其实是一种自欺.

美国演化生物学者崔弗斯(Robert Trivers) 曾经出版过一本以自欺为核心的书，将自欺变成一个科学题材，而不只是人文学者、心理学家感兴趣的人性枷锁. 根据崔弗斯的论述，自欺是欺人的绝招，不出自有意识的努力，而是无意识的心理机制，功能有三：一、有意识地欺骗人，难免泄露意图，自欺就不会；二、既然自己都上当了，当然挥洒自如、行云流水、毫不做作；三、万一被拆穿，也容易卸责，谁会责备受骗的人呢？

自欺的生物根源极为古老，并非人类独有的天赋，从自信到过度自信到自欺，并无明显分际. 只是人能说话，不但方便卖空买空，更容易陷入自欺的漩涡，甚至将旁观者卷入，掀起风潮. 自欺在人类历史中扮演的角色，耐人寻味，但有时也会产生出一些有价值的结果！

刘培杰

2022.4.30

于哈工大

刘培杰数学工作室

已出版(即将出版)图书目录——原版影印

书　　名	出版时间	定　价	编号
数学物理大百科全书.第1卷(英文)	2016－01	418.00	508
数学物理大百科全书.第2卷(英文)	2016－01	408.00	509
数学物理大百科全书.第3卷(英文)	2016－01	396.00	510
数学物理大百科全书.第4卷(英文)	2016－01	408.00	511
数学物理大百科全书.第5卷(英文)	2016－01	368.00	512
zeta 函数,q-zeta 函数,相伴级数与积分(英文)	2015－08	88.00	513
微分形式:理论与练习(英文)	2015－08	58.00	514
离散与微分包含的逼近和优化(英文)	2015－08	58.00	515
艾伦·图灵:他的工作与影响(英文)	2016－01	98.00	560
测度理论概率导论,第2版(英文)	2016－01	88.00	561
带有潜在故障恢复系统的半马尔柯夫模型控制(英文)	2016－01	98.00	562
数学分析原理(英文)	2016－01	88.00	563
随机偏微分方程的有效动力学(英文)	2016－01	88.00	564
图的谱半径(英文)	2016－01	58.00	565
量子机器学习中数据挖掘的量子计算方法(英文)	2016－01	98.00	566
量子物理的非常规方法(英文)	2016－01	118.00	567
运输过程的统一非局部理论:广义波尔兹曼物理动力学,第2版(英文)	2016－01	198.00	568
量子力学与经典力学之间的联系在原子、分子及电动力学系统建模中的应用(英文)	2016－01	58.00	569
算术域(英文)	2018－01	158.00	821
高等数学竞赛:1962—1991年的米洛克斯·史怀哲竞赛(英文)	2018－01	128.00	822
用数学奥林匹克精神解决数论问题(英文)	2018－01	108.00	823
代数几何(德文)	2018－04	68.00	824
丢番图逼近论(英文)	2018－01	78.00	825
代数几何学基础教程(英文)	2018－01	98.00	826
解析数论入门课程(英文)	2018－01	78.00	827
数论中的丢番图问题(英文)	2018－01	78.00	829
数论(梦幻之旅):第五届中日数论研讨会演讲集(英文)	2018－01	68.00	830
数论新应用(英文)	2018－01	68.00	831
数论(英文)	2018－01	78.00	832

刘培杰数学工作室
已出版(即将出版)图书目录——原版影印

书　　名	出版时间	定　价	编号
湍流十讲(英文)	2018－04	108.00	886
无穷维李代数:第3版(英文)	2018－04	98.00	887
等值、不变量和对称性(英文)	2018－04	78.00	888
解析数论(英文)	2018－09	78.00	889
《数学原理》的演化:伯特兰·罗素撰写第二版时的手稿与笔记(英文)	2018－04	108.00	890
哈密尔顿数学论文集(第4卷):几何学、分析学、天文学、概率和有限差分等(英文)	2019－05	108.00	891
偏微分方程全局吸引子的特性(英文)	2018－09	108.00	979
整函数与下调和函数(英文)	2018－09	118.00	980
幂等分析(英文)	2018－09	118.00	981
李群,离散子群与不变量理论(英文)	2018－09	108.00	982
动力系统与统计力学(英文)	2018－09	118.00	983
表示论与动力系统(英文)	2018－09	118.00	984
分析学练习.第1部分(英文)	2021－01	88.00	1247
分析学练习.第2部分,非线性分析(英文)	2021－01	88.00	1248
初级统计学:循序渐进的方法:第10版(英文)	2019－05	68.00	1067
工程师与科学家微分方程用书:第4版(英文)	2019－07	58.00	1068
大学代数与三角学(英文)	2019－06	78.00	1069
培养数学能力的途径(英文)	2019－07	38.00	1070
工程师与科学家统计学:第4版(英文)	2019－06	58.00	1071
贸易与经济中的应用统计学:第6版(英文)	2019－06	58.00	1072
傅立叶级数和边值问题:第8版(英文)	2019－05	48.00	1073
通往天文学的途径:第5版(英文)	2019－05	58.00	1074
拉马努金笔记.第1卷(英文)	2019－06	165.00	1078
拉马努金笔记.第2卷(英文)	2019－06	165.00	1079
拉马努金笔记.第3卷(英文)	2019－06	165.00	1080
拉马努金笔记.第4卷(英文)	2019－06	165.00	1081
拉马努金笔记.第5卷(英文)	2019－06	165.00	1082
拉马努金遗失笔记.第1卷(英文)	2019－06	109.00	1083
拉马努金遗失笔记.第2卷(英文)	2019－06	109.00	1084
拉马努金遗失笔记.第3卷(英文)	2019－06	109.00	1085
拉马努金遗失笔记.第4卷(英文)	2019－06	109.00	1086
数论:1976年纽约洛克菲勒大学数论会议记录(英文)	2020－06	68.00	1145
数论:卡本代尔1979:1979年在南伊利诺伊卡本代尔大学举行的数论会议记录(英文)	2020－06	78.00	1146
数论:诺德韦克豪特1983:1983年在诺德韦克豪特举行的Journees Arithmetiques数论大会会议记录(英文)	2020－06	68.00	1147
数论:1985－1988年在纽约城市大学研究生院和大学中心举办的研讨会(英文)	2020－06	68.00	1148

刘培杰数学工作室

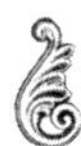

已出版(即将出版)图书目录——原版影印

书　　名	出版时间	定　价	编号
数论:1987 年在乌尔姆举行的 Journees Arithmetiques 数论大会会议记录(英文)	2020－06	68.00	1149
数论:马德拉斯 1987:1987 年在马德拉斯安娜大学举行的国际拉马努金百年纪念大会会议记录(英文)	2020－06	68.00	1150
解析数论:1988 年在东京举行的日法研讨会会议记录(英文)	2020－06	68.00	1151
解析数论:2002 年在意大利切特拉罗举行的 C. I. M. E. 暑期班演讲集(英文)	2020－06	68.00	1152
量子世界中的蝴蝶:最迷人的量子分形故事(英文)	2020－06	118.00	1157
走进量子力学(英文)	2020－06	118.00	1158
计算物理学概论(英文)	2020－06	48.00	1159
物质,空间和时间的理论:量子理论(英文)	2020－10	48.00	1160
物质,空间和时间的理论:经典理论(英文)	2020－10	48.00	1161
量子场理论:解释世界的神秘背景(英文)	2020－07	38.00	1162
计算物理学概论(英文)	2020－06	48.00	1163
行星状星云(英文)	2020－10	38.00	1164
基本宇宙学:从亚里士多德的宇宙到大爆炸(英文)	2020－08	58.00	1165
数学磁流体力学(英文)	2020－07	58.00	1166
计算科学:第 1 卷,计算的科学(日文)	2020－07	88.00	1167
计算科学:第 2 卷,计算与宇宙(日文)	2020－07	88.00	1168
计算科学:第 3 卷,计算与物质(日文)	2020－07	88.00	1169
计算科学:第 4 卷,计算与生命(日文)	2020－07	88.00	1170
计算科学:第 5 卷,计算与地球环境(日文)	2020－07	88.00	1171
计算科学:第 6 卷,计算与社会(日文)	2020－07	88.00	1172
计算科学. 别卷,超级计算机(日文)	2020－07	88.00	1173
多复变函数论(日文)	2022－06	78.00	1518
复变函数入门(日文)	2022－06	78.00	1523
代数与数论:综合方法(英文)	2020－10	78.00	1185
复分析:现代函数理论第一课(英文)	2020－07	58.00	1186
斐波那契数列和卡特兰数:导论(英文)	2020－10	68.00	1187
组合推理:计数艺术介绍(英文)	2020－07	88.00	1188
二次互反律的傅里叶分析证明(英文)	2020－07	48.00	1189
旋瓦兹分布的希尔伯特变换与应用(英文)	2020－07	58.00	1190
泛函分析:巴拿赫空间理论入门(英文)	2020－07	48.00	1191
卡塔兰数入门(英文)	2019－05	68.00	1060
测度与积分(英文)	2019－04	68.00	1059
组合学手册. 第一卷(英文)	2020－06	128.00	1153
*－代数、局部紧群和巴拿赫 *－代数丛的表示. 第一卷,群和代数的基本表示理论(英文)	2020－05	148.00	1154
电磁理论(英文)	2020－08	48.00	1193
连续介质力学中的非线性问题(英文)	2020－09	78.00	1195
多变量数学入门(英文)	2021－05	68.00	1317
偏微分方程入门(英文)	2021－05	88.00	1318
若尔当典范性:理论与实践(英文)	2021－07	68.00	1366
伽罗瓦理论. 第 4 版(英文)	2021－08	88.00	1408

刘培杰数学工作室
已出版(即将出版)图书目录——原版影印

书　　名	出版时间	定　价	编号
典型群,错排与素数(英文)	2020－11	58.00	1204
李代数的表示:通过 gln 进行介绍(英文)	2020－10	38.00	1205
实分析演讲集(英文)	2020－10	38.00	1206
现代分析及其应用的课程(英文)	2020－10	58.00	1207
运动中的抛射物数学(英文)	2020－10	38.00	1208
2－纽结与它们的群(英文)	2020－10	38.00	1209
概率,策略和选择:博弈与选举中的数学(英文)	2020－11	58.00	1210
分析学引论(英文)	2020－11	58.00	1211
量子群:通往流代数的路径(英文)	2020－11	38.00	1212
集合论入门(英文)	2020－10	48.00	1213
酉反射群(英文)	2020－11	58.00	1214
探索数学:吸引人的证明方式(英文)	2020－11	58.00	1215
微分拓扑短期课程(英文)	2020－10	48.00	1216
抽象凸分析(英文)	2020－11	68.00	1222
费马大定理笔记(英文)	2021－03	48.00	1223
高斯与雅可比和(英文)	2021－03	78.00	1224
π与算术几何平均:关于解析数论和计算复杂性的研究(英文)	2021－01	58.00	1225
复分析入门(英文)	2021－03	48.00	1226
爱德华·卢卡斯与素性测定(英文)	2021－03	78.00	1227
通往凸分析及其应用的简单路径(英文)	2021－01	68.00	1229
微分几何的各个方面.第一卷(英文)	2021－01	58.00	1230
微分几何的各个方面.第二卷(英文)	2020－12	58.00	1231
微分几何的各个方面.第三卷(英文)	2020－12	58.00	1232
沃克流形几何学(英文)	2020－11	58.00	1233
彷射和韦尔几何应用(英文)	2020－12	58.00	1234
双曲几何学的旋转向量空间方法(英文)	2021－02	58.00	1235
积分:分析学的关键(英文)	2020－12	48.00	1236
为有天分的新生准备的分析学基础教材(英文)	2020－11	48.00	1237
数学不等式.第一卷.对称多项式不等式(英文)	2021－03	108.00	1273
数学不等式.第二卷.对称有理不等式与对称无理不等式(英文)	2021－03	108.00	1274
数学不等式.第三卷.循环不等式与非循环不等式(英文)	2021－03	108.00	1275
数学不等式.第四卷.Jensen 不等式的扩展与加细(英文)	2021－03	108.00	1276
数学不等式.第五卷.创建不等式与解不等式的其他方法(英文)	2021－04	108.00	1277

刘培杰数学工作室
已出版(即将出版)图书目录——原版影印

书　　名	出版时间	定　价	编号
冯·诺依曼代数中的谱位移函数:半有限冯·诺依曼代数中的谱位移函数与谱流(英文)	2021-06	98.00	1308
链接结构:关于嵌入完全图的直线中链接单形的组合结构(英文)	2021-05	58.00	1309
代数几何方法.第1卷(英文)	2021-06	68.00	1310
代数几何方法.第2卷(英文)	2021-06	68.00	1311
代数几何方法.第3卷(英文)	2021-06	58.00	1312

书　　名	出版时间	定　价	编号
代数、生物信息和机器人技术的算法问题.第四卷,独立恒等式系统(俄文)	2020-08	118.00	1199
代数、生物信息和机器人技术的算法问题.第五卷,相对覆盖性和独立可拆分恒等式系统(俄文)	2020-08	118.00	1200
代数、生物信息和机器人技术的算法问题.第六卷,恒等式和准恒等式的相等 问题、可推导性和可实现性(俄文)	2020-08	128.00	1201
分数阶微积分的应用:非局部动态过程,分数阶导热系数(俄文)	2021-01	68.00	1241
泛函分析问题与练习:第2版(俄文)	2021-01	98.00	1242
集合论、数学逻辑和算法论问题:第5版(俄文)	2021-01	98.00	1243
微分几何和拓扑短期课程(俄文)	2021-01	98.00	1244
素数规律(俄文)	2021-01	88.00	1245
无穷边值问题解的递减:无界域中的拟线性椭圆和抛物方程(俄文)	2021-01	48.00	1246
微分几何讲义(俄文)	2020-12	98.00	1253
二次型和矩阵(俄文)	2021-01	98.00	1255
积分和级数.第2卷,特殊函数(俄文)	2021-01	168.00	1258
积分和级数.第3卷,特殊函数补充:第2版(俄文)	2021-01	178.00	1264
几何图上的微分方程(俄文)	2021-01	138.00	1259
数论教程:第2版(俄文)	2021-01	98.00	1260
非阿基米德分析及其应用(俄文)	2021-03	98.00	1261
古典群和量子群的压缩(俄文)	2021-03	98.00	1263
数学分析习题集.第3卷,多元函数:第3版(俄文)	2021-03	98.00	1266
数学习题:乌拉尔国立大学数学力学系大学生奥林匹克(俄文)	2021-03	98.00	1267
柯西定理和微分方程的特解(俄文)	2021-03	98.00	1268
组合极值问题及其应用:第3版(俄文)	2021-03	98.00	1269
数学词典(俄文)	2021-01	98.00	1271
确定性混沌分析模型(俄文)	2021-06	168.00	1307
精选初等数学习题和定理.立体几何.第3版(俄文)	2021-03	68.00	1316
微分几何习题:第3版(俄文)	2021-05	98.00	1336
精选初等数学习题和定理.平面几何.第4版(俄文)	2021-05	68.00	1335
曲面理论在欧氏空间 En 中的直接表示(俄文)	2022-01	68.00	1444
维纳—霍普夫离散算子和托普利兹算子:某些可数赋范空间中的诺特性和可逆性(俄文)	2022-03	108.00	1496
Maple 中的数论:数论中的计算机计算(俄文)	2022-03	88.00	1497
贝尔曼和克努特问题及其概括:加法运算的复杂性(俄文)	2022-03	138.00	1498

刘培杰数学工作室
已出版(即将出版)图书目录——原版影印

书　　名	出版时间	定　价	编号
复分析:共形映射(俄文)	2022－07	48.00	1542
微积分代数样条和多项式及其在数值方法中的应用(俄文)	2022－08	128.00	1543
蒙特卡罗方法中的随机过程和场模型:算法和应用(俄文)	2022－08	88.00	1544
线性椭圆型方程组:论二阶椭圆型方程的迪利克雷问题(俄文)	2022－08	98.00	1561
动态系统解的增长特性:估值、稳定性、应用(俄文)	2022－08	118.00	1565
群的自由积分解:建立和应用(俄文)	2022－08	78.00	1570

书　　名	出版时间	定　价	编号
狭义相对论与广义相对论:时空与引力导论(英文)	2021－07	88.00	1319
束流物理学和粒子加速器的实践介绍:第2版(英文)	2021－07	88.00	1320
凝聚态物理中的拓扑和微分几何简介(英文)	2021－05	88.00	1321
混沌映射:动力学、分形学和快速涨落(英文)	2021－05	128.00	1322
广义相对论:黑洞、引力波和宇宙学介绍(英文)	2021－06	68.00	1323
现代分析电磁均质化(英文)	2021－06	68.00	1324
为科学家提供的基本流体动力学(英文)	2021－06	88.00	1325
视觉天文学:理解夜空的指南(英文)	2021－06	68.00	1326
物理学中的计算方法(英文)	2021－06	68.00	1327
单星的结构与演化:导论(英文)	2021－06	108.00	1328
超越居里:1903年至1963年物理界四位女性及其著名发现(英文)	2021－06	68.00	1329
范德瓦尔斯流体热力学的进展(英文)	2021－06	68.00	1330
先进的托卡马克稳定性理论(英文)	2021－06	88.00	1331
经典场论导论:基本相互作用的过程(英文)	2021－07	88.00	1332
光致电离量子动力学方法原理(英文)	2021－07	108.00	1333
经典域论和应力:能量张量(英文)	2021－05	88.00	1334
非线性太赫兹光谱的概念与应用(英文)	2021－06	68.00	1337
电磁学中的无穷空间并矢格林函数(英文)	2021－06	88.00	1338
物理科学基础数学.第1卷,齐次边值问题、傅里叶方法和特殊函数(英文)	2021－07	108.00	1339
离散量子力学(英文)	2021－07	68.00	1340
核磁共振的物理学和数学(英文)	2021－07	108.00	1341
分子水平的静电学(英文)	2021－08	68.00	1342
非线性波:理论、计算机模拟、实验(英文)	2021－06	108.00	1343
石墨烯光学:经典问题的电解解决方案(英文)	2021－06	68.00	1344
超材料多元宇宙(英文)	2021－07	68.00	1345
银河系外的天体物理学(英文)	2021－07	68.00	1346
原子物理学(英文)	2021－07	68.00	1347
将光打结:将拓扑学应用于光学(英文)	2021－07	68.00	1348
电磁学:问题与解法(英文)	2021－07	88.00	1364
海浪的原理:介绍量子力学的技巧与应用(英文)	2021－07	108.00	1365
多孔介质中的流体:输运与相变(英文)	2021－07	68.00	1372
洛伦兹群的物理学(英文)	2021－08	68.00	1373
物理导论的数学方法和解决方法手册(英文)	2021－08	68.00	1374
非线性波数学物理学入门(英文)	2021－08	88.00	1376
波:基本原理和动力学(英文)	2021－07	68.00	1377
光电子量子计量学.第1卷,基础(英文)	2021－07	88.00	1383
光电子量子计量学.第2卷,应用与进展(英文)	2021－07	68.00	1384
复杂流的格子玻尔兹曼建模的工程应用(英文)	2021－08	68.00	1393

刘培杰数学工作室
已出版(即将出版)图书目录——原版影印

书　名	出版时间	定　价	编号
电偶极矩挑战(英文)	2021—08	108.00	1394
电动力学:问题与解法(英文)	2021—09	68.00	1395
自由电子激光的经典理论(英文)	2021—08	68.00	1397
曼哈顿计划——核武器物理学简介(英文)	2021—09	68.00	1401
粒子物理学(英文)	2021—09	68.00	1402
引力场中的量子信息(英文)	2021—09	128.00	1403
器件物理学的基本经典力学(英文)	2021—09	68.00	1404
等离子体物理及其空间应用导论.第1卷,基本原理和初步过程(英文)	2021—09	68.00	1405
拓扑与超弦理论焦点问题(英文)	2021—07	58.00	1349
应用数学:理论、方法与实践(英文)	2021—07	78.00	1350
非线性特征值问题:牛顿型方法与非线性瑞利函数(英文)	2021—07	58.00	1351
广义膨胀和齐性:利用齐性构造齐次系统的李雅普诺夫函数和控制律(英文)	2021—06	48.00	1352
解析数论焦点问题(英文)	2021—07	58.00	1353
随机微分方程:动态系统方法(英文)	2021—07	58.00	1354
经典力学与微分几何(英文)	2021—07	58.00	1355
负定相交形式流形上的瞬子模空间几何(英文)	2021—07	68.00	1356
广义卡塔兰轨道分析:广义卡塔兰轨道计算数字的方法(英文)	2021—07	48.00	1367
洛伦兹方法的变分:二维与三维洛伦兹方法(英文)	2021—08	38.00	1378
几何、分析和数论精编(英文)	2021—08	68.00	1380
从一个新角度看数论:通过遗传方法引入现实的概念(英文)	2021—07	58.00	1387
动力系统:短期课程(英文)	2021—08	68.00	1382
几何路径:理论与实践(英文)	2021—08	48.00	1385
论天体力学中某些问题的不可积性(英文)	2021—07	88.00	1396
广义斐波那契数列及其性质(英文)	2021—08	38.00	1386
对称函数和麦克唐纳多项式:余代数结构与Kawanaka恒等式(英文)	2021—09	38.00	1400
杰弗里·英格拉姆·泰勒科学论文集:第1卷.固体力学(英文)	2021—05	78.00	1360
杰弗里·英格拉姆·泰勒科学论文集:第2卷.气象学、海洋学和湍流(英文)	2021—05	68.00	1361
杰弗里·英格拉姆·泰勒科学论文集:第3卷.空气动力学以及落弹数和爆炸的力学(英文)	2021—05	68.00	1362
杰弗里·英格拉姆·泰勒科学论文集:第4卷.有关流体力学(英文)	2021—05	58.00	1363

刘培杰数学工作室
已出版(即将出版)图书目录——原版影印

书　　名	出版时间	定　价	编号
非局域泛函演化方程:积分与分数阶(英文)	2021—08	48.00	1390
理论工作者的高等微分几何:纤维丛、射流流形和拉格朗日理论(英文)	2021—08	68.00	1391
半线性退化椭圆微分方程:局部定理与整体定理(英文)	2021—07	48.00	1392
非交换几何、规范理论和重整化:一般简介与非交换量子场论的重整化(英文)	2021—09	78.00	1406
数论论文集:拉普拉斯变换和带有数论系数的幂级数(俄文)	2021—09	48.00	1407
挠理论专题:相对极大值,单射与扩充模(英文)	2021—09	88.00	1410
强正则图与欧几里得若尔当代数:非通常关系中的启示(英文)	2021—10	48.00	1411
拉格朗日几何和哈密顿几何:力学的应用(英文)	2021—10	48.00	1412
时滞微分方程与差分方程的振动理论:二阶与三阶(英文)	2021—10	98.00	1417
卷积结构与几何函数理论:用以研究特定几何函数理论方向的分数阶微积分算子与卷积结构(英文)	2021—10	48.00	1418
经典数学物理的历史发展(英文)	2021—10	78.00	1419
扩展线性丢番图问题(英文)	2021—10	38.00	1420
一类混沌动力系统的分歧分析与控制:分歧分析与控制(英文)	2021—11	38.00	1421
伽利略空间和伪伽利略空间中一些特殊曲线的几何性质(英文)	2022—01	68.00	1422
一阶偏微分方程:哈密尔顿—雅可比理论(英文)	2021—11	48.00	1424
各向异性黎曼多面体的反问题:分段光滑的各向异性黎曼多面体反边界谱问题:唯一性(英文)	2021—11	38.00	1425
项目反应理论手册.第一卷,模型(英文)	2021—11	138.00	1431
项目反应理论手册.第二卷,统计工具(英文)	2021—11	118.00	1432
项目反应理论手册.第三卷,应用(英文)	2021—11	138.00	1433
二次无理数:经典数论入门(英文)	2022—05	138.00	1434
数,形与对称性:数论,几何和群论导论(英文)	2022—05	128.00	1435
有限域手册(英文)	2021—11	178.00	1436
计算数论(英文)	2021—11	148.00	1437
拟群与其表示简介(英文)	2021—11	88.00	1438
数论与密码学导论:第二版(英文)	2022—01	148.00	1423

刘培杰数学工作室

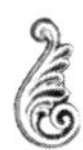

已出版(即将出版)图书目录——原版影印

书名	出版时间	定价	编号
几何分析中的柯西变换与黎兹变换:解析调和容量和李普希兹调和容量、变化和振荡以及一致可求长性(英文)	2021-12	38.00	1465
近似不动点定理及其应用(英文)	2022-05	28.00	1466
局部域的相关内容解析:对局部域的扩展及其伽罗瓦群的研究(英文)	2022-01	38.00	1467
反问题的二进制恢复方法(英文)	2022-03	28.00	1468
对几何函数中某些类的各个方面的研究:复变量理论(英文)	2022-01	38.00	1469
覆盖、对应和非交换几何(英文)	2022-01	28.00	1470
最优控制理论中的随机线性调节器问题:随机最优线性调节器问题(英文)	2022-01	38.00	1473
正交分解法:涡流流体动力学应用的正交分解法(英文)	2022-01	38.00	1475
芬斯勒几何的某些问题(英文)	2022-03	38.00	1476
受限三体问题(英文)	2022-05	38.00	1477
利用马利亚万微积分进行 Greeks 的计算:连续过程、跳跃过程中的马利亚万微积分和金融领域中的 Greeks(英文)	2022-05	48.00	1478
经典分析和泛函分析的应用:分析学的应用(英文)	2022-03	38.00	1479
特殊芬斯勒空间的探究(英文)	2022-03	48.00	1480
某些图形的施泰纳距离的细谷多项式:细谷多项式与图的维纳指数(英文)	2022-05	38.00	1481
图论问题的遗传算法:在新鲜与模糊的环境中(英文)	2022-05	48.00	1482
多项式映射的渐近簇(英文)	2022-05	38.00	1483
一维系统中的混沌:符号动力学,映射序列,一致收敛和沙可夫斯基定理(英文)	2022-05	38.00	1509
多维边界层流动与传热分析:粘性流体流动的数学建模与分析(英文)	2022-05	38.00	1510
演绎理论物理学的原理:一种基于量子力学波函数的逐次置信估计的一般理论的提议(英文)	2022-05	38.00	1511
R^2 和 R^3 中的仿射弹性曲线:概念和方法(英文)	2022-08	38.00	1512
算术数列中除数函数的分布:基本内容、调查、方法、第二矩、新结果(英文)	2022-05	28.00	1513
抛物型狄拉克算子和薛定谔方程:不定常薛定谔方程的抛物型狄拉克算子及其应用(英文)	2022-07	28.00	1514
黎曼-希尔伯特问题与量子场论:可积重正化、戴森-施温格方程(英文)	2022-08	38.00	1515
代数结构和几何结构的形变理论(英文)	2022-08	48.00	1516
概率结构和模糊结构上的不动点:概率结构和直觉模糊度量空间的不动点定理(英文)	2022-08	38.00	1517

刘培杰数学工作室
已出版(即将出版)图书目录——原版影印

书　　名	出版时间	定　价	编号
反若尔当对:简单反若尔当对的自同构(英文)	2022-07	28.00	1533
对某些黎曼-芬斯勒空间变换的研究:芬斯勒几何中的某些变换(英文)	2022-07	38.00	1534
内诣零流形映射的尼尔森数的阿诺索夫关系(英文)	即将出版		1535
与广义积分变换有关的分数次演算:对分数次演算的研究(英文)	即将出版		1536
强子的芬斯勒几何和吕拉几何(宇宙学方面):强子结构的芬斯勒几何和吕拉几何(拓扑缺陷)(英文)	2022-08	38.00	1537
一种基于混沌的非线性最优化问题:作业调度问题(英文)	即将出版		1538
广义概率论发展前景:关于趣味数学与置信函数实际应用的一些原创观点(英文)	即将出版		1539
纽结与物理学:第二版(英文)	2022-09	118.00	1547
正交多项式和q-级数的前沿(英文)	2022-09	98.00	1548
算子理论问题集(英文)	2022-09	108.00	1549
抽象代数:群、环与域的应用导论:第二版(英文)	即将出版		1550
菲尔兹奖得主演讲集:第三版(英文)	即将出版		1551
多元实函数教程(英文)	2022-09	118.00	1552
球面空间形式群的几何学:第二版(英文)	2022-09	98.00	1566

联系地址:哈尔滨市南岗区复华四道街10号　哈尔滨工业大学出版社刘培杰数学工作室
网　　址:http://lpj.hit.edu.cn/
邮　　编:150006
联系电话:0451-86281378　　13904613167
E-mail:lpj1378@163.com